McGRAW-HILL ENCYCLOPEDIA OF
Science & Technology

4

CLI-CYT

McGRAW-HILL ENCYCLOPEDIA OF
Science & Technology

4

CLI-CYT

7th Edition
An international reference work in twenty volumes including an index

McGraw-Hill, Inc.
New York St. Louis San Francisco Auckland Bogotá Caracas Lisbon London Madrid Mexico
Milan Montreal New Delhi Paris San Juan São Paulo Singapore Sydney Tokyo Toronto

1234567890   DOW/DOW   998765432

Library of Congress Cataloging in Publication data

McGraw-Hill encyclopedia of science & technology : an international reference
work in twenty volumes including index. — 7th ed.
    p.   cm.

    Includes bibliographical references and index.
    ISBN 0-07-909206-3
    1. Science—Encyclopedias. 2. Technology—Encyclopedias.
I. Title: McGraw-Hill encyclopedia of science and technology.
II. Title: Encyclopedia of science & technology.
III. Title: Encyclopedia of science and technology.
Q121.M3   1992   503—dc20   91–36349
                                                            CIP

ISBN 0-07-909206-3 {SET}

Organization of the Encyclopedia

This Encyclopedia presents pertinent information in every field of modern science and technology. The 7500 article titles are alphabetically sequenced, so that often the reader can quickly find information by choosing from the 19 text volumes that one bearing the appropriate letter for the subject.

As for the organization of the material throughout the volumes, broad survey articles for each of the disciplines or large subject areas give even the uninitiated reader the basic concepts or rudiments. From the lead article the reader may proceed to other articles that are more restricted in scope by utilizing the cross-referencing system. These cross references are set in small capitals for emphasis and are inserted at the relevant points in the text. In all, some 50,000 cross references are given. In a survey article such as **Petroleum engineering** the reader is directed to other articles such as **Oil and gas well drilling**, **Petroleum reservoir models**, and **Well logging**. The references may lead to subjects that have not occurred to the reader. For example, the article **Chlorine** has such diverse cross references as **Antimicrobial agents**, **Halogen elements**, **Industrial health and safety**, and **Oxidizing agent**.

The pattern of proceeding from the general to the specific has been employed not only in the plan of the Encyclopedia but within the body of the articles. Each article begins with a definition of the subject, followed by sufficient background material to give a frame of reference and permit the reader to move into the detailed text of the article. Within the text are centered heads and two levels of sideheads which outline the article; they are intended to enhance understanding, and can guide the user who prefers to read selectively the sections of a long article.

Alphabetization of article titles is by word, not by letter, with a comma providing a stop in occasional inverted article titles (so that subject matter can be grouped). Two examples of sequence are:

Air	**Earth, age of**
Air-cushion vehicle	**Earth, heat flow in**
Air mass	**Earth crust**
Air-traffic control	**Earth tides**
Aircraft fuel	**Earthquake**

Copious illustrations, both line drawings and halftones, contribute to the utility, clarity, and interest of the text. Each illustration (as well as each table) is called out in boldface at its first mention in the text. This emphasis enables the browsing reader to move from an illustration to the specific point in the text where the illustration is often discussed in detail. Illustrations and tables, insofar as practical, appear on the page spread with their callouts.

Measurements are given in dual systems of units. The U.S. Customary System, continuing in wide application, is used throughout the text. To meet the needs of the Encyclopedia's broad readership, equivalent measurements are given in the International System of Units. In particular cases, such as references to measurements in some illustrations or tables, conversion factors may be given for simplicity.

The contributor's full name appears at the end of an article section or an entire article. Each author is identified in an alphabetical Contributors list in volume 20, which cites the university, laboratory, business, or other organization with which the author is affiliated and the titles of the articles written.

Most of the articles contain bibliographies citing useful sources. The bibliographies are placed at the ends of articles or occasionally at the ends of major sections in long articles. To utilize additional bibliographies, the reader should refer to related articles which are indicated by cross references.

Thus, the alphabetical arrangement of article titles, the text headings, the cross references, and the bibliographies permit the reader to pursue a particular interest by simply taking a volume from the shelf. However, the reader can also find information in the Encyclopedia by using the Analytical Index and the Topical Index in volume 20. The Analytical Index contains each important term, concept, and person—160,000 in all—mentioned throughout the 19 text volumes. It guides the reader to the volume numbers and page numbers concerned with a specific point. The reader wishing to consult everything in the Encyclopedia on a particular aspect of a subject will find that the Analytical Index is the best approach. A broader survey may be made through the Topical Index, which groups all article titles of the Encyclopedia under 79 general headings. For example, under "Geophysics" 59 articles are listed, and under "Organic chemistry," 248. The Topical Index thus enables the reader quickly to identify all articles in the Encyclopedia in a particular subject area.

The Study Guides in volume 20 provide highly structured outlines of six major scientific disciplines (Biology, Chemistry, and so on), and relate groups of Encyclopedia articles to each outline heading. By following a guide, the reader is led through pertinent Encyclopedia articles in a sequence that provides an overall grasp of the discipline.

A useful feature is the section "Scientific Notation" in volume 20. It clarifies usage of symbols, abbreviations, and nomenclature, and is especially valuable in making conversions between International System, U.S. Customary, and metric measurements.

McGRAW-HILL ENCYCLOPEDIA OF
Science &
Technology

4

CLI-CYT

Climate modeling

Construction of a mathematical model of the climate system of the Earth capable of simulating its behavior under present and altered conditions. The Earth's climate is continually changing over time scales ranging from millions of years to a few years. Since the climate is determined by the laws of classical physics, it should be possible in principle to construct such a model. The advent of a worldwide weather observing system capable of gathering data for validation and the development and widespread routine use of digital computers have made this undertaking possible, starting in the mid-1970s. *SEE CLIMATIC CHANGE.*

The first attempts at modeling the planetary climate showed that the Earth's average temperature is determined mainly by the balance of radiant energy absorbed from sunlight and the radiant energy emitted by the Earth system. About 30% of the incoming radiation is reflected directly to space, and 72% of the remainder is absorbed at the surface (**Fig. 1**). The incoming solar radiation is divided among reflection, absorption by the atmospheric constituents, and absorption by the surface of the planet. The outgoing infrared radiation comes from the surface, atmospheric gases, and clouds. In addition, the atmosphere radiates down to the surface, and the surface gives energy to the atmosphere in the forms of latent and sensible heat. The radiation is absorbed unevenly over the Earth, which sets up thermal contrasts that in turn induce convective circulations in the atmosphere and oceans. Climate models attempt to calculate from mathematical algorithms the effects of these contrasts and the resulting motions in order to understand better and perhaps predict future climates in some probabilistic sense. *SEE SOLAR RADIATION; TERRESTRIAL RADIATION.*

Climate models differ in complexity, depending upon the application. The simplest models are intended for describing only the surface thermal field at a fairly coarse resolution. These mainly thermodynamical formulations are successful at describing the seasonal cycle of the present climate, and have been used in some simulations of past climates, for exam-

ple, for different continental arrangements that occurred millions of years ago. At the other end of the spectrum are the most complex climate models, which are extensions of the models used in weather forecasts. These models aim at simulating seasonal and even monthly averages just shortly into the future, based upon conditions such as the temperatures of the tropical-sea surfaces. Intermediate to these extremes are models that attempt to model climate on a decadal basis, and these are used mainly in studies of the impact of hypothesized anthropogenically induced climate change. *SEE CLIMATE MODIFICATION; WEATHER FORECASTING AND PREDICTION.*

Anatomy of models. Since the main interest of climate modelers is in computing the thermal field over the Earth, their primary goal is to represent the conservation of energy at each location in the system. This must include accurate formulations of the absorption and reflection of solar radiation as it passes through the atmosphere and strikes surfaces. It must also include the radiation emitted from each mass element in the Earth–atmosphere system. Conversions of

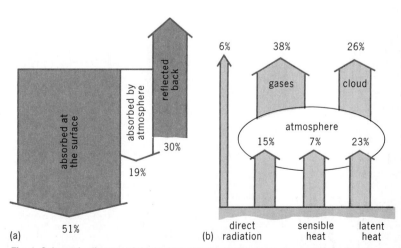

Fig. 1. Schematic diagram of the budget of incoming and outgoing radiation affecting the Earth's climate. (*a*) Solar radiation. (*b*) Terrestrial radiation.

heat from latent to sensible must be taken into account as water changes its phase in the system. Clouds must be included since they participate in the radiation transfer and in the changes of water phase. Similarly, snow and ice cover enter both energy disposal accounts. The thermodynamic expression of the conservation of energy is not complete until allowance is made for the flow of matter of a different temperature into a given region. To include this effect, a model of the circulation must be considered, and this in turn is governed by the same thermal contrasts given by the thermodynamic equation.

The circulation of atmospheric and oceanic material is governed by Newton's second law: local acceleration of a fluid element is proportional to total forces on it. In fluid mechanics this is known as the Navier-Stokes equation. It is a nonlinear partial differential equation that is exceedingly complex. Climate models that include a detailed attempt to solve the fluid dynamics equations must be approached by computer simulation. Even here the solutions are known to differ in detail from nature after only a few weeks at most. However, it is thought that statistics, for example, long-term means for the climate of the numerical model and those of nature, should agree; and this has been borne out in numerous tests. *See Navier-Stokes equations*.

In order to simulate the climate system, the problem must be cast onto a three-dimensional grid in the global ocean-atmosphere volume. The intermediate-sized models typically being used for decadal simulations have a horizontal resolution of about 300 to 600 mi (500 to 1000 km) and vertical resolutions of about 0.6 to 1.8 mi (1 to 3 km). This implies about 25,000 grid points, at each of which the model is keeping track of about six meteorological variables. The system is solved by numerically advancing in time at each grid point, updating at each time step (typically about intervals of 1 h of so-called model time). Simulation of 15 years of model time, which is typical of models that include only simple formulations of the oceanic interaction, may take tens of hours on the fastest computers. About a half dozen models of this type are being investigated around the world. The models as a group simulate the present seasonal cycle of different geographical regions with remarkable fidelity, considering the short time that has passed since the inception of this field. The models are best at simulating the thermal surface field and

weakest in modeling such secondary features as precipitation. *See Simulation*.

Feedback mechanisms and sensitivity. Attempts at modeling climate have demonstrated the extreme complexity and subtlety of the problem. This is due largely to the many feedbacks in the system. One of the simplest and yet most important feedbacks is that due to water vapor. If the Earth is perturbed by an increase in the solar radiation, for example, the first-order response of the system is to increase its temperature. But an increase in air temperature leads to more water vapor evaporating into the air; this in turn leads to increased absorption of space-bound radiation from the ground (greenhouse effect), which leads to an increased equilibrium temperature. This effect, known as a positive feedback mechanism, is illustrated in **Fig. 2**. It roughly doubles the response to most perturbations. Water vapor feedback is not the only amplifier in the system. Another important one is snowcover: a cooler planet leads to more snow and hence more solar radiation reflected to space, since snow is more reflecting of sunlight than soil or vegetation. Other, more subtle mechanisms that are not yet well understood include those involving clouds and the biosphere. *See Greenhouse effect*.

While water vapor and snowcover feedback are fairly straightforward to model, the less understood feedbacks differ in their implementations from one climate model to another. These differences as well as the details of their different numerical formulations have led to slight differences in the sensitivity of the various models to such standard experimental perturbations as doubling carbon dioxide in the atmosphere. All models agree that the planetary average temperature should increase if carbon dioxide concentrations are doubled. However, the predicted response in planetary temperatures ranges, from 4.5 to 9°F (2.5 to 5.0°C). Regional predictions of temperature or precipitation are not reliable enough for detailed response policy formulation. Many of the discrepancies are expected to decrease as model resolution increases (more grid points), since it is easier to include such complicated phenomena as clouds in finer-scale formulations. Similarly, it is anticipated that some observational data (such as rainfall over the oceans) that are needed for validation of the models will soon be available from satellite sensors. *See Meteorological satellites*.

Applications. Climate models are being used in a large variety of applications that aid in the understanding of Earth history. Many simple climate model simulations have been used to sort out the mechanisms responsible for climate change in the past. For example, although not yet fully understood, the astronomical theory of the ice ages states that the waxing and waning of the great continental ice sheets has been forced by the periodic changes in the Earth's elliptical orbit parameters in the past. Similarly, the onset of glaciation in Antarctica and Greenland has been studied by such means.

A problem that has received considerable attention is that of the greenhouse effect. Models are being studied to attempt to achieve better understanding of how the increase of atmospheric carbon dioxide and other trace gases from anthropogenic sources are likely to change the climate in the coming decades. The models are being compared to past climates ranging from the ice ages to the records of the last

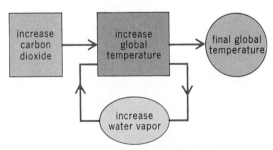

Fig. 2. Diagram illustrating the concept of water vapor feedback in amplifying the sensitivity of the Earth's climate to increases in atmospheric carbon dioxide concentration. Increases in the Earth's temperature cause an increase in water vapor. This in turn causes a warming of the surface because of the greenhouse effect. The net effect is to amplify the response to increases of carbon dioxide.

hundred years, for which an instrumental record exists. *See Climatology; Greenhouse effect.*

Gerald R. North

Bibliography. T. J. Crowley et al., Role of seasonality in the evolution of climate during the last 100 million years, *Science*, 231:579–584, 1986; S. H. Schneider, Climate modeling, *Sci. Amer.*, 256 (5):72–80, 1987; S. H. Schneider and R. E. Dickinson, Climate modeling, *Rev. Geophys. Space Phys.*, 12:447–493, 1974; M. Washington and C. L. Parkinson, *An Introduction to Three-Dimensional Climate Modeling*, 1986.

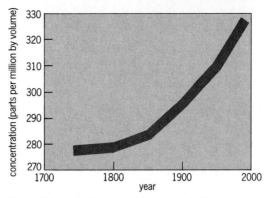

Fig. 1. History of atmospheric CO_2 concentrations as inferred from an analysis of bubbles taken from ice cores in the Antarctic. (*After E. M. Neftel, H. Oeschger, and B. Stauffer, Evidence from polar ice cores for the increase in atmospheric CO_2 in the past two centuries, Nature, 315:45–47, 1985*)

Climate modification

Alteration of the Earth's climate by human activities; humans have the capacity to modify the Earth's climate in several important ways.

Local and regional scale. Conventional agriculture alters the microclimate in the lowest few meters of air, causing changes in the evapotranspiration and local heating characteristics of the air-surface interface. These changes lead to different degrees of air turbulence over the plants and to different moisture and temperature distributions in the local air. *See Evapotranspiration; Irrigation (agriculture).*

Another example of human influence on climate at a larger scale is that the innermost parts of cities are several degrees warmer than the surrounding countryside, and they have slightly more rainfall as well. These changes are brought about by the differing surface features of urban land versus natural countryside and the unique ways that cities dispose of water (for example, storm sewers). The altered urban environment prevents evaporation cooling of surfaces in the city. The modified surface texture of cities (that is, horizontal and vertical planes of buildings and streets versus gently rolling surfaces over natural forest or grassland) leads to a more efficient trapping of solar heating of the near-surface air. The characteristic scales of buildings and other structures also lead to a different pattern of atmospheric boundary-layer turbulence modifying the stirring efficiency of the atmosphere. *See Micrometeorology.*

At the next larger scale, human alteration of regional climates is caused by changes in the Earth's average reflectivity to sunlight. For example the activities of building roads and highways and deforestation change the reflectivity characteristics of the Earth's surface and alter the amount of sunshine that is reflected to space, as opposed to its being absorbed by the surface and thereby heating the air through contact. Such contact heating leads to temperature increases and evaporation of liquid water at the surface. Vapor wakes from jet airplanes are known to block direct solar radiation near busy airports by up to 20%.

Human activities also inject dust, smoke, and other aerosols into the air, causing sunlight to be scattered back to space. Dust particles screen out sunlight before it can enter the lower atmosphere and warm the near-surface air. An extreme case is popularly referred to as the nuclear winter scenario, where a massive injection of smoke particles into the upper atmosphere occurs during a hypothetical exchange of nuclear blasts. The resulting smoke veil theoretically remains in the atmosphere for up to 6 months and leads to a shading of the world and a resulting cooling of continental interiors by as much as 90°F (50°C) for several months. Ocean and coastal regions would experience less than about 18°F (10°C) of cooling. *See Air pollution; Nuclear explosion; Smog.*

Global scale and greenhouse effect. Humans are inadvertently altering the atmospheric chemical composition on a global scale, and this is likely to lead to an unprecedented warming of the global atmosphere during the next generation. It comes about by anthropogenic injection into the atmosphere of relatively inert trace gases that perturb the radiation balance of the globe as a whole. Most of this gaseous waste comes from burning fuels that contain carbon and nitrogen. Other sources include inert gases used in aerosol spray cans and cooling devices. *See Atmospheric chemistry.*

It has been known for over 150 years that some gases act as so-called greenhouse veils and can lead to increased temperatures of the planet if their concentration in the atmosphere is increased. Greenhouse gases such as carbon dioxide have the property of permitting sunlight to pass through volumes containing them but they strongly absorb the surface-originated infrared radiation that would normally pass through a "clean" atmospheric column out to space. The Earth–atmosphere system is balanced on an annual basis between sunlight absorbed by the system and terrestrial infrared radiation emitted to space. When a greenhouse gas is present throughout the column of air over the ground, that gas absorbs some of the surface-originated radiation, and, as a warm body itself, the gas reradiates half of its own emission downward, eventually raising the surface temperature. Increasing the atmospheric concentration of a greenhouse gas invariably raises the temperature of near-surface air and lowers the temperature in the stratosphere. *See Greenhouse effect; Heat balance, terrestrial atmospheric.*

For every carbon atom burned in fossil fuels such as coal, oil, wood, or gasoline, there is one molecule of carbon dioxide (CO_2) released into the global atmospheric system. The relative chemical inertness and water insolubility of CO_2 means that it has a long residence time in the atmosphere. It does eventually react with chemicals and life forms at the Earth's surface and is partially removed by oceanic processes.

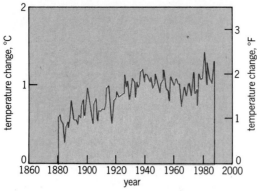

Fig. 2. Global average temperature variations from 1880 to 1988 as inferred from many types of observations. Superimposed on a long-term warming trend are fluctuations that are attributable to volcanic and oceanic activity. The values are relative to an arbitrary baseline.

Currently about 11×10^{12} lb/yr (5.2×10^{12} kg/yr) of carbon are being added to the atmosphere from fossil fuel burning, and another 2×10^{12} lb/yr (10^{12} kg/yr) are being added from the decay of tropical forests that have been cut down and are decaying or being burned. About half of this anthropogenic production remains in the atmosphere as CO_2, with the remainder being taken up by plants and the oceans. This accounts for the observed 0.3% increase in atmospheric CO_2 every year. The 1988 level was 345 parts per million by volume, fully 25% higher than it was in the preindustrial atmosphere as inferred from gas bubbles in ice cores taken from Greenland and Antarctica (**Fig. 1**). A doubling of atmospheric CO_2 is expected to occur by the year 2050. Climate model simulations of the Earth with twice as much atmospheric CO_2 as the present value suggest that the Earth may warm by 5 to 11°F (3 to 6°C) with considerable regional variation. For example, the equatorial regions would warm only by about half of this amount, and the polar regions would warm by two or three times the global average amount. Agriculturally favorable climatic bands might be expected to migrate to new positions. Glacial melting could result in a raising of sea level by as much as 3 ft (1 m). *See Climate modeling.*

An equally important group of greenhouse gases being pumped regularly into the atmosphere are the chlorofluorocarbons, popularly known as freons (CCl_3F and CCl_2F_2); with respect to impact on climate change, one freon molecule is equivalent to 10^4 CO_2 molecules. These gases are produced in industrial processes and are widely used in aerosol cans and air-conditioning equipment. They are relatively more inert and less water-soluble and therefore have an even longer residence time than CO_2. It is believed that they may have a significant potential impact on the ozone layer in the stratosphere. However, they also have an effect on climate, since they are particularly strong greenhouse gases. Two other important greenhouse gases, methane and nitrous oxide, are known to be increasing. Their origins are thought to be connected with anthropogenic activities, although their precise sources are not well understood. The concentrations of these gases increased by 11 and 3.5%, respectively, between 1975 and 1985, and they are projected to continue increasing into the next century. Climate-modeling groups estimate that the contribu-

tions of all these minor trace gases contribute about as much toward anthropogenically induced global climate change as does the most publicized greenhouse gas, CO_2. *See Atmospheric ozone; Freon.*

Since alteration of the atmospheric composition by human activities has taken place for more than 150 years, it is important to ask whether the climate system has started to respond to the forcing. While no definitive answer exists, globally averaged temperature data suggest that on the whole the Earth has been warming over the last 100 years (**Fig. 2**), and the observed warming is consistent with the greenhouse hypothesis. *See Atmosphere; Climatic change; Climatic prediction; Climatology; Weather modification.*

Gerald R. North

Bibliography. R. E. Dickinson and R. J. Cicerone, Future global warming from atmospheric trace gases, *Nature,* 319:109–115, 1986; J. Hansen and S. Lebedeff, Global trends of measured surface air temperature, *J. Geophys. Res.,* D11:13345–13372, 1987; P. D. Jones et al., Evidence for global warming in the past decade, *Nature,* 332:790, 1988; E. M. Neftel, H. Oeschger, and B. Stauffer, Evidence from polar ice cores for the increase in atmospheric CO_2 in the past two centuries, *Nature,* 315:45–47, 1985; V. Ramanathan, The greenhouse theory of climate change: A test by an inadvertent global experiment, *Science,* 240:293–299, 1988.

Climatic change

The long-term fluctuations in precipitation, temperature, wind, and all other aspects of the Earth's climate. The climate, like the Earth itself, has a history extending over several billion years. Climatic fluctuations have occurred at time scales ranging from the longest observable (10^8–10^9 years) to interdecadal variability (10^1 years) and interannual variability (10^0 years). Processes in the atmosphere, oceans, cryosphere (snow cover, sea ice, continental ice sheets), biosphere, and lithosphere, and certain extraterrestrial factors (such as the Sun), are part of the climate system.

The present climate can be described as an ice age climate, since large land surfaces are covered with ice sheets (Antarctica, Greenland). The origins of the present ice age may be traced, at least in part, to movement of the continental plates. With the gradual movement of Antarctica toward its present isolated polar position, ice sheets began to develop about 30,000,000 years ago. Within the past several million years, the Antarctic ice sheet reached approximately its present size, and ice sheets appeared on the lands bordering the northern North Atlantic Ocean. During the past million years of the current ice age, about 10 glacial-interglacial fluctuations have been documented. The most recent glacial period came to an end between about 15,000 and 6000 years ago with the rapid melting of the North American and European ice sheets and the associated rise in sea level. The present climate is described as interglacial. The scope of this article is limited to a discussion of climatic fluctuations within the present interglacial period and, in particular, the climatic fluctuations of the past 100 years—the period of instrumental records. A more complete discussion of past climates is found in other articles. *See Continents, evolution of; Dendrochronology; Geologic thermometry; Glacial epoch;*

Paleoclimatology; Paleogeography; Palynology.

Evidence. Instrumental records of climatic variables such as temperature and precipitation exist for the past 100 years in many locations and for as long as 200 years in a few locations. These records provide evidence of year-to-year and decade-to-decade variability but are completely inadequate for the study of century-to-century and longer-term variability. Even for the study of short-term climatic fluctuations, instrumental records are of limited usefulness, since most observations were made from the continents (only 29% of the Earth's surface area) and limited to the Earth's surface. Aerological observations which permit the study of atmospheric mass, momentum and energy budgets, and the statistical structure of the large-scale circulation are available for only about 20 years. Again there is a bias toward observations over the continents. It is only with the advent of satellites that global monitoring of the components of the Earth's radiation budget (planetary albedo, from which the net incoming solar radiation can be estimated; and the outgoing terrestrial radiation) has begun. *See Heat balance, terrestrial-atmospheric; Meteorological satellites*.

There remain important gaps in the ability to describe the present state of the climate. For example, precipitation estimates, especially over the oceans, are very poor. Oceanic circulation, heat transport, and heat storage are only crudely estimated. Also, the solar irradiance is not being monitored to sufficient accuracy to permit estimation of any variability and evaluation of the possible effect of fluctuations in solar output upon the Earth's climate. Thus, although climatic fluctuations are found in instrumental records, the task of defining the scope of these fluctuations and diagnosing potential causes is at best difficult and at worst impossible.

In spite of the inadequacy of the instrumental records for assessing global climate, there is considerable evidence of regional climatic variations. For example, there is evidence of climatic warming in the polar regions of the Northern Hemisphere during the first four to five decades of the twentieth century. During the 1960s, on the other hand, there is evidence of cooling in the polar and mid-latitude regions of the Northern Hemisphere; and, especially in the early 1970s, there were drier conditions along the northern margin of the monsoon lands of Africa and Asia.

Under the auspices of the World Meteorological Organization and the International Council of Scientific Unions, the Global Atmospheric Research Program (GARP) was charged with developing plans for detailed observation and study of the global climate system—especially the atmosphere, the oceans, the sea ice, and the changeable features of the land surface.

Causes. Many extraterrestrial and terrestrial processes have been hypothesized to be possible causes of climatic fluctuations. A number of these processes are listed and described below.

Solar irradiance. It is possible that variations in total solar irradiance could occur over a wide range of time scales (10^0–10^9 years). If these variations did take place, they would almost certainly have an influence on climate. Radiance variability in limited portions of the solar spectrum has been observed, but not linked clearly to climate variability. *See Solar energy*.

Orbital parameters. Variations of the Earth's orbital parameters (eccentricity of orbit about the Sun, precession, and inclination of the rotational axis to the orbital plane) lead to small but possibly significant variations in incoming solar radiation with regard to seasonal partitioning and latitudinal distribution. These variations occur at times scales of 10^4–10^5 years. *See Earth rotation and orbital motion*.

Lithosphere motions. Sea-floor spreading and continental drift, continental uplift, and mountain building operate over long time scales (10^5–10^9 years) and are almost certainly important factors in long-term climate variation. *See Continental drift; Lithosphere; Marine geology; Orogeny*.

Volcanic activity. Volcanic activity produces gaseous and particulate emissions which lead to the formation of persistent stratospheric aerosol layers. It may be a factor in climatic variations at all time scales. *See Volcano*.

Internal variability of climate system. Components of the climate system (atmosphere, ocean, cryosphere, biosphere, land surface) are interrelated through a variety of feedback processes operating over time scales from, say, 10^0 to 10^9 years. These processes could, in principle, produce fluctuations of sufficient magnitude and variability to explain any observed climate change. For example, atmosphere-ocean interactions may operate over time scales ranging from 10^0 to 10^3 years, and atmosphere-ocean-cryosphere interactions may operate over time scales ranging from 10^0 to 10^5 years. Several hypotheses have been proposed to explain glacial-interglacial fluctuations as complex internal feedbacks among atmosphere, ocean, and cryosphere. (Periodic buildup and surges of the Antarctic ice sheet and periodic fluctuations in sea ice extent and deep ocean circulation provide examples.) Atmosphere-ocean interaction is being studied intensively as a possible cause of short-term climatic variations. It has been observed that anomalous ocean temperature patterns (both equatorial and mid-latitude) are often associated with anomalous atmospheric circulation patterns. Although atmospheric circulation plays a dominant role in establishing a particular ocean temperature pattern (by means of changes in wind-driven currents, upwelling, radiation exchange, evaporation, and so on), the anomalous ocean temperature distribution may then persist for months, seasons, or longer intervals of time because of the large heat capacity of the oceans. These anomalous oceanic heat sources and sinks may, in turn, produce anomalous atmospheric motions.

Human activities. Forest clearing and other large-scale changes in land use, changes in aerosol loading, and the changing CO_2 concentration of the atmosphere are often cited as examples of possible mechanisms through which human activities may influence the large-scale climate. Because of the large observational uncertainties in defining the state of the climate, it has not been possible to establish the relative importance of human activities (as compared to natural processes) in recent climatic fluctuations. There is, however, considerable concern that future human activities may lead to large climatic variations (for example, continued increase in atmospheric CO_2 concentration due to burning of fossil fuels) within the next several decades. *See Air pollution; Atmospheric ozone; Climate modification; Greenhouse effect*.

It is likely that at least several of the above-mentioned processes have played a role in past climatic fluctuations (that is, it is unlikely that all climatic fluctuations are due to one factor). In addition, certain processes may act simultaneously, or in various sequences. Also, the climatic response to some causal process may depend on the particular initial climatic state, which, in turn, depends upon previous climatic states because of the long time constants of oceans and cryosphere. True equilibrium climates may not exist, and the climate system may be in a continual state of transience.

Modeling. Because of the complexity of the real climate system, simplified numerical models of climate are being used to study particular processes and interactions. Some models treat only the global-average conditions, whereas others, particularly the atmospheric models, simulate detailed patterns of climate. These models are still in early stages of development but will undoubtedly be of great importance in attempts to understand climatic processes and to assess the possible effects of human activities on climate. SEE CLIMATOLOGY.

John E. Kutzbach

Ocean-atmosphere interaction. The atmosphere and the oceans have always jointly participated in climatic change, past and contemporary. Some of the contemporary changes can be investigated in the modern records of climatic anomalies in the atmosphere and the oceans.

The most important source of the climatic change surpassing 1-year duration seems to be located along the equatorial zone of the Pacific Ocean. The prevailing winds there are easterly and maintain a westward wind drift of the surface water which diverges, under influence of the Earth's rotation, to the right of the wind direction north of the Equator and the left south of the Equator. The resulting equatorial upwelling of cold water, and subsequent lateral mixing, ordinarily maintains a belt of cold surface water several hundred kilometers wide straddling the Equator from the coast of South America about to the dateline, about one earth quadrant farther west. SEE UPWELLING.

Analogous processes are found in the equatorial belt of the Atlantic, but the upwelling water there covers a much smaller area and is also less cold than in the Pacific. The Indian Ocean has no steady easterlies and thus no equatorial upwelling.

The equatorial process varies in intensity with the equatorial easterlies. Since that wind system is mostly fed by way of the southerlies along the west coasts of South Africa and South America, it is likely that anomalies in the Southern Hemisphere atmospheric circulation frequently are transmitted to the tropical belt. Once an impulse, for example, a strengthening of the Pacific equatorial easterlies, has occurred, the new anomaly has a built-in tendency of self-amplification, because it makes the upwelling strengthen too and thus increases the temperature deficit of the Pacific compared to the persistently warm Indonesian and Indian Ocean tropical waters. This in turn feeds back into further strengthening of the easterlies which started the anomaly in the first place.

The observational proof of this feedback system can be seen in the statistically well-substantiated "southern oscillation," which exhibits opposite contemporaneous anomalies of atmospheric pressure over the tropical parts of the Pacific Ocean on the one side and the Indonesian and Indian Ocean tropical waters on the other (nodal line on the average at 165°E). The

periodicity is rather irregular, so the term oscillation should not be taken too literally. The average length of the cycles is 2–3 years.

The cycles of tropical precipitation of more than a year's length by and large agree with those of pressure wherever special local conditions do not interfere. Satellite photos confirm that in the phase of the southern oscillation with positive pressure anomaly over the Pacific, along with strong equatorial easterlies and strong upwelling, most of the Pacific equatorial belt is arid; while in the opposite phase the western and central part of that belt experiences heavy rainfall. In extreme "El Niño" years this rainfall can also extend to the normally arid coast of northern Peru. SEE TROPICAL METEOROLOGY.

When there is more than normal rainfall at the Equator, the general circulation of the atmosphere is supplied with more-than-normal total heat convertible into kinetic energy. The remote effect of this phenomenon, particularly in the winter hemisphere, is the occurrence of stronger-than-normal tradewinds and midlatitude westerlies. At the opposite extreme, the tradewinds are weak and the westerlies meandering. This produces cold winters in the longitude sectors with winds out of high latitudes and mild winters interspersed at longitudes where wind components from low latitudes prevail. Again, it is in the Pacific longitude sector that these teleconnections are most clearly seen, because the interannual variability of sea temperature up to a range of 5.4°F (3°C) over a large equatorial area is found only in the Pacific sector. SEE ATMOSPHERIC GENERAL CIRCULATION; MARITIME METEOROLOGY.

Jacob Bjerknes

Bibliography. H. P. Berlage, *The Southern Oscillation and World Weather*, Kon. Ned. Meteorol. Inst. Meded. Verh. 88, 1966; J. Bjerknes, A possible response of the Hadley circulation to variations of the heat supply from the equatorial Pacific, *Tellus*, 18:820–829, 1966; H. H. Lamb, *Climate, History, and the Modern World*, 1982; H. H. Lamb, *Climate: Present, Past and Future*, vol. 1: *Fundamentals and Climate Now*, 1972; S. H. Schneider and R. E. Dickinson, Climate modelling, *Rev. Geophys. Space Phys.*, 12:447–493, 1974; H. Shapley (ed.), *Climatic Change*, 1953; B. J. Skinner (ed.), *Climate Past and Present*, 1981; Study of man's impact on climate, in *Inadvertent Climate Modification*, 1971; UNESCO, *Changes of Climate*, Arid Zone Research, 1963; U.S. National Academy of Sciences, *Understanding Climatic Change: A Program for Action*, 1975.

Climatic prediction

Prediction of the response of the Earth-atmosphere system to changes in one of the variables involved, or prediction of future climate from observed present conditions. It is useful to distinguish between these two types of climatic prediction.

First type. Studies of the first type attempt to determine the response of the Earth-atmosphere system to small changes in one of the many variables involved—for example, the solar constant or the atmospheric carbon dioxide content. The most useful results have been obtained by using mathematical models of various degrees of complexity. These range from zero-dimension models, which consider only the radiation balance of the complete global system, to three-dimensional (3-D) models which couple the circulation of the atmosphere and oceans with the hydrologic cycle to reproduce the climatic sequence at several levels in the atmosphere and oceans and at points 130–300 mi (200–500 km) apart.

Sensitivity tests with this hierarchy of models indicate, among other things, that the climate system responds almost identically to a 2% increase in the solar constant and a doubling of the atmospheric carbon dioxide concentration. This is rather surprising considering the differences in the natures of these two forcings. In both cases the globally averaged annual surface temperature rises by 5 to 9°F (3 to 5°C), with increases of almost 36°F (20°C) occurring at high latitudes in both hemispheres in winter. SEE SOLAR CONSTANT.

Various attempts have been made to use paleoclimatic and historical data to predict the regional climatic changes that will occur with changes in the global climate. An interesting approach, used in the Soviet Union is to correlate the local historical temperature record, for example, with that averaged for the hemisphere or globe. The regression coefficient (or slope of the regression line) gives the ratio of the local temperature change to that for the hemisphere or globe. This method could be applied to any variable for which the necessary data are available.

More and more, climatologists realize that the source of climatic variability lies in the tropics, even though the largest climatic changes occur in polar latitudes. The atmosphere is heated primarily by the absorption of infrared radiation emitted by the Earth's surface. However, all of this heat, and more, is radiated away, either back to the surface or to space. This leaves the atmosphere with a net loss of infrared radiation, equivalent to a cooling rate of about 2.3°F (1.3°C) per day. To offset this, the atmosphere is heated by the absorption of solar radiation (0.9°F or 0.5°C/day), the condensation of water vapor (1.3°F or 0.7°C/day), and, to a smaller extent, the convection of heat from the Earth's surface (0.2°F or 0.1°C/day). Between 20°N and 20°S the heat generated by these processes far exceeds the cooling by infrared radiation. The excess heat is transferred to higher latitudes as either potential energy or sensible heat—equatorward of about 30°, by the Hadley cells of the two hemispheres, and poleward of 30°, primarily by moving low- and high-pressure systems. Almost half of the heat added to the atmosphere is added in the 34% of the globe between 20°N and 20°S. Since more than half of this heat is added by condensation, small variations in tropical precipitation, for example, those associated with El Niño, can have a strong impact on the poleward heat transport and, hence, on climatic conditions in middle and high latitudes. SEE GREENHOUSE EFFECT; TERRESTRIAL RADIATION; TROPICAL METEOROLOGY.

Second type. This type of climatic prediction deals with actually predicting the future state of the climate, given the observed present conditions. Forecasts can be made for periods ranging from a month to a millennium or more in the future. Eventually, numerical models may be used for this purpose, but so far this type of prediction, with one exception, involves primarily the application of various statistical techniques.

The exception occurs where the climate-forcing function itself is predictable. A good example is systematic variations in the Earth's orbit about the Sun which produce changes in the seasonal incidence of solar radiation. These changes are completely predictable and can be fed into a climate model to determine the response. SEE EARTH ROTATION AND ORBITAL MOTION.

Extremely popular among the statistical approaches are attempts to isolate climatic cycles, especially those which can be related to solar activity. So far the results have not been especially encouraging, partly because more often than not the phase of a significant cycle changes abruptly and randomly.

Statistically derived interrelationships have been observed between sea-surface temperatures and certain subsequent atmospheric circulation features in the tropics and middle latitudes. These have already been useful in seasonal climatic prediction and may eventually provide a strong base for the numerical prediction of climate. SEE CLIMATIC CHANGE; CLIMATOLOGY; STATISTICS; WEATHER FORECASTING AND PREDICTION.

W. D. Sellers

Bibliography. W. Bach, H.-J. Jung, and H. Knottenberg, *Modeling the Influence of Carbon Dioxide on the Global and Regional Climate*, 1985; S. Manabe (ed.), *Issues in Atmospheric and Oceanic Modeling*: pt. A, *Climate Dynamics*, pt. B, *Weather Dynamics*, 1985; M. E. Schlesinger and J. F. B. Mitchell, Climate model simulations of the equilibrium climatic response to increased carbon dioxide, *Rev. Geophys.*, 25:760–798, 1987.

Climatology

The scientific study of climate. Climate is the expected mean and variability of the weather conditions for a particular location, season, and time of day. The climate is often described in terms of the values of meteorological variables such as temperature, precipitation, wind, humidity, and cloud cover. A complete description also includes the variability of these quantities, and their extreme values. The climate of a region often has regular seasonal and diurnal variations, with the climate for January being very different from that for July at most locations. Climate also exhibits significant year-to-year variability and longer-term changes on both a regional and global basis.

Climate has a central influence on many human needs and activities, such as agriculture, housing, human health, water resources, and energy use. The influence of climate on vegetation and soil type is so strong that the earliest climate classification schemes were often based more on these factors than on the meteorological variables. While technology can be used to mitigate the effects of unfavorable climatic conditions, climate fluctuations that result in significant departures from normal cause serious problems for modern industrialized societies as much as for primitive ones. The goals of climatology are to provide a comprehensive description of the Earth's climate, to understand its features in terms of fundamental physical principles, and to develop models of the Earth's climate that will allow the prediction of future changes that may result from natural and human causes. SEE CLIMATE MODELING; CLIMATIC PREDICTION.

Physical basis of climate. The global mean climate and its regional variations can be explained in terms of physical processes. For example, the temperature is warmer near the Equator than near the poles (**table** and **Figs. 1** and **2**). This is because the source of heat for the Earth is the radiant energy com-

Temperature (T) and precipitation (P) for selected stations in North America, South America, and Antarctica*

Station	Latitude, degrees	T_{annual}	T_{Jan}	T_{July}	P_{annual}	P_{Jan}	P_{July}
Alert	82.50 N	−18.0	−32.1	3.9	156	8	18
Barrow	71.30 N	−12.5	−26.8	3.9	109	5	20
Fairbanks	64.80 N	−3.4	−23.9	15.4	287	23	47
Baker Lake	64.30 N	−12.3	−33.6	10.7	213	7	36
Anchorage	61.17 N	1.8	−10.9	13.9	373	20	47
Juneau	58.37 N	4.5	−3.8	12.9	1288	102	114
Edmonton	53.57 N	2.8	−14.7	17.5	447	25	83
Seattle	47.45 N	10.8	3.9	18.2	980	153	19
Montreal	45.50 N	7.2	−8.9	21.6	999	80	93
Des Moines	41.53 N	10.5	−5.2	25.2	789	32	77
Salt Lake City	40.78 N	10.7	−2.1	24.7	353	34	15
Washington, D.C.	38.85 N	13.9	2.7	25.7	1087	82	107
San Francisco	37.62 N	13.8	9.2	17.1	475	102	t
Nashville	36.12 N	15.6	4.4	26.8	1146	139	94
Los Angeles	33.93 N	18.0	13.2	22.8	373	78	t
Birmingham	33.57 N	17.8	8.1	27.6	1349	128	131
Phoenix	33.43 N	21.4	10.4	32.9	184	19	20
New Orleans	29.95 N	20.0	12.3	27.3	1369	98	171
Havana	23.17 N	24.6	21.8	27.0	1126	54	108
Acapulco	16.83 N	27.6	26.1	28.7	1401	8	230
Caracas	10.60 N	26.1	24.4	26.4	545	42	72
Guayaquil	02.18 S	25.5	26.5	24.2	811	199	0.3
Manaus	03.13 S	27.5	26.7	27.6	2294	279	65
Brazilia	15.78 S	20.4	21.2	18.0	1643	248	6
Rio de Janeiro	22.90 S	23.7	26.1	20.9	1218	211	52
Antofagasta	23.47 S	16.2	19.8	13.2	1.9	0.0	0.3
Santiago	33.45 S	14.4	21.2	8.1	264	0.1	69
Buenos Aires	34.58 S	16.9	23.7	10.6	1029	104	61
Puerto Aisen	45.50 S	9.2	13.9	3.9	3001	203	331
Comodoro Rivadavia	45.78 S	12.6	18.6	6.9	216	16	21
Punta Arenas	53.00 S	6.0	10.4	1.3	362	40	24
Melchior	64.32 S	−3.6	1.0	−9.3	1116	42	90
Byrd Station	80.02 S	−27.9	−14.6	−35.1	39	7	2.5
Amundsen Scott	90.00 S	−49.4	−28.7	−60.3	1.5	t	t

*Temperature is in degrees Celsius; °F = (°C × 1.8) + 32. Precipitation is in millimeters; 1 mm = 0.04 in. t indicates trace of precipitation.

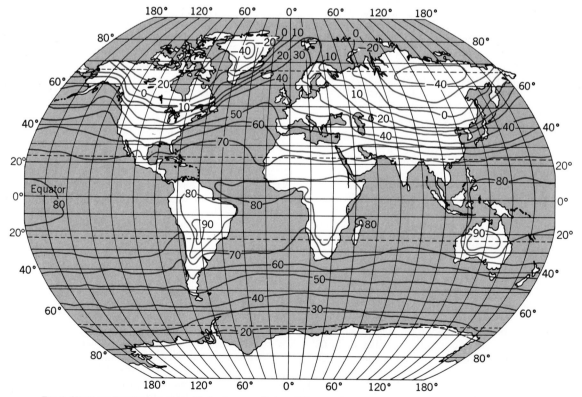

Fig. 1. Mean sea-level temperature, °F, for January. Note cold temperatures near centers of northern land masses. °C = (°F − 32) ÷ 1.8.

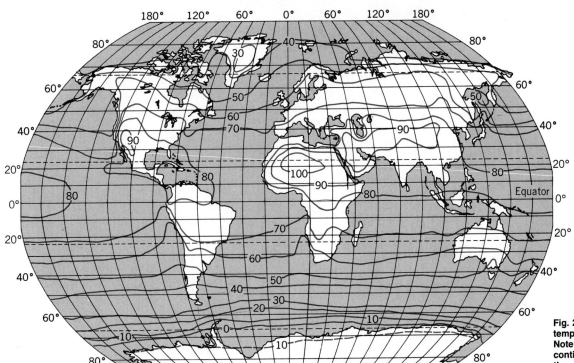

Fig. 2. Mean sea-level temperature, °F, for July. Note that northern continents are warmer than ocean areas at the same latitude. °C = (°F − 32) ÷ 1.8.

ing from the Sun, and the Sun's rays fall more directly on the Equator than on the poles. The circulations in the atmosphere and the oceans transport heat poleward and thereby reduce the Equator-to-pole temperature difference that is continually forced by insolation. The energy released by the rising of warm air in the tropics and the sinking of cold air in high latitudes drives the great wind systems of the atmosphere, such as the trade winds in the tropics and the westerlies of middle latitudes. *SEE ATMOSPHERIC GENERAL CIRCULATION; OCEAN CIRCULATION; WIND.*

The flux of solar energy at the mean distance of the Earth from the Sun is about 1365 W · m^{-2}. The supply of energy per unit of the Earth's surface area is controlled by geometric and astronomical factors. Because the axis of rotation of the Earth is inclined at an oblique angle to the plane of the Earth's orbit, the declination angle of the Sun undergoes a seasonal variation as the Earth makes its annual circuit around it. The declination angle is equivalent to the latitude at which the Sun is directly overhead at noon. The declination angle varies between 23.5° N at northern summer solstice (June 21) and 23.5° S at northern winter solstice (December 22), for the current alignment of the Earth's orbit. The approximate sphericity of the Earth and the annual variation of the declination angle of the Sun cause the incoming solar radiation to be a function of latitude and season (**Fig. 3**). The annual mean insolation is largest at the Equator and decreases toward the poles. The seasonal variation is largest at high latitudes. At the poles 6 months of daylight alternate with 6 months of darkness. The Earth is closer to the Sun during summer in the Southern Hemisphere, so this region receives about 7% more insolation at this time than the Northern Hemisphere. *SEE INSOLATION.*

About half of the energy from the Sun that is inci-

dent at the top of the atmosphere is transmitted through the atmosphere and absorbed at the Earth's surface (**Fig. 4**). About 30% is reflected directly to space, and another 20% is absorbed in the atmosphere. The fraction of the incoming solar radiation that is reflected to space is called the albedo. The albedo increases where clouds or surface ice are present. It increases toward the poles because cloud cover and surface ice increase and because the Sun is closer to the horizon. The albedo of desert areas is generally higher than that of heavily vegetated areas or oceans. The solar energy that reaches the surface may raise the surface temperature, or the energy can be used to evaporate water. The energy that is used to evaporate water is later released into the atmosphere when the water vapor condenses and returns to the surface in the form of precipitation (**Fig. 5**). *SEE ALBEDO; PRECIPITATION (METEOROLOGY).*

Greenhouse effect. In order to achieve an energy balance, the solar energy that is absorbed by the Earth must be returned to space. The Earth emits radiative energy at frequencies that are substantially different from those of the Sun because of the Earth's colder temperature. The Earth emits primarily thermal infrared radiation (wavelengths from 4 to 200 micrometers), whereas most of the energy from the Sun arrives in the form of visible and near-infrared radiation (0.4–4 μm). The atmosphere is much less transparent to thermal radiation than to solar radiation, because water vapor, clouds, and carbon dioxide in the atmosphere absorb thermal radiation. Because the atmosphere prevents thermal radiation emitted from the surface from escaping to space, the surface temperature is warmer than it would be in the absence of the atmosphere. The combination of the relative transparency of the atmosphere to solar radiation and the blanketing effect of the gases in the atmosphere that

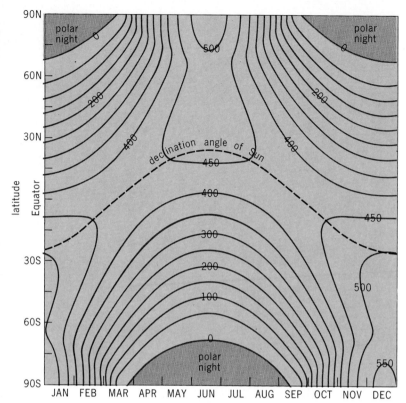

Fig. 3. Daily-average solar energy incident at the top of the atmosphere as a function of latitude and season in watts per square meter of surface area.

the solar energy absorbed is decreased when clouds are present. Since clouds decrease both the solar energy absorbed and the terrestrial energy emitted to space, the net effect of tropical convective clouds on the energy balance of the Earth is less than their individual effects on solar and terrestrial energy. As a result, the distribution of the net radiative energy exchange at the top of the atmosphere does not show the tall cloud signature so clearly. The net radiation shows a strong influx of energy at those latitudes where the insolation is strongest. SEE TERRESTRIAL RADIATION.

A striking feature of the distribution of net radiation during July is the low values over the Sahara and Arabian deserts compared with the surrounding ocean and moist land areas. This is because desert sand has a relatively high albedo, so that less solar radiation is absorbed there. In addition, few clouds and little water vapor are present in the atmosphere over the deserts to absorb the thermal radiation emitted by the very hot surface. Since the emitted thermal radiation is high and the absorbed solar radiation is low, desert areas often show a net loss of radiative energy. This fact plays a central role in the maintenance of desert dryness. The loss of energy requires a continual transport of energy into the desert regions by the atmosphere. This results in downward motion of dry air, which flows outward near the ground and prevents the moist surface air of surrounding regions from reaching the desert interior. SEE DESERT.

Influences of land and ocean. Land and ocean areas have very different seasonal variations because of their different capacities for strong heat. Because the ocean is a fluid, 160–320 ft (50–100 m) of the surface ocean depth are generally in direct thermal communication with the surface. Therefore solar energy incident on the surface can be absorbed in this large heat sink without raising the surface temperature very much. In addition, over the oceans it is possible for solar energy to evaporate water and thus never be realized as heat. In contrast, over land only about the first meter of soil is in thermal contact with the surface, and much less water is available for evapora-

absorb thermal radiation is often referred to as the greenhouse effect. SEE ATMOSPHERE; GREENHOUSE EFFECT.

Clouds affect the energy emission of the Earth. The thermal emission is low over tropical regions where high, cold clouds are present, such as the major precipitation regions of equatorial Africa, South America, southern Asia, Indonesia, and the band of cloudiness along the Equator. Because the albedo is higher,

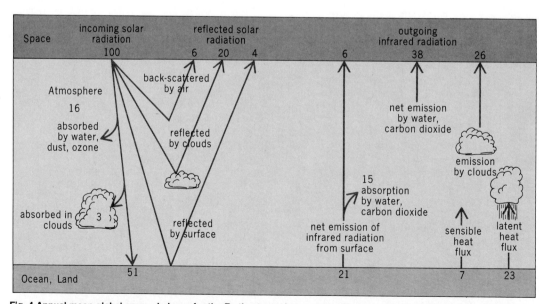

Fig. 4 Annual-mean global energy balance for the Earth–atmosphere system. Numbers given are percentages of the globally averaged solar energy incident upon the top of the atmosphere. (After J. M. Wallace and P. V. Hobbs, Atmospheric Science: An Introductory Survey, Academic Press, 1977)

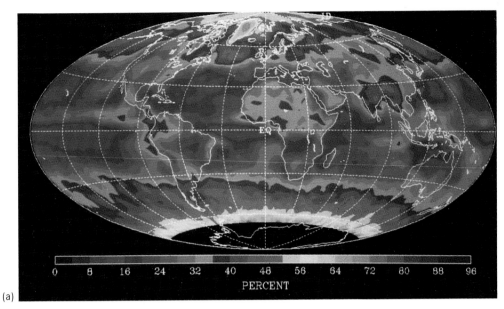

(a)

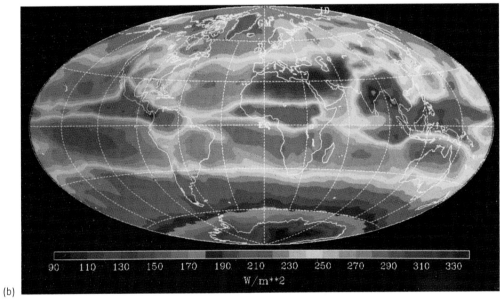

(b)

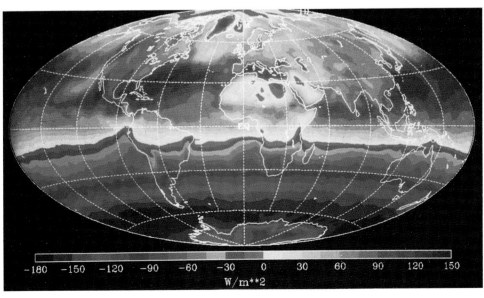

(c)

Radiation balance of Earth for July 1985 measured from space. (*a*) Percentage of incident solar energy reflected to space (albedo). (*b*) Emission of thermal energy by the Earth in watts per square meter of surface area. (*c*) Absorbed solar energy minus emitted thermal energy (net radiation) in watts per square meter of surface area. (*D. L. Hartmann, from Earth Radiation Budget Experiment data, ERBS and NOAA-9 combined data set*)

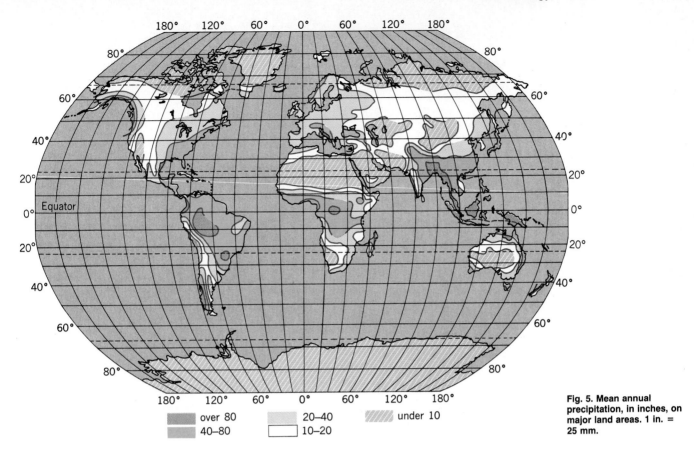

Fig. 5. Mean annual precipitation, in inches, on major land areas. 1 in. = 25 mm.

over 80   20–40   under 10
40–80   10–20

tion. During the winter, the ocean can return heat stored during the summer to the surface, keeping it relatively warm. The land cools off very quickly in winter, warms up rapidly in summer, and experiences large day-night differences. Continental climates in middle latitudes are characterized by hot summers and cold winters, whereas climates that are over or near the oceans are more equable with milder seasons (Figs. 1 and 2).

The distinction between maritime and continental climates can be seen by comparing the annual variations of temperature and precipitation for Tatoosh Island, Washington, with those for Minneapolis, Minnesota (**Fig. 6**). The temperature at Tatoosh Island varies between a mean of 42°F (5.6°C) for January and a mean of 56°F (13.3°C) for August. This narrow temperature range results from the strong influence of the ocean, which is adjacent to and upwind of the Pacific coast and has a small seasonal variation in temperature because of its large capacity to store and release heat. Although it is at nearly the same latitude, Minneapolis has a much larger annual variation of monthly-mean temperatures, with readings of 12°F (−10.9°C) for January and 72°F (22.4°C) for July. Figures 1 and 2 show the large annual variation of temperature near the centers of the continents. SEE CONTINENTALITY (METEOROLOGY); MARINE METEOROLOGY.

The annual variation of precipitation is also very different in the two climate regimes (Fig. 6). At Tatoosh Island the precipitation peaks in the winter season in association with rainfall produced by the cyclones and fronts of wintertime weather. In Minneapolis the precipitation peaks in the summer season. Most of this precipitation is associated with

thunderstorms. Adequate precipitation during the warm summer season is an essential ingredient of the agricultural productivity of the American Midwest.

New York City shows an annual variation of temperature and precipitation that is a combination of maritime and continental. It is near the Atlantic Ocean, but because the prevailing winds are out of the west, it also comes under the influence of air that has been over the continent. It has a fairly large annual variation of temperature (Fig. 6). Monthly-mean precipitation is almost constant through the year. It receives precipitation both from winter storms and summer thunderstorms. Miami, Florida is in the subtropics at 26°N. The annual variation of insolation is weak, so that the seasonal variation of temperature is rather small. The precipitation shows a strong seasonal variation, with maximum precipitation during the summer half-year associated primarily with thunderstorm activity.

General circulation of atmosphere and climate. Many aspects of the Earth's climate are determined by the nature of the circulation that results from the radiative heating of the tropics and cooling of the polar regions. In the belt between the Equator and 30° latitude the bulk of the poleward atmospheric energy transport is carried by a large circulation cell, in which air rises in a narrow band near the Equator and sinks at tropical and subtropical latitudes. The upward motion near the Equator is associated with intense rainfall and wet climates, while the downward motion away from the Equator results in the suppression of rainfall and very dry climates (see Fig. 5). Most of the world's great deserts, including the Sahara, Australian, Arabian, Kalahari, and Atacama,

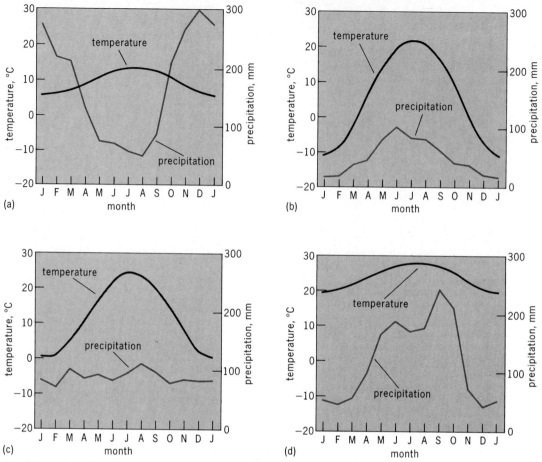

Fig. 6. Annual variation of monthly mean temperature and precipitation at (*a*) Tatoosh Island, Washington; (*b*) Minneapolis, Minnesota; (*c*) New York, New York; and (*d*) Miami, Florida. °F = (°C × 1.8) + 32; 1 mm = 0.04 in.

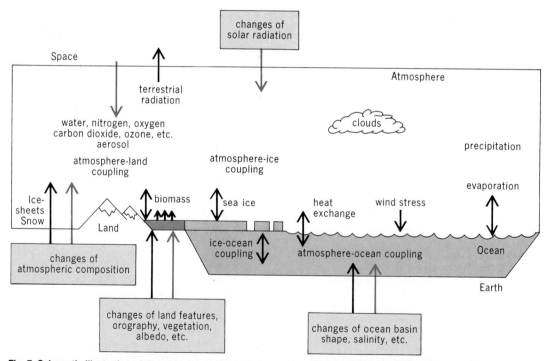

Fig. 7. Schematic illustration of the processes and interactions in the climate system. The colored arrows indicate externally applied conditions, and black arrows show internal processes that act to influence climate change. (*After J. G. Lockwood, Causes of Climate, Halstead Press, 1979*)

are in the belts between 10 and 30° latitude.

In middle latitudes the poleward flow of energy is produced by extratropical cyclones rather than by a mean circulation cell. These storms are thousands of miles across and are characterized by poleward motion of warm, moist air and equatorward motion of cold, dry air. Most of the wintertime precipitation in middle latitudes is associated with weather disturbances of this type.

Surface features are also of importance in determining local climate. Mountain ranges can block the flow of moist air from the oceans, resulting in very low rainfall. The dryness of the Great Basin of North America and the Gobi Desert of Asia is maintained in this way. On the upwind side of mountain ranges, forced ascent of moist air can result in very moist climates; this occurs, for example, on the west slope of the coastal mountains of western America and the south slope of the Himalayas during the summer monsoon. The downwind sides of such mountains are often very dry, since the moisture precipitates out on the upwind side. Two locations which show the effect of topography on local climates are Puerto Aisen, Chile, and Comodoro Rivadavia, Argentina, which are both located near a latitude of 45° (see table). Puerto Aisen is on the westward and upwind side of the Andes mountains in a deep valley that is exposed to the midlatitude westerly winds coming off the Pacific Ocean. It receives more than 10 ft (3 m) of precipitated water annually. Only a few hundred miles downwind of the Andes on the Atlantic seaboard, Comodoro Rivadavia receives only about 8 in. (0.2 m) of precipitation each year.

Complete climate system. The climate of the Earth results from complex interactions among externally applied parameters, like the distribution of insolation, and internal interactions among the atmosphere, the oceans, the ice, and the land (**Fig. 7**). The composition of the atmosphere, which plays a key role in determining the surface temperature through the greenhouse effect, has been radically changed by the life-forms that have developed, and continues to be modified and maintained by them. The atmosphere and the oceans exchange heat, momentum, water, and important constituent gases such as oxygen and carbon dioxide. The exchange of constituent gases is strongly influenced by life in the ocean. The hydrologic cycle of evaporation, cloud formation, and precipitation as rain or snow is intimately connected to the climate through the effects of water vapor, clouds, and surface ice on the radiation balance of the planet. Vegetation interacts strongly with the hydrologic cycle over land to determine the soil moisture, surface albedo, evaporation, precipitation, and surface water runoff. *See* HYDROLOGY.

Evolution of Earth's climate. The climate of the Earth is unique among the planets in the solar system. All of them evolved out of material in the rotating cloud from which the solar system was formed. The subsequent evolution of the planets'' atmospheres depended critically on the mass of each planet and its distance from the Sun.

The mass of the Earth and its distance from the Sun are such that water can exist in liquid form rather than being frozen or escaping to space. The liquid water formed the oceans and led to the development of photosynthetic life, which reduced the carbon dioxide content and increased the molecular oxygen content

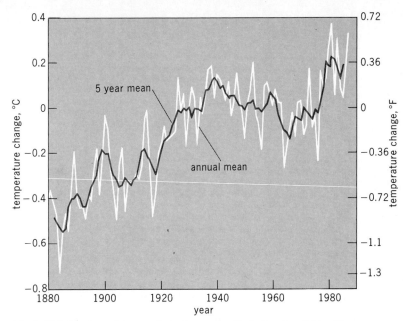

Fig. 8. Global surface air temperature change over the last century inferred from measurements. (*After J. Hansen and S. Lebedeff, Global surface air temperatures:* **Update** *through 1987, Geophys. Res. Lett., 15(4):323, April 1988)*

of the atmosphere. Planets closer to the Sun receive more solar energy per unit area and are thus much hotter than the Earth. The surface of Venus is sufficiently hot that water vapor cannot condense. Venus has a very thick atmosphere composed mostly of carbon dioxide. On Mars, which is farther from the Sun than the Earth and therefore colder, water freezes, leaving a very thin atmosphere of carbon dioxide. Thus the basic climatic conditions of a relatively circular orbit at a favorable distance for the Sun led to a drastically different evolutionary history for the Earth than for the neighboring planets. *See* SOLAR SYSTEM.

History of Earth's climate. Direct measurements allow the estimation of global mean surface temperature of the Earth for only about the last 100 years (**Fig. 8**). Global surface air temperature rose by about 0.9°F (0.5°C) between 1880 and 1940 and then declined slightly in the following three decades. The decade of the 1980s contained some of the warmest years on record.

Evidence of climate variations in prehistoric times must be obtained from proxy indicators in dated sediments, such as pollen spores, the shells of small animals, or isotopic abundances, or from geologic features such as terminal moraines of glaciers or dry lake beds. A wealth of geological evidence exists to indicate that the Earth underwent a great glaciation as recently as 20,000 years ago. During this period, ice sheets nearly 2 mi (3 km) thick covered parts of North America and western Europe. Variations in the relative abundance of oxygen isotopes in deep-sea cores indicate that a succession of these major glaciations separated by relatively warm, ice-free periods called interglacials has occurred during the last million years of Earth history (**Fig. 9**). *See* GLACIAL EPOCH; PALEOCLIMATOLOGY.

Analyses of time series show a relationship between global ice volume and known variations in the Earth's orbit. The parameters of the Earth's orbit include the eccentricity, which measures the departure

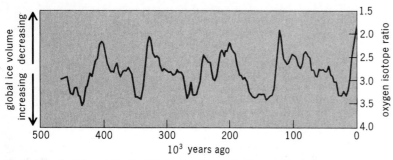

Fig. 9. Climate of the past half-million years. Oxygen isotope ratios are from ocean sediments, which reflect variations in the global volume of ice in glaciers and ice sheets. (*After J. Imbrie and K. P. Imbrie, Ice Ages: Solving the Mystery, Enslow Publishers, 1979*)

from a circular orbit to a more elliptical one; the obliquity, which measures the magnitude of the annual variation in the declination angle of the Sun; and the longitude of perihelion, which measures the season at which the Earth makes its closest approach to the Sun. These parameters control the seasonal and latitudinal distribution of insolation. Because ice sheets form primarily on land and much of the land area is in high northern latitudes, ice sheets form more readily when the summertime insolation is relatively low, which allows some of the winter snow accumulation to survive the summer season. The global climate responds to the presence of the ice sheets through a process called ice-albedo feedback. Ice sheets are more reflective to solar radiation than other surfaces, so that their presence tends to reduce the amount of solar heat absorbed by the Earth. This in turn leads to a cooling of the planet and a greater tendency for the ice sheets to grow. SEE EARTH ROTATION AND ORBITAL MOTION.

Evidence from air bubbles trapped in glacial ice indicates that there were lower atmospheric concentrations of greenhouse gases such as carbon dioxide and methane during past glacial ages than there is at present. The concentration of carbon dioxide 20,000 years ago during the last glacial age was only about 190 parts per million by volume (ppmv), compared with the preindustrial level of 280 ppmv and the current concentration of 350 ppmv. During past glacial periods, the rather low values of carbon dioxide are thought to have been produced by changes in ocean chemistry and biology. Such low greenhouse gas concentrations contribute significantly to the cooling associated with glacial advances.

Future climate changes. Past relationships between the Earth's orbital parameters and global ice volume indicate that the Earth will undergo another major glacial age about 23,000 years from now. Currently, however, the composition of the Earth's atmosphere is changing rapidly as a result of human influences. Of climatic interest are industrial gases that are transparent to solar radiation entering the Earth's atmosphere, but are opaque to the thermal radiation emitted by the Earth. Gases with these characteristics are called greenhouse gases. In order to achieve an energy balance in the presence of increased concentration of greenhouse gases, the Earth's surface must warm, so that its thermal emission will increase to offset increasing downward emission of thermal energy from an atmosphere that is becoming more effective at intercepting outgoing thermal radiation, if other factors remain the same.

The principal greenhouse gases are carbon dioxide, which is increasing principally because of the burning of coal and oil; methane or natural gas, whose increase is related to agriculture and coal and gas mining; nitrous oxide, which is a product of the decay of fixed nitrogen in plants or artificial fertilizer; and chlorofluorocarbons, which are gases used in industry for refrigeration, foam-blowing, and cleaning, and as aerosol propellants. Carbon dioxide is expected to contribute about half of the increase in atmospheric thermal opacity due by the year 2030.

Within the next 100 years, if current trends continue, climate models indicate that human activities will result in a climate that is 4–8°F (2.2–4.4°C) warmer than it is today. The magnitude of this increase is similar to that between the present climate and the last glacial age, and would represent an extremely rapid warming by the standards of natural climate variability and change. Studies also suggest increased probability of drought during the summer growing season in midlatitude agricultural areas and rising sea levels that may be 27 in. (70 cm) higher than present levels by the year 2080. SEE CLIMATE MODIFICATION; CLIMATIC CHANGE; DROUGHT; MESOMETEOROLOGY; METEOROLOGY; MICROMETEOROLOGY.

Dennis L. Hartmann

Bibliography. B. Bolin, B. R. Doos, and J. Jager, *The Greenhouse Effect, Climatic Change and Ecosystems*, 1986; J. F. Griffiths and D. M. Driscoll, *Survey of Climatology*, 1982; J. Imbrie and K. P. Imbrie, *Ice Ages: Solving the Mystery*, 1979; J. G. Lockwood, *Causes of Climate*, 1979; G. T. Trewartha and L. Horn, *An Introduction to Climate*, 1980; W. M. Washington and C. L. Parkinson, *An Introduction to Three-Dimensional Climate Modeling*, 1986.

Clinical microbiology

The adaptation of microbiological techniques to the study of the etiological agents of infectious disease. Some of the most significant scientific advances during the latter half of the nineteenth century and the first few decades of the twentieth century were in medical bacteriology. From the work of Louis Pasteur, Robert Koch, Theobald Smith, and others, it was shown that bacteria were responsible for many diseases of humans. During the early part of the twentieth century, the technology for the growth of bacteria was advanced to the extent that pure cultures of bacteria could be studied from defined environments. Not only did bacteria become models for the study of their own biochemical characteristics, but the same technology enabled bacteriologists to isolate and identify bacteria from patients with infectious disease. This spurred the discovery of most of the bacterial diseases that are currently known.

The role of fungi, protozoa, rickettsias, and viruses in human disease was also elucidated during the first half of the twentieth century. The inhibitory power of materials from a fungus called *Penicillium* was discovered in 1928 by Alexander Fleming. His discovery of penicillin and the contributions of many others marked a great era in the treatment of bacterial diseases. The discovery of penicillin was soon followed by that of streptomycin and tetracycline. For the first time, bacteriologists could detect and identify bacteria and then test in the laboratory the ability of various antibiotics to kill these isolated bacteria. Such was the birth of contemporary clinical microbiology. *See Antibiotic*.

Clinical microbiologists determine the nature of infectious disease and test the ability of various antibiotics to inhibit or kill the isolated microorganisms. A contemporary clinical microbiologist is also responsible for a wide range of microscopic and cultural studies in mycology, parasitology, and virology. The consultative skill of the clinical microbiologist is sought by many clinicians. The clinical microbiologist is often the most competent person available to determine the nature and extent of hospital-acquired infections, as well as public-health problems that affect both the hospital and the community. The clinical microbiologist is a valuable resource to the hospital and community. As clinical microbiology encompasses all aspects of infectious disease, the various specialties will be discussed individually. *See Epidemiology; Hospital infections*.

Bacteriology. Historically, the diagnosis of bacterial disease has been the primary job of clinical microbiology laboratories. Many of the common ailments of humans are bacterial in nature, such as streptococcal sore throat, diphtheria, and pneumococcal pneumonia. Only rarely, however, in recent decades have new bacterial diseases been described. A prime example is the disease legionellosis caused by a hitherto unknown bacterium, *Legionella pneumophila*. *See Legionnaire's disease*.

The bacteriology laboratory plays a major role in the diagnosis of bacterial diseases. It accepts specimens of body fluids, such as sputum, urine, blood, and respiratory or genital secretions, and inoculates the specimens onto various solid and liquid growth media. Following incubation at body temperature, the microbiologist examines these agar plates and tubes and makes a determination as to the relative numbers of organisms growing from the specimen and their importance in the disease process. The microbiologist then identifies these alleged causes of disease and determines their pattern of antibiotic susceptibility to a few chosen agents.

Clinical microbiologists also microscopically examine these body fluids. They report on the presence of bacteria in body fluids and the cellular response to infection, such as the numbers of white blood cells observed in the specimen. Sometimes the kinds of white blood cells observed can provide hints for the identity of the disease. Such microscopic evaluations can provide a rapid aid in the diagnosis of infectious disease, as well as an assessment of the quality of the specimen. For example, a specimen called "pus" which does not contain white blood cells may not be pus. *See Medical bacteriology*.

Mycology. Mycology is the study of fungi. The applications of mycology in the clinical microbiology laboratory include the isolation of fungi which are known to cause disease. In general, diseases caused by fungi are relatively harmless but annoying, such as ringworm and athlete's foot. However, they can be severe, systemic disorders such as blastomycosis, coccidioidomycosis, and histoplasmosis. Fungi can also cause disease in those people whose immunologic processes have been suppressed by drugs or radiation. The diagnosis of fungal infection differs from bacterial infection because of the relatively slower rate of fungal growth. In general, diagnosis of fungal disease is a function of not only isolating and identifying the causative organism and determining which of three or four antifungal antibiotics might be effective, but also determining the patient's antibody response to the disease. *See Medical mycology*.

Parasitology. Parasitic diseases are widespread and of worldwide importance to humans. Such diseases include amebic dysentery, malaria, and Chagas' disease. The clinical microbiologist working in parasitology examines specimens of feces and blood for the presence of parasitic agents. These examinations are primarily microscopic, and the agents are identified on the basis of their characteristic structure. Although antiparasitic agents are available for the treatment of such infection, the parasitologist is usually not involved in the determination of the relative effectiveness of these agents on parasites. *See Medical parasitology*.

Virology. Virology is a specialized part of clinical microbiology. Because of the nature of viruses and their proclivity for growth only in living cell cultures or in eggs, few clinical microbiology laboratories have facilities for the isolation and identification of viruses. Most of these procedures are best done in large hospitals or in state departments of health. There is a growing trend, however, for smaller hospitals to develop screening virology laboratories in which specimens are brought to the laboratory, placed in tissue culture, and preliminary identification steps taken in order to provide a rapid diagnosis of viral disease. Although only a few drugs exist that can be used to treat virus infections, the early recognition of viral disease may allow the clinician to discontinue the use of powerful antibiotics that are effective only for bacterial diseases. Common virus diseases include influenza, herpes, chicken pox, the common cold, and many nonspecific respiratory infections. Much of the viral diagnosis that is done in the laboratory is by the determination of a specific antibody response to viral disease. The routine clinical microbiology laboratory usually does not become involved in such procedures. A special procedure for the detection of viruses and other organisms that affect infants is called the TORCH screen. It is a procedure for the detection of antibodies to the parasitic disease toxoplasmosis and the viral diseases rubella, cytomegalovirus, and herpes. These diseases may cause devastating health problems in babies. *See Animal virus; Disease; Tissue culture; Virus*.

Richard C. Tilton

Technology. Advances in instrumentation and rapid methods in diagnostic microbiology began in the 1970s. The fact that clinical microbiology laboratories have been the last of all the diagnostic laboratories to make use of technological advances can in part be attributed to the nature of the specimens and microorganisms themselves. Unlike specimens submitted for chemical and hematological examination in which a definite substance is to be detected, the living organisms that the microbiologist must analyze take time to grow, and in many instances the microorganism causing infection must be differentiated from a number of other nonpathogenic organisms. *See Clinical pathology*.

Instrumentation. Methods of bacterial identification and susceptibility testing were revolutionized in the 1980s. The advent of microsystems with associated databases for the identification of microorganisms led to the development of automated systems for both the identification and antimicrobial-drug susceptibility testing of microbial organisms. Systems using semiautomated as well as fully automated instrumentation are able to provide results within 3–5 h after initial isolation of the organism. Each of the systems utilizes some type of disposable cassette containing wells with a number of different growth media, biochemicals, or antibiotics. Once the specimen is dispensed into the wells of the machine, photometry is used to recognize microbial growth. A computer sorts the data and reports on identification or susceptibility, either in printed form or on a display terminal.

Although these instruments provide significant savings in labor, as well as clinical data much sooner than would otherwise be possible, the major drawback is still that the microbiological agent responsible for the disease in question must first be isolated from the specimen, a process that may take up to 24–48 h, and with some organisms up to 2 months. Methodologies of molecular genetics and monoclonal antibodies have been applied to the field of diagnostic microbiology, and indeed may revolutionize diagnostic microbiology to the point where it may no longer be necessary to culture, isolate, and identify etiologic agents.

Monoclonal antibodies. Hybridoma technology has had a major impact on the diagnosis of infectious diseases. One of the advantages of monoclonal antibodies is that they can be incorporated into other diagnostic technologies such as enzyme immunoassays, latex agglutination, and fluorescent immunoassays. Presently, a number of immunologic assays use monoclonal antibodies that allow for the direct detection of antigens in clinical specimens in a matter of minutes to a few hours. Diagnostic test kits for some species of bacteria, protozoa, and viruses are already commercially available, and many more potential applications of monoclonal antibodies are being considered.

The use of this technology however, is not without its limitations, such as the narrow range of reactivity. For this and other reasons the overall impact on diagnostics has been less than originally anticipated, especially for direct diagnosis of infectious agents from patient specimens. *SEE IMMUNOASSAY; MONOCLONAL ANTIBODIES.*

DNA probe technology. Deoxyribonucleic acid (DNA) probes are expected to have a major impact on clinical microbiology laboratories. All organisms contain some unique sequence of DNA or ribonucleic acid (RNA) within their genome which distinguishes them from all other organisms. The key to developing a nucleic acid probe is to isolate these sequences, reproduce them in a cloning vector, and attach a labeling molecule to them so that detection can be made during the hybridization process.

Clinical microbiologists believe that DNA probe technology can be very useful in the identification of fastidious microorganisms, such as the mycobateria, at a faster rate and at less cost than biochemical diagnostics. However, a second area of interest, and perhaps the most important, is the use of diagnostic probes for the direct detection of infectious agents in clinical specimens.

There are a number of commercially produced probe kits, including assays, to detect bacteria, protozoa, and viruses, many of which are for research use only. Probes to detect antimicrobial-resistant genes, fungi, and several protozoan and viral pathogens have been described.

Richard Fister

Bibliography. B. J. Howard et al. (eds.), *Clinical and Pathogenic Microbiology,* 1987; H. D. Isenberg, The future of clinical microbiology, *Lab. Med.,* 19:321–323, 1988; E. H. Lennette et al., *Manual of Clinical Microbiology,* 4th ed., 1985; W. J. Payne, Jr., et al., Clinical laboratory application of monoclonal antibodies, *Clin. Microbiol. Rev.,* 1:313–329, 1988; F. C. Tenoven, Deoxyribonucleic acid probes for infectious diseases, *Clin. Microbiol. Rev.,* 1:82–101, 1988; W. A. Volk et al., *Essentials of Medical Microbiology,* 3d ed., 1986.

Clinical pathology

The branch of general pathology (broadly defined as the study of the causes or essential nature of disease) in which disease states are diagnosed and monitored by examining blood, body fluids, and secretions for specific chemical, cytological, microbiological, and immunological abnormalities. In contrast, in anatomical pathology, diagnoses are established by studying the morphology, or structure, of cells, tissues, and organs.

Clinical pathology tests are commonly performed in clinical laboratories within hospitals and ambulatory care facilities. Some of the more routine and simpler laboratory tests are done in physicians' offices. Reference laboratories are available in most large cities to perform tests that are rarely done or are too complex to perform in hospitals and clinics.

Typically, a clinical pathology laboratory is directed by a medical doctor, usually a pathologist. The chief functions of the laboratory director are to oversee the operation of the laboratory and to aid physicians and other medical personnel in choosing appropriate diagnostic tests, interpreting laboratory data, and monitoring therapeutic responses. The day-to-day surveillance of test procedures, the review of the accuracy and precision of test results, and the supervision of the laboratory personnel are commonly performed by clinical scientists or by certified medical technologists. Medical technologists also perform most of the tests in the laboratory. Medical laboratory assistants and clerical personnel assist them in activities such as specimen preparation; quality control; transcription, recording, and filing of test results; and computer operations.

Laboratory operations and function. Clinical pathology laboratories function as an integral part of what is known as the diagnostic cycle. The cycle begins when an individual with signs and symptoms suggestive of disease consults a physician. After obtaining a clinical history and performing a physical examination, the physician may order various laboratory studies on blood, urine, and other body fluids and secretions to establish or confirm a working diagnosis. For many laboratory tests to be accurate, the patient must be properly prepared; for example, before certain specimens are collected, the patient may need to follow a special diet, fast overnight, or avoid ingesting certain chemicals and drugs.

Once the laboratory receives a specimen, the name and location of the patient, the date, the name of the ordering physician, and other information are entered into a logbook or computer record system. The specimen may require specific preparation, such as separating blood cells from serum, making fluid mounts or stained smears for microscopic examination, or mixing the sample with preservatives or with chemicals that produce specific effects. The properly prepared samples are then analyzed; for example, they may be examined visually for color changes or other physical alterations after specific reagents are added, or they may be observed through the microscope to detect elements such as abnormal cells, chemical crystals, or microorganisms. SEE MEDICAL INFORMATION SYSTEMS.

Fluid samples often are analyzed by semiautomated or automated instruments having optical and electronic components that can detect specific constituents. The electronic signals are transferred to recording devices or into computers that can calculate results and print out reports. Blood counts, urine examinations, and many chemical tests can be efficiently performed by such equipment. **Figure 1** illustrates a module by which serum samples are aspirated for analysis in an automated instrument. Small aliquots of the unknown serum samples to be analyzed are contained in plastic capsules that are placed in the slots at the periphery of the wheel. The wheel automatically advances so that each capsule in turn is positioned beneath an aspirator device, where a small amount of sample is mixed with reagents in the instrument and delivered to an electronic reading chamber.

In highly automated laboratories, test results may be transferred directly from the instrument into the hospital-wide computer system, so that clinicians can obtain results from computer terminals present on the wards. The availability of such direct-read instruments and of special chemical and immunologic reagent kits by which specimens can be directly examined permits test results to be obtained within a few minutes rather than hours or days. For example, most cases of bacterial meningitis now can be diagnosed within a few minutes by detecting bacterial antigens in the cerebrospinal fluid; thus, specific therapy can be instituted immediately rather than after the hours or even days required for the causative microorganisms to grow in culture.

Clinical laboratory organization. The clinical pathology laboratory is commonly divided into four major subdisciplines or sections, which in turn may be divided into subsections, as follows:

    Chemical pathology
        General chemistry
        Special chemistry
        Toxicology
        Radioimmunoassay and enzyme-linked
            immunoassay
        Electrophoresis
    Hematology
        Morphologic hematology
        Chemical hematology
        Coagulation
    Blood bank (immunohematology)
        Transfusion service
        Blood donor recruitment
        Atypical antibody detection
        Blood component separation
        Plasmapheresis
    Microbiology
        General aerobic bacteriology
        Anaerobic bacteriology
        Mycobacteriology
        Virology

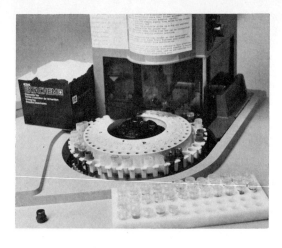

Fig. 1. Sample carousel of an automated chemical analyzer.

        Parasitology
        Mycology
        Immunology
        Antimicrobial susceptibility testing

In addition, clinical laboratories perform urinalysis. The four major areas are briefly described below.

Chemical pathology. In chemical pathology, the chemical constituents of serum, urine, and other body fluids are analyzed. For example, levels of blood glucose may be measured to establish the diagnosis of diabetes, and levels of serum cholesterol may be determined to help assess risk of developing cardiovascular disease. Chemical tests are often clustered into functional groups to help diagnose diseases of the heart, kidney, liver, gastrointestinal tract, and other organ systems. Determining concentrations of electrolytes (sodium, potassium, chloride, and bicarbonate) in the serum can be important when, for instance, patients are receiving fluids intravenously or are taking diuretic drugs that can cause potassium loss. SEE SEROLOGY.

In the toxicology subsection of chemical pathology laboratories, the blood, urine, and other body fluids are analyzed for the presence of drugs and substances of abuse. An equally important application of toxicology testing is to measure the blood levels of therapeutic drugs to assure that concentrations are adequate to treat the disease, but not so high as to be toxic. For example, it can be important to measure levels of the antibiotic gentamicin because the concentrations needed for efficacy can approach those resulting in toxic effects such as hearing loss. An example of the sophisticated equipment required to make these measurements is shown in **Fig. 2.** SEE TOXICOLOGY; URINALYSIS.

Radioimmunoassay is a technique that allows trace quantities of substances to be measured. It involves adding radioactive chemical markers to properly prepared samples and detecting their binding to the substances being measured. Radioactive tritium and iodine (^{125}I) are the radioisotopes commonly used in these assays. Because scintillation counters and gamma radiation counters can detect minute quantities of radioactive material, the interaction of only a few radio-labeled molecules with complementary chemical groups in the unknown sample can be determined. Thus, for example, radioimmunoassay permits levels of thyroid hormone that are present in the blood in minute quantities to be measured. It even permits quantitation of testosterone, which is present in even

Fig. 2. Multiple complex mechanical and electronic modules of a high-pressure liquid chromatography instrument used to measure trace amounts of drugs and other chemical constituents of blood samples.

smaller quantities, measured in nanograms per deciliter (1 nanogram = 10^{-9} gram). *SEE RADIOIMMUNOASSAY*.

Enzyme-linked immunosorbent assay (ELISA) techniques, in which specific enzymes instead of radioisotopes are linked to protein markers, are as sensitive as radioimmunoassay tests. Thus, they can also detect serum substances in trace amounts. ELISA techniques have several advantages over radioimmunoassay procedures. Expensive radiation counters need not be purchased, personnel need not have licenses to use radioactive materials, and there are no radioactive wastes to dispose of. Therefore, ELISA tests have been replacing radioimmunoassay procedures. *SEE IMMUNOASSAY*.

Electrophoresis is a technique used in most chemical pathology laboratories to separate and help identify proteins and lipoproteins. In this technique, a small amount of serum (or concentrated urine or other body fluid) is applied to a filter paper strip, a synthetic gel, or another solid surface containing an electrolyte buffer at specified pH. The strip or gel is then placed in an electric field. Because different proteins and lipoproteins in the sample possess different net surface electric charges, they travel different distances in the electric field and thus are separated. The separated protein species can be stained and quantitated. Applications of electrophoresis include identifying hypogammaglobulinemia (decreased quantities of serum globulins), which is associated with an increased likelihood of developing recurrent infections, helping to diagnose multiple myeloma (a malignancy in which large amounts of a single protein may be produced), and detecting alterations in serum lipid fractions related to heart disease. *SEE ELECTROPHORESIS; IMMUNOELECTROPHORESIS*.

Hematology. The cornerstone of clinical laboratory work in hematology is the assessment of the cellular elements (red blood cells, white blood cells, and platelets) in blood samples. The blood cells may be enumerated, either by manual cell-counting techniques or by automated particle-sensing instruments. In addition, microscopic examinations may be performed on blood smears. These smears are prepared by placing a drop of blood (drawn either from a vein or by finger stick) on a glass slide and spreading it over the surface to form a thin film. This film is then stained with specific dyes and examined under a microscope to determine the types of cells present and their proportions, a procedure called the differential count. *SEE BLOOD; MICROTECHNIQUE*.

Blood smears are particularly useful for detecting cells that appear malignant or are otherwise abnormal. Also, the presence of over 10,000 white cells known as polymorphonuclear leukocytes per cubic millimeter of blood, along with an increased number of young forms (called STAB or juvenile cells), suggests that a patient has an infection. Patients with leukemia often have white blood cell counts over 100,000 mm^3 of blood, and their blood often contains primitive cells called blasts. Laboratory studies also permit anemia to be detected and characterized. Patients with anemia have less than the normal number of red blood cells or a decreased amount of hemoglobin in the blood. A skilled hematologist may be able to judge the type of anemia by observing the size, shape, and hemoglobin content of the red cells in a stained blood smear. Automated cell counters are commonly used to provide a more exact assessment. *SEE ANEMIA; HEMATOLOGIC DISORDERS; LEUKEMIA*.

In complicated cases, where a diagnosis cannot be made by studying the peripheral blood, examining a sample of bone marrow may be necessary. The bone marrow is the "factory" where the peripheral blood cells are made, and thus it can give valuable clues as to why too few or too many cells have been produced. Bone marrow for examination purposes can be obtained by aspiration, or a bone marrow biopsy can be obtained with a large-bore needle. The aspirated cellular material is smeared over a microscope slide by using the technique employed for preparing a blood smear. Bone marrow biopsy specimens are first

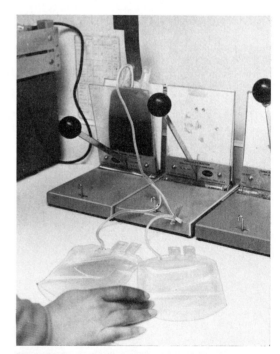

Fig. 3. Mechanical device used in blood banks to separate plasma and cell-rich components from a unit of donor blood.

fixed in formalin to preserve the cells and then placed into molten paraffin. After the paraffin has hardened, thin microscopic sections are cut with a very sharp knife called a microtome. These thin slices are placed on the surface of a microscope slide, stained with special dyes, and examined microscopically for abnormal cells. Special procedures are available by which specific markers on the surfaces of blood cells can be detected by using specific reactants. Thus, leukemias and other blood disorders can be properly categorized. SEE BONE; HEMATOPOIESIS.

Coagulation is an important subdiscipline of hematology involving the study of clotting and bleeding disorders. Women taking oral contraceptives are at increased risk of developing thrombophlebitis. People especially subject to blood clots often receive anticoagulant drugs in order to ''thin'' the blood. The amount of anticoagulant given must be regulated, based on the results of a clotting test known as the prothrombin time, so that the blood is not made too thin. SEE HEMATOLOGIC DISORDERS.

Coagulation tests are also needed to diagnose and manage hemophilia. Blood samples must be obtained and analyzed to determine why the plasma cannot clot. When the nature of the coagulation factor defect is known, it may be possible to provide specific replacement substances. SEE HEMOPHILIA.

The evaluation of platelets in the peripheral blood is also important. Patients with platelet counts under $20,000/\text{mm}^3$ may be in danger of spontaneous bleeding from mucous membranes and into the skin. Platelet function tests are also available to study cases where the total number of platelets is normal or even increased but the patient still bleeds.

Blood bank (immunohematology). The primary function of the blood bank is to ensure that blood taken from a donor will be safe to give to a recipient (that is, it will not produce a transfusion reaction or transmit infectious disease). The surfaces of red blood cells possess chemical markers known as blood group antigens that may be sensed as foreign by another person's immune system. If red blood cells from a donor of one blood type are transfused into a recipient of a different type, antibodies in the plasma may destroy the red cells, and a serious hemolytic reaction may occur. The most important blood groups in this regard are the ABO blood group system (groups O, A, B, and AB) and the Rh system (Rh-positive and Rh-negative). Mismatches of major blood groups can be fatal; mismatches of minor blood groups, white blood cells, platelets, or plasma constituents usually cause less severe reactions.

The chance that a transfused recipient will incur a hemolytic transfusion reaction is decreased to virtually nothing by performing compatibility tests. The most important measure is to give recipients blood of their own ABO and Rh types. To this end, a crossmatch is routinely performed, where red blood cells from the donor are thoroughly mixed on a glass slide or in a test tube with plasma from the recipient. The mixture is carefully examined both visually and microscopically. Any clumping of the cells indicates that the recipient's plasma may destroy donor red blood cells. Blood from the unit showing such clumping cannot be used for that recipient. Also commonly performed is an antibody screen, a test to detect antibodies in the recipient's blood. The patient's plasma is tested against reagent red cells that contain most of the important blood group antigens. If an antibody is found, only red cells lacking the corresponding antigen will be used to transfuse that patient. SEE ANTIBODY; ANTIGEN; BLOOD GROUPS.

Considerable efforts are also made in blood banks to reduce to an absolute minimum the chance for transmission of infectious diseases. The discovery of the Australian antigen has provided a marker by which donor blood can be screened for the presence of the hepatitis B virus, thereby markedly reducing the incidence of transfusion-related hepatitis. Also, remarkable advances in improving the sensitivity and specificity of tests to detect antibodies to the human immunodeficiency virus (HIV) have reduced the chance of transmitting acquired immune deficiency syndrome (AIDS) through blood transfusions to an absolute minimum. The use of volunteer (rather than paid) donors and the careful screening of those donors also have been instrumental in preventing transfusion-related infectious diseases of all types. SEE ACQUIRED IMMUNE DEFICIENCY SYNDROME (AIDS); HEPATITIS.

Transfusion therapy is made more efficient by using blood components rather than whole blood. Blood can be separated into red blood cells, platelets, and fresh frozen plasma. **Figure 3** illustrates a mechanical device by which plasma is separated from the cellular components of blood. After the red cells have been allowed to settle, the blood bag is placed in the separator device, which expresses the plasma through an open port at the top of the bag and into tubing leading to the satellite collection bags.

The cellular components obtained can then be used individually to transfuse patients with specific diseases. For example, red cell transfusions can be given to patients who are severely anemic, and platelet transfusions to patients who are in danger of bleeding because they have insufficient platelets in their blood. Giving only the exact components needed to treat the specific disease process targets therapy specifically to the individual's problem. In addition, red blood cells, platelets, and plasma must be stored at different temperatures (39.2°F or 4°C, 71.6°F or 22°C, −0.4°F or −18°C, respectively) in order to preserve their activity as long as possible. This can be done only if these components are separated.

Donor recruitment programs to provide an adequate supply of blood are maintained in many blood banks. Other services often provided include outpatient therapeutic transfusions; phlebotomy of individuals (withdrawal of blood) who have too many red blood cells (a condition known as polycythemia); and plasmapheresis, a procedure where plasma is separated from the withdrawn blood and the formed elements (red cells and platelets) are immediately retransfused into the donor. Through plasmapheresis, multiple units of certain blood components can be obtained from a willing donor without the danger of inducing anemia or deficiencies of other blood elements. Apheresis instruments are also available that allow blood components such as leukocytes and platelets to be selectively removed from the circulation of a donor without depleting the other elements. Advances in blood banking have made it possible for patients with various diseases to pursue relatively normal lives.

Microbiology. The practice of clinical microbiology is highly complex because there are hundreds of species of medically important microorganisms. They vary in size from the smallest viral particles, which can be seen only by electron microscopy, to intestinal tapeworms that can exceed 40 ft (12 m) in length. Each species of microorganism has an optimal temperature and environment in which it will grow, and some species need specific nutritional factors. Special incubators and a variety of culture media are used in microbiology laboratories to meet these requirements, so that the microorganisms obtained from clinical specimens will grow and can be recovered. *SEE CLINICAL MICROBIOLOGY; CULTURE; MEDICAL BACTERIOLOGY.*

As listed above, the microbiology laboratory has several subsections. Approaches used to recover and identify microorganisms differ within each of these sections, depending on the structure, physiology, and growth characteristics of organisms. For example, most bacteria and fungi grow under aerobic conditions (that is, in the presence of atmospheric oxygen) and can be readily recovered from clinical materials by placing a portion of the sample into an appropriate culture medium and waiting for growth to appear after incubation under specified conditions. Certain species of bacteria (the anaerobes), however, cannot survive in the presence of oxygen. Their cultivation requires the use of transport devices, culture media, and environmental chambers or tents from which oxygen has been excluded. *SEE BACTERIAL PHYSIOLOGY AND METABOLISM.*

Special approaches must also be used to recover and identify viruses. Viruses can live only in viable cells and, for the most part, can survive only briefly outside human or animal hosts. Culture techniques must thus use viable cells; embryonated eggs, cell culture suspensions, thin cell sheets called monolayers, or laboratory animals are usually inoculated with portions of the specimen to be cultured. Species of viruses are identified by observing their ability to produce certain cytopathic effects in the cells where they are growing or to cause recognizable diseases in laboratory animals. The virology laboratory is usually isolated from other sections of the microbiology laboratory to prevent cell cultures from being contaminated by bacteria, fungi, and other microorganisms. *SEE ANIMAL VIRUS.*

Laboratory identification of parasites involves detecting their different forms through direct microscopic examination of body fluids and secretions. Since the parasites generally occur in low densities, attempts are usually made to concentrate the sample to make microscopic examination more fruitful. For example, to examine stool specimens, a small portion of feces is first mixed with an extractant (a mixture of formalin and ethyl acetate) to preserve the parasites and clear out background debris. The mixture is then centrifuged at relatively high speed. The supernatant (overlying liquid) is discarded, and the sediment is examined for the presence of ova and other parasitic forms. Some adult forms, such as the roundworm ascaris and the tapeworm taenia, are visible to the unaided eye and often can be identified by recognizing certain morphologic features. *SEE MEDICAL PARASITOLOGY.*

The primary task of the clinical microbiologist is to identify the organism as soon as possible so that the physician can begin specific treatment. In bacteriology laboratories, this is traditionally done by first isolating on a primary culture plate the pathogenic bacterium from the clinical specimen. Suspicious colonies are first inspected. Then Gram-stained smears are prepared and are viewed microscopically;

the color (blue-staining bacteria are called gram-positive, and red-staining bacteria gram-negative), shape, and arrangement of bacterial cells are assessed. If this does not lead to identification of the species, colonies can be transferred to various test media by which the metabolic and biochemical characteristics of the bacterium can be detected. From the pattern of reactions, a definitive identification can usually be made.

If the organism is considered clinically significant, a susceptibility test can be performed to determine which antibiotics can potentially be used. The disk diffusion (Bauer-Kirby) test is done in most laboratories. This test is performed by covering the entire surface of a special agar medium (Mueller-Hinton agar) with a standardized suspension of organisms obtained from the culture and placing antibiotic-impregnated filter paper disks on it. After the plate is incubated at 95°F (35°C) for 18 h, zones of growth inhibition are visible around the disks of those antibiotics to which the organism is susceptible. The laboratory may also perform broth dilution susceptibility tests, in which the level of antibiotic necessary to inhibit bacterial growth (known as the minimal inhibitory concentration, or MIC) can be determined, thereby giving the physician an indication of the dosage required. *SEE ANTIBIOTIC.*

The approaches used to identify fungi are similar to those used to identify bacteria. Parasites, however, are usually recognized by observing morphologic features of the eggs, larvae, and adult forms in direct mounts or stained smears of various body fluids. Viruses can be detected in smears containing exfoliated cells by looking for specific intranuclear or intracytoplasmic inclusions. Final identifications are usually made by culturing the virus in special cell culture preparations and looking microscopically for specific cytopathic effects.

The availability of monoclonal antibodies and nucleic acid probes, in which a highly specific portion of antigen or a small segment of nucleic acid can be tagged with either a fluorescent or an enzyme-linked detector, has revolutionized the ability to detect specific microbes in biologic specimens and confirm the results of culture. A fluorescent-tagged probe is like a piece in a jigsaw puzzle to which is attached an electric light. As soon as the piece finds its place in the puzzle [the probe finds its complementary sequence in the deoxyribonucleic acid (DNA) molecule], the light can be detected. A probe can be designed to recognize a distinctive marker (an epitope) on the cell surface of the microbe. If the epitope is present, the probe sticks and can be seen under a fluorescent microscope. By using these techniques, exact identifications can often be made within minutes, and the physician need not wait hours or days for a diagnosis while the microorganisms are growing in culture. *SEE FLUORESCENCE MICROSCOPE; MONOCLONAL ANTIBODIES.*

Clinical immunology is that discipline in which infectious diseases are diagnosed by detecting antibodies in serum and other body fluids. Several highly specific techniques (for example, latex agglutination, hemagglutination inhibition, complement fixation, and immunoprecipitation) are used. In practice, immunologic and serologic techniques are used to diagnose an infectious disease when the agent may be difficult to recover in culture. A rise in antibody titer that is higher in a blood sample drawn in the recovery phase of an illness than in a sample drawn at the onset of symptoms (these two samples are usually spaced about 3 weeks apart) is sought. A fourfold increase in titer is usually considered diagnostic of the disease caused by the agent to which the antibiodies have been directed. Techniques using monoclonal antibodies and DNA probes detect antigens more rapidly and specifically than the several conventional test procedures listed above. SEE IMMUNOCHEMISTRY; IMMUNOFLUORESCENCE.

Quality control. A highly sophisticated system of quality control has been implemented in most clinical laboratories to assure the accuracy and precision of the various tests being performed. A quality control sample is run in parallel with the samples being tested each time a test analysis is performed. The results of the quality control sample must fall within a given range of variability for the run to be considered accurate and acceptable. If the quality control result is more than 2 standard deviations outside the range of established values (an event that occurs only 1 in 20 times by chance), the test results cannot be accepted and the run must be repeated, perhaps using a different set of reagents.

In each laboratory, critical and panic values for test results are established, that when exceeded may represent a life-threatening condition that must be brought to the immediate attention of the physician. High-quality laboratories participate in external quality control surveillance programs in which samples provided by an external agency are analyzed and the results compared with those of other participating laboratories. If any laboratory is not getting the correct answers, the supervisor must issue a report indicating what actions have been taken to correct the situation. These programs have contributed greatly to assuring the quality of laboratory testing.

Elmer Koneman

Bibliography. E. W. Bermes, Jr. (ed.), *The Clinical Laboratory in the New Era,* 1985; P. M. Fischer et al., *The Office Laboratory,* 1983; J. A. Halsted and C. H. Halsted, *The Laboratory in Clinical Medicine: Interpretation and Application,* 1981; J. B. Henry (ed.), *Clinical Diagnosis and Management by Laboratory Methods,* 16th ed., 1984; G. D. Lundberg (ed.), *Using the Clinical Laboratory in Medical Decision Making,* 1983; R. Ravel, *Clinical Laboratory Medicine,* 3d ed., 1978; J. R. Snyder, Jr., and A. L. Larsen, *Administration and Supervision in Laboratory Medicine,* 1989; N. W. Tietz, *Textbook of Clinical Chemistry,* 1986; K. M. Trescher, *Clinical Laboratory Tests: Significance and Implications for Nursing,* 1982; W. O., Umiker, *The Effective Laboratory Supervisor,* 1982.

Clinometer

A hand-held surveying device for measuring vertical angles; also called an Abney level. It consists of a sighting tube surmounted by a graduated vertical arc with an attached level bubble. A 45° mirror inside the tube enables the observer to see the bubble at the same time that the observer sights a point or a graduated rod with a horizontal wire. Manipulation of the vertical arc brings the bubble to center in coincidence with the sighted point; the vertical angle is indicated on the arc (see **illus.**).

The clinometer is used mainly to determine slope angles for reduction of measured slope distance to

A typical clinometer.
(*Keuffel and Esser Co.*)

horizontal distance. If set on a sloping surface (such as a street pavement), the clinometer can be used to give the angle of inclination of the surface. With the arc set at 0°, it can be used as a hand level. SEE SURVEYING.

B. Austin Barry

Clipping circuit

An electronic circuit that prevents transmission of any portion of an electrical signal exceeding a prescribed amplitude. The clipping circuit operates, in effect, by disconnecting the transmission path for the portion of the signal to be clipped.

Semiconductor diode as a clipper. A diode may be used as a switch to effect the clipping action indicated in **Fig. 1.** The diode conducts with a very low resistance when the p side is more positive than the n side by the small forward offset voltage V_o of the diode (approximately 0.7 V for a typical silicon diode), and is essentially nonconducting when the p

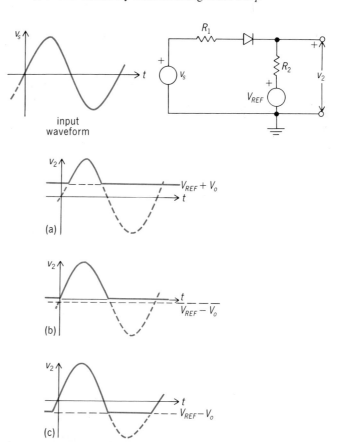

Fig. 1. Series diode clipping with output waveforms for (a) V_{REF} positive; (b) $V_{REF} = (0 - V_o)$; and (c) V_{REF} more negative.

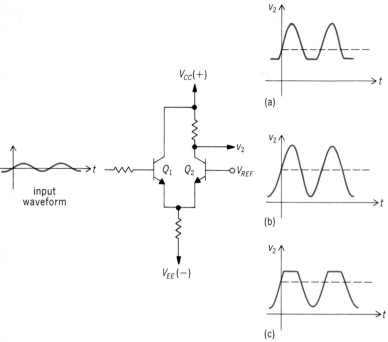

Fig. 2. Differential amplifier as a clipper with output waveforms for (a) V_{REF} positive; (b) V_{REF} = 0; and (c) V_{REF} negative.

input waveform. If V_{REF} is raised, Q_2 conducts more heavily, and the emitter voltage $V_{REF} - V_{EE}$ is raised so that Q_1 becomes nonconducting for part of the most positive portion of the input signal. That portion of the waveform at the output is clipped, and the output is as shown. However, if V_{REF} is lowered, Q_2 is nonconducting during the lower portion of the waveform and that portion is clipped. SEE TRANSISTOR.

Operational amplifier clipper. The circuit of Fig. 2 also represents a crude version of the input stage of a complete operational amplifier which can be used as a clipper, as shown in **Fig. 3.** If the amplifier is overdriven, the output of A_1 will be limited at V_{OH} or V_{OL}, the power supply limits of the operational amplifier. This will, in turn, limit the voltage v_2. Prior to limiting, $v_2 = v_i$. The upper output limit is governed by Eq. (1) and the lower output limit by Eq. (2).

$$V_{2H} = \frac{R_2 V_{REF} + R_1 V_{OH}}{R_1 + R_2} \quad (1)$$

$$V_{2L} = \frac{R_2 V_{REF} + R_1 V_{OL}}{R_1 + R_2} \quad (2)$$

SEE LIMITER CIRCUIT; OPERATIONAL AMPLIFIER.

Glenn M. Glasford

Bibliography. G. Glasford, *Analog Electronic Circuits,* 1986; G. Glasford, *Digital Electronic Circuits,* 1988; J. Millman and A. Grabel, *Microelectronics,* 2d ed., 1987.

side is less than the offset value or negative. Thus for $R_2 \gg R_1$, the diode is conducting for $v_s < (V_{REF} + V_o)$. The output waveforms are shown for values of V_{REF} positive, $V_{REF} = (0 - V_o)$, and V_{REF} more negative. The output levels are slightly attenuated relative to the input signal because of the ratio $R_2/(R_1 + R_2 + r_f)$. Clipping of the opposite polarity can be achieved by reversing the diode connections, and clipping of both positive and negative peaks can be achieved by use of two diodes and a more complicated resistive network. SEE DIODE.

Transistor clipper. Bipolar or field-effect transistors can be used as clipping circuits. For example, the differential amplifier shown in **Fig. 2** may be so designed that for the fixed reference voltage $V_{REF} = 0$, both Q_1 and Q_2 are at the center of their normal operating range, where for very small input signals the output waveform is an amplified reproduction of the

Clock

A device for indicating the passage of time. Most clocks contain a means for producing a regularly recurring action, such as the swing of a pendulum, the oscillation of a spiral spring and balance wheel, the vibration of a tuning fork, the oscillation of a piezoelectric crystal, or the comparison of a high-frequency signal with the radiation from the hyperfine structure of the ground state of atoms. This article describes the refinements of mechanical clocks. SEE ATOMIC CLOCK; QUARTZ CLOCK; WATCH.

Basic movement. The recurring action of a mechanical clock depends on the swing of a pendulum, or the oscillation of a balance wheel and balance

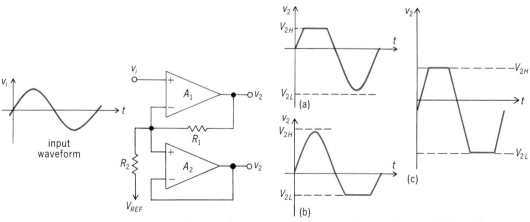

Fig. 3. Operational amplifier as clipper or limiter with output waveform (a) clipped for V_{REF} negative; (b) clipped for V_{REF} positive; and (c) limited at both positive and negative portions for overdriven amplifier because of decreased ratio $R_2/(R_1 + R_2)$.

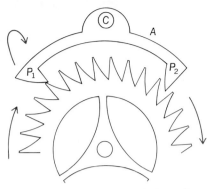

Fig. 1. Anchor escapement. Common in domestic pendulum clocks, this escapement tends to compensate for changes in amplitude of the swing because of irregularly cut gears and varying lubrication.

spring or hairspring, or the vibration of a tuning fork, mechanisms capable of repeating their cyclic movements with great regularity. A counting mechanism, consisting of a gear train with calibrated dial and indicating hands, sometimes with a striking mechanism, marks the number of oscillations that have occurred, although the graduations are in seconds, minutes, and hours. A weight or spring ordinarily supplies power to operate the oscillating and the counting mechanisms. However, temperature changes, accelerations, automatic windings, or electricity may provide the power. (The conventional electric clock is not an independent timepiece but is a repeater whose accuracy depends on a remote primary clock.) Usually an escapement transmits power from the counting mechanism to the oscillating mechanism. The accuracy of a clock depends primarily on the escapement.

Escapements. **Figure 1** depicts an anchor or recoil escapement. Anchor A is connected loosely to the pendulum and swings about C. At some time after midswing, a tooth of the wheel escapes from pallet P_1 or P_2, giving the pallet a push to maintain oscillation. The other pallet then checks another tooth, the curve of the pallet forcing the wheel slightly backward. The reaction helps to reverse the pendulum swing and to correct for circular error.

Figure 2 shows a deadbeat escapement. Faces F are circles about C so that the wheel is stationary ex-

cept near midswing when a thrust on B allows a tooth to escape. **Figure 3** shows a lever escapement in which a balance wheel swings through a much larger angle than does a pendulum. Balance wheel and escapement are connected only while impulse pallet I is in a notch of lever L. As I enters a notch, it pushes the lever and then receives an impulse from it. Disk D and safety pallet S prevent the lever from moving accidentally before I returns; thus the mechanism tolerates appreciable vibration. The escapement pallets are similar to those of Fig. 2, but faces F are drawn or inclined so that the escape wheel moves slightly forward after each tooth locks; this motion pulls S clear of D, leaving the balance wheel completely detached. SEE ESCAPEMENT.

In the escapement of **Fig. 4**, the balance wheel is completely detached except for three brief action pe-

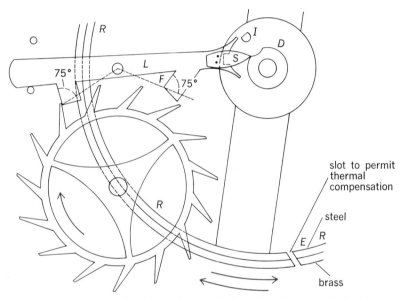

Fig. 3. Lever escapement. Used in good spring clocks and watches, it is detached from the balance wheel during most of a cycle, thus promoting accurate timekeeping.

riods: (1) when pallet P presses detent D to unlock catch L, (2) when an impulse is being given to pallet I, and (3) when P lifts gold spring G on its return half swing. SEE CHRONOMETER.

Power supply. A weight gives uniform power, but a spring develops less force as it runs down. A fuzee by which a cord or chain is unwrapped from an increasing spiral on the first gear provides uniform force from a spring. Alternatively, there may be a remontoir with a small secondary spring driving the clock and being rewound frequently by the mainspring or by an electric motor.

Wind and snow apply large force to the hands of outdoor turret clocks. A gravity escapement isolates the pendulum from such forces. In this design, first used on the Great Westminster Clock (Big Ben), which has run since the mid-1800s with nearly chronometric accuracy, the escape wheel alternately lifts two weighted arms away from the pendulum as it swings out and releases them alternately against the pendulum as it swings back. The maintaining pushes come from the weighted arms rather than directly from the driving power.

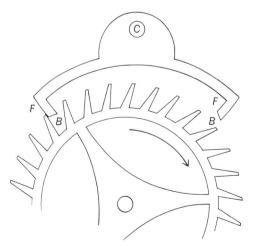

Fig. 2. Deadbeat escapement. Although lacking the corrective feature of the recoil type, this escapement gives better performance if well made.

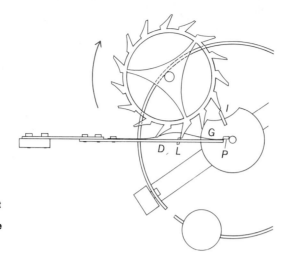

Fig. 4. Escapement used in ships' chronometers. It is sensitive to sudden movements but otherwise can keep within 1 s/day; pallets need no oil.

Temperature compensation. The rate of a pendulum or balance wheel decreases if it expands with rising temperature. Expansion is counteracted by using two metals of different expansion coefficients. In a compensated balance wheel, rim R is of brass and steel fused together and cut (Fig. 3). Small screw weights added near free end E increase the effect; near the spoke they counteract it. Alternatively, the hairspring can be of an alloy that changes in stiffness with temperature.

Shortt clock. For still greater accuracy, in the Shortt or free-pendulum clock, a master pendulum is enclosed in an airtight and nearly evacuated case in a constant-temperature cellar. This pendulum swings freely for about 30 s until wheel W is released by electromagnet E and rolls along arm A attached to the pendulum rod (**Fig. 5**). Near the end of the swing, wheel W rides down slope B, giving the pendulum the slight impulse needed to maintain oscillation for another 30 s. In falling off, wheel W triggers the circuit and is reset by electromagnet F. The trigger pulse also serves as a synchronizing signal to a slave clock which is a complete clock with pendulum and indi-

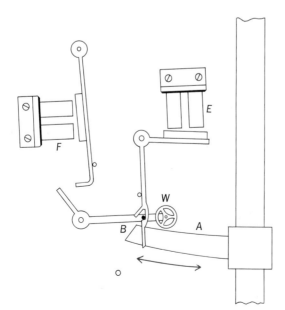

Fig. 5. Shortt free-pendulum clock. The master has only a pendulum and impulse giver; escapement and counting gears are in a separate slave clock (not shown).

cating mechanism. If the slave is slow, the pulse accelerates one swing of its pendulum. SEE HOROLOGY; TIME.

Robert D'E. Atkinson

Tuning-fork mechanism. In the Accutron design the balance wheel and hairspring are replaced by a precision tuning fork which is driven with energy from a tiny battery. A permanent magnet and a surrounding magnetic cup are mounted on each tine of the tuning fork, which vibrates 360 times per second. Fixed coils of wire extend into the area between each magnet and cup, lying inside the magnetic field but not in contact with the moving parts of the tuning fork. Passing a current through one of these coils causes it to become an electromagnet which, depending on the polarities involved, will either repel or attract the corresponding magnet and cup assembly. Conversely, motion of the magnet and cup assembly within its electromagnetic coil causes a voltage to be induced, the polarity being determined by the direction of motion. This electromagnetic system serves three purposes: (1) It drives the tuning fork by converting electrical energy into mechanical impulses, (2) allows a determination of the tuning-fork amplitude by measuring the induced voltage, and (3) permits the accurate determination of the point within each tuning-fork cycle when the driving impulse should be applied. Conversion of the vibrational motion of the tuning fork to rotary motion is achieved by a simple ratchet and pawl mechanism.

Raynor L. Duncombe

Bibliography. F. J. Britten, *Old Clocks and Watches and Their Makers*, 6th ed., 1932, reprint 1971; C. Clutton and G. Daniels, *Watches: A Complete History of the Technical and Decorative Development of the Watch*, 3d ed., 1979; J. E. Haswell, *Horology*, 1928, reprint 1976; H. C. King and J. R. Millburn, *Geared to the Stars: The Evolution of Planetariums, Orreries and Astronomical Clocks*, 1978.

Clock paradox

The paradox produced by the use of incorrect arguments, concepts, and premises in the relativistic twin clock problem in which two initially synchronized perfect clocks, A and B, are first separated and then reunited. It is the true logical contradiction inherent in the assertion that, when reunited, the two clocks, side by side, are slower than each other. Now, two clocks, side by side, cannot be both slower than each other; those who infer this paradox to be a conclusion have, in some way, introduced premises or concepts incompatible with Albert Einstein's theory of relativity. The clock paradox is not a consequence of this theory.

When the relativistic twin clock problem is properly handled, no paradox occurs. Because of the long history of the twin clock paradox problem, some authors refer to it as the clock paradox problem or clock paradox. Authors using the term clock paradox in this sense are not referring to a paradox at all, but to an ordinary relativistic problem for which the correct theoretical prediction is that, upon being reunited, one and only one clock is slower than the other.

In this article, ''time dilation'' and the twin clock problem are distinguished. Next, an example of the twin clock problem is presented. Common conceptual errors that lead to various forms of the clock paradox are discussed, particularly the false notions that (1) nature provides an obvious unique universal rule for determining the simultaneity of distant events, and (2) empty space is to be conceived as an unchangeable, passive constituent of the universe. Finally, brief mention is made of some of the experiments that have empirically confirmed the existence in nature of the twin clock effect, that is, the phenomenon of asymmetric natural aging as a path-dependent function of position and velocity.

Time dilation and twin clock effect. Time dilation is a symmetric artifact of the Lorentz transformations and the Einstein simultaneity convention. The twin clock effect, on the other hand, is an asymmetric physical phenomenon and the physical consequence of the fact that empty space has a metric structure.

Time dilation. Time dilation asserts that clocks at rest in different inertial frames symmetrically observe each other to "run slow." This assertion is no more paradoxical than the assertions that (1) a ship steams past an island and (2) the island floats past the ship. To a viewer on the ship, the island moves; to a viewer on the island, the ship moves. Since the island and the ship use different conventions for the rest, or stationary, state of motion, there is no paradox.

A similar situation arises in time dilation. In each inertial frame the rest clocks do indeed observe the rest clocks of other inertial frames to "run slow," but the Lorentz transformations leading to this result are derived from clocks that are synchronized by a convention that does not have the same meaning in different inertial frames of reference.

Indeed, clocks synchronized within an inertial frame according to the Einstein simultaneity convention are *not* found to be synchronized when they are examined again in other inertial frames according to the "same" convention for distant simultaneity. Given a set of events P_1, P_2, . . ., P_n, . . . , with time and space coordinates (t_1, x_1), (t_2, x_2), . . . , (t_n, x_n) that are simultaneous in one frame, $t_i = t_j$, but spatially separated, $x_i \neq x_j$, then that same set of physical events will, in all other inertial frames, according to the "same" simultaneity convention, have coordinates (t_i', x_i') that are *not* simultaneous, $t_i' \neq t_j'$. This can easily be proved by mere inspection of the Lorentz transformations (**Fig. 1**) or by Lorentz-transforming the coordinates of two point events P_1 and P_2 for which, initially, $t_1 = t_2$ and $x_1 \neq x_2$. SEE LORENTZ TRANSFORMATIONS.

Thus, unless additional information is supplied, the statement that "moving clocks run slow" is an artifact completely without physical significance. Time dilation is not a philosophical statement as to what is or is not real, and it tells nothing about the physical time difference that might exist between spatially separated clocks, were they (as is impossible) magnetically and instantaneously reunited. It is purely a symmetric, definitional artifact of measurement created by the adoption of the Einstein simultaneity convention, a "bookkeeping procedure" adopted to give empirical meaning to the statement that two or more spatially separated events occur simultaneously. This convention depends on the reference frame in the same way that the magnitudes of the components of an ordinary vector depend on the location and orientation of the axes of the reference frame in which the vector is to be resolved into components.

Twin clock effect. The twin clock effect is completely different from time dilation. The twin clock effect is the natural phenomenon that a clock having

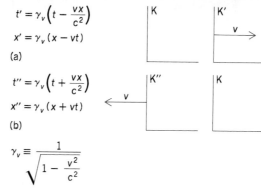

$$t' = \gamma_v\left(t - \frac{vx}{c^2}\right)$$

$$x' = \gamma_v(x - vt)$$

(a)

$$t'' = \gamma_v\left(t + \frac{vx}{c^2}\right)$$

$$x'' = \gamma_v(x + vt)$$

(b)

$$\gamma_v \equiv \frac{1}{\sqrt{1 - \frac{v^2}{c^2}}}$$

Fig. 1. Lorentz transformations. These expressions convert coordinates of events in the K inertial frame into their coordinates in (a) inertial frame K', moving to the right, with respect to frame K, at velocity v, and (b) frame K", moving to the left, with respect to frame K, at velocity v.

a past history of acceleration indicates a total elapsed time less than that indicated by an identical clock having no history of acceleration. It is a physical effect, and the amount by which the accelerated clock is found to be slow is the same regardless of the choice of inertial frame. The physical properties of empty space cause the twin clock effect.

In relativity the concept of empty space is not merely an abstract notion. Empty space is not empty, but has physical content. The physical reality of empty space, or space-time, is, in both the special and general theories, represented by a field whose components are functions of four parameters—three space parameters and one time parameter. In special relativity, the field functions representing the physical properties of empty space are all constant; in general relativity, however, they vary continuously with location and time. In special relativity, the existence of the twin clock effect is implied by the mathematical fact that the metric field functions representing the time-dependent properties of empty space have numerical signs opposite to those representing the spatial physical properties.

In classical mechanics, inertia is the property of massive bodies to resist acceleration. The entity relative to which acceleration occurs is absolute space. By absolute is meant that, as indicated by the fact of inertia, empty space acts upon material objects but these objects do not, in turn, act upon empty space.

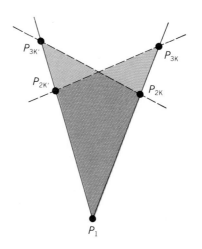

Fig. 2. The five distinct space-time point events that play a role in time dilation.

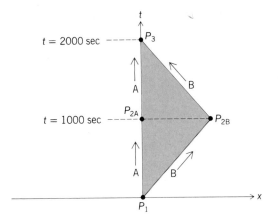

Fig. 3. Space-time diagram of the four events P_1, P_{2A}, P_{2B}, and P_3 in the inertial frame K. Clock A is at rest in this frame.

In the special theory of relativity, empty space not only enables massive objects to resist acceleration but also modifies the rate at which the natural processes of these objects occur. Nevertheless, material objects still do not, in turn, physically influence empty space. In this sense, both acceleration and empty space are absolute in special relativity. *See Kinetics (classical mechanics); Newton's laws of motion.*

However, in general relativity, massive objects warp, or "bend," empty space through the action of their active gravitational masses. To the extent that the metric field functions representing the physical properties of empty space are thereby also made to vary, the timekeeping processes of natural clocks are altered. Gravitational alteration of natural timekeeping processes may, in certain situations, significantly dominate the twin clock effect of special relativity, and it becomes possible for the traveling clock to be, at return, fast (older) rather than, as before, always slower (younger). *See Gravitation.*

In both the general and the special theory the twin clock effect is the consequence of the asymmetric physical metric structure of empty space.

From a purely formal standpoint, time dilation is related to the twin clock effect as a polygon is related to a triangle. Time dilation always involves, as **Fig. 2** illustrates, a minimum of at least five distinct space-time events. Observers at rest in frame K compare space-time intervals P_1P_{2K} and P_1P_{3K}, while observers at rest in frame K' compare intervals $P_1P_{2K'}$ and $P_1P_{3K'}$. Since events P_{2K} and $P_{3K'}$ are simultaneous

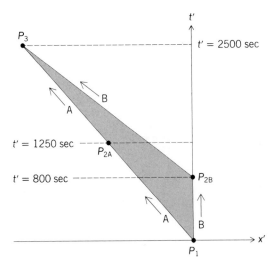

Fig. 4. Space-time diagram of the four events P_1, P_{2A}, P_{2B}, and P_3 in the inertial frame K'.

events in frame K, and events $P_{2K'}$ and P_{3K} are simultaneous in K', rest observers in frames K and K' each conclude that the rest clocks in the other frame are "running slow." On the other hand, the twin clock effect need involve not more than three distinct space-time events.

Twin clock problem. Let A and B be two perfect clocks initially synchronized at a common point of space and time. At some instant, let B depart from A and move along the positive x axis for 1000 s with a constant velocity $v = (3/5)c$, where c is the speed of light (299,792,458 m/s), turn around instantaneously, and take another 1000 s to return to A with $v = -(3/5)c$. There are only three distinct physical events in this problem: B's departure, P_1; B's turnabout, P_{2B}; and B's return, P_3 (**Fig. 3**).

Let K be the inertial frame in which clock A is always at rest. Let K' be an inertial frame moving to the

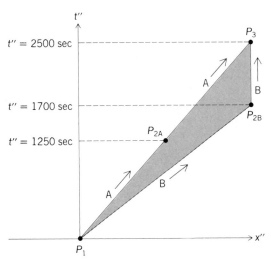

Fig. 5. Space-time diagram of the four events P_1, P_{2A}, P_{2B}, and P_3 in the inertial frame K''.

right with velocity $v = (3/5)c$; and let K'' be an inertial frame moving to the left with velocity $v = -(3/5)c$. For simplicity, the origins of K, K', and K'' are assumed to coincide when B departs from A.

The event P_{2A} is defined to the event at A in K that, according to the Einstein simultaneity convention, is simultaneous with event P_{2B}. The coordinates of the four events P_1, P_{2A}, P_{2B}, and P_3 in frame K are given in the table. By using the Lorentz transformations in Fig. 1, the K frame coordinates of the four events are converted into their corresponding K' and K'' frame coordinates, also given in the **table**. Figure 3 and **Figs. 4** and **5** are space-time diagrams of the four events in frames K, K', and K'', respectively.

Starting with any one of the three frames, coordinates in the other two can be obtained by means of the appropriate Lorentz transformation. For transformations between K' and K'', the relativistic law for composition of velocities must be used to calculate the velocity of K'' with respect to K'.

Clock A's viewpoint. The table indicates that 2000 s elapse between events P_1 and P_3 in frame K. Since clock A is always at rest in K, 2000 s will have elapsed on it when it reaches P_3.

For clock B, however, it is impossible to find an inertial frame in which B is at rest for the entire

Frame K			Frame K′			Frame K″		
Event	t	x	Event	t'	x'	Event	t''	x''
P_1	0 seconds	0 n_c*meters	P_1	0 seconds	0 n_c meters	P_1	0 seconds	0 n_c meters
P_{2A}	+1000	0 n_c	P_{2B}	+800	0 n_c	P_{2A}	+1250	+750 n_c
P_{2B}	+1000	+600 n_c	P_{2A}	+1250	−750 n_c	P_{2B}	+1700	+1500 n_c
P_3	+2000	0 n_c	P_3	+2500	−1500 n_c	P_3	+2500	+1500 n_c

*n_c is a dimensionless constant equal to 299,792,548; it is introduced for calculational convenience.

timelike interval between events P_1 and P_3. Nevertheless, the time elapsed on clock B can be calculated using the fact that in K′, between events P_1 and P_{2B}, clock B is at rest, and in K″, between events P_{2B} and P_3, clock B is again at rest. The table indicates that $800 - 0 = 800$ s elapse between events P_{2B} and P_1 in frame K′ and that $2500 - 1700 = 800$ s elapse between events P_3 and P_{2B} in frame K″. The sum of elapsed times in the various frames in which clock B is at rest is $800 + 800 = 1600$ s. Therefore, 1600 s have elapsed on clock B when it reaches P_3.

Clock B is therefore slower (younger) than clock A by 400 s when they reunite at P_3. This result is independent of the particular inertial frame used to obtain it.

Clock B's viewpoint. In B's view, clock A moves to the left, "turns around," and returns to B. **Figures 6** and **7** show the space-time situation as viewed by B. Figure 6 is identical to Fig. 4 except that all space-time paths occurring after $t' = t'_{2B}$ have been drawn as dashed lines to emphasize that events P_{2A} and P_3 both occur after B departs from frame K′. Even though B is no longer in frame K′ when P_{2A} and P_3 occur, both events still have definite space-time coordinates with frame K′. Figure 7 is identical to Fig. 5 except that all space-time paths occurring before $t'' = t''_{2B}$ have been dashed to indicate that events P_{2A} and P_1 occurred before clock B entered K″. The solid space-time paths in Fig. 6 show what B sees prior to A's turnabout; and the solid space-time paths in Fig. 7 show what B sees after B adopts K″ as its new frame of reference.

As B departs from frame K′ at $t' = t'_{2B}$, event P_{2A}, the event at which A turns around, has yet to occur; but after B has entered frame K″ at $t'' = t''_{2B}$, event P_{2A} has already occurred! The reason is that distant simultaneity does not have the same physical realization in different inertial reference frames. When clock B adopts a new frame of reference, it gains access to an arrangement of physical information that would not have been available to it had B remained at rest in frame K′.

The sudden jump of event P_{2A} from a future event to a past event is not paradoxical; it merely results from the assumption that clock B is subjected to infinite forces when it transfers from frame K′ to K″. If B is subjected to finite, gradual deceleration and acceleration, the time of the distant event P_{2A} changes smoothly and continuously from future event to past event. The "in-between" events which were "missed" because of the mathematical discontinuity originally created by the assumption that B is subject to sharp sudden changes of direction and infinite accelerations will now be observed to occur continuously and rapidly in much the same way the relativistic Doppler shift smoothly changes from a redshift

to a blueshift as B decelerates, stops and reverses direction, and accelerates back to speed. *See Doppler effect.*

The fact that an event, for example, P_{2A}, can be converted, by mere adoption of a new frame of reference, from an event that has yet to occur into one that has already occurred, illustrates why it is impossible to find a common universal time system that all observers could agree to use.

In Fig. 6, $800 - 0 = 800$ s elapse between events P_1 and P_{2B}; in Fig. 7, $2500 - 1700 = 800$ s elapse between events P_{2B} and P_3. Hence in frames K′ and K″, in which clock B is alternately at rest, a total of $800 + 800 = 1600$ s elapse on clock B between events P_1 and P_3.

In frame K′ (Fig. 6), clock A does not "turn around" until event P_{2A}, when $t' = 1250$ s have elapsed in K′. Since clock B leaves K′ at $t' = 800$ s, B misses the 450 s that elapse in K′ between B's departure from K′, event P_{2B}, and the turnaround event P_{2A}. Nevertheless the 1250 s that elapse in frame K′ between event P_1, A's departure from B, and event P_{2A}, A's "turnaround" in K′, still correspond to the elapse on clock A, in the frame K in which A is at rest, of $1000 - 0 = 1000$ s. From B's viewpoint in frame K′, therefore, 1000 s elapse on clock A between A's departure and A's "turnaround."

When B adopts frame K″ (Fig. 7), B discovers that A not only has already turned around but has been returning back toward B for the past $1700 - 1250 = 450$ s. In frame K″, B misses the first 450 s between A's turnaround at $t'' = 1250$ s and A's reunion with B at $t'' = 2500$ s. The $2500 - 1250 = 1250$ s that elapse in frame K″ between events P_{2A} and P_3 corre-

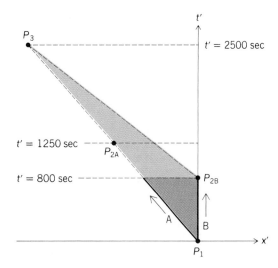

Fig. 6. Space-time diagram of the four events P_1, P_{2A}, P_{2B}, and P_3 in the inertial frame K′ as determined by clock B while B is at rest in K′.

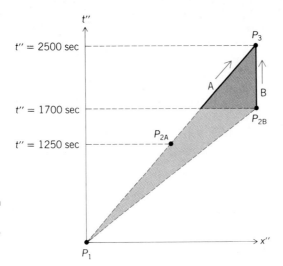

Fig. 7. Space-time diagram of the four events P_1, P_{2A}, P_{2B}, and P_3 in the inertial frame K″ as determined by clock B while B is at rest in K″.

spond, in frame K, in which A is at rest, to the elapse on A of 2000 − 1000 = 1000 s.

As determined by clock B, therefore, 1000 s elapse on clock A on A's outward journey, and another 1000 s on A's return journey, for a total of 1000 + 1000 = 2000 s elapsed in the frame in which A is at rest. Hence, even from clock B's point of view, clock A is still older than clock B when they are reunited at event P_3 by 2000 − 1600 = 400 s.

Significance of motion. One may object that B's viewpoint has not really been adopted, that the argument has been merely repeated from A's point of view. This objection is based on the notion that, in relativity, "only relative motions are significant." This idea is mistaken: in both classical mechanics and special relativity, one cannot, without introducing contradiction (or logical paradox), claim that the same body can move in an inertial, uniform, rectilinear fashion within one inertial frame and in an accelerated, nonuniform, curvilinear fashion in another, and one may not adopt the viewpoint of an accelerated observer for the purpose of pretending that this is possible. Isaac Newton defined an inertial frame of a reference as one in which a body subject to no forces moves in rectilinear path with constant velocity.

Even in general relativity, it is misleading to assert that only purely relative motions are meaningful. Mach's principle does not govern relativity theory in that sense. In general relativity, moving bodies are conceived as being in continuous interaction with empty space, a dynamic, physically active, continuous field endowed with an independent existence of its own. This field is initially created by gravitational mass but possesses thereafter an independent existence, in a manner analogous to the way electromagnetic fields initially created by moving electrons propagate and exist independently.

The relativistic concept of empty space as an existing, dynamic physical field is in common use in modern relativity. It is essential for speculations concerning geometrodynamical interpretations of elementary particles as stable metric "wormhole" structures, and for cosmological speculations concerning massive galactic stars that collapse into black holes. *See Black hole; Gravitational collapse.*

Once one understands that, in relativity, empty space is an existing reality with physical content, and is the immediate physical cause of twin clock effects, then the observable fact that natural clocks, moving

about and interacting with empty space, alter their timekeeping rates should no longer seem enigmatic.

Empirical tests. A. H. Bucherer was the first (1909) to implicity verify the twin clock effect when he experimentally found that high-speed electrons obey relativistic force laws rather than the classical newtonian laws of motion. In 1938 H. E. Ives found that the frequencies of atomic spectral lines of rapidly moving atoms were decreased by an amount in agreement with the decrease predicted on the basis of relativistic time dilation. In 1941 B. Rossi, by counting the rates at which cosmic particles occur at various altitudes in the mountains of Colorado, found that cosmically created mesons survived, on the average, longer than would be possible on the basis of a purely classical newtonian lifetime, and that the excess lifetime was in accord with relativistic time dilation.

During the 1950s many experiments were performed with meson lifetimes. It was repeatedly shown that the number of surviving mesons always exceeded, by a factor in agreement with relativistic time dilation, the number of mesons that were expected to survive on the basis of the classical newtonian lifetime. All these meson experiments represent empirical confirmation of time dilation, and, at least to the extent that meson clocks were accelerated and the laboratory rest clocks were not, also a verification of the twin clock effect.

In 1960 the Mössbauer effect was used by R. V. Pound and G. A. Rebka to verify the general relativistic redshift or, better, blueshift, the relativistic increase in the frequency of an atomic spectral line with the atom's increased distance from a heavy gravitating body. *See Gravitational shift; Mössbauer effect.*

In 1971 J. C. Hafele and R. E. Keating flew four atomically stabilized clocks around the world and directly observed both the special and general relativistic twin clock effects. This experiment excluded theoretical attempts to explain relativistic twin clock effects solely in terms of energy conservation and the principle of equivalence. The Hafele-Keating experiment clearly indicates that both the special and general twin clock effects have as their common physical cause and source the metric structure of the physical field properties of empty space.

The twin clock phenomena of special relativity are well understood, but the twin clock effects of general relativity are just beginning to be explored. Modern navigation systems routinely attain such high worldwide synchronization accuracies that gravitational twin clock effects are significant. Yet it is not known what is meant physically when a clock in Australia is said to be linked with a clock in the United States to an accuracy of 1 nanosecond. The mere fact that the clocks at two such widely separated locations are on opposite sides of the Earth—and therefore alternately farther from and closer to the Sun—leads some scientists to expect that the two clocks have a time difference that varies periodically with a period of 1 day! However, this effect has not been detected. In any case, there is no reason to doubt that asymmetric natural aging as a path-dependent function of position and velocity is an empirical phenomenon exhibited by all known types of stable clocks. *See Relativity; Space-time.*

Richard E. Keating

Bibliography. A. H. Bucherer, The experimental confirmation of the relativity principle, *Ann. Phys.* (Leipzig), 28:513–536, 1909; R. Durbin, H. H. Loar, and W. W. Havens, The lifetime of the π^+ and π^- mesons, *Phys. Rev.*, 88:179–183, 1952; A. Einstein, *The Meaning of Relativity*, 5th ed., 1956; A. Einstein, On the electrodynamics of moving bodies, *Ann. Phys.* (Leipzig), 17:891–921, 1905 (available in a Dover reprint volume, *The Principle of Relativity*); T. T. Frankel, *Gravitational Curvature: An Introduction to Einstein's Theory*, 1979; R. Geroch, *General Relativity from A to B*, 1978; J. C. Graves, *The Conceptual Foundations of Contemporary Relativity Theory*, 1971; J. C. Hafele and R. E. Keating, Around-the-world atomic clocks, *Science*, 177:166–170, 1972; H. E. Ives and G. R. Stilwell, An experimental study of the rate of a moving atomic clock, *J. Opt. Soc. Amer.*, 28:215–266, 1938, and 31:369–374, 1941; L. Marder, *Time and the Space Traveller*, 1974; E. Martinelli and W. K. H. Panofsky, The lifetime of the positive π meson, *Phys. Rev.*, 77:465–469, 1950; C. W. Misner, K. S. Thorne, and J. A. Wheeler, *Gravitation*, 1973; R. V. Pound and G. A. Rebka, Jr., Apparent weight of photons, *Phys. Rev. Lett.*, 4:337–341, 1960; B. Rossi and D. B. Hall, Variation of the rate of mesotrons with momentum, *Phys. Rev.*, 59:223–228, 1941.

Closed-caption television

Method of captioning or subtitling television programs (usually for hearing-impaired or foreign-language viewers) without interfering with the normal television picture, by coding captions as a vertical interval data signal which is decoded at the receiver and superimposed on the television picture. (In open-caption television, captions are part of the normal video picture and are seen by all viewers.) In North America this service is provided primarily by the National Captioning Institute using line 21 of the television picture. In other countries (notably the United Kingdom), closed captions are often provided as part of a more general teletext service. *See* Television; Videotext and Teletext.

Caption origination. Captions are typically prepared at special-purpose workstations optimized for tasks that include preparation and editing of caption texts from the program soundtrack or script; formatting and layout of captions; synchronization of captions to the required video scene (by using auxiliary time-code information recorded with the video as a separate vertical interval or longitudinal data signal); review and correction of the complete captioned program; and storage of caption text and timing information either directly as a vertical interval data signal on the master video or by using auxiliary storage media, such as a floppy disk, for subsequent broadcast transmission.

Caption preparation is labor-intensive, often requiring 10 h or more per hour of program time, and thus more elaborate and specialized techniques are required for captioning live material such as news and sports broadcasts. Two techniques have been used with some success: simultaneous transcription of speech to captions by using machine shorthand (stenograph) input; and summary live captions prepared with conventional keyboards supplemented by short-form dictionaries.

Caption transmission. When captions are transmitted as part of a teletext data transmission, they are stored in the teletext origination computer and transmitted in synchronism with the required video by in-

terrupting the normal cyclic teletext transmissions as necessary to insert a caption page out of the normal page-number sequence. On the other hand, a dedicated caption transmission system (employed, for example, by the National Captioning Institute in North America) uses a much lower data rate, since the caption data are not multiplexed with teletext magazine data; as a result, the data signal is more rugged, can be successfully received under poorer signal conditions, and can be recorded in encoded form on consumer video recorders, whereas teletext data lie outside the bandwidth of such equipment.

Caption reception. Captions can be decoded only by using a teletext television receiver or adaptor, when they are transmitted as part of a teletext service, or by using a special telecaption adaptor (for the National Captioning Institute service). Selecting the caption magazine page associated with the channel being viewed selects and decodes the appropriate vertical interval data, and causes captions to appear superimposed on the picture, generally in a boxed background, as soon as they are received from the transmitter.

A. C. Downton

Bibliography. A. C. Downton et al., Optimal design of broadcast teletext caption preparation systems, *IEEE Trans. Broadcast.*, BC-31 (3):41–50, September 1985.

Closed-circuit television

Television transmitted to a particular audience at specific locations via coaxial cables, telephone wires, fiber-optic strands, microwave radio systems, or communications satellites, as compared with open-circuit television for the public. Television signals are electrical. They combine visual elements, usually generated by the use of video cameras, and related sound elements, usually generated by the use of microphones and audio amplifiers. Television pictures and sounds may also be created synthetically by computer-based equipment. The terms video and television are often used as synonyms. Television signals can be recorded, for example, on magnetic tape. Because television signals have complex electrical characteristics, they usually require a wide-band transmission medium such as coaxial cable or one of the other transmission means noted above. *See* Coaxial cable; Communications cable; Communications satellite; Microwave; Optical communications; Telephone service.

The usual closed-circuit-television picture display device is the television receiver. Most receivers use a cathode-ray tube to produce a visible image, although other devices such as liquid-crystal displays have become more common. *See* Cathode-ray tube; Electronic display; Liquid crystals.

Applications. Much of television can be classified as closed-circuit. Applications include information display, remote monitoring, education, cable television, teleconferences, and special events.

Information display. Flight times in airports are displayed by closed-circuit television, typically on large cathode-ray-tube screens. Similarly, a screen may show the image of a bank teller together with account data during the operation of an automated teller machine. In both cases, closed-circuit television provides information, but different methods are used for generating the video. The airport display usually shows text produced from video signals generated by a computer. The bank machine shows the image of a teller produced by a video camera. These displays exemplify closed-circuit television used in simple form to connect a source of information to a limited number of viewing locations.

Remote monitoring. Closed-circuit television systems produce many kinds of pictures for distant viewing. These include images taken by a video camera on a space probe passing close to a distant planet, or pictures of blast furnace or other industrial operations that could be hazardous to human observers. Parking lots at shopping centers, store cashier locations, bank and hotel lobbies, and other public places are often equipped with closed-circuit television surveillance cameras and displays. These systems may be designed to function in nearly total darkness to detect suspicious activities or hot spots in factory processes. Medical and scientific equipment and industrial robots often incorporate specialized closed-circuit television systems. Modern closed-circuit television remote-monitoring cameras are sensitive, rugged, and relatively inexpensive. Solid-state sensing units in the form of charge-coupled devices, used in many new designs, are durable and have a long service life. Video signals are often recorded on magnetic tape for future use.

Education. For many years, closed-circuit television has been available for educational use. Many universities and metropolitan school systems employ instructional television. Classrooms may be equipped with closed-circuit television cameras, recorders, and receivers. In other cases, under authorization from the Federal Communications Commission, Instructional Television Fixed Service systems transmit lessons to distant classrooms via a form of microwave transmission that uses a special frequency band in the vicinity of 2500 MHz. Receivers with special microwave converters display the lessons in the classrooms. Audio circuits installed at the viewing locations allow two-way conversations between lecturer and students. The general public does not receive Instructional Television Fixed Service transmissions.

Cable television. Cable television is another form of closed-circuit television. Cable systems provide subscribers with a broad selection of broadcast and non-broadcast television signals of high technical quality. They distribute signals from broadcast television stations, some local and some probably distant, and offer closed-circuit television services not available from broadcast sources, such as weather information, satellite-distributed news, music, movies, and community-based programming.

Teleconferences and special events. Closed-circuit television is used to conduct conferences among geographically dispersed groups of people. Signals from television-equipped meeting rooms or studios are distributed, often by communications satellites, to company meeting rooms, school or university classrooms, motels, hotels, or convention centers. In most cases, two-way audio is provided, and conferences can involve many hundreds or thousands of people. Sports or musical events may also be televised and distributed in this fashion to suitably equipped theaters or auditoriums. *See TELECONFERENCING.*

Technology. In contrast to closed-circuit television, broadcast television is limited to a single form of standardized signal for all transmissions. As a result, any television receiver can be used wherever there is a television station. The standardized televi-

sion broadcast system used in the United States, Canada, Mexico, Japan, and some other countries was devised in the 1950s by the National Television Systems Committee (NTSC). *See TELEVISION STANDARDS.*

Closed-circuit television systems are not required to use NTSC signals but many do, mostly for economic reasons. Video cameras, television receivers, videocassette recorders, and accessories are produced in large quantities for NTSC broadcast application. Costs are reduced by large-scale production.

Each second, NTSC television systems display approximately 30 complete television pictures (frames). This figure is referred to as the frame rate and is so high that most images are perceived without flickering or irregular movement. Each frame is composed of 525 horizontal scanning lines arranged from top to bottom. The basic NTSC television signal is monochrome (black and white) and carries all the fine detail conveyed by the system. An auxiliary signal, called the color subcarrier, superimposes color on the monochrome signal.

The NTSC system is well suited for home entertainment, but its signal structure is technically limited by the requirements of simple, inexpensive home receivers and video tape recorders. The development of improved new kinds of television systems is under way in many laboratories. These improvements will probably begin to appear in broadcast television during the mid-1990s and in certain closed-circuit television systems even sooner.

Special requirements. The NTSC signal characteristics are poorly suited to closed-circuit television applications which require high image resolution, high light sensitivity, response to infrared or ultraviolet light, or very rugged components. Closed-circuit television requirements for high resolution of small detail in video images can be satisfied by increasing the number of image scanning lines and the overall video channel bandwidth. Cameras and monitors are now available to meet this need. Industrial standards exist for some high-resolution systems.

In other applications, normal closed-circuit television equipment is satisfactory, but the transmission path to the viewer is of low quality. In these cases, the signal can be adjusted more closely to the needs of the transmission system. For example, channel bandwidth can be reduced without loss of image resolution if the user accepts reduced quality in the representation of image motion. A lower rate than the normal 30 frames per second can be chosen; this is called slow-scan closed-circuit television. Overall transmission bandwidth can then be reduced without loss of resolution. Still-frame video images sometimes suffice, in which case many seconds or even minutes can be assigned to the transmission of a full high-quality still frame, with the picture elements being stored in a memory system and assembled for display only after receipt of all data. Most video pictures from space are transmitted at low frame rates over narrow-bandwidth communications channels; complete reception requires many seconds. As a result, the representation of motion in television signals from spaceships is sometimes crude.

Advanced developments. The digital computer revolution is affecting the design of closed-circuit television equipment. Video signals can be converted into digital form and processed by computer. Special effects often seen on entertainment television, including slow motion, still frames, rotations, inversions, and picture-shape changes, are accomplished by digital processing systems. In closed-circuit television applications, such equipment can produce valuable image enhancement. The excellent pictures from weather and surveillance satellite imaging systems are obtained only after processing and enhancement by special computers. High-performance solid-state camera sensors and digital processing of video signals, taken together, have extended the applications of closed-circuit television systems. SEE IMAGE PROCESSING; REMOTE SENSING; TELEVISION.

Frederick M. Remley, Jr.

Bibliography. K. B. Benson (ed.), *Television Engineering Handbook,* 1986; International Teleproduction Society, *Handbook of Recommended Standards and Procedures,* 1988; E. Jordan (ed.), *Reference Data for Engineers: Radio, Electronics, Computer, and Communications,* 7th ed., 1985.

Cloud

Suspensions of minute droplets or ice crystals produced by the condensation of water vapor. This article presents an outline of cloud formation upon which to base an understanding of cloud classifications. For a more technical consideration of the physical character of atmospheric clouds, including the condensation and precipitation of water vapor, SEE CLOUD PHYSICS; SEE ALSO FOG.

Rudiments of cloud formation. A grasp of a few physical and meteorological relationships aids in an understanding of clouds. First, if water vapor is cooled sufficiently, it becomes saturated and is in equilibrium with a plane surface of liquid water (or ice) at the same temperature. Further cooling in the presence of such a surface causes condensation upon it; in the absence of any surfaces, no condensation occurs until a substantial further cooling provokes condensation upon ions or random large aggregates of water molecules. In the atmosphere, even in the apparent absence of any surfaces, nuclei always exist upon which condensation proceeds at barely appreciable cooling beyond the state of saturation. Consequently, when atmospheric water vapor is cooled sufficiently, condensation nuclei swell into minute water droplets and form a visible cloud. The total concen-

Fig. 2. Cirrus, with trails of slowly falling ice crystals at a high level. (*F. Ellerman, U.S. Weather Bureau*)

tration of liquid in the cloud is controlled by its temperature and the degree of cooling beyond the state in which saturation occurred, and in most clouds approximates to 1 g/m³ (0.001 oz/ft³) of air. The concentration of droplets is controlled by the concentrations and properties of the nuclei and the speed of the cooling at the beginning of the condensation. In the atmosphere these are such that there are usually about 100,000,000 droplets/m³ (2,800,000 droplets/ft³). Because the cloud water is at first fairly evenly shared among them, these droplets are necessarily of microscopic size, and an important part of the study of clouds concerns the ways in which they become aggregated into drops large enough to fall as rain.

The cooling which produces clouds is almost always associated with the upward movements of air which carry heat from the Earth's surface and restore to the atmosphere that heat lost by radiation into space. These movements are most pronounced in storms, which are accompanied by thick, dense clouds, but also take place on a smaller scale in fair weather, producing scattered clouds or dappled skies. SEE STORM.

Rising air cools by several degrees Celsius for each kilometer of ascent, so that even over equatorial re-

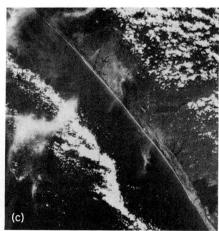

Fig. 1 Clouds as photographed by meteorological satellites. (*a*) A garden of thunderstorms, showing the anvils, forming at the tropopause inversion. (*b*) A hurricane, showing the eye. (c) Convective cloud over land and sea-breeze clouds.

Fig. 3. Small cumulus. (*U.S. Weather Bureau*)

Fig. 4. Overcast of stratus, with some fragments below the hilltops. (*U.S. Weather Bureau*)

gions temperatures below 0°C (32°F) are encountered a few kilometers above the ground, and clouds of frozen particles prevail at higher levels. Of the abundant nuclei which facilitate droplet condensation, very few cause direct condensation into ice crystals or stimulate the freezing of droplets, and especially at temperatures near 0°C (32°F) their numbers may be vanishingly small. Consequently, at these temperatures, clouds of unfrozen droplets are not infrequently encountered (supercooled clouds). In general, however, ice crystals occur in very much smaller concentrations than the droplets of liquid clouds, and may by condensation alone become large enough to fall from their parent cloud. Such ice nuclei are particles of clay, or sometimes bacteria, mixed upward from the Earth's surface. All cloud droplets freeze by self-nucleation (ice crystal formation by random motion of the water molecules) at temperatures below −35 to −40°C (−31 to −40°F). Even small high clouds may produce or become trails of snow crystals, whereas droplet clouds are characteristically compact in appearance with well-defined edges, and produce

rain only when well developed vertically (2 km or 1.5 mi, or more, thick).

Classification of clouds. The contrast in cloud forms mentioned above was recognized in the first widely accepted classification, as well as in several succeeding classifications. The first was that of L. Howard in 1803, recognizing three fundamental types: the stratiform (layer), cumuliform (heap), and cirriform (fibrous). The first two are indeed fundamental, representing clouds formed respectively in stable and in convectively unstable atmospheres, whereas the clouds of the third type are the ice clouds which are in general higher and more tenuous and less clearly reveal the kind of air motion which led to their formation. Succeeding classifications continued to be based upon the visual appearance or form of the clouds, differentiating relatively minor features, but later in the nineteenth century increasing importance was attached to cloud height, because direct measurements of winds above the ground were then very difficult, and it was hoped to obtain wind data on a great scale by combining observations of apparent cloud motion with reasonably accurate estimates of cloud height, based solely on their form. With the advent of satellite and rocket observations in the 1970s, a broader perspective became available for examination of clouds on a global basis. Such pictures readily

Fig. 6. Cirrocumulus, high clouds with a delicate pattern. (*A. A. Lothman, U.S. Weather Bureau*)

Fig. 5. View from Mount Wilson, California. High above is a veil of cirrostratus, and below is the top of a low-level layer cloud. (*F. Ellerman, U.S. Weather Bureau*)

Fig. 7. Altocumulus, which occurs at intermediate levels. (*G. A. Lott, U.S. Weather Bureau*)

Fig. 8. Altostratus, a middle-level layer cloud. Thick layers of such cloud, with bases extending down to low levels, produce prolonged rain or snow, and are then called nimbostratus. (*C. F. Brooks, U.S. Weather Bureau*)

show the distribution of clouds from systems such as a midlatitude cyclone, hurricane, or the intertropical convergence zone; they show the influence of land and ocean in initiating convective motion; and they show dramatically the relationship of lenticular clouds to mountain and island topography (**Fig. 1**). *SEE ME-TEOROLOGICAL ROCKETS; METEOROLOGICAL SATELLITES; SATELLITE METEOROLOGY.*

WMO cloud classification. The World Meteorological Organization (WMO) uses a classification which, with minor modifications, dates from 1894 and represents a choice made at that time from a number of competing classifications. It divides clouds into low-level (base below about 2 km or 1.5 mi), middle-level (about 2 to 7 km or approximately 1 to 4.5 mi), and high-level (between roughly 7 and 14 km or approximately 4 to 8.5 mi) forms within the middle latitudes. The names of the three basic forms of clouds are used in combination to define 10 main characteristic forms, or genera.

1. Cirrus are high white clouds with a silken or fibrous appearance (**Fig. 2**).

2. Cumulus are detached dense clouds which rise in domes or towers from a level low base (**Fig. 3**).

3. Stratus are extensive layers or flat patches of

Fig. 9. Cumulonimbus clouds photographed over the upland adjoining the upper Colorado River valley. (*Lt. B. H. Wyatt, U.S.N., U.S. Weather Bureau*)

low clouds without detail (**Fig. 4**).

4. Cirrostratus is cirrus so abundant as to fuse into a layer (**Fig. 5**).

5. Cirrocumulus is formed of high clouds broken into a delicate wavy or dappled pattern (**Fig. 6**).

6. Stratocumulus is a low-level layer cloud having a dappled, lumpy, or wavy structure. See the foreground of Fig. 5.

7. Altocumulus is similar to stratocumulus but lies at intermediate levels (**Fig. 7**).

8. Altostratus is a thick, extensive, layer cloud at intermediate levels (**Fig. 8**).

Cloud classification based on air motion and associated physical characteristics

Kind of motion	Typical vertical speeds, cm/s*	Kind of cloud	Name	Characteristic dimensions, km†		Characteristic precipitation
				Horizontal	Vertical	
Widespread slow ascent, associated with cyclones (stable atmosphere)	10	Thick layers	Cirrus, later becoming: cirrostratus altostratus altocumulus	10^3	1–2	Snow trails
			nimbostratus	10^3	10	Prolonged moderate rain or snow
Convection, due to passage over warm surface (unstable atmosphere)	10^2	Small heap cloud	Cumulus	1	1	None
	10^3	Shower- and thundercloud	Cumulonimbus	10	10	Intense showers of rain or hail
Irregular stirring causing cooling during passage over cold surface (stable atmosphere)	10	Shallow low layer clouds, fogs	Stratus Stratocumulus	$<10^2$ $<10^3$	<1	None, or slight drizzle or snow

*1 cm = 0.4 in.   †1 km = 0.62 mi.

Fig. 10. A cloud formed by moist air flowing over a mountain. Note the smooth upper profile, indicating low-turbulence flow. (*Courtesy of John Hallet*)

9. Nimbostratus is a dark, widespread cloud with a low base from which prolonged rain or snow falls.

10. Cumulonimbus is a large cumulus which produces a rain or snow shower (**Fig. 9**).

Classification by air motion. Modern studies of clouds have been stimulated by the need to know their composition from the viewpoint of aircraft and rocket penetration, which gives rise to icing, turbulence, and lightning hazards, and by the discovery that seeding of supercooled clouds could, on occasion, give rise to enhanced precipitation. SEE WEATHER MODIFICATION.

These studies show that the external form of clouds gives only indirect and incomplete clues to the physical properties which determine their evolution. Throughout this evolution the most important properties appear to be the air motion and the size-distribution spectrum of all the cloud particles, including the condensation and ice-forming nuclei. These properties vary significantly with time and position within the cloud, so that cloud studies demand the intensive examination of individual clouds with aircraft, radar, and satellites. Nevertheless, the overall cloud shape does give information on the formation process. Lenticular clouds with smooth tops may form from smooth air flow with little environmental mixing (**Fig. 10**); clouds with irregular tops (Fig. 9) show that mixing is strong. The former has nonturbulent, laminar air motions and give a smooth flight; the mixing clouds give modest to strong turbulence and a bumpy flight. An observer of clouds can readily see motions of convective clouds; motions in other systems can be viewed by time-lapse photography at 5–50-s intervals, or sequential satellite pictures at intervals of several hours. From a general meteorological point of view, a classification can be based upon the kind of air motion associated with the cloud, as shown in the **table**.

Frank H. Ludlam; John Hallett

Cloud chamber

A particle detector in which the path of a fast charged particle is made visible by the formation of liquid droplets on the ions left by the particle as it passes through the gas of the chamber. Cloud chambers are used in research in nuclear physics, and can give detailed information on the particle in addition to the simple fact that the particle passed through. They can be operated at pressures below atmospheric, near atmospheric, or up to 50 atm (5000 kilopascals) depending on the application. They vary in size from a few inches in diameter up to large walk-in cham-

bers containing many metal plates. SEE PARTICLE DETECTOR.

Supersaturated vapor is vapor whose pressure exceeds the saturation vapor pressure at the temperature prevailing. It is unstable, and condensation occurs in the presence of suitable nuclei. The interior of a cloud chamber contains a mixture of a gas such as air, argon, or helium and a vapor such as water or alcohol. When such a mixture is made supersaturated in the vapor component by cooling of the gas, the vapor tends to condense preferentially on charged atoms in the gas. Since a fast-moving charged particle produces many such ions in its passage through the gas, condensation of the vapor leaves a trail of droplets to indicate the path of the particle. When properly illuminated, these droplets are visible and may be photographed. A permanent record of the path of the particle may thus be obtained for examination at a later time.

Supersaturation methods. The supersaturation necessary for drop formation may be obtained in two ways. (1) In the Wilson cloud chamber the saturated

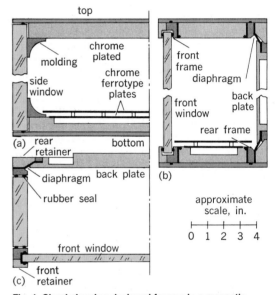

Fig. 1. Cloud chamber designed for use in a magnetic field. (*a*) Vertical section parallel to front. (*b*) Vertical section parallel to side. (*c*) Horizontal section. The back plate moves to produce the expansion. The mechanism for compressing the chamber and producing the expansion is not shown. Photograph in Fig. 2 was taken by using this cloud chamber. 1 in. = 2.5 cm.

vapor-gas mixture is suddenly expanded, causing supersaturation of the vapor at the resulting lower temperature. Usually the expansion is produced by motion of a rubber diaphragm or a piston. The amount of expansion must be carefully controlled; too little expansion will result in insufficient supersaturation to cause drop condensation, while overexpansion will give a dense fog, irrespective of the presence of ions in the gas. (2) The second method for obtaining supersaturation is used in the diffusion cloud chamber. A temperature gradient is maintained in the gas by cooling the bottom of the chamber and warning the top. Vapor introduced at the top diffuses toward the bottom and cools as it goes. A certain region in the chamber becomes supersaturated, and the vapor will condense on ions if they are present. The diffusion chamber is thus continuously sensitive, as contrasted to the Wilson chamber, which is sensitive only for a fraction of a second after an expansion is made.

To overcome the short sensitive time, Wilson cloud chambers are often counter-controlled. Auxiliary particle detectors are used to signal the presence of a fast charged particle and to cause the expansion of the chamber within a few milliseconds after the passage of the particle. Wilson cloud chambers used in cosmic-ray research are usually counter-controlled because of the random time of arrival of cosmic-ray particles.

Information obtained. A cloud chamber can give information on the momentum of the particle if the chamber is placed in a magnetic field (**Figs. 1** and **2**). The curvature of the track can be measured on the photograph, and the value of the curvature, together with the magnitude of the field, gives the momentum. The sign of the charge can usually be determined also, if the direction of motion of the particle is known, since positive and negative particles curve in opposite directions. Another physical quantity that may be determined in a cloud chamber is the velocity

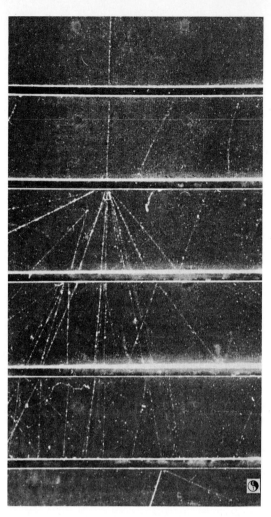

Fig. 3. Nuclear interaction produced by cosmic-ray particle occurring in a lead plate of cloud chamber. Particles produced are mesons, nucleons, and a hyperon, which traverse other lead plates in cloud chamber.

of the particle, since the ionization produced by the charged particle depends on its velocity in a known way. With good photographic technique, individual droplets along the track are visible, and the number of these droplets is closely related to the number of ions along the track. A combination of the measurements of velocity and momentum yields a value for the mass of the particle, thus identifying it.

Cloud chambers may be used to determine the range of a particle, the distance required to bring it to rest. For low-energy particles the energy lost to the gas may be sufficient to bring a particle to rest in the chamber. To stop a high-energy particle, metal plates are inserted in the chamber with spaces between where the tracks may be seen (**Fig. 3**). A particle brought to rest in a given plate will have passed through a certain amount of material, which can be determined to give the range of the particle.

Dead time. Wilson cloud chambers have the disadvantage of a long dead time; it takes at least a minute for an expansion-type chamber to come to equilibrium and be ready for the next expansion. Diffusion chambers have no such dead time and are therefore much more suitable for use near a high-energy particle accelerator, where the particles come out in pulses only a few seconds apart. SEE BUBBLE CHAMBER.

William B. Fretter

Bibliography. American Institute of Physics, *The Cloud Chamber,* 1975; E. Segrè, *Nuclei and Particles,* 2d ed., 1977.

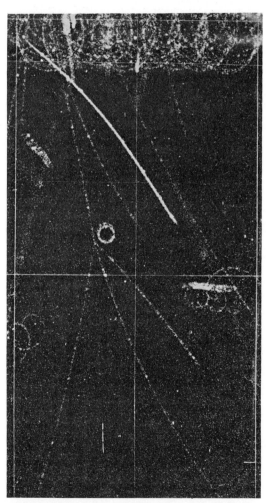

Fig. 2. Nuclear interaction produced by a cosmic-ray particle above chamber analyzed in cloud chamber placed in magnetic field. Circles at top of picture are tracks of electrons spiraling in magnetic field. Heavy track is a proton. Light track which becomes three tracks above center of chamber is a K meson, which disintegrates in flight into three π mesons; $K^+ \rightarrow \pi^+ + \pi^+ + \pi^-$. Tracks are curved because of magnetic field.

Cloud physics

The study of the physical and dynamical processes governing the structure and development of clouds and the release from them of snow, rain, and hail (collectively known as precipitation). *See Precipitation (meteorology)*.

The factors of prime importance are the motion of the air, its water-vapor content, and the number and properties of the particles in the air which act as centers of condensation and freezing. Because of the complexity of atmospheric motions and the enormous variability in vapor and particle content of the air, it seems impossible to construct a detailed, general theory of the manner in which clouds and precipitation develop. However, calculations based on the present conception of laws governing the growth and aggregation of cloud particles and on simple models of air motion provide reasonable explanations for the observed formation of precipitation in different kinds of clouds.

Cloud formation. Clouds are formed by the lifting of damp air which cools by expansion under continuously falling pressure. The relative humidity increases until the air approaches saturation. Then condensation occurs (**Fig. 1**) on some of the wide variety of aerosol particles present; these exist in concentrations ranging from less than 2000 particles/in.^{3} (100/cm^3) in clean, maritime air to perhaps 10^7/in.^{3} (10^6/cm^3) in the highly polluted air of an industrial city. A portion of these particles are hygroscopic and promote condensation at relative humidities below 100%; but for continued condensation leading to the formation of cloud droplets, the air must be slightly supersaturated. Among the highly efficient condensation nuclei are the salt particles produced by the evaporation of sea spray, but it appears that particles produced by human-made fires and by natural combustion (for example, forest fires) also make a major contribution. Condensation onto the nuclei continues as rapidly as the water vapor is made available by cooling of the air and gives rise to droplets of the order of 0.0004 in. (0.01 mm) in diameter. These droplets, usually present in concentrations of several thousand per cubic inch, constitute a nonprecipitating water cloud.

Mechanisms of precipitation release. Growing clouds are sustained by upward air currents, which may vary in strength from about an inch per second to several yards per second. Considerable growth of the cloud droplets (with falling speeds of only about 0.4 in./s or 1 cm/s) is therefore necessary if they are to fall through the cloud, survive evaporation in the unsaturated air beneath, and reach the ground as drizzle or rain. Drizzle drops have radii exceeding 0.004 in. (0.1 mm), while the largest raindrops are about 0.24 in. (6 mm) across and fall at nearly 30 ft/s (10 m/s). The production of a relatively few large particles from a large population of much smaller ones may be achieved in one of two ways.

Coalescence process. Cloud droplets are seldom of uniform size for several reasons. Droplets arise on nuclei of various sizes and grow under slightly different conditions of temperature and supersaturation in different parts of the cloud. Some small drops may remain inside the cloud for longer than others before being carried into the drier air outside.

A droplet appreciably larger than average will fall faster than the smaller ones, and so will collide and fuse (coalesce) with some of those which it overtakes (**Fig. 2**). Calculations show that, in a deep cloud containing strong upward air currents and high concentrations of liquid water, such a droplet will have a sufficiently long journey among its smaller neighbors to grow to raindrop size. This coalescence mechanism is responsible for the showers that fall in tropical and subtropical regions from clouds whose tops do not reach the 32°F (0°C) level and therefore cannot contain ice crystals which are responsible for most precipitation. Radar evidence also suggests that showers in temperate latitudes may sometimes be initiated by the coalescence of waterdrops, although the clouds may later reach to heights at which ice crystals may form in their upper parts.

Initiation of the coalescence mechanism requires the presence of some droplets exceeding 20 micrometers in diameter. Over the oceans and in adjacent land areas they may well be supplied as droplets of sea spray, but in the interiors of continents, where so-called giant salt particles of marine origin are probably scarce, it may be harder for the coalescence mechanism to begin.

Ice crystal process. The second method of releasing precipitation can operate only if the cloud top reaches elevations where temperatures are below 32°F (0°C) and the droplets in the upper cloud regions become supercooled. At temperatures below −40°F (−40°C) the droplets freeze automatically or spontaneously; at higher temperatures they can freeze only if they are infected with special, minute particles called ice nuclei. As the temperature falls below 32°F (0°C), more and more ice nuclei become active, and ice crystals appear in increasing numbers among the supercooled droplets. But such a mixture of supercooled droplets and ice crystals is unstable. The cloudy air, being usually only slightly supersaturated with water vapor as far as the droplets are concerned, is strongly oversaturated for the ice crystals, which therefore grow more rapidly than the droplets. After several minutes the growing crystals will acquire definite falling speeds, and several of them may become joined together to form a snowflake. In falling into the warmer regions of the cloud, however, the snowflake may melt and reach the ground as a raindrop.

Precipitation from layer-cloud systems. The deep, extensive, multilayer-cloud systems, from which pre-

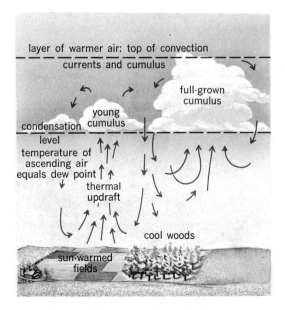

Fig. 1. Conditions leading to birth of a cumulus cloud.

layer of warmer air: top of convection currents and cumulus

full-grown cumulus

young cumulus

condensation level

temperature of ascending air equals dew point

thermal updraft

cool woods

sun-warmed fields

cipitation of a usually widespread, persistent character falls, are generally formed in cyclonic depressions (lows) and near fronts. Such cloud systems are associated with feeble upcurrents of only a few centimeters per second, which last for at least several hours. Although the structure of these great raincloud systems, which are being explored by aircraft and radar, is not yet well understood, it appears that they rarely produce rain as distinct from drizzle, unless their tops are colder than about 10°F (−12°C). This suggests that ice crystals may be responsible. Such a view is supported by the fact that the radar signals from these clouds usually take a characteristic form which has been clearly identified with the melting of snowflakes.

Production of showers. Precipitation from shower clouds and thunderstorms, whether in the form of raindrops, pellets of soft hail, or true hailstones, is generally of greater intensity and shorter duration than that from layer clouds and is usually composed of larger particles. The clouds themselves are characterized by their large vertical depth, strong vertical air currents, and high concentrations of liquid water, all these factors favoring the rapid growth of precipitation elements by accretion.

In a cloud composed wholly of liquid water, raindrops may grow by coalescence with small droplets. For example, a droplet being carried up from the cloud base would grow as it ascends by sweeping up smaller droplets. When it becomes too heavy to be supported by the vertical upcurrents, the droplet will then fall, continuing to grow by the same process on its downward journey. Finally, if the cloud is sufficiently deep, the droplet will emerge from its base as a raindrop.

In a dense, vigorous cloud several kilometers deep, the drop may attain its limiting stable diameter (about 0.2 in. or 5 mm) before reaching the cloud base and thus will break up into several large fragments. Each of these may continue to grow and attain breakup size. The number of raindrops may increase so rapidly in this manner that after a few minutes the accumulated mass of water can no longer be supported by the upcurrents and falls out as a heavy shower. The conditions which favor this rapid multiplication of raindrops occur more readily in tropical regions.

The ice crystals grow initially by sublimation of vapor in much the same way as in layer clouds, but when their diameters exceed about 0.004 in. (0.1 mm), growth by collision with supercooled droplets will usually predominate. At low temperatures the impacting droplets tend to freeze individually and quickly to produce pellets of soft hail. The air spaces between the frozen droplets give the ice a relatively low density; the frozen droplets contain large numbers of tiny air bubbles, which give the pellets an opaque, white appearance. However, when the growing pellet traverses a region of relatively high air temperature or high concentration of liquid water or both, the transfer of latent heat of fusion from the hailstone to the air cannot occur sufficiently rapidly to allow all the deposited water to freeze immediately. There then forms a wet coating of slushy ice, which may later freeze to form a layer of compact, relatively transparent ice. Alternate layers of opaque and clear ice are characteristic of large hailstones, but their formation and detailed structure are determined by many factors such as the number concentration, size and impact velocity of the supercooled cloud droplets, the temperature of the air and hailstone surface, and the size, shape, and aerodynamic behavior of the hailstone.

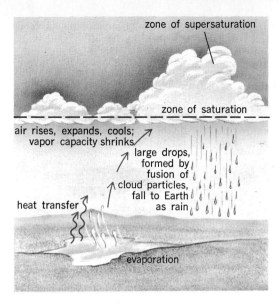

Fig. 2. Diagram of the steps in the formation of rain.

Giant hailstones, up to 4 in. (10 cm) in diameter, which cause enormous damage to crops, buildings, and livestock, most frequently fall not from the large tropical thunderstorms, but from storms in the continental interiors of temperate latitudes. An example is the Nebraska-Wyoming area of the United States, where the organization of larger-scale wind patterns is particularly favorable for the growth of severe storms. *See Hail*.

The development of precipitation in convective clouds is accompanied by electrical effects culminating in lightning. The mechanism by which the electric charge dissipated in lightning flashes is generated and separated within the thunderstorm has been debated for more than 200 years, but there is still no universally accepted theory. However, the majority opinion holds that lightning is closely associated with the appearance of the ice phase, and the most promising theory suggests that the charge is produced by the rebound of a small fraction of the supercooled cloud droplets that collide with the falling hail pellets. *See Lightning*.

Basic aspects of cloud physics. The various stages of the precipitation mechanisms raise a number of interesting and fundamental problems in classical physics. Worthy of mention are the supercooling and freezing of water; the nature, origin, and mode of action of the ice nuclei; and the mechanism of ice-crystal growth which produces the various snow crystal forms.

It has been established how the maximum degree to which a sample of water may be supercooled depends on its purity, volume, and rate of cooling. The freezing temperatures of waterdrops containing foreign particles vary linearly as the logarithm of the droplet volumes for a constant rate of cooling. This relationship, which has been established for drops varying between 0.0004 and 0.4 in. (10 μm and 1 cm) in diameter, characterizes the heterogeneous nucleation of waterdrops and is probably a consequence of the fact that the ice-nucleating ability of atmospheric aerosol increases logarithmically with decreasing temperature.

When extreme precautions are taken to purify the water and to exclude all solid particles, small droplets, about 0.00004 in. (1 μm) in diameter, may be supercooled to −40°F (−40°C) and drops of 0.04 in.

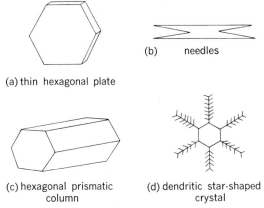

(a) thin hexagonal plate

(b)   needles

(c) hexagonal prismatic column

(d) dendritic star-shaped crystal

Fig. 3. Ice crystal types formed in various temperature ranges: (*a*) 32 to 27°F (0 to −3°C), 18 to 10°F (−8 to −12°C), 3 to −13°F (−16 to −25°C); (*b*) 27 to 23°F (−3 to −5°C); (*c*) 23 to 18°F (−5 to −8°C), below −13°F (−25°C); (*d*) 10 to 3°F (−12 to −16°C).

(1 mm) diameter to −31°F (−35°C). Under these conditions freezing occurs spontaneously without the aid of foreign nuclei.

The nature and origin of the ice nuclei, which are necessary to induce freezing of cloud droplets at temperatures about −40°F (−40°C), are still not clear. Measurements made with large cloud chambers on aircraft indicate that the most efficient nuclei, active at temperatures above 14°F (−10°C), are present in concentrations of only about 30/ft^3 (10/m^3) of air, but as the temperature is lowered, the number of ice crystals increases logarithmically to reach concentrations of about 25/ft^3 (1/liter) at −4°F (−20°C) and 2500/ft^3 (100/liter) at −22°F (−30°C). Since these measured concentrations of nuclei are less than one-hundredth of the number that apparently is consumed in the production of snow, it seems that there must exist processes by which the original number of ice crystals are rapidly multiplied. Laboratory experiments suggest the fragmentation of the delicate snow crystals and the ejection of ice splinters from freezing droplets as probable mechanisms.

The most likely source of atmospheric ice nuclei is provided by the soil and mineral-dust particles carried aloft by the wind. Laboratory tests have shown that, although most common minerals are relatively inactive, a number of silicate minerals of the clay family produce ice crystals in a supercooled cloud at temperatures above −0.4°F (−18°C). A major constituent of some clays, kaolinite, which is active below 16°F (−9°C), is probably the main source of highly efficient nuclei.

The fact that there may often be a deficiency of efficient ice nuclei in the atmosphere has led to a search for artificial nuclei which might be introduced into supercooled clouds in large numbers. Silver iodide is a most effective substance, being active at 25°F (−4°C), while lead iodide and cupric sulfide have threshold temperatures of 21°F (−6°C) for freezing nuclei.

In general, the most effective ice-nucleating substances, both natural and artificial, are hexagonal crystals in which spacings between adjacent rows of atoms differ from those of ice by less than 16%. The detailed surface structure of the nucleus, which is determined only in part by the crystal geometry, is of even greater importance. This is strongly indicated by the discovery that several complex organic sub-

stances, notably steroid compounds, which have apparently little structural resemblance to ice, may act as nucleators for ice at temperatures as high as 30°F (−1°C).

The collection of snow crystals from clouds at different temperatures has revealed their great variety of shape and form. By growing the ice crystals on a fine fiber in a cloud chamber, it has been possible to reproduce all the naturally occurring forms and to show how these are correlated with the temperature and supersaturation of the environment. With the air temperature along the length of a fiber ranging from 32 to −13°F (0 to −25°C), the following clear-cut changes of crystal habit are observed (**Fig. 3**):

hexagonal plates–needles–hollow prisms–
plates–stellar dendrites–plates–prisms

This multiple change of habit over such a small temperature range is remarkable and is thought to be associated with the fact that water molecules apparently migrate between neighboring faces on an ice crystal in a manner which is very sensitive to the temperature. Certainly the temperature rather than the supersaturation of the environment is primarily responsible for determining the basic shape of the crystal, though the supersaturation governs the growth rates of the crystals, the ratio of their linear dimensions, and the development of dendritic forms. *See Snow.*

Artificial stimulation of rain. The presence of either ice crystals or some comparatively large water-droplets (to initiate the coalescence mechanism) appears essential to the natural release of precipitation. Rainmaking experiments are conducted on the assumption that some clouds precipitate inefficiently, or not at all, because they are deficient in natural nuclei; and that this deficiency can be remedied by seeding the clouds artificially with dry ice or silver iodide to produce ice crystals, or by introducing waterdroplets or large hygroscopic nuclei. In the dry-ice method, pellets of about 0.4-in. (1-cm) diameter are dropped from an aircraft into the top of a supercooled cloud. Each pellet chills a thin sheath of air near its surface to well below −40°F (−40°C) and produces perhaps 10^{12} minute ice crystals, which subsequently spread through the cloud, grow, and aggregate into snowflakes. Only a few pounds of dry ice are required to seed a large cumulus cloud. Some hundreds of experiments, carried out mainly in Australia, Canada, South Africa, and the United States, have shown that cumulus clouds in a suitable state of development may be induced to rain by seeding them with dry ice on occasions when neighboring clouds, untreated, do not precipitate. However, the amounts of rain produced have usually been rather small.

For large-scale trials designed to modify the rainfall from widespread cloud systems over large areas, the cost of aircraft is usually prohibitive. The technique in this case is to release a silver iodide smoke from the ground and rely on the air currents to carry it up into the supercooled regions of the cloud. In this method, with no control over the subsequent transport of the smoke, it is not possible to make a reliable estimate of the concentrations of ice nuclei reaching cloud level, nor is it known for how long silver iodide retains its nucleating ability in the atmosphere. It is usually these unknown factors which, together with the impossibility of estimating accurately what would have been the natural rainfall in the absence of seeding activities, make the design and evaluation of a large-scale operation so difficult.

Little convincing evidence can be found that large increases in rainfall have been produced consistently over large areas. Indeed, in temperate latitudes most rain falls from deep layer-cloud systems whose tops usually reach to levels at which there are abundant natural ice nuclei and in which the natural precipitation processes have plenty of time to operate. It is therefore not obvious that seeding of these clouds would produce a significant increase in rainfall, although it is possible that by forestalling natural processes some redistribution might be effected.

Perhaps more promising as additional sources of rain or snow are the persistent supercooled clouds produced by the ascent of damp air over large mountain barriers. The continuous generation of an appropriate concentration of ice crystals near the windward edge might well produce a persistent light snowfall to the leeward, since water vapor is continually being made available for crystal growth by lifting of the air. The condensed water, once converted into snow crystals, has a much greater opportunity of reaching the mountain surface without evaporating, and might accumulate in appreciable amounts if seeding were maintained for many hours.

Trials carried out in favorable locations in the United States and Australia suggest that in some cases seeding has been followed by seasonal precipitation increases of about 10%, but rarely have the effects been reproduced from one season to the next, and overall the evidence for consistent and statistically significant increases of rainfall is not impressive. Indeed, as the experiments have been subjected to stricter statistical design and evaluation, the claims have steadily became more modest, and the difficulty of improving on natural processes has become increasingly apparent.

During the 1960s and 1970s, remaining trials were carried out in some 75 countries, and at one time about 25% of the land area of the United States was being seeded. However, by the late 1980s the activity was much reduced and had virtually ceased in Australia and many other countries. Nevertheless, China and Israel persevered. The Israeli experiments were unique in that they appeared to have produced rainfall increases on the order of 15% year after year for about 15 years. No convincing explanation for this has yet emerged. *See Cloud; Weather modification.*

Basil J. Mason

Bibliography. E. M. Agee and T. Assai (eds.), *Cloud Dynamics*, 1982; L. J. Battan, *Cloud Physics and Cloud Seeding*, 1962, reprint 1979; B. J. Mason, *Clouds, Rain and Rainmaking*, 1975; H. R. Pruppacher and J. D. Klett, *Microphysics of Atmospheric Clouds and Precipitation*, 1980.

Clove

The unopened flower bud (see **illus.**) of a small, conical, symmetrical, evergreen tree, *Eugenia caryophyllata*, of the myrtle family (Myrtaceae). The cloves are picked by hand and dried in the sun or by artificial means. The crop is uncertain and difficult to grow. Cloves, one of the most important and useful spices, are strongly aromatic and have a pungent flavor. They are used as a culinary spice for flavoring pickles, ketchup, and sauces, in medicine, and for perfuming the breath and the air in rooms. The essential oil distilled from cloves by water or steam has even more uses. The chief clove-producing countries are Tanzania

Closed (cloves) and open flower buds on a branch of the evergreen tree *Eugenia caryophyllata*.

with 90% of the total output, Indonesia, Mauritius, and the West Indies. *See Myrtales; Spice and flavoring.*

Perry D. Strausbaugh/Earl L. Core

Clover

A common name used loosely to designate the true clovers, sweet clovers, and other members of the plant family Leguminosa. This article discusses true clovers, sweet clover, and clover diseases.

TRUE CLOVERS

The true clovers are plants of the genus *Trifolium*, order Rosales. There are approximately 250 species in the world. Collectively they represent the most important genus of forage legumes in agriculture. Different species constitute one or more crops on every continent. As indicated by the number of species and their ecotypes (subspecies) that have been collected and described, the center of origin appears to be southeastern Europe and southwestern Asia Minor. Many species are found in countries that border the Mediterranean Sea. About 80 species are native to the United States, most of them occurring in the general regions of the Rocky and Cascade mountains and the Sierra Nevada. Although these native clovers are not used as crop plants, they contribute to the range for grazing and wild hay and also supply nitrogen to associated grasses. Of those that are now named, it is possible that many are variants of other species. Only a few are native to the humid Eastern states. Most clovers are highly palatable and nutritious to livestock. The name clover is often applied to members of legume genera other than *Trifolium*. *See Rosales.*

Characteristics. The true clovers are herbaceous annual or perennial plants. However, many perennial species behave as biennials or annuals under attacks of disease or insects, unfavorable climatic or soil conditions, or improper management. In general, clovers thrive under cool, moist conditions although one native species, desert clover (*T. gymnocarpum*), tolerates semiarid conditions. Annual species usually behave as winter annuals where winter conditions are not severe, and as summer annuals at northern latitudes and high altitudes.

Clovers grow best in areas where adequate supplies of calcium, phosphorus, and potassium are naturally present in the soil or where these elements are applied in limestone and fertilizers. They grow from a few inches to several feet tall, depending on the species and the environmental conditions. The leaves have three to five roundish to spearlike leaflets. The florets, ranging from 5 to 200 in number, are borne in heads. Colors include white, pink, red, purple, and yellow, and various mixtures. The basic haploid chromosome numbers of the species are 6, 7, and 8. Somatic chromosomes range from 12 to about 130, forming diploids, tetraploids, and polyploids. The flowers of some species are self-sterile and must be cross-polli-nated before seed will form; the flowers of other species may be self-fertile and capable of self-pollination; others are self-fertile but do not set seed unless the flowers are tripped and the pollen scattered onto the stigma. Cross-pollination and tripping of flowers are effected principally by bees that visit the flowers for nectar and pollen. Seed color ranges from yellow to deep purple, some being bicolored. There are approximately 60,000–700,000 seeds per pound.

Uses. Clovers are used for hay, pasture, silage, and soil improvement. Certain kinds may be used for

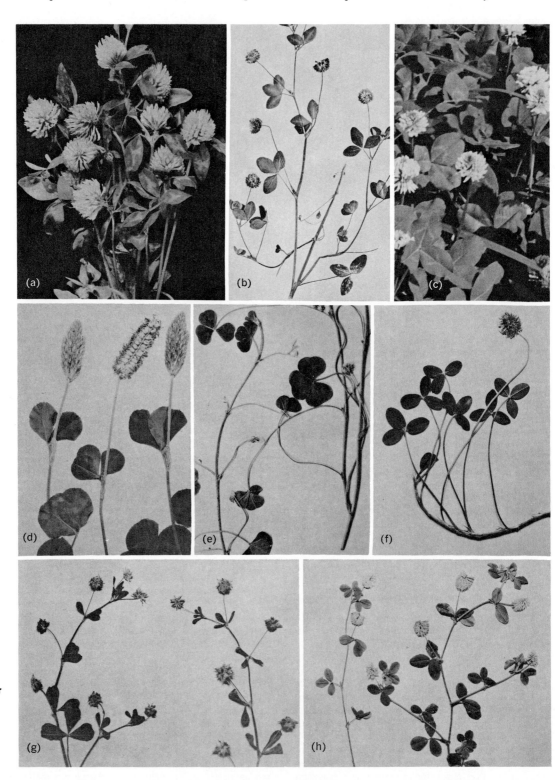

Fig. 1. Important clovers. (a) Red clover (*Trifolium pratense*). (b) Alsike clover (*T. hybridum*). (c) White clover (*T. repens*). (d) Crimson clover (*T. incarnatum*). (e) Subclover (*T. subterraneum*). (f) Strawberry clover (*T. fragiferum*). (g) Persian clover (*T. resupinatum*). (h) Large hop clover (*T. campestre* or *procumbens*). (*USDA*)

all purposes whereas others, because of their low growth, are best suited for grazing. All kinds, when well grown in thick stands, are good for soil improvement. Thoroughly inoculated plants add 50–200 lb of nitrogen per acre (56–222 kg per hectare) when plowed under for soil improvement, the amount added depending on growth, thickness of stand, and length of growing season. Clovers may be grown alone, in combination with grasses and other legumes, or with small grains. In the humid states, or where irrigated, clovers are sown most frequently with small-grain companion crops. In the Corn Belt they are generally spring-seeded, whereas in the Southern states and in the West Coast region they should be seeded in the fall for best results.

Important species. All the clover species of agricultural importance in the United States are introduced (exotic) plants. **Figure 1** shows some of the species most widely used: red clover (*T. pratense*), alsike clover (*T. hybridum*), white clover (*T. repens*), crimson clover (*T. incarnatum*), subclover (*T. subterraneum*), strawberry clover (*T. fragiferum*), persian clover (*T. resupinatum*), and large hop clover (*T. campestre,* or *procumbens*). Other clovers of regional importance, mostly adapted to specific environmental conditions, are rose clover (*T. hirtum*), berseem clover (*T. alexandrinum*), ball clover (*T. nigrescens*), lappa clover (*T. lappaceum*), big-flower clover (*T. michelianum*), and arrowleaf clover (*T. vesiculosum*).

Species characteristics. The following sections discuss the characteristics of the most important species.

Red clover. Red clover is composed of two forms, medium and mammoth, producing two and one hay cuts, respectively. The large purplish-red flower heads are round. An upright-growing perennial, red clover generally persists for 2 years in the northern United States, but behaves as a winter annual in the South. When planted alone or with small grain, seeding rates are 10–15 lb/acre (11–17 kg/hectare). When seeded with grass and other mixtures, 4–8 lb/acre (4.9–9.0 kg/ha) is sufficient. Under favorable growth conditions, seed yields average 70 lb/acre (79 kg/ha); in the West under irrigation they may reach 600–800 lb/acre (675–900 kg/ha). There are several varieties and strains of red clover such as Kenland, Pennscott, Lakeland, Dollard, and Chesapeake. These produce higher yields of forage and are more persistent than common red clover.

Alsike clover. Alsike clover is an upright-growing species that behaves like a biennial. The growth pattern, seeding methods, mixtures, and uses are similar to those of red clover. The flower heads are much like those of white clover in shape and size, but are slightly more pinkish. Alsike clover is more tolerant of wet, poorly drained soils than red clover and occurs widely in mountain meadows of the West.

White clover. White clover, an inhabitant of lawns and closely grazed pastures, is the most important pasture legume in the humid states. The flowers are generally white, but sometimes they are tinged with pink. The stems grow on the soil surface and root at the joints (nodes) as the plant spreads. There are three main types, large, intermediate, and small, with all gradations between. Ladino is the best-known and most widely seeded variety of the large type, and Louisiana Sl represents the intermediate type. The small type is found principally in pastures that are continuously and closely grazed. All types are nutri-

tious and are relished by all classes of livestock and poultry. The protein content ranges from 15 to 30%, depending on plant age and the succulence of growth. Ladino white clover is sometimes used for silage and hay. White clover is mainly grown with low-growing grasses, not being tolerant of the tall-growing kinds. Seeding rates vary from ½ to 2 lb/acre (0.6 to 2.3 kg/ha) depending on the seed mixture. For successful growth, it is important to use the variety or strain best adapted to the particular climatic and soil conditions. Limestone and fertilizer must usually be used to get a thick stand and good growth. Spring seeding is recommended for the Northern states and fall seeding for regions having mild winters. For best growth of the clover, grass-clover mixtures should be grazed or cut frequently.

Crimson clover. Crimson clover is used principally as a winter annual for pasture and as a soil-improving crop from the latitude of the Ohio River southward, and along the West Coast. The yellow seed, approximately 120,000 per pound (260,000 per kilogram), are planted at the rate of 15–20 lb/acre (17–22 kg/ha) during the late summer and fall months in a prepared seedbed or on grass which is closely grazed or cut. The seedlings make a rosette of leafy growth during the winter months. With the advent of spring, growth is rapid; the plant reaches 1½–3 ft (0.4–0.9 m) in height and the flower stems elongate, terminating with pointed crimson flower heads in late spring or early summer. The plant dies when the seed matures. Seed yields average about 250 lb/acre (280 kg/ha) even though large quantities are lost by shattering. Crimson clover is seeded alone, with small grains and grasses, or on grass turf. During the winter it may be grazed when growth reaches 4 in. (10 cm), although if it is too heavily grazed, regrowth is slow. The greatest return for soil improvement is obtained when the largest growth is plowed under. There are several varieties of crimson clover, including Dixie, Auburn, Autauga, Chief, Talladega, and many local strains. When conditions are favorable, all of these will reseed, forming volunteer stands in the fall from seed shattered the previous spring. Common crimson clover does not reseed. When used with Bermuda or other perennial summer-growing grasses, the tall growth of the grass must be closely grazed or clipped.

Subclover. Subclover, a winter annual extensively used for grazing in the coastal sections of the Western states, is the basic pasture crop of the sheep and cattle industry of Australia. The seed are blue-black and the largest of any clover species (about 60,000 per pound or 132,000 per kilogram). The flower heads on decumbent stems are inconspicuous. As the seed develop, fishhook-shaped appendages are produced which, by their twisting action, pull them into the soil, and hence the name subclover. Seed production is large, but the seed are difficult to harvest. When once established, fall stands develop from seed produced the preceding months. The Australian varieties Mount Barker, Tallarook, and Nangeela have proved to be best adapted to most conditions in the United States. Subclover appears to have considerable promise as a pasture legume under many conditions in the southern United States, but better adapted varieties are needed.

Eugene A. Hollowell

Berseem clover. Berseem, or Egyptian clover (*T. alexandrinum*), is used as a cool-season annual forage crop in the states of Washington, Oregon, California, and Arizona and in the southeastern United States, and as a summer annual in northern areas. It is an upright-growing legume with oblong leaflets and hollow stems. It produces yellowish-white florets in heads and has short tap roots. In southern Florida, plants may obtain a height of 18–24 in. (45–60 cm) during a growing season from October to May and heights of 34–42 in. (19–105 cm) in Arizona. Berseem clover's greatest potential is probably as green-chopped forage or pasture. It provides high-quality forage with a crude protein concentration of 20–24% on a dry-weight basis. Berseem clover produces an abundance of seed with yields exceeding 990 lb/acre (1100 kg/ha). Berseem flowers are essentially self-sterile, with cross pollination being accomplished by honeybees.

Cultivars of berseem clover are classified according to their branching behavior; branching influences the number of cuttings that can be made. The cultivars Miscawi and Kahdrawi, which exhibit basal branching, can yield four cuttings per growing season, whereas the Saidi produces both basal and apical branching and can be cut twice. The Fahl cultivar exhibits apical branching and can be cut only once; Nile and Hustler cultivars were developed in the United States and are equal to or superior to Miscawi. A somewhat hardier cultivar Multicut, which tolerates temperatures down to 23°F (-5°C), was released in 1988 for use in parts of California and Mexico; it produces five or six cuttings under irrigation. The release in 1984 of the winter-hardy, reseeding cultivar Bigbee, extended the range of adaptation of berseem clover to Tennessee and Oklahoma and enabled producers to obtain reseeding stands. Prior to the release of Bigbee, berseem clover was considered to be a non-winter-hardy, nonreseeding annual forage legume.

W. E. Knight

Other clovers. Strawberry clover, a perennial, is a nutritious pasture plant similar in growth habit to white clover, for which it is frequently mistaken. The pinkish flower head looks like a strawberry, and hence the common name. It is highly tolerant of the wet, salty soil common to the Western states.

Persian clover is a winter annual used mostly for pasture and soil improvement but is also an excellent silage plant. It is particularly adapted to the wet, heavy soils of the lower southern part of the United States. The flowers are light purple, and the mature seed are enclosed in balloonlike capsules that shatter easily, float on water, and are readily blown about by the wind. Persian clover is relished by all classes of livestock.

Large hop clover and small hop clover, widely distributed as winter annuals in the Southern states, are used in pastures. They are more tolerant of low soil fertility than the other species and less productive, but are highly palatable to livestock. The flower heads are bright yellow and, when mature, look like the flowers of the hop plant. The two species are similar.

Rose clover is a relatively new winter annual and appears to be best adapted to the foothill rangelands of California.

Other species are ball clover, lappa clover, big-flower clover, and arrowleaf clover, all winter annuals. Lappa clover grows best in the heavy dark marl soils of the South, whereas ball, big-flower, and arrowleaf clovers appear to be more widely adapted. All are used for pasture, although big-flower and arrowleaf clovers may also be used for hay.

Seaside clover (*T. willdenovii*), white-tipped clover (*T. variegatum*), and long-stalked clover (*T. longipes*) are the most widely distributed of the native western species. All produce thick stands and good growth under varying conditions. SEE GENETICS; LEGUME FORAGES; REPRODUCTION (PLANT).

SWEET CLOVER

Sweet clover is the common name applied to all but one species of legumes of the genus *Melilotus*, order Rosales. The exception is sour clover (*M. indica*).

Origin and distribution. There are approximately 20 species of sweet clover. Some of the biennial species have an annual form. The roots of the biennials develop crown buds from which the second-year growth arises; the roots of annuals do not develop crown buds. Sweet clovers are native to the Mediterranean region and adjacent countries, but several are widely scattered throughout the world, generally by chance introduction. None is indigenous to the United States. White sweet clover (*M. alba*) and yellow sweet clover (*M. officinalis*) are important forage and soil-improvement plants in the United States and Canada and are found growing along roadsides and in waste places in every state (**Fig. 2**). Sour clover, a yellow-flowered winter annual, is of some value for soil improvement only along the Gulf Coast and in southern New Mexico, Arizona, and California. Improved varieties of white and yellow sweet clovers are available for farm use.

Uses. Sweet clover is used as a field crop in regions of the United States and Canada where the rainfall is 17 in. (42 cm) or more during the growing season, where the soil is neutral, or where limestone and other needed minerals are applied. It is most extensively grown in the Great Plains and Corn Belt, either alone or in rotations with small grains and corn,

Fig. 2. White sweet clover (*Melilotus alba*). (USDA)

and is used for grazing, soil improvement, and hay.

Except for those of certain improved varieties, the plants are somewhat bitter because of the presence of coumarin. External and internal bleeding of animals may result from feeding spoiled sweet-clover hay or improperly preserved silage containing sweet clover, a decomposition product of coumarin (4-hydroxycoumarin, commonly called dicumarol), which develops during spoilage, being the toxic principle. Research has led to its use in medicine and for making warfarin, a rodenticide.   *Eugene A. Hollowell*

DISEASES

Diseases of clovers are caused by bacteria, fungi, nematodes, viruses, a mycoplasma, and air pollutants. These diseases reduce forage yield, quality, and stand. Frequently, perennial clovers are productive for only 1 or 2 years because of diseases. Crop losses are best minimized by growing adapted cultivars that are tolerant of or resistant to one or more of the major diseases. Fungi and viruses are the most important pathogens.

Fungal diseases. Fungi, including *Pythium, Fusarium,* and *Sclerotinia,* cause root and crown rots that reduce stand establishment and persistence (**Fig. 3**). These pathogens can attack clover plants at any

Fig. 4. Typical petiole and stem lesions of northern anthracnose disease of red clover, caused by the fungus *Kabatiella caulivora.*

developmental stage. Plants not killed are weakened, and are thus predisposed to further injury from physical and biological stress.

Fungi, including *Stemphylium, Colletotrichum,* and *Kabatiella,* cause leaf and stem diseases, which usually do not kill the plants but reduce their productive and competitive capabilities (**Fig. 4**).

Viral diseases. Most clovers are susceptible to destructive viruses that cause leaf mottling and distortion and reduced plant vigor. Viruses may be seed-borne or spread by insects and mowing. SEE PLANT PATHOLOGY; PLANT VIRUSES AND VIROIDS.

Kenneth T. Leath

Bibliography. N. C. Brady (ed.), *Adv. Agron.,* 35:165–191, 1982; W. L. Graves et al., Registration of Multicut berseem clover, *Crop Sci.,* 29:235–236, 1989; M. E. Heath, D. S. Metcalfe, and R. F. Barnes (eds.), *Forages: The Science of Agriculture,* 4th ed., 1985; W. E. Knight, Registration of Bigbee berseem clover, *Crop Sci.,* 25:571–572, 1985; N. L. Taylor (ed.), *Agronomy Monograph, no. 24,* ASA-CSSA-SSSA, 1984; J. L. Wheeler and R. D. Mochrie, *Forage Evaluation: Concepts and Techniques,* 1982.

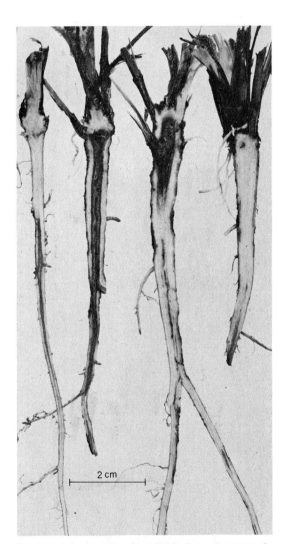

Fig. 3. Red clover roots cut longitudinally to show type of injury caused by crown and root rot organisms.

Clupeiformes

An order of teleost fishes of the subclass Actinopterygii that includes the herrings, sardines, anchovies, and their allies. As presently restricted, the order is much less inclusive than the older Clupeiformes (or Isospondyli), from which the Elopiformes, Osteoglossiformes, Salmoniformes, and Cetomimiformes (in part) have been removed.

The Recent clupeiforms are classified in 2 suborders, 3 (by some authorities as many as 6) families, 71 genera, and about 300 species.

Structure. The Clupeiformes are generalized teleosts. They may be related to the Elopiformes, but alternatively may be independently descended from ancestral early teleosts or advanced holosteans. Clupeiforms are mostly silvery and compressed. A distinctive feature is the extension of the cephalic canal system onto the operculum; the lateral line is undeveloped on the trunk except in *Denticeps.* There are distinctive intracranial swim bladder diverticula encased in bony capsules, and there is an intracranial space into which open the major cephalic sensory ca-

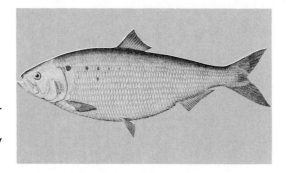

American shad (*Alosa sap-idissima*), which grows to a length of 30 in. (76 cm). (*After G. B. Goode, Fishery Industries of the United States, Sect. 1, 1884*)

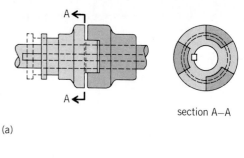

(a)

section A—A

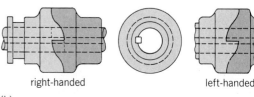

right-handed   left-handed

(b)

Fig. 1. Positive clutches. (a) Square-jaw clutch. (b) Spiral-jaw clutch. (*After E. A. Avallone and T. Baumeister III, eds., Marks' Standard Handbook for Mechanical Engineers, 9th ed., McGraw-Hill, 1987*)

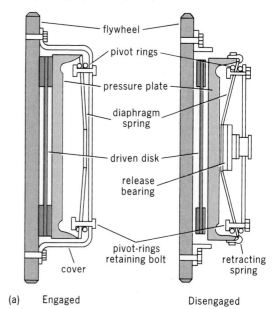

(a)   Engaged   Disengaged

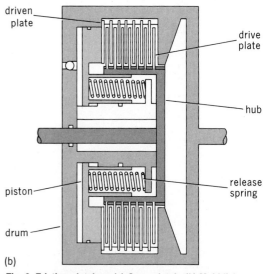

(b)

Fig. 2. Friction clutches. (a) Cone clutch. (b) Multidisk clutch. (*After E. A. Avallone and T. Baumeister III, eds., Marks' Standard Handbook for Mechanical Engineers, 9th ed., McGraw-Hill, 1987*)

nals. They lack fin spines, adipose fin, and gular plate. Branchiostegal rays number 5 to 20. The middle of the belly often bears one or a series of strong scutes. The pelvic fin is abdominal in position, free from the shoulder girdle; the pectoral fin is placed low on the side; and a mesocoracoid arch is present (see **illus**.). The cycloid scales are usually thin and loosely attached. The upper jaw is bordered by premaxillae and maxillae, but the detention is usually feeble. SEE SCALE (ZOOLOGY); SWIM BLADDER.

Distribution and ecology. Clupeiform fishes have left a rich fossil record, especially from the Upper Cretaceous to the early Tertiary, but appeared first in the Upper Jurassic. Some herrings and a few anchovies live in lowland rivers and lakes, and others such as the shad and alewife enter rivers to reproduce; but the majority of clupeiforms occur in bays or shore waters of tropical, temperate, or even northern seas, where they commonly make up enormous schools. None inhabits deep water. Most feed on plankton or other minute organisms, which they capture by straining with the numerous long gill rakers. Thus they are efficient converters from the base of the food chain to fish flesh. The wolf herrings (*Chirocentrus*) of the Indo-Pacific, however, are predatory clupeids provided with strong teeth.

Economic importance. Some of the great fisheries of the world are based on the conversion of plankton by clupeiform fishes; the tremendous anchovy fishery off western South America; the California fisheries for anchovy and (until depleted) Pacific sardine; the menhaden fisheries of the western Atlantic; and the herring and sardine fisheries of northern seas. The catch is processed variously into oil, fertilizer, or fish meal or is prepared directly for human consumption. Indirectly, clupeiforms are of tremendous importance in the food cycles of piscivorous fishes and of some sea birds which depend on them as dietary staples. SEE CETOMIMIFORMES; ELOPIFORMES; MARINE FISHERIES; OSTEOGLOSSIFORMES; SALMONIFORMES.

Reeve M. Bailey

Clutch

A coupling device which permits the engagement and disengagement of coupled shafts during rotation. There are four major types: positive, friction, hydraulic, and electromagnetic.

Positive clutch. This type of clutch is designed to transmit torque without slip. It is the simplest of all shaft connectors, sliding on a keyed shaft section or a splined portion and operating with a shift lever on a collar element. The jaw clutch is the most common type of positive clutch. This is made with square jaws (**Fig. 1***a*) for driving in both directions, or spiral jaws (Fig. 1*b*) for unidirectional drive. Engagement speed should be limited to 10 revolutions per minute (rpm) for a square-jaw clutch and 150 rpm for a spiral-jaw clutch. If desengagement under load is required, the jaws should be finish-machined and lubricated. *See Machine key; Splines.*

Friction clutch. This type of clutch is designed to reduce coupling shock by slipping during the engagement period. It also serves as a safety device by slipping when the torque exceeds its maximum rating. The three common designs for friction clutches are cone, disk, and rim, according to the direction of contact pressure. *See Friction; Torque.*

Cone clutch. This is an axial type clutch. The surfaces of the cone clutch are sections of a pair of cones. This shape uses the wedging action of the mating surfaces under relatively small axial forces to transmit the friction torque. These forces may be established by the compression of axial springs or by the outward displacement of bell-crank levers to apply axial thrust to the conical surfaces.

Disk clutch. This is also an axial-type clutch. The disk clutch may consist of a single plate or multiple disks (**Fig. 2**). With the development of improved friction material, disk clutches have become more common than cone clutches, with wide applications in the industrial and automotive fields. Disk clutches are not subjected to centrifugal effects, present a large friction area in a small space, establish uniform pressure distribution for effective torque transmission, and are capable of effectively dissipating the generated heat to the external housing. The disk clutch may be operated dry, as in most automobile drives, or wet by flooding it with a liquid, as in heavier automotive power equipment and in industrial engines. The advantage of wet operation is the ability to remove heat by circulating the liquid enclosed in the clutch housing. Typical friction materials for clutches and design data are shown in the **table**.

Rim clutch. This is another form of frictional contact clutch that has surface elements applying pressure to the rim either externally or internally. Rim clutches may be subdivided into two groups: those employing either a frictional band or block (**Fig. 3**) to make contact with the rim; and the overrunning clutch (**Fig. 4**), employing the wedging action of a roller or ball. The

Fig. 3. Internal-expanding centrifugal-acting rim clutch. (*Hilliard Corp.*)

clutches of the former type may have hinged shoes connected by an expanding or contracting kinematic mechanism on a hub fixed to one shaft and riding of a drum or rim attached to a hub on a second shaft. They have expansion rings or external bands that, when displaced, are capable of transmitting torque to the clutch rim. The rim may be grooved to increase the surface area; the clutch may have a double grip, internal and external; or pneumatically expanding flexible tubes may set the friction surfaces against the rim (**Fig. 5**). Because the clutch rotates with the shaft when it is engaged, its balancing must be considered.

For overrunning clutches, the driven shaft can run faster than the driving shaft. This action permits freewheeling as the driving shaft slows down or another source of power is applied. Effectively this is a friction pawl-and-ratchet drive, wherein balls or rollers become wedged between the sleeve and recessed pockets machined in the hub. The clutch does not slip when the second shaft is driven, and is released automatically when the second shaft runs faster than the driver. Specially shaped struts or sprags may replace the balls; springs may be used to hold the pawl elements in position. Mechanisms have been devised to reverse the direction of operation. *See Pawl; Ratchet.*

Hydraulic clutch. Clutch action is also produced by hydraulic couplings, with a smoothness not possible

Friction materials for clutches and design data*

Material	Friction coefficient		Maximum temperature		Maximum pressure	
	Wet	Dry	°F	°C	lb/in.²	kPa
Cast iron on cast iron	0.05	0.15–0.20	600	320	150–250	1000–1750
Powdered metal[†] on cast iron	0.05–0.1	0.1–0.4	1000	540	150	1000
Powdered metal[†] on hard steel	0.05–0.1	0.1–0.3	1000	540	300	2100
Wood on steel or cast iron	0.16	0.2–0.35	300	150	60–90	400–620
Leather on steel or cast iron	0.12	0.3–0.5	200	100	10–40	70–280
Cork on steel or cast iron	0.15–0.25	0.3–0.5	200	100	8–14	50–100
Felt on steel or cast iron	0.18	0.22	280	140	5–10	35–70
Woven asbestos[†] on steel or cast iron	0.1–0.2	0.3–0.6	350–500	175–260	50–100	350–700
Molded asbestos[†] on steel or cast iron	0.08–0.12	0.2–0.5	500	260	50–150	350–1000
Impregnated asbestos* on steel or cast iron	0.12	0.32	500–750	260–400	150	1000
Carbon graphite on steel	0.05–0.1	0.25	700–1000	370–540	300	2100

*From J. E. Shigley and L. D. Mitchell, *Mechanical Engineering Design*, p. 742, McGraw-Hill, 1983.
[†]The friction coefficient can be maintained within ± 5% for specific materials in this group.

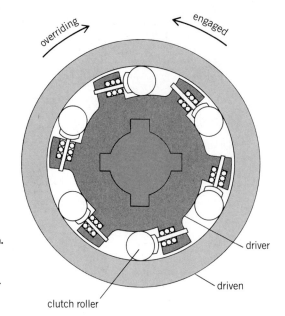

Fig. 4. Overrunning clutch. (*After E. A. Avallone and T. Baumeister III, eds., Marks' Standard Handbook for Mechanical Engineers, 9th ed., McGraw-Hill, 1987*)

be of the dry can type. Upon the application of a magnetic field, the particles tend to coalesce, thereby creating viscous or friction forces between the clutch members. The wet type provides the smoother operation; the dry type is freer from loss of particles. Torque at a fixed moderate excitation is independent of slip.

Eddy-current clutch. The current that is induced by a magnetic field can serve as the torque-transmitting means. In the eddy-current clutch, a coil-carrying direct current in one rotating member establishes poles that generate currents in the adjacent smooth conductive ring of the mating member. The coupling acts as an untuned damper and therefore does not transmit torsional vibration. An eddy-current clutch consists of a separately excited motor and generator that are combined into a single machine. Clutch action can be obtained by electrically connecting a generator on the driving shaft to a motor on the driver. *See Dynamometer; Eddy current; Electric rotating machinery.*

Hysteresis clutch. For low-power applications where continuous control with slip is needed, as in instru-

when using a mechanical clutch. Automatic transmissions in automobiles represent a fundamental use of hydraulic clutches. *See Automotive transmission; Fluid coupling; Hydraulics.*

Electromagnetic clutch. Magnetic coupling between conductors provides a basis for several types of clutches. The magnetic attraction between a current-carrying coil and a ferromagnetic clutch plate serves to actuate a disk-type clutch. Slippage in such a clutch produces heat that must be dissipated, and wear that reduces the life of the clutch plate. Thus the electromagnetically controlled disk clutch is used to engage a load to its driving source. A typical unit 24 in. (61 cm) in diameter and weighing about 600 lb (270 kg) develops 2400 ft-lb (3300 joules) of torque when excited at 2 amperes and 115 dc volts; at a maximum safe speed of 1200 rpm it transmits 540 hp (400 kW). Multiple interleaved disks alternately splined to the driving and driven shafts provide a compact structure (**Fig. 6**). There are three basic types of electromagnetic clutches: magnetic fluid and powder, eddy-current, and hysteresis. *See Electromagnet.*

Magnetic fluid and powder clutch. These may use finely divided magnetic particles in an oil carrier or

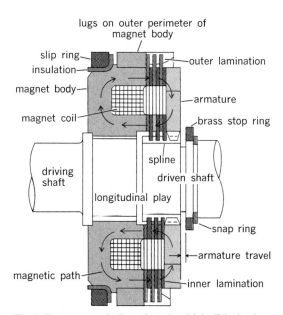

Fig. 6. Electromagnetically activated multiple-disk clutch.

ment servos, a hysteresis clutch may be used. Direct-current excitation of a coil in one part generates a steady flux field that magnetizes an iron ring on the other part, the ring having high magnetic retentivity. The action induced poles in the iron ring. Attraction between the induced poles in the ring and the control field produces a torque that opposes rotation between the two parts of the clutch. Within its load limit the hysteresis clutch transmits rotation without slip. As with the eddy-current clutch, there is no contact surface to wear so that the characteristics are stable and the life is long. *See Hysteresis; Shafting; Torque converter.*

Y. S. Shin

Bibliography. E. A. Avallone and T. Baumeister III (eds.), *Marks' Standard Handbook for Mechanical Engineers,* 9th ed., 1987; J. E. Shigley and L. D. Mitchell, *Mechanical Engineering Design,* 4th ed., 1983.

Fig. 5. External-contracting clutch that is engaged by expanding the flexible tube with compressed air. (*Twin Disc, Inc.*)

Clypeasteroida

An order of exocyclic Euechinoidea in the superorder Neognathostomata which have a monobasal apical system in which all the genital plates fuse together. The ambulacra are petaloid on the aboral side and wider than the interambulacra on the adoral side. The adult has keeled teeth without lat-

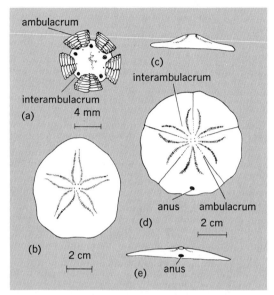

Typical clypeasteroids. *Laganum*, (*a*) monobasal apical system, (*b*) aboral aspect, and (*c*) posterior aspect. *Arachnoides*, (*d*) aboral aspect, and (*e*) posterior aspect.

eral flanges (see **illus.**). The order probably arose in the Late Cretaceous from the Holectypoida. Tertiary fossils are very common, and extant species range the coastlines of tropical and temperate seas. Relatively short time spans of some fossil species make them valuable as index forms in stratigraphy. Many species are gregarious, and occur in closely packed colonies of perhaps millions of individuals. There are 16 families arranged in four suborders. SEE ECHINODERMATA; EUECHINOIDEA; HOLECTYPOIDA; NEOGNATHOSTOMATA.

Howard B. Fell

Bibliography. J. W. Durham, Classification of clypeasteroid echinoids, *Univ. Calif. (Berkeley) Publ. Geol. Sci.*, 31(4):73–198, 1955; D. Nichols, *Echinoderms*, 1962.

Cnidospora

A subphylum of spore-producing Protozoa containing the two classes Myxosporidea and Microsporidea. Myxosporidea contains the three orders Myxosporida, Actinomyxida, and Helicosporida. Microsporida is the only order in the class Microsporidea. Cnidospora are parasites of cells and tissues of invertebrates, fishes, a few amphibians, and turtles. SEE MICROSPORIDEA; MYXOSPORIDEA.

A distinctive structure of this subphylum is the spore, which contains one or more protoplasmic masses called sporoplasms. The spore also contains one or more filaments which lie coiled within the spore proper or within one or more polar capsules.

The spore membrane may consist of a single chitinous piece or two or more parts called valves.

When the spore is ingested by a new host, the filaments are extruded and at the same time the sporoplasm is released. The sporoplasm, now called an amebula, reaches the specific site of infection directly through the intestinal wall or by way of the bloodstream. The amebula becomes a trophozoite, a stage in the life cycle, which feeds and grows at the expense of the host. Asexual reproduction by repeated binary fission, multiple fission, internal and external budding, or plasmotomy results in cells which develop into sporonts and eventually into sporoblasts, producing one or more spores. SEE SPOROZOA.

Because of similarity in multinucleated trophozoites, some protozoologists suggest that the Cnidospora have evolved from "slime mold–like" (Mycetozoia) ancestors. SEE MYCETOZOIA; PROTOZOA.

Ross F. Nigrelli

Coal

A brown to black combustible rock that originated by accumulation and subsequent physical and chemical alteration of plant material over long periods of time, and that on a moisture-free basis contains no more than 50% mineral matter. The initial accumulation of plant debris occurred in various wet environments, commonly peat swamps, where dying plants were largely protected from decay by a high water table and oxygen-deficient water. However, many changes of the original vegetable matter were brought about by bacteria, fungi, and chemical agents. The accumulating spongy, water-saturated, plant-derived organic material known as peat is the precursor of coal. The coalification process transforms peat into lignite or brown coal, subbituminous coal, bituminous coal, and anthracite. This progression is known as a coalification series. Increasingly deeper burial under hundreds to thousands of feet of younger sediments is required to advance coalification to the bituminous coal and anthracite stages. Overburden pressure and heat, as well as length of exposure to them, determine what degree of coalification is reached.

The types of plants contributing to peat accumulations vary greatly, depending on the regional and local environments of deposition (for example, water depths and fluctuations, climate, and flooding by nearby streams or the sea). SEE FOSSIL FUEL; GRAPHITE; LIGNITE.

FORMATION

Economic coal deposits originate from accumulations of plant material (peat) in wetlands commonly referred to as moor, swamp, mire, bog, fen, muskeg, or peatland. Most peat-forming plant material accumulates at or close to the place of growth (autochthonous origin), but some organic material may be transported before deposition, for instance, along beaches or in lakes (allochthonous origin). Autochthonous peats often form in various kinds of regionally coexisting depositional environments; low-lying swamps with forested, marshy, and open water environments (low moors) and raised peat bogs (high moors) are common peat-forming environments. Allochthonous plant material may contribute significantly to otherwise autochthonous peat deposits as open water areas develop and disappear in the evolutionary course of a peat swamp. Coal beds commonly are underlain by seat earths, remnants of ancient soils in which the initial swamp plants were rooted (hence the common occurrence of stigmaria in seat earths). *See Bog; Muskeg; Swamp, marsh, and bog.*

Examples of present-day peat accumulation abound both in humid temperate climates and in subtropical to tropical climates. In North America, peat-forming environments are found in the Everglades of Florida, the Okefenokee Swamp of Georgia and Florida, the Dismal Swamp of Virginia, and in numerous depressions with poor drainage that were created during the ice ages in the plains between the Rocky and Appalachian mountains. In the tropics, thick peat deposits are known in the coastal areas of Sumatra, for instance. Ultimate preservation of deposits of peat depends on long-term subsidence of the area and on burial under younger sediments. *See Depositional systems and environments; Sedimentology.*

The nature of a peat deposit depends on the kind of plants composing the accumulation: sedges, reeds, and other herbaceous vegetation; forest trees and bushes; mosses; and so forth. Such differences create different coal facies that can still be identified in highly coalified coal seams. *See Coal paleobotany.*

Freshly deposited plant material is attacked by microorganisms, fungi, and chemical agents in air and water, enzymes and oxygen in particular. This leads to degradation of the plant material often to the point where the original plant structure is no longer recognizable, and new organic compounds are formed. Cellulose, a major constituent of cell walls, and proteins are much less resistant to decomposition than other chemical plant constituents such as lignin, tannins, fats, waxes, and resins. Humic acids and other humic substances are important decomposition products resulting from the peatification process. Fungi and aerobic bacteria depend on the presence of oxygen to degrade the organic material. During burial, as oxygen is consumed, anaerobic bacteria take over degradation. Woody material passes through a plastic, gellike stage before it solidifies to huminite and eventually vitrinite. As burial depth increases, overburden pressure causes increasing compaction (reduction in porosity and expulsion of moisture), and temperature increases promote chemical changes of the organic components of coal. Peat is transformed into coal through the process of coalification. *See Diagenesis; Humus; Peat.*

ANALYSES

Many kinds of chemical, physical, petrological, and technological analyses are commonly performed on coal samples to ascertain charateristics and origin. The selection of analyses depends on the information needed. Many of these analyses have been standardized at both national and international levels [for example, American Society for Testing and Materials (ASTM) and International Standardization Organization (ISO)].

Megascopic coal description. Megascopic descriptions, often with the help of a hand lens, are routinely performed in the field, in mines, and on diamond drill cores. The detail of description varies between merely distinguishing coal from impure coal, carbonaceous shale, and other rocks, and quite detailed descriptions in terms of coal types or lithotypes, such as banded and nonbanded coal. The latter are further subdivided into bright coal, semisplint (semibright) coal, and

splint (dull) coal for banded coal; cannel and boghead coal for nonbanded coal; or the lithotypes vitrain, attrital coal (clarain, durain), and fusain.

Megascopic descriptions are primarily used for characterizing coal beds in geologic investigations to study lateral variabilities in thickness and structure. Their value in predicting specific coal properties such as ash content or coal rank is limited, and they are generally considered to be only the first step in the characterization of a coal deposit. However, megascopic descriptions are the principal basis for coal resource and reserve determinations.

Proximate analysis. This is probably the most commonly performed type of chemical analysis on coal samples. It involves determination of the ash, moisture, volatile matter, and fixed-carbon contents, which total 100%.

High-temperature ash. Combustion of the organic material at 1290–1380°F (700–750°C) leaves behind the high-temperature ash. At these high temperatures, certain minerals disintegrate, leading to both loss of volatile products and reaction with oxygen from the air to form oxides. Some volatile decomposition products react with the ash before they can escape. Therefore, the weight of the high-temperature ash may differ significantly from the weight of the coal's original mineral matter. High-temperature ash is composed primarily of oxides, altered clay minerals, some carbonates, and sulfates.

Moisture content. This term has different meanings depending on when, how, and on what sample type it is determined, for example, as received, surface, air-dried, bed, inherent, and equilibrium moistures. Alternatively, the determination may be made on raw or washed coal, on various grain sizes, and so forth. The moisture content in the fully saturated stage, but without presence of surface moisture, is used for determining coal rank. Bed, inherent, and equilibrium moistures are representative of this type of moisture content.

Volatile matter and fixed carbon. These are derived from a small-scale retorting test which is performed under rigidly controlled conditions at a temperature of $1740 \pm 36°F$ ($950 \pm 20°C$). The process results in the disintegration of coal into volatile matter, a complex mixture of gases (exclusive of water), and a cokelike residue rich in carbon (fixed carbon).

Mineral matter. This term refers to both inorganic constituents that occur as discrete mineral grains and so-called inorganic elements that are more or less homogeneously distributed through coal. Mineral matter can be calculated from a standard chemical analysis based on a relationship known as the Parr formula, but is often determined directly by low-temperature oxidation of the organic matter of coal in an oxygen-rich plasma at the relatively low temperature range 250–300°F (120–150°C). Minerals remain essentially unaltered, and the resulting low-temperature ash can be subjected to a wide range of analyses (for example, x-ray diffraction) to determine minerals present. *See Mineral.*

Ultimate (elemental) analysis. The elements carbon, hydrogen, oxygen, sulfur, and nitrogen make up essentially all the organic substance of coal and are determined, along with moisture and ash, by the ultimate analysis. The first four of these so-called organic elements are also present in common inorganic components of coal such as moisture, carbonates, sulfides, sulfates, and oxides, making attribution to either the organic or the inorganic constituents of coal potentially difficult. Ultimate analyses are significantly more involved than proximate analyses. Both analyses are now commonly done on computer-controlled sophisticated instruments.

Calorific or heating value. This parameter is determined by observing temperature changes that result from burning a known quantity of coal in an adiabatic calorimeter bomb. In the United States, the calorific value is reported in British thermal units given off per pound of coal burned (Btu/lb); the equivalent SI unit is megajoules per kilogram (MJ/kg) or, until relatively recently, cal/g or kcal/kg (1000 Btu/lb = 2.326 MJ/kg = 560 kcal/kg). The calorific value of a coal is influenced not only by the nature of the organic material of which it is made but also by the proportion of noncombustible material present, primarily mineral matter and moisture. SEE CALORIMETRY; THERMOCHEMISTRY.

Major, minor, and trace elements. Coal contains essentially every element in the periodic table, at least in trace amounts. The concentration of ash-forming elements in whole coal ranges from more than 0.5% for major elements such as aluminum, calcium, iron, and silicon; 0.02–0.5% for minor elements such as potassium, magnesium, sodium, titanium, chlorine, and others; to less than 0.02% (200 parts per million) for trace elements such as arsenic, beryllium, mercury, cadmium, and most other remaining elements of the periodic table. Knowledge of the concentration of these elements is required in the evaluation of coals for combustion and conversion (for example, relative to corrosion, boiler fouling, and poisoning of catalysts) and to assess potential negative environmental impacts resulting from coal use. SEE PERIODIC TABLE.

A multitude of sophisticated instrumental analytical methods for determining trace elements in coal have become available since the 1960s, such as neutron activation analysis, inductively coupled plasma analysis, atomic absorption spectroscopy, x-ray fluorescence spectroscopy, and mass spectroscopy. SEE ACTIVATION ANALYSIS; ATOMIC SPECTROMETRY; MASS SPECTROMETRY; X-RAY FLUORESCENCE ANALYSIS.

Coal macerals. Macerals are the microscopically recognizable constituents of coal. Microscopic analyses are normally performed on representative crushed coal samples embedded in epoxy and polished for observation under incident light, generally with oil immersion lenses. A sufficiently large number of points (500–1000) evenly distributed over the sample is selected for maceral analysis to determine the percent of macerals in the sample by volume. The reflected-light method permits identification of a few nonclay minerals during the course of the maceral analysis. The most accurate determinations of amount and composition of mineral matter are accomplished through other analyses, for example, low-temperature ashing.

Reflectance of vitrinite. The percentage of normally incident light that is reflected from the polished surface of vitrinite (a type of maceral) under oil immersion has become a commonly used measure of coal rank. A reflected-light microscope equipped with a photomultiplier is used. The small size of the measuring spot (about 5 micrometers in diameter) and the large magnification used (500–1000 times) permit selection of highly comparable material for this analysis, in contrast to chemical analyses that test bulk coal with variable mixtures of macerals and minerals. Vitrinite is the preferred material on which to measure reflectance, not only because of its common occurrence in most coals but also because of its homogeneous appearance, its easy recognition under the microscope, and its rapid change in reflectance with increasing coal rank.

Density, porosity, and internal surface area. Because of the large proportion of very small pores (few to few tens of angstroms in diameter), true values for these parameters are difficult to obtain with standard methods. Only very small molecules or atoms (for example, helium) are able to penetrate these micropores. Determinations with gases or liquids whose molecules are of the same order of magnitude as the micropores or larger yield only apparent values. For example, in one experiment the internal surface area of a certain sample of bituminous coal was measured separately with nitrogen and then carbon dioxide; the values obtained were 9 and 205 m^2/g, respectively.

Washability. A large proportion of mined coal is cleaned before it is sold. Washability tests in the laboratory are performed on drill-core and test-pit samples to predict the cleanability of a given coal. Coal samples are crushed to sizes comparable to those that it is anticipated will be produced by a planned mine, and then they are subjected to float-sink tests in heavy liquids. The amounts floating and sinking in each gravity fraction are measured, and their ash and sulfur contents as well as other parameters (for example, heating value) are determined. The data are tabulated and plotted to produce washability curves that characterize a coal's cleanability.

Plastic and coking properties. A major use of coal is coke making, which involves heating finely crushed coal in coke ovens in the absence of air. Certain bituminous coals become plastic or fluid during heating as they disintegrate into gases, liquids, and solids (char, coke). Several empirical tests have been used widely to characterize the behavior of coal during heating.

Free Swelling Index (FSI). This is probably the most widely run test. It requires rapid heating of 1 g (0.35 oz) of finely ground coal to a temperature of $1500 \pm 9°F$ ($820 \pm 5°C$) in a lid-covered crucible. A coke button is formed and its shape is compared to nine standard shapes. The FSI test provides a rapid and inexpensive assessment of the coke-forming behavior of coal.

Plastometer test. By using a device known as the Gieseler plastometer, a small amount of finely crushed coal, in which a small stirrer is submerged, is heated in a retort. A constant torque is applied to the stirrer and its rate of rotation is recorded; the coal becomes plastic while the temperature is raised at a standard rate. The temperatures of initial softening, maximum fluidity, and resolidification, as well as the maximum dial movement in dial divisions per minute (ddpm), are recorded.

Dilatometer test. This test is widely used outside the United States to characterize the behavior of coal during heating in the range 570–1020°F (300–550°C). It measures the volume changes of a sample of coal that is heated at a rate of 5.4°F/min (3°C/min), as it first becomes plastic and then solidifies to coke. A finely ground sample of coal is heated inside a crucible with a piston resting on the sample. The displacement of the piston is recorded as a percentage of the original sample volume.

Other tests. The Gray-King and ROGA tests are commonly performed in several European countries. These essentially standardized laboratory-scale coking tests produce useful index numbers. *See* Coke.

Strength and hardness. Strength and hardness are important coal properties in mine stability, coal winning, and comminution for preparation and utilization. A number of tests are commonly performed in the laboratory to obtain information on a coal's compressive and shearing strengths and hardness; an example is the Hardgrove grindability test. The values thus obtained are only index numbers because of the small size of sample used. They must be translated with caution when full-scale industrial operations are involved.

Ash fusion. During combustion of coal in boilers, the noncombustible portion, such as minerals and inorganic elements, either is volatilized or becomes ash, which forms deposits within the boiler or is removed as fly ash or bottom ash. The ash-fusion test identifies the temperatures at which the ash undergoes major phase changes. In general, for combustion uses of coal, ash that fuses at high temperatures (above 2400°F or 1538°C) is preferred to ash that fuses at low temperatures (below 2000°F or 1093°C). The range of ash-fusion temperatures is about 1600–2800°F (871–1538°C). The properties and relative abundances of minerals present may be critical in the formation of ash deposits in a boiler, as well as in the corrosion of boiler tubes. High alkali and alkali chloride contents are known to cause such problems in boilers. Boilers are designed for certain ash fusion properties.

CLASSIFICATION SCHEMES

The parameters obtained from coal analyses are used to classify coal for general as well as specific purposes. Both national and international standardization organizations have adopted classification schemes that are used in the coal trade, for statistical purposes, and to compile coal resources and reserves by classification category. One of the most widely used classification systems is the American Society for Testing and Materials (ASTM) Standard Classification of Coals by Rank (**Table 1**).

The International Standards Organization (ISO) is considering a new standard that attempts to incorporate both coal rank and coal facies parameters. This standard would replace the United Nations Economic Commission of Europe's classification scheme for hard coal, which combines rank (class), caking (group), and coking (subgroup) parameters into a three-digit code system (**Table 2**).

For the purpose of classification, coal analyses must be standardized, through computation, to a common basis, such as ash-free (af), moisture-free (mf) or dry (d), moisture-and-ash-free (maf) or dry ash-free (daf), mineral-matter-free (mmf), dry mineral-matter-free (dmmf), and so forth. Generally, values are determined on a moisture-free or dry basis and then converted to the desired bases by computation.

COAL FACIES

The coal facies reflects the environments of deposition during accumulation of plant material. The types of plants growing, their preservation, and the kinds and amounts of mineral matter introduced vary with the depositional environments. Coal facies can be recognized both megascopically and microscopically through all or most ranks.

Nonbanded coal. Nonbanded coals are homogeneously fine-granular and devoid of distinct megascopic layers. They have a greasy luster and conchoidal fracture. They formed from subaqueous deposition of finely comminuted plant detritus, commonly with significant amounts of spore or pollen grains, algal remains, wax and resin granules, and various other small plant fragments. Well-known nonbanded coals are cannel coal (composition dominated by spores) and boghead coal (predominance of algae).

Table 1. American Society for Testing and Materials Classification of coals by rank (in box) and other related properties*

ASTM class	ASTM group	1000 Btu/ lb[1]	Agglomer- ating	Volatile matter[2] %	MJ/kg	Maximum reflectance,[3] %	Moisture,[1] %	C,[2] %	O,[2] %	H,[2] %
	Peat	1.0–6.0[†]	No	72–62	2.3–14.0	0.2–0.4	95–50	50–65	42–30	7–5
Lignite	Lignite B	<6.3[†]	No	65–40	<14.6[†]	0.2–0.4	60–40[†]	55–73	35–23	7–5
	Lignite A	6.3–8.3[†]	No	65–40	14.6–19.3[†]	0.2–0.4	50–31[†]	55–73	35–23	7–5
Subbitu- minous	Subbituminous C	8.3–9.5[†]	No	55–35	19.3–22.1[†]	0.3–0.6	38–25[†]	60–80[‡]	28–15[‡]	6.0–4.5
	Subbituminous B	9.5–10.5[†]	No	55–35	22.1–24.4[†]	0.3–0.6	30–20[†]	60–80[‡]	28–15[‡]	6.0–4.5
	Subbituminous A	10.5–11.5[†]	No	55–35	24.4–26.7[†]	0.3–0.7	25–18[†]	60–80[‡]	28–15[‡]	6.0–4.5
Bituminous	High volatile C	10.5–13.0[†]	Yes	55–35	24.4–30.2[†]	0.4–0.7[‡]	25–10[†]	76–83[‡]	18–8[‡]	6.0–4.5
	High volatile B	13.0–14.0[†]	Yes	50–35	30.2–32.5[†]	0.5–0.8[‡]	12–5[†]	77–84[‡]	12–7[‡]	6.0–4.5
	High volatile A	>14.0	Yes	45–31	>32.5	0.6–1.2[‡]	7–1[†]	78–88[‡]	10–6[‡]	6.0–4.5
	Medium volatile	>14.0	Yes	31–22[†]	>32.5	1.0–1.7[†]	<1.5	84–91	9–4	6.0–4.5
	Low volatile	>14.0	Yes	22–14[†]	>32.5	1.5–2.2[†]	<1.5	87–92	5–3	6.0–4.5
Anthracite	Semianthracite	>14.0	No	14–8[†]	>32.5	2.0–3.0[†]	<1.5	89–93[‡]	5–3[‡]	5–3[†]
	Anthracite	>14.0	No	8–2[†]	>32.5[†]	2.6–6.0[†]	0.5–2	90–97[‡]	4–2[‡]	4–2[†]
	Meta-anthracite	>14.0	No	<2	>32.5	>5.5[†]	1–3	>94[‡]	2–1[‡]	2–1[†]

*Modified from Damberger et al., in B. R. Cooper and W. A. Ellingson (eds.), *The Science and Technology of Coal and Coal Utilization*, 1984.
[†]Well suited for rank discrimination in range indicated.
[‡]Moderately well suited for rank discrimination.
[1] Moist, mineral-matter-free.
[2] Dry, mineral-matter-free.
[3] Reflectance of vitrinite under oil immersion.

Table 2. International classification of hard coals

	Value	Volatile matter, %[dry ash-free basis]	Heating value, MJ/kg* %[moist, ash-free basis]
1st digit (coal class);	0	0–3	
	1A	>3–6.5	
	1B	>6.5–10	
	2	>10–14	
	3A	>14–16	
	3B	>16–20	
	4	>20–28	
	5	>28–33	
	6	>33–41[†]	>32.45
	7	>33–44[†]	32.45–30.15
	8	>35–50[†]	30.15–25.55
	9	>42–50[†]	25.55–23.78

	Value	Free Swelling Index	ROGA index
2d digit (coal group):	0	0–0.5	0–5
	1	1–2	>5–20
	2	2.5–4	>20–45
	4	>4	>45

	Value	Gray-King coke type	Dilatometer, maximum dilatation, %
3d digit (coal subgroup):	0	A	Nonsoftening
	1	B–D	Contracting only
	2	E–G	<0
	3	G_1–G_4	>0–50
	4	G_5–G_8	>50–140
	5	>G_8	>140

*1 MJ/kg = 430 Btu/lb.
[†]For information only.

Transitions exist to oil shales. Their proper identification requires microscopic analysis. SEE OIL SHALE; SAPROPEL; TORBANITE.

Nonbanded coals are relatively rare. They may form entire coal seams or layers within banded coal.

Banded coal. This is the most common coal type. In bituminous coal, recognizable bands are composed of lithotypes such as vitrain, attrital coal, and fusain, and of impure coal. The banding is less pronounced in lower rank (subbituminous, lignitic) coals and may be difficult to discern in anthracites. Most bands are between a fraction of a centimeter to several centimeters thick.

Vitrain. This lithotype is highly lustrous in bituminous coals and enhances their banded nature. Vitrain represents coalified, relatively large fragments of wood and the bark of stems, branches, and roots of trees and bushes. In lignite the remains of woody material lack the shiny luster, a characteristic known as previtrain.

Attrital coal. This lithotype represents the fine-grained ground mass of banded coal in which layers of vitrain and fusain are embedded. Attrital coal is highly variable in appearance and composition. It exhibits a finely striated, granulose, or rough texture, much like nonbanded coal. Relatively lustrous attrital coal is often referred to as clarain, dull attrital coal as durain. Dull attrital coal commonly contains relatively high amounts of finely dispersed mineral matter, clays and quartz in particular.

Fusain. In most banded coals fusain represents only a subordinate proportion of the total volume, and it generally occurs in highly lenticular laminae and thin bands that rarely exceed a few centimeters in thickness. Fusain resembles charcoal; it is fibrous, very friable, and has a silky luster. It tends to be concentrated on selected bedding surfaces, forming planes of weakness in coal beds. Its many pores are often filled with secondary mineral matter.

Impure coal. This coal is defined as having 25–50% of ash by weight on the dry basis. The mineral impurities were introduced either during deposition, mostly as clay or as fine-grained detrital mineral matter (bone coal), or later as secondary mineralization (mineralized coal). Bone coal has a dull appearance; it is commonly finely striated. When scratched with a knife or nail, it forms a brown powder; the scratch mark appears shiny. Pyritized coal is the most common mineralized coal.

Coal seams. Minable coal seams occur in many different shapes and compositions. Some coal seams can be traced over tens, even hundreds, of miles in relatively uniform thickness and structure. The extensively mined Herrin Coal of the Illinois Basin and the Pittsburgh Coal of the northern Appalachian Basin are well-known examples. They are 6–8 ft (2–2.5 m) thick over thousands of square miles. These coals originated in peat swamps that developed on vast coastal plains during the Pennsylvanian Period. The German brown coal deposits near Cologne, on the other hand, are characterized by very thick coal deposits (300 ft or 100 m). However, their lateral extent is much more limited than in the two examples from the United States. These peat deposits formed in a gradually subsiding structural graben bounded by major faults. Land lay to the south and the sea to the north. Only a relatively small portion of the subsiding graben block provided optimal conditions for peat accumulation over a long period of time. SEE PENNSYLVANIAN.

Coal seams are commonly composed of a number of benches of alternating coal and shale. The shale

Table 3. Classification of macerals*

Maceral group	Maceral
Vitrinite	Vitrinite
Exinite (or liptinite)	Sporinite
	Cutinite
	Resinite
	Alginite
Inertinite	Fusinite
	Macrinite
	Micrinite
	Sclerotinite
	Semifusinite
	Inertodetrinite

*After American Society for Testing and Materials, 1988.

represents periods of interruption of the peat accumulation by flooding from a river or the sea, or, more rarely, interruption by volcanic ash deposition. The individual benches of a coal bed vary laterally in thickness and composition. The degree of variability is related to the stability of conditions during accumulation. Fluvial and lacustrine environments produce greater lateral variability than deltaic or coastal plain environments. *SEE SHALE.*

Maceral groups. Macerals are the microscopically recognizable organic constituents of coal, equivalent to minerals of other rocks. However, unlike minerals, their chemical and physical properties are not fixed but vary over a wide range depending upon coal rank and facies. Submicroscopic inorganic impurities are considered part of macerals.

Three major groups of macerals are generally recognized: vitrinites, exinites (or liptinites), and inertin-

ites. These maceral groups are divided into maceral subgroups and, depending on the degree of detail desired, even further (**Table 3** and **illustration**). In polished sections, the macerals are distinguished on the basis of their reflectance and morphology, and to some extent on their size and maceral association.

Vitrinite. This is the predominant group of macerals in most coals (commonly 70–85% by volume). The reflectance of vitrinite is intermediate between the darker exinite and brighter inertinite. Vitrinite generally originated from woody plant parts (tree trunks, branches, roots, and bark). The original cellular plant structure is still more or less visible in many vitrinites. The identifiable cell walls are known as telinite, while the structureless cell filling is known as collinite; however, these two are often difficult to distinguish. The reflectance of vitrinite is a good indicator of coal rank, especially in higher-rank coals. Under oil immersion it increases from 0.2–0.3% in low-rank lignites to over 4% in anthracites (Table 1).

Exinite (or liptinite). The macerals of the exinite group reflect light distinctly less than vitrinite (see illus.). Their volatile matter and H contents are significantly higher than those of associated vitrinite. These differences in reflectance, volatile matter, and H contents disappear quickly with increasing rank in coals above about 1% R_o, or about 30% volatile matter (dry ash-free), and converge at about 1.4% R_o, or about 20% volatile matter. The origin of the macerals of the exinite group is clearly implied by their names. They are derived from algae, spores, cuticles, resins, waxes, fats, and oils of plants.

Inertinite. Inertinites reflect light significantly more than associated vitrinites (see illus.), except in high-rank anthracites. Inertinites do not become plastic during heating, such as in coke making; they react as

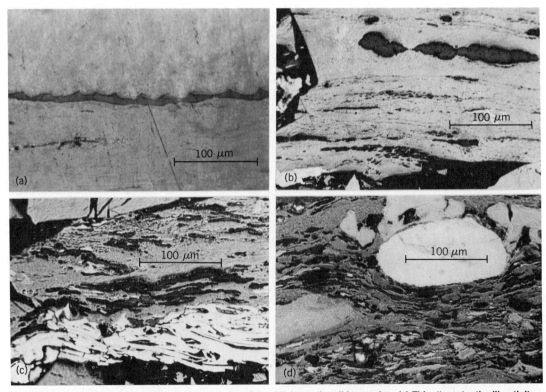

Macerals in polished surfaces of bituminous coals in reflected light, under oil immersion. (*a*) Thin, "saw-toothed" cutinite (dark) enclosed in vitrinite. (*b*) Elongate globs of resinite (dark) and other, smaller exinite macerals (dark) enclosed in vitrinite. (*c*) Sporinite (dark lenses) in vitrinite (gray) with fusinite (white) on lower part of photo. (*d*) Large grains of macrinite (white), smaller inertodetrinite (light gray), and sporinite and other exinites (dark gray) interlayered with vitrinite (gray).

Table 4. Minerals frequently occurring in coals and their stoichiometric compositions, relative abundances, and modes of occurrences*

Mineral	Composition	Common minor- and trace-element associations	Frequency of occurrence in coal seams	Concentration in mineral matter	Chief occurrences Physical[†]	Chief occurrences Genetic[‡]
Clay minerals						
Illite (sericite)	$KAl_2(AlSi_3O_{10})(OH)_2$	Sodium, calcium, iron, lithium, titanium, manganese, fluorine, and other lithophile elements	Common	Abundant	D,L	d,s(?)
Smectite (including mixed layered)	$Al_2Si_4O_{10}(OH)_2 \cdot H_2O$		Common	Abundant	D,L	d,s(?)
Kaolinite group	$Al_2Si_2O_5(OH)_4$		Common	Abundant	L,F	d,e,s(?) / d
Chlorite	$Mg_5Al(AlSi_3O_{10})(OH)_8$		Rare	Moderate	L	
Sulfides						
Pyrite	FeS_2 (isometric)	Arsenic, cobalt, copper, and other chalcophile elements	Common to rare	Variable	D,N,F	s,e
Marcasite	FeS_2 (orthorhombic)		Rare to moderate	Trace	D(?)	s(?)
Pyrrhotite	$Fe_{1-x}S$		Rare	Trace	D	s(?)
Sphalerite	ZnS		Rare	Minor to trace	F	e
Galena/chalcopyrite	$PbS/CuFeS_2$		Rare	Trace	F	e
Carbonates						
Calcite	$CaCO_3$	Manganese, zinc, strontium	Common to rare	Abundant	N,F	e,s
Dolomite (ankerite)	$CaMg(CO_3)_2$		Moderate	Trace	N,L	s,e
Siderite	$FeCO_3$		Rare	Minor	N	s,e
Oxides						
Quartz	SiO_2	—	Common	Abundant	D,L,N	d,s(?)
Magnetite/hematite	Fe_3O_4/Fe_2O_3	Manganese, titanium	Common	Minor to trace	N	s
Rutile and anatase	TiO_2	—	Common	Trace	D	d(?)
Others						
Goethite/limonite	$FeO(OH)$	Manganese, titanium	Common	Trace	N	w
Feldspar	$K(Na)AlSi_3O_8$	Calcium	Moderate	Trace	D,L	d
Zircon	$ZrSiO_4$	—	Moderate	Trace	D	d
Sulfates						
Gypsum	$CaSO_4 \cdot 2H_2O$	—	Moderate	Minor	D,F	w
Barite	$BaSO_4$	Sodium, strontium, lead	Rare	Minor	F	e
Szomolnokite	$FeSO_4 \cdot H_2O$	—	Rare	Trace	D	w
Apatite	$Ca_5(PO_4)_3(F,Cl,OH)$	Manganese, cerium, strontium, uranium	Moderate	Trace	D	d,s(?)
Halite	$NaCl$	Potassium, magnesium	Rare	Trace	D	e

*After R. D. Harvey and R. R. Ruch, Mineral matter in Illinois and other U.S. coals, in K. S. Vorres (ed.), *Mineral Matter and Ash in Coal*, Amer. Chem. Soc. Symp. Ser. 301, pp. 10–40, 1986.
[†]D, disseminated; L, layers (partings), N, nodules; F, fissures (cleat). First listed is the most common occurrence.
[‡]d, detrital; s, syngenetic; e, epigenetic; w, weathering. First listed is the most common occurrence. (?) indicates there is divergence of opinion about the genetic occurrence.

inert materials. Fusinite is characterized by well-preserved cell wall structure (sometimes broken by compaction, a characteristic known as bogenstructure). Cell cavities commonly are filled with minerals. Fusinite's general appearance suggests origin by rapid charring in the peat stage (for example, in a forest fire). On the other hand, macrinite and micrinite lack visible cell structure. Particle size is greater than 10 μm for macrinite and less than 10 μm, commonly about 1 μm, for micrinite. Micrinite is nearly always enclosed within sporinite or vitrinite and apparently is a solid coalification by-product of liptinite macerals. Sclerotinite occurs as high-reflecting round to oval cellular bodies of varying size (20–300 μm) or as interlaced tissues derived from fungal remains. Semifusinite is transitional between fusinite and associated vitrinite in both reflectance and cell structure. Inertodetrinites are small, angular clastic fragments of other inertinites commonly enclosed in vitrinite or associated with mineral matter.

The presence of inertinites indicates oxidative conditions during peat accumulation. Inertinite contents of 5–15% are the rule for many coals of the eastern United States and Europe. But some coals have inertinite contents in the 30–40% range (for example, the High Splint and Winefrede coal seams of eastern Kentucky), indicating strong oxidizing conditions during peat formation. The mostly Permian-age Gondwana coals of Australia, India, South Africa, and South America are well known for their high inertinite contents, up to 85%. These coals were formed in a relatively cold climate with alternating dry and rainy seasons, providing ample opportunity for widespread oxidation of the accumulating plant material. In contrast, the Pennsylvanian-age coals of North America and Europe originated in the humid equatorial zone.

Mineral matter. Whether they occur as discrete mineral grains or are disseminated through coal macerals, the inorganic components are referred to collectively as mineral matter. An ashing technique developed in the mid-1960s has permitted detailed analyses of minerals found in coal (**Table 4**). Mineral matter occurs in coal seams as layers or partings, as nodules, as fissure fillings, as rock fragments, and as small particles finely disseminated throughout. Genetically, minerals can be classified as inherent (plant-derived), detrital (water- or wind-borne during sedimentation), syngenetic (formed early in the peat stage) and epigenetic (formed during and after lithification and during weathering). There is no sharp boundary between syngenetic and epigenetic origins. Minerals, just as macerals, indicate the environments that prevailed during their incorporation into peat or later during coalification. Similarly, the inorganic major, minor, and trace elements of coal were introduced at various stages of coal formation. Multivariate statistical methods permit attribution of most minor and trace elements to specific mineral phases.

Coal Rank

During burial to greater and greater depth, peat is transformed into brown coal or lignite, subbituminous coal, bituminous coal, and anthracite (coalification series) as a result of increasing overburden pressure and rising temperature. Both physical (pressure, heat) and chemical (biochemical, thermochemical) factors are influential in the transformation of peat into the other members of the coalification series. The boundaries between the members of the series are transitional and must be chosen somewhat arbitrarily. The term rank is used to identify the stage of coalification reached in the course of coal metamorphism. Rank is a fundamental property of coal, and its determination is essential in the characterization of a coal. Classification of coal by rank generally is based upon the chemical composition of the coal's ash-free or mineral-matter-free organic substance (Table 1), but parameters derived from empirical tests indicative of technological properties, such as agglomerating characteristics (Table 1) or the Free Swelling Index, as well as the ROGA, Gieseler, and Gray-King tests, are commonly used in several countries outside the United States in addition to coal rank parameters. *See Metamorphism*.

Coal rank increases with depth at differing rates from place to place, depending primarily on the rate of temperature increase with depth (geothermal gradient) at the time of coalification. Coal rank also changes laterally, even in the same coal seam, as former depth of burial and thus exposure to different pressure and temperature vary. Originally established vertical and regional coalification patterns can be significantly altered by various kinds of geologic events, such as the intrusion of magma at depth, or renewed subsidence of a region. Volcanic activity may cause significant local anomalies in coal rank, but rarely leads to regional changes in coalification pattern. *See Magma*.

The chemical parameters customarily used to rank coal may yield misleading results in coals that are not primarily composed of vitrinite. Coals rich in exinite may have significantly higher contents of volatile matter and coals rich in inertinite may have significantly lower contents of volatile matter than associated so-called normal coals. This problem can be minimized by the use of vitrinite reflectance to indicate rank. Reflectance data have the additional advantage of being applicable to small coaly particles that occur dispersed throughout most sediments, permitting rank measurements in strata devoid of coal beds.

As rank increases during maturation, most coal properties change in tandem. This is reflected in the values given in Table 1 and **Table 5**, but can also be represented graphically in correlation charts.

Occurrence and Resources

During geologic history, coal could not form in economic quantities until vascular land plants had evolved during the Devonian Period and conquered the continents during the Carboniferous Period. The Carboniferous (Mississippian and Pennsylvanian of North America) witnessed an explosive development of vast swamp forests along the Euramerican equatorial coal belt, which extended from the North American midcontinent to Russia. The Atlantic Ocean did not exist then, and the North American and European continents were fused together. Another important coal-forming belt stretched over what was then Gondwana of the Southern Hemisphere. Gondwana subsequently drifted north and broke up into South America, Africa, India, and Australia. Most major coal deposits on Gondwana are of Permian age. They formed in a cool, humid climate. Most of the vast coal deposits of the western United States and Canada, along the eastern margin of the Rocky Mountains, originated during the Cretaceous and Tertiary periods. *See Carboniferous; Cretaceous; Permian; Tertiary*.

Not all coal deposits are of economic interest. The governments involved and private industry need to know how much coal exists in the ground (resources) and what portion may be available for development under current economic and legal conditions (reserves). Various classification schemes have been adopted by national and international agencies for determination of coal resources and reserves on comparable bases. However, no universally accepted standard has yet emerged.

Two independent factors generally are considered

Table 5. Proximate and ultimate analyses of samples of each rank of common banded coal in the United States*

Rank	State	Proximate analysis, %				Ultimate analysis, %					Heating value, Btu/lb[†]
		Moisture	Volatile matter	Fixed carbon	Ash	Sulfur	Hydrogen	Carbon	Nitrogen	Oxygen	
Anthracite	Pa.	4.4	4.8	81.8	9.0	0.6	3.4	79.8	1.0	6.2	13,130
Semianthracite	Ark.	2.8	11.9	75.2	10.1	2.2	3.7	78.3	1.7	4.0	13,360
Bituminous coal											
Low-volatile	Md.	2.3	19.6	65.8	12.3	3.1	4.5	74.5	1.4	4.2	13,220
Medium-volatile	Ala	3.1	23.4	63.6	9.9	0.8	4.9	76.7	1.5	6.2	13,530
High-volatile A	Ky.	3.2	36.8	56.4	3.6	0.6	5.6	79.4	1.6	9.2	14,090
High-volatile B	Ohio	5.9	43.8	46.5	3.8	3.0	5.7	72.2	1.3	14.0	13,150
High-volatile C	Ill.	14.8	33.3	39.9	12.0	2.5	5.8	58.8	1.0	19.9	10,550
Subbituminous coal											
Rank A	Wash.	13.9	34.2	41.0	10.9	0.6	6.2	57.5	1.4	23.4	10,330
Rank B	Wyo.	22.2	32.2	40.3	4.3	0.5	6.9	53.9	1.0	33.4	9,610
Rank C	Colo.	25.8	31.1	38.4	4.7	0.3	6.3	50.0	0.6	38.1	8,580
Lignite	N.Dak.	36.8	27.8	30.2	5.2	0.4	6.9	41.2	0.7	45.6	6,960

*After *Technology of Lignitic Coals*, U.S. Bur. Mines Inform. Circ. 769, 1954. Sources of information omitted.
[†]1000 Btu/lb = 2.326 MJ/kg.

Table 6. World coal resources and reserves, based on 1984 survey by World Energy Conference, in 10^9 metric tons

	Proved amount in place	Proved recoverable reserves	Accessible coal in significant coal fields	Estimated additional amount in place
Africa	129	66	35	127
North America	454	273	98	1188
United States	443	264	92	1127
South and Central America	26	16	10	47
Asia and Eastern Europe	1292	348	339	7446
China	737	Unknown	71	2000
Soviet Union	293	245	172	5209
Australia and New Zealand	93	66	56	705
Western Europe	110	69	44	382
West Germany	99	59	25	186
World total	2105	838	581	9895

in these assessments: the degree of economic mina-bility, and the degree of geologic assurance of the existence of a coal deposit of given properties. The World Energy Conference (WEC) has attempted to overcome the lack of standardization between countries by adopting carefully worded definitions of resource and reserve categories, and has asked national organizations to interpret their databases in these terms. **Table 6** is a summary of the worldwide compilation by the World Energy Conference. SEE MINERAL RESOURCE AREAS.

USES

Coal is used primarily for producing steam in electric power plants [87% of coal consumed in the United States in 1987, 64% in the Organization for Economic Cooperation and Development (OECD) in 1983]. Other important uses are for producing steam for industry (9% United States, 13% OECD) and in steel and coke making (4% United States, 11% OECD). Conversion of coal to synthetic liquid or gaseous fuels does not constitute a major use of coal worldwide and in most countries. However, under special circumstances in some countries coal conversion has been and continues to be important. During World War II Germany produced a significant amount of liquid fuels from coal, and South Africa has established a major coal liquefaction industry to avoid dependence on oil imports. One large coal gasification plant in the United States produces synthetic natural gas from North Dakota lignite; it was built with government support. It will take a significant rise in the price of oil and natural gas to make coal conversion competitive in the future. SEE COAL GASIFICATION; COAL LIQUEFACTION.

Heinz H. Damberger

Bibliography. American Society for Testing and Materials, *Annual Book of ASTM Standards*, vol. 05.05, sec. 5: *Petroleum Products, Lubricants, and Fossil Fuels*, 1988; B. R. Cooper and W. A. Ellingson (eds.), *The Science and Technology of Coal and Coal Utilization*, 1984; M. A. Elliott (ed.), *Chemistry of Coal Utilization*, 2d suppl. vol., 1981; G. B. Fettweis, *World Coal Resources: Methods of Assessment and Results*, 1979; R. A. Schmidt, *Coal in America: An Encyclopedia of Reserves, Production and Use*, 1979; E. Stach et al., *Stach's Textbook of Coal Petrology*, 3d ed., 1982; D. W. van Krevelen, *Coal*, 1981; C. R. Ward (ed.), *Coal Geology and Coal Technology*, 1984; World Energy Conference, *Survey of Energy Resources*, sec. 1: *Coal (including Lignite)*, 1986.

Coal balls

Subspherical masses containing mineral matter and embedded plant material, found in coal seams and overlying beds of the late Paleozoic. The mineral content of the balls usually consists of calcium or magnesium carbonate, or sometimes iron pyrites. In some areas coal balls are found only in coal seams overlain by marine limestones. Although some coal balls are barren of recognizable plant remains, others contain fragments showing excellent preservation of internal plant structures (see **illus.**).

All known deposits of coal balls in North America, Europe, and the British Isles are associated with Pennsylvanian and Permian beds of bituminous coal. They usually are present as rounded nodules or irregular masses at the surface of the bed or in the upper part of the coal seam. They vary from small pea-sized nodules to irregularly formed masses weighing tons.

Coal balls are a hindrance to operations when encountered in coal mines. After being sorted from the coal they may be used as ballast for roadbeds, otherwise they are discarded. Some of the best specimens occur in stream beds in which coal seams have been exposed by erosion.

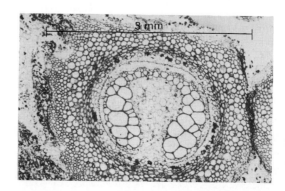

Photomicrograph of the peel made from a coal ball containing the branch of a fern.

To study the embedded plant material, a coal ball is cut with a diamond saw, and serial sections 30–40 micrometers thick are made from the cut surface by the peel technique. Slides are then prepared from the sections of fossilized material. When observed under the microscope, they may show the arrangement of plant tissues, the cell wall structure, and the cytological detail, including nuclei and plastids. This information aids in the reconstruction of ancient plants and the understanding of plant evolution. SEE COAL PALEOBOTANY; CONCRETION; PALEOBOTANY.

Wilson N. Stewart

Coal chemicals

For about 100 years, chemicals obtained as by-products in the primary processing of coal to metallurgical coke were the main source of aromatic compounds used as intermediates in the synthesis of dyes, drugs, antiseptics, and solvents. However, over 97% of the aromatic hydrocarbons, such as benzene, toluene, and xylenes, are now obtained largely from petroleum. Naphthalene, phenanthrene, heterocyclic hydrocarbons such as pyridines and quinolines, and some phenols are still obtained from coal tar.

Coke oven by-products, as a percentage of the coal used, are gas 18.5, light oil 1.0, and tar 3.5. Coke oven gas is a mixture of methane, carbon monoxide, hydrogen, small amounts of higher hydrocarbons, ammonia, and hydrogen sulfide. Most of the ammonia is recovered as about 20 lb (9 kg) of ammonium sulfate per ton (0.9 metric ton) of coal, but only about 10% of the hydrogen sulfide is recovered as elemental sulfur. Most coke oven gas is used as fuel. The composition of a typical coke oven crude light oil is shown in **Table 1**. The unidentified fractions contain very small amounts of a large number of hydrocarbons, and organic compounds containing oxygen and nitrogen.

Although several hundred chemical compounds have been isolated from coal tar, a relatively small number are present in appreciable amounts. These may be grouped as in **Table 2**.

All the compounds in Table 2 except the monomethylnaphthalenes are of some commercial importance. The amounts recovered and sold, however, are only 5–25% of the totals present in the coal tar.

The direct utilization of coal as a source of bulk organic chemicals has been the objective of much research and development. Oxidation of aqueous alkaline slurries of coal with oxygen under pressure yields a mixture of aromatic carboxylic acids. Because of the presence of nitrogen compounds and hydroxy acids, this mixture is difficult to refine. Hydrogenation of coal at elevated temperatures and pressures yields much larger amounts of tar acids and aromatic hydrocarbons of commercial importance than are obtained by carbonization.

Synthetic fuels may be produced from coal by either direct or indirect methods. In direct liquefaction processes, pulverized coal is suspended in a solvent, the resultant slurry is heated and exposed to gaseous hydrogen under pressure, and the liquids produced are separated from the ash and distilled to obtain the fuel fractions. In the indirect process, the coal is decomposed thermally to yield combustible gases, which are synthesized to yield gasoline, alcohols, and waxes. SEE COAL LIQUEFACTION; COKE; DESTRUCTIVE DISTILLATION; ORGANIC CHEMISTRY; PYROLYSIS.

Howard W. Wainwright

Bibliography. L. L. Anderson and D. A. Tillman, *Synthetic Fuels from Coal*, 1979; B. D. Blaustein et al. (eds.), *New Approaches in Coal Chemistry*, 1981; H. H. Lowry (ed.), *Chemistry of Coal Utilization*, 2 vols., 1945, suppl. vol., 1963, suppl. vol. 2, ed. by M. A. Elliot, 1981.

Table 1. Analysis of a typical coke oven crude light oil

Component	% by vol
Forerunnings	
Cyclopentadiene	0.5
Carbon disulfide	0.5
Amylenes, unidentified substances	1.0
Crude benzol	
Benzene	57.0
Thiophene	0.2
Saturated nonaromatic hydrocarbons	3.0
Unsaturates, unidentified substances	3.0
Crude toluol	
Toluene	13.0
Saturated nonaromatic hydrocarbons	0.1
Unsaturates, unidentified substances	1.0
Crude light solvent	
Xylenes	5.0
Ethylbenzene	0.4
Styrene	0.8
Saturated nonaromatic hydrocarbons	0.3
Unsaturates, unidentified substances	1.0
Crude heavy solvent	
Coumarone, indene, dicyclopentadiene	5.0
Polyalkylbenzenes, hydrindene, etc.	4.0
Naphthalene	1.0
Unidentified heavy oils	1.0
Wash oil	
(Used to separate light oil from coke	
oven gas)	5.0

Table 2. Coal tar chemicals

Compound	Fraction of whole tar, %	Use
Naphthalene	10.9	Phthalic acid
Monomethylnaphthalenes	2.5	
Acenaphthenes	1.4	Dye intermediates
Fluorene	1.6	Organic syntheses
Phenanthrene	4.0	Dyes, explosives
Anthracene	1.0	Dye intermediates
Carbazole (and other similar compounds)	2.3	Dye intermediates
Phenol	0.7	Plastics
Cresols and xylenols	1.5	Antiseptics, organic syntheses
Pyridine, picolines, lutidines, quinolines, acridine, and other tar bases	2.3	Drugs, dyes, antioxidants

Coal gasification

The conversion of coal or coal char to gaseous products by reaction with steam, oxygen, air, hydrogen, carbon dioxide, or a mixture of these. Products consist of carbon monoxide, carbon dioxide, hydrogen, methane, and some other gases in proportions dependent upon the specific reactants and conditions (temperatures and pressures) employed within the reactors, and the treatment steps which the gases undergo subsequent to leaving the gasifier. Similar chemistry can also be applied to the gasification of coke derived from petroleum and other sources. The reaction of coal or coal char with air or oxygen to produce heat and carbon dioxide could be called gasification, but it is more properly classified as combustion. The principal purposes of such conversion are the production of synthetic natural gas as a substitute gaseous fuel and synthesis gases for production of chemicals and plastics. *See Combustion.*

Industrial uses. In all cases of commercial interest, gasification with steam, which is endothermic, is an important chemical reaction. The necessary heat input is typically supplied to the gasifier by combusting a portion of the coal with oxygen added along with the steam. From the industrial viewpoint, the final product is either chemical synthesis gas (CSG), medium-Btu gas (MBG), or a substitute natural gas (SNG). Heating values, compositions, and end uses for these gases are compared in the **table**.

Each of the gas types in the table has potential industrial applications. In the chemical industry, synthesis gas from coal is a potential alternative source of hydrogen and carbon monoxide. This mixture is obtained primarily from the steam reforming of natural gas, natural gas liquids, or other petroleum liquids. Fuel users in the industrial sector have studied the feasibility of using medium-Btu gas instead of natural gas or oil for fuel applications. Finally, the natural gas industry is interested in substitute natural gas, which can be distributed in existing pipeline networks. *See Natural gas.*

There has also been some interest by the electric power industry in gasifying coal by using air to provide the necessary heat input. This could produce low-Btu gas (because of the nitrogen present), which can be burned in a combined-cycle power generation system. *See Electric power generation.*

Gasification processes. In nearly all of the processes, the general process-flow diagram is the same (**Fig. 1**). Coal is prepared by crushing and drying, pretreated if necessary to prevent caking, and then gasified with a mixture of air or oxygen and steam. The resulting gas is cooled and cleaned of char fines, hydrogen sulfide, and CO_2 before entering optional processing steps to adjust its composition for the intended end use.

Thermodynamics. In discussions of the thermodynamics of coal gasification, at least one simplifying assumption is usually made, namely, that coal and coal char can be treated as pure carbon. Coal and coal char are really nonhomogeneous solids containing hydrogen, oxygen, sulfur, nitrogen, and mineral matter, but errors associated with this assumption are not likely to be significant.

The basic chemical reactions common to all coal gasification processes are coal and char reactions (1)–

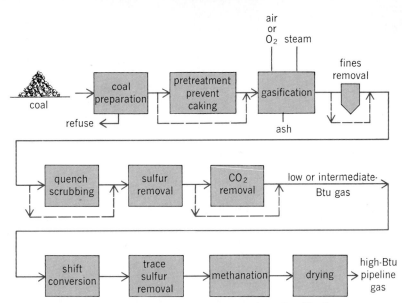

Fig. 1. Schematic representation of the processing steps in coal gasification.

(3) and gaseous reactions (4) and (5). Reaction (2) is highly endothermic, requiring 32 kilocalories/mole

$$\text{Coal} \xrightarrow{\text{heat}} \text{gases (CO, CO}_2\text{, H}_2\text{, CH}_4\text{)} + \text{char} \qquad (1)$$

$$\text{C (char)} + \text{H}_2\text{O} \rightarrow \text{CO} + \text{H}_2 \quad \text{(endothermic)} \qquad (2)$$

$$2\text{C (char)} + \tfrac{3}{2}\text{O}_2 \rightarrow \text{CO}_2 + \text{CO} \quad \text{(exothermic)} \qquad (3)$$

$$\text{CO} + \text{H}_2\text{O} \rightarrow \text{H}_2 + \text{CO}_2 \quad \text{(mildly exothermic)} \qquad (4)$$

$$\text{CO} + 3\text{H}_2 \rightarrow \text{CH}_4 + \text{H}_2\text{O} \quad \text{(exothermic)} \qquad (5)$$

(134 kilojoules/mole). The principal thermodynamic issue in most practical gasification processes is how to supply the heat for this reaction. Several methods have been considered, but generally heat is supplied by allowing exothermic reactions (3) and (5) to occur in the same vessel as reaction (2). The degree to which one reaction or the other is used depends on the process configuration and the desired product. Reactions (4) and (5) are often carried out in downstream processing blocks in order to tailor the gas composition for particular end uses. In these arrangements it is generally not feasible to return the heat to the gasifier, as the downstream reactions are carried out at much lower temperatures. The heat input methods used in the principal commercial and developing processes are given in the process descriptions below.

Processes. The differences in coal gasification processes principally involve the method of heat in-

Coal gasification products

Product	Higher heating value*	Major constituents	Use
Synthesis gas (CSG)	250–400 Btu/SCF† (9–15 MJ/m³)	CO, H_2	Chemical feedstock for synthesis of H_2, NH_3, methanol, hydrocarbons, and so forth
Medium-Btu gas (MBG)	300–500 Btu/SCF (11–19 MJ/m³)	CO, H_2, CH_4	General-purpose fuel for utilities and industries
Substitute natural gas (SNG)	900–1050 Btu/SCF	CH_4	Gaseous fuel

*Based on producing water as a liquid.
†SCF = Standard cubic feet.

put, gasifier type, and ash recovery system. First-generation (commercial) gasifiers include the Lurgi (dry bottom), Koppers-Totzek (K-T), and Winkler gasifiers.

Lurgi. The Lurgi gasifier utilizes a bed of crushed coal moving downward through the reactor with countercurrent flow of gas, and operates at pressures up to 450 lb/in.[2] gage (3100 kilopascals). Crushed coal, screened to remove fines, is fed to the top of the gasifier through a coal lockhopper, and passes downward through the drying pyrolysis, gasification, and combustion zones. Steam and oxygen are admitted through a revolving grate at the bottom of the gasifier which also removes the ash produced. After burning a portion of the coal to provide the heat required, the hot combustion gas passes upward through the zones of the coal bed. Steam is used as a reactant to gasify the coal, and is also added to keep the temperature below the ash melting point. Although the predominant source of methane (CH_4) is devolatilization, some methane comes from coal hydrogenation, releasing heat which is transferred to the coal and minimizing the oxygen requirements. As the coal moves down through the gasifier, the temperature initially rises slowly. Further down in the gasifier, it rises dramatically as the coal approaches the combustion zone where the temperature exceeds 2200°F (1200°C). The Lurgi reactors commonly used are of the general type shown in **Fig. 2.** Because of the countercurrent flow in the gasifier, high carbon utilization and good heat recovery are obtained.

The dry-bottom Lurgi process is generally applicable to subbituminous coals and lignites because the gasifiers operate best with noncaking, nonfriable coals with high reactivity and high ash melting points. The internal design and operating conditions of these reactors may be altered for coals of different caking characteristics and reactivities. Dry-bottom Lurgi gasifiers are used in the Great Plains coal gasification project in North Dakota and also in the world's largest gasification facilities at the SASOL coal-to-oil plants in South Africa.

Koppers-Totzek. The K-T gasifier is an atmospheric, entrained reactor. Dried and pulverized coal is mixed with steam and oxygen and fed to coaxial burners. The pulverized coal is gasified as it is carried in the flow of gas and as a portion of the coal is combusted. Practically every type of coal can be gasified by this technique. The operability of the process is not affected by caking properties of the coal.

The complete entrainment of the feed particles requires high gas velocities to achieve reaction times of only a few seconds; consequently the reaction temperature required is 3300–3500°F (1825–1925°C). At these temperatures, most of the carbon reacts very rapidly with oxygen and steam to produce carbon oxides, hydrogen, and molten slag. Less than 0.1% CH_4 is present in the product gas, and no tars or phenols are produced. More than 50% of the ash flows down the gasifier walls as molten slag into a slag quench tank. The rest leaves as entrained fly ash in the exit gas. The gasifier has a double-walled shell and a thin refractory lining. Water is circulated in the vessel jacket to protect the refractory. Most of the operating plants utilizing K-T gasifiers have been for the production of a H_2-rich gas for NH_3 manufacture.

Winkler. The Winkler gasifier is an atmospheric, fluidized-bed reactor that was developed in Germany in the 1920s. The vessel is cylindrical and refractory-lined, with nozzles for injecting a mixture of steam and oxygen at the bottom of the gasifier and above the top of the char bed to control fines. Crushed, dried coal is fed to the gasifier by means of a screw conveyor. The gasification temperature is controlled around 1800–2000°F (1000–1100°C) by adjusting the ratio of oxygen and steam to coal, and thus the portion of the coal being combusted. Gas is removed overhead, cooled, and dedusted. Product gas is low in CH_4 and contains no tars. Heavy char particles, which contain ash, settle to the bottom of the fluidized bed and are withdrawn from the gasifier by a screw conveyor. Most of the Winkler gasifiers that have been built are in eastern Europe and Asia.

Research and development. Most process research and development is directed toward increasing the efficiency of certain process steps, gasifying a wider range of coals or lowering plant costs. Important programs in the United States include work on partial oxidation, gasification with controlled ash agglomeration, and catalytic gasification. Major European efforts include work on the slagging gasifier and partial oxidation.

Slagging gasifier. The British Gas/Lurgi slagging gasifier is a modification of the commercial dry-bottom Lurgi design. It uses less steam and is less coal-sensitive than the dry-bottom Lurgi. These modifications are achieved by incorporating a high-temperature slagging zone at the bottom of a gasifier. Coal is screened and crushed before being introduced via lockhoppers into the top of the reactor. As the coal moves

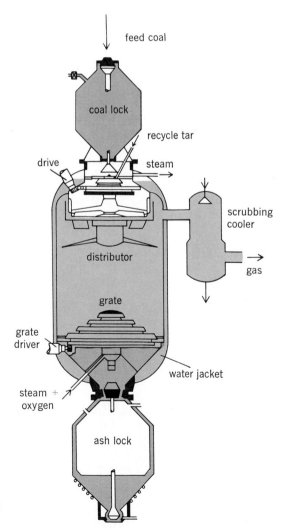

Fig. 2. Diagram of a Lurgi pressure gasifier.

downward through the reactor as a slowly moving bed, it passes through the drying, pyrolysis, gasification, and combustion zones. The coal ash is melted in the combustion zone and collects in a pool, from which it is tapped and quenched with water before being withdrawn from the gasifier. The oxygen and steam for the reaction are supplied through tuyeres in the bottom of the reactor. Tar, oil, and some of the coal fines produced in the crushing step can also be injected through the tuyeres and gasified. Some quantity of dusty tar can also be distributed at the top of the bed. The gas moving upward countercurrently through the moving bed of coal supplies the heat for the various gasification reactions and for drying the coal. The gas exits the reactor at approximately 1000°F (540°C), and is quenched in a scrubber to remove heavy tars and fines before being cooled in a waste heat boiler. This process is considered to be more attractive than the dry-bottom Lurgi for caking coals, such as bituminous coals of the eastern United States.

Partial oxidation. Both Texaco and Shell have pursued partial-oxidation process development. The Texaco gasifier is a pressurized (up to about 1000 lb/in.^{2} absolute or 6800 kPa), entrained reactor. Coal is pulverized in a wet mill and slurried (50–70 wt % solids) in water. The slurry is then pumped to pressure and injected with oxygen and steam into the downflow entrained gasifier. The hot gas and slag exit the bottom of the gasifier and pass through a radiant cooler. Most of the ash is collected at the bottom of the radiant section. The gas that contains some fly ash passes through a convective cooler and a scrubber. Both the radiant and convective coolers generate high-pressure, saturated steam. An alternate design involves cooling the hot gas from the reactor by water quench. In this case, no steam is generated, but the gas produced contains sufficient water for subsequent conversion of the carbon monoxide to hydrogen.

The Shell process is similar, but utilizes a dry feed system rather than a water slurry. In this process, ground coal is metered into an entrained gasifier where it reacts with oxygen and steam.

Gasification with controlled agglomeration. Both Westinghouse and the Institute of Gas Technology have pursued gasification under conditions which result in controlled ash agglomeration. The Westinghouse gasifier is a pressurized (about 250 lb/in.^{2} absolute or 1600 kPa), fluidized-bed reactor. Coal is crushed before pressurizing via lockhoppers. The coal is injected with steam and oxygen into the fluid bed through an axial lift pipe. The higher temperature at the outlet of the lift pipe gives controlled ash agglomeration in the fluid bed. Larger particles of ash concentrate at the bottom of the gasifier and are removed and rejected. Steam and recycled gas are also injected in the lower portion of the reactor in order to fluidize the bed and control ash particle size. The resulting gas product is removed overhead, dedusted by cyclones, and cooled in waste heat boilers. The cooled gas is scrubbed to remove fines, and a portion is recycled to provide fluidization gas.

Catalytic coal gasification. The Exxon catalytic coal gasification process is intended for substitute natural gas production, and differs significantly from other gasification methods in that it uses a potassium catalyst to accomplish steam gasification [reaction (2)] and methanation [reaction (5)] in the same vessel. The catalyst enables the gasifier to operate at mild process conditions of 1300°F (700°C) and 500 lb/in.^{2}

gage (3550 kPa). A major advantage of this "one-step" approach is that the overall reaction is almost thermally neutral. The gasification heat requirement is essentially supplied by the methanation heat release. Therefore, the catalytic coal gasifier does not require a large heat input by means of oxygen or other methods. The gasifier off-gas is treated by commercially available acid-gas–removal technology, and the methane product is cryogenically separated, with H$_2$ and CO recycled to the gasifier. The catalyst-char-ash material which is removed from the gasifier is washed countercurrently with water to recover most of the potassium catalyst for reuse. *SEE HETEROGENEOUS CATALYSIS.*

Underground gasification. Only about one-fifth of the total coal in the ground can be recovered by mining. The remainder is either too deep or in seams that are too thin to be mined. Long-term underground gasification is a promising possibility for utilizing this unminable coal. This process involves drilling wells into coal deposits, preparing a rubbled bed of coal through which gas can flow, and initiating gasification in the bed by using steam and air or oxygen. Such underground field tests have been conducted in the United States, but problems with underground gasification include process control, land subsidence, and groundwater pollution. *SEE COAL MINING.*

Several organizations have been active in this field, emphasizing work on the nature and direction of subsurface fracture systems, the means to calculate fluid movement underground, and environmental impacts. Directional drilling techniques and fracturing of the underground formations with chemical explosives have been studied. *SEE COAL.*

W. R. Epperly

Bibliography. AGA-GRI-DOE-IGU, Proceedings of the 7th, 8th, 9th, and 10th Synthetic Pipeline Gas Symposiums, 1975–1978; M. Elliot (ed.), *Chemistry of Coal Utilization,* 2d suppl. vol., 1981; GRI-AGA-DOE-IGU, Proceedings of the 1st, 2d, and 3d International Gas Research Conferences (1980, 1981, 1983); J. L. Figuerido and J. A. Moulijin (eds.), *Carbon and Coal Gasification: Science and Technology,* 1986; R. L. Hirsch et al., Catalytic coal gasification: An emerging technology, *Science,* 215:121–127, January 8, 1982; J. H. Kolaian and W. G. Schlinger, The Texaco coal gasification process, *Energy Prog.,* 2(4):228–233, December 1982; M. K. Schad and C. F. Hafke, Recent developments in coal gasification, *Chem. Eng. Prog.,* 79(5):45–51, May 1983.

Coal liquefaction

The conversion of most types of coal (with the exception of anthracite) primarily to petroleumlike hydrocarbon liquids which can be substituted for the standard liquid or solid fuels used to meet transportation, residential, commercial, and industrial fuel requirements. Coal liquids contain less sulfur, nitrogen, and ash, and are easier to transport and use than the parent (solid) coal. These liquids are suitable refinery feedstocks for the manufacture of gasoline, heating oil, diesel fuel, jet fuel, turbine fuel, fuel oil, and petrochemicals.

Liquefying coal involves increasing the ratio of hydrogen to carbon atoms (H:C) considerably—from about 0.8 to 1.5–2.0. This can be done in two ways: (1) indirectly, by first gasifying the coal to produce a synthesis gas (carbon monoxide and hydrogen) and then reconstructing liquid molecules by. Fischer-Tropsch or methanol synthesis reactions; or (2) directly, by chemically adding hydrogen to the coal matrix under conditions of high pressure and temperature. In either case (with the exception of methanol synthesis), a wide range of products is obtained, from light hydrocarbon gases to heavy liquids. Even waxes, which are solid at room temperature, may be produced, depending on the specific conditions employed. **Table 1** lists potential uses for coal liquefaction products and indicates the additional processing which would likely be required.

Indirect liquefaction. Indirect coal liquefaction consists of three important steps: first, the coal is gasified; second, the composition of the resulting gas is adjusted (shifted, if necessary, to increase H_2 content; and H_2S and CO_2 are removed); and third, the CO and H_2 in the resulting synthesis gas are catalytically reacted to form the liquids. Commercial processes for catalytically reacting CO and H_2 to form the indirect coal liquids include Fischer-Tropsch synthesis and methanol synthesis. In addition, a Mobil process for conversion of methanol to gasoline is expected to be an important commercial operation.

Fischer-Tropsch synthesis. The synthesis processes based on the Fischer-Tropsch chemistry are represented by reaction (1). This reaction represents a pol-

$$2nH_2 + nCO \rightarrow (-CH_2-)_n + nH_2O \qquad (1)$$

ymeric addition of methylene ($-CH_2-$) groups to form a distribution of linear paraffins. In practice, a broad range of molecular weights is produced, and the product includes branched, olefinic, and oxygenated compounds, depending on reactor type, operating conditions, and catalyst type. The average molecular weight of the product mixture can be adjusted to some extent by the selection of process conditions.

Two synthesis processes (ARGE and Synthol) are used in the commercial coal liquefaction plants of the South African Coal, Oil, and Gas Company (Sasol) in the Republic of South Africa. The ARGE process utilizes a fixed-bed reactor and a precipitated-iron catalyst to produce a high-boiling, linear paraffin product, including a range of waxes. The Synthol process uses an entrained-bed reactor and a fused-iron catalyst to produce a somewhat lower-boiling hydrocarbon product which contains more isoparaffins and olefins.

The Sasol I commercial coal liquefaction plant started operations in 1955 and utilizes both the ARGE fixed-bed and the Synthol entrained-bed processes. Sasol II operation was initiated in 1980, when a site was also prepared for Sasol III. Both Sasol II and Sasol III are modified and enlarged versions of Sasol I. Each is designed to include seven of the Synthol entrained-bed reactors. Production of nonfuel prod-

---

Table 1. Potential uses for coal liquefaction products

Product	Commercial uses	Additional processing required
Light hydrocarbons		
Methane and ethane	Pipeline gas	None
C_3–C_4 hydrocarbons	Liquefied petroleum gas (LPG)	None
C_2–C_4 hydrocarbons	Ethylene and propylene manufacture	Steam cracking
Naphtha		
C_4–160°F (71°C) naphtha	Motor gasoline	Hydrotreating
160–350°F (71–180°C) naphtha	Motor gasoline	Hydrotreating and reforming
	Chemicals (benzene, toluene, xylene)	Extraction
350–430°F (180–220°C) naphtha	Motor gasoline	Hydrotreating
Middle distillate	Turbine fuel	Direct use or hydrotreating
350–650°F (180–340°C) or 430–650°F (220–340°C)	Low- or medium-speed diesel fuel	Direct use
	No. 6 fuel oil blendstock	Direct use
	Home heating oil	Direct use (with burner adjustment) or hydrotreating
	Auto diesel blendstock	Moderate hydrotreating
	Jet, auto diesel, No. 2 fuel oil	Severe hydrotreating or hydroconversion
Heavy distillate		
650–1000°F (340–540°C) vacuum	Fuel oil blendstock	Direct use or hydrotreating
Gas oil	Fluid catalytic cracking or hydroconversion feed	Hydrotreating
Other products		
Solvent refined coal (solid)	Boiler fuel	Direct use
Methanol	Motor fuel, distillate fuel	Direct use (with special distribution system)
Fischer-Tropsch liquids	Gasoline, diesel oil, jet fuel, wax	Hydrotreating, reforming, isomerization, polymerization

Table 2. Leading direct coal liquefaction technologies*

	EDS	H-Coal	SCR II	Modified Bergius-Pier
Demonstration plant	250 tons/day (230 Mg/day), Baytown, Texas	200 tons/day (180 Mg/day), Catlettsburg, Kentucky	35 tons/day (32 Mg/day), Fort Lewis, Washington	220 tons/day (200 Mg/day), Bottrop, West Germany
Recycle solvent	Hydrogenated distillate	Distillate	Product slurry	Distillate
Pressure, atm (MPa)	140–170 (14–17)	210 (21)	140 (14)	300 (30)
Bottoms recycle	Optional	No	Essential	No
Catalyst	Ni-Mo in solvent hydrotreater (none in liquefaction)	Co-Mo (in ebullating bed reactor)	None (relies on catalytic activity in recycled coal ash)	$FeSO_4$ as red mud
Types of coal	Bituminous Subbituminous Lignite	Bituminous Subbituminous	Bituminous	Bituminous

*Mg = megagrams.

ucts will be minimized in Sasol III.

Modified process techniques and conditions can have a significant impact on the Fischer-Tropsch product distribution. New catalyst formulations can improve activity and tailor the product distribution to meet fuel and chemical product demands. Engineering research to define better reactor designs and optimum operating conditions for more economic processes have also been undertaken. SEE FISCHER-TROPSCH PROCESS.

Methanol synthesis. The methanol synthesis processes are based on the chemistry represented by reactions (2) and (3).

$$CO + 2H_2 \rightarrow CH_3OH \qquad (2)$$

$$CO_2 + 3H_2 \rightarrow CH_3OH + H_2O \qquad (3)$$

Commercial processes use copper- and zinc-based catalysts and fixed-bed reactors operating at high pressure (100 atm or 10 megapascals) and low conversion per pass with high recycle ratios. Crude methanol can be utilized in three ways: as a neat fuel in fleet applications, as a gasoline blendstock, or as a feed to the Mobil methanol-to-gasoline process. SEE METHANOL.

Mobil methanol-to-gasoline process. In the Mobil process, a complex sequence of reactions occurs over a zeolite catalyst. Methanol is the first dehydrated to an equilibrium mixture of dimethyl ether, methanol, and water. This mixture is then fed to catalytic reactors where the remaining methanol is dehydrated to dimethyl ether, which reacts to form olefins that polymerize to form aromatic gasoline components. A fixed-bed version of this process is commercially available and has been applied in a New Zealand natural gas-to-gasoline project. A fluid-bed version of the methanol-to-gasoline process is used in a pilot plant in Germany, which started up in late 1982. SEE ZEOLITE.

Direct liquefaction. Direct liquefaction or hydrogenation of coal was developed in Germany in the 1920s by F. Bergius, who received the Nobel Prize in chemistry for his work. This technology was later modified by M. Pier and commercially proved during World War II. However, this technology operated at extremely high pressure (almost 700 atm or 70 MPa) and in small reactors (3-ft or 1-m diameter) and was too costly to be economically justified in peacetime.

Since the war, interest in direct coal liquefaction has periodically been revived, generally in response to projected shortages of petroleum, but no commercial plants have been built. However, four second-generation liquefaction technologies have been demonstrated in large pilot plants, and these are compared in **Table 2**. These technologies were an outgrowth of the Bergius-Pier process, but operate at much lower pressure (140–300 atm or 14–30 MPa). Other technologies have also benefited from major advances in reactor design since the 1940s, which will allow much larger (and therefore more economical) plants to be built. However, notwithstanding the technical improvements which have been made, the basic processing steps in coal liquefaction (see **illus.**) have not changed conceptually.

Coal is first crushed and slurried in a process-derived solvent and then heated and pumped to typical reaction conditions, in the range of 800–930°F (425–500°C) and 140–300 atm (14–30 MPa). Depending on the process, a catalyst may be added to the liquefaction step, or the solvent may be hydrogenated to improve hydrogen transfer to the coal. As the coal is heated to reaction temperatures and held there

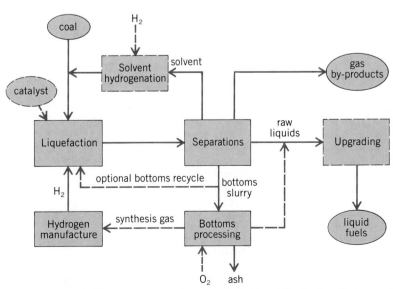

Schematic representation of the processing steps in direct coal liquefaction. Broken lines represent optional steps or flow streams.

for 40–80 min nominal residence time, thermal cracking (thermolysis) occurs. At liquefaction temperature, chemical bonds are broken, that is, cracked, generating free-radical fragments which can participate in secondary reactions—combining with hydrogen to produce stable liquids, reacting further to form lighter liquids, or polymerizing to form high-molecular-weight, high-boiling-point bottoms. Control of these secondary reactions is important in achieving high conversions and good selectivity to distillable liquid products. During liquefaction, heteroatoms in the coal (oxygen, nitrogen, and sulfur) are also liberated, principally as carbon oxides, water, ammonia, and hydrogen sulfide.

After liquefaction, products must be separated from bottoms (ash, heavy coal liquids, and unconverted coal) and solvent recovered. The most common technique for separation is distillation, although hydroclones, which separate by centrifugal force, have been used in Hydrocarbon Research's H-Coal and extraction in Gulf solvent-refined coal (SRC) for removal of ash and coal residues. Gulf's SRC II process, which evolved from SRC, recycles bottoms to increase the ash level in the liquification reactor and improve conversion. The EDS process has operated with and without bottoms recycling, and employs recycling to increase conversion and produce a lower-molecular-weight product slate. The term product slate refers to the end products that are actually sent to the market from the plant. It has been shown that the bottoms contain high-boiling asphaltenes and preasphaltenes which can be readily converted to lighter products through recycling.

The net bottoms product is sent to a bottoms processing step, where the carbon is utilized and coal ash is separated. Carbon utilization can involve recovery of additional liquids by coking or extraction, partial oxidation to produce plant fuel or hydrogen, or combustion to provide plant heat. Hydrogen generation is frequently the preferred bottoms-processing step. In this case, bottoms would be sent to a partial oxidation unit and gasified with oxygen to produce a synthesis gas for hydrogen production. Alternatively, hydrogen could be produced by partial oxidation of coal or by conventional steam reforming of light gases as in present-day petroleum refineries and chemical plants. *SEE COAL GASIFICATION; PETROLEUM PROCESSING; SOLVENT EXTRACTION.*

Upgrading of coal liquids. In most cases, coal liquids require upgrading prior to use. The type and severity of upgrading depends on the initial characteristics of the liquids and the desired product properties (see Table 1). In general, however, the upgrading processes are similar to those employed for petroleum liquids. Hydrotreating (mild hydrogenation) is commonly selected for removal of nitrogen, sulfur, or oxygen in coal liquids. Direct coal liquids are rich in cycloparaffins and aromatics. In the gasoline fraction, the cycloparaffins can be dehydrogenated by catalytic reforming to aromatics, which have a high octane rating. In the middle distillate fraction, the aromatics can be hydrogenated to cycloparaffins, which have improved performance in diesel and jet engines. In addition, catalytic cracking or hydroconversion of heavy distillates can be used to increase the yield of gasoline and middle distillate.

Indirect coal liquids are primarily paraffins and olefins. Polymerization of the propane-butane fraction can be used to increase the yield of gasoline. Reforming of the gasoline fraction increases octane rating by producing aromatics via cyclization of paraffins and subsequent dehydrogenation. Isomerization of linear paraffins also increases octane. Heavy, waxy streams can be cracked to gasoline and middle distillate. *SEE HYDROCRACKING.*

The severity of product upgrading typically decreases with increasing severity of direct liquefaction. The reason for this lies in the basic chemistry of converting high-carbon coal (H:C atomic ratio of about 0.8) into naphtha and light distillate (H:C of 1.2–1.9). Since the overall reaction involves hydrogen addition, it can be seen that a low level of hydrogenation in the liquefaction section would yield a high percentage of high-boiling components which require extensive hydrotreating to produce salable products. On the other hand, a liquefaction process of relatively high severity would yield liquids which require less extensive treatment to achieve the same salable products. As a result, yield comparisons of different coal liquefaction processes are usually closer on a finished product basis than on an intermediate product basis. It is generally agreed that a commercial coal liquefaction plant will produce approximately 3 barrels (0.5 m^3) of liquid product for every ton (0.9 metric ton) of dry, ash-free coal feed.

Continuing research. Research on improving coal liquefaction processes is concentrated in the general areas of staged liquefaction and catalysis. The principal goal of this work is to increase the yield of distillate products while reducing the amount of gas produced. In staged processes, the conditions in each stage are selected to optimize production of the desired liquid products. In catalytic processes, a catalyst which promotes formation of the desired molecular species is selected. Most emerging technologies incorporate elements of both staging and catalysis to obtain the advantages of each approach.

Staged liquefaction processes utilize two or more distinct stages to promote coal conversion or to improve the product slate. Though processes with many stages are theoretically possible, all of the staged liquefaction concepts use only two stages. Their essential features include a coal dissolver section (first stage), a deashing section to remove solids, and a catalytic hydroconversion section (second stage) to upgrade the heavy liquids. The dissolver section could range between a plug-flow reactor and a short-contact-time preheater. Deashing may be accomplished with hydroclones or by using proprietary antisolvent or critical-solvent deashing techniques. The hydroconversion section is typically operated at a lower temperature than the dissolver section (to prevent thermal cracking of liquids into gases) and contains catalyst in either a fixed bed, expanded bed, or ebullating bed.

In addition to these process differences, several variations in processing sequence are possible improvements to the two-stage technology. For example, the degree to which process streams from the hydroconversion section are recycled to the dissolver section may be adjusted to yield an integrated, partially integrated, or nonintegrated operation. Another option is placement of the second stage; since some hydroconversion processes are less sensitive to solids concentration, it may be possible to locate the deashing section downstream of both liquefaction stages. Finally, within the framework of any given processing sequence, the number of recycle streams, the cut points of these streams, and the severity of hydro-treating can also be changed.

The major two-stage liquefaction processes employ catalysts for additional conversion of heavy liquids. The two-stage process of the Electric Power Research Institute and the U.S. Department of Energy and the integrated two-stage liquefaction (ITSL) process of Lummus, both have high-temperature, plug-flow first stages and solvent-extraction deashing followed by lower-temperature, catalytic hydroconversion. In addition, ITSL uses a short-contact time preheater. The Chevron Coal Liquefaction process uses the same process sections but employs solids separation after both of the liquefaction stages. Overall, direct liquefaction is a fertile area for continuing research. *See* Catalysis; Coal; Heterogeneous catalysis.

W. R. Epperly

Bibliography. M. A. Elliott (ed.), *Chemistry of Coal Utilization,* 2d suppl. vol., 1981; T. A. Hendrickson (comp.), *Synthetic Fuels Data Handbook,* Cameron Engineers, Inc., Denver, 1975; R. Meyers, *Handbook for Synfuels Technology,* 1984; Y. T. Shah (ed.), *Reaction Engineering in Direct Coal Liquefaction,* 1981.

Coal mining

The technical and mechanical activities involved in removing coal from the earth and preparing it for market. Coal mining in the industrialized countries is characterized by the integration of a number of complex systems into a production methodology that varies depending upon whether the surface or underground methods are utilized. *See* Coal.

The basic systems to be integrated into the production methodology may be categorized as the following. (1) Extraction systems: the methods and techniques used to break out or "win" the coal. (2) Materials-handling systems: the transport of coal and waste products away from the active production area, and the transport of the necessary materials, equipment, supplies, and workers needed to service the extraction system. (3) Ventilation: the development and operation of an air distribution system which will provide the quantity, quality, and velocity of air, where and when needed, to meet statutory requirements. (4) Ground control: the control of the behavior of underground and surface openings developed by the extraction of coal. (5) Reclamation: the restoration of the mined area to its approximate original state or to an approved state.

To properly plan, design, and engineer a coal mining production method, knowledge of the geology of the deposit and the chemical and physical properties of the coal must be assembled and assessed. Basic information on the geology of the coal deposit is obtained from surface prospecting and mapping, borehole drilling, and general geologic knowledge. This information must be assessed to determine the size and shape of the coal area, the geologic column above and below all minable seams, the continuity and persistence of geologic features throughout the deposit, the presence of water or methane gas, and other special conditions of the deposit which bear on coal production. Proximate chemical analyses are made to determine coal characteristics which affect its utilization in burning. Tests are made to determine the cleaning, grinding, and handling properties of the coal. Ultimate chemical analyses are made to determine the fundamental chemical constituents of the coal. Maps are drawn to summarize this information, and are used for scheduling and sequencing the production from the deposit. *See* Analytical chemistry; Borehole logging; Engineering geology; Prospecting; Rock mechanics; Soil mechanics; Spectroscopy.

SURFACE MINING

When a coal seam lies relatively close to the ground surface, a surface mining production method is indicated. The geographic location and geologic column—in particular, the ratio of the thickness and quality of the overburden and interseam waste to be removed, to the thickness and quality of the coal to be produced—are the initial considerations. Site conditions which affect the production method include types of terrain, temperature range, altitude, and rain- and snowfall. Overburden and interseam waste characteristics affecting the production methods are material weight, stickiness, and swell; ability to form a stable highwall and spoil pile; ratio of rock and other material requiring blasting to soil in the overburden; and need for special handling of waste materials for reclamation purposes. The presence of thicker and good-quality coal that commands higher prices can justify the removal of thick overburden.

Surface coal mines are classified as area or modified open pits, contour and mountaintop removal, and auger mining. Removal of overburden is called stripping and hence the term strip mining is often given to surface coal operations (**Figs. 1** and **2**). Area mining is applicable in relatively flat to gently undulating terrain where coal seams are of considerable area and may be at various dips. Contour mining and mountaintop removal are used in hilly and mountainous country and can be modified to handle coal seams at any dip. Auger mining and punch mining follow the other surface methods when overburden removal becomes uneconomic, and are generally limited to more or less horizontal coal seams. *See* Surface mining.

Area or modified open pit. In a typical area mine the layout is roughly rectangular with adjustments made to fit the boundaries of the property. The length of the pit varies depending on the rate of planned production, but may range to about 2 mi (3 km) or more. Extraction operations involve topsoil and subsoil removal, overburden removal, and coal loading. Drilling and blasting or ripping of the overburden and coal may be required. An intimate relationship exists between the equipment selected for materials handling and the extraction operation.

The equipment used in topsoil and subsoil removal and in stripping overburden is selected based on the characteristics of the soil and rock to be removed. In soft digging, bulldozers, wheel loaders, and elevating scrapers may be used to excavate the soil and overburden and to load it into trucks (**Fig. 3**). In many parts of Europe, where the geologic column consists of soft-digging materials, bucket wheel excavators, used singly and in tandem, excavate and convey the soil and overburden to storage areas. Pan scrapers are found to be useful both in soil and overburden removal and in reclamation. When rock is present, drilling and blasting or ripping are required to loosen and break out the overburden. Draglines (**Fig. 4**) and power shovels (**Fig. 5**) are used to dig the harder and more consolidated materials.

Since outcrops rarely occur where area surface

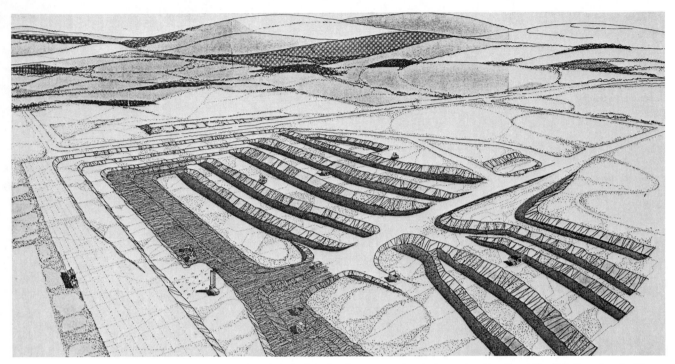

Fig. 1. Diagram of a terrace pit coal mine, showing terrace pit stripping and reclamation. (*After U.S. Department of Energy, The Development of Optimal Terrace Pit Coal Mine Systems, DOE Rep. FE/10023-1, January 1971*)

mining is practiced, mining begins with an initial opening or box cut, with the spoil from the cut being cast to one side or removed to a stockpile, and the uncovered coal is mined. Then an adjoining parallel cut of overburden is removed and deposited in the previous excavation, and the exposed coal is mined. Mining proceeds in this fashion until an economic, technologic, or property limit is reached.

The box cut is usually taken where the coal has the least overburden and in an orientation which allows the pit to advance toward coal with the greatest overburden. In a single-bench operation the box cut is designed to allow for a pit floor width which will ac-

commodate the in-pit equipment and the initial overburden spoil used in backfilling, which must lie at a safe angle of repose. In a dragline or stripping shovel operation the toe of the spoil is adjacent to the coal seam in order to minimize the required operating radius for the equipment. Where multiple benches or terraces are developed for overburden and coal removal, the box cut width varies at each bench or terrace level.

In a single-bench, simple sidecasting dragline operation there is no rehandling of material, and it is therefore a widely practiced and economical method for overburden excavation. Coal is mined by loading

Fig. 2. Diagram of a strip coal mine, showing bucket-wheel excavator prestripping for draglines. (*After U.S. Department of Energy, The Development of Optimal Terrace Pit Coal Mine Systems, DOE Rep. FE/10023-1, January 1971*)

shovels or wheel loaders and is hauled by truck. Sizing of the dragline, which involves determining the operating radius, the dumping height, and the maximum suspended load, is important in a single-seam operation because the depth of the overburden lift which can be sidecasted is determined by the size of the dragline. The dragline may be operated so as to make a rough separation between soil and overburden and hence aid in reclamation of the mined area. Large power shovels are also used for stripping overburden in this type of operation.

At one time either a dragline or a large stripping power shovel was used for overburden excavation, depending on individual site factors and the machine design characteristics. However, draglines have been used exclusively since the 1960s. The superiority of the dragline (**Fig. 6**) are in its long reach and dumping radius, which permits thicker overburden to be mined and the spoil to be placed at a greater distance and in higher piles, and the large capacity of its buckets, which determines the productivity of the machine.

Multiple bench or terraced pits may be developed by using a number of different combinations of equipment, but a shovel-truck combination or a shovel-truck dragline combination is usually high on the list of feasible options. In a typical shovel-truck combination operation, overburden is removed with power shovels, equipped with buckets in the 25–30 yd³ (21–23 m³) range, and loaded into trucks. A bulldozer works with each power shovel for cleanup, leveling, and bench preparation. The overburden benches are about 50 ft (15 m) in height and are mined in 200-ft (61-m) push-backs. Shovels advance the push-back on the top level first and then proceed to the lower benches. Trucks haul the overburden up ramps on the highwall side of the pit, then around the pit on the spoil side, advancing the spoil dump at the same rate as the highwall. Spoil from the box cut is stockpiled for later use in reclamation.

Overburden drilling is done by rotary drills with 6⅞–12½-in.-diameter (17.5–31.8-cm) bits. Coal drilling is done with truck-mounted rotary drills using 6-in.-diameter (15-cm) bits. Blastholes are loaded with ANFO (ammonium nitrate–fuel oil) explosive, usually from bulk mixing trucks, which mix and me-

Fig. 3. Front-end loader.

Fig. 4. Small dragline.

Fig. 5. Power shovel loading coal trucks.

Fig. 6. Dragline working from the surface of the ground. (*Morrison Knudsen Co.*)

ter the amount of explosive for each hole. After blasting, coal is loaded, with either power shovels or wheel loaders, onto bottom dump trucks of 100–150-ton (90–135-metric ton) capacity which are favored for coal hauling because of their low body height. Coal benches are about 30 ft (9 m) high and 200 ft (61 m) wide, and are advanced in 100-ft (30-m) pushbacks. *See Boring and drilling, geotechnical; Explosion and explosive.*

A dragline may be added to the shovel-truck combination after the box cut is excavated and backfilling with spoil has begun. A bench is excavated for the dragline at the appropriate level above the seam. Operating from this bench, the dragline excavates the overburden above the seam and sidecasts it into the pit floor area from which coal has been extracted.

A number of variations on the shovel-truck combinations are possible. For example, an around-the-pit-conveyor transport system may be used to reduce or eliminate truck haulage. Excavated overburden may be loaded directly or may be transported by truck to a hopper equipped with a fixed grizzly screen to reject oversize material. From the hopper, the spoil is fed to a shiftable conveyor system which carries the spoil out of the pit to a stacker-spreader. This in turn spreads the material along a spoil bench in the pit. The conveyor system is shifted forward in 100-ft (30-m) increments as the benches advance.

Bucket wheel excavators are useful in certain mining conditions. Essentially a bucket wheel excavator consists of the following parts: (1) a large wheel with toothed buckets pinned to the circumference and mounted on the end of a maneuverable boom; (2) a discharge boom supporting a belt conveyor; (3) a revolving frame mounted on crawler supports; and (4) electric motors and control devices. Material is excavated by moving the rotating wheel into the overburden or coal and swinging it on the maneuverable boom through an arc. Material is cut, chopped, or scraped into the buckets, and as the wheel continues to rotate, the buckets are emptied. The excavated material is transported away from the wheel by a system

of conveyors and finally moves up the discharge conveyor to the discharge point. The conveyor system may consist of an around-the-pit conveyor incorporating a shiftable conveyor section. The shiftable conveyor must be periodically moved to keep it within the discharge range of the bucket wheel excavator. A mobile transfer conveyer may be added to the system to improve the continuity of material flow. In all cases, however, material flow is more or less continuous.

Bucket wheel excavators are capable of high production under favorable conditions with outputs in the range of 1600–12,000 yd³ (1200–9200 m³) per hour. However, they have difficulty in handling hard, consolidated material even when blasted; alternating hard and soft formations slows production. Also, large boulders, buried vegetation, and sticky materials clog the buckets, transfer points, and machinery. However, they are widely used in Europe for the mining of brown coals. The rock strata, consisting of unconsolidated earth, soft shale, and silt stone overlying thick beds of brown coal, is ideally suited to the capabilities of bucket wheel excavators. The brown coals extend over large areas and may be found in thicknesses of 600 ft (183 m), or more, interbedded with sand and clay zones. Area pit dimensions may be a number of miles, and coal and spoil transport to delivery points is by belt conveyor or rail.

Bucket wheel excavators have found limited application in United States mining conditions. *See Open-pit mining.*

Contour mining and mountaintop removal. This method of surface mining is used in the hilly and mountainous areas of the Appalachians where coal seams are numerous and are generally flat-lying or subhorizontal, outcropping on the steeply sloping valleys and ridges. The geologic column consists of shales, sandstones, and limestones of varying thickness and hardness.

Now illegal, conventional contour mining started with an initial cut along the outcrop line on the mountainside. The overburden was indiscriminately cast over the side of the hill and the coal extracted. This procedure was continued both into the mountainside and along the outcrop line until an economic, technical, or property limit was reached. The resulting excavation left a collar around the mountain consisting of a highwall, bench, and castover spoil bank.

The haulback method of contour mining is an environmentally sound method of coal recovery because it eliminates the casting-over of spoil and provides for concurrent reclamation (**Fig. 7**). The initial cut is made into the mountainside to the highwall limit, and the excavated overburden is hauled to a prepared storage area. The initial cut serves the same purpose as the box cut in area mining. Mining proceeds along the outcrop line with overburden, subsoil, and topsoil being trucked and placed as coal is removed. The method leads to minimum disturbance of the land and allows for contemporaneous reclamation.

In the haulback method the layout of haul roads and the design and construction of the spoil storage area are critical. Haul roads are designed and situated so as to assist in the control of both water flow from the pit and the spillage of spoil on the downslope. Spoil storage areas are generally located in hollow valleys or low areas, and provisions must be made for prevention of contamination of streams by the spoil material and for stability of the material.

The equipment used in the haulback method is

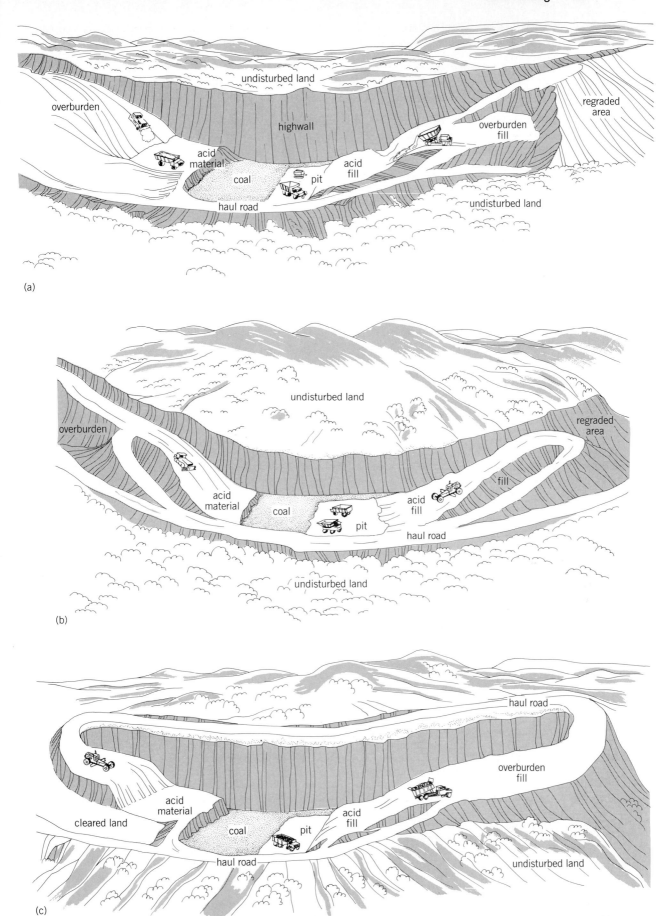

(a)

(b)

(c)

Fig. 7. Typical haulback or lateral movement methods. (*a*) Truck. (*b*) Scraper. (c) Truck and scraper. (*After R. Stefenko, Coal Mining Technology, Theory and Practice, Society of Mining Engineers, AIME, 1983*)

small in comparison to that found in an area mine. Wheel loaders, trucks, dozers, and scrapers are used in various combinations and sometimes power shovels are employed. Drilling and blasting are done with rotary rigs and ANFO. Overburden is drilled and blasted, then loaded by wheel loaders into trucks for haulage to the mined out pit. Subsoil and topsoil that have been removed by dozers are then placed on top of the overburden, and the spoil materials are graded, seeded, and revegetated.

Mountaintop removal might be considered as an extension of area mining techniques to the mountainous and hilly terrains. A box cut is made across the ridge or mountain, and the spoil from this cut is placed in hollow or valley spoil-storage areas. The pit is advanced from the box cut, by excavating parallel slices, with the spoil from the new cut being placed in the adjacent excavated cut. If multiple seams are to be extracted, a bench or terrace pit is developed. The equipment employed is similar to that used in contour mining. The result of mountaintop removal is the transformation of a rugged topography into more level land, but in the process total coal recovery is obtained. Future use of this method will depend upon its perceived advantages in comparison with other techniques.

Auger mining. This method consists of boring large-diameter holes as far as possible into more or less horizontal coal beds, with hole diameter and spacing depending on seam thickness and geologic conditions. The level bench left from a conventional contour mine provides an ideal platform for equipment operation; otherwise a bench must be developed along the coal outcrop line. The auger consists of a cutting head and a set of twisted flights, resembling a carpenter's wood bit or drill. It can be extended by additional flight sections: the drive unit is uncoupled and withdrawn, allowing another flight to be lowered into place for making another advance. As boring proceeds, the coal is moved along the flights and discharged onto an elevating conveyor and then into trucks. Auger mining is attractive because it recovers coal from the highwall side of the contour mine that might otherwise be lost, and it does this with minimal capital and labor cost. Reclamation and environmental problems can significantly affect the cost of auger mining.

Future trends. Surface mining has steadily increased its share of the United States market, and this trend is expected to continue into the foreseeable future. The fundamental extraction and materials-handling systems that have evolved over time present a range of methods that appear capable of coping with future mining problems. Coal lying under deeper and more difficult geologic columns may require some modification in extraction techniques and methods, but it does not appear that a new generation of large equipment needs to be developed. Emphasis will be placed on better planning and managing of the existing methodologies to obtain highly efficient mines. Moreover, the development of highly productive and efficient underground mines will provide an economic limit to surface mines. SEE LAND RECLAMATION.

UNDERGROUND MINING

When a coal seam does not lie close to the surface, it can only be extracted by underground methods. In the design and engineering of these methods, consideration must be given to a set of interacting variables that can be classified as independent or dependent, controllable or uncontrollable, and quantifiable or unquantifiable. The basic uncontrollable variables are geologic environment, coal seam factors, human factors, and government regulations and impositions. From a careful evaluation of these factors plus past experience, the mine designer can select one or several mining methods for consideration. Rating schemes which weigh the importance of specific items in the uncontrollable variables aid in the selection process. Given a mining method, the mine designer can then choose the geometry of the mine layout and the equipment specifications, and thus control the rate at which coal is recovered from the deposit and the total amount or level of recovery from the deposit. For a known set of conditions, the variables will lie within a given range, and the systems performance and sensitivity can be evaluated by a parametric analysis.

Underground coal mining methods may be broadly classified as room-and-pillar, longwall, and others. For each of these methods, modifications to the basic techniques are needed to cope with varying geologic conditions and seam factors. Irregular seam thickness, steep dips, changing rock quality, seam partings, and other geologic and seam factors have a marked influence on the mine geometry and equipment specifications. Seams dipping between 0 and 10° and with regular seam thickness will be assumed in the following discussion of room-and-pillar and longwall mining methods.

Mine development. There are two broad classes of mine openings which must be driven in developing a particular extraction system—development openings and production openings. Development openings provide the primary access to various parts of the coal deposit and are called mains. In room-and-pillar mining, the production openings are the rooms driven in the panel and the extraction cuts in pillar retrieval. In longwall mining, the production openings are the longwall faces. In general, the term development includes all openings and other work which precedes production.

Main access to a horizontal or practically level coal seam is usually by drift entry when the coal seam outcrops on a mountainside; otherwise, slope and shaft entries are used. Slopes are preferred over shafts when the seam is not too deep because conveyor belt haulage can be used. In addition, mining machinery can usually be brought underground with less disassembling. From the foot of these access openings, a set of parallel main entries (usually 7 to 10 individual headings) are driven, and when these have advanced 3000–4000 ft (915–1220 m) a set of submain entries are driven to the right and left of the main entries. This entire set of entries divides the coal property into large blocks, which are further subdivided by panel entries driven off the submains.

Because of ventilation restrictions, mining machinery movements, and ground conditions, connections between entries are made at intervals. These are called crosscuts. The crosscuts divide the coal block between entries into pillars, often called chain pillars, which provide the primary support for the ground overlying the coal deposit. Since main and submain entries must remain open and stable for long periods of time, the design of the pillars and the roof support in the entries is a critical item in mine planning.

Panel entries, up to 4000 ft (1220 m) or longer,

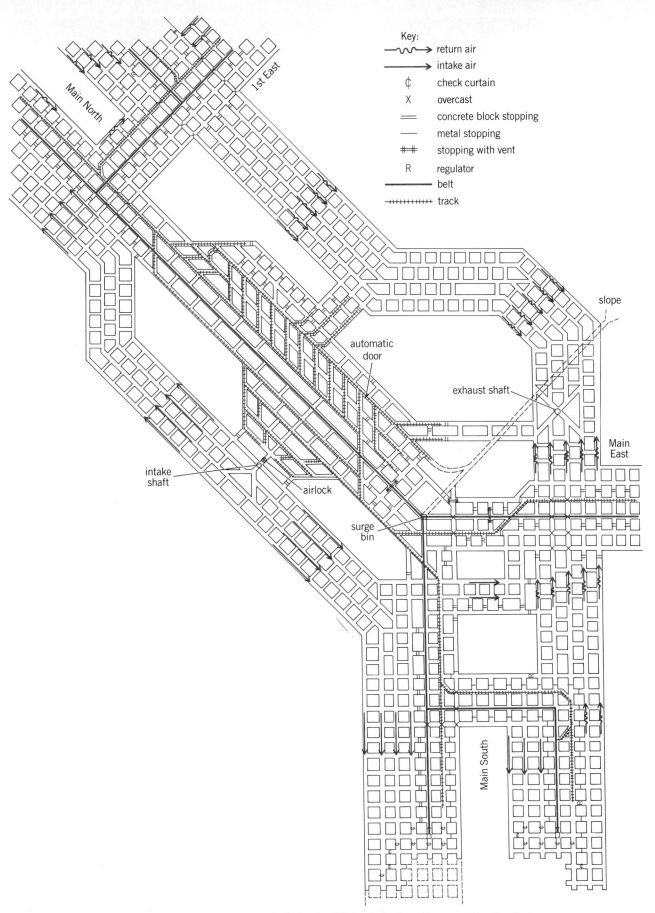

Key:
- return air
- intake air
- ₵ check curtain
- X overcast
- concrete block stopping
- metal stopping
- stopping with vent
- R regulator
- belt
- track

Main North

1st East

automatic door

exhaust shaft

slope

Main East

intake shaft

airlock

surge bin

R

Main South

Fig. 8. Plan of a room-and-pillar mining operation. (*After R. Stefenko, Coal Mining Technology, Theory and Practice, Society of Mining Engineers, AIME, 1983*)

Fig. 9. Single-boom face drill operating in low-seam coal. (*Long-Airdox Co.*)

divide the large coal blocks into rectangular blocks (panels) for production mining. Barrier pillars are left between the mains, the submains, and the rectangular panels in order to minimize the effect of the coal extraction associated with production on the stability of the mains.

Room-and-pillar mining. Coal mining in the room-and-pillar system (**Fig. 8**) may be by either continuous or conventional mining methods. The essential differences between the two methods is in the equipment spectrum and the cycle of face operations that the equipment dictates.

In continuous mining the cycle of operations begins with a continuous miner sumping into the coal face. A shuttle car is positioned behind the miner to receive the cut coal for tramming to the belt feeder. When the shuttle car is filled, it moves out and is replaced by the second shuttle car. This operation continues until the continuous miner operator is located under the last roof support. The continuous miner is then withdrawn, and safety posts or hydraulic jacks are placed in the excavated opening. The miner then widens the room to its final size (usually 18–20 ft or 5.5–6 m) by cutting the adjacent strip. The miner moves onto the next working place—a room or crosscut—and roofbolt support is installed. The face ventilation system is adjusted to compensate for the freshly cut room and the new working location of the continuous miner. Electric cables to the equipment, the power center, the belt feeder, and other items must be adjusted and moved forward as the overall mining section advances.

In conventional mining the breaking out of the coal from the face is done by cutting, drilling, and blasting operations. First, the cutting machine saws a 10–12-ft-deep (3–4-m) slice horizontally across the room width and then moves out to the next place to be cut. Then the drilling machine (**Fig. 9**) moves in and drills holes into the cut face, and this is followed by the blasting operation. After the area is examined for safety, the loading machine and shuttle cars move in and load out the coal. Sufficient working places should be available to minimize the idle time of these various pieces of equipment. Ventilation, electric cables, power center, belt feeder, and other items must be adjusted and moved forward as the overall mining section advances.

Retreat mining. The system of rooms and crosscuts driven in the production panels divides the panel into a series of coal pillars of approximately equal size. These coal pillars are extracted by methods which allow the mining operation to retreat toward the panel entries. As the pillars continue to be extracted, the bridging span of the immediate roof and then of the main roof is exceeded and the ground caves into the opening. This caved rock is called the gob. Since methane gas may accumulate in the gob, a bleeder system of entries is established at the top of the panel to allow a portion of the ventilating air to circulate through the gob.

There are four basic pillar extraction techniques: split-and-fender, pocket-and-wing, outside lift, and open end. The first two are suited for continuous mining methods. Open ending is utilized with conventional mining equipment, and outside lift is primarily used to extract long, thin pillars.

In the split-and-fender method, the pillar is split into two fenders of coal by driving a room parallel to the long side of the pillar. Because of roof control restrictions, which do not allow the splitting of the pillar by a single pass of the continuous miner, two pillars are mined simultaneously. The continuous mining unit, operating in an alternating fashion, mines a cut in each pillar. Rock bolts are used for roof support in the splitting operation, but breaker posts and cribs are used for roof control when the fenders of coal are removed. Check curtains are used to direct the ventilating current to the working faces.

The cuts used in the outside lift method are similar to those used in removing a fender of coal.

In the pocket-and-wing method, two working places can be sequenced in the same pillar. The pillar is alternatingly split transversely, then longitudinally, producing three fenders. These are then removed by

Fig. 10. Remote-controlled continuous miner.

using breaker posts for roof control. Ventilation is controlled by check curtains.

In the open end method of pillar extraction, the sequence of extraction cuts are made adjacent to the gob without a protecting fender of coal.

Equipment. Continuous miners (**Fig. 10**) consist of a cutting head, a coal-gathering system, and a scraper conveyor system; the whole piece of equipment is electrically powered and mounted on a crawler-supported carrier. The cutting heads, of various designs, either rip, bore, or auger the coal from the face. The ripper type of miner is most popular. Rippers have drum cutting heads mounted horizontally and fitted with steel bits. The cutting head, which is supported by arms that allow it to be positioned in a vertical plane, is crowded into the face by moving the carrier forward. Coal is ripped from the face, drops to the floor, and is gathered onto the scraper conveyor by a set of rotating disks or gathering arms. The scraper conveyor transports the coal into the positioned shuttle car. Geologic conditions and seam factors, especially seam height and inseam rock, influence the operation and design of the miner.

A shuttle car is a low-bodied diesel or electrically powered unit (**Figs. 11** and **12**) that carries the coal to a nearby belt conveyor system that takes the coal out of the mine.

Roof bolts are an essential component of both continuous and conventional mining. They are used to provide ground control for the immediate roof strata overlying the mined room, and a means of support that does not obstruct the mine opening. Roof bolters are used to drill the holes and install the bolts. Roof bolting is a prime bottleneck in room-and-pillar mining, and is also a dangerous operation; hence roof bolting equipment must be fast and efficient and provide temporary support to the mine roof while the operator is installing the bolt.

A cutting machine is a large chain saw mounted on a rubber-tired carrier. A universal cutter can be maneuvered to cut a slice at any angle to the horizontal. Blasthole drilling in coal is done with electrically driven augers. Blasting may be done with special coal mine explosives, compressed air, or carbon dioxide cartridges. The loading machine consists of a gathering system which sweeps the coal off the mine floor onto an inclined scraper conveyor for transport to the shuttle car. A belt feeder has a toothed roll crusher for sizing the coal, which is then fed at a uniform rate by a scraper conveyor onto the section belt conveyor.

Ventilation. The primary function of mine ventilation is to dilute, render harmless, and carry away the gas and dust produced in the underground mining of coal. Federal law requires the following conditions to be met: (1) Minimum air velocity in the working place must be 60 ft (18 m) per minute. (2) Minimum air volume at the last open crosscut must by 9000 ft^3 (255 m^3) per minute. (3) Minimum air volume reaching each working face must be 3000 ft^3 (85 m^3) per minute. (4) Respirable dust levels in the mine air cannot exceed an average of 2 mg/m^3 per shift. (5) Air used to ventilate belt conveyors cannot be used to ventilate working faces and must be vented to return airways. (6) Methane accumulations cannot be allowed to exceed 1% in the working area. Additional requirements deal with numerous other situations encountered in underground coal mining.

The mine ventilation system is composed of three subsystems: (1) the primary distribution system,

Fig. 11. Shuttle car for transporting coal within the mine.

which consists of main fans and the main intake and return entries; (2) the face ventilation system (**Fig. 13**), which controls airflow throughout the active production area; and (3) control devices to direct the air through the mine. Main fans are required to be installed on the surface, and usually pull air through the mine. They do this by developing a pressure gradient between the intake air entries and the return air entries in the mine. Hence crosscuts between these two sets of main entries must be closed off by permanent stoppings. Splitting the intake air at each set of submains allows different areas of the mine to have their own independent air, and also serves to minimize power costs. Return air from these splits must be carried over the intake mains by air bridges or overcasts. The flow of air within a split can be controlled by increasing the resistance to flow with regulators placed at strategic locations in various entries and rooms.

The production area of the mine generates the most gas and dust, and hence requires constant examination and modification of the airflow pattern to cope with the changing situation developed by the advancing faces. The blind room headings ahead of the last open crosscut may be ventilated by line brattices (a nylon-plastic curtain which is installed so as to split the blind room heading longitudinally) or auxiliary fans. Air is then directed up one side and down the other. The line brattices are also used to direct the movement of air throughout the production area by acting as check curtains in certain rooms. In order to do an effective job of diluting and sweeping dust and gas from the working face, line brattices must be kept close to the face. Auxiliary fans and tubing in a push-pull arrangement are alternatives to line brattices that

Fig. 12. Diesel scoop in operation. (*WVA Mining Equipment Co.*)

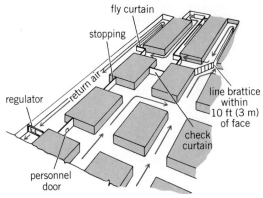

Fig. 13. Diagram of a blower line-brattice face ventilation. Arrows indicate direction of flow of air. (*After R. Stefenko, Coal Mining Technology, Theory and Practice, Society of Mining Engineers, AIME, 1983*)

better diffuse air across the working face but introduce other problems. *See Fan*.

Ground control. The rock strata above the coal seam may be divided into an immediate roof and a main roof. The immediate roof responds quickly to the removal of the coal support and collapses unless some artificial support is installed to prevent and control the deformation. The main roof shows stability for long periods of time. In general, the boundary between these two roofs is formed at the contact zone defined by fairly thick competent rocks, such as sandstones and limestone, and thinly bedded incompetent rocks, such as shale and coal. The function of the pillars of coal developed by driving rooms and crosscuts is to provide support for the total roof structure over both the pillars and the mined openings. The ground load immediately over the pillars is transferred directly to the pillars, but the ground load over the rooms and crosscuts must transfer its load to the pillars. This mechanism of load transfer is called the bridging action of the rocks; competent rocks have good bridging action and incompetent rocks have poor bridging action. Artificial support is used to assist, enhance, or develop the bridging action of the rocks in the immediate roof.

Simple posts with cap pieces were the first roof

supports used. The concept is to decrease the overall bridging span between posts. More posts with larger cap pieces and three-piece sets are added to control less competent roof strata. Posts for roof control are still used in pillar recovery work or retreat mining. However, for the highly mechanized continuous and conventional mining of entries, rooms, and crosscuts, timber posts and three-piece sets cannot be used because they restrict the movement of equipment and may be knocked down.

Prop-free openings with the immediate roof spanning the full entry width can be developed with the aid of roof bolts. The most commonly used roof bolt is an expansion-shell-anchored, ⅝-in.-diameter (1.6-cm), extra-strength, square-head steel bolt. The law specifies that the bolt must be a minimum of 30 in. (76 cm) in length. The function of the bolt is to bind together the various laminae in the immediate roof and make them act as a unit in developing bridging action. The bolts, when anchored in competent rock, also prevent sagging of the immediate roof. The binding and suspension action of the bolt is developed when tension is produced in the bolt by tightening it into the anchoring shell. A number of variations of this basic bolt are in use.

Ground control in main entries or in areas where roof bolts cannot develop the desired support is done by other means. Room and entry intersections often exceed mandated limits on roof bolts, so steel sets or steel beams resting on wood cribs or concrete blocks are used for the support of the intersections. In some main entries, steel arches or yielding steel arches are used. Where rapid temporary support is required, screw jacks or hydraulic props are easy to install and make excellent supports.

In the United States, federal law mandates that each mine operator have an approved roof control plan, a history of all unintentional roof falls, and a systematic evaluation of the roof control system in use. The roof control plan must show the geologic column, and identify the rock type and thickness above and below the coal seam. Further, the plan must describe the sequence of mining and installation of supports, including temporary supports, and must designate the area of the mine covered by the roof control plan.

Longwall mining. Longwall mining (**Fig. 14**) is the basic underground coal mining method practiced in Europe, and it enjoys wide acceptance throughout the world. It has a greater production potential than room-and-pillar systems, and is safer since mining takes place beneath a complete overhead steel canopy (**Fig. 15**) that moves as the face is mined. Longwall methods were tried in the United States for a number of years, but could not compete with the room-and-pillar system until modern longwall mining was introduced in 1960. Longwall mining will be increasingly used in the United States as the exploitation of deeper coal seams under poorer natural conditions becomes necessary.

Longwall systems are classified as retreat or advancing systems based on whether the panel is completely developed by panel entries before panel extraction commences. In the retreat system, entries (gates or roadways) are driven to block out a coal panel 3000–4000 ft (915–1220 m) long by 300–600 ft (91–182 m) wide. Coal is then retreat-mined from the far end of the panel back to the submains. In the advancing system, development entries are kept only

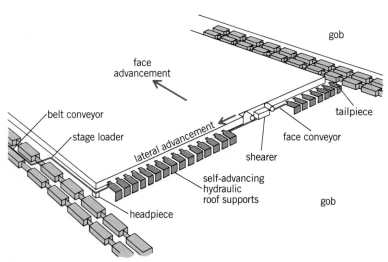

Fig. 14. Diagram showing the basic equipment components on a longwall face. (*After R. Stefenko, Coal Mining Technology, Theory and Practice, Society of Mining Engineers, AIME, 1983*)

a short distance ahead of the working face which advances to the far end of the panel.

Regulatory constraints in the United States require longwall panels to be developed by multiple entries similar to those used in room-and-pillar mining. These entries are usually developed by continuous mining methods. Bleeder entries as well as a longwall equipment setup entry are driven across the top of the panel and connect panel entries. The panel entry containing the beltline, electrical and hydraulic equipment, and other items is called the head gate. Its complement on the other side of the panel is called the tail gate. Panels are arranged so that the previous panel's head gate is the new panel's tail gate.

Single head and tail gate entries are common in Europe. When retreat mining is practiced, full-face tunneling machines are used to drive the entries and rock bolts are used for support. These will be supplemented with steel arches or some other method of support as conditions change due to mining. When advance mining is practiced an in-team miner or stable driving equipment is used. One aspect of longwall mining on the advance is that the head and tail gate must be established and maintained through the gob. This usually means that the roof or floor of these entries must be ripped and the entry cross section re-established and supported.

Three pieces of equipment are fundamental to modern longwall mining—armored face conveyors, powered supports, and the coal cutting machine. The cycle of face operations is based on the movement of this equipment. The armored face conveyor is erected along the coal face, and is connected to powered supports by means of double-acting hydraulic jacks. The cutter-loader usually slides along the top of the armored conveyor and breaks out a strip of coal 20–30 in. (50–75 cm) wide along the face and then pushes the broken coal onto the conveyor. As the cutter-loader progresses, it leaves a web 20–30 in. (50–75 cm) wide into which the armored conveyor is pushed by the hydraulic jacks on the powered supports. The hydraulic jacks supporting the overhead steel canopy are retracted, freeing the powered supports from roof pressure. These supports are then advanced up to the armored conveyor by using the double-acting hydraulic jacks connecting the two and are then reset. When the cutter-loader completes a pass across the face, it is reset to take a cut as it returns across the face and the cycle is repeated. As coal is extracted and the powered supports are moved forward, more and more of the roof is left unsupported. The immediate roof collapses and is followed by the main roof collapsing or settling on the broken rock in the gob.

Equipment. An essential element of the modern longwall system is the armored face conveyor, sometimes called a snaking conveyor because in plan it can be curved around behind the cutter-loader into the web that is created as the cutter-loader advances along the face. The face conveyer is a steel trough running the length of the coal face and strong enough to carry the weight of the cutter-loader. One or more endless chains within the trough pull steel flights (paddles) that drag the coal to the head end where it discharges into the head gate transportation system. The face conveyor may be driven by single or multiple drive units situated at the head and tail gate ends and on either or both sides of the conveyor. Each drive unit consists of a motor, fluid coupling, and reduction gearing. The overall bulk of these units has an influ-

Fig. 15. Longwall shearer and hydraulic roof support.

ence on the width of the head and tail gates. The preferred alternative of having drive units at both ends usually cannot be met because the excessive ground pressure on the tail gate makes it difficult to maintain the necessary room for tail drives. Face lengths are limited to about 600 ft (182 m) by the drive arrangements, the chain pull needed to move the coal, and other factors. The face conveyor does not dump directly onto the belt conveyor taking coal out of the panel. Instead it dumps onto a stage loader which serves as a feeder-breaker system for changing the direction and velocity of coal travel and breaking large lumps of coal and rock.

The good roof control developed by powered supports is a key to successful longwall mining. Powered supports have evolved, with a number of different features being introduced as the need arose. A very popular version in the United States is called a shield support. The unit consists of a set of hydraulic props placed between a steel floor beam and a roof beam with the beams pin-connected to a steel caving shield so that the individual members may rotate with respect to each other. Control valves, hydraulic hoses, and a double-acting jack complete the assembly. The caving shield linking the two beams is arranged so that adjacent supports provide skin-to-skin protection against broken rock in the gob flushing into the traveling way and working area of the face. The roof beam cantilevers over the traveling way and working area of the face, completing the steel canopy protecting workers and giving a propfree front face. The design principles of the shield allow it to cope with both vertical and lateral movements of the immediate roof as it responds to coal extraction. Powered supports have not been designed to work in thin coal seams.

Cutter-loaders are of two types: shearers and plows. A shearer (Fig. 8) consists of a drum fixed to the end of the ranging arm, and the combined unit is connected through gearing to a power unit riding on the conveyor. The drum is a spiral arrangement fitted with picks for breaking out the coal. A cowl is attached to the machine at the back end of the drum, and the combination of spiral drum and cowl acts as a screw conveyor moving the newly mined coal into the face conveyor. A shearer may have two drums mounted at each end of the power unit. Double-drum shearers with cowls that can be repositioned make bidirectional mining feasible as well as allowing the shearer to accomodate to variable seam height. The shearer moves across the face by a rack-and-pinion mechanism between the shearer and the face conveyor.

The coal plow (planer) is pulled back and forth across the face by an endless chain system. It is designed to travel in the space between the face conveyor and the coal, and cuts a layer of coal 2–6 in. (5–15 cm) thick. The coal is deflected or plowed onto the face conveyor. The plow is held against the face by hydraulic jacks located between the conveyor frame and the plow. An operator does not follow the plow along the face, and the plow does not cut as thick a web as the shearer, so planers can probably work under more difficult roof conditions than shearers.

Ventilation and ground control. The speed of coal production and the length of the longwall face give rise to difficulties in face ventilation. Shearers not only produce a fine product and create dust, but also develop a deep web. The net effect of these factors is an increase in methane liberation. Part of the input face ventilation escapes into the gob, and hence is not used to sweep away and dilute these products. Methane that escapes and collects in the gob area may be drained away by a system of boreholes.

Ground control problems usually result from seam factors. Soft floors may have insufficient bearing capacity for support loads. Massive roof strata with long bridging spans will not cave and overload face support. Variations in seam height and uneven floor conditions can unbalance face support systems. Thick seams must be mined by successive lifts. Tail entries are subjected to severe ground pressures and require heavy timbering and hydraulic prop support.

Other methods. Variations on the basic room-and-pillar and longwall methods are numerous. The shortwall and hydraulic mining methods are two variations. Shortwall mining, using equipment from both room-and-pillar and longwall methods, is an attempt to handle the technical and economic alternatives in choosing an extraction system. The shortwall layout is similar to the longwall panels except that the panel width is 150–200 ft (46–61 m) wide. A continuous-miner loading shuttle cars substitutes for the cutter-loader-face conveyor system, and the roof beams of the powered supports need to give cantilever action over a wider working area. Shortwall mining is considered when the coal seam is relatively shallow but is overlain by massive roof strata; when a large, relatively homogeneous and undisturbed area of coal is not available for longwall panels; or when ventilation difficulties from dust preclude face personnel from working downwind from the cutter-loader.

Hydraulic mining uses large quantities of water at high pressure to break and convey the coal from the working area. It is most effectively applied to a steeply dipping, relatively thick coal seam that is bounded by an immediate roof and floor and breaks into large pieces. Jets of water are directed tangentially along the coal face to break off lumps of coal. The coal is washed into a system of troughs for conveying away from the working area. Subsequently the coal may be dewatered and moved by conveyor belt, or it may remain in troughs or be pumped to its final location. Hydraulic mining is successfully applied in China, the Soviet Union, the United States, and Canada.

Future trends. New equipment and equipment modifications needed to meet the challenges of difficult underground mining conditions will dominate the future. How to mine thin seams, thick seams, steeply dipping seams, seams at greater depths, extremely gassy seams, seams under bodies of water, and combinations of these need to be solved. Integration of extraction, haulage, ventilation, ground control, and reclamation systems into an economic and ecologically sound plan will be the challenge for future mining engineers. SEE MINING; UNDERGROUND MINING.

Malcolm T. Wane

Bibliography. N. P. Chironis (ed.), *Coal Age Operating Handbooks of Coal Surface Mining and Reclamation, Underground Coal Mining, Coal Preparation*, 1978; Code of Federal Regulations, Title 30: Mineral resources, 1976; D. F. Crickmer and D. A. Zegeer, *Elements of Pratical Coal Mining*, 2d ed., 1981; R. D. Merritt, *Coal Exploration, Mine Planning, and Development*, 1986; R. Stefanko, *Coal Mining Technology Theory and Practice*, 1983.

Coal paleobotany

A special branch of the paleobotanical sciences concerned with the origin, composition, mode of occurrence, and significance of the fossil plant materials that occur in, or are associated with, coal seams. Information developed in this field of science provides knowledge useful to the biologist in efforts to describe the development of the plant world, aids the geologist in unraveling the complexities of coal measure stratigraphy in order to reconstruct the geography of past ages and to describe ancient climates, and has practical application in the coal, coke, and coal chemical industries. SEE COAL CHEMICALS.

Nature of coal seams. All coal seams consist of countless fragments of fossilized plant material admixed with varying percentages of mineral matter. The organic and inorganic materials initially accumulate in some type of swamp environment. Any chemical or physical alteration experienced by the organic fragments during the course of transportation and deposition is followed by another series of changes effected by the chemical, physical, and microbiological agents characterizing the environment in which the particle comes to rest. Subsequent burial beneath a thick cover of sediment induces further physical and chemical alteration of the particles comprising the coal seam. Usually a consolidated layer of well-bonded fragments results. In some instances, crustal deformation and even volcanism add their modifying effects. Thus, the chemical composition, size, shape, and orientation of the fossilized plant remains are influenced both before and after death by biological processes and by the chemical and physical processes attending their postdepositional history. Even in coal seams that have been metamorphosed to anthracitic rank, certain of the constituents remain recognizable as portions of particular plant organs, tissues, or cells. Such entities are classed as phyterals, and in paleobotanical descriptions of coal seams these are identified as specifically as possible as megaspores, cuticles, periderm, and so on. In seams of peat, lignite, and high-volatile bituminous coal, entities are more readily recognized as particular phyterals than they are in higher-rank deposits. This is partly because fewer are destroyed by the metamorphic processes and partly because of the distinctiveness of the substances composing the entity.

In some instances, the fossilized plant fragments

can be recognized as remnants of a plant of some particular family, genus, or species. When this is possible, information can be obtained on the vegetation extant at the time the source peat was formed, and such data aid greatly in reconstructing paleogeographies and paleoclimatic patterns. Perhaps the most extensive studies of this type are those conducted on the

Tertiary brown coals of the Rhine Valley in Germany. From these, detailed reconstructions have been prepared, describing and illustrating the three major swamp environments that gave rise to the sedimentary layers comprising these spectacular coal seams.

Plant fossils in coal seams. Pollen grains and spores are more adequately preserved in coals than

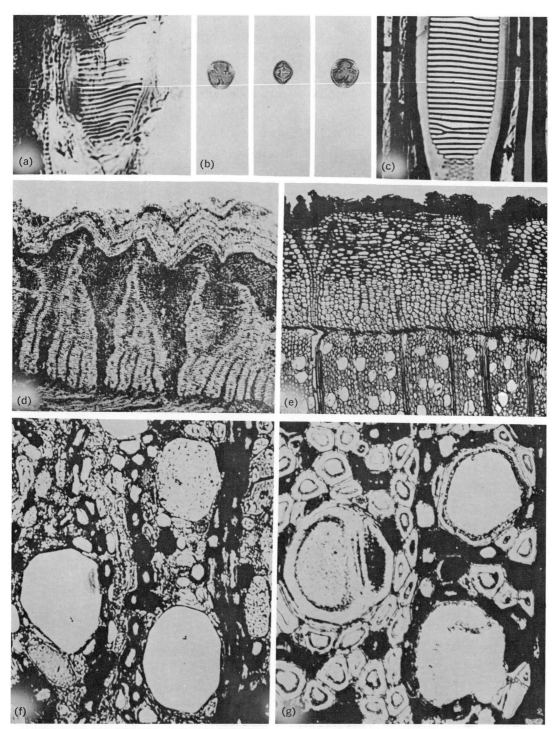

Plant fossils in coal. (a) Coalified remnants of scalariform perforation plate is vessel element of piece of lignitized *Cyrilla* wood from Oligocene lignite deposit near Brandon, Vermont. (b) Coalified *Cyrilla* pollen grains extracted from Brandon lignite. (c) Scalariform perforation plate of vessel in wood of modern *C. racemiflora*. (d) Transverse section of lignitized bark of *Cyrilla* from Brandon lignite. (e) Transverse section of extant *C. racemiflora* bark with portion of secondary xylem shown in lower half of photo. (f) Transverse section of coalified secondary xylem of *Cyrilla* wood from Brandon lignite showing preservation of fine structural detail. (g) Transverse section of secondary xylem of *C. racemiflora* showing the structure of vessels, fiber tracheids, wood parenchyma, and rays. Note the coalification effects by making a comparison with f.

are most other plant parts. Recognition of this fact, coupled with an appreciation of the high degree to which these fossils are diagnostic of floral composition, has led to the rapid development of the paleobotanical subscience of palynology. Fruits, seeds, and identifiable woods also occur as coalified fossils and are deserving of more attention than has been accorded them in the past. Often minute structural details are preserved (see **illus**.). Identifiable leaf fossils are comparatively uncommon, although the coalified cuticles of leaves are frequently encountered and their botanical affinities determined. Occasionally, coal seams contain fossil-rich coal balls. Essentially all types of plant fossil, including entire leaves, cones, and seeds, are encountered in these discrete nodular masses. Within the coal ball, the altered but undistorted plant tissues are thoroughly impregnated with mineral matter, usually with calcite or dolomite. Comparatively few coal seams have been encountered with large concentrations of coal balls, but these seams have provided a wealth of detailed information on the nature of the plants which gave rise to the coals concerned. SEE COAL BALLS; FOSSIL SEEDS AND FRUITS; PALYNOLOGY.

Sedimentary units of coal seams. As implied previously, coal seams generally are composed of several superposed sedimentary layers, each having formed under somewhat different environmental conditions. The coal petrologist and paleobotanist recognize these layers because of their distinctive textural appearance and because each consists of a particular association of organic and inorganic materials. Accordingly, each coal seam usually contains several types of coal. These coal types, or lithotypes, possess characteristic suites of physical properties, and knowledge of these properties is very profitably employed in manipulating coal composition in coal preparation and beneficiation plants.

A given lithotype may form several of the constituent layers of lithobodies of a coal seam. Each lithobody often possesses a characteristic assemblage of botanically identifiable fossil plant fragments rendering the unit recognizable to the coal paleobotanist without reference to its textural or compositional features. Paleobotanical descriptions of coal beds may relate to the entire thickness of the seam without regard for the seam's composite nature, or the fossils may be described in relation to the particular lithobody sequence. The more detailed type of description is required if paleoecological interpretations are to be made or if detailed stratigraphic work is involved.

Interrelations with coal petrology. Another facet of coal paleobotany concerns the nature of the substances which compose the coalified plant fragments. When focusing attention upon such matters, the coal paleobotanist is indistinguishable from the coal petrologist and both begin to encroach on the province of the coal chemist. It is noteworthy, however, that the paleobotanist tends to concentrate attention on the genetic relationships of the various coal substances and hence upon the steps in the derivation and subsequent evolution of maceral materials. The initial stages of coalification tend to be ignored by the coal chemist and petrologist, and the study of these has, quite properly, been thought of as an integral part of coal paleobotany. Thus, through the study of the botanical character of coalified fossils and by means of investigating the substances produced by the coalifi-

cation process, coal paleobotanists extend knowledge of the organically derived sediments of the Earth's crust. SEE COAL.

William Spackman

Coal Sack

An area in one of the brighter regions of the Southern Milky Way which to the naked eye appears entirely devoid of stars and hence dark with respect to the surrounding Milky Way region. Telescopic and spectroscopic observations reveal that the Coal Sack is a cloud of small, solid particles approximately 120 parsecs (400 light-years or 2.3×10^{15} mi or 3.7×10^{15} km) from Earth and 10–15 parsecs ($2-3 \times 10^4$ mi or $3-4.5 \times 10^{14}$ km) in diameter. The cloud not only absorbs about two-thirds of the visible light passing through it but also strongly reddens the transmitted light. The Coal Sack is located just to the southeast of the Southern Cross. SEE CRUX; INTERSTELLAR MATTER.

William Liller

Coastal engineering

A branch of civil engineering concerned with the planning, design, construction, and maintenance of works in the coastal zone. The purposes of these works include control of shoreline erosion; development of navigation channels and harbors; defense against flooding caused by storms, tides, and seismically generated waves (tsunamis); development of coastal recreation; and control of pollution in nearshore waters. Coastal engineering usually involves the construction of structures or the transport and possible stabilization of sand and other coastal sediments.

The successful coastal engineer must have a working knowlege of oceanography and meteorology, hydrodynamics, geomorphology and soil mechanics, statistics, and structural mechanics. Tools that support coastal engineering design include analytical theories of wave motion, wave-structure interaction, diffusion in a turbulent flow field, and so on; numerical and physical hydraulic models; basic experiments in wave and current flumes; and field measurements of basic processes such as beach profile response to wave attack, and the construction of works. Postconstruction monitoring efforts at coastal projects have also contributed greatly to improved design practice.

Environmental forces. The most dominant agent controlling coastal processes and the design of coastal works is usually the waves generated by the wind. Wind waves produce large forces on coastal structures, they generate nearshore currents and the alongshore transport of sediment, and they mold beach profiles. Thus, a primary concern of coastal engineers is to determine the wave climate (statistical distribution of heights, periods, and directions) to be expected at a particular site. This includes the annual average distribution as well as long-term extreme characteristics. In addition, the nearshore effects of wave refraction, diffraction, reflection, breaking, and runup on structures and beaches must be predicted for adequate design. SEE OCEAN WAVES.

Other classes of waves that are of practical importance include the astronomical tide, tsunamis, and waves generated by moving ships. The tide raises and lowers the nearshore water level and thus establishes the range of shoreline over which coastal processes act. It also generates reversing currents in inlets, harbor entrandes, and other locations where water motion is constricted. Tidal currents which often achieve a velocity of 3–6 ft/s (1–2 m/s) can strongly affect navigation, assist with the maintenance of channels by scouring sediments, and dilute polluted waters. *SEE TIDE*.

Tsunamis are quite localized in time and space but can produce devastating effects. Often the only solution is to evacuate tsunami-prone areas or suffer the consequences of a surge that can reach elevations in excess of 33 ft (10 m) above sea level. Some attempts have been made to design structures to withstand tsunami surge or to plant trees and construct offshore works to reduce surge velocities and runup elevations. *SEE TSUNAMI*.

The waves generated by ships can be of greater importance at some locations than are wind-generated waves. Ship waves can cause extensive bank erosion in navigation channels and undesirable disturbance of moored vessels in unprotected marinas.

On coasts having a relatively broad shallow offshore region (such as the Atlantic and Gulf coasts of the United States), the wind and lower pressures in a storm will cause the water level to rise at the shoreline. Hurricanes have been known to cause storm surge elevations of as much as 16 ft (5 m) for periods of several hours to a day or more. Damage is caused primarily by flooding, wave attack at the raised water levels, and high wind speeds. Defense against storm surge usually involves raising the crest elevation of natural dune systems or the construction of a barrier-dike system. *SEE SEA-LEVEL FLUCTUATIONS; STORM SURGE*.

Other environmental forces that impact on coastal works include earthquake disturbances of the sea floor and static and dynamic ice forces. Direct shaking of the ground will cause major structural excitations over a region that can be tens of kilometers wide surrounding the epicenter of a major earthquake. Net dislocation of the ground will modify the effect of active coastal processes and environmental forces on structures. *SEE EARTHQUAKE*.

Ice that is moved by flowing water and wind or raised and lowered by the tide can cause large and often controlling forces on coastal structures. However, shore ice can prevent coastal erosion by keeping wave action from reaching the shore.

Coastal processes. Wind-generated waves are the dominant factor that causes the movement of sand parallel and normal to the shoreline as well as the resulting changes in beach morphology. Thus, structures that modify coastal zone wave activity can strongly influence beach processes and geometry.

Active beach profile zone. **Figure 1** shows typical beach profiles found at a sandy shoreline (which may be backed by a cliff or a dune field). The backshore often has one or two berms with a crest elevation equal to the height of wave runup during high tide. When low swell, common during calm conditions, acts on the beach profile, the beach face is built up by the onshore transport of sand. This accretion of sand adds to the seaward berm. On the other hand, storm waves will attack the beach face, cut back the berm, and carry sand offshore. This active zone of shifting beach profiles occurs primarily landward of the 33-ft (10-m) depth contour. If the storm tide and

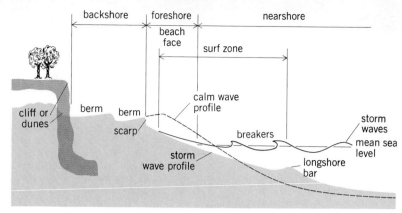

Fig. 1. Typical beach profiles (vertical scale is exaggerated).

waves are sufficiently high, the berms may be eroded away to expose the dunes or cliff to erosion. The beach profile changes shown in Fig. 1 are superimposed on any longer-term advance or retreat of the shoreline caused by a net gain or loss of sand at that location.

Any structure constructed along the shore in the active beach profile zone may retain the sand behind it and thus reduce or prevent erosion. However, wave attack on the seaward face of the structure causes increased turbulence at the base of the structure, and usually increased scour which must be allowed for or prevented, if possible, by the placement of stone or some other protective material.

It is desirable to keep all construction of dwellings, recreational facilities, and such landward of the active beach profile zone, which usually means landward of the frontal dunes or a good distance back from retreating cliffs. It is also desirable to maintain and encourage the growth of the frontal dune system by planting grass or installing sand fencing.

In addition to constructing protective structures and stabilizing the dune system, it is common practice to nourish a beach by placing sand on the beach face and nearshore area. This involves an initial placement of sand to develop the desired profile and periodic replenishment to make up for losses to the profile. A common source for sand, which should be clean and at least as coarse as the native sand, is the offshore area near the nourishment site.

Alongshore current and transport. Waves arriving with their crest oriented at an angle to the shoreline will generate a shore-parallel alongshore current in the nearshore zone. The current flows in the direction of the alongshore component of wave advance and has the highest velocity just inside the breaker line. It may be assisted or hindered by the wind and by tidal action, particularly along sections of the shore adjacent to tidal inlets and harbor entrances. There is a continuous accumulation of flow in the downcoast direction which may be relieved by seaward-flowing jets of water known as rip currents.

The alongshore current transports sand in suspension and as bed load, and is assisted by the breaking waves, which place additional sand in suspension. Also, wave runup and the return flow transports sand particles in a zigzag fashion along the beach face. Coastal structures that obstruct these alongshore transport processes can cause the deposition of sand. They do this by blocking wave action from a section

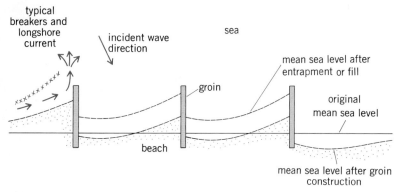

typical breakers and longshore current

sea

incident wave direction

mean sea level after entrapment or fill

groin

original mean sea level

beach

mean sea level after groin construction

Fig. 2. Groin system and beach response.

of the shore and thus removing the wave energy required to maintain the transport system; by interfering with the transport process itself; or by directly shutting off a source of sand that feeds the transport system (such as a structure that protects an eroding shoreline).

The design of most coastal works requires a determination of the volumetric rate of alongshore sand transport at the site—both the gross rate (upcoast plus downcoast transport) and the net rate (upcoast minus downcoast transport). The most reliable method of estimating transport rates is by measuring the rate of erosion or deposition at an artificial or geomorphic structure that interrupts the transport. Also, field studies have developed an approximate relationship between the alongshore transport rate and the alongshore component of incident wave energy per unit time. With this, net and gross transport rates can be estimated if sufficient information is available on the annual wave climate. Typical gross transport rates on exposed ocean shorelines often exceed 500,000 yd^3 (382,000 m^3) per year.

Primary sources of beach sediment include rivers discharging directly to the coast, beach and cliff erosion, and artificial beach nourishment. Sediment transported alongshore from its sources will eventually be deposited at some semipermanent location or sink. Common sinks include harbors and tidal inlets; dune fields; offshore deposition; spits, tombolos, and other geomorphic formations; artificial structures that trap sand; and areas where beach sand is mined.

By evaluating the volumetric transports into and out of a segment of the coast, one can develop a sediment budget for the coastal segment. If the supply exceeds the loss, shoreline accretion will occur, and vice versa. When a coastal project modifies the supply or loss to the segment, geomorphic changes can be expected. For example, when a structure that traps sediment is constructed upcoast of a point of interest, the shoreline at the point of interest can be expected to erode as it resupplies the longshore transport capacity of the waves.

Harbor entrance and tidal inlet control structures built to improve navigation conditions, stabilize navigation channel geometry, and assist with the relief of flood waters will often trap a large portion of the alongshore transport. This can result in undesirable deposition at the harbor or inlet entrance and subsequent downcoast erosion. The solution usually involves designing the entrance structures to trap the sediment at a fixed and acceptable location and to provide protection from wave attack at this location

so a dredge can pump the sand to the downcoast beach. *See Nearshore sedimentary processes*.

Coastal structures. Coastal structures can be classified by the function they serve and by their structural features. Primary functional classes include seawalls, revetments, and bulkheads; groins; jetties; breakwaters; and a group of miscellaneous structures including piers, submerged pipelines, and various harbor and marina structures.

Seawalls, revetments, and bulkheads. These structures are constructed parallel or nearly parallel to the shoreline at the land-sea interface for the purpose of maintaining the shoreline in an advanced position and preventing further shoreline recession. Seawalls are usually massive and rigid, while a revetment is an armoring of the beach face with stone rip-rap or artificial units. A bulkhead acts primarily as a land retaining structure and is found in a more protected environment such as a navigation channel or marina.

A key factor in the design of these structures is that erosion can continue on adjacent shores and flank the structure if it is not tied in at the ends. Erosion on adjacent shores also increases the exposure of the main structure to wave attack. Structures of this class are prone to damage and possible failure caused by wave-induced scour at the toe. In order to prevent this, the toe must be stabilized by driving vertical sheet piling into the beach, laying stone on the beach seaward of the toe, or maintaining a protective beach by artificial nourishment. Revetments that are sufficiently porous will allow leaching of sand from below the structure. This can lead to structure slumping and failure. A proper stone or cloth filter system must be developed to prevent damage to the revetment. *See Revetment*.

Groins. A groin is a structure built perpendicular to the shore and usually extending out through the surf zone under normal wave and surge-level conditions. It functions by trapping sand from the alongshore transport system to widen and protect a beach or by retaining artificially placed sand. The resulting shoreline positions before and after construction of a series of groins and after nourishment is placed between the groins are shown schematically in **Fig. 2**. Typical groin alongshore spacing-to-length ratios vary from 1.5:1 up to 4:1.

There will be erosion downcast of the groin field, the volume of erosion being approximately equal to the volume of sand removed by the groins from the alongshore transport system. Groins must be sufficiently tied into the beach so that downcoast erosion superimposed on seasonal beach profile fluctuations does not flank the landward end of a groin. Even the best-designed groin system will not prevent the loss of sand offshore in time of storms.

Jetties. Jetties are structures built at the entrance to a river or tidal inlet to stabilize the entrance as well as to protect vessels navigating the entrance channel. Stabilization is achieved by eliminating or reducing the deposition of sediment coming from adjacent shores and by confining the river or tidal flow to develop a more uniform and hydraulically efficient channel. Jetties improve navigation conditions by eliminating bothersome crosscurrents and by reducing wave action in the entrance.

At many entrances there are two parallel (or nearly parallel) jetties that extend approximately to the seaward end of the dredged portion of the channel. However, at some locations a single updrift or downdrift jetty has been used, as have other arrangements such as arrowhead jetties (a pair of straight or curved jetties that converge in the seaward direction). Jetty layouts may also be modified to assist sediment-bypassing operations. A unique arrangement is the weir-jetty system in which the updrift jetty has a low section or weir (crest elevation about mean sea level) across the surf zone. This allows sand to move over the weir section and into a deposition basin for subsequent transport to the downcoast shore by dredge and pipeline.

Breakwaters. The primary purpose of a breakwater is to protect a shoreline or harbor anchorage area from wave attack. Breakwaters may be located completely offshore and oriented approximately parallel to shore, or they may be oblique and connected to the shore where they often take on some of the functions of a jetty. At locations where a natural inland site is not available, harbors have been developed by the construction of shore-connected breakwaters that cover two or three sides of the harbor.

Figure 3 shows the breakwater and jetty system at the entrance to Channel Islands Harbor in southern California. The offshore breakwater intercepts incident waves, thus trapping the predominantly southeastern longshore sand transport; it provides a protected area where a dredge can operate to bypass sediment; and it provides protection to the harbor entrance. A series of shore-parallel offshore breakwaters (with or without artificial nourishment) has been used for shore protection at a number of locations. If there is sufficient fill or trapped material, the tombolo formed in the lee of each breakwater may grow until it reaches the breakwater.

Breakwaters are designed to intercept waves, and often extend into relatively deep water, so they tend to be more massive structures than are jetties or groins. Breakwaters constructed to provide a calm anchorage area for ships usually have a high crown elevation to prevent overtopping by incident waves and subsequent regeneration of waves in the lee of the breakwater.

Groins, jetties, and breakwaters are most commonly constructed as rubble mound structures. **Figure 4** shows the cross section of a typical rubble mound breakwater placed on a sand foundation. The breakwater has an outer armor layer consisting of the largest stones or, if sufficiently large stones are not available, the armor units may be molded of concrete with a special shape. Stone sizes decrease toward the

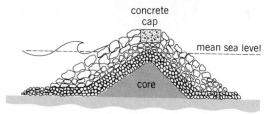

Fig. 4. Rubble mound breakwater cross section.

core and base in order to develop a filter system so that the fine core stone and base sand are not removed by wave and current action. The core made of fine stone sizes is provided to diminish wave transmission through the structure. Jetties and groins have a simpler cross section, consisting typically of only armor and core layers. Breakwaters, groins, and jetties have also been constructed of steel or concrete caissons with sand and gravel fill; wood, steel, and concrete sheet piles; and sand-filled bags.

A different type of breakwater that can be effective where incident wave periods are short and large water-level fluctuations occur (as in a reservoir marina) is the moored floating breakwater. This type has been constructed of hollow concrete prisms, scrap tires, logs, and a variety of other materials.

In an attempt to develop low-cost shore protection, a number of novel materials have been used for shoreline revetments, including cinder blocks, tires, sand-filled rubber tubes, woven-fiber mattresses, and soil-cement paving.

Robert M. Sorensen

Bibliography. American Society of Civil Engineers, *Journal of the Waterway, Port, Coastal and Ocean Engineering Division* and *Proceedings of the International Conferences on Coastal Engineering*; P. Braun, *Design and Construction of Mounds for Breakwater and Coastal Protection,* 1985; B. L. Edge (ed.), *Coastal Engineering: 1982,* 1983; P. D. Komar, *Beach Processes and Sedimentation,* 1976; A. D. Quinn, *Design and Construction of Ports and Marine Structures,* 1972; R. Silvester, *Coastal Engineering,* 2 vols., 1974; U.S. Army Coastal Engineering Research Center, *Shore Protection Manual,* 3 vols., 1977.

Coastal landforms

The characteristic features and patterns of land in a coastal zone subject to processes of erosion and deposition. The interactions of various marine processes with the coast and with terrestrial processes produce a great variety of landforms along the coastal zone. Primary terrestrial processes that influence coastal landforms are those which carry sediment to the coast. These include river movement and mass movement by gravity, such as slumping and slides. Marine processes of waves and currents are predominant in controlling the formation of coastal morphology. Another indirectly important process is change in sea level. A passive yet quite important contributing factor in coastal morphology is the nature of the geology and geomorphology of the area adjacent to the coast.

Landforms associated with, and formed along, wide coastal plains are in striking contrast to those which are developed along high-relief, rocky coasts. A recent classification of coasts based on the modern concepts of global tectonics provides a good frame-

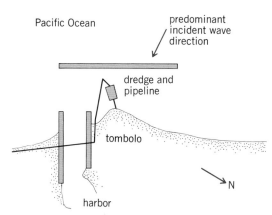

Fig. 3. Overhead view of breakwater and jetty system of Channel Islands Harbor, California.

Fig. 1. Rugged coastal area in Big Sur, California, typical of the morphology associated with leading-edge coasts.

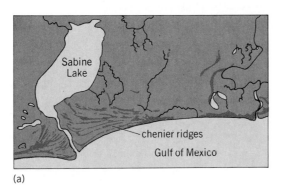

(a)

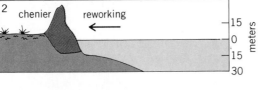

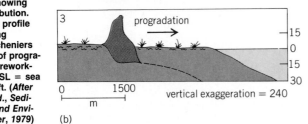

vertical exaggeration = 240

(b)

Fig. 2. Chenier plains along the northern Gulf Coast. (*a*) Map showing geographic distribution. (*b*) Diagrammatic profile sequence showing development of cheniers from alternation of progradation and wave reworking of sediment. SL = sea level. 1 m = 3.3 ft. (*After H. G. Reading, ed., Sedimentary Facies and Environments, Elsevier, 1979*)

work for consideration of coasts. There are essentially two broad types of plate margins: trailing-edge and leading-edge. Although more than one variety of trailing-edge types has been described, the category is characterized by coastal plains and adjacent broad shelves with constructional landforms. The Atlantic coast of the United States south of New York is an example. The leading edge coasts are typically high-relief with bedrock at the coast due to structural deformation. The west coast of the United States is a good example.

The various morphologic entities of the coast will be considered in the two general categories described above from the tectonic-related classification of coasts. As a rule, coastal plain coasts of trailing-edge plates present a greater diversity of depositional environments than high-relief leading-edge coasts. The primary reason is that submarine morphology generally mimics that of the adjacent subaerial coast. That is, trailing-edge plates show broad gently sloping continental shelves upon which sediment accumulates and is distributed by coastal processes. Leading-edge regions show high-relief submarine morphology and a general lack of a continental shelf, resulting in an absence of sites for accumulation of sediment.

Leading-edge coasts. The primary characteristic of leading-edge coasts is the irregular and rugged terrain, typically with cliffs and with deep water adjacent to the shore. Wave-cut features may be drowned or exposed. These include terraces, notches, sea stacks, arches, and caves. Such wave action may develop an irregular coast due to varying rock types or structural control, such as by faults and jointing. Homogeneous bedrock along the coast can result in wave erosion producing a straight coast. All of these specific coastal features combine to produce high-relief morphology (**Fig. 1**).

Trailing-edge coasts. By their very nature, trailing-edge coasts are characterized by a diverse suite of landforms in the transition zone between the terrestrial and the marine environment. This transition zone is characterized by various coastal marine processes, some of which interact with nonmarine processes. Waves, tides, and longshore currents are the predominant marine processes, whereas riverine flow accounts for the dominant terrestrial process. Wind may also be important but is not restricted by or to a coastal location.

Two large-scale types of coastal configurations may characterize a trailing-edge continental mass: a mainland coast and a barrier-island coast. A mainland coast is one along which open coastal marine processes impinge directly on the mainland, whereas on a barrier-island coast there is no direct interaction between the mainland and open marine processes. In the latter, several environments and landforms may be present between the mainland the barrier island. Some large-scale landforms may be common to both coastal types, and others are restricted to one or the other. *See Barrier islands*.

Chenier plains are developed only along mainland coasts, particularly those with much sediment supplied of a generally low coastal energy level. Occasional intense energy events such as tropical storms or hurricanes cause removal of fine sediment while building a sandy ridge parallel to the shore. The resulting coastal morphology is characterized by low-lying muddy areas and sandy, near-parallel ridges (**Fig. 2**). *See Floodplains; Plains*.

Barrier-island coasts have inlets which are characteristic of only this coastal type. All other coastal landform types typical of trailing-edge margins may be found along either mainland or barrier coasts. There are some features which are more common or better developed on one type as opposed to the other. Deltas are a good case in point. They are largest and best developed on mainland coasts. Deltas which form along barrier coasts fill in estuaries behind the barrier and may eventually dominate the coast to the point that the barrier ceases to exist and a mainland coast evolves. *See Delta*.

Independent coastal landforms. Many coastal landform types show no particular association with a specific coastal type. Most, but not all, of these features are relatively small-scale, so-called second-order features. Reefs are probably the most extensive coastal landforms, and their presence has no relation to the type of coast, other than it must lack significant terrigenous runoff. Fiords, like reefs, are climate-related but may develop on a variety of coasts as long as relief is at least modest. These huge glacial excavations are the deepest of the coastal landforms. *See Fiord; Reef*.

Many second-order coastal depositional features can occur along rugged coasts as well as along coastal plains. In general, they are more common along coastal plains because sediment is readily available and the proper site for its accumulation exists. Spits and baymouth bars may develop on rocky coasts, on mainland coasts, or on large deltas due to the combined work of waves and longshore currents. *See Coastal plain; Nearshore sedimentary processes*.

Richard A. Davis, Jr.

Bibliography. A. Clowes and P. Comfort, *Process and Landform: An Outline of Contemporary Geo-morphology*, 1982; J. S. Pethick, *Introduction to Coastal Geomorphology*, 1983.

Coastal plain

A low-relief area which is bounded by the sea on one side and generally by highlands on the landward side. The continental shelf actually represents a drowned portion of the coastal plain and is geologically linked to the coastal plain. The development of coastal plains ranges widely throughout the world, with the Atlantic and Gulf coasts of the United States being among the most extensive (**Fig. 1**). Development of

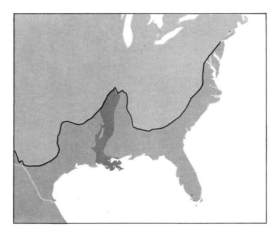

Fig. 1. Diagram showing coastal plains of United States.

coastal plains is associated with trailing-edge coasts in the context of global tectonics. These coasts are relatively stable and structurally simple. *See Continental margin*.

The rise and fall of sea level during the Pleistocene Epoch caused great changes in the areal extent of the coastal plain. During glacial advances and related low stands of sea level, the coastal plains increased their width by as much as 100% as compared to present conditions. Some geomorphic features of the coastal plains are climate-related. High-latitude areas experienced extensive Pleistocene glaciation, which has left a glacial topography overprint on the extensive coastal plains of Arctic regions of North America and Asia. Coastal plains in North Africa and along the Persian Gulf show eolian morphology, and tropical regions may have extensive erosion due to severe weathering and abundant rainfall. *See Fall line*.

Existing coastal plains began their development in the Mesozoic Era and have continued to the present. Sediment accumulations are typically some combination of fluvial, deltaic, and marine shelf depositional environments. There is a general thickening of the stratigraphic section in the seaward direction. Down-to-the-coast and normal faults and intrusion of salt domes represent the most widespread and significant

Fig. 2. Block diagram of belted coastal plain, showing ridges and depressions.

structural features. Surface topography displays low-relief cuestas along the outcrop belt of the more resistant lithofacies (**Fig. 2**). SEE COASTAL LANDFORMS; DEPOSITIONAL SYSTEMS AND ENVIRONMENTS; GEOMORPHOLOGY.

Richard A. Davis, Jr.

Bibliography. J. S. Pethick, *Introduction to Coastal Geomorphology*, 1983.

Coati

The name for three species of carnivorous mammals assigned to the raccoon family Procyonidae. The common or ring-tailed coati (*Nasua nasua*) ranges through the forests of Central and South America (see **illus.**), the mountain coati (*Nasuella ovivacea*) is

Common coati (*Nasua nasua*).

found only in South America, while *Nasua nelsoni* is confined to Cozumel Island in Central America.

The coatis are characterized by their elongated snouts, body, and tail, which is held erect when they walk. They are adept at climbing trees in search of birds and lizards, and their diet also includes fruit, insects, and larvae. *Nasua nasua* roams in bands of females and young during the day in search of food. The males are excluded from the band except during the brief breeding season, when a single male is al-

lowed to join. He is aggressive toward any other male but subordinate to the females. After mating, the male leaves and the female builds an isolated nest in a tree. Here, following a gestation period of 11 weeks, a litter of about five young are born and cared for by the female away from the group. After about 1 month the female and her young rejoin the band. SEE CARNIVORA.

Charles B. Curtin

Coaxial cable

An electrical transmission line comprising an inner, central conductor surrounded by a tubular outer conductor. The two conductors are separated by an electrically insulating medium which supports the inner conductor and keeps it concentric with the outer conductor. One version of coaxial cable is shown in **Fig. 1**, with periodically spaced polyethylene disks supporting the inner conductor. This coaxial is a building block of multicoaxial cables used in L-carrier systems. SEE TRANSMISSION LINES.

The symmetry of the coaxial cable and the fact that the outer conductor surrounds the inner conductor make it a shielded structure. At high frequencies, signal currents concentrate near the inside surface of the outer conductor and the outer surface of the inner conductor. This is called skin effect. The depth to which currents penetrate decreases with increasing frequency. Decreased skin depth improves the cable's self-shielding and increases transmission loss. This loss (expressed in decibels per kilometer) increases approximately as the square root of frequency because of the skin effect. SEE ELECTRICAL SHIELDING; SKIN EFFECT.

Coaxial cables can carry high power without radiating significant electromagnetic energy. In other applications, coaxial cables carry very weak signals and are largely immune to interference from external electromagnetic fields.

A coaxial cable's self-shielding property is vital to successful use in broadband carrier systems, undersea cable systems, radio and TV antenna feeders, and community antenna television (CATV) applications.

Coaxial units are designed for different mechanical behavior depending upon the application. Widely used coaxials are classified as flexible or semirigid.

Materials. Conductors for coaxial cable are generally made of copper or aluminum. The low resistivity of these metals results in lower cable loss. A typical dielectric material is polyethylene; special grades are

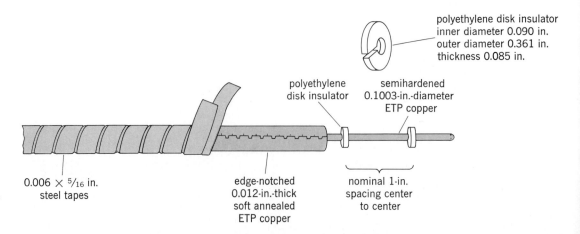

polyethylene disk insulator
inner diameter 0.090 in.
outer diameter 0.361 in.
thickness 0.085 in.

polyethylene
disk insulator

semihardened
0.1003-in.-diameter
ETP copper

0.006 × 5/16 in.
steel tapes

edge-notched
0.012-in.-thick
soft annealed
ETP copper

nominal 1-in.
spacing center
to center

Fig. 1. Air dielectric disk-insulated coaxial cable.
1 in. = 25 mm.

made with very low dielectric loss. Where mechanical factors allow, much of the space between inner and outer conductor is occupied by air. This may be done by using spaced insulating disks (Fig. 1) or expanded or spiraled insulation. SEE ALUMINUM; CONDUCTOR (ELECTRICITY); COPPER; POLYOLEFIN RESINS.

Flexible coaxials. These generally have a braided outer conductor and a stranded inner conductor. Insulation is made of solid or expanded plastic of a low-loss electrical grade. Since a single layer of braid has some air gaps between intersecting wires, a second layer of braid is placed over the first when shielding requirements are severe.

Flexible coaxials are used for high-frequency patch cords and intra- and interbay cabling. Very small-diameter coaxials are used to interconnect components on printed wiring boards. Since runs are short in these applications, loss is not generally important.

Semirigid coaxials. This class of coaxials finds the most extensive use in communications systems, such as cables used in long-distance terrestrial and undersea communications. Also, CATV and closed-circuit television often use semirigid coaxials. To minimize cable loss, the insulation may be of expanded plastic, effectively using air bubbles for up to 80% of the dielectric.

Undersea coaxials. To function reliably in a high-performance system in the ocean environment, undersea coaxials must meet severe requirements. Some of these are (1) stable dimensions and predictable transmission characteristics under pressures of up to 11,000 lb/in.^{2} (76 megapascals); (2) 25-year service lifetime; and (3) ability to withstand handling and scuffing. A solid polyethylene core withstands ocean pressure which would crush a cable with any air fill. The use of inert materials that resist attack by marine organisms assures the desired lifetime. The high-strength (300,000-lb/in.^{2} or 2-gigapascal) steel strand

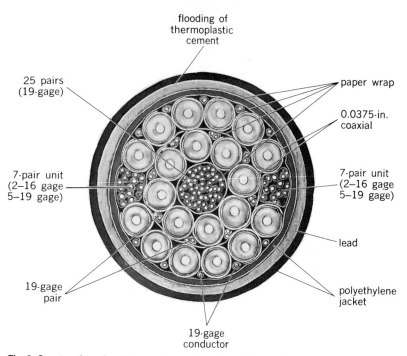

Fig. 3. Construction of multicoaxial transmission line. 1 in. = 25 mm.

and high-density jacket assure survival and continuity despite severe handling. SEE SUBMARINE CABLE.

The type of coaxial cable used in the most advanced analog undersea cable systems is shown in **Fig. 2**. The high-tensile steel strand provides longitudinal strength needed to lay or recover the cable in ocean depths of up to 4 nautical miles (7 km). The inner conductor is a high-conductivity copper tube, 0.48 in. (12 mm) in diameter, which is formed and swaged over the steel strand. Low-loss polyethylene is extruded over the inner conductor assembly to form the 1.7-in.-diameter (43.2-mm) cable core. To achieve an accurate (± 0.001 in. or 0.025 mm) core diameter, the core is extruded slightly oversize and mechanically shaved.

In a further manufacturing operation, the 10-mil-thick (0.010-in. or 0.25-mm) copper outer conductor is formed over the cable core with an overlapped seam. A high-density polyethylene jacket, 2 in. (50 mm) in diameter, is extruded over the outer conductor to complete the cable.

Terrestrial coaxials. A single coaxial unit, which is used in the L3, L4, L5, and L5E carrier systems, is shown in Fig. 1. This unit uses polyethylene disks spaced 1 in. (2.5 cm) apart to keep the inner conductor centered. The outer conductor is then precision-formed around the disk-insulated inner conductor. Two layers of steel tape are overlapped to complete the unit. The tapes give the assembly mechanical stability and provide added shielding at low frequencies.

A coaxial unit such as that of Fig. 1 has low loss and excellent impedance characteristics. In the L5E system a pair of such coaxial units can carry 13,200 telephone calls. A cable carrying 20 of these coaxial units is shown in **Fig. 3**. A 22-unit coaxial version of this cable was used in later L5E installations. Fully equipped, this cable provides 132,000 telephone channel capacity over the 10 working coaxial pairs. The eleventh pair provides service protection for the 10 working pairs.

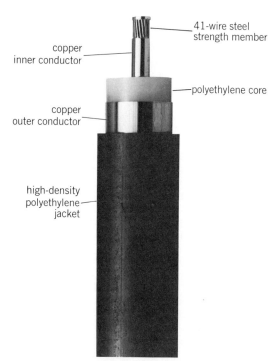

Fig. 2. Deep-water armorless cable used in 4200-channel undersea systems.

Very slight dimensional irregularities can result in small-impedance deviations in the cable. If such deviations are periodically spaced, they can produce reflections which cause large swings in the cable loss versus frequency. Hence, great care is taken in the design, and throughout manufacture. This care extends to raw materials, close dimensional control, careful machinery design and maintenance, precision qualification measurements, and quality control.

Two coaxial cables span the United States from east to west, with many intermediate and shorter installations. Such cables are widely used in North America, Japan, and Europe. Coaxial systems, along with microwave radio, provide most of the long-distance communication in the United States. SEE COMMUNICATIONS CABLE.

S. Theodore Brewer

Cobalt

A lustrous, silvery-blue metallic chemical element, symbol Co, with an atomic number of 27 and an atomic weight of 58.93. Metallic cobalt was isolated in 1735 by the Swedish scientist G. Brandt, who called the impure metal cobalt rex, after the ore from

lanthanide series: 58 Ce 59 Pr 60 Nd 61 Pm 62 Sm 63 Eu 64 Gd 65 Tb 66 Dy 67 Ho 68 Er 69 Tm 70 Yb 71 Lu

actinide series: 90 Th 91 Pa 92 U 93 Np 94 Pu 95 Am 96 Cm 97 Bk 98 Cf 99 Es 100 Fm 101 Md 102 No 103 Lr

which it was extracted. The metal was shown to be a previously unknown element by T. O. Bergman in 1780.

Cobalt is a transition element in the same group as rhodium and iridium. In the periodic table it occupies a position between iron and nickel in the third period. Cobalt resembles iron and nickel in both its free and combined states, possessing similar tensile strength, machinability, thermal properties, and electrochemical behavior. Constituting 0.0029% of the Earth's crust, cobalt is widely distributed in nature, occurring in meteorites, stars, lunar rocks, seawater, fresh water, soils, plants, and animals. SEE PERIODIC TABLE; TRANSITION ELEMENTS.

Cobalt and its alloys resist wear and corrosion even at high temperatures. The most important commercial uses are in making alloys for heavy-wear, high-temperature, and magnetic applications. Small amounts of the element are required by plants and animals. The artificially produced radioactive isotope of cobalt, ^{60}Co, has many medical and industrial applications.

Physical and chemical properties. Cobalt, with a melting point of 1495°C (2723°F) and a boiling point of 3100°C (5612°F), has a density (20°C; 68°F) of 8.90 g·cm^{-3}, an electrical resistivity (20°C) of 6.24 microhm·cm, and a hardness (diamond pyramid, Vickers; 20°C) of 225. It is harder than iron and, although brittle, it can be machined. The latent heat of fusion is 259.4 joules/g, and the latent heat of vaporization is 6276 J/g; the specific heat (15–100°C; 59 = 212°F) is 0.442 J/g·°C. Cobalt is ferromagnetic, with the very high Curie temperature of 1121°C (2050°F). The electronic configuration is $1s^2 2s^2 2p^6 3s^2 3p^6 3d^7 4s^2$. At normal temperatures the stable crystal form of cobalt is hexagonal close-packed, but above 417°C (783°F) face-centered cubic is the stable structure. Although the finely divided metal is pyrophoric in air, cobalt is relatively unreactive and stable to oxygen in the air, unless heated. It is attacked by sulfuric, hydrochloric, and nitric acids, and more slowly by hydrofluoric and phosphoric acids, ammonium hydroxide, and sodium hydroxide. Cobalt reacts when heated with the halogens and other nonmetals such as boron, carbon, phosphorus, arsenic, antimony, and sulfur. Dinitrogen, superoxo, peroxo, and mixed hydride complexes also exist. In its compounds, cobalt exhibits all the oxidation states from −I to IV, the most common being II and III. The highest oxidation state is found in cesium hexa-fluorocobaltate(IV), Cs_2CoF_6, and a few other compounds.

Major compounds. Several oxides of cobalt exist, among them cobalt(II, III) oxides, Co_3O_4, formed when salts are heated in air at temperatures between 400 and 850°C (750 and 1560°F); cobalt(II) oxide, CoO, formed on heating the metal in oxygen at 1100°C (2010°F); and cobalt(III) oxide, Co_2O_3, formed by heating cobalt compounds at low temperatures in an excess of air. The chloride, bromide, nitrate, and sulfate of Co(II) are all formed by the interaction of the metal oxide, hydroxide, or carbonate with the corresponding acid. Cobalt fluoride, CoF_3, is useful in fluorinating hydrocarbons. Most simple cobalt salts contain Co(II). Both Co(II) and Co(III) form coordination complexes, those of the latter being particularly varied and numerous, and extremely diverse in their colors, structures, and stability.

In the late nineteenth century, these so-called complex species that are formed between metals and simple molecules [for example, ammonia (NH_3)] or anions (for example, halides) were the subject of interest and bewilderment, especially since complexes with the same composition (isomers) were often found. The ammines of cobalt, discovered in 1894 by A. Werner, formed the basis for developing the theories of coordination chemistry. Werner postulated (1) that cobalt in this series of compounds exhibited a constant total number (coordination number) of attached molecules and anions equal to six, and (2) that the bonds between the cobalt and the six donor atoms (for example, N in NH_3) had an octahedral arrangement. In all the cases Werner studied, the number of isomers found equaled that expected for octahedral geometry. SEE COORDINATION NUMBER.

The unique role of cobalt in coordination chemistry grows out of the liability of Co(II) complexes, the inertness of Co(III) complexes toward substitution, and the slow electron-transfer rate between Co(II) and Co(III). The important donor atoms are nitrogen, carbon in cyanides (CN), oxygen, sulfur, and the halogens. Cobalt forms cyanide, silyl (SiR_3, SiX_3; R = functional group, X = halide), nitrosyl (NO), nitrile (RCN), cyanate (NCO), and thio- and selenocyanate (SCN and SeCN) complexes, as well as complexes

Cobalt minerals

Name	Formula	Occurrence
Arsenides		
Smaltite, safflorite	$CoAs_2$	Canada, Morocco, U.S.
Skutterudite	$CoAs_3$	Canada, Morocco, U.S., Norway
Cobaltite	CoAsS	Canada, Norway, U.S., Australia
Oxidized minerals		
Asbolite	$CoO \cdot 2MnO_2 \cdot 4H_2O$	New Caledonia
Heterogenite	$CoO \cdot 2Co_2O_3 \cdot 6H_2O$	Zaire, U.S.
Erythrite	$3CoO \cdot As_2O_5 \cdot 8H_2O$	E. Germany, Canada, U.S., Morocco
Sulfides		
Carrolite	$CuS \cdot Co_2S_3$	Zaire, Zambia, U.S., Sweden
Linnaeite	Co_3S_4	U.S., U.S.S.R., Zaire, Germany
Cattierite	CoS_2	Zaire, Zambia
Others		
Cobaltomenite	$CoSeO_3 \cdot 2H_2O$	U. S., Argentina
Sphaerocobaltite	$CoCO_3$	Mexico, Germany, Zaire, Italy

with ureas, amides, and peptides. Complexes with a variety of *N*-heterocyclic bases are known. In oxidation states $-$I to III, cobalt forms a wide variety of phosphine complexes; arsine (AsR_3) complexes are known for both Co(II) and Co(III). Various phosphate complexes of Co(III) are known; complexes of adenosine triphosphate have been shown to mimic many of the biological functions of the magnesium adenosinetriphosphate complex. A great number of Co(III) carboxylate complexes exist, and carbonate complexes are useful in synthesizing other Co(III) complexes. Sulfide and disulfide coordination usually occurs through a bridge between two cobalt centers. Cobalt(II) and, particularly, Co(III) form a variety of thiolates (SR^-) and dithiolates. *SEE COORDINATION CHEMISTRY; COORDINATION COMPLEXES.*

Cobalt also forms organometallic compounds, which have a bond between cobalt and carbon, as in the naturally occurring B_{12} coenzyme (see **illus.**). The first organocobalt compounds discovered were the carbonyls, $[HCo(CO)_4]$ and $[Co_2(CO)_8]$, useful in hydroformylation reactions and as antiknock agents in gasoline. *SEE ANTIKNOCK AGENTS; HYDROFORMYLATION; ORGANOMETALLIC COMPOUND.*

Analytical methods. Cobalt may be determined gravimetrically by precipitation with α-nitroso-β-naphthol and ignition to constant weight as Co_3O_4, or by precipitation and weighing as such complexes as the anthranilate or mercury tetrathiocyanatocobaltate(II), $Hg[Co(SCN)_4]$. Precipitation of cobalt with pyridine in thiocyanate solution and weighing as the complex $[Co(C_5H_5N)_4](SCN)_2$ is a rapid method. Colorimetric methods make use of the blue color developed in the presence of chloride or thiocyanate ions or the red complex formed with nitroso-R-salt. Small amounts of the element may also be analyzed spectrographically or polarographically. Polarographic methods usually depend on the reduction of cobalt(III) ammines, and they can be used to determine the concentration of cobalt in the presence of a 1000-fold excess of nickel. Electrodeposition analysis of platinum gauze electrodes is useful for larger amounts of cobalt. Potentiometric titration with standard potassium hexacyanoferrate(III) in ammonia solution or with standard silver nitrate in aqueous potassium cyanide solutions is a useful volumetric method for intermediate amounts of cobalt, as are atomic absorption spectroscopy or x-ray fluorescence. Neutron activation analysis is used to measure amounts of cobalt in ancient ceramic artifacts. *SEE ACTIVATION ANALYSIS; ANALYTICAL CHEMISTRY; POLAROGRAPHIC ANALYSIS; SPECTROSCOPY.*

Natural occurrence. There are over 200 ores known to contain cobalt; traces of the metal are found in many ores of iron, nickel, copper, silver, manganese, and zinc. However, the commercially important cobalt minerals are the arsenides, oxides, and sulfides (see **table**). Zaire is the chief producer, followed by Zambia. The Soviet Union, Canada, Cuba, Australia, and New Caledonia produce most of the rest. Zaire and Zambia together account for just over 50% of the world's cobalt reserves. Nickel-containing laterites (hydrated iron oxides) found in the soils of the Celebes, Cuba, New Caledonia, and many other tropical areas are being developed as sources of cobalt. The manganese nodules found on the ocean floor are another large potential reserve of cobalt. They are estimated to contain at least 400 times as much cobalt as land-based deposits. *SEE MANGANESE NODULES.*

Metallurgical extraction. Since cobalt production is usually subsidiary to that of copper, nickel, or lead, extraction procedures vary according to which of

Structure of naturally occurring coenzyme B_{12}.

these metals is associated with the cobalt. In general, the ore is roasted to remove stony gangue material as a slag, leaving a speiss of mixed metal and oxides, which is then reduced electrolytically, reduced thermally with aluminum, or leached with sulfuric acid to dissolve iron, cobalt, and nickel, leaving metallic copper behind. Lime is used to precipitate iron, and sodium hypochlorite is used to precipitate cobalt as the hydroxide. The cobalt hydroxide can be heated to give the oxide, which in turn is reduced to the metal by heating with charcoal.

Uses. Cobalt ores have long been used to produce a blue color in pottery, glass, enamels, and glazes. Cobalt is contained in Egyptian pottery dated as early as 2600 B.C. and in the blue and white porcelain ware of the Ming Dynasty in China (1368–1644).

An important modern industrial use involves the addition of small quantities of cobalt oxide during manufacture of ceramic materials to achieve a white color. The cobalt oxide counteracts yellow tints resulting from iron impurities. Cobalt oxide is also used in enamel coatings on steel to improve the adherence of the enamel to the metal. Cobalt arsenates, phosphates, and aluminates are used in artists' pigments, and various cobalt compounds are used in inks for full-color jet printing and in reactive dyes for cotton. Cobalt blue (Thenard's blue), one of the most durable of all blue pigments, is essentially cobalt aluminate. Cobalt linoleates, naphthenates, oleates, and ethylhexoates are used to speed up the drying of paints, lacquers, varnishes, and inks by promoting oxidation. In all, about a third of the world's cobalt production is used to make chemicals for the ceramic and paint industries. *See Ceramics; Dye; Ink; Metal coatings.*

Cobalt catalysts are used throughout the chemical industry for various processes. These include hydrogenations and dehydrogenations, halogenations, aminations, polymerizations (for example, butadiene), oxidation of xylenes to toluic acid, production of hydrogen sulfide and carbon disulfide, carbonylation of methanol to acetic acid, olefin synthesis, denitrogenation and desulfurization of coal tars, reductions with borohydrides, and nitrile syntheses, and such important reactions as the Fisher-Tropsch method for synthesizing liquid fuels and the hydroformylation process. Cobalt catalysts have also been used in the oxidation of poisonous hydrogen cyanide in gas masks and in the oxidation of carbon monoxide in automobile exhausts. *See Catalytic converter.*

Cobalt salts are used as charges for electroplating baths; for example, cobalt sulfate improves the quality of nickel plate. Cobalt carboxylates are used for bonding steel cords to rubber in tire manufacturing. Cobalt salts also act as antifoggants for photographic emulsions and as visual humidity indicators. In the widely used desiccant silica gel, cobalt chloride provides the color change from blue to pink as water is taken up. Cobalt oxides are used in air-deodorizing filters, electrical contacts, thermistors, and varistors. Lasers have been developed from magnesium fluoride doped with cobalt. *See Desiccant.*

Radioactive cobalt (^{60}Co), produced by thermal neutron bombardment of the natural isotope ^{59}Co, has a half-life of 5.271 years and decays by β^- and γ emission to nonradioactive nickel (^{60}Ni). It is used as a concentrated source of γ radiation in cancer therapy and food sterilization, and as a radioactive tracer in biological and industrial applications.

Alloys. Although cobalt was not used in its metallic state until the twentieth century, the principal use of cobalt is as a metal in the production of alloys, chiefly high-temperature and magnetic types. Superalloys needed to stand high stress at high temperatures, as in jet engines and gas turbines, typically contain 20–65% cobalt along with nickel, chromium, molybdenum, tungsten, and other elements.

The ferromagnetism of cobalt makes it useful in the production of alloys for magnets. Perhaps the best-known commercial magnet steels are the alnico magnet alloys, developed in 1935 and used for permanent magnets 25 times more powerful than ordinary steel magnets; these contain 6–12% aluminum, 14–30% nickel, 5–35% cobalt, and the balance iron. Powerful permanent magnets made from alloys of cobalt and rare-earth metals, notably a samarium cobalt alloy, $SmCo^5$, became important in the early 1970s. Cobalt is also used to produce soft-magnet alloys, which retain little magnetism when the applied magnetic field is removed; these are used in generators, motors, and static transformers. The soft-magnet alloys are typically composed of iron, cobalt, and nickel or vanadium. Magnetic and laser recording materials also contain cobalt. *See Ferromagnetism; Magnetic materials.*

Cobalt (3–25%) acts as a matrix, or binder, for tungsten carbide in the manufacture of drill bits and machine tools. The cobalt and tungsten carbide particles are heated, under hydrogen, at 1400–1500°C (2550–2730°F). The carbide particles fuse, and the resulting material has the characteristics of high strength and shock resistance. Cobalt is also used as a binder in polycrystalline sintered diamond tools. An alloy containing 13% cobalt, 84% tungsten, and 3% carbon is one of the hardest alloys known.

Other cobalt alloys have been developed for very specific uses. Alloys approximating 18% cobalt, 28% nickel, and 54% iron are used for glass-to-metal seals. Cobalt-chromium alloys and the stellite alloys of cobalt, chromium, and tungsten are used in high-speed tool steels. Cobalt imparts useful characteristics to electrical resistance alloys, cathode filaments and cores, and semiconductors. An alloy of 36.5% iron, 9.5% chromium, and 54% cobalt has practically zero coefficient of expansion. Alloys of 50–77% gold, 13–45% cobalt, and 4–25% nickel are used in dentistry. The dental and surgical alloy Vitallium (56–68% cobalt, 25–29% chromium, 5–6% molybdenum, 1.8–3.8% nickel, 0–1% manganese, 0–1% silicon, and 0.2–0.3% carbon) is not attacked by body fluids and does not irritate tissues. Similar fracture-resistant alloys are used for prosthetic implants. *See Alloy.*

Biochemistry. In parts of the world where soil and plants are deficient in cobalt, trace amounts of cobalt salts [for example, the chloride and nitrate of Co(II)] are added to livestock feeds and fertilizers to prevent serious wasting diseases of cattle and sheep, such as pining, a debilitating disease especially common in sheep. Symptoms of cobalt deprivation in animals include retarded growth, anemia, loss of appetite, and decreased lactation.

The principal biological role of cobalt involves corrin compounds (porphyrin-like macrocycles). The active forms contain an alkyl group (5′-deoxyadenosine or methyl) attached to the cobalt as well as four nitrogens from the corrin and a nitrogen from a heterocycle, usually 5,6-dimethylbenzimidazole (see illus.). These active forms act in concert with enzymes to catalyze essential reactions in humans. However, the corrin compounds are not synthesized in the body; they must be ingested in very small quantities. Vitamin B_{12}, with cyanide in place of the alkyl, prevents pernicious anemia but is itself inactive. The body metabolizes the vitamin into the active forms. Although the cobalt in corrins is usually Co(III), both Co(II) and Co(I) are involved in enzymic processes. Roughly one-third of all enzymes are metalloenzymes. Cobalt(II) substitutes for zinc in many of these to yield active forms. Such substitution of zinc may account, in part, for the toxicity of cobalt. *SEE ENZYMES; METALS, PATHOLOGY OF; VITAMIN B_{12}.*

Luigi Marzilli; Patricia A. Marzilli

Bibliography. F. A. Cotton and G. Wilkinson, *Advanced Inorganic Chemistry,* 5th ed., 1988; D. Dolphin, *B_{12},* 2 vols, 1982; N. N. Greenwood and A. Earnshaw, *Chemistry of the Elements,* 1984; S. Kirschner (ed.), *Advances in the Chemistry of the Coordination Compounds,* 1961; M. O'Donoghue (ed.), *The Encyclopedia of Minerals and Gemstones,* 1976; A. F. Trotman-Dickenson (ed.), *Comprehensive Inorganic Chemistry,* 1973; U. S. Bureau of Mines, *Minerals Yearbook,* vol. 1: *Metals and Minerals,* 1988; G. Wilkinson (ed.), *Comprehensive Coordination Chemistry: The Synthesis, Reactions, Properties and Applications of Coordination Compounds,* vol. 4, 1987; R. S. Young, *Cobalt,* Amer. Chem. Soc. Monog. 108, 1948.

Cobaltite

A mineral having composition (Co,Fe)AsS. Cobaltite is one of the chief ores of cobalt. It crystallizes in the isometric system, commonly in cubes or pyritohedrons, resembling crystals of pyrite (see **illus.**). There

Cobaltite. Twin and single crystals, found at Håkansboda, Sweden. (*Specimen courtesy of Department of Geology, Bryn Mawr College*)

is perfect cubic cleavage. The luster is metallic and the color silver-white but with a reddish tinge. The hardness is 5.5 (Mohs scale) and the specific gravity is 6.33. Cobaltite is usually found as disseminations in metamorphic rocks or in veins associated with other cobalt and nickel minerals. Notable occurrences are at Skutterud, Norway; Lunaberg, Sweden; Ravensthorpe, Australia; and Cobalt, Ontario, Canada. *SEE COBALT.*

Cornelius S. Hurlbut, Jr.

Coca

Shrubs of the genus *Erythroxylum* in the coca family (Erythroxylaceae). The genus contains over 200 species, most in the New World tropics. Two species, *E. coca* (*see* **illus.**) and *E. novogranatense,* are culti-

Flowering branch of coca (*Erythroxylum coca*).

vated for their content of cocaine and other alkaloids. Leaves of cultivated coca contain 14 alkaloids, chiefly cocaine, and significant amounts of vitamins and minerals. The cocaine content averages 0.5%, rarely exceeding 1.0%.

Millions of South American Indians "chew" coca every day as a mild stumulant by placing dried leaves in the mouth and masticating with various alkalies such as lime or plant ashes to promote the release of the alkaloids, which are swallowed. Coca is important in the folk medicine of Andean Indians, especially as a treatment for gastrointestinal disorders. It may regulate carbohydrate metabolism in a useful way and supplement an otherwise deficient diet. Coca also has great magical and religious significance in Andean societies. Although it has long been the focus of political debate, especially in Peru, ritualized use of the leaf in traditional cultures is not inconsistent with good physical health and social productivity. *SEE COCAINE; COLA; LINALES.*

Andrew T. Weil

Bibliography. R. T. Martin, The role of coca in the history, religion, and medicine of South American Indians, *Econ. Bot.,* 24(4):422–438, 1970; A. T. Weil, Coca leaf as a therapeutic agent, *Amer. J. Drug Alcohol Abuse,* 5(1):75–86, 1978.

Cocaine

The principal alkaloid of coca leaves, a topical anesthetic and stimulant, and popular illicit drug. The molecular formula is given below. Cocaine was first iso-

$$
\begin{array}{c}
CH_2 \!-\! CH \!-\! CH \!-\! COOCH_3 \\
| \qquad | \qquad | \\
\qquad\quad O \\
\qquad\quad \| \\
N \!-\! CH_3 \quad CH \!-\! O \!-\! CC_6H_5 \\
| \qquad\quad | \qquad | \\
CH_2 \!-\! CH \!-\! CH_2
\end{array}
$$

lated by A. Niemann in 1860. In 1884 C. Koller demonstrated its efficacy as an anesthetic in eye surgery, introducing the age of local anesthesia. For the next decade cocaine enjoyed the status of a wonder drug and panacea. It fell into disfavor with increasing reports of acute toxicity and long-term dependence. Today it is used as a topical anesthetic in the eye, nose, mouth, and throat; for injection anesthesia it has been replaced by synthetic drugs with fewer central nervous system effects. *See Anesthesia; Coca.*

Cocaine increases heart rate and blood pressure and causes feelings of alertness and euphoria. It stimulates noradrenergic pathways by blocking reuptake of norepinephrine at the synapse, and also stimulates dopaminergic pathways. In animals it can be shown to be strongly reinforcing. It does not produce physical dependence, as alcohol and opiates do, but many people find it hard to use in a stable and moderate fashion if they have access to it in quantity. Although it is quite active orally, most users of illicit cocaine take it intranasally by snuffing; few inject it intravenously. Aside from local irritation of the nasal membranes, occasional users suffer few adverse effects. The psychological consequences of frequent use can be devastating. The soluble hydrochloride salt is the common form. Insoluble cocaine free base, that is, the free alkaloid form of the drug, may be smoked, a practice that increases the toxicity of the drug and is more likely to be associated with dependence. *See Addictive disorders; Alkaloid; Noradrenergic system.*

Andrew T. Weil

Bibliography. L. Grinspoon and J. Bakalar, *Cocaine: A Drug and Its Social Evolution*, 1976.

Coccidia

A subclass of the class Telosporea. These protozoa are typically intracellular parasites of epithelial tissues in both vertebrates and invertebrates. The group is divided on the basis of life cycles into two orders, the Protococcida (in which there is only sexual reproduction) and the Eucoccida (in which there is both sexual and asexual reproduction). There are only a few species in Protococcida, and all are parasites in marine invertebrates. There are hundreds of species in Eucoccida; these are parasites in both invertebrates and vertebrates. The Eucoccida contains three suborders: Adeleina, which occur mostly in invertebrates and in which there is usually one host for the sexual stages and a different host for the asexual stages (an example is *Hepatozoon muris*, found in rats and transmitted by mites); Haemosporina, which occur mainly in vertebrates and which have their sexual stages in invertebrates (an example is *Plasmodium falciparum*, which causes malaria in humans and is transmitted by mosquitoes); and Eimeriina, which occur mostly in vertebrates and in which the sexual and asexual stages occur in the same host (an example is *Eimeria*, species of which cause coccidiosis in many domestic animals and wildlife, and which are transmitted by filth). The malaria parasites and coccidia are of direct importance to humans, and also cause important losses in chickens, cattle, sheep, and other domesticated animals. *See Eucoccida; Protococcida; Protozoa; Sporozoa; Telosporea.*

Elery R. Becker/Norman D. Levine

Coccolithophorida

A group of unicellular, biflagellate, golden-brown algae characterized by a covering of extremely small (1–35 micrometers) interlocking calcite (the hexagonal form of calcium carbonate) plates called coccoliths. The plates show extinction crosses in a polarizing light microscope; however, detailed study requires the use of electron transmission or electron scanning microscopes. The Coccolithophorida are usually considered plants but possess also some animal characteristics. Botanists assign them to the class Haptophyceae (based on the possession of a haptonema, a threadlike organ of attachment) of the phylum Chry-

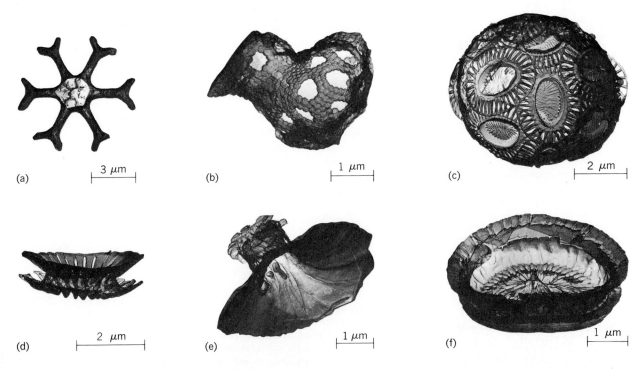

(a)   3 μm

(b)   1 μm

(c)   2 μm

(d)   2 μm

(e)   1 μm

(f)   1 μm

Examples of Coccolithophorida. (*a*) *Discoaster*; (*b*) a holococcolith; (*c*) a coccosphere; (*d*) a coccolith; (*e, f*) heterococcoliths. (*A. McIntyre, Lamont-Doherty Geological Observatory of Columbia University*)

sophyta, and zoologists to the class Phytamastigophorea, superclass Mastigophora, subphylum Sarcomastigophora of the phylum Protozoa. SEE CILIA AND FLAGELLA; PHYTAMASTIGOPHOREA; PROTOZOA.

Classification. The Coccolithophorida are mainly classified by the shape of their coccoliths into two groups: the holococcoliths, with simple rhombic or hexagonal crystals arranged like a mosaic; and the heterococcoliths, with complex crystals arranged into boat, trumpet, basket, or collar-button shapes. Classification is complicated by the presence of polymorphism (more than one type of coccolith per cell) and complex life cycles involving alternation of generations either with different types of coccoliths or with none at all (see **illus.**).

Life cycle. Reproduction occurs mostly by fission. Haploid and diploid phases, motile versus nonmotile phases, resting spores, and ameboid and filamentous stages have been noted but their interrelationships require further study.

Ecology. Chiefly photosynthetic, although epiphytism, phagotrophism, and saprophytism have also been shown, the Coccolithophorida form a significant percentage of the nannoplankton (plankton of less than 74-μm size) of the tropic through subarctic-subantarctic waters of all oceans; a few brackish and fresh-water species also exist. Together with the diatoms they constitute the primary producers of the open ocean food chain. SEE BACILLARIOPHYCEAE; PHYTOPLANKTON.

Coccoliths. Ease of culture makes these organisms a prime choice for studying biologic calcification. Research shows that coccoliths are extruded to the cell surface after being formed intracellularly on an organic matrix located between the nucleus and a "reticular body," which plays some yet undiscovered but essential role, and that the process is light-dependent and linked to photosynthesis. Coccolith formation has been introduced in naked cells by lowering the nitrogen content of the culture medium, and coccolith-bearing cells have lost their covering with the addition of carbon dioxide to their medium.

The function of coccoliths is unknown, but one worker suggests that they serve as a shield against excessive sunlight.

Fossil record. Coccoliths preserve well and have a fossil record dating back into the Jurassic, 180,000,000 years ago. They make up limestone and chalk deposits such as the White Cliffs of Dover on land and up to 30% of the sediments underlying tropical to arctic-antarctic ocean waters.

Their long and involved evolutionary record makes them useful to geologists for dating ancient sediments. The onset of the Pleistocene, or glacial, period is marked by the appearance of new coccolith species and the disappearance of older forms, including the discoasters, a group of star-shaped members prominent during the Tertiary. Extinction of the latter in a sedimentary sequence is a classic marker of the Pliocene-Pleistocene boundary.

Because they live in surface waters, coccoliths are under direct climatic control. Thus many modern species with relatively long fossil records are excellent temperature indicators. Research indicates that these can be used to unravel the climatic variations of the Pleistocene period that evolved into present weather patterns. SEE MICROPALEONTOLOGY.

Andrew McIntyre

Bibliography. M. Black, *Endeavor*, 24:131–137, 1965; W. Fairbridge (ed.), *The Encyclopedia of Oceanography*, 1966; A. McIntyre, *Science*, 158: 1314–1317, 1967; A. McIntyre and A. W. H. Be, *Deep-Sea Res.*, 14:561–597, 1967; R. E. Round, *The Biology of the Algae*, 1965; E. J. F. Wood, *Marine Microbial Ecology*, 1965.

Cockroach

Any of a group of insects (approximately 2500 species) of the family Blattidae, order Dictyoptera. This group is among the most primitive of the living and oldest of the fossilized insects, being found in abundance in the Carboniferous age. In general, cockroaches have a flat oval body, a darkened cuticle, and a thoracic shield, the pronotum, which extends dorsally over the head. A pair of sensory organs, the whiplike antennae, are located on the head, and a much shorter pair of sensory organs, the cerci, are located at the end of the abdomen. Although some species are wingless, most have two pairs of wings but seldom fly. Most species are nocturnal and primarily live in narrow crevices and under stones and debris. The mouthparts, mandibles, are of the biting-chewing type but are relatively weak. Cockroaches are omnivorous, and a few species have adapted to the human environment, becoming household pests since they can chew foodstuffs, clothing, paper, and even plastic insulation. They also can emit a highly disagreeable order.

Females deposit egg cases (oothecae) almost anywhere; each ootheca contains several eggs from which the young nymphs hatch and go through several molts as they grow. At the final molt a winged, sexually mature adult will emerge. The time from egg to adult varies from species to species but can be longer than a year.

There are several well-known household pests in North America. *Periplaneta americana,* the American cockroach or palmetto bug, is a rust-colored insect 1.2–2.4 in. (30–60 mm) long and is found predominantly in the central and southern United States. *Blatta germanica,* the German cockroach or croton bug, and *B. orientalis,* the Asian cockroach, are smaller, darker-brown roaches which infest houses in the eastern and southern regions of the United States.

Other cockroaches of special interest are *Blaberus giganteus* (**illus.** *a*), a South American roach that can grow to 3.2 in. (80 mm) in length; *Cryptocercus punctulatus,* a wood roach which has unicellular or-

Adult cockroaches. (a) *Blaberus giganteus.* (b) *Cryptocercus punctulatus.*

(a)   (b)

ganisms in its gut to digest the consumed wood (illus. *b*); and *Gromphadorhina portentosa*, the Madagascar roach, which can compress its body and force air through modified spiracles (breathing pores), producing a loud hissing sound.

Cockroaches have been claimed as vectors for numerous human diseases. There is little evidence to support this claim; however, there is evidence that they can produce allergenic reactions in humans. Because of their large size, cockroaches have proved to be very useful animals for scientific research on general problems of insect behavior, physiology, and biochemistry.

Charles R. Fourtner

Cocoa powder and chocolate

Products derived from the seeds of the tropical tree *Theobroma cacao,* which grows within a narrow belt along both sides of the Equator. There are three basic varieties: Criollo, which was native to Central and South America; Forastero, which constitutes the bulk of the world supply and is mostly cultivated in West Africa and Brazil; and Trinitario, consisting of various hybrids. *SEE CACAO.*

The seeds, contained in pods which grow on the trunks and lower branches of the trees, are surrounded by a mucilaginous pulp. They are whitish in color and have no normal chocolate flavor. Each mature bean is about ¾–1 in. (2–2.5 cm) in length with an average weight not less than 0.04 oz (1 g). The shell (testa) is about 12% of the total bean weight, the nib (cotyledon) 87%, and germ 1%—the last being the point at which germination would have commenced in the live bean. The main constituents of the dried nib are approximately the following: fat (cocoa butter) 54%; protein 12–14%; starch, other carbohydrates, and mucilage 20%; tannins 6%; theobromine 1.3%; caffeine 0.2%; and moisture 3–5%. Good-quality beans show a brown or purplish-brown color and striations in the sliced cotyledon. Unfermented beans have a dense structure and slaty color. *SEE SEED.*

Processing. To develop a chocolate flavor, the beans are first subjected to a fermentation process. The pods are split open and the seeds and pulp placed in heaps surrounded by plantain leaves or in "sweat boxes." Under these conditions microbiological and enzymatic action occurs, with a rise in temperature to about 120°F (50°C).

After 4–6 days, during which the beans are turned over, they develop a purplish-brown color, and the outer shell and nib become more distinct. The pulp liquefies and drains away, and precursors of the true chocolate flavor are formed. After fermentation the beans are sun-dried in shallow trays that are protected from rainstorms by sliding canopies; in very wet climates, such as that found in Cameroon, various forms of artificial drying are used. After drying, the product is described commercially as cocoa or cocoa beans and should have a moisture content of 6–7%.

A series of well-defined operations are used to produce good-quality cocoa or chocolate.

Cleaning. Machines, by a combination of sieving and air elutriation, remove stones, sand, bag fiber, stalks, and immature beans. Without this treatment, grit would pass to subsequent processes, resulting in machine damage and poor-quality products.

Roasting. In this essential process the chocolate flavor is fully developed from the precursors, the shell loosened from the nib, and the moisture content reduced.

There are many roaster designs. One type is a rotating drum with external heating, which may handle single batches or have a continuous throughput. Other roasters use high-temperature air passing through a continuous stream of beans. Roasting temperatures vary greatly, ranging from 200 to 300°F (95 to 150°C). The temperature used is related to the time of roasting and the type of product required. The shell should never be burned.

In two-stage roasting, the beans are heated at a low temperature in a closed, moist atmosphere, and then a higher temperature and drying follow.

Winnowing. After roasting, the winnowing process separates the shell from the nib. The ideal is complete

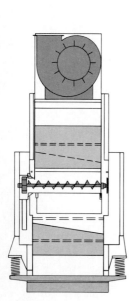
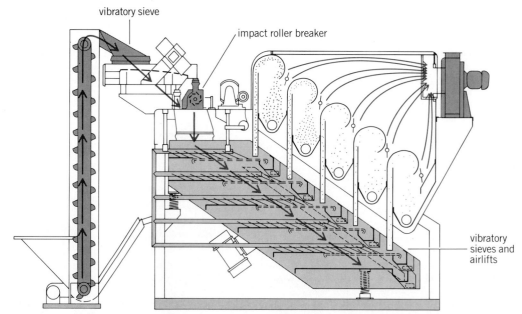

vibratory sieve

impact roller breaker

vibratory sieves and airlifts

Fig. 1. Two views of winnowing machine. (*Bauermeister Maschinen Fabrik, Hamburg, Germany*)

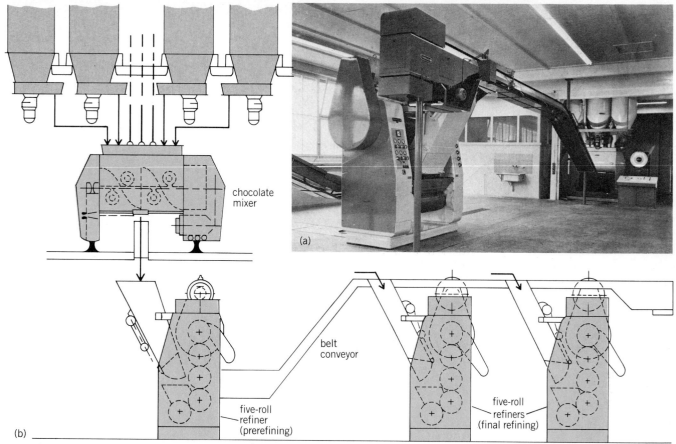

Fig. 2. Modern chocolate-making plant. (*a*) Mixer, conveyor, and refining rolls. (*b*) Diagram of operation of this equipment. (*Buhler-Miag, Uzwil, Switzerland*)

separation with full recovery of the nib, but in practice this is not possible. Winnowing machines have been greatly improved. A modern system uses a bank of vibrating sieves with an air lift to remove shell pieces at the discharge end of each sieve (**Fig. 1**). At the machine entry the beans are broken by hexagonal impact rollers which minimize the production of fine dust. The shell is a waste product that is used as a boiler fuel and an agricultural mulch.

Liquor milling. The roasted nibs are ground at a temperature above the melting point of the cocoa butter constituent (93–95°F or 34–35°C) to produce a dark brown liquid called liquor, mass, or unsweetened chocolate. Finely ground liquor is the basis for manufacture of cocoa powder and chocolate.

Different machines are used for this purpose. Pin and hammer mills in conjunction with roller refiners are frequently used. Other machines include horizontal disk mills with three stages and the vertical ball mill. In the ball mill partially ground liquor is fed through a vertical cylinder containing hard steel or ceramic balls rotated at high speed by means of a central spindle and attached plates.

Cocoa powder manufacture. To manufacture cocoa powder, liquor is subjected to hydraulic pressing, which separates some of the cocoa butter from the solid cocoa matter. The pressing machine consists of a number of pots arranged horizontally, each provided with a filter and linked to a hydraulic ram capable of providing pressures of over 6000 lb/in.^{2} (10^8 pascals). The pots are filled automatically with hot liquor, and when the ram operates liquid cocoa butter is forced

through the filters and into a common channel at the base of the press. The amount of cocoa butter pressed out is determined by control of time and pressure.

At the end of the pressing cycle the ram is reversed, thus opening the pots and releasing compressed cocoa cakes. To make cocoa powder, these cakes are first mechanically broken into small pieces called kibbled cake and then put through pulverizers which, in conjunction with air separators, produce a very fine powder. Such cocoa powder is called natural cocoa.

Additionally, if dark brown or reddish-brown cocoas are required, the liquor, kibbled cake, or preferably the nibs are subjected to a process called alkalization or dutching. This consists of saturating the nibs (or other intermediate) with a solution of potassium or sodium carbonate followed by drying and roasting. In most countries, the amount of added alkali permitted is limited to 3% of the nib.

Color is controlled by the degree of roast of the nib and the quantity and concentration of the alkali used. Alkalization also serves to neutralize the natural acidity of the nib.

Commercial cocoa powders may have a residual cocoa butter content of 10–22%. They are used extensively as ingredients in cookies, biscuits, chocolate syrups and spreads, and vegetable fat coatings.

Chocolate manufacture. Dark, bitter, or sweet chocolate is manufactured from liquor (or nibs), sugar, and cocoa butter—the cocoa butter being obtained from cocoa powder manufacture. Cocoa butter is a very stable, natural fat; it melts just below body temperature and has a narrow melting range. It displays good contraction on solidifying from the liquid state, which enables chocolate to be molded into various shapes. Its melting properties impart good texture in the mouth and flavor release.

Milk chocolate is made from liquor, sugar, milk solids, and cocoa butter. The milk solids are derived from liquid milk, usually by a spray-drying process. Another milk product often used, called milk crumb, is prepared by concentrating and drying liquid milk in the presence of sugar and liquor. This gives a caramelized milk flavor which has proved very popular. *SEE MILK.*

Refining. In the manufacture of chocolate, the solids are mixed mechanically with liquor and some cocoa butter to give a plastic paste. This paste is fed to refiners, which consist of five steel rolls mounted vertically with the bottom feed roll offset to enable the paste to be delivered continuously from the mixers. The rolls work on a system in which the speed of rotation increases from the bottom to the top roll, thereby obtaining a grinding shear between the roll surfaces and enabling the paste to travel upward from one roll to the next. The refined paste is scraped off the top roll by a sharp blade. In some instances refining is done in two stages. **Figure 2** shows a mixing and double refining system.

Conching. The final process in chocolate making is called conching, which removes moisture and volatile substances that would produce acid and harsh flavors. Conching also reduces viscosity and contributes a smoothness to the eating qualities of the chocolate.

At one time conching was carried out in machines which consisted of a heated tank with a flat bottom over which a roller traveled backward and forward through the liquid chocolate. While these were effective, much time was necessary to produce good chocolate; the process also took much energy, and extra cocoa butter was needed to give fluidity.

Later it was shown that with more powerful machinery it was possible to directly conche the friable flake from the refining rolls. This is known as dry conching, a process which removes the volatiles in a much shorter time. Extra fluidity is obtained by mechanical mixing alone. Dry conching is now mostly done in rotary conches (**Fig. 3**), taking about 4 h. Then extra cocoa butter is added and the speed of rotation increased for a further period of 6–12 h de-

pending on the type of chocolate desired.

Besides mechanical improvements, the use of emulsifiers has reduced costs by obtaining the correct fluidity with less cocoa butter. The universally used emulsifier is natural lecithin obtained from the soya bean; in most countries the amount permitted by law is limited to 0.5% of the final chocolate product. About 1 h before the end of the conching period, the emulsifier is added, and the viscosity is checked and adjusted by the addition of more cocoa butter. The final viscosity depends on whether the chocolate is to be used for molding bars or coating confectionery centers.

Tempering. In both refining and conching, the chocolate must be subjected to tempering before it is applied to confectionery centers or deposited into molds. Cocoa butter is polymorphic, which means it crystallizes in several forms on solidifying. Some forms are unstable, and these will transform to the stable form slowly if the chocolate is not tempered, resulting in an unsightly grayish surface film called bloom.

To temper chocolate it is necessary to seed the liquid chocolate with the stable form of crystal. There are several ways of doing this. The most common way is to reduce the temperature of the liquid chocolate from about 100 to 84°F (38 to 29°C), with constant stirring, until cocoa butter seed is formed, and then raising the temperature again to 90–91°F (32–33°C), which will melt out the remaining unstable forms and render the chocolate sufficiently fluid for application. Modern machines called enrobers or coaters carry out this process. *SEE FOOD MANUFACTURING.*

Bernard W. Minifie

Bibliography. L. R. Cook and E. H. Meursing, *Chocolate Production and Use*, 1980; B. W. Minifie, *Chocolate, Cocoa and Confectionery*, 1980.

Coconut

A large palm, *Cocos nucifera*, widely grown throughout the tropics and valuable for its fruit and fiber. Usually found near the seacoast, it requires high humidity, abundant rainfall (60 in. or 1.5 m), and mean annual temperature of about 85°F (31°C). Southern Florida, with mean temperature of 77°F (25°C), is at the limit of successful growth. The origin of the coconut has been in dispute, but strong evidence points to southern Asia with wide dispersal by ocean currents and human migrations. *SEE ARECALES.*

The fruit, 10 in. (25 cm) or more in length, is ovoid and obtusely triangular in cross section (**Fig. 1**). The tough, fibrous outer husk (exocarp) encloses a spherical nut consisting of a hard, bony shell (endocarp) within which is a ½-in. (1.25-cm) layer of fleshy meat or kernel (endosperm). The meat is high in oil and protein and, when dried, is the copra of commerce.

Production and culture. Principal commercial producers are the Philippines, Indonesia, Malaya, Ceylon, and Oceania. Although many trees grow along the seashore and in native villages without special care, the crop lends itself to plantation culture with control of weeds, fertilization, and protection from diseases, insects, and animal pests.

Palms begin to bear nuts the sixth year after planting and reach full bearing about the eighth year. Individual nuts mature about a year after blossoming

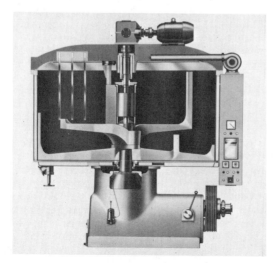

Fig. 3. Modern rotary conche. (*Petzholdt, Frankfurt, Germany*)

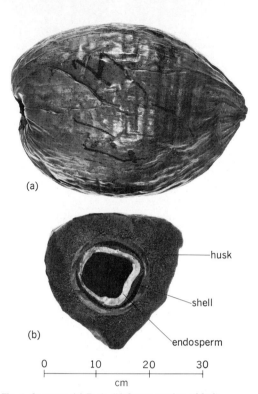

(a)

husk

shell

(b)

endosperm

```
0        10        20        30
|----|----|----|----|----|----|----|
              cm
```

Fig. 1. Coconut. (a) Fruit. (b) Cross section of fruit showing husk, shell, and endosperm (meat). 2.5 cm = 1 in.

and normally fall to the ground. In plantation culture, clusters of mature nuts are cut from the trees with knives on long poles just before maturity. Picking individual nuts by climbing the trees is sometimes practiced.

Processing and products. The thick husks are removed from the nuts, which are then split open with a heavy knife and partially dried to loosen the meat. This is pried out of the shell and dried to about 7% moisture either in the sun, if conditions are favorable, or in kilns with additional heat often furnished by burning the husks and shells.

The oil from the dried coconut meats, or copra, is widely used for margarine, soap, and industrial purposes. High-quality copra may be shredded for confectionery and the baking trade. The residue, after oil removal, is used for animal feed.

Coconut husks are an important source of fiber called coir. In coir production, mostly limited to India, the nuts are harvested about a month before maturity and the husks retted in brackish water for 8–10 days to rot away the soft tissues between the fibers. The fiber is cleaned and washed by hand and dried. Recently the industry has become partially mechanized. The various grades of coir are used for ropes, mats and matting, and upholstery filling. *See* Coir.

In the tropics of both the Orient and the Occident the coco palm is the most useful of all plants to the native population. An important source of food and drink, it also furnishes building material, thatch, hats, dishes, baskets, and many other useful items. *See* Nut crop culture.

Laurence H. MacDaniels

Diseases. Cadang-cadang (**Fig. 2**), a disease confined to certain islands of the Philippines, is caused by a viroid. This same agent is also present in the African oil palm and in the Buri palm. No vector is

yet known. Since the disease affects mature palms, control is based on replanting with early maturing palms.

Red ring is caused by the nematode *Rhadinaphelenchus cocophilus* and is confined to tropical America. Transverse sections of stems from diseased palms show a typical ring. The disease is always lethal, causing death shortly after infection. The nematode is spread by a palm weevil (*Rhyncophorus palmarum*) which in turn is attracted by damaged palm tissues. Control is effected through sanitation and by the application of insecticides.

Heart rot is confined to tropical America and is caused by a trypanosomatid flagellate protozoon of the genus *Phytomonas*. This organism also attacks the African oil palm. Many of the symptoms resemble lethal yellowing. No effective control is known.

Bud rot is worldwide and is caused by the fungus *Phytophthora palmivora*, which attacks the growing point and causes it to rot. There is no effective control.

Kerala wilt is a debilitating disease of coconut palms in southern India. It results in production of

Fig. 2. Midstage symptoms of cadang-cadang. Fronds assume an erect position in the crown, and no nuts are being produced.

fewer buttons and nuts, reduction of nut size, and thinner husks. The cause is unknown.

Blast and dry bud rot affect young palms in the nurseries and right after planting. Both diseases seem most serious in newly released hybrids. Mycoplasma organisms have been suggested as the cause of blast in the Ivory Coast, where the insect *Ricelia mica* is the vector. *See* Plant pathology.

Luigi Chiarappa

Bibliography. T. O. Diener, Viroids: The smallest known agents of infectious diseases, *Annu. Rev. Microbiol.*, 28:23–29, 1974; K. Maramorosch, *A Survey of Coconut Diseases of Unknown Etiology*, FAO,

Rome, 1964; University of Florida, *Proceedings of the 4th Meeting of the International Council on Lethal Yellowing*, 1979.

Codfish

Fish of a subfamily of the Gadidae in the order Gadiformes. Important commercially, these fish are found in cold waters, such as the Baltic and the North Atlantic. *Gadus morhua* is extensively fished off the Newfoundland banks and a circumpolar species, *Boreogadus saida*, the Arctic cod, is found around the ice pack during the summer. *Gadus macrocephalus*, a related species, occurs in the northern Pacific.

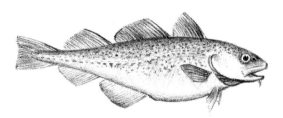

Gadus morhua, a codfish of the northern seas.

The cod is covered with cycloid scales, has a barbel under the chin, and has pelvic fins on the throat; there are two anal and three dorsal fins (see **illus.**). Codfish average about 3 ft (0.9 m) in length and weigh 10–35 lb (4.5–15.7 kg). They live at depths of 100–1500 ft (30–450 m), where they feed on mollusks, small fish, crustaceans, and worms. The cod is a prolific breeder with the female producing as many as 10,000,000 eggs. Spawning occurs from January on to spring. The livers are processed for cod liver oil, which is rich in vitamins, and the swim bladder is made into isinglass. SEE GADIFORMES; SCALE (ZOOLOGY); SWIM BLADDER.

Charles B. Curtin

Coelenterata

That group of the Radiata whose members typically bear tentacles and possess intrinsic nematocysts. The name Cnidaria is also used for this phylum and is preferred by some because the name Coelenterata, as first used, included the sponges (Porifera) and the comb jellies (Ctenophora), as well as the animals called coelenterates. SEE CTENOPHORA; PORIFERA.

TAXONOMY

The coelenterates are mainly marine organisms (**Fig. 1**) and are best known as jellyfish or medusae, sea anemones, corals, the Portuguese man-of-war, small polypoid forms called hydroids, and the freshwater hydras. Taken together, the phylum is divisible into three classes as follows: (1) Hydrozoa, the hydroids, hydras, and hydrozoan or craspedote jellyfish (hydromedusae); (2) Scyphozoa, the acraspedote jellyfish; and (3) Anthozoa, the sea anemones, corals, sea fans, sea pens, and sea pansies. A classification scheme of the phylum including some representative genera and characteristics of the groups is shown in the **table**. See separate articles on each class.

MORPHOLOGY

It is convenient to recognize two basic body forms in this phylum, the polyp and the medusa, into which all coelenterates can be classified. The two, however, have many features in common (**Fig. 2**).

Polyp. This is a radially, biradially, or radiobilaterally symmetrical individual having a longitudinal oral-aboral axis and usually sessile. The mouth is at the free end and is surrounded by one to many whorls or sets of tentacles which may be hollow or solid. The aboral end is commonly developed as an adhesive device for attachment and is conveniently referred to as a base. The polyp is covered by an outer layer of epithelial tissue, the epidermis, lined by a second epithelial tissue layer, the gastrodermis, and the two layers are united by an essentially noncellular layer, the mesoglea. The central body cavity is the gastrovascular cavity also called the enteron or coelenteron.

Medusa. This is a tetramerously or polymerously radial individual and is free-swimming. The body is usually bell- or bowl-shaped with the mouth suspended in the center of the underside of the bell on a stalk or manubrium. Instead of directly surrounding the mouth as in the polyp, the tentacles are located at the margin of the bell. The outer or aboral part of the bell is recognized as the exumbrella and the under or oral part as the subumbrella. The mouth leads to the central stomach which in turn gives rise to four or more radial canals. These radial canals run through the umbrella, on the subumbrellar side, and commonly lead to a ring canal at the margin which is continuous around the margin and may give rise to centripetal and centrifugal canals. The centrifugal canals commonly run from the ring canal throughout the length of the tentacles. Special sensory structures such as eyespots or balance organs (statocysts) may occur along the ring canal, on the tentacle bases or on special marginal structures. The medusa is covered by an outer epidermis; the stomach, radial, ring, and other canals are lined by gastrodermis and between these is a layer poor in cells, the mesoglea. The me-

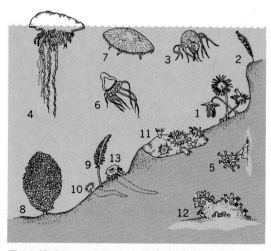

Fig. 1. Marine coelenterates in typical habitats, all reduced but not to same scale: 1, *Tubularia*; 2, *Plumularia*; 3, *Gonionemus*; 4, *Physalia*, Portuguese man-of-war; 5, *Haliclystus*; 6, *Periphylla*; 7, *Aurelia*, common jellyfish; 8, *Gorgonia*, sea fan; 9, *Pennatula*, sea pen; 10, *Edwardsia*; 11, *Epiactis*, sea anemones; 12, *Astrangia*, stony coral; and 13, *Cerianthus*. (After T. I. Storer and R. L. Usinger, *General Zoology*, 3d ed., McGraw-Hill, 1957)

A classification of the phylum Coelenterata

Subdivision	Characteristics	Representative genera
Class Hydrozoa	Majority marine; few species fresh-water	
Order Hydroida	Usually colonial; well-developed polyp stage	
Suborder Anthomedusae	Athecate polyps	*Tubularia; Clava; Pennaria; Hydra*
Suborder Limnomedusae	Several species commensals; hydroids completely naked	*Craspedacusta; Gonionemus*
Suborder Leptomedusae	Hydrotheca present	*Obelia; Sertularia; Plumularia*
Suborder Chondrophora	Polymorphic, colonial, free-floating	*Velella; Porpita*
Order Milleporina	Colonial hydroids; polyps dimorphic	*Millepora*
Order Stylasterina	Dactylozooids lack tentacles	*Stylaster; Allopora*
Order Trachylina	Polypoid generation reduced or absent	
Suborder Trachymedusae	Hydroid generation unknown	*Liriope; Aglantha*
Suborder Narcomedusae	Hydroid generation represented by actinula larva	*Cunina; Aegina*
Suborder Pteromedusae	Modified, bipyramical medusae	*Tetraplatia*
Order Siphonophora	Polymorphic colonies; planktonic	*Physalia; Forskalia; Abyla*
Order Spongiomorphida	Resemble corals; specialized astrorhizal structures	*Spongiomorpha; Heptastylis; Stromatomorpha*
Order Stromatoporoidea	Polymorphic colonies; benthonic	*Actinostroma; Clathrodictyon; Stromatoporella; Hermatostroma; Stromatopora*
Class Scyphozoa	All marine	
Subclass Scyphomedusae	Tetramerous medusae and medusa-like polypoids; reduced marginal tentacles	
Order Stauromedusae	Sessile medusae	*Lucernaria; Haliclystus*
Order Cubomedusae	Velarium present; tropical, subtropical	*Tripedalia; Charybdea*
Order Coronatae	Coronal furrow; lack ocelli	*Nausithoë; Atolla; Linuche; Periphylla*
Order Semaeostomeae	Pedalia and coronal furrow lacking	*Aurelia; Pelagia; Dactylometra*
Order Rhizostomeae	Numerous small mouths; no marginal tentacles	*Rhizostoma; Cassiopea; Cephea*
Subclass Conulata	Tetramerous cone-shaped to elongate pyrimidal; unattached; tentacles on oral margin	
Order Conularida	Characters of subclass	*Conchopelitis; Conularia; Conulariella; Paraconularia*
Class Anthozoa	All marine	
Subclass Alcyonaria	Colonial, eight tentacles; also Octocorallia	
Order Stolonifera	Stolonic network; isolated or compacted skeletal, calcareous spicules	*Tubipora; Clavularia*
Order Telestacea	Calcareous spicules more or less united	*Telesto*
Order Alcyonacea	Soft corals; polyps extend from the coenenchyme	*Alcyonium; Xenia; Anthomastus*
Order Coenothecalia	Blue coral; calcareous skeleton of aragonite	*Heliopora*
Order Gorgonacea	Sea fans; skeleton of gorgonin	*Corallium; Gorgonia; Muricea*
Order Pennatulacea	Sea pens and pansies; colony embedded in coenenchyme	*Pennatula; Renilla; Virgularia*
Order Trachypsammiacea	Colonial; dendroid skeleton	*Trachypsammia*
Subclass Zoantharia	Tentacles and mesenteries hexamerously arranged; Hexacorallia	
Order Actiniaria	Sea anemones; solitary	*Actinia; Metridium; Edwardsia*
Order Ptychodactiaria	Solitary sea anemones; specialized mesenterial filaments absent	*Ptychodactis; Dactylanthus*
Order Corallimorpharia	Resemble corals; lack skeleton; tentacles radially arranged	*Corallimorphus; Corynactis*
Order Rugosa	Solitary or compound corals; calcareous skeleton	*Lambeophyllum; Syringaxon; Petraia*
Order Heterocorallia	Elongate, solitary corals; fibrous calcium carbonate septa and tabulae	*Hexaphyllia; Heterophyllia*
Order Scleractinia	True or stony corals; solitary and colonial; reef formers	*Fungia; Astrangia; Porites*
Order Zoanthidea	No skeleton; solitary or colonial, commonly commensals	*Zoanthus; Palythoa; Parazoanthus*
Order Tabulata	Colonial, calcareous, tabulated corallites; "honeycomb corals"	*Favosites*
Subclass Cerianthipatharia	Includes two related but completely divergent orders	
Order Antipatharia	Thorny or black corals; axial skeleton septa single	*Antipathes; Dendrobranchia*
Order Ceriantharia	Solitary; tube dwellers; lack pedal disk	*Cerianthus*

soglea is much thicker and more resilient in medusae then in polyps and is related to their mode of locomotion. Swimming or locomotion in medusae is accomplished by sudden contractions of the subumbrellar muscles which reduce the volume of the subumbrellar space and thereby force the contained water outward. This propels or jets the medusa in an aboral direction. In medusae of the class Hydrozoa, there is a thin sheet of tissue which arises at the subumbrellar margin of the bell and reduces the effective diameter of the subumbrellar aperture. This structure is called a velum, and medusae possessing a velum are termed craspedote. The resilient and elastic mesoglea is largely responsible for the return of the umbrella to its original form, and the medusa is again ready to repeat the muscular contractions which lead to movement. The contractions occurring in swimming movements are under the control of the nervous system and are regulated in general by the marginal sensory structures. Thus, when the marginal sensory structures are removed, swimming movements stop, unless the medusa is artificially stimulated. When

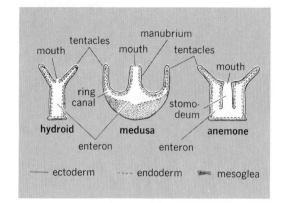

Fig. 2. Comparison of hydroid polyp, medusa (inverted), and anthozoan polyp. (After T. I. Storer and R. L. Usinger, General Zoology, 3d ed., McGraw-Hill, 1957)

only some of the marginal bodies are removed, movement of the medusa is uncoordinated and erratic.

Body form. The body form of the medusa is clearly the by-product of the firm, elastic mesoglea, whereas in the polyp this in only partly so. Small coelenterate polyps depend almost entirely upon the rigidity of their tissues for their form but in all larger polyps, roughly greater than 0.04–0.08 in. (1–2 mm), body form is also dependent upon the amount of water contained within the gastrovascular cavity. Most large polyps possess special ciliated tracts which lead water into the gastrovascular cavity via the mouth. This water, under a very slight positive pressure, literally inflates the polyps as the muscles of the body relax, and once the polyp is expanded this water acts as a type of skeleton. The skeletal nature of this water is best understood if one compares the polyp and its contained water to an ordinary inflated rubber balloon. Thus, if a balloon is squeezed in one fashion, it extends itself in another, and so the polyp by contracting its circular muscles can rapidly extend its length, or by contracting its longitudinal muscles can become shorter and fatter. The water within the body of the polyp can be trapped by closure of the mouth which helps to make the movements described above possible. On strong stimulation leading to a total contraction of a polyp, water is expelled through the mouth, tentacle tips, or special pores in the body wall, and expansion can occur again only after sufficient water has been pumped into the polyp to reinflate it.

Locomotion. Locomotory movements by a creeping of the base of polyps are possible, although they are usually very slow, perhaps a few inches per day. Certain sea anemones such as *Stomphia* and *Boloceroides* can swim in an undirected fashion and some polyps, for example, *Hydra,* can move in a somersault fashion by alternately attaching the base and tentacles to the substrate. Ordinarily, locomotion in polyps is restricted to creeping, although they can also move by using special suctorial devices on the body wall or by merely relasing contact with the substrate and rolling and floating about.

Nematocysts. The unique and most distinctive feature of coelenterates is the possession of intracellular, independent effector organelles called nematocysts, but also known as stinging cells or nettle cells. A coiled thread tube in each cell may be rapidly everted under proper stimulation and used for food gathering and for defense against predators, intruders, or enemies. Nematocysts are produced within cells, called cnidoblasts, and may be produced either in place or

at a location some distance from their definitive site in the coelenterate. The morphologically simplest coelenterates, the Hydrozoa, have nematocysts limited to their outer epidermis whereas the more complex Scyphozoa and Anthozoa bear nematocysts in both the outer epidermis and inner gastrodermis.

There are four structural types of nematocyst found among different species of coelenterates. Although *Hydra* has all four, only three are functionally recognized (**Fig. 3**). These are the volvent, an unarmed, coiled tube, closed at the end, that wraps around the prey; penetrant, with an open tube armed with barbs that penetrates the prey and injects a toxin; and glutinant, an open, sticky tube which is used for anchoring the coelenterate when it is walking on its tentacles.

Nematocysts have been studied and classified by many workers, but it is from the work of Robert Weill that the modern scheme of classification of these structures is taken. Weill described and defined about 17 types of nematocysts. Closely related to nematocysts but differing from them morphologically are spirocysts, another kind of intracellular capsule found in certain of the Zoantharia, such as *Metridium*.

Spirocysts. These structures are thin-walled capsules, the wall of which is probably only a single layer thick, and they contain a long, unarmed, spirally coiled thread of uniform diameter. The reaction of spirocysts to certain dyes differs from that of nematocysts in that spirocysts stain with acid dyes. The nematocysts are, by comparison, double-walled capsules of collagenous proteins containing a thread which may or may not be spirally coiled and of uniform diameter throughout its length. The thread is usually armed with spirally arranged spines. Both spirocysts and nematocysts, on proper stimulation, evert their contained threads. After a nematocyst has been discharged, it cannot be used again. The size of nematocysts varies from a few micrometers to over 0.04 in. (1 mm) in length. Figure 3 illustrates the general morphology and some types of nematocysts.

Classification. Nematocysts may be divided into two broad types, the astomocnidae with threads which have closed ends and the stomocnidae which have open-ended threads. The astomocnidae are either adhesive in function or act as lassos to entangle prey. The stomocnidae, in general, penetrate the prey or foe on discharge and the capsular contents, which are toxic, pour out through the open end of the thread. The toxin or toxins of nematocysts have not been positively identified. They seem to be proteins, protein complexes, or possibly some other substance associated with proteins. Contact with nematocyst-bearing tissues, leading to discharge of nematocysts and the entry of toxins into the contacting organism, has lethal effects on many small invertebrates and fishes. Contact with the Portuguese man-of-war and certain other coelenterates may cause humans extreme pain, vasomotor dysfunction, and even death. *See Toxin.*

Discharge. The discharge of nematocysts is an instantaneous process resulting from contact with the trigger or cnidocil, or from contact between almost any foreign object and the tissues bearing nematocysts. The nervous system of the coelenterate does not seem to be involved in this response. Chemically clean objects do not cause discharge of nematocysts, although contamination of an inert object with organic materials will cause discharge on contact. In life,

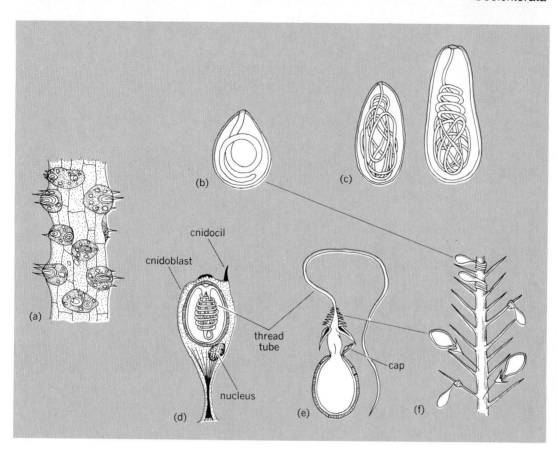

Fig. 3. Nematocysts of *Hydra*. (*a*) Tentacle with nematocyst batteries. (*b*) Volvent type. (*c*) Glutinant type. (*d*) Penetrant type before discharge. (*e*) Penetrant discharged. (*f*) Discharged nematocyst on bristle of a crustacean. (*After T. I. Storer and R. L. Usinger, General Zoology, 3d ed., McGraw-Hill, 1957*)

small organisms which are touched by the nematocyst-laden tentacles of coelenterates cause massive but very local discharges. This is usually sufficient to immobilize the organism being stung, or at least to reduce its activities to such a low level that it cannot free itself from the tentacles. Subsequently, the movements of the tentacles of the coelenterate will carry such a trapped organism to the mouth, where it will be swallowed.

Other organisms. Although intrinsic to coelenterates, nematocysts are known to occur in a variety of other organisms. Certain flatworms and nudibranchs which feed on coelenterates are able to move these structures, without causing their discharge, through their own digestive tract and out into their own tissues. These foreign nematocysts are oriented in the new site so that the eversible end of the capsule is at the surface. Exactly what function these nematocysts now carry out is not clear, although they apparently act as defensive structures for the new host. Nematocysts have been reported in certain dinoflagellates, ctenophores, and larvaceans. Those in dinoflagellates may be intrinsic, although it appears that those in other organisms are of foreign origin. It should be noted also that nematocystlike structures occur in one other group of protists besides the dinoflagellates, the cnidosporidians.

BODY SYSTEMS

The body systems of coelenterates may be divided into the following categories for the purpose of description: epithelial, nervous, muscular, mesogleal, skeletal, digestive, and reproductive (**Fig. 4**). There

is no analog of a circulatory system and the functions of respiration and excretion are carried out by each cell.

Epithelial system. In the coelenterates this system is quite simple. The outer epithelium or epidermis is in general a single layer of cells, varying from a flattened squamous nature to tall and columnar. The epidermis is usually ciliated in polyps so that the net effect of the ciliary beat is to move material away from the mouth toward the margin of the oral disk and toward the tips of the tentacles. In at least one anemone, *Metridium,* the direction of the ciliary beat on the oral disk can be reversed. The cilia on the column move material toward the base. These currents may be interpreted as cleaning currents. Among the epithelial cells may be found numerous specialized types such as mucus-producing cells, sensory cells, undifferentiated interstitial cells, and cnidoblasts. Frequently, some of the epidermal cells have a dual function and the cell base is elongate and possesses contractile fibrils. Such cells are described as epitheliomuscular cells. In simple polyps, such as most hydrozoan polyps, the epitheliomuscular cells of the epidermis and gastrodermis represent the total musculature.

The gastrodermis, or inner epithelial layer, is also abundantly ciliated. These cilia move food materials about within the gastric cavity and circulate the contained water. Among the gastrodermal cells may be found types identifiable as glandular, cnidoblastic, sensory, interstitial, and epitheliomuscular. One of the prime functions of the gastrodermis is digestion, and essentially every cell of the gastrodermis is ca-

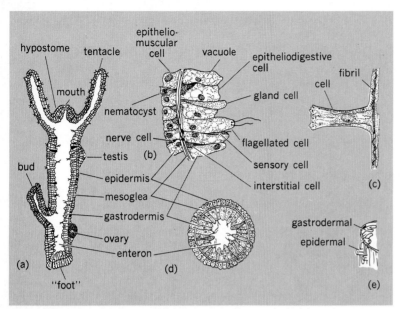

Fig. 4. *Hydra,* in several different microscopic sections. (*a*) Longisection. (*b*) Enlarged cross section. (*c*) Epitheliomuscular cell. (*d*) Cross section. (*e*) Fibril arrangement of the epitheliomuscular cells. (*After T. I. Storer and R. L. Usinger, General Zoology, 3d ed., McGraw-Hill, 1957*)

pable of ingesting small food particles.

Nervous system. The coelenterate nervous system can be defined as consisting basically of an unpolarized network of bipolar and multipolar neurons. The nervous system is subepithelial in location, both subepidermal and subgastrodermal, and may in some forms consist of two networks in only limited contact with one another and with each specialized either for rapid through conduction or for slower, more general spread of conduction. The physiology of coelenterate nervous systems has been much studied, particularly neuromuscular aspects of it. The nervous system is synaptic. Where information is available, action potentials have been demonstrated from single neurons, and the fine morphology of synapses and neuromuscular end plates looks much as it does elsewhere. Cross connections between gastrodermal and epidermal nerve nets occur abundantly in the region of the mouth and may occur elsewhere through the mesoglea, but this latter point needs clarification. In colonial coelenterates the nervous system is continuous from member to member of the whole colony. The response of different kinds of colonies to local stimulation varies from complete contraction of the whole colony to only local contraction in the area of stimulation or various intermediate responses.

In most coelenterates a stimulus at any point causes only a local contraction, whereas succeeding stimuli at the same location cause a greater and greater spread in the response. Thus, with the generally diffuse network, a stimulus at any point can be transmitted via the conducting nervous system to all points in the system. Simple reflexes, such as righting movements, swallowing, and locomotion are possible and are known. The coordination of swimming in medusae is a property of the balance organs at the bell margin. When these controlling structures are removed and a wave of contraction is initiated in a ring-shaped piece of the bell of a medusa, it is possible to cause a trapped circuit wave of neuromuscular activity to travel around and around the ring. One such prepara-

tion made from the medusa *Cassiopea* maintained a circuit wave for 11 days, traveling at 31 in./s (77.5 cm/s) for a total distance of 457 mi (731 km). The coelenterate nervous system is of particular interest because it appears to be the first example of a recognizable nervous system in a metazoan; that is, it is regarded as an example of a primitive nervous system. The system is, however, highly evolved and clearly adapted to coelenterate needs and as such should not be thought of as truly primitive. SEE NERVOUS SYSTEM (INVERTEBRATE).

Muscular system. The muscular system varies considerably among coelenterates. In simple polyps there is an epidermal, longitudinal muscle sheath of epitheliomuscular cells and a gastrodermal, circular sheath. In medusae and in members of the classes Scyphozoa and Anthozoa, the musculature has become separated from the epithelial tissue as a subepithelial system. The cells are elongate and in some medusae are striated, although most coelenterates do not show striated muscle fibers. The muscle cells are closely applied to the mesoglea and the mesoglea is folded to accommodate them. In many anthozoans a large sphincter muscle develops in the gastrodermal surface, at the top of the column, and in some anthozoans this muscle becomes secondarily embedded in the mesoglea. In medusae, where the primary swimming muscle are subumbrellar, large muscle bands, both circular and radial, develop. In large anthozoan polyps the epidermal musculature is usually weak, whereas heavy gastrodermal, circular, and longitudinal muscles are developed. The musculature of tentacles is quite constant throughout the phylum, there being a longitudinal epidermal sheath and a circular gastrodermal sheath.

Mesoglea. The mesogleal system is represented by a layer between epidermis and gastrodermis which varies from a thin structureless cementing layer, the mesolamella of hydrozoan polyps, to a highly complex cellular, fibrous, gelatinous matrix, the collenchyme or mesenchyme of the scyphomedusans and anthozoans. In hydromedusae the mesoglea is essentially free of cells and fibers. Where the mesoglea is well developed, as in medusae and some anthozoans, it serves as a type of internal skeleton, against which muscles may act. This is particularly true in medusae, where the elasticity of the mesoglea returns the bell to its original shape after each pulsation, thus preparing for a new pulsation. In anthozoans the mesoglea serves for insertion and origin of muscle cells; the skeletal function is largely taken over by the water contained in the coelecteron, and the mesoglea is much thinner and less elastic than in medusae.

Skeletal system. Two different types of skeleton occur in the phylum. The first, referred to above, is of an internal nature and is either the mesoglea against which the muscles operate or the contained hydroskeleton of most polyps. Of a quite different nature is the skeleton seen in most hydrozoan polyps, some scyphozoan polyps, and corals. These are exoskeletons, formed as secretions of epidermis, which support and protect the organisms. In the Hydrozoa the skeleton is tubular, usually rigid enough to support many polyps in a colonial system and flexible enough to allow bending in waves or currents. Sometimes the hydrozoan skeleton is elaborately developed with ornamental cups into which each polyp may withdraw, whereas in others the exoskeleton may cover only the column or stalk of the polyp or be variously reduced to the point of being entirely absent. These flexible hydrozoan skeletons are made up of sclerified proteins. In certain colonial pelagic hydrozoans part of the skeleton may be involved in the development of a flotation device, as in *Physalia* or *Velella*. Calcareous exoskeletons occur among both the Hydrozoa and Anthozoa, and the animals possessing them may be referred to as hydrocorals and madreporarian corals, respectively. These skeletons may vary from delicately branched structures to massive encrusting formations, but in general most of the living animal matter is restricted to the outer layer of skeletal material. The polyps themselves occur in cups or depressions in the skeleton, and the individuals of the colonies are connected by sheets of living tissue which cover the whole skeleton except at its attachment to the substrate.

Still another skeletal type known in the coelenterates is an axial skeleton composed of sclerified proteins. It is exceedingly tough and flexible and is characteristic of such anthozoans as gorgonians. In another anthozoan group, the pennatulaceans of sea pens, there is also an axial skeleton which, although flexible up to a point, is also brittle. This latter skeleton is largely calcified. The development of axial skeletons has not been studied sufficiently in detail to state clearly that they are epidermal in origin, although this is assumed to be correct. A number of different coelenterates, such as some hydrozoans and anthozoans, also have spicular calcareous structures which may serve for protection from predators or for the reinforcement of areas on a colony. These structures may be considered skeletal, too, in a broad sense.

Digestive system. The digestive system in coelenterates is developed as a function of the gastrodermis. In hydrozoans, both polyps and medusae, glandular, enzyme-producing cells are abundant throughout the gastrodermis. These cells secrete proteolytic enzymes into the coelenteron which reduce food objects to a fine particulate state. In scyphozoans and anthozoans, although glandular cells are common throughout the gastrodermis, they tend to be most abundant along the free edges of the mesenteries of anthozoans and on the gastric tentacles of scyphozoans. These glandular cells are also the source of proteolytic enzymes which act in the extracellular environment of the coelenteron.

After the preliminary physical breakdown of food to small particles has been accomplished, these particles are engulfed by gastrodermal cells of all types. Food vacuoles form in these cells and definitive digestion of all classes of foods is apparently intracellular. The epidermal cells do not have the ability to engulf food particles, and the nutrition of all tissues of a coelenterate is apparently dependent upon diffusion of foodstuffs from the gastrodermis to other parts of the body. Ameboid wandering cells may play a role as nurse cells, particularly in the development of ova, but whether such cells are involved in the normal nutrition of epidermal tissue is not known.

The phylum is characterized by its carnivorous diet, made possible first by the possession of nematocysts which make the predaceous habit successful. After food has been trapped, movements of the tentacles carry it to the mouth, where with the help of ciliary and muscular devices the food is moved to the coelenteron. Here extracellular proteases prepare the way for final intracellular digestion. No herbivorous coelenterates are known. Certain coelenterates may, however, in a limited sense utilize plant materials as food. It is known, for example, that by-products of photosynthesis of symbiotic algae found in the tissues of coelenterates may be traced from the algae to the tissues of the host. These materials are incorporated into the animal protoplasm, and to this extent the coelenterate may ''feed'' on the plant. Certain anthozoans (some alcyonarians) have never been seen to feed on animal material, and these forms also have extremely abundant symbiotic algae and reduced digestive structures. Moreover, studies on the ecology of coral reefs have suggested that the environment of the reef does not produce enough animal food to support the mass of corals present, and this has led to the conclusion that at least some of the food of reef-forming corals must be obtained directly from their associated algae instead of from the external environment. The degree to which corals are dependent upon their symbiotic algae has not been determined. SEE *FEEDING MECHANISMS (INVERTEBRATE)*.

Reproductive system. The reproductive system of coelenterates consists of specialized areas of epithelia, the gonads, which periodically appear and produce gametes. In the Hydrozoa the gonads are epidermal, whereas in the Scyphozoa and Anthozoa they are gastrodermal. The precise location of gonads is variable in the Hydrozoa, being located either on the manubrium or beneath the radial canals in hydromedusans, or at almost any point on different hydroid polyps. In the Scyphozoa the gonads are located in the floor of the stomach, whereas in the Anthozoa they occur along the sides of the mesenteries which subdivide the coelenteron. There are no ducts for the sex products or any accessory sexual structures. Fertilization usually occurs in the water surrounding the animal, although a few coelenterates have their eggs fertilized in place and may then brood their young. In scyphozoans and anthozoans the sex products are released to the coelenteron and make their way to the outside via the mouth. The sexes are separate in the coelenterates, that is, they are dioecious, with few exceptions.

REPRODUCTION AND DEVELOPMENT

The phenomena of asexual reproduction and regeneration are common throughout the phylum. In all coelenterates the interstitial cells aggregate to form ovaries and testes (not true gonads) from which germ cells originate.

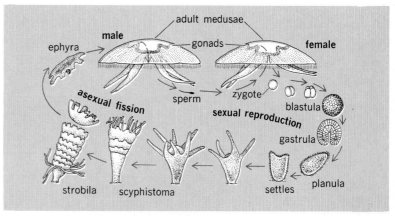

Fig. 5. Life cycle of jellyfish (*Aurelia aurita*), class Scyphozoa. (*After T. I. Storer and R. L. Usinger, General Zoology, 3d ed., McGraw-Hill, 1957*)

Asexual reproduction. This is accomplished in a variety of ways. All colonial coelenterates are the result of the budding of new individuals from existing individuals. Some hydroids, hydromedusae, and anthozoans give rise to free buds, which in turn may bud again to produce still more individuals. Most hydrozoan medusae are budded from hydroid colonies. Small pieces of hydroids may bud from a parent, wander about as undifferentiated, elongated entities called frustules, and in time attach and develop into new individuals or colonies. In sea anemones both transverse and longitudinal fission are known methods of asexual reproduction. Transverse fission is the normal mode of development of the adult medusa from the larval form in some Scyphozoa and is known as a rare event in hydrozoan polyps (**Fig. 5**).

The ability to regenerate lost parts is characteristic of coelenterates. Pieces cut from almost any part of polyps will in time grow into new polyps. The regenerative powers of medusae are much less well developed, and not only will the excised piece not develop but it may not even be replaced by the medusa. Gradients of regenerative ability in polyps exist with the ability for a piece to reconstitute a new whole organism decreasing from the mouth to the base. SEE REGENERATION (BIOLOGY).

Sexual reproduction. The embryology of coelenterates takes slightly differing courses among the many kinds of organisms here involved. After fertilization, cleavage occurs and usually is total, frequently somewhat unequal, and the individual cells are not as a rule definitely arranged. The first stage is a ciliated blastula (Fig. 5). The development of the next stage, the gastrula, follows several courses such as the wandering in of some of the surface cells, the division of the surface layer to produce a second inner layer, or by an infolding of one end of the cell mass. The gastrula is a stereogastrula; that is, it has no cav-

ities. This stage elongates into the typical larval stage of this phylum, the planula larva which is ciliated and free-swimming. The planula tends to have a broadened anterior end, frequently equipped with an elongate tuft of sensory cilia, and is mouthless. This larva, if that of a polyp, will attach by its anterior end and develop a mouth and tentacles at its posterior end. The planula larvae of medusae which have no attached polypoid stage metamorphose to medusae; the anterior end becomes the exumbrellar surface of the bell, and the mouth again develops from the posterior end of the larva. In some hydrozoans the planula larva develops tentacles and a mouth while still free-swimming, and this stage is called an actinula. The actinula of some species may attach and develop into a hydroid, whereas in others it metamorphoses into a medusa. SEE INVERTEBRATE EMBRYOLOGY; REPRODUCTION (ANIMAL). *Cadet Hand*

FOSSILS

The Coelenterata have a long and impressive fossil record stretching from the present into the Precambrian, about 700,000,000 years ago. Thus, the known duration of this phylum equals or exceeds that of other animal phyla. Forms with skeletons, primarily the Conulata, Tabulata, Rugosa, and Scleractinia, have left fairly complete fossil records, whereas softbodied forms, such as most of the Hydrozoa, Scyphomedusae, and Alcyonaria, are represented by very few fossils (**Fig. 6**).

The earliest known coelenterates have been found in Late Precambrian rocks on several continents; an especially well-preserved fauna is known from southern Australia. These early coelenterates are all softbodied, but the fact that the fossils apparently represent all the classes of the Coelenterata indicates that the origin of the phylum was truly ancient. Both free-

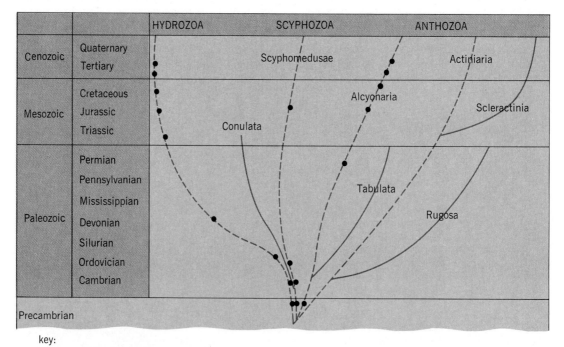

Fig. 6. Geologic distribution of the coelenterates.

key:

—— known record

- - - interpreted path of development

● occurrences of rare groups of fossils

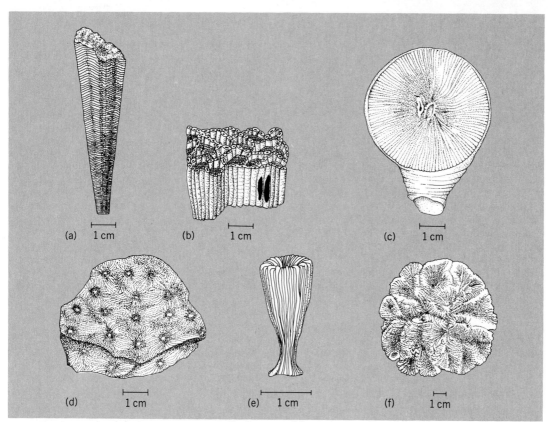

Fig. 7. Coelenterate fossils. (*a*) *Paraconularia*, Middle Paleozoic Conulata. (*b*) *Halysites*, Middle Paleozoic tabulate coral. (c) *Bethanyphyllum*, Devonian solitary rugose coral. (*d*) *Billingsastraea*, Devonian colonial rugose coral. (*e*) *Parasmilia*, Late Mesozoic and Cenozoic solitary scleractinian coral. (*f*) *Isophyllia*, Recent colonial scleractinian coral. (*After R. C. Moore, ed., Treatise on Invertebrate Paleontology, pt. F, Geological Society of America and University of Kansas Press, 1956*)

swimming medusoid forms and bottom-dwelling polypoid forms are represented in these old rocks. Appearing in the succeeding Cambrian strata is the extinct subclass Conulata (**Fig. 7***a*), a group possibly belonging in the Scyphozoa. These forms are represented in the fossil record by an unusual thin, steep-sided, pyramidal, chitino-phosphatic external skeleton. An extinct group of fossils, the Stromatoporoidea, formerly considered coelenterates, also appeared in the Cambrian Period. Work on modern sclerosponges, however, suggests that the stromatoporoids more likely belong in the Porifera (sponges) than in the Coelenterata; thus this group is not included here. SEE SCLEROSPONGIAE.

Two of the three major orders of anthozoans bearing calcium carbonate skeletons, the Rugosa (tetracorals) and the Tabulata (Fig. 7*b*, *c*, and *d*), appear in rocks of the succeeding Ordovician system. These two groups of corals quickly diversified, leaving a vast number of species in the fossil record. Both groups include colonial forms that participated in formation of carbonate banks and reefal buildups. In addition, the rugosans are represented by many solitary horn corals. The rugose corals, especially, were very numerous and widespread in the Early Permian, but in the later part of that period the diversity of forms dwindled, the more complex, bank-forming groups disappearing first. Both the tabulate and rugose corals became extinct during the great wave of Late Permian extinctions, as did a host of other stenohaline marine invertebrates. The succeeding strata of Early Triassic age contain no corals, although the last Conulata have been reported from beds of this age.

In Middle Triassic strata the calcium carbonate skeletons of the Scleractinia (hexacorals) first appear (Fig. 7*e* and *f*). Although it would be convenient to consider that the scleractinians arose from the rugosans, two facts argue against this view: the difference in basic ground plan and the lack of transitional forms in the Early Triassic. Presumably, both the rugosans and scleractinians arose independently (Fig. 6) from the soft-bodied anemones (Actiniaria). By the Jurassic the scleractinians formed reefs similar to the rich, modern reefs of the tropics. Today coelenterates are abundant and widespread in normal marine waters. Solid skeletons that might be preserved as fossils are formed most abundantly and are the most diversified in shallow, tropical waters, but some of the skeleton-bearing scleractinians inhabit deep, cold waters in extratropical regions.

Calvin H. Stevens

Bibliography. R. D. Barnes, *Invertebrate Zoology*, 4th ed., 1980; C. M. Fraser, *Hydroids of the Atlantic Coast of North America*, 1944; W. G. Kükenthal and T. Krumbach, *Handbuch der Zoologie*, vol. 1, 1925; C. E. Lane and E. Dodge, The toxicity of *Physalia* nematocysts, *Biol. Bull.*, 115(2):219–226, 1958; G. O. Mackie (ed.), *Coelenterate Ecology and Behavior*, 1976; A. G. Mayer, *Medusae of the World*, vols. 1–3, Carnegie Inst. Wash. Publ. 109, 1910; R. C. Moore (ed.), *Treatise on Invertebrate Paleontology*, pt. F, 1956; S. P. Parker (ed.), *Synopsis and Classification of Living Organisms*, 2 vols., 1982; F. S. Russell, *The Medusae of the British Isles*, 1953;

T. A. Stephenson, *The British Sea Anemones*, vol. 1, 1928, vol. 2, 1935; A. K. Totton, Siphonophora of the Indian Ocean, *Discovery Rep.*, 27:1–162, 1954; C. M. Yonge, *A Year on the Great Barrier Reef*, 1931.

Coelom

The coelom is the mesodermally lined body cavity of most animals above the flatworms and nonsegmented roundworms. Its manner of origin provides one basis for classifying the major higher groups.

Annelida. Annelids, arthropods, and mollusks have a coelom which develops from solid mesodermal bands. Within the trochophore larva of annelids, a single pole cell proliferates two strips of mesoblast lying on either side of the ventral midline. These bands subdivide transversely into bilateral solid blocks, the somites. Each somite then splits internally to form a hollow vesicle, the cavity of which is the coelom. The somitic vesicles of each side expand until their surfaces become apposed above and below the gut, thus forming its supporting mesenteries. The anteroposterior surfaces of contact between somites form the septa between segments of the body. In some of the annelids which have developed a parasitic mode of life, as the leeches, septa disappear between segments, and the coelom is much reduced by encroachment of mesenchymatous tissue. SEE ANNELIDA.

Mollusca. The mollusks also form bands of mesoderm from a single pole cell, but these bands do not segment. They split internally to form single right and left coelomic sacs, but the cavities are soon reduced and the surrounding mesoblast disperses as separate cells, many of which become muscle. The only remnants of the coelom in the adult are the pericardial cavity and the cavities of the gonads and their ducts. SEE MOLLUSCA.

Arthropoda. In arthropods paired bands of mesoblast may proliferate from a posterior growth center or may separate inward from a blastoderm, a superficial layer of cells, on the ventral surface of the egg. These bands divide into linear series of somites which then hollow out. Their cavities represent the coelom. The upper wall of each somite moves dorsally to meet its partner from either side and fuse into a longitudinal tube, the heart. The outer and inner somitic walls give rise to muscles of the body wall and gut, respectively. The lower wall may form fat body or move into the appendages to form excretory organs or glands. In adult arthropods the coelom persists only as small cavities within the segmental excretory organs or glands and as the cavities of the gonads and gonoducts. SEE ARTHROPODA.

Echinodermata and Chordata. Echinoderms and chordates constitute a second major group, characterized by the origin of the coelom from outpocketings of the primitive gut wall. In echinoderms one pair of bilateral pouches evaginates and separates from the archenteron or primitive digestive cavity. Each pouch constricts into three portions, not homologous to the metameres of other animals. Of these the left middle coelom assumes major importance, giving rise to the water-vascular system and primordia of the radii of the adult. Contributions to the stone canal and axial sinus are made by the left anterior coelom. The rearmost cavities are converted into the main perivisceral coelom of the adult. SEE CHORDATA; ECHINODERMATA.

Protochordata. The protochordates of the groups Hemichordata and Cephalochordata exemplified by *Balanoglossus* and *Branchiostoma*, respectively, have three coelomic pouches formed by separate evaginations of the archenteral roof. The first or head cavity is single and median in origin, the other bilateral and paired. In hemichordates the head cavity remains single as the cavity of the proboscis and has a pore to the exterior on each side. The second pouches form cavities within the collar and also acquire external pores. The third pair is contained within the trunk and forms the major perivisceral cavity.

Cephalochordata. In cephalochordates the head cavity divides into lateral halves. The left side communicates, by a pore, to an ectodermal pit called the wheel organ. The second pair of pouches forms the pair of mesoblastic somites, and the third pouches subdivide transversely to give rise to the remainder of the linear series of somites. The upper or myotomic portion of each somite remains metameric and forms the segmental muscles. As it enlarges, the coelomic space is displaced ventrally and expands above and below the gut to form the perivisceral cavities and mesenteries, as described for annelids. However, the transverse septa perforate and disappear between adjacent somites, so that a continuous anteroposterior body cavity is formed on either side of the gut.

Vertebrata. In vertebrates the mesoderm arises as a solid sheet from surface cells that have been involuted through the blastopore. Lateral to the notochord, beginning at about the level of the ear, the mesoderm subdivides into three parts: (1) the somites; (2) the nephrotomic cord, temporarily segmented in lower vertebrates, which will form excretory organs and ducts; and (3) the unsegmented lateral plate. The coelom arises as a split within the lateral plate. In lower vertebrates the split may continue through the nephrotome into the center of the somite, thus resembling the condition in *Branchiostoma*. Such dorsal extensions are soon lost. The lateral plate expands to meet its partner of the other side, thus forming the body cavity and mesenteries of the gut. SEE GASTRULATION; INVERTEBRATE EMBRYOLOGY.

Howard L. Hamilton

Coelomycetes

Anamorphic (asexual or imperfect) fungi (Deuteromycotina) with sporulation occurring inside fruit bodies (conidiomata) that arise from a thallus consisting of septate hyphae. About 700 genera (with 750 synonyms) containing more than 8500 species are recognized.

The Coelomycetes, like other groups of deuteromycetes, are artificial, comprising almost entirely anamorphic fungi of ascomycete affinity. Some are known anamorphs of Ascomycotina, although there are a few (*Fibulocoela, Cenangiomyces*) with Basidiomycotina affinities because they have clamp connections or dolipore septa. Taxa are referred to as form genera and form species because the absence of a teleomorph (sexual or perfect) state means that they are classified and identified by artificial rather than phylogenetic means. The unifying feature of the group is the production of conidia inside cavities lined by fungal tissue, or by a combination of fungal and host tissue which constitutes the conidioma. SEE ASCOMYCOTINA; BASIDIOMYCOTINA.

Classification and identification. Differences in conidiomatal structure traditionally have been used to separate three orders: the Melanconiales, where conidiomata are subcuticular, epidermal, subepidermal, peridermal, or subperidermal with the basal conidiogenous region formed within the substrate, and dehiscence is by means of rupture of the overlying host tissue; the Sphaeropsidales, where conidiomata are superficial, semi-immersed or immersed, with the conidiogenous region lining the locule, and dehiscence is through a specialized circular or more rarely longitudinal pore, the ostiole; and the Pycnothyriales, where conidiomata are superficial or more rarely subcuticular, flattened, and shield-shaped, with the conidiogenous region restricted to the inner face of the shield or on a lower wall, and dehiscence is through a specialized pore, the ostiole, or stellate fissures, or passively by elevation of the shield at the margin. SEE MELANCONIALES; PYCNOTHYRIALES; SPHAEROPSIDALES.

However, differences in the ways that conidia are produced have been used more recently in classification and identification. Conidiogenesis in Coelomycetes does not show quite the wide range in diversity that is known in Hyphomycetes, but the same principles concerning thallic and blastic types of development apply. Before, during, and after conidial development, the conidiogenous cells proliferate in a variety of different but precise ways so that a plurality of conidia is formed from a single cell, and this to some extent accounts for the copious spore production which characterizes the group. Conidia may be pigmented (*Melanconium*) or hyaline (*Phoma*), single-celled (*Colletotrichum*) or divided into a number of compartments by transverse (*Septoria*) and longitudinal (*Camarosporium*) cross walls of the euseptate (*Diplodia*) or distoseptate (*Coryneum*) type, smooth (*Ascochyta*) or ornamented (*Haplosporella*), and of diverse shape and form from simply globose (*Lasmeniella*) to ellipsoidal (*Ampelomyces*) to filiform (*Phloeospora*), helical (*Helicothyrium*), stellate (*Asterosporium*), and branched (*Furcaspora*). Some conidia are ornamented by cellular (*Pestalotiopsis*) or extracellular (*Neottiospora*) appendages which increase the chances of dispersed conidia becoming trapped on a suitable substratum. Sterile elements resembling paraphyses are occasionally found among conidiophores in some genera (*Phaeocytostroma*). Conidia are often formed in a mucilaginous matrix which oozes out as tendrils or straps or spreads over the surface of the substratum. The matrix inhibits or retards germination until conidia become dispersed, and maintains germinability during periods of environmental stress. SEE HYPHOMYCETES.

Growth media. Many Coelomycetes can be grown in synthetic media in artificial conditions, and, by using a variety of techniques involving the composition of the substratum, light regime, and temperature, may be induced to sporulate. Conidiomatal form, conidial size, shape, and septation, and production of vegetative structures such as chlamydospores and sclerotia become much more variable in such conditions and so are more difficult to apply in classification and identification. Acervular conidiomata in the Melanconiales become indistinguishable from the sporodochial conidiomata of Tuberculariales (Hyphomycetes), so that the traditional separation between these orders breaks down and cannot be applied. Nevertheless, at species level in some important plant pathogenic genera such as *Phoma* and *Colletotrichum*, taxa are now being identified by behavior in artificial culture much in the same way as the hyphomycete genera *Penicillium*, *Aspergillus*, and *Fusarium*.

Biology and distribution. Coelomycetes are known mainly from temperate and tropical regions, and less so from the more extreme climatic areas, although they have been recorded from the arctic and from very arid zones. They grow, reproduce, and survive in a wide range of ecological situations and can be categorized as either a stress-tolerant or combative species. Few ruderal species are known in the Coelomycetes because the very fact that melanized conidiomata are produced means that they are not ephemeral. They are commonly found in and recovered from cultivated and uncultivated soils of different types, leaf litter and other organic debris from both natural and manufactured sources (as biodeteriogens and biodegradative organisms), and saline and fresh water; and on other fungi and lichens. Several are of medical importance, associated with acute conditions in humans and animals, often as opportunistic organisms causing infection in immunocompromised patients. Some mycotoxicoses in cattle and sheep are attributable to animal feed or forage infected with different coelomycetes. Others are entomogenous and, especially on scale insects, show promise as potential biological control agents.

Pathogens. Coelomycetes are consistently isolated from or associated with disease conditions in all types of vascular plants, often in association with other organisms. Without adequate experimental proof it is difficult to determine if they are primary pathogens, are secondary inhabitants, or fulfill a more complex role as one of several factors leading to or actually suppressing the disease condition. Some colonize roots, causing disease, but the majority in soil are benign constituents of the rhizoplane or are latent infections in root systems. Many are primary pathogens of economically important agricultural and forest crops and ornamental plants and weeds, causing a variety of diseases such as leaf, stem, and root necrosis, blights and anthracnoses, cankers, and galls. Others are responsible for flower blights, anther hypertrophy, rot and blemish of growing and harvested fruit, and reduction in viability of seed. A large group of genera is responsible for insidious deteriorative dieback conditions in shrubs and trees. The genera vary in pathogenicity inasmuch as the ability to infect is often dependent on existing or coincident physiological stress or mechanical damage to the host. SEE DEUTEROMYCOTINA; PLANT PATHOLOGY.

B. C. Sutton

Coenopteridales

True ferns which span the Late Devonian through Permian time between the recognizable beginnings of fernlike morphology and the earliest-appearing extant filicalean families (Gleicheniaceae, Osmundaceae). A true fern is a relatively advanced type of vascular land plant with distinct stem, fronds, roots, and foliar-borne annulate sporangia. Coenopterid ferns are mostly small and simple in contrast to late Paleozoic tree ferns of the Marattiales. All well-known genera of the Coenopteridales in the Pennsylvanian Period exhibit fronds with circinate vernation (croziers), laminate vegetative foliage, and annulate sporangia. Their stems are mostly protostelic with a few siphon-

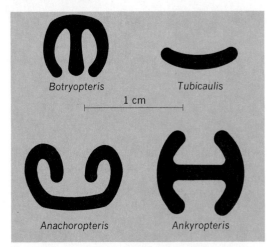

Fig. 1. Diagrammatic cross sections of petiolar vascular strands of representative coenopterid genera, oriented so the parent stem would appear below each diagram.

ostelic species, and one genus (*Zygopteris*) has secondary growth. These ferns are better known from anatomically preserved specimens, especially from peat beds, than from coalified compressions, but they occurred in a variety of habitats mostly known on Northern Hemisphere continents. Their fronds have distinct vascular strands, useful in identifying genera (**Fig. 1**).

There are two major distinct groups. Zygopterid ferns are the most ancient and diverse, and apparently a dead-end evolutionary line; they differ the most from other ferns. *Zygopteris* (Zygopteridaceae: Etapteroideae) had dichotomous rhizomes with distantly spaced erect fronds exhibiting three-dimensionally arranged primary pinnae in four rows. Sessile sori of elongate multiseriate sporangia with paired annuli and with sclerotic nests in their interannular walls were borne abaxially on laminate foliage in *Corynepteris* or trailed from pinnae on slender stalks in *Biscalitheca* (**Fig. 2**). Coenopterid ferns are usually placed in the Anachoropteridaceae and Botryopteridaceae, or in one of several extinct families assigned to the Filicales. Coenopterid ferns *sensu stricto* are probably ancestral to, and consequently form an imperceptible transition with, the Filicales. The best-known genera, *Anachoropteris* and *Botryopteris*, commonly have semierect (like *Osmunda*), protostelic rhizomes with closely spaced fronds bearing two rows of branched pinnae, occasional to frequent buds or vegatative shoots, laminate foliage except on some fertile pinnae, and small to large clusters of uniannulate sporangia. Some of the Pennsylvanian-age fern genera (*Sermaya, Doneggia*) with *Anachoropteris*-like anat-

omy have been assigned to the filicales as members of an extinct family, Sermayaceae.

Evolution of stelar morphology from protostelic to siphonostelic, with or without leaf gaps, is demonstrated in the Anachoropteridaceae and Botryopteridaceae in the Pennsylvanian. Sporangia of various coenopterid genera are osmundoid, gleichenioid, or schizaeoid, generalized and simple morphological types.

The Zygopteridaceae appear in Late Devonian time, the Botryopteridaceae in Visean (Mississippian), the Anachoropteridaceae in the lower Westphalian A (Pennsylvanian), and all families extend into the Permian. SEE PALEOBOTANY.

Tom L. Phillips

Bibliography. R. L. Dennis, Studies of Paleozoic ferns: *Zygopteris* from the middle and late Pennsylvanian of the United States, *Palaeontographica*, 148B:95–136, 1974; D. A. Eggert and T. Delevoryas, Studies of Paleozoic ferns: *Sermaya*, gen. nov., and its bearing on filicalean evolution in the Paleozoic, *Palaeontographica*, 120B:169–180, 1967; J. Galtier and A. C. Scott, Studies of Paleozoic ferns: On the genus *Corynepteris*, a redescription of the type and some other European species, *Palaeontographica*, 170B:81–125, 1979; T. L. Phillips, Evolution of vegetative morphology in coenopterid ferns, *Ann. Mo. Bot. Garden*, 61:427–461, 1974.

Coenothecalia

An order of the class Alcyonaria. The Coenothecalia have no spicules but form colonies with a massive skeleton composed of fibrocrystalline argonite fused into lamellae. The skeleton is perforated by both numerous wide cylindrical cavities occupied by the polyps, and narrow ones containing the solenial systems. In the calyx of the polyp, septalike structures of stony coral, or pseudosepta, are formed. The order includes a few genera, of which *Heliopora*, or the blue coral, is often found on coral reefs. SEE ALCYONARIA; REEF.

Kenji Atoda

Coenurosis

The infection by the larval stage of several species of *Taenia* from canids, such as *T. multiceps, T. serialis,* and *T. brauni*. It is rare in humans. A bladder as large as a hen's egg and containing scolices (protoscolices) forms and daughter bladders bud off either internally or externally, attached by stalks. In subcutaneous or ocular locations the bladders are unilocular, but in the central nervous system they are frequently multilocular (racemose). Symptoms are the result of toxic and allergenic metabolites and pressure effects. Surgery may be attempted; chemotherapy is still ineffective, so the prognosis is poor.

Herbivores feeding on grass contaminated by the parasite's eggs also develop coenurosis. If the central nervous system is involved (as is frequent in sheep) neurological signs appear and the animal staggers and circles. Such an animal is said to be giddy (thus the origin of the term "gid tapeworm"). Economic losses in sheep-raising regions due to coenurosis are considerable. The disease could be controlled by treatment of shepherd dogs. SEE CYCLOPHYLLIDEA.

José F. Maldonado-Moll

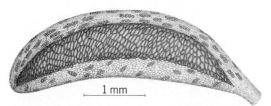

Fig. 2. Side-view reconstruction of *Biscalitheca musata*, a zygopterid sporangium with dorsiventral symmetry, two lateral multiseriate annuli, and small nests of thickened cells in the sporangial wall. (*After S. H. Mamay, Amer. J. Bot., 44(3):236, 1957*)

Coenzyme

An organic cofactor or prosthetic group (nonprotein portion of the enzyme) whose presence is required for the activity of many enzymes. In addition, many enzymes need metal ions, such as copper, manganese, and magnesium, for activation. The prosthetic groups attached to the protein of the enzyme (the apoenzyme) may be regarded as dissociable portions of conjugated proteins. The conenzymes usually contain vitamins as part of their structure. Neither the apoenzyme nor the coenzyme moieties can function singly, since dissociation of the two results in inactivation. In general, the coenzymes function as acceptors of electrons or functional groupings, such as the carboxyl groups in α-keto acids, which are removed from the substrate. Some of the well-known coenzymes are mentioned below. SEE PROTEIN; VITAMIN.

The pyridine nucleotides, nicotinamide adenine dinucleotide (NAD) and nicotinamide adenine dinucleotide phosphate (NADP), function as hydrogen acceptors for a large number of dehydrogenase enzymes. They are also called diphosphopyridine nucleotide (DPN), or coenzyme I, and triphosphopyridine nucleotide (TPN), or coenzyme II, respectively. Thiamine pyrophosphate (TPP), or cocarboxylase, the prosthetic group of carboxylase, is responsible for the decarboxylation of α-keto acids in the body. Flavin mononucleotide (FMN) and flavinadenine dinucleotide (FAD) act as hydrogen acceptors in aerobic dehydrogenases. Pyridoxal phosphate, also known as codecarboxylase and cotransaminase, serves as coenzyme for various racemases in reactions involving the decarboxylation of α-amino acids, and also functions in transamination.

Iron protoporphyrin (hemin) acts as coenzymes of catalase, peroxidases, cytochromes, and hemoglobin. The uridine phosphates are involved in carbohydrate metabolism. Uridinediphosphate (UDP) is the chief transferring coenzyme for carbohydrates. UDP-glucose is responsible for the interconversion of glucose and galactose, is used in the synthesis of glycogen, and can be oxidized to UDP-glucuronate for the synthesis of glucuronides and chrondroitin or converted to UDP-galactose for the synthesis of lactose and chrondroitin.

Adenosinetriphosphate (ATP), adenosinediphosphate (ADP), and adenosinemonophosphate (AMP), the phosphorylated derivatives of adenylic acid, can be active in phosphate transport or transphosphorylation or both. Lipoic acid functions in acyl-generating and in acyl-transfer reactions. In combination with TPP and NAD, it functions in oxidative decarboxylation of α-keto acids. Coenzyme A (CoA), a coenzyme in certain condensing enzymes, acts in acetyl or other acyl group transfer and in fatty acid synthesis and oxidation. Folic acid coenzymes are involved in the metabolism of one carbon unit. Biotin is the coenzyme in a number of carboxylation reactions, where it functions as the actual carrier of carbon dioxide. S-Adenosylmethionine, active methionine, functions in transmethylation reactions. SEE ADENOSINEDIPHOSPHATE (ADP); ADENOSINETRIPHOSPHATE (ATP); BIOCHEMISTRY; BIOLOGICAL OXIDATION; BIOTIN; CARBOHYDRATE METABOLISM; CYTOCHROME; ENZYME; HEMOGLOBIN; NICOTINAMIDE ADENINE DINUCLEOTIDE (NAD); NICOTINAMIDE ADENINE DINUCLEOTIDE PHOSPHATE (NADP); URIDINE DIPHOSPHOGLUCOSE (UDPG).

Mary B. McCann

Coesite

A naturally occurring mineral of wide interest, the high-pressure polymorph of SiO_2. Coesite was first discovered and identified in various meteorite impact craters and in some tektites. Because it requires a unique physical condition, extremely high pressure, for its formation, its occurrence is diagnostic of a special natural phenomenon, in this case, the hypervelocity impact of a meteorite.

Coesite occurs in grains that are usually less than 5 micrometers in size and are generally present in small amounts. The properties of the mineral are known mainly from studies of synthesized crystals. It is colorless with vitreous luster and has no cleavage. It has a specific gravity of 2.915 ± 0.015 and a hardness of about 8 on Mohs scale. It is biaxial positive with $2V$ about $64°$. Its indices of refraction are α 1.5940, β 1.5955, and γ 1.5970 ± 0.0005. Its dispersion is horizontal, with r less than v (weak). The optical orientation is $X = b$, $Z \angle C = 4–6°$, and $\beta = 120°$. Synthetic crystals occur as euhedral to subhedral hexagonal platelets, and laths with positive elongation. Simple contact twins occur with (021) as twin and composition plane.

Natural coesite can be identified from thin sections of rocks by its peculiar habit and high index of refraction. However, its positive identification must rest on the x-ray powder diffraction technique.

Coesite has as yet no evident commercial use and therefore has no obvious economic value. As a stepping-stone in scientific research, it serves in at least two ways: where it occurs naturally, coesite is diagnostic of a past history of high pressure; the natural occurrence of coesite suggests that shock as a process can transform a low-density ordinary substance to one of high density and unique properties.

Coesite has been found in materials ejected from craters formed by the explosion of 500,000 tons (450,000 metric tons) of TNT. The study of coesite and craters is important to the understanding of the impact craters on the Earth as well as those on the Moon.

Edward C. T. Chao

Bibliography. F. R. Boyd and J. L. England, The quartz-coesite transition, *J. Geophys. Res.*, 65(2): 749–756, 1960; E. C. T. Chao, E. M. Shoemaker, and B. M. Madsen, The first natural occurrence of coesite from Meteor Crater, Arizona, *Science*, 132(3421):220–222, 1960; L. S. Walter, Coesite discovered in tektites, *Science*, 147:1029–1032, 1965; T. Zoltai and M. J. Buerger, The crystal structure of coesite, the dense, high pressure form of silica, *Z. Krist.*, 111:129–141, 1959.

Coffee

An evergreen shrub or small tree of the genus *Coffea* (Rubiaceae), a native of northeastern Africa and adjacent southwestern Asia. Commercially, coffee is the most important caffeine beverage plant in the world. There are about 50 species, but only a few of commercial importance. By far the majority of the world's supply of coffee is obtained from the Arabian species, *C. arabica* (**Fig. 1**); another important species is *C. robusta* (*C. canephora*). Of minor importance are Liberian coffee (*C. liberica*) and Zanzibar coffee (*C. zanguebariae*). *Coffea arabica* grows from

Fig. 1. Coffee (*Coffea arabica*) branch.

15 to 30 ft (5 to 10 m) in height but is usually trimmed to about 6 to 10 ft (2 to 3 m) in cultivation. Until about 1850 most coffee grown commercially was *C. arabica,* but the popularity of *C. robusta* increased when it was found to be more resistant to disease. SEE CAFFEINE; RUBIALES.

The small fruit, technically a drupe, is usually red when ripe, with two seeds, the coffee beans. Varieties of *C. arabica* are yellow and white; *C. zanguebariae* has fruits that turn black when fully ripe. The number of genetic varieties is great; environment and methods of cultivation add to the diversity. *Coffea arabica* is usually grown in Latin America, *C. robusta* in Africa, and both are grown in India, Indonesia, and some other countries.

The dozens of different varieties have been given commercial names. Coffee grown in Brazil, for example, is called Brazils, with separate names indicating region of origin, such as Santos, Rio, Victoria, and Parana. Coffee grown elsewhere in Latin America is termed Milds; that originating in Guatemala, Prime Washed; in Costa Rica, Hard Bean; in Colombia, MAMS (from Medellin, Armenia, Manizales, Sevilla, points of origin). Coffee produced in Yemen is called Mocha; grown in Indonesia, it is called Java; and so on.

Cultivation and harvesting. Coffee is strictly a tropical crop, requiring a moderately cool climate that is moist but not wet. A temperature averaging about 70°F (22°C), with 40–70 in. (100–180 mm) of rainfall annually, is most suitable. *Coffea arabica* in Latin America is mostly cultivated at altitudes from 2000 to 6000 ft (600 to 1800 m) above sea level, while *C. robusta* is grown from sea level to 2000 ft (600 m). The plants are killed by prolonged exposure to freezing temperatures. Best crops are normally obtained where there is a dry season of 2 to 3 months each year.

Plantations are usually developed on cleared forest land, on well-drained soil rich in potash; enrichment by organic debris is preferable to commercial fertilizers. Some of the most successful plantations have been located on slopes of extinct volcanoes. The plants may be grown from cuttings, but are grown mostly from seeds in the open field or in seed beds. The plants are set about 15 to 20 ft (5 to 7 m) apart, formerly mostly in light shade but now usually without shade.

The seedlings or cuttings are normally planted at the beginning of the rainy season and begin to produce in 3 or 4 years. Optimum production continues for 15 or 20 years, but berries are produced for many more years.

The fruits are still harvested by hand, no satisfactory mechanical methods having been developed. In Brazil harvest is mostly from May to September, while in Java there may be three crops per year, so that there is virtually a continuous harvest. In Ethiopia there are two harvests annually, from October to March and during May and June.

Brazil is by far the world's leading coffee-producing country, with Colombia second in rank.

Earl L. Core

Diseases. There are more than 350 known diseases of coffee, and new ones are still being discovered on *C. arabica, C. robusta, C. liberica,* and *C. excelsa* in widely separated tropical lands. Probably no crop grows under more varied conditions or is subjected to so many unknown diseases.

Leaf rust (caused by *Hemileia vastatrix*; **Fig. 2**), one of the most important diseases of the crop worldwide, was described as early as 1869 in Ceylon, where there were once flourishing coffee plantations. Rust became so damaging that, in the absence of control measures, coffee production had to be abandoned. In other Eastern and African coffee-producing countries, rust restricted cultivation of coffee to particular altitudes and districts, rendering cultivation more expensive and difficult. As a consequence, the center of the world's potential coffee production shifted to Brazil and other South and Central American countries, where leaf rust was unknown until 1970. Control of the disease can be achieved by fungicidal sprays, but this inevitably increases production costs. Development of resistant varieties is complicated because of the existence of many different races of the coffee rust fungus, but some new introductions show promise.

Other diseases, which are less widespread than leaf rust, have also been extremely damaging. In East Africa, bark disease (caused by *Fusarium stilboides*) destroyed the once-flourishing coffee industry of Ma-

Fig. 2. Coffee leaf rust (*Hemileia vastatrix*). Affected leaves bear conspicuous, orange, spore-bearing pustules and are shed prematurely.

Fig. 3. Coffee berry disease (caused by *Colletotrichum coffeanum*). Berries, which may be infected at any stage of development, are completely destroyed.

lawi and, in the late 1960s, berry disease (caused by *Colletotrichum coffeanum*; **Fig. 3**) devastated the crop in Kenya and threatened the viability of the coffee production. Fungicidal sprays now give control. Berry disease spread from Kenya to all other African coffee-producing countries but is unknown outside the African continent.

American leaf spot (caused by *Mycena citricolor*) is very serious but is found only in the Americas, whereas Cercospora leaf spot and the leaf rot caused by *Pellicularia koleroga* occur in both hemispheres. The last two diseases are controlled by spraying.

Fruit, stem, and branch infections are injurious. Wilts and trunk cankers, caused by *Gibberella*, *Nectria*, and *Ceratocystis*, are troublesome. Probably six *Rosellinias*, a few *Fusaria*, two *Armillarias*, and fungi like *Fomes*, *Ganoderma*, and *Polyporus* attack roots. Root and trunk infections are reduced by good cultural practices. Coffee is also affected by some half dozen viruses, a few bacteria, and numerous parasitic mistletoes. SEE BACTERIA; FUNGI; PLANT PATHOLOGY; PLANT VIRUSES AND VIROIDS.

Ellis Griffiths

Processing. Coffee beverage is prepared by extraction of roasted and ground coffee with hot water. This can be done in the home by the consumer, or by a coffee manufacturer who prepares coffee extract in a concentrated form and removes the water by spray-drying or freeze-drying.

Blending. Green beans are packed and transported in hemp bags. The commercial unit is about 132 lb (60 kg). There has been a significant increase in the usage of containers to transport green coffee. The usual size of containers holds 250 bags. The containers move through the transportation system as a unit, with significant savings in shipping cost compared to loose bulk method. On arrival at the roasting plant, beans from many sources are roasted individually to their optimum flavor development before blending. The beans can also be blended before roasting, but then each source does not reach its best flavor development.

Cleaning. The beans are cleaned both before and after roasting. Pneumatic separators or mechanical devices remove dust, lint, strings, hulls, and other light material from the green beans. Batches of 12 bags, or about 1500 lb (680 kg), are fed through the cleaner, and the beans are completely mixed. A modern unit will handle 5000 lb (2270 kg) an hour.

Roasting. Roasting is the most important operation in coffee processing because it develops the flavor. Modern automatic control makes possible precise transfer of energy from a source to the beans. To roast 1200 lb (545 kg) of beans, approximately 430,000 Btu must be delivered in 11 min.

Systems now employ externally heated gases of combustion, mainly nitrogen, carbon dioxide, and water vapor. In batch roasters, energy at relatively low temperature, 900–1000°F (480–540°C), develops fine flavor and a uniform brown color in about 12 min. In continuous roasters, energy is transferred even more efficiently (**Fig. 4**). Temperature is reduced to about 500°F (260°C) and roasting time to about 5–8 min. Most roasting systems recycle a high percentage of the heated gases for roasting as a means of saving energy.

Light roasts are attained at bean temperatures of 380–390°F (193–199°C); medium roasts, 400–410°F (204–210°C); and dark roasts, 425–430°F (218–221°C). At 450°F (232°C) excessive decomposition of fiber and oils occurs. Continuous roasting systems that reduce roasting times to about 2–3 min by increasing the air-to-bean ratio at these temperatures have been developed.

At the end of batch roasts, water at the rate of 1 gal per 100 lb (8.3 liters per 100 kg) is sprayed onto the beans to check or quench the roast, after which the beans are dumped and rapidly air-cooled.

Continuous roasters that have a wide range of control for rate of throughput, roasting time, and temperatures are also in use. Roasters require afterburners in order to meet environmental requirements.

Stoning. The second cleaning operation is known as stoning. Roasted beans have only half the density of green beans (20 versus 40 lb/ft^3 or 320 versus 640 kg/m^3) and are lifted by air currents away from heavy material, stones, and other debris. The beans are deposited in special bins which prevent segregation as the beans pass to the grinder.

Grinding. There are two types of grinding mills: the plate or compactor, and the roll or granulator. The plate mill tends to crush particles and yields a higher proportion of fine material than does the granulator with its cutting action. Most grinds contain particles ranging in size between 0.023 and 0.055 in. (0.58 and 1.40 mm). Most commercial manufacturers use the granulator mill with cutting or cracking action.

Packaging. Roasted and ground coffee packed in cans with an initial vacuum of 28 in. Hg abs (94.8 kilopascals absolute pressure) or better (relative to a 30-in. barometer or 101.6 kilopascals) maintains

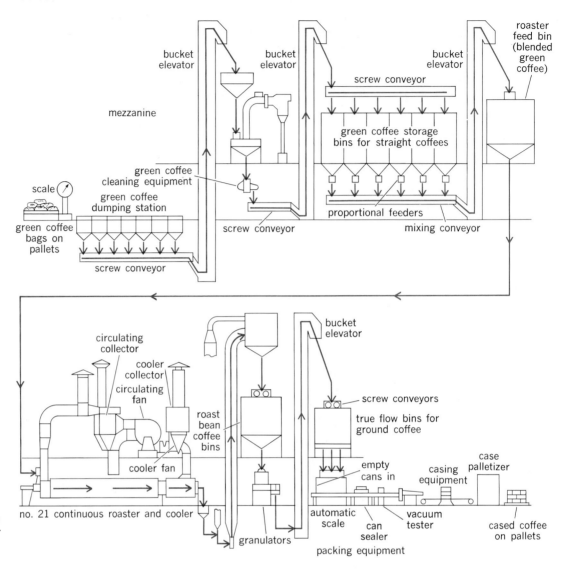

Fig. 4. Modern continuous roasting plant for processing coffee. (*Jabez Burns and Sons, Inc.*)

freshness for at least 2 years. Flexible containers made of paper laminated with plastic film or aluminum foil are not vacuumized and result in shorter shelf life for the coffee.

Many parts of Europe market coffee in a vacuum-packed bag that is made of heavy foil and plastic laminate and can provide a stable fresh product comparable to coffee in a vacuum-packed can. This type of package is also marketed in the United States. There have been significant increases in the introduction of coffees featured as premium as well as roasted whole-bean coffee, which require the use of a grinder in the home.

Instant coffee. Soluble or instant coffee manufacturing utilizes about 20% of the green beans imported. Soluble coffee accounts for about 30% of all coffee consumed in the United States, and about 40% of this is freeze-dried.

Preliminary processing of green beans and the roasting operation are the same as for regular coffee. The roasted beans are equilibrated to a higher moisture content of 7–10% and are ground very coarsely to minimize fines which can interfere with the flow of water in the extraction system.

Coffee extract is made by passing water at temperatures up to 350°F (177°C) through ground coffee in a series of extracting columns, with at least one col-

umn under higher-than-atmospheric pressure. This yields an extract containing 20–30% solids. After filtration, the extract can be concentrated in conventional evaporating equipment, followed by spray-drying to remove most of the water. The process of concentration in an evaporator can also be modified to capture small amounts of flavorful aromatics that come with the water and add this material back to the concentrated extract prior to drying. The spray-dried soluble coffee can be processed further by agglomerating the particles with steam to form a granular-appearing product. The extract also can be frozen and then freeze-dried in a low-temperature vacuum system by sublimation from the ice crystal structure to remove most of the water.

Most manufacturers of soluble coffees return aroma lost during processing to the package to provide the aroma of a roasted and ground coffee. This requires protection in packaging, and most soluble coffees are packed in a low-oxygen atmosphere by using an inert gas such as carbon dioxide or nitrogen. In 1976 there was a significant increase in the use of extenders in coffee such as roasted chicory and roasted wheat. This was prompted by the frost in Brazil in 1975 and the resulting green coffee shortage with increased price at retail. As the green coffee supply was replenished and prices lowered in the late 1970s and early

1980s, many of these extended coffee products were discontinued.

Decaffeinated coffee. Decaffeinated coffee, both regular and instant, is usually prepared from green beans that have been softened by steam, extracted with low-boiling chlorinated solvents to remove the caffeine, then steamed again to remove residual solvent, and dried.

Decaffeinated products prepared with nonchlorinated solvents have appeared on the market. Supercritical liquid carbon dioxide has been identified as a specific caffeine solvent, and a high-pressure system has been developed. The green decaffeinated coffee then can be processed in the manners described above for a regular or a soluble coffee. A number of brands featured as naturally decaffeinated have appeared on the United States market. This refers to the use of a caffeine extracting medium that can be found occurring in nature.

William P. Clinton

Bibliography. R. J. Clarke and R. Macrae, *Coffee Technology*, vol. 2, 1987; M. N. Clifford and K. C. Wilson (eds.), *Coffee: Botany, Biochemistry and Production of Beans and Beverage*, 1985; I. D. Firman and J. M. Waller, *Coffee Berry Disease and Other Colletotrichum Diseases of Coffee*, C.M.I. Phytopathol. Pap. 20, 1977; A. E. Haarer, *Coffee Growing*, 1963; International Coffee Organization, *Coffee Drinking in the United States*, 1982; F. J. Nutman and F. M. Roberts, Coffee leaf rust, *PANS*, 16:606–624, 1970; M. Sivetz, *Coffee Publication*, 1977; M. Sivetz and N. Derosier, *Coffee Technology*, 1979; J. M. Waller, Coffee rust in Latin America, *PANS*, 18:402–408, 1972.

Cofferdam

A temporary, wall-like structure to permit dewatering an area and constructing foundations, bridge piers, dams, dry docks, and like structures in the open air. A dewatered area can be completely surrounded by a cofferdam structure or by a combination of natural earth slopes and cofferdam structure. The type of construction is dependent upon the depth, soil conditions, fluctuations in the water level, availability of materials, working conditions desired inside the cofferdam, and whether the construction is located on land or in water. An important consideration in the design of cofferdams is the hydraulic analysis of seepage conditions, and erosion of the bottom in streams or rivers.

Where the cofferdam structure can be built on a layer of impervious soil (which prevents the passage of water), the area within the cofferdam can be completely sealed off. Where the soils are pervious, the flow of water into the cofferdam cannot be completely stopped economically, and the water must be pumped out periodically and sometimes continuously.

Types. The **illustration** shows the types of cofferdam construction: *a*, *b*, and *e* are used in rectangular form to enclose small areas, such as individual bridge piers where cross bracing, consisting of steel, wood, or concrete beams and struts, to the opposite wall is practical; the other types are for large areas required for dams, river locks, or large buildings, where cross bracing is impractical or undesirable. Cross bracing can be replaced with steel or concrete rings acting in compression when a circular form is used for *a* and *e*.

Sheeted cofferdams are made of sheets of steel, timber, or concrete, driven vertically or placed horizontally. The sheeting may be made reasonably watertight by using interlocking, overlapping, or grouted joints. In gravel or boulders, where sheeting cannot be driven to the required depth, steel H piles are driven vertically into the ground and timber sheeting is placed horizontally between or against the vertical piles (illus. *d*) as the excavation progresses. Watertight cofferdams are designed for a full hydrostatic head of water, plus the pressure of earth when they are on land. Nonwatertight cofferdams have a reduced hydrostatic pressure because of seepage.

Another type of cofferdam suitable for use in gravel or boulders is constructed of cast-in-place concrete. Illustration *b* shows this type in a cylindrical form.

Most sheet-pile cofferdams, with the exception of

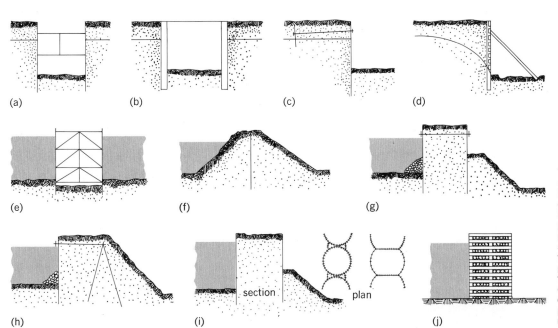

Types of cofferdams. For use on land: (*a*) cross-braced sheet piles; (*b*) cast-in-place concrete cylinder; (*c*) anchored sheet piles; (*d*) braced vertical piles with horizontal sheeting. For use in water: (*e*) cross-braced sheet piles; (*f*) earth dam; (*g*) tied sheet piles; (*h*) anchored sheet piles with earth berm; (*i*) steel sheet-pile cellular cofferdam; (*j*) rock-filled crib.

the cellular type (illus. *i*), require a soil into which the sheets may be embedded to give the walls stability. The cellular cofferdam, consisting of interlocking steel sheet piling driven as a series of interconnecting cells, is independently stable when filled and may be constructed on either rock or soil. The cells of the circular type may be filled individually. The cells of the straight-wall diaphragm type must be filled systematically to avoid a differential height of fill that would deform the diaphragms.

The filled crib (illus. *j*) is an open, boxlike structure of timber, precast concrete, or fabricated steel sections, filled with a suitable material. A rock-filled crib is suitable for construction on a rock bottom in swiftly flowing waters. Sheeting or an impervious core provides the necessary watertightness. Earth dams (illus. *f*) can be constructed of soil or rock, but if pervious, a cutoff is required to control seepage.

Another type for a large open building area has been a slurry wall anchored back to the earth (illus. *c*), constructed by pouring a concrete mix into a trench dug along the proposed wall before excavating the cofferdam area.

Uses. When cofferdams are required for year-round construction in waters subject to flooding, they are made sufficiently high to exclude the flood waters. If overflow is to be permitted, then the cofferdam fill and the bottom of the dewatered area must be protected against scour (eroding action of the water). Scour can also be prevented along the stream side of cofferdams in swiftly flowing waters by protecting the stream bed. The berm (embankment), shown in illus. *g* and *h* on the dewatered side, furnishes stability and reduces the flow of water under the cofferdam.

The cofferdam shown in illus. *b* has been used to form a mine shaft 220 ft (66 m) deep. Type *d* was used extensively in subway construction in New York City.

A nautical application of the term cofferdam is a watertight structure used for making repairs below the waterline of a vessel. The name also is applied to void tanks which protect the buoyancy of a vessel. SEE BRIDGE; DAM; FOUNDATIONS.

Edward J. Quirin

Bibliography. M. Reimbert and A. Reimbert, *Retaining Walls, Anchorages, and Sheet Piling*, pt. 1, 1974; L. White and E. A. Prentis (eds.), *Cofferdams*, 2d rev. ed., 1956.

Cogeneration

The sequential production of electricity and thermal energy in the form of heat or steam, or useful mechanical work, such as shaft power, from the same fuel source.

Basic concept. Cogeneration, a direct approach to conserving fuel by improved energy utilization and higher overall system efficiency, is a proven technology offering energy savings. Cogeneration facilities have been installed and are operating at chemical, petrochemical, refinery, mining and metals, paper and pulp, food-processing, district heating and cooling, and utility complexes. These projects can range in electrical power output from several hundred kilowatts to 800,000 kW.

Cogeneration offers economic, environmental, and social benefits. These may be derived from the improved power-cycle efficiency and associated reduc-

tion in fuel consumption. In addition to the obvious conservation of energy, potential benefits include (1) higher efficiency by utilizing the same fuel to provide electricity and heat, yielding reduced fuel consumption, reduced fuel costs, and most efficient use of capital investment; (2) an increase in the amount of useful energy produced through the recovery of otherwise wasted heat; (3) fewer effects on the environment as a result of efficient fuel use; (4) through onsite generation, possible reduction of energy losses due to transmission and distribution systems; (5) attainment of plant operation in 1–4 years; (6) economic benefits in the form of tax credits and depreciation.

As an energy conservation technology, cogeneration uses oil, gas, and other alternate fuel resources more efficiently than either typical industrial processes or conventional power plants. Experience has demonstrated that cogeneration facilities are capable of operating in excess of 90% availability with system efficiencies up to 80%. Increased efficiency reduces pollution, fuel consumption, and reliance on imported energy sources.

An electric utility requiring additional capacity may look to cogeneration to help defer construction of a new power station. Cogeneration can help a utility to reduce its fuel consumption per kilowatt output, and to count on the availability of cogeneration power during peak-load periods.

Engineering power cycles and applications. Cogeneration projects are typically represented by two basic types of power cycles, topping or bottoming. The topping cycle has the widest industrial application.

The topping cycle (**Fig. 1**) utilizes the primary energy source to generate electrical or mechanical power. Then the rejected heat, in the form of useful thermal energy, is supplied to the process. The cycle consists of a combustion turbine-generator, with the turbine exhaust gases directed into a waste-heat-recovery boiler that converts the exhaust gas heat into steam which drives a steam turbine, extracting steam to the process while driving an electric generator. This cycle is commonly referred to as a combined cycle arrangement. Combustion turbine-generators, steam turbine-generator sets, and reciprocating internal-combustion-engine generators are representative of the major equipment components utilized in a topping cycle. SEE GENERATOR; STEAM TURBINE; TURBINE.

A bottoming cycle (**Fig. 2**) has the primary energy source applied to a useful heating process. The reject heat from the process is then used to generate electrical power. The typical bottoming cycle directs waste heat from a process to a waste-heat recovery boiler that converts this thermal energy to steam which is supplied to a steam turbine, extracting steam to the process and also generating electrical power.

Building and district heating and cooling. Cogeneration for building and district space heating and cooling purposes consists of producing electricity and sequentially utilizing useful energy in the form of steam, hot water, or direct exhaust gases (**Fig. 3**). The two most common heating, ventilating, and air-conditioning cycles are the vapor compression cycle and the absorption cycle. SEE AIR CONDITIONING; CENTRAL HEATING AND COOLING; COMFORT HEATING; DISTRICT HEATING.

Vapor compression cycle. The vapor compression cycle consists of a compressor, condenser, expansion valve, and evaporator. The compressor driver can be an electric motor, steam turbine, or combustion turbine. A single fluid used as the refrigerant vaporizes in the evaporator as a result of heat transfer from the refrigerated space, is compressed by applying work to the compressor, and condenses in the condenser as a result of heat transfer to cooling water or to the surroundings. The refrigerant leaves the condenser as a high-pressure liquid. The pressure of the liquid is decreased as it flows through the expansion valve and some of the liquid flashes into vapor.

Centrifugal-type chillers are the most commonly used chillers for office-building air-conditioning systems; they are simple and clean. This arrangement is essentially the same as the cogeneration arrangement shown in Fig. 3 for the absorption chiller cycle, except that the chiller-equipment-related pumps can be driven by an electric motor, a combustion turbine, a diesel or gas engine, or a steam turbine. The cost of power or energy consumption for an electric-motor-driven centrifugal chiller is approximately 30–50% more than an absorption-type chiller system.

Absorption refrigeration cycle. The absorption refrigeration cycles for cogeneration air-conditioning applications primarily utilize water as the refrigerant and lithium bromide solution as the absorbent. Lithium bromide solution has a strong affinity for water vapor, thus giving the absorbent the ability to compress the refrigerant vapor.

The absorption cycle consists of an absorber generator (concentrator), condenser, expansion valve (pressure-reducing device), evaporator, and pump. The absorber generator functions as a compressor. The pump circulating the absorbent requires less energy than that needed to drive the compressor in the

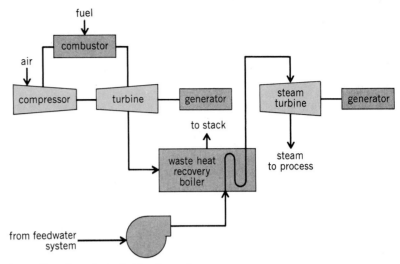

Fig. 1. Flow chart of a typical topping cycle.

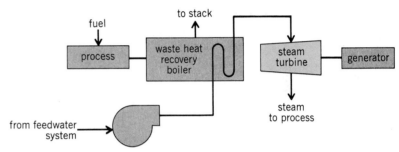

Fig. 2. Flow chart of a typical bottoming cycle.

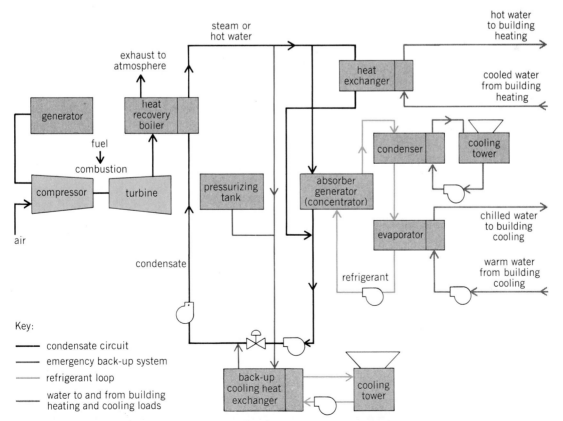

Fig. 3. Schematic of a district heating and cooling cycle utilizing cogeneration absorption.

vapor compression cycle.

This cycle combined with a combustion turbine-generator or internal combustion engine exhausting to a waste-heat-recovery steam or hot-water generator provides a common cogeneration system. SEE REFRIGERATION CYCLE.

Environmental considerations. Air-quality restrictions pose one of the greatest technical problems to cogeneration facilities. Many countries have developed air-pollution control plans to meet ambient-air-quality standards. Burner and combustion control technology can be incorporated into equipment and system design to ensure that air quality does not deteriorate below acceptable levels. Area designations dictate the approach and technology for control of plant air-emission sources. Design departments then undertake achievement of environmental compliance. The next step is to seek approval of permits and applications from the local air-pollution-control districts for authority to construct and operate cogeneration facilities. SEE AIR POLLUTION; ELECTRIC POWER GENERATION.

Charles Butler

Bibliography. C. Butler, *Cogeneration: Engineering, Design, Financing, and Regulatory Compliance*, 1984; D. Hu, *Handbook of Industrial Energy Conservation*, chap. 13, 1983; G. Polimeros, *Energy Cogeneration Handbook*, 1982.

Cognition

The internal structures and processes that are involved in the acquisition and use of knowledge, including sensation, perception, attention, learning, memory, language, thinking, and reasoning. These behaviors are studied by cognitive scientists, a group that includes, but is not limited to, researchers in the areas of cognitive psychology, philosophy, linguistics, computer science, and cognitive neuroscience. Cognitive scientists propose and test theories about the functional components of cognition based on observations of an organism's external behavior in specific situations.

The functional components of cognition take the form of internal states and processes. A model of cognition that details these states and processes is depicted in the **illustration**.

Sensory and short-term memories. When environmental input impinges on an organism's sensors, such as eyes or ears, that sensory information is used to construct an internal representation of the object or event causing the input. This representation undergoes several transformations as more knowledge is activated to make sense of the event. One can think of the organism's knowledge state as undergoing

changes in the amount of information it possesses about the external event over time, with a trade-off occurring between the amount of surface information that is maintained and the amount of meaning that is incorporated into the representation. SEE SENSATION.

Following exposure to a stimulus, a sensory representation (sometimes called an image) is constructed that records nearly all of the surface characteristics of the stimulus (for example, color, shape, location, pitch, and loudness). This information is very short-lived, lasting less than a second. During this time, it is processed for recognizable features; names for the objects comprising these features are also retrieved. Typically, no more than five to nine pieces of information can be identified and maintained. For example, in one experiment, subjects were briefly shown slides of large matrices of letters and were asked to report the contents of each matrix after the slide was extinguished. When asked to report all of the letters in the matrix, they could reliably report only about five or six letters. However, if they were asked to report the letters in some randomly chosen section of the matrix, they could reliably report all of the letters in that section. Since they did not know which area to report until after the slide was extinguished, they must have possessed a memory representation of the entire slide for a brief time following slide extinction.

Long-term memories and knowledge. Aspects of the short-term memory representation that become permanent constitute long-term memory. Typically, these aspects will be ones that are particularly meaningful to the individual or those that are repeatedly encountered (for example, frequently used telephone numbers). These long-term memories constitute an individual's knowledge about the world. They can influence perceptual and short-term memory processes to facilitate interpretation of incoming stimuli. For example, people can detect letters more quickly and reliably if they constitute familiar words rather than random strings.

Memory representations. Long-term memory representations can take the form of images, propositions, or procedures. Propositions are representations of facts. They are made up of a predicate and one or more arguments (for example, nouns). Single-argument propositions typically describe attributes of objects, such as "Apples are red," where apples serve as the argument and "are red" is the predicate. Propositions that contain more than one argument typically describe relationships, such as "John gave the apple to Mary." Here, "gave to" is the predicate, and John, Mary, and the apple are arguments. The same proposition can be expressed in a variety of surface forms. For example, "John gave the apple to Mary" and "John gave Mary the apple" both express the same fact and can be reduced to the same propo-

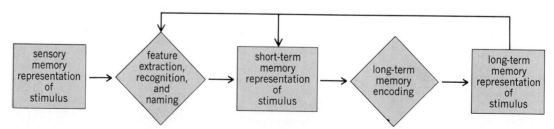

Model detailing the functional components of cognition. The boxes represent internal states, and the diamonds represent processes that produce those states.

sition (GAVE, JOHN, MARY, APPLE). Evidence suggests that memory representations of spoken and written discourse take the form of propositions since people typically remember the gist of what they read or hear and not verbatim words or sentences from the discourse.

Procedures are memory representations of action sequences, such as how to drive a standard transmission or how to make a telephone call. They are typically retrieved as a whole, fluid unit rather than as individual pieces. Often, people find it difficult to decompose procedures into their components, as in trying to explain to a novice driver how to shift gears without stalling the engine. In these cases, it is much easier to do than to tell since the memory representation for the procedure is a tightly woven interconnection of motor commands for coordinated actions that are not readily translated into verbal statements.

Semantic and episodic memories. Long-term memories are sometimes classified as being semantic or episodic in nature. Semantic memories are those that constitute one's knowledge of facts. Episodic memories are those that constitute personal memories about one's own experiences. For example, if a person remembers the date that the *Challenger* space shuttle blew up they experience an instance of semantic memory; if the person remembers where he or she was and the personal reaction to finding out about the tragedy, it is an instance of episodic memory.

Conceptual knowledge. Semantic memories constitute an individual's conceptual knowledge, that is, how the person categorizes objects and events in the world. A category or concept is a cluster of propositions (which may be augmented by images or episodes, for example) about some class of objects or events. Psychological research has indicated that human conceptual memory structures tend to be organized around a prototypical instance of the concept. Category membership is determined by the resemblance between the object to be classified and the prototype. For example, it takes less time to verify that a canary is a bird than to verify that a chicken is a bird because canaries have more features in common with the prototypical bird (that is, it flies, is small, and lives in trees) than do chickens. Similarly, it takes more time to reject the statement ''A bat is a bird'' than ''A collie is a bird'' because bats do have some features in common with birds (for example flight) while collies do not.

Thinking, reasoning, and problem-solving. Because long-term memories tend to be clustered according to shared features, thinking often takes the form of a chain of associations, with one thought (that is, proposition or image) retrieving others to which it is related. Often, problems can be solved by examining the mental structures themselves, such as when one determines how many windows are in one's house by retrieving images of each room and counting. There is a great deal of evidence that people retrieve, construct, and process images to perform certain kinds of tasks, such as mentally rotating images of a geometric figure to draw conclusions about its similarity to another figure.

Other, more difficult problems require that semantic and episodic memories be retrieved and organized into a plan for achieving some goal. When operating in an unfamiliar domain, novices have been observed to employ a strategy called means-ends analysis. In means-ends analysis, one defines the goal, defines the current situation, notes the differences between the two, and works to reduce the differences. Often this requires that subgoals be defined and achieved. For example, when playing chess, novices often attempt dramatic moves that quickly place the opponent's king in check, even though their pieces may be extremely easy to capture in those positions. This is because the goal of the game is to capture the opponent's king, and to do so requires checking the king, that is, reducing the differences between the starting configuration and the ending (checkmate) configuration. Slightly more experienced novices often divide their game-playing strategies into subgoals, such as gaining control of the center of the board and castling to protect their own king.

Experts typically do not use means-ends analysis to solve problems in their domain of expertise. Instead, they retrieve whole plans for achieving goals derived from experience and execute those plans in a straightforward fashion. This means that the path from novice to expert entails moving from backward reasoning (from goal to start state) to forward reasoning (from start state to goal). In chess, for example, experts generally execute plans for developing board position strength early on, responding to familiar board configurations with well-practiced sequences of moves. These differences among novice and expert strategies have been observed in a variety of domains under carefully controlled situations.

Reasoning strategies also tend to be affected strongly by problem format. Novices often are misled by surface characteristics of problems, classifying problems and retrieving knowledge based on surface features. In one series of studies, for example, students who had had only a few courses in physics tended to classify physics problems on the basis of the objects referred to in the problem statement (for example, inclined planes, springs) rather than on the basis of the physics principles and laws required to solve the problem, as experts were found to do. This suggests that the knowledge bases of these novices were organized around surface features found in the domain, and that knowledge of physical principles was associated with these features. Experts' knowledge bases, on the other hand, were organized around physical laws and principles, and it was the physical attributes of, and relationships among, the items referred to in the problem that served as cues for retrieving this information.

Research on reasoning in other domains also suggests that reasoning strategies are determined by the organization of the reasoner's knowledge base. It is not so much what is known as how that knowledge is organized that determines which solution strategies and conclusions a reasoner will produce. In addition, human problem-solving and reasoning performance has been successfully simulated by using computer programs that embody the principles outlined here. *See* Expert systems; Information processing (psychology); Intelligence; Memory; Problem solving (psychology); Psycholinguistics.

Denise Dellarosa Cummins

Bibliography. J. R. Anderson, *Cognitive Psychology and Its Implications*, 2d ed., 1985; M. H. Ashcraft, *Human Memory and Cognition*, 1988; L. E. Bourne, *Cognitive Processes*, 1986; R. S. Cohen, *The Development of Spatial Cognition*, 1985; M. Denis and J. Engelkamp (eds.), *Cognitive and Neurological Approaches to Mental Imagery*, 1988; S. L. Friedman et al. (eds.), *The Brain, Cognition, and Education*, 1986; J. Greene, *Language Understanding: A Cognitive Approach*, 1986; Y. C. Lee, *Evolution, Learning and Cognition*, 1988; N. Stillings et al., *Cognitive Science: An Introduction*, 1987.

Coherence

The attribute of two or more waves, or parts of a wave, whose relative phase is nearly constant during the resolving time of the observer. The concept has been developed most extensively in optics, but is applicable to all wave phenomena.

Coherence of two beams. Consider two waves, with the same mean angular frequency ω, given by Eqs. (1) and (2).

$$\Psi_A(x,t) = A \exp\{i\,[k(\omega)x - \omega t - \delta_A(t)]\} \quad (1)$$

$$\Psi_B(x,t) = B \exp\{i\,[k(\omega)x - \omega t - \delta_B(t)]\} \quad (2)$$

These expressions as they stand could describe de Broglie waves in quantum mechanics. For real waves, such as components of the electric field in light or radio beams, or the pressure oscillations in sound, it is necessary to retain only the real parts of these and subsequent expressions. Assume that the frequency distribution is narrow, in the sense that a Fourier analysis of Eqs. (1) and (2) gives appreciable contributions only for angular frequencies close to ω. This assumption means that, on the average, $\delta_A(t)$ and $\delta_B(t)$ do not change much per period. For any actual wave, however, they must undergo some change; only waves that have existed forever and that fill all of space can have absolutely fixed frequency and phase. *See* ELECTROMAGNETIC RADIATION; QUANTUM MECHANICS; SOUND.

Suppose that the waves are detected by an apparatus with resolving time T; that is, T is the shortest interval between two events for which the events do not seem to be simultaneous. For the human eye and ear, T is about 0.1 s, while a fast electronic device might have a T of 10^{-10} s. If the relative phase $\delta(t)$, given by Eq. (3),

$$\delta(t) = \delta_B(t) - \delta_A(t) \quad (3)$$

does not, on the average, change noticeably during T, then the waves are coherent. If during T there are sufficient random fluctuations for all values of $\delta(t)$,

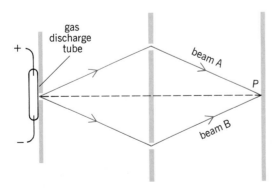

Fig. 1. Young's experiment. If beams *A* and *B* come from the same point source and traverse the same distance, they are coherent around *P* and produce there an interference pattern that has high visibility.

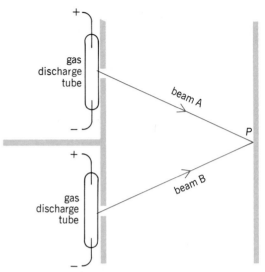

Fig 2. Independent sources that change their phases frequently and randomly during the resolving time *T* are incoherent, and their superposition does not produce a visible interference pattern.

modulus 2π, to be equally probable, then the waves are incoherent. If during T the fluctuations in $\delta(t)$ are noticeable, but not enough to make the waves completely incoherent, then the waves are partially coherent. These distinctions are not useful unless T is specified. Two waves for which the relative phase changes appreciably in 0.01 s would seem incoherent to unaided human perception, but would seem highly coherent to a fast electronic device.

The degree of coherence is related to the interference patterns that can be observed when the two beams are combined. The following variations of Young's two-slit interference experiment illustrate several possibilities. *See* INTERFERENCE OF WAVES.

Young's two-slit experiment. **Figure 1** shows the usual Young's experiment. Near P, for linear wave phenomena, the resultant wave $\Psi(x,t)$ is the sum of $\Psi_A(x,t)$ and $\Psi_B(x,t)$. The observable intensity is proportional to $|\Psi|^2$, the square of the magnitude of Ψ. This observable intensity is the energy density or the mean photon density for electromagnetic waves, the energy density for acoustic waves, and the mean particle density for the wave functions of quantum mechanics. With the wave forms of Eqs. (1) and (2), for real A and B, it is given by Eq. (4).

$$|\Psi|^2 = |\Psi_A + \Psi_B|^2$$
$$= A^2 + B^2 + 2AB \cos\delta(t) \quad (4)$$

The first term gives the intensity of beam A alone, the second gives the intensity of B alone, and the third term depends on the relative phase, given by Eq. (3).

The mean life of a typical excited atomic state is about 10^{-8} s. Collisions and thermal motion reduce the effective time of undisturbed emission to around 10^{-10} or 10^{-11} s in standard discharge tubes. The phase $\delta_A(t)$ of beam A therefore dances about erratically, with substantial changes occurring perhaps 10^{11} times per second. However beam B comes from the same atoms and travels nearly the same distance, and the changes in $\delta_B(t)$ are the same as those in $\delta_A(t)$. The relative phase is always zero at the exact midpoint P and takes on other time-independent values in the neighborhood of P. For each point in that neighborhood, $|\Psi|^2$ has a constant value that satisfies inequality (5).

$$|A - B|^2 \leq |\Psi|^2 \leq |A + B|^2 \qquad (5)$$

A clear interference pattern is observed even with a large-T detector such as a photographic plate.

Independent sources. **Figure 2** shows an arrangement that uses two independent sources to produce the two beams. If the two sources are standard discharge tubes, $\delta_A(t)$ and $\delta_B(t)$ independently change erratically around 10^{11} times per second. Equation (4) is still valid, but the observed quantity is $|\Psi|^2$ averaged over the resolving time T, denoted by $\langle|\Psi|^2\rangle_T$. When the average of Eq. (4) is formed, the last term on the right contributes nothing because $\cos\delta(t)$ randomly takes on values between $+1$ and -1 and thus obeys Eq. (6).

$$\langle\cos\delta(t)\rangle_T = 0 \qquad (6)$$

The two beams are incoherent, the observed intensity is simply the sum of the separate intensities, and the superposition does not produce a visible interference pattern. In general, if incoherent waves Ψ_A, Ψ_B, Ψ_C, . . . , are combined, much the same argument applies because the average over T of each cross term is zero, so that the observed intensity is given by Eq. (7).

$$\langle|\Psi|^2\rangle_T = \langle|\Psi_A|^2\rangle_T + \langle|\Psi_B|^2\rangle_T$$
$$+ \langle|\Psi_C|^2\rangle_T + \cdots \qquad (7)$$

The argument that results in Eq. (6) depends on there being random independent changes in the separate phases during times much shorter than T. If the discharge tubes in Fig. 2 are replaced by very well stabilized lasers, or by loudspeakers or radio transmitters, then the separate phases $\delta_A(t)$ and $\delta_B(t)$ can be made nearly constant during 0.1 s. The relative phase $\delta(t)$ is then also nearly constant, an interference pattern is visible, and the two waves can be called coherent. *SEE LASER.*

Effect of path length difference. **Figure 3**, like Fig. 1, shows two beams that are emitted by the same point source and are therefore coherent at their origin. Here, however, the distance from the source to the point P' of wave addition is larger along beam B than along beam A by an amount l. In Fig. 3, the path length difference is due to the observation point P' not being near the central axis, but it could equally well be caused by the insertion of different optical

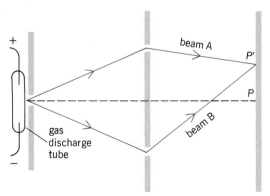

Fig. 3. If the difference between the distances traversed by beams *A* and *B* is of the order of the coherence length, then the beams are partially coherent at *P'*.

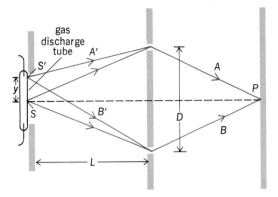

Fig. 4. If the lateral extent *y* of the source is of the order of *Lλ/D*, then the beams are partially coherent at *P*.

devices in the two beams.

The waves that arrive at P' via path B at any instant must have left the source earlier by a time $\tau = l/c$, where c is the wave speed, than those that arrive via path A. Suppose again that the source is a standard discharge tube, so that the phase changes randomly about 10^{11} times per second. For $\tau \ll 10^{-11}$ s, the coherence between A and B at P' is not spoiled by the path length difference, and there is a visible interference pattern around P'. For $\tau \gg 10^{-11}$ s, the light that arrives via A left the source so much later than that which arrives via B at the same instant that the phase changed many times in the interim. The two beams are then incoherent, and their superposition does not produce a visible interference pattern around P'. For $\tau \simeq 10^{-11}$ s, perfect coherence is spoiled but there is less than complete incoherence; there is partial coherence and an interference pattern that is visible but not very sharp. The time difference that gives partial coherence is the coherence time Δt. The coherence time multiplied by the speed c is the coherence length Δl. In this example, Δl is given by Eq. (8).

$$\Delta l \simeq 10^{-11}\ \text{s} \times 3.0 \times 10^{10}\ \text{cm/s} = 0.3\ \text{cm} \qquad (8)$$

Effect of extended source. **Figure 4** differs from Fig. 1 in that it shows a large hole in the collimator in front of the discharge tube, so that there is an extended source rather than a point source. Consider two regions of the source, S near the center and S' near one end, separated by a distance y. During times longer than the coherence time, the difference between the phases of the radiation from the two regions changes randomly. Suppose that, at one moment, the radiation from S happens to have a large amplitude while that from S' happens to be negligible. The waves at the two slits are then in phase because S is equidistant from them, and interference at P gives a maximum. Suppose that a few coherence times later the radiation from S happens to be negligible while that from S' happens to have a large amplitude. If the path length along A' differs from that along B' by half of the wavelength λ, then the waves at the two slits are out of phase. Now interference at P gives a minimum, and the pattern has shifted in a few coherence times. If the resolving time T is much larger than the coherence time Δt, the consequence is that beams A and B are not coherent at P.

The condition that there be a path length difference of $\lambda/2$ between B' and A' is given by Eq. (9),

$$\sqrt{L^2 + (y + D/2)^2}$$
$$- \sqrt{L^2 + (y - D/2)^2} = \lambda/2 \qquad (9)$$

and for D and y both much smaller than L, expansion of the square roots gives Eq. (10).

$$2Dy = L\lambda \qquad (10)$$

This argument is rather rough, but it is clear that a source that satisfies Eq. (11)

$$y \simeq \frac{L\lambda}{D} \qquad (11)$$

is too large to give coherence and too small to give complete incoherence; it gives partial coherence.

Coherence of a single beam. Coherence is also used to describe relations between phases within the same beam. Suppose that a wave represented by Eq. (1) is passing a fixed observer characterized by a resolving time T. The phase δ_A may fluctuate, perhaps because the source of the wave contains many independent radiators. The coherence time Δt_W of the wave is defined to be the average time required for $\delta_A(t)$ to fluctuate appreciably at the position of the observer. If Δt_W is much greater than T, the wave is coherent; if Δt_W is of the order of T, the wave is partially coherent; and if Δt_W is much less than T, the wave is incoherent. These concepts are very close to those developed above. The two beams in Fig. 2 are incoherent with respect to each other if one or both have $\Delta t_W \ll T$, but are coherent with respect to each other if both have $\Delta t_W \gg T$.

The degree of coherence of a beam is of course determined by its source. Discharge tubes that emit beams with Δt_W small compared to any usual resolving time are therefore called incoherent sources, while well-stabilized lasers that give long coherence times are called coherent sources.

Consider two observers that measure the phase of a single wave at the same time. If the observers are close to each other, they will usually measure the same phase. If they are far from each other, the phase difference between them may be entirely random. The observer separation that shows the onset of randomness is defined to be the coherence length of the wave, Δx_W. If this separation is in the direction of propagation, then Δx_W is given by Eq (12),

$$\Delta x_W = c\Delta t_W \qquad (12)$$

where c is the speed of the wave. Of course, the definition of Δx_W is also applicable for observer separations perpendicular to the propagation direction.

The coherence time Δt_W is related to the spectral purity of the beam, as is shown by the following argument. The wave can be viewed as a sequence of packets with spatial lengths around Δx_W. A typical packet requires Δt_W to pass a fixed point. Suppose that n periods of duration $2\pi/\omega$ occur during Δt_W, as expressed in Eq. (13).

$$\Delta t_W \frac{\omega}{2\pi} = n \qquad (13)$$

The packet can be viewed as a superposition of plane waves arranged to cancel each other outside the boundaries of the packet. In order to produce cancellation, waves must be mixed in which are in phase with the wave of angular frequency ω in the middle of the packet and which are half a period out of phase at both ends. In other words, waves must be mixed

in with periods around $2\pi/\omega'$, where Eq. (14) holds.

$$\Delta t_W \frac{\omega'}{2\pi} = n \pm 1 \qquad (14)$$

The difference of Eqs. (13) and (14) gives Eq. (15),

$$\Delta t_W \Delta\omega = 2\pi \qquad (15)$$

where the spread in angular frequencies $|\omega - \omega'|$ is called $\Delta\omega$. This result should be stated as an approximate inequality (16).

$$\Delta t_W \Delta\omega \gtrsim 1 \qquad (16)$$

A large Δt_W permits a small $\Delta\omega$ and therefore a well-defined frequency, while a small Δt_W implies a large $\Delta\omega$ and a poorly defined frequency.

Quantitative definitions. To go beyond qualitative descriptions and order-of-magnitude relations, it is useful to define the fundamental quantities in terms of correlation functions.

Self-coherence function. As discussed above, there are two ways to view the random fluctuations in the phase of a wave. One can discuss splitting the wave into two beams that interfere after a difference τ in traversal time, or one can discuss one wave that is passing over a fixed observer who determines the phase fluctuations. Both approaches concern the correlation between $\Psi(x,t)$ and $\Psi(x, t + \tau)$. This correlation is described by the normalized self-coherence function given by Eq. (17).

$$\gamma(\tau) \equiv \frac{\langle \Psi(x,t + \tau)\Psi^*(x,t)\rangle}{\langle \Psi(x,t)\Psi^*(x,t)\rangle} \qquad (17)$$

The brackets $\langle\ \rangle$ indicate the average of the enclosed quantity over a time which is long compared to the resolving time of the observer. It is assumed here that the statistical character of the wave does not change during the time of interest, so that $\gamma(\tau)$ is not a function of t.

With the aid of the self-coherence function, the coherence time of a wave can be defined quantitatively by Eq. (18).

$$(\Delta t_W)^2 \equiv \frac{\int_{-\infty}^{+\infty} \tau^2 |\gamma(\tau)|^2\, d\tau}{\int_{-\infty}^{+\infty} |\gamma(\tau)|^2\, d\tau} \qquad (18)$$

That is, Δt_W is the root-mean-squared width of $|\gamma(\tau)|^2$. The approximate inequality (16) can also be made precise. The quantity $|g(\omega')|^2$ is proportional to the contribution to Ψ at the angular frequency ω', where $g(\omega')$ is the Fourier transform of Ψ, defined by Eq. (19).

$$g(\omega') \equiv \frac{1}{\sqrt{2\pi}} \int_{-\infty}^{+\infty} \Psi(0,t)e^{-i\omega't}dt \qquad (19)$$

The root-mean-squared width $\Delta\omega$ of this distribution is given by Eq. (20),

$$(\Delta\omega)^2 = \frac{\int_0^\infty (\omega' - \omega)^2 |g(\omega')|^2 d\omega'}{\int_0^\infty |g(\omega')|^2 d\omega'} \qquad (20)$$

where ω is again the mean angular frequency. With these definitions, one can show that inequality (21) is satisfied.

$$\Delta t_W \Delta \omega \geq \tfrac{1}{2} \qquad (21)$$

The symbol $\gtrsim$ has become $\geq$; no situation can lead to a value of $\Delta t_W \Delta \omega$ that is less than $\tfrac{1}{2}$. Relation (21) multiplied by Planck's constant is the Heisenberg uncertainty principle of quantum physics, and can in fact be proved in virtually the same way. *SEE NONRELATIVISTIC QUANTUM THEORY; UNCERTAINTY PRINCIPLE.*

Complex degree of coherence. To describe the coherence between two different waves Ψ_A and Ψ_B, one can use the complex degree of coherence $\gamma_{AB}(\tau)$, defined by Eq. (22),

$$\gamma_{AB}(\tau) \equiv \frac{\langle \Psi_A(t + \tau)\Psi_B{}^*(t)\rangle}{[\langle \Psi_A{}^*(t)\Psi_A(t)\rangle\langle \Psi_B{}^*(t)\Psi_B(t)\rangle]^{1/2}} \qquad (22)$$

where the averaging process $\langle\ \rangle$ and the assumptions are the same as for the definition of $\gamma(\tau)$. If beams A and B are brought together, interference fringes may be formed where the time-averaged resultant intensity $\langle|\Psi|^2\rangle$ has maxima and minima as a function of position. The visibility of these fringes, defined by Eq. (23),

$$V = \frac{\langle|\Psi|^2\rangle_{max} - \langle|\Psi|^2\rangle_{min}}{\langle|\Psi|^2\rangle_{max} + \langle|\Psi|^2\rangle_{min}} \qquad (23)$$

can be shown, for the usual interference studies, to be proportional to $|\gamma_{AB}(\tau)|$.

Examples. The concept of coherence occurs in a great variety of areas. The following are some application and illustrations.

Astronomical applications. As is shown in the discussion of Fig. 4, extended sources give partial coherence and produce interference fringes with visibility V [Eq. (23)] less than unity. A. A. Michelson exploited this fact with his stellar interferometer (**Fig. 5**), a modified double-slit arrangement with movable mirrors that permit adjustment of the effective separation D' of the slits. According to Eq. (11), there is partial coherence if the angular size of the source is about λ/D'. It can be shown that if the source is a uniform disk of angular diameter Θ, then the smallest value of D' that gives zero V is $1.22\lambda/\Theta$. The interferometer can therefore be used in favorable cases to measure stellar diameters. The same approach has also been applied in radio astronomy. A different technique, developed by R. Hanbury Brown and R. Q. Twiss, measures the correlation between the intensities received by separated detectors with fast electronics. *SEE INTERFEROMETRY; RADIO ASTRONOMY.*

Lasers and masers. Because they are highly coherent sources, lasers and masers provide very large intensities per unit frequency, and double-photon absorption experiments have become possible. As a consequence, atomic and molecular spectroscopy were revolutionized. One can selectively excite molecules that have negligible velocity components in the beam direction and thus improve resolution by reduction of Doppler broadening. States that are virtually impossible to excite with single-photon absorption can now be reached. *SEE LASER SPECTROSCOPY; MASER.*

Musical acoustics. Fortunately, for most musical instruments, the coherence time is much less than hu-

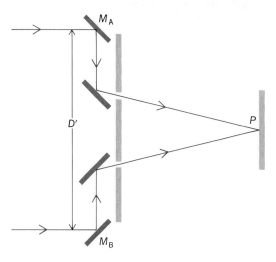

Fig. 5. Michelson's stellar interferometer. The effective slit separation D' can be varied with the movable mirrors M_A and M_B. The angular size of the source can be calculated from the resulting change in the visibility of the interference fringes around P.

man resolving time, so that the separate contributions in an orchestra are perceived incoherently. Equation (7) applies, and 20 violins seem to provide 20 times the intensity of 1 violin. If human resolving time were much shorter, the perceived intensity would drift between 0 and 400 times the intensity of a single violin. *SEE MUSICAL ACOUSTICS.*

Thermodynamic treatment of radiation. Coherence must be considered in thermodynamic treatments of radiation. As one should expect from the connection between entropy and order, entropy is reduced by an increase in coherence. The total entropy of two beams that are coherent with respect to each other is less than the sum of the entropies of the separate beams, because coherence means that the beams are not statistically independent. *SEE ENTROPY; THERMODYNAMIC PRINCIPLES.*

Quantum physics. In quantum physics, a beam of particles can for some purposes be described by Eq. (1) without the phase $\delta_A(t)$. If the wavelength is well defined, measurables such as barrier transmission coefficients and scattering cross sections of sufficiently small targets do not depend on phases and coherence. If, however, the scattering object is large, these aspects may have to be considered.

Electrons emitted by hot filaments show coherence times of roughly 10^{-14} s. In low-energy electron diffraction, where speeds on the order of 10^8 to 10^9 cm/s are used, the coherence length is then on the order of 10^{-5} cm, and the structure can be seen clearly only if the region being studied does not introduce a path length difference greater than this distance. *SEE ELECTRON DIFFRACTION.*

Most nuclear and particle physics experiments deal with energies so high that the wavelengths of the projectiles are less than 10^{-12} cm, far less than the amplitudes of motions of the scattering nuclei. The many nuclei therefore contribute to the scattered beam incoherently. Equation (7) applies, and the intensity of the scattered beam is proportional to the scattering cross section per nucleus times the number of nuclei.

Rolf G. Winter

Bibliography. H. J. Bernstein and F. E. Low, Measurement of longitudinal coherence lengths in particle beams, *Phys. Rev. Lett.*, 59:951–953, 1987; M. Born and E. Wolf, *Principles of Optics*, 6th ed., 1980; F. Haake, L. M. Narducci, and D. Walls (eds.), *Coherence, Cooperation and Fluctuations*, 1986; P. Hariharan, *Optical Interferometry*, 1985; R. G. Winter, *Quantum Physics*, 2d ed., 1986.

Cohesion (physics)

The tendency of atoms or molecules to coalesce into extended condensed states. This tendency is practically universal. In all but exceptional cases, condensation occurs if the temperature is sufficiently low; at higher temperatures, the thermal motions of the constituents increase, and eventually the solid assumes gaseous form. The cohesive energy is the work required to separate the condensed phase into its constituents or, equivalently, the amount by which the energy of the condensed state is lower than that of the isolated constituents. The science of cohesion is the study of the physical origins and manifestations of the forces causing cohesion, as well as those opposing it. It is thus closely related to the science of chemical bonding in molecules, which treats small collections of atoms rather than extended systems. *See Chemical bonding; Intermolecular forces.*

When the constituent atoms or molecules are far separated, an attractive force is present. This force is due, directly or indirectly, to the electrostatic interactions of the negatively charged electrons and the positively charged ions. However, when the distances between the constituents become very short, a strong repulsive force results, which is due to the quantum-mechanical kinetic energy of the electrons. The competition between the attractive and repulsive forces determines the equilibrium density for the condensed state (see **illus.**). A positive pressure at a given volume means that pressure must be applied to the solid to keep it at this volume; thus the forces between the atoms are predominantly repulsive. Similarly, the pressure is negative when the forces are predominantly attractive. Both molybdenum and xenon have attractive interactions at large volumes. As the volume drops, the attractive forces reach a maximum and then begin to be canceled by the repulsive forces. The equilibrium volume of the solid is that for which the cancellation is complete, or the pressure vanishes. The equilibrium point actually occurs when the pressure equals the atmospheric pressure, which is approximately 1 bar (100 kilopascals). However, this pressure is effectively zero on the pressure scale of

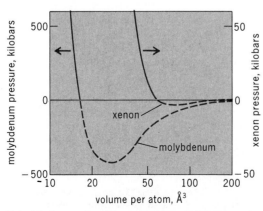

Calculated pressure-volume curves for molybdenum and xenon at absolute zero. Negative-pressure portions of curves are represented by broken lines because they cannot be directly measured. 1 kilobar = 0.1 GPa; 1 Å³ = 0.001 nm³.

the illustration. *See Coulomb's law; Quantum mechanics.*

Attractive forces. The origin and magnitude of the attractive forces depend on the chemical nature of the constituent atoms or molecules. Strong attractive interactions are usually associated with constituents having valence electron shells which are partly filled or open; if the valence electron shells are completely filled or closed, the interactions are weaker. *See Valence.*

Open-shell constituents. This is most easily understood for atomic systems. As the atoms approach, the electron energy levels on different atoms begin to interact, forming a complex of energy levels in the solid. Some of these are below the atomic energy levels and some above. Since the atomic shells are partly filled, the lower energy levels in the solid are filled, but at least some of the higher levels are empty. Thus the average energy of the occupied levels in the solid is lower than that in the isolated atoms, resulting in an attractive force. Bonding in open-shell systems can be approximately divided into three categories, although most cases involve a combination. *See Band theory of solids; Fermi-Dirac statistics; Solid-state physics; Valence band.*

1. Covalent bonding. This type of bonding is most similar to the molecular bond. The electron energy levels in the solid are split into a lower and a higher portion, with the states in the lower one filled and the higher one empty. Covalent bonds are strongly directional, with electron charge accumulating around the bond centers. Materials bonded in this fashion typically form structures with low coordination numbers, prototypical materials elements in group IV of the periodic table, the insulator carbon, and the semiconductors silicon and germanium. *See Periodic table; Semiconductor.*

2. Metallic bonding. In this case, there is no split between the lower and higher states of the electrons in the solid; rather, they occupy levels from the bottom up to a cutoff point known as the Fermi level. For example, in transition metals, the electron states in the solid derived from the atomic *d* orbitals form a complex which is gradually filled with increasing atomic number. The bulk of the cohesive energy is due to this complex. The metallic bond is less directional than the covalent bond, with a more uniform distribution of electronic charge. Metals usually form closely packed structures. *See Fermi surface; Free-electron theory of metals.*

3. Ionic bonding. This occurs in compounds having at least two distinct types of atoms. One or more of the species of atoms (the cations) have only a small number of electrons in their valence shells, whereas at least one species (the anions) has a nearly filled valence shell. As the atoms approach each other, electrons drop from the cation valence states into holes in the anion valence shell, forming a closed-shell configuration in the solid. The different types of atoms in the solid have net charges; a strong attractive force results from the interaction between unlike charges. For example, in sodium chloride (NaCl), the sodium atoms acquire positive charges, and the chlorine atoms acquire negative charges. The closest interatomic separations in the solid are between sodium and chlorine, so that the attractive electrostatic interactions outweigh the repulsive ones. *See Ionic crystals; Solid-state chemistry.*

Closed-shell constituents. In these systems the above

effects are greatly reduced because the atomic or molecular shells are basically inert. The constituents retain their separate identities in the solid environment. If the constituents are atomic, as in rare-gas solids, the cohesion is due to the van der Waals forces. The positions of the electrons in an atom fluctuate over time, and at any given time their distribution is far from spherical. This gives rise to fluctuating long-ranged electric fields, which average zero over time, but can still have appreciable effects on neighboring atoms. The electrons on these atoms move in the direction of the force exerted by the electric field. The net result is that the interactions between unlike charges (electrons and nuclei) are increased in the solid, whereas the interactions between like charges are reduced. Thus the solid has a lower energy than the isolated atoms.

In solids made up of molecules, there are additional electrostatic interactions due to the nonspherical components of the molecular charge density. These interactions are strongest if the molecules are polar. This means that the center of the positive charge on the molecule is at a different point in space from that of the negative charge. Polar molecules, such as water (H_2O), form structures in which the positive charge on a molecule is close to the negative charges of its neighbors. For nonpolar molecules, the electrostatic interactions are usually weaker than the van der Waals forces. The nonspherical interactions in such cases are often so weak that the molecules can rotate freely at elevated temperatures, while the solid is still held together by the van der Waals forces.

Repulsive forces. The repulsive forces in the condensed phase are a dramatic illustration of the combined action of two quantum-mechanical principles, the exclusion principle and the uncertainty principle.

The exclusion principle states that the quantum-mechanical wave function for the electrons in the solid must be antisymmetric under the interchange of the coordinates of any two electrons. Consequently, two electrons of the same spin are forbidden from being very close to each other. *See Exclusion principle*.

The uncertainty principle states that if the motion of an electron is confined, its kinetic energy must rise, resulting in a repulsive force opposing the confinement. The kinetic energy due to the confinement is roughly inversely proportional to the square of the radius of the region of confinement. According to the exclusion principle, the motion of an electron in a solid is partially confined because it is forbidden from closely approaching other electrons of the same spin. Thus the uncertainty principle in turn implies a repulsive force. *See Uncertainty principle*.

The magnitude of the repulsive force always exceeds that of the attractive force when the volume per atom becomes sufficiently small. This is most readily understood in the simple metals, which are, roughly, metals in groups I through IV. In these metals the electrons can be approximately treated as if they were spread uniformly throughout the solid. If the metal is compressed, the repulsive kinetic energy due to the electron confinement is approximately proportional to d^{-2}, where d is the interatomic separation. The energy due to the attractive force, in contrast, is proportional only to d^{-1}. Thus the repulsive force due to the kinetic energy dominates if d is sufficiently small.

Variations between materials. There are enormous variations between materials in the strength of the co-

hesive forces. For example, the maximal attractive force in molybdenum, an open-shell solid, corresponds to a negative pressure of nearly 500,000 bars (50 gigapascals); that in xenon, a closed-shell solid, is roughly 100 times smaller (see illus.) This is reflected in a much larger equilibrium volume per atom for xenon, even though the size of an isolated molybdenum atom exceeds that of a xenon atom. The variations in the cohesive forces are also reflected in the boiling point, which is the highest temperature at which a condensed state exists. This can be over 5000°C (9000°F) in open-shell systems, and is usually higher than 200°C (400°F). In elemental closed-shell systems, in contrast, the boiling point is usually below 0°C (32°F); in helium it is −269°C (−452°F), or 4°C (7°F) above absolute zero. Many other properties correlate with the strength of the cohesive forces, including the melting point, elastic moduli, and the energies required to form defects such as vacancies and surfaces. The variations in these properties from material to material have a large impact on their technological applications. *See Crystal defects*.

Weakening of attractive forces. In analyzing the mechanical properties of solids, it is often conceptually useful to think of the force on a particular atom as being the sum of independent forces due to neighboring atoms. Such a picture is justified for solids consisting of closed-shell atoms or molecules, but is often an inaccurate representation of the forces in open-shell systems, particularly in metals. The attractive force in a solid metal can be several times weaker than that in a dimer molecule, because the atoms interact too strongly with each other to maintain their separate identities; the electrons, instead of being localized on individual atoms, are in states that extend throughout the solid. The weakening of the attractive forces in the solid manifests itself in the atomic geometries around defects. For example, the attractive forces in the vicinity of vacancies or surfaces are intermediate in strength between those in the solid and those in the dimer molecule. This usually results in shorter interatomic separations around these defects, if it is possible for the bonds to contract. In addition, a major contribution to the surface tensions of solids results from changes in the attractive forces near the surface.

Anders E. Carlsson

Bibliography. N. W. Ashcroft and N. D. Mermin, *Solid State Physics*, 1976; P. A. Cox, *The Electronic Structure and Chemistry of Solids*, 1987; W. Harrison, *Electronic Structure and the Properties of Solids: The Physics of the Chemical Bond*, 1980; C. Kittel, *Introduction to Solid State Physics*, 6th ed., 1986.

Coil

One or more turns of wire used to introduce inductance into an electric circuit. At power line and audio frequencies a coil has a large number of turns of insulated wire wound close together on a form made of insulating material, with a closed iron core passing through the center of the coil. This is commonly called a choke and is used to pass direct current while offering high opposition to alternating current.

At higher frequencies a coil may have a powdered iron core or no core at all. The electrical size of a coil is called inductance and is expressed in henries or millihenries. In addition to the resistance of the wire, a coil offers an opposition to alternating current, called reactance, expressed in ohms. The reactance of a coil increases with frequency. *See* INDUCTOR; REACTOR (ELECTRICITY).

John Markus

Coincidence amplifier

An electronic circuit that amplifies only that portion of a signal present when another enabling or controlling signal is simultaneously applied. The controlling signal may be such that the amplifier functions during a specified time interval, called time selection, or such that it functions between specified voltage or current levels, called amplitude selection. For example, the amplifier in a transmission gate operates on the basis of time selection where a control voltage of fixed amplitude is applied during a specific time interval. This voltage is the enabling waveform; it allows the amplifier to function normally only during the time it is applied. *See* GATE CIRCUIT.

A linear amplifier used to amplify a section of a waveform above or below specified voltage or current limits, as determined by a control signal, is an example of a coincidence amplifier that operates by amplitude selection. Limiting and clipping circuits that incorporate the function of amplification may be considered coincidence amplifiers. *See* CLIPPING CIRCUIT; LIMITER CIRCUIT.

A simple example of a coincidence amplifier that functions between two input voltage limits is shown in the **illustration**. This specific coincidence amplifier is sometimes referred to as a slicer or slicer amplifier. The transistor amplifier, in this case an emitter-coupled amplifier, is biased such that, until time T_1, input stage Q_1 functions as an emitter follower, but the voltage at the base of Q_1 is such that no current flows in its collector circuit until input level V_1 is reached. Between levels V_1 and V_2 both transistors are active, and an amplified output signal appears at the collector of Q_2. At time T_2, corresponding to level V_2, the input transistor becomes nonconducting and no further output is produced.

The same operation may be performed using *pnp* transistors with all bias and signal polarities reversed. The same function may also be performed using field-effect transistors or vacuum tubes as source-coupled or cathode-coupled amplifiers.

Such coincidence amplifiers, with their gains sometimes greatly magnified by positive feedback, are often used as elements in linear time-delay circuits. When such an amplifier is used and the interest is primarily in a pulse or other indication when a desired level is reached rather than in preservation of the waveform itself, the term comparator is more often applied. However, the term coincidence circuit is sometimes used interchangeably when the concept of amplification is not specifically considered. In digital logic terminology the AND gate is an example of such a coincidence or comparator circuit. *See* COMPARATOR CIRCUIT.

Glenn M. Glasford

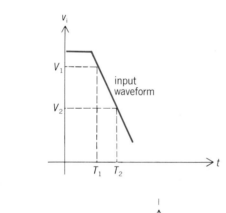

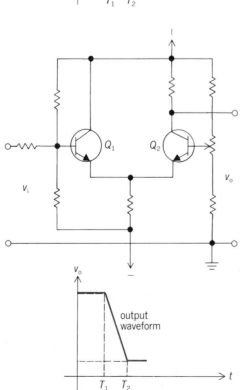

Simple coincidence amplifier.

Coining

A cold metalworking process in a press-type die. Coining is used to produce embossed parts, such as badges and medals, and for minting of coins. It is also used on portions of a blank or workpiece to form corners, indentations, or raised sections, frequently as part of a progressive die operation. The work is subjected to high pressure within the die cavity and thereby forced to flow plastically into the die details. The absence of overflow of excess metal from between the dies is characteristic of coining and is responsible for the fine detail achieved. However, because this action requires highly accurate dies and higher-than-usual die pressures, only essential surfaces are coined. *See* METAL FORMING.

Ralph L. Freeman

Coir

A natural fiber, also known as coco fiber, obtained from the husks of the coconut (*Cocos nucifera*). Although coconuts are grown for copra throughout the tropics and into the subtropics, more than 95% of the coir exports are from Sri Lanka and India. Small amounts are exported from Thailand, Tanzania, Mexico, the Philippines, Malaysia, Kenya, Trinidad, and Tobago. A few other countries produce small amounts of fiber for their own use.

The cells of coir average about 0.028 in. (0.7 mm) in length and about 20 micrometers in width. The cells may have either blunt or pointed ends, and the cell wall has numerous pits. The lumen width and the cell wall thickness are about equal. Coir is a highly lignified fiber and becomes more lignified as it matures.

The outstanding characteristic of coir fiber is its resistance to rot. For example, a coir fiber doormat can stay out in the weather for years. Coir is only intermediate in strength among the vegetable fibers, but its elongation at break is greater than any of the bast or hard fibers.

Types. There are three main types of coir—yarn fiber, bristle fiber, and mattress fiber. Nearly all the yarn fiber comes from southern India; much of the mattress fiber and essentially all the bristle fiber come from Sri Lanka.

Only the finest and longest fiber is suitable for spinning into yarn. It is obtained from the husks of unripe nuts and is the main cash crop rather than a by-product. The bristle fiber and most of the mattress fiber come from mature nuts and are by-products of copra production.

Production. The yarn and bristle fiber and much of the mattress fiber extraction starts with a natural bacteriological process called retting. The loosened fibers are later separated from the nonfibrous material by mechanical action. Immature husks are used for yarn fiber, while mature husks are used to make bristle and mattress fiber.

Yarn fiber. Within a week after harvest the husks are separated from the shell containing the copra and split into four or five pieces which are put into water for retting, to decompose the nonfibrous materials. Retting usually takes 5 to 6 months but sometimes it requires as long as 10 months. The favored retting centers are in the brackish waters of river estuaries, but many husks are retted in tanks, ponds, lakes, and streams. The circulation of water through the husks—either as a result of a tide, the flow of a stream, or changing the water in a tank—improves the quality of the fiber.

After retting, the husks are washed, air-dried, and then beaten with mallets, stones, or sticks to loosen the fiber and remove the nonfiber materials that have decomposed. Some combing is usually done, either by steel combs mounted on wooden wheels or by stationary combs. Then the fiber is spun by hand into single- or two-ply yarn.

Bristle and mattress fiber. For these fibers, the period of retting is much shorter. Before retting, the nuts are frequently allowed to mature further after harvesting so as to even out the quality of the copra produced from them.

The bristle and mattress fiber is usually separated from retted husks by machines, although the separation still requires extensive hand labor. An increasing number of establishments, especially in new areas, are extracting the fiber from green or dry, unretted husks by automatic machines. However, fiber obtained in such a manner is shorter than fiber obtained by traditional processes and is irretrievably mixed so it must be sold as mattress fiber.

Sorting and grading. Coir fiber ranges in color from light tan to dark brown. Bristles are sorted according to length, which ranges from about 6 to 14 in. (15 to 35 cm), and are packed into bundles of about 17.5 oz (500 g) each. Mattress fiber is packed in small bales of about 11 lb (5 kg) each, and yarn is put up in skeins. Before these packages are put into large bales for shipment, the fiber is graded and sometimes resorted. The fiber is graded on the basis of length, stiffness, color, and cleanliness. Long, clean fibers and light color are desirable for any purpose, but stiffness is desirable in bristles and undesirable for yarn.

Uses. In developed countries, yarn fiber is used chiefly in mats and mattings; in the United States, for instance, its best-known use is in light-brown tufted doormats. In Asia it is used extensively for ropes and twines, and locally for hand-made bags. The principal outlet for bristle fiber has been in brush making, but the market has declined and most bristle fiber is now used in upholstery padding. Mattress fiber is used chiefly in innerspring mattresses, though it has found uses as an insulating material. *See Coconut; Natural fiber.*

Elton G. Nelson

Coke

A coherent, cellular, carbonaceous residue remaining from the dry (destructive) distillation of a coking coal. It contains carbon as its principal constituent, together with mineral matter and residual volatile matter. The residue obtained from the carbonization of a noncoking coal, such as subbituminous coal, lignite, or anthracite, is normally called a char. Coke is produced chiefly in chemical-recovery coke ovens (see **illus.**), but a small amount is also produced in beehive or other types of nonrecovery ovens.

Uses and types. Coke is used predominantly as a fuel reductant in the blast furnace, in which it also serves to support the burden. As the fuel, it supplies the heat as well as the gases required for the reduction of the iron ore. It also finds use in other reduction processes, the foundry cupola, and householding. About 91% of the coke made is used in the blast furnace, 4% in the foundry, 1% for water gas, 1% for householding, and 3% for other industries, such as calcium carbide, nonferrous metals, and phosphates. Approximately 1300 lb (560 kg) of coke is consumed per ton of pig iron produced in the modern blast furnace, and about 200 lb (90 kg) of coke is required to melt a ton of pig iron in the cupola.

Coke is classified not only by the oven in which it is made, chemical-recovery or beehive, but also by the temperature at which it is made. High-temperature coke, used mainly for metallurgical purposes, is produced at temperatures of 1650–2100°F (900–1150°C). Medium-temperature coke is produced at 1380–1650°F (750–900°C), and low-temperature coke or char is made at 930–1380°F (500–750°C). The latter cokes are used chiefly for householding, particularly in England. The production of these is rather small as compared to high-temperature coke and usually requires special equipment. Coke is also classified according to its intended use, such as blast-furnace coke, foundry coke, water-gas coke, and domestic coke.

Production. Coke is formed when coal is heated in the absence of air. During the heating in the range of 660–930°F (350–500°C), the coal softens and then fuses into a solid mass. The coal is partially devolatilized in this temperature range, and further heating at temperatures up to 1830–2000°F (1000–1100°C) reduces the volatile matter to less than 1%. The degree of softening attained during heating determines to a large extent the character of the coke produced.

In order to produce coke having desired properties, two or more coals are blended before charging into the coke oven. Although there are some exceptions, high-volatile coals of 32–38% volatile-matter content are generally blended with low-volatile coals of 15–20% volatile matter in blends containing 20–40% low-volatile coal. In some cases, charges consist of blends of high-, medium-, and low-volatile coal. In this way, the desirable properties of each of the coals, whether it be in impurity content or in its contribution to the character of the coke, are utilized. The low-volatile coal is usually added in order to improve the physical properties of the coke, especially its strength and yield. In localities where low-volatile coals are lacking, the high-volatile coal or blends of high-volatile coals are used without the benefit of the low-volatile coal. In some cases, particularly in the manufacture of foundry coke, a small percentage of so-called inert, such as fine anthracite, coke fines, or petroleum coke, is added.

In addition to the types of coals blended, the carbonizing conditions in the coke oven influence the characteristics of the coke produced. Oven temperature is the most important of these and has a significant effect on the size and the strength of the coke. In general, for a given coal, the size and shatter strength of the coke increase with decrease in carbonization temperature. This principle is utilized in the manufacture of foundry coke, where large-sized coke of high shatter strength is required.

Properties. The important properties of coke that are of concern in metallurgical operations are its chemical composition, such as moisture, volatile-matter, ash, and sulfur contents, and its physical character, such as size, strength, and density. For the blast furnace and the foundry cupola, coke of low moisture, volatile-matter, ash, and sulfur content is desired. The moisture and the volatile-matter contents are a function of manner of oven operation and quenching, whereas ash and sulfur contents depend upon the composition of the coal charged. Blast-furnace coke used in the United States normally contains less than 1% volatile matter, 85–90% fixed carbon, 7–12% ash, and 0.5–1.5% sulfur. If the coke is intended for use in the production of Bessemer or acid open-hearth iron, the phosphorus content becomes important and should contain less than 0.01%.

The requirements for foundry coke in analysis are somewhat more exacting than for the blast furnace. The coke should have more than 92% fixed carbon, less than 8% ash content, and less than 0.60% sulfur.

Blast-furnace coke should be uniform in size, about 2½–5 in. (6.4–13 cm), but in order to utilize more of the coke produced in the plant, the practice has been, in some cases, to charge separately into the furnace the smaller sizes after they were closely screened into sizes such as 2½ × 1¾ in. (6.4 × 4.5 cm), 1¾ × 1 in. (4.5 × 2.5 cm), down to about ¾ in. (1.9 cm) in size. There is now a trend, when pelletized ore is used, to crush and screen the coke to a more closely sized material, such as 2½ × ¾ in. (6.4 × 1.9 cm).

The blast-furnace coke should also be uniformly strong so that it will support the column of layers of iron ore, coke, and stone above it in the furnace without degradation. Standard test methods, such as the tumbler test and the shatter test, have been developed by the American Society for Testing and Materials for measuring strength of coke. Good blast-furnace cokes have tumbler test stability factors in the range of 45 to 65 and shatter indices (2-in. or 5-cm sieve) in the range of 70 to 80%. These are not absolute requirements since good blast-furnace performance is obtained in some plants with weaker coke—but with the ore prepared so as to overcome the weakness.

Foundrymen prefer coke that is large and strong. The coke for this purpose ranges in size from 3 to 10

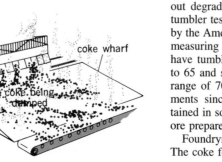

Steps in the production of coke from coal. After the coal has been in the oven for 12–18 h, the doors are removed and a ram mounted on the same machine that operates the leveling bar shoves the coke into a quenching car for cooling. (*American Iron and Steel Institute*)

in. (7.6 to 25 cm) and larger. The size used usually depends upon the size of the cupola and, in general, the maximum size of the coke is about one-twelfth of the diameter of the cupola. High shatter strength is demanded by the cupola operator; shatter indices of at least 97% on 2-in. (5-cm) and 80% on 3-in. (7.6-cm) sieves have been specified.

Other important requirements of foundry coke are high reactivity toward oxygen so that carbon dioxide is produced with high heat evolution, and low reactivity toward carbon dioxide so that the amount of carbon monoxide formed is minimized.

The requirements for coke intended for househeating vary in different localities. Usually the coke is of a narrow size range, and the size used depends upon the size and type of appliance. In general, a high ash-softening temperature is desirable (more than 2500°F or 1370°C).

Coke intended for water-gas generators should be over 2 in. (5 cm) in size, and should have an ash content below 10% and a moderately high ash-softening temperature. SEE CHARCOAL; COAL; COAL CHEMICALS; DESTRUCTIVE DISTILLATION; FOSSIL FUEL; PYROLYSIS.
Michael Perch

Bibliography. American Society for Testing and Materials, *Petroleum Products, Lubricants, and Fossil Fuels,* vol. 0.505, *Gaseous Fuels: Coal and Coke,* 1986; Commission of the European Communities, *Coke Oven Techniques,* 1983; Metal Society Editors (eds.), *Coal, Coke, and the Blast Furnace,* 1978; R. E. Zimmerman, *Evaluating and Testing the Coking Properties of Coal,* 1979.

Coking (petroleum)

A process for thermally converting the heavy residual bottoms of crude oil entirely to lower-boiling petroleum products and by-product petroleum coke. The heavy residual bottoms cannot be catalytically cracked, principally because the metals present are potent catalyst poisons. Traditionally, the residual bottoms have been blended with lighter stocks and marketed as low-quality heavy fuel oils. However, when demands for gasoline, heating oil, and other products increase, refiners resort to deeper vacuum flashing, visbreaking, solvent deasphalting, and coking. The major products from coking are fuel gas,

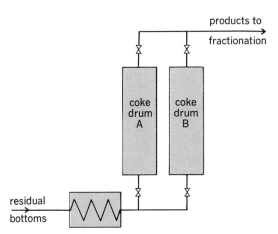

Fig. 1. Delayed coking, in simplified flow plan.

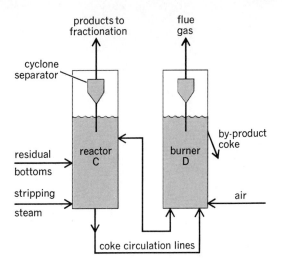

Fig. 2. Fluid coking, in simplified flow plan.

gasoline, gas oil, and petroleum coke. Generally the gasoline is of poor quality and must be further treated before use in premium fuels. The gas oil is commonly used as catalytic-cracking feed stock.

The two coking processes in commercial use are delayed coking and fluid coking. In delayed coking (**Fig. 1**), residual bottoms are heated to temperatures of about 900°F (480°C) in a conventional furnace. The hot oil is charged directly to the bottom of a large coke drum A, where the coking reaction takes place. Coke is deposited in the drum while the lighter products are removed overhead to fractionation. Approximately every 24 h, hot oil is diverted to the second coke drum B, and coke is removed from drum A by hydraulic drilling. In fluid coking (**Fig. 2**), residual bottoms are charged directly to the reactor C containing a bed of fluidized coke particles at about 950°F (510°C). The hot coke particles supply the heat of reaction and also a surface for coke deposition. Converted light products are removed overhead through cyclone separators to fractionation. The coke particles are continuously circulated between the reactor C and the burner D. In the burner D, air is used to burn a portion of the coke produced, thereby maintaining the system in heat balance. Normally, more coke is produced in the reactor than is required for heat balance. The excess is continuously withdrawn from the burner as by-product petroleum coke.

Petroleum coke is used principally as fuel or, after calcining, for carbon electrodes. The crude from which the coke is produced governs its chemical composition (high-sulfur crudes yield high-sulfur cokes). All petroleum cokes are characterized by low total ash contents. SEE CRACKING; PETROLEUM PROCESSING.
J. F. Moser, Jr.

Cola

A tree, *Cola acuminata,* of the family Sterculiaceae and a native of tropical Africa. Its fruit is a star-shaped follicle containing eight hard seeds, the cola nuts of commerce (see **illus.**). These nuts are an important masticatory in many parts of tropical Africa. They have a caffeine content twice that of coffee. The nuts also contain an essential oil and a glucoside, kolanin, which is a heart stimulant. Cola nuts, in com-

Cola (*Cola acuminata*)
branch.

bination with an extract from coca, are used in the manufacture of the beverage Coca-Cola. Cola is now cultivated in West Africa, Jamaica, Brazil, India, and other parts of tropical Asia. See Coca; Malvales.

Perry D. Strausbaugh/Earl L. Core

Colchicine

The major alkaloid obtained from the seed capsules, corms, and bulbs of the meadow saffron (autumn crocus; *Colchicum autumnale*). Colchicine is a fairly abundant alkaloid in dried corms (0.03–0.08%) and seeds (0.6–0.9%). It has been isolated also from other species of *Colchicum* and a few related genera, but they are not important sources of alkaloid. See Alkaloid; Liliales.

Colchicine has an unusual structure (shown below)

containing two seven-membered rings. Ring C is a tropolone system, present also in some fungal metabolites. The molecule has been synthesized in several ways, by constructing the ring system in either the sequence A → AB → ABC or A → A-C → ABC. See Tropolone.

The use of the colchicum plant to relieve the pain of gout has been known since medieval times, and colchicine is still the standard treatment for gout, although it is an extremely toxic substance.

An important property of colchicine is its ability to interrupt the mitotic cycle before cell division occurs. This effect leads to cells with multiple chromosomes, which are of value in plant breeding. The antimitotic effect of colchicine is due to strong binding of the alkaloid to the protein tubulin, thus preventing the assembly of microtubules and formation of the mitotic spindle. See Mitosis; Polyploidy.

James A. Moore

Cold hardiness (plant)

The ability of temperate zone plants to survive subzero temperatures. This characteristic is a predominant factor that determines the geographical distribution of native plant species and the northern limits of cultivation of many important agronomic and horticultural crops. For example, plant hardiness zone maps (**Fig. 1**) are frequently used in recommending the selection of various horticultural crops, especially woody ornamentals (see **table**) and fruit trees, for various regions. Further, freezing injury is a major cause of crop loss resulting from early fall frosts, low midwinter temperatures, or late spring frosts. Problems of cold hardiness are of concern to farmers in diverse areas of agriculture—from coffee growers in Brazil to wheat farmers in Canada; from peach growers in Georgia to apple growers in New York; from tropical foliage plant growers in Florida to producers of woody ornamentals in the northeast; from viticulturists (grape growers) in France to citrus growers in Israel. As a result, the development of varieties of cultivated plants with improved cold hardiness is of long-standing concern.

Within the plant kingdom there is a wide range of diversity in low-temperature tolerance—from low levels of hardiness in herbaceous species such as potatoes (27 to 21°F or −3 to −6°C), to intermediate levels of hardiness for winter annuals such as wheat and rye (−4 to −22°F or −20 to −30°C), to extremely hardy deciduous trees and shrubs such as black locust and red osier dogwood that can withstand temperatures of liquid nitrogen (−321°F or −196°C). Within a given species, the range in hardiness can be substantial. For example, the cold hardiness of different varieties of deciduous fruit trees may vary by 18 to 36°F (10 to 20°C). Within a given plant there is a wide range in the cold hardiness of different tissues and organs. For example, roots are much less tolerant of subzero temperatures than shoots; flower buds are more sensitive than vegetative buds.

Environmental effects. The cold hardiness of a given species is an inherent genetic trait that requires certain environmental cues for its expression. With the shorter days and cooler nights of autumn, temperate zone plants become dormant and increase their cold hardiness. This process is referred to as cold acclimation. In the spring, increasing daylength and warmer temperatures result in the resumed growth and development of the plant and a corresponding decrease in cold hardiness. Thus, hardy species such as winter annuals, biennials, and perennials exhibit an annual periodicity in their low-temperature tolerance (**Fig. 2**). In the spring and summer these species are as susceptible to freezing temperatures as are nonhardy species.

Cold hardiness may be influenced by radiation, temperature, photoperiod, precipitation, and stage of development of the plant, with different optimum conditions for different species or cultivars and ecotypes within a species. The various environmental cues serve to synchronize plant development with the environment. This synchronization has taken centuries to evolve, and freezing injury in cultivated species can result from any factor that disrupts this synchrony. Some varieties may not be responsive to the prevailing environmental cues or may not respond rapidly enough; some may not develop a sufficient degree of hardiness; and some may deacclimate too rapidly. These factors may arise when individual varieties are introduced into areas that are vastly or even slightly different from their natural habitat, where centuries of selection pressures have evolved those individuals most closely synchronized with the prevailing environment.

Temperature is the key environmental parameter for increasing a plant's capacity to withstand freezing temperatures. Low, above-freezing temperatures are conducive to an increase in hardiness in the fall, and warm temperatures are responsible for the decrease in the spring. Generally, it is considered that most plants will acclimate as temperatures are gradually lowered below 50°F (10°C). However, during acclimation, the progressive decline in temperatures (from the relatively warm temperatures in early fall, followed by low, above-freezing temperatures in late fall and early winter, followed by freezing temperatures in winter) is extremely important. The development of cold hardiness may take 4 to 6 weeks.

Photoperiod is the second major factor influencing cold acclimation, but only in those species that are photoperiodically responsive in relation to growth cessation or induction of dormancy (a true physiological rest period). In other species, light is important only in providing sufficient photosynthetic reserves required for the cold acclimation process. In some cases (for example, germinating seeds), sufficient energy reserves are already present and acclimation can occur in the dark. SEE PHOTOPERIODISM.

There are conflicting reports on the role of moisture in relation to cold hardiness. High soil moisture may reduce the degree of cold acclimation; however, severe winter injury of evergreens will occur if soil moisture levels are too low. Most often tissue moisture levels will influence the survival to a given freeze-thaw cycle rather than directly influencing the process of cold acclimation. Thus, whereas temperature and light effects on hardiness are probably mediated through the development of hardiness (cold acclimation), tissue moisture content directly affects the stresses that are incurred during a freeze-thaw cycle.

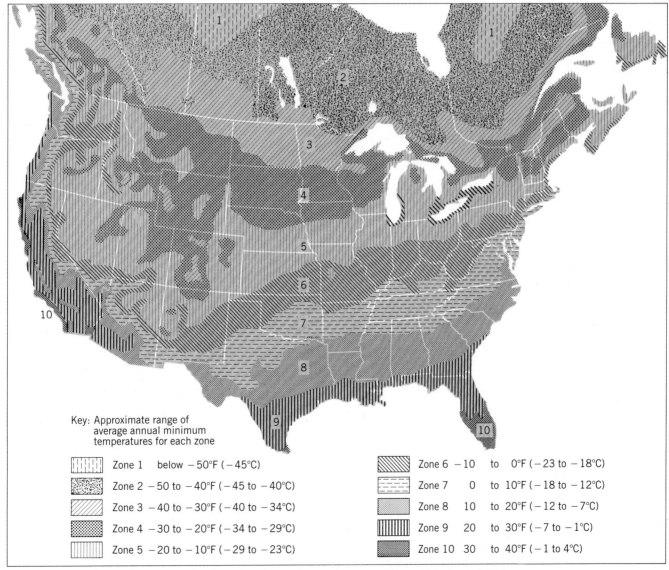

Key: Approximate range of average annual minimum temperatures for each zone

Zone 1	below −50°F (−45°C)
Zone 2	−50 to −40°F (−45 to −40°C)
Zone 3	−40 to −30°F (−40 to −34°C)
Zone 4	−30 to −20°F (−34 to −29°C)
Zone 5	−20 to −10°F (−29 to −23°C)

Zone 6	−10 to 0°F (−23 to −18°C)
Zone 7	0 to 10°F (−18 to −12°C)
Zone 8	10 to 20°F (−12 to −7°C)
Zone 9	20 to 30°F (−7 to −1°C)
Zone 10	30 to 40°F (−1 to 4°C)

Fig. 1. Plant hardiness zone map of the United States based on isotherms of average minimum winter temperatures.

Representative ornamental plants that will survive in the various hardiness zones

Zone	Scientific name	Common name
1 below −50°F (−45°C)	*Betula glandulosa* *Empetrum nigrum* *Populus tremuloides* *Potentilla pennsylvanica* *Rhododendron lapponicum* *Salix reticulata*	Dwarf birch Crowberry Quaking aspen Pennsylvania cinquefoil Lapland rhododendron Netleaf willow
2 −50° to −40°F (−45 to −40°C)	*Betula papyrifera* *Cornus canadensis* *Elaeagnus commutata* *Juniperus communis* *Picea glauca* *Potentilla fruticosa*	Paper birch Bunchberry dogwood Silverberry Common juniper White spruce Bush cinquefoil
3 −40 to −30°F (−40 to −34°C)	*Elaeagnus angustifolia* *Lonicera tatarica* *Malus baccata* *Parthenocissus quinquefolia* *Syringa vulgaris* *Thuja occidentalis*	Russian olive Tatarian honeysuckle Siberian crabapple Virginia creeper Common lilac American arborvitae
4 −30 to −20°C (−34 to −29°C)	*Berberis thunbergi* *Hydrangea paniculata* *Juniperus chinensis* *Ligustrum amurense* *Spiraea vanhouttei* *Taxus cuspidata*	Japanese barberry Panicle hydrangea Chinese juniper Amur River privet Vanhoutte spirea Japanese yew
5 −20 to −10°F (−29 to −23°C)	*Cornus florida* *Deutzia gracilis* *Forsythia ovata* *Parthenocissus tricuspidata* *Rosa multiflora*	Flowering dogwood Slender deutzia Early forsythia Boston ivy Japanese rose
6 −10 to 0°F (−23 to −18°C)	*Acer palmatum* *Buxus sempervirens* *Forsythia suspensa* *Hedera helix* *Ilex opaca* *Ligustrum ovalifolium*	Japanese maple Common boxwood Weeping forsythia English ivy American holly California privet
7 0 to 10°F (−18 to −12°C)	*Azalea Kurume hyb. Hinodegiri* *Cedrus atlantica* *Cercis chinensis* *Chamaecyparis lawsoniana* *Cotoneaster salicifolia* *Ilex aquifolium*	Red Hussar azalea Atlas cedar Chinese redbud Lawson cypress Willowleaf cotoneaster English holly
8 10 to 20°F (−12 to −7°C)	*Arbutus menziesi* *Choisya ternata* *Melia azedarach* *Olearia haasti* *Prunus laurocerasus* *Viburnum tinus*	Pacific madrone Mexican orange Chinaberry New Zealand daisybush Cherry laurel Laurestinus
9 20 to 30°F (−7 to −1°C)	*Arbutus unedo* *Eucalyptus globulus* *Grevillea robusta* *Myrtus communis* *Pittosporum tobira* *Quercus virginiana*	Strawberry tree Tasmanian blue gum Silk oak Myrtle Japanese pittosporum Live oak
10 30° to 40°F (−1 to 4°C)	*Acacia baileyana* *Arecastrum romanozoffianum* *Bougainvillea spectabilis* *Casuarina equisetifolia* *Eucalyptus citriodora* *Ficus macrophylla*	Cootamundra wattle Queen palm Bougainvillea Horsetail beefwood (Australian pine) Lemon eucalyptus Moreton Bay fig

In addition, various cultural practices can influence the cold hardiness of a given plant. For example, late fall applications of fertilizer or improper pruning practices may stimulate flushes of growth that do not have sufficient time to acclimate. Conversely, insufficient mineral nutrition can also impair the development of maximum cold hardiness.

Freezing of plant tissues. Because plants are poikilotherms and assume the temperature of their surroundings, cold hardiness requires that they tolerate ice formation in tissues. When plant tissues freeze, the water they contain changes both its physical state and location. Ice formation may occur within cells (intracellular ice formation), in the intercellular spaces (extracellular ice formation), or in regions that separate different tissues or organs (extraorgan ice formation). In some organs and cells, such as floral buds or xylem ray parenchyma cells of deciduous forest species, ice formation may be precluded entirely. In these cases, the cytoplasm of the cell remains unfrozen below its freezing point (that is, supercooled) down to temperatures as low as $-40°F$ ($-40°C$). The location of ice formation (intracellular versus extracellular) is strongly influenced by the rate of cooling, the minimum temperature attained, and the degree of hardiness of the plant.

During cooling, plant tissues initially supercool; that is, ice formation does not occur at the freezing point of the cytoplasm. Instead, ice nucleation occurs on the external plant surfaces because of the presence of foreign nucleating agents, such as ice crystals in the air or, in some cases, bacteria. Ice then propagates through the extracellular spaces, but the plasma membrane (the outer membrane of the cell) prevents seeding of the cytoplasm by the external ice. Because ice has a lower chemical potential than water at the same temperature, a gradient in water potential exists between the unfrozen cytoplasm and the extracellular ice. Water moves in response to gradients in water potential; hence water will move from the cytoplasm to the extracellular ice and the cell will dehydrate. Because the plasma membrane is a semipermeable membrane that allows for the rapid efflux of water but does not permit the ready efflux of solutes, the cytoplasm becomes more concentrated. With increasing solute concentration, the freezing point of the cytoplasm is progressively lowered and the intracellular solution remains unfrozen. During warming, the sequence of events is reversed. The extracellular ice melts and the extracellular solution becomes less concentrated than the intracellular solution. As a result, water will flow back into the cell. Thus, during a freeze-thaw cycle, the cells behave as osmometers in response to ice formation in the extracellular spaces. SEE CELL MEMBRANES; CELL WALLS (PLANT).

The osmometric behavior of plant cells depends on the semipermeable characteristics of the plasma membrane that allow for the rapid movement of water while retaining solutes within the cell and excluding extracellular ice crystals. This behavior is readily observed microscopically. Although microscopic observations of biological specimens at subzero temperatures have been reported for over a century, there has been a renaissance in cryomicroscopy since the early 1970s. This has occurred because of the development of sophisticated temperature-control systems and advances in video recording and computer-enhanced video image analysis. As a result, cryomicroscopy has emerged as a very powerful technique for quantifying and understanding cellular behavior during a freeze-thaw cycle. Although it has been nearly 100 years since the cooling-rate dependency of intracellular ice formation was observed in higher plants, quantitative studies of the volumetric behavior of higher plant protoplasts are only now being reported. SEE CRYOBIOLOGY.

Isolated protoplasts are obtained from plant tissues by the enzymatic digestion of the cell walls. The liberated protoplasts behave as ideal osmometers. Their spherical shape over a wide range of solute concentrations facilitates quantitative cryomicroscopic studies of their osmometric behavior. Direct measurements of

volumetric behavior during a freeze-thaw cycle (**Fig. 3**) allow for the calculation of water efflux, the extent of supercooling, and water permeability of the plasma membrane. Coupled with a precise time and temperature history, such studies provide considerable insight into the causes and manifestations of freezing injury. Extensive dehydration of the cell occurs at subzero temperatures; following thawing of the suspending medium, the protoplasts rehydrate and osmotically increase in volume. Because the protoplasts in Fig. 3 were isolated from acclimated tissues, they survived the freeze-thaw cycle to a temperature of $14°F$ ($-10°C$).

Intracellular ice formation. Whether ice formation is restricted to the extracellular spaces or also occurs within the cells is of utmost importance because intracellular ice formation is invariably associated with lethal injury. In isolated rye protoplasts subjected to rapid cooling rates ($9°F$ or $5°C$ per minute; **Fig. 4**), intracellular ice formation is manifested as a darkening of the cells (Fig. 4c and d). This optical darkening or flashing is commonly observed, and was long thought to be the result of the diffraction of light by small ice crystals. High-resolution cryomicroscopy of isolated protoplasts reveals that the optical darkening is due to the formation of gas bubbles following intracellular ice formation. Upon melting of the ice, some of the gas bubbles coalesce (Fig. 4e). More importantly, however, intracellular ice formation is associated with lethal injury as manifested by the ruptured

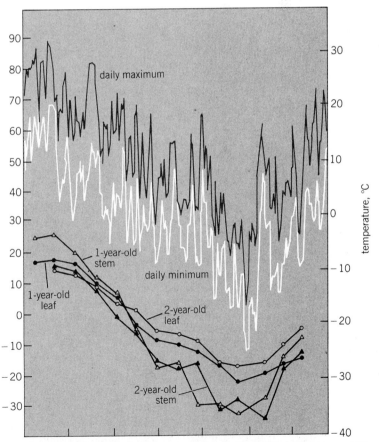

Fig. 2. Development of cold hardiness of *Hedera helix* (English ivy). The curves show killing points for structures at different ages. (*From P. L. Steponkus and F. O. Lanphear, Factors influencing artificial cold acclimation and artificial freezing of Hedera helix "Thorndale," Proc. Amer. Soc. Hort. Sci., 91:735–741, 1967*)

protoplasts (Fig. 4*f*). Intracellular ice formation is associated with early frost injury of herbaceous plants, and may also occur when there are sudden temperature changes in deciduous fruit tree buds.

The incidence of intracellular ice formation is strongly dependent on the rate at which the tissue is cooled and the minimum temperature attained. At

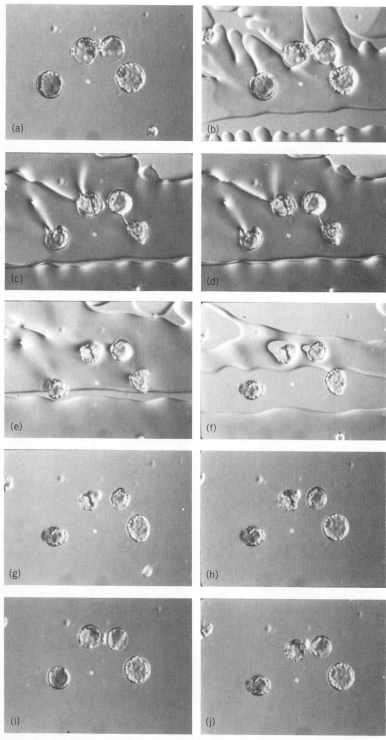

rapid cooling rates, efflux of water from the cell is limited by the water permeability of the plasma membrane or other impediments to water flux. As a result, concentration of the intracellular solutes and depression of the freezing point will not keep pace with the decreasing temperature, and the intracellular solution will become increasingly supercooled during cooling. With suspensions of isolated plant cells or protoplasts where the cells are surrounded by ice and the only barrier to water efflux is the plasma membrane, excessive supercooling will occur at cooling rates greater than 5°F (3°C) per minute. In intact tissues, rates greater than a few degrees per hour are sufficiently rapid to result in supercooling of the intracellular solution. This is because the ice crystals may be far removed from the intracellular solution and water flow is slowed by impediments resulting from tissue organization, contact of cell walls, or other barriers. Supercooling per se does not result in intracellular ice formation; it only predisposes the cell to this possibility. Nucleation or seeding of the supercooled solution is required for intracellular ice formation. This may occur by either homogeneous or heterogeneous nucleation or by seeding by the extracellular ice.

Homogeneous nucleation is a result of the spontaneous aggregation of water molecules to form ice nuclei and occurs only at temperatures below −38°F (−39°C). Thus, floral buds of some species or xylem ray parenchyma of some deciduous species will tolerate temperatures down to −38°F (−39°C) because they remain supercooled. At lower temperatures, however, intracellular ice formation occurs because of homogeneous nucleation. The sudden and rapid formation of ice is invariably lethal. There is a close relationship between the northern limits of many deciduous forest species that exhibit supercooling of the xylem ray parenchyma cells and the occurrence of temperatures below −40°F (−40°C).

Heterogeneous nucleation may occur at higher temperatures, but it requires the presence of foreign materials that are effective ice-nucleating agents. However, there is little evidence that such compounds are present in the cytoplasm. Instead, there is considerable evidence that intracellular ice formation is a consequence of seeding by extracellular ice as a result of perturbation or penetration of the plasma membrane. Direct cryomicroscopic observations of isolated protoplasts reveal that mechanical failure of the plasma membrane occurs immediately before intracellular ice formation. This occurs in the range of 23 to 5°F (−5 to −15°C) in nonacclimated plants. Thus, intracellular ice formation is a consequence of injury to the plasma membrane rather than a cause of injury.

Following cold acclimation, plants are less prone to intracellular ice formation. For many years it was assumed that this was because of an increase in the water permeability of the plasma membrane, which minimized the extent of supercooling. Recently, however, this was shown to be erroneous. Instead, cold acclimation decreases the incidence of intracellular ice formation because the stability of the plasma membrane at lower temperatures is increased and defers seeding by extracellular ice to temperatures as low as −40°F (−40°C).

Freezing injury due to cell dehydration. The visual symptoms of freezing injury—a darkened, water-soaked, flaccid appearance—are apparent immediately following thawing. The rapidity in which injury

Fig. 3. Cryomicroscopic observations of isolated rye (*Secale cereale*) protoplasts during a freeze-thaw cycle. Cooling: (*a*) 32°F (0°C), (*b*) 28°F (−2°C), (*c*) 23°F (−5°C), (*d*) 14°F (−10°C). Warming: (*e*) 17°F (−8°C), (*f*) 25°F (−4°C), (*g*) 32°F (0°C). (*h–j*) Isothermal at 32°F (0°C). Note the extensive dehydration of the cell that occurs at subzero temperatures. (*Courtesy of Peter L. Steponkus*)

is manifested indicates that injury is not the result of metabolic dysfunction. Instead, the primary cause of freezing injury is the disruption of the plasma membrane. Manifestations of injury may range from alterations in the semipermeable characteristics to mechanical rupture, and injury can occur at various times during a freeze-thaw cycle (**Fig. 5**). The various forms of injury observed are (1) expansion-induced lysis during warming and thawing of the suspending medium when the decreasing concentration of the suspending medium results in osmotic expansion of the protoplasts; (2) loss of osmotic responsiveness following cooling at slow rates so that the protoplast is osmotically inactive during warming; and (3) altered osmometric behavior during warming, suggestive of altered semipermeable characteristics or leakiness of the plasma membrane.

These symptoms of injury are not consequences of low temperatures per se because low temperatures in the absence of freezing are not injurious to hardy plants. Thus, thermal stresses that may result in membrane phase transitions are unlikely to contribute to freezing injury. Similarly, there is little evidence that injury occurs because of the presence of ice crystals. Instead, injury is primarily a consequence of cell dehydration, and there is a very good correlation between cold hardiness and the ability to withstand dehydration in the absence of freezing. There are, however, several consequences of cell dehydration that can result in injury to the plasma membrane.

In nonacclimated tissues, cell dehydration results in the deletion of membrane material from the plasma membrane in the form of membrane vesicles (**Fig. 6a**). During contraction of the protoplast, these vesicles are detached from the membrane and become

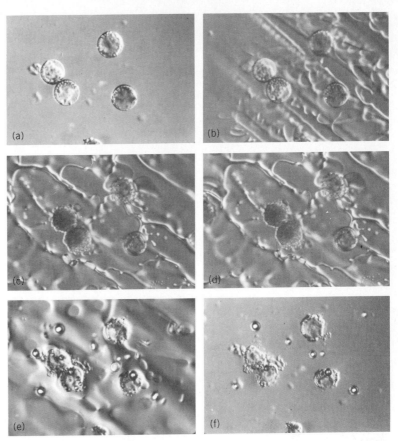

Fig. 4. Intracellular ice formation in isolated rye (*Secale cereale*) protoplasts. Cooling: (*a*) 32°F (0°C), (*b*) 28°F (−2°C), (*c*) 14°F (−10°C), (*d*)14°F (−10°C). Warming: (*e*) 28°F (−2°C), (*f*) 32°F (0°C). (*Courtesy of Peter L. Steponkus*)

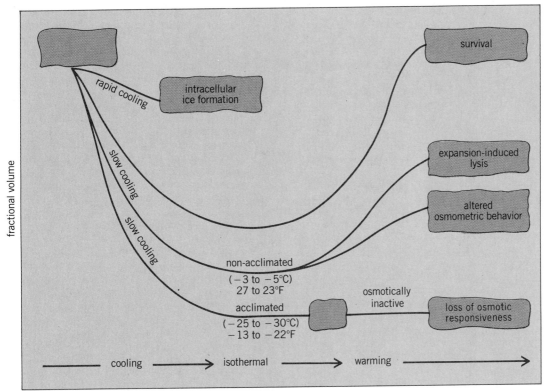

Fig. 5. Manifestations of freezing injury to the plasma membrane.

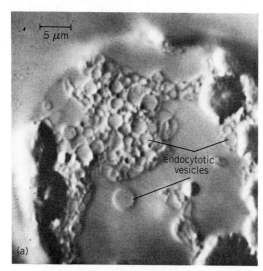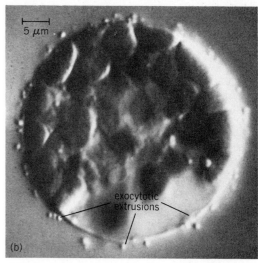

Fig. 6. High-resolution computer-enhanced differential interference contrast micrographs of (a) nonacclimated and (b) acclimated protoplasts subjected to hypertonic solutions. (From P. L. Steponkus et al., Destabilization of the plasma membrane of isolated plant protoplasts during a freeze-thaw cycle: The influence of cold acclimation, Cryobiology, 20:448–465, 1983)

visible in the cytoplasm. Upon thawing, the tissue increases in volume, but the previously deleted membrane material is not reincorporated into the membrane, and rupture of the plasma membrane (lysis) occurs before the cells regain their original size. Hence, this form of injury is referred to as expansion-induced lysis or deplasmolysis injury, and is the predominant form of injury following cell dehydration in nonacclimated herbaceous tissues such as winter cereals (wheat or rye). Injury is largely a consequence of the mechanical stresses imposed on the plasma membrane during cell dehydration and rehydration.

Expansion-induced lysis is not observed in acclimated tissues because membrane deletion does not occur during contraction. Instead, extrusions of the plasma membrane that remain contiguous with the plasma membrane are formed (Fig. 6b). These extrusions are readily reincorporated into the plane of the membrane during expansion. Hence, mechanical stresses of the plasma membrane are not responsible for injury in acclimated tissues. Instead, injury is largely the result of chemical stresses associated with cell dehydration.

Loss of osmotic responsiveness occurs in protoplasts isolated from nonacclimated tissues when the protoplasts are cooled to 14°F (-10°C) or lower. At these temperatures severe freeze-induced dehydration occurs with a loss of nearly 90% of the intracellular water, and with a tenfold increase in the concentration of the intracellular solutes. The semipermeable properties of the plasma membrane are disrupted so that following thawing the protoplasts are osmotically unresponsive. In tissues, the cells appear to be plasmolyzed, and therefore this form of injury has been referred to as frost plasmolysis. However, this term is misleading in that the cells cannot be considered plasmolyzed if they are not osmotically responsive.

Loss of osmotic responsiveness is associated with several changes in the ultrastructure of the plasma membrane, including lateral phase separations and bilayer-to-nonbilayer transitions, such as the formation of inverted cylindrical micelles or hexagonal$_{II}$ phase. These changes occur as a consequence of the close approach (less than 2 to 3 nanometers) of membranes

during cell dehydration. Normally, strongly repulsive hydration forces preclude the close approach of two bilayers. However, the high osmotic pressures that occur during freeze-induced dehydration overcome the hydration forces. Dehydration-induced lateral phase separations and the hexagonal$_{II}$ phase are not observed in protoplasts isolated from cold-acclimated tissues. The increased stability may be a consequence of alterations in the plasma membrane lipid composition that preclude dehydration-induced demixing of the lipids, and the accumulation of cytoplasmic solutes such as sucrose or proline.

Cold acclimation. The process of cold acclimation results in numerous biochemical changes within the plant. These include increases in growth inhibitors and decreases in growth promoters; changes in nucleic acid metabolism; alterations in cellular pigments such as carotenoids and anthocyanins; the accumulation of carbohydrates, amino acids, and water-soluble proteins; increases in fatty acid unsaturation; changes in lipid composition; and the proliferation of cellular membrane systems. Not all of these changes are involved in increased cold hardiness; some are merely changes in response to slower growth rates and decreased photosynthate utilization; others are changes associated with growth at low, above-zero temperatures; and still others are associated with other developmental phenomena, such as vernalization or the induction of dormancy, that also occur during the period of cold acclimation.

As cellular membranes are the primary site of freezing injury, it follows that cold acclimation must involve cellular alterations that allow the membranes to survive lower freezing temperatures. Such alterations may be in the cellular environment so that either the freezing stresses are altered or there is direct protection of the membranes. In addition, cold acclimation may involve changes in the membrane itself, so that its susceptibility to the freezing stresses is decreased.

Large increases in cellular solute concentrations are one of the most universal manifestations of cold acclimation. A doubling of the intracellular solute concentration, most notably sugars, is not uncommon. Such increases have several beneficial effects. First, they serve to depress the freezing point of the intracellular solution. This, however, is of limited advantage because such increases only lower the freezing point to the extent of 1.03°F/osmolal (1.86°C/osmolal).

More important is their effect on the extent of cell dehydration that occurs during freezing. A doubling of the initial intracellular solute concentration will decrease the extent of cell dehydration at any subzero temperature by 50%. An increase in intracellular solutes will also decrease the concentration of toxic solutes at temperatures below 32°F (0°C), because less water will be removed. During dehydration protective compounds, such as sugars, are concentrated along with toxic compounds, such as electrolytes. Because the total concentration of the intracellular solution is a function of the subzero temperature, the toxic compounds will account for only a portion of the total. If the ratio of protective to toxic compounds is initially greater, the final concentration of the toxic compounds at any subzero temperature will be lowered accordingly. In addition, there are some compounds, such as sucrose or proline, that may interact directly with cellular membranes to increase their stability either at low water contents or at high electrolyte concentrations.

Following cold acclimation there are substantial changes in the lipid composition of the plasma membrane. Although there are no lipid molecular species that are unique to the plasma membrane of either nonacclimated or acclimated rye leaves, cold acclimation alters the proportion of virtually every lipid component. This includes an increase in free sterols with corresponding decreases in steryl glucosides and acylated steryl glucosides, a decrease in the glucocerebroside content, and an increase in the phospholipid content. Although the relative proportions of the individual phospholipids (for example, phosphatidylcholine and phosphatidylethanolamine) do not change appreciably, there are substantial differences in the individual lipid molecular species. These include increases in species of phosphatidylcholine and phosphatidylethanolamine that contain two unsaturated fatty acids, whereas species with a single unsaturated fatty acid remain relatively unchanged. The complexity of the plasma membrane lipid composition and the numerous changes that occur during cold acclimation preclude the possibility that any simple correlative analysis of the changes will establish their role in the cold acclimation process; instead, a mechanistic analysis is required.

One approach has been to demonstrate that the differential behavior of the plasma membrane observed in protoplasts isolated from nonacclimated and cold-acclimated leaves (that is, the formation of endocytotic vesicles versus exocytotic extrusions during osmotic contraction and the differential propensity for dehydration-induced bilayer-to-nonbilayer phase transitions) is also observed in liposomes formed from plasma membrane lipid extracts of nonacclimated and acclimated leaves. Thus, the increased cryostability of the plasma membrane following cold acclimation is a consequence of alterations in the lipid composition. However, there are over 100 different lipid molecular species present in the plasma membrane, so that the problem is to determine which of the changes are responsible for the differential cryobehavior. In studies that have used a protoplast–liposome fusion technique

to enrich the plasma membrane with specific lipids, it has been established that increases in the proportion of phosphatidylcholine species composed of one or two unsaturated fatty acids increase the freezing tolerance of protoplasts isolated from nonacclimated rye leaves. The increase in freezing tolerance is the result of a transformation in the cryobehavior of the plasma membrane during osmotic contraction so that exocytotic extrusions are formed rather than endocytotic vesicles. As a result, expansion-induced lysis is precluded. Thus, these studies demonstrate that although there are over 100 lipid molecular species in the plasma membrane, altering the proportion of a single species dramatically alters the cryobehavior of the plasma membrane. Similar membrane engineering studies have been undertaken for the other forms of freezing injury (for example, loss of osmotic responsiveness) and should ultimately provide a molecular understanding of the role of plasma membrane lipid alterations in increasing the cryostability of the plasma membrane. SEE ALTITUDINAL VEGETATION ZONES; CELL, PRESSURE-TEMPERATURE EFFECTS IN; PLANT PHYSIOLOGY; PLANT-WATER RELATIONS.

Peter L. Steponkus

Bibliography. B. W. W. Grout and G. J. Morris (eds.), *The Effect of Low Temperatures on Biological Membranes*, 1987; O. L. Lange et al. (eds.), *Encyclopedia of Plant Physiology*, 1981; J. Levitt, *Responses of Plants to Environmental Stresses*, 2d ed., 1980; P. H. Li and A. Sakai (eds.), *Plant Cold Hardiness and Freezing Stress*, vol. 1, 1978, vol. 2, 1982; D. V. Lynch and P. L. Steponkus, Plasma membrane lipid alterations associated with cold acclimation of winter rye seedlings (*Secale cereale* L. cv Puma), *Plant Physiol.*, 83:761–767, 1987; P. L. Steponkus, Cold hardiness and freezing injury of agronomic crops, *Adv. Agron.*, 30:51–98, 1978; P. L. Steponkus, The role of the plasma membrane in freezing injury and cold acclimation, *Annu. Rev. Plant Physiol.*, vol. 35, 1984; P. L. Steponkus and D. V. Lynch, Freeze/thaw-induced destabilization of the plasma membrane and the effects of cold acclimation, *J. Bioenergetics Biomembranes*, 21:21–24, 1989.

Cold storage

Keeping perishable products at low temperatures in order to extend storage life. Cold storage vastly retards the processes responsible for the natural deterioration of the quality of such products at higher temperatures. Time and temperature are the key factors that determine how well foods, pharmaceuticals, and many manufactured commodities, such as photographic film, can retain properties similar to those they possess at the time of harvest or manufacture.

Food that is placed in cold storage is protected from the degradation that is caused by microorganisms. At 80°F (27°C), bacteria will multiply 3000 times in 12–24 h; at 70°F (21°C), the rate of multiplication is reduced to 15 times; and at 40°F (4°C), it is reduced to 2 times. The lower limit of microbial growth is reached at 14°F (−10°C); microorganisms cannot multiply at or below this temperature. SEE BACTERIAL GROWTH; FOOD MICROBIOLOGY.

Cold storage through refrigeration or freezing makes it possible to extend both the seasons of harvest and the geographic area in which a product is available. In the past, food products were grown locally and had to be marketed within a short period of time. Modern cold storage technology makes virtually any product available year-round on a global basis. Other technologies have been combined with refrigeration to further improve this availability. For example, in controlled-atmosphere storage of apples, controlled temperatures above freezing are maintained in a sealed room where the air is also modified to increase its nitrogen content (20.9% oxygen, 0.03% carbon dioxide, and 78.1% nitrogen with other gases) to keep apples orchard-fresh from one fall harvest through the next.

The cold storage food chain begins at the farm or packing plant where the product is chilled by three principal methods: hydrocooling (immersion in chilled water), forced-air cooling, and vacuum cooling (placing the product in a sealed chamber and creating a vacuum, causing evaporation of some of the water in the product, and subsequent cooling).

If the product is to be frozen, one of five methods is used: (1) air blast freezing (cold air at high velocity is passed over the product); (2) contact freezing (the product is placed in contact with metal plates and heat is drawn off by conduction); (3) immersion freezing (the product is immersed in low-temperature brine); (4) cryogenic freezing [the product is exposed in a chamber to temperatures below $-76°F$ ($-60°C$) by using liquid nitrogen or liquid carbon dioxide]; and (5) liquid refrigerant freezing (the product is immersed and sprayed with a liquid freezant at atmospheric pressure).

The next step in the cold storage food chain is transport by railroad cars, trucks, airplanes, or boats fitted with refrigeration units that maintain temperatures to critical specifications. Electronic technology permits monitoring of temperature and location by satellite transmission from the transport vehicle to a land-based monitoring station.

Refrigerated warehouses and distribution centers maintain the temperatures required to assure continued maintenance of quality before the product makes its last commercial move to supermarkets or to food-service-industry outlets. Typically, a refrigerated warehouse is a fully insulated structure fitted with refrigeration equipment capable of precise maintenance of specific temperatures in rooms holding up to several million pounds of product. For example, individual rooms may be set to maintain refrigerated storage temperatures of $34°F$ ($1°C$) or freezer-room temperatures of $0°F$ ($-18°C$). Some food products are being marketed that may require storage temperatures as low as $-20°F$ ($-29°C$) to maintain texture and to preclude separation of ingredients. *See* REFRIGERATION.

Anhydrous ammonia is the principal refrigerant used in refrigeration systems for cold storage. A number of systems had been introduced that used chlorofluorocarbon refrigerants. However, they are being phased out because of concern about the ozone layer. It is anticipated that such systems will be converted to ammonia or other refrigerants. Most engine rooms are computer-controlled in order to allow optimum use of energy and to transmit alerts if temperatures stray out of the programmed range. *See* GREENHOUSE EFFECT; HALOGENATED HYDROCARBON.

Handling of foods through distribution and cold storage is done by using pallets loaded with 1400–3000 lb (630–1350 kg) of product. Modern freezer rooms are 28–32 ft (8.5–10 m) high for conventional forklift operation, and up to 60 ft (18 m) high when automated or specialized materials-handling equipment is used. *See* FOOD MANUFACTURING; FOOD PRESERVATION; MATERIALS HANDLING.

Michael Shaw

Bibliography. American Society of Heating, Refrigerating, and Air-Conditioning Engineers, *ASHRAE Applications Handbook*, 1982; E. R. Hallowell (ed.), *AVI Cold and Freezer Storage Manual*, 2d ed., 1980; International Institute of Refrigeration, *A History of Refrigeration Throughout the World*, 1979; J. Mogens, *Quality of Frozen Foods*, 1984; A. M. Pearson and T. R. Dutson (eds.), *AVI Advances in Meat Research*, vol. 3, 1987.

Coleoidea

A subclass of the Cephalopoda that appeared in the middle Paleozoic (Early Devonian Period) and presumably evolved from the nautiloid stalk. Of the five orders of fossil coleoids, the belemnites are conspicuous with their fossilized shells termed lightning bolts. All these forms became extinct by the end of the Mesozoic, 65 million years ago. The surviving four orders that developed in the Late Triassic to Late Cretaceous represent all the living cephalopods today, except *Nautilus* (subclass Nautiloidea).

Coleoids are characterized by an internal chitinous or calcareous shell; 8–10 appendages around the mouth (8 arms and 2 tentacles when present) lined with suckers or hooks; one pair of gills; highly developed eyes; a fused, tubelike funnel; an ink sac (lost in some species); chromatophores; and fins on the body (lost in some octopuses).

The living coleoids are the order Sepioidea (cuttlefishes, like *Sepia*; bobtail squids, *Spirula*); the order Teuthoidea (squids—nearshore myopsids, like *Loligo*; open-ocean oegopsids, like *Ommastrephes*); the order

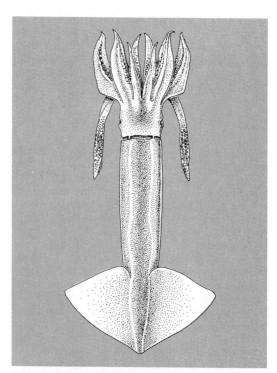

Coleoid cephalopod, the oceanic squid *Ommastrephes bartrami*, from worldwide temperate and tropical waters.

Vampyromorpha (the black, deep-sea vampire squid); and the order Octopoda (octopuses, argonauts, deep-sea finned or cirrate octopods). Living coleoids generally are fast, mobile predators with an advanced brain and central nervous system. SEE BELEMNOIDEA; CEPHALOPODA; MOLLUSCA; NAUTILOIDEA; SEPIOIDEA; TEUTHOIDEA; VAMPYROMORPHA.

Clyde F. E. Roper

Bibliography. M. R. Clarke and E. R. Trueman (ed.), *The Mollusca*, vol. 12: *Paleontology and Neontology of Cephalopods*, 1988; E. R. Trueman and M. R. Clarke (eds.), *The Mollusca*, vol. 11: *Form and Function*, 1988.

Coleoptera

An order of Insecta (generally known as beetles) with a complete metamorphosis (with larva and pupa stages), the forewings forming hard protective elytra under which the hindwings are normally folded in repose, and the indirect flight muscles of the mesothorax having been lost. The head capsule is characterized by a firm ventral (gular) closure, the antennae are basically 11-segmented, and mouthparts are of the biting type with 4-segmented maxillary palpi. The prothorax is large and free. The abdomen has sternite I normally absent, sternite II usually membranous and hidden, and segment X vestigial, with cerci absent. The female has one pair of gonapophyses, on segment IX. There are four or six malpighian tubules, which are often cryptonephric. True labial glands are nearly always absent (**Fig. 1**).

GENERAL BIOLOGICAL FEATURES

The order has well over 250,000 described species, more than any comparable group, showing very great diversity in size, form (**Fig. 2**), color, habits, and physiology. Some Ptiliidae have body lengths as small as 0.01 in. (0.25 mm), and some Cerambycidae may exceed 4 in. (100 mm). Beetles have been found almost everywhere on Earth (except as yet for the Antarctic mainland) where any insects are known, and species of the order exploit almost every habitat and type of food which is used by insects. Many species are economically important.

A common feature is for the exoskeleton to form an unusually hard and close-fitting suit of armor, correlated with a tendency for the adults to be less mobile but longer-lived than those of other orders of higher insects. It seems to be common in a number of groups for adults to survive and breed in 2 or more successive years, a rare occurrence in other Endopterygota. Most beetles, at least in markedly seasonal climates, have a 1-year life cycle, with developmental stages adapted to specific seasons, but many larger species (particularly when larvae burrow in wood or the soil) may take 2 or more years per generation, and in suitable climates some (usually smaller) species may have more than one generation per year. Another feature unusual in Endopterygota is that many beetles have adults and larvae living in the same habitats and feeding on similar food, though many others have adult foods and habitats quite different from the larval ones. Adult beetles tend to be better runners than most other Endopterygota, and some (such as Cicindelidae) are among the fastest of running insects.

Wings and flight. The flight of most beetles is not very ready or frequent, often requiring prior climbing

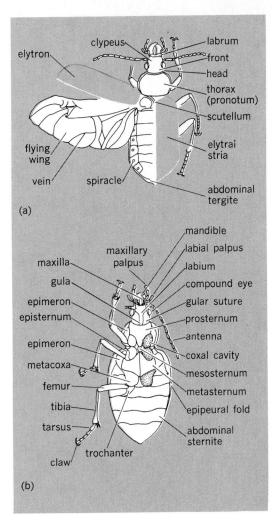

Fig. 1. Diagrammatic sketch of Carabidae member. (*a*) Dorsal view. (*b*) Ventral view.

up onto some eminence, and in many beetles occurs only once or twice in an average life. The elytra may be held in a rigid lateral projection, as in Adephaga, or vibrate through some (usually small) angle in phase with the hindwings, as in many Polyphaga, or be held vertically upward or over the abdomen, almost in the position of repose. In a number of groups, there is evidence that the elytra contribute significantly to lift or to stability in flight. In males of Stylopidae (Strepsiptera) they are reduced to structures resembling (and probably functioning like) the halteres of Diptera. The hindwings have their insertions shifted very far forward, making the pleura of the metathorax very oblique, and their area is expanded in the anal region. Wing loading and wing-beat frequency are usually higher than in most other Pterygota, with the indirect flight muscles of the metathorax of the asynchronous type, contracting several times in response to a single nerve impulse. Wing venation is adapted to the needs of folding as well as of flight. In some groups, the wings unfold automatically if the elytra are lifted (as in Staphylinidae), but folding after flight is a more complex process, involving active participation of the flexible abdomen. In others unfolding requires considerable abdominal activity, but folding is quick and quasi-automatic, resulting from muscle pulls on the axillary sclerites of the wing base. Most beetles seem to be inherently stable (hence not very maneuverable) in flight. In a number of groups (such as Pleocomi-

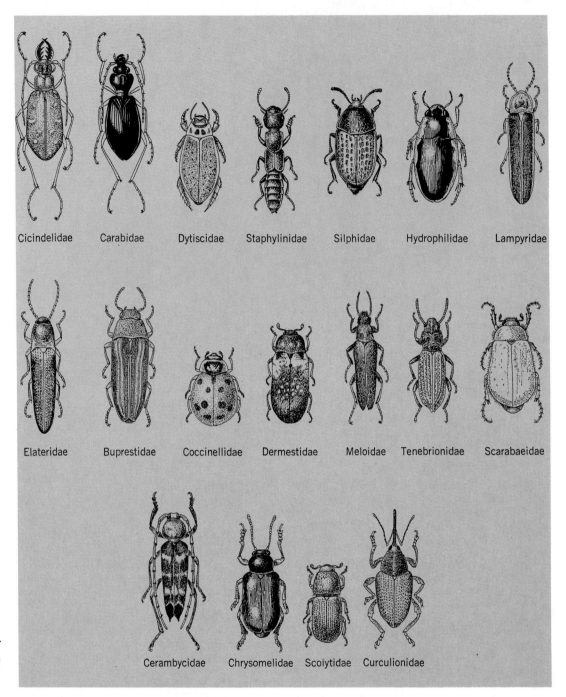

Fig. 2. Representatives of well-known families of Coleoptera. (*After T. I. Storer and R. L. Usinger, General Zoology, 3d ed., McGraw-Hill, 1957*)

Cicindelidae — Carabidae — Dytiscidae — Staphylinidae — Silphidae — Hydrophilidae — Lampyridae

Elateridae — Buprestidae — Coccinellidae — Dermestidae — Meloidae — Tenebrionidae — Scarabaeidae

Cerambycidae — Chrysomelidae — Scoiytidae — Curculionidae

dae, Cebrionidae, Drilidae, Phengodidae, and Stylopidae), flight may be confined to the males, indicating that its prime function is to promote gene dispersal.

Cuticle. Apart from a tendency for the sclerotized layers to be thicker, the cuticle of beetles differs little from that of other insects in constitution or ultrastructure. The epicuticular wax layer shows diversity, being particularly well developed in some of the desert Tenebrionidae. Macrochaetae on the body surface may develop into scales (a trait of many Curculionidae) but are rarely plumose; in larval Dermestidae, very complex defensive hairs are developed.

The black coloring which is prevalent in many groups of beetles is produced by deposition of melanin in the process of cuticle hardening. In many groups, with adults active by day, the black is masked by a metallic structural color produced by interference in the surface cuticle layers. Pigmentary colors, mainly ommochromes, giving reddish to yellowish colors, occur in the cuticle of many species, whereas some groups, notably the tortoise beetles (Cassidinae), may have more unstable greenish pigments in the hypodermal cells beneath the cuticle. Pigmentary colors are often adapted to produce aposematic or pseudaposematic appearances, or cryptic ones. SEE PROTECTIVE COLORATION.

Defense adaptations. Special defenses against predation are common in species with long-lived adults, particularly the ground-living species of Carabidae, Staphylinidae, and Tenebrionidae. These groups commonly possess defensive glands, with reservoirs opening on or near the tip of the abdomen, secreting quinones, unsaturated acids, and similar toxic substances. In some cases, the secretion merely oozes onto the body surface, but in others may be expelled in a jet. In the bombardier beetles (Brachinini), the jet contains quinones and is expelled forcibly at nearly 212°F (100°C) through the agency of hydrogen peroxide in an explosion chamber. In other groups, toxins may be dispersed in the body, as in species of Coccinellidae, Lycidae, and Meloidae, and warning (aposematic) coloration is frequently developed.

Other common defensive adaptations include cryptic and mimetic appearances. The most frequent behavioral adaptations are the drop-off reflex and appendage retraction. Numerous Chrysomelidae, Curculionidae, and other adults will react to visual or tactile stimuli by dropping off the plant foliage on which they occur, falling to the ground with retracted appendages, and lying there for some time before resuming activity. In death feigning, the appendages are tightly retracted, often into grooves of the cuticle, and held thus for some time. In such conditions, beetles often resemble seeds and are difficult for predators to perforate or grasp.

Stridulation. Sound production (stridulation) is widespread in beetles and may be produced by friction between almost any counterposed movable parts of the cuticle. Stridulatory organs may show sex dimorphism, and in some species (such as among Scolytinae) may play an important role in the interrelations of the sexes. Elsewhere, as in species of *Cychrus, Hygrobia,* and *Necrophorus,* stridulation is a common response to molestation. Specialized sound-detecting organs, other than Johnston's organ in the antennal pedicel, are not well documented in Coleoptera.

Cryptonephrism. Cryptonephrism, in which the apical parts of the malpighian tubules are closely applied to the hindgut in the region of the rectal glands, is characteristic of a large section of Polyphaga, including all the "stored product" pests and the dominant desert beetles. This condition, not known in other adult insects, has been shown in *Tenebrio* to economize water by extracting it from the rectal contents to produce very dry feces. In cryptonephric groups, certain more or less aquatic members seem to have lost the condition secondarily.

Symbioses. Of the many types of symbiotic relations into which beetles enter with other organisms, perhaps the most ecologically important are those with fungi, which can be divided into ecto- and endosymbioses. In ectosymbiosis, the beetles carry spores or propagules of the fungi and introduce them into suitable new habitats, and are themselves nutritionally dependent (at least as larvae) on the presence of the right type of fungus in the habitat. Classic examples are in numerous bark beetles (Scolytinae) and Lymexylidae. In endosymbiosis, fungi (or sometimes bacteria) are maintained in the gut (or in mycetomes attached to it), usually of the larvae, and contribute either essential nutrients (such as steroids) or essential enzymes. Endosymbionts are usually passed externally or internally to the eggs by the ovipositing female. In endosymbiosis, it seems that as a rule the symbionts are specific to the beetles concerned, to a rather greater degree than in ectosymbionts.

Genetics. The large majority of beetles have an X-y type of chromosomal sex-determining system, with the y often small (or lost altogether) and forming a "parachute" figure with the X in meiosis. Sex ratios in natural populations are usually 50:50, and sex dimorphism, usually slight, is sometimes very marked.

Thelytokous parthenogenesis, often accompanied by polyploidy, is developed in several groups. A haploid male system, like that of Hymenoptera, has been found in Micromalthidae and a few *Xyleborus* ambrosia beetles. A basic complement of nine pairs of autosomes, plus the sex pair, is maintained in many groups of beetles. *See Genetics.*

Mating and life cycle. Mating in beetles is often a simple process, the male climbing onto the back of the female after a minimum of courtship, but in some groups (such as many Meloidae, Cerambycidae, and Scarabaeidae) more elaborate courtship is the rule. In long-lived species, individuals may copulate several times, often with different partners. Care for the young is commonly limited to the deposition of the eggs in suitable habitats under some sort of cover, but nest construction and provisioning, after the manner of solitary Aculeata Hymenoptera, develop particularly in the dung-feeding Scarabaeinae and Geotrupidae, and in a few Carabidae. Other beetles showing unusual provisions for their offspring include the ambrosia beetles or Attelabidae, some Staphylinidae-Oxytelinae, and some Lamiinae.

Beetle eggs rarely have heavily sculptured chorions or specialized caps for eclosion. Embryonic development is of typically insect types but shows considerable variation within the order. Eggs are often adapted to absorb moisture during embryonic development.

Larvae are very diverse, but usually have a well-developed head capsule with biting mouthparts, often have thoracic legs but very rarely abdominal prolegs (other than a pygopod developed from segment X), and never have true appendages on segment X. The tergite of segment IX often bears paired (sometimes articulated) posterior outgrowths known as urogomphi. First instars often differ considerably from later ones and may have toothlike egg bursters on parts of the body. The number of larval instars usually lies between 3 and 10. Pupae are usually white and exarate, resting on their backs on dorsal spines in some kind of pupal cell. They are never decticous, nor does the adult become active inside the pupal skin.

Hypermetamorphosis, in which two or more sharply different larval forms succeed during individual development, is known in several parasitic or wood-boring groups of beetles. The most common type has an active long-legged first-instar larva which seeks out a suitable feeding site and molts into a grublike type with reduced legs, as in Stylopidae, Rhipiphoridae, and Micromalthidae. The bee-parasitic Meloidae are notable in having two or even three distinct later larval stages.

ADULT ANATOMY

The head of beetles rarely has a marked posterior neck, and the antennal insertions are usually lateral, rather than dorsal as in most other Endopterygota, a feature possibly related to an original habit of creeping under bark. In most Curculionoidea and a few other groups, the head may be drawn out in front of the eyes to form a rostrum. Paired ventral gular sutures (sometimes partially confluent) are usually present, giving rise to a fairly normal tentorium; a frontoclypeal suture (representing an internal ridge) is often present. Two pairs of ventral cervical sclerites are found in Polyphaga only. Dorsal ocelli are rarely present, and never more than two; compound eyes range from occupying most of the head surface (as in male Phengodidae) to total absence (as in many cavernicolous or subterranean forms). Antennal forms are very various, with the number of segments often reduced, but rarely increased, from the basic 11. Mandible forms are very diverse and related to types of food.

Thorax. The prothorax commonly has lateral edges which separate the dorsal part (notum) from the ventral part (hypomeron) of the tergum. The propleuron may be fully exposed (Adephaga, Archostemata) or reduced to an internal sclerite attached to the trochantin (Polyphaga). The front coxal cavities may or may not be closed behind by inward processes from the hypomera or pleura to meet the prosternal process.

The mesothorax is the smallest thoracic segment. Dorsally, its tergum is usually exposed as a small triangular sclerite (scutellum) between the elytral bases, and ventrally, its sternite may have a median pit receiving the tip of the prosternum. The elytra of beetles develop in a different way from those of earwigs, cockroaches, and such. The first stage is the modification of wing venation to form a series of parallel longitudinal ridges connected by a latticework of crossveins (as seen in many modern Archostemata). The veins then expand to obliterate the remaining membrane, often leaving 9 or 10 longitudinal grooves or rows of punctures representing the original wing membrane and having internal pillars (trabeculae) connecting the upper and lower cuticle. There are various types of dovetailing along the sutural margins of the elytra, and their outer margins are often inflexed to form epipleura.

The form of the metathorax is affected by far-forward insertions of the hindwings and far-posterior ones of the hindlegs, so that the pleura are extremely oblique and the sternum usually long. Usually, the episterna are exposed ventrally, but the epimera are covered by the elytral edges. The metasternum commonly has a median longitudinal suture (with internal keel), and primitively, there may be a transverse suture near its hind edge. Internally, the endosternite takes diverse and characteristic forms.

Tarsi. The tarsi of beetles show all gradations from the primitive five-segmented condition to total disappearance. The number of segments may differ between the legs of an individual or between the sexes of a species. A very common feature in the order is the development of ventral lobes, set with specialized adhesive setae, on some tarsal segments. Often, strong lobes of this kind occur on the antepenultimate segments, and the penultimate one is very small and more or less fused to the claw joint. Lobed tarsi are particularly characteristic of species frequenting plant foliage.

Abdomen. Most beetles have the entire abdomen covered dorsally by the elytra in repose, and ventrally only five or six sternites (of segments III to VII or VIII) exposed; segment IX is retracted inside VIII in repose. Tergites I–VI are usually soft and flexible, those of VII and VIII more or less sclerotized. In some groups, the elytra may be truncate, leaving parts of the abdomen exposed dorsally, in which case the exposed tergites are sclerotized. Sternite IX is usually a single (internal) sclerite in males, but divided in females to form a pair of valvifers. Male gonapophyses are typically represented by a median basal piece bearing a pair of parameres (lateral lobes), and those of the female by a pair of coxites and styli attached to the valvifers. In females of many groups, an elongate ovipositor is produced by elongation into rods of sclerites of abdominal segment IX.

Elytra. A feature developed in several groups of Coleoptera, but particularly in Staphylinidae, is the abbreviation of the elytra to leave part of the abdomen uncovered in repose, but usually still covering the folded wings. One advantage of this may be to give greater overall flexibility to the body. This is probably of particular advantage to species living in the litter layer in forests, a habitat in which Staphylinidae are strikingly dominant. Perhaps surprisingly, many beetles with short elytra fly readily.

Internal anatomy. Internally, the gut varies greatly in length and detailed structure. The foregut commonly ends in a distended crop whose posterior part usually has some internal setae and may be developed into a complex proventriculus (Adephaga, many Curculionoidea); the midgut may be partly or wholly covered with small papilae (regenerative crypts) and sometimes has anterior ceca. The four or six malpighian tubules open in various ways into the beginning of the hindgut and in many Polyphaga have their apices attached to it (cryptonephric). The central nervous system shows all degrees of concentration of the ventral chain, from having three thoracic and eight abdominal ganglia distinct (as in some Cantharoidea) to having all of these fused into a single mass (as in some higher Scarabaeidae). There are two main types of testes in the males: Adephaga have each testis as a single, long, tightly coiled tube wrapped around in a membrane; Polyphaga have testes usually of several follicles. In the female, Adephaga have polytrophic ovarioles, while Polyphaga have acrotrophic ones, unlike any others known in Endopterygota.

LARVAL ANATOMY

The head capsule in beetle larvae is usually well developed with fairly typical biting mouthparts, with or without a ventral closure behind the labium. Antennae are basically three-segmented, segment II (pedicel) usually bearing a characteristic sensory appendage beside segment III. In several groups, segment I may be divided to give four-segmented antennae (Carabidae, many Staphylinidae, Scarabaeidae), but division of segment III seems to be peculiar to Helodidae. There are up to six ocelli on each side, variously grouped. The frontal sutures (ecdysial lines) often have a characteristic inverted horseshoe form; a median coronal suture may be present. The mandibles have very diverse forms, related to types of food, and the labrum is often replaced by a toothed nasale, particularly in carnivorous types practicing extraoral digestion. The maxillae also vary considerably. In the more specialized predatory types, the cardo tends to be reduced. In some groups, maxillae and labium are more or less fused together to form a complex resembling the gnathochilarium of Millipedes. In some groups with burrowing or internal-feeding larvae, the head capsule may be largely retracted into the prothorax, with corresponding reductions of sclerotization.

Thorax. In the thorax, the prothorax is usually markedly different from the next two segments. The legs are often well developed, particularly in ground-living predatory types, and in Adephaga and Archostemata may have a separate tarsus with two claws. Reduction or loss of legs is characteristic of many

(a)

(b)

Members of the order Coleoptera. (a) *Lucanus cervus*. (b) *Batocera wallaci*. (c) *Cerapterus lafertei*. (d) *Calosoma sycophanta*. (e) *Tetralobus flabellicornus*. (Editorial Photocolor Archives)

(c)

(d)

(e)

burrowing or internal-feeding larvae. The first pair of spiracles is usually situated close to the front border of the mesothorax, lateroventrally; very rarely, a second thoracic spiracle is distinct.

Abdomen. Ten abdominal segments are usually visible, prolegs are rarely developed on any of the first eight segments, but segment X often forms a proleglike pygopod. The tergite of segment IX is often drawn out into posterior processes known as urogomphi, which may become articulated and movable (as in Hydradephaga, Staphylinoidea). In some aquatic larvae, segments IX and X are reduced, and the spiracles of segment VIII become effectively terminal, as in most Diptera larvae. Larval spiracles take diverse forms, which are systematically important.

Body types. The more specialized burrowers among beetle larvae may be divided into two broad types, the straight- and the curved-bodied (**Fig. 3**). Straight burrowing larvae are normal in Archostemata, Carabidae, Elateroidea, various Heteromera, Cerambycidae, and halticine Chrysomelidae, while curved larvae are the norm in Scarabaeoidea, Bostrychoidea, Chrysomelidae-Eumolpinae, and Curculionoidea. The basic difference is probably in the way in which the larva anchors itself in its burrow, using its mandibles to burrow into a resistant medium. In straight larvae, purchase may be secured by short stout legs (as in Elateridae), by terminal abdominal structures, or by protrusible asperate tuberosities (ampullae) on trunk segments (as is the case with Archostemata and Cerambycidae). In both groups, ocelli rarely number more than one and are often lost, and legs are liable to be reduced or lost. The antennae, which are always reduced in straight burrowers, may be quite long in some curved ones, such as Scarabaeoidea.

PHYSIOLOGY

In the basic life functions, such as digestion, respiration, excretion, circulation, muscle contraction, nerve conduction, sense reception, and growth, most beetles show normal insect features, but certain groups are unusual in one or another of these respects. In digestion, some dermestid larvae can digest keratin, a faculty otherwise known only in some tineoid moth larvae. Some wood-boring larvae, notably in Anobiidae and Cerambycidae, have some ability to digest celluloselike substances, apparently through the aid of microorganisms carried in mycetomes associated with the midgut, and some scarabaeid larvae make similar use of microorganisms in a fermentation chamber in the hindgut. Some beetle larvae are able to live and develop normally in fungus-infested decaying wood that is very low in oxygen and high in CO_2. Some aquatic beetles and their larvae may be found living normally in hot springs at temperatures up to 111°F (44°C). In some adult Scolytinae, it has been found that the indirect flight muscles of the metathorax may be greatly reduced and absorbed during gonad development after an initial nuptial flight, yet redeveloped to become functional again with postreproductive adult feeding. Effective color change in adults, mainly through changes in the hydration of cuticular layers, has been found in certain Scarabaeidae-Dynastinae and Chrysomelidae-Cassidinae. Something approaching homothermy during periods of adult activity has been reported in some of the larger scarabaeoid beetles, utilizing heat generation by mus-

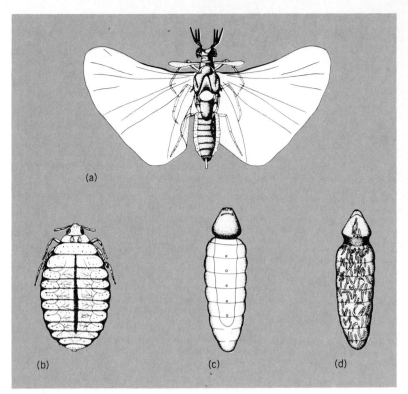

Fig. 3. Stylopidae. (a) *Eoxenos laboulbeni*, adult male and (b) adult female. (c) *Stylops melittae*, adult female (ventral view) and (d) gravid female with larvae. (*After C. P. Clausen, Entomophagous Insects, McGraw-Hill, 1940*)

cles in a preliminary warm-up stage, and air-cooling of the thorax by large tracheae during high activity.

Digestion. The digestive organs and capacities of beetles, and their food requirements, are generally like those of other insects, but some groups have unusual features. In many predatory larvae, also in some adults, extraoral digestion occurs, commonly marked in larvae by the development of a rigid toothed nasale in place of the labrum. In some such larvae, particularly when aquatic, the mandibles may become tubular-channeled. In at least some staphylinine larvae, there appear to be "venom" glands developed in the mandibles.

Among herbivores and wood-borers, capacities to digest celluloses and other "difficult" carbohydrates may be developed, usually and perhaps always by the aid of symbiotic microorganisms producing suitable enzymes. Dermestid larvae have developed the ability to digest keratin, breaking its disulfide bonds in an unusually alkaline midgut and producing free hydrogen sulfide as in some tineoid larvae.

Sense organs. The sense organs of beetles, adult or larval, are generally similar in type and function to those of other insects. For vision, the adult compound eyes show great diversity in form and ommatidial structure. Some types, such as *Cicindela*, are adapted for high visual acuity in bright light, others to high sensitivity in poor light, and many show light-dark adaptations of screening pigment. Color vision seems to be restricted to certain groups in the order, notably some of the floricolous Cetoniinae. In certain beetles, specialized infrared sensitivity is evidenced, notably in the region of the vertex of the head in many Curculionidae, and in complex metasternal organs in certain Buprestidae which are attracted to burning trees in coniferous forests. *See Eye (invertebrate)*.

Antennae. Antennae in beetles are mainly receptors for smell and touch, but detection of aerial vibrations (by Johnston's organ in the pedicel) may be important in some. Some of the long hairlike sensilla may detect movements in the air and serve as "wind socks" when beetles are about to fly. Many of the more actively flying groups have the antennal insertions shifted dorsally from their primitive lateral position. The olfactory sensilla, which commonly are thin-walled multiperforate setae, are often concentrated on expanded apical segments, forming a club. This arrangement probably helps in the detection of odor gradients in still air by crawling beetles. Sex dimorphism in the antennae is not uncommon in groups with short-lived adults, the males being specialized to detect female pheromones. Fairly high olfactory acuities for specific substances have been found, especially in some carrion beetles.

Taste. Taste, or contact chemoreception, is usually mediated by sensilla with a single large apical opening. These sensilla are concentrated on the palpi and other mouthparts and are often found on tibiae or tarsi of the front legs (also probably on the ovipositor in many). In water beetles, the smell-taste distinction becomes obscured.

Sound receptors. Specialized sound receptors, other than Johnston's organ, have not been much studied in Coleoptera, though many species produce sounds and in some cases (in various Scolytinae and other Curculionidae, for example) these play a significant part in courtship. The tibial spurs, in at least some species, serve to detect vibrations in the substratum. *See Animal communication*.

Trichobothria. Very long slender hairs arising from deep cuticular pits occur in many blind cave beetles, along the outer edges of the elytra in most Carabidae, and in larvae of Staphylinidae-Paederinae. These organs are believed to register slight movements in the air, resulting from nearby activity of potential prey or predators.

Gravitational sense. A gravitational sense is evident in many burrowing bettles, which are able to sink accurately vertical burrows despite inclinations of the surface started from. Indications are that such a sense is located in the legs rather than in the antennae, though no beetles are known to have subgenual organs, which may serve this function in other insects.

Temperature sense. A temperature sense is clearly present in some beetles and larvae, but little is known of the sensilla concerned. A type of coeloconic sensillum has been shown to serve this function in palpi of Culicidae, and very similar sensilla have been found in Coleoptera (as on the head capsule of elaterid larvae), which may be sensitive to differences in soil temperature.

Biological clock. What might be called a time sense (otherwise known as a biological clock) clearly operates in many beetles that react specifically to changing daylight lengths marking the seasons in nontropical latitudes. It also manifests itself as a circadian rhythm in "wild" beetles maintained artificially in conditions of uniform temperature, illumination, and humidity. *See Insect physiology; Photoperiodism*.

FOOD SPECIALIZATION

The biting mouthparts of many adult and larval beetles are adaptable to many types of food, and fairly widely polyphagous habits are not uncommon in the order, though most species can be assigned to one or another of a few main food categories: fungivores, carnivores, herbivores, or detritivores. Included in the last category are species feeding on decaying animal or vegetable matter and on dung. Most parasites could be included with carnivores, but those myrmecophiles and termitophiles which are fed by their hosts, and those Meloidae which develop on the food stores of bees, form special categories.

There are reasons for thinking that ancestral Coleoptera fed on fungi on dead wood, and beetles of many groups today are largely or wholly fungus eaters. Many are specialists of particular types of fungi, among whose hosts most major groups of fungi are represented. It could well be that the habit of producing fruit bodies and spores underground (as in truffles, and *Endogone*) originally coevolved in relation to the activities of burrowing beetles (such as Liodini and some Scarabaeoidea), which may still be important spore-dispersal agents for such fungi. The long-lasting fruit bodies of various Polyporaceae on dead wood are important breeding grounds for beetles of a number of families, but few beetles can complete their larval development fast enough to exploit the short-lived types of toadstool. Beetles of several families (including the Scolytinae) may serve as effective vectors, inoculating new habitats with fungi which are significant for the beetles' nutrition.

Predators. Predation, mainly on other insects, is probably the fundamental mode of life in adult and larval Adephaga, and has developed in a number of different lines of Polyphaga. Many carnivorous beetles (such as Carabidae and Staphylinidae) are fairly widely polyphagous and liable to supplement their diets with nonanimal matter to some extent, but more specialized predation also occurs (for example, among Histeridae and various Cleridae). Predacious larvae may give rise to largely herbivorous adults (such as in many Hydrophilidae and Melyridae); the converse relation, though rarer, also occurs (in some Cerambycidae, for example). Predatory beetles often do not require movement of the prey to stimulate attack and feeding behavior, and may feed extensively on eggs or pupae of other insects which escape many other predators. Predation on pupae by beetle larvae may lead to parasitic-type habits, as in Lebiini and Brachinini among Carabidae, or Aleocharini among Staphylinidae.

Herbivores. Herbivorous beetles may eat green plants of all groups, from algae to angiosperms, may attack any part of the plant, and display a great range in host plant specialization. Some, like the Colorado beetle, will feed and develop only on a few closely related species of a single genus. Others, like many of the short-nosed weevils, are highly polyphagous, and are able to live on angiosperms of many different families. Feeding on roots, or on the enclosed seeds of Angiospermae, are mainly larval habits, whereas adults usually eat stems, leaves, or flowers. Many species are injurious to crops (**Fig. 4**), and a few have been used successfully in the biological control of weeds (such as *Chrysolina* spp. against *Hypericum*). A number of beetles are effective transmitters of viruses of field crops (such as *Phyllotreta* spp. on Cruciferae); some may be useful pollinators (such as *Meligethes* spp. on Cruciferae). Many botanists believe that beetle pollination was a feature of the first Angiospermae and that the enclosed carpels were developed to protect ovules against floricolous beetles.

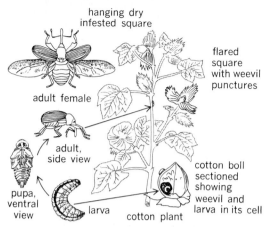

hanging dry
infested square

flared
square
with weevil
punctures

adult female

adult,
side view

cotton boll
sectioned
showing
weevil and
larva in its cell

pupa,
ventral
view

larva   cotton plant

Fig. 4. Cotton boll weevil (*Anthonomus grandis* Boheman).
(*After C. H. Metcalf, W. F. Flint, and R. L. Metcalf,*
***Destructive and Useful Insects, 3d ed., McGraw-Hill, 1951*)**

Adults of a number of families frequent flowers, often eating pollen as well as licking nectar. Botanists distinguish cantharophilous flowers, which depend particularly on beetles for their pollination, and which include what are believed to be some of the most primitive surviving Angiospermae. Characteristic features are white flowers, either large or massed in heads, abundant pollen production, and open shapes, with nectar production often absent.

Larvae developing in flowers or buds are characteristic of Meligethinae and Cateretinae in Nitidulidae. In the latter group, larvae tend to feed on the gynoecium; in the former, on the anthers. Other groups, such as Bruchidae and various weevils, have larvae developing later, in the ripening fruits, and less typically floricolous adults. The tough exines of pollen grains are not easily permeable. For assimilation of the contents, it may be necessary either to crack the outer wall by the mandibles or to cause the grains to begin germination in the crop, as may happen in Oedemeridae and Mordellidae, for example.

WATER BEETLES

Beetles are unusual among the higher insects in that aquatic habits, where present, usually affect adults as well as larvae. The elytra (like the hemielytra of Heteroptera) may have been preadaptive to the invasion of water. Almost any type of fresh or brackish water body is liable to contain some types of water beetles, though very few species can live permanently in full marine salinities. Almost all aquatic beetles maintain an air reservoir, into which the second thoracic and abdominal spiracles open, under their elytra; and some, mainly small, species may have plastrons or physical gills. Aquatic larvae tend to have a last pair of abdominal spiracles that is large and effectively terminal, or tend to develop tracheal gills. Pupae are terrestrial in the large majority of water beetles, and eggs, when deposited underwater, are commonly laid in contact with airspaces in the stems of water plants, or in an air-filled egg cocoon (Hydrophilidae). Most water beetles are confined to shallow waters, and many of them are ready colonists of temporary pools, through adult flight.

Many of the adults, but fewer of the larvae, are active swimmers by means of the legs, while others rely mainly on crawling for locomotion. The whirli-

gig beetles (Gyrinidae) are exceptional in that the adults swim very fast on the water surface, with only the ventral side submerged, by means of highly modified middle and hind legs, and are able to dive when alarmed, while the larvae are fully aquatic with long tracheal gills. The special modifications of the adults include divided compound eyes, one part for aerial, one for underwater vision; short antennae with a highly developed Johnston's organ, serving to pick up vibrations from potential prey or from the movements of other Gyrinidae in the surface film; long raptorial front legs; and a rudderlike end of the abdomen.

Aquatic larvae and more or less terrestrial adults occur in Psephenidae, some Ptilodactylidae, and the Donaciine leaf beetles; the converse relation is known in at least some Hydraenidae.

The foods of water beetles include algae (Myxophaga, Hydraenidae, many Dryopoidea), higher plants (larval Donaciinae, some Curculionidae, adults of many Hydrophilidae), other small animals (most Hydradephaga, Hydrophilid larvae), and mixed diets.

SPECIAL HABITATS AND ADAPTATIONS

More or less unusual adaptations and modes of life are manifest in dung and carrion beetles, ambrosia beetles, cave and subterranean beetles, desert beetles, and luminous beetles. Dung and carrion beetles exploit sporadic and very temporary food resources in which competition (particularly with Diptera larvae) tends to be fierce. Many dung beetles avoid the difficulties by burying dung stores in underground cells where their larvae can develop safely, and some beetles (Necrophorinae) adopt a similar strategy for carrion. They may also develop phoretic and symbiotic relations with specific mites preying on fly eggs.

In ambrosia beetles, the adults usually excavate burrows in wood, carrying with them spores of special fungi, which proceed to develop along the walls of the burrows and are fed on by larvae developing from eggs laid there. This type of relation between beetles, fungi, and trees exists in various forms in a number of families and may be of ecological importance if the fungi concerned are liable to kill trees.

Marginal habitats. More than one family of beetles contains highly adapted types living exclusively in deeper cave systems (cavernicolous) or the deeper layers of the soil (hypogeous). Such species are usually flightless and eyeless, poorly pigmented, and slow-moving, and of very restricted geographical distribution. In the cavernicoles, long trichobothrium-type setae are often developed on the body, and the adults may produce very large eggs singly, with a very abbreviated larval development, approaching the pupiparous condition of some Diptera.

Another marginal habitat in which beetles are the principal insect group represented is the desert, where Tenebrionidae are the dominant group. Here, too, wings are usually lost, but the cuticle tends to be unusually thick, firm, and black. Most desert beetles have burrowing larvae, and the adults also usually take refuge in the soil during the heat of the day. Elaborate adaptations for water economy and dew collection are frequent.

Another marginal habitat in which beetles may be found is hot springs. *Hydroscapha* and certain Hydrophilidae have been found breeding apparently normally in water of temperatures up to 111°F (44°C), which would be lethal for most other animals. A few water beetles are adapted to underground (phreatic) waters, showing parallel features with the cavernic ous and hypogeous terrestrial ones.

Dung and carrion beetles. Dung and carrion provide important but temporary food resources whose insect users are almost exclusively beetles and Diptera. Dung and carrion beetles are almost always good fliers, with strong antennal clubs bearing heavy concentrations of sensilla. May of the adults are burrowers and, particularly among dung beetles, frequently make and provision underground nests, thereby removing the food supply for their larvae from the twin dangers of Diptera and drying up. Necrophorini among carrion beetles shows similar features, with the added feature of actual feeding of the larvae by adults. Another special feature of *Necrophorus* and some dung beetles is an apparently symbiotic relation with *Poecilochirus* and allied mites, which are carried on the bodies of adult beetles and which specifically attack the eggs of competing Diptera in the breeding habitat. A well-known feature of many Scarabaeine dung beetles is the formation of a ball of dung which is rolled to a nest site some distance from the dung patch.

Wood borers. One mode of life in which beetles are undoubtedly the dominant group among insects is wood boring, mainly pursued by larvae, but also by adults in some Bostrychidae, Scolytinae, and Platypodinae. Rather few beetles bore normally into the wood of healthy living trees, which are liable to drown borers in sap, but moribund, dead, or decaying timber is liable to heavy attack. Some borers can develop only in unseasoned wood, requiring remains of cell contents; others develop in seasoned wood; and many require the presence of some type of fungus for successful development. Hardwoods and softwoods have their beetle specialists, and females of some Cerambycidae restrict their oviposition to branches or trunks of some particular girth.

Parasites. More or less parasitic relations to animals of other groups have developed in a number of lines of Coleoptera. The closest parallels to Hymenoptera-Parasitica are to be seen in the families Stylopidae (Strepsiptera of many authors; Fig. 3) and Rhipiphoridae. Stylopid larvae are exclusively endoparasites of other insects, while the adult females are apterous, often legless, and remain in the host's body, while males have large fanlike hindwings and the elytra reduced to haltere-like structures. Adult mouthparts are vestigial in both sexes. The more modified of Rhipiphoridae, parasitizing cockroaches, show close parallels with the more primitive Stylopidae, while more primitive ones, probably parasitizing wood-boring beetles, resemble the allied nonparasitic Mordellidae. Larvae ectoparasitic on lignicolous beetle larvae or pupae are found in Passandridae and some Colydiidae.

A considerable variety of Coleoptera develop normally in the nests of termites and social or solitary Hymenoptera-Aculeata, some being essentially detritivores or scavengers, but most feed either on the young or the food stores of their hosts. Many of the myrmecophiles and termitophiles are highly modified structurally and behaviorally and highly host-specific. Some of these beetles appear to have successfully "broken" the chemical or tactile communication codes of their hosts.

Ectoparasitism on birds or mammals is a rare development in beetles, the best-known examples being in Liodidae-Leptininae, with adults and sometimes larvae living on the bodies of small mammals, and Staphylinidae-Amblyopinini, with apparently similar habits. A few Cryptophagidae and Languriidae-Loberinae have been recorded as adults on bodies of rodents. In the little-known Cavognathidae of the Southern Hemisphere, in at least some species and probably in all of them, the breeding habitat is birds' nests and the larvae feed on nestling birds. SEE POPULATION ECOLOGY.

Luminescence. The beetles are notable in that some of them manifest the highest developments of bioluminescence known in nonmarine animals, the main groups concerned being Phengodidae, Lampyridae, and Elateridae-Pyrophorini, the glowworms and fireflies. In luminescent beetles, the phenomenon always seems to be manifest in the larvae and often in the pupae, but not always in the adults. SEE BIOLUMINESCENCE.

In luminous adult beetles, there is often marked sex dimorphism, and a major function of the lights seems to be the mutual recognition of the sexes of a species. In such cases, there is often a rapid nervous control of the luminescence, and a development of species-specific flash codes. In some cases, notably in adult Pyrophorini and some larval Phengodidae, an individual may carry lights of two different colors.

Another possible function of adult luminosity, and the only one seriously suggested for that of larvae, is as an aposematic signal. There is definite evidence that some adult fireflies are distasteful to some predators, and the luminous larvae of Phengodidae have dorsal glandular openings on the trunk segments that probably have a defensive function.

PHYLOGENETIC HISTORY

Coleoptera are older than the other major endopterygote orders. The earliest fossils showing distinctively beetle features were found rather before the middle of the Permian Period and generally resemble modern Archostemata. By the later Permian, fossils indicate that beetles had become numerous and diverse, and during the Mesozoic Era they appear as a dominant group among insect fossils. Fossils in Triassic deposits have shown features indicative of all four modern suborders, and the Jurassic probably saw the establishment of all modern superfamilies. By early Cretaceous times, it is likely that all "good" modern families had been established as separate lines. In the Baltic Amber fauna, of later Paleogene age (about 40,000,000 years ago), about half the fossil beetle genera appear to be extinct, and the other half still-living representatives (often in remote parts of the world). Beetle fossils in Quaternary (Pleistocene) deposits are very largely of still-living species.

Beetles appear to have been the first insect wood borers, with insect borings in fossil wood of Triassic age (for example, in the Petrified Forest of Arizona) being probably the work of larval Archostemata. Some fossil woods of slightly younger age have shown the oldest indications of a resin-secreting system, probably a defensive reaction against the attacks of wood-boring beetles. Later still, it seems that beetles may have played a major part in the evolution of the angiosperm-type of flower. The Adephaga probably evolved as ground-living predators, and their pygidial defensive glands may well have been a response to amphibian predation, much like the metathoracic glands of the roughly coeval Hemiptera-Heteroptera. Fossils clearly of adephagan water beetles are known from Lower Jurassic deposits, making a further parallel with Heteroptera-Cryptocerata.

CLASSIFICATION

A modern classification of beetles is given below. This system is adapted from that of Crowson (1955) and is intended to be phylogenetic in the sense of Hennig. An asterisk (*) indicates that the family is not known in the fauna of the United States; a dagger (†), that the family is not known from the New World.

Suborder Archostemata
    Superfamily Cupedoidea
        Family: Ommadidae†
            Tetraphaleridae*
            Cupedidae
            Micromalthidae
Suborder Adephaga
    Superfamily Caraboidea
        Family:Paussidae (including Ozaeninae, Metriinae, Sicindisinae)
            Cicindelidae
            Carabidae (including Omophroninae, Rhysodinae)
            Trachypachidae
            Amphizoidae
            Haliplidae
            Hygrobiidae†
            Noteridae
            Dytiscidae
            Gyrinidae
Suborder Myxophaga
    Superfamily Sphaeroidea
        Family: Lepiceridae (Cyathoceridae)*
            Torridincolidae*
            Hydroscaphidae
            Sphaeridae (Sphaeriidae)
Suborder Polyphaga
    Series Staphyliniformia
        Superfamily Hydrophiloidea
            Family: Hydraenidae
                Spercheidae
                Hydrochidae
                Georyssidae
                Hydrophilidae
        Superfamily Histeroidea
            Family: Sphaeritidae
                Synteliidae*
                Histeridae (including Niponiidae)
        Superfamily Staphylinoidea
            Family: Ptiliidae (including Limulodidae)
                Empelidae
                Liodidae (including Leptininae, Catopinae, Coloninae, Catopocerinae)
                Scydmaenidae
                Silphidae
                Dasyceridae
                Micropeplidae
                Staphylinidae (including Scaphidiinae)
                Pselaphidae
    Series Eucinetiformia
        Superfamily Eucinetoidea
            Family: Clambidae
                Eucinetidae
                Helodidae

Series Scarabaeiformia
    Superfamily Scarabaeoidea
        Family: Pleocomidae
            Geotrupidae
            Passalidae
            Lucanidae
            Trogidae
            Acanthoceridae
            Hybosoridae
            Glaphyridae
            Scarabaeidae
    Superfamily Dascilloidea
        Family: Dascillidae
            Karumiidae*
            Rhipiceridae (Sandalidae)
Series Elateriformia
    Superfamily Byrrhoidea
        Family Byrrhidae
    Superfamily Dryopoidea
        Family: Eulichadidae
            Ptilodactylidae
            Chelonariidae
            Psephenidae
            Elmidae (Elminthidae)
            Lutrochidae
            Dryopidae
            Limnichidae
            Heteroceridae
    Superfamily Buprestoidea
        Family Buprestidae (including Schizopidae)
    Superfamily Artematopoidea
        Family: Artematopidae
            Callirhipidae
    Superfamily Elateroidea
        Family: Cebrionidae
            Elateridae
            Throscidae
            Cerophytidae
            Perothopidae
            Phylloceridae†
            Eucnemidae
    Superfamily Cantharoidea
        Family: Brachypsectridae
            Cneoglossidae*
            Plastoceridae†
            Homalisidae†
            Lycidae
            Drilidae†
            Phengodidae
            Telegeusidae
            Lampyridae
            Omethidae
            Cantharidae
Series Bostrychiformia
    Superfamily Dermestoidea
        Family: Derodontidae
            Nosodendridae
            Dermestidae
            Thorictidae†
            Jacobsoniidae (Sarothriidae)*
    Superfamily Bostrychoidea
        Family: Bostrychidae (including Lyctinae)
            Anobiidae
            Ptinidae (including Ectrephini, Gnostini)
Series Cucujiformia
    Superfamily Cleroidea
        Family: Phloiophilidae†

Series: Peltidae
    Family: Lophocateridae
        Trogossitidae
        Chaetosomatidae[†]
        Cleridae
        Acanthocnemidae
        Phycosecidae[†]
        Melyridae (including Malachiinae)
    Superfamily Lymexyloidea
    Family: Lymexylidae
        Stylopidae (Strepsiptera, Stylopoidea)
    Superfamily Cucujoidea
    Family: Nitidulidae
        Rhizophagidae (including Monotominae)
        Protocucujidae*
        Sphindidae
        Boganiidae[†]
        Cucujidae
        Laemophloeidae
        Passandridae
        Phalacridae
        Phloeostichidae*
        Silvanidae
        Cavognathidae*
        Cryptophagidae (including Hypocoprinae)
        Lamingtoniidae[†]
        Helotidae[†]
        Languriidae
        Erotylidae (including Pharaxonothinae)
        Biphyllidae
        Cryptophilidae (including Propalticinae)[†]
        Cerylonidae
        Corylophidae
        Alexiidae (Sphaerosomatidae)[†]
        Endomychidae
        Coccinellidae
        Discolomidae
        Merophysiidae
        Lathridiidae
        Byturidae
        Mycetophagidae
        Cisidae
        Pterogeniidae[†]
        Tetratomidae
        Melandryidae
        Mordellidae
        Rhipiphoridae
        Merycidae[†]
        Colydiidae
        Synchroidae
        Cephaloidae
        Pythidae
        Pyrochroidae
        Anthicidae
        Meloidae
        Aderidae
        Scraptiidae (including Anaspidinae)
        Oedemeridae
        Cononotidae
        Othniidae
        Salpingidae
        Inopeplidae
        Mycteridae (including Hemipeplinae)
        Monommidae

    Family: Zopheridae
        Tenebrionidae (including Lagriinae, Nilioninae, Alleculinae)
    Superfamily Chrysomeloidea
    Family: Disteniidae (including Oxypeltinae, Philinae,[†] Vesperinae)
        Cerambycidae
        Megalopodidae (including Zeugophorinae)
        Bruchidae (including Sagrinae)
        Chrysomelidae
    Superfamily Curculionoidea
    Family: Nemonychidae
        Anthribidae (including Bruchelinae)
        Belidae*
        Oxypeltidae*
        Aglycyderidae (Proterhinidae)[†]
        Allocorynidae
        Attelabidae
        Apionidae (including Antliarrhininae,[†] Ithycerinae, Nanophyinae)
        Brenthidae
        Curculionidae (including Scolytinae and Platypodinae)

This classification is based on a great variety of characteristics, including internal ones and those of immature stages, and takes account of the fossil evidence. Important characteristics include the structure of the propleura, the cervical sclerites, the ovarioles, larval leg segments, form of testes, wing venation and folding, number and arrangement of malpighian tubules, forms of the female and male external genitalia, presence or absence of a free larval labrum, larval spiracular structures, mode of formation of adult midgut in the pupa, form of the metendosternite, number of female accessory glands, form of adult antennae, tarsal segmentation, presence of pupal "gin traps," number of pupal spiracles, number of adult abdominal spiracles, presence of defensive glands in adults or larvae, egg bursters of the first-instar larva, form of adult tentorium and gular sutures, forms of adult coxal cavities, and internal structure of adult ommatidia.

GEOGRAPHICAL DISTRIBUTION

Almost every type of continuous or discontinuous distribution pattern which is known in any animal group could be matched in some taxon of Coleoptera, and every significant zoogeographical region or area could be characterized by endemic taxa of beetles. In flightless taxa, distributional areas are generally more limited than those of comparable winged taxa. Distinct distributional categories can be seen in those small, readily flying groups (in Staphylinidae and Nitidulidae, for example) which are liable to form part of "aerial plankton," and in those wood borers which may survive for extended periods in sea-drifted logs, both of which are liable to occur in oceanic islands beyond the ranges of most other beetle taxa. Climatic factors often seem to impose limits on the spread of beetle species (and sometimes of genera or families), and beetle remains in peats have been found to be sensitive indicators of climatic changes in glacial and postglacial times. SEE INSECTA; SOCIAL INSECTS.

Roy A. Crowson

Bibliography. R. A. Crowson, *The Natural Classification of the Families of Coleoptera*, 1955; R. A. Crowson, The natural classification of the families of

Coleoptera: Addenda and corrigenda, *Entomol. Mon. Mag.*, 103:209–214, 1967; S. P. Parker (ed.), *Synopsis and Classification of Living Organisms*, 2 vols., 1982.

Coliiformes

A small order of birds containing only the family Coliidae with six species (the mousebirds) restricted to Africa. The mousebirds are small, grayish to brownish, with a long tail. The legs are short and the feet strong, with the four toes movable into many positions from all four pointing forward to two reversed backward. Mousebirds perch, climb, crawl, and scramble agilely in bushes and trees. They are largely vegetarian but eat some insects. They are nonmigratory and gregarious, and sleep in clusters, but they are monogamous as breeders. The nest is an open cup in a tree or bush, and the two to four young remain in the nest, cared for by both adults until they can fly. The relationships of the mousebirds to other birds are obscure. Mousebirds have a surprisingly good fossil record from the Miocene of France and Germany. SEE AVES.

Walter J. Bock

Coliphage

Any bacteriophage able to infect the bacterium *Escherichia coli*. Many are phages able to attack more than one strain of this organism. The T series of phage (T1–T7), propagated on a special culture of *E. coli*, strain B, have been used in extensive studies from which most of the knowledge of phages is derived.

A typical phage particle is shaped something like a spermatozoon. The outer skin, or shell, consists of protein and encloses contents of deoxyribonucleic acid (DNA). This DNA can be separated from the protein by osmotic shock, and it is then found that the protein shell alone can cause death of a bacterium when adsorbed to the cell surface. This does not constitute a true infection, however, since no new phage is formed.

The usual lytic infection of a *coli* B cell by a T phage proceeds as follows (see **illus.**). The phage is adsorbed, by its tail, to the cell surface. Following this adsorption, the phage DNA gains access to the cell. The protein skin has now served its purpose and, if desired, can be removed from the cell surface without affecting subsequent events. Phage in an infectious state no longer exists; it is now spoken of as vegetative phage and will remain in this state during its life cycle until the production of new infective particles occurs. Immediately following the injection of the DNA, the metabolic activities of the cell are profoundly altered. Normal growth ceases abruptly, although respiration continues at a steady rate. Cell enzymes still function, but their activity is now redirected and controlled by "information" supplied by the phage DNA. New protein synthesis can be detected at once; this is not bacterial protein, however, but phage protein which is to serve later as part of the new phage particles. Synthesis of bacterial ribonucleic acid (RNA) ceases. Bacterial DNA breaks down, and after about 10 min, phage DNA begins to make its appearance, formed in part from the bacterial

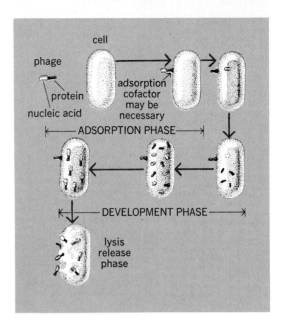

Steps in the two phases of a lytic infection by phage. (*After E. L. Oginsky and W. W. Umbreit, An Introduction to Bacterial Physiology, Freeman, 1954*).

DNA and in part from new synthesis. This new DNA is incorporated with the phage protein to make a new phage which can now be detected as an infectious unit. In another 10 min or so, lysis occurs after more phage has accumulated within the cell. The time will vary with different phages. The infectious phage, released in numbers of 100 or more particles, is able to infect susceptible bacteria. SEE BACTERIOPHAGE; LYSOGENY; LYTIC INFECTION; VIRUS.

Philip B. Cowles

Collagen

The major fibrous protein of many animals. Located in the extracellular connective tissue, collagen is probably the most abundant animal protein in nature. It is estimated that collagen accounts for about 30% of the total human body protein. It is present in all types of multicellular animals, both invertebrate and vertebrate, although in insects and crustaceans chitin replaces collagen as the fibrous supporting matrix in the exoskeleton. Collagen constitutes the fibrillar component of the soft connective tissues (for example skin, ligament, and tendon) and is the major component of the organic matrix of the calcified tissues such as bone and dentine. Collagen is synthesized predominantly by mesenchymal cells such as fibroblasts, chondroblasts, osteoblasts, and odontoblasts. Besides its structural significance, collagen plays an important role in such events as development and wound healing, and has been implicated in aging and in a number of disease processes.

Derivatives and related proteins. In mammals, the thick inner skin or corium under the epidermis is chiefly collagen. Corium is the raw material in the manufacture of leather by tanning, a process known since antiquity in which the hide is protected against bacterial decay. During the tanning process, tannins and other agents form an irreversible chemical combination with the collagen of the corium, making cross-links and converting rawhide into leather. SEE INTEGUMENT; LEATHER AND FUR PROCESSING.

Upon treatment with heat or certain chemical

agents, collagen can be denatured and partially degraded to gelatin. Prolonged boiling of hides and bones produces glue, an extract from impure gelatin. Bone gelatin is also called ossein. The distinction between gelatin, glue, and size is based only on their relative purity. The first extracts, which are mainly free of extraneous materials, are called gelatin, and the later, less pure extracts are called glue or size. Fish gelatin and glue are made from the collagen of fish, present in bones, skin, swim bladders, and offal. SEE GELATIN.

The collagen of fish swim-bladder undertunics is called ichthyocol. The protein present in the fins of sharks and other elasmobranchs is called elastoidin and is related to collagen. Elastic tissues such as lungs, large blood vessels (for example, aorta), ligaments, and certain elastic cartilages contain, in addition to collagen, the fibrous protein elastin, which is quite different from collagen in composition and properties.

Structure and properties. At least four genetically distinct collagens have been described. The most familiar is called type I and is composed of three α-polypeptide chains, each of about 95,000 mol wt. Two chains are identical and are called α1(I); the other is called α2. Type I collagen forms the major portion of the collagen of both soft (skin, tendon, fascia, and so on) and hard (bone and dentine) connective tissue. Type II collagen is the major collagen of cartilage and is composed of three α1(II) chains. Type III collagen is composed of three α1(III) chains and is found in blood vessels, wounds, and certain tumors and appears to be elevated in the tissues of young animals. Reticulin fibers appear to be identified with type III collagen. Basement membrane collagens have been classified as type IV, and other collagen types have been identified.

Collagen is recognized by its relative insolubility under mild conditions, by its resistance to most of the proteolytic enzymes of vertebrates, and by various histologic staining properties. Its appearance in the electron microscope is unique, and it has a characteristic banded or striated appearance with a 64-nanometer spacing. On examination by x-ray diffraction, it shows a wide-angle pattern with a meridional 0.286-nm spacing related to the three-stranded left-handed helical conformation of its constituent polypeptide chains.

Collagen has a characteristic amino acid composition, containing 33% glycine and about 25% proline and hydroxyproline. The amino acid δ-hydroxy-L-lysine and the amino acid 4-hydroxy-L-proline are almost unique to collagen among the proteins of the vertebrates and invertebrates (see **illus**.). Almost all four α chains of types I, II, and III collagen have been completely sequenced. Each α1(I) chain contains 1052 residues. Residues 17−1027 comprise sequences starting with a glycyl residue at every third position. The cross-links which impart stability and insolubility to tissue collagen involve residues derived from lysine or hydroxylysine residues and are primarily located at the extremities of the α chains. SEE HYDROXYPROLINE.

Biochemistry and disease. Once the α chains have been synthesized, a number of modifications occur, leading to the formation of a collagen which is tailored to meet its particular functional role. The maturation of collagen in nature involves the intramolecular and intermolecular cross-linking of the polypeptide chains in a three-stranded molecule called tropocollagen, which is initially free of covalent cross-links. This cross-linking process leads to increases in the stability of collagen. In addition to specific lysine and hydroxylysine residues which are modified to aldehydic compounds in the maturation process, residues of unmodified lysine, hydroxylysine, and histidine are involved in a complex process leading to the biosynthesis of a wide variety of cross-linking compounds. Certain agents such as β-aminopropionitrile which are present in the sweet pea (*Lathyrus odoratus*) and related plants inhibit the maturation and produce a serious disorder called osteolathyrism in animals which ingest these sweet peas. This disorder is characterized by skeletal deformities and loss of connective tissue strength.

Several heritable disorders of connective tissue have been elucidated at the molecular level. The basic defect has been determined in four types of the

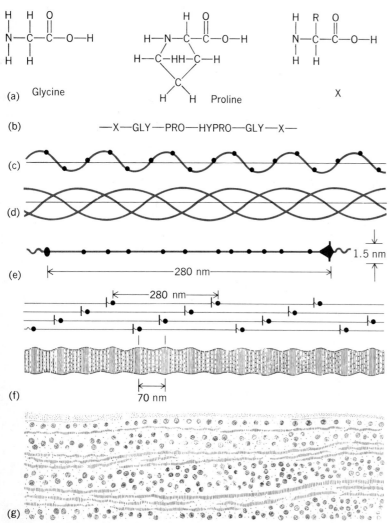

Formation of collagen, visualized in seven steps. (a) Starting materials are free amino acids; the letter R in amino acid X represents any of some 20 different side chains. (b) Hypro stands for hydroxyproline, created from proline after molecular chain has been formed. (c) The chain twists itself into a left-handed helix (×17,500,000). (d) Three chains that intertwine to form a right-handed superhelix (×17,500,000), the tropocollagen molecule. (e) Many molecules line up in staggered fashion (×330,000), (f) overlapping by one-quarter of their length to form a fibril (×120,000). (g) Fibrils in connective tissues (×50,000) are often stacked in layers with fibrils aligned at right angles. (*After J. Gross, Collagen, Sci. Amer., 204:121, 1961*)

Ehlers-Danlos syndrome. People with this condition have "human pretzel" or "India rubber man" characteristics, with hyperextensible skin, hypermobile joints, fragile tissues, and a bleeding diathesis. In three varieties of the syndrome an enzyme defect has been identified, and in the fourth variety it is suggested that the type III collagen is lacking. A disease of cattle called dermatosporaxis mimics the human condition. Several types of osteogenesis imperfecta have also been identified. The individuals exhibit fragile bones, clear or blue sclera, deafness, and loose ligaments. It appears that the bone collagen fails to mature, and there is evidence that the ratio of type I/type III collagen may be abnormal. SEE CONNECTIVE TISSUE; FIBROUS PROTEIN.

David A. Keith

Bibliography. P. M. Gallop and M. A. Paz, Post-translational protein modifications with special attention to collagen and elastin, *Physiol. Rev.*, 55(3):418–487, 1975; L. Goldstein and D. M. Prescott (eds.), *Cell Biology: A Comprehensive Treatise*, vol. 4, 1980; K. I. Kivirikko and L. Risteli, Biosynthesis of collagen and its alterations in pathological states, *Med. Biol.*, 54:159–186, 1976; D. J. Prockop et al., The biosynthesis of collagen and its disorders, *New Engl. J. Med.*, 301(1):13–23 and (2):77–85, 1979; G. N. Ramachandran and A. H. Reddi, *Biochemistry of Collagen*, 1976.

Collard

A cool-season biennial crucifer, *Brassica oleracea* var. *acephala*, similar to nonheading cabbage. Collard is of Mediterranean origin and is grown for its rosette of leaves which are cooked fresh as a vegetable (see **illus.**). Kale and collard differ only in the form of their leaves; both have been referred to as coleworts, a name taken from the Anglo-Saxon term meaning cabbage plants.

Propagation is by seed. Cultural practices are similar to those used for cabbage; however, collards are more tolerant of high temperatures. Georgia and Vates are popular varieties (cultivars). Collard is moderately tolerant of acid soils. Harvesting is usually 75 days after planting. Important production centers are in the southern United States, where collards

Collard (*Brassica oleracea* var. *acephala*).

are an important nutritious green, especially during winter months. SEE CABBAGE; CAPPARALES; KALE; VEGETABLE GROWING.

H. John Carew

Collembola

An order of primitive insects, commonly called springtails, belonging to the subclass Apterygota. These tiny insects do not undergo a metamorphosis. They have six abdominal segments, some of which may be ankylosed to give an apparently smaller number (see **illus.**). The tarsi are united with the tibiae,

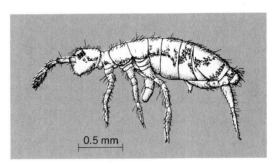

0.5 mm

A collembolan, *Entomobrya cubensis*. (After J. W. Folsom, Proc. U.S. Nat. Mus., 72(6), plate 6, 1927).

and a pretarsus is present, which bears a dorsal median claw and usually a ventral clawlike appendage. The ventral tube always occurs on the first abdominal segment and connects with the mouthparts through a ventral canal consisting of two integumentary folds which overlap. Appended to the fourth abdominal segment in most of the species is a spring, or furcula. This consists of a basal piece and two apical parallel structures which end in hooks. Another pair of structures, united at the base, occur ventrally on the third abdominal segment. These bear teeth which engage with the furcula and hold it beneath the body. When released, the spring flies backward, catapulting the insect. Two families, Sminthuridae and Actaletidae, have simple tracheal systems, while the others breathe through the integument. Collembola live in humid places, often in leaf mold. Some species are active at cool temperatures, hence another common name, snowflea, is used. The earliest known insect fossil, *Rhyniella praecursor*, is thought to belong to this order.

Harlow B. Mills

Bibliography. H. B. Mills, *A Monograph of the Collembola of Iowa*, Iowa State Coll. Div. Ind. Sci. Monogr. 3, 1934.

Collenchyma

A primary, or early differentiated, supporting tissue of young shoot parts appearing while these parts are still elongating. It is located near the surface, usually just under the epidermis (see **illus.** *a*). When observed in transverse sections, it is characterized structurally by cell walls that are intermittently thickened, generally in the corners or places of juncture of three or more cells (see illus. *b* and *c*). Collenchyma is typically formed in the petioles and vein ribs of leaves, the elongating zone of young stems, and the pedicels

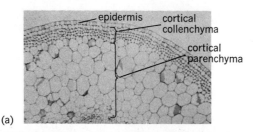

(a)

(b)   (c)

Transverse sections of Jimsonweed (*Datura stramonium*), progressively magnified. (*a*) Stem section. (*b*) Stem section, showing the corner thickenings of collenchyma cell walls and air spaces surrounded by wall thickenings. (*c*) Collenchyma section, showing lamellation of the cell wall forming the corner thickenings. (*Courtesy of W. S. Walker*)

of flowers. SEE CELL WALLS (PLANT).

As in parenchyma, the cells in collenchyma are living and may contain chloroplasts and starch grains. The cell wall of a collenchyma cell is its most striking feature structurally and functionally. It is composed of cellulose and pectic compounds plus a very high proportion of water. In some studies collenchyma cell walls were found to contain 50–60% water by fresh weight in a variety of species examined. The cellulose and pectic compounds are present in alternating layers. A striking feature of collenchyma cell walls is their plasticity. They are capable of great elongation during the period of growth in length of the plant. Despite this elongation, the characteristic thickening of the wall continues to increase because new wall material is added during growth. The plasticity of collenchyma is associated with a tensile strength comparable to that shown by fibers of sclerenchyma. The combination of strength and plasticity makes the collenchyma effective as a strengthening tissue in developing stems and leaves having no other supporting tissue at that time. SEE CELLULOSE; CHLOROPHYLL; EPIDERMIS (PLANT); FLOWER; LEAF; PARENCHYMA; PECTIN; STEM.

Robert L. Hulbary

Bibliography. J. O. Mauseth, *Plant Anatomy*, 1988.

Collision (physics)

Any interaction between particles, aggregates of particles, or rigid bodies in which they come near enough to exert a mutual influence, generally with exchange of energy. The term collision, as used in physics, does not necessarily imply actual contact.

In classical mechanics, collision problems are concerned with the relation of the magnitudes and direction of the velocities of colliding bodies after collision to the velocity vectors of the bodies before collision. When the only forces on the colliding bodies are those exerted by the bodies themselves, the principle

of conservation of momentum states that the total momentum of the system is unchanged in the collision process. This result is particularly useful when the forces between the colliding bodies act only during the instant of collision. The velocities can then change only during the collision process, which takes place in a short time interval. Under these conditions the forces can be treated as impulsive forces, the effects of which can be expressed in terms of an experimental parameter known as the coefficient of restitution, which is discussed later. SEE CONSERVATION OF MOMENTUM; IMPACT.

The study of collisions of molecules, atoms, and nuclear particles is an important field of physics. Here the object is usually to obtain information about the forces acting between the particles. The velocities of the particles are measured before and after collision. Although quantum mechanics instead of classical mechanics should be used to describe the motion of the particles, many of the conclusions of classical collision theory are valid. SEE SCATTERING EXPERIMENTS (ATOMS AND MOLECULES); SCATTERING EXPERIMENTS (NUCLEI).

Classification. Collisions can be classed as elastic and inelastic. In an elastic collision, mechanical energy is conserved; that is, the total kinetic energy of the system of particles after collision equals the total kinetic energy before collision. For inelastic collisions, however, the total kinetic energy after collision is different from the initial total kinetic energy.

In classical mechanics the total mechanical energy after an inelastic collision is ordinarily less than the initial total mechanical energy, and the mechanical energy which is lost is converted into heat. However, an inelastic collision in which the total energy after collision is greater than the initial total energy sometimes can occur in classical mechanics. For example, a collision can cause an explosion which converts chemical energy into mechanical energy. In molecular, atomic, and nuclear systems, which are governed by quantum mechanics, the energy levels of the particles can be changed during collisions. Thus these inelastic collisions can involve either a gain or loss in mechanical energy.

Consider a one-dimensional collision of two particles in which the particles have masses m_1 and m_2 and initial velocities u_1 and u_2. If they interact only during the collision, an application of the principle of conservation of momentum yields Eq. (1), where v_1 and

$$m_1u_1 + m_2u_2 = m_1v_1 + m_2v_2 \qquad (1)$$

v_2 are the velocities of m_1 and m_2, respectively, after collision.

Coefficient of restitution. It has been found experimentally that in collision processes Eq. (2) holds,

$$e = \frac{v_2 - v_1}{u_1 - u_2} \qquad (2)$$

where e is a constant known as the coefficient of restitution, the value of which depends on the properties of the colliding bodies. The magnitude of e varies from 0 to 1. A coefficient of restitution equal to 1 can be shown to be equivalent to an elastic collision, while a coefficient of restitution of zero is equivalent to what is sometimes called a perfectly inelastic collision. From the definition of e one can show that in a perfectly inelastic collision the colliding bodies stick together after collision, as two colliding balls of putty or a bullet fired into a wooden block would do. Equations (1) and (2) can be solved for the unknown velocities v_2 and v_1 in the one-dimensional collision of two particles.

The concept of coefficient of restitution can be generalized to treat collisions involving the plane motion of smooth bodies—both of particles and larger bodies for which rotation effects must be considered. For these collisions, experiments show that the velocity components to be used in Eq. (2) for e are the components along the common normal to the surfaces of the bodies at the point where they make contact in the collision. For smooth bodies the velocity components perpendicular to this direction are unchanged. Use of this result and the principle of conservation of momentum is sufficient to solve two-dimensional collision problems of smooth bodies. For collisions of smooth spheres the velocity components to be used in Eq. (2) for e are those on the line joining the centers of the spheres. Velocity components perpendicular to this direction are unchanged.

Center-of-mass coordinates. A simplification of the description of both classical and quantum mechanical collisions can be obtained by using a coordinate system which moves with the velocity of the center of mass before collision. (Since for an isolated system the center of mass of the system can be shown to be unaccelerated at all times, the velocity of the center of mass of the system of particles does not change during collision.) The coordinate system which moves with the center of mass is called the center-of-mass system, while the stationary system is the laboratory system.

The description of a collision in the center-of-mass system is simplified because in this coordinate system the total momentum is equal to zero, both before and after collision. In the case of a two-particle collision the particles therefore must be oppositely directed after collision, and the magnitude of one of the velocities in the center-of-mass system can be determined if the other magnitude is known. SEE CENTER OF MASS.

Paul W. Schmidt

Bibliography. D. Halliday and R. Resnick, *Physics*, 3d ed., 1978; F. W. Sears et al., *University Physics*, 5th ed., 1976.

Colloid

A state of matter characterized by large specific surface areas, that is, large surfaces per unit volume or unit mass. The term colloid refers to any matter, regardless of chemical composition, structure (crystalline or amorphous), geometric form, or degree of condensation (solid, liquid, or gas), as long as at least one of the dimensions is less than approximately 1 micrometer but larger than about 1 nanometer. Thus, it is possible to distinguish films (for example, oil slick), fibers (spider web), or colloidal particles (fog) if one, two, or three dimensions, respectively, are within the submicrometer range.

A colloid consists of dispersed matter in a given medium. In the case of finely subdivided particles, classification of a number of systems is possible, as given in the **table**. In addition to the colloids listed in the table, there are systems that do not fit into any of the listed categories. Among these are gels, which consist of a network-type internal structure loaded with larger or smaller amounts of fluid. Some gels

Types of colloid dispersions

Medium	Dispersed matter	Technical name	Examples
Gas	Liquid	Aerosol	Fog, sprays
	Solid	Aerosol	Smoke, atmospheric or interstellar dust
Liquid	Gas	Foam	Head on beer, lather
	Liquid	Emulsion	Milk, cosmetic lotions
	Solid	Sol	Paints, muddy water
Solid	Gas	Solid foam	Foam rubber
	Liquid	Solid emulsion	Opal
	Solid	Solid sol	Steel

may have the consistency of a solid, while others are truly elastic bodies that can reversibly deform. Another colloid system that may occur is termed coacervate, and is identified as a liquid phase separated on coagulation of hydrophilic colloids, such as proteins. SEE GEL.

It is customary to distinguish between hydrophobic and hydrophilic colloids. The former are assumed to be solvent-repellent, while the latter are solvent-attractant (dispersed matter is said to be solvated). In reality there are various degrees of hydrophilicity for which the degree of solvation cannot be determined quantitatively.

Properties. Certain properties of matter are greatly enhanced in the colloidal state due to the large specific surface area. Thus, finely dispersed particles are excellent adsorbents; that is, they can bind various molecules or ions on their surfaces. This property may be used for removal of toxic gases from the atmosphere (in gas masks), for elimination of soluble contaminants in purification of water, or decolorization of sugar, to give just a few examples. SEE ADSORPTION.

Colloids are too small to be seen by the naked eye or in optical microscopes. However, they can be observed and photographed in transmission or scanning electron microscopes. Owing to their small size, they cannot be separated from the medium (liquid or gas) by simple filtration or normal centrifugation. Special membranes with exceedingly small pores, known as ultrafilters, can be used for collection of such finely dispersed particles. The ultracentrifuge, which spins at very high velocities, can also be employed to promote colloid settling. SEE ELECTRON MICROSCOPE; SCANNING ELECTRON MICROSCOPE; ULTRACENTRIFUGE; ULTRAFILTRATION.

Colloids show characteristic optical properties. They strongly scatter light, causing turbidity such as in fog, milk, or muddy water. Scattering of light (recognized by the Tyndall beam) can be used for the observation of tiny particles in the ultramicroscope. Colloidal state of silica is also responsible for iridescence, the beautiful effect observed with opals. SEE SCATTERING OF ELECTROMAGNETIC RADIATION; TYNDALL EFFECT.

Preparation. Since the characteristic dimensions of colloids fall between those of simple ions or molecules and those of coarse systems, there are in prin-

Examples of colloid particles. (a) Zinc sulfide (sphalerite). (b) Cadmium carbonate; at right is an enlargement of the boxed section in the left photograph (*courtesy of Egon Matijević*).

ciple two sets of techniques available for their preparation: dispersion and condensation. In dispersion methods the starting materials consist of coarse units which are broken down into finely dispersed particles, drawn into fibers, or flattened into films. For example, colloid mills grind solids to colloid sizes, nebulizers can produce finely dispersed droplets from bulk liquids, and blenders are used to prepare emulsions from two immiscible liquids (such as oil and water). *See Emulsion*.

In condensation methods, ions or molecules are aggregated to give colloidal particles, fibers, or films. Thus, insoluble monolayer films can be developed by spreading onto the surface of water a long-chain fatty acid (for example, stearic acid) from a solution in an organic liquid (such as benzene or ethyl ether). Colloidal aggregates of detergents (micelles) form by dissolving the surface-active material in an aqueous solution in amounts that exceed the critical micelle concentration. *See Micelle; Monomolecular film*.

The most common procedure to prepare sols is by homogeneous precipitation of electrolytes. Thus, if aqueous silver nitrate and potassium bromide solutions are mixed in proper concentrations, colloidal dispersions of silver bromide will form, which may remain stable for a long time. Major efforts have focused on preparation of monodispersed sols, which consist of colloidal particles that are uniform in size, shape, and composition (see **illus.**). *See Precipitation (chemistry)*.

Dispersion stability. Dispersions of colloids (sols) are inherently unstable because they represent systems of high free energy. Particles may remain in suspended state as long as they are small enough for gravity to be compensated by the kinetic energy

(brownian movement). Consequently, particle aggregation on collision must be prevented, which can be achieved by various means. *See Brownian movement*.

The most common cause for sol stability is electrostatic repulsion, which results from the charge on colloidal particles. The latter can be due to adsorption of excess ions (such as of bromide ions on silver bromide particles), surface acid or base reactions (for example, removal or addition of protons on the surface of metal hydroxides), or other interfacial chemical reactions. Alternately, sol stability can be induced by adsorption of polymers or by solvation. Stability of colloid dispersions that occur naturally may represent an undesirable state, as is the case with muddy waters whose turbidity is due to the presence of tiny particles (clays, iron oxides, and so forth). When needed, stability can be artificially induced, as in the production of paints. *See Solvation*.

Coagulation or flocculation. Processes by which a large number of finely dispersed particles is aggregated into larger units, known as coagula or flocs, are known as coagulation or flocculation; the distinction between the two is rather subtle. The concepts most commonly refer to colloidal solids in liquids (sols), although analogous principles can be employed in explaining aggregation of solid particles or liquid droplets in gases (aerosols).

Coagulation of stable sols can be accomplished by various additives (electrolytes, surfactants, polyelectrolytes, or other colloids) whose efficiency depends on their charge and adsorptivity, as well as on the nature of suspended particles to be treated. A common example of applications of coagulation is in water purification in which particulate colloid contaminants must be aggregated in order to be removed. Coagulation is also used to describe solidification of polymers (for example, proteins), which can be achieved by heating, as exemplified by boiling an egg. *See Emulsion; Flocculation; Foam; Fog; Polymer; Smoke*.

Egon Matijević

Bibliography. P. C. Hiemenz, *Principles of Colloid and Surface Chemistry,* 2d ed., 1986; R. D. Vold and M. J. Vold, *Colloid and Interface Chemistry,* 1983.

Colloidal crystals

Periodic arrays of suspended colloidal particles. Common colloidal suspensions (colloids) such as milk, blood, or latex are polydisperse; that is, the suspended particles have a distribution of sizes and shapes. However, suspensions of particles of identical size, shape, and interaction, the so-called monodisperse colloids, do occur. In such suspensions, a new phenomenon that is not found in polydisperse systems, colloidal crystallization, appears: under appropriate conditions, the particles can spontaneously arrange themselves into spatially periodic structures. This ordering is analogous to that of identical atoms or molecules into periodic arrays to form atomic or molecular crystals. However, colloidal crystals are distinguished from molecular crystals, such as those formed by very large protein molecules, in that the individual particles do not have precisely identical internal atomic or molecular arrangements. On the other hand, they are distinguished from periodic stackings of macroscopic objects like cannonballs in that the periodic ordering is spontaneously adopted by the

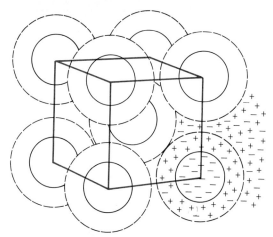

Fig. 1. Arrangement of particles and distribution of electrical charges in a charged stabilized body-centered cubic colloidal crystal. Spheres indicated by broken lines show extent of positive counterion clouds.

system through the thermal agitation (brownian motion) of the particles. These conditions limit the sizes of particles which can form colloidal crystals to the range from about 0.01 to about 5 micrometers. *SEE BROWNIAN MOVEMENT; KINETIC THEORY OF MATTER.*

The most spectacular evidence for colloidal crystallization is the existence of naturally occurring opals. The ideal opal structure is a periodic close-packed three-dimensional array of silica microspheres with hydrated silica filling the spaces not occupied by particles. Opals are the fossilized remains of an earlier colloidal crystal suspension. Another important class of naturally occurring colloidal crystals are found in concentrated suspensions of nearly spherical virus particles, such as *Tipula* iridescent virus and tomato bushy stunt virus. Colloidal crystals can also be made from the synthetic monodisperse colloids, suspensions of plastic (organic polymer) microspheres. Such suspensions have become important systems for the study of colloidal crystals, by virtue of the controllability of the particle size and interaction. *SEE OPAL; VIRUS.*

Crystal structure and properties. In order for colloidal crystals to form, the forces between the particles must be repulsive, since if the forces are attractive the particles will aggregate into noncrystalline clusters. The repulsion can be electrostatic, if the particles are charged (charge-stabilized), or can come from interparticle collisions (steric-stabilized). In the latter case the particle surface must be treated with an amphiphilic or polymer layer to prevent the particles from sticking together.

In colloidal crystals of spherical or nearly spherical particles, the crystal structures adopted are identical to those found in crystals formed by spherical atoms, such as the noble gases and alkali metals. The phase behavior is also quite similar. Consider, for example, a suspension in water of polymer microspheres. These particles are readily prepared with acid surface groups which dissociate in water, leaving the particle with a large negative charge (as large as several thousand electron charges). This charge attracts positively charged small ions, which form clouds around the particles (**Fig. 1**). The spheres will repel each other if they come close enough so that their ion clouds overlap, much the same as noble gas atoms repel if their

electron clouds overlap. At low concentration, the particles rarely encounter one another, executing independent brownian motion and forming the colloidal gas phase. As the concentration is increased to where the average interparticle separation is comparable to the overlap separation, the particles become strongly confined by their neighbors and form the colloidal liquid phase. If the concentration is further increased, an abrupt freezing transition to the ordered colloidal crystal phase will occur. The colloidal crystal structure found near the freezing transition is the body-centered cubic (Fig. 1), as is often the case in atomic systems. At higher concentrations, the more densely packed, in fact closest-packed, face-centered cubic and hexagonal close-packed structures are found. *SEE GAS; LIQUID.*

Despite these structural similarities, atomic and colloidal crystals are characterized by vastly different length, stiffness, and time scales. First, particles in atomic crystals are separated by less than a nanometer compared to micrometers in colloidal crystals. This means that colloidal crystal structure can be determined by visible-light techniques, as opposed to x-rays, which are required in the atomic crystal case. The spectacular iridescent light reflection by opal and other colloidal crystals (**Fig. 2**) is a direct result of the periodic lattice structure. Second, colloidal atomic crystals are about 10^8 times weaker than atomic crystals; that is, their elastic constants are 10^8 times smaller. Finally, significant atomic motions, such as the relative vibration of adjacent particles, occur in 10^{-12} s, whereas the equivalent time in a colloidal system is 10^{-2} s. Hence, it is possible to watch in detail through a microscope the motions of particles in a colloidal lattice. The distinctive properties of colloidal crystals thus provide a new avenue to the understanding of many phenomena found in atomic systems, such as lattice vibrations (sound waves), crystal dislocation and other defects, the response of crystals to large stresses (plastic flow), and crystallization itself. *SEE CRYSTAL DEFECTS; CRYSTAL GROWTH; LATTICE VIBRATIONS; OPALESCENCE; PLASTICITY.*

Complex systems. In addition to the simple colloidal crystals found in bulk (three-dimensional) suspensions of spherical particles, there are a number of more exotic colloidal systems. Colloidal alloys can be formed in suspensions containing particles of several different sizes. Binary mixtures form mixed crystals, sometimes with approximately 100 particles per unit cell, analogous to the alloy structures found in binary

Fig. 2. Reflection of white light from a polycrystalline colloidal crystal suspension. The crystals are body-centered cubic, with a particle diameter of 0.1 μm and interparticle spacing of 1.5 μm.

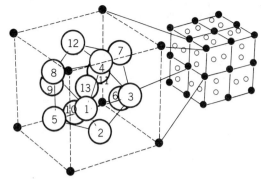

Fig. 3. Crystal structure found in a binary mixture of polystyrene microspheres of two different sizes: 550 nm in diameter (open circles) and 270 nm in diameter (closed circles). The colloidal alloy has the crystal structure found in atomic binary alloys such as NaZn₁₃. (After S. Hachisu and S. Yoshimura, Optical demonstration of crystalline superstructures in binary mixtures of latex globules, Nature, 283:188–189, 1980)

atomic metal mixtures (**Fig. 3**). Other systems include colloidal glasses, which are binary or multicomponent suspensions that exhibit shear rigidity without periodic structure; two-dimensional colloidal crystals, which are obtained by trapping a monolayer or several layers of particles; and colloidal liquid crystals, which are found in suspensions of interacting nonspherical (rod- or disk-shaped) particles such as tobacco mosaic virus. SEE ALLOY STRUCTURES; AMORPHOUS SOLID; COLLOID; CRYSTAL STRUCTURE; LIQUID CRYSTALS.

Noel A. Clark

Bibliography. N. A. Clark, A. J. Hurd, and B. J. Ackerson, Single colloidal crystals, *Nature,* 281:57–60, 1979; D. J. Darragh, A. J. Gaskin, and J. V. Sanders, Opals, *Sci. Amer.,* 234(4):84–95, April 1976; R. M. Fitch (ed.), *Polymer Colloids,* vols. 1 and 2, 1978; S. Hachisu and S. Yoshimura, Optical demonstration of crystalline superstructures in binary mixtures of latex globules, *Nature,* 283:188–189, 1980; P. Pieranski, Colloidal crystals, *Contemp. Phys.,* 24:25–73, 1983; P. N. Pusey and W. van Megan, Observation of a glass transition in suspensions of spherical colloidal particles, *Phys. Rev. Lett.,* 59:2083–2086, 1987.

Colon

The portion of the intestine that runs from the cecum to the rectum; in some mammals, it may be separated from the small intestine by an ileocecal valve. It is also known as the large intestine. The colon is usually divided into ascending, transverse, and descending portions. In the humans a fourth section, the sigmoid, is found. The colon is longer in herbivores and shorter in carnivores, and is about 4 to 6 ft (1 to 2 m) long in humans. Although no digestive enzymes are secreted in the colon, digestion is assisted by an alkaline fluid. Much digestion (for example, all breakdown of cellulose) occurs by bacteria, of which *Escherichia coli* is the most common. Most of the fluid added to the food during digestion is reabsorbed into the body in the colon. All digestive action, water absorption, and so on, is completed before the food materials pass out of the colon into the rectum. SEE DIGESTIVE SYSTEM.

Walter J. Bock

Color

That aspect of visual sensation enabling a human observer to distinguish differences between two structure-free fields of light having the same size, shape, and duration. Although luminance differences alone permit such discriminations to be made, the term color is usually restricted to a class of differences still perceived at equal luminance. These depend upon physical differences in the spectral compositions of the two fields, usually revealed to the observer as differences of hue or saturation.

Photoreceptors. Color discriminations are possible because the human eye contains three classes of cone photoreceptors that differ in the photopigments they contain and in their neural connections. Two of these, the R and G cones, are sensitive to all wavelengths of the visible spectrum from 380 to 700 nanometers. (Even longer or shorter wavelengths may be effective if sufficient energy is available.) R cones are maximally sensitive at about 570 nm, G cones at about 540 nm. The ratio R/G of cone sensitivities is minimal at 465 nm and increases monotonically for wavelengths both shorter and longer than this. This ratio is independent of intensity, and the red-green dimension of color variation is encoded in terms of it. The B cones, whose sensitivity peaks at about 440 nm, are not appreciably excited by wavelengths longer than 540 nm. The perception of blueness and yellowness depends upon the level of excitation of B cones in relation to that of R and G cones. No two wavelengths of light can produce equal excitations in all three kinds of cones. It follows that, provided they are sufficiently different to be discriminable, no two wavelengths can give rise to identical sensations.

The foregoing is not true for the comparison of two different complex spectral distributions. These usually, but not always, look different. Suitable amounts of short-, middle-, and long-wavelength lights, if additively mixed, can for example excite the R, G, and B cones exactly as does a light containing equal energy at all wavelengths. As a result, both stimuli look the same. This is an extreme example of the subjective identity of physically different stimuli known as chromatic metamerism. Additive mixture is achievable by optical superposition, rapid alternation at frequencies too high for the visual system to follow, or (as in color television) by the juxtaposition of very small elements which make up a field structure so fine as to exceed the limits of visual acuity. The integration of light takes place within each receptor, where photons are individually absorbed by single photopigment molecules, leading to receptor potentials that carry no information about the wavelength of the absorbed photons. SEE EYE (VERTEBRATE); LIGHT.

Colorimetry. Although colors are often defined by appeal to standard samples, the trivariant nature of color vision permits their specification in terms of three values. Ideally these might be the relative excitations of the R, G, and B cones. Because too little was known about cone action spectra in 1931, the International Commission on Illumination (CIE) adopted at that time a different but related system for the prediction of metamers (the CIE system of colorimetry). This widely used system permits the specification of tristimulus values X, Y, and Z, which make almost the same predictions about color matches as do calculations based upon cone action spectra. If, for fields 1 and 2, $X_1 = X_2$, $Y_1 = Y_2$, and $Z_1 = Z_2$, then the two stimuli are said to match (and therefore have the same color) whether they are physically the same (isometric) or different (metameric).

The use of the CIE system may be illustrated by a sample problem. Suppose it is necessary to describe quantitatively the color of a certain paint when viewed under illumination by a tungsten lamp of known color temperature. The first step is to measure the reflectance of the paint continuously across the visible spectrum with a spectrophotometer. The reflectance at a given wavelength is symbolized as R_λ. The next step is to multiply R_λ by the relative amount of light E_λ emitted by the lamp at the same wavelength. The product $E_\lambda R_\lambda$ describes the amount of light reflected from the paint at wavelength λ. Next, $E_\lambda R_\lambda$ is multiplied by a value $\bar{x}_\lambda$, which is taken from a table of X tristimulus values for an equal-energy spectrum. The integral $\int E_\lambda R_\lambda \bar{x}_\lambda d_\lambda$ gives the tristimulus value X for all of the light reflected from the paint. Similar computations using $\bar{y}_\lambda$ and $\bar{z}_\lambda$ yield tristimulus values Y and Z.

Tables of $\bar{x}_\lambda$, $\bar{y}_\lambda$, and $\bar{z}_\lambda$ are by convention carried to more decimal places than are warranted by the precision of the color matching data upon which they are based. As a result, colorimetric calculations of the type just described will almost never yield identical values, even for two physically different fields that are identical in appearance. For this and other reasons it is necessary to specify tolerances for color differences. Such differential colorimetry is primarily based upon experiments in which observers attempted color matches repeatedly, with the standard deviations of many matches being taken as the discrimination unit.

Chromaticity diagram. Colors are often specified in a two-dimensional chart known as the CIE chromaticity diagram, which shows the relations among tristimulus values independently of luminance. In this plane, y is by convention plotted as a function of x, where $y = Y/(X + Y + Z)$ and $x = X/(X + Y + Z)$. [The value $z = Z/(X + Y + Z)$ also equals $1 - (x + y)$ and therefore carries no additional information.] Such a diagram is shown in the **illustration**, in which the continuous locus of spectrum colors is represented by the outermost contour. All nonspectral colors are contained within an area defined by this boundary and a straight line running from red to violet. The diagram also shows discrimination data for 25 regions, which plot as ellipses represented at 10 times their actual size. A discrimination unit is one-tenth the distance from the ellipse's center to its perimeter. Predictive schemes for interpolation to other regions of the CIE diagram have been worked out.

If discrimination ellipses were all circles of equal size, then a discrimination unit would be represented by the same distance in any direction anywhere in the chart. Because this is dramatically untrue, other chromaticity diagrams have been developed as linear projections of the CIE chart. These represent discrimination units in a relatively more uniform way, but never perfectly so.

A chromaticity diagram has some very convenient properties. Chief among them is the fact that additive mixtures of colors plot along straight lines connecting the chromaticities of the colors being mixed. Although it is sometimes convenient to visualize colors in terms of the chromaticity chart, it is important to realize that this is not a psychological color diagram. Rather, the chromaticity diagram makes a statement about the results of metameric color matches, in the

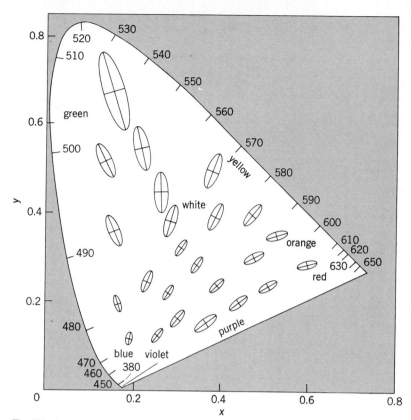

The 1931 CIE chromaticity diagram showing MacAdam's ellipses 10 times enlarged. *(After G. W. Wyszecki and W. S. Stiles, Color Science, John Wiley and Sons, 1967)*

sense that a given point on the diagram represents the locus of all possible metamers plotting at chromaticity coordinates x, y. However, this does not specify the appearance of the color, which be dramatically altered by preexposing the eye to colored lights (chromatic adaptation) or, in the complex scenes of real life, by other colors present in nearby or remote areas (color contrast and assimilation). Nevertheless, within limits, metamers whose color appearance is thereby changed continue to match.

For simple, directly fixated, and unstructured fields presented in an otherwise dark environment, there are consistent relations between the chromaticity coordinates of a color and the color sensations that are elicited. Therefore, regions of the chromaticity diagram are often denoted by color names, as shown in the illustration.

Although the CIE system works rather well in practice, there are important limitations. Normal human observers do not agree exactly about their color matches, chiefly because of the differential absorption of light by inert pigments in front of the photoreceptors. Much larger individual differences exist for differential colorimetry, and the system is overall inappropriate for the 4% of the population (mostly males) whose color vision is abnormal. The system works only for an intermediate range of luminances, below which rods (the receptors of night vision) intrude, and above which the bleaching of visual photopigments significantly alters the absorption spectra of the cones. *See* COLOR VISION.

Robert M. Boynton

Bibliography. F. W. Billmeyer and M. Saltzman, *Principles of Color Technology,* 1981; R. M. Boynton, *Human Color Vision,* 1979; R. W. Burnham,

R. M. Hanes, and C. J. Bartleson, *Color: A Guide to Basic Facts and Concepts*, 1963; D. B. Judd and G. W. Wyszecki, *Color in Business, Science, and Industry*, 3d ed., 1975; K. Nassau, *The Physics and Chemistry of Color*, 1983; G. W. Wyszecki and W. S. Stiles, *Color Science*, 2d ed., 1982.

Color (quantum mechanics)

A term used to describe a hypothetical quantum number carried by the quarks which are thought to make up the strongly interacting elementary particles. It has nothing to do with the ordinary, visual use of the word color.

The quarks which are thought to make up the strongly interacting particles have a spin angular momentum of one-half unit of $\hbar$ (Planck's constant). According to a fundamental theorem of relativity combined with quantum mechanics, they must therefore obey Fermi-Dirac statistics and be subject to the Pauli exclusion principle. No two quarks within a particular system can have exactly the same quantum numbers. SEE EXCLUSION PRINCIPLE; FERMI-DIRAC STATISTICS.

However, in making up a baryon, it often seemed necessary to violate this principle. The omega particle, for example, is made of three strange quarks, and all three had to be in exactly the same state. O. W. Greenberg was responsible for the essential idea for the solution to this paradox. In 1964 he suggested that each quark type (*u, d,* and *s*) comes in three varieties identical in all measurable qualities but different in an additional property, which has come to be known as color. The exclusion principle could then be satisfied and quarks could remain fermions, because the quarks in the baryon would not all have the same quantum numbers. They would differ in color even if they were the same in all other respects.

The color hypothesis triples the number of quarks but does not increase the number of baryons and mesons. The rules for assembling them ensures this. Tripling the number of quarks does, however, have at least two experimental consequences. It triples the rate at which the neutral π meson decays into two photons and brings the predicted rate into agreement with the observed rate.

The total production cross section for baryons and mesons in electron-positron annihilation is also tripled. The experimental result at energies between 2 and 3 GeV is in reasonable agreement with the color hypothesis and completely incompatible with the simple quark model without color. SEE BARYON; ELEMENTARY PARTICLE; GLUONS; MESON; QUARKS.

Thomas Appelquist

Bibliography. H. Georgi, A unified theory of elementary particles and forces, *Sci. Amer.*, 244(4):48–63, 1981; S. L. Glashow, Quarks with color and flavor, *Sci. Amer.*, 233(4):38–50, 1975; O. W. Greenberg, Spin and unitary spin independence in a paraquark model of baryons and mesons, *Phys. Rev. Lett.*, 13:598–602, 1964; G. K. O'Neill and D. Cheng, *Elementary Particle Physics*, 1979.

Color centers

Atomic and electronic defects of various types in solids which produce optical absorption bands in otherwise transparent crystals such as the alkali halides, alkaline earth fluorides, or metal oxides. They are general phenomena found in a wide range of materials. Color centers are produced by gamma radiation or x-radiation, by addition of impurities or excess constituents, and sometimes through electrolysis. A well-known example is that of the *F*-center in alkali halides such as sodium chloride, NaCl. The designation *F*-center comes from the German word *Farbe*, which means color. *F*-centers in NaCl produce a band of optical absorption toward the blue end of the visible spectrum; thus the colored crystal appears yellow under transmitted light. On the other hand, KCl with *F*-centers appears magenta, and KBr appears blue. SEE CRYSTAL DEFECTS.

Theoretical studies guided by detailed experimental work have yielded a deep understanding of specific color centers. The crystals in which color centers appear tend to be transparent to light and to other forms of electromagnetic radiation, such as microwaves. Consequently, experiments which can be carried out include optical spectroscopy, luminescence and Raman scattering, magnetic circular dichroism, magnetic resonance, and electromodulation. Color centers find practical application in radiation dosimeters, schemes have been proposed to use color centers in high-density memory devices, and tunable lasers have been made from crystals containing color centers. SEE ABSORPTION OF ELECTROMAGNETIC RADIATION.

Origin. Figure 1 shows the absorption bands due to color centers produced in potassium bromide by exposure of the crystal at the temperature of liquid nitrogen (81 K) to intense penetrating x-rays. Several prominent bands appear as a result of the irradiation. The *F*-band appears at 600 nanometers and the so-called *V*-bands appear in the ultraviolet. Uncolored alkali halide crystals may be grown readily from a melt with as few imperfections as one part in 10^5. In the unirradiated and therefore uncolored state, they show no appreciable absorption from the far-infrared through the visible, to the region of characteristic absorption in the far-ultraviolet (beginning at about 200 nm in Fig. 1). Color centers may be introduced into such crystals by means other than exposure to penetrating radiation. The most important of these is a method known as additive coloration, in which the composition of the crystal is made to deviate from stoichiometric proportions by heating in the presence of the alkali metal vapor.

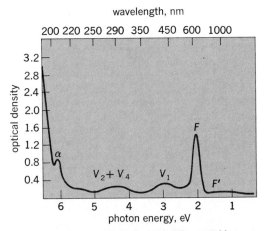

Fig. 1. Absorption bands produced in KBr crystal by exposure to x-rays at 81 K. Optical density is equal to $\log_{10} (I_0/I)$, where I_0 is the intensity of incident light and I the intensity of transmitted light. Steep rise in optical density at far left is due to intrinsic absorption in crystal (modified slightly by existence of *F*-centers).

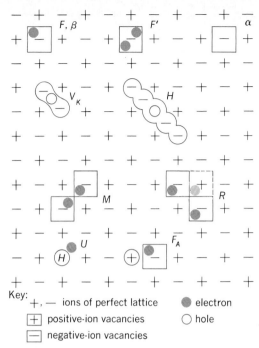

Key:
+, — ions of perfect lattice   ● electron
⊞ positive-ion vacancies   ○ hole
☐ negative-ion vacancies

Fig. 2. Well-established models for several color centers in ionic crystal of NaCl structure. Designation β for *F*-center signifies disturbing influence of *F*-centers on tail of fundamental absorption band; characteristic absorption is also influenced by presence of vacancies, or α-centers (see Fig. 1). The *V*ₖ-center can be described as a self-trapped hole forming a Cl₂-molecule-ion, and the *H*-center can be described as a hole trapped at an interstitial halide ion. Both are stable at low temperature only. The *R*-center contains three negative-ion vacancies, one of which is in an adjacent lattice plane. *H* in circle at lower left represents hydrogen impurity.

Color bands such as the *F*-band and the *V*-band arise because of light absorption at defects dispersed throughout the lattice. This absorption is caused by electronic transitions at the centers. On the other hand, colloidal particles, each consisting of many atoms, dispersed through an optical medium also produce color bands. In this case, if the particles are large enough, the extinction of light is due to both light scattering and light absorption. Colloidal gold is responsible for the color of some types of ruby glass. Colloids may also form in alkali halide crystals—for example, during heat treatment of an additively colored crystal which contains an excess of alkali metal.

Atomically dispersed centers such as *F*-centers are part of the general phenomena of trapped electrons and holes in solids. The accepted model of the *F*-center is an electron trapped at a negative ion vacancy. Many other combinations of electrons, holes, and clusters of lattice vacancies have been used to explain the various absorption bands observed in ionic crystals. The centers and models shown in **Fig. 2** have been positively identified either by electron spin resonance or by a combination of other types of experimental evidence.

Impurities can play an important role in color-center phenomena. Certain impurities in ionic crystals produce color bands characteristic of the foreign ion. For example, hydrogen can be incorporated into the alkali halides as H⁻ ions substituting for halogen ions with resultant appearance of an absorption band (the *U*-band) in the ultraviolet. In this case, the *U*-centers interact with other defects; for example, they trap holes to become hydrogen atoms. The rate at which

F-centers are produced by x-irradiation is greatly increased by the incorporation of hydrogen, the *U*-centers being converted into *F*-centers with high efficiency.

F-centers. *F*-centers may be produced in uncolored crystals by irradiation with ultraviolet, x-rays, gamma rays, and high-speed particles, and also by electrolysis. However, one of the most convenient methods is that of additive coloration. Alkali halide crystals may be additively colored by heating to several hundred degrees Celsius in the presence of the alkali metal vapor, then cooling rapidly to room temperature. The *F*-band that results is the same as that produced by irradiation and is dependent upon the particular alkali halide used, not upon the alkali metal vapor. For example, the coloration produced in KCl is the same whether the crystal is heated in potassium or in sodium vapor. This and other evidence indicate that excess alkali atoms enter the crystal as normal lattice ions by donating an electron to accompanying negative-ion vacancies to form *F*-centers. **Figure 3** shows the position of the *F*-band in several alkali halides.

Both the width and the exact position of the *F*-band change with temperature in the manner shown in **Fig. 4**. This behavior can be explained in terms of the thermal motion of the ions surrounding the center. Thermal motion is most important at high temperatures, whereas at low temperatures the width of the absorption band becomes less and approaches a constant value which prevails down to the very lowest temperatures.

Concentration of centers. The height of the absorption maximum and the width at half maximum may be used to calculate the concentration N of absorbing centers. For KCl, classical theory gives the concentration as $N = 1.3 \times 10^{16}\, AH\, \mathrm{cm}^{-3}$, where A is the maximum absorption in cm^{-1}, and H is the half width of the absorption band in electronvolts. The numerical factor in this equation depends upon the material under consideration and is given for KCl. The general relation between absorption characteristics and defect concentration is slightly more complicated but is nevertheless useful to determine concentrations below the limit of detection by chemical means.

Shape of F-band. Refined measurements of the *F*-band at low temperatures show that it is not a simple bell-shaped curve indicated in Figs. 3 and 4. In fact, it is found to contain a shoulder on the short-wavelength side, which is referred to as the *K*-band. It has been proposed that the *K*-band is in agreement with a model for the *F*-center in which an electron is trapped in a vacancy surrounded by point ions which produce the cubic crystal field of the lattice. An electron in the vicinity of a negative-ion vacancy should have a ground state and one or more excited states below the bottom of the conduction band. The main absorption peak corresponds to excitation of the electron from the 1*s*-like ground state to an excited 2*p*-like state. The *K*-band is thought to be due to transitions to the 3*p* and higher states which finally merge with the conduction band itself. *F*-center photoconductivity measurements tend to confirm the idea of discrete ex-

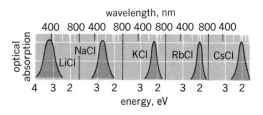

Fig. 3. F-bands in different alkali halide crystals.

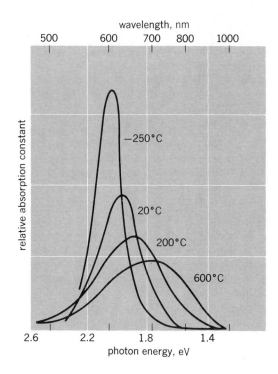

Fig. 4. Variation with temperature of the width of the *F*-band in a crystal of potassium bromide. °F = (°C × 1.8) + 32°.

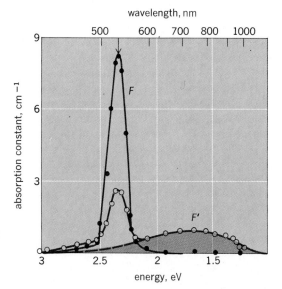

Fig. 5. *F*- and *F'*-bands in KCl at −235°C (−391°F). *F'*-band was produced by irradiating crystal with light of wavelength shown by arrow at center of *F*-band. Such light causes the *F*-band to bleach as *F'*-band grows.

cited states, such as the 2*p*, which lie below the conduction band or continuum. *See Photoconductivity.*

Magnetic experiments. Some of the most detailed information on *F*-centers comes from electron spin resonance and from electron-nuclear double resonance. From these experiments the extent and character of the wave function of the electron trapped at the halogen-ion vacancy can be obtained. The electron even in the ground state is not highly localized but is spread out over the six neighboring alkali ions and to some extent over more distant neighbors. Faraday rotation studies indicated fine structure in the excited state of the *F*-center due to spin-orbit interaction. The spin-orbit splitting of the *F*-center is negative, in agreement with the theoretical prediction for an electron trapped at a vacancy. *See Magnetic resonance.*

Lifetime of excited F-center. At low temperature an excited *F*-center decays spontaneously to the ground state with the emission of an infrared luminescent quantum. The radiative lifetime has been measured and found to be unexpectedly long, 0.57 microsecond in KCl and as long as 2.0 μs in KI. There are indications that this is due to a drastic change in the wave function following excitation and lattice relaxation. Lattice polarization must be taken into account in order to calculate the effect. The relaxed excited state is very spread out in the lattice, whereas the ground state is somewhat more localized.

F'-centers. When a crystal containing *F*-centers is irradiated at low temperature with light in the *F*-band itself, the *F*-absorption decreases and a new band, known as the *F'* band, appears (**Fig. 5**). Experiments have shown that for every *F'*-center created, two *F*-centers are destroyed. The *F'*-center consists of an *F*-center which has captured an extra electron. Important photoelectric effects occur during conversion of *F*- to *F'*-centers and during irradiation of alkali halides with *F*-band light.

V-centers. As Fig. 1 shows, absorption bands arise in the ultraviolet as well as in the visible portion of the spectrum when KBr is irradiated while cold. These *V*-bands are apparently not present in crystals additively colored by the alkali metal vapor. Irradiation of a crystal produces both electrons and holes so that bands associated with trapped positive charges are to be expected in addition to electrons trapped at negative-ion vacancies (*F*-centers). *V*-bands can also be produced, at least in the case of KBr and KI, by introduction of an excess of halogen into the crystal.

Crystals containing *V*-centers have been studied by spin resonance techniques with the result that a center not previously observed optically has been identified. This center, sometimes referred to as the V_K-center (Fig. 2), is due to a hole trapped in the vicinity of two halogen ions which have become displaced from their normal lattice position to form a Cl_2-molecule-ion (in the case of KCl). It has been found in several different alkali halides, and its optical absorption and energy level structure have been correlated. The V_K-center is known to play an important role in the intrinsic luminescence of ionic crystals.

Other centers. Only the simpler defects in pure crystals, such as the alkali halides, are discussed above. Aggregate centers and centers associated with impurities may be important. The *M*- and *R*-centers shown in Fig. 2 are examples of the former; *U*-centers and F_A-centers, also shown in Fig. 2, are important examples of the latter. The alkali halide lattice is the classic matrix for color centers, which frequently can be readily formed by irradiation. An efficient photolytic process is involved, which follows recombination of self-trapped ions. Analogous phenomena occur in crystals with the fluorite (CaF_2) structure and in oxides such as CaO and MgO. In these cases, the photolytic processes are different and apparently not as efficient as in an alkali halide. Crystal purity is also frequently a problem. *See Photolysis.*

Lasers. Color-center lasers are sources of tunable coherent radiation in the near-infrared. These devices make use of *F*-centers with nearby impurities as well as foreign atoms such as thallium in ionic crystals. *See Laser.*

Frederick C. Brown

Bibliography. N. W. Ashcroft and N. D. Mermin, *Solid State Physics,* 1976; F. C. Brown, *The Physics of Solids,* 1967; W. B. Fowler, *The Physics of Color Centers,* 1968; W. Hayes, *Crystals with the Fluorite Structure,* 1974; H. Herman (ed.), *Treatise on Mate-*

rial Science and Technology, vol. 5, 1974; A. E. Hughes and B. Henderson, in J. Crawford and L. Slifkin (eds.), Point Defects in Solids, vol. 1, 1972; L. F. Mollenauer, H. D. Vieira, and L. Szeto, Mode locking by synchronous pumping using a gain medium with microsecond decay times, Opt. Lett., 7:414–416, 1982; I. Schneider and N. C. Moss, Color center laser continuously tunable from 1.67 to 2.46 micrometers, Opt. Lett., 8:7–8, 1983.

Color filter

An optical element that partially absorbs incident radiation, often called an absorption filter. The absorption is selective with respect to wavelength, or color, limiting the colors that are transmitted by limiting those that are absorbed. Color filters absorb all the colors not transmitted. They are used in photography, optical instruments, and illuminating devices to control the amount and spectral composition of the light.

Color filters are made of glass for maximum permanence, of liquid solutions in cells with transparent faces for flexibility of control, and of dyed gelatin or plastic (usually cellulose acetate) for economy, convenience, and flexibility. The plastic filters are often of satisfactory permanence, but they are sometimes cemented between glass plates for greater toughness and scratch resistance. They do not have as good quality as gelatin filters.

Color filters are sometimes classified according to their type of spectral absorption: short-wavelength pass, long-wavelength pass or band-pass; diffuse or sharp-cutting; monochromatic or conversion. The short-wavelength pass transmits all wavelengths up to the specified one and then absorbs. The long-wavelength pass is the opposite. Every filter is a band-pass filter when considered generally. Even an ordinary piece of glass does not transmit in the ultraviolet or infrared parts of the spectrum. Color filters, however, are usually discussed in terms of the portion of the visible part of the spectrum. Sharp and diffuse denote the sharpness of the edges of the filter band pass. Monochromatic filters are very narrow band-pass filters. Conversion filters alter the spectral response or distribution of one selective detector or source to that of another, for example, from that of a light bulb to that of the Sun (see **illus.**).

Neutral-density filters transmit a constant fraction of the light across the spectrum considered. They are made either by including an appropriate absorbent like carbon that is spectrally nonselective, or by a thin, partially transparent reflective layer, often of aluminum.

The transmittance of a filter is the ratio of the transmitted flux to the incident flux, expressed as either a ratio or a percentage. The density of a filter is the logarithm to base 10 of the reciprocal of the transmittance. For example, a filter with a 1% transmittance has a density of 2.

Other, closely related filters are the Christiansen, Lyot or birefringent, residual-ray, and interference filters. They use the phenomena of refraction, polarization, reflection, and interference. SEE ABSORPTION OF ELECTROMAGNETIC RADIATION; COLOR; INTERFERENCE FILTER; REFLECTION OF ELECTROMAGNETIC RADIATION; SUN.

William L. Wolfe

Color vision

The ability to discriminate light on the basis of wavelength composition. It is found in humans, in other primates, and in certain species of birds, fishes, reptiles, and insects. These animals have visual receptors that respond differentially to the various wavelengths of visible light. Each type of receptor is especially sensitive to light of a particular wavelength composition. Evidence indicates that primates, including humans, possess three types of cone receptor, and that the cones of each type possess a pigment that selectively absorbs light from a particular region of the visible spectrum.

If the wavelength composition of the light is known, its color can be specified. However, the reverse statement cannot be made. A given color may usually be produced by any one of an infinite number of combinations of wavelength. This supports the conclusion that there are not many different types of color receptor. Each type is capable of being stimulated by light from a considerable region of the spectrum, rather than being narrowly tuned to a single wavelength of light. The trichromatic system of colorimetry, using only three primary colors, is based on the concept of cone receptors with sensitivities having their peaks, respectively, in the long, middle, and short wavelengths of the spectrum. The number of such curves and their possible shapes have long been subjects for study, but not until the 1960s were direct measurements made of the spectral sensitivities of individual cone receptors in humans and various animals. SEE COLOR; COLORIMETRY.

Color recognition. Color is usually presented to the individual by the surfaces of objects on which a more or less white light is falling. A red surface, for example, is one that absorbs most of the short-wave light and reflects the long-wave light to the eye. The surface colors are easily described by reference to the color solid shown in **Fig. 1**. The central axis defines the lightness or darkness of the surface as determined by its overall reflectance of white light, the lowest reflectance being called black and the highest, white. The circumference denotes hue, related primarily to the selective reflectance of the surface for particular wavelengths of light. The color solid is pointed at its top and bottom to represent the fact that as colors become whiter or blacker they lose hue. The distance from central axis to periphery indicates saturation, a characteristic that depends chiefly on the narrowness, or purity, of the band of wavelengths reflected by the surface. At the center of the figure is a medium-gray surface, one which exhibits a moderate amount of reflectance for all wavelengths of light. Colors that are opposite one another are called complementaries, for example, yellow and blue, red and blue-green, purple

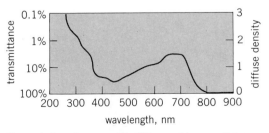

Transmission of a conversion filter, used to convert the spectral distribution of a light bulb (color temperature of 2360 K or 3790°F) to that of the Sun (color temperature of 5500 K or 9440°F). (After Eastman Kodak Co., Kodak Filters for Scientific and Technical Uses, 1972)

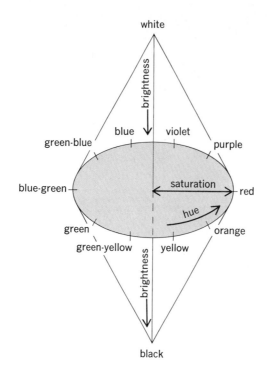

and green, and white and black.

Color mixture and contrast. Two complementaries, when added together in the proper proportions, can oppose or neutralize one another to produce a colorless white or gray. Various contrast effects also attest to the opposition of the complementaries. Staring at a bright-red surface against a gray background results in the appearance of a blue-green border around the red; this is an example of simultaneous contrast. Similarly, when a red light is turned off, there is frequently a negative afterimage that appears to have the opposite color, blue-green.

A set of primary colors can be chosen so that any other color can be produced from additive mixtures of the primaries in the proper proportions. Thus, red, green, and blue lights can be added together in various proportions to produce white, purple, yellow, or any of the various intermediate colors. Three-color printing, color photography, and color television are examples of the use of primaries to produce plausible imitations of colors of the original objects. *SEE PHO-TOGRAPHY; PRINTING.*

Achromatic colors. Colors lying along a continuum from white to black are known as the gray, or achromatic, colors. They have no particular hue, and are therefore represented by the central axis of the color diagram in Fig. 1. White is shown at the top of the diagram since it represents the high-brightness extreme of the series of achromatic colors. With respect to the surface of an object, the diffuse, uniform reflectance of all wavelengths characterizes this series from black through the grays to white in order of increasing reflectance. Whiteness is a relative term; white paper, paint, and snow reflect some 80% or more of the light of all visible wavelengths, while black surfaces typically reflect less than 10% of the light. The term white is also applied to a luminous object, such as a gas or solid, at a temperature high enough to emit fairly uniformly light of all visible wavelengths. In the same connotation, a sound is described as white noise if its energy is nearly the same at all audible frequencies.

Gray. Gray is the term applied to all the intermediate colors in the series of achromatic colors. Gray may result from a mixture of two complementary colors, from a mixture of all primary colors, or from a fairly uniform mixture of lights of all wavelengths throughout the visible spectrum. Grayness is relative; a light gray is an achromatic color that is lighter than its surroundings, while a dark gray is so called because it is darker than its surroundings. Thus, the same surface may be called light or dark gray when carried from one situation to another. A gray color is one of minimum saturation; it corresponds to zero on a scale of excitation purity. At the other extreme on this scale is the color evoked by pure monochromatic light.

Black. The opposite extreme from white in the series of achromatic colors is black. Blackness is a relative term applied to surfaces that uniformly absorb large percentages of light of all visible wavelengths. A black object in sunlight absorbs a large percentage of the light, but it may reflect a larger absolute quantity of light to the eye than does a white object in the shade. Black may also be used to refer to invisible light; ultraviolet rays, for example, may be called black light if they fall on fluorescent materials that thereby emit visible light.

Color blindness. Color blindness is a condition of faulty color vision. It appears to be the normal state of animals that are active only at night. It is also characteristic of human vision when the level of illumination is quite low or when objects are seen only at the periphery of the retina. Under these conditions, vision is mediated not by cone receptors but by rods, which respond to low intensities of light. In rare individuals, known as monochromats, there is total color blindness even at high light levels. Such persons are typically deficient or lacking in cone receptors, so that their form vision is also poor.

Dichromats are partially color-blind individuals whose vision appears to be based on two primaries rather than the normal three. Dichromatism occurs more often in men than in women because it is a sex-linked, recessive hereditary condition. One form of dichromatism is protanopia, in which there appears to be a lack of normal red-sensitive receptors. Red lights appear dim to protanopes and cannot be distinguished from dim yellow or green lights. A second form is deuteranopia, in which there is no marked reduction in the brightness of any color, but again there is a confusion of the colors normally described as red, yellow, and green. A third and much rarer form is tritanopia, which involves a confusion among the greens and blues. *SEE HUMAN GENETICS.*

Many so-called color-blind individuals might better be called color-weak. They are classified as anomalous trichromats because they have trichromatic vision of a sort, but fail to agree with normal subjects with respect to color matching or discrimination tests. Protanomaly is a case of this type, in which there is subnormal discrimination of red from green, with some darkening of the red end of the spectrum. Deuteranomaly is a mild form of red-green confusion with no marked brightness loss. Nearly 8% of human males have some degree of either anomalous trichromatism or dichromatism as a result of hereditary factors; less than 1% of females are color-defective. A few forms of color defect result from abnormal conditions of the visual system brought on by poisoning, drugs, or disease.

Color blindness is most commonly tested by the

use of color plates in which various dots of color define a figure against a background of other dots. The normal eye readily distinguishes the figure, but the colors are so chosen that even the milder forms of color anomaly cause the figure to be indistinguishable from its background. Other tests involve the ability to mix or distinguish colored lights, or the ability to sort colored objects according to hue.

Theories. Theories of color vision are faced with the task of accounting for the facts of color mixture, contrast, and color blindness. The schema shown in **Fig. 2** may be useful in considering the various theories of color vision. It has no necessary resemblance to the structures or functions that are actually present in the visual system.

Young-Helmholtz theory. Thomas Young, early in the nineteenth century, realized that color is not merely a property of objects or surfaces. Rather, it involves a sensory experience, the characteristics of which are determined by the nature of the visual receptors and the way in which the signals that they generate are processed within the nervous system. Young was thus led to a trichromatic theory of color vision, because he found that only three types of excitation were needed to account for the range of perceived colors. Hermann von Helmholtz clarified the theory and provided it with the idea of differential absorption of light of various wavelengths by each of three types of receptor. Hence, the theory, commonly known as the Young-Helmholtz theory, is concerned with the reception phase in the color vision schema shown. This theory assumes the existence of three primary types of color receptor that respond to short, medium, and long waves of visible light, respectively. Primary colors are those that stimulate most successfully the three types of receptor; a mixture of all three primaries is seen as white or gray. A mixture of two complementary colors also stimulates all three. Blue light, for example, stimulates the blue-sensitive receptors; hence, its complementary color is yellow, which stimulates both the red receptors and the green. Similarly, the complement of red is blue-green and the complement of green is purple (blue and red). Protanopia is explained as a condition in which no long-wave sensitive receptors are present. Deuteranopia results from loss of the green receptor.

Hering theory. The theory formulated by Ewald Hering is an opponent-colors theory. In its simplest form, it states that three qualitatively different processes are present in the visual system, and that each of the three is capable of responding in two opposite ways. Thus, this theory is concerned with the processing phase (Fig. 2), in which there is a yellow-blue process, a red-green process, and a black-white process. The theory obviously was inspired by the opposition of colors that is exhibited during simultaneous contrast, afterimages, and complementary colors. It is a theory best related to the phenomenal and the neural-processing aspects of color rather than to the action of light on photopigments. This theory gains credibility when it is realized that opponent responses are common enough in the central nervous system. Well-known examples are polarization and depolarization of cell membranes, excitation and inhibition of nerve impulses, and reciprocal innervation of antagonistic response mechanism. The theory also maintains that there are four primary colors, since yellow appears to be just as unique perceptually, as red, green, or blue. However, it does not necessarily assume the existence of a separate yellow receptor.

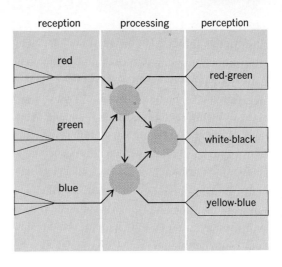

reception   processing   perception

red

green

blue

red-green

white-black

yellow-blue

Fig. 2. Schematic representation of the three phases of color vision which can be generally used to describe the various theories of color vision.

Other theories. Various forms of stage theory or zone theory of color vision have been proposed. One form combines the principles of the Young-Helmholtz and Hering theories. It proposes that (1) there are three types of cone receptor; (2) the responses of red-sensitive, green-sensitive, and blue-sensitive receptors (and possibly others) are conducted to the higher visual centers; (3) at the same time, interactions are occurring at some stage along these separate conducting paths so that strong activity in a red response path inhibits, for example, the activity of the other response paths; and (4) this inhibiting effect is specific to the time and place of the strong red activity. The last point means that in neighboring regions and at subsequent times the blue and green mechanisms are less inhibited, so that blue-green afterimages and borders are commonly experienced as an opponent reaction to strong red stimulation.

Experimental evidence. Techniques of microspectrophotometry have been used to measure the absorption of light by single cone receptors from the eyes of primates, including humans. The results confirm the major assumption of the Young-Helmholtz theory, namely, that three types of visual receptor are specialized to absorb light over characteristic ranges of wavelength, with a maximum absorption at particular regions of the spectrum. Wide differences are found, from one receptor to another, in the wavelength at which this maximum absorption of light occurs. The three types of human cone receptors absorb maximally at about 420, 530, and 560 nanometers. In addition there are rod receptors sensitive to low intensities of light over a broad range of wavelengths peaking at about 500 nm. In each of the four types of receptor there is a photosensitive pigment that is distinguished by a particular protein molecule. This determines the range and spectral location of the light which it absorbs.

In other experiments, a microelectrode has been used to penetrate individual cone receptors in living retinal tissue of certain species of fish. Electrical responses are thus recorded as the receptor cell is stimulated by a beam of light. The wavelength of this test beam is varied from one end of the spectrum to the other. The result is that each cone shows an electrical response to light which presumably initiates impulses in the nerve cells of the visual system and is qualitatively the same regardless of the wavelength of the stimulating light. Quantitatively, however, the cones are found to differ widely in the size of response potential that they generate at each wavelength. In fact, they fall into three classes, responding maximally at about 460, 530, and 610 nm. This confirms, at least in the case of certain species of fish, the trichromatic aspect of the Young-Helmholtz theory.

The Hering theory has also received powerful support from electrophysiological experiments. A microelectrode may be used to record from retinal nerve cells that are close to the cone receptors described above in the fish. Electrical responses from some of these cells are again found to depend on the wavelength of the stimulating light. But, unlike the receptors themselves, these cells often have qualitatively different responses depending on wavelength of stimulation. Thus, a given cell may exhibit a positive response potential over one region of the spectrum and a negative one over another region. A retinal basis thus exists in these cells for the opponent process aspect of the Hering theory.

In monkeys having visual systems similar to that of humans, experiments have also shown opponent-color modes of responding. Cells in certain lower brain centers, for example, exhibit nerve impulses more or less continuously. Some of these cells are found to increase their activity in response to red light, for example, but decrease it in response to green. Other cells have the opposite reaction, and still others show blue-yellow opposition of the sort that is also demanded by the Hering theory.

Central nervous system factors are evident from the fact that a red light in one eye and a green light in the other can sometimes be fused to produce an appearance of yellow. Further evidence of central effects is found in the phenomenon of color constancy. The greenness of trees or the redness of bricks is only slightly less vivid when both are viewed through dark or yellow goggles; a black cat in the sunshine reflects more light to the eye than a white cat seen by indoor illumination. Color vision, like other forms of perception, is highly dependent on the experience of the observer and on the context in which the object is perceived. SEE EYE (INVERTEBRATE); EYE (VERTEBRATE); NERVOUS SYSTEM (VERTEBRATE); PERCEPTION; PHOTORECEPTION; VISION.   Lorrin A. Riggs

Bibliography. R. M. Boynton, *Human Color Vision,* 1979; R. Fletcher and J. Voke, *Defective Color Vision,* 1985; C. H. Graham (ed.), *Vision and Visual Perception,* 1965; L. M. Hurvich, *Color Vision,* 1981; J. Mollon and T. Sharpe (eds.), *Color Vision: Physiology and Psychophysics,* 1983; W. S. Stiles, *Mechanisms of Color Vision,* 1978.

Colorimetry

Any technique by which an unknown color is evaluated in terms of known colors. Colorimetry may be visual, photoelectric, or indirect by means of spectrophotometry. These techniques are widely used in scientific studies involving the appearance of objects and lights, but are of greatest importance in the color specification of the raw materials and finished products of industry.

Visual colorimetry. In this type, the unknown color is presented beside a comparison field into which may be introduced any one or a range of known colors from which the operator chooses the one matching the unknown. To be generally applicable, the comparison field must not only cover a suf-

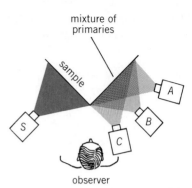

The tricolorimeter, in which the given color in *S* is matched by the addition of any three primaries in *A, B,* and *C*.

ficient color range but must also be continuously adjustable in color in three independent ways: (1) by superposition of light primaries (for example, red, green, and blue) in any required proportions (additive colorimeters, see **illus.**); (2) by successive transmission through primary filters (for example, yellow, cyan, and magenta) of adjustable thickness (subtractive colorimeters); or (3) by a rotating sectored disk whose four differently colored sectors are adjustable in relative area (disk colorimeters). Many colorimeters, however, provide but a single series of comparison colors and are limited to specimens known in advance to have colors identical or nearly identical to one or another of this single series of colors.

Indirect colorimetry. In this type, the light leaving the unknown specimen is split into its component spectral parts by means of a prism or diffraction grating, and the amount of each component part is separately measured by a photometer. The quantity evaluated is spectral radiance of a light source, spectral transmittance of a filter (glass, plastic, gelatin, or liquid), or spectral reflectance of an opaque body. SEE SPECTROPHOTOMETRIC ANALYSIS.

From spectrophotometric data such as these, it is possible to predict the amounts of known light primaries (red, green, and blue) required to produce the color of the specimen by superposition. This prediction is made by using the color-mixture functions (CIE standard observer) recommended for this purpose in 1931 by the International Commission on Illumination (CIE). These functions give the amounts of the light primaries required by the standard observer to match unit radiance of each part of the spectrum in turn. The prediction is found by adding up for the whole visible spectrum the amounts of red primary appropriate to the unknown specimen, and then doing the same thing for the appropriate amounts of green and blue primaries. SEE COLOR.

Photoelectric colorimetry. In this type, the light leaving the specimen is measured separately by three photocells. The spectral sensitivity of these photocells is adjusted, usually by color filters, to conform as closely as possible to the three color-mixture functions for the average normal human eye (CIE standard observer). The responses of the photocells give directly the amounts of red, green, and blue primaries required to produce the color of the unknown specimen for the kind of vision represented by the three photocells. Photoelectric colorimetry is used primarily for the control of the colors of manufactured articles to a preestablished tolerance.

Metamers. If two objects have the same color because the light leaving one of them toward the eye is spectrally identical to that leaving the other, any type of colorimetry serves reliably to establish the fact of color match. If, however, the two lights are spectrally dissimilar, they may still color-match for any one observer; such pairs of lights are called metamers. Normal color vision differs sufficiently from person to person so that a metameric color match for one observer may be seriously mismatched for another. On this account, the question of color match of spectrally dissimilar lights can be reliably settled only by the indirect method which uses spectrophotometry combined with a precisely defined standard observer. Examples are the color matching of a fluorescent lamp to an incandescent lamp or the matching of a specimen colored by one set of dyes or pigments to a specimen colored by a different set. SEE COLOR FILTER; COLOR VISION.

Colorimetric analysis. By strict definition, the term colorimetric analysis is limited to the techniques for visual identification and comparison of colored solutions. By common usage among analytical chemists, it has become a generic term for all types of analysis involving colored solutions.

Identification of a substance by the hue of its solution is termed qualitative analysis. Determination of its concentration in a solution by comparison of the intensity of its color to color intensity standards is termed quantitative analysis. When the human eye is used as the detector, quantitative colorimetric methods have relatively poor precision. Moreover, the precision varies with the color because of the varying response of the eye to different colors.

Most visual color-comparison methods utilize incident white light containing all wavelengths in the visible region. Filters or monochromators are seldom used, because colorimetric methods are advantageous principally for their simplicity, speed, and low cost, and for their modest demands for skill and training of the operator. Color intensities of solutions are usually compared to a set of permanent color standards. These standards may be solutions containing the same substance of similar hue, or colored glass.

One technique for the comparison of sample solutions to standard solutions is the use of long test tubes with flat bottoms called Nessler tubes, which may contain as much as 100 ml of solution. These tubes permit comparisons through depths of as much as 300 mm of solution, which increases the color intensities of dilute solutions. The sample solution is matched to one, or placed between two, of a series of standard solutions of graduated concentrations. A more sophisticated technique, now largely obsolete due to the availability of spectrophotometers, is based on the Duboscq colorimeter, in which one standard solution is used. With this instrument, the depths of the sample and standard solution are made continuously variable by means of transparent glass plungers. The color intensity of the sample solution is matched to that of the standard solution through the use of a split-field optical system, and the sample concentration is then determined by calculations based on the Beer-Lambert law.

A simple device used in water analysis for pH, chlorine, ammonia, iron, and phosphate is the Hellige comparator, which utilizes a series of colored glass standards mounted on a rotatable disk. The sample solution, after appropriate treatment with a reagent, is matched to a colored glass standard. With one device, the sample solution is viewed through a long Nessler tube; with a pocket-type comparator, the solution is viewed transversely in a square glass cell.

Deane B. Judd/Jack L. Lambert

Bibliography. G. Charlot, *Colorimetric Determination of Elements: Principles and Methods*, 1964; F. W. Clulow, *Color: Its Principles and Their Applications*, 1972; *Industrial Color Technology*, American Chemical Society Advances in Chemistry Series 107, 1972; F. D. Snell and C. T. Snell, *Colorimetric Methods of Analysis*, 3d ed., 1948–1971; L. C. Thomas and G. J. Chamberlin, *Colorimetric Chemical Analytical Methods*, 9th ed., 1980.

Columbiformes

An order of birds containing three families, the largest of which is the worldwide pigeons and doves (Columbidae). The relationships of the sandgrouse (Pteroclidae) are disputed, and many workers argue that these birds are members of the Charadriiformes. The members of this order are characterized by an ability to drink water by sucking instead of the sip-and-tilt method of most birds. However, some other groups of birds are able to suck water by various methods. SEE CHARADRIIFORMES.

Classification. The order Columbiformes is divided into the suborder Pterocletes, with the single family Pteroclidae (sandgrouse; 16 species), and the suborder Columbae with the families Raphidae (dodos; 3 species) and Columbidae (pigeons and doves; 303 species). The pigeons are divided into several subfamilies: the worldwide true pigeons (Columbinae), the pheasant-pigeon of New Guinea (Otidiphapinae), the large, crowned pigeons of New Guinea (Gourinae), the tooth-billed pigeon of Samoa (Didunculinae), and the fruit pigeons of the Old World tropics (Treroninae). Relationships of Columbiformes appear to be to the Charadriiformes in one direction and to the Psittaciformes in another. Possibly the Columbiformes are a central stock in the evolution of birds. SEE PSITTACIFORMES.

Fossil record. The fossil record of the Columbiformes is meager, as would be expected for these terrestrial, mainly woodland and grassland birds. The extinct dodos (Raphidae), from the Mascarene Islands, are known mainly from subfossil remains.

Columbidae. Pigeons are found mainly in the tropics, but a number of species are common in temperate regions. They have a sleek plumage ranging from browns and grays to the brilliant greens, yellows, and reds of the tropical fruit pigeons. They feed on seeds, fruit, and other vegetarian food. Most pigeons live in flocks but breed solitarily. Almost all are nonmigratory. The one or two young remain in the nest where they are fed by both parents, first on milk produced in the crop and then on seeds mixed with this milk. The extinct North American passenger pigeon lived and bred obligatorily in vast flocks, feeding largely on acorns and other mast. Although the passenger pigeon was once an abundant bird, its numbers dwindled because of overhunting and reduction of the extensive forests on which it depended for food and breeding; the last individual died in captivity in 1914.

Pigeons are important game birds throughout the world. Two species of Old World origin, the rock dove (*Columba livia*) and the ring dove (*Streptopelia risoria*) have been domesticated; both have escaped and exist as feral birds throughout the world. Numerous breeds of the rock dove have been developed, and these served as the primary example of C. Darwin in introducing his idea of evolution by natural selection. One breed of rock doves, the carrier pigeons, have served for centuries to transport messages back to their home loft. Most species of pigeons are kept and bred by aviculturists. Both domesticated species have important roles in biological research, such as studies of orientation and navigation, including work on the ''magnetic compass'' possessed by birds. In spite of hunting pressures, most species of pigeons are still in good numbers. A few, however, are endangered. SEE MIGRATORY BEHAVIOR.

Pteroclidae. The sandgrouse live in flocks in dry grasslands and deserts of the Old World, but depend on ponds of water to which flocks come in large numbers. The nest is generally a long distance from water, which is transported to the young by soaking specialized belly feathers. The young drink only from these moistened feathers, and will refuse water in pans placed before them.

Raphidae. The three species of dodos and solitaires were found only in the several Macarene Islands and were completely exterminated by the seventeenth century by sailors and released pigs. One of the common forest trees of these islands has almost become extinct as a result, because its fruits were eaten by dodos and the hard seeds were adapted to the grinding action of the bird's stomach; the hard seed coat had to be cracked in this way so that water could reach the ovary of the excreted seed and start germination. Seeds can now be planted after cracking the seed coat artificially. SEE AVES.

Walter J. Bock

Bibliography. D. Goodwin, *Pigeons and Doves of the World*, 1970.

Columbite

A mineral with composition $(Fe,Mn)Nb_2O_6$. Tantalum substitutes for niobium (columbium) in all proportions, and a complete series extends to tantalite

muscovite

columbite

3 cm

Columbite associated with muscovite from Minas Gerais, Brazil. (*American Museum of Natural History*)

$(Fe,Mn)Ta_2O_6$. Columbite without some tantalum is rare; iron and manganese also vary considerably in their relative amounts. It crystallizes in the orthorhombic system in short prismatic crystals. There is perfect side pinacoid cleavage. The hardness is 6 (Mohs scale), and the specific gravity is 5.2 for pure columbite; both increase with increased amounts of tantalum. The luster is submetallic and the color iron-black (see **illus.**). Columbite is found in granite pegmatites and in detrital deposits derived by their disintegration.

Columbite is the chief ore mineral of niobium and is found in small amounts in many pegmatites through the world. The chief producers are the Congo region and Nigeria, where it is recovered from decomposed granite. SEE NIOBIUM; TANTALITE; TANTALUM.

Cornelius S. Hurlbut, Jr.

Column

A slender compression member which fails by instability when the applied axial load reaches a critical value. Instability or buckling is said to occur when the system cannot attain an equilibrium condition and the load causes a deformation of an indeterminate amount. The collapse load is a function of the end restraints, the ratio of length to the least radius of gyration (L/r, the slenderness ratio), and the combined action of axial and lateral loads applied to the member. Columns are present in all types of structures, such as bridges, buildings, machinery, cranes, and airplanes. Different materials may be used in the various structures, such as steel, concrete, wood, aluminum, and alloy steels of all types. SEE LOADS, TRANSVERSE.

Column design. A column is designed to support a required load in compression. Columns of different slenderness ratio behave differently under load and thus the criteria for maximum capacity differ. Long columns (those with high slenderness ratio) become unstable at a critical load, usually the maximum resistance, with compressive stresses less than the elastic limit, so that failure is a phenomenon of elastic instability. The column buckles with a lateral deflection or with a twist. Columns of intermediate length, with slenderness ratios between long columns and short blocks, develop inelastic stresses with ultimate failure by inelastic buckling. Short blocks have such dimensions that lateral deflections can be neglected, failure being determined by the yield strength of the material.

The load capacity of columns is further influenced by the properties of the material, end conditions, initial crookedness, defects, and residual stresses, particularly in the intermediate range of slenderness.

Elastic buckling. An abrupt increase of lateral deflection of a compression member at a critical load while the stresses are wholly elastic constitutes elastic buckling. An initially straight column under axial load remains straight until a condition of neutral equilibrium is reached, when the column will continue to support the full load with considerable lateral deflection. A small increase of load above this condition causes a large increase in deflection, leading to collapse.

Leonhard Euler's expression for the critical buckling load P_{cr} of an ideally straight, axially loaded column with no rational restraint at the ends is $P_{cr} =$

$\pi^2 EI/L^2$ (I is the moment of inertia), from which the critical average stress f_{cr} is defined by Eq. (1), where

$$f_{cr} = \frac{\pi^2 E}{(L/r)^2} \qquad (1)$$

E is modulus of elasticity of the perfectly elastic material, L/r is the slenderness ratio, and r is the least radius of gyration of the section.

Euler's formula is directly applicable to the design of long columns whose safe load is Euler's predicted critical load divided by a factor of safety. Elastic buckling loads for columns with other restraining conditions modifying the elastic curve are found by using the length between points of contraflexure of the deflected member as the effective length KL, where $K = 1$ for rotationally free ends, $K = 0.5$ for both ends fixed, $K = 2$ for one end fixed with the other free to rotate and translate, and $K = 0.70$ for one end rotationally fixed with the other only fixed in position. The buckling load depends only on the stiffness E and the slenderness ratio KL/r, which is determined by end conditions. In design, end conditions are designated restraint coefficients, $C = 1/k^2$, with $P_{cr} = C\pi^2 EI/L^2$. The coefficient is taken to provide partial restraint for various types of end connections.

Eccentrically loaded columns. Off-axis loads subject a column to combined bending because of end moments and axial load. Deflections are produced immediately on application of the load and increase rapidly until the member yields under compression because of the combined loading, with inelastic buckling ultimately producing the collapse. Buckling may be accompanied by twisting, depending upon the shape of the section, lateral bracing, and slenderness ratio.

An estimate of the load capacity can be made according to either initial yield, or relations involving ratios of actual to ultimate thrust and moment, called interaction formulas. When initial yield under the combined stress is assumed to be the critical condition leading to collapse, the corresponding load is taken as a minimum value of the collapse load. The limiting average stress for design is defined by the secant formula, derived for free-end columns of material having a distinct yield point, loaded with equal end moments. As used, actual load eccentricity is increased to account for crookedness, defects, accidental eccentricity, and the effect of residual stresses. This formula is prescribed in bridge specifications for steel, with modifications to provide for effective length and double eccentricity. *SEE BEAM COLUMN.*

Interaction formulas are more convenient for general application. They depend upon the formulas adopted for predicted column and bending resistance.

Inelastic buckling. When the compressive stress in the column reaches the elastic limit before elastic buckling develops, any sudden increase of deflection or twist leads to collapse. The behavior of columns under inelastic stress depends on the shape of the stress-strain curve as determined by testing a short length of the actual column section. Because of residual stresses caused by unequal cooling of rolled sections, structural rolled steel has a proportional limit approximately half the yield point for annealed material, and the stress-strain curve has a gradual knee similar to alloy steels, aluminum, or magnesium alloys. Euler's formula is not applicable to these materials because the modulus of elasticity E is not con-

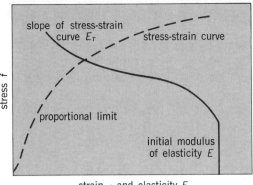

Fig. 1. Stress-strain curve for an aluminum alloy.

stant. An analytical approach, called the tangent modulus theory, leads to prediction of the buckling load. *SEE STRESS AND STRAIN.*

The tangent modulus formula (Engesser formula) is a modification of Euler's formula and is shown in Eq. (2), where E_T is the slope of the stress-strain curve at

$$f_{cr} = \frac{\pi^2 E_T}{(KL/r)^2} \qquad (2)$$

the critical stress (**Fig. 1**).

Values of E_T and stress f are interdependent. Solution is accomplished by trial or graphically for a given value of KL/r. Also, values of $\pi\sqrt{E_T/f}$ are consistent with particular values of KL/r, and a curve can be constructed to represent column resistance (**Fig. 2**). The resistance curve in the inelastic range can be represented by a parabola. For columns of wide-flange steel sections, in which residual stress is an important factor, a recommended column strength equation is Eq. (3), where F_y is the stress of the material at its yield point.

$$f_{cr} = F_y - \frac{F_y^2}{4\pi^2 E}\left(\frac{KL}{r}\right)^2 \qquad (3)$$

Local buckling. Relatively thin elements of a column section may buckle locally in a series of waves or wrinkles before the member as a whole buckles elastically. The outstanding legs of thin angles, thin flanges of channels, or thin plates connecting parts of the section are susceptible to local buckling. Thin-wall tubes buckle in a similar manner. Usually, local buckling reduces the load capacity of the member, but redistribution of stresses may produce conditions favorable to increased column loads. The likelihood of local buckling can be predicted by the theory of flat

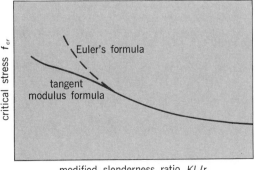

Fig. 2. Column resistance to inelastic buckling.

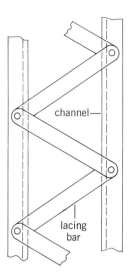

Fig. 3. Laced channel section.

The unsupported length between batten plates is limited to provide smaller slenderness in the flange elements than in the overall column, to avoid local buckling.

Torsional buckling. The dimensions and shape of a column section should provide a large radius of gyration to reduce KL/r and increase the allowable stress. A large radius of gyration requires that the area be spread far from the centroid, and for a given area the elements of the section become thin. Columns with open sections composed of thin elements, particularly angle and T sections, are susceptible to failure either by local buckling or by twisting.

A short compression member may fail by twisting of the section while the axis remains straight. This is called primary torsional buckling, in which the member becomes torsionally unstable at a condition of neutral equilibrium similar to a long column in bending. The critical load may be determined by energy methods. Columns which bend upon approaching maximum load are subjected to transverse shear, which may also produce twist buckling if the shear center does not coincide with the centroid. Local buckling strength determined by plate theory approximates the twist buckling load.

Column baseplates. Column loads and moments are transferred by baseplates to footings and foundations. The plate must be large enough to distribute the load without excessive bearing stresses. The column is connected to the plate by welded or bolted connections. Where no moment is involved, the connection only maintains alignment. The pressure of the plate on the foundation is nonuniformly distributed and depends on relative stiffnesses.

An approximate design procedure assumes uniform pressure and cantilever action of the plate projecting beyond the column boundaries. Anchorage is not required for axially loaded columns or when the resultant pressure due to moment and load falls within the mid-third of the base width. However, bolts are used to position the column. Where lateral forces, such as wind pressures, are involved, the resultant moment requires anchor bolts.

John B. Scalzi

Bibliography. E. H. Gaylord and C. N. Gaylord, *Design of Steel Structures*, 2d ed., 1972; E. H. Gaylord and C. N. Gaylord, *Structural Engineering Handbook*, 3d ed., 1989; B. G. Johnston (ed.), *Guide to Stability Design Criteria for Metal Structures*, 3d ed., 1976; W. McGuire, *Steel Structures*, 1968; C. G. Salmon and J. E. Johnson, *Steel Structures: Design and Behavior*, 1980.

plates, which involves the thickness-to-width ratio and edge supporting conditions. *See* Structural plate.

Shear is produced in columns by variation of bending moment associated with initial and buckling curvatures causing a shear force, lateral external forces, and lateral forces transmitted by distortion of the connecting members of the frame of which the column forms a part. An allowance for shear is taken as 2–2.5% of the axial load for columns composed of aluminum alloys or structural steel. In rolled sections the web provides shear resistance, whereas in open-web members resistance is provided by lacing or batten plates (**Fig. 3**).

Lacing. A system of diagonal bars connected to the elements of an open-web column forms a truss capable of resisting transverse shear (Fig. 3). These bars, tying the separated elements together, are usually inclined at 45 or 60° and are designed to act in either tension or compression. The slenderness ratio of the flange elements between lacing-bar connections is limited by specification to avoid local buckling.

Batten plates. Alternatively, flat plates can be attached transversely to the flange of an open-web column, at intervals along the length (**Fig. 4**). These batten plates serve as ties holding the elements in position, and, together with the column flanges, form a continuous frame resisting transverse shear. The flanges and plates are subject to bending and shear.

Combinatorial theory

The branch of mathematics which studies arrangements of elements (usually a finite number) into sets under certain prescribed constraints. Problems combinatorialists attempt to solve include the enumeration problem (how many such arrangements are there?), the structure problem (what are the properties of these arrangements and how efficiently can associated calculations be made?), and, when the constraints become more subtle, the existence problem (is there such an arrangement?).

Other names for the field include combinatorial mathematics, combinatorics, and combinatorial analysis, the last term having been coined by G. W. Leib-

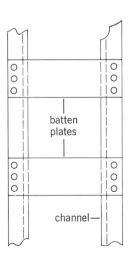

Fig. 4. Batten plates system.

erties) t. Then clearly $g(\phi) = 52$, $g(\{H\}) = 13$, $g(\{P\}) = 12$, and $g(\{P,H\}) = 3$, and g and f stand in the relation given by Eq. (12). For boolean algebras the Möbius function is given as follows: $\mu(s,t) = 0$ if s is not a subset of t, and $\mu(s,t) = (-1)^p$ otherwise, where p is the number of elements in t but not in s. [Thus, for all $s \subseteq t$, $\mu(s,t) = 1$ if the difference in sizes of s and t is even and $\mu(s,t) = -1$ if it is odd.] In our example, $f(\phi) = g(\phi) - g(\{H\}) - g(\{P\}) + g(\{P,H\}) = 30$ cards which are neither hearts nor picture cards. *See Boolean algebra*.

For a general boolean algebra, the Möbius inversion formula for $f(\phi)$ is more often called the principle of inclusion-exclusion, or sieve formula, because at various stages of the summation the correct number is alternately overcounted and then undercounted. Assume S has $n = g(0)$ elements each of which may have some of the properties $P_1, \ldots, P_r$. Further, let $g(i_1, i_2, \ldots, i_k)$ be the number of elements with (at least) properties $P_{i_1}, P_{i_2}, \ldots, P_{i_k}$. Again Eq. (12) is satisfied, so by Eq. (13), one arrives at formula (14) for $f(0) = f(\phi)$.

$$
\begin{aligned}
f(0) = n - \sum_{1 \le i_1 \le n} g(i_1) + \cdots \\
+ (-1)^p \sum_{1 \le i_1 < \cdots < i_p \le n} g(i_1, \ldots, i_p) + \cdots \quad (14)
\end{aligned}
$$

Derangements. The derangement numbers D_n count complete permutations (without repetition) subject to the additional constraint that every element be "wrongly labeled." That is, if $S = \{1, \ldots, n\}$, then D_n is the number of ways to order S into sequences $a_1, \ldots, a_n$ such that $a_i \ne i$ for $i = 1, \ldots, n$. If the property P_i of a permutation is that $a_i = i$, then the sieve formula may be applied to prove Eq. (15).

$$
\begin{aligned}
D_n &= P(n,n) - C(n,1)P(n-1,n-1) \\
&\qquad + C(n,2)P(n-2,n-2) - \cdots \\
&= n!\left(1 - 1 + \frac{1}{2!} - \frac{1}{3!} + \cdots \right. \\
&\qquad \left. + \frac{(-1)^n}{n}\right) \approx \frac{n!}{e} \quad (15)
\end{aligned}
$$

In probabilistic terms, Eq. (15) implies that no matter what the size of the decks is, the probability is approximately $1/e$ that two identical and randomly shuffled decks of cards will never have the same card in the same place.

Gaussian coefficients. The Möbius function has been computed for other lattices which arise in enumeration, such as the lattice of all positive divisors of a given integer n (with elements ordered by divisibility), a lattice of integer partitions, the lattice of partitions of an n-element set (ordered by refinement: $s \le t$ if every block of s is contained in a block of t), and the lattice of all subspaces of a vector space of dimension n over a finite field with q elements. The number of subspaces of dimension k in such a vector space is expressed in formula (16) for gaussian coef-

$$
\begin{bmatrix} n \\ k \end{bmatrix}_q = \frac{(q^n - 1)\,(q^{n-1} - 1) \cdots (q^{n-k+1} - 1)}{(q^k - 1)\,(q^{k-1} - 1) \cdots (q - 1)} \quad (16)
$$

ficients. This apparent rational function of q is, in fact, a polynomial which is the generating function for the numbers $p_{m,k}(n)$ of integer partitions of n into at most m parts whose largest part is at most k, Eq. (17). An easy fact, established from the Ferrer's dia-

$$
\begin{bmatrix} m + k \\ k \end{bmatrix}_q = \sum_{q=0}^{mk} p_{m,k}(n)q^n \quad (17)
$$

gram, is that $p_{m,k}(n) = p_{m,k}(mk - n)$. A much harder result is that the sequence of coefficients is unimodal: $P_{m,k}(n + 1) \ge p_{m,k}(n)$ for all $n < mk/2$. *See Linear algebra*.

The lattices for boolean algebras, subspaces, and set-partitions show striking similarities [such as the similarity between Eqs. (2) and (16)], which are studied in the general theory of geometric lattices or combinatorial geometries. For example, the methods applied to set partitions to study the four-color theorem of graph theory, when applied to the lattice of subspaces, provide a way to attack the problem of finding efficient error-correcting codes. *See Information theory*.

Magic squares. A (generalized) magic square is an n by n matrix of (not necessarily distinct) nonnegative integers all of whose row and column sums are equal to a prescribed number x. Thus $x = 15$ in the 3×3 magic square given earlier. The number of such magic squares is 1 when $n = 1$, $x + 1$ when $n = 2$, and in general is a polynomial of the form $a_m x^m + a_{m-1} x^{m-1} + \cdots + a_1 x + 1$, where $m = (n - 1)^2$. Although the coefficients a_i are not given explicitly in this formula, for each n they may be computed by the method of undetermined coefficients, so that a computer search of magic squares for small values of x will also determine the number of such squares for all values of x. *See Matrix theory*.

Pólya counting formula. If the number of ways to paint the faces of a cube with red and blue is to be computed, formula (3) would seem to give $2^6 = 64$; however, it is reasonable to assume that all six arrangements in which exactly one face is blue are identical, since for all such arrangements a rotation of the cube will make the top face blue. Under the condition that two coloring patterns are to be assumed equivalent if one can be transformed into the other by a rotation, the number of configurations is 10. This number can be derived by exhausting all cases or by the Pólya counting formula. The Pólya formula counts the number of functions from a set D (in the example, the faces of the cube) to a set R (the colors red and blue), with two functions f and g assumed to be the same if some element of a fixed group G of (complete) permutations of D (in this case, the 24 rotations of a cube) takes f into g. Actually, the Pólya formula provides a way to compute the generating function, with variables $x_1, \ldots, x_k$ corresponding to the elements of R, for which the coefficient of $x_1^{a_1} \cdots x_k^{a_k}$ is the number of functions which use each x_i value exactly a_i times. (For the painted cube, if $x_i =$ red and $x_2 =$ blue, then the generating function is given by $x_1^6 + x_1^5 x_2 + 2x_1^4 x_2^2 + 2x_1^3 x_2^3 + 2x_1^2 x_2^4 + x_1 x_2^5 + x_2^6$.) The proof of the Pólya theorem depends on a special case proved earlier by W. Burnside: if G is a group consisting of m (complete) permutations of S, then the number of inequivalent permutations of S is

$$
\frac{1}{m} \sum_{\sigma \in G} f(\sigma)
$$

Here two permutations π_1 and π_2 are regarded as equivalent if there is some σ in G such that the relabeling that σ effects on the sequence π_1 gives the sequence π_2; and for each σ in G, $f(\sigma)$ denotes the number of permutations of S which do not change when relabeled by σ. SEE GROUP THEORY.

Application of Lefschetz theorem. A wide range of techniques is employed in enumeration theory. For example, the hard Lefschetz theorem from algebraic geometry was used to show that if A is any set of n distinct positive real numbers, and if for any subset B of A, $s(B)$ is the sum of its members, then $s(B)$ is constant on at most $g(n)$ subsets B, where $g(n) = \overline{p}_n([n + 1)(n)/4])$, the number of partitions of $[(n + 1)(n)/4]$ into distinct parts whose largest part is at most n. Thus $g(n)$ is the middle coefficient of the polynomial $(1 + x)(1 + x^2)(1 + x^3) \cdots (1 + x^n)$, and $g(n)$ is achieved by letting A be the arithmetic progression $\{1,2, \ldots, n\}$ and setting the sum equal to $[(n + 1)(n)/4]$.

PROPERTIES OF ARRANGEMENTS

An arrangement is primarily a family of subsets of a set S. An alternate way of formulating this concept is as an incidence system, also called a relation, which specifies when a particular element s of S is in a subset S_i of the family, for example, by means of the incidence matrix $M = [m_{ij}]$ of the family: the rows of the incidence matrix are indexed by the elements $s_1, \ldots, s_n$ of the set S, the columns of the matrix are indexed by the subsets $S_1, \ldots, S_m$ in the family, $m_{ij} = 1$ if s_i is in S_j, and $m_{ij} = 0$ otherwise. For example, if $S = \{s_1,s_2,s_3,s_4\}$, $S_1 = \{s_1,s_2\}$, $S_2 = S_4 = \{s_1,s_3,s_4\}$, and $S_3 = \{s_2\}$, the incidence of matrix of this arrangement is

$$\begin{bmatrix} 1 & 1 & 0 & 1 \\ 1 & 0 & 1 & 0 \\ 0 & 1 & 0 & 1 \\ 0 & 1 & 0 & 1 \end{bmatrix}$$

Systems of distinct representatives. If $n = m$ in an incidence system, a system of distinct representatives for the sets is a permutation $a_1, \ldots, a_n$ of S such that a_i is in S_i for $i = 1,2, \ldots, n$. (In a congressional committee S, where each S_j might denote a particular cause or coalition, it might be desirable to find a system of distinct representatives so that each cause or coalition would have a different proponent.) In the above example, s_1,s_3,s_2,s_4 is such a system.

Marriage theorem. A system of distinct representations certainly cannot exist unless any k of the subsets S_i together contain at least k distinct elements, and in fact it can be shown that this condition is also sufficient. This theorem is whimsically called the marriage theorem, since if S consists of a collection of n boys and each subset S_i consists of the boys acquainted with a particular girl g_i, then the conditions of the theorem guarantee that each girl could marry a boy of her acquaintance.

Algorithms for finding systems. Algorithms for computers exist which find systems of distinct representatives with reasonable efficiency: as n increases, the time a computer needs to find a system increases as the square of n. SEE DIGITAL COMPUTER.

Permanent. To enumerate the systems of distinct representatives, it is enough to compute the permanent of the incidence matrix defined by $\Sigma m_{1,a_1} m_{2,a_2}$ $\cdots m_{n,a_n}$, where the sum is taken over all $n!$ permutations $(a_1, a_2, \ldots, a_n)$ of the set of column indices.

Maximal systems. When a complete system of distinct representatives does not exist, the size of a largest possible system is equal to the smallest possible value of the function $n - k + s_{i_1, \ldots, i_k}$ for all possible combinations $S_{i_1}, \ldots, S_{i_k}$ of subsets in the family, where $s_{i_1, \ldots, i_k}$ is the number of elements in $S_{i_1} \cup \cdots \cup S_{i_k}$. A related theorem asserts that the minimum number of blocks into which a partially ordered set may be partitioned subject to the constraint that any pair of elements x and y in the same block be related ($x \le y$ or $y \le x$) is equal to the maximum possible size of a set S of pairwise unrelated elements. It is always possible to find such a set S with the further property that any (complete) permutation π which preserves the partial order [if $x \le y$, then $\pi(x) \le \pi(y)$] must take members of S to other members of S. This fact can be used to show that in a boolean algebra the maximum size of S is $C(n,[n/2])$. This number is the middle (largest) binomial coefficient and corresponds to the set S of all subsets of size $[n/2]$.

Assignment problem. Suppose it is desired to assign n workers to do n jobs, and entry a_{ij} of a matrix A measures how well the ith worker does the jth job. An assignment would be a permutation $j_1, \ldots, j_n$ of the jobs so that the ith worker performs the job j_i. The assignment problem requires that one find an assignment maximizing the total utility T, given by Eq. (18).

$$T = \sum_{i=1}^{n} a_{ij_i} \tag{18}$$

The problem may be solved by introducing auxiliary row and column numbers $u_1, \ldots, u_n$ and $v_1, \ldots, v_n$ subject to the condition $u_i + v_j \ge a_{ij}$ for all i and j. Then Eq. (19) is valid for any assignment,

$$T = \sum_{i=1}^{n} a_{ij_i} \le \sum_{i=1}^{n} u_i + \sum_{j=1}^{n} v_j = R \tag{19}$$

since in an assignment each row and column is used exactly once. The marriage theorem shows that the maximum utility M of T is equal to the minimum m of R. Further, an algorithm is known which systematically changes the u's and v's to reduce R, arriving at numbers u_i and v_j for which the set of a_{ij}'s such that $a_{ij} = u_i + v_j$ admits a complete system of distinct representatives. For an assignment derived from such a system, $T = R$ and so M is achieved for this T. This method is practical in that the algorithm leads directly to a solution; and since the number of possible assignments is usually very large (for example if $n = 25$, this number is $25! > 10^{25}$), without some guidance there is no practical way to search for the optimal assignment or to recognize it when it is found. Another application of the marriage theorem is to prove topological fixed-point theorems. SEE TOPOLOGY.

Doubly stochastic matrices. A doubly stochastic matrix, a matrix of nonnegative real numbers such that every row sum and every column sum is equal to one, is a real-number analog of a magic square. The set of all n by n doubly stochastic matrices is convex in the sense that if $M_1, \ldots, M_k$ are all doubly stochastic and $a_1, \ldots, a_k$ are nonnegative numbers whose sum is one, then the convex combination

$$\sum_{i=1}^{k} a_i M_i$$

is also doubly stochastic. These matrices in fact form a convex polytope (a generalization of the notion of convex polygon) in euclidean space $\mathbf{R}^{n^2}$, in that their entries are bounded and there are a finite number of vertices (a minimal set of matrices such that every doubly stochastic matrix is a convex combination of elements of this set). The marriage theorem can be used to prove that the $n!$ permutation matrices with exactly one 1 in each row and column are the vertices of the convex polytope of doubly stochastic matrices.

By a theorem in linear programming, linear functions are optimized on the vertices of a convex polytope. For the assignment problem this means that for a more general assignment in which the ith worker spends a fraction m_{ij} of his time on the jth job and each job is always being worked on (so that the matrix $M = [m_{ij}]$ is doubly stochastic), a maximum utility is obtained when each worker spends all his time at one job. *SEE LINEAR PROGRAMMING*.

Upper-bound problem. In a typical problem in linear programming, polytopes in $\mathbf{R}^m$ occur which are defined by n linear constraints, such as those given by (20), where the x_j's are the variables (corresponding

$$\sum_{j=1}^{m} a_{ij} x_j \leq b_i \qquad i = 1, \ldots, n \qquad (20)$$

to vectors in $\mathbf{R}^m$). Algorithms to maximize linear functions on these polytopes usually involve proceeding from one vertex to another at which the function is greater. [An algorithm was given in 1979 using different (nonlinear) techniques, which guarantees that the number of steps needed to solve any such linear program is no more than a polynomial function of mn, but for most practical applications it has been found that a vertex-to-vertex method such as the simplex algorithm gives a faster answer.] It is important when allocating computer time to place an upper bound on the number of vertices. In 1970 it was shown that the number of vertices of a polytope in $\mathbf{R}^m$ defined by n linear constraints is at most $M(m,n)$, given by Eq. (21).

$$M(m,n) = \begin{pmatrix} n - \left[\dfrac{m+1}{2}\right] \\ n - m \end{pmatrix}$$
$$+ \begin{pmatrix} n - \left[\dfrac{m+2}{2}\right] \\ n - m \end{pmatrix} \qquad (21)$$

EXISTENCE AND CONSTRUCTION

In the 1920s R. A. Fisher noted that "the design and analysis of statistical experiments requires the construction of orthogonal latin squares and balanced incomplete block design." *SEE STATISTICS*.

Orthogonal Latin squares. A Latin square is a square n by n matrix with entries from the set $N = \{0,1, \ldots, n-1\}$ so that each number occurs exactly once in each row and exactly once in each column. Two Latin squares are said to be orthogonal if

when they are superposed the n^2 cells contain each of the n^2 pairs of numbers from N exactly once. For $n = 3$ one has the superposed orthogonal squares

00	11	22
12	20	01
21	02	10

where the first digits form the first square, and the second digits the second square. This square may be used to design an agricultural experiment which tests the interaction of three varieties of grain with three types of fertilizers. A field is divided into nine plots in each of which the choice of grain is made according to the first digit and the choice of fertilizer according to the second digit. The Latin squares assure the even distribution of the varieties of grain and fertilizer in both directions, so that effects such as the variation of the soil are mimimized, and the orthogonality allows the experimenter to try each fertilizer with each variety of grain.

For $n \leq 10$ each of these arrays can be viewed as a particular type of magic square whose entries are distinct two-digit numbers and such that each row and column sum is $11n(n-1)/2$ (33 for $n = 3$ and 495 for $n = 10$).

If n is odd, it is easy to construct two orthogonal squares A and B. One takes $A = [a_{ij}]$ and $B = [b_{ij}]$ where $a_{ij} = i + j$, $b_{ij} = i + 2j$, reducing these sums by n or $2n$ if necessary to put them in the range $0, \ldots, n-1$. If n is a multiple of 4, it is not much more difficult to make a construction. But for n of the form $4m + 2$ it is considerably harder. It is clearly impossible when $n = 2$, and, by an exhaustive trial, G. Tarry showed it to be also impossible for $n = 6$. Euler had conjectured in 1782 that no pair of orthogonal Latin squares existed for any n of the form $4m + 2$, but in 1959 orthogonal Latin squares were found for every $4m + 2$ greater than or equal to 10 (**Fig. 1**).

Block designs. T. Kirkman posed the following puzzle in the 1850s: Is it possible for 15 schoolgirls to go for walks in 5 groups of 3 every afternoon, so that in seven afternoons every girl shall have walked with every other girl? This arrangement is indeed possible and the following is a solution, the girls being represented by letters $a, \ldots, o$:

Sun.	Mon.	Tue.	Wed.	Thur.	Fri.	Sat.
abi	*acj*	*adk*	*ael*	*afm*	*agn*	*aho*
cdf	*deg*	*efh*	*fgb*	*ghc*	*hbd*	*bce*
gjo	*hki*	*blj*	*cmk*	*dnl*	*eom*	*fin*
ekn	*flo*	*gmi*	*hnj*	*bok*	*cil*	*djm*
hlm	*bmn*	*cno*	*doi*	*eij*	*fjk*	*glk*

The cited design is a special case of a balanced incomplete block design or (b,v,r,k,λ)-design: an arrangement of v elements or treatments x_i, $i = 1, \ldots, v$, into b subsets or blocks B_j, $j = 1, \ldots, b$, so that each B_j contains exactly k distinct elements, each element occurs in r blocks, and every combination of two elements $x_u x_v$ occurs together in exactly λ blocks.

Counting the total number of plots (incidences of x_i in B_j or symbols in the above arrangement) in two ways—first summing over the blocks and then summing over the treatments, the constraint (22) is ob-

$$bk = vr \qquad (22)$$

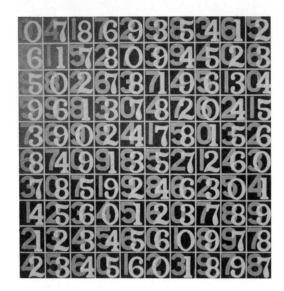

Fig. 1. Two orthogonal Latin squares for *n* = 10. (*Courtesy of Thomas Brylawski; Karl Petersen, photographer*)

tained. By considering all two-element subsets which contain *x*, constraint (23) is obtained. The two technical constraints (24) imply that the blocks are incom-

$$r(k - 1) = \lambda(v - 1) \qquad (23)$$

nical constraints (24) imply that the blocks are incom-

$$2 < k < v - 1, \lambda > 0 \qquad (24)$$

plete in that none of the blocks contains all or all but one of the treatments and none consists of a single treatment.

Steiner triple systems. When $\lambda = 1$ and $k = 3$, such (b,v,r,k,λ)-designs are known as Steiner triple systems, and in this case constraints (22) and (23) imply that *v* is of the form $6m + 1$ or $6m + 3$. Kirkman himself proved that triple systems exist for all such *v*.

The schoolgirl arrangement has the additional property that the blocks themselves are partitioned into *r* families called parallel classes of *v/k* blocks each such that every element occurs in exactly one block of each parallel class. Such designs are termed resolvable. A resolvable Steiner triple system is called a Kirkman triple system, and clearly in this case *v* must be divisible by three. In 1968 Kirkman triple systems (resolvable balanced incomplete block designs with block size three) were shown to exist for all $v = 6m + 3$.

An application of Steiner triple systems might arise in testing taste preferences among *v* products if every pair of products is to be compared, but each taster can efficiently taste only three products per day. A Kirkman system will, in addition, give an efficient schedule in which each product is used exactly once each day.

Fisher's inequality. Fisher's inequality states that $b \geq v$ in every design. If $b = v$, then necessarily $k = r$ and the design is called a symmetric (v,k,λ)-design. It is a property of these designs that every two blocks intersect in λ elements, and hence the transpose of the incidence matrix of a (v,k,λ)-design is itself a (v,k,λ)-design. Elementary matrix theory may be used to show that if *v* is even, then a symmetric design with parameters v,k,λ exists only if $k - \lambda$ is a perfect square; and deep results in number theory have been used to show that when *v* is odd, there

must exist integers *c*, *y* and *z* not all zero such that Eq. (25) holds. *See Number theory*.

$$x^2 = (k - \lambda)y^2 + (-1)^{(v - 1)/2}\lambda z^2 \qquad (25)$$

Existence for fixed *k*. If *k* is fixed, then, for all but finitely many sets of parameters b,v,r,k,λ satisfying Eqs. (22), (23), and (24), there exists a block design having those parameters. When $k \leq 5$, all sets of parameters for which a block design exists have been determined. (A design is simple when $\lambda = 1$.)

Projective planes. When a symmetric design has $\lambda = 1$, condition (23) shows that if $k = n + 1$, then $v = n^2 + n + 1$. In analogy with the points (treatments) and lines (blocks) of a plane in projective geometry, such configurations are called projective planes. In this case, condition (25) states that if the parameter *n*, called the order of the plane, is of the form $4m + 1$ or $4m + 2$, then integers *u* and *v* exist with $n = u^2 + v^2$. Thus an infinite number of possible values of *n* are excluded, beginning with 6, 14, and 21. Whenever *n* is a power of a prime (and thus the size of a finite field), a plane exists, namely, the projective plane coordinatized over the field (but these are not the only planes which exist for prime-power orders). In the early part of the twentieth century, planes were constructed over many algebraic systems in which linear equations could be solved, but which were not associative. Planes coordinatized over such systems are different from those coordinatized over fields. The smallest of these planes has order 9 and thus has 91 points. *See Algebra; Projective geometry*.

Non-Desarguesian planes. The theorem of G. Desargues (which states that if corresponding vertices of two triangles lie on three concurrent lines, then the three points of intersection of respective sides of those triangles will lie on a line) holds in general only for those planes which arise from fields, and hence the term non-Desarguesian plane is used for the others. No (necessarily non-Desarguesian) projective plane has been found for any *n* that is not a prime power [and no *n* has been excluded except those not satisfying condition (19)]. Whether a projective plane exists for an *n* that is not a power of a prime is an important unsolved problem in existence theory and projective geometry. Orthogonal Latin squares and projective planes are related in that the existence of a projective plane of order *n* is equivalent to the existence of a set of $(n - 1)$ *n* by *n* Latin squares every pair of which is orthogonal. This would mean that for the first unresolved case, $n = 10$, nine pairwise orthogonal Latin squares must be found to produce a projective plane of order 10. No one has found more than two such squares.

Friendship theorem. Only for trivial projective planes can all the points be paired with all the lines in such a way that the incidence system is dualized, and each line is paired with a point not on that line. Stated differently, this is the friendship theorem: Among *n* people, if every pair of people has exactly one common friend, then there is someone who knows everyone else.

t-Designs. The concept of a block design in which every pair of elements is in the same number of blocks can be generalized to *t*-designs with parameters $t - (v,k,\lambda)$, in which each *t*-element subset of a *v*-element set is in a fixed number λ of *k*-element blocks. Every *t*-design is an *s*-design for all *s* such that $0 \leq s \leq t$ where the λ_s are given by Eq. (26),

$$\lambda_s = \frac{\lambda P(v - s, t - s)}{P(k - s, t - s)} \qquad (26)$$

and these numbers must all be integers. Formulas (22) and (23) are special cases of Eq. (26), with $\lambda_0 = b$, the number of blocks, and $\lambda_1 = r$, the replication number.

Designs are only known for relatively small values of t. For $t = 3$, there are the inversive planes which, like finite projective planes, combine classical geometry and finite fields. Let P be the points of a finite affine plane coordinatized by a field F with q elements. These points are in correspondence with the elements of an extension field F' with q^2 elements, where each point (r,s) is paired with the element $r + sx$, with x the root of an irreducible quadratic equation (like $\sqrt{-1}$ over the real numbers). An ideal point at infinity, ∞, is added giving $v = q^2 + 1$ points. The blocks are then the images of the set B_0 consisting of the line $\{(s,0)\colon s$ element of $F\}$ along with ∞ under all linear fractional transformations, given by Eq. (27).

$$y \rightarrow \frac{ay + b}{cy + d} \qquad (27)$$
$$(a,b,c,d \text{ elements of } F', \ ad \neq bc)$$

Here ∞ acts like $1/0$ so that Eq. (28) holds. The

$$\frac{a^\infty + b}{c^\infty + d} = \frac{a}{c} \qquad (28)$$

blocks (all with $q + 1$ elements) are then called circles (with the original lines in the affine plane corresponding to circles through ∞). Circles have the property that every triple of points determines a unique circle ($\lambda = 1$), and further that if C is a circle which contains p but not q, there is a unique circle which contains q and is "tangent" to C at p.

The construction of t-designs has been closely related to the study of multiply transitive groups. The Mathieu groups give two designs for $t = 5$. One has parameters $5 - (12,6,1)$, and the other has parameters $5 - (24,8,1)$. In the first case, the blocks which contain a fixed three-element subset of points form a design isomorphic to the affine plane of order three, and in the second the blocks through three fixed points form a projective plane of order four.

At present no simple designs are known for $t > 5$. Steiner quadruple systems [designs with parameters $3 - (v,4,1)$] have been constructed for all v congruent to 2 or 4 modulo 6.

Error-correcting codes. The following matrix is the incidence system of a projective plane of order 2 (and symmetric Steiner triple system):

$$
\begin{array}{ccccccc}
1 & 0 & 0 & 0 & 1 & 0 & 1 \\
1 & 1 & 0 & 0 & 0 & 1 & 0 \\
0 & 1 & 1 & 0 & 0 & 0 & 1 \\
1 & 0 & 1 & 1 & 0 & 0 & 0 \\
0 & 1 & 0 & 1 & 1 & 0 & 0 \\
0 & 0 & 1 & 0 & 1 & 1 & 0 \\
0 & 0 & 0 & 1 & 0 & 1 & 1 \\
\end{array}
$$

Since $k = 3$ and $\lambda = 1$, each row has exactly three 1's and any two rows overlap in only one of these columns. Thus, any two rows differ in four of their entries. This matrix is an example of a binary block code of distance $m = 4$ and length $n = 7$: a set of codewords, each consisting of a sequence of n binary digits (zeros and ones), such that any two codewords differ in at least m places. If seven messages are coded by rows of the above matrix and one is transmitted to a receiver, the original message can be decoded—even if during transmission one of the 1's were changed to a 0 or conversely—by choosing the codeword closest to the received sequence. For example, if 1001011 were received, the last codeword was transmitted. On the other hand, if two digits were changed, resulting for example in a received message 1000110, the receiver could not tell whether the intended message corresponded to the first codeword or the second. Such a code will then "correct" one transmission error and decide if an intended codeword with up to three such errors has been correctly transmitted.

In general, a block code will correct up to $[(m - 1)/2]$ errors. The fundamental problem in coding theory is to construct codes with a large number of words and large distance but small length. In the example, the size of the code may be increased without decreasing the distance by adding a 1 to the end of all seven codewords, adjoining a new word of eight 1's, and then forming the eight additional words constructed from each of the others by interchanging all 0's and 1's.

Subdivision of square. An elegant combinatorial application of the theories of electrical networks and three dimensional convex polytopes led in the 1940s to the construction of a square subdivided into n smaller squares no two of which are the same size (**Fig. 2**, where $n = 26$). The least number of unequal squares into which a square can be divided is 21, and that, up to symmetry, there is only the following subdivision. A square of length 112 is partitioned into squares of length 50, 35, 27, 8, 19, 15, 17, 11, 6, 24, 29, 25, 9, 2, 7, 18, 16, 42, 4, 37, and 33, respectively, where each square enters the big square as far north as possible and then as far west as possible.

Pigeonhole principle. The pigeonhole (or Dirichlet drawer) principle states that: if a very large set of elements is partitioned into a small number of blocks, then at least one block contains a rather large number

Fig. 2. Subdivision of a square into 26 smaller squares; no two of which are the same size. (*Courtesy of Thomas Brylawski; Karl Petersen, photographer*)

of elements. From this result it follows, for example, that any permutation of the numbers from 1 to $mn + 1$ must contain either an increasing subsequence of length $m + 1$ or a decreasing subsequence of length $n + 1$. To see this, associate with each number in the permutation P the pair of numbers (i,d), where i (respectively d) is the length of the longest increasing (respectively decreasing) subsequence in which it is the first element. Then no two numbers in P are associated with the same pair (since if p precedes q, p initiates a longer increasing subsequence if $p < q$ and a longer decreasing subsequence if $p > q$), so that not every number can be associated with one of the mn pairs (i,d), $i = 1, \ldots, m$; $d = 1, \ldots, n$. One can do no better than $mn + 1$, since if the numbers from 1 to mn are arranged sequentially in an n by m array and read off from left to right starting at the bottom, then the longest increasing subsequence has length m and the longest decreasing subsequence has length n. This construction belongs to the field of extremal combinatorial theory, a branch of combinatorics which constructs counterexamples of size n for theorems which hold for all $k \geq n + 1$.

Ramsey's theorem. Ramsey's theorem, proved in the 1930s, generalizes the pigeonhole principle as follows: For any parameters $(r; r_1, \ldots, r_k)$ with each $r_i \geq r$, there exists a number N such that for all $n \geq N$, if S is an n-element set all

$$\binom{n}{r}$$

of whose r-element subsets are partitioned into blocks $A_1, \ldots, A_k$, then there is some r_i-element subset S' of S such that all

$$\binom{r_i}{r}$$

of the r-element subsets of S' are in the block A_i. For a given set of parameters, one then asks for the Ramsey number, or smallest N for which the theorem is true. When $r = 1$, by the pigeonhole principle $N = r_1 + \cdots + r_k - k + 1$. For the parameters $(2;3,3)$, the theorem can be interpreted as follows: When the edges and diagonals of a regular polygon with a sufficiently large number N of vertices are all colored red or blue, there are three vertices of the polygon which are also the vertices of a triangle all of whose edges are the same color. One can always find a monochromatic triangle in a hexagon, and the extremal configuration consisting of a pentagon whose edges are red and whose five diagonals are blue shows that the Ramsey number for these parameters is 6.

Analogs of Ramsey's theorem have been proved for graphs, vector spaces, configurations in the Euclidean plane, and (with infinite cardinals as parameters) in set theory. However, very few Ramsey numbers have been computed. For example, when the edges and diagonals of a polygon are colored with four colors, the smallest N which guarantees a monochromatic triangle [that is for $(2;3,3,3,3)$] is known only to be between 51 and 63.

Many asymptotic results are known and have interpretations in graph theory. For example, the Ramsey number $(2;3,t)$ is known to be at most $ct^2/\log t$, where c is an absolute constant not depending on t. This means that in any graph with $ct^2/\log t$ vertices and no

triangles, there is a subset of t vertices, no two of which are connected by an edge.

Thomas Brylawski

Bibliography. M. Aigner, *Combinatorial Theory*, 1979; I. Anderson, *A First Course in Combinatorial Mathematics*, 1979; E. F. Beckenbach (ed.), *Applied Combinatorial Mathematics*, 1964, reprint 1981; C. Berge, *Principles of Combinatorics*, 1971; K. P. Bogart, *Introductory Combinatorics*, 1983; R. Bose and B. Manvil, *Introduction to Combinatorial Theory*, 1984; R. A. Brualdi, *Introductory Combinatorics*, 1977; L. Comtet, *Advanced Combinatorics*, 1974; C. L. Liu, *Topics in Combinatorial Mathematics*, 1972; J. Riordan, *An Introduction to Combinatorial Analysis*, 1980; G.-C. Rota (ed.), *Studies in Combinatorics*, 1979.

Combining volumes, law of

The principle that when gases take part in chemical reactions the volumes of the reacting gases and those of the products, if gaseous, are in the ratio of small whole numbers, provided that all measurements are made at the same temperature and pressure. The law is illustrated by the following reactions:

1. One volume of chlorine and one volume of hydrogen chloride combines to give two volumes of hydrogen chloride.
2. Two volumes of hydrogen and one volume of oxygen combine to give two volumes of steam.
3. One volume of ammonia and one volume of hydrogen chloride combine to give solid ammonium chloride.
4. One volume of oxygen when heated with solid carbon gives one volume of carbon dioxide.

It should be noted that the law applies to all reactions in which gases take part, even though solids or liquids are also reactants or products.

The law of combining volumes was put forward on the basis of experimental evidence, and was first explained by Avogadro's hypothesis that equal volumes of all gases and vapors under the same conditions of temperature and pressure contain identical numbers of molecules. SEE AVOGADRO'S NUMBER.

The law of combining volumes is similar to the other gas laws in that it is strictly true only for an ideal gas, though most gases obey it closely at room temperatures and atmospheric pressure. Under high pressures used in many large-scale industrial operations, such as the manufacture of ammonia from hydrogen and nitrogen, the law ceases to be even approximately true. SEE GAS.

Thomas C. Waddington

Combustion

The burning of any substance, whether it be gaseous, liquid, or solid. In combustion, a fuel is oxidized, evolving heat and often light. The oxidizer need not be oxygen per se. The oxygen may be a part of a chemical compound, such as nitric acid (HNO_3) or ammonium perchlorate (NH_4ClO_4), and become available to burn the fuel during a complex series of chemical steps. The oxidizer may even be a non-oxygen-containing material. Fluorine is such a substance. It combines with the fuel hydrogen, liberating light and heat. In the strictest sense, a single chemical substance can undergo combustion by decomposition, with emission of heat and light. Acetylene, ozone, and hydrogen peroxide are examples. The products of their decomposition are carbon and hydrogen for acetylene, oxygen for ozone, and water and oxygen for hydrogen peroxide.

Solids and liquids. The combustion of solids such as coal and wood occurs in stages. First, volatile matter is driven out of the solid by thermal decomposition of the fuel and burns in the air. At usual combustion temperatures, the burning of the hot, solid residue is controlled by the rate at which oxygen of the air diffuses to its surface. If the residue is cooled by radiation of heat, combustion ceases.

The first product of combustion at the surface of char, or coke, is carbon monoxide. This gas burns to carbon dioxide in the air surrounding the solid, unless it is chilled by some surface. Carbon monoxide is a poison and it is particularly dangerous because it is odorless. Its release from poorly designed, or malfunctioning, open heaters constitutes a serious hazard to human health.

Liquid fuels do not burn as liquids but as vapors above the liquid surface. The heat evolved evaporates more liquid, and the vapor combines with the oxygen of the air.

Spontaneous combustion. This occurs when certain materials are stored in bulk. The oxidizing action of microorganisms often produces the initial heat.

As the temperature increases, the air trapped in the material takes over the oxidation process, liberating more heat. Because the heat cannot be dissipated to the surroundings, the temperature of the material rises still more and the rate of oxidation increases. Eventually the material reaches an ignition point and bursts into flame. Coal is subject to spontaneous combustion and is generally stored in shallow piles to allow the heat of oxidation to dissipate.

Gases. At ordinary temperatures, molecular collisions do not usually cause combustion. At elevated temperatures the collisions of the thermally agitated molecules are more frequent. More important as a cause of chemical reaction is the greater energy involved in the collisions. Moreover, it has been reasonably well established that there is very little combustion attributable to direct reaction between the molecules. Instead, a high-energy collision dissociates a molecule into atoms, or free radicals. These molecular fragments react with greater ease, and the combustion process proceeds generally by a chain reaction involving these fragments. An illustration will make this clear. The combustion of hydrogen and oxygen to form water does not occur in a single step, reaction (1). In this seemingly simple case, some

$$2H + O_2 \rightarrow 2H_2O \qquad (1)$$

fourteen reactions have been identified. A hydrogen atom is first formed by collision; it then reacts with oxygen molecules, reaction (2), forming an OH radi-

$$H + O_2 \rightarrow OH + O \qquad (2)$$

cal. The latter in turn reacts with a hydrogen molecule, reaction (3), forming water and regenerating the

$$OH + H_2 \rightarrow H_2O + H \qquad (3)$$

H atom which repeats the process. This sequence of reactions constitutes a chain reaction. Sometimes the O atom reacts with a hydrogen molecule to form an OH radical and another H atom, reaction (4). Thus a

$$O + H_2 \rightarrow OH + H \qquad (4)$$

single H atom can form a new H atom in addition to regenerating itself. This process constitutes a branched-chain reaction. Atoms and radicals recombine with each other to form a neutral molecule, either in the gas space or at a surface after being adsorbed. Thus, chain reactions may be suppressed by proximity of surfaces; and the number and length of the chains may be controlled by regulating the temperature, the composition of the mixtures, and other conditions.

Under certain conditions, where the rate of chain branching equals or exceeds the rate at which chains are terminated, the combustion process speeds up to explosive proportions; because of the rapidity of molecular events, a large number of chains are formed in a short time so that essentially all of the gas undergoes reaction at the same time; that is, an explosion results. The branched-chain type of explosion is similar in principle to atomic explosions of the fission type, where more than one neutron is generated by the reaction between a neutron and a uranium nucleus. Another cause of explosion in gaseous combustion arises when the rate at which heat is liberated in the reaction is greater than the rate at which the heat dissipates to the surroundings. The temperature increases, accelerates the reaction rate, liberates more heat, and so on, until the entire gas mixture reacts in a very short time. This type of explosion is known as a thermal explosion. There are cases intermediate between branched-chain and thermal explosions which depend upon the type and proportion of gases mixed, the temperature, and the density.

In slow combustion, intermediate products can be isolated. Aldehydes, acids, and peroxides are formed in the slow combustion of hydrocarbons, and hydrogen peroxide in the slow combustion of hydrogen and oxygen. At the relatively low temperature of combustion of paraffin hydrocarbons (propane, butane, ethers) a bluish glow is seen. This light from activated formaldehyde formed in the process is called a cool flame.

In the gaseous combustion and explosive reactions described above, the processes proceed simultaneously throughout the vessel. The gas mixture in a vessel may also be consumed by a combustion wave which, when initiated locally by a spark or a small flame, travels as a narrow intense reaction zone through the explosive mixture. The gasoline engine operates on this principle. Such combustion waves travel with moderate velocity, ranging from 1 ft/s (0.3 m/s) in hydrocarbons and air to 20–30 ft/s (6–9 m/s) in hydrogen and air. The introduction of turbulence or agitation accelerates the combustion wave. The accelerating wave sends out compression or shock waves which are reflected back and forth in the vessel. Under certain conditions these waves coalesce and change from a slow combustion wave to a high-velocity detonation wave. In hydrogen and oxygen mixtures, the speed is almost 2 mi/s (3.2 km/s). The pressure created by detonation can be very high and dangerous.

Combustion mixtures can be made to react at lower temperatures by employing a catalyst. The molecules are adsorbed on the catalyst, where they may be dissociated into atoms or radicals, and thus brought to reaction condition. An example is the catalytic combination of hydrogen and oxygen at ordinary temperatures on the surface of platinum. The platinum glows as a result of the heat liberated in the surface combustion. See Chain reaction (chemistry); Chemical dynamics; Explosion and explosive; Flame; Free radical.

Bernard Lewis

Spectroscopy. The spectroscopy of combustion is an experimental technique for obtaining data from flames without interfering with the combustion process. The spectrum of light emitted or absorbed in a flame is a physical property of the materials present in the flame. See Spectroscopy.

Flame spectra are used in interpreting combustion mechanisms and in determining flame temperatures. Various spectrographic techniques are used depending on the type of measurement desired. The method of line reversal, in which the radiation intensity from thermally excited metal atoms is compared to a blackbody lamp filament of controllable brightness, is one technique that gives a measure of temperature; it can be handled by simple equipment with only filters to isolate radiation. An example is the addition of small amounts of sodium salts to gaseous or liquid fuels—a technique known as sodium D-line reversal.

Band spectra from reactive combustion intermediates are usually studied with a spectroscope of either the prism or grating type; for extremely fine resolution, an interferometer is sometimes used. The dispersed radiation is detected either photographically or by photomultiplier tubes. Photography is of value in recording spectra over a range of wavelengths simultaneously. If only specific lines are of interest, two or more photomultiplier tubes can be used to record relative line brightnesses. A continuous scan of a spectrum can be achieved by moving one phototube and slit along the focus of the spectral radiation and displaying the output on an oscilloscope.

The band spectra from flames has been associated with the quantized energy changes in molecules due to rotation and vibration. Each spectral line in a band spectrum represents a discrete energy level. Under equilibrium temperature conditions, there is ideally a Maxwell-Boltzmann distribution of molecules in different energy states. According to the kinetic theory of gases, this is a dynamic equilibrium with molecules having their energy distributed in a specific way over these energy states. The radiation of light is a measure of the number of molecules in the process of changing energy levels. From these measurements, on band spectra, flame temperatures and reaction intermediates can be determined. See Burning velocity measurement; Kinetic theory of matter.

Roderick S. Spindt

Bibliography. N. A. Chigier (ed.), *Progress in Energy and Combustion*, vols. 1–10, 1976–1986; A. G. Gaydon and H. G. Wolfhard, *Flames: Their Structure, Radiation, and Temperature*, 4th ed., 1979; B. Lewis and G. von Elbe, *Combustion, Flames and Explosions of Gases*, 3d ed., 1987; H. B. Palmer and J. M. Beer (eds.), *Combustion Technology: Some Modern Developments*, 1974; M. L. Parsons et al., *Handbook of Flame Spectroscopy*, 1975; M. L. Parsons and P. M. McElfresh, *Flame Spectroscopy: Atlas of Spectral Lines*, 1971; F. J. Weinberg, *Advanced Combustion Methods*, 1986; F. A. Williams, *Combustion Theory*, 2d ed., 1985.

Combustion chamber

The space at the head end of an internal combustion engine cylinder where most of the combustion takes place. See Combustion.

Spark-ignition engine. In the spark-ignition engine, combustion is initiated in the mixture of fuel and air by an electrical discharge. The resulting reaction moves radially across the combustion space as a zone of active burning, known as the flame front. The velocity of the flame increases nearly in proportion to engine speed so that the distance the engine shaft turns during the burning process is not seriously affected by changes in speed. See Internal combustion engine; Spark plug.

For high efficiency the mixture should burn as quickly as possible while the piston is near top center position (constant-volume combustion). Short burning time is achieved by locating the spark plug in a central position to minimize the distance the flame front must travel. The chamber itself is made as compact as possible for the required volume, to keep the flame paths short (**Fig. 1**). A compact chamber presents less wall area to the enclosed volume, thereby reducing the heat loss from the flame front. Because motion of the flame depends upon transferring heat from the flame to the adjacent layers of unburned mixture, the reduced heat loss increases the flame velocity. One method for increasing flame velocity is to provide small-scale turbulence in the cylinder charge, often by designing the chamber so that part of the piston head comes close to the cylinder head at top center position (Fig. 1a). The squish that results forces the mixture in this region into the rest of the combustion chamber at high velocity. The turbulence so produced increases the rate of heat transfer from the flame to the unburned mixture, greatly increasing the flame velocity. Other engines may use high swirl in the intake port and combustion chamber or an auxiliary or precombustion chamber to accomplish the same result (Fig. 1e). When a prechamber is used, combustion begins in the richer mixture around the spark plug in the prechamber. The burning mixture then enters the main chamber, which is filled with a leaner mixture, where combustion is completed. This process is known as stratified charge.

Occasionally a high burning rate, or too rapid change in burning rate, gives rise to unusual noise and vibration called engine roughness. Roughness may be reduced by using less squish or by shaping the combustion chamber to control the area of the flame front. A short burning time is helpful in eliminating knock because the last part of the charge is burned by the flame before it has time to ignite spontaneously. See Spark knock.

For high power output, the shape of the combustion chamber must permit the use of large valves with space around them for unrestricted gas flow. In many engine designs, three and four valves per cylinder are used to provide higher volumetric efficiency. See Valve train.

Pockets or narrowed sections in the combustion chamber that trap thin layers of combustible mixture

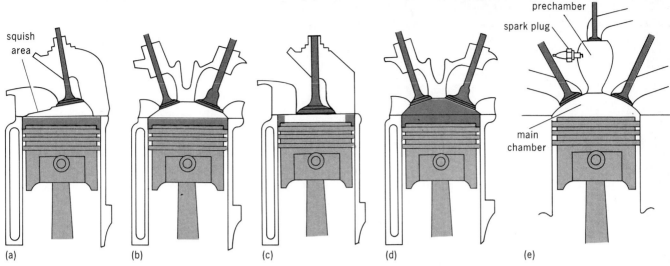

Fig. 1. Various combustion chambers used in spark-ignition engines. (a) Wedge. (b) Hemispheric (open). (c) Bowl-in-piston (cup). (d) Crescent (pent-roof). (e) Prechamber. (*Automotive and Technical Writing, Charlottesville***)**

between adjacent surfaces interfere with combustion and increase the unburned fuel hydrocarbons normally present in the exhaust gases. Because unburned hydrocarbons are one of the constituents of smog-forming air pollution, the newer combustion chambers have less squish and less surface area per unit of volume. *SEE AIR POLLUTION; SMOG*.

Compression-ignition engine. In compression-ignition (diesel) engines, the fuel is injected late in the compression stroke into highly compressed air. Mixing must take place quickly, especially in smaller

high-speed engines, if the fuel is to find oxygen and burn while the piston remains near top center. After a short delay, the injected fuel ignites from contact with the hot air in the cylinder. There is no flame front travel to limit the combustion rate. Injection may be into either an undivided combustion chamber (direct injection) or a divided combustion chamber (indirect injection) [**Fig. 2**].

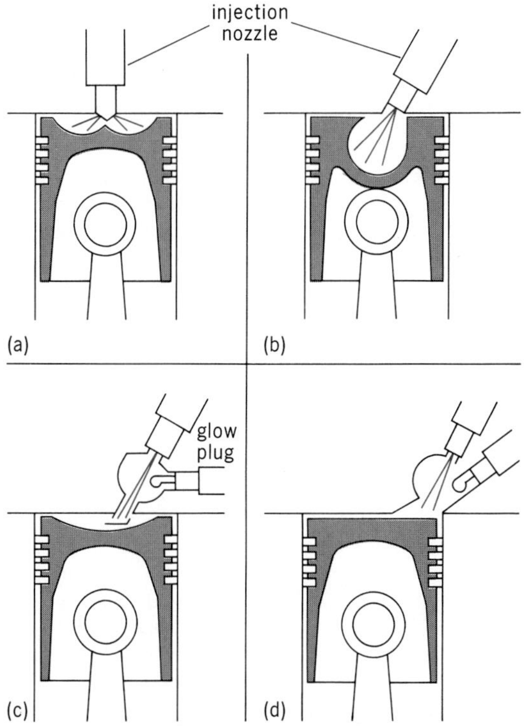

During direct injection into an undivided chamber, mixing can be improved by producing a swirl in the intake air so that it is moving relative to the fuel spray (Fig. 2a). Swirl velocity increases with higher inlet velocities present at high engine speeds. Therefore, mixing becomes more rapid as the time available for mixing becomes less. Another type of undivided combustion chamber is formed by a narrow cavity in the piston head (Fig. 2b). For rapid vaporization, the fuel is sprayed against the hot wall of the combustion chamber in the direction of the swirling air. Smooth combustion with relatively little exhaust smoke results.

Most high-speed diesel engines have indirect injection, in which the combustion chamber is divided into a main and a connected auxiliary chamber. Fuel is sprayed into the auxiliary chamber, where rapid mixing and combustion takes place (Fig. 2c and d). The burning mixture then issues into the main chamber, where any remaining fuel is burned.

If mixing of fuel and air is too thorough by the end of the delay period, high rates of pressure rise result, and the operation of the engine is rough and noisy. To avoid this condition, the auxiliary chamber is most compression-ignition engines operates at high temperature so that the fuel ignites soon after injection begins. This reduces the amount of fuel present and the degree of mixing at the time that ignition takes place.

High rates of pressure rise can also be reduced by keeping most of the fuel separated from the chamber air until the end of the delay period. Rapid mixing must then take place to ensure efficient burning of the fuel while the piston is near top center. *SEE DIESEL ENGINE*.

Augustus R. Rogowski/Donald L. Anglin
Bibliography. *Bosch Automotive Handbook*, 1986;

Fig. 2. Various combustion chambers used in compression-ignition (diesel) engines. (a) Direct injection. (b) MAN M system (German manufacturer). (c) Prechamber. (d) Swirl chamber. (*Robert Bosch Corp.***)**

W. H. Crouse and D. L. Anglin, *Automotive Engines*, 1986; J. B. Heywood, *Internal Combustion Engine Fundamentals*, 1988.

Combustion wave measurement

Observation of the transient phenomena accompanying combustion. When an explosive mixture is ignited, a zone of reacting material propagates through the mixture. The flow of heat and chemical reactants from the ignition source into adjacent layers of the mixture causes a self-sustaining process to occur, usually manifest as a flame; this phenomenon is the combustion wave. Since combustion can occur under diverse conditions, many types of measurement have been applied to the combustion wave.

The rate at which the wave moves through the mixture is the flame speed. SEE BURNING VELOCITY MEASUREMENT.

Measurement of the intermediate reactants and the products of combustion is needed for an understanding of the chemical reactions that sustain combustion waves. For example, a mixture of hydrogen and oxygen will support a combustion wave in which the product of combustion is water. The intermediate reactants include hydrogen and oxygen atoms and hydroxyl radicals. Evidence for hydroxyl radicals is from spectroscopic measurements.

The temperature rise that accompanies a combustion wave can be measured by various techniques based on radiation and absorption properties of gases; temperature-sensitive probes; velocity of sound through gases; critical flow velocities of gases; and dissociation reactions of certain gases.

By the usual definition the temperature of a gas refers to molecules in thermodynamic equilibrium. In a system undergoing active combustion, molecules can gain excess energy in their vibrational and rotational degrees of freedom. Under these nonequilibrium conditions it is not possible to define a unique temperature.

Spectroscopic measurement of band spectra can give temperatures based on vibrational and rotational motion. Other spectroscopic methods include line reversal and two-color pyrometry.

Temperature-sensitive probes include the familiar thermocouple. By the use of adequate shielding and corrections, temperatures of low-velocity flames can be accurately measured; however, these devices are not entirely satisfactory for complex flames having a large temperature distribution. Compensated hot wires have also been used with transparent flames in regions where combustion is complete. SEE THERMOCOUPLE.

The flow of gases through the combustion wave is an important aid to understanding the shape and structure of the wave. Illumination of fine particles of magnesium oxide flowing through the flame gives information on direction and velocity of gas flow. SEE COMBUSTION.

Roderick S. Spindt

Comesomatoidea

A superfamily of nematodes in the order Chromadorida and containing only one family, the Comesomatidae. These nematodes are distinguished by their wide multispiral amphids that make at least two complete turns. The stoma is shallow and bowl-shaped, generally containing small denticles. The body cuticle is smooth but ornamented with punctations that simulate annuli. In males the spicules may be greatly elongate or short and complex, but they are always accompanied by a gubernaculum. All known species are found in marine habitats, but their feeding habits are unknown. SEE NEMATA.

Armand R. Maggenti

Comet

One of the major types of objects that move in closed orbits around the Sun. Compared to the orbits of planets and asteroids, comet orbits are more eccentric and have a much greater range of inclinations to the ecliptic (the plane of the Earth's orbit). Physically, a comet is a small, solid body which is roughly a half mile in diameter, contains a high fraction of icy substances, and shows a complex morphology, often including the production of an extensive atmosphere and tail, as it approaches the Sun. SEE ASTEROID; PLANET.

About 10 comets are discovered or rediscovered each year. On the average, one per year is a bright comet, visible to the unaided eye and generating much interest among the public as well as among comet workers.

Astronomers consider comets to be worthy of detailed study for several reasons: (1) They are intrinsically interesting, involving a large range of physical and chemical processes. (2) They are valuable tools for probing the solar wind. (3) They are considered to be remnants of the solar system's original material and, hence, prime objects to be studied for clues about the nature of the solar system in the distant past. (4) Comets may be required to explain other solar system phenomena.

Appearance. As seen from Earth, comets are nebulous in appearance, and the tail is usually the most visually striking feature. This tail can in some cases stretch along a substantial arc in the sky. An example is given in **Fig. 1**, which shows Comet West dominating the eastern sky on March 9, 1976, with tails some 30° in length. Some fainter comets, however, have little or no tail.

The coma or head of a comet is seen as the ball of light from which the tail or tails emanate. Within the coma is the nucleus, the origin of the material in the tail and coma.

Discovery and designation. Comets are discovered by both amateur and professional astronomers. The fainter ones are often discovered by professionals on wide-field photographic plates taken for other purposes. Amateurs usually carry out systematic searches of the sky using wide-field binoculars or telescopes. Discoveries are communicated to the Bureau for Astronomical Telegrams, Smithsonian Astrophysical Observatory, Cambridge, Massachusetts, and are then announced by the International Astronomical Union.

Normally, comets are named after their discoverers, and up to three independent codiscoverers are allowed. An example is Comet Kobayashi-Berger-Milon. The need for such a rule is easily understood when one realizes that some bright comets have been discovered almost simultaneously by dozens of individuals. Comets are numbered in two ways. The first comet discovered in 1990 would be 1990a, the second

1990b, and so on. After the orbits have been calculated, they are assigned a roman numeral in order of perihelion (point of closest approach to the Sun) passage. The third comet to pass perihelion in 1990 would be 1990 III. Halley's Comet at its last appearance was first comet 1982i and then comet 1986 III. Halley's Comet is also designated P/Halley, the P indicating a periodic comet (a comet with period less than 200 years).

Occasionally, a comet has been named after the person who computed its orbit. Examples are Halley's Comet and Encke's Comet.

Oort Cloud and comet evolution. A major step in understanding the origin of comets was taken in 1950 by J. Oort. He developed the idea that comets are in effect "stored" in a Sun-centered spherical cloud of radius approximately 50,000 AU from the Sun (1 AU is the distance from the Sun to the Earth, 9.3×10^7 mi or 1.50×10^8 km). Occasionally, gravitational perturbations by passing stars send comets into the inner solar system, where they are discovered and their phenomena are observed. This process, of course, leads to the eventual destruction of the comet, because passages near the Sun cause a loss of material which is not recovered by the comet. Very rough estimates indicate that a comet loses 1% of its mass at each perihelion passage. Thus, a comet with radius of 0.6 mi (1 km) would lose a layer approximately 10 ft (3 m) thick on each passage. The origin and evolution of comets are discussed in more detail below.

The current view considers the Oort Cloud as a steady-state reservoir which loses comets as just described and gains them from an inner cloud of comets located between the orbit of Neptune and the traditional Oort Cloud. Ideas that periodic comet showers could result from a tenth planet in the inner cloud or a faint solar companion, called Nemesis, are novel but not generally accepted.

Observations. Observations of comets run the gamut of modern observing techniques. These include photographs in visual and ultraviolet wavelengths, photometry (accurate brightness measurements) in visual and infrared wavelengths, spectral scans in many wavelength regions, radio observations, and observations in extreme ultraviolet wavelengths from rockets and orbiting spacecraft above the Earth's atmosphere. A sample spectrum in the visual wavelength range

Fig. 1. Comet West as photographed on March 9, 1976, showing the general appearance of a bright comet. The fan-shaped structure emanating from the head or coma is the dust tail, while the single straight structure is the plasma tail. (*S. M. Larson, Lunar and Planetary Laboratory, University of Arizona*)

with some constituents identified is shown in **Fig. 2**. Visual wavelength spectra of comets show considerable variation. Ultraviolet spectra, obtained regularly with the *International Ultraviolet Explorer* satellite, have been found to be rather similar. A sample spectrum in ultraviolet wavelengths is shown in **Fig. 3**. Observations of meteors also contribute to the understanding of comets. SEE ASTRONOMICAL SPECTROSCOPY.

Measurements. The advent of space missions to comets has added an entirely new dimension to obtaining cometary information. A wide variety of direct measurement techniques have already been applied. Mass spectrometers have determined the properties of neutral gases, plasmas, and dust particles (via impact ionization). Dust counters have recorded the fluxes of dust and determined the mass distribution. Magnetometers have measured the

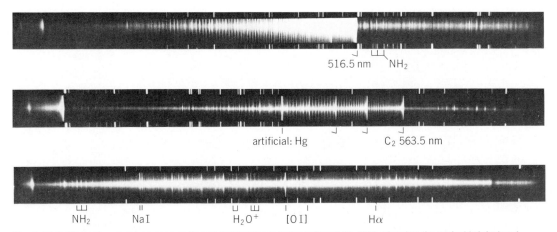

Fig. 2. High-dispersion spectrograms of Comet Kohoutek on January 9 and 11, 1974, showing the typical bright-band structure in the visible region of the spectrum. The bands are produced by molecular species; the amine radical (NH_2), diatomic carbon (C_2), and ionized water (H_2O^+) are marked. The lines of the atomic species sodium (NaI), oxygen ([OI]), and hydrogen ($H\alpha$) are also from the comet, but the mercury line (Hg) is from street lights. (*Lick Observatory, University of California*)

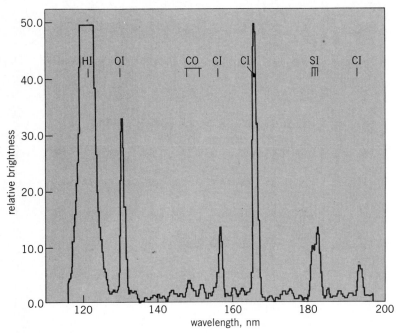

Fig. 3. Ultraviolet spectrum of Comet Bradfield obtained by the *International Ultraviolet Explorer* satellite in January 1980. The spectrum shows the hydrogen line (HI) from the hydrogen cloud as well as lines of oxygen (OI), carbon monoxide (CO), carbon (CI), and sulfur (SI). (*After P. D. Feldman et al., IUE observations of the UV spectrum of Comet Bradfield, Nature, 236, 132–135, 1980*)

Sun and a few tenths of an AU from Earth. Two specific exceptions are discussed here.

While comets have been observed at distances from the Sun of less than 0.1 AU for decades, only in 1979 was it learned that comets occasionally impact the Sun (**Fig. 6**). This fact was determined from a satellite (*SOLRAD*) used to monitor the Sun's corona; hence, the cometary discovery was not planned. The instrument carries an occulting disk which blocks out the bright glare of the Sun's surface, producing the dark circle with an apparent diameter of 5 solar radii in Fig. 6. Comet Howard-Koomen-Michels hit the Sun on August 30, 1979; the estimated time of impact was 22^h29^m Greenwich mean time. Figure 6 clearly shows the comet approaching the Sun, but its reappearance, which would have produced a cometary image on the frames shown, did not occur. Instead, a brightening of a major portion of the corona was observed which then slowly faded away. The brightening was presumably caused by cometary debris ultimately blown away by the Sun's radiation pressure following the disintegration of the nucleus.

Analysis of the data from this solar monitoring satellite has turned up several more cometary impacts with the Sun. These results imply that solar impacts by relatively small comets are quite frequent.

Comets very near the Earth are rare, but provide unique opportunities for new discoveries. Comet IRAS-Araki-Alcock was another example of a comet being first discovered by a satellite, in this case the *Infrared Astronomical Satellite* (*IRAS*), which also discovered additional comets. This comet passed within 0.031 AU of Earth on May 11, 1983. This distance was only about 12 times the average distance from the Earth to the Moon. Photographs of the comet showed a faint tail on May 9 (**Fig. 7a**); but by May 12 the tail was not visible (Fig. 7b) and the coma clearly displayed a distinct dual structure with bright inner and faint outer components.

The small distance to Comet IRAS-Araki-Alcock also provided a perfect opportunity to bounce radar waves off the nucleus. The experiment was successful, and the echoes provided evidence for the existence of a solid cometary nucleus with diameter of 0.6 mi (1 km).

strength and orientation of the magnetic field. Further plasma properties have been determined by plasma wave detectors.

As illustrations of this vast data source, **Fig. 4** shows the measurements of plasma wave activity around Comet Giacobini-Zinner, and **Fig. 5** shows the measurements of the densities of water (H_2O) and carbon dioxide (CO_2) in Halley's coma.

In addition, detailed imaging of the nucleus was made possible by the space missions. SEE *HALLEY'S COMET.*

Near-Sun and near-Earth comets. Most comets are observed under so-called normal circumstances; that is, the comet is between 0.5 and 1.5 AU from the

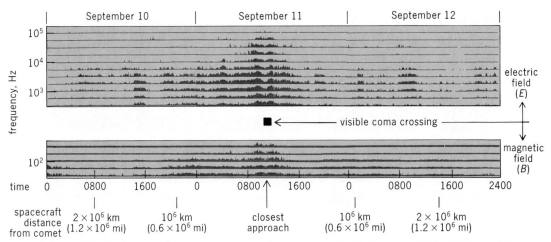

Fig. 4. Observations of plasma waves, produced when cometary ions interact with the solar wind, for a 3-day period centered on the time of closest approach of the *International Cometary Explorer* spacecraft on September 11, 1985. The peak amplitudes for the different frequency ranges are shown on a logarithmic scale. The region of intense plasma wave activity is much larger than the visible coma. (*F. L. Scarf, TRW Space and Technology Group*)

The discovery of small comets hitting the Sun and the comets discovered by *IRAS* (including Comet IRAS-Araki-Alcock) imply that small comets are more numerous than had been generally believed. As the details for the origin of comets are worked out, it

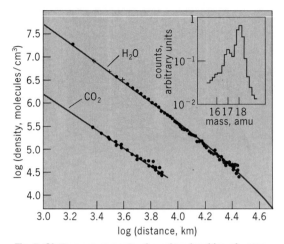

Fig. 5. *Giotto* measurements of number densities of water vapor (H_2O) and carbon dioxide (CO_2) molecules as a function of distance from the nucleus of Halley's Comet. The inset shows measurements in the mass range including 16, 17, and 18 amu and clearly indicates that H_2O (mass = 18 amu) is the dominant species. 1 km = 0.6 mi. 1 molecule/cm³ = 16.4 molecules/in.³ (*After D. Krankowsky et al., In situ gas and ion measurements at Comet Halley, Nature, 321:326–329, 1986*)

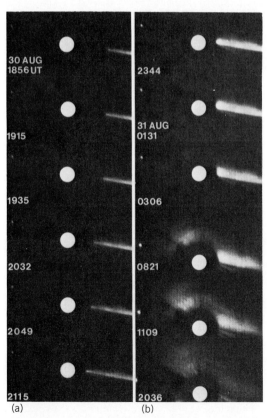

Fig. 6. Sequence of photographs obtained from a *SOLRAD* satellite showing Comet Howard-Koomen-Michels hitting the Sun in August 1979. Captions give date and universal time (Greenwich mean time) of each photograph. (*a*) Before impact. (*b*) After impact. (*Naval Research Laboratory, Washington, D.C.*)

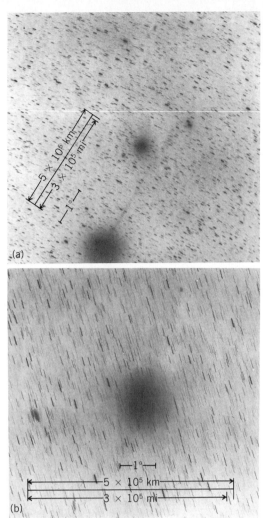

Fig. 7. Comet IRAS-Araki-Alcock. (*a*) May 9, 1983. (*b*) May 12, 1983. Scales are marked on the photograph. (*Joint Observatory for Cometary Research, operated by NASA–Goddard Space Flight Center and New Mexico Institute of Mining and Technology*)

will be necessary to account for this fact. There has been increasing interest in small comets or cometlike-objects, and the number of these objects in the inner solar system may be much larger than previously thought. Such objects might be responsible for brightness decreases in the Earth's upper atmosphere and for a component of Lyman-alpha emission. In the past, these objects may have been important in lunar cratering, and the total population of comets might have been an important source of atmospheres for the terrestrial planets as well as a possible source of moderately complex organic molecules necessary for the initial development of life on Earth.

Orbits. The first closed comet orbit to be calculated was Edmond Halley's elliptical orbit for the comet of 1680. This work indicated that comet orbits were ellipses with the Sun at one focus. In subsequent work, Halley noticed the striking similarity of the orbits of what were thought to be three different comets observed in 1531, 1607, and 1682. In 1705 he concluded that these were the same comet with a period of 75 or 76 years, and predicted its return in 1758. This comet is the one bearing Halley's name. It made an appearance in 1910 (**Fig. 8**), and again in 1985 and 1986.

The second comet to have its return successfully predicted was named after J. F. Encke. This comet has the shortest known period, 3.3 years. At its 1838

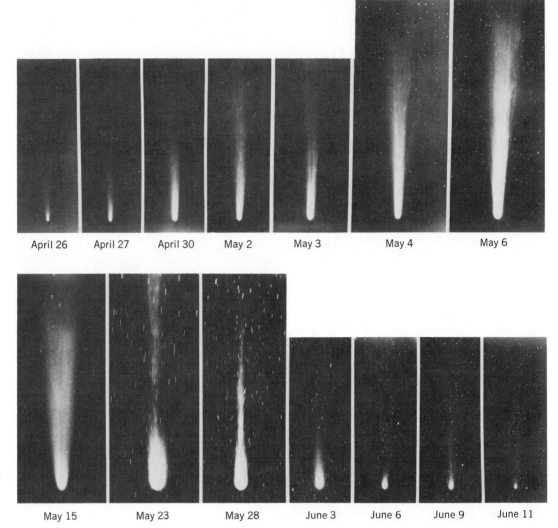

| April 26 | April 27 | April 30 | May 2 | May 3 | May 4 | May 6 |

Fig. 8. Sequence of photographs showing the changing appearance of Halley's Comet in 1910. (*Hale Observatories*)

| May 15 | May 23 | May 28 | June 3 | June 6 | June 9 | June 11 |

return, Encke's Comet showed another common property of comets, that is, a steadily changing period. This phenomenon is now known to result from the so-called nongravitational forces which must be explained by any successful comet model.

Six parameters are necessary to describe completely the orbit of a comet. They specify the orientation of the orbital plane in space, the orientation of the orbit in this plane, the size and shape of the ellipse, and the position of the comet along the orbit. In principle, three observations of position on the celestial sphere are sufficient, because each observation consists of independent measures of two coordinates.

Atomic and molecular species observed in comets*

Neutrals	Ions
H, OH, O, S, H_2O, H_2CO C, C_2, C_3, CH, CN, CO, CS, S_2 HCN, CH_3CN, NH NH_2, Na, Fe, K Ca, V, Cr, Mn, Co, Ni, Cu $(H_2CO)_n$	CO^+, CO_2^+, H_2O^+, OH^+, H_3O^+ CH^+, N_2^+, Ca^+, C^+, CN^+

*Species observed in coma and tail in spectroscopic studies and direct spacecraft measurements.

In practice, definitive orbits are derived from many observations, often hundreds.

While comet orbits are represented by ellipses to a good approximation, there are departures caused by the nongravitational forces and by the gravitational perturbations of the planets. When orbital parameters are listed for a comet, they refer to an ellipse which exactly matches the comet's position and motion at a specific time. Such an osculating orbit, as it is called, forms the starting point for studies of orbital evolution and for accurate predictions of the time and location of a comet's appearance in the sky. *See* CELESTIAL MECHANICS.

Orbital parameters have been determined for over 750 individual comets. Over 600 are classified as long-period comets, that is, comets with orbital periods greater than 200 years. The orbital planes of the long-period comets have approximately random inclinations with respect to the ecliptic. This means that there are as many comets with direct orbits (revolving around the Sun in the same sense as the planets) as with retrograde orbits (revolving in the sense opposed to the planets' motion). Careful examination of the original (that is, inbound) orbits of the long-period comets shows none that are hyperbolas; that is, no interstellar comets have yet been observed. This fact strongly implies that the cloud of comets is gravitationally bound to the Sun and therefore is a part of the solar system.

About 150 short-period comets arc mostly in direct orbits with inclinations of less than 30°. The distribution of periods shows a peak between 7 and 8 years, and the majority have an aphelion (point of greatest distance from the Sun) near the orbit of Jupiter.

Composition. The results of many spectroscopic studies and direct spacecraft measurements on the composition of comets are summarized in the **table**, where the atoms, molecules, ions, and classes of substances that have been observed are listed. Fairly complex molecules such as polymerized formaldehyde [$(H_2CO)_2$] and methyl cyanide (CH_3CN) are present.

Observations of the gas composition in Halley's coma have found these values by number of molecules: water, approximately 80%; carbon monoxide, roughly 10%; carbon dioxide, approximately 3.5%; complex organic compounds such as polymerized formaldehyde, a few percent; and trace substances, the remainder. These values were obtained from direct measurements, except for that of carbon monoxide, which was determined from a rocket experiment. These values are for the outer layers of a comet which has passed through the inner solar system many times. The values of a new comet or the interior of Halley may be different with a higher fraction of carbon dioxide likely.

Measurements of dust composition were made by instruments on the spacecraft sent to Halley's Comet. Some of the particles contain essentially only the atoms hydrogen, carbon, nitrogen, and oxygen; these are the "CHON" particles. Other particles have a silicate composition, that is, they resemble rocks found throughout the solar system. The dominant dust composition is a mixture of these two and resembles carbonaceous chondrites enriched in the light elements hydrogen, carbon, nitrogen, and oxygen.

Structure. The main components of a comet are the nucleus, coma, hydrogen cloud, and tails.

Nucleus. Strong evidence points to the existence of a central nuclear body or nucleus for all comets from which all cometary material, both gas and dust, originates. In the early 1950s, F. L. Whipple proposed an icy conglomerate model of the nucleus. In this model, the nucleus is not a cloud of particles as had previously been thought (the "sandbank" model), but a single mass of ice with embedded dust particles, commonly called a dirty snowball. Such a nucleus could supply an adequate amount of gas to explain cometary phenomena and last through many apparitions because only a relatively thin surface layer would be eroded away (by sublimation) during each passage near the Sun.

The nucleus of Halley's Comet has been directly observed. Its shape was highly irregular, the surface exhibited features and structure, and the albedo (fraction of reflected light) was very low. While the confirmation of the existence of a single, nuclear body is important, Halley is just one comet and generalizations from it may not be valid. SEE ALBEDO.

A schematic drawing of a cometary nucleus, embodying the best available information, is given in **Fig. 9**. The generic body is probably irregular, with radius ranging from around 1000 ft (300 m) to 12 mi (20 km). Masses would range in order of magnitude from 10^{11} to 10^{16} kg (10^{11} to 10^{16} lbm), with roughly

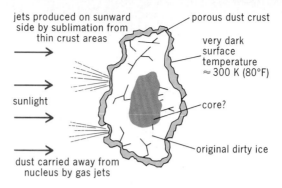

Fig. 9. Schematic model of the cometary nucleus.

equal parts of ice and dust. The average density would be in the range 0.2 to 2 g/cm³ (0.2 to 2 times that of water). Nuclei of comets sometimes split into two or more pieces; Comet West (1976 VI) is an example.

Coma. The coma is observed as an essentially spherical cloud of gas and dust surrounding the nucleus. The principal gaseous constituents are the neutral molecules listed in the table, and the dust composition has been described above. The coma can extend as far as 10^5 to 10^6 km (10^5 to 10^6 mi) from the nucleus, and the material is flowing away from the nucleus at a typical speed of 0.6 mi/s (1.0 km/s). As the gas flows away from the nucleus, the dust particles are dragged along. For comets at heliocentric distances greater than 2.5 to 3 AU, the coma is not normally visible and is presumed not to be present.

Hydrogen cloud. In 1970, observations of Comet Tago-Sato-Kosaka (1969g) and Comet Bennett (1969i) from orbiting spacecraft showed that these comets were surrounded by a giant hydrogen cloud that extends to distances on the order of 10^7 km (10^7 mi), or a size larger than the Sun. The observations were made in the resonance line of atomic hydrogen at 121.6 nm. Hydrogen clouds have since been observed for many other comets. A photograph of Comet West's hydrogen cloud is given in **Fig. 10**, along with a visual photograph to the same scale for comparison. Fairly bright comets (such as Bennett) have a hydrogen production rate by sublimation from

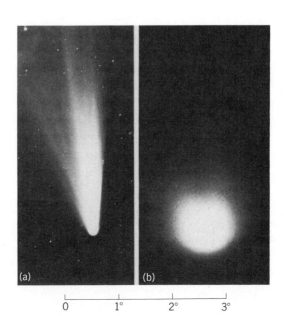

Fig. 10. Comparison of photographs of Comet West, obtained from a rocket on March 5, 1976, and printed to same scale. (*a*) Visual photograph (*P. D. Feldman, Johns Hopkins University*). (*b*) Ultraviolet photograph showing hydrogen cloud (*C. B. Opal and G. R. Carruthers, Naval Research Laboratory*).

Fig. 11. Halley's Comet as recorded by the United Kingdom Schmidt Telescope in Australia on February 22, 1986. The image shows dust tail structures (above, right), plasma tail (right, below), and an antitail (left, above). (*Royal Observatory, Edinburgh*)

in 1951 to postulate the existence of a continuous outflow of ionized material from the Sun, which he called the solar corpuscular radiation, now called the solar wind. The interaction of the solar wind and its magnetic field, as suggested by H. Alfvén in 1957, plays an important role in cometary physics and in the formation of plasma tails. The direct measurements on the sunward side by the spacecraft sent to Halley's Comet and on the tailward side by the *International Cometary Explorer* mission to Comet Giacobini-Zinner in 1985 have confirmed these views. *SEE SOLAR WIND.*

The plasma tails are generally straight and have lengths that range in order of magnitude from 10^7 to 10^8 km (10^7 to 10^8 mi). The Great Comet of 1843 had a plasma tail extending over 2 AU in length. The plasma in these tails is composed of electrons and molecular ions. The dominant visible ion is CO^+, and the other ions known to be present are listed in the table. The zone of production for the molecular plasma appears to be in the coma near the sunward side of the nucleus. The material in the plasma tails is concentrated into thin bundles or streamers, and additional structure is found in the tail in the form of knots and kinks. These features appear to move along the tail away from the head at speeds of 6 mi/s (10 km/s) to 120 mi/s (200 km/s). Plasma tails are generally not observed beyond heliocentric distances of 1.5 to 2 AU; an exception is Comet Humason, which showed a spectacular, disturbed plasma tail well beyond normal distances.

The dust tails are usually curved and have lengths that range in order of magnitude from 10^6 to 10^7 km (10^6 to 10^7 mi). Normally, the dust tails are relatively homogeneous; an exception, Comet West, is shown in Fig. 1. Observations indicate that the dust particles are typically 1 micrometer in diameter and are probably silicate in composition. Occasionally dust tails are seen which appear to point in the sunward direction, the so-called antitails. Examples are Comet Arend-Roland in 1957 and Halley's Comet in 1986 (**Fig. 11**). These are not truly sunward appendages but are the result of projection effects.

The structure and dimensions of the constituent

the nucleus at heliocentric distance of 1 AU in the range 3 to 8×10^{29} atoms/s. The size of the hydrogen cloud depends on the velocity of the outflowing hydrogen atoms, and 5 mi/s (8 km/s) has been derived. This velocity would arise (from energy balance considerations) if most of the hydrogen in the cloud were produced by the photodissociation of the hydroxyl radical (OH).

Tails. Photographs of bright comets generally show two distinct types of tails (Fig. 1): the dust tails and the plasma tails. They can exist separately or together in the same comet. In a color photograph, the dust tails appear yellow because the light is reflected sunlight, and the plasma tails appear blue from emission due to ionized carbon monoxide, CO^+.

Studies of features in plasma tails led L. Biermann

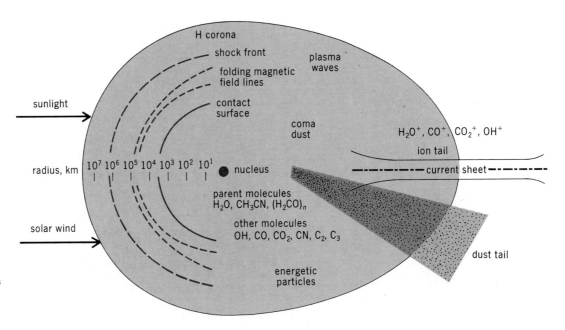

Fig. 12. Cometary features diagrammed on a logarithmic scale. 1 km = 0.6 mi.

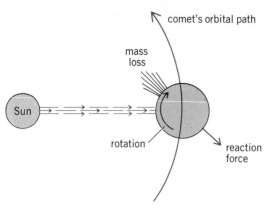

Fig. 13. Mechanism by which mass loss can produce a reaction force on the nucleus by the rocket effect.

parts of comets—nucleus, coma, hydrogen cloud, and tails—are summarized schematically in **Fig. 12**.

Modern theory. The goal of modern comet theory is to explain the facts about comets outlined above. The broad, theoretical approach appears to be in reasonably good shape and has survived the tests of direct exploration by spacecraft.

Sublimation from nucleus. The cornerstone of the current best ideas is F. L. Whipple's icy conglomerate model of the nucleus as further developed by A. Delsemme. As a comet approaches the Sun on its orbit, sunlight supplies radiant energy to the surface of the nucleus. The energy received heats the nucleus when it is far from the Sun. As the comet continues toward the Sun, the temperature of the surface layers increases to a value, determined by the thermodynamic properties of the ice, where sublimation (passage from the solid state directly to the gaseous state) occurs. Then, most of the incident energy goes to the sublimation of ices. The situation has at least one additional complexity. Sublimation of ice from a dust–ice mixture probably leaves a dust crust which is heated to temperatures higher than the sublimation temperature of water ice. The heat is conducted inward to the ice layers, and the sublimated gases then pass outward through the dust crust. The jets of material seen on Halley's Comet originate in areas of minimal dust crust.

The onset of activity in comets at 2.5 to 3 AU is entirely consistent with water ice as the dominant ice constituent of the nucleus. But a problem still exists. Most other possible constituents are predicted from theory to begin sublimation much farther from the Sun than water ice, but, for example, molecular emissions from CN are visible essentially at the onset of activity.

The accepted solution to the problem is to assume that comets are made up of the type of ice called a clathrate hydrate. Very simply, ice in the crystalline form has cavities which are formed by the bonds that hold the crystal together and into which other substances can be trapped. Thus, the release of most minor constituents is controlled by the thermodynamic properties of water ice and not by the properties of the minor constituent. If all the available cavities are filled, the minor constituents can amount to 17% of the icy lattice material by number of atoms. If more minor constituents, such as carbon dioxide (CO_2), are present in the nucleus, they would be vaporized well beyond 3 AU on the comet's first approach to the

Sun. This phenomenon could explain comets found to be anomalously bright at large distances from the Sun and the dimming of some comets after their first perihelion passage.

Origin of nongravitational forces. The nongravitational forces have a ready explanation if the cometary nucleus is rotating. The result of radiant energy producing sublimation of ices in an ice-dust mixture is to leave a crust of dusty material. Thus, there is a time lag while the heat traverses the dust layer between maximum solar energy received and maximum loss through sublimation. If the nucleus were not rotating, the maximum mass loss would be directly toward the Sun. For a rotating nucleus, the mass loss occurs away from the sunward direction, toward the afternoon side of the comet. The analogous situation on Earth (that is, the time lag between cause and effect) produces the warmest time of day in the afternoon, not at noon. The mass loss under these circumstances produces a force on the nucleus via the rocket effect (reaction force), and this force can accelerate or retard the motion of the comet in its orbit, as shown schematically in **Fig. 13**. Detailed studies of the nongrav-

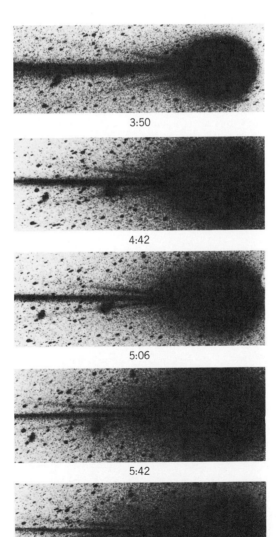

3:50

4:42

5:06

5:42

6:12

|10 arc-minutes

Fig. 14. Sequence of photographs of Comet Kobayashi-Berger-Milon on July 31, 1975, showing the capture of magnetic field from the solar wind. The dominant pair of tail streamers visible on either side of the main tail lengthen and turn toward the tail axis in this sequence. (*Joint Observatory for Cometary Research, operated by NASA–Goddard Space Flight Center and New Mexico Institute of Mining and Technology*)

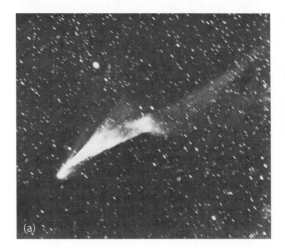

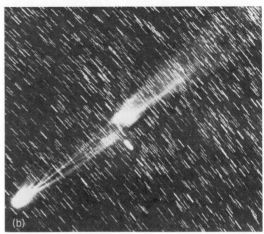

Fig. 15. Comet More-house. (a) Beginning of separation of tail from comet's head on September 30, 1908. (b) Tail widely separated from the head on October 1, 1908. (*Yerkes Observatory, University of Chicago*)

itational forces in comets show that they are entirely consistent with water ice as the controlling substance.

Formation of coma. The sublimated gases, mostly neutral molecules, flow away from the nucleus, dragging some of the dust particles with them to form the coma. Close to the nucleus, the densities are high enough that chemical reactions can occur between molecular species. Photodissociation is also important. Thus, the molecules observed spectroscopically far from the nucleus often are not the same as the initial composition.

Formation of dust tails. The dust particles carried away from the nucleus by the flow of coma gases are blown in the antisolar direction by the Sun's radiation pressure to form the dust tails. The general theory has been developed by M. L. Finson and R. F. Probstein, and good agreement can be obtained with the observed shapes and sizes of the tails if the emission of dust from the nucleus has a peak before perihelion. The larger particles liberated from the nucleus can or-

bit the Sun and reflect sunlight to produce the zodiacal light.

Formation of plasma tails. The gas flowing away from the nucleus has a more involved fate. Under normal circumstances, when a comet's heliocentric distance is about 1.5 to 2 AU, significant ionization of the coma molecules occurs (probably by solar radiation), and this triggers a reaction with the solar wind. At the Earth, the solar wind, a fully ionized proton-electron gas, flows away from the Sun at 250–310 mi/s (400–500 km/s) and has an embedded magnetic field. Because of the magnetic field, the ionized cometary molecules cause the solar-wind field lines to slow down in the vicinity of the comet while proceeding at the full solar-wind speed away from the comet. This situation causes the field lines and the trapped plasma to wrap around the nucleus like a folding umbrella, to form the plasma tail. This picture has been completely confirmed by the spacecraft measurements. In addition, the folding can be seen and photographed because of the emission from the trapped ions (such as CO^+) which serve as tracers of the field lines. A photographic sequence showing this phenomenon is given in **Fig. 14**. Thus, while the ionized molecules are indeed swept in the antisolar direction by the solar wind, the plasma tail should be thought of as a part of the comet attached to the near-nuclear region by the magnetic field captured from the solar wind.

Exceptions occur at times apparently when the polarity of the solar-wind magnetic field changes. This can disrupt the magnetic connection to the near-nuclear region and literally causes the old plasma tail to disconnect while the new tail is forming. This process is quite common. **Figure 15** shows an example from Comet Morehouse in 1908, and **Fig. 16** from Comet Halley in 1986. The physics of this phenomenon is complex and not generally understood.

Fate of comets. When the process of sublimation has been carried out over an extensive period of time, as would be the case for the short-period comets, the ices will be exhausted and the inactive or "dead" comet should consist of dust particles and larger-sized rocky material. These remnants are dispersed along the comet's orbit by perturbations, and are the particles responsible for producing meteors when they enter the Earth's atmosphere. Small particles, very probably of cometary origin, have been collected by high-flying aircraft. *See* METEOR.

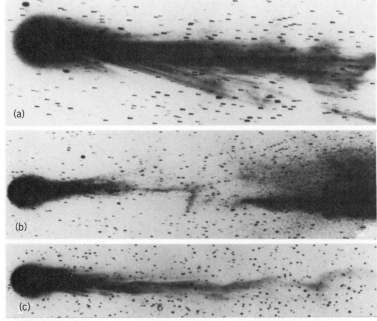

Fig. 16. Images of Halley's Comet clearly showing a disconnection event, taken on (a) January 9 and (b) January 10, 1986, at the Calar Alto Observatory of the Max-Planck Institute für Astronomy, and (c) January 11, 1986, at the Haute-Provence Observatory.

A detailed summary of the physical processes in comets is given in **Fig. 17**.

Origin. For years, it seemed likely that the population of short-period comets could be produced from the population of the long-period comets—those in Oort's Cloud—by their gravitational interaction with Jupiter. It is now understood that they originate in a ring of comets in the vicinity of the orbits of Neptune and Uranus.

The Oort Cloud is the source of the long-period comets. The evidence from the statistics of cometary orbits indicates an essentially spherical cloud of comets with dimensions in the range 10^4 to 10^5 AU. Gravitational perturbations from passing stars have several effects on the cloud. They limit its size and tend to make the orbits random (as observed). Most importantly, the perturbations continually send new comets from the cloud into the inner solar system, where they are observed. Thus, the Oort Cloud can

be considered as a steady-state reservoir for new comets. Evidence is mounting for the view that the Oort Cloud is supplied by an inner cloud of comets.

The current consensus on the origin of comets holds that they condensed from the solar nebula at the same time as the formation of the Sun and planets. In other words, although the details are sketchy, comets are probably a natural by-product of the solar system's origin. They may be remnants of the formation process, and their material may be little altered from the era of condensation to the present time. *SEE SOLAR SYSTEM*.

The consensus scenario for the origin is as follows. Generally accepted models of the solar nebula have temperature and density conditions suitable for the condensation of cometary materials at solar distances around the orbits of Uranus and Neptune. The condensate could coalesce to produce a ring of comets or cometesimals. Gravitational perturbations by the ma-

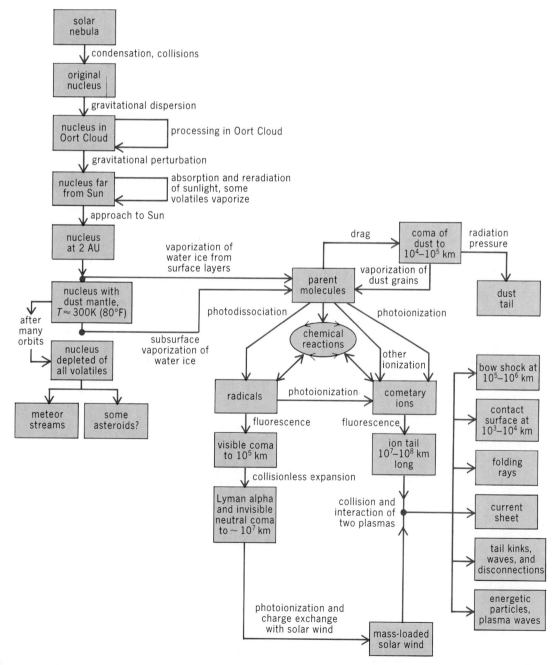

Fig. 17. Summary of processes involved in the formation of comets and the interaction with sunlight and the solar wind. 1 km = 0.6 mi.

jor planets would disperse the ring of comets, sending some into the inner cloud or Oort's Cloud and sending some inward to become short-period comets.

Some scenarios for formation and evolution assign additional roles to the comets. They may have been a major source of the atmospheres of the terrestrial planets and might even have provided the organic molecules necessary for the evolution of life on Earth. While some of these ideas are speculative, they help show why comets are objects of keen interest and active study, and why several space agencies have launched deep-space probes to comets during the 1980s.

Space missions to comets. Interest in sending a spacecraft to a comet was heightened by the appearance of Halley's Comet in 1985 and 1986. The first mission was the diversion of the third *International Sun-Earth Explorer* (*ISEE 3*) from Earth orbit to pass through the tail of Comet Giacobini-Zinner in September 1985. The spacecraft, renamed *International Cometary Explorer* (*ICE*), measured the properties of the tail, including the magnetic field, and the general plasma environment.

For a discussion of the missions to Halley's Comet SEE HALLEY'S COMET. Some of these spacecraft may be diverted to new targets.

Even as the data from the first space missions are under detailed analysis, plans are being made for the next generation of missions. All of the first missions were flybys, and the next step should be rendezvous missions carried out hopefully by the year 2000. Beyond this time period, missions could involve the return to Earth of cometary material. SEE SPACE PROBE.

Comets and the public. There is no scientific evidence for the belief that comets are harbingers, omens, or actual producers of evil, disasters, or natural calamities. Some facts about comets can be distorted to provide an apparent basis for some fears. There are toxic gases in comets, but the amount is very small and, moreover, the cometary gas would not penetrate the Earth's atmosphere. The only real problem would be an impact of a comet on the Earth's surface. The odds of an impact are very small, and studies indicate that the somewhat fragile cometary nucleus would break up high in the Earth's atmosphere. Thus, the Tunguska event, which took place in central Siberia on June 30, 1908, and which produced substantial devastation, is no longer believed to have a cometary origin.

The opportunities for much of the public to view bright comets are becoming increasingly rare due to the combination of atmospheric pollution and bright city lights. In January 1974 there was general public disappointment over the poor visibility of Comet Kohoutek, particularly as viewed from cities in the populous northeastern United States, although it was a conspicuous object to astronomers at mountaintop observatories in the southwestern United States. Halley's Comet was not a spectacular object for public viewing in 1985 and 1986. In part, this was due to the circumstances of the comet's orbit, which led to the comet's times of greatest intrinsic brightness occurring well away from the Earth in contrast to the situation in 1910. The march of light pollution is inexorable, and the interested viewer of comets will require an observing site well away from urban population centers.

John C. Brandt

Bibliography. J. C. Brandt, *Comets: A Scientific American Reader*, 1981; J. C. Brandt and R. D. Chapman, *Introduction to Comets*, 1981; R. D. Chapman and J. C. Brandt, *The Comet Book*, 1984; M. Grewing, F. Praderie, and R. Reinhard (eds.), *Exploration of Halley's Comet*, 1988; The *International Cometary Explorer* Mission to Comet Giacobini-Zinner, *Science*, 232:353–385, 1986; Voyages to Comet Halley, *Nature*, 321:259–365, 1986; F. L. Whipple, *The Mystery of Comets*, 1985; L. L. Wilkening, *Comets*, 1982.

Comfort heating

The maintenance of the temperature in a closed volume, such as a home, office, or factory, at a comfortable level during periods of low outside temperature. Two principal factors determine the amount of heat required to maintain a comfortable inside temperature: the difference between inside and outside temperatures and the ease with which heat can flow out through the enclosure. SEE CONDUCTION (HEAT).

Heating load. The first step in planning a heating system is to estimate the heating requirements. This involves calculating heat loss from the space, which in turn depends upon the difference between outside and inside space temperatures and upon the heat transfer coefficients of the surrounding structural members.

Outside and inside design temperatures are first selected. Ideally, a heating system should maintain the desired inside temperature under the most severe weather conditions. Economically, however, the lowest outside temperature on record for a locality is seldom used. The design temperature selected depends upon the heat capacity of the structure, amount of insulation, wind exposure, proportion of heat loss due to infiltration or ventilation, nature and time of occupancy or use of the space, difference between daily maximum and minimum temperatures, and other factors. Usually the outside design temperature used is the median of extreme temperatures.

The selected inside design temperature depends upon the use and occupancy of the space. Generally it is between 66 and 75°F (19 and 24°C).

The total heat loss from a space consists of losses through windows and doors, walls or partitions, ceiling or roof, and floor, plus air leakage or ventilation. All items but the last are calculated from $H_l = UA(t_i - t_o)$, where heat loss H_l is in British thermal units per hour (or in watts), U is overall coefficient of heat transmission from inside to outside air in $Btu/(h)(ft^2)(°F)$ (or $J/s \cdot m^2 \cdot °C$), A is inside surface area in square feet (or square meters), t_i is inside design temperature, and t_o is outside design temperature in °F (or °C).

Values for U can be calculated from heat transfer coefficients of air films and heat conductivities for building materials or obtained directly for various materials and types of construction from heating guides and handbooks.

The heating engineer should work with the architect and building engineer on the economics of the completed structure. Consideration should be given to the use of double glass or storm sash in areas where outside design temperature is 10°F (-12°C) or lower. Heat loss through windows and doors can be more than halved and comfort considerably improved with double glazing. Insulation in exposed walls, ceilings, and around the edges of the ground slab can usually reduce local heat loss by 50–75%. **Table 1** compares

Table 1. Effectiveness of double glass and insulation*

Heat-loss members	Area, ft² (m²)	Heat loss, Btu/h[†]	
		With single-glass weather-stripped windows and doors	With double-glass windows, storm doors, and 2-in. (5.1-cm) wall insulation
Windows and doors	439 (40)	39,600	15,800
Walls	1952 (181)	32,800	14,100
Ceiling	900 (84)	5,800	5,800
Infiltration		20,800	20,800
Total heat loss		99,000	56,500
Duct loss in basement and walls (20% of total loss)		19,800	11,300
Total required furnace output		118,800	67,800

*Data are for two-story house with basement in St. Louis, Missouri. Walls are frame with brick veneer and 25/32-in. (2.0-cm) insulation plus gypsum lath and plaster. Attic floor has 3-in. (7.5-cm) fibrous insulation or its equivalent. Infiltration of outside air is taken as a 1-h air change in the 14,400 ft³ (408 m³) of heating space. Outside design temperature is −5°F (−21°C); inside temperature is selected as 75°F (24°C).
[†] 1 Btu/h = 0.293 W.

two typical dwellings. The 43% reduction in heat loss of the insulated house produces a worthwhile decrease in the cost of the heating plant and its operation. Building the house tight reduces the large heat loss due to infiltration of outside air. High heating-energy costs may now warrant 4 in. (10 cm) of insulation in the walls and 8 in. (20 cm) or more in the ceiling.

Humidification. In localities where outdoor temperatures are often below 36°F (2°C), it is advisable to provide means for adding moisture in heated spaces to improve comfort. The colder the outside air is, the less moisture it can hold. When it is heated to room temperature, the relative humidity in the space becomes low enough to dry out nasal membranes, furniture, and other hygroscopic materials. This results in discomfort as well as deterioration of physical products.

Various types of humidifiers are available. The most satisfactory type provides for the evaporation of the water to take place on a mold-resistant treated material which can be easily washed to get rid of the resultant deposits. When a higher relative humidity is maintained in a room, a lower dry-bulb temperature or thermostat setting will provide an equal sensation of warmth. This does not mean, however, that there is a saving in heating fuel, because heat from some source is required to evaporate the moisture.

Some humidifiers operate whenever the furnace fan runs, and usually are fed water through a float-controlled valve. With radiation heating, a unitary humidifier located in the room and controlled by a humidistat can be used.

Insulation and vapor barrier. Good insulating material has air cells or several reflective surfaces. A good vapor barrier should be used with or in addition to insulation, or serious trouble may result. Outdoor air or any air at subfreezing temperatures is comparatively dry, and the colder it is the drier it can be. Air inside a space in which moisture has been added from cooking, washing, drying, or humidifying has a much higher vapor pressure than cold outdoor air. Therefore, moisture in vapor form passes from the high vapor pressure space to the lower pressure space and will readily pass through most building materials. When this moisture reaches a subfreezing temperature in the structure, it may condense and freeze. When the structure is later warmed, this moisture will thaw and soak the building material, which may be harmful. For example, in a house that has 4 in. (10 cm) or more of mineral wool insulation in the attic floor,

moisture can penetrate up through the second floor ceiling and freeze in the attic when the temperature there is below freezing. When a warm day comes, the ice will melt and can ruin the second floor ceiling. Ventilating the attic helps because the dry outdoor air readily absorbs the moisture before it condenses on the surfaces. Installing a vapor barrier in insulated outside walls is recommended, preferably on the room side of the insulation. Good vapor barriers include asphalt-impregnated paper, metal foil, and some plastic-coated papers. The joints should be sealed to be most effective. SEE HEAT INSULATION.

Thermography. Remote heat-sensing techniques evolved from space technology developments related to weather satellites can be used to detect comparative heat energy losses from roofs, walls, windows, and so on. A method called thermography is defined as

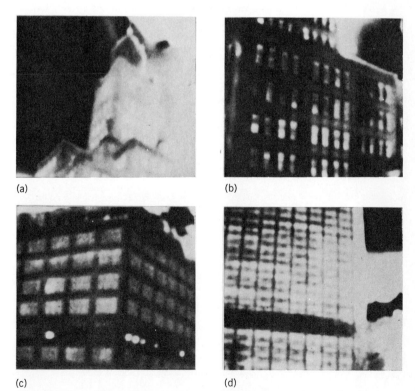

(a)   (b)

(c)   (d)

Thermograms of building structures: (*a-c*) masonry buildings; (*d*) glass-faced building. Black indicates negligible heat loss; gray, partial loss; and white, excessive loss. (*Courtesy of A. P. Pontello*)

the conversion of a temperature pattern detected on a surface by contrast into an image called a thermogram (see **illus.**). Thermovision is defined as the technique of utilizing the infrared radiation from a surface, which varies with the surface temperatures, to produce a thermal picture or thermogram. A camera can scan the area in question and focus the radiation on a sensitive detector which in turn converts it to an electronic signal. The signal can be amplified and displayed on a cathode-ray tube as a thermogram.

Normally the relative temperature gradients will vary from white through gray to black. Temperatures from -22 to $3540°F$ (-30 to $2000°C$) can be measured. Color cathode-ray tubes may be used to display color-coded thermograms showing as many as 10 different isotherms. Permanent records are possible by using photos or magnetic tape.

Infrared thermography is used to point out where energy can be saved, and comparative insulation installations and practices can be evaluated. Thermograms of roofs are also used to indicate areas of wet insulation caused by leaks in the roof.

Infiltration. In **Table 2**, the loss due to infiltration is large. It is the most difficult item to estimate accurately and depends upon how well the house is built. If a masonry or brick-veneer house is not well caulked or if the windows are not tightly fitted and weather-stripped, this loss can be quite large. Sometimes, infiltration is estimated more accurately by measuring the length of crack around windows and doors. Illustrative quantities of air leakage for various types of window construction are shown in Table 2. The figures given are in cubic feet of air per foot of crack per hour.

Design. Before a heating system can be designed, it is necessary to estimate the heating load for each room so that the proper amount of radiation or the proper size of supply air outlets can be selected and the connecting pipe or duct work designed. SEE AIR REGISTER; CENTRAL HEATING AND COOLING; HOT WATER HEATING SYSTEM; OIL BURNER; RADIATOR; STEAM HEATING; WARM AIR HEATING SYSTEM.

Heat is released into the space by electric lights and equipment, by machines, and by people. Credit to these in reducing the size of the heating system can be given only to the extent that the equipment is in use continuously or if forced ventilation, which may be a big heat load factor, is not used when these items are not giving off heat, as in a factory. When these internal heat gain items are large, it may be advisable to estimate the heat requirements at different times during a design day under different load conditions to maintain inside temperatures at the desired level.

Cost of operation. Design and selection of a heating system should include operating costs. The quantity of fuel required for an average heating season may be calculated from

$$F = \frac{Q \times 24 \times DD}{(t_i - t_o) \times Eff \times H}$$

where F = annual fuel quantity, same units as H
Q = total heat loss, Btu/h (or J/s)
t_i = inside design temperature, °F (or °C)
t_o = outside design temperature, °F (or °C)
Eff = efficiency of total heating system (not just the furnace) as a decimal
H = heating value of fuel
DD = degree-days for the locality for 65°F (19.3°C) base, which is the sum of 65 (19.3) minus each day's mean temperature in °F (or °C) for all the days of the year

If a gas furnace is used for the insulated house of Table 1, the annual fuel consumption would be

$$F = \frac{56,500 \times 24 \times 4699}{[75 - (-5)] \times 0.80 \times 1050}$$
$$= 94,800 \text{ ft}^3 \ (2684 \text{ m}^3)$$

For a 5°F (3°C), 6–8-h night setback, this consumption would be reduced by about 5%. SEE THERMOSTAT.

Gayle B. Priester

Bibliography. American Society of Heating, Refrigerating, and Air Conditioning Engineers, *Handbook of Fundamentals*, 1977; A. P. Pontello, Thermography: Bringing energy waste to light, *Heat. Piping Air Condit.*, 50(3):55–61, 1978.

Comfort temperatures

Air temperatures adjusted to represent human comfort or discomfort under prevailing conditions of temperature, humidity, radiation, and wind. Theoretical formulas attempt to compare the rate of heat loss to surroundings with rate of heat production by work and metabolism. Most modern empirical relations, based on relative comfort expressed by human subjects under differing atmospheric combinations, attempt to indicate the temperature at which air at some standard humidity, air motion, and radiation load would be just as uncomfortable (or comfortable). Many former indices, however, had arbitrary scales.

Heat is produced constantly by the human body at a rate depending on muscular activity. For body heat balance to be maintained, this heat must be dissipated by conduction to cooler air, by evaporation of perspiration into unsaturated air, and by radiative exchange with surroundings. Air motion (wind) affects the rate of conductive and evaporative cooling of skin, but not of lungs; radiative losses occur only from bare skin or clothing, and depend on its temperature and that of surroundings, as well as sunshine intensity.

Table 2. Infiltration loss with 15-mi/h (24 km/h) outside wind

Building item	Infiltration, ft³/(ft)(h)
Double-hung unlocked wood sash windows of average tightness, non-weather-stripped including wood frame leakage	39
Same window, weather-stripped	24
Same window poorly fitted, non-weather-stripped	111
Same window poorly fitted, weather-stripped	34
Double-hung metal windows unlocked, non-weather-stripped	74
Same window, weather-stripped	32
Residential metal casement, 1/64-in. (0.4-mm) crack	33
Residential metal casement, 1/32-in. (0.8-mm) crack	52

As air temperatures approach body temperature, conductive heat loss decreases and evaporative loss increases in importance. Hence, at warmer temperature, humidity is the second-most important atmospheric property controlling heat loss and hence comfort, and the various sensible temperature formulas incorporate some humidity measure. *See* BIOMETEOROLOGY.

Temperature and humidity. Almost a century ago, wet-bulb temperature was thought to indicate human comfort under warm conditions, especially in deep mines, ship engine rooms, and steel mills and other factories, on the assumption that sweaty skin approximated a wet bulb. It was superseded after 1923 by the effective temperature, which combined dry- and wet-bulb temperatures, at constant low-ventilation rate, as evaluated by 300 trained subjects walking between controlled-environment rooms. In the 1950s this was approximated by the discomfort index, soon renamed temperature-humidity index (THI) and used by the Weather Bureau (now National Weather Service) for several years. In Canada it was called humidity index. It is calculated by either Eq. (1) or Eq.

$$THI = 15 - 0.4\,(t_a + t_w) \qquad (1)$$

(2), where the dry-bulb (air) and wet-bulb tempera-

$$THI = t_a - 0.55(1 - RH)(t_a - 58) \qquad (2)$$

tures—t_a and t_w respectively—are expressed in °F, and relative humidity (RH) is expressed as a decimal fraction. Except in saturated air (RH = 1.0, that is, 100%), the temperature-humidity index is usually lower than air temperature. A different temperature-humidity index was used in 1980 by another National Weather Service office to compute a heat wave index, ranging from 0 (mild) to 15 (extreme).

A simple average of air temperature in °F and relative humidity in percent was called humiture in 1937, changed in 1959 to a simple average of dry- and wet-bulb temperatures in °F. Later, this was further changed to an average of air temperature in °F and excess vapor pressure e, above 10 millibars, with the saturation vapor pressure at 45°F. This index is called humidex in Canada. The originator of humidex opposed raising the threshold to 21 mb (saturation at 65°F), which was being used by at least one television weatherperson. No two of these formulations are equivalent, but each is usually less than air temperature. *See* AIR TEMPERATURE; HUMIDITY; PSYCHROMETRICS.

Wind and radiation. Meanwhile, more refined measurements led to more elaborate formulas involving all aspects of heat loss. Operative temperature modifies effective temperature to include radiation at constant wind speed, and standard operative temperature considers also temperature and heat conduction of skin. The 1955 heat stress index, essentially the ratio of evaporation needed to maintain a stable body temperature to the evaporation possible under existing conditions and working rate, has been evaluated by a detailed computer program. The 1979 apparent temperature, a very complete formulation, is calculated by use of a nomograph or computer as the heat index of the National Oceanic and Atmospheric Administration (see **illus.**). It is a nonlinear combination of air temperature and relative humidity, adjusted for wind.

A less detailed graphic computation yields a similar but not equivalent value known as humisery. Another

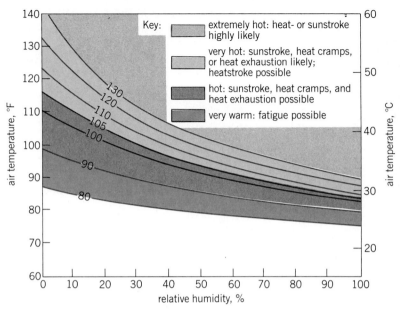

Heat index used by the National Weather Service, National Oceanographic and Atmospheric Administration. The numerical values on each curve represent a heat index, that is, apparent temperature in °F.

approach uses the shaded comfort-factor thermometer, which is a large bulb half-covered by a wetted cloth so that air temperature, wet-bulb temperature, wind, and radiation are integrated into a single number similar to a Fahrenheit temperature; most people are said to be comfortable at values from 63 to 72.

Cold conditions. Under cold conditions, atmospheric moisture is negligible and wind becomes important in heat removal. Wind chill, given by Eq. (3), where H represents heat loss in kcal/(m²)(h²), does not estimate sensible temperature as such, but rather heat loss in kcal/(m²)(h) for wind speed v in m/s and air temperature t in °C.

$$H = (10.45 + 10\sqrt{v} - v)(33 - t) \qquad (3)$$

The formula was based on the rate of freezing, at various temperature-wind combinations during the Antarctic winter, of water in a plastic cylinder, not on responses of human subjects. It has never been validated by the latter approach. Nevertheless, the temperature of which the estimated heat loss at some reference wind speed equals that at the existing temperature-wind combination is the wind chill temperature. The reference wind speed is variously 3 mi/h, 3 knots, or 4 mi/h, and in a revised formulation is 0.

Arnold Court

Bibliography. P. O. Fanger, *Thermal Comfort*, 1972, reprint 1982; H. Landsberg (ed.), *World Survey of Climatology*, vol. 3: *General Climatology*, 1981.

Commelinales

An order of flowering plants, division Magnoliophyta (Angiospermae), which is included in the subclass Commelinidae in the class Liliopsida (monocotyledons). It consists of 4 families and about 1000 species, the bulk of which is in the families Commelinaceae (about 700 species) and Xyridaceae (about 200

species). The order is marked by its flowers that ordinarily have the perianth well differentiated into sepals and petals but do not have nectaries or nectar. The ovary is consistently superior and the fruit capsular. The wandering Jew (species of *Tradescantia* and *Zebrina* in the Commelinaceae) belongs to this order. *See* COMMELINIDAE; LILIOPSIDA; MAGNOLIOPHYTA; PLANT KINGDOM.

Arthur Cronquist; T. M. Barkley

Commelinidae

A subclass of the class Liliopsida (monocotyledons) of the division Magnoliophyta (Angiospermae), the flowering plants, consisting of 7 orders, 16 families, and nearly 15,000 species. The orders include Commelinales, Eriocaulales, Restionales, Juncales, Cyperales, Hydatellales, and Typhales. For further information see separate articles on each order.

These monocotyledons are syncarpous (the carpels are united in a compound ovary) or pseudomonomerous (reduced to a single carpel from a syncarpous ancestry). The endosperm is usually starchy, and the perianth is either well differentiated into sepals and petals or more or less reduced and not petallike. The stomates have two or more subsidiary cells, the pollen is either binucleate or more often trinucleate, and the endosperm may be nuclear. Many of the families have well-developed vessels in all vegetative organs. Several of the orders of Commelinidae have often been treated as a single order Farinosae or Farinales. *See* LILIOPSIDA; MAGNOLIOPHYTA; PLANT KINGDOM.

Arthur Cronquist; T. M. Barkley

Common cold

A viral disease of humans, most frequently caused by the rhinoviruses, a subgroup of the picornavirus group. However, other viruses (including parainfluenza, influenza, and respiratory syncytial viruses and members of the adenovirus, enterovirus, and reovirus groups) also may cause common cold symptoms. *See* ANIMAL VIRUS.

The virus enters via the nose or mouth. After an incubation period of 2–5 days, typical cold symptoms and histopathology of mucous membranes develop, and continue for 4–6 days. Experiments under controlled conditions have not shown that chilling, or wearing wet socks, produces the cold or increases susceptibility to the virus. Chilliness is an early symptom of the common cold.

The common cold is a worldwide disease. The average individual in temperate zones has two to four attacks per year, usually during the winter or spring. Persons in isolated areas have fewer colds, but have more severe illness and a higher incidence of infection when a cold is introduced than do those in areas where these viruses circulate regularly. Bacterial vaccines, antihistamines, and attempts at air sterilization have not been successful in preventing colds. *See* ADENOVIRIDAE; ENTEROVIRUS; PARAINFLUENZA VIRUS; RHINOVIRUS.

Joseph L. Melnick

Bibliography. C. H. Andrewes and D. A. J. Tyrrell, Rhinoviruses, in F. L. Horsfall, Jr., and I. Tamm (eds.), *Viral and Rickettsial Infections of Man*, 4th ed., 1965; G. J. Galasso et al. (eds.), *Antivirals and Virus Diseases of Man*, 1979; D. Hamre, Rhinoviruses, *Monographs in Virology*, vol. 1, 1968; S. E. Luria et al., *General Virology*, 3d ed., 1978; E. Lycke and E. Norrby (eds.), *Textbook of Medical Virology*, 1983.

Communications cable

A cable that transmits information signals between geographically separated points. The heart of a communications cable is the transmission medium, which may be optical fibers, coaxial conductors, or twisted wire pairs. A mechanical structure protects the heart of the cable against handling forces and the external environment. The structure of a cable depends on the application.

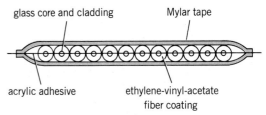

Fig. 1. Cross section of optical fiber ribbon (12 fibers).

Optical communications cables. Figures 1 and **2** show an optical communications cable for terrestrial use, where they may be installed aerially, or by direct burial, or in protective ducts. The cable used in the SL Undersea System is illustrated in **Fig. 3.** These cables were designed by Bell Laboratories for use in local and long-distance communications.

The terrestrial and undersea cables differ in several respects. First, the terrestrial cable requires only enough longitudinal strength to allow installers to pull

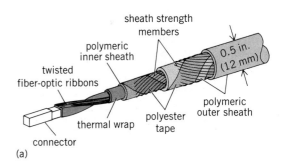

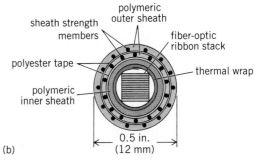

Fig. 2. Multifiber cable for terrestrial use (144 fibers maximum). (*a*) Cutaway view. (*b*) Cross section.

the cable into ducts, or lay it in a trench. For the undersea cable, the high-strength steel strand allows it to be laid and recovered in ocean depths up to 4.5 mi (7315 m). A second difference is that the undersea cable must deliver as much as 1.6 amperes of direct current to the spaced, series-connected repeaters. Repeaters for the terrestrial system are generally located in buildings, and are powered from locally available sources. A third difference is due to the environment. Terrestrial cables are often installed in ducts. Hence, only limited protection is required against local hazards. The undersea cable, on the other hand, must withstand moderately severe handling in laying, and it sustains the full pressure of ocean depths. At 4.5 mi (7315 m) depth, this pressure is 10,700 lb/in.^{2} (73.8 megapascals). To withstand this pressure, the cable is a filled design (no voids). The thick polyethylene outer layer provides electrical insulation for the power-carrying copper conductor, and is tough enough to resist severe handling. In addition to providing longitudinal strength, the steel strand overlaid with copper acts as a pressure cage to protect the core and fibers. *See Electrical insulation; Polyolefin resins; Submarine cable.*

Optical fibers. Signals in these cables are carried by light pulses which are guided down the optical fiber. In most applications, two fibers make up a complete two-way signal channel. The guiding effect of the fiber confines light to the core of the glass fiber and prevents interference between signals being carried on different fibers. The guiding effect also delivers the strongest possible signal to the far end of the cable. Exceptionally pure silica glass in the fiber minimizes light loss for signals passing longitudinally through the glass fiber. Developers have realized fiber losses of 0.40 dB or less per statute mile (0.25 dB per kilometer), at a wavelength of 1550 nanometers. For such a fiber, the signal can travel 7.5 mi (12 km) before losing half its power, and can travel 50 mi (80 km) or more before requiring electronic regeneration. Coaxial cable systems require regeneration at much closer intervals. In the L5 and T4M Coaxial Cable systems, for example, signals are amplified or regenerated every mile (every 1.6 km). *See Optical fibers.*

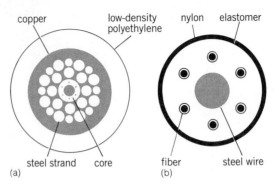

Fig. 3. Optical-fiber cable for undersea systems. (*a*) Cross section. (*b*) Detail of cross section of core.

Signal regeneration. Optical cable systems are usually digital. Thus, information is coded into a train of off-or-on light pulses. These are detected by a photodetector at the far end of a cable span and converted into electronic pulses which are amplified, retimed, recognized in a decision circuit, and finally used to drive an optical transmitter. In the transmitter, a laser or light-emitting diode converts the electric signals back into a train of light pulses which are strong enough to traverse another cable span. By placing many spans in tandem, optical cable systems can carry signals faithfully for thousands of miles. *See Light-emitting diode; Photodetector.*

Figure 4 shows an electronic regenerator for the SL Undersea Cable System. This regenerator functions in TAT-8 (the eighth United States–Europe transatlantic cable), which entered service in 1988. Six regenerators are mounted in a repeater housing. Thus, the repeater regenerates signals in three pairs of fibers. **Table 1** presents a summary of the characteristics of the TAT-8 cable system.

Optical communications cables are often used to carry input and output data to computers, or to carry such data from one computer to another. Then they are generally referred to as optical data links. The links are often short enough that intermediate regeneration of the signals is not needed.

Repeaterless undersea cables. The trend toward lower fiber loss will allow the linking of mainland to offshore islands without undersea electronics and op-

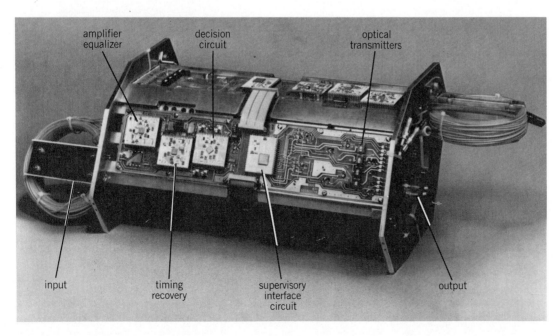

Fig. 4. Regenerator for SL optical undersea cable communications system.

Table 1. Principal characteristics of the eighth United States-to-Europe undersea cable system

Characteristic	Value
Line rate	295.6 megabits/s
Number of service fibers	4
Number of protection fibers	2
Regenerators per repeater	6
Number of voice channels	8000
Number of conversations (with digital circuit multiplication equipment)	40,000
Optical wavelength	1310 nm

Table 2. Repeaterless cable system

Characteristic	Value
System length	93 mi (150 km) maximum
Fiber type	SL, single-mode
Fiber proof stress	200,000 lb/in.^{2} (1.38 GPa)
Cable design	SL, embedded core
Optical wavelength	1.55 μm
Optical source	Buried heterostructure laser
Optical detector	*pin* diode
Line data rate	3.088 megabits/s
Line coding	Dipulse

tics. Cable systems of up to 93 mi (150 km) are feasible, with future capability trending toward 125–150 mi (200–250 km). Parameters for a 93-mi (150-km) repeaterless cable system are shown in **Table 2**.

Advantages. Optical communications cables have four key advantages over earlier metallic cables: much greater repeater spacings; enormous information capacity (many individual voice, data, or television channels); immunity from outside interference (other signals or noise); and smaller size and lower weight.

Because of these advantages and because optical cable technology is in its infancy, many additional applications can be expected to develop, and the cost of such communications cables can be expected to continue to decrease. Such communications cables require the conversion of the signal from optical to electronic mode and from electronic to optical at each regenerator. As the cost of lasers and optical detectors decreases, this complication will present less of a cost penalty. It also may become possible to achieve regenerators which process the signals optically. Such regenerators would eliminate the need for optoelectronic conversion, except at the ends of the entire link. *See* OPTICAL COMMUNICATIONS.

Coaxial cable communications systems. Coaxial communication systems evolved before optical systems. Most of these systems are analog in nature. Signals are represented by the amplitude of a wave representing the signal to be transmitted. In a multichannel system, each voice, data, or picture signal occupies its unique portion of a broadband signal which is carried on a shared coaxial conductor or "pipe." In the transmitting terminal, various signals are combined in the frequency-division transmitting multiplex equipment. At the receiving end of a link, signals are separated in the receiving demultiplex equipment. This combining and separation operates much as broadcast radio and television do, and the principles are identical. *See* AMPLITUDE MODULATION; ELECTRICAL COMMUNICATIONS; FREQUENCY MODULATION.

In order to carry as many channels as possible, the multiplexing and demultiplexing use single-sideband (SSB) transmission. The fixed carrier signals, which do not convey useful information, are suppressed in the transmitting multiplex and reintroduced in the receiving multiplex. This avoids unnecessary load on the system. *See* SINGLE SIDEBAND.

Terrestrial cables. The channel capacity of terrestrial coaxial systems has increased dramatically over the

Fig. 5. Multicoaxial cable used in the L5E communications system.

Table 3. L-system characteristics

Characteristic	L1	L3	L4	L5	L5E
Service date	1946	1953	1967	1974	1976
Repeater spacing, mi (km)	8 (12.9)	4 (6.4)	2 (3.2)	1 (1.6)	1 (1.6)
Voice-band channel capacity per ⅜-in. (9.5-mm) coaxial pair	600	1860	3600	10,800	13,200
Coaxial pairs	4	6	10	11	11
Working pairs	3	5	9	10	10
Channel capacity	1800	9300	32,400	108,000	132,000

years to meet increased long-distance calling. This has resulted in decreased cost per channel, which has stimulated still further demand. While demand was the driving force, it is improving technology which has allowed such tremendous increases in capacity. This historical evolution is summarized in **Table 3**.

The L1 and L3 systems used electron tubes as the amplifying device, while all later systems used transistors. The L5 system benefited by the use of thin-film integrated circuits. These advances were responsible for the 22-fold increase in the number of channels per coaxial pair, as shown in Table 3. Because the number of coaxial pairs per cable was also increased, the total route capacity was increased by a factor of 73 over the 30-year period covered in Table 3. **Figure 5** shows the 11-coaxial-pair cable used in the L5 and L5E systems. The overall diameter of this cable is 3 in. (76 mm). SEE ELECTRON TUBE; INTEGRATED CIRCUIT; TRANSISTOR.

Table 3 shows how repeater spacing decreases as the capacity rises. Multiplying channel capacity by 22 has meant going to higher frequencies, since each additional channel requires another 4 kHz of spectrum. At higher frequencies the cable has higher loss, and therefore such signals must be amplified at closer intervals.

In all of the above L-carrier systems, one of the coaxial pairs is reserved for protection of the working pairs. Since transfer to the protection pair is automatic upon failure of a working pair, this arrangement provides exceptionally high system reliability.

Since the various generations of L-carrier systems use the same basic coaxial unit, replacement of earlier electronics with the latest design can yield dramatic increases in channel capacity. As an example, if a cable with three working pairs of coaxials is upgraded from L-3 to L-5E, the capacity increases from 5580 to 39,600 channels. Since the cable and right of way are already established, this is a very economical method of enhancing the capacity of a route.

Undersea cables. The history of undersea communications goes back to 1858, when the first transatlantic telegraph cable was completed by C. W. Field. This cable had no repeaters and thus was capable of carrying only a single slow-speed telegraph signal. Nevertheless, it allowed news to be communicated between North America and Europe without the long delay of surface mail. While telegraph cables were improved over the years, the next major breakthrough occurred in 1956, when the first transatlantic telephone cable was completed from Nova Scotia to Newfoundland, and then across the Atlantic to Scotland.

With the completion of TAT-1 (Transatlantic Telephone-1), callers could for the first time talk across the Atlantic Ocean as easily as they could within the continents. The response was a dramatic increase in use of overseas calling. Since 1956, cables have been installed throughout the world. England, France, Japan, and the United States have been active in designing and installing undersea communications cables.

Low delay is one respect in which undersea cables have a significant edge over satellites in a geosynchronous orbit. Delay is usually evaluated in terms of round-trip message time. For a telephone conversation, it is the extra time added between question and answer because of the transmission time. For the satellite connection, this delay is about 0.5 s. For a

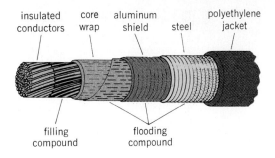

insulated conductors · core wrap · aluminum shield · steel · polyethylene jacket · filling compound · flooding compound

Fig. 6. Waterproof cable design.

4000-statute-mile (6400-km) undersea cable link, delay is only 0.06 s. Longer delay has two effects: it makes echo more bothersome to voice communications, and it requires more complex and less efficient protocols for data communications.

Traffic growth on the Atlantic route has exceeded 20% per year from the time of the first cable. In response to this rapid growth, AT&T Bell Laboratories designed five generations of communications cables. A summary of this technological evolution is presented in **Table 4**. SEE COAXIAL CABLE.

Multipair communications cables. These cables may contain from 6 to 4200 insulated twisted wire pairs in a protective sheath. Earlier cable designs used extruded lead sheaths. In new installations, sheaths using composites of metal and extruded plastic dominate the market. The wire pairs, which were insulated with paper pulp in early practice, are now more commonly insulated with extruded plastic. The conductors are most frequently copper, although aluminum is sometimes used for cost reasons. SEE CONDUCTOR (ELECTRICITY).

Development. Following the invention of the telephone in 1876, the first telephone systems used open wire lines. These lines differed from the earlier tele-

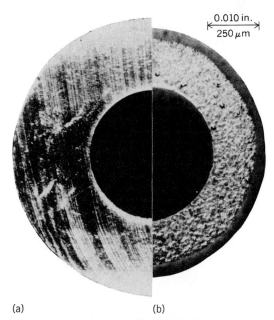

(a)   (b)

Fig. 7. Comparison of (a) solid plastic insulation with (b) dual-layer expanded insulation (0.002-in. or 50-µm solid skin over 45% expanded core) showing both the structure and the smaller diameter of the dual-layer insulation compared with the solid insulation. Transmission properties are equivalent. (Bell Telephone Laboratories)

Table 4. Undersea cable systems from the United States to Europe				
System designation	First service	Number of channels	Top frequency	Active devices
TAT-1	1956	36	0.164 MHz	Electron tubes
TAT-3	1963	138	1.1 MHz	Electron tubes
TAT-5	1970	840	6.0 MHz	Germanium transistors
TAT-6	1976	4200	30.0 MHz	Silicon transistors
TAT-8*	1988	8000*	295.6 megabits/s	Lasers, *pin* diodes, silicon integrated circuits

*With digital circuit multiplication equipment, capability is 40,000 conversations. This is the first digital transatlantic light-wave system using optical-fiber communications cable.

graph systems, which used a single wire with an earth return to complete the circuit. Such an arrangement would have produced intolerable noise and crosstalk at the higher frequencies (3500 Hz) used for telephony. Hence, open wire lines with a pair of spaced conductors for each circuit were employed. The two conductors were transposed periodically to further reduce noise and crosstalk. As the telephone grew in popularity, telephone poles became taller and carried more crossarms to hold the large number of open wire pairs. *SEE CROSSTALK; ELECTRICAL NOISE.*

At this point, the multipair cable was introduced, providing a flexible means of carrying large numbers of circuits. Within the cables, wires are grouped in bundles of 25 pairs. Each pair and bundle is individually color-coded.

Types and uses. Paired cables are often classified by application into three classes: exchange area cable, toll cable, and building or switchboard cable.

Exchange area cable connects the telephone subscriber to the central office, or one central office to another in the same area. Pulp paper insulation used in older cables has given way to plastic or expanded plastic insulation. Most new cables are buried rather than pole-mounted. Within the exchange area, subscriber loop carrier (SLC) systems are being used to expand the capacity of existing cables and to avoid the need for installing new cables and ducts.

Toll cable connects central offices in different areas. As indicated above, optical communications and such means as L-carrier and microwave radio now carry a higher fraction of intercity traffic. In addition, T-carrier systems have been applied to existing multipair cables. Application of T-1, for example, provides 24 circuits over two pairs of conductors. Where carrier systems are applied, requirements of crosstalk, balance, and circuit impedance are more severe than for voice-frequency service.

Building or switchboard cable is for inside use, where fire resistance is of great importance. Because of this, polyvinyl chloride (PVC) is used for pair insulation and sheath. From a transmission point of view, PVC is inferior to paper or polyethylene, because its higher dielectric constant results in a cable with greater loss per unit length. This higher loss is generally not important because cable runs are relatively short. *SEE POLYVINYL RESINS.*

Pressure maintenance and waterproof cable. Lead-sheathed and other nonwaterproof cables are usually maintained by placing the entire inside of the cable under 6–10 lb/in.^{2} (40–70 kilopascals) dry gas pressure. Each section may have a flow meter or pressure sensor whose information appears in the central office. This system quickly identifies a leak in the pressure-filled section before an electrical failure occurs. Maintenance personnel can pinpoint the sheath failure and repair it before moisture damages the pairs.

To further ease maintenance, waterproof cables evolved (**Fig. 6**). The cable interstices are flooded with "petroleum-based grease" which prevents water penetration, even in the face of a damaged sheath. These waterproof cables do not require gas pressure and have greatly reduced problems due to moisture.

As cables have gone underground, they have been better protected against storm damage but more continuously exposed to moisture. Hence, there is a strong preference for filled waterproof cable.

Wire pair insulation. Early cables used paper pulp for insulation. To be effective, this insulation had to be kept dry. If the lead sheath or end seals were damaged, moisture could penetrate, causing noise and ultimately failure of the cable.

In the 1950s, high-speed plastic extrusion was first used to apply wire insulation. Thus, pulp gave way to polyethylene as an insulating material. Polyethylene has the advantage of excellent dielectric properties (high insulation strength and low dielectric constant).

The "grease" with which interstices of waterproof

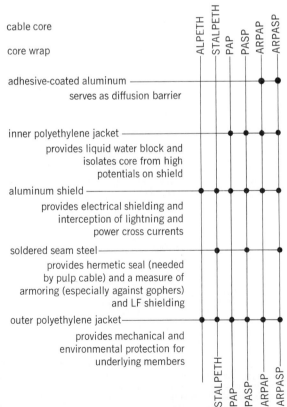

cable core

core wrap

adhesive-coated aluminum —
 serves as diffusion barrier

inner polyethylene jacket —
 provides liquid water block and
 isolates core from high
 potentials on shield

aluminum shield —
 provides electrical shielding and
 interception of lightning and
 power cross currents

soldered seam steel —
 provides hermetic seal (needed
 by pulp cable) and a measure of
 armoring (especially against gophers)
 and LF shielding

outer polyethylene jacket —
 provides mechanical and
 environmental protection for
 underlying members

ALPETH STALPETH PAP PASP ARPAP ARPASP

Fig. 8. Representative sheath designs. The choice of layers incorporated in a particular design depends upon the type of cable core and environmental requirements.

cables are filled has a dielectric constant of about 2.3. This grease replaces the air, which has a dielectric constant of 1.0. The result is an increase in the total capacitance of each pair, and hence an undersirable increase of cable loss. To overcome this effect, composite wire insulations were developed (**Fig. 7**). This design uses expanded plastic (about 50% air bubbles) to replace most of the insulation. A second, 50-μm-thick (0.002-in.) solid plastic layer gives moisutre protection and provides a carrier which can be color-coded for easy pair identification. The two layers are applied by tandem extrusion in a single insulation line. The dual-layer insulation allows the waterproof cable to have the same low-loss characteristics as the earlier air-filled cable, with no increase in outside cable diameter and minimal change in the amount of raw materials required.

Sheaths. Lead is the sheath material of all early cables. It provides an excellent moisture barrier and good corrosion resistance. Since it can be extruded at low temperatures, it can be placed over the paper-wrapped core of pairs without damaging them. However, the ends of a lead-sheathed cable must be sealed.

The development of newer materials, plus lead shortages, ultimately led to the evolution of composite sheaths, in which two or more layers provide the cable protection. One such design is shown in Fig. 6; the aluminum shield, the corrugated steel layer, and the polyethylene jacket replace the lead jacket.

The use of composite layers, which can be applied in a single in-line operation, gives great flexibility. Cables can be fabricated to meet widely differing environments. **Figure 8** represents some of the designs which have been manufactured. The particular designs are indicated by the column headings—acronyms which indicate the layers included in the particular design. Where there is a dot at a horizontal intersection, it indicates that this particular layer is included in that design. For example, the design shown as waterproof cable in Fig. 6 is a STALPETH design: It includes layers of steel, aluminum, and polyethylene.

Installation methods. Earlier cables were usually installed aerially on poles, except in cities, where they were pulled into ducts. The vast majority of new cables are placed underground. In cities and larger towns, duct installation dominates. In smaller communities, cable may be installed in ducts or buried directly. SEE TELEPHONE SERVICE; TELEPHONE SYSTEMS CONSTRUCTION.

S. T. Brewer

Bibliography. M. K. Barnoski (ed.), *Fundamentals of Optical Fiber Communications*, 1976; Bell Telephone Laboratories, *Transmission Systems for Communications*, 5th ed., 1982; R. D. Ehrbar, Undersea cables for telephony, *Commun. Mag.*, 21(5):18–27, August 1983; S. E. Miller and A. G. Chynoweth (eds.), *Optical Fiber Telecommunications*, 1979; *1980 World's Submarine Telephone Cable Systems*, U.S. Department of Commerce, NTIA-CR-80-6.

Communications satellite

A spacecraft placed in orbit around the Earth to receive and retransmit radio signals. Communications satellites amplify and sort or route these signals. Their function is similar to that of a ground microwave repeater but with greatly increased coverage. Whereas a ground repeater relays signals between two fixed locations, a communications satellite interconnects many locations, fixed and mobile, over a wide area.

Development. In 1945, advances in rocket and microwave engineering which had occurred during World War II led Arthur C. Clarke to propose the use of radio repeaters (receivers and transmitters) on artificial satellites in geosynchronous equatorial (geostationary) orbit. By 1954, both passive and active communications satellites had been studied in some detail. The space age began on October 4, 1957, with *Sputnik 1* (strictly speaking, not a communications satellite), whose transmissions at 20 and 40 MHz were received on Earth. In December 1958, a human voice was heard from space for the first time. President Eisenhower's Christmas message, transmitted by an earth station and received by the *SCORE* satellite, was recorded on board and played back. From 1960 to 1964, several communications satellites, both passive (*Echo 1* in 1960 and *Echo 2* in 1964, which were metallized Mylar balloons) and active (*Courier* in 1960, *Telstar* in 1962 and 1963, and *Relay* in 1962 and 1964), were placed in low-altitude (less than 600 mi or 1000 km) or medium-altitude orbits (less than 6000 mi or 10,000 km). Satellite orbital height and payload mass were limited by the modest capabilities of the early rockets. Passive satellites that scatter back to Earth only a minute fraction of the energy radiated by earth stations provide very limited communications capacities. Thus, after reliable long-life space-qualified receivers and transmitters were developed, only active satellites, which can provide much higher communications capacity, were used.

After the first unsuccessful attempt to place a communications satellite in geostationary orbit (*Syncom 1* in February 1963), the second and third launches (*Syncom 2* in July 1963 and *Syncom 3* in August 1964) were successful. In April 1965, INTELSAT I (*Early Bird*), the first geostationary commercial communications satellite, was launched. It provided 240 high-quality telephone circuits and the capability of television program exchange between western Europe and northeastern North America.

Uses. Geostationary commercial communications satellites carry about three-fourths of the long-distance international telephone traffic. In addition to telephone services (which constitute the bulk of the traffic), television, data, facsimile, and electronic mail services are offered. Worldwide television would be impossible without satellites, because no other wideband transmission system exists. SEE DATA COMMUNICATIONS; ELECTRONIC MAIL; FACSIMILE; TELEPHONE SERVICE; TELEVISION; TELEVISION NETWORKS.

In addition to fixed-point services, satellites provide worldwide ship-to-shore commercial maritime communications of high quality and reliability. The use of satellites in other mobile communications systems, such as ground-to-aircraft and land vehicles, is being developed.

Broadcasting via satellites to individual homes and community antenna television (CATV) cable heads has been in use since 1983 by medium-power satellites; high-power satellites are creating a new industry. SEE DIRECT BROADCAST SATELLITE SYSTEMS.

Satellites are also widely used for military communications between fixed stations and mobile terminals on ships, airplanes, and land vehicles. SEE MILITARY SATELLITES.

Basic configuration. Satellites represent the most recent and significant step in the evolution of radio

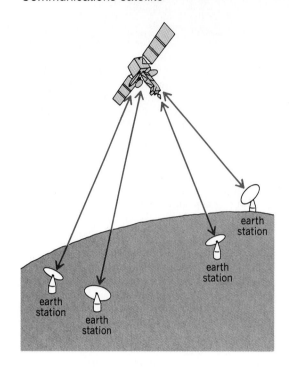

Fig. 1. Configuration of a satellite communications system.

of an active communications satellite. The spacecraft carries the power subsystem, station-keeping and orientation devices, and the useful payload, which is the communications subsystem. Wideband linear receivers amplify the uplink signals. After a process of frequency conversion, the signals are further amplified in separate channels (to minimize intermodulation noise) and fed to the downlinks. Transmitter power output of individual channels falls in the range 5–20 W in satellites for fixed services. Traveling-wave tubes are mostly used as power amplifiers, but solid-state power amplifiers based upon field-effect transistors are competitive at 4 GHz and can be expected to become competitive at 12 GHz. Broadcasting satellites, serving a multitude of users with small, inexpensive receivers, require power up to a few hundred watts provided by traveling-wave-tube amplifiers. *See* AMPLIFIERS; MICROWAVE SOLID-STATE DEVICES; TRANSISTOR; TRAVELING-WAVE TUBE.

Satellites and earth stations are interactive elements tied together by the transmission system (multiplexing, modulation and coding, and multiple access) in use. The type of traffic, topology of the network, spectrum allocations, transmission methods, characteristics of the devices used, and economic considerations are important factors in the overall system design.

Antennas. Antennas play an important role in satellite communications systems, both on the ground and on the spacecraft. With the early small satellites, which were very power-limited and used simple low-gain antennas, large antennas (up to 100-ft- or 30-m-diameter paraboloids) were necessary at the earth stations.

Spacecraft antennas were small and exhibited little directivity. In the case of dipoles aligned with the spacecraft spin axis, a toroidal radiation pattern was produced, with the resultant waste of energy outside the 18° angle subtending the Earth from geostationary orbit. In 1968, mechanical despun antennas, consisting of a conical horn and a reflector, providing about 19 dB of gain, were introduced to provide Earth coverage. Progress has continued with parabolic reflector antennas providing spot beam coverage down to a few degrees. Through frequency reuse, the allocated frequency bands are utilized many times over by means of orthogonal polarizations (vertical–horizontal or clockwise–counterclockwise) or by means of spatially separated beams. Beams have been synthesized to follow the contours of geographical areas such as continents and national or regional boundaries. By using suitably arranged feed horns, each excited with proper amplitude and phase, the radiated energy impinges upon a reflector and illuminates the desired areas on Earth.

In contour-shaped beams and frequency reuse systems, adequate isolation up to 30 dB must be maintained between dually polarized and spatially separate beams. Until the 1980s, parabolic reflectors with offset feed assemblies in the focal region were adequate. However, when more beams and higher frequency reuse factors are desired, the single-offset reflector design becomes inadequate. Dual-offset reflectors with Gregorian or Cassegrainian feeds provide better control of the illumination and permit the use of larger reflectors without the need for excessive focal lengths. Lens-type antennas or combination of lenses and reflectors or phased arrays and reflectors can also be used. Ultimately, size and mass limitations are encountered. The shape of the payload limits the size of

communications systems, whose progress can be largely attributed to the use of ever-higher carrier frequencies and, consequently, wider signaling bandwidths. Terrestrial ultrahigh-frequency (UHF) and superhigh-frequency (SHF) radio relay systems have high communications capacity, but are range-limited because the quasioptical behavior of microwaves requires line-of-sight visibility between stations. Long overland distances require cascaded links with receivers and transmitters placed on high towers. Clearly, when terminals are separated by oceans upon which towers cannot be erected, the placement of electronic equipment on an orbiting spacecraft provides the solution. *See* BANDWIDTH REQUIREMENTS (COMMUNICATIONS); MICROWAVE.

Figure 1 shows the basic configuration of a satellite communications system, and **Fig. 2** the elements

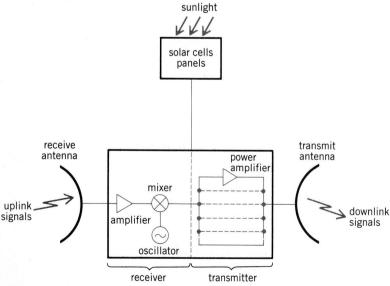

Fig. 2. Elements of an active communications satellite.

fixed antennas. This limitation can be circumvented by using deployable apertures, although structural problems arise in regard to surface tolerances.

The combination of sophisticated spacecraft antennas and the larger amounts of radio-frequency (rf) power available from bigger satellites has permitted the size of the earth station antennas to be reduced, with considerable savings in cost. SEE ANTENNA (ELECTROMAGNETISM).

Transponders. Transponders are microwave repeaters carried by communications satellites. For the configuration shown in Fig. 2, transponders are designated as transparent because any signal whose format can fit in the transponder bandwidth can be handled. No signal processing occurs other than that of heterodyning (frequency changing) the uplink frequency bands to those of the downlinks. Such a satellite communications system is referred to as a bent-pipe system. Connectivity among earth stations, which is maximum with global-coverage antennas, is reduced when multiple narrow beans are used. Hence, the evolution proceeded from the transparent transponder of Fig. 2 to transponders that can perform signal switching and format processing as shown in **Figs. 3** and **4**, respectively.

Choice of orbit. Figure 5 illustrates the relationship between orbital height, period, and great-circle-arc distance which can be covered. Satellites in low- and medium-altitude orbits must be tracked by the transmit-and-receive stations on Earth. The duration of an interconnection is limited to the interval of satellite visibility by the stations to be connected. For continuous communications, several satellites must be orbited so that before one sets on the horizon another rises. To transfer the traffic from one satellite to another, two antennas and additional equipment are required at each earth station. Prior to the advent of geostationary satellites, studies of medium-altitude communications satellite systems showed that 30–50 spacecraft would be needed to provide continuous communications between the United States and Europe.

At an altitude of 22,280 mi (35,860 km) the orbital period corresponds to a sidereal day (23 h 56 min 4 s), and if the plane of the orbit coincides with the equatorial plane the satellite appears geostationary. It hovers at a fixed point with respect to the rotating Earth, being supported by a virtual keplerian–newtonian tower under the balance of gravitational and centrifugal forces. The angle subtended by the Earth, as seen from a geostationary satellite, is about 17°, and the maximum great-circle-arc distance is about 11,000 mi (18,000 km). Thus, with coverage of about two-fifths of the entire Earth's surface from a single satellite; three geostationary spacecraft could, in principle, provide worldwide coverage. SEE ORBITAL MOTION.

For n earth stations visible from its orbital position, the satellite becomes a communications node capable of simultaneously establishing as many as $n(n - 1)/2$ connections. This multiple-access capability is a unique characteristic of satellite systems. The accesses are kept separate in a physical domain such as frequency (frequency-division multiple access, FDMA) or time (time-division multiple access, TDMA), or by coding techniques (code-division multiple access, CDMA). SEE ELECTRICAL COMMUNICATIONS.

Numerous other advantages are inherent in the geostationary orbit. High-energy particles are trapped

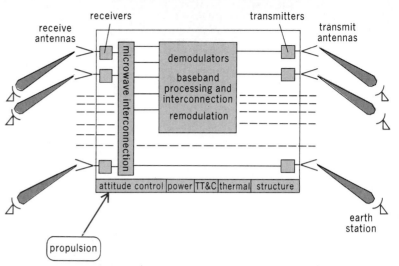

Fig. 3. Satellite transponder with switching at radio frequency. TT&C = telemetry, telecommunications, and control.

by the Earth's magnetic field in the Van Allen belts. As the inner belt extends from 750 to 5500 mi (1200 to 9000 km) and the outer belt from 5500 to about 37,000 mi (9000 to 60,000 km) with peak intensity around 10,000 mi (16,000 km), geostationary satellites are only mildly affected by radiation. Additional advantages of the geostationary orbit are zero Doppler effect, fewer eclipse-induced thermal stresses, reduced energy drain from batteries during eclipses, and reduced perturbations by the magnetic field of the Earth. SEE VAN ALLEN RADIATION.

However, extreme latitudes cannot be covered from the geostationary orbit, a restriction that is not too serious in view of the distribution of the world's population. To cover extreme latitudes, inclined orbits are used. For its domestic system initiated in 1965, the Soviet Union chose orbits inclined 63.5° with respect to the equatorial plane, with perigee at 300 mi (500 km), apogee at 25,000 mi (40,000 km), and orbital period of 12 h (although their satellite communications system includes geostationary satellites as well). For the above-mentioned orbit inclination, no rotation of the line of the apsides (otherwise induced

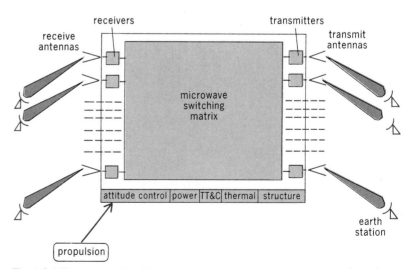

Fig. 4. Satellite transponder with switching at baseband. TT&C = telemetry, telecommunications, and control. (After C. E. Mahle, G. Hyde, and T. Inukai, Satellite scenarios and technologies for the 1990's, IEEE J. Selected Areas Commun., SAC-5:556–570, 1987)

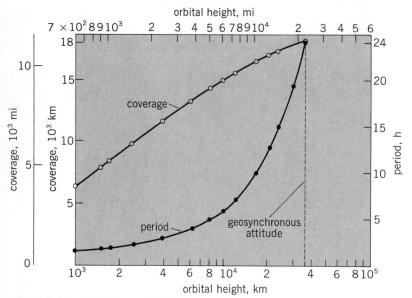

Fig. 5. Relationship between orbital height, period, and coverage (maximum great-circle-arc distance between earth stations) of communications satellites.

by the Earth's oblateness) occurs, and the need for orbit maneuvers and corrections is reduced. It is necessary, however, to track the satellites, and several are required for continuous communications.

The 0.25-s transmission delay encountered with geostationary satellites was regarded by some as a fundamental obstacle, especially in the presence of echo conditions. Echoes are not due to satellites, but are generated in the land-based telephone plant at the points where long-distance four-wire circuits are con-

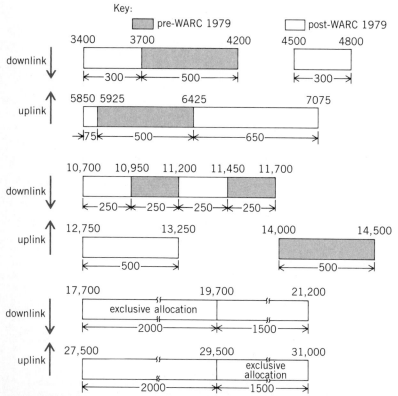

Fig. 6. Fixed-services satellite frequency allocations, in MHz.

nected to local two-wire circuits. When the echo exceeds a certain threshold, it becomes very annoying to the user, and for a given intensity the annoyance is proportional to the transmission delay.

These difficulties were somewhat ameliorated at first by improved voice-activated devices (echo suppressors) used for many decades in long-distance terrestrial telephone lines. They have been virtually replaced with echo cancelers, which generate a negative replica of the echo, resulting in cancellation of the echo as soon as a conversation begins.

Placement in orbit. With disposable rockets, injection of a spacecraft in geostationary orbit typically involves two-stage rockets which place the satellite and an attached third-stage rocket motor into a transfer elliptical orbit with apogee at geosynchronous altitude and a relatively low perigee. The inclination of this orbit with respect to the equatorial plane depends on the latitude of the launch site. The orbit is circularized by firing the third stage at an apogee passage, and the orbital plane is changed by firing on-board thrusters. *See Rocket staging.*

With the United States Space Transportation System (STS, that is, the shuttle), the payload with an attached upper-stage module is first released at an altitude of about 200 mi (350 km). Firing of the upper-stage perigee motor places the payload in an elliptical transfer orbit with apogee at geosynchronous altitude. The orbit is eventually circularized by firing the apogee motor, and the orbital plane is changed to coincide with that of the Equator. These complex maneuvers are executed with the help of high-speed computers and tracking, telemetry, command, and control communications links between earth stations and the spacecraft. *See Space shuttle.*

Position and orientation control. A geostationary satellite must be kept in position, as it is subjected to perturbing forces. Some of these are gravitational (oblateness of the Earth, ellipticity of the Equator, and third-body effects due to the Moon and the Sun), whereas the principal nongravitational perturbation is solar radiation pressure. To counteract these forces, orbit control (station-keeping) is exercised by activating on-board auxilary propulsion devices upon command from Earth. Hydrazine has been the most commonly used propellant. Ideally, longitudinal drift and orbit inclination, that is, east-west and north-south station-keeping, can be controlled independently. With advancing technology, tighter position control has been enforced. Most communications satellites are kept within ±0.1° of their nominal position, although position tightness can be reduced to 0.03°; this contributes to more efficient use of the geostationary orbit, which is a unique and finite resource. Position is controlled by firing auxiliary jets at intervals of 2–5 weeks. The amount of on-board propellant is one of the elements that determines the lifetime of a satellite in orbit, which is 10–15 years.

In addition to orbital position control, communications satellites need orientation control to keep their highly directive antennas pointed toward the Earth. Spacecraft orientation is determined by on-board optical Earth, Sun, and star sensors, as well as by infrared and rf sensors. The most common method is by spinning the main body of the spacecraft. Gyroscopic stiffness provides the necessary stabilization while the antennas are despun toward the Earth. Alternatively, the spacecraft is stabilized by reaction wheels and gyroscopes sensitive to perturbations and is capable of generating correction torques by various means. *SEE SPACE TECHNOLOGY; SPACECRAFT STRUCTURE.*

Transmission-systems design. The rate of information transmission, R (bits per second), between two points in free space separated by a distance r can be expressed by Eq. (1),

$$R = \frac{P_t G_t A_r}{4\pi r^2 \beta N_0} \qquad (1)$$

where P_t = transmit power (in watts)

G_t = transmit antenna gain

$P_t G_t$ = effective radiated power in the direction of maximum antenna gain (eirp)

A_r = receive-antenna effective area (in square meters)

β = E/N_0, ratio of the energy per information bit to noise power density

N_0 = kT, noise power density (in W/Hz); k = Boltzmann's constant (1.38 $\times 10^{-23}$ joule/K) and T = noise temperature in kelvins

As the quantity N_0 is frequency-dependent, spectral regions yielding minimum noise are desirable. A satellite system comprises an uplink, active elements in the on-board transponders, and a downlink; therefore it is important to minimize all noise contributions. At the Earth's surface in the range 0.1–100 GHz, the external noise has two components, one galactic and the other atmospheric; the latter is also a function of the elevation angle. It is preferable to operate communications satellite systems in the minimum-noise spectral region between 1 and 10 GHz. High-gain, low-noise linear amplifiers are available in this range, and precautions must be taken to minimize the intermodulation noise produced by nonlinear power amplifiers. *SEE DISTORTION (ELECTRONIC CIRCUITS); ELECTRICAL NOISE.*

In 1963, as portions of the above-mentioned spectral region were already occupied by land-based communications and radar systems, satellites were assigned shared allocations agreed upon at international conferences. Mutual interference is kept under control by enforcing adequate geometric separation (distance and angle) or limitations on transmitter power. **Figure 6** shows the spectral regions allocated to fixed-services satellite systems at the World Administrative Radio Conference of 1979 (WARC-1979); earlier allocations are also shown. *SEE RADIO SPECTRUM ALLOCATIONS.*

Frequencies above 10 GHz, although not as favorable in terms of atmospheric noise, are used because the lower-frequency bands have become saturated in certain parts of the world. In the presence of rain, atmospheric noise is higher and the signals are attenuated (other hydrometeors, such as snow, ice, and hail, tend to produce scattering and depolarization). At 14/11 GHz, links are designed with added power margins up to a factor around 10. (In this article, the notation 14/11 GHz means an uplink frequency of 14 GHz and a downlink frequency of 11 GHz.) At 20 and 30 GHz, signals at times can be almost obliterated by rain; a solution can be found by using space-diversity techniques. With two or more earth stations separated by distances of 9–18 mi (15–30 km), the probability that all paths are simultaneously blocked by rain is markedly reduced. Use of the higher fre-

quencies is, however, advantageous because smaller earth stations can be installed at the user's premises, avoiding the additional cost of land interconnections.

Figure 7 shows the noise temperature observable under clear sky conditions at the surface of the Earth. At frequencies below 5 GHz, where low-noise receivers are used, nonatmospheric noise effects may be more severe than the noise arising from attenuation in the troposphere.

The effect of varying the antenna-beam elevation angle is also shown in Fig. 7. At low elevation angles, larger amounts of noise are collected by the antenna main beam in the atmosphere; in addition, the antenna sidelobes pick up noise radiated by the surface of the Earth as well as more atmospheric noise.

Link budgets. The planning of a satellite communications system involves complex trade-offs which require careful design of up- and downlinks, consideration of the transponder characteristics, the original signals to be transmitted, and the multiplexing, modulation, coding, and multiple-access methods. When the spacecraft functions as a relay station in space, the information flow is the same on both the uplink and the downlink. Messages generated by individual sources and converted into electrical signals in the terrestrial network are routed to satellite earth stations. After suitable signal processing, modulated rf carriers are beamed to the satellite by a high-gain antenna. Conversely, the signals received from the satellite are processed, fed into the terrestrial network, and eventually distributed to the intended destinations.

Although the information transmission rate can be obtained in terms of Eq. (1) for the up- or downlink or for their combination, it is convenient to use separate relationships for the signal-to-noise power ratios in the up- and downlinks. In this manner, individual contributions to the link budgets can be identified. In analog transmission systems, overall signal-to-noise power ratio is calculated by including additional noise contributions, such as interference and intermodulation. Finally, the number of channels (telephone, television, or other) is computed for specific combina-

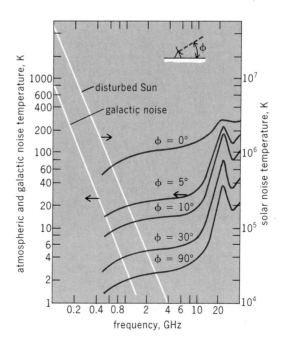

Fig. 7. Noise temperature versus frequency, observable under clear sky conditions at the surface of the Earth. (*After E. C. Jordan, ed., Reference Data for Engineers: Radio, Electronics, Computer and Communications, 7th ed., Chap. 27, Howard W. Sams, 1986*)

tions of multiplexing, modulation, and multiple-access techniques. *SEE ELECTRICAL INTERFERENCE.*

In the case of digital transmission systems, Eq. (2)

$$\frac{C}{N} = \frac{E}{N_0}\frac{R}{B} \qquad (2)$$

where C = carrier power
N = noise power
E = energy per bit
N_0 = noise power density
R = transmission rate
B = channel bandwidth

applies, and the transmission rate is limited by power or bandwidth. Message bit error probability can be computed in terms of the C/N ratio for a given modulation format.

In power-limited situations, forward error control can be helpful when the desired transmission rate cannot be attained with a prescribed value of bit error probability. Error control in the channel requires redundant bits to be added to the information bits and, hence, an increase of the overall rate and of the required bandwidth. The error probability of the recovered message can be reduced in terms of the added redundancy.

Various codes can be used in satellite systems; block and convolutional codes predominate. Although the former may be attractive in certain cases, the latter have been given preference on account of their easier implementation and the availability of efficient decoding schemes (Viterbi and sequential). *SEE INFORMATION THEORY.*

Advances in technology. Improved performance and lower cost to the user arise from advances in the technology of satellite communications systems. Advances are found in communications payload technology, earth station technology, and transmission methodology and hardware.

As mentioned above, at low power levels, solid-state power amplifiers using field-effect transistors have begun to replace traveling-wave tube amplifiers at the C-band (6/4 GHz) and the K_u-band (14/11 GHz). Gallium arsenide (GaAs) monolithic microwave integrated circuits (MMICs) are now ready for application. Low-noise amplifiers, microwave switch matrices (for switching at radio frequency), buffer amplifiers, digital variable phase shifters, and digital variable attenuators (for beam-forming networks behind array antennas) have all been developed and implemented in the laboratory and are ready for flight implementation. Advanced regenerative repeaters, bulk demodulators, and on-board baseband switches will be available for satellites to be deployed in the late 1990s. These and the array antennas implicit in the beam-forming network technology suggest a significant increase in computing power, control intelligence, and autonomy, and will be included in such satellites, most likely using very large-scale integrated circuits in silicon. *SEE INTEGRATED CIRCUITS.*

Antennas have increased in size, gain, and beam-shaping capability to the point where arrays are practical, and may be necessary in some cases, to provide multiple scanning or hopping beams and reconfigurability. Microwave filter technology advances continue to reduce filter size and weight. The use of dielectric loading (with low-loss dielectrics having relative permittivities greater than 30) permits significant reduction in the size of high-Q cavities. In another technique the orthogonality in electromagnetic modes has been exploited to obtain fourfold reuse of a single cavity; an eight-pole filter has been built reusing two cavities, each four times. *SEE CAVITY RESONATOR; DIELECTRIC MATERIALS.*

The energy storage of the satellite electrical power systems has been significantly improved through the use of nickel-hydrogen battery systems, which permit higher power density, greater depth of discharge, and longer life. In addition, solar-cell technology has advanced further. Silicon solar cells have end-of-life (after 10 years in orbit) efficiencies of 12% or more. Gallium arsenide solar cells combined with solar energy collectors are being developed which could double both the conversion and area efficiency. *SEE SPACE POWER SYSTEMS.*

The significance of the higher-power, higher-gain satellites is that the system balance has changed, whereby the requirements and costs of earth stations can be drastically lowered. This has given rise to very small-aperture terminals (VSATs) and direct broadcast service (DBS) television to 15 in. $\times$ 23 in (0.4 m $\times$ 0.6 m) K_u-band flat-plate antennas (K_u-band planar arrays of electromagnetically coupled elements). It has also promoted the rise of private networks through international business services (IBS) terminals at teleports and even terminals at customer premises.

Such developments require efficient transmission through satellites. This has led to the creation of protocols for satellite data-transmission systems, such as the random access with notification (RAN) protocol for very small-aperture terminals. It has also required modifications to existing protocols to permit very high efficiency throughput (more than 95%) in data transmission over satellite international circuits.

To make satellite voice transmission more efficient, a 32-kilobit/s adaptive differential pulse-code-modulation (ADPCM) digital voice has been developed, and a 16-kbit/s digital voice will follow. When combined with digital speech interpolation (DSI), this approach will allow digital voice circuit multiplication of a single bearer circuit by factors of up to 5 and 10. The end result will be that a single satellite of the INTELSAT VI class, if loaded optimally and solely with voice channels, could theoretically carry over 100,000 of them. Circuit multiplication methods have been also developed for television which make it possible to send three television channels through a single 36-MHz transponder. *SEE MODULATION; PULSE MODULATION.*

Coding and modulation methods have been optimized, and efficiencies of almost 2 bits/Hz have been obtained. In 1988, tests were conducted between the United States and France over a 72-MHz transponder in an INTELSAT V, for a data stream at 140 megabits/s. The transponder employed coded octal phase-shift keying (COPSK) modulation and rate 7/9 coding, that is, a 180-Mbit/s raw bit rate at 60 megasymbols per second. It achieved a bit error rate (over the 140-Mbit/s information rate channel) of 10^{-9} over one channel. Work on even higher transmission rates, while maintaining 10^{-9} error rates, has been undertaken.

Thus, communications satellites will have to handle everything from a few voice and data circuits from customers' premises or developing countries' earth terminals to heavy trunk routes operating at 45–155

Mbit/s. This multiplicity of demands over wide ranges of bit rates makes on-board processing at baseband (bulk demodulation, baseband switching, routing, rate changing) very useful to interconnect different beams and service rates where necessary. The result will be highly efficient communications satellites, providing many voice channels per kilogram and requiring smaller, less expensive earth stations.

Regional and domestic systems. Countries having vast territories with geographical characteristics unfavorable to the establishment of land links (for example, mountain ranges, islands, or jungles) greatly benefit from satellite systems for internal (domestic) communications.

After the establishment of the Soviet Union's dedicated domestic system in 1965 with nonsynchronous Molniya satellites, Canada was the first country to establish a geosynchronous domestic satellite system (TELESAT) in 1972. The three Anik A satellites, launched in 1972, 1973, and 1975, provided telephone, television, data, and facsimile services throughout the country to a variety of earth stations. Each spacecraft, with 655-lb (297-kg) in orbit weight and 12 transponders operating at 6/4 GHz, provided a communications capacity up to 7000 voice circuits. In 1975, TELESAT was the first commercial satellite system to use time-division multiple access techniques operationally. The system expanded with the introduction of three new series of spacecraft: Anik B (1978), Anik C (1980), and Anik D (1982). The bands at 14/11 GHz have been extensively used in addition to those at 6/4 GHz.

Indonesia started its Palapa A system in 1976 with two satellites similar to those of the Anik A type and 40 earth stations scattered over its many islands. In 1982, Palapa B satellites were added to the system, which has been continuously expanding and extending its services to neighboring countries.

In 1982, India launched its first domestic satellite, *INSAT 1A,* which developed a malfunction; a second satellite, *INSAT 2B,* was launched in 1983. These satellites provide three distinct services: fixed services, television broadcasting, and meterology.

In the United States, a domestic satellite system did not become operative until 1974, partially because of the existence of a well-developed terrestrial communications network. After the WESTAR I system of Western Union (1974), other systems, such as the RCA Satcom system (1975–1976), the COMSAT General Corporation (for AT&T and GTE) COMSTAR system (1976), the leased satellite system of the American Satellite Corporation, and the Satellite Business System (SBS, 1980), became operational. Subsequent systems include those of AT&T (TELSTAR 3), the Hughes Aircraft Company (GALAXY), and GTE (G-STAR).

United States domestic satellites have been found highly cost-effective, flexible, and efficient for relaying or distributing television programs within networks and to users. Several hundred thousand receive-only stations are in service.

The availability of powerful satellites in the 14/11-GHz bands triggered the development, first in the United States but later in other countries, of very small-aperture terminal networks. This approach (**Fig. 8**) is based on a star topology with a hub central station, allowing the efficient interconnection of very small-antenna (3–8-ft or 1–2.4-m paraboloid) terminals. With terminals located on the customers' prem-

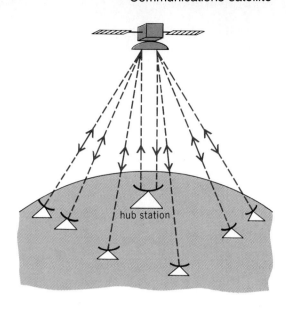

Fig. 8. Very small-aperture terminal system with central hub station.

ises, the high cost of interconnecting land lines is eliminated. The hub station provides the sensitivity and power required to operate the satellite transponders in an efficient manner. This approach has received widespread acceptance also because of advantages gained by the full control of the network by its owner.

Outside the United States, as many as 40 countries have established or plan to establish their own domestic services via satellite by leasing INTELSAT transponders and building their own earth stations. Countries which have their own systems or are planning or constructing domestic satellite systems include France, Australia, Italy, the Scandinavian countries, the 22 countries of the Arab League, Columbia, Mexico, Brazil, and China.

The first *European Communications Satellite* (*ECS*) of the European Telecommunications Satellite Organization (EUTELSAT) was launched in 1983. EUTELSAT satellites operate at 14/11 GHz. Experimental satellites preceding domestic operational systems have been orbited by Japan and Italy. Satellites to provide a variety of domestic satellite services, including fixed, mobile, and television to home, were launched by Japan in 1987 and 1988, including the JCSAT, BS, GMS, SCS, and Superbird series, which provide operational services.

France launched its first TELECOM IA satellite in 1984 as a multimission communications spacecraft (civilian and military) covering France and its overseas territories. The system was expanded in 1988 with the addition of TELECOM I.

The Australian system (AUSSAT) consists of two satellites operating at 14/11 GHz launched in 1985. Public switched telephone services are provided with coverage over Australia and Papua, New Guinea. The radiated power is adequate also to provide direct-to-home television broadcasting with receive antennas 5 ft (1.5 m) in diameter.

The 22 Arab countries, spanning the vast area from Morocco to Saudi Arabia, operate the ARABSAT system; their two launches occurred in 1984 and 1985. The spacecraft are body-stabilized and carry 25 transponders at 6/4 GHz plus a direct broadcasting transmitter at 2.5 GHz capable of serving receive-only earth terminals 10 ft (3 m) in diameter.

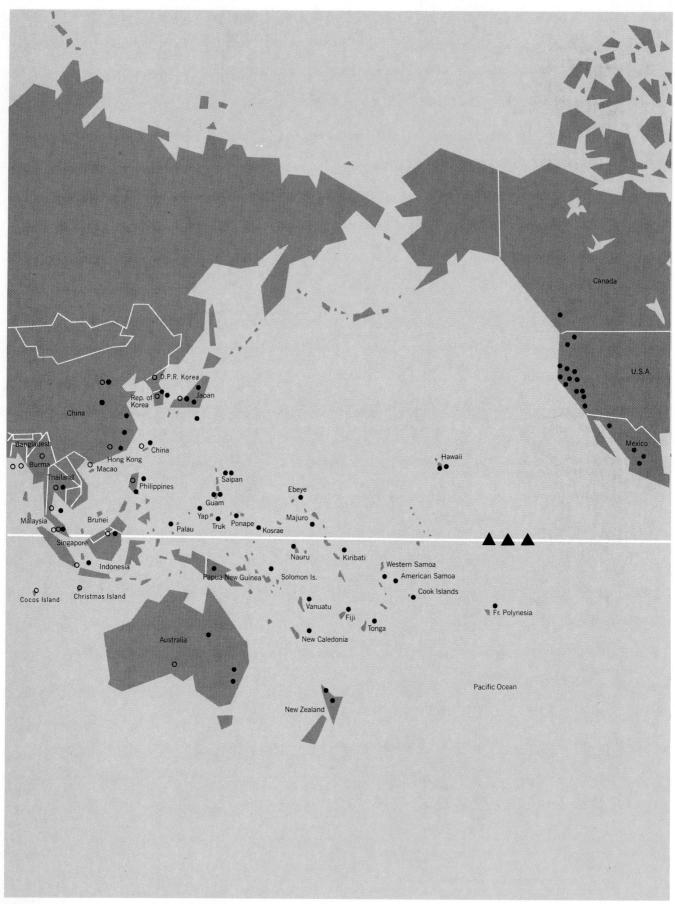

Fig. 9. Worldwide deployment of the INTELSAT system.

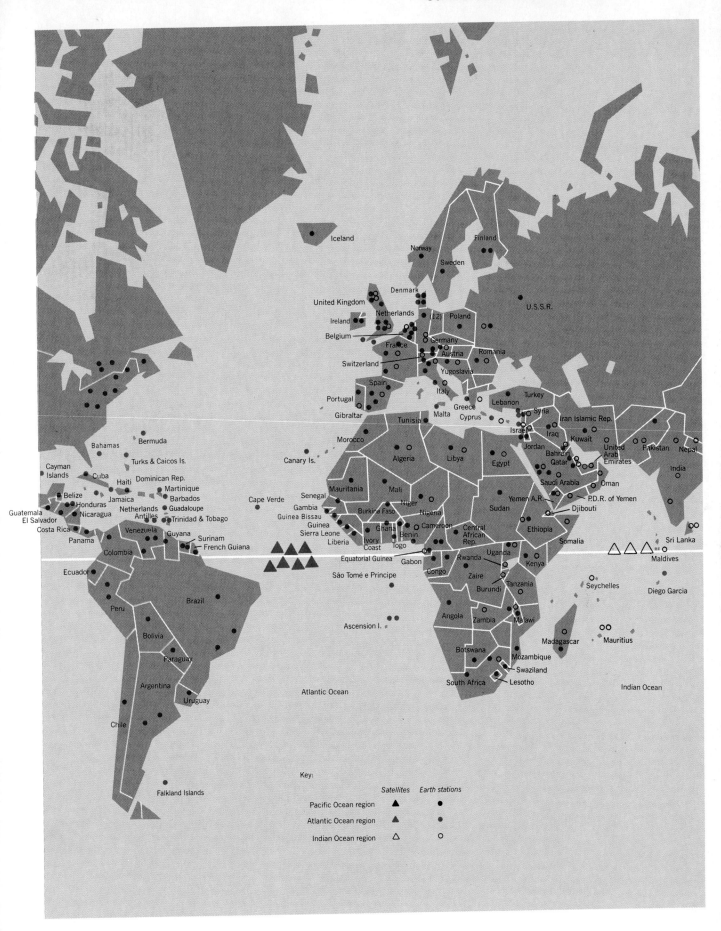

Iceland

Norway

Finland

Sweden

Denmark

United Kingdom

Netherlands

Ireland

Belgium

(12)   Poland

U.S.S.R.

France   Germany

Switzerland   Austria   Romania

Yugoslavia

Spain   Italy

Portugal   Greece   Lebanon   Turkey

Gibraltar   Malta   Cyprus   Syria   Iran Islamic Rep.

Tunisia   Israel   Iraq   Kuwait

Bahamas   Bermuda   Morocco   Jordan   Bahrain   United   Pakistan   Nepal

Algeria   Libya   Egypt   Qatar   Arab   India

Turks & Caicos Is.   Saudi Arabia   Emirates

Canary Is.   Oman

Cayman   Cuba   P.D.R. of Yemen

Islands   Haiti   Dominican Rep.   Mauritania   Mali   Yemen A.R.

Belize   Jamaica   Martinique   Cape Verde   Senegal   Niger   Sudan   Djibouti

Honduras   Barbados   Gambia   Burkina Faso   Nigeria   Ethiopia   Somalia   Sri Lanka

Guatemala   Nicaragua   Netherlands   Guadeloupe   Guinea Bissau   Cameroon

El Salvador   Antilles   Trinidad & Tobago   Guinea   Ghana   Central   Maldives

Costa Rica   Sierra Leone   Benin   African   Uganda   Kenya

Venezuela   Guyana   Liberia   Ivory   Togo   Rep.   Seychelles   Diego Garcia

Panama   Surinam   Coast   Rwanda

Colombia   French Guiana   Equatorial Guinea   Gabon   Congo   Burundi   Tanzania

Ecuador   São Tomé e Principe   Zaire   Mauritius

Peru   Brazil   Angola   Zambia   Malawi   Madagascar

Bolivia   Ascension I.   Botswana   Mozambique

Paraguay   South Africa   Swaziland   Indian Ocean

Chile   Lesotho

Argentina   Uruguay   Atlantic Ocean

Falkland Islands

Key:

Satellites   *Earth stations*

Pacific Ocean region   ▲   ●

Atlantic Ocean region   ▲   ·

Indian Ocean region   △   ○

Table 1. Principal characteristics of INTELSAT satellites

	INTELSAT I	INTELSAT II	INTELSAT III	INTELSAT IV	INTELSAT IVA	INTELSAT V	INTELSAT VI
Year of first launch	1965	1967	1968	1971	1975	1980	1989
Dimensions, ft (m): diameter	2.36 (0.72)	4.66 (1.42)	4.66 (1.42)	7.81 (2.38	7.81 (2.38)	6.56 (2.00)	11.81 (3.60)
height	1.94 (0.59)	2.20 (0.67)	3.41 (1.04)	17.32 (5.28)	19.36 (5.90)	51.5 (15.70)	17.39 (5.30)
In-orbit mass, lb (kg)	84 (38)	190 (86)	335 (152)	1543 (700)	1742 (790)	2286 (1037)	4938 2240)
Launch vehicle	Thor-Delta	Thor-Delta	Thor-Delta	Atlas/Centaur	Atlas/Centaur	Atlas/Centaur-STS-Ariane	STS-Ariane 4
Primary power, watts	40	75	120	400	500	1200	2300
Number of transponders	2	1	2	12	20	27	50
Usable bandwidth, MHz	50	130	450	500	800	2300	3262
eirp, dBW*	11.5	15.5	23	22.5 (global) 33.7 (spot beam)	22 (global) 29 (spot beam)	22–29 (4 GHz) 44 (11 GHz)	26–30 (4 GHz) 45 (11 GHz)
Number of telephone circuits	240†	240	1200	4000	6000	12,000	40,000
Lifetime, years	1.5	3	5	7	7	7+	10

*Effective radiated power in the direction of maximum antenna gain, decibels above 1 W.
†No multiple access.

Mexico started operation of its MORELOS satellite system in 1985. Two satellites operating primarily at 6/4 GHz have been launched which have contributed to a substantial improvement of the country's internal telephone network. A second series, MORELOS second generation, has been proposed for the early 1990s.

Brazil established the 6/4-GHz Brazilsat SBTS system in orbit with two satellites in 1985 and 1986, providing a variety of fixed services; extensions are planned for the 1990s.

INTELSAT. The Communications Satellite Act of 1962 led to INTELSAT, an international organization established in 1964 in Washington, D.C., by a group of 14 nations, under the leadership of COMSAT, the United States signatory to INTELSAT under this act. In 1988, this organization linked via satellite 115 member nations and 57 nonmembers users. **Figure 9** illustrates the system's deployment in the areas surrounding the Atlantic, Indian, and Pacific oceans.

Table 1 lists the major characteristics of seven generations of INTELSAT satellites launched since 1965. Technological advances have led to a 60-fold increase in the in-orbit mass and available prime power, a 65-fold increase in usable bandwidth, and a 165-fold increase in communications capacity available in a single spacecraft. Operational lifetime up to 10 years or more and the above-mentioned increase in communications capacity have greatly reduced the per-unit cost of the space segment.

At 6/4 GHz, mainly two types of earth stations are used. Standard A stations at terminals with heavy traffic have steerable parabolic antennas 90–100 ft (27–30 m) in diameter and low-noise receivers that yield a G/T ratio of about 41 dB/K, where G is the antenna gain and T is the noise temperature in kelvins. Standard B stations at terminals with light and medium traffic use parabolas 39–43 ft (12–13 m) in diameter and low-noise receivers to yield a G/T ratio of 31 dB/K. Traveling-wave tubes with a power output of 1–12 kW are used in the earth station transmitters. At 14/11 GHz, standard C stations are used for trunk telephone traffic with parabolas of 39-ft (12-m) diameter and low-noise receivers to yield a G/T ratio of 39 dB/K. Transmitter power is up to 5 kW. For business (IBS) services at 14/11 GHz, antennas of 5–8-m diameter have recently been introduced.

The bulk of the INTELSAT traffic is carried by frequency-division multiplex, frequency-modulation, frequency-division multiplex-access (FDM-FM-FDMA) analog transmission techniques. For thin routes, single-channel-per-carrier (SCPC) and demand-assigned multiple-access techniques are used. Since 1985, pulse-code-modulation, time-division-multiplex, quadrature-phase-shift keying, time-division multiple-access (PCM-TDM-QPSK-TDMA) digital techniques have been widely used. *See Frequency-modulation radio.*

Among the numerous advances which have characterized this family of satellites, the following ones are noteworthy: multiple access capabilities became available for the first time in the INTELSAT II series, Earth-coverage antennas occurred first in the INTELSAT III series, and increasingly sophisticated transmit and receive antenna systems were used in the INTELSAT IV, IVA, and V series.

The time-division multiple-access techniques, which had been experimented with since 1969 with INTELSAT III, IV, and IVA series, were introduced operationally on the INTELSAT V satellites in 1984 and 1985 in the Atlantic and Indian ocean regions. By coupling these techniques with digital speech interpolation, capacities in excess of 3200 voice channels per 72-MHz-bandwidth transponder were realized (a voice channel corresponding to its digital equivalent at 64 kbit/s). Digital speech interpolation takes advantage of the intermittent nature of speech to assign channels to active speakers. Since, in a two-way conversation, the average speaker is active only about 40% of the time, with a large traffic pool, a 2.5-fold capacity increase can be achieved. In addition, the use of digital circuit multiplication equipment (DCME), which allows transmission at 32 kbit/s with digital speech interpolation, can further double the capacity to 6400 voice channels per transponder.

In INTELSAT V, time-division multiple-access at 120 Mbit/s is employed in conjunction with a static interconnection of the receive and transmit antenna beams to provide wideband high-capacity international digital communications. The beam-to-beam connections remain fixed over relatively long periods of time, changed only as required by modified traffic patterns or to serve different needs if the satellite is moved to another orbital position. Other digital services initiated at lower bit rates include the IBS (INTELSAT business services) and the IDR (international data rate).

In a satellite with a large number of beams, such

as INTELSAT VI, flexible (or dynamic) interconnections among them are essential if the potential transponder capacities available from the use of time-division multiple-access and digital speech interpolation are to be realized. Dynamic interconnection is achieved on INTELSAT VI by means of an on-board time-division multiple-access switch. The transponders in the transponder banks designated (1-2) and (3-4) are connected to the on-board time-division multiple-access control switch. Interconnections are varied cyclically according to a predetermined sequence or a 2-ms switch frame period. The same sequence of interconnections is repeated in each switch frame. The interconnection between the transponders in either of the two banks is accomplished by switching units, each of which comprises a microwave switching matrix and a distribution and control unit. Normally, the microwave matrix provides the rf interconnections among the receiving and transmitting beams, while the distribution and control unit stores the sequence of required beam interconnections and implements this by controlling the connection points of the microwave switching matrix. The interconnection established within each of the two banks of transponders corresponding to channels (1-2) and (3-4) is accomplished by separating switching units controlled by a common timing source. No interruptions in the beam-to-beam connection occurs with this mode of operation either during or between frames.

A bypass static switch matrix is also available which can make interconnections in the static mode if the satellite-switched time-division multiple-access facility is not required on some transponders, or if the satellite-switched time-division multiple-access switch should suffer a catastrophic failure. A combination of dynamic and static switch operation is possible by ground command. In either the dynamic or the static mode, broadcast interconnections can be established whereby a single receiving beam is connected to two to six transmitting beams.

Characteristics of INTELSAT VII, a series of satellites currently under development, are shown in **Table 2**.

Private international systems. Until 1988, INTELSAT was the only entity providing international telecommunications services via satellite on the basis of well-established national and international agreements. Following the breakup of the Bell Telephone System monopoly in the United States, the arena became open to competitive innovative satellite services at the national as well as the international level. Between 1983 and 1986, several companies filed applications with the Federal Communications Commission (FCC) requesting authorization to offer and operate specialized business communications services, other than telephony, via satellite. Originally, six companies petitioned the FCC: Orion Satellite Corporation, International Satellite, Inc., Cygnus Corporation, Pan-American Satellite Corporation, RCA, and Financial Satellite corporation. However, as a result of mergers and withdrawals, four companies were left to compete in providing specialized business-oriented communications services between the United States and Europe, South America, Australia, and the Far East.

In September 1985, the FCC issued a separate satellite system order giving the applicants authority to enter the marketplace under the condition of forbidding them to compete against INTELSAT in the area of public-switched telephone services, thus leaving competition open for data transmission, television (regular and slow-scan or encoded), teleconferencing, electronic mail, and facsimile. The separate systems applications are essentially based on the proposed use of high-power satellites (mostly at 14/11 GHz, but in some cases also at 6/4 GHz) which could permit the installation at customer premises of smaller and simpler, and hence more economical, earth stations of a kind previously not acceptable within the standards imposed by INTELSAT upon all earth terminals in its systems.

In June 1988, the Pan-American Satellite Corporation placed its first satellite in orbit (orbital position 315°E longitude; 18 transponders at C-band and 6 transponders at K_u-band). Business traffic is exchanged between Peru and the United States with extensions foreseen at K_u-band to European countries and for inter–South American traffic. In response to the competition, INTELSAT established in 1987 its business service (IBS) and also requested modifications to three of its satellites under construction in order to offer competitive data services by means of spot beams providing higher eirp, thus allowing the use of smaller earth terminals.

Soviet satellite systems. In 1965, the Soviet Union began launching satellites of the Molniya type in highly elliptical, 63.5°-inclined, 12-h orbits with perigee at 300 mi (500 km) and apogee at 25,000 mi (40,000 km). *Molniya 1* satellites operated at both 0.8–1.0 GHz and 6/4 GHz; the *Molniya 2* (1971) and

Table 2. INTELSAT VII preliminary design

Characteristic	At C-band	At K_u-band
Number of transponders	26	10
Transponders and bandwidths	16 with 80-MHz bandwidth for hemi-zone coverage	2 of 120-MHz bandwidth for spot coverage
	4 with 40-MHz bandwidth for hemi-zone coverage	8 of 80-MHz bandwidth for spot coverage
	2 with 40-MHz bandwidth for global TV coverage (Channel 12)	
Coverage	Hemi-zone, global	Spot
eirp	Hemi-zone ≈ 33 dBW	Spot 41.4–47.8 dBW
	Global ≈ 26–29 dBW	
Number of telephone circuits	Full-bearer circuits ≈ 36,000 maximum* (two ½-bearer circuits)	
Lifetime	10–15 years	

*A bearer circuit is a noninterpolated 64,000-bit/s, end-to-end circuit. Such circuits, using digital circuit multiplication equipment, may carry up to five interpolated telephone voice circuits.

Molniya 3 (1974) series operated only at 6/4 GHz. These satellites were used in the Orbita system for communications within the Soviet Union. By the 1970s, the system was extended to several eastern European countries, Mongolia, Vietnam, and Cuba, and became known as the International INTERSPUT-NIK System.

As geostationary satellites cannot provide efficient coverage of high-latitude regions because of the low elevation angle of the slant transmission path above the horizon, inclined Molniya-type orbits were initially favored by the Soviet Union, with its populated areas at high latitudes. Both the Molniya and the Orbita systems require several satellites with tracking and handover equipment at earth stations to switch traffic from one satellite to another. In 1975, the Soviet Union began launching geostationary satellites; two of these, *Statsionar 4* and *5*, located at 14°W and 5°E longitude, respectively, operating in the 6/4-GHz bands, carry most of the traffic of the INTERSPUT-NIK system. The earth stations have a *G/T* ratio of 31 dB/K. Maximum radiated power is 55 dBW for frequency-modulation or pulse-code-modulation–phase-shift-keying telephony and 85 dBW for frequency-modulation television transmissions. The geostationary satellites of the Soviet Union include various programs designated as Statsionar, Raduga, Ekran, Gorizont, Kosmos, Molniya, Luch, Gals, Potok, and Volna. Services offered include point-to-point telephony, television distribution and data transmission, television direct broadcasting, data relay, and mobile communications. The combination of satellites in low-altitude orbits (500–900 mi or 800–1500 km) Molniya-type orbits, and geostationary and quasigeostationary orbits provide communications for civilian, government, and military traffic.

Mobile maritime satellites. The first operational maritime system, called MARISAT, was established in 1976 by the United States. Three satellites, each with an in-orbit mass of 719 lb (326 kg), were positioned in geostationary orbit at 15°W, 176.5°E, and 73°E longitude to serve the Atlantic, Indian, and Pacific ocean regions, respectively. Shore-to-satellite links operated at 6 and 4 GHz (up- and downlinks). Ship-to-satellite links operated in the allocated bands at about 1.6 and 1.5 GHz (up- and downlinks), Telex, telephone, data, and facsimile services were offered. Transmission methods such as companded frequency-modulation single-channel-per-carrier multiple-access for voice and time-division multiplex and multiple-access for telegraphy were used. Shore stations at Southbury, Connecticut, and Santa Paula, California, linked the United States with the Atlantic and Pacific MARISAT satellites. A third station became operative at Yamaguchi, Japan, in 1978.

Efforts to organize an international maritime satellite communications system initiated in 1972 led to the establishment of the INMARSAT consortium with headquarters in London. Agreements were reached to begin an ordered transition from MARISAT to IN-MARSAT in 1982. INMARSAT used the MARISAT satellites, then leased four INTELSAT V satellites modified to include maritime services capabilities, and finally used two European-built satellites. The first of these, *Marecs A,* was launched in 1982. Communications capacity was increased from the 10-voice channels of the MARISAT satellite to 30-voice channels in the modified INTELSAT V maritime subsystem and 60-voice channels in the Marecs satellites.

Three second-generation INMARSAT II satellites are designed to support at least 200 voice channels per spacecraft. There are 20 operational coastal earth stations (CES) and over 6000 terminals on ships, oil platforms, and other seagoing equipment. The IN-MARSAT satellites have the capability to serve aircraft as well as ships. Such aircraft services have begun, mainly with European signatories and Japan, in an exploratory fashion. There is a great deal of activity in the development of mobile terminals for aircraft, which have stringent requirements. The aeronautical mobile satellite services (AMSS) can be expected to grow as quickly as the maritime mobile satellite services (MMSS) did in the past.

Land mobile satellite service. Geostationary satellites can substantially enhance and augment the capability of land mobile radio systems. The land mobile satellite service (LMSS) has been assigned (in the United States) the uplink band of 1646.5–1660.5 MHz and the downlink of 1545–1559 MHz for mobile communications. The feeder links between satellites and fixed earth stations (gateways) are at 14–14.5 GHz for uplinks and 11.7–12.2 GHz for downlinks. As the interconnection with the public-switched telephone network occurs at the gateway stations, telephone service will be provided for trucks, automobiles, trains, and other moving vehicles. Data transmission will also be available for a broad category of digital dispatch services, such as one-way paging, two-way alphanumeric messaging, monitoring, telemetering, remote control, alarm and security signaling, facsimile, and low-speed television. Land mobile satellite service also has great potential to serve lightly populated rural areas unlikely to be covered for reasons of cost by terrestrial mobile radio, which serves urban and suburban areas with denser populations. The combination of land mobile satellite service and terrestrial mobile radio will ultimately result in complete coverage of national territories as the rulings of the FCC allow the intermingling of the two services.

Taking into account the low directivity of the antennas carried by moving terrestrial vehicles, a fairly high radiated power is required on the satellites and, as a consequence of the chosen frequency bands, large antennas will be needed on the spacecraft. Plans are under way in the United States and Canada for an operational land mobile satellite service system in the 1990s. A consortium has been formed in the United States among the former applicants for land mobile satellite service; in Canada the project is being promoted by the Department of Transportation.

Direct broadcast service. In 1976, the experimental Canada–United States *Communications Technology Satellite* (*CTS* or *Hermes*) operating at 14/12 GHz with transmit power up to 230 W made possible the reception of quality color television programs with small (1.6–3.3-ft or 0.5–1.0-m) antennas. During the 4-year lifetime of this satellite, many experiments in Canada, the United States, and South America confirmed the potential of direct broadcast service. Small antennas mounted on the rooftops of individual homes or apartment houses combined with appropriate amplification and frequency conversion equipment provide signals which can be fed to standard very high-frequency–ultrahigh-frequency television receivers. The reception is especially good as there is a simple path from the spacecraft to the receive antennas, with the elimination of multipath reception (ghosts and multiple images) which frequently occurs with terrestrial television systems.

At the 1983 World Administrative Radio Conference, frequency bands in the range 11.7–12.75 GHz were assigned to direct broadcast service with feeder links (uplinks) in the band 17.3–18.1 GHz from central earth stations to the satellites. Subsequently, considerable interest in these systems was generated in the United States, Canada, Europe, and Japan.

In the United States, notwithstanding several applications submitted to the FCC, no commercial service is yet available. In Europe, as a result of full governmental support, direct broadcast service systems have been developed in Germany, France, the United Kingdom, and Luxembourg, all of which either have launched satellites or are scheduled to do so. Japan has a medium-power quasioperational direct broadcast service satellite in orbit.

Undoubtedly, other countries will develop direct broadcast service systems. Important for this development is the availability of low-noise field-effect transistor preamplifiers and flat-plate antennas to be offered in home receiver equipment. Even with medium-power satellites currently operational in the United States, it is possible to receive television programs from space with earth station antennas 7–10 ft (2–3 m) in diameter. The cost of this equipment has dropped sharply; television reception via satellite is now commonplace in the United States, especially in rural areas where ordinary television coverage is not available. It is estimated that more than 1,500,000 television sets are equipped for such reception in the United States.

The Soviet Union uses the inclined-orbit geosynchronous Ekran system (at 0.7 GHz and about 200 W, 1 channel), the geostationary Gorizont/Moskva system (at about 3.8 GHz and 40 W, 5 channels), and the Molniya 3/Orbita system (at about 3.8 GHz and 40 W, 5 channels) to achieve very widespread television distribution through local rediffusion, most people having access to two or more channels. A direct-to-home broadcast system, at 12 GHz, *STV-12*, has been developed.

Pier L. Bargellini; Geoffrey Hyde

Bibliography. COMSAT Technical Review, Communications Satellite Corporation, biannually; D. M. Jansky and M. C. Jeruchim, *Communications Satellites in the Geostationary Orbit*, 2d ed., 1987; E. C. Jordan (ed.), *Reference Data for Engineers: Radio, Electronics, Computer and Communications*, Chap. 27, 7th ed., 1986; L. Ya. Kantor (ed.), *Handbook of Satellite Telecommunication and Broadcasting*, 1987; C. E. Mahle, G. Hyde, and T. Inukai, Satellite scenarios and technologies for the 1990's, *IEEE J. Selected Areas Commun.*, SAC-5:556–570, 1987; W. L. Pritchard and J. A. Sciulli, *Satellite Communications Engineering*, 1986; Special issue on broadcasting satellites, *IEEE J. Selected Areas Commun.*, vol. SAC-3, no. 1, January 1985.

Communications scrambling

The methods for ensuring the privacy of voice, data, and video transmissions. Various techniques are commonly utilized to perform such functions.

Voice. Analog methods typically involve splitting the voice frequency spectrum into a number of sections by means of a filter bank and then shifting or reversing the sections for transmission in a manner determined by switch settings similar to those of a combination lock; the reverse process takes places at the receive end. The strength of these methods depends on the number of switch settings available (often called the key size) and the rate at which the switch settings are changed. *See Electric filter.*

Digital methods first convert the analog voice to digital form by a number of digital coding techniques such as pulse code modulation (PCM), delta modulation, or linear predictive coding (LPC). The digital voice data are then scrambled or encrypted by one of the many methods discussed below and are transmitted digitally. After the data are received, they are descrambled or deciphered and then converted back to analog form by the appropriate voice decoder. *See Analog-to-digital converter; Digital-to-analog converter; Modulation; Pulse modulation.*

Data. A simple data-scrambling method involves the addition of a pseudorandom number sequence to the data at the transmit end. The same sequence must be subtracted at the receive end to recover the data. The strength of the method depends on the period of the sequence and the way the sequence is generated, as well as the number of different pseudorandom number sequences available. Devices using this method are known as stream ciphers. Typically, the pseudorandom number sequences are in binary form; and the addition and subtraction are performed in binary form also.

A second method partitions the data into blocks. Data within a block may be permutated bit by bit or substituted by some other data in a manner determined by the switch setting, which is often called a key. At the receive side, the reverse permutation or substitution is performed to recover the data. Devices using this method are known as block ciphers. An example of block ciphers is the Data Encryption Standard (DES), accepted as a standard by the National Bureau of Standards in 1977. DES, which uses both permutation and substitution iteratively, is approved for nonmilitary, nondiplomatic governmental use.

The National Security Agency established a commercial communication security endorsement program under which encryption devices suitable for different applications are developed and certified by the government. *See Cryptography.*

Video. Typical video scrambling devices used for cable television applications involve modifying the amplitude or polarity of the synchronization signals, thereby preventing the normal receiver from detecting the synchronization signals. A device capable of undoing the modifications is used to recover the signal. Some devices also invert the polarity of the video signal so that the black and white levels are reversed. *See Closed-circuit television.*

A more sophisticated technique, used in satellite transmission, introduces a random delay to the active video signal on a line-by-line basis which is determined by a pseudorandom number sequence. Knowing the sequence, the receiver can compensate for the delay and recover the signal. An even more advanced technique called cut-and-rotate has been proposed. This method selects a random point of time in a scan line and cuts the line into two sections. The second section of the line is transmitted first, followed by the first section. By knowing exactly where the cut takes place, the receive side can perform the reverse process and reconstruct the signal.

Video signals can also be digitized by a number of coding techniques and then scrambled by any of the data-scrambling techniques discussed above to achieve high security. *See* Television.

Lin-Nan Lee

Communications systems protection

The protection of wire and optical communications systems equipment and service from electrical disturbances. This includes the electrical protection of lines, terminal equipment, and switching centers, and inductive coordination, or the protection against interference from nearby electric power lines.

Sources of disturbance. The principal sources of destructive electrical disturbances on wire communications systems are lightning and accidental energization by power lines. Lightning that directly strikes aerial or buried communication cable may cause localized thermal damage and crushing. Simultaneously, it energizes the communication line with a high-level voltage transient that is conducted to terminal equipment. Nearby strikes are more common than direct strikes, and although they normally do not produce mechanical damage, they cause the propagation of electrical transients along the line. The rate of occurrence of these transients is geographically dependent, increasing with the level of thunderstorm activity (keraunic level) and earth resistivity. The distance over which a buried cable collects a lightning stroke to the nearby earth, and the resulting voltage produced within a cable both increase as resistivity of the earth increases. The duration of the lightning-caused voltage transient on shielded communication cable is typically less than 1 or 2 milliseconds for each lightning strike. The magnitude of the voltage between conductors and shield of a cable can reach several thousand volts. In regions of high thunderstorm activity and earth resistivity, a communication line several miles long may experience annually more than 1000 voltage transients of magnitude greater than 250 V. The accompanying current surge usually is less than 100 A, but much higher currents can result from a direct strike to the cable. *See* Lightning.

During fault conditions on commercial power lines, high voltages may occur in nearby communication lines by several mechanisms. The most common mechanism is magnetically induced voltage caused by the high unbalanced currents of a phase-to-ground power-line fault. During such an event, an aerial or buried communication line that is parallel to the faulted power line intercepts its time-varying magnetic field, incurring a high longitudinal voltage. In effect, the system is a one-turn to one-turn air-core transformer. This inductively coupled voltage may be several hundred volts, with current limited by the impedance of the communication line and its terminations. Another source of high voltage is a damaged aerial power line that falls and directly contacts an aerial communication cable, or a digging machine that simultaneously contacts power and communication cable jointly buried in a common trench. In these cases, currents conductively enter the communication lines and are limited in magnitude only by fusing capabilities. Finally, communication lines near a power station may be subjected to high voltages from the rise in ground potential as power fault currents return through the ground mat of the power station. In each of these mechanisms, power-line fault-clearing devices are intended to limit the duration of high current conduction to a few seconds. However, if the power-line fault impedance is high, the automatic clearing devices may not operate, and the overvoltage can last indefinitely. *See* Electric power systems; Electromagnetic induction; Grounding; Transformer; Transmission lines.

In city centers, the diversion of lightning strikes by steel-framed buildings, and the shielding effect of the many underground metallic utility systems, considerably reduce the probability of high-voltage transients from lightning or induction from power lines. Since the communication and power facilities are routed in separate conduits, the possibility of a power contact is remote.

Electrical protection. The components of a communications system requiring electrical protection include both metallic and optical-fiber lines, terminal equipment, and switching centers.

Metallic lines. Metallic communication lines often are made up of many closely spaced pairs of wires arranged as a cable. A grounded circumferential metallic shield on the cable reduces the magnitude of electrical transients from nearby lightning strikes, and can intercept a low-current direct strike, minimizing damage to the internal conductors. The effectiveness of the shield is improved if its resistance per unit length is low and if the dielectric strength of the insulation between the shield and the internal conductors is high. Frequent grounding of the cable shield eases the return of lightning currents to the earth, thereby reducing voltages in the cable. Voltage breakdown from transients within the cable can be avoided by installing voltage protective devices between the pairs and shield. The devices may be air gaps consisting of carbon electrodes separated by a few thousandths of an inch (about 0.1 mm), or gas tube surge arresters, which are metal electrodes separated by a gap in an enclosed discharge medium. These devices limit voltages at their terminals to the capabilities of the cable insulation. They are used principally in areas of high lightning activity, especially where the cable insulation has low dielectric strength. *See* Communications cable; Electrical shielding; Gas tube; Lightning and surge protection; Surge suppressor.

The localized damage caused by a direct strike to buried cable can be reduced or eliminated by the burial of one or more wires at least 1 ft (0.3 m) above the cable. These wires divert all or much of the lightning current from the cable.

Minimization of damage to cable plant from a power-line contact is usually effected by providing frequent bonds between the cable shield or support strand and the neutral conductor of the power line. These bonds create low resistance paths for fault currents returning to the neutral, and serve to hasten deenergization of the fault by automatic protective devices. Their close spacing limits the length of communication cable that is damaged by the contact.

The outer metal shield of belowground cables may be attacked by electrolytic action and require protective measures against corrosion. Electrolysis occurs where direct current leaves the shield. Severe sources of electrolytic currents include stray direct current from electrified railways and from industries using large amounts of direct current in grounded circuits. An effective remedial measure is to install a gap, known as an insulating joint, in the cable shield to interrupt the current. This gap can hinder the electrical protection and noise mitigation function of the shield unless it is bridged with a capacitor that provides a bypass for lightning and power-frequency currents. SEE CORROSION.

Optical-fiber lines. Optical-fiber communication lines are placed in cables that may contain metallic components to provide mechanical strength, waterproofing, local communications, or a rodent barrier. Though the cables would otherwise be immune to the effects of lightning or nearby power lines, such metallic components introduce a measure of susceptibility that is made all the more important by the high information load often carried by the fibers. A direct lightning strike to a metallic component of buried optical-fiber cable can cause localized thermal damage, arcing, and crushing that together may damage the optical fibers. Electrical protection is provided by cable designs that withstand these effects and that avoid arcing paths that intercept and damage fibers. As with wire-pair cables, the probability of this damage can be reduced by burying one or more shield wires at least 1 ft (0.3 m) above the cable.

Alternating currents may be conducted on the metallic sheath components of optical-fiber cables during an accidental contact with power-line conductors. Cable damage is minimized in extent by bonding the sheath to the neutral conductor of the power line, and by providing enough conductivity to carry the currents without damage to the fibers. SEE OPTICAL COMMUNICATIONS; OPTICAL FIBERS.

Terminal equipment. To protect the users, their premises, and terminal equipment, communication lines that are exposed to lightning or contacts with power lines are usually provided with protectors. Article 800 of the National Electrical Code requires that communication lines exposed to contact with power lines of voltages of greater than 300 V be equipped with a protector at the entrance to the served premises. It is also common practice to minimize equipment damage by so equipping communication lines in an area of significant lightning exposure even if there is no power-contact hazard. For the most part, the protector contains a voltage-limiting device, such as a carbon block or gas tube surge arrester, connected between each wire of the line and a local ground electrode. The voltage surge on the wires thereby is limited with respect to local ground. Voltages between the communication lines and other grounded systems within the premises are limited by bonds between the protector ground, the power neutral, and the metallic water piping. SEE ELECTRICAL CODES.

Sparking between communication lines and nearby objects, and electric shock hazard to the user, are minimized if the insulation of the terminal equipment withstands voltages at least as great as the limiting voltage of the protector. Service interruptions are avoided if electronic circuits within the equipment are able to withstand these surge voltages. In some equipment, surge immunity is achieved by applying additional voltage limiters, such as Zener diodes or special low-voltage gas tubes within the equipment itself. SEE ZENER DIODE.

Sustained currents from power-line contacts can overheat the exterior wires connected to the protector unless interrupted by a current-limiting device. A short length of 24- or 26-gage wire in series with the communication line, or a fuse at the protector, is usu-

ally installed for that purpose. If the terminal equipment contains an internal voltage limiter, an additional fuse may be needed to protect the premises wire connecting the equipment to the protector. SEE FUSE (ELECTRICITY).

Switching centers. Switching centers contain electronic equipment that routes communications to their proper destinations. This equipment is protected from the effects of electrical transients appearing at interfaces with external communication lines in the same way as described for terminal equipment. In addition to these effects at external interfaces, communication equipment within buildings can be subjected to transient voltages resulting from lightning strikes to nearby objects or by direct strikes to the building. Such events cause rapidly varying currents in metallic structural members and in the grounding system. These currents can be resistively and magnetically coupled to the wires interconnecting equipment in the building. Depending on the relative paths taken by the lightning currents and the wiring, transient differences in voltage may appear between nearby ground and conductors that extend between internal locations.

One method of controlling these voltages is to bond, as often as possible, all metallic equipment frame members and grounded circuits to the floor ground of the building. This technique, known as an integrated ground, reduces potential differences, but permits surge currents to circulate throughout the system. Sensitive modern electronics can be damaged by the local potential differences caused by these currents, so many designs now avoid circulating currents by isolating the entire system from ground except at one location, which is deliberately grounded. This system, known as the isolated ground plane, has found effective use in many modern switching centers. SEE ELECTRONIC EQUIPMENT GROUNDING; SWITCHING SYSTEMS (COMMUNICATIONS).

Inductive coordination. Inductive coordination refers to measures that reduce the magnitudes and effects of steady-state potentials and currents induced in metallic communication lines from paralleling power facilities. Where power and communication lines are parallel for long distances, normal-load power currents returning to the power generator via the neutral conductor or earth induce voltages into the communication lines. Malfunction of terminal equipment may result. These voltages normally appear equally in each wire of a pair; electrical asymmetries in the lines or terminal equipment convert them to a voltage between the wires of a pair and may cause background noise or data errors. Remedial measures that reduce the induction effect are (1) balancing the power-line currents to reduce the neutral component returning to the generator; (2) using a grounded metallic shield on the communication cable so that shield currents partly cancel the field from the disturbing power line; (3) using well-balanced communication line and terminal equipment to minimize the conversion to voltages between wires; (4) installing reactive components in the communication line to reduce induced currents; and (5) using terminal equipment that is tolerant of induced voltages. SEE ELECTRIC PROTECTIVE DEVICES; ELECTRICAL INTERFERENCE; ELECTRICAL NOISE; ELECTROMAGNETIC COMPATIBILITY.

Michael Parente

Bibliography. R. L. Carroll, Loop transient measurements in Cleveland, South Carolina, *Bell Sys. Tech. J.*, 59:1645–1680, 1980; R. H. Golde, *Light-*

ning, vol. 2, 1977; *Guide for the Application of Gas Tube Arrester Low-Voltage Surge Protective Devices*, ANSI/IEEE C62.42-1987, 1987; *National Electrical Code*, ANSI/NFPA 70, Article 800, National Fire Protection Ass., 1987; E. D. Sunde, *Earth Conduction Effects in Transmission Systems*, 1968.

Communications traffic-network engineering

The engineering of voice and data networks where customers have a common set of communications paths, involving the determination of the number of paths needed to maintain delays below a specified level. The task of making this determination is known as the sizing problem.

Single-link example. The sizing problem is based on a branch of applied probability theory known as queueing or traffic theory. Consider a simple telephone network (**Fig. 1**) that enables any one of the 10,000 telephones in region A to be temporarily connected to any of the 8000 telephones in region B. The permanent connection between each telephone and its local end office is called a loop; a single talking path between end offices A and B is a trunk. When, for example, customer A_1 calls customer B_1, the switching equipment at A first transmits a dial tone to A_1 to acknowledge that A is ready to receive the called number; second, A connects A_1's loop to a trunk; and third, A transmits the dialed number to end office B. Finally, B connects B_1's loop to that same trunk and rings B_1's phone. When A_1 hangs up, the connection is taken down. *See Queueing theory; Switching systems (communications); Switching theory.*

The link sizing problem here is to determine the amount of switching equipment to be provided at A and B, and the number of trunks to be provided between A and B.

To guarantee immediate service to these subscribers would require the equivalent of 10,000 operators at A (in the unlikely event that everyone picks up their telephone at nearly the same instant), 8000 operators at B, and 8000 trunks between A and B (since there can, at most, be 8000 conversations between A and B).

Clearly, such a solution would be prohibitively expensive. If less than this amount of equipment is provided, however, the possibility exists that the network will be unable to set up a call immediately. Any customer denied immediate service is said to be blocked.

When a dial tone is not received immediately, the customer waits for it. The blocked call is then said to be delayed. When all trunks are busy, however, the customer is asked (by a reorder, or fast-busy, tone) to hang up and try the call at a later time. In this case, the blocked call is said to be cleared.

To provide an acceptable grade of service, the end office is sized so that only a small percentage (typically, 1.5% during the busy hour) of the customers will have to wait more than a small interval (for example, 3 s) to receive a dial tone. Trunk groups are sized so that no more than a small percentage (typically, 1% during the busy hour) of call attempts will be blocked and cleared from the system.

To illustrate, suppose that 300 call attempts per hour, each of 2 min duration, are placed between end offices A and B. If each call has the same origination time, then 297 trunks would be required to achieve 1% blocking. At the other extreme, since one trunk can carry at most 30 2-min calls per hour, 10 trunks are sufficient if the calls arrive back to back in stacks of 10. Thus, no matter how the calls arrive, the required number of trunks should be between 10 and 297.

Binomial formula. To describe a more realistic case, suppose that call origination times occur at random during the hour. For the moment, assume that enough trunks are provided to carry all calls. The probability that exactly x trunks are busy at some time, t_0, is then equal to the probability that exactly x calls arrive in the 2-min interval prior to t_0. Since the probability that a particular subscriber places a call in this interval is 2/60, the probability, p_x, that exactly x trunks are busy is given by the binomial distribution, Eq. (1). In general, if n calls each of duration h are

$$p_x = \binom{300}{x}\left(\frac{2}{60}\right)^x\left(1 - \frac{2}{60}\right)^{300-x} \tag{1}$$

placed in a time interval of length T, the probability that x trunks are busy is given by Eq. (2).

$$p_x = \binom{n}{x}\left(\frac{h}{T}\right)^x\left(1 - \frac{h}{T}\right)^{n-x} \tag{2}$$

The individual and cumulative frequency distributions, for $n = 300$ and $h/T = 1/30$, are shown in **Fig. 2**. The most probable number of busy trunks is $x = 10$, which, independent of how the calls arrive, is also the average number of busy trunks over the hour. The probability of 18 or more busy trunks is about 0.01, which suggests that 18 trunks should provide immediate service to all but about 1 out of 100 customers (compared with the 300 required to guarantee immediate service to all customers).

The binomial distribution can be used to estimate the blocking probability for a group of s trunks under an artificial assumption intermediate between the blocked-calls-cleared and blocked-calls-delayed assumptions. Specifically, blocked calls are assumed to be held in the system, either waiting or holding a trunk if one becomes available, for their normal call-holding times. Under this assumption, if s trunks are provided to serve $n + 1$ subscribers, the probability of blocking is given by Eq. (3).

$$P(s,n,h/T) = \sum_{x=s}^{n} \binom{n}{x}\left(\frac{h}{T}\right)^x\left(1 - \frac{h}{T}\right)^{n-x} \tag{3}$$

See Binomial distribution; Combinatorial theory.

Poisson formula. The binomial formula depends upon both n and h/T. However, for most applications (n large and h/T small), the binomial formula is well

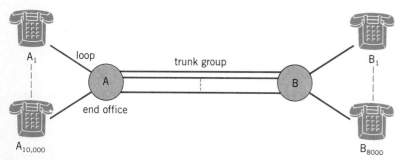

Fig. 1. A simple telephone network.

Fig. 2. Direct-current generator showing the typical brush assembly. (*Allis-Chalmers*)

windings characteristically have two parallel paths between any two commutator bars separated 180 electrical degrees. Machines employing commutators are usually restricted to voltages below the kilovolt range because flashover between the closely spaced commutator segments may result from deposition of carbon dust from the brushes and from uneven voltage distribution between commutator bars under heavy loading. For machines using commutators SEE DIRECT-CURRENT GENERATOR; DIRECT-CURRENT MOTOR; REPULSION MOTOR; UNIVERSAL MOTOR. SEE ALSO WINDINGS IN ELECTRIC MACHINERY.

Arthur R. Eckels

Compact disk

A system of audio or video recording or data storage in which digitally encoded information in the form of microscopic pits on a rotating disk is accessed by optical readout. The implementation of these ideas in the audio compact disk will be described, along with their extension to current developments in video disks and optical computer storage.

Optical readout. The acoustic signal waveform is recorded on the disk in the form of a binary code, as a series of 0s and 1s. This is done by forming pits along spiral tracks on a transparent plastic disk, overlaying this with a reflective coating, and then covering this coating with a protective layer. The light from a semiconductor laser is focused onto the pits from below (**Fig. 1**), and the reflected light, picked up by a photodetector, converts the presence or absence of the pits into a binary electrical signal. SEE LASER.

With the light focused at the level of the pits, the surface of the transparent disk is well out of focus; hence the presence of dust, finger marks, and minor scratches on this surface is so blurred as not to affect the detection process. Even more important, the use of an optical pickup means that there is no mechanical contact and hence no wear.

Because the focused laser spot size is so small, the amount of information that can be packed onto the surface of the disk is very great. Adjacent tracks of the spiral of pits need be only 1.6 μm apart, and hence 20,000 such tracks are available on a 120-mm-

diameter (5-in.) audio compact disk. The pits, and the separations between pits, vary in length from 0.9 to 3.3 μm (**Fig. 2**). The track is optically scanned at a constant linear velocity of 1.25 m/s or 4.1 ft/s.

Digitally encoded waveform. The human ear is insensitive to sounds whose frequency is higher than about 20,000 Hz. C. Shannon's sampling theorem guarantees that a waveform can be reproduced exactly from samples taken at a uniform rate of at least twice that of the highest frequency. To allow a small margin of error, the standard sampling rate for audio compact disks is 44.1 kHz. The amplitude of each of these samples is then approximated to the nearest of 2^{16} (65,536) equally spaced levels (a process known as quantization). Hence each sample amplitude can be represented by a 16-bit binary word. With two separate stereo channels, the number of bits per second required to be stored on the disk is given by Eq. (1).

$$2 \times 16 \times 44,100 = 1,411,200 \text{ bits/s} \qquad (1)$$

SEE INFORMATION THEORY; PULSE MODULATION.

Two important performance figures are the dynamic range and signal-to-noise ratio. If it is assumed that all unwanted noise and interference have been eliminated, the remaining noise will be that due to the process of quantization, with amplitude of the order of the difference in amplitude between adjacent quantization levels. Hence, both the dynamic range, defined as the ratio of the amplitude of the largest undistorted signal to the amplitude of the noise, and the signal-to-noise ratio are just 2^{16}, and are given in decibels by Eq. (2). This is very close to the accepted

$$\text{Dynamic range} = 10 \log_{10}(2^{16})^2 = 92 \text{ dB} \qquad (2)$$

figure of 100 dB for the dynamic range of a live orchestra. As a comparison, the best figure of dynamic range for stylus disks and professional magnetic tapes is about 70 dB, and acceptable signal-to-noise ratios for conventional audio systems are in the region of 50–60 dB.

Writing to and reading from the disk. Figure 3 shows the disk mastering and replication process. A glass disk is covered by a uniform coating of photoresist material (Fig. 3*a*) upon which a laser is shone

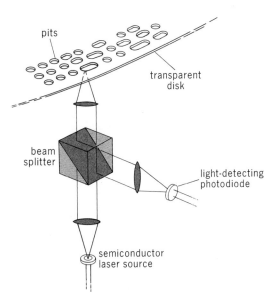

Fig. 1. Optical readout system for audio compact disk.

Fig. 2. Pits on an audio compact disk. (*From M. G. Carasso et al., Compact disc: Digital audio, Philips Tech. Rev., 40(6):151–180, 1982*)

where the pits are to be formed. The photoresist is then developed and washed, leaving the master disk (Fig. 3b). A nickel mother is derived from this master (Fig. 3c) and is then used to stamp out multiple copies of the disk in transparent plastic material (Fig. 3d). Each of these is then coated with a thin metallic reflecting layer, with a protective plastic coating on top of that (Fig. 3e).

The correspondence between the channel bit stream (the electrical signal) and the cross section of an audio disk is shown in Fig. 3e and f. A "1" appears in the time slot immediately following both the beginning and end of a pit; otherwise the signal is "0."

Electromechanical control. Realization of advantages of the compact disk, with its high density and near-incorruptibility of stored data, requires the design of advanced control systems in the compact-disk player.

The lens which focuses the laser onto the pit track has a depth of focus of 2 μm. But, since the vertical movement of the disk is more than a hundred times this, automatic focusing must be provided. This is done by inserting a cylindrical lens in the path of the reflected beam and allowing this to impinge on a four-quadrant segmented photodetector. The changing ellipticity of the focused spot is registered as a differential signal between the two opposite pairs of photodetector segments, which provides the control signal for the focusing servomechanism. *See Servomechanism.*

Lateral tracking control is achieved by providing two extra laser spots, with one straddling each edge of the track when the main spot is properly centered. The differential signal from two photodetectors positioned to pick up the reflection from the two extra spots provides the lateral position control signal.

The constant linear velocity of the beam along the track is ensured by synchronizing the data read off the disk with a quartz-crystal oscillator. *See Oscillator.*

Signal processing. Since the information that is read off the disk is in digital form, as a sequence of 0s and 1s, it can be processed in many more ways than were possible with analog systems. To enable this, information can be stored for as long as is desired, and then output at a rate that is controlled by the player's quartz-crystal (oscillator) clock, hence eliminating the wow and flutter of conventional systems entirely.

Error correction. Errors in the information read off the disk fall broadly into two categories. A random error in an individual bit may be due to an imperfectly

formed pit or imperfection formed in the plastic disk during the stamping process. A burstlike error produced by severe scratches or dirt on the outer layer of the transparent disk affects many adjacent tracks.

The detection and correction of single-bit errors is achieved by the addition of extra check bits known as parity bits. The bit stream can be divided into blocks and the bits in a block arranged in the form of a rectangular grid or matrix. If a parity bit is assigned to every row and every column, then, on the assumption that only one bit per block is in error, it is possible to identify precisely which bit is in error, and hence to correct it. Extensions of this same idea are used in the detection and possible correction of bursts of errors, but implementation is very complex and is ultimately realized in an integrated circuit of 36,000 transistors.

Conversion of signal to analog form. The mathematical theory of sampling is misleading as regards its practical implementation, since it is impossible to make filters required by the theory to convert the digital signal back to analog form. The filters should have a flat amplitude and linear phase with frequency out to a little beyond 20 kHz, at which point the amplitude should suddenly fall to zero.

This difficulty is surmounted by applying the technique of oversampling, in which the sampling rate is

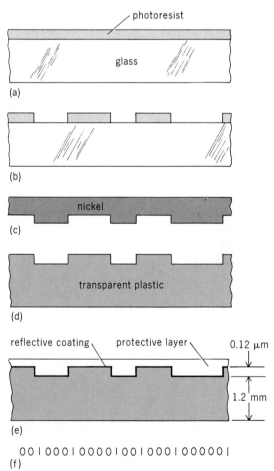

Fig. 3. Mastering and replication process for an audio compact disk. (a) Glass disk covered by uniform layer of photoresist. (b) Pits formed by a focused laser in the photoresist, which is developed and washed away, giving the master. (c) Nickel mother, formed from master. (d) Transparent plastic disk stamped from nickel mother. (e) Disk with reflective coating and protective layer added. (f) Associated binary electrical signal (channel bit stream).

increased by a factor of, say, 4, from 44.1 to 176.4 kHz. This technique has the effect of requiring filter bandwidths of a little over 80 kHz while the audio bandwidth remains limited to 20 kHz. Filters having the necessary linearity in both amplitude and phase over this 0–20 kHz range, but falling off in response at much higher frequencies, are relatively easily implemented.

Video disks and computer storage. The video disk and the compact disk for computer storage have been slow to develop for both economic and technical reasons. A major difficulty in video recording is the large bandwidths required. Conventional television requires a bandwidth of 5 MHz, while the more advanced high-definition television (HDTV) will require about 30 MHz. (The corresponding rates are 10 and 60 megabits per second.)

The main difficulty in using video and compact disks for recording computer data involves the allowable error rate. For data this must be effectively zero, whereas for both audio and video recording a small but finite error rate is tolerable.

Current methods of writing to video disks for permanent recording are by melting pits or by the process of photopolymerization. Both sides of video disks are used, and 8000 megabits can be stored per side.

Erasable compact disks are also produced. The information is stored magnetooptically. A laser heats the surface of a special magnetic material, and a local magnetic field is applied to reverse the magnetization, thus forming a bit. A lower-power laser is used to read this bit by being focused on the spot, and the change in polarization of the reflected light indicates the presence of the bit. The bit is erased by using a higher-power laser and reversing the direction of magnetization. *See* Computer storage technology; Disk recording; Video disk recording.

Richard H. Clarke

Bibliography. M. G. Carasso et al., Compact disc: Digital audio, *Philips Tech. Rev.*, 40(6):151–180, 1982; S. Miyaoka, Digital audio is compact and rugged, *IEEE Spectrum*, 21(3):35–39, 1984; J. B. H. Peek, Communications aspects of the compact disc digital audio system, *IEEE Commun. Mag.*, 23(2):7–15, 1985.

Comparator circuit

An electronic circuit that produces an output voltage or current whenever two input levels simultaneously satisfy predetermined amplitude requirements. A comparator circuit may be designed to respond to continuously varying (analog) or discrete (digital) signals, and its output may be in the form of signaling pulses which occur at the comparison point or in the form of discrete direct-current (dc) levels.

Linear or analog comparator. A linear comparator operates on continuous, or nondiscrete, waveforms. Most often one voltage, referred to as the reference voltage, is a variable dc or level-setting voltage and the other is a time-varying waveform. When the signal voltage becomes equal to the reference voltage, a discrete output level is obtained. If the time-varying (signal) voltage approaches the reference voltage from a more negative level the output voltage is of one polarity; if it approaches the reference from a more positive value the output is of the opposite polarity.

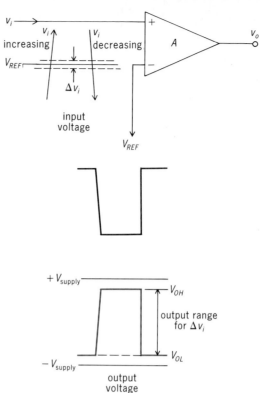

Fig. 1. High-gain nonregenerative operational amplifier as voltage comparator.

Operational-amplifier voltage comparator. A very high-gain operational amplifier whose output is inverting with respect to one input terminal and noninverting with respect to the other and whose output voltage is limited at upper and lower levels (usually at voltages near the supply voltage levels, $+V_{\text{supply}}$) may be used as a voltage comparator (**Fig. 1**). The upper and lower output levels are designated as V_{OH} and V_{OL}, respectively. When the input signal v_i equals V_{REF}, the output is approximately in the center of this range. The change in input voltage around V_{REF} required to drive the output to its respective limits is

$$\Delta v_i = \frac{V_{OH} - V_{OL}}{A} \quad (1)$$

given by Eq. (1), where A is the voltage gain of the amplifier. If A is very large, Δv_i approaches 0; as v_i approaches V_{REF} from the negative side, the output switches abruptly from V_{OL} to V_{OH} as soon as v_i becomes bigger than V_{REF}; and as v_i approaches V_{REF} from the positive side, the output switches abruptly to V_{OL} as soon as v_i becomes less than V_{REF}. If v_i and V_{REF} are interchanged, the output polarity shifts are reversed.

Comparators are often designed with a built-in offset voltage; that is V_{OH} and V_{OL} need not be symmetrical about V_{REF}. For example, high and low output levels may be designed as $V_{OH} = +5$ V and $V_{OL} = 0$, which would be compatible with standard saturated logic levels; or as $V_{OH} = 0$ and $V_{OL} = -5$ V, for outputs compatible with standard emitter-coupled logic levels. *See* Logic circuits.

The high-gain operational-amplifier comparator used in the open-loop mode is designated as a nonregenerative comparator because there is no positive feedback path from the output back to the input.

An operational-amplifier comparator connected in a positive-feedback mode (**Fig. 2**) is referred to as a

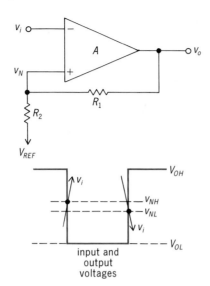

Fig. 2. Regenerative operational amplifier as comparator (Schmitt trigger). The input and output voltages are superimposed, showing hysteresis.

regenerative comparator. If the voltage gain A is very high, then the feedback voltage v_N is given by Eq. (2)

$$v_N = v_{NH} = \frac{R_2 V_{OH} + R_1 V_{REF}}{R_1 + R_2} \qquad (2)$$

when the output voltage is $v_o = V_{OH}$; and when $v_o = V_{OL}$ the feedback voltage is given by Eq. (3). For this

$$v_N = v_{NL} = \frac{R_2 V_{OL} + R_1 V_{REF}}{R_1 + R_2} \qquad (3)$$

circuit there is a difference between the input levels which will cause the output to switch, given by Eq. (4). This difference in input levels is referred to as

$$v_{NH} - v_{NL} = \frac{R_2}{R_1 + R_2}(V_{OH} - V_{OL}) \qquad (4)$$

the hysteresis of the circuit. Such regenerative comparators are generically referred to as Schmitt triggers. Historically, they were first implemented as discrete two-vacuum-tube devices, which were later replaced by transistors and then by the operational amplifier feedback circuit. *See* AMPLIFIER; FEEDBACK CIRCUIT; HYSTERESIS; OPERATIONAL AMPLIFIER.

Digital comparator. The term digital comparator has historically been used when the comparator circuit is specifically designed to respond to a combination of discrete level (digital) signals, for example, when one or more such input signals simultaneously reach the reference level which causes the change of state of the output. Among other applications, such comparators perform the function of the logic gate such as the AND, OR, NOR, and NAND functions. More often, the term digital comparator describes an array of logic gates designed specifically to determine whether one binary number is less than or greater than another binary number. Such digital comparators are sometimes called magnitude or binary comparators.

Applications. Comparators may take many forms and can find many uses, in addition to those which have been discussed. For example, the electronically regulated dc voltage supply uses a circuit which compares the dc output voltage with a fixed reference level. The resulting difference signal controls an am-

plifier which in turn changes the output to the desired level. In a radio receiver the automatic gain control circuit may be thought of broadly as a comparator; it measures the short-term average of the signal at the output of the detector, compares this output with a desired bias level on the radio-frequency amplifier stages, and changes that bias to maintain a constant average level output from the detector. *See* AUTOMATIC GAIN CONTROL (AGC).

Regenerative comparators (Schmitt triggers) are often used as the key element in the voltage-controlled oscillator (VCO), which in turn is the control element in the phase-locked loop (PLL). Nonregenerative comparators are often used in parallel analog-to-digital converters where a number of comparators (equal to the desired bit rate) are used at a succession of higher V_{REF} levels. *See* ANALOG-TO-DIGITAL CONVERTER; PHASE-LOCKED LOOPS.

Glenn M. Glasford

Bibliography. G. Glasford, *Analog Electronic Circuits*, 1986; G. Glasford, *Digital Electronic Circuits*, 1988; J. Millman and A. Grabel, *Microelectronics*, 2d ed., 1987; J. Millman and H. Taub, *Pulse Digital and Switching Waveforms*, 1965; H. Taub and D. Schilling, *Digital Intergrated Electronics*, 1977.

Complement

A group of proteins in the blood and body fluids that are sequentially activated by limited proteolysis and play an important role in humoral immunity and the generation of inflammation. When activated by antigen–antibody complexes, or by other agents such as proteolytic enzymes (for example, plasmin), complement kills bacteria and other microorganisms. In addition, complement activation results in the release of peptides that enhance vascular permeability, release histamine, and attract white blood cells (chemotaxis). The binding of complement to target cells also enhances their phagocytosis by white blood cells. The most important step in complement system function is the activation of the third component of complement (C3), which is the most abundant of these proteins in the blood. *See* IMMUNITY.

There are two mechanisms for the activation of C3, the classical and the alternative pathways. The classical pathway requires calcium and is activated when the C1q component binds to immunoglobulin antibodies IgG or IgM. In contrast, activation of the alternative pathway requires magnesium and is not dependent on antibodies. The activation of C3 by either pathway results in the cleavage of C3 to a small fragment, C3a, and a large fragment, C3b. The latter binds to a bimolecular enzyme, and a new trimolecular complex is formed which cleaves C5 into a small fragment, C5a, and a large fragment, C5b. This is the final enzymatic step in complement activation; the rest of the activation process is nonenzymatic. An amphiphilic macromolecular complex is formed by the binding of one molecule of C5b to one molecule each of C6, C7, and C8, and 12 or more molecules of C9. This so-called membrane attack complex invades the membrane of target cells and kills them. At first these complexes become embedded in target-cell membranes without inflicting any damage. After a molecule of C8 binds to the C5b,6,7 complex, a transmembrane channel (pore) is formed, and the complex can destroy target cells, although very slowly. The process is much accelerated by polymerization of C9 into the complexes, forming a cylindrical structure. The top of the cylinder is hydrophilic and capped by a ring of annulus that remains outside the target cell. The other end is lipophilic; it becomes embedded in the cell membrane and forms large transmembrane channels which coalesce into so-called leaky patches. The target cell loses potassium, imbibes sodium and water, swells, and bursts. This is called cell lysis or hemolysis in the case of a red blood cell.

As mentioned earlier, several small fragments are enzymatically generated during complement activation, and these act to release histamine from mast cells. The C5a fragment is potent in chemotaxis and in activating phagocytes. The C2b fragment contains a peptide that, upon release by further proteolysis, markedly enhances vascular permeability. Phagocytes and red blood cells have a receptor for fragments C4b and C3b, called complement receptor (CR1), which appears to play an important role in clearing immune complexes from blood. Phagocytes also have a complement receptor 3 (CR3) which, when engaged, stimulates phagocytes to produce superoxide and enhances phagocytic killing of ingested bacteria. A third complement receptor (CR2) is found only on B lymphocytes. SEE ALLERGY; CELLULAR IMMUNOLOGY; HISTAMINE.

Genetic deficiencies of certain complement subcomponents, including properdin, have been found in humans, rabbits, guinea pigs, and mice. All are inherited as autosomal recessives, except for properdin deficiency, which is X-linked recessive, and C1 inhibitor deficiency, which is autosomal dominant. Certain deficiencies lead to immune-complex diseases, such as systemic lupus erythematosus; other deficiencies result in increased susceptibility to bacterial infections, particularly those of the genus *Neisseria* (for example, gonorrhea and meningococcal meningitis), and hereditary angioneurotic edema. SEE COMPLEMENT-FIXATION TEST.

Fred S. Rosen

Bibliography. R. D. Campbell et al., Structure, organization, and regulations of the complement genes, *Annu. Rev. Immunol.*, 6:161–196, 1988; H. J. Muller-Eberhard, The membrane attack complex of complement, *Annu. Rev. Immunol.*, 4:503–528, 1986.

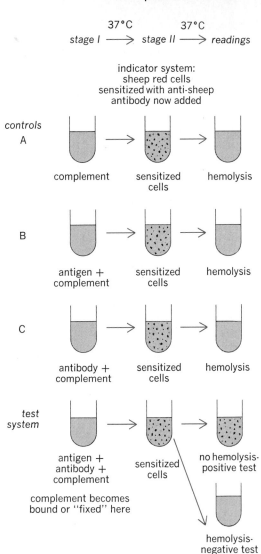

Diagram of the complement-fixation reaction. 37°C = 98.6°F

Complement-fixation test

A sensitive reaction used in serology for the detection of either antigen or antibody, as in the diagnosis of many bacterial, viral, and other diseases, including syphilis. It involves two stages: state I is the binding or fixation of complement if certain antigen-antibody reactions occur, and stage II is detection of residual unbound complement, if any, by its hemolytic action on the sensitized erythrocytes subsequently added. This is shown in the **illustration**. SEE ANTIBODY; ANTIGEN; COMPLEMENT.

In the first stage either the antigen or the antibody must be supplied as a reagent, with the other of the pair as the test unknown. Fresh guinea pig serum is normally used as a complement source. Sheep erythrocytes which are coated with their corresponding antibody (amboceptor or hemolysin) are used in the second stage.

The controls A, B, and C in the illustration demonstrate that sufficient complement is present to effect hemolysis of the sensitized indicator cells and that neither antigen nor antibody added alone will interfere with this by binding complement. In the test system the combination of a suitable antigen and antibody in the presence of complement will bind the complement to the complex so that the complement becomes unavailable for the hemolysis of the indicator cells added in stage II. If either antigen or antibody is added as a reagent in stage I, then the presence of the other in the test unknown that is added can be detected through its ability to complete the antigen-antibody system. A lack of hemolysis denotes that the complement is bound. This result is termed a positive complement-fixation reaction and, if the system is controlled against false reactions, implies that the antigen (or antibody) assumed to be in the test unknown was actually present. Conversely, complete hemolysis of the sensitized erythrocytes implies, ultimately, absence of antigen or antibody and is termed a negative complement-fixation reaction. In its most sensitive form the microcomplement-fixation test can distinguish the minute antigenic difference that results from the shift in a single amino acid between hemoglobins A and S. SEE IMMUNOLOGY; LYTIC REACTION; SEROLOGY.

Henry P. Treffers

Bibliography. W. R. Clark, *The Experimental Foundations of Modern Immunology*, 1986; S. B. Pal (ed.), *Reviews on Immunoassay Technology*, vol. 1, 1988.

Complementation (genetics)

The complementary action of different genetic factors. The term usually implies two homologous chromosomes or chromosome sets, each defective because of mutation and unable by itself to promote the normal development or metabolism of the organism, but able to do so jointly when brought together in the same cell. *See* Chromosome; Mutation.

Methods for testing. In a diploid organism, such as a flowering plant or an animal, in which there are two sets of chromosomes in each cell nucleus, one can test for complementation by crossing mutant strains and observing whether or not the hybrid progeny are mutant in character. In fungi, which are usually haploid (one set of chromosomes in each nucleus), complementation tests can often be made by forming heterokaryons, in which two different kinds of nuclei are present together in the same cell. In bacteria, which are haploid with only one chromosome, there are techniques for introducing an additional chromosome fragment into a cell in order to test the fragment's ability to complement the chromosome already present.

Test analysis. Complementation tests are used to investigate the relationships between different genetic mutations which individually cause various functional defects in the organism such as failures of growth or development. If mutations a and b are at different sites in the same chromosome, a^+ and b^+ being the corresponding nonmutant sites, a complementation test involves making the chromosome combination ab^+/a^+b. This is called the trans configuration. In principle it can be compared to the situation in which both mutations are present in one chromosome and both normal sites in the other (the cis configuration) ab/a^+b^+. If a^+ and b^+ act independently, the cis and trans arrangements should both give a nondefective organism, since both contain both a^+ and b^+. If, on the contrary, a and b are within a functionally indivisible unit, cis will be normal but trans mutant (no complementation).

Noncomplementary units. S. Benzer proposed the term cistron for the unit within which mutants do not complement each other. The word gene is often used in the same sense. The usual biochemical function of a cistron, or gene, is to determine the structure of a specific polypeptide component of a protein. Full complementation between different genes is the rule except when, as sometimes in bacteria, the genes form part of a functionally coordinated complex (operon). Allelic mutants (mutants within one gene) show limited complementation in some cases, for example, when certain pairs of mutant polypeptides correct each other's defects through coaggregation in a complex protein. *See* Genetics; Operon.

J. R. S. Fincham

Bibliography. R. E. Langman, *The Immune System*, 1989; S. K. A. Law and K. B. M. Reid, *Complement*, 1988; G. D. Ross, *Immunobiology of the Complement System*, 1986; K. Rother and G. O. Till (eds.), *The Complement Systems*, 1988.

Complex numbers and complex variables

A natural and extremely useful extension of the familiar real numbers. They can be introduced formally as follows. Consider the two-dimensional real vector space consisting of all ordered pairs (a_1,a_2) of real numbers. Geometrically this space can be identified with the ordinary euclidean plane, viewing the real numbers a_1,a_2 as the coordinates of a point in the plane.

The addition of vectors is defined by $(a_1,a_2) + (b_1,b_2) = (a_1 + b_1, a_2 + b_2)$ and is just the usual addition of vectors by the parallelogram law (**Fig. 1**). The multiplication of a vector (a_1,a_2) by a real number c is defined by $c(a_1,a_2) = (ca_1,ca_2)$, and is just the uniform dilation of the plane by the factor c. It may be asked whether it is possible to define a multiplication of one vector by another in such a manner that this multiplication is linear and satisfies the same formal rules as multiplication of real numbers. That is to say, it may be asked whether it is possible to associate to any two vectors (a_1,a_2) and (b_1,b_2) a product vector $(a_1,a_2) \cdot (b_1,b_2)$ in such a manner that:

(1) The multiplication is linear in the sense that

$$(a_1,a_2) \cdot [(b_1,b_2) + (c_1,c_2)]$$
$$= (a_1,a_2) \cdot (b_1,b_2) + (a_1,a_2) \cdot (c_1,c_2)$$

(2) It is associative,

$$(a_1,a_2) \cdot [(b_1,b_2) \cdot (c_1,c_2)]$$
$$= [(a_1,a_2) \cdot (b_1,b_2)] \cdot (c_1,c_2)$$

(3) It is commutative,

$$(a_1,a_2) \cdot (b_1,b_2) = (b_1,b_2) \cdot (a_1,a_2)$$

(4) There is an identity vector (e_1,e_2) such that $(e_1,e_2) \cdot (a_1,a_2) = (a_1,a_2)$ for all vectors (a_1,a_2).

(5) For each vector (a_1,a_2) other than $(0,0)$ there is a vector (b_1,b_2) for which $(b_1,b_2) \cdot (a_1,a_2) = (e_1,e_2)$.

There does exist such a multiplication, given by Eq. (1), and it is essentially unique in the sense that

$$(a_1,a_2) \cdot (b_1,b_2) = (a_1b_1 - a_2b_2, a_1b_2 + a_2b_1) \quad (1)$$

any multiplication satisfying all the desired properties can be reduced to Eq. (1) by a suitable choice of co-

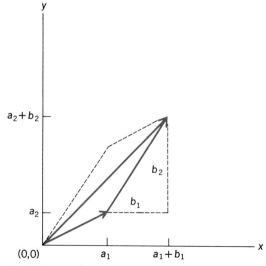

Fig. 1. Vector addition of two complex numbers by the parallelogram law.

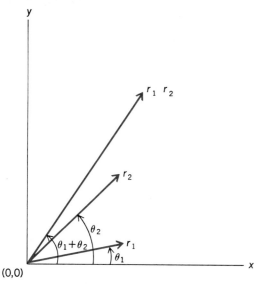

Fig. 2. Multiplication of complex numbers.

ordinates in the plane. The plane with the ordinary addition and scalar multiplication of vectors and with the vector multiplication in Eq. (1) is the complex number system. It is not possible to introduce a corresponding multiplication on vector spaces of higher dimensions, although, for example, if the commutativity condition is dropped, there is a multiplication satisfying the remaining conditions on the four-dimensional vector space (the quaternions). *See* Quaternions.

Geometric interpretation. The multiplication in Eq. (1) can most easily be interpreted geometrically by introducing polar coordinates in the plane and writing $(a_1,a_2) = (r \cos \theta, r \sin \theta)$, where r is the length of the vector (a_1,a_2), or the modulus of the complex number (a_1,a_2), defined by $r^2 = a_1^2 + a_2^2$, and θ is the angle between the vector (a_1,a_2) and the first coordinate axis, or the argument of the complex number (a_1,a_2), defined by $\tan \theta = a_2/a_1$. It follows immediately from Eq. (1), by using the familiar addition formulas for the sine and cosine functions that

$$(r_1 \cos \theta_1, r_1 \sin \theta_1) \cdot (r_2 \cos \theta_2, r_2 \sin \theta_2)$$
$$= [r_1r_2 \cos (\theta_1 + \theta_2), r_1r_2 \sin (\theta_1 + \theta_2)]$$

Thus multiplication of complex numbers amounts to multiplying their moduli and adding their arguments (**Fig. 2**). *See* Plane trigonometry.

Real and imaginary parts. Introducing the basis vectors $l = (1,0)$ and $i = (0,1)$, any vector (x,y) can be written uniquely as the sum $(x,y) = xl + yi$. Since the multiplication given by Eq. (1) is linear, the product of any two vectors is determined by giving the multiplication table for the basis vectors l and i. It follows readily from Eq. (1) that $l \cdot l = l$, $l \cdot i = i \cdot l = i$, and $i \cdot i = -l$, so that in particular the vector l is the identity element for the multiplication of complex numbers. Furthermore, the addition and multiplication of vectors of the form xl for all real numbers x reduce to the ordinary addition and multiplication of real numbers. There is consequently a natural imbedding of the real numbers into the complex numbers by associating to any real number x the complex number xl. The complex number l can be identified with the ordinary real number 1, and re-

flecting this identification the notation can be simplified by writing a complex number $z = (x,y) = xl + yi$ merely as $z = x + iy$. The component x is called the real part of the complex number z, and the component y is called the imaginary part. The modulus of the complex number z is denoted by $|z|$.

Complex conjugates. The vectors l and $-i$ are also a basis and satisfy exactly the same multiplication table as the vectors l and i, so the mapping which associates to any complex number $z = x + iy$ the complex number $\bar{z} = x - iy$ is a one-to-one mapping preserving the algebraic structure of the complex number system; thus $\overline{(z_1 + z_2)} = \bar{z_1} + \bar{z_2}$ and $\overline{(z_1 \cdot z_2)} = \bar{z_1} \cdot \bar{z_2}$. The complex number $\bar{z}$ is called the conjugate of z. It has the properties that $z = \bar{z}$ precisely when z is a real number, that $z \cdot \bar{z} = x^2 + y^2 = |z|^2$, and that in polar coordinates $\bar{z}$ is a complex number with the same modulus as z but with the negative argument (**Fig. 3**).

Polynomials. An ordinary real polynomial function $f(x) = x_n + a_1x^{n-1} + \cdots + a_n$ can be extended to a function of a complex variable z in the obvious manner, merely setting $f(z) = z^n + a_1z^{n-1} + \cdots + a_n$ and using the algebraic operations as defined for complex numbers. The advantage of this extension is that the fundamental theorem of algebra then holds: a nontrivial polynomial function over the complex numbers always has a root α, a complex value for which $f(\alpha) = 0$. The same result is true for polynomials with complex coefficients $a_1, \ldots, a_n$. A simple consequence of this result is that any polynomial function of degree n can be written as a product $f(z) = (z - \alpha_1)(z - \alpha_2) \cdots (z - \alpha_n)$ where the complex numbers $\alpha_1, \alpha_2, \ldots, \alpha_n$ are precisely the roots of this polynomial repeated according to multiplicity. Thus passing from the real to the complex number system simplifies the analysis of polynomial functions and clarifies many of their properties. This was in a sense the original motivation for introducing the complex number system.

Power series. In a somewhat similar manner, passing from the real to the complex number system simplifies and clarifies the analysis of more general functions than polynomials. Many of the familiar functions treated in elementary calculus have convergent Taylor series expansions and so can be written

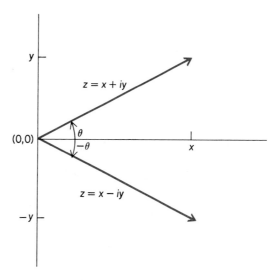

Fig. 3. Complex conjugate numbers.

for all real numbers x sufficiently near a real number a in the form of a convergent infinite series

$$f(x) = \sum_{n=0}^{\infty} a_n(x - a)^n$$

where $n!a_n = f^{(n)}(a)$. If this series converges when the real number x is replaced by nearby complex numbers z, there results a natural extension of the function f to a complex-valued function of the complex variable z.

The question of convergence can be handled rather easily. A series of the form

$$\sum_{n=0}^{\infty} a_n(z - a)^n$$

where a_n, a, z are complex numbers is called a complex power series centered at the point a. For any such series there is a value r, a nonnegative real number or ∞, such that the power series converges for all complex numbers z satisfying $|z - a| < r$ and diverges for all complex numbers z satisfying $|z - a| > r$; r is called the radius of convergence of the power series. If $r = 0$, the power series diverges for all complex numbers z other than $z = a$, while if $r = \infty$, the power series converges for all complex numbers. Otherwise, the series converges for all complex numbers lying within a circle of radius r centered at the point a and diverges for all complex numbers outside that circle. The series may or may not converge at complex numbers on the circle itself. The radius of convergence can be determined by Hadamard's formula: $1/r = \lim \sup_{n\to\infty} |a_n|^{1/n}$. Alternatively and generally more simply the radius of convergence satisfies

$$\lim \inf_{n\to\infty}|a_n|/|a_{n+1}| \le r \le \lim \sup_{n\to\infty}|a_n|/|a_{n+1}|$$

If r is the radius of convergence of a power series, then that series converges absolutely and uniformly on the set of all complex numbers z satisfying $|z - a| \le \rho$ whenever $\rho < r$.

Using these observations it is easy to see, for example, that the exponential function can be extended to a complex-valued function of the complex variable z defined for any z by

$$e^z = \sum_{n=0}^{\infty} z^n/n!$$

Among the trigonometric functions the sine and cosine functions can also be defined for arbitrary complex numbers z by their Taylorseries. Moreover, these functions satisfy the identity $e^{iz} = \cos z + i \sin z$ for all complex numbers z, an interesting relation among the elementary functions only apparent when passing to complex numbers. SEE SERIES.

Analytic functions. A complex-valued function f of a complex variable defined on some region Ω in the complex plane is analytic or holomorphic in Ω if for each point a in Ω the values of the function at all points z sufficiently near a are given by a power series expansion

$$f(z) = \sum_{n=0}^{\infty} a_n(z - a)^n$$

This power series must then be the Taylor series expansion of the function f at the point a and must con-

verge in at least the largest circle centered at a and contained in Ω. An analytic function in a region Ω is necessarily a continuous function in Ω; moreover, as a function of the two real variables (x,y), an analytic function has partial derivatives of all orders.

Characterization by differentiability. There are a number of alternative characterizations of analytic functions, all of which are useful. The most common and perhaps the most surprising characterization, which is often taken as the definition of an analytic function, is that a function f is analytic in Ω precisely when the limit given by expression (2) exists at all points a in

$$\lim_{z\to a}\frac{f(z) - f(a)}{z - a} \tag{2}$$

Ω. This limit is the complex analog of the ordinary derivative and is therefore called the complex derivative and written $f'(a)$. If

$$f(z) = \sum_{n=0}^{\infty} a_n(z - a)^n$$

where the power series converges in a disc Δ centered at a, then

$$f'(z) = \sum_{n=0}^{\infty} na_n(z - a)^{n-1}$$

and this series also converges in the disc Δ. SEE CALCULUS.

Cauchy-Riemann equation. If f is analytic near a point a, then the complex derivative (2) can be calculated by letting z approach a through a sequence of points having the same imaginary value as a, hence $f'(a) = \partial f/\partial x$. Alternatively the complex derivative can be calculated by letting z approach a through a sequence of points having the same real value as a, hence $f'(a) = -i\,\partial f/\partial y$. Upon comparing these two results it follows that an analytic function f must satisfy the Cauchy-Riemann equation $\partial f/\partial x + i\,\partial f/\partial y = 0$, which is sometimes written $\partial f/\partial \bar{z} = 0$. Conversely if f is a complex-valued function defined in a region Ω of the complex plane, and if f has continuous first partial derivatives with respect to the real coordinates x and y, and if they satisfy the Cauchy-Riemann equation at all points of Ω, then f is analytic in Ω. It is not enough merely to require the existence of the first partial derivatives, and so this is not such a clean characterization of analytic functions as the preceding one, but it is nonetheless very useful. Weaker regularity conditions than the continuity of the first partial derivatives are possible. For instance, if the function f has first partial derivatives in the sense of distributions and if they satisfy the Cauchy-Riemann equation, then f is analytic.

Relation to harmonic functions. Setting $f(x,y) = u(x,y) + iv(x,y)$, where u and v are real-valued functions, the Cauchy-Riemann equation can be written equivalently as the pair of equations $\partial u/\partial x = \partial v/\partial y$, $\partial u/\partial y = -\partial v/\partial x$. It follows immediately from this that $\partial^2 u/\partial x^2 + \partial^2 v/\partial y^2 = 0$ and similarly for v, so that the real and imaginary parts of an analytic function are harmonic functions. Conversely if u is harmonic in a simply connected region Ω of the complex plane, the pair of real Cauchy-Riemann equations determines a harmonic function v such that $u + iv$ is analytic in Ω, and v is unique up to an additive constant; if Ω is not simply connected, the function v may not be single-valued. SEE DIFFERENTIAL EQUATION;

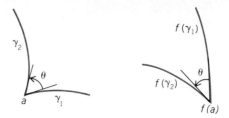

Fig. 4. A conformal mapping that preserves angles at *a*.

Laplace's differential equation.

Morera's theorem. If f is an analytic function in a region Ω of the complex plane, then it follows from the Cauchy-Riemann equation that, using the terminology and notation of differential forms, the differential form $fdz = fdx + ifdy$ is closed, since

$$d(fdx + ifdy) = (i\partial f/\partial x - \partial f/\partial y)dx \wedge dy = 0$$

Therefore

$$\int_\gamma fdz = 0$$

for any closed path γ homologous to zero in Ω. Conversely Morera's theorem asserts that if f is a continuous complex-valued function in Ω and if

$$\int_\gamma fdz = 0$$

for all closed paths γ homologous to zero in Ω, then f is necessarily analytic in Ω. From this characterization of analyticity it follows immediately that the limit of a uniformly convergent sequence of analytic functions is also analytic.

Conformal mapping. An analytic function f in a region Ω of the complex plane can be viewed as a mapping from the plane region Ω into the complex plane. If $f(z) = u(z) + iv(z)$, this mapping is described in real terms by the pair of functions $u(x,y)$, $v(x,y)$. It follows readily from the Cauchy-Riemann equation that the jacobian determinant of this real mapping is given by Eq. (3). Since the jacobian determinant is

$$\det\begin{pmatrix} u_x & u_y \\ v_x & v_y \end{pmatrix} = u_x^2 + v_x^2 = |f'(z)|^2 \qquad (3)$$

thus always nonnegative, the mapping always preserves orientation. Moreover, the mapping is locally one-to-one whenever $f'(z) \neq 0$, and its inverse is readily seen also to satisfy the Cauchy-Riemann equation and therefore to be analytic. By analyzing further the condition imposed by the Cauchy-Riemann equation it follows that if $f'(a) \neq 0$, if γ_1 and γ_2 are two differentiable paths through the point a, and if the tangent vectors to γ_1 and γ_2 at the point a are at an angle θ apart, then the tangent vectors to the paths $f(\gamma_1)$ and $f(\gamma_2)$ at the point $f(a)$ are also at an angle θ apart. The mapping f thus preserves angles at a, and is therefore said to be conformal at a (**Fig. 4**). Conversely any continuously differentiable conformal mapping is an analytic function with a nonzero complex derivative, thus providing a very geometric characterization of analytic functions. *See Conformal Mapping; Partial Differentiation.*

For completeness, something should be said about the mapping property of an analytic function at a point at which the complex derivative is zero. The analytic function $f(z) = z^n$, for example, has a zero derivative at $z = 0$ whenever $n > 1$. The mapping is not conformal but increases angles at the origin by a factor of n (**Fig. 5**) and so can be viewed as a mapping which wraps each circle about the origin in its domain n times about the origin in its range. Now if f is any analytic function at the origin with $f(0) = f'(0) = 0$, then this function can be written in the form $f(z) = g(z)^n$ for some integer $n > 1$, where g is analytic at the origin, $g(0) = 0$ but $g'(0) \neq 0$. Therefore the change of variable $w = g(z)$ reduces the function f to the form $f(w) = w^n$ just considered. It is clear from this that an analytic function takes open sets to open sets even when its derivative is zero and that any analytic function satisfies the maximum mod-

ulus theorem: if f is analytic at a and $|f(a)| \geq |f(z)|$ for all points z in some neighborhood of a, then f must actually be a constant. This seemingly simple property has a considerable range of applications.

Cauchy integral formula. An integral representation formula due to A. Cauchy plays a fundamental role in complex analysis and is traditionally the principal tool used in proving the equivalence of some of the alternative characterizations of analytic functions given above. The formula asserts that if γ is the smooth boundary curve of a finite region Ω in the complex plane (**Fig. 6**) and if f is analytic in an open neighborhood of the closed region $\Omega \cup \gamma$ (the union of Ω and γ), then Eq. (4) is valid. This shows, for

$$\frac{1}{2\pi i}\int_\gamma \frac{f(\zeta)d\zeta}{\zeta - z} = \begin{cases} f(z) & \text{whenever } z \text{ is a} \\ & \text{member of } \Omega \\ 0 & \text{whenever } z \text{ is not a} \\ & \text{member of } \Omega \cup \gamma \end{cases} \qquad (4)$$

example, that the values of the analytic function f in Ω are completely determined by the values of that function on the boundary of Ω. Equation (4) can be differentiated under the integral sign to yield the companion formula (5) for the derivatives of the function

$$f^{(n)}(z) = \frac{n!}{2\pi i}\int_\gamma \frac{f(\zeta)\,d\zeta}{(\zeta - z)^{n+1}} \qquad (5)$$

f at any point z in Ω. In particular, if Ω is a disc of radius r centered at a point a and if $|f(z)| \leq M$ on the boundary γ of Ω, then (5) implies that $|f^{(n)}(a)| \leq Mr^{-n}n!$. Thus the derivatives of an analytic function cannot grow faster than $r^{-n}n!$ for some r as n tends to infinity. An immediate consequence of this estimate is Liouville's theorem, the assertion that the only functions analytic and bounded in the entire plane are the constant functions.

Relation to Green's integral theorem. The Cauchy integral formula (4) can be considered as a special case of a more general integral representation formula, following fairly directly from Green's integral theorem. If Ω and γ are as before and if f is any continuously differentiable complex-valued function of the real variables (x,y) in an open neighborhood of the closed region $\Omega \cup \gamma$, then Eq. (6) is valid for all z in Ω. Since analytic functions are characterized by the Cauchy-Riemann equation $\partial f/\partial \bar{z} = 0$, it is evident that (6)

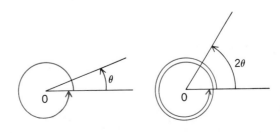

Fig. 5. The mapping z^2 at the origin.

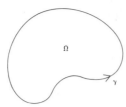

Fig. 6. The geometric situation in which the Cauchy integral formula holds.

$$f(z) = \frac{1}{2\pi i} \int_{\gamma} f(\zeta) \frac{d\zeta}{\zeta - z}$$
$$+ \frac{1}{2\pi i} \int \int_{\gamma} \frac{\partial f(\zeta)}{\partial \bar{\zeta}} \frac{d\zeta \wedge d\bar{\zeta}}{\zeta - z} \quad (6)$$

reduces to (4) for analytic functions. The more general formula (6) is useful in investigating solutions of the partial differential equation $\partial f / \partial \bar{z} = g$ for a given function g. This equation has many applications in complex analysis.

Relation to Poisson integral formula. The Cauchy integral formula (4) is also closely related to the Poisson integral formula for harmonic functions, which can be derived from Eq. (4). The Poisson formula is useful in solving boundary-value problems for harmonic functions. For any continuous function on the smooth boundary γ of a bounded region Ω, there exists a harmonic function u in Ω with the given boundary values. If Ω is simply connected, the function u is the real part of an analytic function in Ω, so there thus exists a holomorphic function in Ω with an arbitrarily specified real part on γ. The imaginary part of this analytic function is determined uniquely up to an additive constant by the real part, so there cannot exist an analytic function in Ω with arbitrarily specified values on the boundary curve γ. On the other hand, if g is any sufficiently smooth function on γ, the integral

$$\int_{\gamma} g(\zeta) d\zeta / (\zeta - z)$$

represents an analytic function $f_i(z)$ at all points z in Ω and another analytic function $f_0(z)$ at all points outside $\Omega \cup \gamma$. These functions have well-defined boundary values on γ, and $f_i(z) - f_0(z) = 2\pi i g(z)$ whenever z is on γ; the function $f_0(z)$ vanishes only when g is the boundary value on γ of an analytic function in Ω.

Isolated singularities. If a function f is analytic at all points of a disc $\triangle$ centered at a, except perhaps at the point a itself, a is called an isolated singularity of the function f. There are three mutually exclusive possibilities for the nature of such a singularity.

(1) It may happen that $|f(z)|$ remains bounded near the point a. In that case the function can be extended to an analytic function even at the point a by suitably choosing the value of $f(a)$; the point a is then called a removable singularity.

(2) It may happen that $\lim_{z \to a} |f(z)| = \infty$. In that case a is a removable singularity for the function $g(z) = 1/f(z)$, so that $f(z) = 1/g(z)$ where $g(z)$ is an analytic function vanishing at the point a; and if $g(z) = (z - a)^k g_0(z)$ where $g_0(z)$ is analytic and $g_0(a) \neq 0$, then $f(z) = (z - a)^{-k} f_0(z)$ where $k > 0$ and $f_0(z)$ is analytic and $f_0(a) \neq 0$. The function $f(z)$ is said to have a pole of order k at the point a. The function

$f_0(z)$ has a Taylor expansion at the point a as usual, so the function $f(z)$ has an expansion given by Eq. (7).

$$f(z) = a_{-k}(z - a)^{-k} + \cdots + a_{-1}(z - a)^{-1}$$
$$+ a_0 + a_1(z - a) + a_2(z - a)^2 + \cdots \quad (7)$$

The portion

$$\sum_{n=-k}^{-1} a_n (z - a)_n$$

of this expansion is called the principal part or singular part of the function $f(z)$ at the point a, while the remainder of the series represents an analytic function near a.

(3) If neither of the two preceding cases arises, the function $f(z)$ is said to have an essential singularity at the point a. A function having an essential singularity admits a Laurent series expansion near the point a of the form (8), in which infinitely many terms involving

$$f(z) = \sum_{n=-\infty}^{+\infty} a_n(z - a)^n \quad (8)$$

negative powers of the variable $(z - a)$ appear. Singularities of this type can occur and do so frequently. For example, whenever $f(z)$ has a pole at a point a, then $e^{f(z)}$ has an essential singularity at the point a. The behavior of a function near an essential singularity is rather complicated. A theorem of K. Weierstrass asserts that in any neighborhood of an essential singularity the values taken by the function come arbitrarily near any complex value. A much deeper theorem of E. Picard asserts that in any neighborhood of an isolated singularity a function actually takes all complex values with at most one possible exception.

Global properties. The zeros of an analytic function are isolated, in the sense that if f is analytic near a point a and $f(a) = 0$, then f has no other zeros in some neighborhood of a. That leads quite easily to the identity theorem: if $f(z)$ and $g(z)$ are analytic in a connected region Ω and if $f(a_n) = g(a_n)$ for an infinite sequence of points a_n having a limit point inside Ω, then $f(z) = g(z)$ at all points z in Ω. On the other hand, if a_n is any sequence of points in Ω having no limit point inside Ω, there are analytic functions in Ω having zeros precisely at the points a_n. This result is particularly easy in case that Ω is the entire complex plane. Indeed, in that case any analytic function having zeros at the points a_n can be written as a product given by Eq. (9), where $h(z)$ is an analytic function

$$f(z) = z^m e^{h(z)} \sum_{n} \left(1 - \frac{z}{a_n}\right) e^{P_n(z)} \quad (9)$$

in the entire plane and $P_n(z)$ are suitably chosen polynomials ensuring that the product converges when there are infinitely many points a_n. There is a good deal of quite detailed information relating the growth of the function $f(z)$, the distribution of the zeros a_n, and the canonical product formula (9). This area of investigation is called the theory of entire functions. The infinite product expansions of even the elementary functions are quite interesting. For example, the expansion of $\sin \pi z$ is given by Eq. (10).

$$\sin \pi z = \pi z \sum_{n=1}^{\infty} \left(1 - \frac{z^2}{n^2}\right) \quad (10)$$

A function analytic in a region Ω of the complex

plane, except possibly for poles at some points of Ω, is called a meromorphic function in Ω. The points at which the function has poles can of course have no limit point inside Ω. Near any pole a function can be written as the quotient of two analytic functions, as already noted. There is an analogous global result: any meromorphic function in Ω can be written as the quotient of two functions each analytic in Ω. If a_1, a_2, . . . is any sequence of points in Ω having no limit point inside Ω and if $f_1(z), f_2(z)$, . . . are principal parts of poles at these points, so that

$$f_v = \sum_{n=-k_v}^{-1} a_{v,n}(z - a_v)^n$$

then there exists a meromorphic function $f(z)$ in Ω having poles with the specified principal parts at the specified points and no other singularities. This is the analog of the familiar partial fraction expansion for quotients of polynomials. The resulting expansions of even the elementary functions are quite interesting, as for example the expansion given by Eq. (11).

$$\pi \cot \pi z = \frac{1}{z} + \sum_{n=1}^{\infty} \frac{2z}{z^2 - n^2} \qquad (11)$$

Analytic functions of several variables. A complex-valued function depending on several complex variables $f(z_1, . . . , z_n)$ is called analytic at a point $(a_1, . . . , a_n)$ if the values of that function at all points z_j sufficiently near a_j are given by a multiple power series expansion (12). If this series converges

$$f(z_1, . . . , z_n) =$$
$$\sum_{i_1=0}^{\infty} \cdots \sum_{i_n=0}^{\infty} a_{i_1 \ldots i_n} (z - a_1)^{i_1} \cdots (z - a_n)^{i_n} \qquad (12)$$

at a point $(z_1^0, . . . , z_n^0)$ with $|z_i^0 - a_i| = r_i > 0$, then it is absolutely and uniformly convergent on the set of all points $(z_1, . . . , z_n)$ for which $|z_i - a_i| \geq \rho_i$ whenever ρ_i are any positive constants such that $\rho_i < r_i$. This series thus converges to the same value for any ordering of the terms. For example, it can be rewritten as the power series (13)

$$f(z_1, . . . , z_n) =$$
$$\sum_{i_n=0}^{\infty} a_{i_0} (z_1, . . . , z_{n-1}) \cdot (z_n - a_n)^{i_n} \qquad (13)$$

in the variable z_n with coefficients which are analytic functions of the variables $z_1, . . . , z_{n-1}$. Hence holding $z_1, . . . , z_{n-1}$ constant, the function $f(z_1, . . . , z_n)$ is analytic in the variable z_n alone, or more generally the function $f(z_1, . . . , z_n)$ is analytic in each variable separately. Conversely a surprisingly nontrivial theorem of F. Hartogs asserts that a function $f(z_i, . . . , z_n)$ that is analytic in each variable separately at all points of a region Ω is an analytic function of all n variables $z_1, . . . , z_n$ in Ω.

Extension of results from one variable. Some results extend quite directly from one to n complex variables: the uniform limit of analytic functions is analytic; the maximum modulus theorem holds; and there is an extension of the Cauchy integral formula, asserting that if $f(z_1, . . . , z_n)$ is analytic whenever z_j is a member of $\Omega_j \cup \gamma_j$ where Ω_j is a plane domain with smooth boundary γ_j, then Eq. (14) holds whenever z_j is a member of Ω_j for all indices j. However, for the most part, the theory of analytic functions of several com-

$$f(z_1, . . . , z_n) = \left(\frac{1}{2\pi i}\right)^n \int_{\zeta_1 \epsilon \gamma_1} \cdots \int_{\zeta_n \epsilon \gamma_n}$$
$$\frac{f(\zeta_1, . . . , \zeta_n)}{(\zeta_1 - z_1) \cdots (\zeta_n - z_n)} d\zeta_1 \cdots d\zeta_n \qquad (14)$$

plex variables is far from being merely an extension of the standard results of the classical theory of functions of a single complex variable, by considering one variable at a time. Indeed, the theory of functions of several complex variables has quite a distinctive character with a considerable variety of results which perhaps are not the expected generalizations of the classical results but which shed new light on the familiar theory of analytic functions of a single variable.

Analytic continuation. An examination of the extended Cauchy integral formula (14) indicates some of the differences between the cases $n = 1$ and $n > 1$. When $n = 1$, either formula (14) or (4) expresses the values of an analytic function inside a domain Ω in terms of the values of the analytic function on the boundary of Ω, whereas when $n > 1$, formula (14) expresses the values of an analytic function inside a domain $\Omega_1 \times \cdots \times \Omega_n$ in terms of the values of the analytic function on a very small piece of the boundary. For example, when $n = 2$ the domain $\Omega_1 \times \Omega_2$ can be viewed as an open set of the four-dimensional euclidean space, its boundary is the three-dimensional subset $(\Omega_1 \times \gamma_2) \cup (\gamma_1 \times \Omega_2)$, but the integration in (14) is extended across the two-dimensional part $\gamma_1 \times \gamma_2$ of the boundary. This difference is not superficial; it reflects the greater possibilities for analytic continuation of functions of $n > 1$ variables than of functions of a single variable.

For any domain Ω in the complex plane there are functions which are analytic in that domain but which cannot be extended as analytic functions across any part of the boundary of Ω. For example, there exist nontrivial analytic functions vanishing at any infinite sequence of points a_1, a_2, . . . of Ω having no limit point inside Ω, and if every boundary point of Ω is a limit of a subsequence of these points a_1, a_2, . . . then such an analytic function cannot be continued across any boundary point. However, there are domains Ω in the space of $n > 1$ complex variables such that any function analytic in Ω extends to an analytic function in a properly larger domain. For example, it was shown by Hartogs that if Ω is an open neighborhood of the boundary of a compact subset K in the space $n > 1$ complex variables, then every function analytic in Ω extends to a function analytic in the union of K and Ω (**Fig. 7**).

Thus in the study of functions of $n > 1$ complex variables there arise the problems of characterizing the natural domains of existence of analytic functions (called domains of holomorphy) and of determining for those domains which are not domains of holomor-

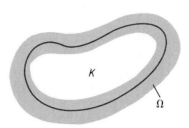

Fig. 7. Analytic continuation according to the theorem of Hartogs.

phy the largest domains to which all analytic functions can be extended (called the envelopes of holomorphy). Both problems have been extensively studied, the first one particularly so. Perhaps the most surprising result is that the domains of holomorphy can be characterized purely locally: a domain Ω in the space of n complex variables is a domain of holomorphy if and only if each point on the boundary of Ω has an open neighborhood $\cup$ such that the intersection of $\cup$ and Ω is a domain of holomorphy. The determination of the envelope of holomorphy of a domain is generally rather difficult, the situation being complicated by the fact that the envelope of holomorphy may not be realizable as a domain in the space of n complex variables.

Zeros and singularities. Another and not entirely unrelated difference between functions of $n > 1$ and functions of one complex variable is that in the case $n > 1$ the zeros and singularities of an analytic function are never isolated. The set of zeros of an analytic function of $n > 1$ variables is a set of topological dimension $2n - 2$ and is in most places locally euclidean but can have very complicated singularities. Strange spaces, such as the exotic seven spheres, manifolds which are topologically but not differentiably equivalent to the ordinary seven-dimensional sphere, arise in this context. Even the purely local study of functions of several complex variables thus presents a considerable challenge. SEE TOPOLOGY.

Robert C. Gunning

Bibliography. L. V. Ahlfors, *Complex Analysis*, 3d ed., 1979; H. Behnke and P. Thullen, *Theorie der Funktionen mehrer komplexer Veränderlichen*, 1970; J. B. Conway, *Functions of One Complex Variable*, 2d ed., 1979; E. Hille, *Analytic Function Theory*, 2d ed., 1973; L. Hormander, *An Introduction to Complex Analysis in Several Variables*, 2d ed., 1973; R. Nevanlinna and V. Paatero, *Introduction to Complex Analysis*, 2d ed., 1982.

Composite beam

Composite beams are so called because they provide beam action of two materials joined together in such a way as to act as a unit. In civil engineering the term is most commonly given to the beam action developed by a concrete slab resting on a steel beam, usually an I or a wide flange shape, the slab and the beam being made to act together by shear connectors. SEE CONCRETE SLAB.

In noncomposite design the structural function of the concrete slab is only to span transversely across a series of parallel steel beams. In composite design the concrete slab performs its normal function and in ad-

Fig. 1. Steel part of composite beam; spiral shear connectors welded to I beams. (*Porete Manufacturing Co.*)

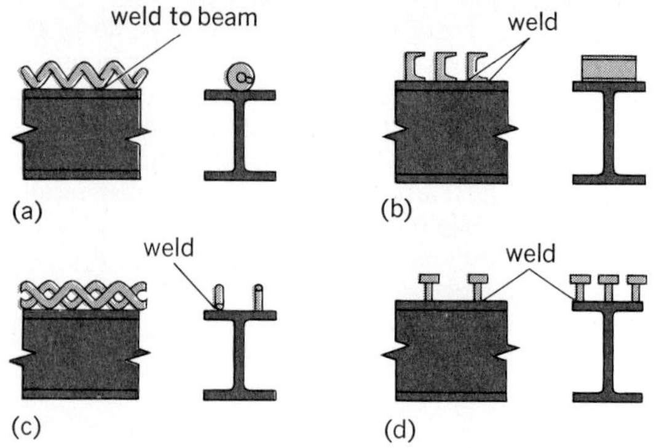

Fig. 2. Typical shear connectors. (*a*) Spiral shear connector. (*b*) Channel shear connector. (*c*) Serpentine shear connector. (*d*) Stud bolt connector.

dition increases to a considerable extent the capacity of the structure to resist bending in the direction of the steel beams. The unit stresses in the concrete slab are not added algebraically from the dual function the slab performs, because the flexural stresses are in mutually perpendicular directions. Thus the concrete slab thickness and area of reinforcing steel are not necessarily greater for this type of design than for the noncomposite. SEE STRESS AND STRAIN.

Shear connectors are devices that provide unified action between the concrete slab and the steel beam by resisting the horizontal shear developed in beam bending (**Fig. 1**). They take the form of steel projections welded to the top flange of the steel beam and around which the concrete slab is poured. The connectors perform a necessary function in the development of composite beam action, for without them the two elements, the concrete slab and the steel beam, would act separately as two beams and the beam action would be analogous to two unconnected wooden planks placed one on top of the other and slipping on each other when bent under a transverse load. There are many types of shear connectors, but three commonly used types are stud bolts, a helical coil of steel rod, and short lengths of a standard channel shape. **Figure 2** shows several types of shear connectors. A typical stud bolt might be ½ in. (1.2 cm) in diameter and 5 in. (12.5 cm) in length embedded in a 6-in.-thick (15-cm) concrete slab. The bolt would have a cylindrical head ¾ in. (1.8 cm) in diameter and perhaps ½ in. (1.2 cm) high. All shear connectors must have some provision to prevent the concrete slab from lifting vertically from the steel beam when bent under load. In the case of the stud bolt the head provides this restraint. The bolts are welded to the beam flange, several in a grouping, and spaced longitudinally along the beam in accordance with the design based on the intensity of the horizontal shearing stress at each position along the beam. A short channel length welded across the beam flange provides the same shear connection function as the stud bolts, or a helical coil of rod with a pitch of 6 or 8 in. (16 or 20 cm) and a diameter of 4 or 5 in. (10 or 12.5 cm) can connect the two elements. A typical rod diameter is ½ in. (1.2 cm). SEE SHEAR.

Composite beams are used both in building and highway bridge construction and can be used in any type of construction where it is appropriate to use a concrete slab resting on steel beams or stringers. Definite and considerable economy is achieved in composite construction when the spans are of sufficient length to offset the increased cost of fabricating the shear connectors with the saving from the lesser beam size required. Composite construction is generally not considered economically feasible for spans under 35 or 40 ft (10 or 12 m) but is used on simply supported spans up to 100–120 ft (30–36 m) or longer with specially fabricated steel beams. In the longer spans the size of the steel stringers can be reduced at least to the next lower standard rolled steel section. In most size ranges the beams are listed in groupings of beam depths in 3-in. (7.5-cm) multiples. For example, a beam design for noncomposite action which required a depth of 36 in. (0.9 m) could be expected to be reduced to a depth of 30–33 in. (0.7–0.8 m) if composite action is provided. *See* Beam.

Henry L. Kinnier

Bibliography. P. M. Ferguson, *Reinforced Concrete Fundamentals*, 1982; E. H. Gaylord and C. N. Gaylord, *Design of Steel Structures*, 2d ed., 1972; E. H. Gaylord and C. N. Gaylord, *Structural Engineering Handbook*, 3d ed., 1989; C. K. Wang and C. G. Salmon, *Reinforced Concrete Design*, 4th ed., 1984; G. Winter and A. H. Nilson, *Design of Concrete Structures*, 9th ed., 1979.

Composite material

A material that results when two or more materials, each having its own, usually different characteristics, are combined to provide the composite with useful properties for specific applications. Each input material should serve a specific function in the composite, which in turn should show distinctive new or improved characteristics (**Table 1**). As knowledge about the behavior and preparation of these materials grows, the scope of their useful applications will expand. Advanced composites are a class of structural high-performance, fiber-reinforced plastic or metallic materials.

The development and use of composite materials goes back to antiquity. Combining two or more high- and low-carbon irons or steels by skillfully hand-forging them together into a samurai sword with a lightweight, tough blade which could keep its edge is a typical example. Modern technology applies them to a growing variety of functions from fishing poles to lightweight automobiles or aircraft, and from machine tool bits to the thermal protection systems for reentry from space. The expanding activity in composites is stimulated both by the demands of such applications and by the potential embodied in the outstanding properties which have been achieved in substances such as high-strength glass filaments or stiff, refractory graphite filaments. The challenge with such filaments, for example, is to combine them with a suitable supporting, load-transferring, environmental protective material such as an epoxy polymer or a graphitizable pitch to produce efficient structural or refractory composite materials.

Attempts are often made to differentiate composite materials from conventional materials on the basis of the heterogeneous or multiphase nature of the former, but many materials which are considered conventional by generally accepted usage, such as high-strength structural alloys, are also multiphase. The range of anisotropy of properties which can result from different arrangements of the components of a composite has also been taken to be a characteristic feature, but many conventional materials can also exhibit considerable anisotropy. For some materials, such as oxide dispersion–strengthened alloys or directionally solidified eutectics, the usage seems to depend upon a particular author. Similarly, there is considerable lack of consensus in differentiations between, and the use of, terms such as composite, mixture, agglomerate, laminate, and compound. Such materials as concrete, plywood, carbon black–reinforced rubber, or metal foil–foamed plastic sandwiches may or may not be considered composites. *See* Concrete; Plywood; Rubber.

CLASSIFICATION

This discussion will present several possible classifications or typologies of composite materials of current interest.

Functional classification. This classification considers that composites obtain their useful characteristics either from the unique properties of at least one of the components or from the unique arrangement of the components. Examples of the former are the utilization of the very high strength-to-weight ratio of chopped glass fibers in bulk or sheet molding compounds (**Fig. 1**) for fabrication to lightweight automobile parts, or the hardness of carbide particles in

Table 1. A conceptual framework for evaluating the status of knowledge about metal matrix composites

Synthesis	←——→	Structure	←——→	Properties	←——→	Use
Filament surface preparation		Filament diameter and length		Modulus of elasticity		Wing skin
Molten metal dip or infiltration		Filament spacing and orientation		Tensile strength		Compressor blade
Sintering or hot pressing powder		Matrix composition		Impact resistance		Reentry body
Diffusion bonding		Filament-matrix interaction products		Anisotropy		Submarine hull
Rolling or extruding		Matrix grain size and shape		Oxidation resistance		Engine shaft
Annealing		Cracks or voids		Thermal stability		Bridge truss

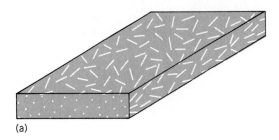

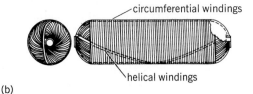

Fig. 1. Glass-plastic composites. (a) High-strength, lightweight sheet-molding compound. (b) Very high-strength, lightweight, filament-wound pressure vessel. (*After M. W. Riley et al., Filament wound reinforced plastics: State of the art, Mater. Des. Eng., pp. 127–146, August 1960*)

metal-bonded carbide tools. In each case the matrix, polymer or metal, also serves important functions such as mechanical support, and environmental protection or heat transfer. An example of the latter is the use of polymer-bonded continuous glass filaments in filament-wound rocket motor cases (Fig. 1b) or other applications where the winding patterns are designed to orient the filaments so that in service they are stressed as purely in tension as possible. This example also utilizes the extremely high unidirectional tensile strength offered by the glass filaments and attempts to minimize their deficiencies in compression. It combines the unique property of a component with the opportunity of designing an optimum composite material for a specific use. An individual component can also fulfill several functions. In glass-reinforced phenolic ablative heat shields, glass filaments provide strength and influence the ablative characteristics.

The functional classification emphasizes the contribution of each component to the new or improved characteristics of a composite. Each of the examples given above is appropriately labeled a composite in this typology. However, when a second component merely serves the function of a cost-reducing filler or extender, as in some plastics, the resulting mixture or compound would not be called a composite.

Geometric classification. In the simplest typology, particles (all dimensions small), discontinuous fibers, or continuous filaments (two dimensions small), and platelets or films (one dimension small) are bonded into a composite by a suitable matrix (**Fig. 2**). They may be randomly dispersed or deliberately arranged in the matrix. Many composite materials fall into one of these simple categories, probably because it is pos-

sible to achieve unique properties in a form of material such as a film or filament where at least one dimension is small. A material such as a filament has only limited applicability (as in cables or woven goods) unless its unique properties can, in part at least, be transferred to a composite material. The attempts to do this have led to a wide variety of reinforced materials, or polymer-, metal-, or ceramic-bonded filaments using glass, boron, graphite, metal, or aramid ordered polymer filaments. (If the filaments predominate they are said to be bonded; if the matrix predominates it is said to be reinforced.) Utilizing the extremely high strength (approaching theoretical limits) of fine, short fiber whiskers in appropriate composites is another possibility. Examples of particles

Fig. 3. General appearance of three types of silicon carbide used to reinforce aluminum alloys. (a) Standard SiC powder. (b) Submicrometer SiC powder. (c) Submicrometer whiskers. (*ARCO Metals Co.*)

Fig. 2. Composites illustrating, (a) particles, (b) fibers, and (c) platelets in a matrix.

and whiskers being evaluated as reinforcements for metals are shown in **Fig. 3**.

More complex geometries are, of course, both possible and useful (**Fig. 4**). The dimensions of the particles, fibers, or films may become similar to or large with respect to the interparticle spacing, as in concrete or some cermets. Instead of discrete particles in a continuous matrix, there may be two interpenetrating continuous phases as in some self-lubricated bearings or refractory composites. Very often, filaments or films are combined with a matrix into plies which are then stacked and bonded together into a laminate. Three-dimensional filament arrays to provide improved multiaxial strength are also used, as in modern carbon-carbon composites. Also, filaments need not be uniformly dispersed through a matrix; they can be used selectively to stiffen or strengthen a selected volume of a component. In these latter examples, the emphasis and opportunity are once again to design an optimum composite material for a specific application. The common examples of coated or surface-altered materials, as in a painted or electroplated part or an induction hardened steel, might also be considered as composite materials, but this usage is rare. SEE CERMET.

Dimensional classification. The dimensions of some of the components of composite materials vary widely and overlap the dimensions of the microstructural features of common conventional materials (**Fig. 5**). They range from extremely small particles or fine whiskers to the large aggregate particles or rods in reinforced concrete. Although in modern usage all of these are called composites, there is some tendency to refrain from doing so at the two ends of the scale. However, complex components such as an automobile tire or a foam core sandwich building panel are also sometimes labeled as composite materials.

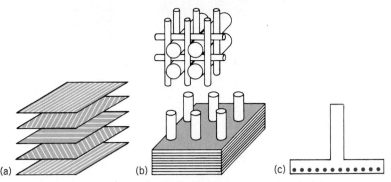

Fig. 4. Different geometrical arrangements. (a) Lay-up of plies. (b) Typical three-dimensional constructions (after R. Stedfeld, Carbon, the lightweight composite for the hot spots, Mater. Eng., 9:61–63, January 1980). (c) Selective reinforcement.

Preparatory method classification. The most common usage considers composite materials to be prepared by mechanically combining the individual components. Thus, it is primarily the mechanical placement of the filament that determines its orientation and spacing in a diffusion-bonded, silicon carbide–reinforced titanium fan blade. The creation of a multiphase microstructure by the familiar processes of solidification, phase transformation, precipitation, and so on, is commonly felt to be in the domain of conventional materials. However, the real situation is not this simple. In all except the most elementary composites, complex adsorption, interdiffusion, chemical reaction, and phase transformation processes can occur at the interface between the components. In fact, the detailed behavior of the composites is a function of the characteristics of the resulting interphase regions. On the other hand, a quite compositelike microstructure can be created by the controlled solidifi-

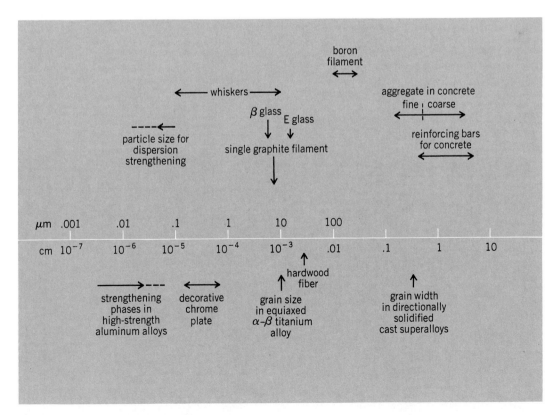

Fig. 5. Dimensional range of microstructural features in composite and conventional materials. Filament and fiber dimensions are diameters. 1 cm = 0.39 in.

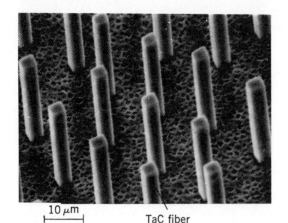

Fig. 6. Microstructure of directionally solidified alloy with tantalum carbide fibers. (*General Electric Co., Cincinnati*)

10 μm

TaC fiber

cle can be examined from similar viewpoints as synthetic composites. The principles established for the latter can help explain the behavior of the former, while the existence of unique composite structures in nature can provide ideas for synthetic substances.

Harris M. Burte

ADVANCED COMPOSITES

Advanced composites comprise structural materials that have been developed for high-technology applications, such as airframe structures, for which other materials are not sufficiently stiff. In these materials, extremely stiff and strong continuous or discontinuous fibers, whiskers, or small particles are dispersed in the matrix. A number of matrix materials are available, including carbon, ceramics, glasses, metals, and polymers. Advanced composites possess enhanced stiffness and lower density compared to fiber-glass and conventional monolithic materials. While composite strength is primarily a function of the reinforcement, the ability of the matrix to support the fibers or particles and to transfer load to the reinforcement is of no less importance. Also, the matrix frequently dictates service conditions, for example, the upper temperature limit of the composite.

Reinforcements. There are several elements and compounds that exhibit high specific stiffness, that is, the ratio of Young's modulus to density. In bulk form, however, these solids are typically weak and brittle. The weakening effect is significantly reduced if the material exists in filamentary form, such as a fiber (0.1–0.2 mm or 0.04–0.08 in. diameter) or whisker (1–25 micrometers or $3.9–98 \times 10^{-5}$ in. diameter). *SEE YOUNG'S MODULUS.*

Continuous filamentary materials being used as the reinforcing constituents in advanced composites are carbonaceous fibers, organic fibers, inorganic fibers, ceramic fibers, and metal wires. Inorganic materials are also used in the form of discontinuous fibers and whiskers. The characteristics and properties of selected reinforcements are summarized in **Table 2.** *SEE STRENGTH OF MATERIALS.*

Carbonaceous fibers. Carbon and graphite fibers offer high modulus and the highest strength of all reinforcing fibers. The fibers are produced by heating and controlling the temperature, tension, and atmosphere of an organic precursor fiber in a pyrolysis chamber. Three different precursor materials are used to pro-

cation of a eutectic (**Fig. 6**). The term in-situ composite has been used for these unique materials; these directionally solidified eutectics may be labeled as composites as often as not.

In another example, there was some tendency to label dispersion-strengthened high-temperature alloys as composites whether they were initially formed by mixing two powders or by a chemical or solid-state process such as internal oxidation. In this case, the tendency diminished probably because of an enhanced understanding that the desirable properties of dispersion-strengthened materials are as much a result of subsequent thermomechanical processing by familiar metalworking procedures as of the initial particle-matrix microstructure. In general, as the diversity of methods for synthesizing controlled microstructures grows, so will differences and conflicts in usage of what is labeled a conventional material and what is a composite material. *SEE ALLOY; EUTECTICS.*

The domain of naturally occurring biological materials is also one where usage of the term composites is nonuniform and sporadic. Some authors arbitrarily hold that the term composites should be reserved for manufactured materials. On the other hand, as a result of the heterogeneous structure of a cell in which components are combined for specific functions as a result of evolution, a literal interpretation of the introductory definition might classify all biological materials as composites. Such an extreme is rare, but naturally occurring materials such as bone, bamboo, and mus-

Table 2. Properties of reinforcements used in advanced composites*

Materials	Form	Size, μm	Density, $g \cdot cm^{-3}$	Tensile modulus, GPa	Tensile strength, GPa	Specific stiffness, $GPa/(g \cdot cm^{-3})$
Carbon (polyacrylonitrile precursor)	Continuous fiber	4	1.8	320	5.5	178
Para-Aramid Kevlar 149	Continuous fiber	12	1.47	186	3.4	127
Boron	Continuous fiber	140	2.49	400	3.6	161
Silicon carbide (SiC)	Continuous fiber	140	3.0	400	3.9	133
Silicon carbide (SiC)	Whisker	0.1–0.5	3.19	400–700	3–14	125–219
Silicon nitride (Si_3N_4)	Whisker	0.1–1.6	3.2	550	5	172
Polycrystalline aluminum oxide	Continuous fiber	20	3.95	379	1.38	96
Tungsten alloy (W-Re-Hf-C)	Continuous fiber (wire)	380	19.4	~400	2.17	21

*1 μm = 3.9×10^{-5} in.   1 $g \cdot cm^{-3}$ = 0.58 oz in.$^{-3}$ 1 GPa = 1.45×10^5 lb/in.^{2}

Table 3. Properties of advanced composites at room temperature*

Composite system	Form of reinforcement	Vol % of reinforcement	Density, g · cm^{-3}	Tensile modulus, GPa	Tensile strength GPa	Specific stiffness, GPa/(g · cm^{-3})
Carbon-epoxy resin	Continuous fiber	60	1.62	220	1.4	135
Kevlar 49– epoxy resin	Continuous fiber	63	1.25	80	1.4	64
Al 6061 (T6)– boron[†]	Continuous fiber	51	2.7	232	1.4	86
Al 6061 (T6)– SiC[†]	Continuous fiber	47	2.85	207	1.4	73
Al 6061 (T6)– SiC[†]	Discontinuous fiber	16	2.86	104	0.3	36
Al 2124 (T6)– SiC[†]	Whisker	20	2.86	127	0.86	44
Glass ceramic– silicon carbide	Continuous fiber	50	—	140	0.6	—
Alumina– silicon carbide	Continuous fiber	30	—	375	0.65	—

*1 g · cm^{-3} = 0.58 oz in.$^{-3}$ 1 GPa = 1.45 × 10^5 lb/in.^{2}
[†]Aluminum alloy-silicon carbide.

duce carbon fibers commercially: rayon, polyacrylonitrile (PAN), and pitch. Modulus is determined by the alignment of the graphite structure in the filament. High-modulus carbon fibers are available in an array of yarns and tows (a bundle of continuous filaments) with differing moduli, strengths, cross-sectional areas, twists, and plies. SEE CARBON; GRAPHITE.

Organic fibers. Almost any polymer fiber can be used in a composite structure, but the first one with a high enough tensile modulus and strength to be used as a reinforcement in advanced composites was an aramid or aromatic polyamide fiber. Aramid fibers are the predominant organic reinforcing fiber; a high-strength high-modulus polyethylene fiber is also available. Aramid fibers are produced by extrusion and spinning processes which lead to alignment of the stiff polymer molecules. SEE MANUFACTURED FIBER; POLYMER.

Inorganic fibers. The most important inorganic continuous fibers for advanced composites are boron and silicon carbide (SiC). Both exhibit high stiffness, high strength, and low density. Continuous fibers of silicon carbide are made by chemical vapor deposition of microcrystalline or amorphous boron from boron trichloride onto a heated tungsten-core filament. Similarly, silicon carbide fiber is produced in a chemical vapor deposition process that involves the reaction of hydrogen with a mixture of chlorinated alkyl silanes at the surface of a heated spun-carbon monofilament substrate which is coated with a thin layer of pyrolitic graphite. Stiff, strong discontinuous fibers that predominate as reinforcements for metal matrix composites are silicon carbide, alumina, and aluminosilicates. Discontinuous oxide fibers are made from chemicals or sol-gels. SEE BORON; METAL MATRIX COMPOSITES; VAPOR DEPOSITION.

Ceramic fibers. An example of a commercial continuous fiber is polycrystalline aluminum oxide (Al_2O_3), which exhibits high stiffness, high strength, a high melting point, and exceptional resistance to corrosive environments. The fibers are produced by the dry-spinning technique from various solutions, followed by heat treatment. An alternate form utilizes a coating of silica (SiO_2) on the continuous polycrystalline fiber, and other continuous polycrystalline oxide fibers

are available. SEE CERAMICS.

Continuous metal fibers. The primary use of continuous metal fibers is in the reinforcement of matrices of nickel and iron-base alloys for high-temperature applications. Strong, stiff, creep-resistant fibers are prepared from tungsten and tungsten alloys. SEE HIGH-TEMPERATURE MATERIALS; TUNGSTEN.

Whiskers. These are single crystals that exhibit fibrous characteristics. Compared to continuous or discontinuous polycrystalline fibers, whiskers exhibit exceptionally high strength and stiffness. Silicon carbide whiskers are prepared by chemical reaction processes or by the pyrolysis of rice hulls. Whiskers of silicon nitride (Si_3N_4) are also available.

Composite systems. In the composite, the fibers are bonded and held in position by the matrix. In many advanced composites the matrix is organic, but the use of metal matrices is increasing. Organic matrix materials are lighter than metals, adhere better to the fibers, and offer more flexibility in shaping and forming. Ceramic matrix composites have begun to find application where organic or metal matrix systems are unsuitable.

Representative property levels for a group of advanced composite systems in which the fibers are aligned in one direction only are given in **Table 3.** For purposes of comparison, the corresponding properties of some conventional (monolithic) engineering materials are given in **Table 4.** Tensile modulus and strength are measured parallel to the direction of fiber reinforcement. When all the fibers in the composite are aligned in the same direction, the material exhibits high stiffness and strength in this direction. At right angles to the fibers, the stiffness and strength are low. Multidirectional fiber arrangements can be used to achieve intermediate levels of stiffness and strength. The effect is illustrated in **Fig. 7** for graphite fibers in an epoxy resin matrix.

Organic matrix. Epoxy resins are used extensively as the matrix material in advanced structural composites. Alternatively, bismalemide resins (BMIs) and polyimide resins are being developed to enhance in-service temperatures. Thermoplastic resins with improved hot and wet properties and impact resistance are in limited use; examples are polyether ether ke-

Table 4. Properties of some monolithic engineering materials at room temperature*

Composite system	Vol % of reinforcement	Density, g · cm⁻³	Tensile modulus, GPa	Tensile strength, GPa	Specific stiffness, GPa/(g · cm⁻³)
Monolithic, aluminum-zinc-magnesium (Al-Zn-Mg) alloy	0	2.8	72	0.5	26
Monolithic low alloy steel (quench and temper)	0	7.85	207	0.6–2.0	26
Monolithic nickel-base superalloy	0	8.2	204	1.2	25
Monolithic epoxy resin	0	1.1–1.4	3-6	0.035–0.1	2.1–5.5

*1 g · cm⁻³ = 0.58 oz in.⁻³   1 GPa = 1.4 × 10⁵ lb/in.²

tone (PEEK) and polyphenylene sulfide (PPS).

The continuous reinforcing fibers for organic matrices are available in several product forms: monofilaments, multifilament fiber bundles, unidirectional ribbons, and single-layer and multilayer fabric mats. Frequently, the continuous reinforcing fibers and matrix resins are combined into a nonfinal form called a prepreg. Product forms for discontinuous fiber-reinforced materials include sheet molding compounds, bulk molding compounds, injection molding compounds, and dry preforms for use in resin transfer processes.

Many processes are available for the fabrication of the final composite organic matrix composite. The first step involves fiber placement in order to orient the unidirectional layers at discrete angles to one another (rather like plywood) in order to achieve the desired load distribution. Hand or machine lay-up techniques and filament winding are in common use. In lay-up, the prepreg is cut and laid up, layer by layer, to produce a laminate of the desired thickness, number of plies, and ply orientations. In filament winding, a fiber bundle or ribbon is impregnated with resin and wound on a mandrel to produce simple or complex shapes. Consolidation into the final composite structure is usually achieved by means of heat and pressure. *See POLYMERIC COMPOSITE.*

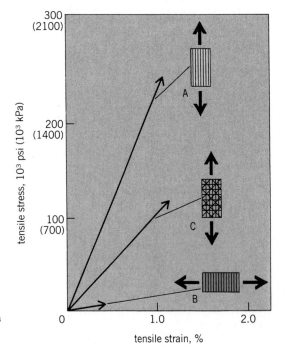

Fig. 7. Stress–strain curves for simple laminates made of graphite fibers in a matrix of epoxy resin. Laminate A, all fibers at 0° unidirectional. B, all fibers at 90°. C, fibers at 0°, +45°, −45°, and 90°.

Metal matrix. Boron-aluminum composites are available commercially. They are fabricated by high-pressure diffusion bonding of a preform consisting of continuous boron fibers in aluminum foil. Aluminum (Al) and titanium (Ti) matrix composites reinforced with continuous fibers of silicon carbide are also in use. Processes suitable for making Al-SiC are investment casting and low-pressure hot molding. Superplastic forming with diffusion bonding and hot isostatic pressing are processes commonly used to fabricate complex shapes of Ti-SiC. To illustrate the distribution of the fiber reinforcement in a metal matrix, the fracture surface of a composite consisting of continuous SiC fibers in a titanium alloy matrix is shown in **Fig. 8**; the composite was fabricated by diffusion bonding.

Continuous graphite-fiber–reinforced metal matrix composites make up a class of extremely high-performance materials. First a composite precursor is fabricated by infiltrating liquid metal [aluminum or magnesium (Mg)] into the multifilament graphite tow. Precursor layers are then diffusion-bonded together under pressure at elevated temperature. Alternatively, lay-ups of Al-C and Mg-C can be consolidated by hot isothermal drawing into shapes such as rods and tubing. Complex-shaped continuous graphite-fiber metal matrix composites can be cast without the use of a preform. The fibers are first formed into the desired configuration in the mold, and molten metal is poured in to form a shaped composite. Aluminum oxide-fiber–reinforced metal matrix composites are suitable for use in weight-sensitive elevated-temperature applications. A number of matrices have been evaluated, in particular Al and Mg. Shaped metal components are usually fabricated by casting. *See ALUMINUM ALLOYS; METAL CASTING.*

The use of continuous metal fibers as a reinforcement in a metal matrix is best illustrated by the tungsten-reinforced superalloys. The tungsten or tungsten alloy fibers can be incorporated into a nickel or iron-base alloy matrix by one of several techniques: liquid metal infiltration, investment casting, powder metallurgy processing, or diffusion bonding. Fabrication methods for the incorporation of discontinuous fibers or whiskers into a metal matrix are similar to those described for continuous reinforcement. Powder metallurgy, liquid metal infiltration and casting techniques predominate. *See METAL MATRIX COMPOSITES; POWDER METALLURGY.*

Ceramic matrix. With the promise of increased toughness and reliability, ceramic matrix composites have demonstrated that they have the potential for taking advantage of properties such as high-temperature strength and wear resistance. Material forms suitable for reinforcement include continuous fibers, whiskers, platelets, and particles. Examples are carbon fibers, silicon carbide fibers, alumina fibers, and silicon carbide whiskers. The matrix is usually a glass or glass ceramic (a ceramic produced by fine-scale crystallization from a glass). Candidate matrices are borosilicate glass, lithium alumino silicate glass ceramic, alumina, and silicon nitride.

For continuous fiber reinforcement, ceramic matrix composites are made by slurry mixing and infiltration, sol-gel or polymer pyrolysis, in-situ growth or chemical reaction, and transfer molding. For chopped fibers or whiskers, fabrication involves mixing of the reinforcement in a slurry, drying, hot pressing or spray drying, forming, and activated sintering. *See* *Sintering*.

Applications. The use of fiber-reinforced materials in engineering applications has grown rapidly. This is due primarily to the replacement of traditional materials, in particular, metals and alloys. Selection of a composite rather than a monolithic material for a given application is dictated by its properties. Frequently, the high values of specific stiffness and specific strength will be the determining factor. In other applications, the selection of a composite material reflects unique matrix characteristics in combination with the properties of the reinforcement, for example, advanced composites that exhibit wear resistance or strength retention at elevated temperatures.

The first applications for advanced composite materials were in aerospace structures and sports and leisure equipment. Applications have grown to include artificial joints and organs, reinforced building materials, automotive components, precision audio equipment, and marine structures.

Aerospace. Components fabricated from advanced organic-matrix fiber-reinforced composites are used extensively on commercial production aircraft such as the Boeing 737-300, 757 and 767, and the Airbus A310 (**Fig. 9**). Most of these composite components are of a honeycomb sandwich construction.

Advanced carbon-reinforced composites have found wide acceptance in military aerospace applications. About 26% of the structural weight of the U.S. Navy's AV-8B, a vertical takeoff and landing aircraft, is a carbon fiber–epoxy composite. On the F-18 aircraft, carbon fiber–reinforced composites make up about 10% of the structural weight and more than 50% of the plane's surface area. Uses include wing skins, horizontal and vertical tail boxes, and doors. The B-18 bomber employs a number of composite structural components such as weapons and equipment bay doors, and flaps. The V-22 Osprey tilt-rotor aircraft combines the advantages of vertical takeoff and landing of a helicopter with the smooth, high-speed cruise and extended range of a fixed-wing airplane. Wing panels have integrally stiffened laminate skins, and the fuselage of the V-22 utilizes composite materials that make up 50% of its structural weight. In the future, the use of carbon fiber composites could reach 75–80% of the structure, equivalent to a 26% weight saving on the all-up aircraft weight. Composite brake materials consisting of carbon fibers embedded in a carbonaceous matrix are used in military and certain commercial (Airbus and Concorde) aircraft. The carbon-carbon composite exhibits low weight, excellent friction behavior, and wear resistance. *See* *Aircraft*; *Convertiplane*; *Fuselage*; *Helicopter*; *Mil-*

Fig. 8. The fracture surface of a composite consisting of continuous silicon carbide (SiC) fibers in a titanium alloy metal matrix. (*M. J. Koczak, Drexel University*)

itary aircraft; *Vertical takeoff and landing (VTOL)*; *Wing*.

In the jet engine (gas turbine), metal matrix composites have been considered as a replacement for several components. These include low- and high-pressure compressor casings (aluminum matrix), the low- to medium-temperature compressor blades (titanium matrix), and the high-temperature turbine blades (tungsten-reinforced iron-base superalloy). Ceramic matrix composites have the potential for application in hot-section engine components.

Automotive. Because of the need to build lighter, quieter, and more fuel-efficient engines, consideration has been given to the use of advanced composites in several components. One example is the use of metal matrix composite to selectively reinforce the crown and ring groove in diesel engine pistons to improve thermal fatigue and wear. The composite consists of short alumina fibers in an aluminum alloy matrix. Metal matrix composite connecting rods have been developed; materials are aluminum reinforced with stainless steel fibers and aluminum reinforced with polycrystalline alumina fibers. Ceramic matrix composites are the next materials for applications in turbocharged diesel engines; examples are valve seats and inserts, piston rings, liners, and exhaust mani-

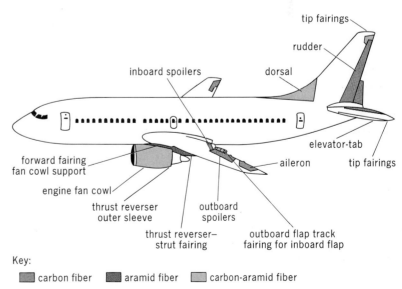

Key:
⬛ carbon fiber   ⬛ aramid fiber   ⬜ carbon-aramid fiber

Fig. 9. Diagram showing the applications of advanced composites on the Boeing 737-300. (*Engineered Materials Handbook, vol. 1; Composites, ASM International , 1987*)

folds. Candidate composites are silicon carbide–reinforced lithium aluminum silicate glass or aluminum silicate glass. *See Motor vehicle*.

Leisure and sporting products. Boron fiber– or carbon fiber–reinforced epoxies are used in the frames of rackets for tennis, squash, racquetball, and badminton. Other products utilizing composites are fishing rods, skis, and golf club shafts. Golf club heads of hot forged Al-SiC whisker composite are also available. High-performance sailing craft fabricated from honeycomb sandwich panels of carbon fiber–reinforced plastic have strong, rigid, lightweight hulls. Similarly, many racing cars make use of a composite chassis consisting of carbon fiber–reinforced plastic skins bonded to a honeycomb core. The use of computer-controlled fiber placement to fabricate the preform of a monocoque car body is illustrated in **Fig. 10**. The development of this advanced braiding approach resulted in a 30:1 reduction in manufacturing steps compared to conventional car manufacturing methods. Bicycle frames have been fabricated from boron-aluminum composites, and extruded bicycle rims have been fabricated from an aluminum–aluminum oxide composite. The composite rim is stiffer and lighter than a rim fashioned from a monolithic aluminum alloy. *See Epoxy resins*.

Electrical, electronics, and communications. Metal heat sinks to remove heat from integrated circuits have a coefficient of thermal expansion different from that of the circuit, and this leads to failure of the device by thermal fatigue. However, heat sinks fabricated from metal matrix composites with exactly matched thermal properties have solved this problem. Precision radar systems require antennas with low inertial and structural rigidity. Hence metal matrix composites with high specific stiffness and strength and a low coefficient of thermal expansion are being evaluated. A monolithic alloy of iron, nickel, and cobalt is being replaced by graphite-reinforced copper in some microwave circuits. *See Integrated circuits; Microwave; Radar*.

Other applications. Alumina reinforced with silicon carbide whiskers has become a state-of-the-art cut-ting-tool material for rough-machining superalloys. The whisker reinforcement enhances toughness compared to monolithic alumina. Other advantages include higher speeds and feeds and improved reliability. In the machining of cast irons, silicon nitride reinforced with silicon carbide whiskers is beginning to displace cemented carbide and alumina-base cutting tool materials. *See Machining operations*.

Ceramic matrix composite laser mirrors are in use in outer space applications. The borosilicate glass matrix is reinforced with graphite fibers. Distortion and weight are significantly reduced compared to conventional molybdenum mirrors. Metal matrix composites consisting of discontinuous silicon carbide reinforcement in aluminum offer advantages in optical-instrument applications, for example, optical-grade mirrors.

Advanced composites have made it possible to mimic the properties of human bone; a composite prosthesis consisting of a three-dimensional reinforcement in a suitable matrix has been developed. One example of a biocompatible composite considered for bone replacement and joint implants is a three-dimensional structure of graphite fibers infiltrated in a hydrophilic matrix material. Carbon-carbon composites consisting of carbon fibers embedded in a carbonaceous matrix are biocompatible with living tissue, and the modulus is close to that of bone. As a result, these advanced composites are finding applications in medicine. *See Prosthesis*.

Alan Lawley

Bibliography. American Society for Metals, International Handbook Committee, *Engineered Materials Handbook*, vol. 1: *Composites*, 1987; J. Delmonte, *Technology of Carbon and Graphite Fiber Composites*, 1981; J. E. Gordon, *The Science of Structures and Materials*, 1988; D. Hull, *An Introduction to Composite Materials*, 1987; A. Kelly, *Strong Solids*, 2d ed., 1973; T. Richardson, *Composites: A Design Guide*, 1987; S. A. Wainwright and W. D. Biggs, *Mechanical Design in Organisms*, 1982.

Composition board

A wood product in which the grain structure of the original wood is drastically altered. Composition board may be divided into several types. When wood serves as the raw material for chemical processing, the resultant product may be insulation board, hardboard, or other pulp product. When the wood is broken down only by mechanical means, the resultant product is particle board. Because composition board can use waste products of established wood industries and because there is a need to find marketable uses for young trees, manufacture of composition board is one of the most rapidly developing portions of the wood industry. *See Paper*.

Fiberboard. One form of fiberboard is produced by loading a batch of wood chips into a chamber which is then heated and pressurized by steam. After about 2 min, the 1000-lb/in.² (6.9-megapascal) pressure is abruptly released to hydrolyze and fluff the chips into a brown fiber. The fiber is refined, washed, and felted into a mat on a wire conveyor so that some of the water can drain out, and then the mat is cut to length for loading on a screen into a press. At controlled temperature in the press, the lignin rebonds the material while water is driven off as steam through the

Fig. 10. Example of large-scale manufacturing: the fiber preform for a composite car. (*F. Ko, Drexel University*)

screen. The finished reconstituted wood product is a hard isotropic board as a consequence of the felting of the fibers and the ligneous bonding, possibly augmented by synthetic adhesive.

Alternatively, a similar board is produced by a continuous process. A screw feed delivers wood chips from a hopper to a steam preheater where the chips partially hydrolyze in the vicinity of 150 lb/in.^{2} (0.11 MPa). The hot chips pass between grinding disks to discharge as pulp, which is then formed into sheets essentially as described above. The wood chips may also be processed entirely by grinding. A further variation is to deliver the pulp slurry into a deckle box, in which case most of the water is removed by suction applied below the box before the mat is compressed into the finished sheet.

Instead of conveying the wet fiber or pulp to the mat, it can be dried sufficiently to be conveyed by air; it is then introduced between chrome-plated cauls and pressed. Because of the low-moisture content, no screen is needed; the finished sheet is smooth on both sides. Most new plants use variants of the dry process.

In any of the foregoing methods synthetic resin may be added as a binder before the board is formed. After the board is formed, it may be impregnated with drying oils and heated in a kiln at 300°F (150°C) until the oils are completely polymerized to produced tempered board. *See Drying oil; Resin.*

If insulating board is required instead of hardboard, the material is less compacted, the degree of compaction being described by the specific gravity of the finished board.

Particle board. When formed from wood particles that retain their woody structure, the product is termed particle board. Properties of such boards depend on the size and orientation of the particles, which may be dimensioned flakes, random-sized shavings, or splinters. After the particular type of particles are produced, they are screened to remove fines and to return oversizes for further reduction. Graded particles are dried, mixed with synthetic adhesive and other additives such as preservatives, and delivered to the board-forming machine.

Forming machines are basically of two types. In one type mixed materials are metered by weight or volume onto an open flat tray, prepressed at room temperature, and finally consolidated under heat at high pressure to the finished dimension. Whether carried out as a batch or a continuous process, this flat-press method tends to orient the particles with their long dimensions approximately parallel to the face of the board. In the second type of machine the mixed materials are extruded under pressure between heated platens. Thick boards made this way may have hollow cores produced by rods fixed in the aperture ahead of the platens. Particles in extruded boards are randomly oriented. Boards formed by either method may be layered, with different-sized particles at the surfaces from those in the center, which are also usually coarser.

Characteristics. Hardboards produced by a continuous process can be formed to great length, although a finished length of 16 ft (5 m) is usual stock maximum. Standard thicknesses range from ⅛ to 5/16 in. (3.2 to 7.9 mm); standard densities range from a specific gravity of 0.80 to 1.15, with boards up to 1.45 available if needed.

Tensile strength parallel to the face ranges from 1000 lb/in.^{2} (6.9 MPa) for untreated board to 5500 lb/in.^{2} (37.9 MPa) for tempered hardboard; perpendicular to the face it ranges from 125 to 400 lb/in.^{2} (0.86 to 2.76 MPa). Compression strength parallel to the face ranges from 2000 to 5300 lb/in.^{2} (13.8 to 36.5 MPa).

Lengths and widths of particle boards are similar to those of hardboards; thicknesses range from ¼ to 1 in. (6 to 25 mm); thicker sheets are most economically produced by laminating sheets of standard thicknesses. Properties of flat-pressed particle board vary more than do those of hardboard. They have low density: 0.4–1.2 specific gravity. Tensile strength parallel to their faces ranges from 500 to 5000 lb/in.^{2} (3.4 to 34.5 MPa) and perpendicular from 40 to 400 lb/in.^{2} (276 kPa to 2.76 MPa). Extruded particle board has low bending strength; it is used chiefly as the core of plywood-type panel. Development of adhesives specifically for composition boards has extended their utilitarian value, and variety of textures has increased their esthetic appeal. *See Wood products.*

Frank H. Rockett

Compressible flow

Flow in which density change is not negligible. Compressibility is a thermodynamic property of all substances. If a fluid element of volume v is subject to a pressure change Δp, its volume experiences a corresponding change Δv, and the compressibility τ is defined as the fractional change in volume per unit change in pressure, given by Eq. (1). The minus sign

$$\tau = \frac{-(\Delta v/v)}{\Delta p} \qquad (1)$$

accounts for a decrease in volume corresponding to an increase in pressure. All substances have a finite τ, but τ for gases is orders of magnitude larger than that for liquids and solids. If the element has a unit mass, v is the volume per unit mass. Density ρ is the mass per unit volume; hence Eq. (2) holds. In turn,

$$v = \frac{1}{\rho} \qquad (2)$$

the compressibility is given by Eq. (3). Alternatively, Eq. (4) is valid.

$$\tau = \frac{\Delta\rho/\rho}{\Delta p} \qquad (3)$$

$$\frac{\Delta\rho}{\rho} = \tau\Delta p \qquad (4)$$

This last equation is the essence of the definition of compressible flow. By virtue of moving in a flow, the fluid element experiences a change in pressure, Δp; this change in pressure in part provides the force for accelerating and decelerating the element in the flow. This causes a corresponding change in density given by Eq. (4). For a flow in which the product $\tau\Delta p$ is very small, the corresponding change in density, $\Delta\rho$, is negligible. Such constant-density flows are defined as incompressible flows. In contrast, for a flow in which the product $\tau \Delta p$ is moderate or large, the change in density is not negligible. Such flows with variable density are defined as compressible flows. *SEE INCOMPRESSIBLE FLOW.*

Strictly speaking, all flows are compressible because τ is finite for all substances, and hence all flows experience a finite change in density, $\Delta \rho$. In this sense, incompressible flow does not really occur. However, as a rule of thumb, a flow can be considered incompressible where $\Delta \rho / \rho < 0.05$ and compressible where $\Delta \rho / \rho \geq 0.05$; thus, if the density changes throughout by 5% or more, the flow should be treated as compressible.

An important parameter that governs compressible flow is the Mach number. If a denotes the speed of sound in a substance and V denotes the velocity of the flow, the Mach number M by definition is given by Eq. (5). Both V and a can vary throughout the flow;

$$M = \frac{V}{a} \qquad (5)$$

hence M is also variable. For most low-speed flows where $M < 0.3$, $\Delta \rho / \rho$ is negligibly small, and hence the flow can be considered incompressible. For higher-speed flows where $M > 0.3$, $\Delta \rho / \rho$ is not negligible, and flow must be considered compressible.

The Mach number is also useful for defining several different forms of compressible flow: (1) flow where $M < 1$ everywhere is called subsonic flow; (2) flow where $M = 1$ is called sonic flow; (3) flow with regions where M varies slightly below and slightly above 1 (for example, $0.8 < M < 1.2$) is called transonic flow; (4) flow where $M > 1$ everywhere is called supersonic flow; (5) a special type of high supersonic flow, where $M > 5$, is called hypersonic flow. *SEE FLUID FLOW; MACH NUMBER.*

John D. Anderson, Jr.

Bibliography. J. D. Anderson, Jr., *Fundamentals of Aerodynamics*, 1984; J. D. Anderson, Jr., *Introduction to Flight*, 3d ed., 1989; J. D. Anderson, Jr., *Modern Compressible Flow: With Historical Perspective*, 2d ed., 1990.

Compression ratio

In a cylinder, the piston displacement plus clearance volume, divided by the clearance volume. This is the nominal compression ratio determined by cylinder geometry alone. In practice, the actual compression ratio is appreciably less than the nominal value because the volumetric efficiency of an unsupercharged engine is less than 100%, partly because of late intake valve closing. In spark ignition engines the allowable compression ratio is limited by incipient knock at wide-open throttle. Knock in turn depends on the molecular structure of the fuel and on such engine features as the temperature of the combustible mixture prior to ignition, the geometry and size of the combustion space, and the ignition timing. For example, isooctane, benzene, and alcohol can be burned at much higher compression ratios than *n*-heptane. In compression ignition engines critical compression ratio is that necessary to ignite the fuel and depends on fuel and cylinder geometry. *SEE COMBUSTION CHAMBER; DIESEL ENGINE; INTERNAL COMBUSTION ENGINE; SPARK KNOCK; VOLUMETRIC EFFICIENCY.*

Neil MacCoull

Compressor

A machine that increases the pressure of a gas or vapor (typically air), or mixture of gases and vapors. The pressure of the fluid is increased by reducing the fluid specific volume during passage of the fluid through the compressor. When compared with centrifugal or axial-flow fans on the basis of discharge pressure, compressors are generally classed as high-pressure and fans as low-pressure machines.

Compressors are used to increase the pressure of a wide variety of gases and vapors for a multitude of purposes (**Fig. 1**). A common application is the air compressor used to supply high-pressure air for conveying, paint spraying, tire inflating, cleaning, pneumatic tools, and rock drills. The refrigeration compressor is used to compress the gas formed in the evaporator. Other applications of compressors include chemical processing, gas transmission, gas turbines, and construction. *SEE GAS TURBINE; REFRIGERATION.*

Characteristics. Compressor displacement is the volume displaced by the compressing element per unit of time and is usually expressed in cubic feet per minute (cfm). Where the fluid being compressed flows in series through more than one separate compressing element (as a cylinder), the displacement of the compressor equals that of the first element. Compressor capacity is the actual quantity of fluid compressed and delivered, expressed in cubic feet per minute at the conditions of total temperature, total pressure, and composition prevailing at the compressor inlet. The capacity is always expressed in terms of air or gas at intake (ambient) conditions rather than in terms of arbitrarily selected standard conditions.

Air compressors often have their displacement and capacity expressed in terms of free air. Free air is air at atmospheric conditions at any specific location. Since the altitude, barometer, and temperature may vary from one location to another, this term does not mean air under uniform or standard conditions. Standard air is at 68°F (20°C), 14.7 lb/in.^{2} (101.3 kilopascals absolute pressure), and a relative humidity of 36%. Gas industries usually consider 60°F (15.6°C) air as standard.

Types. Compressors can be classified as reciprocating, rotary, jet, centrifugal, or axial-flow, depend-

Fig. 1. Mobile air compressor unit which is driven by engine in the forward compartment. (*Ingersoll-Rand*)

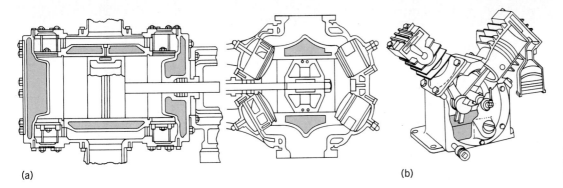

Fig. 2. Compressor cylinders, (a) water-cooled and (b) air-cooled.

ing on the mechanical means used to produce compression of the fluid, or as positive-displacement or dynamic-type, depending on how the mechanical elements act on the fluid to be compressed. Positive-displacement compressors confine successive volumes of fluid within a closed space in which the pressure of the fluid is increased as the volume of the closed space is decreased. Dynamic-type compressors use rotating vanes or impellers to impart velocity and pressure to the fluid.

Reciprocating compressors are positive-displacement types having one or more cylinders, each fitted with a piston driven by a crankshaft through a connecting rod. Each cylinder also has intake and discharge valves, and a means for cooling the mechanical parts (**Fig. 2**). Fluid is drawn into the cylinder during the suction stroke. At the end of the suction stroke, motion of the piston reverses and the fluid is compressed and expelled during the discharge stroke. When only one end of the piston acts on the fluid, the compressor is termed a single-acting unit. When both ends of the piston act on the fluid, the compressor is double-acting. The double-acting compressor discharges about twice as much fluid per cylinder per cycle as the single-acting (**Fig. 3**).

Single-stage compressors raise the fluid pressure from inlet to discharge on each working stroke of the piston in each cylinder. Two-stage compressors use one cylinder to compress the fluid to an intermediate pressure and another cylinder to raise it to final discharge pressure. When more than two stages are used, the compressor is called a multistage unit.

Vertical and horizontal compressors may be single-cylinder or multicylinder units. The angle type is multicylinder with one or more horizontal and vertical compressing elements (**Fig. 4**). Single-frame (straight-line) units are horizontal or vertical, double-acting, with one or more cylinders in line with a single frame having one crank throw, connecting rod, and crosshead. The V or Y type is a two-cylinder, vertical, double-acting machine with cylinders usually at a 45° angle with the vertical. A single crank is used. Semiradial compressors are similar to V or Y type, but have horizontal double-acting cylinders on each side. Duplex compressors have cylinders on two parallel frames connected through a common crankshaft. In duplex-tandem steam-driven units, steam cylinders are in line with air cylinders. Duplex four-cornered steam-driven compressors have one or more compressing cylinders on each end of the frame and one or more steam cylinders on the opposite end. In four-cornered motor-driven units, the motor is on a shaft between compressor frames.

Reciprocating compressors are built to handle fluid

capacities ranging to 100,000 ft³/min (2800 m³/min); pressures range to over 35,000 lb/in.² (240 megapascals). Special units can be built for larger capacities or higher pressures. Water is the usual coolant for cylinders, intercoolers, and aftercoolers, but other liquids, including refrigerants, may also be used.

Compressor thermodynamics. Compression efficiency in any compressor is compared against two theoretical standards—isothermal and adiabatic. Neither occurs in an actual compressor because of unavoidable losses. A plot of the compression process on a pressure–volume diagram shows that an actual unit works between these two standards (**Fig. 5**). Isothermal compression has perfect cooling—air remains at inlet temperature while being compressed. The work input to the compressor, measured by area *ABCD*, is the least possible. Adiabatic compression has no cooling; the temperature rises steadily during compression. Discharge pressure is reached sooner than with isothermal compression. Since air pressure is higher during every part of the piston's stroke, more work input is needed, as shown by area *ABCE*. SEE THERMODYNAMIC PROCESSES.

If compression is divided into two or more steps or stages, air can be cooled between stages. This intercooling brings the actual compression line closer to the isothermal line. Area *BCDE* shows the power saved.

Water vapor in air entering the compressor leaves as superheated vapor because its temperature is in excess of that corresponding to its pressure. It can be converted to water only by cooling to a temperature below the saturation temperature corresponding to its

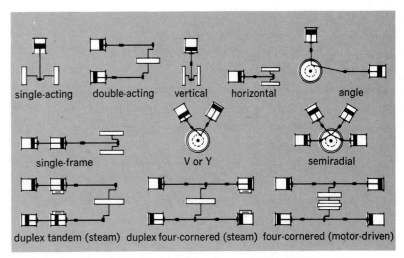

single-acting   double-acting   vertical   horizontal   angle

single-frame   V or Y   semiradial

duplex tandem (steam)   duplex four-cornered (steam)   four-cornered (motor-driven)

Fig. 3. Frame arrangements of positive-displacement piston compressors.

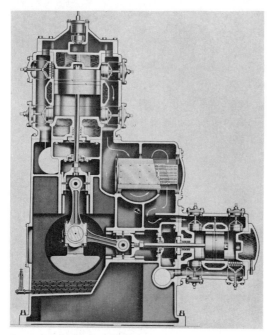

Fig. 4. Cross section of stationary angle-type compressor with first stage vertical and second stage horizontal.

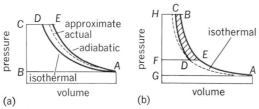

Fig. 5. Compression curves for (a) theoretical and actual units and (b) effect of intercooling.

pressure. Immediate cooling of air after it leaves the compressor proves best, because this prevents vapor from reaching the distribution system. Either air- or water-cooled heat exchangers, called aftercoolers, are used for this.

The usual single-stage reciprocating compressor is built for pressures to about 150 lb/in.^{2} (1.0 MPa); two-stage types for up to about 500 lb/in.^{2} (3.4 MPa); three-stage for up to about 2500 lb/in.^{2} (17 MPa); four- and five-stage for up to 15,000 lb/in.^{2} (100 MPa), and higher.

Rotary compressors. Other forms of positive-displacement compressor are the rotary types.

Sliding-vane type. In the sliding-vane rotary compressor, fluid is trapped between vanes as the rotor passes the inlet opening (**Fig. 6**). Further rotation of

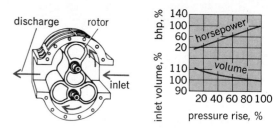

Fig. 7. Two-lobe compressor and its performance curves at constant speed.

the rotor reduces the volume of the space in which the fluid is trapped. Fluid pressure rises until the discharge port is reached, when discharge occurs.

Depending on design, the compressor may be cooled with atmospheric air, oil, or water. For low pressures one stage of compression is used for higher pressures, two stages for higher pressures. Capacities range to about 5000 ft^3/min (140 m^3/min). Some compressors of this type have rings around the vanes to keep them from bearing on cylinder walls. Others have vanes bearing against the cylinder.

Lobe type. In the lobed rotary compressor, fluid is trapped between two or more rotors held in fixed relationship to each other (**Fig. 7**). Rotation of the impellers reduces the volume in which the fluid is trapped, producing a pressure rise. Fluid is discharged when the rotors pass the outlet port. Either two or three rotors are used. Capacities range from about 5 to 50,000 ft^3/min (0.1–1400 m^3/min). Pressures above about 15 lb/in.^{2} (100 kPa) are obtained by operating two or more lobe-type compressors in series. Rotors are straight, as shown, or slightly twisted.

Liquid-piston type. In the liquid-piston rotary compressor, a round multiblade rotor revolves in an elliptical casing partly filled with liquid, usually water. When the rotor turns, it carries liquid around with it, the blades forming a series of buckets (**Fig. 8**). Because the liquid follows the casing contours, it alternately leaves and returns to the space between blades (twice each revolution). As the liquid leaves the bucket, the fluid to be compressed is drawn in. When the liquid returns, it compresses the fluid to discharge pressure.

Liquid-piston compressors handle up to about 5000 ft^3/min (40 m^3/min). Single-stage units can develop pressures to about 75 lb/in.^{2} (520 kPa); multistage designs are used for higher pressures. Water or almost any other low-viscosity liquid serves as the compressant. For exacting services, the compressor may be sealed with chilled water to prevent condensation in lines.

Radial-flow units. Dynamic-type centrifugal compressors use rotating elements to accelerate the fluid radially. By diffusing action, velocity is converted to static pressure. Thus, the static pressure is higher in the enlarged section. Centrifugal compressors usually take in fluid at the impeller eye (or central inlet of the circular impeller) and accelerate it radially outward (**Fig. 9**). Some static-pressure rise occurs in the impeller, but most of the pressure rise occurs in the diffuser section of the casing, where velocity is converted to static pressure. Each impeller-diffuser set is a stage of the compressor. Centrifugal compressors are built with from 1 to 12 or more stages, depending on the final pressure desired and the volume of fluid to be handled. The pressure ratio, or compression ra-

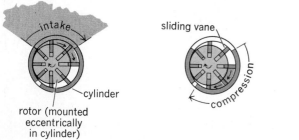

Fig. 6. Operation of sliding-vane rotary compressor.

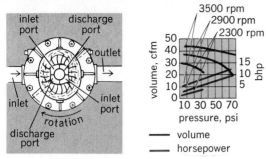

Fig. 8. Liquid-piston rotary compressor and its constant-speed performance curves. (*Compressed Air and Gas Institute*)

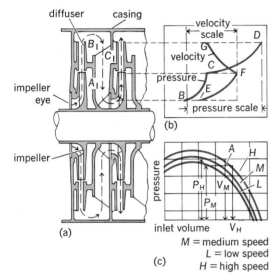

Fig. 9. Performance curves for radial-flow or centrifugal compressor. (*a*) Air flow through centrifugal compressor following paths shown by arrows. (*Compressed Air and Gas Institute*). (*b*) Pressure and velocity relationships of air in a typical centrifugal compressor. (*c*) Volume and pressure relationships of a centrifugal compressor at various speeds.

tio, of a compressor is the ratio of the absolute discharge pressure to the absolute inlet pressure. SEE BERNOULLI'S THEOREM; DIFFUSER.

In the typical compressor (Fig. 9), centrifugal action of the impeller produces pressure rise *BC* and a large increase in air velocity *EF*. In the diffuser, velocity energy is converted to static pressure. Velocity falls from *F* to *G*, pressure rises from *C* to *D*.

Pressure-volume curves for a centrifugal compressor operating at different speeds show that, for example, at speed *M* the unit delivers volume V_M at pressure P_M (point *A* on the pressure-volume diagram, Fig. 9). Increasing speed to *H* raises volume to V_H at P_M, or old volume V_M can be delivered at higher pressure P_H.

Pressure delivered by a centrifugal compressor is practically constant over a relatively wide range of capacities.

Pumping limit, also called surge point or pulsation point, is the lower limit of stable operation. Percentage stability is 100 minus the pumping limit in percent of design capacity.

Multistage centrifugal compressors handle 500 to more than 150,000 ft³/min (14 to 4300 m³/m) at pres-

sures as high as 5000 lb/in.² (35 MPa), but are limited to compression ratios in the order of 10.

Axial-flow units. Compressors that accelerate the fluid in a direction generally parallel to the rotating shaft consist of pairs of moving and stationary blade-rows, each pair forming a stage. The pressure rise per stage is small, compared with a radial-flow unit. Hence the usual axial-flow compressor has more stages than a centrifugal compressor working through the same pressure range. Single-stage axial compressors have capacities from a few to more than 100,000 ft³/min (2800 m³/m) at pressures from less than 1 to several pounds per square inch. Multistage axial-flow compressors compress air to 150 lb/in.²(1 MPa) or more. Some special machines handle over 2,000,000 ft³/min (60,000 m³/m). Pressure rise per stage is generally relatively small, so units for higher pressures frequently have a considerable number of stages—20 or more (**Fig. 10**).

While centrifugal machines deliver practically constant pressure over a considerable range of capacities, axials have a substantially constant delivery at variable pressures. In general, centrifugals have a wider stable operating range than axials. Because of their more or less straight-through flow, axials tend to be smaller in diameter than centrifugals and are apt to be longer. Efficiency of axials usually runs slightly higher.

To prevent surging at extremely low loads, large-capacity axials are sometimes fitted with a blowoff system that discharges excess air to the atmosphere. Then there is always enough air passing through the machine to keep it in its stable range.

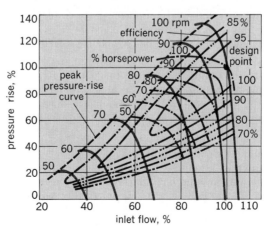

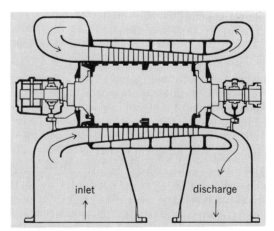

Fig. 10. Axial-flow compressor and performance curves of its constant-speed operation. (*Compressed Air and Gas Institute*)

Because of difficulty in accurately predicting the performance curves of centrifugal and axial compressors, only one capacity and one discharge-pressure rating, together with corresponding power input, are normally guaranteed. Shape of the curve may be indicated but is never guaranteed. SEE FAN; PUMPING MACHINERY.

Tyler G. Hicks; Donald L. Anglin

Bibliography. Business Trend Analysts, *Pumps and Compressors*, 1982; Compressed Gas Association, *Handbook of Compressed Gases*, 2d ed., 1981; *Compressor Handbook for the Hydrocarbon Processing Industries*, 1979; L. Sheel, *Gas Machinery*, 1972.

Compton effect

The increase in wavelength of electromagnetic radiation, observed mainly in the x-ray and gamma-ray region, on being scattered by material objects. This increase in wavelength, $\lambda_2 - \lambda_1 = \Delta\lambda$, of the scattered radiation, which is caused by the interaction of the radiation with the weakly bound electrons in the matter in which the scattering takes place, is given to good approximation by Eq. (1). Here λ_1 is the wave-

$$\lambda_2 - \lambda_1 = \Delta\lambda = \frac{h}{m_0 c}(1 - \cos\phi) \qquad (1)$$

length of the incident radiation, λ_2 is the wavelength of the radiation scattered at the angle ϕ, h is Planck's constant, m_0 is the rest mass of the electron, c is the speed of light, and ϕ is the angle that the direction of the scattered radiation makes with the direction of the incident radiation.

The Compton effect illustrates one of the most fundamental interactions between radiation and matter and displays in a very graphic way the true quantum nature of electromagnetic radiation. Together with the laws of atomic spectra, the photoelectric effect, and pair production, the Compton effect has provided the experimental basis for the quantum theory of electromagnetic radiation. For information on these and related topics SEE ANGULAR MOMENTUM; ATOMIC STRUCTURE AND SPECTRA; ELECTRON-POSITRON PAIR PRODUCTION; LIGHT; NONRELATIVISTIC QUANTUM THEORY; PHOTOEMISSION; QUANTUM MECHANICS; UNCERTAINTY PRINCIPLE.

The Compton effect represents a great departure from earlier ideas concerning electromagnetic radiation. According to the original theory for the scattering of electromagnetic radiation by electrons in matter, which was developed by J. J. Thomson about 1900, the scattered radiation should have exactly the same wavelength as the incident radiation. This theory considers the incident radiation to have an oscillating electric field and shows that an electron would be forced by this electric field to oscillate with the same frequency as the field. The theory of electromagnetic radiation developed in the latter part of the nineteenth century by James Clerk Maxwell predicts that a point charge, such as an electron, when oscillating with a given frequency, will itself emit in all directions waves of electromagnetic radiation of exactly the same frequency. Therefore an increase in the wavelength, corresponding to a decrease in the frequency, of the scattered radiation is not to be expected if the scattering of x-rays takes place according to Thomson's theory. SEE ELECTROMAGNETIC RADIATION.

In a series of experiments, beginning in 1922, A. H. Compton confirmed the earlier conclusions of other scientists that the wavelengths of scattered x-rays increase, depending on the angle of scattering, a result in direct conflict with Thomson's theory. This discovery, along with its subsequent explanation by Compton, is regarded as one of the most significant contributions in physics. Compton showed that a beam of x-rays is composed of individual particles, called photons, each of which carries the energy $h\nu$ and also the linear momentum $h\nu/c$ in the direction of the beam, where h is Planck's constant, ν is the frequency of the radiation, and c is the speed of light. Moreover, the photon can impart energy and linear momentum to an individual electron in an elastic collision with the electron.

Experimental results. Using a crystal spectrometer, Compton made careful measurements of the wavelength spectrum of molybdenum K x-rays after they had been scattered at different angles from graphite. A diagram of the experimental apparatus for these measurements is shown in **Fig. 1a**. The spectrum of wavelengths of the molybdenum K x-rays after scattering from graphite at various angles is shown in Fig. 1b. While the incident radiation before scattering consists mainly of a fairly narrow range of wavelengths (that is, the molybdenum K line), the spectrum observed after scattering consists of two peaks. One of these peaks P has essentially the same wavelength as the molybdenum K line, but the second peak M has a longer wavelength. The wavelength of this second peak depends on the scattering angle, and is longer for larger scattering angles (Fig. 1b).

The dependence of the wavelength shift $\lambda_2 - \lambda_1 = \Delta\lambda$ on scattering angle ϕ is plotted in Fig. 1c. The range of wavelengths due to the inhomogeneity in the scattering angle required by these experiments is also shown for the angles 45, 90, and 135°, and labeled $\delta\lambda$ in the figure. The first peak in Fig. 1b, which has the same wavelength as the incident radiation, is called the unmodified line. The longer-wavelength peak is called the modified line and is clearly due to a different type of scattering than Thomson predicted, since the wavelengths of the x-rays are increased in the scattering process. This is the type of scattering which yields the longer wavelengths in Compton scattering.

Theoretical explanation. Compton's explanation of the observed wavelength shift in x-ray scattering is based on developments which took place early in the twentieth century. Max Planck provided an explanation for the observed intensity distribution with wavelength of electromagnetic radiation from a blackbody by introducing the idea that energy could be emitted or absorbed in the blackbody only in discrete amounts equal to $h\nu$, where h is Planck's constant and ν is the frequency of the radiation. Planck's discovery, which marked the birth of quantum theory, was followed by Albert Einstein's explanation for photoemission. When light falls upon a surface, energetic electrons are emitted, and the energy of the electrons is found to be independent of the intensity but dependent on the frequency of the incident light which liberates the electrons. Einstein supplied the solution for this puzzle by postulating that the radiation field, that is, the beam of light, consists of particles called photons, each having energy $h\nu$, where h and ν are as Planck defined them. The photoelectric effect takes place when a photon is absorbed by an electron, so that the photon disappears and the electron assumes all of the energy $h\nu$ of the photon, less the energy required to

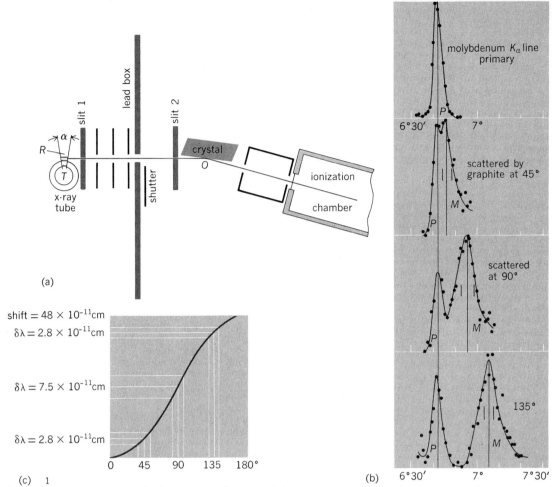

shift = 48×10^{-11}cm

$\delta\lambda = 2.8 \times 10^{-11}$cm

$\delta\lambda = 7.5 \times 10^{-11}$cm

$\delta\lambda = 2.8 \times 10^{-11}$cm

(a)

(c) 1

(b)

molybdenum K_α line primary

6°30′ P 7°

scattered by graphite at 45°

M

P

scattered at 90°

M

P

135°

P M

6°30′ 7° 7°30′

Fig. 1. Compton's experiment. (*a*) Characteristic *K*-lines from molybdenum target *T* of x-ray tube (shown in cross section) fall on graphite scatterer *R*. Scattered radiation is passed through slits 1 and 2, and analyzed spectrally by slow rotation of the calcite crystal and ionization chamber around pivot *O*. Longer wavelengths are observed by Bragg diffraction for larger angles of the calcite crystal with respect to the scattered beam defined by slits 1 and 2. Spread, or inhomogeneity, in scattering angle φ due to width of scatterer *R* is denoted by α. (*b*) Resulting spectra at three scattering angles. The ordinate is the intensity of the beam detected with ionization chamber and the abscissa is the angle that diffracting planes of calcite crystal make with scattered beam defined by slits 1 and 2. Larger angles of the calcite crystal spectrometer correspond to longer wavelengths in the spectrum of scattered radiation. (*c*) Graph showing dependence of shift $\lambda_2 - \lambda_1$ on φ. Effect of inhomogeneity α, about 0.31 radian, in scattering angle φ is shown in the graph. This spread gives rise to apparent spread δλ in wavelength shift due to thickness of scatterer *R*, as shown on the ordinate of the graph.

bind the electron in its medium. *See Blackbody; Heat radiation.*

Compton's explanation of the wavelength shift assumes that, since x-rays are electromagnetic radiation, the picture given by Einstein of the quantum nature of the radiation field in the photoelectric effect describes the incident beam of x-rays. The scattering of the x-rays which gives rise to the peak of the modified line in the spectrum of the scattered x-rays (Fig. 1*b*) is due to photons or quanta of electromagnetic radiation scattering like material particles in collision with free or loosely bound electrons in the scattering material. Further, a photon, in addition to the energy $h\nu$, has a linear momentum $h\nu/c$ in the direction of travel of the beam. This momentum corresponds to that possessed by a massless particle having energy $h\nu$ and moving with the speed of light c, a fact which follows from Einstein's theory of special relativity. *See Relativity.*

In other words, the x-ray scattering process, which gives rise to the increase in wavelength, is a process in which an x-ray photon of energy $h\nu$ and linear mo-

mentum $h\nu/c$ scatters as a mechanical particle would in colliding elastically with an electron which is at rest. In this type of collision, the x-ray particle transmits some of its energy and linear momentum to the electron which recoils. Therefore the x-ray photon after the collision has less energy $h\nu_2$ than before the collision. Since $\nu_2 = c/\lambda_2$ for a wave traveling with speed c, the wavelength λ_2 for the photon after the collision will be longer than λ_1, the wavelength it had before the scattering.

An important property of the elastic collision between two mechanical particles is that the energy and linear momentum will be conserved in the process. These two principles were used by Compton to calculate the increase in wavelength of the x-rays after scattering through a definite angle.

Scattering process. **Figure 2** shows the Compton scattering process. The incident photon with linear momentum p_1 and wavelength λ_1 collides with the electron, which is initially at rest. After the collision, the photon scatters off at an angle φ with respect to the incident direction. The electron recoils and moves

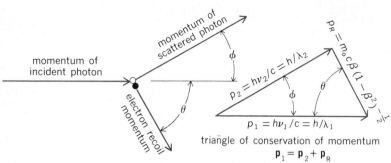

Fig. 2. Diagram for derivation of wavelength shift, electron recoil energy, and angular correlation between scattering angle ϕ and electron recoil angle θ.

off at an angle θ with respect to the incident direction. A triangle resulting from the conservation of linear momentum of the particles involved in the collision is also shown in Fig. 2. This triangle shows the angular relationship which must be satisfied by ϕ and θ.

The principle of conservation of energy gives Eq. (2), where $h\nu_1$ is the energy of the incident photon,

$$hv_1 = hv_2 + m_0c^2\left(\frac{1}{\sqrt{1 - \beta^2}} - 1\right) \quad (2)$$

$h\nu_2$ is the energy of the scattered photon, and the second term on the right-hand side of the equation is the relativistic form of the kinetic energy of the recoiling electron. *See Conservation of energy.*

In applying the principles of conservation of linear momentum, two additional equations are obtained. Conservation of the horizontal component of linear momentum (Fig. 2) gives Eq. (3), while conservation

$$\frac{hv_1}{c} = \frac{hv_2}{c}\cos\phi + \frac{m_0V}{\sqrt{1 - \beta^2}}\cos\theta \quad (3)$$

of the vertical component yields Eq. (4). In these

$$0 = \frac{hv_2}{c}\sin\phi - \frac{m_0V}{\sqrt{1 - \beta^2}}\sin\theta \quad (4)$$

equations, V is the velocity of the recoiling electron and β is V/c.

After making the substitution $\nu = c/\lambda$, squaring the equations, and combining the results, Eq. (1) is obtained. The quantity in Eq. (1), $h/m_0c = 24.26 \times 10^{-11}$ cm, is called the Compton wavelength of a free electron, and is the wavelength of a photon having energy $h\nu = m_0c^2$, the rest energy of the electron. The Compton wavelength is the shift in the wavelength of a photon which is scattered through 90°, as seen in Eq. (1).

Two important physical results follow from Eq. (1). First, the wavelength shift $\Delta\lambda$ is independent of the wavelength of the incident photons and is, therefore, independent of the photon energy. Second, the shift $\Delta\lambda$ is independent of the type of material of the scatterer.

From Eq. (1) it follows that the energy of the scattered photon can be expressed as shown in Eq. (5). It

$$hv_2 = \frac{hv_1}{1 + \frac{hv_1}{m_0c^2}(1 - \cos\phi)} \quad (5)$$

then follows from Eqs. (2) and (5) that the kinetic

energy E_R of the recoil electron can be expressed as in Eq. (6).

$$E_R = hv_1 - hv_2$$

$$= hv_1 \frac{\frac{hv_1}{m_0c^2}(1 - \cos\phi)}{1 + \frac{hv_1}{m_0c^2}(1 - \cos\phi)} \quad (6)$$

From the geometry of the momentum triangle of Fig. 2, Eq. (7) can be obtained. This is the required

$$\cot\theta = -\left(1 + \frac{hv_1}{m_0c^2}\right)\tan\frac{\phi}{2} \quad (7)$$

relationship for the angular correlation between the scattered photon and the recoil electron.

Compton-Debye effect. Peter Debye knew of Compton's published measurements of the wavelength shift of scattered x-rays, and independently developed the same theoretical explanation as Compton. His results were published at about the same time as Compton's; in Europe the effect has been known as the Compton-Debye effect.

Experimental verification. Verification of Compton's ideas appeared soon after his theoretical explanation for the shift. The recoil electrons, which are predicted by Compton's scattering theory, were detected independently by C. T. R. Wilson and Walther Bothe in Wilson cloud chambers. The momenta of the recoil electrons were measured by magnetic deflection and found to agree with Compton's theory by A. A. Bless. Bothe and Hans Geiger demonstrated the simultaneity of the appearance of the scattered photon and recoil electron. Then Compton and A. W. Simon, using a cloud chamber with partitions, were able to show that the predicted correlation in Eq. (7) between the direction of the scattered photon and that of the recoil electron is satisfied in individual Compton scattering processes. Later experiments with improved techniques showed the correctness of Compton's theory with greater accuracy. *See Cloud chamber.*

The results of all these experiments led to the definite conclusion that electromagnetic radiation scattering from an electron, instead of spreading out in all directions around the scattering center as waves are expected to do, takes a definite direction in each individual process as a particle would.

Intensity distribution. Thomson's theory for the scattering of electromagnetic radiation predicts that the relative intensity of scattered radiation is symmetrical about a 90° scattering angle and proportional to the factor $1 + \cos^2\phi$, where ϕ is the scattering angle. The angular distribution of scattered electromagnetic radiation as given by Thomson's theory is written as Eq. (8). Here I_ϕ/I_0 is the ratio of the radiation

$$I_\phi/I_0 = \frac{ne^4}{2r^2m_0^2c^4}(1 + \cos^2\phi) \quad (8)$$

scattered at an angle ϕ to the incident radiation intensity, n is the effective number of independently scattering electrons, e is the charge on the electron, and r is the distance from the scatterer.

Thomson's theory did not take into account the effect of the magnetic vector of the radiation. Including the effect of the magnetic vector would have provided the electron with a recoil in the scattering of radiation from the effect known as radiation pressure. Consideration of recoil of the electron from radiation pressure would have led to the prediction of an increase in the wavelength of the scattered radiation from the Doppler shift. Paul Dirac, Gregory Breit, and others took into account the effect of the relativistic Doppler shift of the photon in scattering from the electron, and found that the intensity distribution from the Thomson theory was to be multiplied by the Breit-Dirac recoil factor R, given by Eq. (9).

$$R = \left[1 + \frac{h\nu_1}{m_0 c^2} (1 - \cos \phi) \right]^{-3} \qquad (9)$$

See Doppler effect; Radiation pressure.

In addition to their charge and mass, electrons also have a quantized spin or angular momentum and an associated magnetic moment. Using Dirac's relativistic quantum mechanics and taking into account the interaction of the spin and magnetic moment of the electron with the electromagnetic radiation, O. Klein and Y. Nishina obtained a further factor which modifies the angular distribution of scattered x-rays. This factor can be written as notation (10). Combining this

$$1 + \frac{\left(\dfrac{h\nu_1}{m_0 c^2}\right)^2 (1 - \cos \phi)^2}{(1 + \cos^2 \phi) \left[1 + \dfrac{h\nu_1}{m_0 c^2} (1 - \cos \phi) \right]} \qquad (10)$$

factor and the Breit-Dirac recoil factor with the Thomson formula for the scattered x-ray angular distribution, the Klein-Nishina formula is obtained. This is written as Eq. (11). The quantities on the left-hand

$$I_\phi/I_0 = \frac{ne^4}{2r^2 m_0^2 c^4} \cdot \frac{1 + \cos^2 \phi}{\left[1 + \dfrac{h\nu_1}{m_0 c^2} (1 - \cos \phi) \right]^3}$$

$$\cdot \left\{ 1 + \frac{\left(\dfrac{h\nu_1}{m_0 c^2}\right)^2 (1 - \cos \phi)^2}{(1 + \cos^2 \phi) \left[1 + \dfrac{h\nu_1}{m_0 c^2} (1 - \cos \phi) \right]} \right\}$$

$$\qquad (11)$$

side of the formula represent the same quantities as in the Thomson formula, Eq. (8). For the limiting case of low-energy x-rays ($h\nu_1 \rightarrow 0$), the Klein-Nishina formula reduces to the Thomson scattering law. But at photon energies above a few hundred kilovolts, the angular distribution departs significantly from the $1 + \cos^2 \phi$ distribution of the Thomson law. *See Electron spin; Relativistic quantum mechanics.*

Figure 3 shows the predicted results of the x-ray scattering angular distribution, comparing the Klein-Nishina formula with the Thomson formula at various energies. The full lines in Fig. 3 are the theoretical predictions, and the points are experimental measurements.

If the modified line of Fig. 1b is due to scattering of an x-ray photon from an electron which recoils, the presence of the unmodified line in the spectrum of scattered radiation must be due to some other type of scattering. The fact that the wavelength changes in Compton scattering, giving rise to the modified line

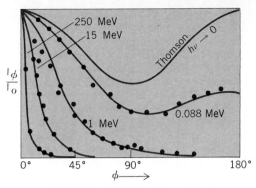

Fig. 3. Angular distribution of Compton scattered gamma rays. I_ϕ represents intensity of gamma rays scattered at angle ϕ with respect to incident beam, while I_0 represents intensity of incident beam before scattering. (*After R. S. Shankland, Atomic and Nuclear Physics, 2d ed., 1960*)

of Fig. 1b, is due to the decrease in energy of the scattered photon in imparting recoil energy to the electron. The unmodified line is due to photons that scatter from electrons which are too tightly held in the atom to recoil from the impact of the photon. That is, these interactions correspond to photon-electron interactions for which the conservation of momentum and energy would require the free-electron recoil energy to be comparable to, or less than, the electron binding energy in the atom. The momentum change accompanying scattering is communicated in such cases to the atom as a whole. Because of the atom's larger mass when compared to that of an electron, the photon transmits only a small contribution of linear momentum to the atom. Therefore the photon itself loses only a negligible amount of energy. The unmodified line is more prominent in scattering from atoms of higher atomic number because of the greater number of tightly bound electrons in these atoms. The theory giving the ratio of intensities of modified and unmodified scattering has been developed by I. Waller and D. R. Hartree.

There are many ways in which photons can interact with matter. In these interactions, usually photons either disappear completely or are scattered out of the initial beam of photons, so that the intensity of the beam is diminished as it moves through the scattering material. Of the many possible modes of interaction of photons with matter, there are four which predominate: Compton scattering (with a change in wavelength of the photon giving rise to the modified line in the scattered spectrum), elastic scattering (with no change in wavelength giving rise to the unmodified line), photoelectric effect or photoemission (when the visible and ultraviolet regions are most often involved), and electron-positron pair production (for photon energies above 1 MeV).

The Klein-Nishina formula, Eq. (11), can be integrated over the angle ϕ to obtain a formula which gives the relative intensity of photons scattered by the Compton effect compared with the initial intensity of the incident beam in the scattering material. This formula is related to the probability that Compton scattering will occur to a photon of a given energy $h\nu$. Formulas have been obtained giving the probability that the other three processes can occur. It has been demonstrated that, while all these processes can happen to some extent at most energies (except for electron-positron pair production which has a threshold at

1 MeV below which it cannot occur), the Compton effect is the predominant interaction for a large range of atomic numbers for photon energies between 1 and approximately 10 MeV. *See Gamma rays.*

Applications in science. The Compton effect has played a significant role in several diverse scientific areas. Compton scattering (often referred to as incoherent scattering, in contrast to Thomson scattering and also Rayleigh scattering, which are called coherent scattering) is important in nuclear engineering (radiation shielding), experimental and theoretical nuclear physics, atomic physics, plasma physics, x-ray crystallography, elementary particle physics, and astrophysics, to mention some of these areas.

In addition the Compton effect provides an important research tool in some branches of medicine, in molecular chemistry and solid-state physics, and in the use of high-energy electron accelerators and charged particle storage rings. Some examples of the role of the Compton effect in a few of these fields will be described below.

Attenuation of electromagnetic radiation. As mentioned above, if a beam of electromagnetic radiation traverses matter it is reduced in intensity by the absorption of photons and by the scattering of photons out of the beam by the matter. The relative importance of the principal processes in attenuating the beam depends in part on the energies of the photons in the beam. Considerable effort has been made to determine as accurately as possible the parameters determining the rate of attenuation of electromagnetic radiation as a function of energy through Compton scattering and through the other processes. For example, the contribution of Compton scattering (incoherent scattering) is an important correction in crystallographic studies through coherent x-ray scattering. *See X-ray crystallography.*

Information on the attenuation properties of various materials is obviously important in the design of radiation shielding. *See Radiation shielding.*

Medical applications. In addition to its role in the interpretation of x-ray photographs, the Compton effect has been used directly in the diagnoses of human diseases. One application is the use of Compton scattering at 90° of gamma rays from various radioactive sources for the early detection of the bone disease osteoporosis. Since the intensity of the Compton-scattered gamma rays is proportional to n, the effective number of independently scattering electrons [see Eq. (11)], and since the number of effective scattering electrons can be closely related to the density of the scattering material, the intensity of the Compton-scattered photons can be used as a measure of the density of the scattering material. Changes in density of bone are an indication of osteoporosis. Similar techniques have been developed to determine the density of lung tissues in the body to assist in the diagnoses of disease.

Study of electrons in matter. The Compton effect has been used to study the electronic structures of molecules and solid crystals. It was realized by J. W. M. DuMond that in addition to the Doppler shift from the recoil of the electron in the Compton-scattering process, there is also an additional Doppler shift of the photons which is due to the fact that the electrons from which the photons are scattered are not really at rest but are rapidly moving in their atoms. This second Doppler shift adds to or subtracts from the Compton shift, depending on whether the electron is moving toward or away from the incident photon before scattering. The result is a broadening of the modified line. Physicists and chemists have made extensive use of this Doppler broadening to study the momenta of electrons in atoms, molecules, and conducting and semiconducting crystals.

Inverse Compton effect. In 1948 E. Feenberg and H. Primakoff studied theoretically the possibility that the inverse Compton effect takes place in interstellar space and that it may be an astrophysical phenomenon important in the depletion of high-energy electrons. In the inverse Compton effect energetic electrons undergo elastic collisions with low-energy photons so that the electrons lose energy and the photons gain energy in the collision. The wavelength shift for the photon is to shorter wavelength. Feenberg and Primakoff suggested that energetic electrons would interact through the inverse Compton effect with starlight to reduce the energy of the electrons. In so doing, of course, the photons would often receive high energies and would then become energetic gamma rays.

Such a mechanism has been extensively considered as a source for the production of high-energy gamma rays observed in cosmic rays. The universal microwave radiation, discovered in 1965 by A. Penzias and R. Wilson, has been considered as the source of initial photons for this process, since it provides much higher density of photons in space than light could. The nature and source of the primary high-energy electrons which produce the gamma rays can be studied from the observed gamma-ray spectrum. *See Cosmic background radiation; Gamma-ray astronomy.*

The inverse Compton effect can be used as a method for producing high-energy polarized gamma rays, by directing a homogeneous beam of photons of light from a laser into the energetic electron beam of an electron accelerator. **Figure 4** shows the arrangement used by Richard Milburn and his colleagues at the Cambridge electron accelerator when this technique was being developed. In that arrangement the photons which are backscattered through the inverse Compton effect are increased uniformly in energy so that high-energy gamma rays are emitted from the accelerator and can be detected by a Cerenkov counter or some other type of gamma-ray detector. If photons are circularly polarized, the polarization state will be maintained and the resulting gamma rays will be circularly polarized in the same sense as the original laser photons. *See Particle accelerator; Polarized light.*

The inverse Compton backscatter technique using photons from lasers is a useful method for determining the degree of spin polarization of the circulating electron and positron beams in high-energy storage rings. Under certain conditions there is a gradual buildup of this polarization through the effects of the synchrotron radiation emitted by the circulating beams. The calculation described earlier for the intensity of Compton scattered photons, the Klein-Nishina formula (11), applies in the case of unpolarized electrons and unpolarized photons. When the calculation is carried out for the case of spin-polarized electrons and circularly polarized photons, it is found that there will be an asymmetry in the scattering depending on the degree of polarization of the electrons. This effect has been used to detect circularly polarized gamma rays scattered from polarized electrons in magnetized iron, in studying nonconservation of parity in beta decay. Thus, circularly polarized photons from a

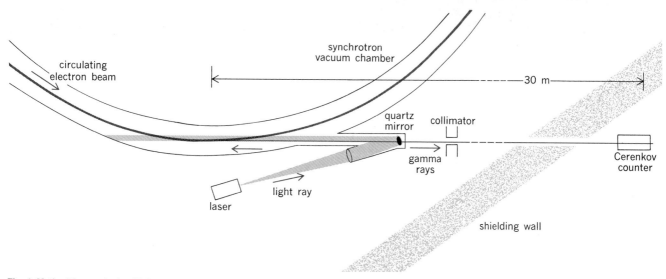

Fig. 4. Method for producing high-energy gamma rays at the Cambridge electron accelerator. Light from the ruby laser is reflected at the quartz mirror to meet 6-GeV electrons in the straight section of the machine. Backscattered gammas pass through the mirror and are counted in the total-absorption Cerenkov counter. (*After Polarized gamma rays made by Compton scattering, Phys. Today, 21(5):77–79, 1968*)

laser can be used to determine the extent of the polarization of the beams of circulating electrons and positrons in storage rings by examining the scattering asymmetry. *See Electron spin; Parity (quantum mechanics).*

Another application of the inverse Compton effect using photons from a laser is to monitor the intensity of the charged particle beams, such as a proton beam, in high-energy accelerators. By monitoring the intensity of high-energy gamma rays produced through the inverse Compton backscattering from the beam, the intensity of the beam can be determined once the system has been calibrated. *See Charged particle beams.*

Compton-like scattering from other particles. The Compton effect is usually associated with the elastic collision of a photon with an electron. However, a similar process occurs when a photon is scattered by a proton or some other elementary particle, such as a pion. Consideration has also been given to Compton-like scattering of neutrinos by nuclear matter in interstellar space.

It is believed that research on Compton scattering of high-energy gamma-ray photons from protons will provide details of the internal structure of the proton and possibly also of the photon itself. Quarks, believed to be the constituent particles of the proton, can be observed through their internal motions which give rise to a Doppler shift, similar to that observed in experiments on the motion of electrons in atoms. *See Quarks.*

Conclusion. During the past two centuries an overwhelming amount of evidence has been obtained which shows that electromagnetic radiation, including light, x-rays, and gamma rays, has the nature of waves moving through space. The phenomena of diffraction and interference, which are exhibited by x-rays and gamma rays as well as by light, can be explained only if the radiation has a wave character.

Experiments involving the polarization of the electric vector of x-rays and gamma rays from scattering have shown that the waves are transverse vibrations, and that x-rays and gamma rays as well as light are linearly polarized with the electric vector of the beam

predominantly in a given direction after scattering in matter. If the radiation is scattered at 90° with respect to its incident direction, the scattered radiation is completely linearly polarized. This statement is found to be true for both the unmodified and the Compton-scattered-modified line.

Perhaps the greatest significance of the Compton effect is that it demonstrates directly and clearly that in addition to its wave nature with transverse oscillations, electromagnetic radiation has a particle nature and that these particles, the photons, behave quite like material particles in collisions with electrons. This discovery by Compton and Debye was followed by the formulation of quantum mechanics by W. Heisenberg and E. Schrödinger and provided the basis for the beginning of the theory of quantum electrodynamics, the theory of the interactions of electrons with the electromagnetic field. *See Quantum electrodynamics.*

A finding also of fundamental importance is that material particles have a wave nature and exhibit interference effects. This dual nature of both electromagnetic radiation and material particles, like electrons and protons, lies at the very heart of modern quantum theory and is called the wave-particle duality. A striking example of this dual nature of material particles is that an electron will undergo a Compton-like effect, that is, there will be an increase in its de Broglie wavelength, when it scatters, say, from a proton, since some of the electron's momentum will be transmitted to the proton. However, the formula for the wavelength increase will be different from Eq. (1), since the electron is not a massless particle. *See De Broglie wavelength; Electron diffraction.*

The discovery and successful explanation of the Compton effect and subsequent experiments have provided the resolution of an important scientific controversy of 200 years' duration: does electromagnetic radiation consist of waves or particles? The astonishing answer is that electromagnetic radiation is both. *See Scattering of electromagnetic radiation.*

Eastman N. Hatch

Bibliography. A. H. Compton, The scattering of x-rays as particles, *Amer. J. Phys.*, 29:817–820, 1961;

R. D. Evans, The Compton effect, *Handbuch der Physik*, vol. 34, pp. 218–298, 1958; G. Hazan et al., The early detection of osteoporosis by Compton gamma ray spectroscopy, *Phys. Med. Biol.*, 22:1073–1084, 1977; J. H. Hubbell et al., Pair, triplet and total atomic cross sections and mass attenuation coefficients for 1 MeV–100 GeV photons in elements Z = 1 to 100, *J. Phys. Chem. Ref. Data*, 9(4):1023–47, 1980; I. Kaplan, *Nuclear Physics*, 2d ed., 1962; W. Niemann and K. O. Thielheim, Universal microwave radiation and extragalactic γ-radiation, *Nature*, 282:48–50, 1979; P. J. Schinder and S. L. Shapiro, Neutral currents and neutrino comptonization in high-temperature, nuclear matter, *Astrophys. J.*, 233:961–973, 1979; R. F. Schwitters, Experimental review of beam polarization in high-energy e^+e^- storage rings, *American Institute of Physics Conference Proceedings*, no. 51, pp. 91–108, 1979; R. S. Shankland, *Atomic and Nuclear Physics*, 2d ed., 1960; R. H. Stuewer, *The Compton Effect: Turning Point in Physics*, 1975; C. E. Webber and G. Coates, A clinical system for the in vivo measurement of lung density, *Med. Phys.*, 9(4):473–477, 1982; B. G. Williams (ed.), *Compton Scattering*, 1976; R. Wilson, From the Compton effect to quarks and asymptotic freedom, *Amer. J. Phys.*, 45:1139–1147, 1977.

Compton wavelength

A convenient unit of length that is characteristic of any particle. By definition, the Compton wavelength λ_c of a particle of rest mass m is $\lambda_c = h/mc$, where h is Planck's constant and c is the velocity of light; this definition is analogous to that of the quantum-mechanical de Broglie wavelength λ_B, defined as $\lambda_B = h/p$, where p is the momentum classically given by $p = mv$, with m and v the particle mass and velocity respectively. *SEE DE BROGLIE WAVELENGTH; PLANCK'S CONSTANT.*

The Compton wavelength of the electron is 2.42631×10^{-12} m; of the proton 1.32141×10^{-15} m; of the muon 1.1734×10^{-14} m; and of the pion 8.883×10^{-15} m. The last is, by definition, the range of the strong nuclear force; in general, the Compton wavelength provides a convenient scale length in any given quantum-mechanical situation. *SEE ELEMENTARY PARTICLE; QUANTUM FIELD THEORY.*

The so-called reduced Compton wavelength, $\lambda_c = \lambda_c/2\pi$, is very frequently used instead of the Compton wavelength itself. *SEE QUANTUM MECHANICS.*

D. Allan Bromley

Computational chemistry

A branch of theoretical chemistry that uses a digital computer to model systems of chemical interest. In this discipline, the computer itself is the primary instrument of research. The use of computers for analysis of experimental data, and for the storage and display of results obtained with other tools, is distinct from computational chemistry. The latter permits calculation of quantities which can be measured experimentally, such as molecular geometries of ground and excited states, heats of formation, and ionization potentials. Alternatively, quantities not readily accessible by existing experimental techniques, such as geometries of transition states and detailed structure of

liquids, may be evaluated. *SEE DIGITAL COMPUTER.*

Because of the increasing power and availability of computers, and the simultaneous development of well-tested and reliable theoretical methods, the use of computational chemistry as an adjunct to experimental research has increased rapidly. Calculations ranging from a few seconds to many hours of computer time can serve as a guide to exclude less favorable reactions or unstable products, or to select several more fruitful procedures from the many possible ones. In addition, modeling of chemical systems with a computer enables the researcher to examine them on a scale of space or time as yet unmeasurable by experimental techniques. This can give insight into a chemical system beyond that provided by experiment. Examples are examination of the dynamics of a chemical reaction or of detailed changes in conformation of a polymer in solution. Examination of the molecular orbitals occupied by the electrons of the molecule can provide insight into chemical bonding and the electronic interactions which determine specific geometric configurations. Thus computational chemistry can yield information which may not be experimentally available. *SEE CHEMICAL DYNAMICS.*

Computational chemistry may be the application of existing theory and numerical methods to new molecules, or it may be the development of new computational methods. The latter may include incorporating more physics into the mathematical model in order to provide a better theoretical description of the system being studied, for example, inclusion of interactions between individual electrons in molecular orbital calculations. These more complete studies, for "large" molecules, usually need to be performed on a supercomputer, or on a mid-size computer with an array processor. Another approach is to devise simpler methods which will approximate the accuracy of more complex calculations. These include semiempirical methods in which values of hard-to-calculate terms are derived from experiment. Such an approach allows fruitful work with a mid-size or even a desktop computer, avoiding the need for expensive computer resources. *SEE MICROCOMPUTER.*

Before a computational study is begun, the suitable theoretical method must be determined. Some methods are known to yield more accurate results for systems of a certain type, for example, simple organic molecules. The preferred method must be balanced against the computing facilities available and the amount of computer time to be allocated.

Molecular orbital theory. The most active area of computational chemistry involves molecular orbital calculations which yield a description of the electronic distribution in molecules. In this method, a solution is sought for Eq. (1), the Schrödinger equation,

$$H\psi = E\psi \tag{1}$$

where H is the hamiltonian operator, ψ is the electronic wave function, and E is the energy of the wave function. The equation is solved self-consistently until there is no change in ψ or E. A molecular orbital is composed of linear combinations of atomic orbitals. These are expressed as Slater functions, Eq. (2),

$$\eta = r^{n-1}e^{-\alpha r} \tag{2}$$

where η is the quantum number, or more often gaussian functions or linear combinations of gaussians,

Eq. (3), and p, q, and s are integers. In Eqs. (2) and

$$\eta = \sum_i c_i x^p y^q z^s e^{-\alpha_i r^2} \qquad (3)$$

(3), $r = \sqrt{x^2 + y^2 + z^2}$ is the distance from the atom and the c's and α's are coefficients and exponents which define the basis function. Gaussian functions are used most often, as the necessary integrals may be evaluated analytically.

Molecular orbital computational methods vary in complexity from extended Hückel theory (which deals only with the valence electrons) to ab initio methods which calculate everything from first principles. In between are a number of semiempirical methods [complete neglect of differential overlap (CNDO), intermediate neglect of differential (INDO), and modified neglect of differential overlap)]. Many of the programs that have been developed have provision for geometry optimization using the gradient of the energy to determine a minimum-energy geometry.

Molecular orbital calculations can take from a few seconds of computer time for extended Hückel calculations to many hours on the largest supercomputers for detailed ab initio calculations. The time for any method usually increases as N^4, where N is the number of atomic orbitals.

An example of molecular orbital techniques applied to novel, possibly speculative, systems can be seen in the results of ab initio computations completed by P. v. R. Schleyer and coworkers. They concluded that the carbon-lithium compounds CLi_5 and CLi_6 may be stable and thus violate the octet rule. The computed C-Li (**Fig. 1**) distances are similar for these compounds and CLi_4. The pattern of electron density indicates that the extra electrons are involved in Li-Li bonding. This group cites experimental results which support the possible existence of these hypervalent species. *SEE MOLECULAR ORBITAL THEORY; QUANTUM CHEMISTRY.*

Molecular dynamics. Molecular dynamics calculations simulate the behavior of physical systems by integration of Newton's equations of motion. This method has been applied to studies of liquid water, deposition of molecules on surfaces, cluster formation, and conformational changes of alkanes and polymer chains, among others. As the time step of the integration must be small relative to the vibrations of the particles in order for the numerical integration to be stable, the method requires vast amounts of computer time. Because of this, it is usually not practical to include more than several hundred particles. However, the system may be considered infinite because of the imposition of periodic boundary conditions. This means that the box containing the particles is considered to be replicated in all dimensions, in the manner that strips of wallpaper continue a pattern around a room.

To decrease the amount of computer time needed for these simulations, a family of related stochastic (using random forces) methods has been developed. These range from Monte Carlo methods, where each particle movement is determined randomly, to Langevin dynamics methods, which treat forces between nearby particles exactly and approximate those due to remote particles stochastically. *SEE MONTE CARLO METHOD; STOCHASTIC PROCESS.*

This area of computational chemistry has grown rapidly. Research activity includes application of the method to more complex physical systems, such as

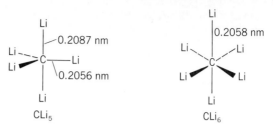

Fig. 1. Computed geometries of CLi_5 and CLi_6. (*After P. v. R. Schleyer et al., CLi_5, CLi_6, and the related effectively hypervalent first-row molecules, $CLi_{5-n}H_n$ and $CLi_{6-n}H_n$, J. Amer. Chem. Soc., 105:5930–5932, 1983*)

freezing, melting, and phase separations, as well as comparisons of the accuracy and costs of the various related methods.

Polymer relaxation processes have been studied by a number of experimental techniques such as nuclear magnetic resonance, dielectric relaxation, ultrasonic attenuation, dynamical light scattering, fluorescence depolarization, and excimer fluorescence. While these studies detect transitions between trans and gauche conformations, they provide little information about the detailed process by which a bond undergoes a transition.

Most of the experiments detect an activation energy of a single barrier height (between trans and gauche states), indicating that transitions occur independently. However, a single transition in a polymer chain would necessitate a wide swing by the rest of the chain, which is unlikely to occur. To investigate the detailed mechanisms of these transitions, E. Helfand and coworkers performed brownian dynamics simulations (solution of the Langevin equations) of polymer conformational transitions. These studies confirmed an activation energy of a single barrier height. But in addition, they indicated a highly enhanced rate of transition of a bond whose second-nearest neighbor has recently undergone a transition. These favorable pairs are separated by a trans central bond. A clockwise transition of a bond on one side would be associated with a counterclockwise rotation on the other (**Fig. 2**). This pair of changes leads to lateral motions of the tails, rather than large swings.

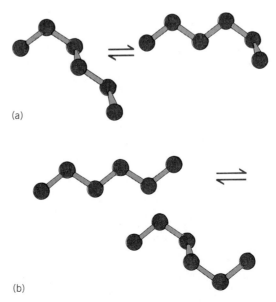

Fig. 2. Two pairs of transitions which result in translation of the polymer tails. (*a*) Gauche migration gtt ⇌ ttg, transitions on second-nearest neighbors which result in a shift of the gauche bond from one end to the other. (*b*) Pair gauche production ttt ⇌ g^+tg^-, transitions on second-nearest neighbors which transform an all trans chain to one with a gauche at either end. (*After E. Helfand, Theory of the kinetics of conformational transitions in polymers, J. Chem. Phys., 54:4651–4661, 1971*)

Evidence for these pairs of transitions was subsequently detected experimentally. *See* BROWNIAN MOVEMENT; CONFORMATIONAL ANALYSIS.

Molecular mechanics. This method has been in use since the early 1950s, particularly for small organic molecules. The molecular mechanics model expresses the energy of a system as potential functions of the bond lengths, bond angles, dihedral angles, and van der Waals interactions of the atoms in the system. Using the gradient of the energy, the programs adjust the coordinates of the atoms until a configuration of minimum energy is determined. The potential functions, or force fields, contain parameters which are optimized to produce results which agree with a set of experimental measurements. The assumption is that these parameter values will be equally valid for other molecules of similar structure which have not been studied experimentally.

While the existing programs contain parameters which apply to a wide range of molecules, difficulties arise when attempts are made to use them for calculations on novel systems with unusual interactions or geometric structures. Such parameter values may not be valid, or they may not be known. Research has continued to widen the class of molecules for which the method may be expected to yield reliable results.

When the relative stability of chemical structures is under consideration, molecular mechanics is an appealing technique. Calculations take several orders of magnitude less computer time than molecular orbital calculations, and the method provides the only feasible approach for large biological molecules. *See* BOND ANGLE AND DISTANCE; CHEMICAL BONDING.

Interactive computer graphics. This field owes its existence to the rapid development of the capabilities of interactive computer graphic devices. The chemist is essentially working with a sophisticated set of molecular models which are stored in the computer memory and displayed on a screen. Moving a joystick or turning a dial moves atoms about on the screen as easily as tangible models can be moved physically. Stereoscopic views are produced either by placing left- and right-eye images next to each other on the screen and using a viewer to merge the images, or by alternately blinking left- and right-eye images and looking through a viewer which presents the correct image to each eye.

This technique is most useful when studying interactions between large molecules, and it is often used to fit potential drug molecules into the active sites of biomolecules. The intuition of the chemist takes the place of theoretical calculations in determining geometric configurations, and the most common application is in establishing starting geometries for subsequent molecular mechanics refinement. When faster computers become available, it may be combined with energetic calculations performed in real time.

P. A. Kollman and coworkers modeled the stereoselective hydrolysis of peptides catalyzed by enzymes. They used molecular mechanics calculations to simulate interactions of the enzyme α-chymotrypsin with both the substrate L-*N*-acetyltryptophanamide and the inhibiting D form. Considering both an initial nonbonded complex and the covalently bonded transition state, the calculations confirm the high stereoselectivity of the enzyme. The selectivity is not associated with the initial state but only with the bonded complex, and relaxation of the enzyme is essential for stereoselectivity. *See* STEREOCHEMISTRY.

The Kollman group used interactive graphic techniques to position the molecules for subsequent energy refinement. The efficiency of the graphics procedure allowed the sampling of large areas of the potential energy surfaces needed to locate the energy minima characteristic of the complexes. *See* COMPUTER GRAPHICS; SIMULATION.

Zelda R. Wasserman

Bibliography. U. Burkert and N. Allinger, *Molecular Mechanics*, American Chemical Society, 1982; W. H. Miller et al. (eds.), *Modern Theoretical Chemistry Series*, vols. 1–8, 1976–1977.

Computer

A device that receives, processes, and presents information. The two basic types of computers are analog and digital. Although generally not regarded as such, the most prevalent computer is the simple mechanical analog computer, in which gears, levers, ratchets, and pawls perform mathematical operations—for example, the speedometer and the watt-hour meter (used to measure accumulated electrical usage). The general public has become much more aware of the digital computer with the rapid proliferation of the hand-held calculator and a large variety of intelligent devices, ranging from typewriters to washing machines.

Analog computer. An analog computer uses inputs that are proportional to the instantaneous value of variable quantities, combines these inputs in a predetermined way, and produces outputs that are a continuously varying function of the inputs and the processing. These outputs are then displayed or connected to another device to cause action, as in the case of a speed governor or other control device.

The electronic analog computer is often used for the solution of complex dynamic problems. Electrical circuits, usually transistorized, perform the processing. Electronic amplifiers allow signals to be impressed upon cascaded circuits without significant electrical loss of attenuation through loading of prior stages, a feature absent in purely mechanical computers. Friction in a mechanical analog computer builds up and limits the complexity of the device.

Small electronic analog computers are frequently used as components in control systems. Inputs come from measuring devices which output an electrical signal (transducers). These electrical signals are presented to the analog computer, which processes them and provides a series of electronic outputs that are then displayed on a meter for observation by a human operator or connected to an electrical action device to ring a bell, flash a light, or adjust a remotely controlled valve to change the flow in a pipeline system. If the analog computer is built solely for one purpose, it is termed a special-purpose electronic analog computer. *See* CONTROL SYSTEMS.

General-purpose electronic analog computers are used by scientists and engineers for analyzing dynamic problems. A general-purpose analog computer receives its degree of flexibility through the use of removable control panels, each of which carries a series of mating plugs. Outputs from one component are routed to the input of another component by connecting an electrical conductor from one mating plug on the removable board (output) to another plug on the removable board (input). This process is called patching, and the removable panel is frequently called a patch board.

Thus, in any analog computer the key concepts involve special versus general-purpose computer designs, and the technology utilized to construct the computer itself, mechanical or electronic. In any case, an analog computer receives inputs that are instantaneous representations of variable quantities and produces output results dynamically to a graphical display device, a visual display device, or in the case of a control system, a device which causes mechanical motion. SEE ANALOG COMPUTER.

Digital computer. In contrast, a digital computer uses symbolic representations of its variables. The arithmetic unit is constructed to follow the rules of one (or more) number systems. Further, the digital computer uses individual discrete states to represent the digits of the number system chosen.

Electronic versus mechanical computers. The most prevalent special-purpose mechanical digital computers have been the supermarket cash register, the office adding machine, and the desk calculator. Each of these is being widely replaced by electronic devices allowing much greater logical decision making and greatly increased speed. For example, most products now carry a bar code, the Universal Product Code (UPC); in suitably equipped supermarkets, the code is scanned by a light-sensitive device, bringing information about each product into the point-of-sale (POS) terminal that has replaced the mechanical cash register. The POS terminal then computes total charges and provides a receipt for the customer. It may also communicate with a centralized computer system that controls inventory, accounts payable, salaries and commissions, and so on. While a mechanical cash register could carry out only a small number of operations each minute, and some electromechanical devices might handle several hundred operations per second, even a small general-purpose electronic computer can carry out its computations at a million operations per second or more. SEE CALCULATORS.

Stored program operation. A digital computer works with a symbolic representation of variables; consequently, it can easily store and manipulate numbers, letters, or graphical information represented by a symbolic code. Typically, a general-purpose electronic digital computer operates on numbers by using both decimal and binary number systems, and on symbolic data expressed in an alphabet. It contains both an arithmetic unit and a storage unit. As the digital computer processes its input, it proceeds through a series of discrete steps called a program. The storage unit serves to retain both the values of the variables and the program to process those variables. The arithmetic unit may operate on either variables or coded program instructions interchangeably, since both are usually retained in the storage unit in the same form. Thus, the digital computer has the capability to be adaptive, because processing can be determined by the previously prepared program, by the data values supplied as input to the computation, and by the values generated during the course of the computation. Through the use of the stored program, the digital computer achieves a degree of flexibility unequaled by any other computing or data-processing devices. SEE DIGITAL COMPUTER PROGRAMMING.

Applications. In the past, most digital computers were confined to standard applications, such as bookkeeping, accounting, engineering design, and test data reduction. However, the advent of the relatively inexpensive and readily available personal computer has dramatically altered that pattern. The most common application now is probably text and word processing, followed by electronic mail. Ready access to national networks has fostered the growth of communities of computer users who communicate easily and effectively through their computers. The ability of even modest computer systems to store, organize, and retrieve very large amounts of information has brought about radical changes in the very nature of many business offices. SEE ELECTRONIC MAIL; MICROCOMPUTER; WIDE-AREA NETWORKS; WORD PROCESSING.

Personal workstations, used in engineering design and other applications requiring intense computation and sophisticated graphics, have become more powerful than earlier, very large computer systems. This has moved computation, and even the storage of many large data files, directly onto each person's desk. In addition, applications that were once considered esoteric have led to industrial applications, such as the use of robots on manufacturing assembly lines. Many of the heuristic techniques employed by these robots are based on algorithms developed in such artificial-intelligence applications as chess playing and remotely controlled sensing devices. SEE ARTIFICIAL INTELLIGENCE; COMPUTER-AIDED DESIGN AND MANUFACTURING; COMPUTER-AIDED ENGINEERING; COMPUTER GRAPHICS; DATA PROCESSING; DATABASE MANAGEMENT SYSTEMS; DIGITAL COMPUTER; ROBOTICS.

Bernard A. Galler

User interfaces. The user interface to a computer system defines the communication between the computer and the user. It includes input devices, to transform users' physical actions to a form that the computer can process; output devices, to present messages from the computer system in a form which users can assimilate; a specification for the computer actions to respond to user inputs and to generate output presentations; and implementation of the computer actions in software and hardware.

Human–computer interaction. User interfaces are the basis for the process of human–computer interaction. In addition to the observable events defined in the user interface, the interaction includes the processing by which the person and the computer interpret and generate inputs and outputs.

On the user side, interface processing begins with the user's goals in interacting with the computer. These can be refined into a sequence of actions to be taken by the computer, a sequence of inputs to the computer to cause these actions to take place, and physical actions to communicate the inputs. The outputs generated by the computer in response must then be perceived and interpreted. This interpretation may cause the user to continue the planned sequence of inputs or to bring it into better conformity with the goals.

The details of this process are the subject of research in human–computer interaction. Some of the cognitive phenomena that influence the process include models of the tasks to be accomplished, models of the computer system, models of the user interface events, and goals and plans for the interaction process. An understanding of these phenomena relies on a knowledge of the underlying cognitive architecture. SEE COGNITION.

The computer side of the interaction is implemented in the underlying hardware and systems software. For a specific user interface, the implementer designs representations for the interaction and algorithms for processing. Research in computing science has improved these components to allow development of more effective user interfaces.

Styles. Interaction with a computer can resemble a variety of everyday experiences, depending on the user interface. If the interaction is modeled on dialog between people, the interface may appear to the user as a command language for giving instructions. If the interaction is modeled on using a tool, the user interface appears to offer direct manipulation of the objects in the task.

Each user interface style reflects a different assignment of roles between user and computer. The set of roles suitable for a particular situation depends on the information and processing available on each side of the interaction. For example, a menu-based user-interface style presents the user with sets of choices for actions. This style is often chosen for users having limited computer experience.

For an infrequent user of an interactive system, it is essential that the system provide ongoing information about the options available. The mechanics of use must be kept simple. More regular users work effectively with increasingly complex interfaces, to exercise greater discretion over operations. Frequent users of several different user interfaces may require some measure of integration among them, to give patterns of actions consistent effects.

Effective interface styles therefore provide a growth path as knowledge increases. Experienced users are expected to progress from an early focus on the ease of learning a system to a concern for power and ease in their tasks. Later mastery will make efficiency of use a major interest. Finally, expert users are most concerned with the ease of extending and modifying the user interface. This learning sequence requires a corresponding sequence of interaction styles.

Effective user interfaces anticipate user problems and provide mechanisms for error recovery and assistance. In addition to stating problem symptoms, an effective interface reports the current status of suspended activities, a diagnosis of causes for the problem, and a set of possible corrective actions. Simple errors like those in spelling can often be corrected automatically. When users ask for help, an effective interface provides specific information about present conditions.

Development. Developing a user interface requires design of the processes on both sides of the interaction. The materials of the user interface, such as devices and processing specifications, are familiar to computer scientists as a design medium. However, the need to design user behavior as well implies additional knowledge and skills in applied cognitive science. Other contributing disciplines include ergonomics and graphics design. SEE HUMAN-FACTORS ENGINEERING.

Just as user interfaces form one component of human–computer interaction, the interaction is itself a component of a wider system which determines its purpose and constraints. This task environment includes processes to be controlled, jobs to be performed, and other people who are involved.

The requirements of the interaction process are determined by this context and by characteristics of target users. Interface designers must allow for differences in the duration and purpose of work, the degree of control users have over work performance, the amount of satisfaction derived from it and the collaboration of other people. When these system requirements are defined, the interaction process and the user interface can be designed, implemented, and evaluated.

Design. To select the appropriate user-interface styles, the designers consider the cognitive models and processing required by the users. The design must also balance the costs and benefits of the devices, processing, and implementation on the computer side.

The user-interface design is often specified as a series of levels of action. At the lowest level are definitions of individual actions, like typing a character for a command language style. At the next level, characters are aggregated into words and sentences with syntactic properties. Sentences form sequences of actions with semantic properties. The design consists of rules for events at each level and transformations between levels.

There are similar design levels for other styles. Designs for menu-based interfaces specify mechanisms for selecting menu items, the words or graphic images (icons) to represent the items, how users request displays of the menus, and the position and sequence of the menus. Direct manipulation designs must consider how to represent objects and processing to create the feeling of working ''hands-on.''

Implementation. Advances in computing science have made it easier to implement advanced user-interface functions by incorporating improved tools for user-interface software into the programming environment. These include window managers to provide multiple screen views; animation kits for programming dynamic displays; tool kits of interaction objects, like menus, which can be adapted to particular applications; and design tools for rapid prototyping and evaluation. Implementation also includes design and development of training materials and system documentation. Users must be instructed in system operation.

Evaluation. The evaluation has to consider both the user–interface behavior (human and machine) and the effectiveness of the interaction in the overall system. Since much of the human behavior is not directly observable, testing techniques use methods from the behavioral and social sciences. User interfaces often evolve through a series of prototypes which are refined by iterative designing and testing. SEE HUMAN FACTORS ENGINEERING.

Evolution. When processor cycles were scarce, interactive users typed one-line statements at a keyboard and received a one-line response. With increasing power available, user-interface devices now offer increased communication bandwidth between the person and the system. User-interface devices may contain several display areas (windows) as views of different ongoing activities. Improvements in input techniques have made it easier to select items from displays, often using a hand-controlled pointing device (mouse). Future interfaces will make increased use of graphic presentations, including animations and interactive video.

The techniques used to implement user interfaces have improved. Although natural-language input and output are often featured in fictional descriptions of future computers, progress in implementation in this area has been slow (but steady). Yet it is now evident that emulating other aspects of human behavior is also important for user interfaces. For example, representations of the user's cognitive models allows the computer system to respond more intelligently to user actions. This includes questioning actions which have severe consequences, and analyzing patterns of user behavior to support better advising and error correction.

Tom Carey

Bibliography. R. Baecker and W. Buxton (ed.), *Readings in Human–Computer Interaction*, 1988; J. G. Brookshear, *Computer Science: An Overview*, 1988; R. Rubinstein and H. Hersh, *User Interface Design*, 1984; J. E. Savage, S. Magidson, and A. M. Stein, *The Mystical Machine*, 1986; B. Shneiderman, *Designing the User Interface*, 1986.

Computer-aided design and manufacturing

The application of digital computers in engineering design and production. Computer-aided design (CAD) refers to the use of computers in converting the initial idea for a product into a detailed engineering design. The evolution of a design typically involves the creation of geometric models of the product, which can be manipulated, analyzed, and refined. In CAD, computer graphics replace the sketches and engineering drawings traditionally used to visualize products and communicate design information.

Engineers also use computer programs to estimate the performance and cost of design prototypes and to calculate the optimal values for design parameters. These programs supplement and extend traditional hand calculations and physical tests. When combined with CAD, these automated analysis and optimization capabilities are called computer-aided engineering (CAE). *See* Computer-aided engineering; Optimization.

Computer-aided manufacturing (CAM) refers to the use of computers in converting engineering designs into finished products. Production requires the creation of process plans and production schedules, which explain how the product will be made, what resources will be required, and when and where these resources will be deployed. Production also requires the control and coordination of the necessary physical processes, equipment, materials, and labor. In CAM, computers assist managers, manufacturing engineers, and production workers by automating many production tasks. Computers help to develop process plans, order and track materials, and monitor production schedules. They also help to control the machines, industrial robots, test equipment, and systems which move and store materials in the factory.

Advantages of automation. CAD/CAM is more expensive than traditional manufacturing technology and requires a more highly trained work force. Wisely applied, however, CAD/CAM can improve productivity, product quality, and profitability. Computers can eliminate redundant design and production tasks, improve the efficiency of workers, increase the utilization of equipment, reduce inventories, waste, and scrap, decrease the time required to design and make a product, and improve the ability of the factory to produce different products. Today most manufacturers employ CAM/CAM to varying degrees. *See* Productivity.

Although CAD/CAM technology is comparatively new, two factors account for its rapid development since the 1960s. First, digital computers are becoming faster, more capable, smaller, and less expensive. Together with improved techniques for software engineering, this makes CAD/CAM increasingly cost-effective even for small companies. Second, world markets for manufactured goods are more competitive. This frequently makes CAD/CAM a key tool for economic success. *See* Software engineering.

Computer-integrated manufacturing. The deployment of CAD/CAM in the manufacturing industry is helping to shape new ideas about organization and production. Traditional manufacturing companies, especially large companies, are organized in functional units such as marketing, design, engineering, and production. These units are coordinated by higher-level management without a great deal of direct communication among the units themselves. The introduction of CAD/CAM technology results in so-called islands of automation.

The fact that CAD, CAE, and CAM work best together has led to the breakdown of many of the traditional barriers between functional and manufacturing units. The goal of computer-integrated manufacturing (CIM) is a database, created and maintained on a factory-wide computer network, that will be used for design, analysis, optimization, process planning, production scheduling, robot programming, materials handling, inventory control, maintenance, and marketing. Although many technical and managerial obstacles must be overcome, computer-integrated manufacturing appears to be the future of CAD/CAM. *See* Computer-integrated manufacturing; Database management systems.

CAD graphics. A CAD workstation (**Fig. 1**) consists of a graphics display terminal (a cathode-ray tube) and various input devices, such as a keyboard, joystick, mouse, light pen, and digitizing tablet. One or more workstations are linked to a computer, which runs the CAD software. Peripheral devices, such as printers, plotters, and mass storage devices, are also linked to the computer. The CAD system allows the designer to create a model of the product design in the computer memory, display different views of the model on the terminal, modify it as desired, save different models in a database for later recall, and make printed copies of design notes and engineering drawings as needed.

The computer model used depends on the product. In the design of printed wiring assemblies, for example, two-dimensional models are used to lay out circuits and electronic components on the face of a circuit board. In mechanical and civil engineering design, three-dimensional, geometric models repre-

Fig. 1. Computer-aided design workstation where the designer creates and refines a geometric model of a product. (*Center for Computer-Aided Engineering, University of Virginia*)

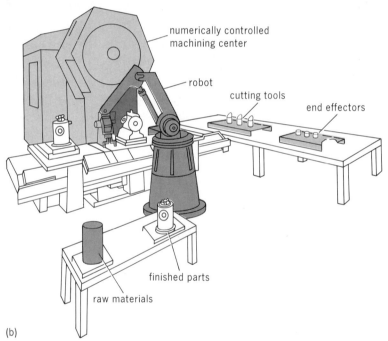

Fig. 2. Horizontal machining workstation. (*a*) Workstation with an industrial robot that loads and unloads parts processed on a numerically controlled machining center. (*b*) Computer simulation display used to verify the robot and part programs which control the workstation. (*Automated Manufacturing Research Facility, National Institute of Standards and Technology*)

sent parts, mechanical assemblies, and engineering structures.

Several types of geometric models are possible, including solid models, wire-frame (line) drawings, and surface representations, such as planes, patches, and sculptured contours. Solids modeling is the basis for most current CAD systems, because the information in the model database represents a distinct solid object, completely and unambiguously. Complicated solid objects are built up by adding and subtracting simpler geometric shapes (cubes, slabs, cylinders, and cones) called primitives.

Most CAD systems are currently used as electronic

drafting devices. This has many time- and labor-saving advantages. The traditional drafting person needs to make several views of a single object to capture its geometry, whereas the CAD drafting person creates a single solid model from which any particular view can be derived automatically. As the design evolves, each change does not require an entirely new model. In the future, as CAD, CAE, and CAM are integrated in CIM, the CAD model will become part of a common database for the entire design and manufacturing cycle. *SEE COMPUTER GRAPHICS; DRAFTING; ENGINEERING DRAWING.*

CAM and numerically controlled machines. The origin of automated manufacturing predates CAM and is usually associated with the introduction of numerically controlled machine tools in the mid-1950s. Machine tools, such as drills, lathes, saws, and milling machines, are used to cut and shape parts from metal stock. Numerically controlled machine tools use precoded information to operate the machine, rather than a human machinist. *SEE MACHINE TOOLS; NUMERICAL CONTROL.*

The precoded information, called a program of instructions, contains a detailed set of directions, in symbolic form, that instructs the machine which operations to perform. The program may be punched on paper tape, recorded on magnetic tape, or manually input to the controller unit by a human operator. The more sophisticated numerically controlled systems used in CAM employ direct numerical control, where the program is transmitted electronically to the controller unit over a direct link with a computer.

The numerical controller unit reads the program, interprets the symbols, and sends the appropriate control signals to servomotors and other physical controls on the machine itself. Various sensors send information back to the controller unit. The purpose of this closed-loop or feedback control is to ensure that the actual process is following the program. Most modern numerical control systems are computer–numerically controlled, and the control unit is based on a microcomputer. Microcomputers are also used to monitor and control other types of equipment and processes. The microprocessor control units are called programmable controllers. *SEE MICROCOMPUTER; MICROPROCESSOR; PROGRAMMABLE CONTROLLERS; TOOLING.*

Machining centers and cells. Numerically controlled machines range in complexity from simple, open-loop, tape-controlled drill presses to versatile, multifunction machining centers. A horizontal numerical control machining center is shown in **Fig. 2**, as one component of a still more complex and versatile group of automated machines called a machining workstation or cell. The machining center can perform a variety of functions, such as drilling, boring, reaming, tapping, and milling. The tools required for these operations are stored in a circular tool drum. When the controller calls for a new tool, an automated tool changer removes the current tool from the spindle chuck and returns it to the drum. The drum rotates the new tool to the position in front of the changer, and the changer inserts the new tool in the chuck. The machining center also controls the position and orientation of the part being made, so that it can be machined on all sides.

Automated part programming. Preparing numerically controlled programs is called part programming. Manual part programming is difficult and tedious, because the programmer must calculate and record a long list of the relative positions of tools and parts for each movement along the tool path. Computer-assisted part programming simplifies this task. The programmer specifies the part geometry and machining information (tool sizes, speed, feed rates, and tolerances) in a high-level programming language. The module creates the part program. **Figure 3** shows the display on a numerically controlled machine tool creating a part program for a connecting rod; **Fig. 4** shows this rod being cut from stock. Automation significantly reduces the time required for part programming and program verification, especially for complex parts with many machining steps. In integrated CAD/CAM systems, the part geometry is specified in the CAD database and does not need to be reentered during programming.

Robotics. Industrial robots are general-purpose, programmable manipulators that perform a variety of production tasks. Robots can move material, load and unload parts on other machines, assemble components, spray paint, and spot-weld. The first industrial robot was installed in a United States automotive die-casting facility in 1961. Figure 2 shows a type of robot that loads and unloads parts and changes tools in the tool drum at a machining center. To spare humans the tasks, robots can do dangerous, unpleasant, fatiguing, or boring jobs, such as working in areas that are contaminated by fumes, acid, or radiation.

Robots can be reprogrammed for different tasks as needed. In walk-through and lead-through training, the robot arm and end effector (hand) are moved through the desired sequence. The program is created by recording this sequence in the robot's memory. With a more sophisticated technique, the robot pro-

Fig. 4 Numerically controlled machine tool in profiling operation using data from the part program. (*Control Data Corp.*)

gram is created on an off-line computer by using a high-level programming language. The program is loaded into the robot memory when needed. This eliminates the loss of production time during robot training, and allows a repertoire of programs to be stored and used interchangeably.

Robots also can be equipped with humanlike senses (vision, touch, and hearing) to perform more complicated tasks. For example, a video camera and a computer equipped with image-processing hardware and software are used for robot vision. This allows the robot to discern objects in the camera field and recognize the position and orientation of familiar objects. In this way the robot can retrieve randomly oriented parts, recognize specific parts in the presence of other objects, perform visual inspections, and assemble products requiring precise alignment. Similarly, touch, force (stress), and proximity sensors let the robot sense contact with another object, gage and control the force of contact (for example, its grip strength), and sense the presence of nearby objects it has not touched. *See* Computer vision; Robotics.

Automated materials handling. Materials handling refers to the movement, storage, retrieval, and tracking of raw materials, work in process, and completed products. Conveyor belts, hand trucks, fork lifts, and cranes are familiar equipment used to move materials. Shelves, bins, pallets, and silos are used for storage. Inventory sheets, traveler cards, parts lists, and fixed storage locations aid in tracking and retrieving materials.

In CAM, the goal is to automate all materials handling under computer control. Mobile robots, called automated guided vehicles, follow routes on the factory floor defined by wires or tape. Combined with automated pickup and delivery stands, automated guided vehicles move bins or trays of material without human intervention. A typical automated storage and retrieval system uses conveyors and lifts to move bins into and out of specialized inventory shelving. The computer automatically assigns entering bins to empty shelf locations, recording the bin number and shelf location for future retrieval. The bins are labeled with bar codes and can be identified automatically by optical scanners. Large-scale software systems, such

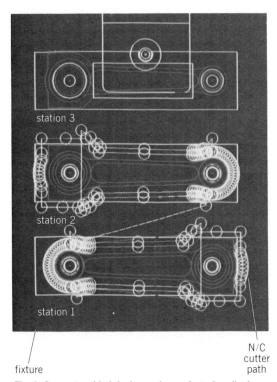

Fig. 3. Computer-aided design and manufacturing display on a system that allows the part programmer to create a cutter location file and preview the cutter path before machining. (*Control Data Corp.*)

as manufacturing resource planning, assist managers in determining what raw materials and components will be required during the current production period. These systems also help the managers to decide when materials and components must be ordered or manufactured to maintain production schedules. *See* MATERIALS HANDLING.

Flexible manufacturing systems. Flexible manufacturing systems are a form of computer-integrated manufacturing used to make small to moderate-sized batches of parts. A typical flexible manufacturing system comprises a cluster of numerical control machine tools, served by an automated materials-handling system and controlled by a central computer. A typical flexible manufacturing system can make from 4 to 100 different kinds of parts in volumes of 40–2000 parts per year, all with minimal human intervention. Such a system is efficient because it is highly automated and requires little downtime for changeovers for making batches of different parts. The flexible manufacturing system is in contrast to the familiar assembly line, which depends for its efficiency on making a single product in very large numbers. *See* AUTOFACTURING; DIGITAL COMPUTER; DIGITAL COMPUTER PROGRAMMING; ENGINEERING DESIGN; FLEXIBLE MANUFACTURING SYSTEM.

Bibliography. M. P. Groover and E. W. Zimmers, Jr., *CAD/CAM: Computer-Aided Design and Manufacturing*, 1984; H. F. Jackson and A. W. T. Jones, An architecture for decision making in the factory of the future, *Interfaces*, 17:15–28, 1987; National Research Council, *Toward a New Era in U.S. Manufacturing: The Need for a National Vision*, 1986.

K. Preston White, Jr.

Computer-aided engineering

Any use of computer software to solve engineering problems. With the improvement of graphics displays, engineering workstations, and graphics standards, computer-aided engineering (CAE) has come to mean the computer solution of engineering problems with the assistance of interactive computer graphics. *See* COMPUTER GRAPHICS.

CAE software is used on various types of computers, such as mainframes and superminis, engineering workstations, and even personal computers. The choice of a computer system is frequently dictated by the computing power required for the CAE application or the level (and speed) of graphics interaction desired. The trend is toward more use of engineering workstations, especially a new type known as supergraphics workstations. *See* DIGITAL COMPUTER; MICROCOMPUTER.

Design engineers use a variety of CAE tools, including large, general-purpose commercial programs and many specialized programs written in-house or elsewhere in the industry. Solution of a single engineering problem frequently requires the application of several CAE tools. Communication of data between these software tools presents a challenge for most applications. Data are usually passed through proprietary neutral file formats, data interchange standards, or a system database.

A typical CAE program is made up of a number of mathematical models encoded by algorithms written in a programming language. The natural phenomena being analyzed are represented by an engineering model. The physical configuration is described by a geometric model. The results, together with the geometry, are made visible via a user interface on the display device and a rendering model (graphics image). *See* ALGORITHM; DIGITAL COMPUTER PROGRAMMING; PROGRAMMING LANGUAGES.

The development of international graphics standards and data interchange standards for product models is simplifying the task of creating specialized CAE software and improving its transferability from one computer system to another.

CAE allows for many more iterations of the analysis-design cycle than was possible by hand computation. The benefits are translated into improved productivity and quality of design.

Program structure. A CAE program usually consists of a series of mathematical models and a data structure. **Figure 1** illustrates a simplified view of a typical CAE program operating in an engineering workstation environment. First, a mathematical description of the physical phenomena being analyzed is written. This engineering model may consist of equations such as Newton's second law to describe the dynamics of a system or the Navier-Stokes equations to analyze a fluid flow field. Next, a model of the physical configuration is created. This geometric model may consist of two- or three-dimensional (2D or 3D) curves, surfaces, faceted approximations to surfaces, or solid elements. The results of the engineering analysis are frequently displayed on the geometric model by color fringing to show the variation of a scalar parameter. Large amounts of data are created during this modeling phase, and the need for a data structure to store and retrieve them is greater than for the engineering model.

Although the engineering and the geometry are fully described, they cannot be viewed on the display until a model for rendering has been formulated and coded. A mathematical description of the lighting conditions, the approximate intensities of light reflected by the nodes of the geometric model, and the corresponding shades of color provide the graphics data necessary for the model to be realistically shaded. Before viewing, the graphics data are transformed from geometric model coordinates to a normalized coordinate system. Parameters are set for an orthographic or perspective projection from 3D model coordinates to 2D coordinates for final transformation to the display screen in device coordinates. The CAE applications programmer accomplishes most of these tasks with the help of precoded algorithms provided by graphics support software. Interactive communication with the graphics image and the geometric and engineering models occurs through a user interface written with the assistance of the graphics software or windowing software provided with the workstation.

In the past, CAE programs have been predominantly coded in the FORTRAN language. The present trend is toward the C programming language together with UNIX operating systems.

Graphics standards. Portability, the ability to move programs easily from one computer to another, is important for CAE software. Although it has long been possible to make a computational program code portable by using standard programming languages, this was not previously possible for CAE software because of the lack of graphics standards. A proposed 3D graphics standard (CORE), introduced to the American National Standards Institute (ANSI), was superseded by the adoption in 1985 of the 2D Graphical Kernel System (GKS) as an international standard by the International Standards Organization (ISO).

A new 3D device-independent graphics standard, the Programmer's Hierarchical Interactive Graphics System (PHIGS), was proposed by ANSI in 1985, and adopted as an international standard by ISO in 1988. Where possible, the concepts and nomenclature of GKS have been used in PHIGS.

The principal limitation to CAE software portability has been the wide variety of graphics hardware and the direct dependence of CAE software on this hardware. Both GKS and PHIGS give the programmer device-independent graphics primitives and coordinate systems as well as a set of logical graphics input devices to replace the wide variety of input hardware. For example, the pick action (selection of a graphics entity on the screen) could be physically accomplished by a light pen, a cursor and tablet, or a mouse. Using graphics standards, the CAE programmer will always specify a logical pick device regardless of the physical device used to achieve the pick.

PHIGS has several important advantages over GKS for CAE software. It is a full 3D system for viewing and modeling transformations, and allows both graphics and nongraphics data to be stored in its data structure. The data structure can invoke other structures and store transformations as attributes. The result is a hierarchical graphics data structure well suited for animation and representation of entities with multiple components. Extensions to PHIGS, called PHIGS-plus (PHIGS +), have been considered by standards committees. PHIGS + provides support for most of the rendering model and some of the geometric model in a CAE program. PHIGS + routines address lighting, shading, hidden surface elimination, transparency, and also nonuniform, rational B-spline curves and surfaces. These PHIGS + features are directed at the new breed of CAE engineering workstations.

Hardware. All sizes of computer systems are used for CAE software. Industries such as aerospace and automobile manufacture make wide use of very large mainframe computers to support computer-intensive CAE software such as finite element analysis and CAD/CAM systems (especially solid modelers). These mainframes are connected by high-speed data transmission lines to networks of from 10 to 100 graphics display terminals with local controllers (**Fig. 2**). Although these large time-sharing systems are easier to administer than an equivalent number of engineering workstations, when a large number of CAE users invoke computer-intensive programs, a significant delay in response usually occurs. *SEE MULTIACCESS COMPUTER.*

For this reason, together with cost considerations, many companies have chosen engineering workstations. Many small CAE applications programs are written for personal computers, of which most engineering companies have a wide array. However, most large, commercial CAE programs do not perform effectively or at all on personal computers. Engineering workstations, on the other hand, have computer power from 1 to 10 times as high as the common time-sharing superminicomputers available in 1986. These workstations usually run one CAE program at a time and serve one user. Engineering workstations are frequently connected by local-area networks (such as Ethernet) across which they share interactive com-

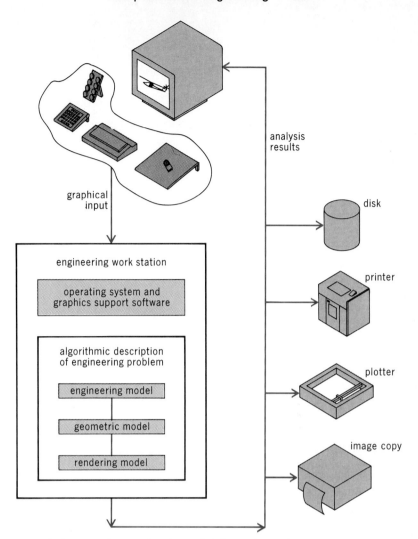

Fig. 1. CAE program in an engineering workstation environment.

munication and disk resources. When they also share processor resources across the network, transparent to the user, they will act as a true distributed computer system. As a result of this independence, a demanding CAE code at one workstation node in a network will not affect the response time for other users. *SEE DISTRIBUTED SYSTEMS (COMPUTERS); LOCAL-AREA NETWORKS.*

Two advances in engineering workstations that have made a significant impact on CAE are supergraphics and supercomputing workstations. Supergraphics workstations have computer engines that run at speeds of 3–15 million instructions per second (MIPS). They also have a number of specialized high-speed, parallel graphics engines, display list memory, image memory, buffer memory, and alpha memory to permit the real-time generation of color-shaded images desired for CAE applications. The graphics engines perform graphics modeling and viewing transformations, lighting and shading computations, and scan conversion of models for raster-scan display. Display list memory retains the rendering model structure, or in the case of PHIGS, the hierarchical PHIGS structure. The z-buffer provides fast hardware support for hidden surface elimination, whereas alpha memory allows fast hardware-assisted transparency computations. The final rasterized image is stored in one of two image memories for fast transmission to

Fig. 2. Solid modeling on a mainframe CAE display station.

the display screen. Most supergraphics workstations support the PHIGS graphics standard. Some support PHIGS+, and at least one uses a PHIGS+ native instruction set for advanced rendering operations.

A supercomputing workstation may be a workstation engine augmented by array processors or a smaller version of a supercomputer architecture. Although these systems do not provide superfast rendering features, they allow execution of computer-intensive CAE codes at very high speeds [30–60 million floating-point operations per second (MFLOPS)]. A trend is the combination of supergraphics with supercomputing workstations. SEE SUPERCOMPUTER.

A wide variety of output devices are available for CAE images. The most popular are film recorders for color slides, ink-jet and thermal printers for color images on paper and Mylar, and electrostatic and pen plotters for line drawings. Many types of graphics input devices are available for workstations and mainframe displays. Cursor and tablet, mouse, function key box, and dial box are the most widely used. Light pens are generally in use on older-technology displays.

Commercial CAE software. Most companies use a combination of a few commercial CAE programs and a number of smaller, specialized CAE applications programs typically written in-house or elsewhere in the industry. There are several classes of commercial CAE software. The first large commercial programs used for CAE were dynamic simulation systems.

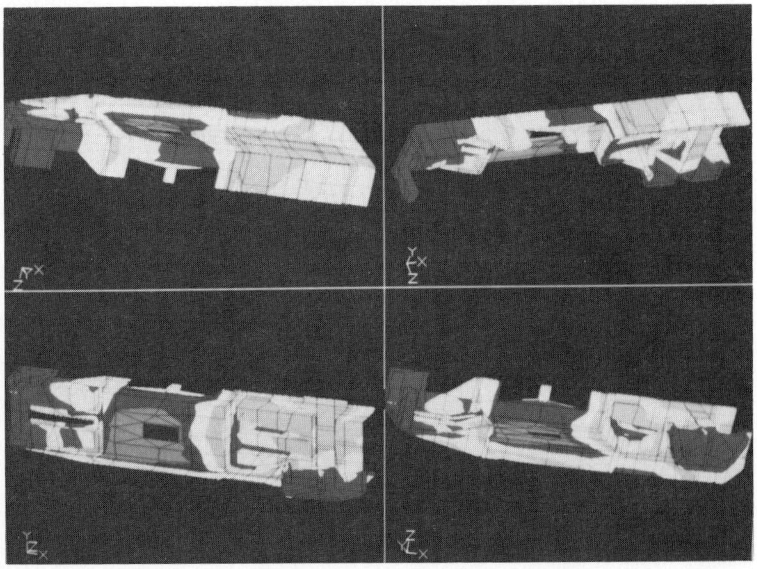

These systems usually had no graphical output or at best a printer–plotter output of graphs. They required the engineer to give input in the form of ordinary differential equations or building-block elements to describe the dynamic behavior of a real system, in many ways similar to the patching (programming) of analog computers. These systems numerically integrate the set of coupled differential equations, thereby simulating the dynamic behavior of the particular system. Simulation systems are still widely used, and although the graphics output has improved considerably, the description of the input has not changed greatly. SEE ANALOG COMPUTER; NUMERICAL ANALYSIS; SIMULATION.

A need to solve detailed problems, such as stress or deformation analysis, for components of systems led to the development of the finite element method or finite element analysis. In this method, a complex object is broken down into simpler elements. With these, a set of equations is formulated which, when solved, predict the behavior of the object as modeled by the set of elements. The modeling of the object is known as finite element preprocessing, and the validity of the solution is to a great extent dependent on the skill of the modeling engineer. Although very advanced methods of interactive graphics are used in finite element preprocessing, it is still a very tedious and expensive operation. The solution of the equations is highly compute-intensive. Finite element methods are applicable for a wide variety of physical phenomena, including mechanical stress and strain, fluid flow, acoustics, heat transfer, and electrical fields. **Figure 3** illustrates the solution of a finite element stress analysis problem with a commercial CAE code running on a mainframe computer. SEE FINITE ELEMENT METHOD.

Computer-aided design and manufacturing. CAD/CAM systems were created by the aerospace industry in the early 1960s to assist with the massive design and documentation tasks associated with producing airplanes. By the late 1970s, these codes were being distributed to other industries. CAD/CAM systems have been used primarily for detail design and drafting along with the generation of numerical control instructions for manufacturing. Gradually, more CAE functions are being added to CAD/CAM systems. A trend toward open architecture with flexible geometry interfaces is stimulating the addition of more analysis and manufacturing functions. Modeling with CAD/CAM systems has become fairly sophisticated. Most popular commercial systems support 2D and 3D wireframe, surface models and solid models. Rendered surface models differ from solid models in that the latter have full informaton about the interior of the object. For solid models a combination of three types of representation is commonly used: constructive solid geometry, boundary representation, and sweep representation. Although CAD/CAM solid models are improving, so'' modeling is not yet a production design tool. Complex systems require significant amounts of processing power, primary memory space, and disk space. SEE COMPUTER-AIDED DESIGN AND MANUFACTURING.

Integration of CAE software. Frequently, all of these CAE tools are needed together with specialized CAE applications programs to solve a single engineering problem. The integration of these tools or the communication of data between them is a challenging problem. To successfully integrate these programs, a

Fig. 3. Finite element stress analysis using commercial CAE software.

centralized database is necessary. Data from the various CAE tools are processed into and retrieved from this database through a database management system. If the CAE software is proprietary and the data structure not easily accessible, the proprietary programming interfaces may be used or, with a sacrifice of functionality and interactivity, a data interchange standard may be used. SEE DATABASE MANAGEMENT SYSTEMS.

Initial Graphics Exchange Specification (IGES) was developed under the leadership of the National Bureau of Standards and was accepted as a standard by ANSI in 1981. The goal of IGES was to allow the transfer of product data between dissimilar CAD systems. A future alternative to IGES is the Product Data Exchange Specification (PDES); parts will be defined based on solids using feature descriptors. Instead of dimensions, PDES will define a tolerance envelope for the part. Additional nongeometric data will be included, such as material information and manufacturing process data.

Applications. The CAE methods for electrical and electronics engineering are well developed. The geometry is generally two-dimensional, and the problems are primarily linear or can be linearized with sufficient accuracy. Chemical engineering makes extensive use of CAE with process simulation and control software. The fields of civil, architectural, and construction engineering have CAE interests similar to mechanical CAE with emphasis on structures. Aerospace, mechanical, industrial, and manufacturing engineering all make use of mechanical CAE software together with specialized software.

An example of CAE is the design of an aircraft landing-gear mechanism. The first step is definition of the problem and creation of a set of performance specifications. Next, the conceptual design phase may be aided by specialized programs to determine a size estimate of the landing gear based on specified loads and deflections. Commercial CAE programs are available for kinematic synthesis of mechanisms based on specified motion requirements, but this landing-gear mechanism will be designed with an in-house program written for the purpose. SEE LANDING GEAR.

The next phase is preliminary design. An applications program will be used to analyze the deflection and response of a shock-absorbing, energy-dissipating strut. Dynamic analysis of the guiding mechanism and complete assembly will be determined by a commercial code. When the dynamic loads are determined, a finite element stress analysis of each link of the mechanism will be done by using commercial finite element–method software. Following the stress analysis, some links will be changed in size and the dynamic analysis repeated to determine new loads. Using the new loading, another iteration of the finite element-method software will be made to verify that the stresses fall below the strength limits.

The next phase is the final design. All components of the assembly will be drawn in 2D on a CAD/CAM system and detailed, giving dimensions, material specifications, and other instructions. An assembly drawing will be created from the components, and mating of components will be verified. An alternative approach would be to create a solid model of each component, assemble the solid components, and run an automatic interference-clearance check; 2D drop-offs are then automatically made of each component and manually detailed (at the workstation) by a drafter. From the final part geometry, instructions for numerically controlled machine tools are generated to produce the part. Some systems may support tooling design and process planning. Finally, a design release is made to the manufacturing department. SEE TOOLING.

Arvid Myklebust

Bibliography. K. J. Bathe, *Finite Element Procedures in Engineering Analysis*, 1982; M. D. Brown, *Understanding PHIGS, The Hierarchical Graphics Standard*, 1985; J. D. Foley and A. Van Dam, *Fundamentals of Interactive Computer Graphics*, 1982; M. K. Gillenson, *Database, Step-by-Step*, 1985; F. R. A. Hopgood et al., *Introduction to the Graphical Kernel System (GKS)*, 1986; J. Krouse, *What Every Engineer Should Know about CAD/CAM*, 1982; M. E. Mortenson, *Geometric Modeling*, 1985; D. F. Rogers, *Procedural Elements for Computer Graphics*, 1985.

Computer graphics

Communication between people and computers using imagery as opposed to text and numbers. Computer graphics enables people to channel communication with computers into pictures, thus employing the highly evolved human ability to recognize patterns. Trends portrayed in a graph or the relationship between objects shown on a map are absorbed much faster and more firmly than the same information presented in numbers or words.

Graphics increasingly permeate the entire user interface with computers. Instructions are given by devices pointing to pictograms or icons on the computer screen rather than typing commands in text form (**Fig. 1**). SEE COMPUTER.

Computers can generate all types of images, from simple line drawings to realistically shaded pictures that resemble photographs. These can be totally synthesized by the computer or can be enhanced or manipulated natural images. SEE IMAGE PROCESSING.

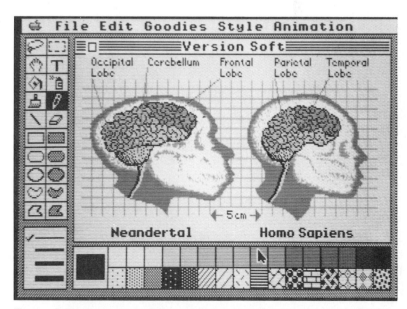

Fig. 1.   Raster display of a personal computer composed of 512 pixels across and 370 down, in which text and graphics are fully integrated as a bit map in the computer's memory. The user interface is based on graphical icons (left column), selected by moving a pointer with a mouse. (*Apple Computer, Inc.*)

Graphics-display hardware. Computers must be equipped with special hardware in order to display images. There are two types of graphics; vector and raster.

Vector images are specified as a series of lines. The computer stores a list of coordinates for the starting and end points of each line, called a display list. To display the image, the computer passes through the list, redrawing each line on a cathode-ray tube from 10 to 30 times a second, depending on the number of vectors in the image. Vector graphics became a practical medium in the 1960s (**Fig. 2**), but the hardware was costly because it had to work at high speed to continuously renew the displayed image. In 1965, with the invention of the direct-view storage tube, vector graphics became much more widespread. The direct-view storage tube consists of a cathode-ray tube with an internal grid near the phosphor which employs a capacitive effect to retain an image indefinitely. Such tubes eliminated the need for image-renewing hardware, reducing the cost of a computer graphics system by a factor of almost 10. SEE CATHODE-RAY TUBE; STORAGE TUBE.

Raster images are continuous smoothly shaded pictures, composed of rows of square or rectangular picture elements, called pixels. The quality of a raster image is determined by its spatial and contrast resolution. The spatial resolution, or sharpness, is determined by the number of rows and columns of pixels. A low-resolution image might consist of 300 by 200 pixels, whereas a high-resolution image would be composed of, say, 1280 by 1024 pixels. The contrast resolution refers to the number of different shades of gray (in a monochrome image) or different colors that a pixel can assume. Inexpensive personal computers may offer only eight colors, whereas 24 bits per pixel, offering 2^{24} or 16,777,216 different colors, are required for smooth tones in realistic images. Higher resolution, both spatial and contrast, requires more data to represent an image and more memory to store it. With the price of integrated circuit memory falling by a factor of 2 every 2 years, affordable image quality has increased steadily. A raster image is stored as a bit map in a special, high-speed video memory within the computer. The memory contents are read out and passed to the video display 30 times a second, fast enough to create a flicker-free image. Raster graphics has largely supplanted the line drawings of vector graphics because it offers smoothly shaded images in continuous tones. In addition, raster displays (of a given resolution) can handle complex images without incurring any extra burden, whereas the vector-display refresh rate drops with increasing number of displayed vectors. SEE INTEGRATED CIRCUIT; SEMICONDUCTOR MEMORIES.

The most common display device for raster images is the cathode-ray tube. Monochrome graphics displays use the same type of tube as home black-and-white television receivers, whereas color displays use shadow-mask cathode-ray tubes, also similar to home color television receivers. However, the monitors required to display high-resolution images are manufactured with much finer-spaced masks, with triads of red, green, and blue holes spaced on centers about 0.01 in. (0.3 mm) apart, twice the resolution of home television. SEE TELEVISION RECEIVER.

Another raster display device is the plasma panel. It consists of an array of cells containing gas which, like an array of neon lights, emits an orange light under an applied voltage. The screens are flat, flicker-free, and rugged, making them suitable for military and industrial use. Another raster flat-screen technology is the liquid-crystal display, a large version of the system first employed in digital watches and pocket

Fig. 2. Early vector refresh graphics used in the SAGE system, deployed by the U.S. Air Force, starting in 1958, to collect and display radar data as well as to guide intercepting aircraft to radar-detected targets. The light gun was pointed at the screen to select a particular object for identification. (*MITRE Corp.*)

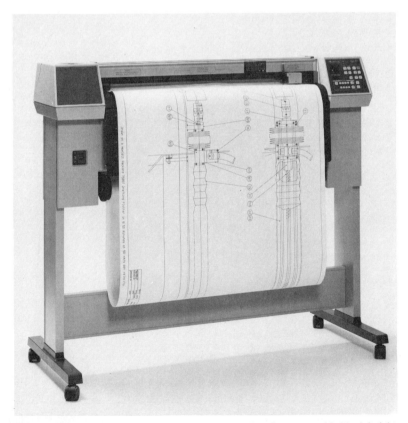

Fig. 3. High-precision drafting plotter. The two axes of motion are provided by left-right movement of the pen (visible at top right) and vertical movement of the paper. (*Hewlett-Packard*)

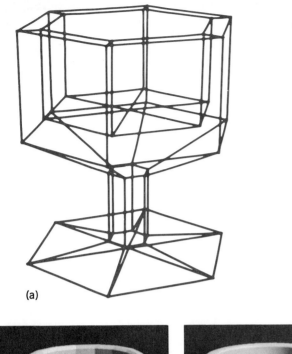

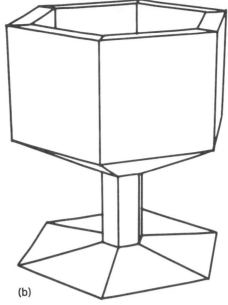

(a)   (b)

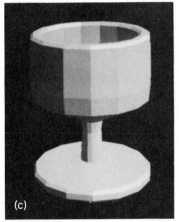

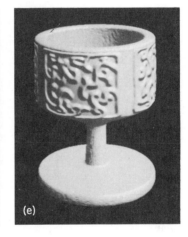

(c)   (d)   (e)

Fig. 4. Rendering a three-dimensional object. (*a*) Wire-frame rendering of a polygon mesh. (*b*) Wire-frame with hidden-edge suppression. (*c*) Uniform shading for each face. (*d*) Use of curved surface patches instead of polygons. (*e*) Perturbing the surface normal, a technique called bump mapping, to introduce the effect of surface relief. (*James F. Blinn*)

calculators. Plasma and liquid crystal are commonly used in lap-top computers. *See* Electronic display; Liquid crystals.

Hardcopy output. Computer-generated images can be recorded on paper in different ways. Vector drawings can guide pen plotters, in which electric motors, under computer control, move pens over paper. Extremely precise plotters are suitable for the creation of engineering drawings, integrated circuit masks, and maps (**Fig. 3**). Intelligent plotters compute pen movements that effect hatched shading. *See* Drafting.

Raster images can be recorded on film by cameras that photograph images displayed on a small, extremely high-resolution monitor. Raster images can also be printed by such devices as electrostatic plotters and ink-jet and dot matrix printers. Electrostatic plotters use a technology similar to that found in photocopy machines. To produce an image, paper is passed over a row of tiny, electrically charged wires. The wires produce charged spots on the paper, which, in turn, attract particles of toner to the paper which stick, forming a dot. For each row, the appropriate pattern of wires is charged, building up an image row by row. Electrostatic plotters are faster than pen plotters and are approaching the same degree of preci-

sion. *See* Data-processing systems; Photocopying processes.

Input devices. Line drawings can be entered into a computer by tracing them with a special stylus or cursor over a tablet embedded with a fine grid of wires. Pulses sent through the grid are detected by the stylus. The time lapse between issuing and detecting a pulse is used to compute the location of the stylus.

Entire images can be input at a stroke by means of such devices as scanning digitizers and charge-coupled-device video cameras. In a scanner, an image is placed flat or in some cases wrapped around a cylindrical drum; a beam of light scans the image and a detector measures the reflected brightness. The scanner has higher resolution and accuracy than a video camera, but is slower and more expensive.

Interactive graphics. Interactive graphics enables users to create, modify, and respond to graphical objects in real time. Graphical commands are mediated by a pointing or locator device, such as a mouse, trackball, joystick, or stylus with tablet. These devices convert the physical movement of the user's hand into the movement of a cursor or an object on the computer display screen. The mouse, which is the most widely used device, is a hand-held box with a ball on the underside coupled to rollers that respond to movement in two orthogonal directions. Moving the mouse over a flat surface turns the rollers, which are connected to detectors that sense the movement. *See Potentiometer*.

The stylus with tablet is used for precise sketching, being easier to control for fine movements than the mouse. In a paint program, the movement of the stylus simulates the effect of a pencil or paintbrush on the computer display. Color palettes, cut-and-paste commands, and other special functions such as automatic filling of outlines with patterns are selected by pointing the stylus to special locations around the perimeter of the tablet. The joystick is used for coarser movements and is popular for controlling motion in video games. *See Video games*.

Image synthesis. A collection of surfaces can be used to model three-dimensional objects. This is called geometric or shape modeling. One commonly used method is to create a mesh of polygons, a connected set of flat surfaces. This works best for objects that have flat sides, such as buildings or furniture, but can also be used to approximate curved surfaces. The smaller the polygons, the more accurately a curve can be represented by flat segments. However, larger numbers of small polygons are required to cover a surface, with a corresponding increase in storage space and execution time. Another technique approximates the surface of objects with a set of patches, each of which is described by a set of three bicubic equations. one each for the *x, y,* and *z* axes. Fewer bicubic patches are required to represent a curved surface to a given accuracy, but the bicubic equations involve more computation than the planes that represent polygons.

Unlike geometric modeling, solid modeling uses three-dimensional solid primitives for representing objects. Complex shapes are built up by combining cubes, cones, spheres, ellipsoids, and cylinders. *See Computer-aided engineering*.

The creation of a geometric or solid model is only the first step toward the realistic rendering of an object. Merely drawing all the lines that bound the polygons in a mesh, or the cubic curves that bound bicubic patches, will often produce a confused jumble of lines. The first improvement that can be made is to provide some depth cuing by eliminating all edges or surfaces that would be obscured by visible surfaces. This step usually improves the realism of the model greatly.

Shading. The next step in producing realistic images is to shade the visible surfaces, taking into account the light sources, surface properties, and positions and orientations of the surfaces. To model diffuse reflection from dull surfaces, the brightness seen from any direction depends only on the orientation and position of the surface with respect to the light source, and not on the viewer's position. However, shiny surfaces exhibit highly directional specular reflection with highlights; here viewer position must be taken into account. Differing degrees of shine can be modeled by adjusting the rate at which reflected light falls off around the specular angle, for which the angle of incidence equals the angle of reflection. The simplest way of applying a shading model to an object defined by a polygon mesh is to assign a single constant shade to each polygon. Unfortunately, even small changes in intensity between adjacent facets are exaggerated by the human eye; consequently, some form of intensity interpolation is almost always used. Intensity interpolation shading, or Gouraud shading, eliminates the discontinuities by interpolating the intensities across a polygon between values determined for each vertex. (The vertex value is computed for a normal that is the average of the adjacent surface normals.) A more sophisticated

Fig. 5. Image of an airport generated in real time by a flight simulator. (*Rediffusion Simulation, Inc.*)

Synthetic image, each of whose elements was created by a special program: the rocks, road, lake, hills, fence, rainbows, grass, bushes, and even the ripples on the puddles were rendered using distinct techniques. The final image was made by careful compositing of the constituent elements. (© *Pixar*)

(*Right*) Synthetic image of a steel mill, created by a technique which models the thermodynamics of light emission and propagation. The procedure provides for interreflections, color bleeding, shadows, and penumbra. (*Cornell Program of Computer Graphics*)

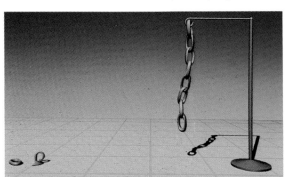

Simulation of balls in motion on a pool table. Realistic reflections of a pool room were rendered on the balls using a technique in which the paths of millions of imaginary rays of light were traced. The simulation of motion blur helps provide the illusion of smooth motion in an animated movie that uses a succession of such images. (© *Pixar*)

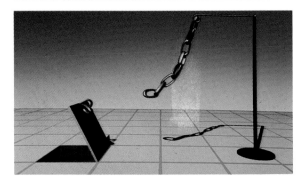

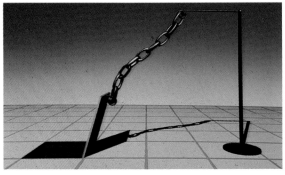

Time series (from bottom to top) based on computer program that incorporates newtonian physics to enhance the realism of the synthetic images. The series was modeled by providing a computer with information describing the constraints and the force of gravity acting on the chain. The program calculated the position of each link. (*Caltech Computer Graphics Group*)

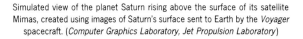

Simulated view of the planet Saturn rising above the surface of its satellite Mimas, created using images of Saturn's surface sent to Earth by the *Voyager* spacecraft. (*Computer Graphics Laboratory, Jet Propulsion Laboratory*)

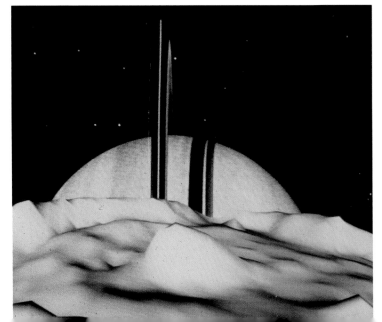

method, known as normal-vector interpolation or Phong shading, has the additional advantage of preserving specular highlights. The surface normal is interpolated between the vertices, and the corresponding intensity is computed for each point (**Fig. 4**).

Further refinements include the modeling of shadows, the refraction of light by transparent objects, and the interplay between objects and their environments. The most realistic and costly technique of modeling shadows, reflections, and refractions is ray tracing, in which a separate light ray for each pixel is traced backward from the viewpoint through the pixel to its origin. The color of this end point determines the color of the pixel. Ray tracing can model only the effects of geometric optics. To include the effects of diffuse illumination, a procedure based on the computation of total luminosity is used.

The computation of shading tends to be very demanding of computer time, and increases linearly with the number of pixels in the image. Many graphics-display systems incorporate special hardware for shading. For example, a high-end graphics workstation introduced in 1988 uses dedicated hardware to compute and display 150,000 Gouraud-shaded polygons per second.

Models of nature. Realism in computer image synthesis can depend on the convincing generation of natural structures. The models are usually grounded in physical or biological concepts, but simplify or alter reality for computational convenience. Clouds, plants, and mountain ranges have been simulated by using fractals. These are mathematical objects that, when approximated by computer algorithms, produce complex shapes with a natural-looking balance between small-scale detail and large-scale form. The surface of water has been modeled by solving the wave equation in two dimensions. The appearance of different materials, ranging from wood to marble, is simulated by wrapping a surface texture map onto the desired shape and selecting an appropriate light reflection law. SEE FRACTALS.

Applications. In most bit-mapped computer displays, the distinction between text and graphics is not carried through to the display hardware (Fig. 1). In publishing, camera-ready art is prepared by compositing text and graphics on a high-resolution screen. The decreasing cost of such systems has led to desktop publishing, in which the traditional functions of typesetting, graphic art preparation, and page layout are all carried out with graphics-based software.

The flexible creation of synthetic images enables designers and engineers from a wide range of disciplines to explore designs interactively on a screen without a physical model. Computer graphics is an indispensable part of the design process for products as diverse as integrated circuits, automobiles, running shoes, or skyscrapers. SEE COMPUTER-AIDED DESIGN AND MANUFACTURING.

Computer graphics is used to present the results of scientific or engineering computations in ways that offer insights that would be difficult or impossible to extract from numerical results alone. Pharmaceutical firms, for example, use computer graphic renderings of molecular models to design drugs and simulate their effects, bypassing weeks of laboratory investigation. Aircraft designers view graphic simulations of the airflow over a wing, and can see the results of complex aerodynamic calculations in an intuitively comprehensible fashion. Astrophysicists experiment

Fig. 6. Single frame from a computer-animated film. The characters are animated by using highly flexible computer-based models with over 100 degrees of freedom. The synthetic trees, grass, and bushes are generated by recording the forking paths of clouds of simulated particles, shaded to simulate sunlight with shadows. (© *Pixar*)

with artificial galaxies composed of artificial stars, trying a range of models to select that which best matches the real universe. SEE SIMULATION.

An important use of real-time computer graphic simulations is the training of aircraft pilots. Sophisticated flight simulators build high-speed computer-image-generating systems into mock-ups of a cockpit to simulate the view seen through the cockpit window (**Fig. 5**). SEE AIRCRAFT TESTING.

Another application of computer graphics is the creation of computer animated films (**Fig. 6**). Here the computer can be directed not only to create the key frames, that is, the images at the significant junctures in an animated film, but also to interpolate between them, to generate smooth motion. Computer animation is still costly to create, as 30 images are needed for each second of the film. Most computer animation has been created for television advertising, but the medium has also proved effective in educational television and motion pictures. SEE DIGITAL COMPUTER.

Oliver B. R. Strimpel

Bibliography. J. C. Beatty and K. S. Booth, *Tutorial: Computer Graphics,* 1982; D. R. Clark (ed.), *Computers for Imagemaking,* 1981; J. D. Foley and A. VanDam, *Fundamentals of Interactive Computer Graphics,* 1982; D. P. Greenberg and A. Marcus, *The Computer Image: Applications of Computer Graphics,* 1982; Time-Life Books, *Computer Images,* 1986.

Computer-integrated manufacturing

A system in which individual engineering, production, and marketing and support functions of a manufacturing enterprise are organized into a computer-integrated system. Functional areas such as design, analysis, planning, purchasing, cost accounting, inventory control, and distribution are linked through the computer with factory floor functions such as materials handling and management, providing direct control and monitoring of all process operations.

Computer-integrated manufacturing (CIM) may be viewed as the successor technology which links computer-aided design (CAD), computer-aided manufacturing (CAM), robotics, numerically controlled machine tools (NCMT), automatic storage and retrieval systems (AS/RS), flexible manufacturing systems (FMS), and other computer-based manufacturing technology. Computer-integrated manufacturing is also known as integrated computer-aided manufacturing (ICAM). Autofacturing includes computer-integrated manufacturing, but also includes conventional machinery, human operators, and their relationships within a total system. *SEE AUTOFACTURING; COMPUTER-AIDED DESIGN AND MANUFACTURING; FLEXIBLE MANUFACTURING SYSTEM; ROBOTICS; TOOLING.*

Factory of the future. The term computer-integrated manufacturing was first popularized by Joseph Harrington in 1975. He predicted that future manufacturing success would be based more on effective management of data and information flow rather than on how efficiently either piece parts or finished products were manufactured. He postulated that in the CIM factory of the future there would be areas of "departmentalized" decision making. "Departments" would be defined as being process ("hard technology") based, that is, involving processes such as milling, drilling, routing, and grinding. "Nondepartments" are defined as being information ("soft technology") based, such as design engineering, process planning, and inventory management, the basic premise being that they are generic throughout an organization. For example, process planning considered as the function "planning" applies equally to all processes in the factory, whether they be machining of metal or lay-up of composite material. Investment made to computerize the function of process planning based on the assumption that it is generic will result in reuse of this investment many times throughout the organization. The computer-based process planning function will not change at all; but the data on which it operates will be specific to the process of the moment. Computer-aided design systems are the best example of the concept.

The CIM factory of the future will integrate the traditional hard or process-based technology factory (departments) with emerging software of systems-based technology (nondepartments). **Figure 1** shows such a factory concept. The goal of this factory is to be efficiently integrated and continuously flexible and economical in the face of change.

Effectiveness, efficiency, and economy. Efficiency is most often associated with transfer-line or assembly-line technology. The best-known developer of this technology was Henry Ford. His Model T factory could produce cars at an unprecedented rate, but the consumer had no choice at all about the specifications of the product. Flexibility is most often associated with robotics and other reprogrammable automation. Effectiveness is usually associated with the soft technology and organizational support structure of the enterprise.

Optimization of the CIM factory is based on specific criteria. By considering efficiency and flexibility and effectiveness, the CIM factory enables economy

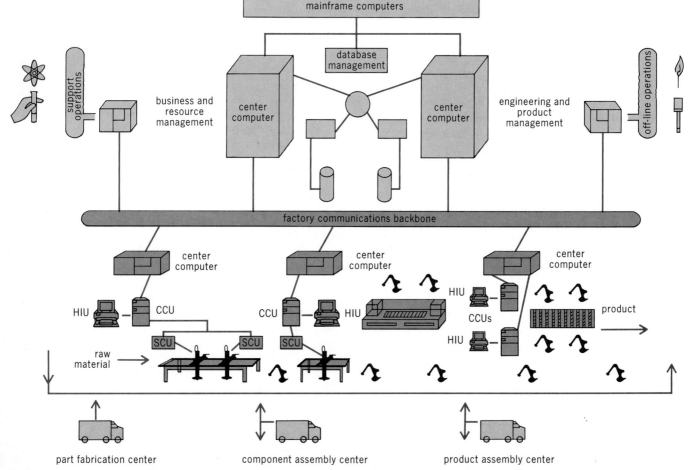

Fig. 1. Factory-of-the-future concept. CCU = cell control unit; SCU = station control unit; HIU = human interface unit.

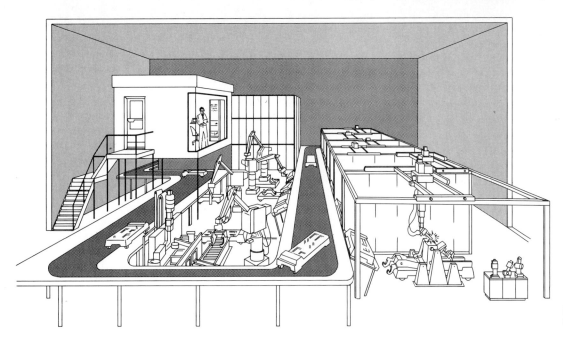

Fig. 2. Typical CIM work cell.

of scope. In economy of scope, flexibility is at least as important as efficiency, because the factory is seen to be an ever-changing dynamic environment which must always respond quickly to the needs of the marketplace, if not actually lead the need. The traditional factory with both rigid transfer-line technology and a rigid organizational structure cannot satisfy this objective; it strives for maximum volume of each product. In the ideal case, the CIM factory provides for profit with an order quantity of one, that is, when no two products are identical.

Organization. The CIM factory concept is illustrated in Fig. 1: soft technology is shown in the upper half and hard technology in the lower half. Soft technology can be thought of as the intellect or brains of the factory, and hard technology as the muscles of the factory. The type of hard technology employed depends upon the products or family of products made by the factory. For metalworking, typical processes would include milling, turning, forming, casting, grinding, forging, drilling, routing, inspecting, coating, moving, positioning, assembling, and packaging. More important than the list of processes is their organization.

The CIM factory is made up of a part fabrication center, a component assembly center, and a product assembly center (Fig. 1). Centers are subdivided into work cells, cells into stations, and stations into processes. Processes comprise the basic transformations of raw materials into parts which will be assembled into products. In order for the factory to achieve maximum efficiency, raw material must come into the factory at the left end and move smoothly and continuously through the factory to emerge as a product at the right end. No part must ever be standing; each part is either being worked on or is on its way to the next work station. In conventional factories, a typical part is worked on about 5% of the time.

In the part fabrication center, raw material is transformed into piece parts. Some piece parts move by robot carrier or automatic guided vehicle (AGV) to the component fabrication center. Other piece parts (excess capacity) move out of the factory to sister factories for assembly. There is no storage of work in process and no warehousing in the CIM factory of the future. To accomplish this objective, part movement is handled by robots of various types. These robots serve as the focus or controlling element of work cells (**Fig. 2**) and workstations (**Fig. 3**). Each work cell contains a number of workstations. The station is where the piece part transformation occurs—from a raw material to a part, after being worked on by a particular process.

Components, also known as subassemblies, are created in the component assembly center. Here robots of various types, and other reprogrammable automation, put piece parts together. Components may then be transferred to the product assembly center, or out of the factory (excess capacity) to sister factories for final assembly operations there. Parts from other

Fig. 3. Typical CIM workstation.

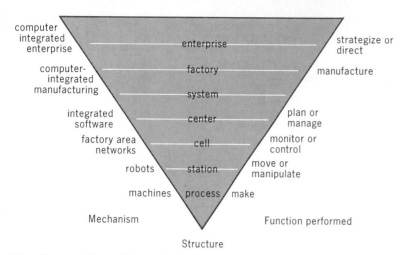

computer integrated enterprise

computer-integrated manufacturing

integrated software

factory area networks

robots

machines

enterprise

factory

system

center

cell

station

process

strategize or direct

manufacture

plan or manage

monitor or control

move or manipulate

make

Mechanism

Structure

Function performed

Fig. 4. Diagram of the architecture of manufacturing.

factories may come into the component assembly center of this factory, and components from other factories may come into the product assembly center of this factory. The final product moves out of the product assembly center to the end user. SEE AUTOMATION.

Integration. The premise of CIM is that a network is created in which every part of the enterprise works for the maximum benefit of the whole enterprise. Independent of the degree of automation employed, whether it is robotic or not, the organization of computer hardware and software is essential (upper half of Fig. 1). The particular processes (lower half of Fig. 1) employed by the factory are specific to the product being made, but the functions performed (upper half of Fig. 1) can be virtually unchanged in the CIM factory of the future no matter what the product. These typical functions include forecasting, designing, predicting, controlling, inventorying, grouping, monitoring, releasing, planning, scheduling, ordering, changing, communicating, and analyzing.

It must be recognized that independent, optimum performance of the individual functions is not as important as their integration with one another, or their integration with the factory floor itself. This integra-

tion is brought about in two ways. The first is through an architecture of manufacturing which specifies precisely how and when each function is integrated with the other (**Fig. 4**). The second is through a cell controller and cell network.

The cell controller ensures integration of data between each machine, robot, automatic guided vehicle, and so forth, of the cell. The cell network (**Fig. 5**) actually performs this communication within the cell, and connects cells together into centers. In the cell network, communication is possible between each machine of the network and between different networks. The cell controller itself serves to perform such tasks as downloading part programs from the CAD system to each machine in the cell, monitoring actual performance of each machine and comparing this performance to the plan, selecting alternate routing for a part if a machine is not operable, notifying operators of pending out-of-tolerance conditions, archiving historical performance of the cell, and transmitting to the center level on an exception basis the cell performance compared to plan. The center level of the CIM factory of the future is all of the policies and procedures that run the factory. These are embodied in computer software, which is in turn based upon an overintegration plan or architecture of manufacturing.

The purpose for the architecture of manufacturing is to provide a blueprint for employing CIM. While there are as many variations of the details of the architecture of the CIM factory of the future as there are factories, the structure of the architecture is generic. The focus is the flexible manufacturing system (FMS), which is made up of cells, stations, and processes. The information to be fed back to functions where decisions are made is more important than the product itself. In this example, the functions are defined very broadly and are used only to illustrate the absolute necessity for interaction between functions if the CIM factory is to be made to work efficiently, flexibly, and effectively.

In the ideal CIM factory of the future, before any real decision is made, it is tested on a computerized model of the architecture to determine its feasibility. Thereby, any mistakes are made in theory, not in practice. Several simulation software packages are

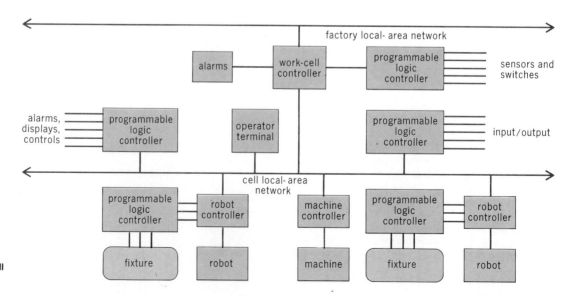

Fig. 5. CIM production cell network.

routinely used for this purpose. For the true CIM factory of the future to be realized, it will be necessary to build a complete simulation model.

Dennis Wisnosky

Bibliography. S. S. Cohen and J. Zysman, *Manufacturing Matters*, 1987; W. Deming, *Out of the Crisis*, 2d ed., 1986; J. D. Goldhar and M. Jelinek, Plan for economy of scope, *Harvard Bus. Rev.*, 6/(6):141–148, November/December 1983; Y. Monden, *Applying Just in Time: The American/Japanese Experience*, 1986; R. J. Schonberger, *Japanese Manufacturing Techniques*, 1982; D. L. Shunk, Group technology provides organized approach to realizing benefits of CIMS, *Ind. Eng.*, 24(4):74–80, April 1985; D. E. Wisnosky, *USAF ICAM Perspectus*, 1977.

Computer security

The process of ensuring confidentiality, integrity, and availability of computers, their programs, hardware devices, and data. Lack of security results from a failure of one of these three properties. A variety of causes—human, mechanical, environmental; malicious, unintentional, unavoidable; internal to the computing system, external to it—lead to a lack of security.

The lack of confidentiality is unauthorized disclosure of data or unauthorized access to a computing system or a program. Disclosure of data, also called leakage or breach of secrecy, is a serious threat because of the large number of computers that contain sensitive or personal information about individuals, proprietary information about businesses, or classified information about governments. A failure of integrity results from unauthorized modification of data or damage to a computing system or program. Computer users expect that saved data will not be changed—either maliciously or inadvertently—from one use to the next, even though the data may be stored without access for years. A lack of availability of computing resources results in what is called denial of service.

An act or event that has the potential to cause a failure of computer security is called a threat. Some threats are effectively deflected by countermeasures called controls. For example, a human guard and a locked door are generally effective controls against the theft of computing hardware. Kinds of controls are physical (for example, locked doors that prevent direct access), administrative (procedures that require an account number to be issued before access is permitted), logical (internal restrictions within a program that allow the execution of that program by only certain individuals), cryptographic (encoding data to make it unintelligible to unauthorized persons), legal (such as laws making unauthorized access a crime), and ethical (moral positions discouraging acts such as looking at another person's private computer records). Threats that are not countered by controls are called vulnerabilities.

Encryption. Encryption is a very effective technique for preserving the secrecy of computer data, and in some cases it can also be employed to ensure integrity and availability. An encrypted message is converted to a form presumed unrecognizable to unauthorized individuals. A message to be encrypted is called plaintext, and its encrypted equivalent is called ciphertext; the encryption process is denoted $C = E(P)$, where P is a body of plaintext, and C is its resulting ciphertext.

The principal advantage of encryption is that it renders interception useless. An encrypted message must be intelligible to intended receivers, and so an encryption algorithm has an inverse function, called decryption, by which ciphertext is converted back to the plaintext from which it was produced. Decryption is denoted $P = D(C) = D[E(P)]$.

The security of encryption is enhanced by the use of an encryption key, a parameter that modifies the effect of the algorithm. Keyed encryption is denoted $C = E(P,k)$. The key must be supplied to encrypt or to decrypt. Thus, a single algorithm can represent many different encryptions with different keys.

For further discussion of encryption, including public-key systems, the Data Encryption System, digital signatures, and the RSA algorithm, *SEE* CRYPTOGRAPHY.

Access control. Computer security implies that access be limited to authorized users. Therefore, techniques are required to control access and to securely identify users. Access controls are typically logical controls designed into the hardware and software of a computing system. Identification is accomplished both under program control and by using physical controls.

Identification and authentication. Computing systems must identify and authenticate users in order to enforce access control restrictions appropriately. Each user must first state an identity (a name or a prearranged log-in name). Then, so that the system can be certain that the user is authentic (that is, one user is not trying to impersonate another), it will require further proof of identity. Users can prove their identity based on (1) what they know, such as a password; (2) who they are, such as a physical characteristic that can be sensed (for example, handwriting or the pattern of blood vessels in the eye); or (3) what they possess, such as a token or an identifying card, or some physical device.

Control mechanisms. Typically, access within a computing system is limited by an access control matrix (see **illus.**) administered by the operating system or a processing program. All users are represented as

Objects Subjects	File-1	File-2	•••	SORT ROUTINE	•••	File-*n*
SYSADMIN	– –	– –		OWN READ EXECUTE		– –
USER 1	OWN READ	OWN READ WRITE		EXECUTE		– –
SORT ROUTINE	READ	WRITE		– –		– –
•••						
USER 17	– –	READ		– –		OWN READ WRITE

Access matrix for representing discretionary authorization information.

subjects by programs executing on behalf of the users; the resources, called the objects of a computing system, consist of files, programs, devices, and other items to which users' accesses are to be controlled. The matrix specifies for each subject the objects that can be accessed and the kinds of access that are allowed.

Although an access control matrix is the most straightforward means of describing allowable accesses, it is very inefficient in its use of space. A system with 200 users, each of whom has 50 files, require $200 \cdot 50 = 10,000$ table entries. If most users' files are private, most cells in the access control matrix are empty.

An access control list is effectively a column of an access control matrix. For each object there is one list that includes only those users who can access the object and the type of access allowed. For public objects, such as compilers and shared data files, a "wild card" subject can be specified; for example, the system programmer may have read and deleted access, while all other users are allowed read access. Access control lists are especially effective for denoting single users who have specific types of access to certain objects.

Alternatively, a capability list, which corresponds to a row of the access control matrix, can be maintained to control access. The capability list indicates the objects to which each subject is allowed access. Other types of access control mechanisms are capabilities, which are effectively tokens or tickets that a user must possess in order to access an object, and group authorizations, in which subjects are allowed access to objects based on defined membership in a group, such as all employees of a single department, or the collaborators on a particular project.

Discretionary and mandatory control. Access control as described above relates to individual permissions. Typically, such access is called discretionary access control because the control is applied at the discretion of the object's owner or someone else with permission. With a second type of access control, called mandatory access control, each object in the system is assigned a sensitivity level, which is a rating of how serious would be the consequences if the object were lost, modified, or disclosed. Also, each subject is assigned a level of trust. For example, a company might have three classes of data: public, company confidential, and personal. Anyone can be trusted not to improperly modify or disclose public data, but only certain people can be trusted to access company confidential data, and only a few of these people are also allowed access to personal data.

Models. In mandatory access control, a decision is made on each attempted access, based on a comparison of the requesting subject's trust level with the object's sensitivity level. These rules have been formalized and extended in an access control model known as the Bell–La Padula model. If the subject's level is at least as high as the object's level, the subject is allowed to view (read) the object. However, to prevent a malicious subject from writing a copy of a sensitive object with a lower sensitivity level (which could then be read by other subjects who did not have trust levels as high as the original), a subject can write an object only if his or her trust level is no higher than that of the object to be written. By a combination of these two rules, a subject can read and modify only objects at the subject's level. These access rules are similar to the military regulations for handling of classified information.

An alternative model of access control is called the information flow model. In this model, a program statement such as

$$a := b$$

causes information to flow from b to a. However, information can also flow in other, less direct means. The statement

$$\text{if } a = 0 \text{ then } c := 1 \text{ else } c := 2$$

causes information to flow from a to c (it is possible to determine whether a was zero by testing whether c is 1 or 2), even though there is no direct assignment from a to c. This indirect information flow represents a possible path of disclosure. Similar analysis is performed on loops and input/output instructions to determine what data potentially pass from a sensitive source to any output (in order to determine what sensitive data may be disclosed by a program). Information flow analysis can be performed as a program is compiled, and a program can then be certified as safe or unsafe from information leakage.

Pitfalls. Access control is not necessarily as direct as just described. Unauthorized access can occur through a covert channel. Two students taking a test can communicate by working out a scheme of coughs, pencil taps, and nods; this communication uses a covert or hidden medium of exchange. In computing systems, one process can signal something to another by opening and closing files, creating records, causing a device to be busy, or changing the size of an object. All of these are acceptable actions, and so their use for covert communication is essentially impossible to detect, let alone prevent.

Another issue in access control is granularity, or the fineness of data over which control is exercised. In some applications, a single piece of data would be treated as sensitive. However, the bookkeeping to enforce access control at this level of granularity is often prohibitive. Thus, larger bodies of data, such as a file, or an entire piece of output, are treated as a single unit to which access is either permitted or denied.

Security of programs. Computer programs are the first line of defense in computer security, since programs provide logical controls. Computer programs perform user identification and authentication, make access control decisions, and maintain the tables of data to support these other functions. Programs, however, are subject to error, which can affect computer security.

Correctness, completeness, and exactness. A computer program is correct if it meets the requirements for which it was designed. A program is complete if it meets all requirements. Finally, a program is exact if it performs only those operations specified by requirements. For example, a program to compute square roots would be correct if it returned an accurate square root for every positive integer supplied. However, if there are additional requirements, such as returning both positive and negative roots, or giving an appropriate error indication if the argument is negative, the procedure is incomplete if it fails to meet these additional requirements as well. If the program also performs something else, such as recording the name of its caller, it is inexact.

Correctness and completeness can sometimes be determined by exhaustive testing, although with large, complex systems it may be infeasible to test all possible inputs in all possible situations. Determining exactness is even more difficult. Every program performs some additional actions, such as assigning values to temporary variables, closing files, and testing for debugging inputs. Deciding which of these additional actions are security-relevant is not trivial.

Program errors. Simple programmer errors are the cause of most program failures. Fortunately, with careful and thorough design techniques, including structured design, program review, and team programming, many program flaws are eliminated before a program is completed. Program testing removes still more errors. Thus, the quality of software produced under rigorous design and production standards is likely to be quite high. However, a programmer who intends to create a faulty program can do so, in spite of all of these development controls. *SEE SOFTWARE ENGINEERING.*

A salami attack is a method in which an accounting program reduces some accounts by a small amount, while increasing one other account by the sum of the amounts subtracted. The amount reduced is expected to be insignificant; yet, the net amount summed over all accounts is much larger. Because there is no change to the total, the accounts may seem to be in balance.

Some programs have intentional trapdoors, additional undocumented entry points. For example, programmers often include an unknown debugging entry, so that test data can be supplied directly to a module being developed or so that errors can be diagnosed once a system is operational. If these trapdoors remain in operational systems, they can be used illicitly by the programmer or discovered accidentally by others.

A Trojan horse is an intentional program error by which a program performs some function in addition to its advertised use. For example, a program that ostensibly produces a formatted listing of stored files may write copies of those files on a second device to which a malicious programmer has access. Or, a program to produce paychecks may reduce one employee's check by an amount and add a similar amount to the programmer's check.

A program virus is a particular type of Trojan horse that is self-replicating. In addition to performing some illicit act, the program creates a copy of itself which it then embeds in other, innocent programs. Each time the innocent program is run, the attached virus code is activated as well; the virus can then replicate and spread itself to other, uninfected programs.

Trojan horses and viruses are serious computer security threats for several reasons. First, they can be very small, so that their presence will be difficult if not impossible to detect in a large source program. Second, there is no known feasible countermeasure to halt or even detect their presence. Finally, these malicious techniques have the potential to cause serious harm to computing resources by changing or destroying data, affecting hardware, and denying service to legitimate users.

Security of operating systems. Operating systems are the heart of computer security enforcement. Originally, operating systems were designed to provide services to users and then to promote equitable shar-ing of computer resources. As multiprogramming grew, operating systems were required to protect each user from the other users, as well as to establish a central mechanism for controlled access to shared resources. Currently, operating systems perform most access control mediation, most identification and authentication, and most assurance of data and program integrity and continuity of service.

System design. Operating systems structured specifically for security are built in a kernelized manner, embodying the reference monitor concept. A kernelized operating system is designed in layers. The innermost layer provides direct access to the hardware facilities of the computing system and exports very primitive abstract objects to the next layer. These objects may be things such as messages, units of memory, and input/output primitives. From these objects, the next layer builds more complex objects, such as files and virtual memory space. These objects are then exported to the next layer, which may build even more complex objects, such as an access control matrix.

The reference monitor is effectively a gate between subjects and objects; each time the subject seeks to access the object, the reference monitor is invoked to determine whether the access is to be permitted or not.

System evaluation. The U.S. National Computer Security Center (NCSC) has developed a set of criteria by which computer (operating) systems can be evaluated for security purposes. The criteria, called the Trusted Computer System Evaluation Criteria (TCSEC or Orange Book), specify both security-relevant features that a computing system must provide and assurances that must be included to convince someone that the system operates as it is supposed to. The NCSC evaluates systems based on the criteria and assigns ratings. *SEE OPERATING SYSTEM.*

Security of databases. A database is a collection of records containing fields, organized in such a way that a single user can be allowed access to none, some, or all of the data. Typically the data are shared among several users, although not every user will have access to every item of data. A database is accessed by a database management system that performs the user interface to the database.

Integrity is a much more encompassing issue for databases than for general applications programs, because of the shared nature of the data. Integrity has many interpretations, such as assurance that data are not inadvertently overwritten, lost, or scrambled; that data are changed only by authorized individuals; that when authorized individuals change data, they do so correctly; that if several people access data at a time, their uses will not conflict (for example, two people cannot write a single data item at the same time); and that if data are somehow damaged, they can be recovered.

A database management system is an application program that does not necessarily have a reliable interface to the operating system, and the operating system cannot necessarily be highly trusted. Therefore, a database management system typically maintains its own lists of users and their permitted actions, and the database management system may perform its own identification and authentication of users, quite independent from that performed by the operating system.

Inference and aggregation. Database systems are especially prone to inference and aggregation. Through inference, a user may be able to derive a sensitive or prohibited piece of information by deduction from nonsensitive results without accessing the sensitive information itself. For example, knowledge of any individual's salary may be restricted, but statistical measures (such as mean, median, maximum, and minimum) of the general salary pool may be freely released. If a user can formulate a database query that selects exactly one record (for example, ''select all employees who are 6′2″ tall and live in Glenwood, Maryland''), the user can then determine a single individual's salary by asking for the mean salary for the select set (of one employee). Various statistical methods make it very difficult to prevent inference.

A related problem is aggregation, the ability of two or more separate data items to be more (or less) sensitive together than separately. For example, neither the longitude nor the latitude of the location of a secret gold mine is sensitive by itself, but these two values together pinpoint the mine. It is infeasible to release one value to some people and the second value to others (even though no single person is told both) because two people could pool their knowledge. Aggregation is extremely difficult to prevent, since users can access great volumes of data from a database over long periods of time and then correlate the data independently.

Database controls. Encryption is sometimes used in database management systems both for confidentiality and integrity. A device called an integrity lock is used to ensure that the contents of a field are not modified. The field may be stored in plaintext, but a cryptographic checksum of the field, the integrity lock, is also stored. Each time the field is changed, a new integrity lock is computed and stored; each time the contents of the field are retrieved (in response to a query), a lock is computed on the value of the field and compared with the stored lock. If the two values do not match, the field has been modified without authorization. To guard against swapping data and their locks, the encrypted data may comprise the actual data of the field together with the record identifier and the name of the field. *See* Database management systems.

Security of networks. As computing needs expand, users interconnect computers. Network connectivity, however, increases the security risks in computing.

Impediments to security. Users of one machine are protected by some physical controls; proximity (need to be close to the machine, and therefore observed), physical access limitations (such as locked doors), and familiarity (being known to other users, and therefore recognized) complement the normal identification and authentication functions. However, with network access, a user can easily be thousands of miles from the actual computer.

In networks, a user's interaction is broken into messages, and a routing is established by which the messages travel from the user's machine to the destination machine. This routing may involve many intermediate machines, called hosts, each of which is a possible point where the message can be modified or deleted, or a new message fabricated. A serious threat is the possibility of one machine's impersonating another on a network in order to be able to intercept communications passing through the impersonated machine.

Another impediment to security is the use of ex-posed communication media, typically telephone lines. While a user of a single machine may be protected by a physical enclosure (the building, an office, or the machine room in which the computer is kept), the physical protection is lost when a network message leaves a building.

Finally, a network is a larger, more complex structure than a single computer. The reliability of single resources or vital links has a direct effect on the availability of the network resources, and hence, on the many users of the network.

Network security controls. The principal method for improving security of communications within a network is encryption. Messages can be encrypted link or end to end. With link encryption, the message is decrypted at each intermediate host and reencrypted before being transmitted to the next host. Management of keys within a network is a complex problem, but the problem is reduced if one host communicates only with a small number of nearest neighbors (hosts that are connected by a single link). Furthermore, link encryption can be performed by a hardware device between the computer and the communication line. End-to-end encryption is applied by the originator of a message and removed only by the ultimate recipient. Therefore, the contents of the message are not exposed during transmission between hosts. However, this encryption protects only the message, not any header or routing information, which must be available while the message is en route. Worst, end-to-end encryption must be applied before the message is being transmitted, and so it is typically performed by software; such encryption can be slower and more error-prone than hardware encryption. Neither link nor end-to-end encryption is, thus, ideal. *See* Local-area networks; Wide-area networks.

Charles P. Pfleeger

Bibliography. D. Denning, *Cryptography and Data Security,* 1982; M. Gasser, *Building a Secure System,* 1988; C. Pfleeger, *Security in Computing,* 1989.

Computer storage technology

The techniques, equipment, and organization for providing the memory capability required by computers in order to store instructions and data for processing at high electronic speeds. In early computer systems, memory technology was very limited in speed and high in cost. Since the mid-1970s, the advent of high-density, high-speed random-access memory (RAM) chips has reduced the cost of computer main memory by more than two orders of magnitude. Chips are no larger than ¼ in. (6 mm) square and contain all the essential electronics to store tens to hundreds of thousands of bits of data or instructions. An analogous increase in magnetic recording density has increased the capacity and reduced the cost per bit of secondary memory. Traditionally, computer stroage has consisted of a hierarchy of three or more types of memory storage devices (for example, RAM chips, disks, and magnetic tape units). *See* Bit.

MEMORY HIERARCHY

Memory hierarchy refers to the different types of memory devices and equipment configured into an operational computer system to provide the necessary attributes of storage capacity, speed, access time, and cost to make a cost-effective practical system. The fastest-access memory in any hierarchy is the main memory in the computer. In most computers manufactured after the late 1970s, RAM chips are used because of their high speed and low cost. Magnetic core memories were the predominant main-memory technology in the 1960s and early 1970s prior to the RAM chip. The secondary storage in the hierarchy usually consists of disks. Significant density improvements have been achieved in disk technology, so that disk capacity has doubled every 3 to 4 years. Between main-memory and secondary-memory hierarchy levels, however, there has always been the "memory gap."

The memory gap noted by J. P. Eckert (a developer of the ENIAC computer in the 1940s) still presents a problem for the data-processing system designer. The memory gap for access time is bounded on one side by a 100-nanosecond typical computer main memory cycle time and on the other by the 30-millisecond typical disk-drive access time. Capacity is bounded on the one side by a typical mainframe memory capacity

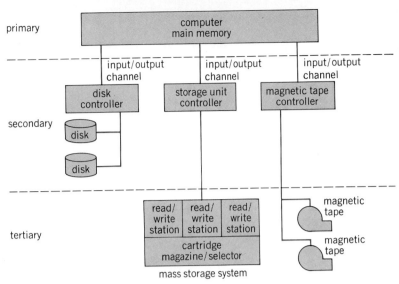

Fig. 1. Memory hierarchy levels and equipment types.

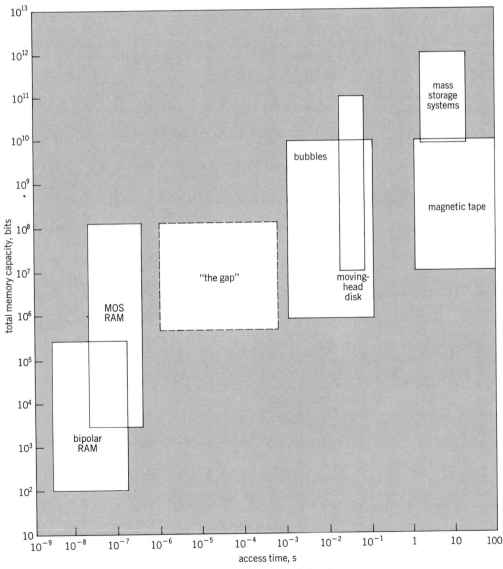

Fig. 2. Memory capacity versus access time for various storage technologies.

of 256 megabytes and on the other by multiple disks, each with a capacity of from 100 to over 1 gigabyte.

The last, or bottom, level (sometimes called the tertiary level) of storage hierarchy is made up of magnetic tape transports and mass-storage tape systems (**Fig. 1**). Performance (**Fig. 2**) is usually measured by two parameters: capacity and access time. (Speed or data rate is a third parameter, but it is not so much a function of the device itself as of the overall memory design.) Capacity refers to the maximum on-line user capacity of a single connectable memory unit. Access time is the time required to obtain the first byte of a randomly located set of data.

MEMORY ORGANIZATION

The efficient combination of memory devices from the various hierarchy levels must be integrated with the central processor and input/output equipment, making this the real challenge to successful computer design. The resulting system should operate at the speed of the fastest element, provide the bulk of its capacity at the cost of its least expensive element, and provide sufficiently short access time to retain these attributes in its application environment. Another key ingredient of a successful computer system is an operating system (that is, software) that allows the user to execute jobs on the hardware efficiently. Operating systems are available which achieve this objective reasonably well. SEE COMPUTER SYSTEMS ARCHITECTURE.

The computer system hardware and the operating system software must work integrally as one resource. In many computer systems, the manufacturer provides a virtual memory system. It gives each programmer automatic access to the total capacity of the memory hierarchy without specifically moving data up and down the hierarchy and to and from the central processing unit (CPU). During the early years of computing, each programmer had to incorporate storage allocation procedures by determining at each moment of time how information would be distributed among the different hierarchic levels of memory, whenever the totality of information was expected to exceed the size of main memory. These procedures involved dividing the program into segments which would overlay one another in main memory. The programmer was intimately familiar with both the details of the machine and the application algorithms of the program. This all changed in the 1970s and was much improved in the 1980s, when sophisticated higher-level program languages and database management software became well established, to provide significantly greater problem-solving capability. Thus manufacturer-supported operating systems evolved, with complete built-in virtual memory support capabilities, which made it possible for the user to ignore the details of memory hierarchy internal software and hardware operations. SEE DATABASE MANAGEMENT SYSTEMS; DIGITAL COMPUTER PROGRAMMING; OPERATING SYSTEM; PROGRAMMING LANGUAGES.

In the area of memory organization, two types of memory augmentation have been employed in the more widely used computers to enhance the overall computer performance capabilities. These two memory organization techniques are the cache memory, which speeds up the flow of instructions and data into the central processing unit from main memory, and an intelligent disk controller memory that is used as a staging buffer memory to queue up instructions and data for more rapid access into main memory.

Cache memory. A cache memory is a small, fast buffer located between the processor and the main system memory. Data and instructions in current use are moved into cache, producing two benefits. First, the average access time for the processor's memory requests is reduced, increasing the processor's throughput. Second, the processor's utilization of the available memory bandwidth is thereby reduced, allowing other devices on the system bus to use the memory without interfering with the processor. Cache memory is thus used to speed up the flow of instructions and data into the central processing unit from main memory. This cache function is important because the main memory cycle time is typically slower than the central processing unit clocking rates. To achieve this rapid data transfer, cache memories are usually built from the faster bipolar RAM devices rather than the slower metal oxide semiconductor (MOS) RAM devices.

Performance of mainframe systems is very dependent upon achieving a high percentage of accesses from the cache rather than main memory. In typical systems 80–95% of accesses are to cache memory, with a typical cache size of 8K bytes. (1K bytes is equal to 1024 bytes.) Since the cache contents are a duplicate copy of information in main memory, writes (instructions to enter data) to the cache must eventually be made to the same data in main memory. This is done in two ways: write-through cache, in which write is made to the corresponding data in both cache and main memory; and write-back cache, in which main memory is not updated until the cache page is returned to main memory, at which time main memory is overwritten.

Intelligent disk controller memory. This is used as a cache memory between disk and main memories. Typically it consists of MOS RAM chips which overlap the disk operations, with their longer access time, to mask out the disk access delays so that main memory can execute subsequent tasks more rapidly and efficiently. Intelligent disk controllers provide the latest techniques to fill the memory gap with the best practical memory organization techniques for high performance. Microprocessors are also an integral part of intelligent disk controllers, and carry out many central processing unit operating system functions necessary for disk operations. This off-loads the mainframe computer from doing this kind of overhead processing to a large extent. SEE MICROPROCESSOR.

MAIN SEMICONDUCTOR MEMORY

The rapid growth in high-density very large-scale integrated (VLSI) circuits has advanced to a point where only a few applications require the tens to hundreds of thousands of transistors that can now be placed on a chip. One obvious exception is computer main memory, in which there is a continual demand for higher and higher capacity at lower cost. SEE INTEGRATED CIRCUITS.

In the 1960s and early 1970s, magnetic core memories dominated computer main-memory technology, but these have been completely replaced by semiconductor RAM chip devices of ever-increasing density. This transition started with the introduction of the first MOS 1K-bit RAM memory in 1971. This was followed with the 4K-bit RAM chip in 1973, the 16K-bit chip in 1976, the 64K-bit chip in 1979, the 256K-bit chip in 1982, and the 1M-bit chip in 1985. (A 1M-bit chip has 1024×1024 or 1,048,576 bits.) **Figure**

3 shows the progression of RAM chips, which follows the "rule of four," according to which the cost of development of a new RAM device generation can be justified only by a factor-of-four increase in capacity. SEE SEMICONDUCTOR MEMORIES.

1M-bit RAM chips. The 1M-bit MOS RAM has continued to push photolithographic fabrication techniques with feature sizes of 1 to 2 micrometers or less. For RAM chip densities of 1M bits or more per device, the integrated circuit industry has had to make a twofold improvement to maintain volume production and device reliability: better means, such as x-ray step-and-repeat equipment, to achieve features under 2 μm; and plasma-etching or reactive-ion-etching machines to achieve vertical profiles needed as horizontal dimensions decrease. Considerable progress has been made in the use of both x-ray and electron-beam techniques to achieve the submicrometer-size features needed to make even higher-density RAM chips. Most 1M-bit RAMs are dynamic RAMs. The production techniques for dynamic and static RAM chips are identical. Therefore, the cost per unit quickly becomes the cost for mass-producing one chip. Since building costs per chip are about the same whether they store 4K bits or 1M bits, higher densities lead to lower costs per bit.

RAM chip types and technologies. RAM chips come in a wide variety of organizations and types. Computer main memories are organized into random addressable words in which the word length is fixed to some power-of-2 bits (for example, 4, 8, 16, 32, or 64 bits). But there are exceptions, such as 12-, 18-, 24-, 48-, and 60-bit word-length machines. Usually RAMs contain NK · 1 (for example, 64K · 1) bits, so the main memory design consists of a stack of chips in parallel with the number of chips corresponding to that machine's word length. There are two basic types of RAMs, static and dynamic. The differences are significant. Dynamic RAMs are those which require their contents to be refreshed periodically. They require supplementary circuits on-chip to do the refreshing and to assure that conflicts do not occur between refreshing and normal read-write operations. Even with those extra circuits, dynamic RAMs still require fewer on-chip components per bit than do static RAMs (which do not require refreshing).

Static RAMs are easier to design, and compete well in applications in which less memory is to be provided, since their higher cost then becomes less important. They are often chosen for minicomputer memory, or especially for microcomputers. Because they require more components per chip, making higher bit densities more difficult to achieve, the introduction of static RAMs of any given density occurs

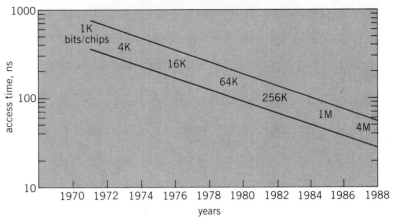

Fig. 3. Trends in performance and density (in bits per chip) of dynamic RAM chips. 1K = 1024 bits; 1M = 1024 × 1024 bits = 1,048,576 bits.

behind that of dynamic versions.

There is another trade-off to be made with semiconductor RAMs in addition to the choice between static and dynamic types, namely that between MOS and bipolar chips. Biopolar devices are faster, but have not yet achieved the higher densities (and hence the lower costs) of MOS. Within each basic technology, MOS and bipolar, there are several methods of constructing devices, and these variations achieve a variety of memory speeds and access times, as well as power consumption and price differences. Within the basic MOS technologies there are several types, such as the n-channel MOS referred to as NMOS and the complementary MOS solid-state structure referred to as CMOS. For bipolar there are several types such as transistor-to-transistor logic (TTL) and the emitter-coupled logic (ECL). **Table 1** provides a sampling of the many different memory chip types, where both the basic silicon technologies, MOS and bipolar semiconductor, are represented, and their varying characteristics. SEE LOGIC CIRCUITS.

Memory cycle and access times. The RAM memory cycle time in a computer is defined as that time interval required to read or write one word of data. Access time is defined as the time interval in which the data are available after the initiation of the read cycle. From the user's point of view, cycle time is also an important characteristic because it has more impact on overall computational speed of the system. A new data read-write cycle does not begin until the previous read cycle is completed. The specified timing, signal overlap, and tolerances allowed vary with each RAM main memory system.

Error checking and correction. Most computer main

Characteristics	Dynamic RAMs	Static RAMs	Static RAMs	ROMs	PROMs	PROMs	EPROMs	EEPROMs (EAROMs)
Chip technology	NMOS CMOS	NMOS CMOS	Bipolar (ECL)	CMOS	Bipolar	CMOS	MOS (UVPROMs)	MOS
Speed (access time)	100 ns	50–100 ns	1–5 ns	100 ns	5–100 ns	50–200 ns	50–300 ns	50–300 ns
Volatility	Volatile	Volatile	Volatile	Nonvolatile	Nonvolatile	Nonvolatile	Nonvolatile	Nonvolatile
Programmability	Easy	Easy	Easy	During manufacture	Once only	Once only	Many times, long procedure	Many times, easier and faster than EPROMs
Power dissipation	Low	Low	High	Low	High	Low	Medium	Medium
Chip density of currently available devices	1M	256K	4K, 16K	256K, 1M	16K, 64K	16K, 64K	256K, 1M	64K, 256K

Table 1. Summary of characteristics of various memory chip types

memories have a memory fault control consisting of a memory controller and a fault control subsystem which automatically corrects all single-bit errors and detects and reports all double-bit or three-bit errors. Each bit in a word is read from a separate RAM device. The fault control permits computer operation to continue even if a memory module is malfunctioning, and thus the computer will operate even if a memory chip is removed from the board. Failures are pinpointed to the specific chip by fault-indicating light-emitting diodes built into each array board. *See* Light-emitting diode.

FIFO and LIFO chips. As memory chips increased in storage density and designers attempted to minimize interconnection circuitry, specialty memory function organizations such as first-in first-out (FIFO) buffers and last-in first-out (LIFO) buffers became available as chips. These kinds of chips are readily used in memory buffering applications to accommodate varying data rates between one element and another. The other complementary feature of FIFO and LIFO chips is the dual data paths (called ports) to allow simultaneous transfer of data into and out of these memory buffer chips. These chip types are commonly used on microcomputer boards and in the computer's peripheral interface and controller electronics.

Gallium arsenide memory chips. All the memory chips discussed above are based on the predominant silicon technology. Reliable gallium arsenide (GaAs) integrated circuits have been produced: 4K-bit and 16K-bit static RAM (SRAM) devices with 1-nanosecond access times. Development of 1M-bit SRAMs has been undertaken for military applications. Considering the need for and emphasis on increasingly rapid computers, there is definite potential for high-speed gallium arsenide logic and memory devices. *See* Gallium.

ROMs, PROMs, and EPROMs. Microcomputers have evolved their own special set of semiconductor memory chips (see Table 1) to suit their application needs. Whereas large, medium-size, and minicomputers primarily use only RAMs that have read-write capability, microcomputers have found significant use for read-only memory (ROM) chips and programmable read-only memory (PROM) chips. Data and program storage in most microcomputer applications is separately allocated between RAMs and ROMs. ROMs provide protection, since the contents are fixed or hardwired and the chips are completely nonvolatile. During microcomputer program development, PROMs are typically used. A PROM can be written only once and involves the irreversible process of blowing polysilicon fuse links. These PROMs are neither erasable nor reprogrammable. Another kind of device, called an erasable PROM (EPROM), is cleared or rewritten by putting the EPROM chip under an ultraviolet light to zero out its contents and then using a PROM programmer to write in the new bit pattern for the modified program. After the microcomputer application program has completed final tests and acceptance, the final bit pattern in the EPROM can be put into ROMs or PROMs (using a chip mask) to facilitate quantity production. *See* Microcomputer; Microcomputer development system.

EEPROMs. There are also memory chip devices called electrically erasable programmable read-only memories (EEPROMs). They have an internal switch on the chip to permit a user to electrically rewrite new contents into them. This means EEPROMs do not have to be removed from the circuit in order to clear their contents to put in the new or modified bit pattern representing another program of instructions. Thus EEPROMs have met the need in microcomputer systems for a nonvolatile in-circuit reprogrammable random word access memory device.

EEPROMs use two distinctively different technologies. The more mature technology is metal nitride oxide semiconductor (MNOS), which is a very different gate insulator process technology from that used for MOS. The MNOS technology previously used the generic acronym EAROM (electrically alterable read-only memories) until EEPROMs became common in usage for both technologies. The other technology is floating-gate MOS (also called FAMOS). FAMOS technology was used for ultraviolet erasable EPROMs (UVPROMs) and was subsequently refined to provide an electrically erasable technology. Both technologies rely on Fowler-Nordheim tunneling to move the charge to be stored from the substrate to the gate. FAMOS stores the charge on the gate itself, whereas MNOS devices trap the charge in the nitride layer under the gate.

SECONDARY MEMORY

High-capacity, slower-speed memory consists of two major functional types: random-access, which has been provided primarily by disk drives, and sequential-access, which has been provided primarily by tape drives. Since tape drives provide removability of the medium from the computer, tape is used for the majority of off-line, archival storage, although some disks are removable also. The on-line random-access disk devices are classed as secondary, and tape-based systems are classed as tertiary.

Over the history of electronic computers, while the technology for processors and main memory has been evolving from vacuum tubes to transistors to very large-scale integration (VLSI) chips, the predominant technology for secondary memory has continued to be magnetic recording on tape and disk. This has not been due so much to the absence of competing alternatives as to the continuous and rapid progression in magnetic recording capability. The increase in recording density on disks is shown in **Fig. 4**. The scale of the figure is logarithmic, so the rate of improvement has been exponential.

The current magnetic-disk technology will be discussed, and also the potential of new magnetic recording techniques such as thin-film, vertical recording, and bubble-memory devices; and a nonmagnetic technology, optical recording.

Magnetic disk storage. Conventional magnetic-disk memories consist of units which vary in capacity from the small 360-kilobyte floppy disks (used with microcomputers) to 1-gigabyte disk drives used with large-scale computers. Hard-disk memories are characterized by access times in the 20–80-ms region (versus access times of tens to hundreds of nanoseconds for RAMs). **Figure 5** shows the historical progression in several performance factors: capacity, transfer rate, and access time.

Capacity. The major area of development in disks has been the progressive and even spectacular increases in capacity per drive, particularly in terms of price per byte. There is a substantial economy of scale in storage capacity, in that the cost per byte goes down as the drive capacity goes up. Between small disks and top-end drives, the differences in access time and transfer rate may be only a factor of two to five, while the capacity difference is a factor of as much as 2000, that is, from 5 megabytes to several gigabytes.

Transfer rate. The rotation rate of the disk platter, like the action of the head actuators, is limited by considerations of physical dynamics. Increases in transfer rate will come from greater linear bit density around the track. In an absolute sense, transfer rates are not as often a system performance bottleneck as is access time.

Average access time. Since the arm supporting the read-write head is one of the few moving mechanical parts in a computer system, it has not shown the multiple-orders-of-magnitude performance improvement over time that the electronic technologies have. Also, average access times for the highest-priced and lowest-prices disks do not vary by an enormous factor, being in the range of 10 to 80 ms for a typical microcomputer hard (as opposed to floppy) disk, and 10 to 30 ms for large disks on mainframe computers. There is nothing which indicates a breakthrough in the technology governing access time, even though greater capacities and transfer rates are still coming. Therefore, for random-access bottlenecks, the solution in the future will be the same as it is now: multiple smaller drives to allow overlapped seeks instead of a single large-capacity drive.

Basic disk technology. Many significant advances in the capabilities of commercially available disk memory technology were made during the late 1970s. One of the major trends was toward larger-capacity disks, progressing from 100 megabytes to several gigabytes per disk drive. The majority of technology improvements to provide higher capacities, lower costs, and better reliabilities resulted from the Winchester technology. Prior to this technology, removable disk packs established themselves as the most advanced technology for large-capacity disks. The Winchester technology is characterized by nonremovable, sealed disk packs, which inherently provide more reliability due to less handling and make more stringent alignment tolerances practical.

The head, or read-write transducer, has undergone substantial refinement. Central to the advance in increasing disk-packing density was the reduction in the flying height of the read-write head. Because flux spreads with distance, reduced separation between the read-write head and the magnetic surface will decrease the area occupied by an information bit. This obviously limits the number of bits that can be defined along an inch of track and increases the minimum spacing between tracks. The ideal would be a direct contact between head and surface, which is the case with magnetic tape. But this is not possible with a rigid aluminum disk whose surface is traveling at rates that can exceed 100 mi/h (45 m/s). By applying a load on the head assembly, proper spacing is maintained. All movable media memories utilize an air-bearing mechanism known as the slider or flying head to space the transducer in proximity to the relatively moving media. The slider, when forced toward the moving surface of the medium, rests on a wedge of air formed between it and the medium by virtue of their relative motion. The thickness of the wedge can be controlled by the applied force, and the wedge in effect is an extremely stiff pneumatic spring. Head-to-medium spacings were well under 10 μin. (0.25 μm) in 1982. SEE TRANSDUCER.

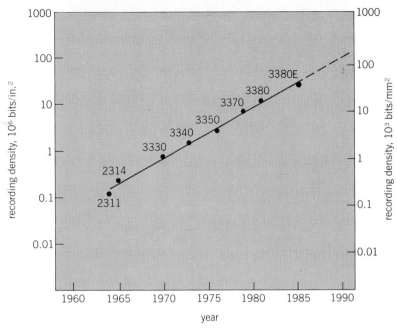

Fig. 4. Increase in recording density on magnetic disk storage. Data points represent IBM models, and model numbers are indicated.

Some standards have been developed for interfaces between the drive and the controller, to permit compatibility among drives of various capacities and manufacturers. Among these are the Storage Module Device (SMD) interface, Enhanced Small Device Interface (ESDI), Small Computer System Interface (SCSI), and Intelligent Peripheral Interface (IPI).

SMD drives can run at 3 megabytes per second. ESDI drives can also attain that rate. The IPI Level II drives can operate at 10 megabytes per second. The transfer rates of SCSI vary. As an intelligent interface between peripherals and computer systems, SCSI allows system designers to avoid device-level idiosyncrasies that make quick integration of new peripherals difficult. It is suitable for use on small personal computers with standard asynchronous transfer rates of 1.5 megabytes per second. It can also be used with

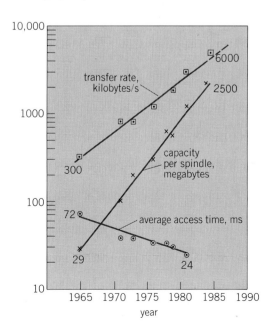

Fig. 5. Development of magnetic disk performance factors.

supermicrocomputers and minicomputers with an optional synchronous transfer rate of 4 megabytes per second. Work has been done toward raising the transfer rate to 8 megabytes per second by tightening the timing specifications of the SCSI interface. Although initial versions of SCSI were designed for small computers, SCSI can meet the high-performance and high-capacity requirements of larger computers. For example, a single SCSI bus can support six SCSI-to-ESDI controllers with 24 disks, providing 7.2 gigabytes of storage, in addition to tape drives and other devices.

In addition to fixed Winchester-type disk units of various capacities, disk storage is used in several removable forms.

Cartridge disks. Removable storage cartridges containing single-disk packs are used extensively with minicomputer systems. The cartridge disk-drive unit has capacities ranging from 1 to 10 megabytes, depending on the recording density. Average access times range from 35 to 75 ms, and data transfer rates from 200 to 300 kilobytes per second. Physically, these plastic-encased cartridges are either top-loaded or front-loaded units. These units, besides having a removable disk, also have a fixed disk. The performance features of cartridge disks compete favorably with the larger-capacity disk pack storage equipment. A disadvantage of removable disk cartridges or disk packs is reduced reliability compared to fixed units.

Floppy disks. Floppy disks have become the widely used random-access secondary memory for microcomputer systems. The floppy disk was originally developed by IBM in 1965 for internal use on its large 370 computers. The disk was designed to be a permanent nonvolatile memory storage device, written at the factory. However, large numbers of new and varied applications evolved for these relatively small, compact disk systems. Floppy disks derive their name from the recording medium, an oxide-coated flexible disk (similar to, but more flexible than, plastic 45-rpm music records) enclosed within a protective plastic envelope. The Mylar flexible disk is 7.875 in. (200 mm) in diameter, 0.003 in. (0.076 mm) thick, and records 3200 bits per inch (125 bits per millimeter) and 48 tracks per inch (1.9 tracks per millimeter). The disk is permanently enclosed in a protective envelope, and this package is 8 in. (203 mm) square, hence the common reference to the size as 8-in. The protective envelope contains alignment holes for spindle loading, head contact, sector-index detection, and write-protect detection. The capacity of this easily transportable medium is 8 megabits, commonly laid out on 77 tracks, each divided into 26 sectors.

The flexible-disk drive subassembly consists of the metal base frame, the disk-drive motor that rotates the spindle (360 rpm for 8-in. or 203-mm, 300 rpm for 5¼-in. or 133-mm) through a belt drive, and a stepping motor with an integral lead screw to form the head-positioning actuator. The read-write head is mounted on a carriage that the lead screw moves. Head load is achieved by moving a load pad against the disk and restraining the disk between the head and the load pad to accomplish the head-to-disk compliance.

The removable floppy disk, called a diskette, was originally 8 in. (203 mm), in both single and double densities, and single- and double-sided. A variety of incompatible formats exist, particularly in double density, as the majority of suppliers use a common (IBM) format in single density. The most conspicuous format difference is in sector boundaries, hard-sector referring to boundaries of physical holes in the diskette, and soft-sector to boundaries that are software-formatted. As recording density increased and packaging considerations for microcomputers became important, a smaller, 5¼-in. (133-mm) diskette was introduced in 1976 and has become dominant, with a similar proliferation of formats as in 8-in. (203-mm). Newer systems use yet smaller sizes such as 3, 3¼, and 3½ in. (76, 83, and 90 mm). The 3½-in. (90-mm) floppy diskettes are widely used with lap-top computers and newer versions of personal computers. Standard storage capacities range from 720 kilobytes to 1.44 megabytes, and higher densities have been investigated.

Thin-film and vertical recording. Several technologies are in development for disk storage.

The recording technology for conventional magnetic disk recording has been based upon ferrite heads and a recording surface consisting of a coating embedded with ferrous oxide particles. One of the limiting factors is the thickness of the recording layer, which is 20 to 50 μin. (0.5 to 1.25 μm). With current products pushing the limits of recording density achievable with that approach, a successor technology has been developed, called thin-film.

The thin-film technique uses a continuous magnetic film of cobalt or nickel, applied by electroplating to a controlled thickness of as little as 1 μin. (25 nanometers). Either the head or the recording surface or both can use thin-film technology. Most of the major conventional large disks use thin-film heads and an oxide recording surface. Advanced models are designed to use a thin-film recording surface. Thin-film technology is the source for current progress in disk capacity and transfer rate. It was first introduced in small disks, partly due to greater ability to withstand the more portable environment of the small computer. The hardness of the medium which provides this resiliency also reduces both the likelihood and consequences of head crashes.

Another magnetic recording technique is vertical, or perpendicular, recording. In this approach the miniature magnets which are formed in the recording media are oriented downward into the surface rather than longitudinally along the surface. This allows closer packing of the magnetized spots, not just because of spatial compression, but also because of improved magnetic interaction between adjacent spots. The adjacent spots repel each other in the longitudinal case but attract in the vertical case. This technique offers the potential for significantly extending recording density and the accompanying performance parameters.

Recording density is the product of the bit density along the recording track, in bits per inch (or bits per millimeter), and the number of tracks per inch (or tracks per millimeter) across the recording surface. For a given rotation rate of the disk, the bit density along the track determines the transfer rate. A comparison of these characteristics for various recording techniques is given in **Table 2**.

The point at which vertical recording becomes attractive to develop into commercial products depends in part upon the level of capability which can ultimately be achieved with thin-film technology. The

Table 2. Storage densities for various recording technologies

Technology	Bit density along track		Track density		Recording density	
	Bits/in.	Bits/mm	Tracks/in.	Tracks/mm	10^6 bits/in.^{2}	10^3 bits/mm^2
Magnetic recording						
Conventional (IBM 3380)	15,000	600	800	32	12	19
Thin-film*	25,000	1000	1,200	47	30	47
Vertical*	100,000	4000	1,200	47	120	190
Optical recording	15,000	600	18,750	710	225	350

*Figures are projections.

density level achieved by vertical recording technology may represent a limit for magnetic recording. If so, one possible answer to the question of how further levels of performance can be obtained is optical recording.

Optical recording. Optical recording is a nonmagnetic disk technology that uses a laser beam to burn pits in the recording medium to represent the bits of information, and a lower-power laser to sense the presence or absence of pits for reading. This technology has the potential for higher ultimate recording density than magnetic recording. The medium is removable and relatively inexpensive. The removability is significant in light of the nonremovability of most current and projected high-capacity magnetic disks. The recording density increase is apparent when it is noted that 4-gigabyte capacity is achieved on a single disk, whereas current-model large-capacity (2.5-gigabyte) magnetic drives typically consist of a cabinet containing two disk assemblies, each assembly having 16 platters.

One of the first optical disk mass-storage devices has 4 gigabytes on a removable, nonerasable, 14-in. (356-mm) disk. The transfer rate of 1.5 megabytes per second buffered or 3.0 megabytes streaming is similar to that of conventional magnetic disks (3.0 for the IBM 3380). The average access time is slower than that of high-performance magnetic disks at 84.7 milliseconds, versus the 3380's 24 milliseconds, but track-to track access is a comparatively fast 1.0 millisecond. The corrected bit error rate is 1 in 10^{13}, which is compatible with magnetic disk standards.

In addition to the 14-in. (356-mm) disk for large-capacity systems, 12-in. (305-mm) optical disk systems in the 1-gigabyte capacity range have been produced for use with 16- or 32-bit microcomputer and minicomputer systems.

Disadvantages. Optical technology has several disadvantages. First, the medium is not inherently erasable, although systems which can erase and rewrite are also in development. The primary problem in producing a commercial optical disk system has been getting the bit error rate down to an acceptable level, with sufficient recording density to be competitive with magnetic recording. To be competitive with magnetic tape, a bit error rate, after error detect and correct, of 1 in 10^{10} to 10^{11} is needed, and to compete with magnetic disk, 1 in 10^{13} to 10^{14}. Through error detect and correct techniques, the 1 in 10^{13} level has been attained.

The bit error rate problem is far less critical for storage of documents and images, since bit dropout is seldom noticeable, in contrast with financial information, where it could be disastrous. This is the reason that optical disk recording was first used successfully in consumer and industrial systems for video-tape image information, and for document storage and retrieval systems. SEE VIDEO DISK RECORDING.

Applications. The nonerasability of the technology can be an advantage for some applications. Examples are data which must be kept for historical or audit-trail purposes, and applications where large amounts of data must be shipped among different locations.

When this technology matures, it will be an event of major significance, being a removable, large-capacity, random-access device using a recording technology with large potential for further growth. The potential for development of yet higher recording densities is perhaps the most significant factor, but in addition, having large capacity on a removable platter would allow some other kinds of system development. One possible system, referred to as a jukebox, would store on the order of 100 disks in an automatic retrieval and mounting device, so that a given disk could be fetched and accessed in around 5 seconds giving an on-line storage in the range of 500 gigabytes. Previous systems for very large on-line storage which have been delivered in significant quantity are the cartridge tape systems from IBM and CDC and automated tape libraries discussed below. The cartridge tape systems have capacities to 50 megabytes (IBM) and 8 megabytes (CDC) per cartridge; hence the optical system would provide a great improvement in the capacity available on the basic storage unit, and the automated tape libraries of course have the limitation of sequential data access. The 3½-in. (90-mm) floppy diskettes are widely used with lap-top computers and newer versions of personal computers. Standard storage capacities range from 720 kilobytes to 1.44 megabytes, and higher densities have been investigated.

CD-ROM and WORM. CD-ROM (compact disk–read-only memory) and WORM (write once, read many) are special types of optical disks. CD-ROM resembles the related audio compact disk technology in that users of CD-ROM can read only pre-recorded data on the disk. A 5-in. (127-mm) CD-ROM can hold 500 to 600 megabytes, which is equivalent to 1400 to 1700 (360-kilobyte) diskettes. For example, the text of a 20-volume encyclopedia could be easily stored on a CD-ROM disk. CD-ROM has become an add-on peripheral device for some personal computers and workstations.

The CD-ROM disk and its drive are physically similar to the audio compact disk and its player, but there are several key differences: (1) A CD-ROM drive does not have a digital-to-analog converter. (2) It does have more powerful error correction than the audio system, because computer data requires a lower error rate, typically one error every 10^{15} bits, while the ratio of one error in 10^8 is acceptable in an audio compact-disk system. (3) The CD-ROM drives employ data scrambling to protect the servo control circuits from harmful dc components.

In CD-ROM, data are stored digitally on the surface of the substrate as a series of pits and lands of variable lengths. A land is the surface of the substrate between pits. Instead of using pits and lands to represent ones and zeros, the data in CD-ROM is represented in terms of the transitions from a pit to a land and vice versa (ones) and by the lack of such transitions (zeros). This arrangement, together with a modulation scheme, enhances the resolution with which data are read off the disk from about 1 μm—the diameter of the laser beam—to about 0.6 μm. The result is a 60% increase in the density of the data on the disk.

The pits on the disk surface can be arranged in either concentric circles, like tracks on magnetic disks, or in a spiral track, as on vinyl phonograph records. One common data format uses concentric tracks with the disk rotating at a constant angular speed. This simplifies the servomechanism but wastes space on the outer part of the disk because the bits in the outer tracks are separated from each other by more space than that necessary for adequate resolution. With the spiral track, the disk can be rotated so that the track passes the read head at a constant linear speed. This means that the disk must rotate more slowly when the laser beam is focused on the outer tracks and more quickly when the beam is on the inner tracks. Whether the tracks are concentric or spiral, the inter-track distance is about 1.6 μm, and the width of the pits is about 0.6 μm.

While CDD-ROM technology is fairly well established, the lack of established standards for WORM technology could delay its acceptance. WORM drives come in different sizes: 3½, 5¼, 8, 12, and 14 (90, 133, 203, 305, and 356 mm). Capacities of 5¼ in. (133 mm) WORM drives range from 200 bytes to 800 megabytes, while 12 in. (305 mm) WORM drives are in the 2–3-gigabyte range.

Bubble memory devices. Bubble memories are chips rather than disks, but are different from semiconductor memories in that they are magnetic devices, in which the absence or presence of a magnetic domain is the basis for a binary 1 to 0. The performance characteristics of these devices makes them competitive as small-capacity secondary storage. A magnetic bubble is in reality a cylindrical magnetic domain with a polarization opposite to that of the magnetic film in which it is contained. These cylinders appear under a microscope as small circles (which give them their name) with diameters from 80 to 800 μin. (2 to 20 μm). The size of a bubble is determined by the material characteristics and thickness of the magnetic film.

Bubbles are moved or circulated by establishing a magnetic field through a separate conductor mounted adjacent to the bubble chip. A large portion of the bubble chip must be given over to circuitry for generating, detecting, and annihilating the bubbles. Magnetic-domain bubble devices typically operate in an endless serial loop fashion or in an organization with serial minor loops feeding major loop registers. Bubble memories are particularly well suited to applications such as portable recorders because of their phys-

ical advantages (low power requirements and light weight) and speed advantage over electromechanical devices such as cassettes and floppy disks. Like their electromechanical counterparts, bubble devices are nonvolatile; that is, they retain their contents when the power goes off.

The controller is central to any bubble memory system, which it serves as the interface between the bubble chip and the system bus. It generates all of the system timing and control functions and supervises the handshaking operations required to access and transfer data between the system bus and the bubble memory module. Usually a controller can operate more than one bubble chip module, in most cases up to eight modules. The function-driver integrated circuit produces the control currents required to generate and input or ouput tracks on the chip. The sense-amplifier integrated circuit converts the analog output signals produced by the bubble detector into a transmittable data stream that passes to the controller. The coil-driver integrated circuits excite the x and y coils of the bubble chip package with out-of-phase signals to produce the rotating field that moves the bubbles.

Several companies are producing 1-megabit and 4-megabit bubble memory chips, and 16-megabit bubble memory chips are in the development stages. Under a cross-license and alternate source agreement, these companies have agreed to jointly adopt a low-height leaded package for complete component level interchangeability. Each company developed its peripheral chip set for the 4-megabit magnetic bubble module that will be compatible with the 1-megabit units.

For portable and other special applications, bubbles have definite advantages such as nonvolatility, low power, and high compactness. Performance capabilities relative to floppy disks are 100 kilobits per second for bubbles versus 200–250 kilobits per second for floppies, and 40 milliseconds average access time for bubbles versus 200–250 milliseconds for floppies. *See* MAGNETIC BUBBLE MEMORY.

General trends. The technologies which will provide the growth in performance in disk systems will also allow improved price-performance, so that the cost per byte on-line can be expected to continue to drop. Although this disscussion has concentrated on high-end disks in terms of performance parameters, small- and medium-range disks, including floppies, will benefit from the same technological and price-performance progression. Some of these technological developments in fact can appear first in the lower-end products, one such possibility being vertical recording in floppy diskettes.

In both hard disks and floppies, the higher recording densities permit progressively more compact packaging for microcomputers, but the constant change has aggravated the existing difficulties with lack of standardization, which is unfortunate for the potentially convenient interchange medium which the floppy diskette could provide. Although such possibilities as a 40-megabyte floppy diskette are impressive, the prospects for standards permitting interchangeability are becoming less, with the proliferation of sizes and incompatible formats discussed above.

MAGNETIC TAPE UNITS

In magnetic tape units, the tape maintains physical contact with the fixed head while in motion, allowing high-density recording. The long access times to find user data on the tape are strictly due to the fact that all intervening data have to be searched until the desired data are found. This is not true of rotating disk memories or RAM word-addressable main memories. The primary use of tape storage is for seldom-used data files and as back-up storage for disk data files.

Half-inch tapes. Half-inch (12.7-mm) tape has been the industry standard since it was first used commercially in 1953. Half-inch magnetic tape drive transports are reel-to-reel recorders with extremely high tape speeds (up to 200 in. or 5 m per second), and fast start, stop (on the order of 1 millisecond), reverse, and rewind times.

Performance and data capacity of magnetic tape have improved by orders of magnitude, as shown in **Table 3**. Just prior to the 1970s came the single-capstan tape drive, which improved access times to a few milliseconds. These vacuum-column tape drives have such features as automatic hub engagement and disengagement, cartridge loading, and automatic thread operation. There are two primary recording techniques, namely, nonreturn-to-zero-inverted (NRZI) and phase-encoded (PE), used with packing densities of 800 and 1600 bytes per inch (31.5 and 63 bytes per millimeter). These typically use a 0.6-in. (15-mm) interrecord gap. In phase encoding, a logical ''one'' is defined as the transition from one magnetic polarity to another positioned at the center of the bit cell. ''Zero'' is defined as the transition in the opposite direction also at the center of the bit cell, whereas NRZI would involve only one polarity. The advantage of the phase-encoding scheme over NRZI is that there is always one or more transitions per bit cell, which gives phase encoding a self-clocking capability, alleviating the need for an external clock. The disadvantage of phase encoding over NRZI is that at certain bit patterns the system must be able to record at twice the transition density of NRZI.

Computer-compatible magnetic tape units are available with tape drives of 6250 bytes per inch (246 bytes per millimeter). They have nine tracks, where eight bits are data and one bit is parity. Each track is recorded by a technique called group-coded record (GCR), which uses a 0.3-in. (7.6-mm) record gap. Every eighth byte in conjunction with the parity track is used for detection and correction of all single-bit errors. Thus, GCR offers inherently greater reliability because of its coding scheme and error-correction capability. It does, however, involve much more complex circuitry.

The on-line capacity is strictly a function of the number of tape drives that one controller can handle (eight is typical). A 2400-ft (730-m) nine-track magnetic tape at 6250 bytes per inch (246 bytes per mil-

limeter) provides a capacity of approximately 10^9 bits, depending on record and block sizes.

Quarter-inch tapes The installation of ¼-in. (6.35-mm) streaming tape drives has increased very rapidly. There are several recording-format standards for ¼-in. (6.35-mm) tape drives; the most common one is called QIC (quarter-inch compatibility)-24, which connotes a full-size cartridge class with formatted storage capacity of 60 megabytes on nine tracks, using 600 ft (183 m) of tape. The device interface for a QIC-24 drive is either an intelligent interface such as QIC-02; the SCSI; or a basic drive interface, QIC-36. Another recording format, QIC-120, provides 125 megabytes on 15 tracks.

Cassettes and cartridges. The most frequently used magnetic tape memory devices for microcomputer systems are cassette and cartridge tape units. Both provide very low-cost storage, although their access times are long and their overall throughput performance is not as great as that of floppy disks. Cassette and cartridge units both use ¼-in. (6.35-mm) magnetic tape.

The digital cassette transport was orginally an outgrowth of an audio cassette unit. Unfortunately, the very low-cost audio-designed transport did not meet the endurance needs of true digital computer applications. There are two basic design approaches to digital cassette transports: capstan drive and reel-to-reel drive. Capstan tape drives are better for maintaining long-term and short-term tape-speed accuracy, while reel-to-reel transports have better mechanical reliability.

During the 1960s the first true digital tape cartridge was developed. With the cassette, a capstan must penetrate the plastic case containing the tape in order to make contact. There is no such penetration system with the cartridge because the transport capstan simply drives an elastomer belt by pressing against the rim of the belt's drive wheel. This simplicity eliminates a major source of tape damage and oxide flaking.

Cassettes have undergone evolution as a digital medium, and capacity has increased to 10 megabytes, at a density of 5120 bits per inch (202 bits per millimeter). Transfer rates are 24 kilobits per second at 30 in. (0.76 m) per second, or 72 kilobits per second at 90 in. (2.29 m) per second. The ¼-in. (6.35-mm) streaming tape cartridge has a typical capacity range of 10 to 30 megabytes, with some available with over 50 megabytes. Transfer rates are 30 to 90 kilobits per second, at densities of from 6400 to 10,000 bits per inch (252 to 394 bits per millimeter).

A typical mainframe tape cartridge can store data at 38,000 bits per inch (1496 bits per millimeter). At typical blocking factors, a 550-ft (168-m) tape car-

Table 3. Standard recording densities for magnetic tape					
		Bit density along track		Area density	
Year	Number of tracks	bits/in.	bits/mm	bits/in.²	bits/mm²
1953	7	100	4	1,400	2
1955	7	200	8	2,800	4
1959	7	556	22	7,784	12
1962	7	800	32	11,200	17
1963	9	800	32	14,400	22
1965	9	1600	63	28,800	45
1973	9	6250	246	112,500	174

tridge contains about the same amount of data as a 2400 ft (730-m) reel of tape at 6250 bits per inch (246 bits per millimeter), but it requires less storage space. The data transfer rate between the tape cartridge drive and the process is 4.5 megabytes per second.

Half-inch (12.7-mm) tape is also used for back-up for microcomputer hard disk systems. Configured to standard microcomputer packaging sizes, these give 50-megabyte capacity in 5¼-in. (133-mm) form, and 300 megabytes in 8-in. (203-mm) form.

Mass Storage Tape Systems

With the gradual acceptance of virtual memory and sophisticated operating systems, a significant operational problem arose with computer systems, particularly the large-scale installations. The expense and attendant delays and errors of humans storing, mounting, and demounting tape reels at the command of the operating system began to become a problem. Cartridge storage facilities are designed to alleviate this problem.

Their common attributes are: capacity large enough to accommodate a very large database on-line; access times between those of movable-head disks and tapes; and operability, without human intervention, under the strict control of the operating system. The cartridge storage facility is included within the virtual address range. All such configurations mechanically extract from a bin, mount on some sort of tape transport, and replace in a bin, following reading or writing, a reel or cartridge of magnetic tape.

Cartridge storage systems are hardware devices that need operating system and database software in order to produce a truly integrated, practical hardware-software system. Users require fast access to their files, and thus there is a definite need to queue up (stage) files from the cartridge storage device onto the disks. The database software must function efficiently to make this happen. In general, users base their storage device selection on the file sizes involved and the number of accesses per month. Magnetic tape units are used for very large files accessed seldom or infrequently. Mass-storage devices are for intermediate file sizes and access frequencies. Disk units are used for small files or those which are accessed often.

In practice, most users have been satisfied with tapes and disks, and have not chosen to install mass storage systems. During the 1970s, two basic kinds were delivered, although some others were built and installed in the earlier years. In different ways, these units combine the low cost of tape as a storage medium with the operating advantages of on-line access. One is a mechanical selection and mounting unit to load and unload tape reels onto and off standard magnetic tape units; another is a mechanical selection and accessing unit to operate with special honeycomb short-tape units. The first type (sometimes called automated tape libraries) eliminated manual tape-mounting operations. The main objective of the short-tape honeycomb cartridge system is to improve access time. The shorter tape (770 in. or 19.6 m, versus 2400 ft or 730 m) results in better access time. Operationally, the honeycomb cartridge tape system and the mechanical standard tape-mounting units are capable of handling 100 to 300 cartridge or tape loads per hour.

The first type of fully automated tape library uses standard magnetic tape (½ in. or 12.7 mm wide). Under computer control, this equipment automatically brings the tapes from storage, mounts them on tape drives, dismounts the tapes when the job is completed, and returns them to storage. Accessing up to 150 reels per hour, this unit can store up to 7000 standard tapes or 8000 thin-line reels in a lockable self-contained library that can service up to 32 tape drives and that can interface with up to four computers.

The honeycomb cartridge storage system uses a storage component called a data cartridge. Housed in honeycomb storage compartments, these 2 × 4 in. (24 × 50 mm) plastic cartridges can each hold up to 50,000,000 bytes of information on 770 in. (19.6 m) of magnetic tape approximately 3 in. (76 mm) wide. Whenever information from a cartridge is needed by the computer, a mechanism selects the desired cartridge and transports it to one of up to eight reading stations. There the data are read out and transferred to the staging disk drives. See Computer; Data-processing systems; Digital computer.

Michael Plesset; Douglas Theis; Peter P. Chen

Bibliography. R. Brechtlein, Comparing disc technologies, *Datamation*, 24(1):130–150, January 1978; P. P. Chen, The compact disk ROM: How it works, *IEEE Spectrum*, 23(4):44–49, April 1986; S. Chi, Advances in computer mass storage technology, *IEEE Comput.*, 19(5):60–74, May 1982; M. Elphick, Disk and tape memory systems, *Comput. Des.*, 22(1): 85–126, January 1983; A. H. Eschenfelder, *Magnetic Bubble Technology,* 1980; T. Moran, New developments in floppy disks, *Byte Mag.*, 8(3):68–82, March 1983; M. Plesset, Future developments in disc storage, *Data Process.*, 25(8):28–30, October 1983; R. M. White, Disk-storage technology, *Sci. Amer.*, 2432:(1)138–148, August 1980.

Computer systems architecture

The discipline that defines the conceptual structure and functional behavior of a computer system. It is analogous to the architecture of a building, determining the overall organization, the attributes of the component parts, and how these parts are combined. It is related to, but different from, computer implementation. Architecture consists of those characteristics which affect the design and development of software programs, whereas implementation focuses on those characteristics which determine the relative cost and performance of the system. This division is necessary because of the existence of computer families. A family is a series of several computers (usually from the same manufacturer) that offers a variety of cost and

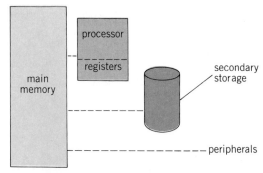

Fig. 1. Overview of a computer system. Storage is made up of registers, main memory, and secondary storage. Broken lines indicate input/output.

Table 1. Storage characteristics of typical computers

Storage type	Capacity (words)	Access time, s	Typical uses
Registers	10^0 to 10^3	10^{-8} to 10^{-6}	Computation
Main memory	10^3 to 10^8	10^{-7} to 10^{-5}	Storage of programs and data
Secondary storage	10^5 to 10^{12}	10^{-2} to 10^1	Archival or long-term data and program storage

performance options (different implementations) but can run the same software programs (that is, all have the same architecture).

The architect's main goal has long been to produce a computer with the highest performance possible, within a given set of cost constraints. This requires a strong background in electrical engineering. Over the years, other goals have been added, such as making it easier to run multiple programs concurrently or improving the performance of programs written in higher-level languages. Thus, the architect also needs a strong background in the software (programming) aspect of computing. *See* DIGITAL COMPUTER PROGRAMMING; PROGRAMMING LANGUAGES; SOFTWARE ENGINEERING.

A computer system consists of four major components (**Fig. 1**): storage, processor, peripherals, and input/output (communication). The storage system is used to keep data and programs; the processor is the unit that controls the operation of the system and carries out various computations; the peripheral devices are used to communicate with the outside world; and the input/output system allows the other components to communicate with one another.

Storage. The storage or memory of a computer system holds the data that the computer will process and the instructions that indicate what processing is to be done. In a digital computer, these are stored in a form known as binary, which means that each datum or instruction is represented by a series of bits. Bits are conceptually combined into larger units called bytes (usually 8 bits each) and words (usually 8 to 64 bits each). *See* BIT.

A computer will generally have several different kinds of storage devices, each organized to hold one or more words of data. These types include registers, main memory, and secondary or auxiliary storage (**Table 1**).

Registers. These are the fastest and most costly storage units in a computer (**Table 2**). Normally contained within the processing unit, registers hold data that are involved with the computation currently being performed. Registers are also used to hold information describing the current state of the computing pro-

cess. Implementation of registers may vary with different members of the same architecture family.

Main memory. This device holds the data to be processed and the instructions that specify what processing is to be done. Main memory consists of a sequence of words or bytes, each individually addressable and each capable of being read or written to. Two good measures of computer system performance are how much main memory it can have and how fast that memory can be accessed. A major goal of the computer architect is to increase the effective speed and size of a memory system without incurring a large cost penalty. Two prevalent techniques for increasing effective speed are interleaving and cacheing, while virtual memory is a popular way to increase the effective size.

Interleaving involves the use of two or more independent memory systems, combined in a way that makes them appear to be a single, faster system. In one approach, all words with even addresses come from one memory system and all words with odd addresses come from the other. When an even-numbered word is fetched, the next-higher odd-numbered word is fetched at the same time. If that odd-numbered word is requested next (a situation that occurs often), it has already been fetched and thus has an access time of zero. This nearly doubles the average access speed.

With cacheing, a small, fast memory system contains the most frequently used words from a slower, larger main memory. With careful design, cacheing can yield considerably improved average memory speeds.

Virtual memory is a technique whereby the programmer is given the illusion of a very large main memory, when in fact it has only a modest size. This is achieved by placing the contents of the large, "virtual" memory on a large but slow auxiliary storage device, and bringing portions of it into main memory, as required by the programs, in a way that is transparent to the programmer.

Auxiliary memory. This portion of memory (sometimes called secondary storage) is the slowest, lowest-

Table 2. Registers commonly found in computers

Register	Other names	Function
Instruction register	I register	To hold the instruction currently being executed by the processor
Accumulator	Arithmetic register	To hold the results of computation
Status register	State register	To indicate the results of tests, occurrence of unusual conditions, and status of certain activities
Program counter	Instruction pointer	To hold the address in main memory of the next instruction to be executed
Index register	Counter, B register	For counting (for example, number of times through a repeated computation)
Base register	Pointer register, address register	For pointing to (holding the address of) something in main memory

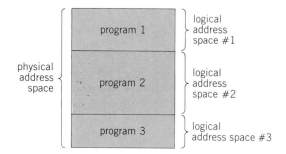

Fig. 2. Several logical address spaces within a physical address space.

cost, and highest-capacity computer storage area. Programs and data are kept in auxiliary memory when not in immediate use, so that auxiliary memory is essentially a long-term storage medium.

There are two basic types of secondary storage: sequential and direct-access. Sequential-access secondary storage devices, of which magnetic tape is the most common, permit data to be accessed in a linear sequence. Thus, in order to access the 100th datum on such a device, the first 99 must be "passed over." For processes where sequential access is suitable, such as printing a list of names or merging two files of insurance records, sequential-access devices are very cost-effective.

A direct-access device is one whose data may be accessed in any order. Disks and drums are the most commonly encountered devices of this type, although certain low-speed semiconductor devices, such as bubble memory, are also used. Direct-access devices permit high-speed access to any block of data, regardless of its location on the device, and thus are well suited to situations where sequential access is not convenient, such as obtaining individual insurance records, at random, from a large file. *SEE MAGNETIC BUBBLE MEMORY; SEMICONDUCTOR MEMORIES*.

Memory mapping. This is one of the most important aspects of modern computer memory designs. In order to understand its function, the concept of an address space must be considered. When a program resides in a computer's main memory, there is a set of memory cells assigned to the program and its data. This is known as the program's logical address space.

If several programs reside in memory at the same time, each has its own logical address space.

The computer's physical address space is the set of memory cells actually comprising the main memory. In the simplest case, where only one program resides in main memory, the logical and physical address spaces may be identical. A more common situation (**Fig. 2**) is where several programs, and thus several logical address spaces, reside within the same physical address space. In such a case, it is desirable to design the system so that the program does not need to know in which part of the physical address space it resides; this anonymity permits the program to be placed anywhere. It is also desirable for the computer to prevent one program from accidentally accessing data in another program's memory. These goals are achieved by means of memory mapping.

Memory mapping is simply the method by which the computer translates between the computer's logical and physical address spaces. The most straightforward mapping scheme involves use of a bias register (**Fig. 3**). Suppose the physical address space begins at location 0, but the program is placed in memory at location 10,000. With the bias scheme, the program is written as though it began at location 0, and the number 10,000 is placed in the bias register. Each time memory is referenced by the program, the contents of the bias register are added to the memory address, thus offsetting the program addresses by the bias amount. Assignment of a different bias value to each program in memory enables the programs to coexist without interference.

Another strategy for mapping is known as paging. This technique involves dividing both logical and physical address spaces into equal-sized blocks called pages. Mapping is achieved by means of a page map, which can be thought of as a series of bias registers (**Fig. 4**). Each logical address is divided into a page number and an offset; the offset is combined with the page map value to yield a physical address.

A simple extension of paging, known as demand paging, allows implementation of virtual memory. In this scheme, the page map includes a "presence" bit for each page, indicating whether that page actually resides in main memory. For a page that is located in main memory, the operation is identical to that of simple paging. If, however, the presence bit indicates that the page is not in main memory, a page fault occurs, causing the computer to obtain the desired page from secondary storage and copy it into main memory. The above procedure is carried out by the operating system and is essentially unseen by the application program. *SEE COMPUTER STORAGE TECHNOLOGY*.

Processing. A computer's processor (processing unit) consists of a control unit, which directs the operation of the system, and an arithmetic and logic unit (ALU), which performs computational operations. The design of a processing unit involves selection of a register set, communication paths between these registers, and a means of directing and controlling how these operate. Normally, a processor is directed by a program, which consists of a series of instructions that are kept in main memory. Each instruction is a group of bits, usually one or more words in length, specifying an operation to be carried out by the processing unit. The basic cycle of a processing unit consists of the following steps: (1) fetch an instruction from main memory into an instruction register; (2) decode the instruction (determine what it in-

(a)

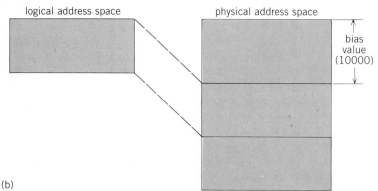

(b)

Fig. 3. Memory mapping by means of a bias register. (a) Translation formula. (b) Memory structure.

dicates should be done); (3) carry out the operation specified by the instruction; and (4) determine where the next instruction is. Normally, the next instruction is the one immediately following the current one.

Each instruction must indicate an operation to be performed and the data to which the operation should be applied. The assembly language reference manual for any computer system will contain a good illustration of its repertoire of instructions, including both their formats and their functions. There is considerable variation between instruction sets from different computers, and selection of a set of instructions is a major job of the computer architect.

Microprogramming. In the above discussion, it has been assumed that the process of decoding and executing instructions is carried out by logic circuitry. Although such is often the case, the complexity of instruction sets can lead to very large and cumbersome circuits for this purpose. To alleviate this problem, a technique known as microprogramming was developed. With microprogramming, each instruction is actually a macrocommand that is carried out by a microprogram, written in a microinstruction language. The microinstructions are very simple, directing data to flow between registers, memories, and arithmetic units. The microprogramming cycle is similar to that described above for macroinstructions, except that the microinstructions are stored in a special microstore memory that cannot be used for storing regular programs and data.

The advantage of microprogramming is that the architect can create very complicated macroinstructions by simply increasing the complexity of the microprograms. The logic circuitry need not be made more complex, and the only cost is in providing a larger microstore to hold the increasingly complex microprograms.

It should be noted that microprogramming has nothing to do with microprocessors. A microprocessor is a processor implemented through a single, highly integrated circuit. *See Microprocessor.*

Parallelism. There are limits to the speed of circuitry, determined by the state of technology and ultimately by the laws of physics. Efforts to design a faster computer often depend on parallelism. Because most techniques used to exploit parallelism belong to the area of implementation, they will be discussed only briefly.

Parallelism can occur in many ways. Parallel computer architectures have been categorized into various groups, depending on whether the instructions or the data are being handled in parallel. An example of the former is a system that fetches one instruction before it finishes with the previous one. An example of the latter is a system whose instructions specify groups of data to be processed at the same time (as in "add this vector to that one"). Perhaps the most widespread technique for parallelism, called pipelining, is similar to an assembly line in which the execution of an instruction is divided into stages, and there is a different instruction at each stage. *See Concurrent processing; Dataflow systems.*

RISC instruction sets. It has long been known that a typical computer program will make extensive use of some instructions and little or no use of others. In-depth studies of use patterns have led to the concept of a reduced instruction set computer (RISC), in which instruction sets are minimized and simplified so they can be executed faster and with less complex circuit-

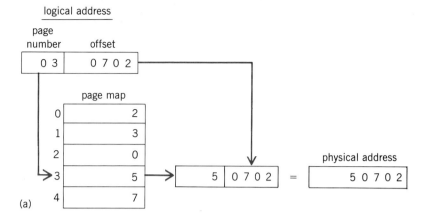

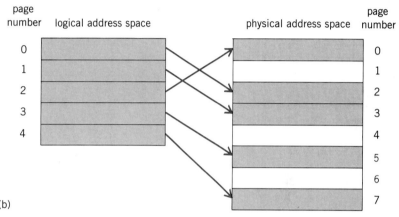

Fig. 4. Memory mapping by means of a page table. (*a*) Translation formula. (*b*) Memory structure.

ry. (The resulting instructions are still more complex than microinstructions, however.) There is always a trade-off in computer architecture: the price paid for the speed of a RISC design is a requirement for more registers and main memory because it takes more (small) steps to perform the equivalent of conventional (larger step) instructions. This is often a worthwhile trade because memory sizes and costs have improved significantly as compared with relatively modest improvements in circuit speed.

Peripherals and communication. A typical computer system includes a variety of peripheral devices such as printers, keyboards, and displays. These devices translate electronic signals into mechanical motion or light (or vice versa) so as to communicate with people. For example, as a person strikes a key (say, the letter K), a particular sequence of signals is transmitted to the computer. This sequence is interpreted as a string of bits, which the computer sees as the letter K. The details of how a peripheral device works are in the realm of mechanical and electrical engineering. The computer architect is more concerned with how the electrical signals are communicated to the computer. *See Electronic display.*

There are two common approaches for connecting peripherals and secondary storage devices to the rest of the computer: the channel and the bus. A channel is essentially a wire or group of wires between a peripheral device and a memory device (either main memory or a register). An input operation occurs when the memory receives data from the peripheral device, and an output operation occurs when the contents of the memory are transmitted to the device. A multiplexed channel allows several devices to be connected to the same wire.

A bus is a form of multiplexed channel that can be shared by a large number of devices. The overhead of sharing many devices means that the bus has lower peak performance than a channel; but for a system with many peripherals, the bus is more economical than a large number of channels.

A computer controls the flow of data across buses or channels by means of special instructions and other mechanisms. In addition to the wires used to move data, a channel or bus has one or more wires that control what happens when. The simplest scheme is known as program-controlled input/output (I/O). In this approach, the computer has an instruction such as "read from channel n" which causes the processor to send a control signal across a particular channel. That control signal is interpreted by the device as a sign to transmit a word or byte of data across the channel. The data will be deposited in some register, and subsequent instructions will process that data or store the data into main memory for later use.

Direct memory access (DMA) I/O is a technique by which the computer signals the device to transmit a block of data, and the data are transmitted directly to memory, without the processor needing to wait. This is also a form of parallelism, for the processor may carry out other functions while the data transfer is taking place. Memory mapped I/O is a similar technique, except that the computer places commands to the device in a designated portion of memory rather than executing a special instruction.

Interrupts are a form of signal by which a peripheral device notifies a processor that it has completed transmitting data. This is very helpful in a direct memory access scheme, for the processor cannot always predict how long it will take to transmit a block of data. Architects often design elaborate interrupt schemes to simplify the situation where several peripherals are active simultaneously. For example, there might be a separate interrupt signal for each peripheral, or a priority scheme so that, if two peripherals send interrupts at the same time, the more important one is transmitted to the processor first. SEE COMPUTER; DATA PROCESSING SYSTEMS; DIGITAL COMPUTER.

Dennis J. Frailey

Bibliography. P. J. Denning, Third generation computer systems, *ACM Computing Surveys*, 3(4):175–216, December 1971; K. Hwang and F. A. Briggs, *Computer Architecture and Parallel Processing*, 1984; G. J. Myers, *Advances in Computer Architecture*, 2d ed., 1982; D. Patterson, Reduced instruction set computers, *Commun. ACM*, 28(1):8–21, January 1985; Special issue on computer architecture, *Communications of the ACM*, vol. 21, no. 1, January 1978.

Computer vision

The use of digital computer techniques for extracting, characterizing, and interpreting information in visual images of a three-dimensional world; it is also known as machine vision. Visual sensing technology is receiving increased attention as a means to endow machines with the capability of exhibiting in a greater degree of "intelligence" in dealing with their environment. Thus, a robot or other machine that can "see" and "feel" should be easier to train in the performance of complex tasks while requiring less stringent control mechanisms than preprogrammed machines. A sensory, trainable system is also adaptable to a much larger variety of tasks, thus achieving a degree of universality that ultimately translates into lower production and maintenance costs.

The computer vision process may be divided into five principal areas: sensing, segmentation, description, recognition, and interpretation. These categories are suggested to a large extent by the way in which computer vision systems are generally implemented. It is not implied that human vision and reasoning can be neatly subdivided, nor that these processes are carried out independently of each other. For instance, it is logical to assume that recognition and interpretation are highly interrelated functions in a human. These relationships, however, are not yet understood to the point where they can be modeled analytically. Thus, the subdivision of functions discussed below may be viewed as a practical (albeit limited) approach for im-

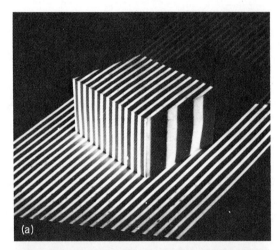

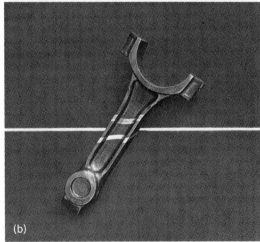

Fig. 1. Two examples of scene illumination by structured lighting. (a) Illumination of a block by a series of parallel strips of light (*from W. Myers, Industry begins to use visual pattern recognition, Computer, 13(5):21–31, 1980*). (b) Illumination of a workpiece by two strips of light which coincide at a background surface (*from F. Rocher and E. Keissling, Methods for analyzing three-dimensional scenes, Proceedings of the 1975 International Conference on Artificial Intelligence, American Association for Artificial Intelligence, pp. 669–673*).

plementing computer vision systems, given the level of understanding and the analytical tools presently available.

Visual sensing. Visual information is converted to electrical signals by the use of visual sensors. The most commonly used visual sensors are vidicon cameras and solid-state diode arrays. Vidicons are the usual vacuum-tube cameras used as television imaging devices. An input video signal is digitized and transferred to a computer in an image of a size ranging typically from 256 × 256 to 1024 × 1024 discrete image elements, depending on the resolution requirements of a given application. *See* Television Camera.

Solid-state devices are available as linear and area arrays. If the scene to be imaged is in continuous, uniform motion (as in belt conveyors), a linear array can be used to scan a line across the conveyor, and the motion of an object in the direction perpendicular to scan produces the desired two-dimensional image. In a one-dimensional array each element is read every N time intervals, where N is the number of image elements in the line, whereas in a two-dimensional array each element is read every N^2 time intervals; therefore, the two-dimensional array maintains a higher output data rate while allowing a long integration time for noise reduction. Linear arrays with resolution exceeding 2048 elements are available.

In order for a vision system to be able to interact with its environment, a geometric relationship between the real world and the images in the picture seen by the system must be established. This relationship is a transformation of measurements from a three-dimensional coordinate system to the image coordinate system, or vice versa. Essential to the derivation of the object-image relationship is a precise mathematical description for the camera (that is, a camera model). The use of two or more cameras and their mathematical models allows extraction of three-dimensional information, such as depth.

Illumination of a scene is an important factor affecting the complexity of vision algorithms. Arbitrary lighting of the environment often results in low-contrast images, specular reflections, shadows, and extraneous details. When control of the illumination is possible, the lighting system should illuminate the scene so that the complexity of the resulting image is minimized, while the information required for analysis is enhanced. For instance, the so-called structured lighting approach used in some industrial vision systems projects points, stripes, or grids onto the scene. The way in which these features are distorted by the presence of an object simplifies the computer interpretation of the scene. Two examples of this approach are shown in **Fig. 1**.

Segmentation. Segmentation is the process that breaks up a sensed scene into its constituent parts or objects. Hundreds of segmentation algorithms have been proposed since about 1970. This is still an active area of research because of its importance as the first processing step in any practical computer vision application. Although image segmentation has proved to be a difficult task in unconstrained situations such as automatic target detection, the problems encountered in industrial applications can be considerably simplified by special lighting techniques such as those discussed above.

Segmentation algorithms are generally based on one of two basic principles: discontinuity and similar-

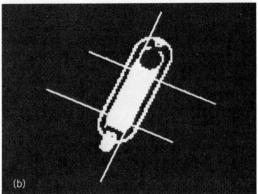

Fig. 2. Description of an object by extraction of its boundary and principal directional axes. (*a*) Bin of workpieces. (*b*) Description of one of the segmented objects. (*From J. Birk, R. Kelley, and H. Martins, An orienting robot for feeding workpieces stored in bins, IEEE Trans. SMC, SMC-11:151–160, 1981*)

ity. The principal approach in the first category is edge detection. The principal approaches in the second category are thresholding and region growing. Most edge-detection techniques for industrial applications are based on the use of spatial convolution masks in order to reduce processing time. The idea is to move a mask over the entire image area, one image element location at a time and, at each location, to compute a measure proportional to discontinuity (for example, the gradient) in the image area directly under the mask. Thresholding is by far the most widely used approach for segmentation in industrial applications of computer vision. There are two reasons. First, thresholding techniques (in their simpler forms) are fast, and in addition, they are quite straightforward to implement in hardware. Second, the lighting environment is usually a controllable factor in industrial application; this results in images that often readily lend themselves to a thresholding approach for object extraction. Region-growing techniques are applicable in situations where objects cannot be differentiated from each other or the background by thresholding or edge detection. Although region growing has been used extensively in scene analysis, it has not found wide applicability in industrial applications because this method is usually impractical from a computational or hardware implementation point of view, and many of the problems which would require region growing for segmentation can usually be handled by special lighting or other enhancement techniques. A considerable amount of work has dealt with techniques that attempt to incorporate contextual information in the segmentation process. This includes the use of relaxation, plan-guided analysis, and the use of semantic information.

Description. The description problem in computer vision is one of extracting features from an object for the purpose of recognition. Ideally, these features should be independent of object location and orientation and should contain enough discriminatory information to uniquely identify between objects. Descriptors for computer vision are based primarily on shape

and amplitude (for example, intensity) information. Shape descriptors attempt to capture invariant geometrical properties of an object. Approaches for shape analysis and description are generally either global-oriented (region-oriented) or boundary-oriented. Global techniques include principal-axes analysis, texture, two- and three-dimensional moment invariants, geometrical descriptors such as p^2/A (where p is the length of the perimeter and A is the area) and the extrema of a region, topological properties, and decomposition into primary convex subsets. Boundary-oriented techniques include Fourier descriptors, chain codes, graph representations, of which strings

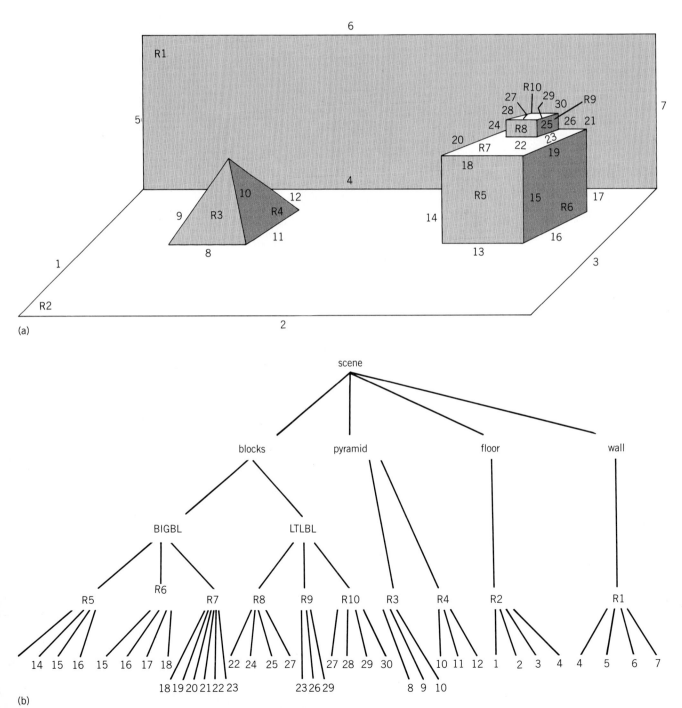

Fig. 3. Hierarchical representation of a scene. (a) Scene with blocks, pyramid, floor, and wall. (b) Hierarchical representation. (*After M. G. Thomason and R. C. Gonzalez, Database Representation in Hierarchical Scene Analysis, in L. Kanal and A. Rosenfeld, eds., Progress in Pattern Recognition, Elsevier-North Holland, 1982*)

and trees are special cases, and shape numbers. Boundary feature extraction is often preceded by linking procedures which fit straight-line segments or polynomials to the edge points resulting from segmentation. An example of object description by extraction of its boundary and principal directional axes is shown in **Fig. 2**.

Recognition. Recognition is basically a labeling process; that is, the function of recognition algorithms is to identify each segmented object in a scene and to assign a label (for example, wrench, seal, or bolt) to that object. Recognition approaches presently in use may be subdivided into two principal categories: decision-theoretic and structural. Decision-theoretic techniques are based on the use of decision (discriminant) functions. Given M object classes, ω_1, ω_2, . . . , ω_M, the basic problem in decision-theoretic pattern recognition is to identify M decision functions, $d_1(x)$, $d_2(x)$, . . . , $d_M(x)$, with the property that, for any pattern x^* from class ω_i, $d_i(x^*) > d_j(x^*)$, for $j = 1, 2, . . . , M, j \neq i$. The objective is to find M decision functions such that this condition holds for all classes with minimum error in classification. *SEE DECISION THEORY.*

Decision-theoretic methods deal with patterns on a quantitative basis and largely ignore structural interrelationships among pattern primitives. Structural methods of pattern recognition attempt to describe fundamental relationships among pattern primitives via discrete mathematical modes. Here, the most widely used method is syntactic pattern recognition in which concepts and results from formal language theory provide the basic mechanisms for handling structural descriptions. The existence of a recognizable and finitely describable structure is essential in the success of the syntactic approach. Basically, a formal grammar is developed to generate elements of a language that defines a pattern class, and an automaton (or equivalently, a parsing algorithm) is developed to recognize precisely that same language. Such a language may consist of strings of primitives and relational operators (for example, directed line segments along the boundary of a polygonal representation of a workpiece) or of higher-order data structures, such as trees, graphs, and webs.

One of the most significant recent extensions of syntactic techniques has been the inclusion of semantic evaluations simultaneously with syntactic analysis by means of attributed grammars. In this approach, a pattern primitive is defined by two components: a token or symbol from a finite alphabet, and an associated list of attributes consisting of logical, numerical, or vector values. The syntactic rules provide the basic structural description, while the semantic rules assign meaning to that description.

Interpretation and models. In this discussion, interpretation is viewed as the process which endows a vision system with a higher level of conception about its environment. Sensed information, tasks to be performed, and types of parts to be handled are all essential items in establishing the level of competence and adaptability of a vision system. Given the limited state of development in "truly intelligent" vision systems, careful definition of a constrained set of operating conditions is essential. This is usually accomplished via models.

The structure and complexity of a model depends on the stage of visual processing in which it is used. Computer vision techniques may be divided into three basic levels of processing: low-, medium-, and high-

level vision. Although this division is somewhat arbitrary, it does provide a convenient method for categorizing the various processes that are inherent components of a computer vision system.

Low-level vision techniques attempt to extract "primitive" information from a scene. Examples of the use of models in low-level vision procedures range from modeling the characteristics of incident and reflected light properties of a body, to the detection of edge segments in a scene by modeling an edge as an abrupt change in intensity amenable to detection by gradient operators. Medium-level vision refers to procedures which use the results from low-level vision to produce structures that somehow carry more meaning than the elements extracted by the low-level vision process. Medium-level vision processes include edge linking, segmentation, description, and recognition of individual objects. High-level computer vision may be viewed as the process that attempts to emulate cognition. At this level of processing, the present knowledge and understanding of a suitable model is considerably more vague and speculative. While models for low- and medium-level vision tend to be rather specific in nature, a model for high-level vision encompasses a considerably broader spectrum of processing functions, ranging from the actual formation of a digital scene through interpretation of interrelationships between objects in a scene. **Figure 3** illustrates modeling of a scene by decomposing it into successively simpler elements. The simplest element considered in this case is an edge. Thus, regions are composed of edges, objects are composed of regions, and the scene is composed of objects. *SEE ARTIFICIAL INTELLIGENCE; CHARACTER RECOGNITION; COMPUTER GRAPHICS; ROBOTICS.*

R. C. Gonzalez

Bibliography. H. G. Barrow and J. M. Tenebaum, Computational vision, *Proc. IEEE*, 69:572–595, 1981; G. G. Dodd and L. Rossol (eds.), *Computer Vision and Sensor-Based Robots*, 1979; K. S. Fu, R. C. Gonzalez, and C. S. G. Lee, *Introduction to Robotics*, 1984; R. C. Gonzalez and R. Safabakhsh, Computer vision techniques for industrial applications and robot control, *Computer*, 15(12):17–32, 1982.

Computerized tomography

An imaging technique which uses an array of detectors to collect information from a beam that has passed through an object (for example, a portion of the human body). The information collected is then used by a computer to reconstruct the internal structures, and the resulting image can be displayed—for example, on a television screen.

Technique. Wave phenomena can penetrate into regions where it is impossible or undesirable to introduce ordinary probes—regions as diverse as the ocean, the human body, and the deep interior of the Earth. Since wave attenuation and velocity are affected by material properties, such as temperature and density, information concerning these properties is accumulated by a wave packet as it travels along its path. Thus, the intensities and travel times of such pulses measured at a receiver represent such accumulations. Although a single such datum does not allow point-to-point variation to be inferred, when many rays crisscross a region the corresponding interrelated data contain constraints that allow an inversion to determine interior structures. Considerable computational power and mathematical sophistication are needed in dealing with indirect, incomplete, and imperfect measurements. The relevant branch of mathematics is called inverse theory. The term computerized tomography for this type of remote sensing arose in medicine, where x-ray attenuation data along many straight paths confined to a plane were used to obtain a map of density structure in a body section. Previous use of the term tomography referred to methods that use analog techniques rather than computation. *See* Inverse scattering theory.

Examining a person or object in the laboratory requires the use of a gantry, composed of an x-ray tube, an array of detectors opposite the tube, and a central aperture in which the person or object is placed. The rigid gantry maintains the proper alignment between the x-ray tube and the detectors. A conventional x-ray tube with a rotating anode and a small focal spot is used. X-rays are generated in short bursts, usually lasting 2–3 ms. The thickness of the section to be examined can vary from 1.5 to 13 mm and is determined by collimating the x-ray beam. Detectors pick up the "invisible image" of the internal structures revealed by the exiting x-ray beam and feed it into a computer.

Two types of detectors are in use in the laboratory: scintillation crystals (sodium iodide, calcium fluoride, and bismuth germanate) coupled with a photomultiplier tube; and an ionization chamber filled with xenon gas. X-rays emitted from different positions are needed to gather information. Early scanners used linear and rotational motions of both x-ray tubes and detectors. One early unit, for example, had a pencil-shaped x-ray beam and a single detector, and the time needed for a single scan was between 4.5 and 5 min. By adopting a fan-shaped beam with an array of detectors (as many as 30) and by using a greater rotatory increment (30° instead of 1°), the time required for a tomographic scan was reduced to 10–90 s. Newer machines, which use more detectors (up to 300) and a wider fan-shaped beam, can complete a scan in 2–10 s. *See* Ionization chamber; Scintillation counter.

The computer reconstructs the image from the information collected by the detectors in the following way. The cross section of the object to be reconstructed is divided into tiny blocks, called voxels, giving a square matrix. Matrices can be constructed having voxels of different sizes; for example, 80 × 80, 160 × 160, or 320 × 320 voxels. In the 320 × 320 matrix, the voxels are simply one-fourth of the volume of those of an 80 × 80 matrix, and therefore the resolution is greater. The computer assigns each voxel a number proportional to the degree that the voxel has attenuated the beam passing through it. Once the voxel has received a computer number, it is called a pixel (picture element). The linear attenuation coefficient (a measure of the quantity of radiation attenuated by each centimeter of absorber) is used to quantify the attenuation of the beam. Each voxel is "examined" by the beam incident from several positions, and this generates many "positional" equations. The computer must solve many thousands of equations to determine the linear attenuation coefficient for a single voxel. In order to obtain enough information to calculate one image, the newer scanners can take as many as 90,000 readings (300 pulses and 300 detectors).

Applications. The ability of computerized tomography to explore the internal structure of objects has made it a valuable tool in medicine and geophysics.

Medical tomography. In medicine, computerized tomography, introduced by G. N. Hounsfield in 1972, represents a noninvasive way of seeing internal structures. In the brain, for example, computerized tomography can readily locate tumors and hemorrhages, thereby providing immediate information for evaluating neurological emergencies. Equally important is the fact that, in a great majority of cases, computerized tomography offers an alternative approach to the more problematic invasive procedures such as pneumoencephalography and angiography. Reductions in scanning time have made it possible to do whole-body scans, and many abdominal tumors, hemorrhages, and infections can now be detected without the need of exploratory surgery. Other medical imaging techniques that make use of computerized tomographic methods include magnetic resonance imaging, positron emission tomography, and single-photon emission tomography. *See* Medical imaging; Radiography.

Nicole-Fr. Bolender

Geophysical tomography. After the success of computerized tomography in medicine, its possibilities in other fields were quickly realized. In the earth, atmospheric, and ocean sciences it has supplemented, but by no means replaced, older methods of remote sensing. Seismic tomography is now an important tool for investigating the deep structure of the Earth, testing theories such as plate tectonics, and exploring for oil. Ocean acoustic tomography is applied to physical oceanography, climatology, and antisubmarine warfare. Atmospheric tomography finds applications to weather, climate, and the environment.

1. Seismic tomography. The data for tomographic analysis are the earthquake signal arrival times monitored by the global network of seismic stations. These signals are propagated as elastodynamic waves that can be classified as compressional or shear and as body or surface waves. Compressional waves such as acoustic waves consist of alternating compressions and rarefactions caused by the oscillations of material particles in the direction of wave propagation. Shear waves are of a different (slower) type in which the motion of material particles is at right angles to the wave motion. Surface waves are guided by the Earth's surface; they travel along great circle paths and penetrate the upper mantle, but not deeper. Body waves, on the other hand, are not confined to the surface and can penetrate deeply, even reaching the Earth's core. They can be of the compressional or shear type. There are also two kinds of surface waves: Rayleigh waves, composed of a vertical shear component and a compressional component, and Love waves, which are pure horizontal shear waves. *See* Seismology.

The diversity of wave phenomena allows various physical variables to be computed from wave velocities. Temperature, density, and crystalline structure of rock are the most important of these. Body-wave ray paths are not straight as in medical tomography but are curved by refraction in the vertical due to variation in the Earth's mechanical properties. Body and surface ray paths and travel times are calculated from arrival time data at many stations by means of a standard spherically symmetric reference model of the Earth. The travel time associated with thousands of crossing paths are treated as tomographic data and used to draw maps of wave velocity anomalies (deviations from the reference model). These maps reveal much about internal structure and dynamics of the Earth. Since velocity decreases as temperature in-

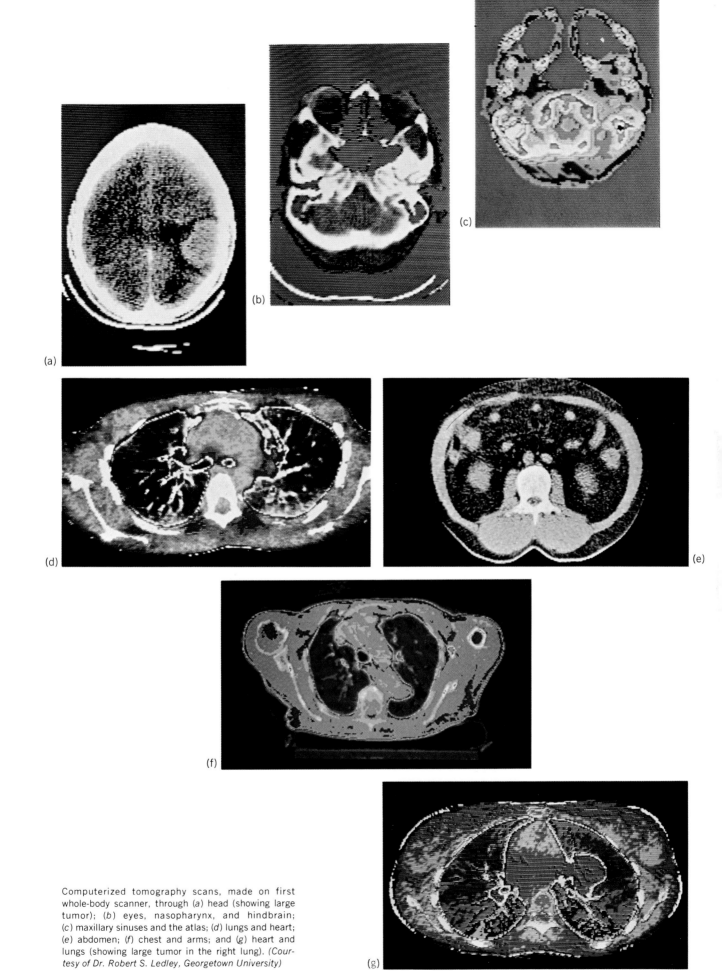

Computerized tomography scans, made on first whole-body scanner, through (a) head (showing large tumor); (b) eyes, nasopharynx, and hindbrain; (c) maxillary sinuses and the atlas; (d) lungs and heart; (e) abdomen; (f) chest and arms; and (g) heart and lungs (showing large tumor in the right lung). (Courtesy of Dr. Robert S. Ledley, Georgetown University)

creases, an anomalously slow velocity may point, for example, to upwelling hot material from the mantle toward the surface. An opposite anomaly may represent subduction of cool surface matter as one tectonic plate is thrust beneath another. Large anomalies near the core have revealed a previously unknown aspherical (bumpy) structure. *See Plate tectonics.*

Smaller-scale seismic tomography serves as a tool for oil exploration. In cross-borehole tomography a seismic source such as a small explosive or impact device creates a mechanical disturbance at a number of different depths in a shaft. A linear array of receivers in a second borehole measures signal arrival times. The propagation paths connecting each source position with each receiver form a tomographic network. Maps of propagation velocity in the region between boreholes are used to detect changes in rock strata. *See Seismic exploration for oil and gas.*

2. Ocean acoustic tomography. A tomographic array in the ocean consists of a number of acoustic sources and receivers that span an ocean region of interest. Each source transmits a pulse that is detected at each receiver. Because of refraction due to temperature variations and other (smaller) effects, the pulses follow curved paths, which usually reverse direction in the vertical at a number of upper and lower turning points. These paths are determined mainly by the temperature stratification of the ocean and tend to be restricted to a vertical plane because of strong vertical stratification. The elements of each source–receiver pair are connected by many acoustic paths (called multipaths). Since a system typically contains many source–receiver pairs, the ocean volume within the tomographic array is penetrated by hundreds of acoustic paths. Pulse travel time along these paths are the measured data. They are determined by the variation of sound speed and currents along the respective paths. To the first order, sound speed in the ocean is determined by temperature, and the effect of currents is negligible. Thus, the information contained in the data is related mainly to the temperature variation in the ocean, and the monitoring of this distribution has been the primary objective in the development of this technology. However, monitoring of currents is also possible by the use of reciprocal (two-way) transmissions. By subtracting reciprocal travel times, temperature effects are canceled out and the effect of currents becomes dominant.

The operation of a system begins with initialization of the array, referring to the fact that a knowledge of the distribution of ocean variables (mainly temperature) is required at the initial moment of operation in order to identify the acoustic paths that connect sources and receivers. Thereafter, as these paths change position, they can be tracked by the system. In early experiments the necessary initial information was supplied by ship surveys of vertical temperature profiles.

It is not possible in the ocean environment to achieve the extremely fine spatial resolution obtained in medical scanning; a prohibitively large number of sources and receivers would be required. However, future ocean monitoring systems will include satellite observations, providing global coverage of the ocean surface with excellent horizontal resolution but no significant vertical penetration. Acoustic tomography provides vertical penetration but poorer resolution. These two techniques can play complementary roles in ocean monitoring. *See Applications satellites; Oceanography; Underwater sound.*

3. Atmospheric tomography. Electromagnetic waves interact with the atmosphere in two ways useful for remote sensing. Parts of the atmosphere can reflect incident radiation directly to a detector or can absorb energy and reradiate it. Lidar, a form of radar that uses laser beams, is a suitable source for active tomographic systems because it confines incident energy to narrow beams and because optical frequencies interact strongly with the atmosphere. Alternatively, a system can be based on passive detection of emitted energy from the atmosphere by means of the radiometer, an instrument that measures the intensity of waves arriving from a particular direction. For active systems, measured data are the attenuation of lidar pulses along straight path segments from source to point of reflection to receiving telescope. For passive systems, a measured datum is the total intensity of radiation emitted toward the radiometer by all sources on its line of sight. *See Lidar.*

A simple example of a configuration that provides the network of crossing paths required for tomography consists of two radiometers in fan beam mode: Each instrument moves its line of sight up and down, forming a fan of receiving beams. The intersection of the two fans forms a network of paths for tomography; this configuration has been used to map the liquid water content of clouds. Other tomographic configurations have been obtained from instruments mounted on satellites, aircraft, surface vehicles, and the ground.

As a characteristic tomography, the data represents not the value of physical variables at a point but the integration of variables along a line. Point-to-point mapping is accomplished by inversion algorithms similar to those used in other fields. Although attenuation and emissivity are often of interest in themselves, they can also be related to other physical variables of interest. Thus, atmospheric applications also include mapping of visible airglow, temperature, and properties of aerosols. *See Airglow.*

Thomas J. Eisler; Ronald New

Bibliography. D. L. Anderson and A. M. Dziewonski, Seismic tomography, *Sci. Amer.*, 251:60–68, 1984; A. M. Dziewonski and J. H. Woodhouse, Global images of the Earth's interior, *Science*, 236:47–48, 1987; W. R. Hendee, *The Physical Principles of Computed Tomography*, 1983; W. Menke, *Geophysical Data Analysis: Discrete Inverse Theory*, 1984; W. H. Munk and P. F. Worcester, Ocean acoustic tomography, *Oceanography*, 1:8–10, 1988; F. Natterer, *The Mathematics of Computerized Tomography*, 1986.

Concentration scales

Numerical systems defining the quantitative relations of the components of mixtures. In solutions the concentration is expressed as the mass, volume, or number of moles of solute present in proportion to the amount of solvent or of total solution. Each scale of concentration has significant features of experimental simplicity or of theoretically significant relationship.

Percentage. The simplest scale to measure is percentage; hence it is often used for medicinal or household solutions. Weight percent is the number of parts of weight of solute per hundred parts of weight of solution (total). For example, a 10% saline solution contains 10 g of salt in 90 g of water, that is, 100 g total weight. Gaseous mixtures, being difficult to weigh, are often expressed as volume percent. Thus, air is said to contain 78% nitrogen by volume. Solutions of liquids in liquids (say, alcohol in water) may also be expressed in volume percent.

Molarity. To the chemist, the number of moles of solute is of more significance than the number of grams. The molarity (abbreviated M) is the number of moles of solute per liter of total solution. Thus, 12 M HCl means that the solution contains 12 formula weights (12 × 36.5), or 438 g, of hydrochloric acid per liter. Concentration is an intensive, rather than an extensive, property of solutions. Thus, 0.5 liter of the acid just mentioned is still 12 M, although it contains only 6 moles of solute. As used here, the mole is an amount of substance whose weight in grams is numerically the same as the molecular weight. Chemical engineers sometimes use a mole which is a pound-molecular weight.

Molality. Certain solution properties, for example, the lowering of the freezing point of water by addition of salt, require the use of a concentration scale which relates the number of moles of solute to the weight of solvent rather than to the volume of solution. This scale, called molality (abbreviated m), indicates the number of moles of solute per 1000 g of solvent. Thus 34.2 g of sucrose ($C_{12}H_{22}O_{11}$, mol wt 342), if dissolved in 200 g of water, has the concentration of 0.5 mole of sucrose per 1000 g of water, and hence is 0.5 m. For dilute aqueous solutions, the molality is essentially identical with the molarity. In solutions with densities other than unity, the two scales differ. Molality may be computed from molarity, as shown in the equation below, by first subtracting from the

$$m = M\left(\frac{1000}{1000d - M \times \text{mol wt}}\right)$$

weight of 1 liter (1000 × density d) the weight of solute, and then scaling the result by proportion to the number of moles per 1000 g, where the molecular weight is that of the solute.

Normality. Molarity does not represent reactive capacity for solutes which possess more than one active unit per molecule. Since sulfuric acid (H_2SO_4) molecules yield twice as many hydrogen ions (H^+) as do those of hydrochloric acid, a liter of 0.1 M sulfuric acid will neutralize twice as much base as will a liter of 0.1 M hydrochloric acid. When it is important to know the reactive capacities of reagents, as in volumetric analysis, the normality scale is used. Normality (abbreviated N) is found by multiplying molarity by the number of active units in the formula.

In metathesis (double decomposition) reactions, normality may be an ambiguous concept unless referred to a specific reaction. Phosphoric acid (H_3PO_4) may be reacted with sodium hydroxide (NaOH) to yield NaH_2PO_4, Na_2HPO_4, or Na_3PO_4; thus, phosphoric acid may have three normalities, depending on whether the reaction involves replacing one, two, or three hydrogen atoms. In oxidation-reduction reactions, there is a change in oxidation number, and normality must be calculated on the basis of this change. SEE OXIDATION-REDUCTION.

For reaction in a solution, the product of normality times the volume is the same for both reactants; for a solution of normality N_1, determined by titrating a volume V_1 with volume V_2 of a known solution of normality N_2, this equivalence can be expressed as $V_2N_2 = N_1V_1$. If the volume is expressed in milliliters, the products are in milliequivalents; if in liters, in equivalents.

Formality. The molarity scale is ambiguous when applied to ionic reactions. For example, 1 mole of Na_2HPO_4 in a liter of solution can yield 2 moles of sodium ion (Na^+), 1 mole of H^+, and 1 mole of phosphate ion (PO_4^{3-}), leaving zero mole of undissociated Na_2HPO_4. What the proper molarity is may not be clear. To obviate the confusion, it is possible to disregard moles and to use instead the number of formula weights of solute per liter of solution. This scale is designated the formality scale. In the example given, the solution would be 1 F Na_2HPO_4. This identifies the amount of total solute per liter and removes ambiguity in making up a solution, the reactive capability of the solution under various conditions is not considered.

Other scales. For very dilute solutions, for example, hard water, it is useful to express concentration in parts per million (ppm). Parts per billion are given in nanograms (10^{-9} g) per liter. With modern analytical techniques, quantities can be expressed in picogram (10^{-12} g) and femtogram (10^{-15} g) ranges. Particular applications often require special scales; for example, radon concentration in the atmosphere is given in units of picocuries per liter.

Mole percent (mole fraction). Many properties of solutions (for example, vapor pressure of one component) are dependent on the ratio of the number of moles of solute to the number of moles of solvent, rather than on the ratios of respective volumes or masses. The mole fraction (abbreviated N_A or X_A for component A) is the ratio of the number of moles of solute to the total number of moles of all components. Thus for 16 g of methanol (0.5 mole) dissolved in 18 g of water (1 mole), the mole fraction of methanol is 0.5/1.5, or 1/3; the mole percent is 33.3. For gases the mole percent is identical with the volume percent. SEE GRAM-EQUIVALENT WEIGHT; GRAM-MOLECULAR WEIGHT; SOLUTION; STOICHIOMETRY; TITRATION.

Allen L. Hanson

Bibliography. S. W. Benson, *Chemical Calculations*, 3d ed., 1971; J. Brady and J. Holum, *Fundamentals of Chemistry*, 1988; R. DeLorenzo, *Problem Solving in General Chemistry*, 1981.

Concrete

Any of several manufactured, stonelike materials composed of particles, called aggregates, that are selected and graded into specified sizes for construction purposes, usually with a substantial portion retained on a No. 4 (0.19-in. or 4.75-mm) sieve, and that are bonded together by one or more cementitious materials into a solid mass.

Composition. The term concrete, when used without a modifying adjective, ordinarily is intended to indicate the product formed from a mix of portland cement, sand, gravel or crushed stone, and water. There are, however, many different types of concrete. The names of some are distinguished by the types, sizes, and densities of aggregates—for example, wood-fiber, lightweight, normal-weight, or heavyweight concrete. The names of others may indicate the type of binder used—for example, blended-hydraulic cement, natural-cement, polymer, or bituminous (asphaltic) concrete.

Concretes are similar in composition to mortars,

Fig. 1. Cross section through hardened ordinary concrete shows the random distribution of coarse aggregates, such as gravel (largest particles), and fine aggregates, such as sand, in a matrix of portland cement.

which are used to bond unit masonry. Mortars, however, are normally made with sand as the sole aggregate, whereas concretes contain much larger aggregates and thus usually have greater strength. As a result, concretes have a much wider range of structural applications, including pavements, footings, pipes, unit masonry, floor slabs, beams, columns, walls, dams, and tanks. *SEE CONCRETE BEAM; CONCRETE COLUMN; CONCRETE SLAB; MASONRY; MORTAR.*

Characteristics. Design of a concrete mix (**Fig. 1**) specifies ingredients to achieve specific objectives, such as strength, durability, abrasion resistance, low volume change, and minimum cost. The ingredients are mixed together to ensure that coarse, or large-size, aggregates are uniformly dispersed, that fine aggregates fill the gaps between the larger ones, and that all aggregates are coated with the cementitious materials. Before the cementing action commences, the mix is plastic and can be rolled or molded in forms into desired shapes. Recommended practices for measuring, mixing, transporting, placing, and testing concretes are promulgated by such organizations as the American Concrete Institute (ACI) and the American Association of State Highway and Transportation Officials (AASHTO).

Classification. Concretes may be classified as flexible or rigid. These characteristics are determined mainly by the cementitious materials used to bond the aggregates. Flexible concretes tend to deform plastically under heavy loads or when heated. Rigid concretes are considerably stronger in compression than in tension and tend to be brittle. To overcome this deficiency, strong reinforcement may be incorporated in the concrete, or prestress may be applied to keep the concrete under compression.

Flexible concretes. Usually, bituminous, or asphaltic, concretes are used when a flexible concrete is desired. The main use of such concretes is for pavements. *SEE PAVEMENT.*

The aggregates generally used are sand, gravel or crushed stone, and mineral dust; the binder is asphalt cement, an asphalt specially refined for the purpose. A semisolid at normal temperatures, the asphalt cement may be heated until liquefied for bonding the aggregates. Ingredients usually are mixed mechanically in a pug mill, which has pairs of blades revolving in opposite directions. While the mix is still hot and plastic, it can be spread to a specified thickness and shaped with a paving machine and compacted with a roller or by tamping to a desired density. When the mix cools, it hardens sufficiently to withstand heavy loads.

Sulfur, rubber, or hydrated lime may be added to an asphalt-concrete mix to improve the performance of the product.

Rigid concretes. Ordinary rigid concretes are made with portland cement, sand and stone, or crushed gravel. The mixes incorporate water to hydrate the cement to bond the aggregates into a solid mass. These concretes meet the requirements of standard specifications developed by the American Society for Testing and Materials (ASTM). Substances called admixtures may be added to the mix to impart specific properties to it or to the hardened concrete. The ACI publishes a recommended practice for measuring, mixing, transporting, and placing concrete.

Other types of rigid concretes include nailable concretes; insulating concretes; heavyweight concretes; lightweight concretes; fiber-reinforced concretes, in which short steel or glass fibers are embedded for resistance to tensile stresses; polymer and pozzolan concretes, which exhibit improvement in several properties; and silica-fume concretes, which possess high strength. Air-entrained concrete formulations, in which tiny air bubbles have been incorporated, may be considered variations of ordinary concrete if in conformance with ASTM specifications.

Stress and reinforcement. Because ordinary concrete is much weaker in tension than in compression, it is usually reinforced or prestressed with a much stronger material, such as steel, to resist tension. Use of plain, or unreinforced, concrete is restricted to structures in which tensile stresses will be small, such as massive dams, heavy foundations, and unit-masonry walls. For reinforcement of other types of structures, steel bars (**Fig. 2**) or structural-steel shapes may be incorporated in the concrete. Prestress to offset tensile stresses may be applied at specific locations

Fig. 2. Large steel bucket, supported by a crane, transports concrete for a floor slab to formwork on which steel reinforcement bars for the slab have previously been positioned. (*From Engineering News-Record, pp. 40FC–40GC, June 16, 1988*)

by permanently installed compressing jacks, high-strength steel bars, or steel strands. Alternately, pre-stress may be distributed throughout a concrete component by embedded pretensioned steel elements. Another option is use of a cement that tends to expand concrete while enclosures prevent that action, thus imposing compression on the concrete. *See Pre-stressed concrete; Reinforced concrete*.

Cementitious materials. By definition, ordinary concrete is made with portland cement, which usually is specified to conform with the ASTM Standard Specification for Portland Cement. This specification details requirements for general-purpose cements, cements modified to achieve low heat of hydration and to produce a concrete resistant to sulfate attack, high-early-strength cement, and air-entraining cement. *See Air-entraining portland cement; Cement; Portland cement*.

Several other types of cement are sometimes used instead of portland cement for specific applications—for example, hydraulic cements, which can set and harden underwater. Included in this category, in addition to portland cement, are aluminous, natural, and white portland cements, and blends, such as portland blast-furnace slag, portland-pozzolan, and slag cements.

Other cementitious materials, such as polymers and silica fume, may be used as replacements for cement. Polymers are plastics with long-chain molecules; concretes made with them have many qualities much superior to those of ordinary concrete. Silica fume, also known as microsilica, is a waste product of electric-arc furnaces. The silica reacts with lime in concrete to form a cementitious material. A fume particle has a diameter only 1% of that of a cement particle. *See Polymer*.

Aggregates for ordinary concrete. Aggregates should be inert, dimensionally stable particles, preferably hard, tough, and round or cubical. They should be free of clay, silt, organic matter, and salts.

Coarse aggregates are retained on a No. 4 sieve. Fine aggregates are retained on a No. 200 (75-micrometer) sieve. Sand generally is used as the fine

aggregate for ordinary concrete. The coarse aggregate generally is gravel or crushed stone. A lighter-weight alternative, however, is blast-furnace slag, which sometimes is used when it is economically available. Aggregates that cause a deleterious alkali reaction with the cement are avoided, if feasible.

Grading and maximum size of aggregate influence selection of the relative proportions of a mix. Particle-size distribution is determined by separation of the aggregates with a series of standard sieves. Coarse aggregate usually is graded up to the largest size practical for a project; the normal upper limit is a maximum dimension of 6 in. (150 mm).

Mix design and concrete properties. To achieve the normally desired properties of ordinary concrete, mix design usually specifies the required ratios, by weight, of water to cement, of cement to aggregates, and of fine to coarse aggregates. Also, the design may specify the maximum size of coarse aggregate and the type or brand of cement, as well as admixtures to be incorporated to attain desired results at minimum cost.

Water. Mixing water for concrete should be clear and visibly clean. If there is any doubt about an available water source, information concerning its potential effects on concrete should be obtained from tests in which mortars made with that water and with water of known acceptable quality are compared.

Density. Concrete mixes made with conventional aggregates usually weigh about 150 lb/ft^3 (2400 kg/m^3). This value generally is assumed also for the density of hardened concrete and is used in the design of formwork and structural members. Values differ for lighter- and heavier-weight concretes.

Strength. The prime objective of a mix design is to obtain a workable concrete with the specified compressive strength, measured in pounds per square inch or megapascals. Many properties of hardened concrete are significantly influenced by this value. Because the strength of concrete increases with age, compressive strength at 28 days is conventionally specified for design purposes.

The principal factor determining strength of concrete is the water-cement ratio, expressed either by relative weights of water and cement or by gallons of water per sack of cement. [A sack contains 1 ft^3 (0.03 m^3) of cement and weighs 94 lb (43 kg).] The smaller the water-cement ratio, the higher the strength.

It is noteworthy that the larger the cement content, the greater is the strength, whereas the more water in the mix, the lower the strength. Excessive water, in addition, may leave voids and cause undesirable volume changes and cracks in the concrete.

Usually, more water is incorporated in a mix than is necessary for complete hydration of the cement and development of full strength. The role of the additional water is to make the mix easier to place in the forms and around the material used for reinforcement and to ensure good workability. To reduce water requirements so as to enhance strength and yet provide acceptable workability, water-reducing admixtures may be added to the mix. As an alternative, excess water may be sucked from concrete with vacuum pads after it has been placed in the forms; the process also reduces voids in the concrete. The resulting product, often referred to as vacuum concrete, has much greater strength than would have been attained by the initially produced concrete.

The short-term strength of ordinary concrete may be increased by using type III (high-early-strength) portland cement or accelerating admixtures, such as calcium chloride, or by increasing curing temperatures, but long-term strengths may not be affected.

Consistency. This is an indication of workability, or ability to flow without segregation of components in the initial plastic state of a mix. Workability is of importance in placing and shaping the mix.

Consistency of a mix usually is measured by slump in a test with a standard, open-ended, truncated cone. The cone is filled with the mix, set on a flat surface with the small end up, and then removed, in accordance with a standard procedure. Slump is the sag of the mix after removal of the cone.

Volume change. The dimensional change of ordinary concrete after setting occurs is of major concern in concrete structures. As concrete dries, it shrinks. Allowance must be made for this in constructing beams and columns of tall buildings and large floor slabs. Expansion of concrete also should be provided for in construction. The thermal coefficient of expansion of concrete averages about 5.5×10^{-6} in./in. · °F $(10^{-5}$ mm/mm · °C). *See Thermal expansion*.

Creep. Under a constant long-time load, deformations known as creep, or plastic flow, occur, which increases gradually with time. *See Creep (materials)*.

Modulus of elasticity. For concrete under a gradually increasing load, the stress–strain relationship plots as a curve. Nevertheless, the portion of the curve below about 40% of ultimate load is sufficiently straight that it can be approximated by a secant. The slope of the secant, accordingly, may be taken as the modulus of elasticity; that is, stress may be taken as the product of the secant slope and the strain. The modulus, however, like strength, increases with the age of the concrete. *See Elasticity; Stress and strain*.

Admixtures. These are substances other than portland cement, aggregates, and water that are added to a concrete mix to modify its properties.

Types. The ASTM specifies several types of admixtures as described below.

1. Water reducers (type A) decrease the quantity of mixing water that otherwise would be required to produce concrete with a specific strength and consistency. Hence, for a given cement content and consistency, a stronger concrete results. Water reducers, for example, can increase slump of ordinary concrete from 2 to 4 in. (50 to 100 mm) without addition of water and thus produce a stronger concrete.

2. Retarders (type B) slow the setting of concrete; such admixtures are useful for working with concrete in hot weather.

3. Accelerators (type C) hasten the setting and promote early development of strength in the concrete; with these admixtures, 28-day strengths of concrete can develop in about 7 days.

4. Water-reducing and retarding agents (type D) both decrease water requirements and delay setting of the concrete.

5. Water-reducing and accelerating agents (type E) both decrease water requirements and hasten setting of concrete.

6. Water-reducing, high-range agents (type-F) reduce by 12% or more the quantity of mixing water needed to produce a concrete of specific consistency.

7. Water-reducing, high-range, and retarding agents (type G) act much like type F but in addition retard setting of concrete.

Special admixtures. Several admixtures have been developed for specific applications, including air-entraining compositions, waterproofing compositions, pozzolans, superplasticizers, and pigments.

1. Air-entraining admixtures. These admixtures are an alternative to air-entraining cement. Their use permits better control of air content with changes in sand, temperature, or job requirements.

2. Waterproofing admixtures. Admixtures such as stearates and oils are sometimes employed to reduce permeability of concrete. Other alternatives, however, are generally preferable, for example, surface coatings or high-strength concrete.

3. Pozzolans. These are sometimes used as admixtures to prevent alkali–aggregate reactions or to decrease cement requirements.

4. Superplasticizers. These are water-reducing admixtures, such as sulfonated melamine-formaldehyde condensates and sulfonated napththalene-formaldehyde condensates or other polymers. They provide considerable increases in concrete strength for a specific consistency, and they are also useful for producing fluid mixes for long-distance pumping from a mixer to form work. For example, an initial slump of 3–4 in. (75–100 mm) for an ordinary-concrete mix may be increased to 7–8 in. (175–200 mm) by addition of a superplasticizer without addition of water and with no decrease in strength.

5. Pigments. These materials may be added to a mix to produce colored concrete. They usually are used with white portland cement to attain the full coloring value of the pigment. *See Pigment (material)*.

Lightweight concretes. Appreciable savings in weight, with associated reductions in costs of structural supports, may be achieved with concretes that are lighter in weight than ordinary concrete. Also, they may offer better thermal and sound insulation, improved fire resistance and nailability, and may be used as lightweight fill in construction. These concretes are usually made with lightweight aggregates, but they also may be produced by expanding or foaming concrete mixes, usually by incorporation of an admixture.

Heavyweight concrete. This type of concrete is used for shielding of nuclear reactors. For this purpose, heavyweight aggregates such as barite, limonite, magnetite, steel punchings, sheared steel bars, and steel shot (fine aggregate) are used. With grading and mix proportions similar to those for ordinary concrete, heavyweight concrete weighing up to about 385 lb/ft^3 (6170 kg/m^3) can be produced.

Polymer concretes. Polymers are used in several ways to improve concrete properties: they are impregnated in hardened concrete, are incorporated in a mix, or replace portland cement.

Impregnation is sometimes used for concrete road surfaces. It can more than double compressive strength and elastic modulus, decrease creep, and improve resistance to freezing and thawing.

Monomers and polymers added as admixtures are used for restoring and resurfacing deteriorated roads. The concrete hardens more rapidly than ordinary concrete, enabling faster return of roads to service.

Polymer concretes in which a polymer replaces portland cement possess strength and other properties similar to those of impregnated concrete. After curing for a relatively short time, for example, overnight at room temperatures, polymer concretes are ready for use; in comparison, ordinary concrete may have to

Fig. 3. Building under construction is enclosed by large precast-concrete wall panels installed with a crane. (*From Engineering News-Record, p. 23, June 9, 1988*)

cure for a week or more before exposure to service loads.

High-tensile-strength concrete. This can be made with an organic admixture that forms a slurry that fills the pores, or voids, in concrete. As a result, the concrete develops about 20 times the tensile strength of ordinary concrete.

Silica-fume concretes. Formulations incorporating silica fume, or microsilica, as a partial replacement for portland cement exhibit increased compressive strength. Without other admixtures, strength increases of about 25% may result. With suitable proportioning of concrete mixes and addition of admixtures, strengths in the range of 20,000 lb/in.^{2} (140 MPa) can be produced.

Roller-compacted concrete. This type of concrete is formulated with low contents of portland cement and water and is compacted in layers in final position, with vibratory rollers, to maximum density.

Whereas ordinary concrete is made with five to six bags of cement per cubic yard of concrete, only three to five bags are used for roller-compacted concrete. Water content is so low [only 18–24 gallons/yd^3 (90–120 kg/m^3)] that the mix looks like damp gravel. The mix can be transported by dump trucks or loaders and spread with bulldozers or graders. Roller-compacted concrete is used for pavements and dams. For such applications, it is relatively low in cost, can employ a broad range of aggregate gradations, can facilitate a high rate of concrete production, and is strong and durable. For pavements, roller-compacted concrete can be sized and shaped with conventional paving machines. Construction of gravity dams using roller-compacted concrete employs methods and equipment similar to those used for earth dams. This type of construction avoids the undesirable conditions created by the high heat of hydration of mass concrete. *See Dam.*

Casting. There are various methods employed for casting ordinary concrete. For very small projects, sacks of prepared mixes may be purchased and mixed on the site with water, usually in a drum-type, portable, mechanical mixer. For large projects, mix ingredients are weighed separately and deposited in a stationary batch mixer, a truck mixer, or a continuous mixer. Concrete mixed or agitated in a truck is called ready-mixed concrete. Paving mixers are self-propelled, track-mounted machines that move at about 1 mi/h (0.45 m/s) while mixing.

At the site, concrete may be conveyed to the forms in wheelbarrows, carts, trucks, chutes, conveyor belts, buckets, or pipelines. Where chutes or belts are used, the flow of material is maintained as continuous as possible. On large projects, concrete may be delivered to the formwork in dump buckets by cableways or by cranes. In many instances, direct pumping through piplines is used to convey concrete from the mixer or hopper to the point of placement.

At times, concrete must be placed under water. The conveyance used, called a tremie, consists of a large metal tube with a hopper at the top end and a valve arrangement at the bottom, submerged end of the tube. While the tremie method is not as satisfactory as working in the dry, good results can be achieved.

In general, concrete is placed and consolidated in forms by hand tamping or puddling around reinforcing steel or by spading at or near vertical surfaces. Another technique, vibration or mechanical puddling, is the most satisfactory one for achieving proper consolidation.

In the Gunite process, consolidation is achieved by blowing a mixture of cement, sand, and water through a nozzle onto a formed surface or a prepared concrete surface. Depth is built up in a series of passes over the surface. This technique is particularly useful for repairing portions of existing structures, lining reservoirs, and encasing steel for fireproofing.

In the Prepakt process, consolidation is accomplished by packing the form with the aggregate and then filling the voids with grout pumped into the mass under pressure. The grout usually consists of cement, water, a workability admixture, and a retarder to ensure that the grout remains fluid until filling is completed.

An alternative to delivering plastic concrete mixes to a site is precasting the concrete components at a convenient location, for rapid, controlled fabrication, and transporting the hardened components to the site. For economy and fast construction, precast concrete often is used for wall panels (**Fig. 3**), floor and roof slabs, and structural members. *See Buildings; Floor construction; Roof construction.*

Finishing. Finishes for exposed concrete surfaces are obtained in a number of ways. Surfaces cast against forms can be given textures by using patterned form liners or by treating the surface after forms are removed, for instance, by brushing, scrubbing, floating, rubbing, or plastering. After the surface is thoroughly hardened, other textures can be achieved by grinding, chipping, bush-hammering, or sandblasting.

Unformed surfaces, such as the top of pavement slabs or floor slabs, may be either broomed or smoothed with a trowel. Brooming or dragging burlap over the surface produces scoring, which reduces skidding when the pavement is wet.

Curing. Adequate curing is essential to bring the concrete to required strength and quality. The aim of curing is to promote the hydration of the cementing material. This is accomplished by preventing moisture loss and, when necessary, by controlling temperature.

Moisture is a necessary ingredient in the curing process, since hydration is a chemical reaction between the water and the cementing material. Unformed surfaces are protected against moisture loss immediately after final finishing by means of wet burlap, soaked cotton mats, wet earth or sand, sprayed-on sealing compounds, waterproof paper, or waterproof plastic sheets. Formed surfaces, particularly vertical surfaces, may be protected against moisture loss by leaving the forms on as long as possible, covering with wet canvas or burlap, spraying a small stream of water over the surface, or applying sprayed-on sealing compounds.

The length of the curing period depends upon the properties desired and upon atmospheric conditions, such as temperature, humidity, and wind velocity, during this period. Short curing periods are used in fabricating concrete products such as block or precast structural elements. Curing time is shortened by the use of elevated temperatures.

Frederick S. Merritt

Bibliography. American Concrete Institute, *ACI Manual of Concrete Practice,* annually; F. Kong and R. H. Evans, *Handbook of Structural Concrete,* 1983; F. R. McMillan and L. H. Tuthill, *Concrete Primer,* 1973; F. S. Merritt, *Building Design and Construction Handbook,* 4th ed., 1982; F. S. Merritt, *Standard Handbook for Civil Engineers,* 3d ed., 1983; J. C. Ropke, *Concrete Problems: Causes and Cures,* 1982.

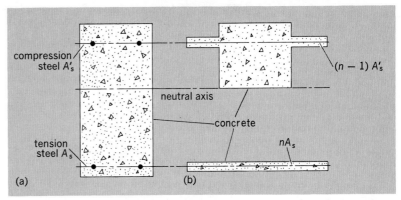

Fig. 1. Sketch of concrete beam. (*a*) Actual beam section. (*b*) Same section transformed. The use of the term $(n - 1)A'_s$ prevents compression steel area from being included twice when computing moment of inertia of section; $(n - 1)A'_s = nA'_s - A'_s$, because computations require the equivalent steel area to be added to the original concrete area above the neutral axis.

Concrete beam

A structural member of reinforced concrete placed horizontally to carry loads over openings.

Because both bending and shear in such beams induce tensile stresses, steel reinforcing tremendously increases beam strength. Usually, beams are designed under the assumption that tensile stresses have cracked the concrete and the steel reinforcing is carrying all the tension. SEE STRESS AND STRAIN.

Two design theories are used, elastic design and ultimate-load design.

Elastic design. The following assumptions are made for elastic design:

1. Plane sections remain plane after bending and are perpendicular to the longitudinal fibers.
2. The stress-strain curve is a straight line.
3. The ratio n of the modulus of elasticity of steel E_s to that of concrete E_c is a constant $n = E_s/E_c$.
4. The concrete does not carry tensile stress.

Transformed section. One approach to elastic design of reinforced concrete beams is to convert the steel to concrete. Because the steel and concrete are assumed to be firmly bonded and thus strained the same amount, the stress in the steel is n times the concrete stress. Hence the steel area may be replaced by an equivalent concrete area which is n times as large (**Fig. 1**).

If the equivalent area is placed at the same level as the steel and the moment of inertia I computed for the transformed section, the bending stresses can be computed from the simple flexural formula $f = Mc/I$, where M is the bending moment and c the distance from the neutral axis to the level at which stresses are to be computed.

Rectangular beams. The following formulas can be derived from the basic assumptions of elastic theory. Equation (1) locates the neutral axis, given steel and

$$\frac{nf_c}{f_s} = \frac{k}{1 - k} \qquad (1)$$

concrete extreme stresses, where n is the ratio of the modulus of elasticity of steel to that of concrete, f_c is the stress in the extreme fiber of the concrete, f_s is the stress in the steel, and k is the ratio of the distance between the top of the beam and the neutral axis to

the distance between the top of the beam and the steel (**Fig. 2**).

The design equation for equal moment resistance of concrete and steel (balanced design) is Eq. (2). Equa-

$$k = \frac{1}{1 + f_s/nf_c} \qquad (2)$$

tions (3)–(6) are review equations.

$$k = \sqrt{2np - (np)^2} - np \qquad (3)$$

$$j = \frac{1 - k}{3} \qquad (4)$$

$$M_c = \tfrac{1}{2}f_c kjbd^2 \qquad (5)$$

$$M_s = f_s A_s jd \qquad (6)$$

Here p is the ratio of effective area of tension reinforcement to effective area of concrete in beams, j is the ratio of lever arm of resisting couple to depth d, M_c is the moment resistance of the concrete, b is the width of the rectangular beam or width of flange of the T beam, d is the depth from the compressive surface of beam or slab to center of longitudinal tension reinforcement, M_s is the moment resistance of the steel, and A_s is the steel area.

When M_s is less than M_c, the capacity of the steel determines the maximum moment that the beam will carry. The beam is called underreinforced. If the beam is loaded to failure, the steel rather than the

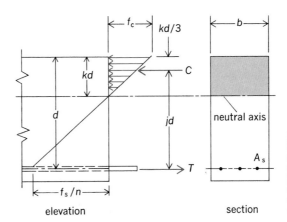

Fig. 2. Stress distribution in a rectangular beam designed according to elastic theory. *C* is total compressive force in concrete and *T* is total tensile stress in longitudinal reinforcement.

concrete determines the maximum load that is sustained.

When M_c is less than M_s, the concrete determines the maximum moment the beam will carry. The beam is said to be overreinforced. Usually, overreinforced beams are avoided because they are not considered economical and failure may occur without warning.

To design a rectangular beam by the elastic theory: (1) Select allowable unit stresses and determine k from the formula for balanced design; (2) compute j; (3) assuming the resisting moment of the concrete equal to the bending moment on the section, calculate bd^2 and select values for b and d; and (4) determine the steel area by equating the resisting moment of the steel to the bending moment on the section. This equation indicates that the amount of steel required can be reduced by increasing the depth. If the depth is fixed and an overreinforced beam results, it may be economical to use compression steel.

Compression-reinforced rectangular beams. Reinforcement may be added in the compression zone of concrete beams when the resisting moment of the concrete is less than the bending moment on the section or to avoid an overreinforced design. The compression steel is assumed to act as in plastic design; the Building Code Requirements of the American Concrete Institute allow the steel to take twice the stress given by elastic-theory formulas, provided the allowable tensile stress of the steel is not exceeded. Formulas for design of beams with both tension and compression reinforcement can be derived from the elastic theory in the same manner as for beams with tension reinforcement only, but they are too complicated for ordinary use. Equations (7)–(12) are either exact or a close approximation.

$$k = \frac{1}{1 + f_s/nf_c} \tag{7}$$

$$f'_s = nf_c\frac{kd - d'}{d - kd} \tag{8}$$

$$A_s = \frac{M}{f_s jd} \tag{9}$$

$$M_c = \tfrac{1}{2}f_c bkd\left(d - \frac{kd}{3}\right) \tag{10}$$

$$M'_s = M - M_c \tag{11}$$

$$A'_s = \frac{M'_s}{2f'_s(d - d')} \tag{12}$$

k = ratio of distance between top of beam and neutral axis to the distance between top of beam and tension steel

f_s = allowable steel stress

f_c = allowable concrete stress

n = ratio of modulus of elasticity of steel to that of concrete

f'_s = stress in compression steel computed from elastic theory formula

d' = distance from top of beam to compression steel

A_s = area of tension steel

M = bending moment on the section

jd = moment arm of the tensile reinforcement (assumed)

M_c = resisting moment of the concrete

b = width of beam

d = distance from top of beam to tension steel

M'_s = resisting moment of the compression steel

A'_s = area of compression steel

Concrete T beams. A slab cast integrally with a rectangular concrete beam usually is assumed to assist the beam in carrying loads. In regions of positive bending moment, the two act together as a T beam. In regions of negative moment, the beam is designed as a rectangular section because the slab is in tension and is assumed not to be able to resist such stresses, which must be taken by the reinforcing. SEE CONCRETE; CONCRETE SLAB.

If the neutral axis of a T beam is within the slab, it is designed as a rectangular beam, with width b the same as that of the flange, to resist bending moments. For shear, however, only the width of stem b' can be assumed to be effective.

If the neutral axis falls within the stem, the section can be designed as a T beam. However, the compression in the stem is negligible and can be ignored to simplify computations.

The American Concrete Institute's code recommends the following limits for the part of the slab that can be considered effective as the flange: (1) b shall be less than one-fourth the span length, (2) the overhanging width of flange shall not exceed eight times the slab thickness, and (3) the overhanging width shall not exceed one-half the clear distance between beams.

If t is the flange thickness and the remaining symbols are the same as for elastic-theory design of rectangular beams, Eqs. (13) through (20) can be used

$$k = \frac{1}{1 + f_s/nf_c} \tag{13}$$

$$\frac{f_s}{f_c} = \frac{bt(2kd - t)}{2A_s kd} \tag{14}$$

$$kd = \frac{2ndA_s + bt^2}{2nA_s + 2bt} \tag{15}$$

$$z = \frac{t(3kd - 2t)}{3(2kd - t)} \tag{16}$$

$$jd = d - z \tag{17}$$

$$M_s = A_s f_s jd \tag{18}$$

$$M_c = \frac{f_c btjd}{2kd}(2kd - t) \tag{19}$$

$$f_c = \frac{f_s}{n}\left(\frac{kd}{d - kd}\right) \tag{20}$$

for T beams when stem compression is neglected.

The shear unit stress in T beams is computed from Eq. (21), where V is the shear on the section.

$$v = \frac{V}{b'd} \tag{21}$$

SEE ELASTICITY; HOOKE'S LAW; YOUNG'S MODULUS.

Ultimate-load design. Other names for ultimate-load design are ultimate-strength design, limit-load design, and the plastic design. Design assumptions differ from those of the elastic theory principally in that the stress-strain curve is not a straight line. Instead of being designed to carry allowable unit stresses, as in the elastic theory, beams are proportioned to carry at ultimate capacity the design load multiplied by a safety factor.

Among the reasons for using ultimate-load design are the following:

1. The elastic theory is not corroborated with sufficiently great accuracy by beam tests.

2. It is logical to use different safety factors for live and dead loads. Different factors can easily be used with ultimate-load design but not with elastic design.

3. Column design is based on a modified ultimate-load theory. To avoid inconsistency, structural members subjected to both bending and compressive stress should also be designed by ultimate-load theory.

4. The ultimate strength of beams carrying both bending and axial compression, as determined in tests, conforms closely with ultimate-load theory.

Rectangular beams. The compressive stress distribution may be assumed to be a rectangle, parabola, trapezoid, or any other shape that conforms to test data. Maximum concrete stress is assumed to be $0.85f'_c$, where f'_c is the compressive strength of a standard-test concrete cylinder at 28 days (**Fig. 3**).

When the moment resistance of the steel is less than that of the concrete, the bending moment that a rectangular beam with only tension reinforcement can sustain under ultimate load is Eq. (22).

$$M_u = \phi[A_s f_y(d - a/2)]$$
$$= \phi[bd^2 f'_c q(1 - 0.59q)] \quad (22)$$

A_s = area of tension reinforcement
f_y = yield point stress of steel
f'_c = compressive strength of standard-test concrete cylinder at 28 days
ϕ = reduction factor = 0.90 for this type of beam in flexure
b = width of beam
d = distance from top of beam to centroid of the steel
a = depth of rectangular stress block
q = $A_s f_y/bdf'_c$

Reinforcement ratio $p = A_s/bd$ should not exceed $0.75p_b$, where p_b is given by Eq. (23), and $k_1 = 0.85$

$$p_b = \frac{0.85k_1 f'_c}{f_y}\left(\frac{87,000}{87,000 + f_y}\right) \quad (23)$$

for values of f'_c up to 4000 lb/in.², 0.80 for 5000 lb/in.², and 0.75 for 6000 lb/in.².

Compression-reinforced rectangular beams. Based on a nonlinear stress-strain relation, the bending moment that a rectangular beam with both compression and tension reinforcement can sustain under ultimate loads is shown in Eq. (24).

$$M_u = \phi[(A_s - A'_s)f_y(d - a/2) + A'_s f_y(d - d')] \quad (24)$$

A_s = area of tension steel
A'_s = area of compression steel
f_y = yield point stress of the steel

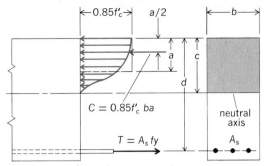

Fig. 3. Stress distribution in a rectangular beam under ultimate load; _T_ and _C_ are as in Fig. 2.

d = distance from top of beam to centroid of tension steel
d' = distance from top of beam to centroid of compression steel
a = $(A_s - A'_s)f_y/0.85f'_c b$
b = width of beam
f'_c = compressive strength of standard-test concrete cylinder at 28 days

Equation (24) holds only if Eq. (25) is true, where

$$(p - p') \geq 0.85k_1 \frac{f'_c d'}{f_y d}\left(\frac{87,000}{87,000 - f_y}\right) \leq 0.75p_b \quad (25)$$

$p = A_s/bd$; $p' = A'_s/bd$; and p_b and k_1 are the same as for beams with tension steel only.

Concrete T beams. Two cases should be considered, one with a relatively thick slab and one with a thin slab. If the flange thickness exceeds $1.18qd/k_1$, where $q = pf_y/f'_c$, the bending moment under ultimate load may be taken to be the same as that for a rectangular beam with tension reinforcement only. The value of p used in computing q should be A_s/bd, with b the width of the flange. For thinner flanges, use Eq. (26)

$$M_u = \phi[(A_s - A_{sf})f_y(d - a/2) + A_{sf}f_y(d - 0.5t)] \quad (26)$$

A_{sf} = $0.85f'_c(b - b')t/f_y$
f_y = yield point stress of the steel
t = flange thickness
b = width of flange
b' = width of stem
d = distance from top of beam to centroid of tension steel
a = $(A_s - A_{sf})f_y/0.85f'_c b'$

to compute bending moment.

Equation (26) holds only if $p_w - p_f$ does not exceed $0.75p_b$, where $p_w = A_s/bd$, $p_f = A_{sf}/bd$, and p_b is the same as for beams with tension reinforcement only, as in Eq. (23).

Shear and bond. Maximum unit shear stress v acting on a section of a beam subjected to total shear V is given by Eq. (27), where b is the beam width, and

$$v = \frac{V}{bd} \quad (27)$$

d is depth from top of beam to the tensile steel. The bond stress can be computed from Eq. (28), where

$$u = \frac{V}{\Sigma_o jd} \qquad (28)$$

Σ_o is the sum of the bar perimeters. This formula applies only to tension steel. Bond stresses will be at a maximum where shear is a maximum and the steel is in the tension side of the concrete.

To develop a given bar stress through bond, a bar should be embedded a length L at least equal to Eq. (29), where a is the side of a square bar or the di-

$$L = \frac{f_s a}{4u} \qquad (29)$$

ameter of a round bar. This length of embedment is called anchorage. Usually, bars are extended 10 diameters past the section where they are no longer required for bending stress.

Shear in itself is not as important in the design of a concrete beam as the tensile stresses that accompany it on a diagonal plane. To resist these stresses, concrete beams should be reinforced with bent-up bars or with stirrups. The latter are bars placed vertically or on an incline in a beam. They may be the legs of a single U-shaped bar or the sides of a rectangle.

The shear V' taken by stirrups is assumed to be the total shear on the section at which the stirrups are to be placed, less the shear taken by the concrete, vbd. The cross-sectional area of the stirrups needed at a section is given by Eq. (30), where f_v is the allowable

$$A_v = \frac{V's}{f_v d} \qquad (30)$$

tensile stress of the steel, and s is the stirrup spacing.

If the stirrups, instead of being vertical, are laid at an angle α with the horizontal, greater steel area is required, as in Eq. (31). Every potential 45° crack

$$A_v = \frac{V's}{d(\sin \alpha + \cos \alpha)f_v} \qquad (31)$$

should be crossed by at least one line of reinforcement.

If the area of the stirrups is given, the spacing can be computed from the two formulas above. The first stirrup is usually placed as close to the support as practical, generally 2 in. (5 cm). Stirrups should be placed throughout a beam, even if theoretically they are not needed. They serve also as supports for the longitudinal steel. When not required for diagonal tension, stirrups should be placed at most 18 in. (45 cm) apart.

It is common practice to cut off bars or bend them up where they are no longer needed to resist tension at the bottom of the beam. The bent bars serve as diagonal-tension reinforcement and tensile reinforcement at the top of the beam over the support.

Frederick S. Merritt

Bibliography. American Concrete Institute, *Building Code Requirements for Reinforced Concrete,* 1977; C. Davies, *Steel-Concrete Composite Beams for Building,* 1975; K. Leet, *Reinforced Concrete Design,* 1982; G. Winter and A. H. Nilson, *Design of Concrete Structures,* 9th ed., 1979.

Concrete column

A structural member subjected principally to compressive stresses. Concrete columns may be unreinforced, or they may be reinforced with longitudinal

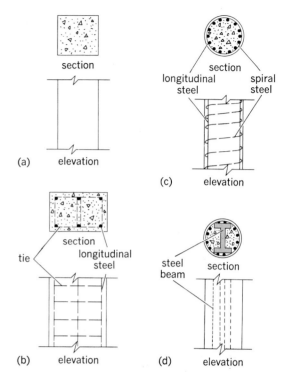

Column types. (*a*) Plain concrete. (*b*) Tied column. (*c*) Spiral-reinforced column. (*d*) Composite column.

bars and ties (tied columns) or with longitudinal bars and spiral steel (spiral-reinforced columns). Sometimes the columns may be a composite of structural steel of cast iron and concrete (see **illus.**).

Plain concrete columns. Unreinforced concrete columns are seldom used because of transverse tensile stresses and the possibility of longitudinal tensile stresses being induced by bucking or unanticipated bending. Because concrete is weak in tension, such stresses are generally avoided.

When plain concrete columns are used, they usually are limited in height to five or six times the least thickness. Under axial loading, the load divided by the cross-sectional area of the concrete should not exceed the allowable unit compressive stress for the concrete.

Axially loaded reinforced columns. Reinforced concrete columns are designed by ultimate-load theory. Two types of column are considered: short and long. Those whose length is three to ten times their least lateral dimension are called short columns.

For spiral-reinforced short columns, the American Concrete Institute code gives the formula for the allowable load in pounds shown in Eq. (1), where f'_c is

$$P = 0.25f'_c A_g + f_s A_s \qquad (1)$$

the 28-day compressive strength of a standard concrete test cylinder in pounds per square inch (lb/in.²), A_g the gross area of section in square inches, f_s the allowable unit stress for steel in pounds per square inch, and A_s the area of reinforcing steel in square inches.

For a tied column, the allowable load is 85% of that for a spiral-reinforced column. Spiral-reinforced columns are stronger because columns tend to fail by a lateral bursting of the concrete as the longitudinal bars bend outward, and spiral reinforcing is more effective than ties in restraining the concrete. If spirals are used, the longitudinal bars should be arranged in a circle.

For long columns, the allowable load is reduced from that permitted for short columns because of the possibility of buckling. When h/d, the ratio of unsupported length of column to least lateral dimension, is equal to or greater than 10, one should use the appropriate reduction factor given in Sec. 916 of the Building Code Requirements for Reinforced Concrete (ACI 318–63).

Spiral reinforcement is determined from Eq. (2),

$$p' \geqq \frac{0.45(R - 1) f'_c}{f'_s} \qquad (2)$$

where p' is the ratio of the volume of spiral reinforcement to the volume of the spiral core (out-to-out of the spiral), R is the ratio of gross area to core area, and f'_s is the yield-point stress of spiral steel, with a maximum of 60,000 lb/in.^{2}

Composite columns. A concrete compression member having a structural steel or cast iron core with a cross-sectional area not exceeding 20% of the gross area of the column is called a composite column. Spiral and longitudinal reinforcing also may be incorporated in the concrete.

The allowable load in pounds on a composite column is given by Eq. (3), where A_c is the net area of

$$P = 0.225 A_c f'_c + f_r A_r + f_s A_s \qquad (3)$$

the concrete section $(A_g - A_r - A_s)$, A_s is the area of longitudinal reinforcement other than the metal core, A_r is the cross-sectional area of the structural steel or cast iron core, f_r is the allowable unit stress for the core, and f_s is the allowable unit stress for the longitudinal reinforcement.

If the core area is 20% or more of the gross area and the concrete is at least 2 in. (5 cm) thick over all metal except rivet heads, the members are called a combination column. If the concrete is to be allowed to share the load with the core, the concrete must be reinforced; usually wire fabric is wrapped completely around the core.

Combined bending and axial load. Because columns are designed by ultimate-load theory and this theroy has also been developed for beams, it is logical to design columns subjected to both bending and axial loads by ultimate-load theory. The ultimate load is given by Eqs. (4) and (5), where f'_c is the compressive

$$P_u = \phi(0.85 f'_c ba + A'_s f_y - A_s f_s) \qquad (4)$$

$$P_u e = \phi[0.85 f'_c ba(d - a/2) + A'_s f_y (d - d')] \qquad (5)$$

strength of concrete at 28 days; f_y is the yield strength reinforcement; f_s is the calculated stress in reinforcement when less than yield strength; A'_s is the area of compression reinforcement; A_s is the area of tension reinforcement; e is the eccentricity of axial load at the end of the member measured from centroid of tension reinforcement; a is the depth of equivalent rectangular compression-stress block; b is the width of compression flange; d is the distance from extreme compression fiber to centroid of tension steel; d' is the distance from extreme compression fiber to centroid of compression steel; and ϕ is a capacity reduction factor. It equals 0.75 for spiral-reinforced members and 0.70 for tied members.

For strength reduction factors when length-thick-

ness ratio exceeds 10, one should consult Sec. 916 of the Building Code Requirements for Reinforced Concrete (ACI 318–63). SEE CONCRETE SLAB; REINFORCED CONCRETE.

Frederick S. Merritt

Concrete slab

A shallow, reinforced-concrete structural member that is very wide compared with depth. Spanning between beams, girders, or columns, slabs are used for floors, roofs, and bridge decks. If they are cast integrally with beams or girders, they may be considered the top flange of those members and act with them as a T beam. SEE CONCRETE; CONCRETE BEAM.

One-way slab. A slab supported on four sides but with a much larger span in one direction than in the other may be assumed to be supported only along its long sides. It may be designed as a beam spanning in the short direction. For this purpose a 1-ft width can be chosen and the depth of slab and reinforcing determined for this unit.

Some steel is also placed in the long direction to resist temperature stresses and distribute concentrated loads. The area of the steel generally is at least 0.20% of the concrete area.

Two-way slab. A slab supported on four sides and with reinforcing steel perpendicular to all sides is called a two-way slab. Such slabs generally are designed by empirical methods. A two-way slab is divided into strips for design purposes.

The American Concrete Institute (ACI) code recommends the following design method: Divide the slab in both directions into a column strip and middle strip (**Fig. 1**). If the ratio S/L of the short span to the long span is equal to or greater than 0.5, the width of the middle strip extending in the short direction equals L/2, as shown. If S/L is less than 0.5, the width of the middle strip in the short direction is $L - S$; the remaining width is divided equally between the two column strips. However, when S/L is less than 0.5, most of the load would be carried in the short direction, and it would be desirable to design the slab as a one-way slab.

A table in the ACI code gives coefficients for calculation of the bending moments in the middle strip for different values of S/L and different types of panels. The moment in the column strip is assumed to be two-thirds that in the middle strip. The reinforcing steel area is determined from $A_s = M/f_s jd$, where M is the bending moment, f_s the tensile unit stress in

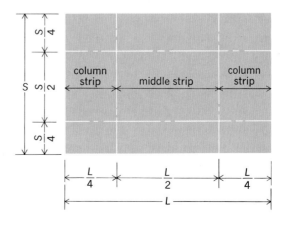

Fig. 1. A two-way slab in strips for design purposes.

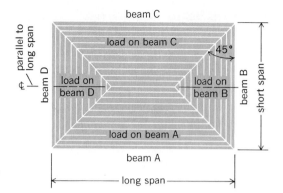
load on beam C

45°

load on beam D

load on beam B

load on beam A

beam C

beam A

beam D

beam B

short span

long span

parallel to long span

Fig. 2. Load distribution assumed for beams supporting a uniformly loaded two-way slab.

longitudinal reinforcement, and j the ratio of lever arm of resisting couple to d, the depth from compressive surface of beam or slab to center of longitudinal tension reinforcement.

In the design of the beams that support a two-way slab along its sides, the load is assumed to be uniform over the slab and distributed to the beams as shown in **Fig. 2**. If w is the slab load in lb/ft^2, moments in the beams may be approximated by assuming the beams loaded with the following equivalent uniform loads: for the short span, $wS/3$; for the long span, $wS(3 - S^2/L^2)/6$.

Flat slabs. When a slab is supported directly on columns, without beams and girders, it is called a flat plate or flat slab (**Fig. 3**).

A flat plate generally is of uniform thickness throughout. Usually, the columns also have constant dimensions throughout the story height below the slab (excluding the portion forming capitals, which are discussed later). In designing such construction, it is necessary to investigate shear and diagonal tension in the vicinity of the columns; frequently, reinforcements which are known as shear heads are embedded at the column tops.

In flat-slab construction it is customary to flare out the columns at the top to form capitals so as to give the slab-column junction greater rigidity. The capital is usually sloped at 45°. For exterior columns the capital is sometimes only a bracket that projects inward.

Flat slabs generally are thickened in the region around the columns. The thickened portion, called a drop panel, may be extended until it reaches from column to column, forming shallow beams and giving the effect of a paneled ceiling.

Although thicker and more heavily reinforced than slabs in beam-and-girder construction, flat slabs are advantageous because they offer no obstruction to passage of light (as beam construction does); savings in story height and in the simpler formwork involved; less danger of collapse due to overload; and better fire protection with a sprinkler system because the spray is not obstructed by beams.

Flat slabs may be reinforced in several ways:

1. Two-way system. When the columns are arranged to form rectangular bays, the reinforcing steel may be placed in two directions, perpendicular to the column lines. Design is based on column strips and middle strips, similar to that for a two-way slab.

2. Four-way system. Column strip steel is similar to that for a two-way slab. But middle-strip reinforcing consists of diagonal bands of steel extending over the columns.

3. Circumferential system. Bars are placed in the top of the slab in concentric rings around the columns, also radially. Similar radial and circular bars are placed in the bottom of the slab in the central portion.

4. Three-way system. When columns are arranged so that lines joining them would form triangles, the reinforcing steel may be laid parallel to the column lines.

Design of flat slabs and flat plates is based on empirical formulas. The ACI code presents design rules that are widely used. SEE CONCRETE COLUMN; REINFORCED CONCRETE.

Frederick S. Merritt

Concretion

A loosely defined term used for a sedimentary mineral segregation that may range in size from inches to many feet. Concretions are usually distinguished from the sedimentary matrix enclosing them by a difference in mineralogy, color, hardness, and weathering characteristics. Some concretions show definite sharp boundaries with the matrix, while others have gradational boundaries. Most concretions are composed dominantly of calcium carbonate, with or without an admixture of various amounts of silt, clay, or organic material. Less common are the clay-ironstone concretions characteristic of the Carboniferous coal measures in many parts of the world. The latter are mixtures of iron carbonate minerals and iron silicate minerals. Coal balls are calcareous concretions, found in or immediately above coal beds, in which there may be a high percentage of original plant organic matter, showing wonderfully preserved plant fossils in a noncompressed condition. Concretions are normally spherical or ellipsoidal; some are flattened to disklike shapes. Frequently a concretion is dumbbell-shaped, indicating that two separate concretionary centers have grown together. SEE COAL BALLS; SEDIMENTARY ROCKS.

Raymond Siever

Concurrent processing

The simultaneous execution of several interrelated computer programs. A sequential computer program consists of a series of instructions to be executed one

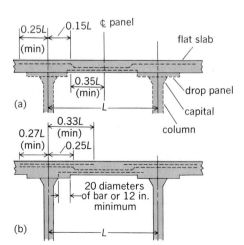

0.25L (min)

0.15L

¢ panel

flat slab

0.35L (min)

L

drop panel

capital

column

(a)

0.27L (min)

0.33L (min)

0.25L

20 diameters of bar or 12 in. minimum

L

(b)

Fig. 3. Details of a flat slab drawn in cross section at two points. (a) Middle strip. (b) Column strip. L is the center-to-center distance between columns.

after another. A concurrent program consists of several sequential programs to be executed in parallel. Each of the concurrently executing sequential programs is called a process. Process execution, although concurrent, is usually not independent. Processes may affect each other's behavior through shared data, shared resources, communication, and synchronization.

Concurrent programs can be executed in several ways. Multiprogramming systems have one processing unit and one memory bank. Concurrent process execution is simulated by randomly interleaving instructions of the sequential programs. All processes have access to a common pool of data. In contrast, multiprocessing systems have several processing units and one memory bank. Processes are executed in parallel on the separate processing units while sharing common data. In distributed systems, or computer networks, each process is executed on its own processor with its own memory bank. Interaction between processes occurs by transmission of data from one process to another along a communication channel. SEE DISTRIBUTED SYSTEMS (COMPUTERS); MULTIPROCESSING.

Uses and examples Concurrent programs are inherently more complex to create and to execute than sequential programs. However, concurrent programs are often more appropriate than sequential programs for a particular computing environment. Additionally, concurrent programs often perform the same tasks much more efficiently than sequential programs.

Historically, one of the first uses of concurrent processing was in operating systems. The operating system of a computer is the software that provides an interface between the user, the computer, and input/output devices such as terminals, printers, and disk drives. If the computer is to support a multiuser environment, the operating system must employ concurrent programming techniques to allow several users to access the computer simultaneously. The operating system should also permit several input/output devices to be used simultaneously, again utilizing concurrent processing. SEE MULTIACCESS COMPUTER; OPERATING SYSTEM.

Concurrent programming is also used when several computers are joined in a network. An airline reservation system is one example of concurrent processing on a distributed network of computers. Agents at many different locations on different computers simultaneously book passenger reservations. The concurrent program controlling the system allows this simultaneity while ensuring that the parallel processes do not conflict, for example, that two agents do not book the same seat on the same flight for two different passengers. SEE LOCAL-AREA NETWORKS; WIDE-AREA NETWORKS.

A simple example of a task that can be performed more efficiently by concurrent processing is a program to calculate the sum of a large list of numbers. A sequential program must iterate through each element of the list, accumulating the sum. In a concurrent program, several processes can simultaneously compute the sum of a subset of the list, after which these sums are added to produce the final total. Depending on the size of the list and the number of processes, the concurrent program may perform the computation in only a small fraction of the time required by the sequential program.

Process interaction. The processes of a concurrent program are usually required to interact and cooperate. An operating system must ensure that processes coordinate the use of shared resources such as printers, memory, and processors. An airline reservation system must ensure that conflicting reservations are not made at different sites. The processes of a concurrent program to sum a list of numbers must cooperate to combine their subtotals into a final total. Many tasks also require concurrent processes to synchronize certain actions. For example, a process controlling a sensor that produces data at a varying rate must be synchronized with a process that reads and analyzes these data.

In multiprogramming and multiprocessing systems, communication and synchronization are achieved through shared variables. Certain memory locations contain data to be read and modified by several processes, such as the number of disk drives available, the next process to send a file to the printer, the number of elements in a buffer, or the total of a sublist of numbers. Reading and writing of shared variable values must be synchronized to avoid undesirable situations. For example, one process may read a shared value in order to update it, and a second process may change the value before the first process is completed. The change in the second process is lost when the first process finally performs its update. This is known as the lost update problem, and it is prevented by ensuring that processes have the ability to gain exclusive access to shared variables for the duration of an operation.

Distributed systems have no shared memory; communication and synchronization are achieved by transmitting messages along shared communication links. Communication can be either synchronous or asynchronous. With synchronous communication, both the sending and receiving processes must be ready to communicate before a message is actually transmitted. With asynchronous communication, the sending process may transmit messages whether or not the receiving process is prepared to receive them. The sent messages are held in a queue until requested by the receiving process. Synchronous communication is used when coordination between processes may be required, as in a series of computers controlling an automated assembly line. Asynchronous communication is used when coordination between processes is not required, as in the transmission of electronic mail messages from one computer to another. SEE DATA COMMUNICATIONS; ELECTRONIC MAIL.

Creating concurrent programs. Concurrent programs can be created explicitly or implicitly. Explicit concurrent programs are written in a programming language designed for specifying processes to be executed concurrently. Implicit concurrent programs are created by a compiler that automatically translates programs written in a sequential programming language into programs with several components to be executed in parallel.

Many programming languages provide facilities for creating concurrent programs. These languages typically include a component for specifying sequential programs and a component for combining and coordinating several sequential programs into one concurrent program. Examples of such languages are Ada, Communicating Sequential Processes (CSP), Concurrent Pascal, Mesa, Modula, and Synchronizing Resources (SR).

Compilers that translate sequential programs into concurrent programs are called parallelizing compilers. They employ a technique called dependency analysis to identify regions of a sequential program that are independent and may therefore be executed in parallel. The regions may be as large as several hundred lines of code or as small as a single arithmetic operation. The object of a parallelizing compiler is to introduce as much parallelism as possible in order to increase the speed of the program. SEE DATA-FLOW SYSTEMS; PROGRAMMING LANGUAGES.

Jennifer Widom

Bibliography. G. R. Andrews and F. B. Schneider, Concepts and notations for concurrent programming, *Comput. Surv.*, 15(1):3–43, March 1983; M. Ben-Ari, *Principles of Concurrent Programming*, 1982; P. Brinch Hansen, *The Architecture of Concurrent Programs*, 1977; J. Peterson and A. Silberschatz, *Operating System Concepts*, 1983.

Concussion

A state following injury in which there is temporary functional impairment without physical evidence of damage to the impaired tissues. The term usually refers to cerebral concussion produced by any type of trauma.

From a clinical point of view cerebral concussion is produced by a head injury which causes temporary unconsciousness but with complete recovery within 24 h. This temporary alteration may affect reflexes and other nervous system components, and is believed to result from one of several mechanisms. In all of these a sudden acceleration or deceleration appears to be a prerequisite. The sudden movement is thought to cause an unequal shifting of tissues of different specific gravities within the skull, between skull and brain, or between different brain tissues. Because no permanent detectable lesion is produced, the exact mechanisms remain speculative.

There are increasing orders of damage which pass beyond concussion states. Clinically it is difficult to judge whether a brief period of unconsciousness is the result of concussion or is an early sign of laceration, contusion, or hemorrhage of brain tissues and related structures. Therefore a guarded prognosis is made and careful watch is maintained, often for many weeks, after a head injury has been sustained. SEE BRAIN.

Edward G. Stuart/N. Karle Mottet

Condensation reaction

One of a class of chemical reactions involving a combination between molecules or between parts of the same molecule. A relatively small molecule such as water or alcohol is often eliminated in the process. The conversions of oxygen to ozone and of diethyl pimelate to 2-carbethoxycyclohexanone are examples, as seen in reactions (1) and (2).

$$3O_2 \rightarrow 2O_3 \qquad (1)$$

$$C_2H_5O_2C(CH_2)_5CO_2C_2H_5 \rightarrow$$

$$\underset{\substack{|\\ CH_2CH_2CHCO_2C_2H_5}}{CH_2CH_2C{=}O} + C_2H_5OH \qquad (2)$$

The term condensation reaction is used very loosely by chemists. Some authors define the term to include all reactions which result in formation of carbon-carbon bonds. General usage, however, does not classify all reactions resulting in formation of carbon-carbon bonds as condensation reactions. Furthermore, the reaction commonly referred to as condensation polymerization involves formation of carbon-oxygen or carbon-nitrogen bonds. SEE POLYMERIZATION.

The following reactions of organic compounds are usually described as condensation reactions.

Aldol condensations. The reactions involve a carbonyl component and an α-hydrogen component, as shown in reaction (3). The carbonyl component is

most often an aldehyde (RCHO), but may also be a ketone ($R_2C{=}O$). The α-hydrogen component may be an aldehyde, ketone, simple ester (RCH_2CO_2R), succinate ester, $RO_2CCH_2CH_2CO_2R$ (the Stobbe condensation), carboxylic acid anhydride, $(RCH_2CO)_2O$ (the Perkin reaction), or compounds containing an activated methylene of the type Y—CH₂—Z, where Y and Z may be groups such as —CO₂H, —CO₂R, —CO—R, —CN, —CONH₂, or —NO₂ (the Knoevenagel reaction).

Aldol condensations are catalyzed by bases such as sodium hydroxide, sodium ethoxide, or amines. Catalysis by acids is feasible in certain cases.

The α-hydroxycarbonyl compounds initially formed in aldol condensations can sometimes be isolated, but they often eliminate water under the experimental conditions and form α,β-unsaturated carbonyl compounds.

At least two α-hydrogen atoms are present in the α-hydrogen component of most aldol condensations, but the first stage of the reaction sometimes involves a single α-hydrogen atom.

The Darzens condensation of aldehydes and ketones with α-haloesters to produce glycidic esters is also an aldol-type condensation, as shown in reactions (4) and (5).

Simple aldol:

The Stobbe, Perkin, Knoevenagel, and Darzens reactions are represented by reactions (6)–(9).

Stobbe reaction:

$(C_6H_5)_2C{=}O + CH_3O{-}\overset{O}{\overset{\|}{C}}CH_2CH_2\overset{O}{\overset{\|}{C}}{-}OCH_3$

Benzophenone   Dimethyl succinate

$+ NaOCH_3 \rightarrow (C_6H_5)_2C{=}\underset{CH_2{-}CO_2Na}{C}{-}CO_2CH_3 + 2CH_3OH$ (6)

Sodium
methoxide

Perkin reaction:

Salicyl aldehyde   Acetic anhydride

Coumarin   Acetic acid   (7)

Knoevenagel reaction:

Cyclo-   Ethyl
hexanone   cyanoacetate

(8)

Darzens reaction:

$C_6H_5{-}\overset{O}{\overset{\|}{C}}{-}CH_3 + ClCH_2CO_2C_2H_5 + NaNH_2 \rightarrow$

Acetophenone   Ethyl   Sodium
               chloroacetate amide

$C_6H_5{-}\underset{CH_3}{C}{-}CHCO_2C_2H_5 + NaCl + NH_3$ (9)

Claisen condensations. The carbonyl component in a Claisen condensation is an ester. Esters of oxalic acid (RO_2CCO_2R) are particularly reactive as carbonyl components. The α-hydrogen component may be an ester or ketone. A generalized reaction is shown by reaction (10).

$-\overset{O}{\overset{\|}{C}}{-}OR + {-}CH_2{-}\overset{O}{\overset{\|}{C}}{-} \rightleftharpoons$

Carbonyl   α-Hydrogen
component   component

$-\overset{O}{\overset{\|}{C}}{-}\underset{H}{C}{-}\overset{O}{\overset{\|}{C}}{-} + HOR$ (10)

Claisen condensations involving esters as both components are known as acetoacetic ester condensations and have as products β-ketoesters $(RCO{-}CHRCO_2R)$. This type of condensation reaction is represented by reaction (11).

$2CH_3CO_2C_2H_5 + NaOC_2H_5 \rightarrow$

Ethyl   Sodium
acetate   ethoxide

$Na(CH_3CO\,CHCO_2C_2H_5) + C_2H_5OH$

$\downarrow$ HCl

$CH_3{-}\overset{O}{\overset{\|}{C}}{-}CH_2{-}\overset{O}{\overset{\|}{C}}{-}OC_2H_5 + NaCl$ (11)

Simple esters react with ketones to produce β-diketones $(RCO{-}CHR{-}CO{-}R)$, as shown in reaction (12), and oxalic esters react with ketones, as

$C_6H_5CO_2C_2H_5 + C_6H_5{-}\overset{O}{\overset{\|}{C}}{-}CH_3 \xrightarrow[\text{(2) HCl}]{\text{(1) NaOC}_2\text{H}_5}$

$C_6H_5{-}\overset{O}{\overset{\|}{C}}{-}CH_2{-}\overset{O}{\overset{\|}{C}}{-}C_6H_5$ (12)

shown in reaction (13), to give oxalyl esters

$\xrightarrow[\text{(2) HCl}]{\text{(1) NaOC}_2\text{H}_5}$

(13)

$(R{-}CO{-}CHR{-}CO{-}CO_2R)$ as products. Condensations of esters of dicarboxylic acids which produce cyclic β-ketoesters are known as Dieckman condensations.

Claisen condensations are brought about by strong bases (sodium alkoxides, sodium amide, sodium hydride). At least 1 mole of base for each mole of α-hydrogen component is required.

α-Hydrogen components with one α-hydrogen atom may be utilized in Claisen condensations if very strong bases such as sodium triphenylmethyl are used.

Michael condensations. Compounds containing an active methylene group often add to the carbon-carbon double bond of α,β-unsaturated carbonyl compounds in the presence of basic catalysts. This is represented by reactions (14) and (15).

$-\underset{|}{C}{=}\underset{|}{C}{-}\underset{|}{C}{=}O + X{-}CH_2{-}Y \rightleftharpoons$

$X{-}\underset{\overset{|}{Y}}{CH}{-}\underset{\overset{|}{H}}{C}{-}\underset{|}{C}{-}C{=}O$ (14)

$H_2C{=}CHCO_2C_2H_5 + H_2C(CO_2C_2H;)_2 \xrightarrow{\text{NaOC}_2\text{H}_5}$

Ethyl acrylate   Diethyl malonate

$\underset{CH(CO_2C_2H_5)_2}{CH_2{-}CH_2{-}CO_2C_2H_5}$ (15)

Benzoin condensations. Aromatic aldehydes in the presence of catalytic amounts of potassium cyanide are converted to hydroxy ketones of the type Ar—CHOH—CO—Ar (benzoins) reaction (16).

$$2C_6H_5\overset{\overset{\displaystyle O}{\|}}{C}-H \xrightarrow{\text{KCN}} C_6H_5-\overset{\overset{\displaystyle OH}{|}}{C}H-\overset{\overset{\displaystyle O}{\|}}{C}-C_6H_5 \quad (16)$$

Acyloin condensations. Aliphatic esters react with metallic sodium to produce intermediates which are converted by hydrolysis into aliphatic α-hydroxyketones, called acyloins, as shown in reaction (17).

$$2C_3H_7CO_2C_2H_5 \xrightarrow[\text{(2) H}_2\text{O}]{\text{(1) Na}}$$

$$C_3H_7-\overset{\overset{\displaystyle OH}{|}}{C}H-\overset{\overset{\displaystyle O}{\|}}{C}-C_3H_7 + 2C_2H_5OH + 4NaOH \quad (17)$$

Mannich reaction. The most common application involves a ketone, formaldehyde, and dimethylamine, as in reaction (18). This results in replacement

$$CH_3-\overset{\overset{\displaystyle O}{\|}}{C}-CH_3 + CH_2O + (CH_3)_2NH_2 \rightarrow$$

$$CH_3-\overset{\overset{\displaystyle O}{\|}}{C}-CH_2CH_2N(CH_3)_2 + H_2O \quad (18)$$

of α-hydrogen atoms with one or more dimethylaminomethyl groups. Other secondary amines may be used.

Cyanoethylation. Acrylonitrile adds a variety of weakly acidic compounds. This is shown in reactions (19)–(22).

$$CH_3COCH_3 + 3CH_2{=}CHCN \rightarrow$$
$$CH_3-COC(CH_2CH_2CN)_3 \quad (19)$$

$$(CH_3)_2CH\overset{\overset{\displaystyle H}{|}}{C}{=}O + CH_2{=}CHCN \rightarrow$$
$$NCCH_2CH_2C(CH_3)_2-\overset{\overset{\displaystyle H}{|}}{C}{=}O \quad (20)$$

$$CH_3CH_2OH + CH_2{=}CHCN \rightarrow$$
$$CH_3CH_2OCH_2CH_2CN \quad (21)$$

$$CH_3CH_2NH_2 + 2CH_2{=}CHCN \rightarrow$$
$$CH_3CH_2N(CH_2CH_2CN)_2 \quad (22)$$

The reaction is catalyzed by strong bases—for example, potassium hydroxide. *See Cyanoethylation.*

Pinacol formation. Ketones are converted to tetrasubstituted ethylene glycols (called pinacols) by certain reducing agents, as in reactions (23) and (24).

$$2CH_3COCH_3 + Mg \rightarrow \begin{matrix} (CH_3)_2C-O \\ | \quad\quad\; Mg \\ (CH_3)_2C-O \end{matrix} \xrightarrow{\text{HCl}}$$

$$\begin{matrix} (CH_3)_2C-OH \\ | \\ (CH_3)_2C-OH \end{matrix} + MgCl_2 \quad (23)$$

$$2(C_6H_5)_2CO + (CH_3)_2CHOH \xrightarrow{\text{light}}$$

$$\begin{matrix} (C_6H_5)_2C-OH \\ | \\ (C_6H_5)_2C-OH \end{matrix} + CH_3COCH_3 \quad (24)$$

Reformatsky reaction. Aldehydes and ketones react with α-haloesters in the presence of zinc. Treatment of the reaction mixture with a dilute acid produces β-hydroxy esters, as shown in reactions (25) and (26).

$$R-\overset{\overset{\displaystyle O}{\|}}{C}- + X-\overset{|}{\underset{|}{C}}-CO_2C_2H_5 + Zn \rightarrow$$

$$R-\overset{\overset{\displaystyle XZnO}{|}}{\underset{|}{C}}-\overset{|}{\underset{|}{C}}-CO_2C_2H_5$$

$$\downarrow \text{HX}$$

$$R-\overset{\overset{\displaystyle OH}{|}}{\underset{|}{C}}-\overset{|}{\underset{|}{C}}-CO_2C_2H_5 + ZnX_2 \quad (25)$$

$$C_6H_5COCH_3 + CH_3CHBrCO_2C_2H_5 \xrightarrow[\text{(2) acid}]{\text{(1) Zn}}$$

$$C_6H_5-\overset{\overset{\displaystyle CH_3}{|}}{C}OH-\overset{\overset{\displaystyle CH_3}{|}}{C}H-CO_2C_2H_5 \quad (26)$$

See Reformatsky reaction.

Additions of Grignard reagents. Grignard reagents (RMgX) add to carbonyl groups and to nitriles. Alcohols, imines, and ketones may be prepared by this method, as shown in reactions (27)–(30).

$$R-\overset{\overset{\displaystyle O}{\|}}{C}- + R'-MgX \rightarrow$$

$$R-\overset{\overset{\displaystyle OMgX}{|}}{\underset{|}{C}}-R' \xrightarrow{\text{acid}} R-\overset{\overset{\displaystyle OH}{|}}{\underset{|}{C}}-R' \quad (27)$$

$$CH_3COCH_3 \xrightarrow[\text{(2) acid}]{\text{(1) CH}_3\text{CH}_2\text{MgBr}} CH_3-\overset{\overset{\displaystyle OH}{|}}{\underset{\underset{\displaystyle CH_3}{|}}{C}}-CH_2CH_3 \quad (28)$$

$$C_6H_5-\overset{\overset{\displaystyle O}{\|}}{C}-OCH_3 \xrightarrow[\text{(2) acid}]{\text{(1)C}_6\text{H}_5\text{MgBr}} (C_6H_5)_3COH \quad (29)$$

$$C_6H_5CN \xrightarrow[\text{(2) NH}_4\text{Cl}]{\text{(1) CH}_3\text{MgBr}} C_6H_5\overset{\overset{\displaystyle N-H}{\|}}{C}-CH_3 \xrightarrow{\text{H}_2\text{O}}$$

$$C_6H_5-\overset{\overset{\displaystyle O}{\|}}{C}-CH_3 + NH_3 \quad (30)$$

Organolithium compounds (RLi) and organozinc compounds (R_2Zn) add to carbonyl groups in the same way that Grignard reagents react. *See Grignard reaction.*

Friedel-Crafts alkylations and acylations. Alkyl and acyl groups may be substituted for hydrogen atoms in aromatic compounds. Aluminum chloride is the usual catalyst for the reaction, but other strong acids may be used. Alkyl and acyl halides are often the substituting reagents, but alcohols, olefins, anhydrides, organic acids, aldehydes, ketones, ethers, and esters may be employed. Reactions (31)–(34) represent Friedel-Crafts alkylations and acylations.

$$C_6H_6 + CH_3CH_2Br \xrightarrow{AlCl_3} C_6H_5CH_2CH_3 + HBr \quad (31)$$

$$C_6H_6 + CH_3-\overset{\overset{\displaystyle O}{\|}}{C}-Cl \xrightarrow{AlCl_3}$$
$$C_6H_5\overset{\overset{\displaystyle O}{\|}}{C}-CH_3 + HCl \quad (32)$$

$$2C_6H_6 + Cl_3C\overset{\overset{\displaystyle O}{\|}}{C}-H \xrightarrow{H_2SO_4}$$
$$(C_6H_5)_2CHCCl_3 + H_2O \quad (33)$$

$$+ H_2O \quad (34)$$

See FRIEDEL-CRAFTS REACTION.

Alkylations of organometallic compounds. Compounds of the type RMe, where Me represents a metal such as Na, K, Li, or MgX, react with many alkyl halides to eliminate MeX and replace the metal with an alkyl group, as in reaction (35). The availability of

$$RMe + R'X \rightarrow R-R' + MeX \quad (35)$$

the compounds RMe and the reactivity of the halides R'X are the limiting factors.

Grignard reagents are readily available from halides and magnesium metal. These reagents are alkylated by reactive halides such as allyl bromide, as in reaction (36). Compounds of the type Y—CH₂—Z where

$$CH_3CH_2MgBr + BrCH_2CH=CH_2 \rightarrow$$
$$CH_3(CH_2)_2CH=CH_2 + MgBr_2 \quad (36)$$

Y and Z are activating groups are converted to metallocompounds by treatment with sodium or potassium alkoxides, and may then be alkylated by allyl, primary, or secondary halides. This is shown by reactions (37) and (38).

$$H_2CYZ + MeOR \rightarrow MeHCYZ + HOR \quad (37)$$

$$MeHCYZ + R'X \rightarrow R'HCYZ + MeX \quad (38)$$

A second alkyl group may be introduced by repeating the process, as in reaction (39). Alkylations of

$$R'HCYZ \xrightarrow[\text{(2) R''X}]{\text{(1) MeOR}} R'R''CYZ \quad (39)$$

acetoacetic and malonic esters are classical examples of this procedure. These alkylations are represented by reactions (40) and (41).

$$CH_3\overset{\overset{\displaystyle O}{\|}}{C}CH_2CO_2C_2H_5 \xrightarrow[\text{(2) CH}_3\text{I}]{\text{(1) NaOC}_2\text{H}_5}$$
$$CH_3\overset{\overset{\displaystyle O}{\|}}{C}CHCO_2C_2H_5 \quad (40)$$
$$\overset{\displaystyle |}{CH_3}$$

$$CH_3CH(CO_2C_2H_5)_2 \xrightarrow[\text{(2) CH}_3\text{CH}_2\text{Br}]{\text{(1) NaOC}_2\text{H}_5}$$
$$CH_3CH_2\overset{\overset{\displaystyle CH_3}{|}}{C}(CO_2C_2H_5)_2 \quad (41)$$

Ketones with at least one α-hydrogen atom may usually be metallated with sodium amide and subsequently alkylated, as in reaction (42).

$$C_6H_6COCH_3 \xrightarrow{NaNH_2} C_6H_5COCH_2Na \xrightarrow{CH_3I}$$
$$C_6H_5COCH_2CH_3 \quad (42)$$

See ORGANIC REACTION MECHANISM; ORGANOMETALLIC COMPOUND.

William B. Renfrow, Jr.

Bibliography. R. T. Morrison and R. N. Boyd, *Organic Chemistry*, 5th ed., 1987; J. D. Roberts and M. C. Caserio, *Basic Principles of Organic Chemistry*, 2d ed., 1977.

Condensed-matter physics

The fundamental science of very large numbers of strongly interacting particles. It includes the familiar solid and liquid states but covers as well dense plasmas, liquid crystals, glasses, polymers, gels, and so forth. It may be regarded as generalizing the more traditional field of solid-state physics in several directions. Whereas solid-state physics has provided a quantitative and predictive understanding of the structures and properties of well-ordered (crystalline), homogeneous systems in three dimensions under nearly equilibrium conditions, condensed-matter physics also treats disorder in both space and time, and heterogeneous compositions in both one and two, as well as three dimensions. It seeks to provide an understanding of macroscopic physical (for example, electric, magnetic, mechanical, and optical) properties in terms of microscopic interactions and phenomena at the atomic level. This requires a thorough understanding and innovative application of the laws and equations of quantum mechanics. Many of the phenomena in condensed-matter physics, such as superconductivity, magnetism, and superfluidity, cannot be understood on the basis of classical physics alone. *See* AMORPHOUS SOLID; COLLOID; GEL; GLASS; LIQUID CRYSTALS; MAGNETISM; PLASMA PHYSICS; POLYMER; QUANTUM MECHANICS; SUPERCONDUCTIVITY; SUPERFLUIDITY.

Systems of many particles, especially under strongly nonequilibrium conditions, exhibit complex dynamical behavior which cannot be simply inferred from the behavior of a few particles. Thus condensed-matter physics encompasses such subjects as chaos and turbulence, microstructures of alloys, glassy superconductors, hot electron devices, nonlinear optical phenomena, and random magnetic systems. *See* ALLOY STRUCTURES; CHAOS; NONLINEAR OPTICS; TURBULENT FLOW.

Condensed-matter physics is one of the most active fields of research. Both its theoretical and experimental methods continue to strongly influence other fields of science as well as a broad range of technologies. Of all the subfields of physics, condensed-matter physics has had the greatest impact on everyday life through the technologies it has generated. Among these technologies are the electronics industry based on the transistor and the integrated circuit or semiconductor chip, which has revolutionized computing, communications, and entertainment; the optical fiber and the miniature semiconductor laser, which are rapidly bringing about the photonic age; superconducting magnets, which have made magnetic-resonance imaging a practical and powerful tool for sophisticated medical diagnostics; and composite materials with extraordinary strength or other useful properties. *See* Composite material; Electronics; Integrated circuits; Laser; Medical imaging; Optical communications; Optical fibers; Semiconductor; Transistor.

Paul A. Fleury

Bibliography. N. W. Ashcroft and N. D. Mermin, *Solid State Physics*, 1976; National Academy of Sciences, *Physics Through the 1990's: Condensed Matter*, 1986; F. Seitz et al. (eds.), *Solid State Physics: Advances in Research and Applications*, vols. 1–40, 1955–1987.

Conditioned reflex

A learned response performed by a trained animal to a signal that was previously associated with an event of consequence for that animal. Conditioned reflex (CR) was first used by the Russian physiologist I. P. Pavlov to denote the criterion measure of a behavioral element of learning, that is, a new association between the signal and the consequential event, referred to as the conditioned stimulus (CS) and unconditioned stimulus (US), respectively. In Pavlov's classic experiment, the conditioned stimulus was a bell and the unconditioned stimulus was sour fluid delivered into the mouth of a dog restrained by harness; the conditioned stimulus was followed by the unconditioned stimulus regardless of the dog's response. After training, the conditioned reflex is manifested when the dog salivates to the sound of the bell.

Classical conditioning. Ideally, certain conditions must be met to demonstrate the establishment of a conditioned reflex according to Pavlov's classical conditioning method. Before conditioning, the bell conditioned stimulus should attract the dog's attention or elicit the orienting reflex (OR), but it should not elicit salivation, the response to be conditioned. That response should be specifically and reflexively elicited by the sour unconditioned stimulus, thus establishing its unlearned or unconditioned status. After conditioned pairings of the conditioned stimulus and the unconditioned stimulus, salivation is manifested prior to the delivery of the sour unconditioned stimulus. Salivation in response to the auditory conditional stimulus is now a "psychic secretion" or the conditioned reflex.

This simple conditioned reflex was important to Pavlov because it presumably demonstrated concretely a new temporary connection in the brain, a neural connection established by training between the auditory apparatus and salivary mechanisms. It provided him with an opportunity to study the psyche objectively. "Here we have exact and constant facts," Pavlov said, "facts that seem to imply intelligence." He had already won the Nobel Prize in physiology for his studies of digestion and its associated secretory reflexes. He knew that the food stimuli contracted oral receptors, and how the impulses were transmitted via centripetal nerves to a specific area of the brain stem. He knew that this reflexive mechanism controlled salivation by centrifugal nerves, so as to match the quality and quantity of saliva precisely with the nature of oral stimulation in accordance with the "wisdom of the body." He understood how the auditory conditioned stimulus might control the head-turning orienting reflex in the dog. But the final mystery was how the auditory conditioned stimulus gained control over the salivary reflex through conditioning or, in other words, how the bell became a stimulus substitute for food. Pavlov's goal was to elucidate these new pathways of control in the brain, and that task was not completed. But to this day, Pavlov's methods provide important guidelines for basic research upon brain mechanisms in learning and memory.

By 1909, other Russian workers had extended the study of the conditioned reflex to other stimulus-reflex combinations in dogs and humans, and since that time scientists all over the world have paired a vast array of stimuli with an enormous repertoire of reflexes to test conditioned reflexes in representative species of almost all phyla, classes, and orders of animals. As a result, classical conditioning is now considered a general biological or psychobiological phenomenon which promotes adaptive functioning in a wide variety of physiological systems in various phylogenetic settings.

Selective learning and homeostatic feedback. The unconditioned stimulus is usually contact with a object or event that is significant for survival, evoking pleasure or pain. This is only a rough guide because some "pleasures" (for example, drugs and alcohol) can be ultimately detrimental. Food, the most popular unconditioned stimulus, is vital for survival. Consummation is followed by digestion which provides feedback (FB) to adjust the palatability of food commensurate with its homeostatic effect. If food is nutritious, it tastes "good" next time. If food is followed by nausea induced by injection, conditioned food aversion is acquired and the food tastes "bad." Therefore the entire conditioning sequence is composed of two different associative processes denoted as CS-US—FB. The first CS-US link and the second US—FB link differ in many ways. For rapid one-trial learning: (1) The CU-US interval must be a matter of seconds, but the US—FB interval can be a matter of hours. (2) The unconditioned stimulus must be presented to an alert subject, but the feedback is completely effective when delivered to a subject asleep under anesthesia. (3) After CS-US training, the subject gains a precise appreciation of both stimuli and their association in time and space and modifies its behavior accordingly. But after conditioned food aversion training the hedonic value of the unconditional stimulus is modified even in the absence of any memory of its association with the nauseous feedback. The subject often approaches the unconditioned stimulus again and rejects it in surprise and disgust.

Furthermore, while many different conditioned stimuli can easily be linked with food unconditioned stimulus, there is a strong selective affinity of the food feedback for taste. In the complete absence of taste, even an aversion to odor paired with poison is absent or weak. With repeated trials, the aversion spreads to other cues associated with feeding. In descending order of effectiveness, these cues are oral stimuli other than taste or odor, visual and tactual aspects of the food, visual and tactual aspects of the food locus, and auditory stimulation paired with food. All these cues are potentiated when accompanied by taste. Natural selection resulted in this specialized US—FB learning system to protect the gut from natural toxins. Charles Darwin was concerned with the mechanisms by which natural toxins protected poisonous insect larvae from predators. Subsequent research revealed that the predators acquired a conditional food aversion for such insects and their mimics; thus the US—FB association has been called Darwinian conditioning.

A brief electric shock impinging on the skin, also a well-known unconditioned stimulus, engages another specialized CS-US—FB system designed by natural selection to protect the skin from predatory attacks. Since the attack is usually swift and preceded by vibrations in the substrate, an immediate noise-shock combination is about as effective for conditioned fear as a delayed taste-toxin combination is for conditioned food aversion. Conversely, taste-shock and noise-toxin are relatively ineffective combinations for rapid conditioning because these stimuli impinge upon two different integrative systems. The shock unconditioned stimulus is followed by an endogenous analgesic feedback that raises the pain threshold. This feedback action is of long duration and is also effective in subjects sleeping under anesthesia. US—FB relationships have not be elucidated for other unconditoned stimulus events, but there is evidence that they exist. Core temperature of the body is apparently feedback for thermal skin receptors. When core temperature is low, a warm unconditioned stimulus is pleasant to the skin. When the core is warmed, the same unconditioned stimulus is unpleasant. In mating behavior of mammals, reproductive success seems to make odors associated with mates more attractive.

Instrumental conditioning. In this procedure the experimenter selects a response (R) emitted by the animal and reinforces it with an unconditioned stimulus. R-US methodology is favored by B. F. Skinner. For example, consider an operant conditioning situation in which a pigeon is placed in the Skinner box with a key or target connected to a food delivery mechanism. Whenever the key illuminates, the pigeon can obtain food by depressing it. Skinner's procedure differs from Pavlov's situation in that the subject is unrestrained, and food is contingent upon depression of the key whenever the light is on. No key press, no food; thus the birds appeared to be working for the pay of food in economic fashion. Skinner, avoiding physiological mechanisms, defined his conditioned reflex as a depression of the key and called it an operant; his pigeon was operating the key to get food. However, high-speed photographs indiated that the pigeon pecked at the target in the same way it pecked at grain. When depressing the lighted key to get water, the pigeon worked the key with the same action it used to drink water. For other birds, the key was disconnected from the food delivery mechanism and seeds were delivered regardless of the pigeon's response, a procedure called autoshaping. These pigeons also learned to peck at the lighted key as if it were a substitute for food. The pigeon's peck, like the dog's salivation, is a Pavlovian conditioned reflex that is characteristic of the consummatory phase of learning.

True instrumental learning is displayed by subjects long before they gain their consummatory objective. For example, E. L. Thorndike confined hungry cats in a puzzle box and placed the food (unconditioned stimulus) outside. Manipulation of a trigger (the re-sponse) released the cat, giving access to food. Another procedure is to place a hungry rat in the starting box and to provide food at the end of a maze with cul-de-sacs along the way. With repeated trials in such apparatus, animals will reduce extraneous responses and decrease time from start to finish. With prolonged training under controlled conditions, subjects can combine CS-US and R-US units into long instrumental sequences. However, behavioral units and sequences do not provide a complete answer to the question "What is learned?" Laboratory studies and detailed observations under natural conditions indicate that animals often forgo satisfaction of needs to spontaneously explore their surroundings, acquiring memorial representation or a cognitive map of the spatial and temporal relationships among many stimuli. With this cognitive map well in hand, they will exhibit novel shortcuts, detours, strategies, and expectations, modifying instrumental means to achieve desired ends. After consumption, these incentives to action are automatically reevaluted in acordance with their homeostatic utility. SEE COGNITION; MEMORY.

John Garcia

Bibliography. D. L. Alkon and J. Farley (eds.), *Primary Neural Substrates of Learning and Behavioral Change*, 1984; N. Bravemen and P. Bronstein (eds.), *Experimental Assessments and Clinical Applications of Conditioned Taste Aversions*, 1985; M. Cabanac, Sensory pleasure, *Quart. Rev. Biol.*, 54:1–29, 1979; J. Garcia and M. D. Holder, Time, space and value, *Human Neurobiol.*, 4:81–89, 1985; R. A. Hinde and J. Stevenson-Hinde (eds.), *Constraints on Learning: Limitations and Predispositions*, 1973; N. Mackintosh, *The Psychology of Animal Learning*, 1974; A. Trevor and L-G. Nilsson (eds.), *Aversion, Avoidance, and Anxiety: Perspectives in Aversively Motivated Behavior*, 1989.

Conductance

The real part of the admittance of an alternating-current circuit. The admittance Y of an alternating-current circuit is a complex number given by Eq. (1).

$$Y = G + jB \qquad (1)$$

The real part G is the conductance. The units of conductance, like those of admittance, are called siemens or mhos. Conductance is a positive quantity. The conductance of a resistor R is given by Eq. (2).

$$G = \frac{1}{R} \qquad (2)$$

In general the conductance of a circuit may depend on the capacitors and inductors in the circuit as well as on the resistors. For example, the circuit in the

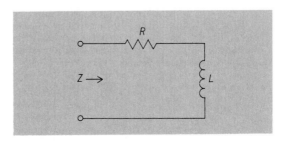

Circuit with a resistor and inductor in series.

illustration has impedance at frequency ω given by Eq. (3) and admittance given by Eq. (4), so that the

$$Z = R + jL\omega \tag{3}$$

$$Y = \frac{1}{R + jL\omega} \tag{4}$$

conductance, given by Eq. (5), depends on the induc-

$$G = \frac{R}{R^2 + L^2\omega^2} \tag{5}$$

tance L as well as the resistance R. SEE ADMITTANCE; ALTERNATING-CURRENT CIRCUIT THEORY; ELECTRICAL IMPEDANCE.

J. O. Scanlan

Conduction (electricity)

The passage of electric charges due to a force exerted on them by an electric field. Conductivity is the measure of the ability of a conductor to carry electric current; it is defined as the ratio of the amount of charge passing through unit area of the conductor (perpendicular to the current direction) per second divided by the electric field intensity (the force on a unit charge). Conductivity is the reciprocal of resistivity and is therefore commonly expressed in units of siemens per meter. SEE ELECTRICAL RESISTIVITY.

The magnitude of the conductivity of a material or system is determined by two properties: the number n of its charges in unit volume that are free to move in the field, and how effectively the field can move them. Acted on by the field, which accelerates them, and by the resistance of the material to their motion, the free charges achieve a drift velocity that is proportional to the electric field intensity. The mobility μ of the charges is defined as the ratio of this drift velocity to the field. From the above definition of the conductivity σ, it follows directly that σ = $ne\mu$, where e is the charge of a free carrier and n is the number of free carriers per unit volume. In metals and semiconductors (such as germanium and silicon) the charges that are responsible for current are free electrons and holes (which, as missing electrons, act like positive charges). These are electrons or holes not bound to any particular atom and therefore able to move freely in the field. Conductivity due to electrons is known as n-type conductivity; that due to holes is known as p-type. SEE HOLE STATES IN SOLIDS; SEMICONDUCTOR.

The conductivity of metals is much higher than that of semiconductors because they have many more free electrons or holes. In semiconductors the number of free electrons or holes is limited by the necessity for them to be excited across an energy gap; this requires energy that is usually provided by raising the temperature of the semiconductor. Thus the number of free electrons or holes, that is, free carriers, and the resulting conductivity usually increase rapidly with temperature. In metals the number of free carriers does not vary with temperature. Increasing temperature, however, causes the lattice atoms to vibrate more strongly, impeding the motion of the free carriers in the field and causing the conductivity to decrease. This effect also occurs in semiconductors, but the increase in number of free carriers with tempera-

ture is usually a stronger effect. At low temperatures the thermal vibrations are weak, and the impediment to the motion of free carriers in the field comes from imperfections and impurities, which in metals usually does not vary with temperature. At the lowest temperatures, close to absolute zero, certain metals become superconductors, possessing infinite conductivity. SEE ELECTRICAL CONDUCTIVITY OF METALS; FREE ELECTRON THEORY OF METALS; SUPERCONDUCTIVITY.

Electrolytes conduct electricity by means of the positive and negative ions in solution. In ionic crystals, conduction may also take place by the motion of ions. This motion is much affected by the presence of lattice defects such as interstitial ions, vacancies, and foreign ions. SEE ELECTROLYTIC CONDUCTANCE; IONIC CRYSTALS.

Ionic conduction can also take place in a gas by using a strong electric field to ionize the gas molecules. If enough ions are formed, there may be a spark. SEE ELECTRIC SPARK; ELECTRICAL CONDUCTION IN GASES.

Electric current can flow through an evacuated region if electrons or ions are supplied. In a vacuum tube the current carriers are electrons emitted by a heated filament. The conductivity is low because only a small number of electrons can be "boiled off" at the normal temperatures of electron-emitting filaments. SEE ELECTRON EMISSION; ELECTRON MOTION IN VACUUM.

Esther M. Conwell

Bibliography. C. Kittel, *Introduction to Solid State Physics*, 5th ed., 1976; L. B. Loeb, *Fundamental Processes of Electrical Discharge in Gases*, 1939; W. Shockley, *Electrons and Holes in Semiconductors*, 1950; P. Vashishta, J. N. Mundy, and G. K. Shenoy (eds.), Fast ion transport in solids, *Proceedings of the International Conference at Lake Geneva, Wisconsin*, May 21–25, 1979.

Conduction (heat)

The flow of thermal energy through a substance from a higher- to a lower-temperature region. Heat conduction occurs by atomic or molecular interactions. Conduction is one of the three basic methods of heat transfer, the other two being convection and radiation. SEE CONVECTION (HEAT); HEAT RADIATION; HEAT TRANSFER.

Steady-state conduction is said to exist when the temperature at all locations in a substance is constant with time, as in the case of heat flow through a uniform wall. Examples of essentially pure transient or periodic heat conduction and simple or complex combinations of the two are encountered in the heat-treating of metals, air conditioning, food processing, and the pouring and curing of large concrete structures. Also, the daily and yearly temperature variations near the surface of the Earth can be predicted reasonably well by assuming a simple sinusoidal temperature variation at the surface and treating the Earth as a semi-infinite solid. The widespread importance of transient heat flow in particular has stimulated the development of a large variety of analytical solutions to many problems. The use of many of these has been facilitated by presentation in graphical form.

For an example of the conduction process, consider a gas such as nitrogen which normally consists of diatomic molecules. The temperature at any location can be interpreted as a quantitative specification of the mean kinetic and potential energy stored in the molecules or atoms at this location. This stored energy will be partly kinetic because of the random translational and rotational velocities of the molecules, partly potential because of internal vibrations, and partly ionic if the temperature (energy) level is high enough to cause dissociation. The flow of energy results from the random travel of high-temperature molecules into low-temperature regions and vice versa. In colliding with molecules in the low-temperature region, the high-temperature molecules give up some of their energy. The reverse occurs in the high-temperature region. These processes take place almost instantaneously in infinitesimal distances, the result being a quasi-equilibrium state with energy transfer. The mechanism for energy flow in liquids and solids is similar to that in gases in principle, but different in detail.

Fourier equation. The mathematical theory as well as the practical calculation of heat conduction is based on a macroscopic interpretation, as contrasted to the basic microscopic mechanism just described. From a physical point of view, it is reasoned that the steady heat flow from a surface (**Fig. 1**) at temperature t_1 to a parallel surface at t_2 is directly proportional to $(t_1 - t_2)$, the area A normal to the direction of flow, and the time of flow τ, and inversely proportional to the distance l between the two planes. These factors are modified by a coefficient κ accounting for the heat-conducting nature of the particular substance between the two planes. Thus, the heat flow Q (in British thermal units, for example) is given by Eq. (1).

$$Q = \kappa A \frac{t_1 - t_2}{l} \tau \qquad (1)$$

In terms of the time rate of flow $q = Q/\tau$ through an infinitesimally thin layer dx, in which the temperature change is dt, this becomes Eq. (2). The minus sign is

$$q = -\kappa A \frac{dt}{dx} \qquad (2)$$

conventionally included to make q positive when heat flows in the increasing direction of x, since dt/dx is then negative. Although this equation was first proposed by J. Biot, it is named after J. Fourier in honor of Fourier's extensive contributions to the theory of heat conduction.

Thermal conductivity. The coefficient κ in Eqs. (1) and (2), called the thermal conductivity, is an important property of matter. It accounts for the heat-conducting ability of a substance, and depends not only on the particular substance involved, but also on the state of that substance. The Fourier equation is essentially a definition of κ which (Fig. 1) can be interpreted as the rate of heat flow per unit of area normal to the direction of flow when a unit temperature difference exists in unit length. Thus from Eq. (2) is derived Eq. (3), where κ is seen to have the dimen-

$$\kappa = \frac{-q/A}{dt/dx} = \frac{q/A}{(t_1 - t_2)/l} \qquad (3)$$

sions of a heat rate per unit area and per unit of temperature gradient. In the cgs system, it can be expressed in cal(s)(cm²)(°C/cm), which is equivalent to cal/(s)(cm)(°C). In engineering, the units most frequently used are Btu/(h)(ft)(°F).

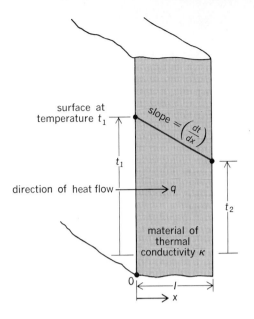

Fig. 1. Heat flow by conduction.

Considerable progress has been made in the interpretation of thermal properties from theories of matter. This is particularly true for gases, where theory involving intermolecular forces has yielded very accurate results. The process of heat conduction in liquids is believed to be similar to that of sound transmission. In dielectric solids, energy is transmitted primarily by means of waves traveling through the atomic lattice; in metals, the electrons behave like an electron gas and provide for energy transfer as well as electrical conduction. This is the basis of the Widemann-Franz law, which states that $\kappa/\sigma T = $ constant (σ is the electrical conductivity and T the absolute temperature). *See* INTERMOLECULAR FORCES; KINETIC THEORY OF MATTER; THERMAL CONDUCTION IN SOLIDS.

For materials occurring as crystalline or amorphous solids, the general trend of κ at atmospheric pressure throughout the three physical states is as shown in **Fig. 2**. The numerical value at the maximum, which occurs near absolute zero in crystalline substances, is comparatively high. For example, the κ of a copper crystal at 20 K has been found to be 7050 Btu/(h)(ft)(°F)—more than 30 times its value at room temperature. Thermal conductivity of solids is discussed in a later section.

Because of the complexity and incomplete understanding of the mechanisms responsible for heat conduction, values of κ are usually determined experimentally. Results for typical gases, liquids, and solids in appropriate temperature ranges are shown in **Fig. 3**. The

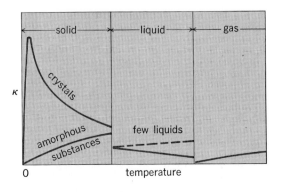

Fig. 2. General variation of thermal conductivity with temperature throughout the three physical states. (*After L. S. Kowalczyk, Trans. ASME, 77:1021–1035, 1955*)

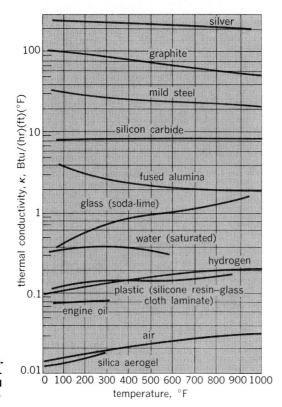

Fig. 3. Thermal conductivities of some typical examples of gases, liquids, and solids. °C = (°F − 32)/1.8.

effect of pressure is significant primarily in gases.

Differential equation of conduction. The evidence of heat flow by conduction through a substance is the variation of the temperature with location and time. If the temperature as a function of the space coordinates and time is known or can be determined, the heat flow at any location and in any direction can be specified by appropriate differentiation. A given problem is normally attacked by solving the differential equation governing the temperature distribution in a homogeneous substance and making this solution fit the prescribed initial or boundary conditions. This differential equation, essentially an expression of the first law of thermodynamics applied to the heat flow, is derived by making a heat balance on an elemental volume in a medium (**Fig. 4**).

Considering first the x direction, the net heat flow into the element in time $\Delta\tau$ is the difference between the flowing in on the left minus that flowing out on the right; or, applying Eq. (1), Eq. (4) is obtained.

$$\Delta Q_x = -\kappa\,\Delta y\,\Delta z\left(\frac{\partial t}{\partial x}\right)_x \Delta\tau$$
$$-\left[-\kappa\,\Delta y\,\Delta z\left(\frac{\partial t}{\partial x}\right)_{x+\Delta x}\Delta\tau\right] \quad (4)$$

Accounting for the variation of κ with temperature is extremely difficult and may make the analytical solution of a problem impossible. Because of this, it is customary to use an appropriate average value which is regarded as constant. Equation (4) can then be rearranged to read as Eq. (5a), which, as $\Delta x \to 0$, becomes Eq. (5b). Similar expressions apply for the y

$$\Delta Q_x = \kappa\,\Delta y\,\Delta z\,\Delta x\,\frac{\left(\frac{\partial t}{\partial x}\right)_{x+\Delta x} - \left(\frac{\partial t}{\partial x}\right)_x}{\Delta x}\Delta\tau \quad (5a)$$

$$\Delta Q_x = \kappa\,\Delta x\,\Delta y\,\Delta z\,\Delta\tau\,\frac{\partial^2 t}{\partial x^2} \quad (5b)$$

and z directions. Heat generated within the element at a uniform rate G per unit volume and time would add an amount $G\,\Delta x\,\Delta y\,\Delta z\,\Delta\tau$.

The net heat flow into the element would be manifest as stored energy and would be equal to notation (6), where w is the specific weight (weight per unit

$$\Delta x\,\Delta y\,\Delta z\,wc\,\Delta t \quad (6)$$

volume) of the medium, c its specific heat, and Δt the temperature rise in the time increment $\Delta\tau$. Equating the net flow into the element to that stored and letting Δx, Δy, Δz, and $\Delta\tau \to 0$ leads to Eq. (7). When no

$$\kappa\left(\frac{\partial^2 t}{\partial x^2} + \frac{\partial^2 t}{\partial y^2} + \frac{\partial^2 t}{\partial z^2}\right) + G = wc\,\frac{\partial t}{\partial\tau} \quad (7)$$

heat source is present, Eq. (7) becomes Eq. (8). The

$$\left(\frac{\partial^2 t}{\partial x^2} + \frac{\partial^2 t}{\partial y^2} + \frac{\partial^2 t}{\partial z^2}\right) = \frac{1}{\kappa/wc}\frac{\partial t}{\partial\tau} = \frac{1}{\alpha}\frac{\partial t}{\partial\tau} \quad (8)$$

ratio $\kappa/wc = \alpha$ is defined as the thermal diffusivity, and is the significant thermal property of a material for transient heat flow (**Fig. 5**). Equation (8) will be recognized as the equation governing a potential field written in terms of temperature. Similar equations are satisfied by other potential field phenomena, such as electricity, magnetism, diffusion, and ideal fluid flow. Because of this, solutions to problems in one field are applicable to analogous systems in the others. Also, an experimental solution of a problem in one field may be obtained from an analogous system in another. SEE POTENTIALS.

Steady-state conduction. When the temperature at all locations is constant with time or, mathematically, when $\partial t/\partial\tau = 0$, steady-state conduction exists. Many important practical problems fall in this category, the most familiar being heat flow through a wall and a hollow cylinder. Referring to Fig. 1 and assuming no heat generation within the wall, Eq. (8) reduces to Eq. (9). The terms $\partial^2 t/\partial y^2$ and $\partial^2 t/\partial z^2$ are eliminated,

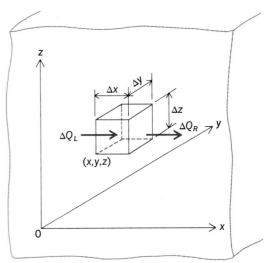

Fig. 4. Diagram of the heat flow through an elemental volume in a homogeneous medium.

$$\frac{\partial^2 t}{\partial x^2} = 0 \qquad (9)$$

since t is considered to vary only with x. The desired temperature distribution is obtained by integrating Eq. (9) twice and evaluating the two constants from the known temperatures at $x = 0$ and $x = l$. This leads to Eq. (10). Application of Eq. (2) to obtain q yields Eq. (11).

$$t = t_1 - (t_1 - t_2)\frac{x}{l} \qquad (10)$$

$$q = \kappa A \frac{(t_1 - t_2)}{l} \qquad (11)$$

In the case of steady radial heat flow through a cylindrical wall (**Fig. 6**), Eq. (2) is applicable to any imaginary thin annular ring in the wall; thus Eq. (12)

$$q = -\kappa A \frac{dt}{dr} = -\kappa 2\pi r l \frac{dt}{dr} \qquad (12)$$

is obtained. Integration from $t = t_1$ to t, and r_1 to r leads to Eq. (13), which indicates that the effect of

$$t = t_1 - q \frac{\ln (r/r_1)}{2\pi \kappa l} \qquad (13)$$

the increasing area for heat flow is to produce a logarithmic variation in the temperature, as shown by the curve in the left part of Fig. 6.

Heat flow through a cylindrical wall is usually expressed in the same form as Eq. (11), written as Eq. (14). Solving Eq. (13) for q and substituting in Eq.

$$q = \kappa A_m \frac{t_1 - t_2}{r_2 - r_1} \qquad (14)$$

(14) shows that the appropriate value for A_m must be as given in Eq. (15), which is called the logarithmic-mean area.

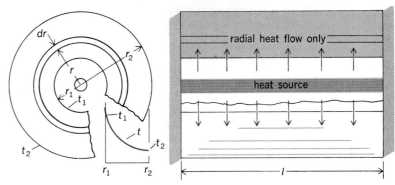

Fig. 6. Steady-state conduction. Diagrams show the temperature distribution for steady radial heat flow through a circular cylinder wall.

$$A_m = \frac{A_2 - A_1}{\ln (A_2/A_1)} \qquad (15)$$

If the rate of heat flow and the inner and outer temperatures of a plane wall or hollow cylinder are measured during steady heat conduction, values of κ can be determined from Eq. (11) or Eq. (14). Because of the simplicity of the equations and the physical systems, most devices for measuring thermal conductivity are based on these types of heat flow.

Interface resistance. Now consider steady-state heat flow through a wall composed of two or more layers of material, each with different uniform thermal properties. If the surfaces of the various layers are very smooth and in very good contact with each other, the temperature distribution will be continuous. At any interface (**Fig. 7a**), since q is constant, Eq. (16) holds. This shows that there is a discontinuity in

$$\kappa_1 \left(\frac{dt}{dx}\right)_1 = \kappa_2 \left(\frac{dt}{dx}\right)_2 \qquad (16)$$

the temperature gradient due to the change in κ.

Actual surfaces, even polished ones, are not smooth but have small projections and depressions. Consequently, when two surfaces are brought together, contact occurs primarily at projecting spots, as illustrated in Fig. 7b. The resulting contact area is only a small fraction of the nominal contact area. Plastic deformation at the points of contact of one or both materials usually occurs with the application of force to hold them together. Heat flows through both the small contact areas and the substance (usually a gas or liquid) filling the voids between the contacting protuberances.

The impairment to the heat-flow path caused by this imperfect contact is referred to as contact resistance. This is determined by extrapolating the measured temperature distribution in each material to the apparent interface location (Fig. 7b). The quotient of the resulting temperature difference Δt_i thus determined and the heat flux defines an interface resistance $R_i = \Delta t_i/q$. R_i depends on the roughnesses of the surfaces, the gas or liquid filling the voids, and the contact pressure. In general, the effect of contact resistance is significant only at low (for example, below 100 lb/in.^{2} or 700 kPa) interface pressures.

Internally generated heat. Conduction of heat generated internally occurs, for example, in the fuel elements used in nuclear reactors. Many of these elements are essentially long, flat plates over which a coolant flows. It is usually necessary to clad the fis-

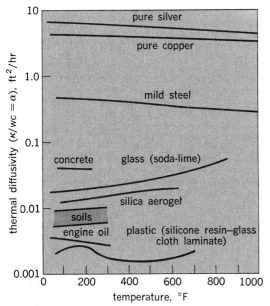

Fig. 5. Thermal diffusivities of some materials. °C = (°F − 32)/1.8.

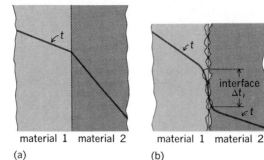

Fig. 7. Temperature distribution through composite wall. (a) With perfect interface contact. (b) For typical actual surfaces.

material 1   material 2
(a)

material 1   material 2
(b)

sionable material to prevent corrosion and keep radioactive particles from entering the coolant (**Fig. 8**). This cladding is undesirable from a heat-transfer standpoint, and is made as thin as possible and of the best adaptable heat-conducting material. Consequently, the temperature drop through it is usually small. *See Nuclear reactor*.

To illustrate the effect of the heat generation, assume that it is uniform with space and time in the radioactive material of Fig. 8 [that is, G in Eq. (7) is constant]. Heat flow from the ends of the elements and parallel to the flow direction of the coolant is negligible.

Therefore, Eq. (7) reduces to Eq. (17). A solution

$$\frac{d^2t}{dx^2} + \frac{G}{\kappa} = 0 \qquad (17)$$

for t is obtained by integrating twice and applying the boundary conditions (considering cooling to be the same on each side) $dt/dx = 0$ at $x = 0$ and $t = t_0$ at $x = l$. The result shows the temperature distribution to be parabolic, as in Eq. (18). The rate of heat trans-

$$t = t_0 + \frac{G}{2\kappa}(l^2 - x^2) \qquad (18)$$

fer to the coolant is obtained by differentiating Eq. (18) with respect to x, evaluating at $x = l$, and substituting in Eq. (2).

Since the heat generation rate is frequently dependent on the temperature, G will be more complex, possibly making an analytical solution impossible.

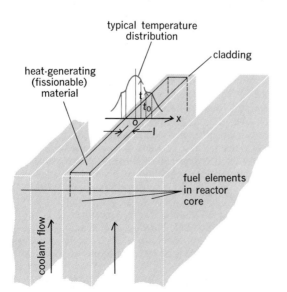

typical temperature distribution

cladding

heat-generating (fissionable) material

fuel elements in reactor core

coolant flow

Fig. 8. Temperature distribution in reactor fuel plate.

Numerical methods of solution are then employed.

Periodic and transient conduction. These are the two kinds of non-steady-state heat flow. Periodic means a quasi-steady-state condition in which the temperature and heat flow at any location in a body vary continuously with time, but pass through the same series of values in a definite period of time, τ_0. A transient state results when the heat flow at any location is momentarily or permanently changed. The duration of the transient period is the time required for the system to return to its original or a new steady-state condition. A transient change may be superimposed on a periodic variation.

Restricting consideration to examples in which no heat generation is present, the fundamental differential equation to be satisfied for either periodic or transient heat flow is Eq. (8). An interesting application is the temperature variation in the Earth due to the diurnal temperature variation, or to a sudden change in surface temperature. Equation (8) reduces in this case to Eq. (19). Assume the existence of a mean

$$\alpha \frac{\partial^2 t}{\partial x^2} = \frac{\partial t}{\partial \tau} \qquad (19)$$

temperature of the Earth which is invariable with depth and that its surface temperature has been varying in a steady periodic manner long enough so that the original transient state due to starting the cyclic surface temperature has reached a steady periodic condition. If the surface temperature variation is given by Eq. (20), the appropriate solution to Eq.

$$t = t_0 \cos 2\pi \frac{\tau}{\tau_0} \qquad (20)$$

(19) is given by Eq. (21). This result shows that the

$$t = t_0 e^{-\sqrt{\pi/2\alpha}\, x} \cos\left(2\pi\frac{\tau}{\tau_0} = \sqrt{\pi/2\alpha}\, x\right) \qquad (21)$$

temperature distribution looks like a wave traveling into the medium with the amplitude decreasing as the factor $e^{-\sqrt{\pi/2\alpha}\, x}$. Computation of the heat flow by determining $(\partial t/\partial t)_{x=0}$ from Eq. (21) indicates that heat flows in during one-half of the period τ_0 and out during the other half.

Calculations of the temperature distribution in the Earth over an 8-h interval, using Eq. (21), are shown in **Fig. 9**. In this case the surface temperature varied from 29 to 41°F (-1.6 to 5.0°C) over a 24-h period; the diffusivity of the soil was 0.0065 ft²/h (6.0 × 10^{-4} m²/h).

An illustration of a transient state is the temperature variation resulting from a sudden change of magnitude t_0 in the temperature at the surface of the Earth. When the initial temperature throughout is uniform and taken as the datum, the applicable solution to Eq. (19) is given by Eq. (22). If a body of soil in the

$$t = \frac{2t_0}{\sqrt{\pi}} \int_0^{x/2\sqrt{\alpha\tau}} e^{-\beta^2}\, d\beta \qquad (22)$$

Earth's surface is at a uniform temperature of 40°F (4.4°C) and the surface temperature suddenly drops to 20°F (-6.7°C), Eq. (22) can be applied to determine the depth at which the temperature will have dropped to freezing (32°F or 0°C) in 12 h. Taking $\alpha = 0.0065$ ft²/h, $t = 32$°F after 12 h is found to occur at $x = 0.27$ ft = 3.2 in. = 8.2 cm.

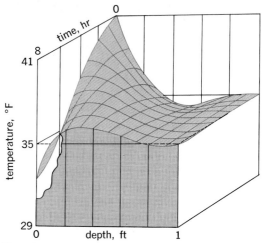

Fig. 9. Temperature distribution in the ground for a daily periodic surface variation. °C = (°F − 32)/1.8.

Variable thermal conductivity. Most materials are sufficiently homogeneous so that their thermal properties are independent of position. Assuming that they vary only with temperature, the heat balance on the element of Fig. 4 leads to Eq. (23). Upon carrying out the indicated differentiation, Eq. (23) becomes Eq. (24). By introducing a new variable θ called the

$$\frac{\partial}{\partial x}\left(\kappa \frac{\partial t}{\partial x}\right) + \frac{\partial}{\partial y}\left(\kappa \frac{\partial t}{\partial y}\right) + \frac{\partial}{\partial z}\left(\kappa \frac{\partial t}{\partial z}\right) + G = wc\frac{\partial t}{\partial \tau} \quad (23)$$

out the indicated differentiation, Eq. (23) becomes Eq. (24). By introducing a new variable θ called the

$$\kappa\left(\frac{\partial^2 t}{\partial x^2} + \frac{\partial^2 t}{\partial y^2} + \frac{\partial^2 t}{\partial z^2}\right)$$
$$+ \frac{\partial \kappa}{\partial t}\left[\left(\frac{\partial t}{\partial x}\right)^2 + \left(\frac{\partial t}{\partial y}\right)^2 + \left(\frac{\partial t}{\partial z}\right)^2\right]$$
$$+ G = wc\frac{\partial t}{\partial \tau} \quad (24)$$

conductivity potential, notation (25), Eq. (24) can be simplified to Eq. (26).

$$\theta = \int_0^t \kappa \, dt \quad (25)$$

$$\frac{\partial^2 \theta}{\partial x^2} + \frac{\partial^2 \theta}{\partial y^2} + \frac{\partial^2 \theta}{\partial z^2} + G = \frac{1}{\alpha}\frac{\partial \theta}{\partial \tau} \quad (26)$$

Equation (26) has exactly the same form as Eq. (7). In many cases the variation of α with temperature (Fig. 5) is less than that of κ so that a mean constant value may be selected. This is true, for example, of metals at temperatures near absolute zero. In such cases, if G is not a function of t (or no heat source is present), solutions for constant α will apply with θ replacing t, provided that the boundary conditions are specified in terms of t or $\kappa(\partial t/\partial n)$ where n represents the variables x, y, z.

The utility of the conductivity potential is demonstrated in the calculation of the heat leak to a liquid-nitrogen tank through a support rod. Consider steady-state conditions with the exposed end of the rod at

300 K and the other at 77.3 K. Also assume there are no losses from the side of the rod which is 1 cm² in cross section, is 15 cm long, and is made of stainless steel (for which κ decreases from 0.15 at 300 K to 0.08 watt/cmK at 77.3 K). At any location q is given by Eq. (27). Since q is constant, Eq. (28) holds. Then Eqs. (29) follows.

$$q = -A\,\kappa\,\frac{dt}{dx} \quad (27)$$

$$\frac{q}{A}\int_0^L dx = -\int_{t_1}^{t_2}\kappa\,dt = \int_0^{t_1}\kappa\,dt - \int_0^{t_2}\kappa\,dt \quad (28)$$

$$\frac{qL}{A} = \theta_1 - \theta_2$$
$$q = A\frac{\theta_1 - \theta_2}{L} = 1\frac{30 - 3.3}{15} = 1.78 \text{ watts} \quad (29)$$

Warren H. Giedt

Bibliography. V. S. Arpaci, *Conduction Heat Transfer*, 1966; R. Berman, *Thermal Conduction in Solids*, 1976; H. S. Carslaw and J. C. Jaeger, *Conduction of Heat in Solids*, 2d ed., 1959; V. Grigull and H. Sandner, *Heat Conduction*, 1983; J. P. Holman, *Heat Transfer*, 5th ed., 1981; M. N. Ozisik, *Heat Conduction*, 1980; W. M. Rohsenow and J. P. Hartnett (eds.), *Handbook of Heat Transfer*, 1973.

Conduction band

The electronic energy band of a crystalline solid which is partially occupied by electrons. The electrons in this energy band can increase their energies by going to higher energy levels within the band when an electric field is applied to accelerate them or when the temperature of the crystal is raised. These electrons are called conduction electrons, as distinct from the electrons in filled energy bands which, as a whole, do not contribute to electrical and thermal conduction. In metallic conductors the conduction electrons correspond to the valence electrons (or a portion of the valence electrons) of the constituent atoms. In semiconductors and insulators at sufficiently low temperatures, the conduction band is empty of electrons. Conduction electrons come from thermal excitation of electrons from a lower energy band or from impurity atoms in the crystal. SEE BAND THEORY OF SOLIDS; ELECTRIC INSULATOR; ELECTRICAL CONDUCTIVITY OF METALS; SEMICONDUCTOR; VALENCE BAND.

H. Y. Fan

Conductor (electricity)

Metal wires, cables, rods, tubes, and bus-bars used for the purpose of carrying electric current. Although any metal assembly or structure can conduct electricity, the term conductor usually refers to the component parts of the current-carrying circuit or system.

Types of conductor. The most common forms of conductors are wires, cables, and bus-bars.

Wires. Wires employed as electrical conductors are slender rods or filaments of metal, usually soft and flexible. They may be bare or covered by some form of flexible insulating material. They are usually circular in cross section; for special purposes they may be drawn in square, rectangular, ribbon, or other shapes. Conductors may be solid or stranded, that is,

Fig. 1. End views of
stranded round conduc-
tor.

19-strand   7-strand

37-strand

built up by a helical lay or assembly of smaller solid conductors (**Fig. 1**).

Cables. Insulated stranded conductors in the larger sizes are called cables. Small, flexible, insulated cables are called cords. Assemblies of two or more insulated wires or cables within a common jacket or sheath are called multiconductor cables.

Bus-bars. Bus-bars are rigid, solid conductors and are made in various shapes, including rectangular, rods, tubes, and hollow squares. Bus-bars may be applied as single conductors, one bus-bar per phase, or as multiple conductors, two or more bus-bars per phase. The individual conductors of a multiple-conductor installation are identical. *See Bus-bar.*

Sizes. Most round conductors less than ½ in. (1.3 cm) in diameter are sized according to the American wire gage (AWG)—also known as the Brown & Sharpe gage. AWG sizes are based on a simple mathematical law in which intermediate wire sizes between no. 36 (0.0050-in. or 0.127-mm diameter) and no. 0000 (0.4600-in. or 11.684-mm diameter) are formed in geometrical progression. There are 38 sizes between these two diameters. An increase of three gage sizes (for example, from no. 10 to no. 7) doubles the cross-sectional area, and an increase of six gage sizes doubles the diameter of the wire.

Sizes of conductors greater than no. 0000 are usually measured in terms of cross-sectional area. Circular mil (cmil) is usually used to define cross-sectional area and is a unit of area equal to the area of a circle 1 mil (0.001 in. or 25.4 micrometers) in diameter.

Wire lengths are usually expressed in units of feet or miles in the United States. Bus-bar sizes are usually defined by their physical dimensions—height and width in inches or fractions of an inch, and length in feet.

Materials. Most conductors are metals, but there is a wide range of conductivities (as much as a 70-to-1 ratio between different metals). Conductors are usually classified as good conductors, such as copper, aluminum, silver; and poor conductors, such as iron; alloys of nickel, iron, copper, and chromium; and carbon products. In 1914, the U.S. Bureau of Standards made measurements of a large number of representative samples of copper and established standard values of resistivity and temperature coefficients, which have been adopted by the International Electrotechnical Commission. *See Electrical conductivity of metals; Electrical resistivity; Metal.*

Most wires, cables, and bus-bars are made from either copper or aluminum. Copper, of all the metals except silver, offers the least resistance to the flow of electric current. Both copper and aluminum may be bent and formed readily and have good flexibility in small sizes and in stranded constructions. Typical conductors are shown in Fig. 1.

Aluminum, because of its higher resistance, has less current-carrying capacity than copper for a given cross-sectional area. However, its low cost and light weight (only 30% that of the same volume of copper) permit wide use of aluminum for bus-bars, transmission lines, and large insulated-cable installations. *See Aluminum; Copper.*

Metallic sodium conductors were used in 1965 on a trial basis for underground distribution insulated for both primary and secondary voltages. Sodium cable offered light weight and low cost for equivalent current-carrying rating compared with other conductor metals. Because of marketing problems and a few safety problems—the metal is reactive with water—the use of this cable was abandoned. *See Sodium.*

For overhead transmission lines where superior strength is required, special conductor constructions are used. Typical of these are aluminum conductors, steel reinforced (ACSR), a composite construction of electrical-grade aluminum strands surrounding a stranded steel core (**Fig. 2**). Other constructions include stranded, high-strength aluminum alloy and a composite construction of aluminum strands around a stranded high-strength aluminum alloy core (ACAR). *See Transmission lines.*

For extrahigh-voltage (EHV) transmission lines, conductor size is often established by corona performance rather than current-carrying capacity. Thus special "expanded" constructions are used to provide a large circumference without excessive weight. Typical constructions use helical lays of widely spaced aluminum strands around a stranded steel core. The space between the expanding strands is filled with paper twine, and outer layers of conventional aluminum strands are applied. In another construction the outer conductor stranding is applied directly over lays of widely separated helical expanding strands, without filler, leaving substantial voids between the stranded steel core and the closely spaced outer conductor layers. Diameters of 1.6 to 2.5 in. (41 to 63 mm) are typical. For lower reactance, conductors are "bundled," spaced 6–18 in. (15–45 cm) apart, and paralleled in groups of two, four, or more per phase. Figure 2 shows views of typical aluminum-conductor–steel-reinforced and expanded constructions.

Materials research has focused on amorphous or glassy materials, which include amorphous silicon-based semiconductors and metal alloys that totally lack crystalline structure. *See Amorphous solid.*

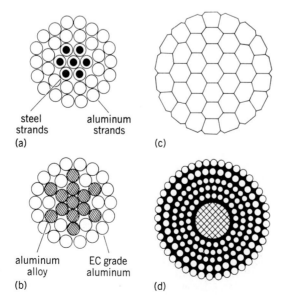

steel   aluminum
strands   strands
(a)   (c)

aluminum   EC grade
alloy   aluminum
(b)   (d)

Fig. 2. Aluminum
conductors. (a) ACSR. (b)
ACAR. (c) Compact
concentric stranded
conductor. (d) Expanded-
core concentric stranded
conductor.

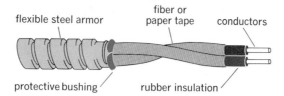

Fig. 4. Two-conductor armored cable.

Bare conductors. Bare wires and cables are used almost exclusively in outdoor power transmission and distribution lines. Conductors are supported on or from insulators, usually porcelain, of various designs and constructions, depending upon the voltage of the line and the mechanical considerations involved. Voltages as high as 765 kV are in use, and research has been undertaken into the use of EHV transmission lines, with voltages as high as 1500 kV.

Bare bus-bars are used extensively in outdoor substation construction, in switchboards, and for feeders and connections to electrolytic and electroplating processes. Where dangerously high voltages are carried, the use of bare bus-bars is usually restricted to areas accessible only to authorized personnel. Bare bus-bars are supported on insulators which have a design suitable for the voltage being carried.

Insulated conductors. Insulated electric conductors are provided with a continuous covering of flexible insulating material. A great variety of insulating materials and constructions has been developed to serve particular needs and applications. The selection of an appropriate insulation depends upon the voltage of the circuit, the operating temperature, the handling and abrasion likely to be encountered in installation and operation, environmental considerations such as exposure to moisture, oils, or chemicals, and applicable codes and standards. SEE ELECTRICAL INSULATION.

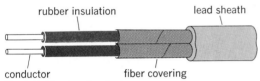

Fig. 3. Rubber-insulated, fiber-covered, lead-sheathed cable, useful for installation in wet places.

Magnet wires, used in the windings of motors, solenoids, transformers, and other electromagnetic devices, have relatively thin insulations, usually of enamel or cotton or both. Magnet wire is manufactured for use at temperatures ranging from 220 to 390°F (105 to 200°C). SEE MAGNET WIRE.

Conductors in buildings. Building wires and cables are used in electrical systems in buildings to transmit electric power from the point of electric service (where the system is connected to the utility lines) to the various outlets, fixtures, and utilization devices. Building wires are designed for 600-V operation but are commonly used at utilization voltages substantially below that value, typically 120, 240, or 480 V. Insulations commonly used include thermoplastic, natural rubber, synthetic rubber, and rubberlike compounds. Rubber insulations are usually covered with an additional jacket, such as fibrous braid or polyvinyl chloride, to resist abrasion. Building wires are grouped by type in several application classifications in the National Electrical Code. SEE ELECTRICAL CODES.

Classification is by a letter which usually designates the kind of insulation and, often, its application characteristics. For example, type R indicates rubber or rubberlike insulation. TW indicates a thermoplastic, moisture-resistant insulation suitable for use in dry or wet locations; THW indicates a thermoplastic insulation with moisture and heat resistance. Other insulations in commercial use include silicone, fluorinated ethylene, propylene, varnished cambric, asbes-

tos, polyethylene, and combinations of these.

Building wires and cables are also available in duplex and multiple-conductor assemblies; the individual insulated conductors are covered by a common jacket. For installation in wet locations, wires and cables are often provided with a lead sheath (**Fig. 3**).

For residential wiring, the common constructions used are nonmetallic-sheath cables, twin- and multiconductor assemblies in a tough abrasion-resistant jacket; and armored cable with twin- and multiconductor assemblies encased in a helical, flexible steel armor as in **Fig. 4**.

Power cables. Power cables are a class of electrical conductors used by utility systems for the distribution of electricity. They are usually installed in underground ducts and conduits. Power cables are also used in the electric power systems of industrial plants and large buildings.

Power-cable insulations in common use include rubber, paper, varnished cambric, asbestos, and thermoplastic. Cables insulated with rubber (**Fig. 5**), polyethylene, and varnished cambric (**Fig. 6**) are used up through 69 kV, and impregnated paper to 138 kV. The type and thickness of the insulation for various voltages and applications are specified by the Insulated Power Cable Engineer Association (IPCEA).

Spaced aerial cable. These cable systems are used for pole-line distribution at, typically, 5–15 kV, three-phase. Insulated conductors are suspended from a messenger, which may also serve as a neutral conductor, with ceramic or plastic insulating spacers, usually of diamond configuration. For 15 kV, a typical system may have conductors with 10/64-in. (4-mm) polyethylene insulation, 9-in. (23-cm) spacing between conductors, and 20 ft (6 m) between spacers.

High-voltage cable. High-voltage cable constructions and standards for installation and application are described by IPCEA. Insulations include (1) paper, solid type; (2) paper, low-pressure, gas-filled; (3) paper, low-pressure, oil-filled; (4) pipe cable, fully impregnated, oil-pressure; (5) pipe cable, gas-filled, gas-pressure; (6) rubber or plastic with neoprene or plastic jacket; (7) varnished cloth; (8) AVA and AVL (asbestos-varnished cloth).

Underground transmission cables are in service at voltages through 345 kV, and trial installations have been tested at 500 kV. Research has also been undertaken to develop cryoresistive and superconducting cables for transmitting power at high density.

Superconducting cables. It may be possible to apply the phenomenon of superconductivity to help pro-

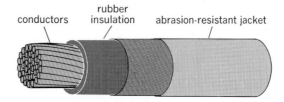

Fig. 5. Rubber-insulated power cable.

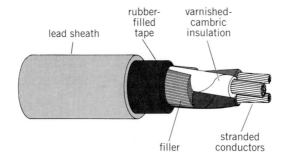

Fig. 6. Varnished cambric-insulated, lead-sheathed power cable, for voltages up to 28 kV.

lead sheath · rubber-filled tape · varnished-cambric insulation · filler · stranded conductors

vide more efficient and more compact electrical systems. In the area of electric power transmission, generation, and storage, superconductivity could have a major impact by increasing energy conservation. With the advent, in 1986, of superconducting materials with transition temperatures as high as 125 K ($-235°$F), there is a possibility of designing superconducting transmission lines that would carry as much as 1 GVA (10^9 W) of power. SEE SUPERCONDUCTING MATERIALS; SUPERCONDUCTIVITY.

Busways. Enclosed bus-bar assemblies, or busways, are extensively used for service conductors and feeders in the electrical distribution systems of industrial plants and commercial buildings. They consist of prefabricated assemblies in standard lengths of busbars rigidly supported by solid insulation and enclosed in a sheet-metal housing.

Busways are made in two general types, feeder and plug-in. Feeder busways have no provision for taps or connections between the ends of the assembly. Low-reactance feeder busways are so constructed that conductors of different phases are in close proximity to minimize inductive reactance. Plug-in busways have provisions at intervals along the length of the assembly for the insertion of bus plugs.

Voltage drop in conductors. In electric circuits the resistance and (in ac circuits) the reactance of the circuit conductors result in a reduction in the voltage available at the load (except for capacitive loads). Since the line and load resistances are in series, the source voltage is divided proportionally. The difference between the source voltage and the voltage at the load is called voltage drop.

Electrical utilization devices (such as motors, household appliances, and electronic equipment) are designed to operate at a particular voltage or within a narrow range of voltages around a design value. The performance and efficiency of these devices are adversely affected if they are operated at a significantly lower voltage. Incandescent-lamp light output is lowered; fluorescent-lamp light output is lowered and starting becomes slow and erratic; the starting and pull-out torque of motors is seriously reduced.

Voltage drop in electric circuits caused by line resistance also represents a loss in power which appears as heat in the conductors. In excessive cases, the heat may rapidly age or destroy the insulation. Power loss also appears as a component of total energy use and cost. SEE COPPER LOSS.

Thus, conductors of electric power systems must be large enough to keep voltage drop at an acceptable value, or power-factor corrective devices—such as capacitors or synchronous condensers—must be installed. A typical maximum for a building wiring system for light and power is 3% voltage drop from the utility connection to any outlet under full-load condition. SEE WIRING.

H. Wayne Beaty

Bibliography. S. A. Boggs, F. Y. Chu, and N. Fujimoto (eds.), *Gas-Insulated Substations: Technology and Practice, Proceedings of the International Symposium*, 1986; J. Douglas, Pursuing the promise of superconductivity, *EPRI J.*, 12(6):4–15, September 1987; D. Fink and H. W. Beaty (eds.), *Standard Handbook for Electrical Engineers*, 12th ed., 1987; Frontiers in chemistry: Materials science, *Science*, 235:953, 997–1035, 1987; Institute of Electrical and Electronics Engineers, *Fiber Optic Applications in Electrical Utilities*, Publ. 84TH0119-8-PWR, 1984; R. I. Jaffee, Materials and electricity, *Metallurg. Trans.*, 17A:755–775, 1986; J. McPartland, *McGraw-Hill's National Electrical Code Handbook*, 19th ed., 1987; M. Rabionowitz, Superconductivity, *Elec. Eng. Rep.*, 2(1):1–14, 1988.

Condylarthra

A mammalian order of extinct, primitive hoofed herbivores with five-toed plantigrade to semidigitigrade feet. Condylarths are not far removed from carnivorous ancestors. The astragalus is carnivorelike and known skulls are carnivorelike in outline, although the dentition has been modified for an omnivorous or herbivorous diet. Inflated auditory bullae are not present. While the teeth are usually low-crowned and superficially piglike, one group possesses selenodont teeth. Condylarths are best represented in the early Cenozoic deposits of Europe and of North and South America, appearing first in the North American Cretaceous. The order is near the ancestry of all living hoofed mammals. SEE CARNIVORA; DENTITION; TOOTH.

Condylarthra is subdivided into seven families. The Arctocyonidae (*Protungulatum, Arctocyon, Deltatherium,* and *Triisodon*) form the central and most carnivorelike group of condylarths. They also appear in the record earlier than other members of the order.

The Hyopsodontidae (*Mioclaenus, Oxyacodon, Protoselene,* and *Hyopsodus*), represented abundantly in the Paleocene and Eocene of North America but also known from Europe and possibly southern Asia, are seldom larger than rabbits. They are descendants of early arctocyonids and may be ancestral to artiodactyls. SEE ARTIODACTYLA.

Phenacodontidae (*Tetraclaenodon, Ectocion,* and *Phenacodus*), abundantly represented in the early Paleocene to early Eocene of North America and in the early and middle Eocene of Europe, are large herbivores similar in body proportions to carnivores (see

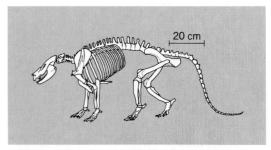

20 cm

Skeleton of *Ectoconus*, an early Paleocene phenacodont condylarth.

illus.). Their teeth are much like those of arctocyonids but are lower-crowned. Like the Arctocyonidae they show affinity with South American didolodontids and may be near the ancestry of the Proboscidea, Desmostylia, Tubulidentata, Sirenia, and Perissodactyla. SEE DESMOSTYLIA; PERISSODACTYLA; PROBOSCIDEA; SIRENIA; TUBULIDENTATA.

Didolodontidae (*Didolodus* and *Asmithwoodwardia*), found in late Paleocene and Eocene of Argentina and Brazil but surviving until late Miocene in Columbia (*Megadolodus*), are generally medium-sized herbivores. They show possible relationships to the earliest notoungulates and astrapotheres and gave rise to the litopterns. SEE ASTRAPOTHERIA; LITOPTERNA; NOTOUNGULATA.

Periptychidae (*Anisonchus*, *Ectoconus*, and *Periptychus*), of the Paleocene of North America, were a short-lived but highly successful group of herbivores with specialized, fluted teeth. They are most closely related to arctocyonids, hyopsodontids, and phenacodontids.

Meniscotheriidae (*Meniscotherium*, *Orthaspidotherium*, and *Pleuraspidotherium*), late Paleocene of France and late Paleocene and early Eocene of North America, possess selenodont teeth and molarized premolars. They are not closely related to other condylarths but have been thought by some to be related to the Hyracoidea. SEE HYRACOIDEA.

The Mesonychidae, formerly considered to be members of the order Carnivora, are now placed in the Condylarthra from one of whose subfamilies (Triisodontinae) they arose. Mesonychids never developed the type of shearing teeth possessed by members of the Carnivora, Deltatheridia, or Insectivora, but instead developed peculiar carnivorous adaptations of their own. One group of mesonychids appears to be close to the ancestry of the earliest whales. SEE DELTATHERIDIA; INSECTIVORA; MAMMALIA.

Malcolm C. McKenna

Cone

The solid of revolution obtained by revolving a right triangle about one of its shorter sides is called a cone, or more precisely a right circular cone (see **illus.**). More generally, the term cone is used in solid geometry to describe a solid bounded by a plane and a portion of one nappe of a conical surface. In analytic geometry, however, the term cone refers not to a solid but to a conical surface. This is a surface generated by a straight line which moves so that it always intersects a given plane curve, called the directrix, and passes through a point, called the vertex, not in the plane of the directrix. The generating line in each of its positions is called an element of the cone. The vertex divides the surface into two parts, called nappes, that are congruent to each other in the extended sense (under rotation and reflection combined) but may not be superposable by a rigid motion in three-dimensional space.

If the elements of a cone make equal angles with a line through the vertex, the cone is called a cone of revolution, the line is called the axis of revolution, and the angle is called the semivertical angle ϵ. Plane sections of a cone of revolution are called conic sections.

If the plane makes an angle θ with the axis and

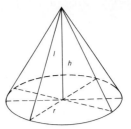

Right circular cone.

does not contain the vertex, the resulting conic section is called nondegenerate and is a circle if $\theta = 90°$, an ellipse if $\epsilon < \theta < 90°$, a parabola if $\theta = \epsilon$, or a hyperbola if $\theta < \epsilon$. In the latter case the plane cuts both nappes of the cone, and the conic section consists of two separate congruent arcs called the branches of the hyperbola.

If the cutting plane contains the vertex of the cone, the conic section is degenerate and consists of a point, a line, or a pair of intersecting lines. Cones whose directrices are conic sections are called quadric cones.

The volume of a solid cone is $V = Bh/3$, where B is the base area included within the directrix, and h is the altitude (or height) measured from the vertex to the plane of the base. The volume V and surface area S of a right circular cone are $V = \pi r^2 h/3$ and $S = \pi r l + \pi r^2$, where r denotes the radius of the base and $l = \sqrt{r^2 + h^2}$ denotes the slant height, measured from vertex to base along an element of the cone.

A frustum of a cone is a solid bounded by two parallel planes and a portion of a conical surface. The plane boundaries are called bases. If the bases are circles of radii r and R, the volume of the frustum is $V = (\pi h/3)(r^2 + rR + R^2)$. SEE CONIC SECTION; EUCLIDEAN GEOMETRY; SURFACE AND SOLID OF REVOLUTION.

J. Sutherland Frame

Confocal microscopy

A technique that creates high-resolution images of very small objects but differs from conventional optical microscopy in that it uses a condenser lens to focus the illuminating light from a point source into a very small, diffraction-limited spot within the specimen, and an objective lens to focus the light emitted from that spot onto a small pinhole in an opaque screen (see **illus.**). Located behind the screen is a detector capable of quantifying how much light passes through the hole at any instant. Because only light from within the illuminated spot is properly focused to pass through the pinhole and reach the detector, any stray light from structures above, below, or to the side of the spot is filtered out. The image quality is therefore greatly enhanced.

Only the smallest possible spot is illuminated at any one time, and so a coherent image must be built up by scanning point by point over the desired field of view and recording the intensity of the light emitted from each spot. The size of the spot is equal to the ultimate resolution of the instrument and is typically about 0.25 micrometer in diameter and about 0.5 μm deep, although the dimensions vary with the wavelength of the light and the lens system used.

Scanning can be accomplished in several ways, but

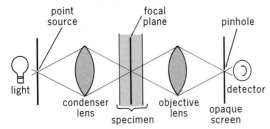

Schematic diagram of a confocal microscope. Light passes through a point source and a condenser lens to a point on the specimen. Only light coming from that spot is focused again by the objective lens onto a pinhole, through which it passes to reach the detector. Light from structures above or below the focal plane is too diffuse to pass through, and light from structures beside the spot is directed to the side of the pinhole. (*After R. J. Taylor, Confocal microscopy sheds new light on the dynamics of living cells, J. NIH Res., 1(1):113–115, 1989*)

the most common system is laser scanning. This system uses a pair of mirrors, vibrating at precisely controlled frequencies, which are placed in the path of a laser beam and direct it in a sweeping pattern over the specimen. The intensity of light emitted from each spot is digitally recorded, and the complete set of intensities from each point is displayed on a video screen, allowing the total image to be seen at one time. Because the information is already in a digital form, it can easily be passed to a computer for storage or further analysis.

Although confocal microscopy was first described in 1961, instruments were not available commercially until the late 1980s. They have found applications in several fields, including materials science and the semiconductor industry, but they have a number of advantages that make them particularly useful in the biological sciences. Many dyes used in biological studies are fluorescent and emit light at a wavelength different from the illuminating light. The amount of fluorescence from different parts of a cell, which is often directly proportional to the concentration of a specific intracellular constituent, can be measured far more precisely with a confocal microscope, because of the rejection of interference from other parts of a specimen. Haze from out-of-focus structures above the focal plane is eliminated, and so it is possible to see very clear images from fairly deep within a relatively thick slice of tissue without having to disrupt its three-dimensional organization. A stack of images from successive focal planes can be obtained simply by moving the narrow focal plane through the specimen. That technique makes it relatively easy to build up complex, three-dimensional views of cellular interconnections or subcellular structures. SEE FLUORES-CENCE MICROSCOPE.

Robert J. Taylor

Bibliography. H. Shuman, J. M. Murray, and C. DiLullo, Confocal microscopy: An overview, *Bio-Techniques,* 7(2):154–163, 1989; R. J. Taylor, Confocal microscopy sheds new light on the dynamics of living cells, *J. NIH Res.,* 1(1):113–115, 1989; J. G. White and W. B. Amos, Confocal microscopy comes of age, *Nature,* 328:183–184, 1987; J. G. White, W. B. Amos, and M. Fordham, An evaluation of confocal versus conventional imaging of biological structures by fluorescence light microscopy, *J. Cell Biol.,* 105:41–48, 1987.

Conformal mapping

A special operation in mathematics in which a set of points in one coordinate system is mapped or transformed into a corresponding set in another coordinate system, preserving the angle of intersection between pairs of curves.

A mapping or transformation of a set E of points in the xy plane onto a set F in the uv plane is a correspondence, Eqs. (1), that is defined for each point

$$u = \varphi(x,y) \qquad v = \psi(x,y) \qquad (1)$$

(x,y) in E and sends it to a point (u,v) in F, so that each point in F is the image of some point in E. A mapping is one to one if distinct points in E are transformed to distinct points in F. A mapping is conformal if it is one to one and it preserves the magnitudes and orientations of the angles between curves (**Fig. 1**). Conformal mappings preserve the shape but not the size of small figures.

Relation to analytic functions. If the points (x,y) and (u,v) are viewed as the complex numbers $z = x + iy$ and $w = u + iv$, the mapping becomes a function of a complex variable: $w = f(z)$. It is an important fact that a one-to-one mapping is conformal if and only if the function f is analytic and its derivative $f'(z)$ is never equal to zero. This can be seen by considering a function f that is analytic near $z_0 = x_0 + iy_0$, with $f'(z_0) \neq 0$, and mapping a curve C passing through z_0 to a curve Γ passing through $w_0 = f(z_0)$. Then, as a point z tends to z_0 along C, the angle $\arg\{z - z_0\}$ tends to the angle α between C and the line $y = y_0$ (**Fig. 2**). But by assumption, expression (2) is valid. In particular, $\arg\{f(z) - w_0\} - \arg\{z -$

$$\frac{f(z) - w_0}{z - z_0} \rightarrow f'(z_0) \qquad (2)$$

$z_0\}$ tends to a limit $\gamma = \arg\{f'(z_0)\}$. In other words, $\arg\{w - w_0\} \rightarrow \beta$ as $w \rightarrow w_0$ along Γ, where $\beta = \alpha + \gamma$. But any other curve through z_0 is rotated by the same angle γ, so that the mapping preserves the angle between the two curves. In a similar manner, it can be shown that a conformal mapping is necessarily given by an analytic function. SEE COMPLEX NUMBERS AND COMPLEX VARIABLES.

Examples. Some examples of functions that provide conformal mappings (and one that is not conformal) will now be given.

1. The function $w = (z - 1)/(z + 1)$ maps the right half-plane, defined by $\text{Re}\{z\} > 0$, conformally onto the unit disk, defined by $|w| < 1$.

2. The function $w = \log z$ maps the right half-plane conformally onto the horizontal strip defined by $-\pi/2 < \text{Im}\{w\} < \pi/2$.

3. The function $w = z^2$ maps the upper semidisk,

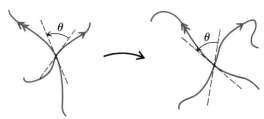

Fig. 1. Angle-preserving property of a conformal mapping.

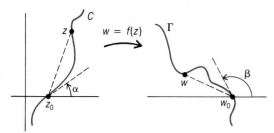

Fig. 2. Illustration for proof that analytic functions preserve angles.

defined by $\text{Im}\{z\} > 0$ and $|z| < 1$, conformally onto the unit disk $|w| < 1$ with the segment $0 \le u < 1$ removed. It doubles angles at the origin, but this is a boundary point which does not lie in the semidisk.

4. The function $w = \bar{z}$ of complex conjugation preserves the magnitudes but not the orientations of angles between curves. It is nowhere conformal.

5. The function $w = z + 1/z$ maps the unit disk $|z| < 1$ conformally onto the extended complex plane (including the point at infinity) with the line segment $-2 \le u \le 2$ removed.

6. The Koebe function $k(z) = z(1 - z)^{-2}$ maps the unit disk conformally onto the complex plane with the half-line $-\infty < u \le -1/4$ removed.

Linear fractional transformations. A linear fractional transformation is a function of the form given by Eq. (3), where a, b, c, and d are complex con-

$$w = \frac{az + b}{cz + d} \qquad ad - bc \neq 0 \qquad (3)$$

stants. It is also known as a Möbius function. One example is the function $w = (z - 1)/(z + 1)$, discussed above. Simpler examples are magnifications, given by Eq. (4), rotations, given by Eq. (5), and inversion, given by Eq. (6). Every linear fractional

$$w = az \qquad a > 0 \qquad (4)$$

$$w = az \qquad |a| = 1 \qquad (5)$$

$$w = 1/z \qquad (6)$$

transformation is a composition of linear fractional transformations of these three special types. Thus each linear fractional transformation provides a conformal mapping of the extended complex plane onto itself, and in fact the linear fractional transformations are the only such mappings. There is a unique linear fractional transformation which carries three prescribed (distinct) points z_1, z_2, z_3 to prescribed images w_1, w_2, w_3. The most general conformal mapping of the unit disk onto itself is a linear fractional transformation of the form given by Eq. (7).

$$w = \frac{z - z_0}{1 - \bar{z}_0 z} \qquad |z_0| < 1 \qquad (7)$$

Each linear fractional transformation carries circles to circles and symmetric points to symmetric points. Here a circle means a circle or a line. Two points are said to be symmetric with respect to a circle if they lie on the same ray from the center and the product of their distances from the center is equal to the square of the radius. Two points are symmetric with respect to a line if the line is the perpendicular bisector of the segment joining the two points. As an in-

stance of this general property of linear fractional transformations, the mapping $w = (z - 1)/(z + 1)$ sends the family of circles of Apollonius with symmetric points 1 and -1 (defined by requiring that on each circle the quotient of the distances from 1 and -1 be constant) onto the family of all circles centered at the origin. It carries the orthogonal family of curves, consisting of all circles through the points 1 and -1, onto the family of all lines through the origin (**Fig. 3**).

Applications. Conformal mappings are important in two-dimensional problems of fluid flow, heat conduction, and potential theory. They provide suitable changes of coordinates for the analysis of difficult problems. For example, the problem of finding the steady-state distribution of temperature in a conducting plate requires the calculation of a harmonic function with prescribed boundary values. If the region can be mapped conformally onto the unit disk, the transformed problem is readily solved by the Poisson integral formula, and the required solution is the composition of the resulting harmonic function with the conformal mapping. The method works because a harmonic function of an analytic function is always harmonic. SEE CONDUCTION (HEAT); FLUID-FLOW PRINCIPLES; LAPLACE'S DIFFERENTIAL EQUATION; POTENTIALS.

The term conformal applies in a more general context to the mapping of any surface onto another. A problem of great importance for navigation is to produce conformal mappings of a portion of the Earth's surface onto a portion of the plane. The Mercator and stereographic projections are conformal in this sense. SEE MAP PROJECTIONS.

Riemann mapping theorem. In 1851, G. F. B. Riemann enunciated the theorem that every open simply connected region in the complex plane except for the whole plane can be mapped conformally onto the unit disk. Riemann's proof was incomplete; the first valid proof was given by W. F. Osgood in 1900. Most proofs exhibit the required mapping as the solution of an extremal problem over an appropriate family of analytic univalent functions. (A univalent function is simply a one-to-one mapping.) For instance, the Riemann mapping maximizes $|f'(z_0)|$ among all analytic univalent functions which map the given region into the unit disk: $|f(z)| < 1$. Here z_0 is chosen arbitrarily in the region, and the Riemann mapping has $f(z_0) = 0$.

Multiply connected regions. There is no exact analog of the Riemann mapping theorem for multiply

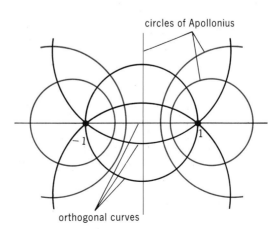

Fig. 3. Circles of Apollonius and members of the orthogonal family of curves.

connected regions. For instance, two annuli, $r_1 < |z| < r_2$ and $R_1 < |w| < R_2$, are conformally equivalent if and only if $r_2/r_1 = R_2/R_1$. Any doubly connected region can be mapped conformally onto a (possibly degenerate) annulus. Any finitely connected region (other than a punctured plane) can be mapped conformally onto the unit disk minus a system of concentric circular arcs, or onto the whole plane minus a system of parallel segments, or radial segments, or concentric circular arcs. Other canonical regions are bounded by arcs of lemniscates or logarithmic spirals, or by full circles.

Distortion theorems. Conformal mappings are often studied by considering the class S of functions $f(z)$ which are analytic and univalent in the unit disk and have the normalizing properties $f(0) = 0$ and $f'(0) = 1$. Alternatively, the class S may be defined as the class of all univalent power series of the form $f(z) = z + a_2 z^2 + a_3 z^3 + \cdots$ that are convergent for $|z| < 1$. The Koebe distortion theorem gives the sharp bounds, Eqs. (8), for all functions f in S and all points

$$r(1 + r)^{-2} \le |f(z)| \le r(1 - r)^{-2} \quad r = |z|$$
$$(1 - r)(1 + r)^{-3} \le |f'(z)| \le (1 + r)(1 - r)^{-3} \tag{8}$$

z in the disk. The Koebe 1/4-theorem asserts that each function f in S includes the full disk $|w| < 1/4$ in its range. Suitable rotations of the Koebe function (discussed above) show that each of these bounds is the best possible. All of these statements can be deduced from a theorem of L. Bieberbach (1916) that $|a_2| \le 2$ for all functions f in S, with equality occurring only for the Koebe function, given by Eq. (9),

$$k(z) = z(1 - z)^{-2} = z + 2z^2 + 3z^3 + \cdots \tag{9}$$

or one of its rotations, given by $e^{-i\theta}k(e^{i\theta}z)$ for some value of θ. Bieberbach conjectured that in general $|a_n| \le n$ for all n. For many years, the Bieberbach conjecture stood as a challenge and inspired the development of powerful methods in geometric function theory. After a long series of advances by many mathematicians, the final step in the proof of the Bieberbach conjecture was taken by L. de Branges in 1984.

Boundary correspondence. The open region inside a simple closed curve is called a Jordan region. Riemann's theorem ensures the existence of a conformal mapping of one Jordan region onto another. C. Carathéodory proved in 1913 that such a mapping can always be extended to the boundary and the extended mapping is a homemorphism, or a bicontinuous one-to-one mapping, between the closures of the two regions. In fact, Carathéodory proved a much more general theorem which admits inaccessible boundary points and establishes a homeomorphic correspondence between "clusters" of boundary points, known as prime ends. For a conformal mapping of the unit disk onto a Jordan region with a rectifiable boundary, the Carathéodory extension preserves sets of measure zero (or zero "length") on the boundary. It is also angle-preserving at almost every boundary point, that is, except for a set of points of measure zero. If the boundary of the Jordan region has a smoothly turning tangent direction, the derivative of the mapping function can also be extended continuously to the boundary. SEE MEASURE THEORY.

Quasiconformal mappings. A theory of generalized conformal mappings, known as quasiconformal

mappings, has evolved. Roughly speaking, a univalent function $w = f(z)$ is said to be K-quasiconformal in a certain region if it maps infinitesimal circles to infinitesimal ellipses in which the ratios of major to minor axes are bounded above by a constant $K \ge 1$. Equivalently, a mapping is quasiconformal if it distorts angles by no more than a fixed ratio. The simplest examples are linear mappings of the form $u = ax + by$, $v = cx + dy$, $ad - bc \ne 0$, where a, b, c, and d are real constants. The 1-quasiconformal mappings are simply the conformal mappings. The notion of a quasiconformal mapping is readily extended to higher dimensions.

Peter L. Duren

Bibliography. L. Bieberbach, *Conformal Mapping*, 1952; R. V. Churchill and J. W. Brown, *Complex Variables and Applications*, 1984; P. L. Duren, *Univalent Functions*, 1983; H. Kober, *Dictionary of Conformal Representations*, 1957; Z. Nehari, *Conformal Mapping*, 1975; Ch. Pommerenke, *Univalent Functions*, 1975.

Conformational analysis

The study of the energies and structures of conformations of organic molecules and their chemical and physical properties. Organic molecules are not static entities. The constituent atoms vibrate and groups rotate about the bond axes.

Linear structures. Rotation about the C—C bond in ethane, for example, results in an infinite number of slightly different structures called conformations, two of these are indicated below.

In conformation (I) of ethane, the pairs of C—H bonds on the two carbon atoms reside in a plane, and are termed eclipsed. (The circle represents the frontmost carbon atom, and the long bonds those from that carbon atom to the hydrogen atoms. The bonds from the circle represent those bonds from the rearward carbon to its hydrogen atoms.) Conformation (I) is called the eclipsed conformation. In conformation (II), the C—H bonds on one carbon reside between the C—H bonds on the other carbon. Conformation (II) is called the staggered conformation. An infinite number of conformations between (I) and (II) are possible; however, conformations (I) and (II) are of greatest interest because they are the maximum and minimum energy structures. In conformation (I), the electrons in the C—H bonds and the nuclear charges of the hydrogen atoms repel each other, resulting in a higher energy state. In conformation (II), the electrons and hydrogen atoms are at their greatest possible separation and the repulsion is at a minimum. The conversion of (II) to (III) by rotation of one of the methyl groups by 120° requires the input of energy, approximately 3.0 kcal (13 kilojoules) per mole, in order to pass through the eclipsed conformation. This amount is small compared to the thermal energy at room temperature, and the rotation about the C—C bond in ethane occurs at about 10^9–10^{10} times per second.

Butane provides a more complicated case. There are two different eclipsed (IV and VI) and two staggered (V and VII) conformations which are designated by the names below the structures. As the methyl group is larger than a hydrogen atom, (IV) is higher in energy than (VI), and (V) is higher in en-

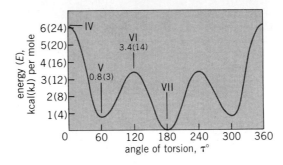

(I)

(II) (III)

Potential energy diagram for butane. (After E. L. Eliel et al., Conformational Analysis, Interscience Publishers, 1965)

ergy than (VII) [see **illus.**]. As the molecular weight of an alkane increases, the number of molecular conformations (combinations of all possible individual bond conformations) increases dramatically, although the antiperiplanar conformations are always favored.

(IV)
synperiplanar

(V)
synclinal

(VI)
anticlinal

(VII)
antiperiplanar

1,2-Dichloroethane also exists in conformations similar to those of butane (see interconverting conformations below). In the gas phase, the antiperiplanar

synclinal   anitperiplanar

1,2-Dicholorethane

conformation predominates. This is not due to a steric repulsion in the synclinal conformation, but to C—Cl bond dipole repulsion. (The bond dipole moment arises from the fact that chlorine is more electronegative than carbon.) In a polar solvent, the energy difference between the antiperiplanar and synclinal conformations is decreased due to stabilization of the synclinal form. In pure liquid 1,2-dichloroethane, the enthalpy difference is zero. Such solvent effects are not observed with nonpolar molecules such as the alkanes.

Cyclic structures. These also exist in various conformations. Cyclopropane, a planar structure, can exist in only one conformation. In cyclobutane, slight twisting about the C—C bonds can occur which relieves some of the C—H eclipsing strain energy. Cyclobutane exists in rapidly interconverting so-called butterfly conformations (VIII). Cyclopentane exists in two types of nonplanar conformations: the envelope conformation (IX) and the half-chair (or twist) conformation (X). The flap atom of (IX), the atom out of

(VIII)

(IX)   (X)

the plane of the other four, can migrate around the ring, as can also the twist in (X). These motions, called pseudorotation, require very little energy, and cyclopentane presents a very complex conformational system.

Cyclohexane exists predominantly in the chair conformation (XI), in which all C—C bond conformations are of the staggered type. Interconversion between chair conformations occurs rapidly at room temperature, passing through the high-energy, boat conformation (XII). In the chair conformation, there

(XI)   (XII)

are two distinctly different types of hydrogen atoms; one set oriented perpendicular to the general plane of the ring, called axial (a) hydrogens, and one set oriented parallel to the plane of the ring, called equatorial (e) hydrogens. These hydrogens interchange orientation on chair interconversion. Axial and equatorial hydrogens possess different chemical and physical properties, although at room temperature the interconversion occurs very rapidly ($\sim 10^5$ per second), and it is in general not possible to detect the differences. It is easy to design molecules in which the interconversion is not possible and the differences in properties become readily apparent.

A substituted cyclohexane, for example methylcyclohexane, exists in two different chair conformations which are in equilibrium. In the axial conformation (XIII), the methyl group is in proximity with the other two axial hydrogens on the same side of the ring. This steric repulsion disfavors the axial conformation and favors equatorial conformation (XIV). The mea-

(XIII)   (XIV)

surement of conformational equilibrium constants has allowed the construction of a relative size scale for organic functional groups.

Rotation about C=C bond. This requires a very large amount of energy, and is rarely observed except under very forcing conditions. Thus, although *cis*- and *trans*-2-butene can be considered to be conformational isomers, practically this does not occur.

An illustration of how the conformation of a molecule affects a chemical reaction is provided by the elimination of hydrogen bromide from 2-bromobutane, which can exist in either of the two staggered conformations (XV) and (XVI). The elimination of hydrogen bromide requires that the hydrogen and bromine be antiperiplanar. Thus, the elimination from (XV) produces *trans*-2-butene [reaction (1)], and from XVI, *cis*-2-butene [reaction (2)].

trans-2-Butene   (1)

(XV)

cis-2-Butene   (2)

(XVI)

Conformational analysis represents a very large area of study and has provided important information on the structures and reactivities of complex natural products such as steroids and alkaloids. SEE MOLECULAR ISOMERISM; STEREOCHEMISTRY.

Daniel J. Pasto

Bibliography. E. L. Eliel, *Stereochemistry of Carbon Compounds*, 1962; E. L. Eliel et al., *Conformational Analysis*, 1965; B. O. Ramsay, *Stereochemistry*, 1982.

Congenital anomalies

Structural abnormalities of the body that develop during embryogenesis and the fetal period; also called birth defects. Children with significant birth defects need more medical care than other children do, require more frequent hospitalizations, need community support services, and often require special education programs. Among children, over half of all visits to subspecialty medical clinics and admissions to hospitals are for treatment of disorders resulting from errors in embryonic development, chromosomal abnormalities, and genetic and familial disorders. Two-thirds of the deaths of infants and children in pediatric hospitals in developed countries are caused by underlying congenital anomalies.

Incidence. Although any particular birth defect may be rare, birth defects as a group are common. They are identified in 2–3% of newborns and in about an equal percentage later, in the first year of life, as structural defects and development delays become evident.

The overall incidence of congenital anomalies is about the same throughout the world, but certain geographic and racial groups have higher rates of certain disorders. For example, those developmental defects of the brain and spinal cord known as neural tube disorders are much more common in the British Isles than in North America, occurring in 10 per 1000 live births in Ireland as compared with 1–2 per 1000 in Canada and the United States.

Screening. Screening programs are available to identify fetuses and newborns likely to have disorders such as congenital malformations and genetic diseases. Examples of screening include amniocentesis to detect fetal chromosomal abnormalities in mothers over 35 years of age; measurement of maternal serum alpha-fetoprotein levels at 15–16 weeks' gestation to help identify fetuses with certain malformations; and screening of newborns for phenylketonuria, sickle cell disease, thalassemia, galactosemia, and congenital hypothyroidism. SEE ALPHA FETOPROTEIN; PHENYLKETONURIA; SICKLE CELL DISEASE.

Causes. Birth defects can be caused by genetic factors, exposure to malformation-causing agents (teratogens), or a combination of both. It is estimated that single-gene disorders cause 7.5% of birth defects identified in newborns and structural abnormalities of chromosomes cause 6%. Maternal infections that have been transmitted to the embryo or fetus are responsible for 2–3% of congenital anomalies. Maternal diabetes is associated with 1.5% and other maternal diseases with less than 1.5%. Maternal use of drugs and alcohol is responsible in about 1–2% of cases. In more than 50% of newborns with congenital anomalies, however, the specific etiology is not known. SEE DIABETES, FETAL ALCOHOL SYNDROME; HUMAN GENETICS.

Dysmorphology is the area of medicine and science concerned with the cause of congenital anomalies resulting from errors in embryonic development (dysmorphogenesis). The study of normal and abnormal embryonic development allows identification of the latest time in embryogenesis when a malformation could occur. Examples of such times include 28 days of gestation for neural tube defects, 36 days for cleft lip, and 10 weeks for cleft palate. Because the heart is an embryologically complicated organ, different congenital heart defects develop at different times in embryogenesis. Abnormalities of digestive tract formation can occur as early as 30 days of gestation and as late as 10 weeks.

Classification. To facilitate the study, birth defects are divided into malformations, disruptions deformations, and dysplasias.

Malformations. Malformations are structural defects that are caused by primary errors in morphogenesis. They are classified as major and minor. Major malformations require medical or surgical intervention or are of substantial cosmetic importance. Various defects that were mentioned above are among such malformations. Minor malformations do not require such treatment or do not greatly affect appearance; examples include protruding ears, wide-set eyes, curvature of the fifth finger, webbing of the toes, and many others. Most individuals have two or three minor malformations, and the same minor malformations are frequently found in various family members. An excess of minor malformations is often associated with major malformations.

Disruptions. These are structural defects resulting from interruption of normal morphogenesis, with consequent destruction of previously existing structures and incomplete development of tissues. For example, if the fetus becomes entangled in the amnion (the inner lining of the placental sac), ring constrictions of the limbs can occur.

Deformations. These are congenital anomalies resulting from external compression of a normally formed part of the fetus. The most common deformations are bowing of the lower legs and in-toeing of the feet. In otherwise normal infants, minor deformations usually improve without treatment. If deformations are major, splints and special shoes may be required for correction.

Dysplasias. These are disorders that result from an abnormal organization of cells into tissues; the morphological result is called dyshistogenesis. Examples of such disorders include hemangiomas and lymphangiomas, which result from overgrowth of blood vessels and lymphatic vessels, respectively. Neurofibromatosis is an autosomal dominant genetic disorder that results in overgrowth of neural tissue.

Syndromes. Some individuals have various major and minor congenital anomalies that together form a recognizable pattern, called a syndrome. The medical diagnosis of a specific congenital syndrome in a child aids in determining etiology, identifying associated anomalies that previously were unrecognized, and establishing prognosis. Often, further diagnostic studies are required to confirm the diagnosis of a syndrome. Syndromes can have both genetic and environmental causes.

Teratology. Tetratology is the study of the effect of environmental agents on the developing embryo and fetus. Teratogens are agents that interfere with normal embryonic development. They can cause miscarriages, retard prenatal growth, and produce congenital anomalies or mental retardation.

There are five general groups of teratogens: (1) infectious diseases and agents, including rubella, toxoplasmosis, syphilis, cytomegalovirus, chickenpox, Venezuelan equine encephalitis, and human immunodeficiency virus; (2) physical agents, such as radiation, maternal hyperthermia, and abnormalities of the intrauterine environment; (3) drugs and chemical agents; (4) maternal metabolic and genetic factors, such as diabetes, epilepsy, smoking, and nutritional deficiencies; and (5) paternal factors, although rare. (Paternal exposures contribute mainly to decreased fertility rather than birth defects.)

Treatment, detection, and prevention. Treatment of congenital anomalies is specific for each individual. Individuals with severe or numerous abnormalities usually require multidisciplinary treatment, including such measures as medical management, surgical correction, nursing care, special diets, rehabilitation, prosthetic devices, special education, and community support.

If a couple has had a child with a congenital anomaly, tests may be done in subsequent pregnancies to help detect recurrence. Prenatal tests that are sometimes performed include chorionic villus sampling and amniocentesis to detect chromosomal and some metabolic defects, detailed ultrasound imaging of the fetus to identify significant structural malformations, and fetal echocardiograms to detect congenital heart disease.

Measures that help reduce the risk of having a child with congenital anomalies include avoidance of teratogenic exposures, medical treatment of maternal illnesses, good nutrition, and routine obstetrical care. However, much remains to be learned about the cause, detection, and prevention of congenital anomalies. SEE PREGNANCY.

Margot I. Van Allen

Bibliography. K. L. Jones, *Smith's Recognizable Patterns of Human Malformation*, 4th ed., 1988; T. H. Shepard, *Catalog of Teratogenic Agents*, 5th ed., 1986; J. S. Thompson and M. W. Thompson, *Genetics in Medicine*, 1986.

Conglomerate

The consolidated equivalent of gravel. Conglomerates are aggregates of more or less rounded particles greater than 0.08 in. (2 mm) in diameter. Frequently they are divided on the basis of size of particles into pebble (fine), cobble (medium), and boulder (coarse) conglomerates. The common admixture of sand-sized and gravel-sized particles in the same deposit leads to further subdivisions, into conglomerates (50% or more pebbles), sandy conglomerates (25–50% pebbles), and pebbly or conglomeratic sandstones (less than 25% pebbles). The pebbles of conglomerates are always somewhat rounded, giving evidence of abrasion during transportation; this distinguishes them from some tillites and from breccias, whose particles are sharp and angular (see **illus.**).

Conglomerates fall into two general classes, based on the range of lithologic types represented by the pebbles, and on the degree of sorting and amount of matrix present. The well-sorted, matrix-poor conglomerates with homogeneous pebble lithology are one type, and the poorly sorted, matrix-rich conglomerates with heterogeneous pebble lithology are the other. The well-sorted class includes quartz-pebbles, chert-pebble, and limestone-pebble conglomerates. The quartz pebbles were derived from long, continued erosion of the source-rock terrain that resulted in the disappearance of all unstable minerals, such as feldspars and ferromagnesian minerals, that make up most igneous and metamorphic rocks. Typically the quartz-pebble conglomerates are thin sheets that overlie an unconformity and are basal to a series of overlapping marine beds. Chert pebble conglomerates are derived from weathered limestone terrains. Limestone pebble conglomerates seemingly are the result of special conditions involving rapid mechanical erosion and short transport distances; otherwise the lime-

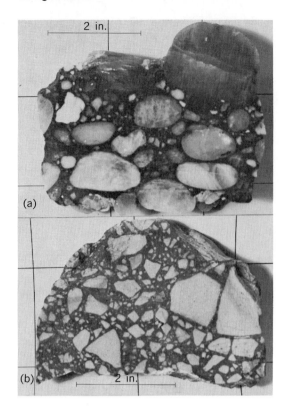

(a)

(b)

Lithified gravels. (a) Conglomerate, composed of rounded pebbles. (b) Breccia, containing many angular fragments. 2 in. = 5 cm. (Specimens from Princeton University Museum of Natural History; photo by Willard Starks)

of these pebbles, they have often been called breccias. SEE BRECCIA; GRAVEL; GRAYWACKE; TILL.

Raymond Siever

Conglutination

A term used in serology to describe the completion or enhancement of an incomplete agglutinating system by the addition of certain substances. Some bacteria or erythrocyte suspensions do not exhibit the visible agglutination ordinarily expected after they have been coated with their specific antibodies and complement. Further addition of a conglutinating agent—normal bovine serum—however, initiates visible agglutination. Complement is an essential component of the system. The conglutination reagent is itself without effect in the absence of antibody and complement. Similar overall actions, such as agglutination of Rh$^+$ cells, occur when human serum, gelatin, or bovine serum albumin are added to the incomplete Rh antibodies found in some sera, but since this enhancing effect occurs also in the absence of complement, these reactions probably are to be distinguished in mechanism from that of the traditional conglutination. SEE BLOOD GROUPS.

Like the related complement-fixation reaction, the conglutination reaction can be used for a variety of serological diagnostic reactions in bacterial and viral infections. SEE AGGLUTINATION REACTION; COMPLEMENT-FIXATION TEST; SEROLOGY.

Henry P. Treffers

Bibliography. C. W. Bennett, *Clinical Serology*, 1977; N. J. Bryant, *Laboratory Immunology and Serology*, 1978; R. R. A. Coombs, A. M. Coombs, and D. G. Ingram, *The Serology of Conglutination and Its Relation to Disease*, 1961.

Conic section

One of the class of curves in which a plane may cut a cone (surface) of revolution. They were extensively studied by the ancient Greeks. The section is a parabola if the plane is parallel to an element of the cone, an ellipse or circle if the plane cuts all elements of one nappe (but does not go through the apex), and a hyperbola if the plane cuts elements of both nappes (for example, the plane parallel to the cone's axis of revolution) and does not go through the apex (see **illus.**). If a line g intersects the cone's axis perpendicularly at a point distinct from the apex, a plane that revolves about g will cut all three kinds of conics from the cone. If a plane contains the cone's axis, it intersects the cone in a pair of lines (a degenerate hyperbola); if it goes through the apex but does not contain the axis, it has in common with the cone either the apex alone (degenerate ellipse) or an element of the cone (degenerate parabola).

Algebraic definitions. After the advent of analytic geometry, the synthetic method used by the Greeks to develop the properties of conics gave way to algebraic procedures which, using other definitions, divorced these curves from their original relationship to cones and regarded them as graphs of second-degree equations in cartesian coordinates x, y. Let plane π cut all the elements of one nappe of a cone C (π not perpendicular to the cone's axis and not passing through the apex), and let E denote the curve of intersection of π

stone would quickly become degraded to finer sizes and dissolve. The well-sorted conglomerates tend to be distributed in thin, widespread sheets, normally interbedded with well-sorted, quartzose sandstones. SEE UNCONFORMITY.

The poorly sorted, lithologically heterogeneous conglomerates include many different types, all related in having very large amounts of sandy or clayey matrix and pebbles of many different rock classes. The graywacke conglomerates are the outstanding representatives. They are composed of pebbles, many times only slightly rounded, or many different kinds of igneous and metamorphic rock as well as sedimentary rocks, bound together by a matrix that is a mixture of sand, silt, and clay. They seem to have been formed from the products of rigorous mechanical erosion of highlands and transported in large part by turbidity or density currents. Another representative of the poorly sorted class is fanglomerate, a conglomerate formed on an alluvial fan. Fanglomerates in many cases are much better sorted than the graywacke conglomerates but have heterogeneous pebble composition. A tillite is another representative of this class of conglomerate. All of these poorly sorted conglomerates tend to occur in fairly thick sequences, and some of them, typically the fanglomerates, are wedge-shaped accumulations. SEE SEDIMENTARY ROCKS; TURBIDITY CURRENT.

Special types of conglomerates, such as volcanic conglomerates and agglomerates and some intraformational conglomerates composed of shale pebbles or deformed limestone pebbles, do not seem to fall easily into either class. Some of the shale and limestone intraformational conglomerates have formed from the tearing up of previously deposited beds at the sea bottom while they were still relatively soft and only partially consolidated. Because of the angularity of some

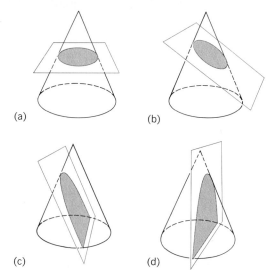

Four conic sections. (*a*) Circle. (*b*) Ellipse. (*c*) Parabola. (*d*) Hyperbola.

and C. To show that E is an ellipse, consider the spheres S_1 and S_2 inside C that are tangent to π and to all the elements of C and that lie on the opposite sides of π. The contact points of S_1 and S_2 with C form circles whose planes π_1 and π_2 are perpendicular to the axis of C. Let F_1 and F_2 denote the contact points of S_1 and S_2, respectively, with π. If P is any point of E, then Eqs. (1) holds, where Q_1, Q_2 are

$$\text{dist } (P,F_1) = \text{dist } (P,Q_1) \qquad (1a)$$

$$\text{dist } (P,F_2) = \text{dist } (P,Q_2) \qquad (1b)$$

those contact points of S_1, S_2, respectively, that lie on the element of C through P. Since dist (P,Q_1) + dist (P,Q_2) is the (constant) distance between the parallel planes π_1, π_2, then Eq. (2) holds, and so E is an

$$\text{dist } (P,F_1) + \text{dist } (P,F_2) = \text{constant} \qquad (2)$$

ellipse. The directrices of E are the lines in which π_1, π_2 intersect π. *SEE ANALYTIC GEOMETRY.*

Conics are basic in astronomy, where it is shown that the newtonian inverse-square law of gravitation implies that the orbits of the planets are ellipses, with the Sun at one focus. Other heavenly bodies, certain comets, for example, have parabolic and hyperbolic paths.

Projective geometry. The third phase in the study of conic sections began with the development of projective geometry. In that subject it is shown that the locus of the points of intersection of two projective (nonperspective), coplanar (nonconcentric) pencils of lines is a nondegenerate (point) conic. All nondegenerate conics are projectively equivalent; that is to say, if C_1, C_2 are any two such conics, there is a projective transformation of their plane into itself that transforms C_1 into C_2. Conics may be classified into types by their behavior with respect to an arbitrarily selected line g, say the ideal line of the plane. In the geometry (affine) obtained by considering only those projective transformations that transform g into itself, a conic is a hyperbola, a parabola, or an ellipse accordingly as the number of points that the conic has in common with g is two, one, or none, respectively.

Dual to point conics are line conics, the locus of all lines joining corresponding points of two projective (not perspective), coplanar (not collinear) point ranges. The tangents to a point conic form a line conic, and the contact points of a line conic form a point conic. One of the most famous theorems concerning conics was proved by the French philosopher-mathematician Blaise Pascal in 1640, when only 16 years old. The theorem asserts that if six points A_i, with $i = 1, 2, \ldots, 6$, are on a conic, then the intersections P,Q,R of the lines (A_1,A_2) and (A_4,A_5), (A_2,A_3) and (A_5,A_6), (A_3,A_4) and (A_6,A_1), respectively, lie on a line. This theorem, together with its converse (which is also valid) had numerous interesting consequences. The Brianchon theorem, proved 166 years later, is its dual. It states that if a hexagon circumscribes a conic, the three lines joining the three pairs of opposite vertices are concurrent. *SEE CONE; ELLIPSE; HYPERBOLA; PARABOLA; PROJECTIVE GEOMETRY.*

Leonard M. Blumenthal

Bibliography. C. C. Carico and I. Drooyan, *Analytic Geometry*, 1980; W. J. Fishback, *Projective and Euclidean Geometry*, 1962; G. Fuller, *Analytic Geometry*, 5th ed., 1979; L. E. Garner, *Outline of Projective Geometry*, 1981; R. R. Middlemiss, *Analytic Geometry*, 3d ed., 1968; B. Spain, *Analytic Conics*, 1957.

Conjugation and hyperconjugation

A higher-order bonding interaction between electron orbitals on three or more contiguous atoms in a molecule, which leads to characteristic changes in physical properties and chemical reactivity. One participant in this interaction can be the electron pair in the π-orbital of a multiple (that is, double or triple) bond between two atoms, or a single electron or electron pair or electron vacancy on a single atom. The second component will be the pair of π-electrons in an adjacent multiple bond in the case of conjugation, and in the case of hyperconjugation it will be the pair of electrons in an adjacent polarized σ-bond (that is, a bond where the electrons are held closer to one atom than the other due to electronegativity differences between the two atoms). *SEE CHEMICAL BONDING.*

Conjugation. The conjugated orbitals reside on atoms that are separated by a single bond in the classical valence-bond molecular model, and the conjugation effect is at a maximum when the axes of the component orbitals are aligned in a parallel fashion because this allows maximum orbital overlap. Conjugation thus has a stereoelectronic requirement, or a restriction on how the participating orbitals must be oriented with respect to each other. Two simple examples are shown in **Fig. 1**; in 1,3-butadiene ($H_2C{=}HC{-}CH{=}CH_2$) conjugation occurs between the p orbitals (π-bonds) of the two double bonds, and in methyl vinyl ether ($H_2C{=}HC{-}O{-}CH_3$) the nonbonding sp^3 orbital on oxygen is conjugated with the p orbitals of the double bond. This interaction is manifest in an effective bond order between single and double for the underlined single bond.

If the component orbitals are orthogonal (at right angles) to each other, or if the atoms bearing these orbitals are separated by more than one single bond or by no bond, the molecule is not conjugated. Two examples of molecules which do not exhibit conjugation are 1,4-pentadiene ($H_2C{=}CH{-}CH_2{-}CH{=}CH_2$) and allene ($H_2C{=}C{=}CH_2$), the former because there is more than

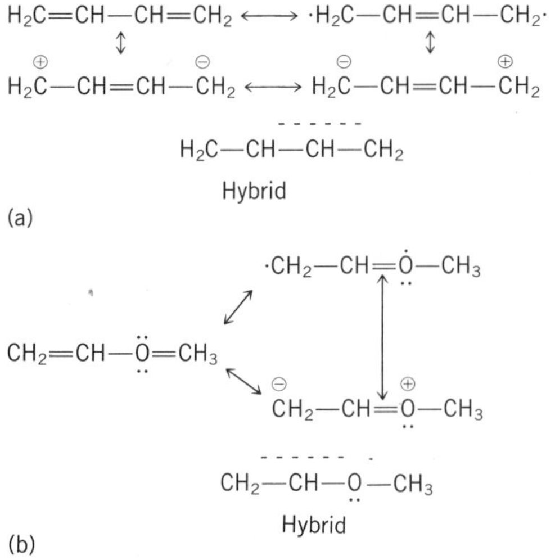

$$H_2C\!\!=\!\!CH\!\!-\!\!CH\!\!=\!\!CH_2 \longleftrightarrow \cdot H_2C\!\!-\!\!CH\!\!=\!\!CH\!\!-\!\!CH_2\cdot$$

$$\overset{\oplus}{H_2C}\!\!-\!\!CH\!\!=\!\!CH\!\!-\!\!\overset{\ominus}{CH_2} \longleftrightarrow \overset{\ominus}{H_2C}\!\!-\!\!CH\!\!=\!\!CH\!\!-\!\!\overset{\oplus}{CH_2}$$

$$H_2C\!\!-\!\!CH\!\!-\!\!CH\!\!-\!\!CH_2$$

Hybrid

(a)

$$\cdot CH_2\!\!-\!\!CH\!\!=\!\!\overset{\cdot}{\underset{\cdot\cdot}{O}}\!\!-\!\!CH_3$$

$$CH_2\!\!=\!\!CH\!\!-\!\!\overset{\cdot\cdot}{\underset{\cdot\cdot}{O}}\!\!=\!\!CH_3$$

$$\overset{\ominus}{CH_2}\!\!-\!\!CH\!\!=\!\!\overset{\oplus}{\underset{\cdot\cdot}{O}}\!\!-\!\!CH_3$$

$$CH_2\!\!-\!\!CH\!\!-\!\!O\!\!-\!\!CH_3$$

Hybrid

(b)

Fig. 1. Conjugated molecules. Broken overbars indicate the effects of conjugation. (*a*) 1,3-Butadiene. (*b*) Methyl vinyl ether.

one single bond between the component orbitals and the latter because the component orbitals are orthogonal to each other and are not separated by a single bond.

Chemists have used the conjugation rationale since the midnineteenth century, and it has become well accepted. The valence-bond model of conjugation invokes delocalization: the participating electrons are no longer localized or fixed on a particular atom or between a pair of atoms, but are shared throughout the conjugated orbitals. This increases their entropy, which contributes to a lower overall energy and generally results in greater stabilization for the molecule. From the valence-bond point of view, conjugated molecules are viewed as a weighted average hybrid of two or more valence-bond structures as shown in Fig. 1 for the two examples. Thus butadiene has a major contribution from the standard valence-bond structure, $H_2C\!\!=\!\!CH\!\!-\!\!CH\!\!=\!\!CH_2$, but the diradicals or the structures with separation of charge are valid alternative representations, and the same is true for methyl vinyl ether. The effects of conjugation are suggested by the broken lines on the hybrid structures in Fig. 1. When more orbital units are linked together or conjugated, more resonance structures can be drawn, and the molecule will be more stable and behave less according to the expectation based on the simple structure drawn from the classical valence-bond model. *See* CHEMICAL STRUCTURES; DELOCALIZATION; ENTROPY; RESONANCE (MOLECULAR STRUCTURE).

Three of the four resonance contributors for butadiene are drawn with only one multiple bond compared to the two multiple bonds in the standard valence-bond structure (Fig. 1*a*); this is known as sacrificial conjugation, in that some bonds in the standard depiction are sacrificed. In methyl vinyl ether (Fig. 1*b*) and in other molecules or chemical intermediates, the different pictures drawn for the resonance contributors have the same number of multiple bonds; this is known as isovalent conjugation and results in a greater resonance energy stabilization than in the case of sacrificial conjugation. Other examples of isovalent conjugation include the allyl cation ($CH_2\!\!=\!\!CH\!\!-\!\!CH_2{}^+$ and $^+CH_2\!\!-\!\!CH\!\!=\!\!CH_2$) and benzene.

Conjugation is better understood by considering the molecular orbitals which can be approximated by the mathematical combination of the wave functions for the constituent atomic *s* and *p* orbitals (and *d* orbitals for elements beneath the first row of the periodic table). **Figure 2** shows simple pictorial results of complex molecular orbital (MO) calculations for the π-electrons in butadiene; here the σ-bonds are neglected for clarity. The combination of four atomic *p* orbitals (two each for each of the double bonds) leads to four molecular orbitals. Each of the molecular orbitals is a quartet of "*p*-like" orbitals of different size (determined by calculation) situated on the four carbon atoms of butadiene (C^1–C^4). The four π-electrons are placed in these orbitals from the bottom up, so that the two lowest-energy orbitals, MO #1 and MO #2, both contain paired electrons. These two molecular orbitals thus account for the π-bonding in butadiene. In MO #1 the continuous bonding across all four lobes can be seen, with the highest orbital density between C^2 and C^3. In MO #2 there is a phase change between C^2 and C^3, with consequent repulsion between the adjacent lobes, but this repulsion is relatively less important because the electron orbital density (depicted as orbital size) is much lower here. In

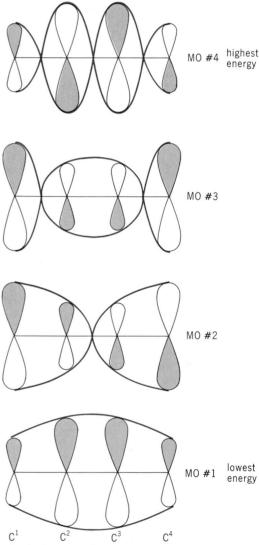

MO #4   highest energy

MO #3

MO #2

MO #1   lowest energy

C^1 C^2 C^3 C^4

Fig. 2. Molecular orbital (MO) diagram for butadiene as determined by calculation. Algebraic phase is indicated by the presence or absence of tint. Orbital density is depicted as orbital size.

both of these molecular orbitals, the bonding interactions between C^1—C^2 and C^3—C^4 are in phase and add up to strong π-bonds. Thus, the π-delocalization interaction across C^2—C^3 that characterizes conjugation appears very naturally from the molecular orbital description. *SEE MOLECULAR ORBITAL THEORY.*

Hyperconjugation. This concept posits the conjugation of polarized σ-bonds with adjacent π-orbitals, and was introduced in the late 1930s by R. S. Mulliken. This rationale was used to explain successfully a wide variety of chemical phenomena; however, the confusing adaptation of the valence-bond model necessary to depict it and its inappropriate extension to some phenomena led to difficulties. The advent of the molecular orbital treatment has eliminated many of these difficulties.

Early on, hyperconjugation was used to explain the stabilization by alkyl groups of carbocations, or positively charged trivalent carbon. **Figure 3** shows how the orbitals of the σ-bonds of a methyl group (H_3C) exert a hyperconjugative stabilizing effect upon a neighboring p orbital on a methylene group (CH_2). The bonding molecular orbitals of the methyl group utilize aspects of its carbon p orbitals; in both of the conformations or rotations the methyl group has such an orbital which overlaps in phase with the neighboring p orbital on CH_2. As hydrogen is an electropositive atom, the methyl group acts as an electron donor to stabilize an empty p orbital and to destabilize a filled p orbital on the right-hand carbon. This is in accord with empirical observations. If H^1 is replaced with some other atom X, the hyperconjugative effect of X will be at a maximum in conformation A, where there is orbital density on X, and it will be at a minimum in conformation B, where X is in the nodal or null plane. If X is a more electropositive atom such as silicon, the C—X bond is a donor bond and will stabilize an adjacent empty p orbital, such as in a cation. If X is a more electronegative element such as fluorine, the C—X bond is an acceptor bond and will stabilize an adjacent filled p orbital such as in an anion; this result accords with experiment as well. The donor and acceptor effects are also important where the single p orbital is replaced by a multiple bond. *SEE REACTIVE INTERMEDIATES.*

Physical properties and chemical reactivity. The most notable differences in physical properties attributed to conjugation are as follows: (1) The single bonds lying between the two orbital components are shortened. For example, the length of the C^2—C^3 bond in butadiene is 0.148 nanometer, as compared to 0.154 nm for the carbon-carbon single bond in ethane. This shortening can also be explained by the changes in orbital hybridization at these carbons. (2) Electronic absorption spectra begin at significantly longer wavelengths for conjugated than for related unconjugated molecules. (3) Ionization potentials are lower than for isomeric unconjugated molecules. (4) Related to points 2 and 3 is the fact that polarizabilities are larger for conjugated than for the corresponding unconjugated dienes. Polarizability is the ease with which the electron orbitals (electron clouds) of the molecule can be distorted through dipole-dipole interactions with other molecules, ions, atoms, or electrons; it is generally a measure of how loosely the electrons are held by the nuclei. (5) As compared with predictions from formulas for unconjugated molecules, conjugated molecules show lower energies, for example, about 6 kcal/mol (25 kilopascals/mol)

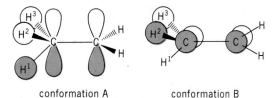

conformation A   conformation B

Fig. 3. Diagram showing hyperconjugative effect of orbitals of the σ-bonds of a methyl group (H_3C) on a neighboring p orbital of a methylene group (CH_2). Algebraic phrase is indicated by the presence or absence of tint.

lower for 1,3-butadiene than for unconjugated dienes; the excess is describable theoretically as resonance energy (valence-bond picture) or delocalization (molecular orbital theory picture). (6) The magnitude and orientation of dipole moments can be derived from hyperconjugation arguments, although the aforementioned differences in hybridization can also explain this. *SEE BOND ANGLE AND DISTANCE; DIPOLE MOMENT.*

In terms of changes in chemical reactivity, the following are exemplary: (1) The acidity of carboxylic acids stems from the conjugation stabilization of their ionized forms. (2) The addition of many chemical reactants takes place across the conjugated system (conjugate addition) rather than at the isolated double bonds, for example, the addition of bromine to butadiene to give a mixture of 1,4-dibromo-2-butene and 1,2-dibromo-3-butene. (3) The well-recognized orientation rules for the addition of chemical reagents to substituted benzenes and other aromatic systems can be rationalized through conjugated intermediates. (4) The facile formation of an anion from cyclopentadiene and of a cation from cycloheptatriene derives from the conjugation stabilization for those ions. (5) The chemistry of unsaturated organosilicon compounds can be explained only by hyperconjugation effects. (6) Certain changes in chemical reaction rates that occur when hydrogen atoms are substituted by deuterium atoms are clearly hyperconjugative effects.

A number of scientific phenomena depend on the properties of conjugated systems; these include vision (the highly tuned photoreceptors are triggered by molecules with extended conjugation), electrical conduction (organic semiconductors such as polyacetylenes are extended conjugated systems), color (most dyes are conjugated molecules designed to absorb particular wavelengths of light), and medicine (a number of antibiotics and cancer chemotherapy agents contain conjugated systems which trap enzyme sulfhydryl groups by conjugate addition). *SEE VALENCE.*

Matthew F. Schlecht

Bibliography. M. J. S. Dewar, *Hyperconjugation,* 1962; I. Fleming, *Frontier Orbitals and Organic Chemical Reactions,* 1976; J. March, *Advanced Organic Chemistry,* 3d ed., 1985; R. S. Mulliken, C. A. Rieke, and W. G. Brown, Hyperconjugation, *J. Amer. Chem. Soc.,* 63:41, 1941; V. J. Shiner, Jr., and E. Campaigne (eds.), Papers from the Conference on Hyperconjugation at Indiana University, June, 1958, *Tetrahedron,* 5:105–274, 1959.

Connecting rod

A link in several kinds of mechanisms. Usually one end of a connecting rod is intended to follow a circular path, while the other end follows a path along a

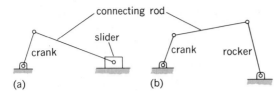

Fig. 1. Connecting rod in (a) slider-crank mechanism and (b) crank-and-rocker mechanism.

straight line or a curve of large radius. The term is sometimes applied, however, to any straight link that transmits motion or power from one linkage to another within a mechanism. **Figure 1** shows conventional arrangements of connecting rods in typical mechanisms. In some applications (for example, the connecting rod between the crank and an overhead oscillating member or walking beam in a well-drilling rig, or between the steering column and cross-links in an automobile) the connecting rod is called a pitman. SEE AUTOMOTIVE STEERING.

The connecting rod of the four-bar linkage, often called the coupler, has special significance. The motion of its plane can now be synthesized to furnish desired paths for points not necessarily on the straight line *AB* (coupler-point paths), or desired positions of the entire connecting-rod plane. The connecting rod is then not primarily used for transmission of force or motion from input to output crank, but the entire mechanism is employed to impart to the connecting-rod plane certain displacements, and sometimes velocities and accelerations. **Figure 2** illustrates paths

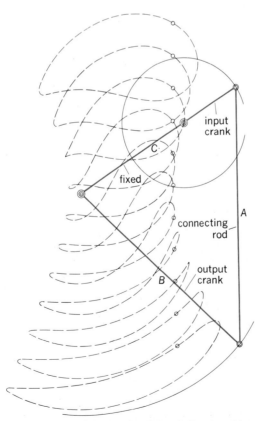

Fig. 2. Paths of points on connecting-rod plane traced by four-bar crank-and-rocker mechanisms for *A* = 4.0, *B* = 3.5, *C* = 2.0. *(After J. A. Hrones and G. L. Nelson, Analysis of Four-Bar Linkage, John Wiley and Sons, 1951)*

traced by various points in the plane of the rod, also revealing some velocities and accelerations there. SEE FOUR-BAR LINKAGE; MECHANISM.

Douglas P. Adams

Bibliography. E. A. Dijksman, *Motion Geometry of Mechanisms*, 1976; G. H. Martin, *Kinematics and Dynamics of Machines*, 2d ed., 1982; A. Ramous, *Applied Kinematics*, 1972.

Connective tissue

One of the four primary tissues of the body. It differs from the other three tissues in that the extracellular components (fibers and intercellular substances) are abundant. It cannot be sharply delimited from the blood, whose cells may give rise to connective tissue cells, and whose plasma components continually interchange with and augment the ground substance of connective tissue. Bone and cartilage are special kinds of connective tissue. SEE BLOOD.

Functions. The functions of connective tissues are varied. They are largely responsible for the cohesion of the body as an organism, of organs as functioning units, and of tissues as structural systems. This cohesive function is achieved through their permeation of other tissues of the body. The connective tissues are essential for the protection of the body both in the elaborate defense mechanisms against infection and in repair from chemical or physical injuries. Nutrition of nearly all cells of the body and the removal of their waste products are both mediated through the connective tissues.

Connective tissues are important in the development and growth of many structures. Constituting the major environment of most cells, they are probably the major contributor to the homeostatic mechanisms of the body so far as salts and water are concerned. They act as the great storehouse for the body of salts and minerals, as well as of fat. The connective tissues determine in most cases the pigmentation of the body. Finally, the skeletal system (cartilage and bones) plus other kinds of connective tissue (tendons, ligaments, fasciae, and others) make motion possible.

Cellular components. The connective tissues consist of cells and extracellular or intercellular substance (see **illus.**). The cells include many varieties, of which the following are the most important: (1) fibroblasts—variable in shape, frequently outstretched to form many fine processes, some as small as 20–30 nanometers; (2) macrophages (histiocytes)—variable in shape, strongly motile or outstretched, frequently with numerous minute cell processes, highly phagocytic (the ability to take particulate material into the cytoplasm); (3) mast cells—frequently large, ameboid, nucleus centrally located, cytoplasm very often packed with granules about 0.4 micrometer in diameter; (4) plasma cells—generally ovoid, weakly ameboid, nucleus eccentric, granular cytoplasm staining strongly with basic dyes (toluidine blue, methylene blue, hematoxylin); (5) melanocytes—cytoplasm of highly branched cells filled with minute granules which appear yellow or brown because of the melanin they contain; and (6) fat cells of two kinds: (a) yellow fat cells—very large (80 μm or more), generally spherical, with a thin shell of protoplasm (3.0–0.1 μm or less) enclosing a single enlarged fat droplet which appears yellowish; and (b) brown fat cells—moderately large, generally spherical, with small droplets of variable size (about 2 μm or less), scattered in the cytoplasm. In addition, connective tissues may contain, especially in the vicinity of small blood vessels, undifferentiated mesenchymal cells which may be considered as potentially embryonic, and blood cells which have wandered through the walls of blood vessels, including lymphocytes and polymorphonuclear granular leukocytes. It is especially important to realize that most of the cells of the connective tissue are developmentally related even in the adult; for example, fibroblasts may be developed from histiocytes or from undifferentiated mesenchymal cells.

Extracellular components. The extracellular components of connective tissues may be fibrillar or nonfibrillar. The fibrillar components are (1) reticular fibers which may be stained differentially with certain silver stains; (2) collagenous fibers which have a longitudinally striated appearance with the light microscope and are readily stainable with the acid dyes cosin and aniline blue; and (3) elastic fibers which are highly refractile, appear slightly yellowish, and may be stained differentially by certain dyes like orcein. With the higher resolution achieved with the electron microscope, both reticular and collagenous fibers are seen to be made up of smaller fibrils of about 20–100 nm in diameter, each with an internal cross-banded structure which is repeated on an average of every 64 nm. The unit of structure is tropocollagen. This is an elongated protein molecule about 300 nm long and about 1.3 nm in diameter, and is made up of three subunits. These are wound spirally around each other in very specific manners. The elastic fibers are also resolvable into finer fibrils which may be 8 nm or larger in diameter. All fibers visible with the light microscope are thus composed of bundles of submicroscopic fibrils visible only with the electron microscope. In addition, there are many collagenous and elastic fibers composed of the same fibrillar constituents which are too fine to be visible with the light microscope (fibers that are less than 0.2 μm in diameter). *See* Collagen.

The nonfibrillar component of connective tissues appears amorphous with the light microscope and is the matrix in which cells and fibers are embedded. It consists of two groups of substances: (1) those probably derived from secretory activity of connective tissue cells including mucoproteins, protein-polysaccharide complexes, tropocollagen, and antibodies; and (2) those probably derived from the blood plasma, including albumin, globulins, inorganic and organic anions and cations, and water. In addition, the ground substance contains metabolites derived from, or destined for, the blood.

The ground substance of local origin and the tissue fluid of hematogenous origin are not only coextensive but are dynamically related in a two-phase system. The two phases of this thermodynamic system are probably recognizable in the electron microscope as a series of small, watery vacuoles (about 60 nm in diameter) embedded in a denser matrix. The basement membrane is a specially differentiated sheetlike part of the ground substance (10–50 nm thick) which underlies directly many epithelia, and encloses muscle fibers and fat cells. It is denser than the adjacent ground substance, with which it is continuous, and may contain fibrils. There is some evidence that epithelial cells may contribute some constituents to the basement membrane.

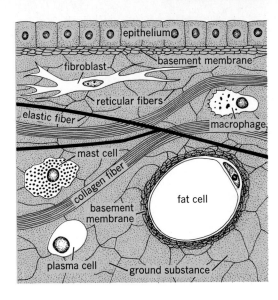

Components of connective tissue.

Classification. All the manifold varieties of connective tissue may contain all the cells and fibers discussed above in addition to ground substance. They differ from each other in the relative occurrence of one or another cell type, in the relative proportions of cells and fibers, in the preponderance and arrangement of one or another fiber, and in the relative amount and chemical composition of ground substance. In the classification which follows, some of these differences are pointed out.

1. Irregularly arranged connective tissue—which may be loose (subcutaneous connective tissue) or dense (dermis). The dominant fiber type is collagen, though others may be present. The fibers are irregularly interwoven and the cells irregularly distributed. The looser type has more ground substance than the denser type.

2. Regularly arranged connective tissue—primarily collagenous—with the fibers arranged in certain patterns depending on whether they occur in tendons or as membranes (dura matter, capsules, fasciae, aponeuroses, or ligaments). *See* Tendon.

3. Mucous connective tissue—ground substance especially prominent (umbilical cord).

4. Elastic connective tissue—predominance of elastic fibers or bands (ligamentum nuchae) or lamellae (aorta). *See* Ligament.

5. Reticular connective tissue—fibers mostly reticular, moderately rich in ground substance, frequently numerous undifferentiated mesenchymal cells. In contrast with collagenous fibers which are unbranched, reticular fibers are branched and may be selectively stained by silver methods. These branched fibers, however, consist of submicroscopic fibrils which have the same submicrostructure as unbranched collagen fibrils.

6. Adipose connective tissue—yellow or brown fat cells constituting chief cell type, reticular fibers most numerous. *See* Adipose tissue.

7. Pigment tissue—melanocytes numerous.

8. Cartilage—cells exclusively of one type, derived from mesenchymal cells. The latter persist even in the adult in the perichondrium which encloses cartilage. The extracellular matrix is of three types: (1) hyaline (joint cartilage), containing no bundles of fibers; (2) fibrocartilage (patella), containing numerous coarse bundles of collagen fibers; and (3) elastic car-

tilage (external ear), containing elastic fibers. Hyaline cartilage does, however, contain collagen fibrils of submicroscopic dimensions. In all three types of cartilage, the fibrils and fibers are embedded in a homogeneous matrix, which is almost entirely composed of protein-polysaccharide complex, whose predominant polysaccharide is chondroitin sulfate. SEE CARTILAGE.

9. Bone—cells predominantly osteocytes, but also include fibroblasts, mesenchymal cells, endothelial cells, and osteoclasts. The fibers are almost entirely collagenous in nature, and may be highly organized into osteones. The fibers are embedded in a ground substance matrix. The outstanding property of bone is its hardness, due to deposition in the extracellular regions of many minute crystals of a calcium mineral, apatite. SEE BONE.

In some forms of connective tissue (fat) blood capillaries are extremely numerous. In others (cartilage and dentine) blood capillaries are lacking. Tendon is nearly avascular, and bone is also poorly supplied with blood capillaries. But all these different kinds of connective tissue must be supplied with nutrients and oxygen and must be drained of metabolic waste products in order to maintain the cells, however sparse, in a functional condition, and to effect turnover of extracellular components. How an adequate nutritive state is maintained is a great puzzle, and is attracting the attention of research workers. SEE HISTOLOGY.

Isidore Gersh

Bibliography. G. Bevelander and J. A. Ramaley, *Essentials of Histology*, 8th ed., 1981; L. Weiss and L. Lansing (eds.), *Histology: Cell and Tissue Biology*, 5th ed., 1983.

Connective tissue disease

Any of a group of diseases involving connective tissue; formerly termed collagen vascular disease. These diseases are clinically and pathologically discrete from each other but have overlapping features. The group includes lupus erythematosus, systemic vasculitis, scleroderma (and systemic sclerosis), polymyositis, and Sjögren's syndrome. Each of the diseases can involve multiple organ systems and is often coupled with various immunologic abnormalities. SEE CONNECTIVE TISSUE.

Lupus erythematosus. This set of diseases includes limited, primarily cutaneous disorders (discoid lupus and subacute cutaneous lupus) and a diffuse systemic illness (systemic lupus erythematosus), all of which are of unknown cause. Also, a lupuslike reaction known as drug-induced lupus can be caused by certain therapeutic agents.

Systemic lupus erythematosus. This is an inflammatory, multisystem disorder in which tissue injury is mediated by immune complexes. In this incompletely understood disorder, lymphocytes known as T cells lose some of their control over the antibody production by those known as B cells. Thus, in most individuals with systemic lupus, antibodies are found in the blood serum that are not normally present.

The immunologic events of systemic lupus are influenced by multiple factors. In some families, a genetic predisposition seems to be present. That the disease sometimes flares up after excessive exposure to ultraviolet light emphasizes the importance of environmental factors. Finally, hormonal factors also seem to play a contributing role; eight to nine times as many young women as men have systemic lupus,

and in the animal model of lupus, femaleness is associated with more severe disease. SEE CELLULAR IMMUNOLOGY.

Although systemic lupus can be an acute and fulminating illness, it is far more often a chronic disorder with long periods (years, sometimes even decades) of remission and good health and with infrequent flare-ups. With the recognition of the immune markers (that is, autoantibodies) in serum, milder forms of the disorder can now be identified. SEE IMMUNOPATHOLOGY.

The manifestations and course of systemic lupus vary greatly from one individual to another and even in one person over time. Fever and other constitutional symptoms (for example, fatiguability, lack of appetite, and weight loss) are prominent. Skin rashes are common, and the so-called butterfly rash over the nose and cheeks is most characteristic. Joint pain and inflammation that is, arthritis, are among the common features and often are the first manifestations of the disease. Many other parts of the body—such as the lungs, heart, skeletal muscles, kidneys, and nervous system—can be involved. Small blood vessels at any site may be inflamed. Hematologic abnormalities also can occur when antibodies reactive with cell-membrane antigens are present. Anemia and low levels of white blood cells and platelets are common. A chronic false-positive test for syphilis is possible and is often associated with the presence of a circulating anticoagulant in serum. SEE ARTHRITIS; HEMATOLOGIC DISORDERS.

The diagnosis of systemic lupus is based on a combination of clinical and laboratory findings; no single clinical feature or laboratory test makes the diagnosis. Even the characteristic immune markers in serum (that is, the auto-reactive antibodies) are not specific. For example, the routine screening test for antinuclear antibodies (ANA), which is positive in 95% of patients with systemic lupus erythematosus, can also be positive in those with the other connective tissue disorders and certain other diseases, as well as some healthy family members of patients with systemic lupus.

Discoid lupus and subacute cutaneous lupus are variants, primarily with skin involvement. Photosensitivity is prominent. Systemic features are infrequent and milder than in systemic lupus erythematosus, and kidney and nervous systems are usually spared. ANA positivity is occasionally present.

Drug-induced lupus. This form subsides when the offending agent is discontinued. Many drugs that are chemically unrelated to each other can produce the disorder. Genetic factors play a role in susceptibility to induction of lupus by at least some drugs.

Cause. Except for drug-related lupus, the cause of systemic lupus and its variants is not known; thus, a rational basis for cure does not exist. However, the clinical manifestations of systemic lupus can generally be managed effectively for long years or even decades. Prognosis is steadily improving. Drugs that are used to control the disease by suppressing inflammation and altering the immune system include aspirin and other antiinflammatory drugs, and antimalarial agent (hydroxychloroquine), corticosteroids and cytotoxic agents. Sound general medical management and vigorous treatment of associated illnesses (especially infections) are essential.

Systemic vasculitis. A series of different clinical illnesses that are all characterized by intense inflammation in the walls of blood vessels, especially arteries (that is, arteritis). These illnesses, some of which are discussed below, differ clinically, pathologically, and with respect to therapeutic responsiveness and outcome. Even the age and sex of the typical affected person varies from one to another.

The clinical pattern of illness varies with the sites of blood vessel involvement and the character of vascular injury. Tissue death in the vessel wall can lead to narrowing, or even blockage, of the vessel or weakening of the wall with formation of an aneurysm. Impaired circulation and altered blood flow result. Symptoms, therefore, can arise not only from the tissue inflammation itself but also from the lack of adequate blood supply. Vasculitis can be a systemic condition or can be confined to a single organ or to the skin. *SEE ANEURYSM; CIRCULATION DISORDERS.*

Polyarteritis. Formerly called polyarteritis nodosa, polyarteritis most commonly occurs in young adult men. In most instances, the cause of the inflammatory reaction, which affects medium-sized and small arteries, is not known. Sometimes, however, the condition appears to be associated with a viral infection (that is, hepatitis B) or to be induced by a drug.

The clinical features of polyarteritis include both nonspecific manifestations of diffuse inflammation and reflections of arterial involvement at specific sites. Fever, fatiguability, and loss of appetite and weight are prominent. Arteritis may occur in the kidney (often associated with high blood pressure), the heart, the liver, and the circulation to the bowel. A peripheral nervous system disease is more frequent than involvement of the central nervous system. Joint pain and arthritis, muscle pain, and vasculitic skin lesions are also common. Polyarteritis varies greatly from one person to another; any one of these features, or any combination, can be present and cause clinical problems.

The laboratory manifestations are mostly nonspecific reflections of tissue inflammation, with elevations of the white blood cell count and platelets. The blood sedimentation rate is a standard test for the presence of an inflammatory process of some type; it is elevated well above the normal in most individuals with polyarteritis. There is no diagnostic blood test specifically for polyarteritis; the diagnosis is established by either a biopsy of an involved site (for example, muscle or peripheral nerve) or demonstration of structural changes in the arteries by x-ray visualization after injection of dye—an angiogram.

Optimal therapy is yet to be defined, but disease control and prognosis have progressively improved with, first, the administration of corticosteroids and then their combination with a cytotoxic drug.

Granulomatous arteritis. This category encompasses several rare disorders, including Churg-Strauss syndrome, Wegener's granulomatosis, aortic arch arteritis, giant cell arteritis, and central nervous system arteritis. These disorders cause very different clinical problems, but all are characterized by the presence of granulomas and giant cells in the affected sites. The causes and mechanisms are unknown. These disorders differ from each other in age group and sex predominantly affected, proper therapy, and prognosis. Early recognition and prompt treatment are key to an optimal outcome. Indeed, without treatment these disorders tend to have a dismal prognosis.

Churg-Strauss syndrome (allergic granulomatosis) is rare. It involves small and medium-sized arteries as well as the smallest arterioles and venules. Pulmonary involvement predominates. A characteristic feature of this syndrome is a marked increase in the number of white blood cells known as eosinophils, both in peripheral blood and in the affected tissue.

Wegener's granulomatosis is characterized by inflammation of small vessels in the respiratory tract, in the kidneys, and elsewhere. Both the upper respiratory tract (nasopharynx and sinuses) and lungs are involved; biopsy of lung tissue best establishes the diagnosis. Addition of a cytotoxic drug, especially cyclophosphamide, to corticosteroid therapy can reverse the pulmonary process and prevent kidney failure.

Aortic arch arteritis, also called Takayasu's arteritis, occurs almost exclusively in young women. Large and medium-sized arteries that branch off the aorta are involved, with consequent narrowing or even occlusion. Impaired blood flow results, with diminished or absent pulses in the upper extremities and neck; thus this condition is sometimes called the pulseless syndrome. Decreased circulation to the head can impair brain function, with recurrent lightheadedness and fainting. Similarly, impaired coronary artery flow may result in heart failure and the pain of angina. These major complications occur late and can be prevented if the process is recognized and treated early. Diagnosis is established by intra-arterial dye studies that demonstrate characteristic structural changes in the large vessels.

Giant cell arteritis occurs mainly in women who are over 50 years of age and especially in those who are elderly. Aching pain and severe stiffness in the proximal limb-girdle regions (that is, shoulder/arm, pelvis/thigh) are common early manifestations. Later, with temporal artery involvement (in the head), symptoms such as headaches, scalp tenderness, and visual abnormalities can occur. A tender, dilated temporal artery may be found. Visual loss is the most feared complication, and is preventable by corticosteroid therapy. The musculoskeletal pain syndrome, which can be progressive and in capacitating, also promptly responds to steroids.

Central nervous system arteritis usually affects the elderly. It is manifested mainly by a neuropsychiatric syndrome. Headaches, altered mentation and confusion, decreased level of consciousness, and even coma are common; infrequently, unexplained seizures can occur. Although thought to be a discrete entity, the condition may be an unusual expression of giant cell arteritis.

Hypersensitivity angitis. In this disease there is an intense inflammation of the smaller vessels, especially venules, that represents an adverse immunologic reaction to an offending agent. The agent could be a drug, a foreign protein, or an infectious agent. Skin lesions occur in all cases; they vary in type but typically include localized areas of bleeding into the skin from injured vessels over the feet and ankles. Systemic features also can be present. Henoch-Schönlein purpura is a designation applied to one subset of this disorder.

In all types of hypersensitivity vasculitis, the diagnosis in clinical since there are no specific blood tests. A skin biopsy may support the diagnosis if inflammation of small vessels is found. Treatment focuses on withdrawal of the offending antigen, if known, or treatment of any causal infection, if present. Occasionally, use of corticosteroids may be necessary.

Vasculitis associated with specific diseases. Vasculitis can also be associated with other connective tissue disorders (for example, systemic lupus, polymyositis, and Sjögren's syndrome) and with rheumatoid arthritis, infections and malignant neoplasms. The existence of this wide array of underlying disorders suggests that various immunologic pathways can lead to vascular injury. Although the appropriate therapy is generally determined by the associated disease process, emergence of vasculitis may indicate the need to administer, or increase the dosage of, corticosteroids. SEE VASCULAR DISORDERS.

Polymyositis. Polymyositis involves intense inflammation in skeletal muscle that, if untreated, can lead to destruction of muscle fibers. When typical skin lesions are present, the term dermatomyositis is applied.

The cause of polymyositis, or dermatomyositis, is not known in most cases. These conditions can occur alone or in association with other connective tissue disorders (that is, systemic lupus, scleroderma, and Sjögren's syndrome). In older persons, they sometimes accompany malignant tumors.

Polymyositis may be abrupt or insidious in onset. The dominant manifestation is progressive weakness of the proximal limb-girdle muscle groups. Difficulty rising from a chair, in climbing stairs, lifting modest weights, and keeping the arms elevated are common symptoms. Despite intense inflammation in the skeletal musculature, pain and tenderness of the involved muscles are infrequent. Most serious are the potential involvement of pharyngeal musculature which may impair swallowing; of the chest muscles, which may impair breathing; and the cardiac muscle, which gives rise to altered heart rate and rhythm.

Destruction of muscle fibers causes release into the circulation of various enzymes normally contained in muscle; measuring levels of these enzymes in blood thus aids in diagnosis. Also, inflamed muscle is abnormal in its electrical activity; therefore the test known as an electromyogram yields abnormal results. Finally, muscle biopsy firmly establishes the diagnosis; it can be important in distinguishing polymyositis from noninflammatory disorders affecting muscle. Although polymyositis is presumed to be an immune-mediated process, the mechanisms are obscure. Antinuclear antibodies may be found in the serum. Other less frequent but more myositis-specific autoantibodies may also be present.

The inflammatory process is generally responsive to corticosteroid therapy, but, in some cases, the addition of a cytotoxic drug, especially methotrexate, is required. In the few individuals with tumor-related polymyositis, appropriate treatment of the malignancy is essential; and when polymyositis occurs together with other connective tissue disorders, the additional features of these illnesses must also be controlled.

If diagnosed early, before loss of muscle fibers and replacement by fibrous scarring is extensive, myositis is reversible; recovery with minimal residual loss of strength is to be expected in almost all patients.

Scleroderma. The term scleroderma means "hard skin," designating a disorder in which increased deposition of collagen fibers in the deeper dermis leads to thickened, leathery, bound-down skin. This deposition is variable in extent and degree. When organ involvement is associated with such skin changes, the term systemic sclerosis is used.

Raynaud's phenomenon, a hyperreactivity to cold exposure with blanching and discoloration of the fingers and toes, is the common initial manifestation and may precede sclerodermatous skin changes by decades. Only a small fraction of those with Raynaud's eventually develop scleroderma or systemic illness; but of those who do have scleroderma, at least 80% will have Raynaud's. Skin changes may be confined to the distal digits or become generalized.

Systemically, the gastrointestinal tract is most commonly involved. Difficulty in swallowing and midchest discomfort result from loss of peristalsis in the esophagus. Diarrhea, inability to absorb nutrients, abdominal pain, and distention can occur with bowel involvement and lead to weight loss and wasting. Scarring of the lungs and decreased pulmonary function occur frequently. The most serious lesions are those of the heart and kidney, which may result from vascular abnormalities related to Raynaud's phenomenon. Joint inflammation is mild and uncommon; but advanced skin changes, especially in the hands, may restrict joint mobility. Polymyositis is occasionally present and is the major indication for corticosteroid therapy in this disorder.

There is a variant called the CRST syndrome (calcinosis, Raynaud's, sclerodactyl- or scleroderma-like finger, and telangiectasia) which is important because of its mildness and slow progression.

Multiple autoantibodies (including ANA) may be found in serum. A skin biopsy is diagnostic, with increased collagen fibers in the deeper dermis without any inflammatory reaction.

Management involves treatment of the high blood pressure that can arise because of kidney involvement, use of agents to improve circulation by reducing small-vessel spasm, and reducing gastrointestinal complications with special diets and sometimes antibiotics. The drug D-penicillamine may produce improvement in the skin and possibly systemic features. In a few cases, skin changes may subside spontaneously.

Sjögren's syndrome. This condition is characterized primarily by dryness of the membranes due to excretory gland failure, especially dryness of the eyes (xerophthalmia) and mouth (xerostomia) from loss of tears and salvia, respectively. This sicca (dryness) syndrome reflects the infiltration of lacrimal and salivary glands by immunologically competent cells (lymphocytes).

Sjögren's syndrome may be primary or secondary; that is, superimposed on another disorder like rheumatoid arthritis or systemic lupus. The typical individual with primary Sjögren's is a woman later in life. In most cases, the sicca complex dominates, but there can be loss of secretions in the skin, bronchial tree, esophagus and stomach, and vaginal vault. In some people with primary Sjögren's, systemic features may evolve. Diagnosis is primarily clinical, but ocular dryness can be quantitated. Lip biopsy of the minor salivary glands can confirm the infiltration by immune cells without the hazard of biopsying the parotid salivary glands. Noninvasive scanning of the parotids can confirm the loss of functional glandular structure. Autoantibodies in the blood are common, but in no instance is a blood test diagnostic of either Sjögren's or any companion disorder.

Management involves artificially keeping membranes moist (for example, with artificial tears). Extraglandular inflammatory lesions and associated disorders are treated with anti-inflammatory agents and occasionally cytotoxic agents. SEE IMMUNOLOGY; INFLAMMATION; STEROID.

Mary Betty Stevens

Conodont

A group of extinct marine animals that are often abundant in strata of Late Cambrian to Late Triassic age, a time span of about 300 million years. Only the mineralized elements, which are usually 0.2 to 2 mm (0.008 to 0.08 in.) in dimension (the largest known reach 14 mm or 0.6 in.), are normally preserved. They are routinely extracted as isolated discrete specimens by chemical degradation of the rock in which they occur. The apatite (calcium phosphate) of which conodont elements are composed is laid down as lamellae. In the earliest euconodonts ("true" conodonts, as opposed to the more primitive, and possibly unrelated, protoconodonts and paraconodonts), the elements comprise an upper crown and a basal body. The basal body occupies a cavity in the base of the crown, but is not present in the majority of post-Devonian species. In advanced conodonts the crown incorporates regular patches of opaque, finely crystalline, white matter.

For many years, conodont taxonomists treated individual element types as separate species. There are three major shape categories, coniform, ramiform, and pectiniform (**Fig. 1**). Coniform elements were dominant in the Cambrian to Early Ordovician and common until the Devonian. They vary in curvature and in cross section, including the development of costae. Ramiform (comblike) elements extend into elongate processes with various arrangements of denticles. Pentiniform elements include straight and arched blades, and may be expanded laterally to form a platform.

Skeletal apparatus. In the 1930s the discovery of "natural assemblages" on bedding surfaces provided direct evidence that each animal possessed a skeletal apparatus consisting of several different kinds of elements. Several hundred such associations have now been found, each representing the skeletal remains of an individual animal. Most are from shales, as their preservation requires that the soft parts of the animal decay undisturbed by currents, so that the skeleton remains intact. Some elements may also become fused into clusters during the diagenesis of the sediments in which they occur, and these, too, give an indication of the configuration of the conodont apparatus. Although most conodont species are still known only from elements scattered in the sediment, it is now conventional practice in conodont taxonomy to reconstruct their skeletal apparatus by using morphological and statistical criteria.

The evidence of bedding-plane assemblages shows that most of the elements in an apparatus were symmetrically paired across the midline. A common type of apparatus in the Upper Paleozoic (Fig. 1i) consisted of a set of 7 to 11 ramiform elements (one of which was an unpaired median, bilaterally symmetrical form), followed by a pair of arched blades and, posterior of these, a pair of platform pectiniform elements. Variations in this arrangement occur; early Paleozoic apparatus often consisted of a series of coniform elements alone.

Soft parts. In the absence of preserved soft parts, the nature of the affinities of conodonts was the subject of considerable speculation and debate. Since the first discovery of isolated elements in 1856, conodonts have been variously aligned with algae, higher plants, several wormlike phyla, mollusks, arthropods, lophophorates, chaetognaths, and chordates, or have

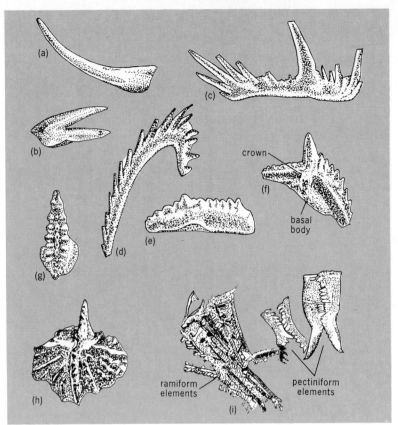

Fig. 1. Conodont elements: (*a, b*) coniform elements, (*c, d*) ramiform elements, (*e, f*) pectiniform blade elements, (*g, h*) pectiniform platform elements, (*i*) bedding-plane assemblage.

been assigned to a separate phylum, Conodonta. It was not until 1983 that evidence of the soft parts was described by D. E. G. Briggs, E. N. K. Clarkson, and R. J. Aldridge, on the basis of the first of several specimens discovered in lower Carboniferous rocks near Edinburgh, Scotland.

Features of the soft parts are preserved as a calcium phosphate film (**Fig. 2**). They show that conodonts were elongate animals (40–60 mm or 1.6–2.4 in.). The apparatus lies in the head region, behind a bilobed structure which is preserved as a film darker than the rest of the body (**Fig. 3**). All of the Carboniferous examples discovered to date are compacted in lateral view, and the apparatus lies to one side, in some cases beyond the apparent outline of the body. This suggests that the soft tissue which supported the elements is not preserved. The trunk shows clear evidence of V-shaped structures, the apex of the V's pointing anteriorly. These structures are interpreted as muscle blocks, but the nature of two parallel lines which run the length of the trunk is less certain. Fin rays indicate the presence of an asymmetrical fin around the margin of the tail. A single poorly preserved specimen from the Silurian of Wisconsin, the only other example known which shows traces of the soft parts, suggests that the body of some conodonts may have been broader than that in the Scottish specimens.

Biological affinities. The evidence of the soft-part morphology indicates that the conodonts belong within the chordates; it is no longer possible to justify their separation as a phylum, Conodonta. The lack of jaws and of a bony skeleton suggests that they repre-

Fig. 2. The first specimen of the conodont animal discovered in Lower Carboniferous rocks in Edinburgh. (*a*) Photograph of specimen. (*b*) Outline of the specimen for comparison. (*From D. E. G. Briggs, E. N. K. Clarkson, and R. J. Aldridge, The conodont animal, Lethaia, 16:1–14, 1983)

sent a primitive group of jawless craniates, perhaps close to the Myxinoidea (hagfishes). The apparatus is thought to have functioned in food capture. The ramiform elements (or coniform elements in earlier conodonts) appear to be adapted for grasping prey. In later conodonts this would have been processed by the shearing and grinding action of the paired pectiniform elements. *See* Chordata.

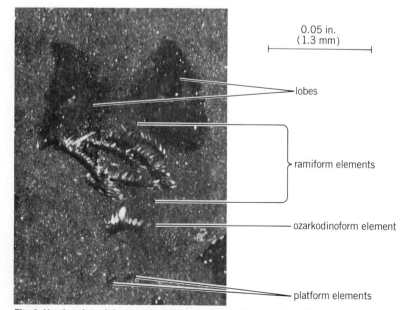

Fig. 3. Head region of the conodont animal, showing the conodont elements on the side of the slab opposing that illustrated in Fig. 2. (*From D. E. G. Briggs, E. N. K. Clarkson, and R. J. Aldridge, The conodont animal, Lethaia, 16:1–14, 1983)

Geological applications. Although the biological affinities of conodonts and the function of the elements were essentially unknown until recently, they have nonetheless been extensively studied because of their important geological applications. Most significant of these is the use of conodont elements in biostratigraphy. Biozonal schemes have been erected for all the systems through which conodonts range, Cambrian to Triassic; the most detailed subdivision is in Upper Devonian strata (conodont subbiozones representing, on average, 0.5 million years). Some conodonts are restricted in their distribution to certain sedimentary environments and can be used to define biofacies. Most conodont biofacies are, at least indirectly, related to depth, as reflected in the contrasts between nearshore and offshore assemblages.

Conodont elements can also be used to determine the thermal history of strata. The color of elements alters from pale yellow through brown to black as their carbon content is fixed during organic metamorphism. This is the basis for a conodont Color Alteration Index (CAI, from 1 to 5) which can be applied in assessing oil and gas potential, and in wider studies of the thermal history of sedimentary basins.

The apatite of which conodont elements are formed is stable and resistant to diagenetic alteration. The trace-element and isotopic-ratio characteristics of the seawater in which the organism lived were retained in the apatite, which can be analyzed to yield significant information about the temperature and chemistry of ancient oceans. Thus an assemblage of conodont elements extracted from a rock sample not only may allow biostratigraphic correlation, but also may give an indication of the paleoenvironment of the locality; the temperature, geochemistry, and oxygen content of the seawater; and the subsequent thermal history of the sedimentary basin. SEE STRATIGRAPHY.

Richard J. Aldridge; Derek E. G. Briggs

Bibliography. R. J. Aldridge (ed.), *Palaeobiology of Conodonts*, 1987; R. J. Aldridge et al., The affinities of conodonts: New evidence from the Carboniferous of Edinburgh, Scotland, *Lethaia*, 19:279–291, 1986; D. E. G. Briggs, E. N. K. Clarkson, and R. J. Aldridge, The conodont animal, *Lethaia*, 16:1–14, 1983; R. A. Robison (ed.), *Treatise on Invertebrate Paleontology*, W, suppl. 2, *Conodonta*, 1981.

Consciousness

Accounting for the sensations or "raw feels" which constitute conscious experience has been a persistent problem for scientific approaches to psychology. In addressing this problem, each new scientific approach has put forth a new account of the relationship between those conscious sensations, as they are privately experienced by individuals, and the nervous system, as it is experienced by scientists.

Fechner on consciousness. In 1860 Gustav Fechner published his *Elemente der Psychophysik*, wherein he developed scientific techniques to measure the magnitudes of conscious sensations such as brightness and loudness, and then derived a mathematical equation to describe the relationship between each sensory magnitude and the physical magnitude which stimulated it. The resulting psychophysical equation is known as Fechner's law: $S = k \log R$, where S is the magnitude of the conscious sensation, k is a constant which takes on specific values for specific sensory magnitudes like brightness and heaviness, and R is the magnitude of the physical stimulus evoking the sensation.

Fechner derived this psychophysical equation as a scientific solution to the mind-body problem. He interpreted it as support for panpsychism, the metaphysical position that all matter is conscious of other matter. His interpretation was based on the equation's implication that a very small physical magnitude stimulates a negative sensation: the consciousness of a magnitude beneath the threshold at which neural substances produce awareness of each other's magnitudes, a conscious sensation such as that experienced by flowers and other nonneural substances.

Inasmuch as the mathematical relationship between psychological sensations and physical stimuli must be derivable from psychoneural and neurophysical equations, Fechner's panpsychist interpretation of his psychophysical equation assumed that the underlying psychoneural equation was a logarithmic equation yielding the negative sensations and that the neurophysical equation was linear. However, scientific critics of Fechner's metaphysics and philosophic critics of his scientific psychology rejected the latter assumptions. They argued that the logarithmic relationship reflected the neurophysical relationship between the magnitude of the brain state evoked by the stimulus and the magnitude of the physical stimulus itself, both of which are physically and scientifically measureable magnitudes. Despite the ingenuity and rigor of his scientific approach to consciousness, Fechner failed to rebut the age-old argument that science is sufficient only to deal with the outer world of physical reality, and not to deal with the inner world of consciousness. SEE PSYCHOPHYSICAL METHODS.

Wundt on consciousness. Historically, Wilhelm Wundt convinced the scientific establishment that the methods of science are sufficient to study the sensory data of consciousness. Rather than defend Fechner's contention that science applies both to outer reality and to inner reality, Wundt took the offensive. He argued that the inner aspects of conscious experience, the data of psychology, are the only immediately known data of science: that the outer world of physical reality has to be inferred from the data of scientific psychology and, thus, is known only mediately. Accordingly, a psychological science with sensory data had to be sufficient because such a science was necessary for there to be any sciences with mediately known data, such as external stimuli and brains.

Consistent with the methodological justifications above, Wundt's scientific studies of perceptual consciousness confirmed that perceptual experience provides no immediate knowledge of external reality. For instance, introspection—the scientific observation of consciousness—reveals that an observer's experience of a dimly illuminated sheet of white paper and his or her experience of a brightly illuminated sheet of white paper do not contain any of the same sensations of whiteness, even though both sheets are "known" to be of the same whiteness. Such knowledge of whiteness must, according to Wundt, be an illusory belief resulting from the previous association of the observer's two experiences of white paper.

In logically and experimentally supporting his thesis that the data of scientific psychology constitute the foundation for all sciences, Wundt sought to reduce physical reality to a subset of consciousness itself. Three founders of subsequent approaches to psychology were not satisfied with this effort to rewrite psychophysical and psychoneural equations in purely mental terms. One of them—John Watson, who founded behaviorism around 1913—sought, instead, to reduce mental terms to physical terms. The other two—Sigmund Freud, who founded psychoanalysis around 1900, and Max Wertheimer, who founded Gestalt psychology around 1912—sought to put mental terms on the psychological sides of the psychoneural and psychophysical equations and to put physical terms on the opposite sides.

Freudian reaction to Wundt. By arguing for the existence of unconscious mental events, Freud took issue with both the metaphysical and the methodological underpinnings of Wundt's approach to the psychology of consciousness. If unconscious mental events existed, then much of the mind itself—much of the data of scientific psychology—was no more immediately knowable than the inferred data of physical sciences. Moreover, to the extent that the unconscious mind was actually part of physical reality, Wundt was not justified in treating physical reality as a subset of the mind. In the light of Freud's analysis of the unconscious mind, the psychophysical and psychoneural relationships reemerged as unsolved and problematical relationships.

Behaviorist reaction to Wundt and Freud. In creating and developing behaviorism, Watson introduced both a new methodological approach to psychology and a new metaphysical approach to consciousness. In many ways, Watson's behavioral approaches were parallel but antithetical to Wundt's introspective approaches. Both Wundt's scientific methods and his metaphysical positions were addressed to the psychology of the self. The starting point for introspective psychology was the scientist's immediate experience of his or her own consciousness; metaphysical inferences about physical realities outside of consciousness were based on illusory associations. Watson's scientific methods and metaphysical positions were addressed to the psychology of the other. The starting point for behaviorism was the scientist's immediate experience of another organism's behavioral response to a stimulus; metaphysical inferences about conscious events within that other organism were deemed to be unnecessary and unwarranted.

From the standpoint of behaviorism, all scientific conclusions regarding the psychophysical and psychoneural relationships—including Fechner's panpsychist interpretation of his psychophysical equation and Freud's distinction between conscious and unconscious mental events—were meaningless. Such conclusions were deemed to be neither true nor false but to be meaningless, because their premises contained both a mental term referring to the scientist's experience of other organisms and other entities—terms which, according to many turn-of-the-century philosophers, came from logically incompatible categories. This did not necessarily imply that mental terms such as consciousness were meaningless. It did imply that the behavioral scientist's physical descriptions of other entities like brains (including the physical description of his or her own brain) could not meaningfully be related to mental descriptions of his or her own consciousness.

From behaviorism to computer models. At the beginning of the twentieth century, Ivan Pavlov and Edward Thorndike initiated psychological research on animals, and Watson's psychology of the other organism provided the rationale for viewing other animals and other people as scientifically equivalent subjects of psychological study. Since the late 1960s, animal-behavior models of psychological processes have been replaced by computer models, and Watson's animal behaviorism has been transformed into a computer behaviorism or a psychology of the other machine. The starting point for computer behaviorism is the scientist's immediate experience of another machine's behavioral output to controlled input. The goal of computer behaviorism is to program computers to produce output responses which observers experience as indistinguishable from human beings' responses. If and when this goal is achieved, computer models of psychological processes will provide logically sufficient explanations of the psychological processes of other human beings. The critical question is whether such computer models of psychological processes would also provide sufficient explanations of the psychological processes of the scientist who consciously experiences the other machine. Two different answers to this question have been offered by computer behaviorists.

One set of computer behaviorists has offered an answer consistent with animal behaviorists' position that the scientist's conscious experience has no causal effect on his or her own behavior. This first set of computer behaviorists argues that neither the scientist's conscious experience nor the computer's lack thereof has any causal effect on behavior. However, a subset of these computer behaviorists prefers not to concede that consciousness distinguishes human beings from machines. Instead, this subset further qualifies the be-

haviorist argument against the causal efficacy of consciousness by adopting an epiphenomenalist argument: that any conscious psychological process in the scientist's computerlike brain can actually occur with or without consciousness of that process. Such an argument implies that the scientist's psychological processes are, in principle, not different from the computer's. It also implies that the scientist's brain contains a metaphysical subject which can consciously inspect or not inspect the same neural process. Traditionally, however, epiphenomenalism has presupposed this metaphysical subject or homunculus to be a person's soul; computer behaviorism presupposes no alternative definitions of a metaphysical subject inhabiting the scientist's brain.

Rather than argue that the scientist is reducible to an unconscious machine or concede that consciousness distinguishes human beings from machines, the other set of computer behaviorists adopts the functionalist argument that sensations like blueness and other privately experienced phenomena will "emerge" from any functionally equivalent psychological processes present either in the scientist's computerlike brain or in the other computer studied by the scientist. This "emergence" of sensory qualities from the functionally defined compounding of computer circuits is deemed to be analogous to the emergence of water from the compounding of hydrogen and oxygen. The problem with this analogy is that the emergence of a conscious whole from unconscious parts represents a qualitative change, whereas the emergence of water from gaseous elements represents merely a quantitative change. The greater number of chemical bonds in the water molecule produces a boiling point that is quantitatively higher than the boiling points that is produced by the lesser numbers of chemical bonds in hydrogen alone and oxygen alone.

Metaphysically, computer behaviorism presupposes that the conscious mind is a machine; methodologically, it starts with the scientist's conscious experience of other machines. Another metaphysical and methodological approach, the approach of Gestalt psychology, presupposed also that the mind is a machine, but started instead with the scientist's phenomenal experience of his or her own machine.

Gestalt reaction to Wundt and behaviorism. While the Gestalt psychologist Max Wertheimer rejected Watson's scientific psychology of the other organism, he also rejected Wundt's contention that nothing in the scientist's immediately known experience corresponds directly with the mediately known physical world. Wundt's scientific observations of his own conscious experience had revealed that his experience of a dimly illuminated sheet of white paper and his experience of a brightly illuminated sheet of white paper did not contain any of the same sensations of whiteness. Gestalt psychologists objected that Wundt had restricted his observations to elementary sensations and had not observed the Gestalt, the holistic relationship among the immediately known sensations. Thus, according to this Gestalt amelioration of the introspective method, dimly illuminated white paper against a dimly illuminated gray background and brightly illuminated white paper against a brightly illuminated gray background resulted immediately and directly in a constant experience of relative whiteness because the ratio of paper brightness to background brightness was constant. According to Gestalt psychology, there was a direct psychophysical correspondence between perceptual experience and physical stimulation.

From Gestalt to mind-brain identity theory. In his effort to rebut Wundt's contention that any psychophysical correspondence is a necessarily indirect, arbitrarily learned, and basically illusory correspondence, Wertheimer adopted the scientist's conscious experience as his starting point and looked for phenomenal evidence to support a direct psychophysical relationship. In so making his case for a direct psychophysical correspondence, he set the stage for the Gestalt psychologist Wolfgang Köhler and other mind-brain identity theorists to argue for a direct psychoneural relationship.

According to mind-brain identity theorists, the scientist's brain contains no homunculus, no soul, no inner person who is "conscious of" his or her sensations. There is never any "consciousness of" anything else. Rather, the scientist's sensations are qualities of particular brain structures in their excited state. Thus, the scientist is "conscious of" neither sensations nor the brain; the scientist is the sensory qualities of his or her brain states. The scientist immediately experiences nothing besides the qualities of the brain states which he or she is; and the scientist's experiential qualities change over time, simply because the patterning of excited structures in his or her brain changes. Whereas the epiphenomenalism that is espoused by some computer behaviorists implies that exactly the same brain state may sometimes correspond to a conscious experience and at other times correspond to no conscious experience, mind-brain identity theory implies that the same conscious (or unconscious) quality is always a quality of a particular brain structure in a given state of excitation.

According to mind-brain identity theory, a scientist could simultaneously experience his or her own brain both immediately and mediately, by visually experiencing the exposed brain on a television screen. In spite of the fact that such a mediately experienced brain would be a subset of the immediately experienced brain, the scientist could look for mediately experienced brain states which correlate perfectly with immediately experienced sensory qualities. With regard to brain states which the scientist immediately experiences as blue in quality, the scientist's mediate experience of those same brain states need not also be blue in quality. Brain states with blue qualities need not reflect 475-nanometer waves of light and, thus, need not be mediately experienced as blue. Still, to the extent that physical brain states and the light patterns reflected by those states and the mediate experiences induced by those patterns correspond directly with each other, as Gestalt psychologists argue that they do, mind-brain identity theory predicts that psychoneural correspondences can be discovered. Scientists adopting a mind-brain identity position have not yet discovered any structural differences in various parts of the brain, much less discovered any correspondences between conscious qualities and brain structures. SEE PSYCHOLOGY, PHYSIOLOGICAL AND EXPERIMENTAL.

Robert G. Kunzendorf

Bibliography. J. Brzezinski (ed.), *Consciousness: Methodological and Psychological Approaches*, 1985; R. W. Coan, *Human Consciousness and its Evolution*, 1987; R. Ellis, *An Ontology of Consciousness*, 1986; A. J. Hobson, *States of Brain and Mind*, 1988; A. J. Marcel and E. Bisiach (eds.), *Consciousness in Contemporary Science*, 1988.

Conservation of energy

The principle of conservation of energy states that energy cannot be created or destroyed, although it can be changed from one form to another. Thus in any isolated or closed system, the sum of all forms of energy remains constant. The energy of the system may be interconverted among many different forms—mechanical, electrical, magnetic, thermal, chemical, nuclear, and so on—and as time progresses, it tends to become less and less available; but within the limits of small experimental uncertainty, no change in total amount of energy has been observed in any situation in which it has been possible to ensure that energy has not entered or left the system in the form of work or heat. For a system that is both gaining and losing energy in the form of work and heat, as is true of any machine in operation, the energy principle asserts that the net gain of energy is equal to the total change of the system's internal energy. SEE THERMODYNAMIC PRINCIPLES.

Application to life processes. The energy principle as applied to life processes has also been studied. For instance, the quantity of heat obtained by burning food equivalent to the daily food intake of an animal is found to be equal to the daily amount of energy released by the animal in the forms of heat, work done, and energy in the waste products. (It is assumed that the animal is not gaining or losing weight.) Studies with similar results have also been made of photosynthesis, the process upon which the existence of practically all plant and animal life ultimately depends. SEE METABOLISM; PHOTOSYNTHESIS.

Conservation of mechanical energy. There are many other ways in which the principle of conservation of energy may be stated, depending on the intended application. Examples are the various methods of stating the first law of thermodynamics, the work-kinetic energy theorem, and the assertion that perpetual motion of the first kind is impossible. Of particular interest is the special form of the principle known as the principle of conservation of mechanical energy (kinetic E_k plus potential E_p) of any system of bodies connected together in any way is conserved, provided that the system is free of all frictional forces, including internal friction that could arise during collisions of the bodies of the system. Although frictional or other nonconservative forces are always present in any actual situation, their effects in many cases are so small that the principle of conservation of mechanical energy is a very useful approximation. Thus for a missile or satellite traveling high in space, the dissipative effects arising from such sources as the residual air and meteoric dust are so exceedingly small that the loss of mechanical energy $E_k + E_p$ of the body as it proceeds along its trajectory may, for many purposes, be disregarded. SEE ENERGY; PERPETUAL MOTION.

Mechanical equivalent of heat. The mechanical energy principle is very old, being directly derivable as a theorem from Newton's law of motion. Also very old are the notions that the disappearance of mechanical energy in actual situations is always accompanied by the production of heat and that heat itself is to be ascribed to the random motions of the particles of which matter is composed. But a really clear conception of heat as a form of energy came only near the middle of the nineteenth century, when J. P. Joule and others demonstrated the equivalence of heat and work by showing experimentally that for every definite amount of work done against friction there always appears a definite quantity of heat. The experiments usually were so arranged that the heat generated was absorbed by a given quantity of water, and it was observed that a given expenditure of mechanical energy always produced the same rise of temperature in the water. The resulting numerical relation between quantities of mechanical energy and heat is called the Joule equivalent, or mechanical equivalent of heat. The present accepted value is one 15° calorie = 4.1855 ± 0.0004 joules.

Conservation of mass-energy. In view of the principle of equivalence of mass and energy in the restricted theory of relativity, the classical principle of conservation of energy must be regarded as a special case of the principle of conservation of mass-energy. However, this more general principle need be invoked only when dealing with certain nuclear phenomena or when speeds comparable with the speed of light (1.86×10^5 mi/s or 3.00×10^5 km/s) are involved. *SEE RELATIVITY.*

If the mass-energy relation, $E = mc^2$, where c is the speed of light, is considered as providing an equivalence between energy E and mass m in the same sense as the Joule equivalent provides an equivalence between mechanical energy and heat, there results the relation, $1 \text{ kg} = 9 \times 10^{16}$ joules.

Laws of motion. The law of conservation of energy has been established by thousands of meticulous measurements of gains and losses of all known forms of energy. It is now known that the total energy of a properly isolated system remains constant. Some parts or particles of the system may gain energy but others must lose just as much. The actual behavior of all the particles, and thus of the whole system, obeys certain laws of motion. These laws of motion must therefore be such that the energy of the total system is not changed by collisions or other interactions of its parts. It is a remarkable fact that one can test for this property of the laws of motion by a simple mathematical manipulation that is the same for all known laws: classical, relativistic, and quantum mechanical.

The mathematical test is as follows. Replace the variable t, which stands for time, by $t + a$, where a is a constant. If the equations of motion are not changed by such a substitution, it can be proved that the energy of any system governed by these equations is conserved. For example, if the only expression containing time is $t_2 - t_1$, changing t_2 to $t_2 + a$ and t_1 to $t_1 + a$ leaves the expression unchanged. Such expressions are said to be invariant under time displacement. When daylight-saving time goes into effect, every t is changed to $t + 1$ h. It is unnecessary to make this substitution in any known laws of nature, which are all invariant under time displacement. Without such invariance laws of nature would change with the passage of time, and repeating an experiment would have no clear-cut meaning. In fact, science, as it is known today, would not exist.

Duane E. Roller/Leo Nedelsky

Bibliography. K. R. Atkins, *Physics,* 3d ed., 1976; D. Halliday and R. Resnick, *Physics,* 3d ed., 1978; G. Laundry et al., *Physics: An Energy Introduction,* 1979; F. W. Sears et al., *University Physics,* 7th ed., 1987; E. P. Wigner, Symmetry and conservation laws, *Phys. Today,* 17(3):34–40, March 1964.

Conservation of mass

The notion that mass, or matter, can be neither created nor destroyed. According to conservation of mass, reactions and interactions which change the properties of substances leave unchanged their total mass; for instance, when charcoal burns, the mass of all of the products of combustion, such as ashes, soot, and gases, equals the original mass of charcoal and the oxygen with which it reacted.

The special theory of relativity of Albert Einstein, which has been verified by experiment, has shown, however, that the mass of a body changes as the energy possessed by the body changes. Such changes in mass are too small to be detected except in subatomic phenomena. Futhermore, matter may be created, for instance, by the materialization of a photon (quantum of electromagnetic energy) into an electron-positron pair; or it may be destroyed, by the annihilation of this pair of elementary particles to produce a pair of photons. *SEE ELECTRON-POSITRON PAIR PRODUCTION; LIGHT; MASS; RELATIVITY.*

Leo Nedelsky

Conservation of momentum

The principle that, when a system of masses is subject only to forces that masses of the system exert on one another, the total vector momentum of the system is constant. Since vector momentum is conserved, in problems involving more than one dimension the component of momentum in any direction will remain constant. The principle of conservation of momentum holds generally and is applicable in all fields of physics. In particular, momentum is conserved even if the particles of a system exert forces on one another or if the total mechanical energy is not conserved.

Use of the principle of conservation of momentum is fundamental in the solution of collision problems. *SEE COLLISION (PHYSICS).*

If a person standing on a well-lubricated cart steps forward, the cart moves backward. One can explain this result by momentum conservation, considering the system to consist of cart and human. If both person and cart are originally at rest, the momentum of the system is zero. If the person then acquires forward momentum by stepping forward, the cart must receive a backward momentum of equal magnitude in order for the system to retain a total momentum of zero.

When the principle of conservation of momentum is applied, care must be taken that the system under consideration is really isolated. For example, when a rough rock rolls down a hill, the isolated system would have to consist of the rock plus the earth, and not the rock alone, since momentum exchanges between the rock and the earth cannot be neglected.

Rocket propulsion. The propulsion of a rocket through space can be explained in terms of momentum conservation. Hot gases produced by the combustion of the fuel are expelled at high speed from the rear of the rocket. Although the total mass of these hot gases may not be large, the gases move with such a high velocity that the total momentum associated with them is appreciable. The momentum of the gases is directed backward. For momentum to be conserved, the rocket must acquire an equal momentum in the forward direction. If the rocket carries all the materials needed for the combustion of its fuel, its propulsion does not require air, and it can move through empty space. SEE PROPULSION; ROCKET.

Exploding bomb. An exploding bomb gives another application of the conservation of momentum. The total resultant vector momentum of all the pieces of the bomb immediately after explosion must equal the momentum of the unexploded bomb just before the explosion. SEE EXPLOSION AND EXPLOSIVE.

Proof of principle. The principle of conservation of momentum follows directly from Newton's second and third laws. While the principle will be proved here only for the straight-line motion of a two-particle system, it can be generalized to systems containing any number of particles. A particle is a mass with dimensions so small that rotational effects are negligible. Momentum will also be conserved for rigid bodies large enough that rotation must be considered, since rigid bodies can be treated as assemblies of many particles.

For the one-dimensional motion of an isolated two-particle system, Newton's third law states that the force F_{12} that particle 1 exerts on particle 2 is equal in magnitude and opposite in direction to the force F_{21} that particle 2 exerts on particle 1. Thus Eq. (1) holds.

$$F_{21} = -F_{12} \qquad (1)$$

By use of Newton's second law this equation can be expressed in terms of the momenta $m_1 v_1$ and $m_2 v_2$ of particles 1 and 2, respectively, where m_1, m_2, v_1, and v_2 are the masses and velocities of particles 1 and 2, respectively. Then Eq. (2) holds. Integration gives

$$m_1 \frac{dv_1}{dt} = -m_2 \frac{dv_2}{dt} \qquad (2)$$

Eq. (3), where c is a constant. This equation expresses

$$m_1 v_1 + m_2 v_2 = c \qquad (3)$$

the conversation of momentum for two particles moving in the same straight line.

Finally it should be mentioned that angular and linear momentum are independent quantities. A complete description of a system must include both quantities. The angular momentum of a system is conserved under quite general conditions. SEE ANGULAR MOMENTUM; CONSERVATION OF ENERGY; MOMENTUM.

Paul W. Schmidt

Conservation of resources

Management of natural resources to prevent overexploitation or destruction. Natural resources (sunlight, water, plants, animals, soil, and minerals) help support life. Many natural resources must be utilized to maintain the quality of human life, yet some are needlessly wasted or destroyed, diminishing the heritage of future generations. Direct effects on humans are not the only or the highest standard for conservation decisions, but actions affecting any natural system eventually affect human systems because of the interrelations of all parts of the biosphere. Renewable natural resources, such as forests and fisheries, are to be preserved while providing the maximum benefit. Others, such as fossil fuels, are nonrenewable except over geologic time scales and should be used cautiously with a view to eventual replacement by other resources. Conservation sometimes requires unconditional protection, in the belief that natural systems are important in themselves rather than for exploitation. SEE BIOSPHERE.

Historically, conservation measures came into being after an adequate supply of material goods was available to an organized population; until that time, the meeting of material needs took precedence, and the long-term health of the environment was considered secondary to human needs.

The emerging recognition among ecologists that seemingly local environmental concerns interact with larger-scale problems has created an awareness that decisions affecting localized systems will eventually affect other systems in a regional or global sense. For example, air pollution from fossil fuel–burning electric power plants was at first considered a normal and acceptable by-product of a growing economy; health effects, scenic damage, or objections to odors were consigned to local nuisance lawsuits. Eventually government regulations forced the development of modern pollution-control technologies. Within a few years it was discovered that air pollution was damaging water bodies hundreds of miles from the source through acid deposition. These effects have become international problems. Today the scale is global, as climate alterations caused by gases, partly produced by power-generating plants, are challenging human societies to alter basic technologies. It is rare to find an environmental action affecting only local concerns. SEE ACID RAIN; AIR POLLUTION.

The purpose of natural resource conservation is to maintain valuable and essential resources in a condition of good health that will allow development of the necessary resources for maximum long-term benefits. The human responsibility is to manage the environment in such a way as to ensure that there will be adequate supplies of natural resources for future generations.

An ecological perspective on the conservation of natural resources looks to a sustained optimum yield of the resources that can be continually utilized without depletion. This is not necessarily the same as maximum yield. Additionally, this perspective recognizes that there are certain sensitive systems and species from which humans must no longer seek direct benefit, either to avoid destruction of the resource or to allow time for recovery from past use. Historically, the pattern was to deplete local resources, then move to more bountiful areas that had yet to be developed. This allowed depleted areas adequate time for natural recovery. This process is still practiced in some undeveloped countries, but with the rapid growth of human population in these areas, adequate recovery is no longer possible. Moreover, intensive use over a short time can create soil erosion and plant destruction. Problems of uneven distribution of natural resources exacerbate problems of overutilization; when a scarce resource is available in a growing economy, efforts to restrict use are difficult to establish and enforce. SEE HUMAN ECOLOGY.

Wildlife conservation. The first conservation efforts in the United States focused on forestry and wildlife preservation and management. The term wildlife includes all undomesticated animals, birds, and fish living in their natural environments. Wildlife populations historically were regulated by natural processes, including catastrophes such as fire, flood, and drought, and biological controls such as predation, disease, and food limitations. The increasing influence of human populations has caused extensive wildlife species extinctions, which came about first through overhunting and later through agricultural and building practices and environmental pollution. In the past, uncontrolled hunting was the principal single cause of species loss (for example, the extinction of the passenger pigeon, which once covered North America with a population in the billions). Today, hunting laws protect rare or endangered species. With the exception of threats to the whales from hunting, habitat destruction is the largest single cause of species loss and endangerment today. Human activities that cause habitat degradation or destruction include the development of land for human habitation, transportation, or farming, leading to deforestation and wetlands draining; sewage and solid-waste disposal; decreased streamside vegetative cover and stream channelization; clear-cutting of forested areas; improper applications of pesticides and fertilizers; increased sedimentation in waters; and air and water pollution.

Food, water, and shelter requirements vary greatly among species, even in the same ecosystem. Actions affecting one part of an ecosystem can have unforeseen and deleterious effects in another part. The classic example is the application of DDT for mosquito control, which led to eggshell thinning and population reductions of pelicans and other carnivorous birds. The disturbed animals cannot simply move to a less affected area. Yet often the degradation is too rapid for species to adapt. Over 90% of the original wetlands in the United States have disappeared because of human activities. Some elements of an ecosystem can remain stable only if large contiguous land areas are undisturbed; this is the reason for the decline in grizzly bear populations outside Alaska. Other types of habitats are disturbed by activities upstream, as rivers and estuaries receive pollutants transported by water. Upwind activities harm forests and lakes with acid precipitation or air pollution.

Wildlife has greatly decreased in importance as a food and clothing source even in undeveloped countries. Consequently, hunting and fishing control must be based on current information. Properly controlled hunting benefits game species, especially large animals, keeping them within the carrying capacity of the area. Hunting and fishing are normally regulated through licenses, with limits on open seasons and numbers taken; ecologists urge the harvesting of wildlife populations that exceed the carrying capacity of the natural area to prevent damage to habitats from excess population pressures and to avoid disease and starvation.

Other wildlife habitat conservation activities encourage wild strips left along cultivated fields; clean farming practices remove food and shelter used by indigenous species that are often beneficial for the farmers, especially in pest control. Proper use of fertilizers and pesticides avoids problems of chemical persistence and runoff to water sources. Wildlife conservation encourages good forest management practices such as selective cutting and underbrush clearing.

Rangeland conservation. Grass-covered lands, also known as grasslands, grazing lands, and rangelands, represent the optimum growth in areas where low to moderate precipitation is too little to support farming or forests. Rangelands provide food and shelter for a wide variety of wildlife and forage for livestock (cattle, horses, and sheep). The native nonwoody plants enhance soil stability, maintain soil productivity, and assure replenishment of underground water sources. The balance between soil stability and plant cover in rangeland areas is delicate, although rangelands seem to be boundless, hardy, and easily replaceable. They often recover rapidly from fire, even with improved productivity; however, they are very sensitive to overgrazing, which can cause rapid deterioration.

Grazing domestic animals on rangelands was an early practice. Some areas are still grazed in a fully nomadic manner, though this is very rare in the United States. Under these circumstances, there is little incentive for conservation, since the users do not own or permanently occupy the land. As the human population grew, overgrazing became an increasing problem. When the number of animals exceeds the carrying capacity of the ecosystem, desirable grasses give way to weeds and brush, which are less nutritious and provide less protection to soils against water erosion and wind damage. Bare soil areas destroy habitats of indigenous species. Overgrazed slopes cannot hold moisture, and the soil-laden runoff pollutes streams and reservoirs and aggravates flooding. Sedimentation reduces the storage capacity of reservoirs. *See* Erosion.

When vast areas of the rangelands began to be farmed, plows turned over the grasses, eventually creating near-desert conditions because the soil's ability to hold moisture and resist wind erosion was destroyed and the fields were left barren after crops were harvested. The dust bowl seen in the 1930s in the midwestern United States is being repeated today in marginal agricultural areas in sub-Saharan Africa. In both cases, the combination of overexploitation of the environment and extreme climatic conditions was much worse than either alone. The combined adverse effects of human and natural stresses on the environment will be felt increasingly as the population grows and the Earth's climate changes.

Overgrazing damage is reduced by properly located watering facilities to decrease daily travel by livestock. Rotation of grazing areas allows time for recovery of grass. Proper positioning of fences and salt lures also reduces damage. Rangelands can be restored, as was done in the late 1930s in response to the dust bowl conditions in the United States. Reseeding of grasses and the installation of irrigation or water storage tanks were of value. Grazing was avoided in marginal areas. Some land can be easily restored if grazing is allowed only during one season; some rangelands should be grazed only during nongrowth seasons. Some advocate the use of herbicides and burning to destroy weeds and poisonous plants. The goal of rangeland conservation is to rehabilitate mismanaged areas and adopt sound practices.

Most rangeland in the United States is owned by the federal government and administered by the U.S. Bureau of Land Management or the U.S. Forest Service. Ranchers are issued permits to use public grazing lands; these permits limit the number of animals that can be grazed and prescribe their movements. The management of rangelands is intended to protect wildlife, the economic value of grazing animals, and recreational uses.

Plant and animal conservation   Natural resource conservation requires the preservation of species and the genetic diversity of life. The growth of human populations assures that, despite all measures taken, some species will be lost. Scientists do not know how many species exist on the Earth today; estimates range from 2 to 30 million or more. It is agreed that species are being lost at alarming rates, especially in the tropics as a result of large-scale deforestation. At these rates, virtually all tropical forests will be lost in a few decades, and with them millions or tens of millions of species. But loss of diversity is not limited to the tropics, and temperate regions of the world have already been remade almost totally into human-affected environments.

For most species that become extinct, the costs of this loss will never be assessed, as their possible value cannot be estimated. The loss of so-called keystone species can result in a cascading effect, in which many other species disappear because of associated changes in the food chains that define ecosystems. Other species or groups of species perform critical functions such as nutrient cycling. The loss of these functional groups of species can undermine entire ecosystems. *See Ecosystem; Food web.*

Some species that were almost eliminated through intensive hunting and fishing or through habitat destruction have survived in zoos, fish hatcheries, and captive breeding programs (California condor, bison, and wildebeest). However, management of species and habitats is much preferable, benefiting many more species for the same effort and cost. National parks, nature reserves, national forests, wildlife refuges, biosphere reserves, and similar areas have been

established in many parts of the world. They create opportunities for scientific research and recreational activities. Some communities should be left undisturbed to act as genetic reserves, where ecological knowledge can be gained through the study of undisturbed systems.

Biological reserves can provide for the preservation of seeds, seedlings, and shoots that could be used in the future as lost habitats are reestablished. The purpose of natural resource conservation should be to maintain the biosphere in a healthy condition. Ensuring the long-term security of natural resources in an increasingly stressed global environment will secure a vital heritage for future generations. *See Fishery conservation; Forest management; Land-use planning; Mineral resources; Soil conservation; Water conservation.*

Christine C. Harwell; Mark A. Harwell

Constellation

One of the 88 areas into which the sky is divided. Each constellation has a name that reflects its earliest recognition. Though pictures are associated with the constellations, they have no official status, and constellations have been depicted differently by different artists.

The identifications of the constellations are lost in antiquity. No doubt ancient peoples associated myths with the heavens and imagined pictures connecting or surrounding the bright stars. The names of some constellations have been handed down by the Chaldeans or Egyptians, but most of those that can be seen from midnorthern latitudes have Greek or Roman origins. Star maps are found in various cultures, and constel-

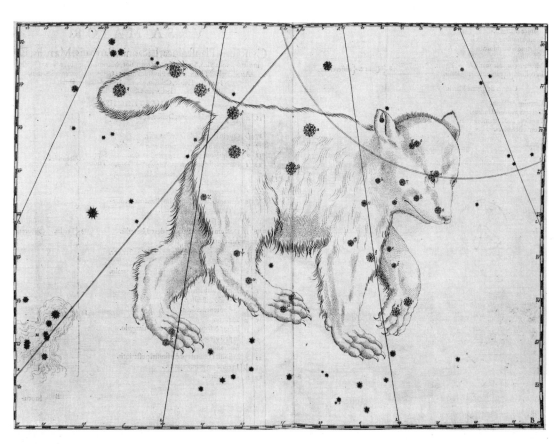

Fig. 1. The constellation Ursa Major from Bayer's star atlas.

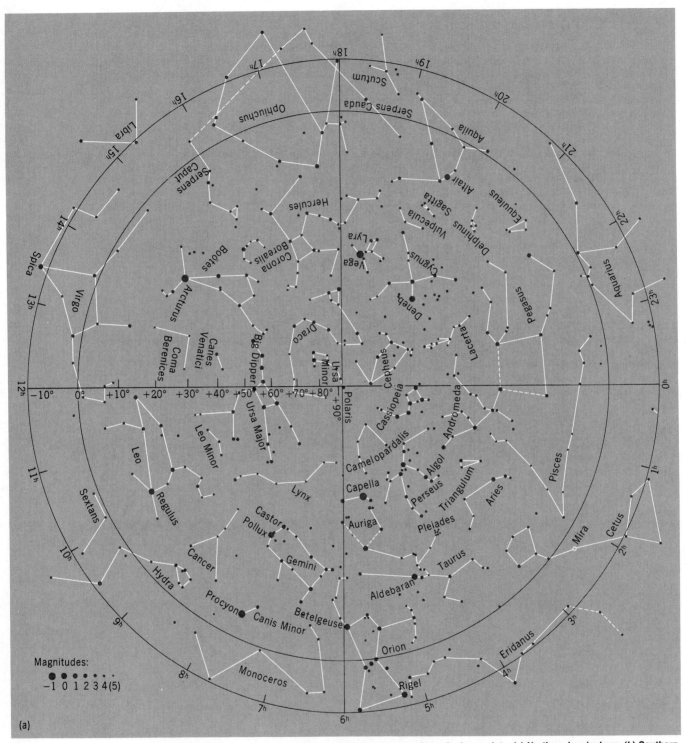

Fig. 2. Universally accepted constellations; brighter stars are shown by larger dots. (a) Northern hemisphere. (b) Southern hemisphere. (*After J. M. Pasachoff and D. H. Menzel, A Field Guide to the Stars and Planets, Houghton Mifflin, 2d ed., 1984*)

lation figures and myths usually differ from those found in Greek and Roman sources.

Star catalogs and atlases. The catalog of Ptolemy, in Hellenic Alexandria in the second century of the Christian Era, included over 1000 stars grouped into 48 constellations. Many of the images of constellations are derived from the beautiful engravings by Alexander Mair in Johann Bayer's *Uranometria* (1603) [**Fig. 1**]. Bayer included the constellations listed by

Ptolemy and also named 12 new ones containing stars observed on expeditions to the Southern Hemisphere. Bayer originated the scheme of labeling individual stars in constellations with Greek and other letters, roughly in order of brightness, and the genitive form of the constellation name. For example, the bright star Betelgeuse (the second brightest in the constellation Orion) is alpha Orionis (alpha of Orion), and Sirius, the brightest star in the sky, is alpha Canis Ma-

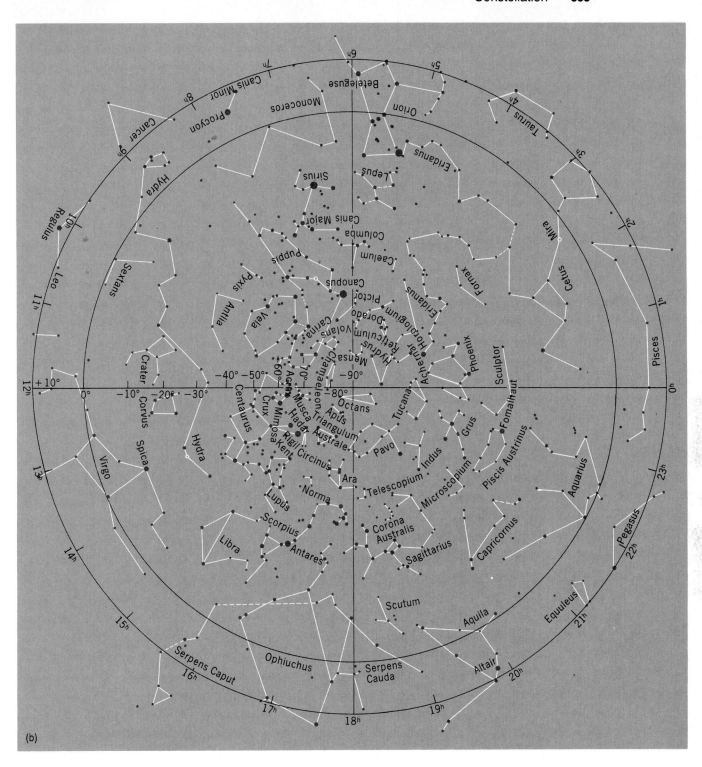

(b)

joris (alpha of the Big Dog). In some cases, Bayer labeled stars in order around figures in the sky, as for the Big Dipper. Bayer also used some uppercase (as in the star P Cygni) and lowercase Latin letters. In 1624, Jakob Bartsch placed three new constellations in gaps in Bayer's atlas and separated the southern constellation Crux from Centaurus. Coma Berenices (Berenice's Hair) was also added at about this time, reportedly by Tycho Brahe.

Johannes Hevelius added nine more southern constellations in his 1690 star atlas, *Firmamentum Sobiescianum sive Uranographia*. His figures represent the view of a celestial globe from the outside and thus are reversed from Bayer's. Nicolas Louis de Lacaille added 14 constellations in 1763 from his expedition to the Cape of Good Hope, using names reflecting the mechanical age, such as air pump, microscope, and telescope.

The constellations*

Latin name	Genitive	Abbreviation	English translation
Andromeda	Andromedae	And	Andromeda[†]
Antlia	Antliae	Ant	Pump
Apus	Apodis	Aps	Bird of Paradise
Aquarius	Aquarii	Aqr	Water Bearer
Aquila	Aquilae	Aql	Eagle
Ara	Arae	Ara	Altar
Aries	Arietis	Ari	Ram
Auriga	Aurigae	Aur	Charioteer
Boötes	Boötis	Boo	Herdsman
Caelum	Caeli	Cae	Chisel
Camelopardalis	Camelopardalis	Cam	Giraffe
Cancer	Cancri	Cnc	Crab
Canes Venatici	Canum Venaticorum	CVn	Hunting Dogs
Canis Major	Canis Majoris	CMa	Big Dog
Canis Minor	Canis Minoris	CMi	Little Dog
Capricornus	Capricorni	Cap	Goat
Carina	Carinae	Car	Ship's Keel[‡]
Cassiopeia	Cassiopeiae	Cas	Cassiopeia[†]
Centaurus	Centauri	Cen	Centaur[†]
Cepheus	Cephei	Cep	Cepheus[†]
Cetus	Ceti	Cet	Whale
Chamaeleon	Chamaeleonis	Cha	Chameleon
Circinus	Circini	Cir	Compass
Columba	Columbae	Col	Dove
Coma Berenices	Comae Berenices	Com	Berenice's Hair[†]
Corona Australis	Coronae Australis	CrA	Southern Crown
Corona Borealis	Coronae Borealis	CrB	Northern Crown
Corvus	Corvi	Crv	Crow
Crater	Crateris	Crt	Cup
Crux	Crucis	Cru	Southern Cross
Cygnus	Cygni	Gyg	Swan
Delphinus	Delphini	Del	Dolphin
Dorado	Doradus	Dor	Swordfish
Draco	Draconis	Dra	Dragon
Equuleus	Equulei	Equ	Little Horse
Eridanus	Eridani	Eri	River Eridanus[†]
Fornax	Fornacis	For	Furnace
Gemini	Geminorum	Gem	Twins
Grus	Gruis	Gru	Crane
Hercules	Herculis	Her	Hercules[†]
Horologium	Horologii	Hor	Clock
Hydra	Hydrae	Hya	Hydra[†] (water monster)
Hydrus	Hydri	Hyi	Sea Serpent
Indus	Indi	Ind	Indian
Lacerta	Lacertae	Lac	Lizard
Leo	Leonis	Leo	Lion
Leo Minor	Leonis Minoris	LMi	Little Lion
Lepus	Leporis	Lep	Hare
Libra	Librae	Lib	Scales
Lupus	Lupi	Lup	Wolf
Lynx	Lyncis	Lyn	Lynx
Lyra	Lyrae	Lyr	Harp
Mensa	Mensae	Men	Table (mountain)
Microscopium	Microscopii	Mic	Microscope
Monoceros	Monocerotis	Mon	Unicorn
Musca	Muscae	Mus	Fly
Norma	Normae	Nor	Level (square)
Octans	Octantis	Oct	Octant
Ophiuchus	Ophiuchi	Oph	Ophiuchus[†] (serpent bearer)
Orion	Orionis	Ori	Orion[†]
Pavo	Pavonis	Pav	Peacock
Pegasus	Pegasi	Peg	Pegasus[†] (winged horse)
Perseus	Persei	Per	Perseus[†]
Phoenix	Phoenicis	Phe	Phoenix
Pictor	Pictoris	Pic	Easel
Pisces	Piscium	Psc	Fish
Piscis Austrinus	Piscis Austrini	PsA	Southern Fish
Puppis	Puppis	Pup	Ship's Stern[‡]
Pyxis	Pyxidis	Pyx	Ship's Compass[‡]
Reticulum	Reticuli	Ret	Net
Sagitta	Sagittae	Sge	Arrow
Sagittarius	Sagittarii	Sgr	Archer
Scorpius	Scorpii	Sco	Scorpion
Sculptor	Sculptoris	Scl	Sculptor
Scutum	Scuti	Sct	Shield
Serpens	Serpentis	Ser	Serpent
Sextans	Sextantis	Sex	Sextant
Taurus	Tauri	Tau	Bull
Telescopium	Telescopii	Tel	Telescope

The constellations* (continued)

Latin name	Genitive	Abbreviation	English translation
Triangulum	Trianguli	Tri	Triangle
Triangulum Australe	Trianguli Australis	TrA	Southern Triangle
Tucana	Tucanae	Tuc	Toucan
Ursa Major	Ursae Majoris	UMa	Big Bear
Ursa Minor	Ursae Minoris	UMi	Little Bear
Vela	Velorum	Vel	Ship's Sails‡
Virgo	Virginis	Vir	Virgin
Volans	Volantis	Vol	Flying Fish
Vulpecula	Vulpeculae	Vul	Little Fox

*After J. M. Pasachoff, *Contemporary Astronomy*, 4th ed., 1989.
†Proper names.
‡Formerly formed the consellation Argo Navis, the Argonauts' Ship.

John Flamsteed (1729) and Johan Elert Bode (1801) produced other elegant star atlases, with engravings of constellations. Since the mid-1800s, Ptolemy's largest constellation, Argo Navis, has been divided into Carina, the keel; Puppis, the stern; and Vela, the sails.

Modern star atlases, from *Uranometria 2000.0* to the *Guide Star Catalogue* compiled in the late 1980s for the use of observers with the Hubble Space Telescope, do not usually show the figures of tradition. *See* ASTRONOMICAL ATLASES; ASTRONOMICAL CATALOGS.

International agreement. In 1928, the International Astronomical Union formally accepted the division of the sky into 88 constellations (see **table**), with the final list provided 2 years later; each star now falls in only one constellation (**Fig. 2**). The boundaries follow north-south or east-west celestial coordinates (right ascension and declination lines) from the year 1875; because of precession, the current boundaries do not match rounded values of celestial coordinates. *See* ASTRONOMICAL COORDINATE SYSTEMS; PRECESSION OF EQUINOXES.

Asterisms. Some of the most familiar patterns in the sky are asterisms rather than constellations. For example, the asterism known as the Big Dipper is part of the constellation Ursa Major. The asterism known as the Great Square of Pegasus has three of its corners in Pegasus but the fourth in Andromeda. The Northern Cross is made of stars in Cygnus.

Zodiac. The Sun, the Moon, and the planets move through a band in the sky known as the ecliptic. The constellations that fall close to the ecliptic through which the Sun traditionally moves are known as the zodiac. Actually, the Sun moves through 13 constellations, and precession has changed the dates at which the Sun passes through the zodiacal constellations. The intersection of the ecliptic and the celestial equator (the extension of the Earth's Equator into the sky) is known as the vernal equinox, or the first point of Aries. It has been used to mark the zero point of the celestial coordinate system. Westward from Aries, the zodiacal constellations are Taurus, the Bull; Gemini, the Twins; Cancer, the Crab; Leo, the Lion; Virgo, the Virgin; Libra, the Scales; Scorpius, the Scorpion; Sagittarius, the Archer; Capricornus, the Seat Goat; Aquarius, the Water Bearer; and Pisces, the Fish. *See* ECLIPTIC; EQUINOX; ZODIAC.

With the advent of computer control of telescopes, astronomers no longer use the constellations to locate objects in the sky. Even amateur telescopes are often equipped with computer locators.

Jay M. Pasachoff

Bibliography. J. M. Pasachoff, *Peterson's First Guide to Astronomy*, 1988; J. M. Pasachoff and D. H. Menzel, *A Field Guide to the Stars and Planets*, 2d ed., 1983; H. A. Rey, *The Stars*, rev. ed., 1988; W. Tirion, B. Rappaport, and G. Lovi, *Uranometria 2000.0*, 1987; D. J. Warner, *The Sky Explored: Celestial Cartography 1500–1800*, 1979.

Constraint

A restriction on the natural degrees of freedom of a system. If n and m are the numbers of the natural and actual degrees of freedom, the difference $n - m$ is the number of constraints. In principle $n = 3N$, where N is the number of particles, for example, atoms. In practice n is determined by the number of effectively rigid components. *See* DEGREE OF FREEDOM (MECHANICS).

A holonomic system is one in which the n original coordinates can be expressed in terms of m independent coordinates and possibly also the time. It is characterized by frictionless contacts and inextensible linkages. The new coordinates, $q_1, q_2, \ldots, q_m$, are called generalized coordinates. The equations of equilibrium and of motion may be expressed in terms of these coordinates. *See* LAGRANGE'S EQUATIONS.

Nonholonomic systems cannot be reduced to independent coordinates because the constraints are not on the n coordinate values themselves but on their possible changes. For example, an ice skate may point in all directions but at each position it must point along its path. This is a condition between changes (dx,dy) in the two position coordinates (x,y) and the direction angle θ, as in Eq. (1). This cannot be put in an inte-

$$dy = \tan \theta \, dx \qquad (1)$$

grated form, as in Eq. (2), since a skater can pass in

$$f(x,y,\theta) = 0 \qquad (2)$$

different directions repeatedly over a point.

The static equilibrium conditions of a constrained system under impressed forces F_1, F_2, and so on are contained in a general statement, the principle of virtual work or virtual displacement. *See* STATICS; VIRTUAL WORK PRINCIPLE.

A moving constraint is one which changes with time, as in the case of a system on a moving platform. Moving constraints differ from the stationary constraints of static equilibrium in being able to do work. For example, a mass constrained to move on the floor of a rising elevator is carried also in the direction of the (vertical) force of constraint which does work on it.

Bernard Goodman

Bibliography. H. C. Corben and P. Stehle, *Classical Mechanics,* 2d ed., 1974; H. Goldstein, *Classical Mechanics,* 2d ed., 1980; D. T. Greenwood, *Classical Dynamics,* 1977; L. D. Landau and E. M. Lifshitz, *Mechanics,* 3d ed., 1976; K. R. Symon, *Mechanics,* 1971.

Construction engineering

A specialized branch of civil engineering concerned with the planning, execution, and control of construction operations for such projects as highways, buildings, dams, airports, and utility lines.

Planning consists of scheduling the work to be done and selecting the most suitable construction methods and equipment for the project. Execution requires the timely mobilization of all drawings, layouts, and materials on the job to prevent delays to the work. Control consists of analyzing progress and cost to ensure that the project will be done on schedule and within the estimated cost.

Planning. The planning phase starts with a detailed study of construction plans and specifications. From this study a list of all items of work is prepared, and related items are then grouped together for listing on a master schedule. A sequence of construction and the time to be allotted for each item is then indicated. The method of operation and the equipment to be used for the individual work items are selected to satisfy the schedule and the character of the project at the lowest possible cost. SEE CONSTRUCTION EQUIPMENT; CONSTRUCTION METHODS.

The amount of time allotted for a certain operation and the selection of methods of operation and equipment are generally determined by the equipment that is readily available to the contractor. After the master or general construction schedule has been drawn up, subsidiary detailed schedules or forecasts are prepared from the master schedule. These include individual schedules for procurement of material, equipment, and labor, as well as forecasts of cost and income.

Execution. The speedy execution of the project requires the ready supply of all materials, equipment, and labor when needed. The construction engineer is generally responsible for initiating the purchase of most construction materials and expediting their delivery to the project. Some materials, such as structural steel and mechanical equipment, require partial or complete fabrication by a supplier. For these fabricated materials the engineer must prepare or check all fabrication drawings for accuracy and ease of assembly and often inspect the supplier's fabrication.

Other construction engineering duties are the layout of the work by surveying methods, the preparation of detail drawings to clarify the design engineer's drawings for the construction crews, and the inspection of the work to ensure that it complies with plans and specifications.

On most large projects it is necessary to design and prepare construction drawings for temporary construction facilities, such as drainage structures, access roads, office and storage buildings, formwork, and cofferdams. Other problems are the selection of electrical and mechanical equipment and the design of structural features for concrete material processing and mixing plants and for compressed air, water, and electrical distribution systems.

Control. Progress control is obtained by comparing actual performance on the work against the desired performance set up on the master or detailed schedules. Since delay on one feature of the project could easily affect the entire job, it is often necessary to add equipment or crews to speed up the work. SEE PERT.

Cost control is obtained by comparing actual unit costs for individual work items against estimated or budgeted unit costs, which are set up at the beginning of the work. A unit cost is obtained by dividing the total cost of an operation by the number of units in that operation.

Typical units are cubic yards for excavation or concrete work and tons for structural steel. The actual unit cost for any item at any time is obtained by dividing the accumulated costs charged to that item by the accumulated units of work performed.

Individual work item costs are obtained by periodically distributing job costs, such as payroll and invoices to the various work item accounts. Payroll and equipment rental charges are distributed with the aid of time cards prepared by crew forepersons. The cards indicate the time spent by the job crews and equipment on the different elements of the work. The allocation of material costs is based on the quantity of each type of material used for each specific item.

When the comparison of actual and estimated unit costs indicates an overrun, an analysis is made to pinpoint the cause. If the overrun is in equipment costs, it may be that the equipment has insufficient capacity or that it is not working properly. If the overrun is in labor costs, it may be that the crews have too many workers, lack of proper supervision, or are being delayed for lack of materials or layout. In such cases time studies are invaluable in analyzing productivity.

William Hershleder

Bibliography. D. W. Halpin *Financial and Cost Concepts for Construction Management,* 1985; A. L. Iannone and A. M. Civitello, *Construction Scheduling Simplified,* 1985; T. C. Kavanagh, F. Muller, and J. J. O'Brien, *Construction Management: A Professional Approach,* 1978; R. L. Peurifoy, *Construction Planning, Equipment, and Methods,* 4th ed., 1985.

Construction equipment

A wide variety of relatively heavy machines which perform specific construction (or demolition) functions under power. The power plant (which is treated in later paragraphs) is commonly an integral part of an individual machine, although in some cases it is contained in a separate prime mover, for example, a towed wagon or roller. It is customary to classify construction machines in accordance with their functions such as hoisting, excavating, hauling, grading, paving, drilling, or pile driving. There have been few changes for many years in the basic types of machines available for specific jobs, and few in the basic configurations of those that have long been available. Design emphasis for new machines is on modifications that increase speed, efficiency, and accuracy (particularly through more sophisticated controls); that improve operator comfort and safety; and that protect the public through sound attenuation and emission control. The selection of a machine for a specific job is mainly a question of economics and depends primarily on the ability of the machine to complete the job efficiently, and secondarily on its availability.

Hoisting equipment. This class of equipment is used to raise or lower materials from one elevation to another or to move them from one point to another over an obstruction. The main types of hoisting equipment are derricks, cableways, cranes, elevators, and conveyors. *See Bulk-handling machines; Hoisting machines.*

Derricks. The two main types of derricks are the guy and the stiff-leg. The former has a mast that is held in a vertical position by guy wires and a boom that can rotate with the mast 360°. In the stiff-leg the mast is tied to two or more rigid structural members, and the rotation of the boom is limited by the position of these members. The derrick is practical only where little mobility is required as in some types of steel erection, excavation of a shaft, or hoisting of materials through a shaft. *See Derrick.*

Cableway. This is a combination hoist and tram system comprising a trolley that runs on main load cables stretched between two or more towers which may be stationary or tiltable. Since even with towers that can be tilted slightly from side to side the cableway is limited almost entirely to linear movement of materials, its use in construction is restricted to dams, to some dredging and drag-line operations, and in special cases to bridges.

Elevator. The construction elevator, like the passenger elevator, consists of a car or platform that operates within a structural framework and is raised and lowered by cables. Because the elevator can move materials only in a vertical linear direction and because the materials so moved must subsequently be moved horizontally by other means, the elevator is limited to moving relatively light materials on jobs of small areas. *See Elevating machines.*

Crane. A crane is basically a fast-moving boom mounted on a frame containing the power supply and mechanisms for moving the boom and for raising and lowering the load-bearing cables that run through sheaves at the top of the boom. Once all cranes in the United States were mobile; that is, they were attached to a chassis mounted on pneumatic tires, crawler treads, or flanged (railroad) wheels. This type of machine is still predominant in all operations where mobility is a must. **Figure 1** shows a hydraulic truck crane. However, European and Australian tower cranes and climbing cranes have become popular throughout the United States and the world for high-rise building construction where mobility requirements are negligible. These machines, which carry horizontal or diagonal masts or booms, can ''grow'' with the construction by climbing with it or by adding sections to their own towers. They have the fast boom and line action of the mobile crane but are not restricted in boom reach.

Conveyor. Occasionally a conveyor is used to raise materials needed for construction of low buildings. This machine, which has a movable endless belt mounted on a frame, can carry only light loads such as bricks and cement sacks and is not practical for buildings of more than one or two floors. A conveyor carrying rock over a railroad yard is shown in **Fig. 2.**

Excavating equipment. This type of equipment is divided into two main classes: standard land excavators and marine dredges; each has many variations.

Fig. 1. Hydraulic truck crane.

Standard land excavator. This comprises machines that merely dig earth and rock and place it in separate hauling units, as well as those that pick up and transport the materials. Among the former are power shovels, draglines, backhoes, cranes with a variety of buckets, front-end loaders, excavating belt loaders, trenchers, and the continuous bucket excavator. The second group includes such machines as bulldozers, scrapers of various types, and sometimes the front-end loader.

Power shovel. This has for years been regarded as the most efficient machine for digging into vertical banks and for handling heavy rock. Generally mounted on crawler treads, the power shovel carries a short boom on which rides a movable dipper stick carrying an open-topped bucket. The bucket digs in

Fig. 2. Conveyor carrying rock over a railroad yard.

Fig. 3. Power shovels loading over-the-road rear-dump trucks.

an upward direction away from the machine and dumps its load by lowering the front of its hinged bottom. **Figure 3** shows power shovels in the process of loading dump trucks.

Front-end loader. This has made heavy inroads into the domain of the power shovel and in time may well replace it in all but the biggest operations. Loaders are built that can handle buckets having 12, 14, and 24 yd³ (9, 11, and 18 m³) capacities. The loader, basically an articulated bucket mounted on a series of movable arms at the front of a crawler or rubber-tired tractor, has the advantage of mobility, speed, economy of cost and operation, and light weight. A loader of any given capacity can cost as little as one-half or even one-third as much as a power shovel of the same capacity and requires only one person to run it as opposed to the two or more normally required for the power shovel. **Figure 4** shows a loader operating in a swamp.

Backhoe. This is fundamentally an upside-down power shovel. Its bucket, mounted on a hinged boom, digs upward toward the machine and unloads by

being inverted over the dumping point. Backhoes are manufactured as individual machines but more often are attachments mounted on a crawler crane, a rubber-tired truck crane, or a tractor (which usually has a front-end loader bucket at the other end). It is particularly well adapted to the digging of deep trenches.

Dragline. This is a four-sided bucket that is used mainly on soil too wet to support an excavating machine. It is usually carried on a mobile machine mounting a crane boom, but it can also be worked from a cableway where excavating distances are great. The bucket is carried or cast to a point ahead of its support and dropped to the ground. The dragline is designed to excavate and to fill itself as it is drawn across the ground. It empties when its front end is lowered.

Clamshell. A clamshell is a two-sided bucket that can dig only in a vertical direction. The bucket is dropped while its leaves are open and digs as they close. Formerly, clamshells were suspended from cables on cranes or gantries and worked only under their own weight. Therefore they were imprecise in dig-

Fig. 4. A loader operating in a swamp.

Fig. 5. A scraper in operation, as it is pulled by a tractor.

ging and were practical only in relatively soft soil or loose rock. Presently many clamshell buckets are attached directly to powered booms, are closed hydraulically, and can work with greater precision in dense soils.

Orange peel. This is a multileaved bucket, generally round in configuration. It is normally cable-supported and, like the clamshell, works under its own weight. Like the clamshell, however, it too is being mounted on powered booms and closed hydraulically. One of the commonest uses of the orange peel is the cleaning of small shafts, sewers, and storm drains. Larger sizes are often used for handling broken rock.

Grapple. This group includes a wide variety of special-purpose tined grabs that work on the principle of the clamshell and orange peel. The grapple is used mainly for handling rock, pipe, and logs.

Excavating belt loader. This type of machine is used mainly for loading granular materials from a stockpile, or as a vertical or inclined belt, or as a pair of chains on which are mounted small buckets. Some belt loaders are mobile and can move into a bank under their own power as the material is moved. Others are fixed and are fed by bulldozers which push material onto the belt from above. Their advantage over most excavators lies in their continuous operation.

Trenching machine. This equipment for digging trenches ranges in size from small hand-pushed units used for getting small pipes from streets to private homes to monsters of many tons capable of cutting a 4-ft-wide (1.2-m) swath for 2000 mi (3200 km) of transcontinental pipelines. All, however, work on basically the same principle: A series of buckets, mounted on a chair or a wheel, lift dirt from the ground and deposit it alongside the trench being dug.

Continuous bucket excavator. This excavator works like a trenching machine but is designed to remove earth and loose rock from a wider area and at shallower depths. These machines are self-propelled; they can load trucks in a continuous operation and can switch from a filled to an empty truck in less than 2 s. The continuity of loading and the absence of in-

termittent activities, such as those necessary in shovel or loader operations, make the continuous bucket excavator a fast, economical machine for large excavating jobs.

Combination excavators and haulers. This classification of construction equipment includes bulldozers, carrying scrapers, self-loading or elevating scrapers, and sometimes front-end loaders.

Bulldozer. This is a curved blade mounted on the front of a crawler or rubber-tired tractor for the purpose of digging and pushing earth and broken rock from one place to another. It is often used in conjunction with a ripper, a heavy tooth or series of teeth mounted on the rear of a tractor to break up rock so that it can be handled by the dozer. Because the amount of dirt the dozer can move at one time is limited by the size of its blade, it is not economical to use the dozer for moving dirt more than a few hundred feet.

Front-end loader. This is sometimes used for moving dirt in the same manner as the dozer, but with the same limitations.

Scraper. This is the most economical of the combination excavating-hauling units for hauls of over a few hundred feet. A wheel-mounted, open-top box or "bowl," the scraper has a hinged bottom that is lowered to scoop up dirt as the machine moves forward. When the bowl is full, the bottom is closed and the unit is moved to the dumping area, where the bottom is lowered and a hydraulic pushing unit in the back of the bowl moves the dirt out through the opening in the bottom, to spread the dirt as needed. Scrapers are pulled by tractors, most of which are an integral part of the unit. A scraper is shown in **Fig. 5**.

Self-loading or elevating scraper. This group of machines does the same work as the conventional scraper, but in front of the bowl is a series of horizontal plates mounted on moving chains. As the chains move, the plates lift the dirt into the bucket in the same manner as a bucket loader. The advantage of the elevating scraper is that it needs less tractor power during the loading operation.

Marine excavator. Usually called a dredge, the marine excavator is an excavating machine mounted on a barge or boat. Two common types are similar to land excavators, the clamshell and the bucket excavator. The suction dredge is different; it comprises a movable suction pipe which can be lowered to the bottom, usually with a fast-moving cutter head at the bottom end. The cutter churns up the bottom so that the pumps on the barge or boat suck up water and the earth suspended in it. When practical, the pumped material is then piped to land, where the solids settle out, allowing the liquids to run back to the body of water being dredged. When distance to land is too great or where there is no empty land on which to discharge the material, the effluent is pumped into barges in which the settled-out solids are towed to a dump area at a distant point.

Hauling equipment. Excavated materials are moved great distances by a wide variety of conveyances. The most common of these are the self-propelled rubber-tired rear-dump trucks, which are classed as over-the-road or off-the-road trucks. The main difference between these two is weight and carrying capacity. Many states restrict the total weight of highway-using trucks (truck and load) to as little as 30 tons (27 metric tons). For jobs where trucks remain on site, such as at dams and some highway construction areas, trucks with a carrying capacity of 50 tons (45 metric tons) are common and some capable of carrying loads in excess of 100 tons (90 metric tons) are in use. A 100-ton (90-metric-ton) rear-dump truck is shown in **Fig. 6**. Wagons towed by a rubber-tired prime mover are also used for hauling dirt. These commonly have bottom dumps which permit spreading dirt as the vehicle moves. In special cases side-dump trucks are also used.

Conveyors, while not commonly used on construction jobs for hauling earth and rock great distances, have been used to good advantage on large jobs where obstructions make impractical the passage of trucks. One such conveyor which was utilized on a New Jersey highway job exceeded 2 mi (3.2 km) in length.

Graders. Graders are high-bodied, wheeled vehicles that mount a leveling blade between the front and rear wheels. The principal use is for fine-grading relatively loose and level earth. Commonly considered a maintenance unit, the grader is still an important machine in highway construction. While the configura-

Fig. 6. A 100-ton (90-metric-ton) rear-dump truck.

Fig. 7. Crawler-mounted drill. (*Gardner-Denver*)

tion of graders has not changed in many years, some models have been built with articulated main frames that permit faster turns of shorter radii and safer operation at the edges of steep slopes.

Pavers. These place, smooth, and compact paving materials. They may be mounted on rubber tires, endless tracks, or flanged wheels. Asphalt pavers embody tamping pads that consolidate the material; concrete pavers use vibrators for the same purpose. Concrete pavers had been required to use forms to contain the material until the advent of the slip-form pavers. These drag their own short forms behind them and consolidate the concrete sufficiently to stand without slumping after removal of the forms. Many high-speed conventional pavers have been produced with highly sophisticated automated control devices that provide extremely accurate and consistent surface elevations.

Drilling equipment. Holes are drilled in rock for wells and for blasting, grouting, and exploring. Drills are classified according to the way in which they penetrate rock, namely, percussion, rotary percussion, and rotary. The first two types are the most common for holes up to 6 in. (15 cm) in diameter and under 40 ft (12 m) in depth. In smaller sizes these drills may be hand-held, but for production work they are mounted on masts which are supported by trucks or, as shown in **Fig. 7**, special tracks mounted on drill rigs. In tunnel work the drills are commonly mounted on platform-supported movable arms or posts that are hydraulically or pneumatically controlled to permit drilling in horizontal or angled positions, with a minimum of effort by the drill operator. The rotary drill is most common in larger sizes, but drills as small as 3 in. (7.5 cm) in diameter are used for coring rock to obtain samples.

Specialized equipment. Among the most common specialized construction equipment are augers, compactors, and pile hammers.

Augers. These are used for drilling wells, dewatering purposes, and cutting holes that can be filled with concrete for foundations. Augers up to 6 ft (1.8 m) in diameter are common.

Compactors. These machines are designed solely to consolidate earth and paving materials to sustain loads greater than those sustained in the uncompacted state. They range in size from small pneumatic hand-held tampers to multiwheeled machines weighing more than 60 tons (50 metric tons). Actual compaction may be achieved by heavily loaded rubber tires or steel rollers. The steel rollers may be solid cylinders, or have separate pads or grids, or contain protrusions (sheep-foot roller). Many machines induce a vibratory action into the compacting units so that compaction is achieved by impact force rather than sheer weight.

Pile hammers. These machines are used to install bearing piles for foundations, or sheet piles for cofferdams and retaining walls. The prehistoric pile hammer consisted of a heavy weight lifted by a rope and allowed to drop on top of a pile. This drop hammer is still in use, basically unchanged. More efficient modern hammers are true machines activated by steam, air, oil, hydraulic fluid, or electricty. Conventional hammers contain pistons that are raised by one of these means and then either allowed to fall freely because of gravity (single-acting) or to be driven downward by the same means (double-acting) to impart an impact to the pile. Vibratory hammers have been developed which use electrically activated eccentric cams to vibrate piles into place. Relatively rare are machines that use hydraulic action to install piles.

Road planer. This machine looks like a conventional motor grader, but instead of a blade it carries near its center a horizontal drum that has on its periphery many rows of replaceable hardened steel teeth. As the drum revolves, the teeth pulverize cracked asphalt or decayed concrete pavement to depths of up to 4 in. (10 cm). This permits the placement of a smooth new riding surface on a sound base without raising the road surface elevation, having to raise manhole covers, and reducing the clearances under bridges.

Bore tunneling machine. This is a machine capable of placing water, sewer, or utility pipes accurately for great distances underground. Key to the operation is a series of intermediate hydraulic jacking rings that expand or contract as controlled to permit the movement of any section of pipe independent of the sections before or behind. This independent movement of short sections reduces the friction on the line and therefore the amount of pressure that would be required if the entire pipe were moved from a single push point.

Power plants. Steam is seldom used as a source of power today except in some marine applications and rarely for pile driving. Gasoline and diesel engines of the piston type are the most common source of power for construction machines, with the diesel attaining an increasingly greater prominence for at least two major reasons. First, the size of many construction machines has increased to proportions formerly thought impractical. With this increase in size, the economy of operation and maintenance of the diesel has come to far outweigh its greater initial cost. Second, the increased popularity of the diesel lies in the fact that manufacturers have been able to build lightweight economical diesel engines in very small sizes, thus making practical their application in small compressors, pumps, portable electric power plants, and so on. SEE DIESEL ENGINE.

Electricity is sometimes, though not often, used as a primary source of power for some construction machines, but usually only at the site of large dams and strip-mining operations where mobility is not a primary requirement. Electricity is making its appearance as a secondary source of power on several types of mobile construction machinery, such as trucks and scrapers, which use diesel or gas engines to turn generators that provide power for electric wheels.

A more common type of secondary power, however, is the hydraulic motor which, activated by a gasoline or diesel-powered hydraulic pump, is being used to provide direct power for virtually every movement, including travel, of construction machines.

Some manufacturers have experimented with turbine engines for construction applications. While they are being used with success as primary power sources for large stationary electric power plants on job sites, their practical use in mobile equipment is not evident. SEE CONSTRUCTION ENGINEERING.

Edward M. Young

Bibliography. J. Russel, *Construction Equipment,* 1985; M. J. Tomlinson, *Pile Design and Construction Practice,* 1987.

Construction methods

The procedures and techniques utilized during construction. Construction operations are generally classified according to specialized fields. These include preparation of the project site, earthmoving, foundation treatment, steel erection, concrete placement, asphalt paving, and electrical and mechanical installations. Procedures for each of these fields are generally the same, even when applied to different projects, such as buildings, dams, or airports. However, the relative importance of each field is not the same in all cases. For a description of tunnel construction, which involves different procedures, SEE TUNNEL.

Preparation of site. This consists of the removal and clearing of all surface structures and growth from the site of the proposed structure. A bulldozer is used for small structures and trees. Larger structures must be dismantled.

Earthmoving. This includes excavation and the placement of earth fill. Excavation follows preparation of the site, and is performed when the existing grade must be brought down to a new elevation. Excavation generally starts with the separate stripping of the organic topsoil, which is later reused for landscaping around the new structure. This also prevents contamination of the nonorganic material which is below the topsoil and which may be required for fill. Excavation may be done by any of several excavators, such as shovels, draglines, clamshells, cranes, and scrapers. For a discussion of their application SEE CONSTRUCTION EQUIPMENT.

Efficient excavation on land requires a dry excavation area, because many soils are unstable when wet and cannot support excavating and hauling equipment. Dewatering becomes a major operation when the excavation lies below the natural water table and intercepts the groundwater flow. When this occurs, dewatering and stabilizing of the soil may be accomplished by trenches, which conduct seepage to a sump from which the water is pumped out. Dewatering and stabilizing of the soil may in other cases be accomplished by wellpoints and electroosmosis. *SEE WELL-POINT SYSTEMS*.

Some materials, such as rock, cemented gravels, and hard clays, require blasting to loosen or fragment the material. Blast holes are drilled in the material; explosives are then placed in the blast holes and detonated. The quantity of explosives and the blast-hole spacing are dependent upon the type and structure of the rock and the diameter and depth of the blast holes.

After placement of the earth fill, it is almost always compacted to prevent subsequent settlement. Compaction is generally done with sheep's-foot, grid,

pneumatic-tired, and vibratory-type rollers, which are towed by tractors over the fill as it is being placed. Hand-held, gasoline-driven rammers are used for compaction close to structures where there is no room for rollers to operate.

Foundation treatment. When subsurface investigation reveals structural defects in the foundation area to be used for a structure, the foundation must be strengthened. Water passages, cavities, fissures, faults, and other defects are filled and strengthened by grouting. Grouting consists of injection of fluid mixtures under pressure. The fluids subsequently solidify in the voids of the strata. Most grouting is done with cement and water mixtures, but other mixture ingre-

Fig. 1. Steel erection with guy derricks on a high structure. (*Bethlehem Steel Co.*)

dients are asphalt, cement and clay, and precipitating chemicals. *See Foundations*.

Steel erection. The construction of a steel structure consists of the assembly at the site of mill-rolled or shop-fabricated steel sections. The steel sections may consist of beams, columns, or small trusses which are joined together by riveting, bolting, or welding. It is more economical to assemble sections of the structure at a fabricating shop rather than in the field, but the size of preassembled units is limited by the capacity of transportation and erection equipment. The crane is the most common type of erection equipment, but when a structure is too high or extensive in area to be erected by a crane, it is necessary to place one or more derricks on the structure to handle the steel (**Fig. 1**). In high structures the derrick must be constantly dismantled and reerected to successively higher levels to raise the structure. For river bridges the steel may be handled by cranes on barges, or, if the bridge is too high, by traveling derricks which ride on the bridge being erected. Cables for long suspension bridges are assembled in place by special equipment that pulls the wire from a reel, set up at one anchorage, across to the opposite anchorage, repeating the operation until the bundle of wires is of the required size.

Concrete construction. Concrete construction consists of several operations: forming, concrete production, placement, and curing. Forming is required to contain and support the fluid concrete within its desired final outline until it solidifies and can support itself. The form is made of timber or steel sections or a combination of both and is held together during the concrete placing by external bracing or internal ties (**Fig. 2**). The forms and ties are designed to withstand the temporary fluid pressure of the concrete.

The usual practice for vertical walls is to leave the forms in position for at least a day after the concrete is placed. They are removed when the concrete has solidified or set. Slip-forming is a method where the form is constantly in motion, just ahead of the level of fresh concrete. The form is lifted upward by means of jacks which are mounted on vertical rods embedded in the concrete and are spaced along the perimeter of the structure. Slip forms are used for high structures such as silos, tanks, or chimneys.

Concrete may be obtained from commercial batch plants which deliver it in mix trucks if the job is close to such a plant, or it may be produced at the job site. Concrete production at the job site requires the erection of a mixing plant, and of cement and aggregate receiving and handling plants. Aggregates are sometimes produced at or near the job site. This requires opening a quarry and erecting processing equipment such as crushers and screens.

Concrete is placed by chuting directly from the mix truck, where possible, or from buckets handled by means of cranes or cableways, or it can be pumped into place by special concrete pumps.

Curing of exposed surfaces is required to prevent evaporation of mix water or to replace moisture that does evaporate. The proper balance of water and cement is required to develop full design strength. *See Concrete*.

Concrete paving for airports and highways is a fully mechanized operation. Batches of concrete are placed between the road forms from a mix truck or a movable paver, which is a combination mixer and placer. A series of specialized pieces of equipment,

Fig. 2. Concrete form. (*Superior Concrete Accessories*)

which ride on the forms, follow to spread and vibrate the concrete, smooth its surface, cut contraction joints, and apply a curing compound. *See Highway engineering; Pavement; Runway*.

Asphalt paving. This is an amalgam of crushed aggregate and a bituminous binder. It may be placed on the roadbed in separate operations or mixed in a mix plant and spread at one time on the roadbed. Then the pavement is compacted by rollers.

William Hershleder

Contact aureole

The zone of alteration surrounding a body of igneous rock caused by heat and volatiles given off as the magma crystallized. Changes can be in mineralogy, texture, or elemental and isotopic composition of the original enclosing (country or wall) rocks, and progressively increase closer to the igneous contact. The contact aureole is the shell of metamorphosed or metasomatized rock enveloping the igneous body (**Fig. 1**). The ideal contact aureole forms locally around a single magma after it is emplaced. Metamorphism over a much larger area can result from coalescing of several contact aureoles. This is termed a contact-regional metamorphic aureole and is thought responsible for the regional metamorphism of several mountain areas. Other contact aureoles develop at greater depths and may be physically emplaced to shallower levels along with the igneous body. These are termed

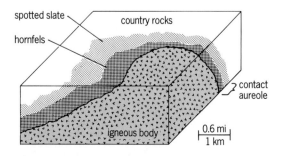

Fig. 1. Zoned contact aureole developed around an igneous body.

dynamothermal aureoles. *See Igneous rocks; Magma; Metamorphic rocks; Metamorphism; Metasomatism; Pluton.*

Extent. The aureole extends from the igneous contact, where the metamorphic effects are the greatest, out into the country rocks to where the temperature or heat energy is insufficient to effect any changes. This temperature lies between 400 and 750°F (200 and 400°C), and actual widths of contact aureoles range from several inches to miles.

Most important in determining the metamorphic intensity and the extent of the aureole are the heat content of the magma and the temperature contrast between the magma and country rock. Initial temperature of the magma depends on its type; metamorphic effects around gabbro magma (~2200°F or 1200°C) are generally more extensive and intensive than those around granite (~1400°F or 750° C). Better developed aureoles are found around larger igneous bodies, magmas which were not extensively crystallized before emplacement, and convecting magmas. Under the right circumstances, the highest temperature in the aureole can approach that of the magma. At deeper crustal levels, country rock temperatures increasingly coincide with those of common magmas, and the contact effects are correspondingly reduced until they are indistinguishable from metamorphic changes brought about by normal burial. Thus contact aureoles are most commonly found around shallow igneous bodies where magma temperatures exceed those of the country rock. They are considered to have low-pressure metamorphism, occurring at pressures less than 5 kilobars (500 megapascals), but usually much less. This corresponds to depths of less than 12 mi (20 km).

Factors of heat transfer also affect the extent and intensity of the metamorphism in the contact aureole. Geometry and orientation of the igneous body are important; intense metamorphism occurs at acute corners of the body and wider aureoles occur at the roof of an igneous body, where most heat is escaping, as opposed to the floor. Wider aureoles and less intense metamorphism occur in country rocks with good thermal conductivity. Heat transfer by conduction produces narrower, more regular, aureoles in contrast to heat transfer by fluid convection. Convective heat transport can produce wide or very narrow aureoles depending on the fluid content of the magma, fluid content of the wall rock, permeability of the system, and direction of fluid flow. Dry magmas and impermeable systems tend to have narrow aureoles. Cool fluid from the wall rocks can flow into the magma, cooling it and preventing development of the aureole. Conversely, hot fluid from the magma or wall rock fluid heated by the magma flowing outward can produce an extensive aureole.

Aureoles are most pronounced where wall rocks are easily affected by metamorphism because of their composition, texture, or previous metamorphism. Shale and limestone generally show a more conspicuous metamorphism than does sandstone. Aureoles are best developed in fine-grained and previously unmetamorphosed or weakly metamorphosed rocks; they may not be detected in intensely metamorphosed rocks because minerals of the wall rocks are stable at the conditions imposed upon them by the magma. Heat absorption by energy-consuming reactions will lessen the intensity of metamorphism elsewhere in the aureole. If the rate of heating and subsequent cooling exceeds the rate of recrystallization or reaction in the aureole, metamorphic effects will be reduced.

Rocks. Weakly contact-metamorphosed wall rocks may be only discolored or indurated. Contact aureoles with more intense metamorphism are progressively zoned, with appearance of new textures and minerals toward the igneous body (**Fig. 2**). At the outer edge of the aureole, textural changes include development of scattered aggregates of existing or new minerals, which are larger than the matrix minerals. If the host rock is shale, these recrystallized rocks become spotted slates. Further inward there is sequential appearance of new minerals, more extensive recrystallization, and color changes. Average grain size may increase but is still small, usually less than 0.08 in. (2 mm). Fine grain size in contact metamorphism results from short-duration heating and lack of deformation. Well within the aureole, where new mineral growth and recrystallization is complete, older structures (bedding or schistosity) may be eliminated. Mineral grains tend to become equidimensional with little or no preferred orientation and lack their crystal form in an interlocking mosaic (granoblastic texture); the rock is called a hornfels. The term hornfels is applied in a wide sense to any contact-metamorphosed

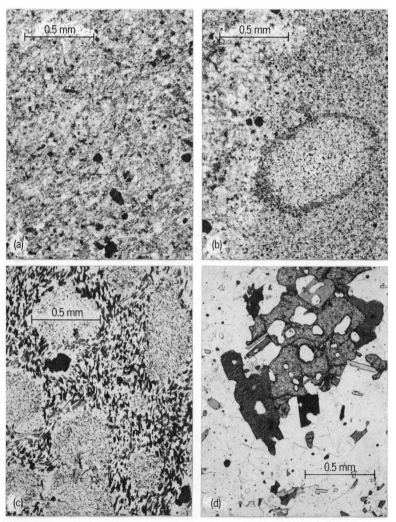

Fig. 2. Microscopic appearance of progressive metamorphism in a contact aureole. (*a*) Original slate country rock comprising chlorite, muscovite, albite, and quartz. (*b*) Spotted slate with the same mineralogy and texture as country rock but with the addition of cordierite plus biotite spots. (c) Hornfels with poikiloblastic porphyroblasts of cordierite in a matrix of biotite, muscovite, quartz, and plagioclase. (*d*) Granoblastic hornfels with garnet, biotite, cordierite, alkali feldspar, plagioclase, and quartz.

rock and made specific by mineral or rock compositional modifiers. The massive texture of hornfelses results from the lithostatic or hydrostatic nature of the pressure; directed pressure as in regional metamorphism is absent. However, some hornfelses develop a preferred alignment because of intrusion of the adjacent magma. Hornfelses typically have a few larger mineral grains (porphyroblasts) which can contain abundant inclusions of the smaller matrix minerals (poikiloblastic texture). Adjacent to the igneous contact, melting of the wall rocks can occur. SEE HORN-FELS; PORPHYROBLAST; ROCK; SCHISTOSITY.

Mineral facies. Contact metamorphism can occur over a wide range of temperatures, pressures, or chemical gradients in rocks of any composition. Thus any mineral assemblage or facies of metamorphic rocks an be found. However, the nature of contact aureoles results in minerals characteristic of low to moderate pressures and moderate to high temperatures usually in common rock types: shales, basalt, limestone, and sandstone. Characteristic minerals developed in shales are andalusite, sillimanite, cordierite, biotite, orthopyroxene, and garnet. At the highest temperatures, tridymite, sanidine, mullite, and pigeonite form; whereas in limestone unusual calcium silicates form, including tilleyite, spurrite, rankinite, larnite, merwinite, akermanite, monticellite, and melilite. SEE FACIES (GEOLOGY).

Chemical alterations. Compositional changes in a contact aureole range from none to great, but as a rule, contact metamorphism entails relatively little change in bulk rock composition. Because metamorphic changes are largely brought about by heat, contact aureoles are often termed thermal aureoles. However, there is a tendency for volatiles (water, carbon dioxide, oxygen) and alkalies (sodium, potassium) to be lost from rocks in the aureole. Stable isotope compositions (oxygen, sulfur) change in response to the thermal gradient and flow of fluids through the rocks. In some cases, volatiles (boron, fluorine, and chlorine) and other elements from the crystallizing magma are gained.

Some wall rock compositions, such as limestone, can be greatly changed and form rocks termed skarn. These contact aureoles are economically important because they often contain ore deposits of iron, copper, tungsten, graphite, zinc, lead, molybdenum, and tin. Conversely, the magma can incorporate material from the wall rocks by assimilation or mixing with any partial melts formed. Mixing results in elemental and isotopic contamination of the magma, crystallization of different minerals from the melt, and hybrid rock types at the margin of the igneous body. SEE ORE AND MINERAL DEPOSITS; PNEUMATOLYSIS; SKARN.

J. Alexander Speer

Bibliography. C. Gillen, *Metamorphic Geology*, 1982; A. Miyashiro, *Metamorphism and Metamorphic Belts*, 1975; V. V. Reverdatto, *The Facies of Contact Metamorphism*, 1970; A. Spry, *Metamorphic Textures*, 1969; F. J. Turner, *Metamorphic Petrology*, 2d ed., 1980; H. G. F. Winkler, *Petrogenesis of Metamorphic Rocks*, 4th ed., 1976.

Contact condenser

A device in which a vapor is brought into direct contact with a cooling liquid and condensed by giving up its latent heat to the liquid. In almost all cases the cooling liquid is water, and the condensing vapor is

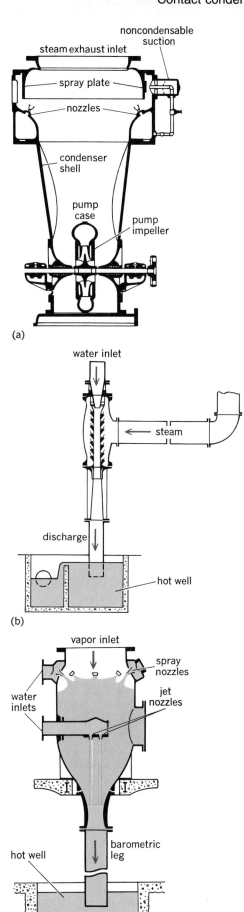

Three basic types of contact condenser: (a) low-level jet condenser (*C. H. Wheeler Manufacturing Co.*); (b) single-jet ejector condenser; (c) multijet barometric condenser (*Schutte and Koerting Co.*).

steam. Contact condensers are classified as jet, barometric, and ejector condensers (see **illus.**). In all three types the steam and cooling water are mixed in a condensing chamber and withdrawn together. Noncondensable gases are removed separately from the jet condenser, entrained in the cooling water of the ejector condenser, and removed either separately or entrained in the barometric condenser. The jet condenser requires a pump to remove the mixture of condensate and cooling water and a vacuum breaker to avoid accidental flooding. The barometric condenser is self-draining. The ejector condenser converts the energy of high-velocity injection water to pressure in order to discharge the water, condensate, and noncondensables at atmospheric pressure. SEE VAPOR CONDENSER.

Joseph F. Sebald

Contact potential difference

The potential difference that exists across the space between two electrically connected metals; also, the potential difference between the bulk regions of a junction of two semiconductors. In **Fig. 1a** consider two metals M_1 and M_2 with work functions ϕ_1 and ϕ_2, respectively. When the metals are brought in contact, the Fermi levels, that is, the levels corresponding to the most energetic electrons, must coincide, as in Fig. 1b. Consequently, if $\phi_1 > \phi_2$, metal M_1 will acquire a negative and metal M_2 a positive surface charge at the contact area. In the circuit indicated in **Fig. 2** there will thus exist a potential difference between the plates such that M_1 is negative with respect

to M_2; the magnitude of the contact potential is equal to $(\phi_1 - \phi_2)/e$ (of the order of a few volts or less) where $e = 1.6 \times 10^{-19}$ coulomb. Note that the contact potential is independent of the number of different materials used in the circuit external to the plates in vacuum. The contact potential difference may be measured by placing a variable electromotive force in the external circuit, which can be balanced to make the potential difference between the plates equal to zero. Contact potentials must be taken into account in analyzing physical electronics experiments. SEE ELECTRONIC WORK FUNCTION.

A. J. Dekker

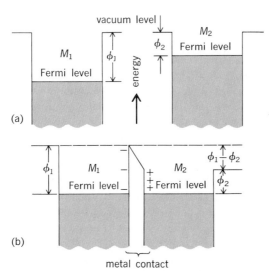

Fig. 1. Energy-level diagram for conduction electrons of two metals. (a) Before contact. (b) After contact.

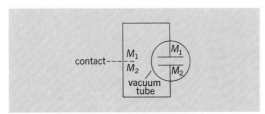

Fig. 2. Two metallic plates in vacuum between which a contact potential difference may be measured.

Continent

A protuberance of the Earth's crustal shell with an area of several million square miles and with sufficient elevation above neighboring depressions (the ocean basins) so that much of it is above sea level.

The term was originally applied to the most extensive continuous land areas of the globe. The great majority of maps now in use imply that the boundaries of continents are their shorelines. From the geological point of view, however, the line of demarcation between a continent and an adjacent ocean basin lies offshore, at distances ranging from a few to several hundred miles, where the gentle slope of the continental shelf changes somewhat abruptly to a steeper declivity. This change occurs at depths ranging from a few to several hundred fathoms at different places around the periphery of various continents. SEE CONTINENTAL MARGIN.

On such a basis, numerous offshore islands, including the British Isles, Greenland, Borneo, Sumatra, Java, Taiwan (formerly Formosa), Japan, and Sri Lanka (formerly Ceylon), are parts of the nearby continent. Thus, there are six continents: Eurasia (Europe, China, and India are parts of this largest continent), Africa, North America, South America, Australasia (including Australia and New Guinea), and Antarctica.

All continents have similar structural features but display great variety in detail. Each includes a basement complex (shield) of metamorphosed sedimentary and volcanic rocks of Precambrian age, with associated igneous rocks, mainly granite. Originally formed at considerable depths below the surface, this shield was later exposed by extensive erosion, then largely covered by sediments of Paleozoic, Mesozoic, and Cenozoic age, chiefly marine limestones, shales, and sandstones. In at least one area on each continent, these basement rocks are now at the surface (an example is the Canadian Shield of North America). In some places they have disappeared beneath a sedimentary platform occupying a large fraction of the area of each continent, such as the area in the broad lowland drained by the Mississippi River in the United States. In each continent there are long belts of mountains in which thick masses of sedimentary rocks have been compressed into folds and broken by faults. SEE BASEMENT ROCKS; CONTINENTS, EVOLUTION OF.

Presumably because of the large volume of acidic aluminum-rich granitic rocks beneath the surface of each continent, the velocity of transmission of earthquake vibrations in continental segments of the Earth's crust is slower than in oceanic segments, in which alkaline, iron-rich basaltic rather than granitic rocks are the dominant type. For the same reason the average density of the rocks composing continents is less than that of the rocks beneath the deep-sea floors. SEE EARTHQUAKE.

Continents are the less dense, subaerially exposed portion of the large and not-so-large plates that make up the Earth's lithosphere, or outer shell of rigid rock material. As such, continents together with part of the ocean's floor are intimately joined portions of the lithospheric plates. As plates rip apart and migrate horizontally over the Earth's surface, so too do continents rip apart and migrate, sometimes colliding with other continental segments scores of millions of years later. Mountain systems, such as the Appalachian-Ouachita, the Arbuckle-Wichita, and the Urals systems, are now believed to represent the sutures of former continents attached to their respective plates which collided long ago. The Red Sea and the linear volcano-lakes district of Africa are also believed by many to manifest continental ripping and early continental drifting. Such continental collision and accretion are believed to have occurred throughout most of the Earth's history. SEE AFRICA; ANTARCTICA; ASIA; AUSTRALIA AND NEW ZEALAND; EARTH, GRAVITY FIELD OF; EUROPE; ISOSTASY; LITHOSPHERE; NORTH AMERICA; PLATE TECTONICS; SEISMOLOGY; SOUTH AMERICA.

Donald Lee Johnson

Continental drift

The concept that the world's continents once formed part of a single mass and have since drifted into their present positions. Although it was outlined by Alfred Wegener in 1912, the idea was not particularly new. Paleontological studies had already demonstrated such strong similarities between the flora and fauna of the southern continents between 300,000,000 and 150,000,000 years ago that a huge supercontinent, Gondwana, containing South America, Africa, India, Australia, and Antarctica, had been proposed. However, Gondwana was thought to be the southern continents linked by land bridges, rather than contiguous units.

Wegener's ideas were almost universally rejected at a meeting of the American Association of Petroleum Geologists in 1928. His evidence and timing were undoubtedly wrong in many instances, and only a few scientists accepted his general concepts and remained sympathetic to the ideas. The fundamental objection was the lack of a suitable mechanism. Studies of the passage of earthquake waves through the Earth had shown that, whereas the core is liquid, the Earth's mantle and crust are solid. It was therefore difficult to find sufficient force to move the continents through the mantle and oceanic crust, and even more difficult to picture them moving without leaving obvious evidence of their passage.

Almost simultaneously with the temporary eclipse of Wegener's theory, Arthur Holmes was considering a mechanism that is still widely accepted. With the knowledge that the presence of radioactivity within the Earth could result in an internal zone of slippage, Holmes conceived the idea of convective currents within the Earth's mantle which were driven by the radiogenic heat produced by radioactive minerals within the mantle. Since the rates of motion were slow, but vast and inexorable, they were not evident during the rapid passage of seismic waves. This ac-

counted for the apparent discrepancy between the seismic observations and the drift of continents. At that time, Holmes's ideas, like those of Wegener, were largely ignored, as it was no longer thought that continental drift was worthy of further consideration. Nonetheless, several geologists, particularly those living in the Southern Hemisphere, continued to believe the theory and accumulate more data in its support.

By the 1950s, convincing evidence had accumulated, with studies of the magnetization of rocks, paleomagnetism, beginning to provide numerical parameters on the past latitude and orientation of the continental blocks. Early work in North America and Europe clearly indicated how these continents had once been contiguous and had since separated. The discovery of the mid-oceanic ridge system also provided more evidence for the geometric matching of continental edges, but the discovery of magnetic anomalies parallel to these ridges and their interpretation in terms of sea-floor spreading finally led to almost universal acceptance of continental drift as a reality. In the 1970s and 1980s, the interest changed from proving the reality of the concept to applying it to the geologic record, leading to a greater understanding of how the Earth has evolved through time. SEE CONTINENTS, EVOLUTION OF; EARTH, CONVECTION IN; PALEOMAGNETISM; PLATE TECTONICS; TECTONOPHYSICS.

D. H. Tarling

Bibliography. P. A. Davies and S. K. Runcorn (eds.), *Mechanisms of Continental Drift and Plate Tectonics*, 1981; F. Press and R. Siever, *Earth*, 4th ed., 1986; W. Sullivan, *Continents in Motion*, 1974; D. H. Tarling and M. P. Tarling, *Continental Drift*, 2d ed., 1975; T. H. Van Andel, *New Views on an Old Planet: Continental Drift and the History of the Earth*, 1985.

Continental margin

The submerged fringes of a continent, comprising three characteristic zones: the continental shelf, the continental slope, and the continental rise. The continental shelf is the zone around the continent, extending from the low-water line to the depth at which there is a marked increase in slope to greater depth. The continental slope is the declivity from the edge of the shelf extending down to great ocean depths. The shelf and slope comprise the continental terrace, which is the submerged fringe of the continent, connecting the shoreline with the 2½-mi-deep (4-km) abyssal ocean floor (**Fig. 1**). The value of oil from the continental shelves around the United States exceeds that of fisheries. SEE OIL AND GAS, OFFSHORE.

Continental shelf. This comparatively featureless plain, with an average width of 45 mi (72 km), slopes gently seaward at about 10 ft/mi (1.9 m/km). At a depth of about 70 fathoms (128 m) there generally is an abrupt increase in declivity called the shelf break, or the shelf edge. This break marks the limit of the shelf, the top of the continental slope, and the brink of the deep sea. However, some shelves are as deep as 200–300 fathoms (360–540 m), especially in past or presently glaciated regions. For some purposes, especially legal, the 100-fathom line or 180-m line is conventionally taken as the limit of the shelf. Characteristically, the shelves are thinly veneered with clastic sands, silts, and silty muds, which are patchily distributed. Geologically, the shelf is an extension of,

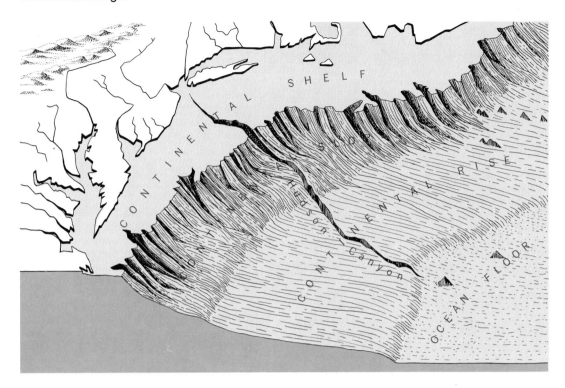

Fig. 1. Continental margin off northeastern United States.

and in unity with, the adjacent coastal plain. The position of the shoreline is geologically ephemeral, being subject to constant prograding and retrograding, so that its precise position at any particular time is not important. Genetically, the origin of the shelf seems to be primarily related to shallow wave cutting [waves cut effectively as breakers and surf only down to about 5 fathoms (9 m), the depth of vigorous abrasion], shoreline deposition, and oscillations of sea level, which have been especially strong during the Pleistocene and Recent. Although worldwide in distribution and comprising 5% of the area of the Earth, shelves differ considerably in width. Off the east coast of the United States the shelf is about 75 mi (120 km) wide, while off the west coast it is about 20 mi (32 km) wide. Especially broad shelves fringe northern Australia, Argentina, and the Arctic Ocean. As along the eastern United States, continental shelves commonly acquire a prism of sediments as the continental margin downflexes. Such capping prisms appear to be nascent miogeoclines.

Continental slope and rise. The drowned edges of the low-density ''granitic'' or sialic continental masses are the continental slopes. The continental plateaus float like icebergs in the Earth's mantle with the slopes marking the transition between the low-density continents and the heavier oceanic segments on the Earth's crust. Averaging 2½ mi (4 km) high and in some places attaining 6 mi (10 km), the continental slopes are the most imposing escarpments on the Earth. The slope is comparatively steep with an average declivity of 4.25° for the upper 1000 fathoms (1.8 km). Most slopes resemble a straight mountain front but are highly irregular in detail; in places they are deeply incised by submarine canyons, some of which cut deeply into the shelf. Usually the slope does not connect directly with the sea floor; instead there is a transitional area, the continental rise, or

apron, built by the shedding of sediments from the continental block. SEE MARINE GEOLOGY; SUBMARINE CANYON.

Robert S. Dietz

Shelf circulation. Shelf circulation is the pattern of flow over continental shelves. An important part of this pattern is any exchange of water with the deep ocean across the shelf-break and with estuaries or marginal seas at the coast. The circulation transports and distributes materials dissolved or suspended in the water, such as nutrients for marine life, fresh-water and fine sediments originating in rivers, and domestic and industrial waste.

Water movements over continental shelves are extremely complex in detail, but may be thought of as consisting of eddies of relatively small size and short lifetime, tidal motions, wind-driven currents, and long-term mean circulation. Tidal currents reverse every half tidal cycle (about 6 h for the usually dominant semidiurnal tide), and water particles undergo back-and-forth excursions due to tides amounting to a few miles. Storms generate currents which persist typically for several days and lead to particle excursions of tens of kilometers. A pattern of long-term mean circulation is characteristic of each major subdivision of the continental shelves. Off the North American east coast, for example, the region of the shelf between the Grand Banks of Newfoundland and Cape Hatteras is characterized by a persistent southwestward drift, always present over the outer portions of the shelf (in water at least 150 ft or 50 m deep). The mean shelf circulation in this area is one leg of a larger oceanic gyre lying offshore (**Fig. 2**). Water particles are eventually swept off the shelf and into a deepwater current adjacent to the Gulf Stream. However, the speed of this mean flow is low, typically less than 0.3 ft/s (0.1 m/s).

Important for the exchange of water between a

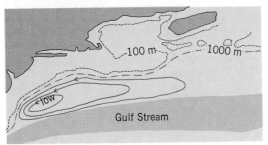

Fig. 2. Deepwater gyre between the Gulf Stream and the North American continent, influencing continental shelf circulation. Georges Bank is at the center of the illustration, separating the Gulf of Maine from deep water, and is subject to intense tidal flow. 1 m = 3.3 ft.

shelf region and the deep ocean are differences in motion between surface and bottom layers. At the ocean surface, the wind is a dominant influence, while at depths greater than about 30 ft (10 m) internal forces determine the intensity and direction of motion. In the east coast example quoted before, the surface layers move significantly offshore over the long term, a consequence of westerly winds, in addition to their general southwestward drift. Over shorter periods they move in whatever direction the wind carries them. The bottom layers move shoreward, to replace the water driven offshore by the wind, transporting saltier, nutrient-rich water from the deep ocean onto the shelf.

Tidal currents are mainly important for shelf circulation on account of the mixing they cause. While tidal currents are typically 0.6 ft/s (0.2 m/s) in amplitude, they are much stronger in some locations where shallow depth constricts the flow, for example, over Georges Bank where tidal currents are up to 3 ft/s (1 m/s). Currents of this intensity cause strong bottom friction, turbulence, and mixing. Tides on continental shelves are not caused directly by the pull of the Moon and Sun, but propagate in from the deep ocean, where the gravitational attraction of the heavenly bod-

ies acts over a much larger water mass.

The inflow of fresh water from land also contributes to shelf circulation, because such water would tend to spread out on the surface on account of its low density. Rapid nearshore mixing reduces the density contrast, and the Earth's rotation deflects the offshore flow into a shore-parallel direction, leaving the coast to the right. A compensating shoreward flow at depth is deflected in the opposite direction, adding to the complexity of shelf circulation. SEE NEARSHORE SEDIMENTARY PROCESSES; OCEAN CIRCULATION.

G. T. Csanady

Bibliography. J. S. Allen, Models of wind-driven currents on the continental shelf, *Annu. Rev. Fluid Mech.*, 12:389–433, 1980; R. C. Beardsley, W. C. Boicourt, and D. V. Hansen, Physical oceanography of the Middle Atlantic Bight, *American Society for Limnology and Oceanography, Special Symposium*, 2:20–34, 1976; D. F. Bumpus, A description of the circulation on the continental shelf of the east coast of the United States, in *Progress in Oceanography*, vol. 6, pp. 111–159, 1973; C. Burk and C. Drake, *The Geology of Continental Margins*, 1974; G. T. Csanady, Circulation in the coastal ocean, *Adv. Geophys.*, vol. 23, 1981; F. P. Shepard, *Submarine Geology*, 3d ed., 1973.

Continentality (meteorology)

The attributes of notably greater temperature ranges with associated weather characteristics that make inland and continental regions distinctly different from marine areas or areas bordering large bodies of water. Unlike the latitudinal uniformity of climate which might result from exposure of a uniform surface by the Earth to the Sun's rays, some of the most outstanding differences are those between continental and oceanic areas, particularly in the middle latitudes. These latitudes, designated temperate zones by the Greeks of the classical period, actually contain some of the least temperate weather and climate of the

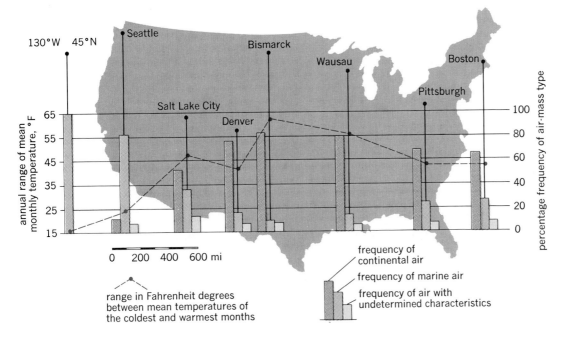

Amplitude of the annual range in temperature in the United States related to continental location and to frequencies of the marine and the continental air masses. 1 mi = 1.6 km. °C = (°F − 32)/1.8.

world over continental areas, as well as truly temperate moderation in maritime provinces.

The reasons are associated with the greatly different thermal behavior of water and land surfaces. The soil surfaces of the continents can be heated or cooled comparatively rapidly because the thermal effects are limited to a shallow surface layer. The heat received at the ground is used to warm the top layers, which consequently become hot in summer. In winter, by contrast, the surface becomes notably cold since there is little conduction of heat from subsurface depths. As a result of the responsiveness of the surface air temperature to the temperature of the underlying soil surface, the air in contact with the ground undergoes temperature fluctuations of large amplitude, lagging slightly behind the annual and the diurnal periods of insolation (see **illus.**). *See Maritime ecology.*

On the other hand, the water surfaces which make up the oceans are heated or cooled much less rapidly, partly because the thermal effects extend to greater depths and partly because much of the solar energy absorbed at the sea surface is utilized for evaporation rather than heating. As a result, the upper layers of the ocean are not heated as strongly as the land surfaces in summer or cooled as strongly by radiation in winter. It follows that large annual and diurnal ranges in temperature with hot summers and cold winters are characteristic of the continental interiors of middle and high latitudes. In contrast, the summers over the oceans are cool, while the winters are comparatively mild.

Because the increase in the annual range in temperature inland over the continents is the most striking effect of the continental surface on climate, climatologists have taken the annual range in temperature (ΔT) as a measure of the continentality. The most frequently used formula is the one shown here, advanced

$$ K = \frac{1.6 \, \Delta T}{\sin \phi} - 14 $$

by O. V. Johansson. Here the index of continentality is K (in percent, a purely oceanic climate presenting a value of 0% and a completely continental climate of 100%), ΔT is annual range of monthly mean temperature in °C, and ϕ is latitude. Note that this equation is not to be applied near the Equator, where $\sin \phi = 0$. W. Gorczynski and D. Brunt presented similar formulas, although the one by Brunt, while of greater physical significance than the other two, is somewhat difficult to evaluate because it involves the use of solar radiation data as well as temperature data.

Woodrow C. Jacobs

Bibliography. G. T. Trewartha and L. H. Horn, *An Introduction to Climate,* 5th ed., 1980.

Continents, evolution of

The processes that led to the formation of the Earth's continents. Many theories concerning the evolution of the continents have been proposed and then replaced by newer models. As recently as the 1950s, elementary school children were taught that the Earth's crust formed as the Earth cooled from a liquid ball and that mountains were formed in response to the shrinking of the Earth and its crust by cooling. The mountains, and indeed the continents, were thought to be merely

wrinkles on the Earth's surface, much as a prune is a wrinkled plum shrunk by drying.

Early in the twentieth century, seismic methods led to the discovery of the Mohorovičić discontinuity (Moho), which separates the crust of the Earth from the mantle beneath. The Moho is marked by a rather abrupt change in seismic P-wave velocity from 4.2–4.5 mi/s (6.7–7.2 km/s) in the lower crust to 4.7–5.2 mi/s (7.6–8.3 km/s) in the upper mantle. In the early 1900s earthquake seismologists discovered the Earth's liquid core; by the 1930s they had elucidated the presence of a central solid core. *See Moho (Mohorovičić discontinuity); Seismology.*

In the 1960s the concept of plate tectonics (then called continental drift) was accepted by leading geoscientists. The discussion that follows includes a brief review of the tenets of plate tectonics, and then an examination of the evolution of the continents in those terms. This evolution has had major impact on the presence and concentration of natural resources and on the occurrence of many natural disasters. Future exploitation of resources will be enhanced by new knowledge about the continents; in addition, this knowledge will be essential in planning for protection from natural disasters. *See Continental drift; Plate tectonics.*

The dynamic Earth. Prior to about 1960, earth scientists thought the Earth was made up of concentric spherical shells of material of particular compositions and densities (**Fig. 1**). Although it was revealed that the density of the shells generally increased with depth, little was known about chemical compositions of rocks below the base of the Earth's crust.

However, since 1960 the concepts of a brittle lithosphere about 80–120 km (50–75 mi) thick and the soft asthenosphere between that depth and 400 km (250 mi) have been added to the model of the Earth. The lithosphere includes the crust and the brittle part of the upper mantle above the asthenosphere. Geologists and geophysicists now perceive the Earth as a dynamic system. Although it had been observed for centuries that earthquakes and volcanoes often accompany one another, the reasons for their close relationship became apparent only with the develop-

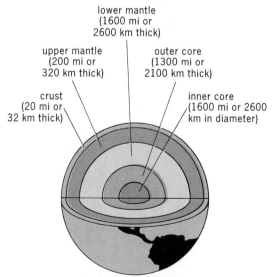

Fig. 1. Cross section of the model of the Earth as a series of concentric spherical shells. (*After D. W. Steeples, Earthquakes, Kansas Survey Educational Series, 1979***)**

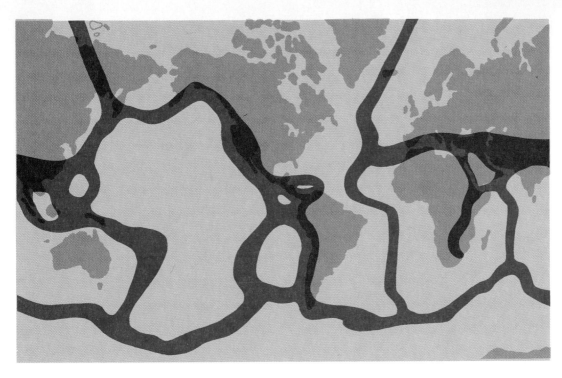

Fig. 2. Map showing the principal earthquake zones of the Earth, an area that includes both subduction zones and spreading ridges. (*After D. W. Steeples, Earthquakes, Kansas Survey Educational Series, 1979*)

ment of the concept of plate tectonics. Over 98% of the world's earthquakes and a similar percentage of volcanoes can be explained in terms of plate tectonics (see **Fig. 2**).

Gravity and magnetic fields. Variations in the Earth's gravity and magnetic fields were important pieces of evidence in the development of plate tectonic theory. Magnetic maps of the sea floor made during World War II showed variations occurring in a striped pattern. In the 1960s it was discovered that these patterns in the oceanic crust were caused by the reversing of the polarity of the Earth's magnetic field every few million years. Younger stripes located near the oceanic ridges were discovered by water-depth soundings made during World War II. The older stripes were rafted away from the ridges by lithospheric plate motions. *SEE ROCK MAGNETISM*.

Gravity is unusually strong around the rim of the Pacific Ocean, which is also a zone of unusually large numbers of earthquakes and volcanoes. In fact, about 75% of the world's earthquake energy originates around the Pacific rim.

Plate tectonic theory is based partly on strong gravitational, seismic, and volcanic evidence that the Earth's outer rigid shell (the lithosphere) is sinking gradually into the soft underlying shell (the asthenosphere) around the rim of the Pacific (**Fig. 3**). This happens because the lithospheric slabs are cooler and denser. As they sink into the warmer and lighter asthenosphere, gravity highs are created above them. *SEE EARTH, GRAVITY FIELD OF*.

Lithosphere. The slabs of lithosphere contain ocean water and sediments along their upper surfaces. The presence of water lowers the melting temperature of rocks, and as the cool, dense lithosphere descends into the hot asthenosphere, a process known as subduction, the water-laden materials with lower melting points are turned into molten rock. These materials are propelled upward by pressure from steam and other volatile components, resulting in volcanic activity above the descending lithosphere. The hundreds of

active volcanoes around the rim of the Pacific have their origin in this process. *SEE VOLCANO*.

Lithospheric material is disappearing around the Pacific rim, and only a part of it returns to the Earth's surface through volcanic melting of its upper layers. New lithosphere is created at spreading oceanic ridges, such as the Mid-Atlantic Ridge, a prominent linear submarine topographic high (buried mountain range) near the middle of the Atlantic Ocean. New lithosphere appears to be created at the approximate rate that old lithosphere is consumed. *SEE ASTHENOSPHERE; LITHOSPHERE*.

The cool, dense lithosphere is also involved in the earthquake processes around the Pacific rim. The warm, soft asthenosphere is a low-strength material, not capable of sustaining the shear stress necessary to produce large earthquakes. The descending lithosphere, however, is cool and brittle enough to sustain earthquake-producing stresses to depths of 430 mi (700 km). Earthquakes rarely occur deeper than 45 mi (75 km) in regions not showing a descending lithospheric slab. *SEE EARTHQUAKE*.

Sea-floor spreading. The geology of Iceland illustrates spreading processes that are occurring at oceanic ridges. The volcanoes in Iceland produce dense, dark rock known as basalt; these basalts are produced by convective ascension of asthenospheric

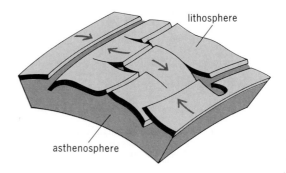

Fig. 3. Diagram showing lithospheric plates moving (direction indicated by arrows) on top of the asthenosphere.

material. As the material approaches the Earth's surface, the decreased pressure and the presence of small amounts of seawater circulating deep into the oceanic lithosphere lower the melting temperature. The result is volcanic activity at the midocean ridge, in this case, on Iceland. *See Basalt.*

The oceanic ridge volcanoes produce new lithospheric material that is gradually rafted away from the ridge at the rate of a few centimeters per year. As the basalt solidifies from magma and cools, it becomes permanently magnetized in a direction parallel to the magnetic field at the place and time of cooling. Because the Earth's magnetic field reverses occasionally, some basalt deposits are magnetized in a normal direction, and some are magnetized in a reverse direction. The result is a sea floor that has striped magnetic patterns. The stripes are parallel to the ridge from which the basalt was erupted, and younger rocks are represented by stripes nearer the ridge.

The spreading can be observed directly by measuring the steadily increasing distance between the east and west coasts of Iceland. As new basalt is emplaced near the center of Iceland, the eastern half of the island moves east and the western half moves west to make room for it. *See Marine geology.*

Earthquakes and volcanoes. Worldwide earthquake activity, ranging from minor to moderate, neatly outlines the midocean ridges (Fig. 2 and **Fig. 4**). Such activity is concentrated at the spreading ridge and at the collision zone at the edge of the continent. It is significant that the molten rock erupted at spreading ridges cools to form basalt, whereas volcanoes around the Pacific rim, which contain more silica from the melted sediments in the upper lithospheric layer, form andesite upon cooling. The high silica content produces molten rocks that are much stiffer and require higher pressures for volcanic activity. Volcanic eruptions around the Pacific rim often are initiated by huge volcanic explosions, such as the one that occurred at Mount St. Helens in Washington in 1980; volcanic activity at ocean ridges is quite mild by comparison. *See Andesite; Volcanology.*

Ancient continents. Approximately 200 million years ago, the present major continents were united in one landmass known as Pangea (**Fig. 5**). Another continent, Pacifica, may also have been present at some place on the Earth's surface when Pangea existed. Evidence suggests that Pacifica was broken up by one or more spreading ridges (continental rifts) and plastered onto continents around the edges of what is now the Pacific Ocean. Possible remnants of Pacifica have been found in exotic rock assemblages in Alaska and other parts of western North America since the mid-1970s. *See Earth, convection in.*

Age of continents and ocean basins. The age of the Earth was a point of major controversy early in the twentieth century. Prior to the discovery of radioactivity, age estimates of the Earth were of the magnitude of a few hundred thousand years. The discovery of radioactivity led to the realization that the Sun and possibly the planets could be much older than previously estimated. *See Radioactivity.*

In the early 1900s a method was developed for determination of the ages of some types of igneous rocks, including most granites. These calculations relied on the precise measurement of very small amounts of radioactive elements and isotopes produced by radioactive decay of some elements in the rock materials. These studies demonstrated that the Earth was billions rather than thousands or millions of years old. Rocks as old as 3.8 billion years have been found in parts of Greenland and Africa. The oldest rocks brought back from the Moon are about 4.6 billion years old. *See Moon; Rock age determination.*

In the 1960s it was discovered that the floors of the ocean basins were almost all less than 200 million years old. This implied that the major masses of the continents were geologically very old and the ocean basins were geologically very young. Thus there must be very different processes of formation and consumption of the two types of geologic terranes. *See Earth, age of; Oceanic islands.*

Changing sizes of continents. There are several ways by which material is added to or subtracted from the continental crust. In some cases, material is added immediately to the areal extent of a continent. Sometimes the addition to the continental area comes later in the form of increased sedimentation rates at the continent's edge. In other cases, material is removed from the areal extent of one or more continents. Over geologic time, however, the continents are probably increasing gradually in total area. Material can be added by the collision of continents with island arcs at subduction zones. For example, at the present time rapid uplift is occurring as a result of the impending collision of the island of Timor with northern Australia. Over the long term this collision will provide an addition to the continental area of Australia.

When continents collide, the net effect is a decrease in continental area. The continental crust, which is really the upper part of the lithosphere, is of lower density (about 2.9–3.0 g/cm^3) than the lower part of the lithosphere, which is composed of upper-mantle material (density of about 3.3 g/cm^3). As a result, the lower-density material is scraped off and remains at or near the Earth's surface. The Himalayas are located where continental collision is affecting the area of the continents. The Earth's crust is becoming thicker at that location; and the continental area is decreasing as a result, even though uplift is causing high rates of sedimentation along southern Asia. *See Orogeny.*

Another way occurs when a continent is pulled

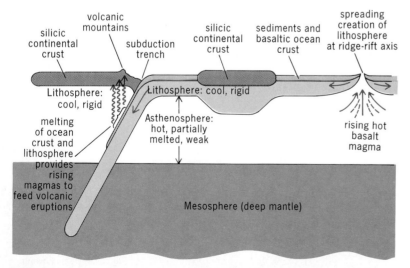

Fig. 4. Cross section of the upper mantle and two continental masses encased in the lithosphere. The continental masses are complex mixtures of silicate rocks of varying ages and compositions. The complex boundary between the crust and upper mantle is contained within the lithosphere.

apart by rifts that abort before they become ocean basins. A present-day example is the East African Rift, which contains large continental lakes that are gradually being filled with a mixture of continental sediments and volcanic rocks that are erupted to the surface. Another example, which occurred about 1 billion years ago in the central part of the United States, is the Midcontinent Rift System extending from Lake Superior in Minnesota southwestward to central Kansas (**Fig. 6**). The increased crustal area is the central part of the rift, filled mostly with basalt. SEE EARTH CRUST.

Sampling the continents. There are several methods for sampling the continents. Some involve direct sampling of the rocks; some involve indirect geophysical and geochemical methods to infer the presence of bodies and their mineral compositions and physical properties.

Drilling. One of the most obvious direct sampling methods is drilling. During drilling, chips of rock (drill cuttings) are brought to the Earth's surface for physical and chemical examination. Alternatively, and at greater expense, cylindrical-shaped samples of intact rock a few inches in diameter and several feet long can be obtained by coring. Both types of samples give a direct reading of rocks at depths that are precisely and accurately known. SEE BORING AND DRILLING, GEOTECHNICAL.

In the 1950s, a group of earth scientists proposed Project Mohole, whose purpose would have been to drill through the Earth's crust to sample the upper mantle directly. They believed that obtaining such samples would help in developing geophysical and geochemical models of the Earth's deep interior below the Moho. However, this project was not undertaken. By the late 1960s, geologists realized that plate tectonic processes had left pieces of mantle material exposed at the Earth's surface in many places on several continents. At the same time, deep seismic-refraction studies showed that the seismic P-wave and S-wave velocities in the upper mantle varied by as much as 10% from place to place. This implied that a single set of samples from a Mohole would not necessarily characterize the Earth's upper mantle very well, especially since for technical reasons the drilling would have been done in a place where the Moho was unusually shallow and possibly atypical. SEE EARTH INTERIOR.

Also in the late 1960s, certain igneous rocks were found to have originated in the upper mantle at depths of 60 mi (100 km) or more. Furthermore, they contained samples (xenoliths) of the rock layers they passed through on the way to the surface. It was demonstrated that the statistical size distribution of these xenoliths was related to the depth at which the particular types of xenoliths originated. This allowed geochemists to construct models of rock type versus depth for the upper mantle at many places on the Earth.

Chief among these rock types were kimberlites. Their presence showed that in many places the lower crust was composed of gabbro and granodiorite. Gabbro is chemically similar to basalt but is not erupted at the Earth's surface. Granodiorite is much like granite but contains less silica. The upper mantle is composed of rocks such as eclogite and peridotite. Eclogite is composed of garnets and other silicates, such as sodic pyroxene, that form at high pressure and high temperature. Peridotite is an igneous rock composed

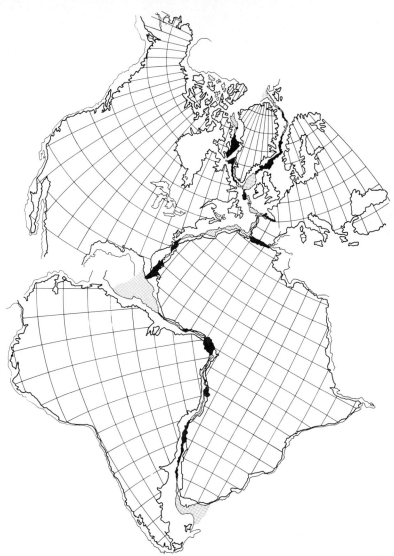

Fig. 5. Map showing how South America, North America, Greenland, Europe, and Africa can be fitted together to form part of Pangea. The fit is made along the continental slope at the 500-fathom (914-m) contour line. Regions where the continents overlap are shown in black. Areas without continental crust are indicated by stippled pattern.

chiefly of olivine, a mineral common in low-silica igneous rocks. SEE IGNEOUS ROCKS.

Laboratory experiments in which granodiorites, gabbros, eclogites, and peridotites were subjected to High pressures and high temperatures have shown that the Mohole project was unnecessary. In addition to chemical analyses, these rock samples have been measured for mass density, magnetic and electrical properties, and seismic properties. These experiments have led to increased understanding of the Earth's physical and chemical structure to depths of at least 125 mi (200 km).

Geophysical methods. Geophysical methods are also used to sample the continents. Maps of the Earth's gravity and magnetic fields are even more useful for studying continents than for studying ocean basins. Because of its antiquity and the complex processes by which it was formed, the continental crust is generally much more complicated than the oceanic crust. Therefore gravity and magnetic maps contain more information about continents than about oceans. This information can be enhanced by modern digital processing techniques; these include filtering, which can

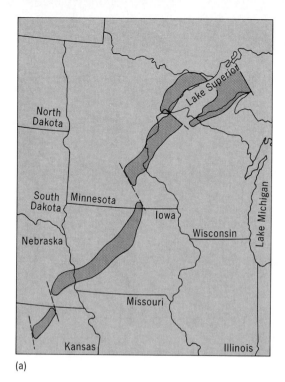

(a)

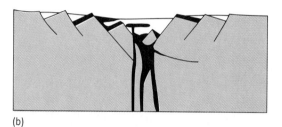

(b)

Fig. 6. Midcontinent rift system. (a) Map showing the areal extent; the shaded area represents crust that was added about 1 billion years ago. (b) Schematic cross section of the system occurring in Kansas. Area shown in black represents addition to the crust by intrusion of basalt about 1 billion years ago. Crustal blocks were formed by faulting as the crust was stretched and extended. The shaded area represents sediments that have been deposited since the rifting ended, obscuring the rift from view.

emphasize particularly interesting features such as size, depth, azimuthal direction of anomalous rock bodies, and geologic faults. SEE GEOPHYSICAL EXPLORATION.

When features of interest are being mapped by geologic, gravity, and magnetic methods, they are sometimes also investigated by deep seismic soundings. Results of such investigations have drastically changed the way geoscientists view the deep crust and upper mantle beneath the continents. Since the Moho was discovered early in the twentieth century, the assumption from earthquake studies and from deep crustal seismic-refraction studies was that the Moho was a discrete boundary between the lower crust and the upper mantle. However, it has been shown that this is rarely the case, and that the Moho boundary is usually diffuse and penetrated by intrusions of igneous rocks that complicate it.

At many locations, seismic refraction and earthquake studies had indicated the presence of a seismic-wave velocity boundary in the deep crust known as the Conrad discontinuity. This was assumed to be caused by distinct layering in the mid-to-lower crust. It is now known that the Conrad discontinuity is highly variable and often absent. Where present, it usually represents the top of a huge intrusion or intrusions of igneous material at depths of 6–18 mi (10–30 km) in the crust.

Before North America separated from Europe with the breakup of Pangea, a collision occurred between these two continents to help form Pangea. As a result of this collision, a crustal layer a few kilometers thick was thrust from the collision boundary westward about 120 mi (200 km). Thus, the Appalachian Mountains in the eastern United States are a belt of rocks about 4 mi (6 km) thick that have been thrust on top of preexisting rocks; a suite of rocks formerly at the Earth's surface is now buried deep beneath this thrust sheet. Prior to this discovery, geoscientists had not realized that horizontal thrusting could occur over such great distances with such a thin sheet of rock.

In many other parts of the United States, data show much greater complexity in the crust than had been expected. From these and other types of data, geoscientists have determined that many cycles of plate tectonic activity have built the continents over the past 4 billion years. Many of these cycles left their mark as mountain ranges, such as the Alps and Urals in Europe, while others have left exotic accretion terranes, such as those found in Alaska.

Resources and disasters. Almost all of the Earth's energy, fresh water, and mineral resources exist on the continents and continental shelves. Also, most devastating earthquakes, volcanic eruptions, floods, and landslides occur on the continental masses. The evolutionary processes of formation of the continents have strongly controlled the deposition and degree of concentration of these resources and these disasters.

The occurrence of faulting is important in the presence of many mineral deposits of economic interest, and this also is the cause of earthquakes not caused by volcanic eruptions. Minerals of economic value, such as gold, silver, and copper, are concentrated into ores by fluids moving upward from deep in the Earth along faults. The hot fluids containing high concentrations of these and other elements often are the last remaining liquids from a dying magma chamber, particularly in magma chambers that do not result in surface eruptions. As the liquids cool, they deposit concentrations of their dissolved metals and other minerals. SEE EARTH; FAULT AND FAULT STRUCTURES; MAGMA; ORE AND MINERAL DEPOSITS.

Donald Wallace Steeples

Bibliography. A. Cox and R. B. Hart, *Plate Tectonics: How It Works,* 1986; F. Press, and R. Siever, *Earth,* 1986; D. G. Smith (ed.), *The Cambridge Encyclopedia of Earth Sciences,* 1981; D. H. Tarling (ed.), *Evolution f the Earth's Crust,* 1978; B. F. Windley, *The Evolving Continents,* 1977.

Continuous-wave radar

A radar in which the transmitter output is uninterrupted, in contrast to pulse radar, where the output consists of short pulses. Among the advantages of continuous-wave (CW) radar is its ability to measure velocity with extreme accuracy by means of the Doppler shift in the frequency of the echo. The detected, reflected wave is shifted in frequency by an amount which is a function of the relative velocity between the target and the transmitter-receiver. Range data are extracted from the change in Doppler frequency with time. *See Doppler radar.*

In order to measure the range of targets, some form of frequency modulation (FM) of the continuous-wave output must be used. In one very effective form of modulation, the carrier frequency of the transmitted signal is varied at a uniform rate. Range is determined by comparing the frequency of the echo with that of the transmitter, the difference being proportional to the range of the target that produced the echo. Systems in which this is done are known as FM-CW radars. *See Frequency modulation.*

A modified form of FM-CW radar employs long, but not continuous, transmission. This might be regarded as the same as transmitting extremely long pulses on an FM carrier. Systems of this type are referred to as pulse compression radars.

An FM-CW radar used in the radio-proximity fuses of missiles generates functions of the Doppler signal envelope that have identifiable characteristics at certain ranges. The amplitude of the received FM-CW signal is used to determine when the missile has attained the proper altitude. Then the generated functions tell when the missile has arrived at the desired point of burst. Three triangular-wave modulators can be used to enhance the accuracy of an FM-CW radar fusing system.

In advanced short-range air defense radar (ASHORAD) a coded continuous-wave signal reflected from the target is fed to a tapped delay line, where a predetermined delay between adjacent taps is used to establish the range.

Design objectives. Design objectives for continuous-wave radar include: protecting the receiver from the transmitter output and close-by return echoes when a single antenna is used; resolving side-lobe ambiguity; resolving range ambiguity; distinguishing between approaching and receding targets; eliminating noise and clutter; simultaneously measuring both target range and velocity; determining the shapes of targets; handling multiple targets; measuring target acceleration; communicating with targets; and increasing the received signal-to-noise ratio.

Receiver protection. One disadvantage of continuous-wave radar is that when a single antenna is used for both reception and transmission it is difficult to protect the receiver against the transmitter because, in contrast to pulse radar, both are on all the time. Use of isolation circuitry gives the receiver protection from a transmitter output up to 200 W. Use of magnetically biased yttrium-iron-garnet (YIG) provides three tuned, tandem-connected power-limiter stages in one X-band FM-CW radar. *See Ferrimagnetic garnets.*

In continuous-wave surveillance radars where wide bandwidth is required to meet range and range-rate measurement requirements despite clutter and other interfering environments, the transmitter output can be coded by pseudorandom binary signals that vary with range. This is an ideal waveform for permitting range measurements, retaining a capability for range-rate measurements, and rejecting the transmitter output and other nearby echoes.

Side-lobe ambiguity. The problem of distinguishing between echoes from the main lobe and those from the side lobes has been addressed by providing a demodulated range response that has one polarity for the main-lobe echoes and the opposite polarity for side-lobe echoes.

Side-lobe ambiguity is removed in an FM-CW radar fusing system in which the reflected signal is compared with a sample of the continuous-wave output in a pair of FM discriminators displaced from each other in the frequency domain. Thresholding circuits determine in which half of which discriminator the difference frequency signal is concentrated.

Side-lobe ambiguity in the range response of an FM-CW radar can be resolved when a periodic modulation frequency is reflected from a target and combined with a sample of the transmitted signal. Selected harmonics of the combination signal are processed to develop a main-lobe response of one polarity with all side lobes of the opposite polarity.

Range ambiguity. This ambiguity results from reception of echoes from targets beyond the range of interest and of second-time-around echoes. It can be resolved by range gates that make the radar insensitive to targets beyond the range of interest or by filters that put limits on range. Resolution of range ambiguity is very important in radars for personnel surveillance.

Directional ambiguity. To distinguish between approaching and receding targets, a continuous wave beamed by a radar is sampled continuously but delivers an output voltage only when the distance to the target is changing significantly. The voltage is positive if the target is closing, and negative if it is receding.

Noise and clutter. Suppression of noise, clutter, chaff, and jamming has received a great deal of attention. In one approach, a fixed pair of amplitude-modulation (AM) sidebands is added and used to null out the noise.

Echo signal analyzers can be selectively adjusted for a certain moving-target size so that small targets are not detected. A series of digital range cells can be defined such that each cell represents the minimum allowable distance between two reflective points on a target. The radar also can be adaptively adjusted to attend to a desired average signal level.

Use of digitally controlled voltage-tunable radio-frequency generators permit frequency jumping to avoid jamming or spoofing. Pseudorandom-noise-modulated continuous-wave radar can be used to establish perimeter surveillance in a heavily foliaged environment. Pseudorandomly coded continuous-wave radar can also be used to detect targets in the presence of clutter. *See Electrical noise; Electronic warfare.*

Range and velocity sensing. Several techniques are used to obtain simultaneous measurements of range and velocity. One involves transmitting sidebands of linearly increasing and decreasing frequency or using triangular modulation, and comparing the difference between transmitted and received frequencies with the difference between the received frequency and the transmitted frequency delayed by a period of time proportional to the range of the target. Phase-shift keying by a pseudorandom code also permits simultaneous measurement of range and velocity. *See Modulation.*

A double-sideband linear FM radar obtains simultaneous range and Doppler measurements. Waveforms of both increasing and decreasing frequency are transmitted simultaneously and combined coherently to give an output that is a measure of the target's radial velocity. The received signals are processed separately to obtain range information.

Target shape. Millimeter-wave FM-CW radar can be used to determine the shape of a target. These radars operate at frequencies in excess of 90 GHz. They often use Gunn oscillators as the source. *See Microwave solid-state devices.*

French and Soviet research suggests that in some applications it may be necessary to use space diversity in which two beams illuminate the target in order to compensate for nonlinearities in frequency scanning.

The dependence of radar cross section on transmitted frequency over a wide bandwidth can be used to make the signal amplitude scintillate so that target characteristics can be determined.

Rapid discontinuous fluctuations in the scattering from a moving multielement target is known as the intermittent-contact RADAM (radar detection of agitated metals) effect. This effect is observed in the contact between the drive sprocket and the tread of a tank and can be used to define signatures of various kinds of fighting vehicles.

Multiple targets. Discrete Fourier analysis of the video heterodyne (beat) signals using a digital computer permits multiple-target handling and achieves good resolution between approaching and receding targets. *See Spectrum analyzer.*

Target acceleration. An indication of the acceleration of a target can be obtained by double subtraction and delay of the returned signal.

Communicating with the target. In radar missile guidance, homing beacons, and identification–friend-or-foe, it is necessary to communicate with the target. This can be done by interrupting the range beat frequency at short intervals near its peak excursion and transmitting AM signals to the illuminated target.

Signal-to-noise ratio. The signal-to-noise ratio can be improved by digitizing the beat frequency signals. Moreover, compressing the received signals to a sine-squared shape can increase the signal level more than 20 decibels. *See Signal-to-noise ratio.*

Applications. Applications of continuous-wave radar include missile guidance; detection of hostile targets; terrain clearance indication and ground surveillance; laser radar systems; atmospheric studies; automobile safety; surveillance of personnel; ice studies; remote sensing; and reproduction of the shape of a patient's pulse.

Missile guidance. The principal use of continuous-wave radar is in short-range missile guidance. Typically the missile's course is tracked from the ground while the missile is simultaneously illuminated. Continuous-wave radar, in some cases the same radar, can be used for both tracking and illumination, although it is more common for pulse Doppler radar to be used for tracking.

Advances in the design of phased-array antennas have led to pulse Doppler radar becoming more attractive than continuous-wave radar even for illumination. However, at least one system allows the operator to select either continuous-wave or pulse Doppler radar for illumination.

Illumination radar is for target acquisition. Two signals are of interest. One, the directly received illumination of the missile, is called the rear signal. The other, the signal reflected from the target, is called the front signal. The front signal is shifted in frequency as the missile closes with the target. This shift occurs because of the Doppler effect. The shift, and therefore the range to the target, is obtained by coherently detecting the front signal against the rear signal. In an active guidance system, detection is performed on board the missile. Semiactive missile guidance is more frequently used, and in this system the signals are relayed over a data link for processing at the ground station.

A late version of the MPQ-34 tracking and acquisition radar for the Hawk ground-to-air missile incorporates the choice of either continuous-wave or pulse Doppler radar illumination. The ability of continuous-wave radar to discriminate against clutter made it attractive in this system, which was intended to be used against low-flying aircraft.

Pulsed tracking radar with continuous-wave acquisition radar is used with the Sparrow air-to-air missile system and with the Tartar shipborne missile system.

A miss-distance indicator (MDI) is a special application of continuous-wave radar. It uses radio-frequency signals to determine the relative velocity between two objects, the time of intercept when the distance will be a minimum, and the actual value of that minimum distance. With the availability of real-time intercept information, a prior determination of guidance parameters that will produce a direct hit can be made. *See Guidance systems; Guided missile.*

Detection of hostile targets. Modulated continuous-wave radar has been used to detect hostile military vehicles and personnel.

The Navy MK-92 fire-control system uses a linear FM-CW radar with multiple resonant filters. *See Fire-control systems.*

An FM-CW naval radar operating at 95.6 GHz uses pseudorandom-coded phase-shift keying, digital signal processing, and microprocessor-controlled azimuth and elevation scanning. It can display the shape, range, velocity, and direction of vessels. *See Naval armament.*

In the United States, Canada, and the United Kingdom, there has been interest in continuous-wave radar for detecting nonmetallic mines. One radar operates between 2 and 4 GHz and can detect plastic objects buried in 10 in. (25 cm) of wet sand.

Terrain surveillance. One of the first uses of FM continuous-wave radar was in terrain clearance indicators or radar altimeters. Today intruder aircraft and air-breathing missiles obtain continuous guidance from that kind of equipment. *See Altimeter.*

The ability of side-looking radar on reconnaissance aircraft and Earth-observation satellites to resolve targets has been enhanced by the use of synthetic-aperture radar. Although this technique usually requires a huge data rate and great processing complexity, special-purpose processors have been used to extract useful data economically. Such a processor on the *Seasat 1* was able to determine the direction of ocean wave movement from 800 mi (1300 km) up. *See Applications satellites; Remote sensing.*

Earth surveillance has been hampered by some radar reflection phenomena, notably foreshadowing, which distorts distance; layover, which tends to fill in depressions; and shadow, which obscures features. These problems place limiting requirements on the size and contrast of images needed for successful interpretation. These limitations can be partially overcome by training radargrammetrists with simulators.

Laser radar systems. An experimental modulated continuous-wave laser radar was built to track low-flying airborne targets from the ground. The system can determine both range and the rate at which the range is changing. An advantage of this system is its extreme precision. The divergence of the beam is less than 1 milliradian. The transmitter is a carbon dioxide laser operating at a wavelength of 10.6 micrometers. The infrared carrier is frequency-modulated with 10-MHz excursions occurring at a frequency of 1 kHz. *See* Laser.

Atmospheric studies. The ability of FM-CW radar to measure distance and velocity despite smoke, dust, and thermal gradients makes it attractive for many industrial automatic control applications. The Naval Ocean Systems Center uses S-band continuous-wave radar to update refractive index profiles of the troposphere. This work forms the basis for radio interception and over-the-horizon targeting. In France, continuous-wave radar has been used to measure the electron temperatures in the ionosphere. The National Oceanic and Atmospheric Administration (NOAA) has used continuous-wave radar to measure wind in clear air.

Automobile safety. An experimental continuous-wave radar which exploits the Doppler principle has been developed for use in automobiles. The radar can anticipate a crash when an obstacle is 30 ft (10 m) away in order to deploy air-cushion-type passive restraints. It can sense obstacles 500 ft (150 m) away to govern automatic braking and headway control. The source of carrier power is an X-band Gunn-type solid-state oscillator.

Surveillance of personnel. A variant of continuous-wave radar is used to illuminate persons under surveillance by techniques employing semiconductor tracer diodes. These devices are secreted on or implanted in the subject's body without his or her knowledge, or else concealed in objects that are to be protected against theft. Despite the fact that pulsed X-band sources could provide the needed power levels more conveniently, the requirement to reduce clutter makes it desirable to use continuous-wave power.

These techniques are also used to sweep premises to discover clandestine listening devices and to guard protected locations against the introduction of concealed tape recorders.

Implementation of tracer-diode surveillance utilizes a single carrier frequency and looks for reflections of the third harmonic as evidence that the person sought has been acquired by the beam. The surveillance operations have been carried out from fixed posts, vans, and low-flying aircraft.

An improved version makes use of two carrier frequencies for illumination and depends upon coherent detection of the third and fourth harmonics of the sum and difference frequencies to signify acquisition of the target. *See* Electronic listening devices.

Other applications. In Finland, a continuous-wave radar operating between 1 and 1.8 GHz has been used to measure ice and frost thickness in lakes and bogs. Continuous-wave radars operating at 1.5 and from 8 to 18 GHz have been used to discriminate among different types of Arctic Sea ice.

Continuous-wave radar has been used in the Netherlands to discriminate among different types of vegetation, crops, and bare soils.

With its antenna placed over a patient's wrist artery, continuous-wave radar has been shown to be able to reproduce the shape of the arterial pulse. *See* Radar.

John M. Carroll

Bibliography. J. L. Eaves and E. K. Reedy, *Principles of Modern Radar*, 1987; P. J. Martin, Direct determination of the two-dimensional image spectrum from raw synthetic aperture radar data, *IEEE Transactions on Geoscience and Remote Sensing*, GE-19:194–203, 1981; H. E. Penrose and R. S. H. Boulding, *Principles and Practice of Radar*, 6th ed., 1959; *Proceedings of the National Aerospace and Electronics Conference (NAECON)*, 1983; A. B. Przedpelski, CW finds new uses in the '80s, *Microwaves*, 21(4):89–94, 1982; J. A. Stiles et al., The recognition of extended targets: SAR images for level and hilly terrain, *IEEE Transactions on Geoscience and Remote Sensing*, GE-20:205–211, 1982.

Contour

The locus of points of equal elevation used in topographic mapping. Contour lines represent a uniform series of elevations, the difference in elevation between adjacent lines being the contour interval of the given map. Thus, contours represent the shape of terrain on the flat map surface (see **illus.**). Closely

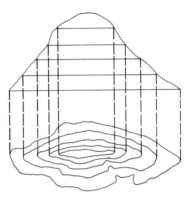

Contour representation; vertical projection indicates the profile of the elevation as it relates to the spacing of the contours.

spaced contours indicate steep ground; sparseness or absence of contours indicates gentle slope or flat ground. Contours do not cross each other unless there is an overhang. *See* Topographic surveying and mapping.

In ground-survey procedures the horizontal positions and elevation of numerous points are observed. Points on the elevations chosen for contouring are then located by interpolation, and each contour is sketched by connecting the points found for its elevation. *See* Surveying.

In photogrammetric mapping with a stereoplotting instrument, the operator sets the plotting table at the chosen contour elevation; the elevation is "hunted" by shifting the table until a floating dot appears to touch the stereoscopic image; the operator then lowers the plotting pencil and traces the contour by keeping the floating dot in apparent touch with the image. *See* Cartography; Photogrammetry.

Robert H. Dodds

Control chart

A chart for the analysis and presentation of data obtained from production processes and research investigations. It is also used to display data in other commercial applications, such as labor turnover, office

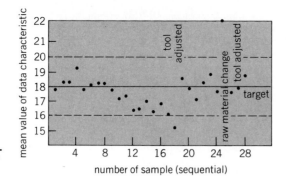

vide an invaluable diagnostic tool to production supervision.

Consider, for example, the chart shown in **Fig. 1**. The solid horizontal line near the center of the chart represents the design target value for the characteristic in question. The broken horizontal lines represent values which the average of the process must not exceed. The first 9 points plotted on this chart are each the average of 5 units drawn at random from the process at hourly intervals, and they will be seen to be grouped closely around the target value. Points 10 to 17 indicate a downward drift in the process, and point 18 actually falls outside the lower broken line. The process was adjusted at this point and brought back on target. It then ran satisfactorily for another 6 h, and then the average increased suddenly to a point above the upper line. Reference to production records

efficiency, costs, and accident rates. In the simplest charts, results are plotted in sequence over time, so that trends and changes may be identified and acted upon. In a more advanced form, limits calculated by statistical methods are placed on the chart. Provided that results stay within these limits, no action is taken, but if a result or series of results fall outside the limits, some action is indicated. Such statistical control charts are known as Shewhart charts, after W. A. Shewhart who initiated them in the 1930s.

Due to natural variations in raw materials and processes, it is rare to find any two manufactured units which are exactly alike. There is also a tendency for industrial processes to change over time (due to tool wear and changes in external conditions, for example). In order to detect such changes, quality-control personnel take samples from the process output and make measurements on important characteristics of the units in the sample. In isolation the results of individual samples are not very informative, but when plotted on a control chart they provide an effective cumulative picture of process behavior. Together with a history of changes made to the process, they pro-

Factors for computing control chart lines				
	Chart for averages		Chart for ranges	
	Factor for control limits	Factor for central line	Factors for control limits	
Number of observations in sample, n	A_2	d_2	D_3	D_4
2	1.88	1.13	0	3.27
3	1.02	1.69	0	2.58
4	0.73	2.06	0	2.28
5	0.58	2.33	0	2.12
6	0.48	2.53	0	2.00
7	0.42	2.70	0.08	1.92
8	0.37	2.85	0.14	1.86
9	0.34	2.97	0.18	1.82
10	0.31	3.08	0.22	1.78

show that the first change occurred due to a tool setting drifting out of specification, and the second occurred when a new batch of raw material was used.

It will be noticed that even when the process was running satisfactorily there was still some variation in results about the target value. Such variations are due to a system of chance causes, and provided they can be tolerated in the finished product, it is unnecessary and usually uneconomical to try and reduce them. Under these conditions the process is said to be in a state of statistical control. However, changes which move the process average away from its target, such as those shown in Fig. 1, must be identified and dealt with. These changes are due to assignable causes, as distinct from chance causes. The object of Shewhart control charts is to provide rules for determining whether changes are more likely due to assignable causes or chance variation.

The statistical theory on which Shewhart charts are based depends on the assumption that the underlying data are normally distributed, and this must be kept in mind when interpreting charts which are constructed by the methods given below. Misleading indications can result from the assumptions being violated, and therefore a thorough analysis of the data by skilled statisticians may be necessary. Custom-designed charts may be required in some cases.

Average and range. The charts most frequently used by the quality control engineer are the $\bar{X}$ (X bar) and R charts. These charts display the average char-

Sample no.	Observations or measurements				Total	Average ($\bar{X}$)	Largest value	Smallest value	Range (R)
1	223	207	229	229	888	222	229	207	22
2	248	248	248	235	979	245	248	235	13
3	248	248	228	228	952	238	248	228	20
4	201	248	197	217	863	216	248	197	51
5	229	207	241	241	918	229	241	207	34
6	235	228	217	217	897	224	235	217	18
7	228	217	192	217	854	213	228	192	36
8	192	212	212	207	823	206	212	192	20
9	207	207	241	223	878	219	241	207	34
10	228	228	229	241	926	231	241	228	13

Section: _W-5_

Machine: _HEAT TREAT #3_

Date: _MARCH 10-14, 1982_

Dimension controlled: _BRINELL HARDNESS_

acteristic $\overline{X}$ of production units and the range R of the characteristic. The first step toward $\overline{X}$ and R charts is to collect the data and make the calculations. In **Fig. 2** four measurements were made on each lot sample, and $\overline{X}$ and R computed for each sample. The next step is to calculate for the 10 samples grand average $\overline{\overline{X}}$ and average range $\overline{R}$ by using Eqs. (1) and (2). From

$$\overline{\overline{X}} = \sum_{i=1}^{i=10} \frac{\overline{X}_i}{10} = \frac{2243}{10} = 224 \qquad (1)$$

$$\overline{R} = \sum_{i=1}^{i=10} \frac{\overline{R}_i}{10} = \frac{261}{10} = 26 \qquad (2)$$

these data control limits are calculated based on the normal distribution curve. Tabulated chart factors simplify such calculations (see **table**). The control limits for X are shown by Eqs. (3) and (4), and for R, by Eqs. (5) and (6).

$$\begin{aligned} \text{Upper limit} &= \overline{\overline{X}} + A_2\overline{R} \\ &= 224 + (0.73 \times 26) = 243 \qquad (3) \end{aligned}$$

$$\begin{aligned} \text{Lower limit} &= \overline{\overline{X}} - A_2\overline{R} \\ &= 224 - (0.73 \times 26) = 205 \qquad (4) \end{aligned}$$

$$\text{Upper limit} = D_4\overline{R} = 2.28 \times 26 = 59 \qquad (5)$$

$$\text{Lower limit} = D_3\overline{R} = 0 \times 26 = 0 \qquad (6)$$

Finally an $\overline{X}$ and an R chart are constructed from the calculations (**Fig. 3**). With the control limits thus established, subsequent lots are inspected and $\overline{X}$ and R (broken lines in Fig. 3) are plotted. A point outside the control limits, such as for lot 2 (on $\overline{X}$ chart), indicates need for a change in the manufacturing process.

Fraction defective. In some situations the quality characteristic being inspected is judged as acceptable or not acceptable. This is inspection by attributes. The inspected unit is either passed or rejected without

reference to any degree of goodness. The product may be inspected for one characteristic or for several (**Fig. 4**).

The average fraction defective is represented by $\overline{p}$ and is the ratio of the number of rejected units to the

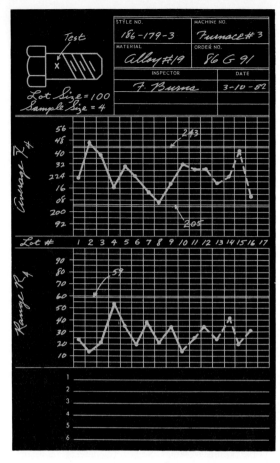

Fig. 3. $\overline{X}$ *and* R chart constructed from the data and calculations which are given in Fig. 2.

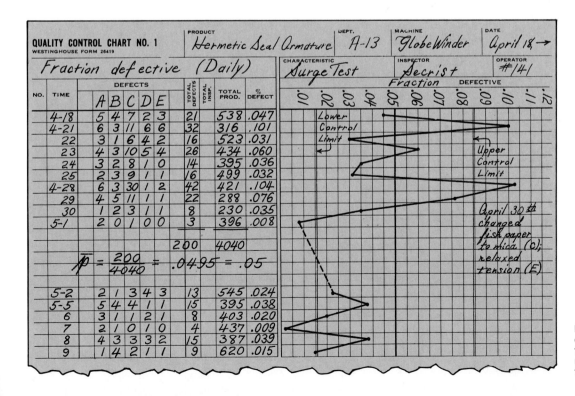

Fig. 4. Work sheet and $\overline{p}$ chart. Defects are classified to assist in determination of corrective action.

number of units inspected. Control limit lines are established by expression (7), where $\bar{p}$ and $\bar{n}$ are defined

$$\bar{p} \pm 3 \sqrt{\frac{\bar{p}(1 - \bar{p})}{\bar{n}}} \qquad (7)$$

by Eqs. (8) and (9), respectively. The cause of points

$$\bar{p} = \frac{\text{total units rejected}}{\text{total units inspected}} \qquad (8)$$

$$\bar{n} = \frac{\text{total units inspected}}{\text{number lots inspected}} \qquad (9)$$

falling outside control limit lines should be determined, and action should be taken to correct the situation.

Defects per unit. The c chart, generally known as a defects-per-unit chart, is used to show the number of defects found in one unit of product. For this purpose the item is inspected, and the number of defects found is recorded. One or more items in a lot may be inspected, but a separate record is made of the number of defects found on each unit, and this number is plotted on the chart (**Fig. 5**). Control limits for a c chart are shown by expression (10). Here $\bar{c}$ is the av-

$$\bar{c} \pm 3\sqrt{\bar{c}} \qquad (10)$$

erage number of defects per unit, which is calculated from the inspection of a number of units, as determined by Eq. (11).

$$\bar{c} = \frac{\text{total number of defects}}{\text{total number of units inspected}} \qquad (11)$$

Recording different classes of defects facilitates corrective action when required. In Fig. 5, 64 defects were found in 20 units inspected. Therefore $\bar{c} = 64/20 = 3.2$, giving an upper control limit of 8.6 and a lower control limit of 0, because a negative control is meaningless in this case.

Average number of defects per unit. The $\bar{c}$ chart is used to show the average number of defects per unit

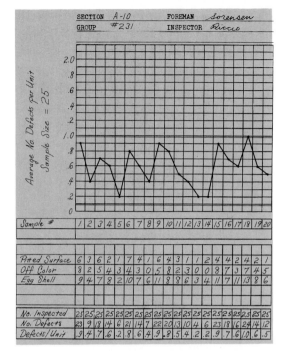

Fig. 6. Control chart for recording sample size and average number of defects per unit in each sample.

| SECTION A-10 | FOREMAN Sorensen |
| GROUP #231 | INSPECTOR Riccio |

Sample #	1	2	3	4	5	6	7	8	9	10	11	12	13	14	15	16	17	18	19	20
Pitted Surface	6	3	6	2	1		7	4		6	4	3	1	1	2	4	4	2	4	2	1
Off Color	8	2	5	4	3	4	3	0	5	8	2	3	0	0	8	7	3	7	4	5
Egg Shell	9	4	7	8	2	10	7	6	11	8	8	6	3	4	11	7	11	13	8	6
No. Inspected	25	25	25	25	25	25	25	25	25	25	25	25	25	25	25	25	25	25	25	25
No. Defects	23	9	18	14	6	21	14	7	22	20	13	10	4	6	23	18	16	24	14	12
Defects/Unit	.9	.4	.7	.6	.2	.8	.6	.4	.9	.8	.5	.4	.2	.2	.9	.7	.6	1.0	.6	.5

in a sample. For this purpose a number of items, generally four or five, are inspected, and a record is made of the number of defects found in all units in the sample. Then the number of defects is divided by the number of units in the sample, as shown in Eq. (12),

$$\bar{c} = \frac{\text{number of defects found in sample}}{\text{number units in sample}} \qquad (12)$$

and this average number is plotted on the $\bar{c}$ chart. Control limits are shown by expression (13) where n

$$\bar{\bar{c}} \pm 3\sqrt{\bar{\bar{c}}/n} \qquad (13)$$

is the number of units in the sample, and $\bar{\bar{c}}$ is the average number of defects in all units in a number of samples, usually 10 or more, as shown in Eq. (14).

$$\bar{\bar{c}} = \frac{\text{total number of defects}}{\text{total number of units inspected}} \qquad (14)$$

Figure 6 shows a simple form of $\bar{c}$ chart, which provides for recording the size of sample inspected and the number of defects found in each sample. In this case, 294 defects were found in 500 units. Therefore $\bar{\bar{c}} = 294/500 = 0.6$, giving upper and lower control limits of 1.06 and 0.14, respectively.

The quality-control engineer finds many uses for control charts. The $\bar{X}$ and R charts are widely used. The $\bar{c}$ chart and c chart are as informative and could be used more widely. For unusual situations the engineer may have to design special charts. SEE CONTROL SYSTEMS; GAGES; INDUSTRIAL COST CONTROL; INSPECTION AND TESTING; QUALITY CONTROL.

John A. Clements

Bibliography. I. W. Burr, *Statistical Quality Control Methods*, 1976; A. J. Duncan, *Quality Control and Industrial Statistics*, 5th ed., 1986; E. L. Grant and R. S. Leavenworth, *Statistical Quality Control*, 5th ed., 1980; *Journal of Quality Technology, Quality Progress, Technometrics*, periodically; J. M. Juran and F. M. Gryna, *Juran's Quality Control Handbook*, 4th ed., 1988.

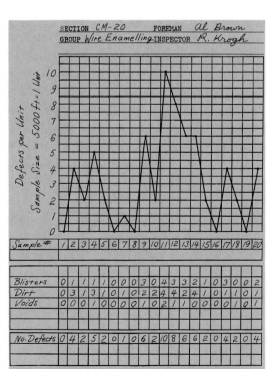

Fig. 5. Control chart for defects per unit.

| SECTION CM-20 | FOREMAN Al Brown |
| GROUP Wire Enamelling | INSPECTOR R. Krogh |

Sample #	1	2	3	4	5	6	7	8	9	10	11	12	13	14	15	16	17	18	19	20
Blisters	0	1	1	1	1	0	0	0	3	0	4	3	3	2	1	0	3	0	0	2
Dirt	0	3	1	3	1	0	1	0	2	2	4	4	2	4	1	0	1	1	0	1
Voids	0	0	0	1	0	0	0	0	1	0	2	1	1	0	0	0	0	1	0	1
No. Defects	0	4	2	5	2	0	1	0	6	2	10	8	6	6	2	0	4	2	0	4

Control system stability

The well-behavedness of the output of a control system in response to either external inputs or internal excitations. Several definitions of stability are in use, each being appropriate to a different application. The objective of stability theory is to obtain information concerning the behavior of a control system without actually solving the equations that describe the system.

Stability theory can be applied in a wide variety of disciplines, such as engineering, mathematics, physics, biology, and economics.

A few examples illustrate such stability theory:

1. The classical problem studied by the Soviet mathematician A. M. Lyapunov was the stability of the solar system. The objective was to determine the ultimate behavior of the trajectories of the various planets, in particular whether they would all spiral into the Sun, fly off into space, or do something else.

2. In digital filters, it is sometimes possible to encounter limit cycles, that is, round off errors that do not decay, perhaps due to the finite word length of the operations. It has been experimentally observed that certain configurations and certain coefficient values never exhibit limit cycles, and a mathematical explanation of this phenomenon would be useful. *See* ELECTRIC FILTER.

3. A voltage source is connected to an electronic circuit containing resistors, capacitors, and operational amplifiers. For a broad but prespecified class of voltage excitations, it would be useful to know whether the resulting current will eventually decay to zero.

4. In a plasma laboratory, a scheme is proposed for confining the plasma to a specified cylindrical region; it would be useful to know whether the proposed scheme will work. *See* NUCLEAR FUSION.

5. A model is developed for the spread of an infectious disease within a population, and it would be useful to be able to determine how the number of survivors is affected by various parameters of the model.

As shown by the discussion below, stability theory encompasses systems described by a wide variety of models. Thus the equations governing the system may be linear or nonlinear; they may be difference equations or differential equations; they may be ordinary, partial, or functional differential equations; finally, they may be difference equations in more than one parameter. This article deals chiefly with differential equations; the theory for other types of equations is similar in concept, though not in detail.

Routh, Hurwitz, and related tests. Consider a system whose transfer function is of the form given by Eq. (1), where $\hat{n}$ and $\hat{d}$ are polynomials and s denotes

$$\hat{h}(s) = \frac{\hat{n}(s)}{\hat{d}(s)} \qquad (1)$$

the variable of Laplace transformation. Such a system exhibits a bounded response to every bounded excitation if and only if the following two conditions hold: (1) $\hat{h}$ is proper, that is, the degree of the polynomial $\hat{n}$ is less than or equal to that of $\hat{d}$; (2) all poles of $\hat{h}$, that is, all zeros of the polynomial $\hat{d}$, must have negative real parts. (A polynomial satisfying these conditions is called a strictly Hurwitz polynomial.) Given a transfer function $\hat{h}$, inspection will verify whether or not the first condition is satisfied, but the

second condition is harder to verify. In particular, it would be useful to determine whether or not all zeros of $\hat{d}$ have negative real parts (that is, whether or not $\hat{d}$ is strictly Hurwitz) without actually computing these zeros. The various tests described below are addressed to this problem. *See* LAPLACE TRANSFORM.

Routh test. Suppose the polynomial $\hat{d}$ is of the form given by Eq. (2), where it may be supposed, without

$$\hat{d}(s) = d_m s^m + d_{m-1} s^{m-1} + \cdots + d_1 s + d_0 \qquad (2)$$

loss of generality, that $d_m > 0$. The Routh test is based on forming the ratio of the even part of $\hat{d}$ over the odd part of $\hat{d}$ (or vice versa), and then carrying out a continued fraction expansion. To illustrate, suppose $m = 2k$ is an even number; then $\hat{d}_{even}(s)$ and $d_{odd}(s)$ are defined by Eqs. (3) and (4). Then the ratio

$$d_{even}(s) = \sum_{i=0}^{k} d_{2i} s^{2i} \qquad (3)$$

$$d_{odd}(s) = \sum_{i=0}^{k-1} d_{2i+1} s^{2i+1} \qquad (4)$$

$\hat{d}_{even}(s)/\hat{d}_{odd}(s)$ is expressed in the form given by Eq. (5). [If m is odd, $\hat{d}_{odd}(s)/\hat{d}_{even}(s)$ is expressed in

$$\frac{\hat{d}_{even}(s)}{\hat{d}_{odd}(s)} = b_m s + \cfrac{1}{b_{m-1}s + \cfrac{1}{\cdots + \cfrac{1}{b_1 s}}} \qquad (5)$$

this form.] The basis of the Routh test is in the following fact: $\hat{d}$ is a strictly Hurwitz polynomial if and only if (1) it is possible to completely carry out the continued fraction expansion in Eq. (5), and (2) all the constants $b_m, \ldots, b_1$ are positive. The test also furnishes one additional piece of information. Suppose that the expansion in Eq. (5) is completed, but not all the b_i's are positive. (Then of course $\hat{d}$ is not strictly Hurwitz.) Under these conditions, if none of the b_i's is zero, then the number of zeros of $\hat{d}$ with positive real part is equal to the number of negative b_i's.

Though the above is the theoretical basis of the Routh test, this is not the manner in which the test is actually carried out. To perform the test in an efficient manner, an array is set up, called the Routh array, as shown in Eq. (6), for the case that m is even, and the

$$
\begin{array}{c|cccccc}
s^m & d_m & d_{m-2} & \cdots & d_2 & & d_0 \\
s^{m-1} & d_{m-1} & d_{m-3} & \cdots & d_1 & \\
s^{m-2} & a_{m-2,1} & a_{m-2,2} & \cdots & a_{m-2,k} & \\
\cdot & & & & & \\
\cdot & & & & & \\
\cdot & & & & & \\
s^1 & a_{1,1} & & & & \\
s^0 & a_{0,1} & & & & \\
\end{array}
\qquad (6)
$$

entries in the array are recursively defined by Eq. (7).

$$a_{i,j} = a_{i+2,j+1} - \frac{a_{i+2,1} \cdot a_{i+1,j+1}}{a_{i+1,1}} \qquad (7)$$

With reference to Eq. (5), it can be shown that the b_i's are given by Eq. (8). Hence the Routh test can

$$b_i = a_{i,1}/a_{i-1,1} \qquad i = m, \ldots, 1 \qquad (8)$$

be concisely stated as follows: The polynomial $\hat{d}$ is strictly Hurwitz if and only if all the entries in the first column of the Routh array are positive. Now

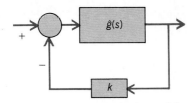

Fig. 1. Feedback system with feedback containing a constant gain _k_.

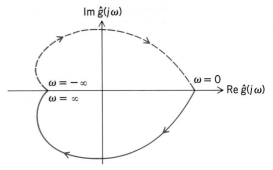

Fig. 2. Nyquist plot of a transfer function _ĝ_.

suppose d is not strictly Hurwitz, but none of the entries in the first column of the Routh array is zero. Then the number of zeros of $\hat{d}$ with positive real part is equal to the number of sign changes in the first column of the Routh array.

Hurwitz test. The Routh test is frequently misnamed the Routh-Hurwitz test. Actually the Hurwitz test is different from the Routh test, though they are equivalent in the sense that they both give the same information. The Hurwitz test is as follows: Given a polynomial $\hat{d}$ of the form in Eq. (2), construct the $m \times m$ matrix $\mathbf{H}$ given by Eq. (9), and let $\Delta_1, \ldots, \Delta_m$

$$\mathbf{H} = \begin{bmatrix} d_{m-1} & d_{m-3} & \cdots & 0 & 0 & \cdots & 0 \\ d_m & d_{m-2} & \cdots & & & & \\ 0 & d_{m-1} & d_{m-3} & \cdots & & & \\ 0 & d_m & d_{m-2} & \cdots & & & \\ \cdot & & & & & & \\ \cdot & & & & & & \\ \cdot & & & & & & \\ 0 & 0 & \cdots & & & d_3 & d_1 \\ 0 & 0 & \cdots & & & d_2 & d_0 \end{bmatrix} \quad (9)$$

denote the leading principal minors of $\mathbf{H}$. (That is, Δ_i is the determinant of the matrix consisting of the entries in $\mathbf{H}$ that belong to both the first i rows and the first i columns.) Then $\hat{d}$ is a strictly Hurwitz polynomial if and only if $\Delta_i > 0$ for all i. If $\hat{d}$ is not strictly Hurwitz but none of the Δ_i's is zero, then the number of zeros of $\hat{d}$ with positive real part is equal to the number of negative Δ_i's. _See_ DETERMINANT.

The relationship between the Hurwitz determinants and the first column of the Routh array is given by Eq. (10), where $\Delta_0 = 1$.

$$a_{i,1} = \Delta_{m-i}/\Delta_{m-i-1} \quad (10)$$
$$i = 1, \ldots, m-1$$

Liénard-Chipart test. Given a polynomial $\hat{d}$ of the form in Eq. (5), a necessary but by no means sufficient condition for $\hat{d}$ to be strictly Hurwitz is that all the coefficients d_i be positive. In this case, it is possible to somewhat simplify the Hurwitz test. In particular, if $d_i > 0$ for all i, then $\hat{d}$ is strictly Hurwitz if and only if either of the equivalent conditions (11) and

$$\Delta_1 > 0, \Delta_3 > 0, \Delta_5 > 0, \ldots \quad (11)$$

and (12) is satisfied. This simplification is known as the Liénard-Chipart test.

$$\Delta_2 > 0, \Delta_4 > 0, \Delta_6 > 0, \ldots \quad (12)$$

Tests for zeros all of magnitude less than 1. In analyzing the stability of sampled-data control systems, digital filters, and so forth, it is desirable to determine not whether all zeros of $\hat{d}$ have negative real parts, but whether all zeros of $\hat{d}$ have magnitude less than 1. One means of doing so is by using the bilinear transformation defined by Eq. (13), which maps the

$$s = \frac{v+1}{v-1} \quad (13)$$

interior of the unit circle in the s plane into the left half of the v plane. Thus, all zeros of $\hat{d}$ in the form (2) have magnitude less than 1 if and only if the associated polynomial given by Eq. (14) is strictly Hur-

$$p(v) = \sum_{i=0}^{m} d_i(v+1)^i(v-1)^{m-i} \quad (14)$$

witz. This can be determined using the tests described above.

However, there are many tests that can be applied directly to the polynomial $\hat{d}$ without requiring any transformation. One of these is the Schur-Cohn test: Let $\hat{d}(s)$ be a polynomial of the form in Eq. (2), and define the determinants Δ_i by Eq. (15). Δ_i is the determinant of a $2i \times 2i$ matrix. The Schur-Cohn test states that all zeros of $\hat{d}$ have magnitude less than 1 if and only if $(-1)^i \Delta_i > 0$ for all i, that is, Δ_i is positive for even i and negative for odd i.

The theoretical advantage of the Schur-Cohn test is that it can be used (with slight modifications) even if the coefficients d_i are complex. However, in practical situations, the coefficients d_i are all real, in which case it is possible to simplify the Schur-Cohn test.

$$\Delta_i = \left| \begin{array}{ccccc|cccc} d_0 & 0 & 0 & \cdots & 0 & d_m & d_{m-1} & \cdots & d_{m-i+1} \\ d_1 & d_0 & 0 & \cdots & 0 & 0 & d_m & \cdots & d_{m-i+2} \\ d_2 & d_1 & d_0 & \cdots & 0 & 0 & 0 & \cdots & d_{m-i+3} \\ \cdot & & & & & \cdot & & & \\ \cdot & & & & & \cdot & & & \\ \cdot & & & & & \cdot & & & \\ d_{i-1} & d_{i-2} & d_{i-3} & \cdots & d_0 & 0 & 0 & \cdots & d_m \\ \hline d_m & 0 & 0 & \cdots & 0 & d_0 & d_1 & \cdots & d_{i-1} \\ d_{m-1} & d_m & 0 & \cdots & 0 & 0 & d_0 & \cdots & d_{i-2} \\ \cdot & & & & & \cdot & & & \\ \cdot & & & & & \cdot & & & \\ \cdot & & & & & \cdot & & & \\ d_{m-i+1} & d_{m-i+2} & & \cdots & d_m & 0 & 0 & \cdots & d_0 \end{array} \right| \quad (15)$$

The Jury-Blanchard test is as follows: Given a polynomial $\hat{d}$ in the form of Eq. (2), define an array of elements $a_{i,j}$, $i = 0$ to $m = 2$, $j = 0$ to $(m - i)$, by the rules of Eqs. (16) and (17). With these defini-

$$a_{0,j} = d_j \qquad j = 0, \ldots, m \qquad (16)$$

$$a_{i,j} = \begin{vmatrix} a_{i-1,0} & a_{i-1,m-i-j} \\ a_{i-1,m-i} & a_{i-1,j} \end{vmatrix} \qquad (17)$$

tions, all zeros of $\hat{d}$ have magnitude less than 1 if and only if conditions (18) and (19) are satisfied.

$$\hat{d}(1) > 0 \quad (-1)^m \, \hat{d}(-1) > 0 \quad |d_0| > |d_m| \qquad (18)$$

$$|a_{i,0}| > |a_{i,m-i}| \text{ for } i = 1 \text{ to } m - 2 \qquad (19)$$

Nyquist, circle, and Popov criteria. The Nyquist criterion provides a necessary and sufficient condition for the stability of a broad class of linear, time-invariant systems. The criterion can be stated at various levels of generality, and what is given below is one of the simplest versions.

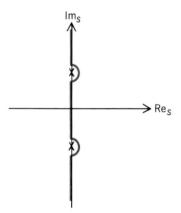

Fig. 3. Modification of Laplace transformation variable to avoid poles of the transfer function $\hat{g}$.

Nyquist plot. Consider a feedback system of the form shown in **Fig. 1**, where the forward element has a transfer function $\hat{g}(s)$ and the feedback contains a constant gain k. It is assumed that the transfer function consists of the sum of three terms, in the form of Eq. (20), where g_0 is a constant; $\hat{g}_a(s)$ is the Laplace

$$\hat{g}(s) = g_0 + \hat{g}_a(s) + \hat{g}_b(s) \qquad (20)$$

transform of an absolutely integrable function, that is, Eq. (21) is valid, and $\hat{g}_b(s)$ is a proper rational func-

$$\hat{g}_a(s) = \mathcal{L}[g_a(t)] \quad \left[\text{where } \int_0^\infty |g_a(t)| \, dt \text{ is finite} \right] \qquad (21)$$

tion. Under these conditions, it is possible to determine whether or not the closed-loop system is stable by examining only the behavior of $\hat{g}(j\omega)$, where $j = \sqrt{-1}$ and the frequency parameter ω is allowed to vary along the real axis from $-\infty$ to ∞.

The plot of the real part of $\hat{g}(j\omega)$ versus the imaginary part of $\hat{g}(j\omega)$, as the frequency parameter ω varies from $-\infty$ to ∞, is called the Nyquist plot of $\hat{g}$ (**Fig. 2**). Since $\hat{g}(-j\omega)$ is the complex conjugate of $\hat{g}(j\omega)$, the plot as ω varies from 0 to $-\infty$ is the mirror image, about the real axis, of the plot as ω varies from 0 to ∞.

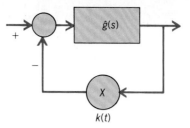

Fig. 4. Feedback system with feedback containing a time-varying gain $k(t)$.

It sometimes happens that $\hat{g}$ has some poles on the imaginary axis, in which case $\hat{g}(j\omega)$ may be undefined for certain values of ω. In such cases, it is necessary to indent the $j\omega$ axis around the poles of $\hat{g}(j\omega)$, as shown in **Fig. 3**.

Nyquist criterion. With these conventions, the Nyquist criterion can be stated as follows: The system of Fig. 1 is stable if and only if (1) the Nyquist plot of $\hat{g}(j\omega)$ does not intersect the critical point $-(1/k) + j0$, and (2) as ω increases from $-\infty$ to ∞, the Nyquist plot encircles the critical point exactly μ times in the counterclockwise direction, where μ is the number of poles of $\hat{g}$ with positive real parts.

Several advantages of the Nyquist criterion are immediately apparent. (1) It is based on the behavior of $\hat{g}(j\omega)$, which can be readily obtained by frequency response measurements. (2) Since $\hat{g}$ is not required to be a rational function of s (only $\hat{g}_b$ is required to be rational), the criterion is applicable to systems containing transmission lines, time delays, and so forth. (3) For a given transfer function $\hat{g}$, one can readily calculate the range of values of k that result in a stable closed-loop system. (4) Even if the system of Fig. 1 is unstable, it can easily be determined what sort of ''compensation'' is needed to make the system stable.

Circle criterion. Now consider the system of **Fig. 4**, which is similar to that of Fig. 1, except that the constant gain k has been replaced by a time-varying gain $k(t)$. The assumptions on $\hat{g}(s)$ are the same as before. The circle criterion gives sufficient conditions for the stability of such a system, and can be stated as follows: Suppose the gain $k(t)$ varies in the range $[\alpha, \beta]$, that is, $\alpha \le k(t) \le \beta$ for all t. Let $D(\alpha, \beta)$ denote the critical disk, which passes through $-1/\alpha + j0$ and $-1/\beta + j0$, and is centered on the real axis. Under these conditions, the system of Fig. 4 is stable if one of the following conditions, as appropriate, holds:

Case 1: $0 < \alpha < \beta$. The Nyquist plot of $\hat{g}(j\omega)$ does not intersect the critical disk $D(\alpha, \beta)$ and encircles it in the counterclockwise direction exactly μ times (**Fig. 5**).

Case 2: $0 = \alpha < \beta$. The Nyquist plot of $\hat{g}(j\omega)$ lies in the half-plane Re $\hat{g}(j\omega) < 1/\beta$.

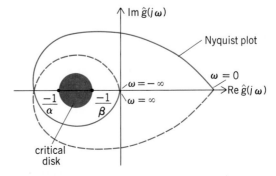

Fig. 5. Circle criterion for $0 < \alpha < \beta$. The Nyquist plot does not intersect the critical disk and encircles it in the counterclockwise direction exactly μ times, where μ is the number of poles of $\hat{g}$ with positive real parts.

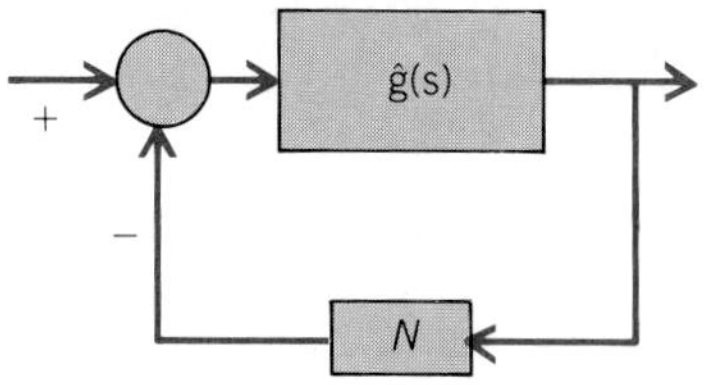

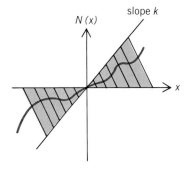

Fig. 6. Feedback system with feedback containing a memoryless nonlinear element N.

Case 3: $\alpha < 0 < \beta$. The Nyquist plot of $\hat{g}(j\omega)$ is contained within the critical disk $D(\alpha,\beta)$.

Case 4: $\alpha < \beta < 0$. Replace $\hat{g}$ by $-\hat{g}$, α by $-\beta$, β by $-\alpha$, and apply case 1 (since $0 < -\beta < -\alpha$).

The circle criterion shares the same advantages as the Nyquist criterion. In addition, if $\alpha \to \beta$, so that the time-varying gain $k(t)$ approaches a constant gain β, then the critical disk in the circle criterion shrinks to the critical point of the Nyquist criterion.

Popov criterion. Finally, consider the system shown in **Fig. 6**, wherein the feedback contains a memoryless nonlinear element N. Suppose the graph of the nonlinearity lies in the sector $[0,k]$ that is, between the horizontal axis and a line of slope k through the origin (**Fig. 7**). Finally, suppose $\hat{g}(s)$ is strictly proper and that all poles of $\hat{g}$ have negative real parts. The Popov criterion gives a sufficient condition for the stability of such a system. It states that the system of Fig. 6 is stable, provided one can find a number $q \geq 0$ and a number $\delta > 0$ such that inequality (22)

$$\text{Re}\,[(1 + j\omega q)\hat{g}(j\omega)] + 1/k \geq \delta > 0 \text{ for all } \omega \quad (22)$$

is satisfied. The criterion (22) can also be given a graphical interpretation by means of a plot of Re $\hat{g}(j\omega)$ versus ω Im $\hat{g}(j\omega)$, called a Popov plot (**Fig. 8**). It is only necessary to do the plot for $\omega \geq 0$, since both Re $\hat{g}(j\omega)$ and ω Im $\hat{g}(j\omega)$ are even functions of ω. The inequality (22) states that it is possible to draw a line of slope $1/q$ through the point $-1/k + j0$ in such a way that the Popov plot of $\hat{g}(j\omega)$ lies strictly to the right of this line (Fig. 8).

Lyapunov theory. This can be used to analyze the behavior of the solutions of the vector differential equation (23), corresponding to various initial conditions $\mathbf{x}(t_0) = \mathbf{x}_0$ without actually solving the differ-

$$\dot{\mathbf{x}}(t) = \mathbf{f}(t,\mathbf{x}(t)) \quad (23)$$

ential equation (23). The form of Eq. (23) is quite general, and is referred to as the set of state equations for a system; it is distinguished by the fact that the system under study is described by a set of first-order differential equations. The form of Eq. (23) is sufficiently general that any set of coupled ordinary dif-

ferential equations, of any order, can be equivalently expressed in this form. To simplify the exposition, only the case where the time t does not explicitly appear in the state equations will be discussed, that is, where the state equations are of the form of Eq. (24).

$$\dot{\mathbf{x}}(t) = \mathbf{f}(\mathbf{x}(t)) \quad (24)$$

SEE DIFFERENTIAL EQUATION; LINEAR SYSTEM ANALYSIS.

Equilibrium. An important concept for the system of Eq. (23) is that of an equilibrium. A vector $\mathbf{x}_0$ is called an equilibrium of the system of Eq. (24) if Eq. (25) is satisfied. Equivalently, if $\mathbf{x}_0$ is an equilibrium

$$\mathbf{f}(\mathbf{x}_0) = \mathbf{0} \quad (25)$$

of Eq. (24), then the solution of Eq. (24) corresponding to the initial condition $\mathbf{x}(0) = \mathbf{x}_0$ is $\mathbf{x}(t) = \mathbf{x_0}$ for all $t \geq 0$.

Asymptotic stability. There are many distinct concepts of "stability" in Lyapunov theory, each of them suitable for a different application, and only one of them will be discussed here. Suppose $\mathbf{x} = \mathbf{0}$ is an equilibrium of Eq. (24) [that is, $\mathbf{f}(\mathbf{0}) = \mathbf{0}$], and let $\mathbf{x}(t,\mathbf{x}_0)$ denote the solution of Eq. (24), corresponding to the initial condition $\mathbf{x}(0) = \mathbf{x_0}$. Finally, let the norm of the vector $\mathbf{x}$ be defined by Eq. (26). Then $\mathbf{0}$

$$|\mathbf{x}| = \left(\sum_i x_i^2\right)^{1/2} \quad (26)$$

is called an asymptotically stable equilibrium of Eq. (24) if the following two conditions are satisfied: (1) There is a number $\delta_0 > 0$ such that $|\mathbf{x}(t,\mathbf{x}_0)|$ is bounded as a function of t whenever $|x_0| < \delta_0$. Moreover, the function ϕ defined by Eq. (27) is continuous

$$\phi(r) = \sup_{|\mathbf{x}_0| \leq r} \sup_{t \geq 0} |\mathbf{x}(t,\mathbf{x}_0)| \quad (27)$$

at $r = 0$. (2) There is a number $\delta_1 > 0$ such that $|\mathbf{x}(t,\mathbf{x}_0)| \to 0$ as $t \to \infty$, whenever $|\mathbf{x}_0| < \phi_1$.

Basically, condition (1) states that, whenever the magnitude of the initial condition is sufficiently small ($|\mathbf{x}_0| < \delta_0$), small perturbations in the initial condition produce small perturbations in the corresponding solution. Condition (2) states that there is a domain of attraction (the sphere of radius δ_1) such that, whenever the initial condition belongs to this domain, the corresponding solution approaches $\mathbf{0}$ as $t \to \infty$.

Linear systems. In the case of a linear system, Eq. (28), where $\mathbf{x}(t)$ is an $m \times 1$ vector and $\mathbf{a}$ is an

$$\dot{\mathbf{x}}(t) = \mathbf{Ax}(t) \quad (28)$$

$m \times m$ matrix, the equilibrium $\mathbf{x} = \mathbf{0}$ is asymptotically stable if and only if all eigenvalues of $\mathbf{A}$ have negative real parts, that is, the characteristic polynomial $\phi(s) = \det(s\mathbf{I} - \mathbf{A})$ is strictly Hurwitz. Moreover, if $\mathbf{x} = \mathbf{0}$ is asymptotically stable, then the domain of attraction is the entire state space, that is, $|\mathbf{x}(t,\mathbf{x}_0)| \to 0$ as $t \to \infty$ for every $\mathbf{x}_0$.

A test that can be used to determine whether or not all eigenvalues of $\mathbf{A}$ have negative real parts (that is, whether $\mathbf{A}$ is a Hurwitz matrix), without actually computing them, is summarized in the following theorem:

Given an $m \times m$ matrix $\mathbf{A}$, the following three conditions are equivalent: (1) All eigenvalues of $\mathbf{A}$ have negative real parts. (2) For some symmetric positive definite matrix $\mathbf{Q}$, Eq. (29) has a unique solution

$$\mathbf{A}^\mathrm{T}\mathbf{P} + \mathbf{PA} = -\mathbf{Q} \quad (29)$$

slope k

$N(x)$

x

Fig. 7. Graph of nonlinear element N.

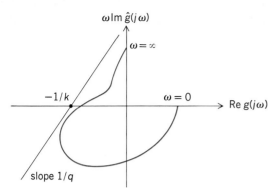

Fig. 8. Popov plot of a transfer function $\hat{g}$.

Bibliography. C. A. Desoer and M. Vidyasagar, *Feedback Systems: Input-Output Properties*, 1975; E. I. Jury, *Inners and Stability of Dynamic Systems*, 2d ed., 1982; K. S. Narendra and J. H. Taylor, *Frequency Domain Criteria for Absolute Stability*, 1973; M. Vidyasagar, *Nonlinear Systems Analysis*, 1978.

Control systems

Interconnections of components forming system configurations which will provide a desired system response as time progresses. The steering of an automobile is a familiar example. The driver observes the position of the car relative to the desired location and makes corrections by turning the steering wheel. The car responds by changing direction, and the driver attempts to decrease the error between the desired and actual course of travel. In this case, the controlled output is the automobile's direction of travel, and the control system includes the driver, the automobile, and the road surface. The control engineer attempts to design a steering control mechanism which will provide a desired response for the automobile's direction control. Different steering designs and automobile designs result in rapid responses, as in the case of sports cars, or relatively slow and comfortable responses, as in the case of large autos with power steering.

Open- and closed-loop control. The basis for analysis of a control system is the foundation provided by linear system theory, which assumes a cause-effect relationship for the components of a system. A component or process to be controlled can be represented by a block. Each block possesses an input (cause) and output (effect). The input-output relation represents the cause-and-effect relationship of the process, which in turn represents a processing of the input signal to provide an output signal variable, often with power amplification. An open-loop control system utilizes a controller or control actuator in order to obtain the desired response (**Fig. 1**). *See* BLOCK DIAGRAM.

A closed-loop control system utilizes an additional measure of the actual output in order to compare the actual output with the desired output response (**Fig. 2**). A standard definition of a feedback control system is a control system which tends to maintain a prescribed relationship of one system variable to another by comparing functions of these variables and using the difference as a means of control. In the case of the driver steering an automobile, the driver visually measures and compares the actual location of the car with the desired location. The driver then serves as the controller, turning the steering wheel. The process represents the dynamics of the steering mechanism and the automobile response.

A feedback control system often uses a function of a prescribed relationship between the output and reference input to control the process. Often, the difference between the output of the process under control

P; moreover, this solution is symmetric and positive definite. (3) For every symmetric positive definite matrix **Q**, Eq. (29) has a unique solution for **P**, and this solution is positive definite.

Thus the testing procedure is as follows: Given a matrix **A**, select a symmetric positive definite matrix **Q** arbitrarily, and solve Eq. (29) [known as the Lyapunov matrix equation] for **P**. If a unique **P** cannot be found, or if the unique **P** is not positive definite, then **A** is not a Hurwitz matrix; if there is a unique **P** satisfying Eq. (29) and this **P** is positive definite, the **A** is a Hurwitz matrix. The test also provides additional information: If Eq. (29) has a unique solution for **P**, then the number of negative eigenvalues of **P** equals the number of eigenvalues of **A** having positive real part. *See* EIGENFUNCTION; MATRIX THEORY.

Nonlinear systems. To analyze the nonlinear system of Eq. (24), two general methods are available, namely the direct and indirect methods of Lyapunov. The indirect method is based on expanding $\mathbf{f}(\mathbf{x})$ in a Taylor series around the point $\mathbf{x} = \mathbf{0}$. Specifically, let **A** be defined by Eq. (30). Since $\mathbf{f}(\mathbf{0}) = \mathbf{0}$ by assump-

$$\mathbf{A} = \left.\frac{\partial \mathbf{f}}{\partial \mathbf{x}}\right|_{\mathbf{x}=\mathbf{0}} \quad (30)$$

tion, Eq. (31) follows. Lyapunov's indirect method is

$$\mathbf{f}(\mathbf{x}) = \mathbf{A}\mathbf{x} + \text{higher-order terms in } \mathbf{x} \quad (31)$$

based on the following fact: If **A** is strictly Hurwitz, then $\mathbf{x} = \mathbf{0}$ is an asymptotically stable equilibrium of the nonlinear system of Eq. (24); if **A** has some eigenvalues with positive real parts, then $\mathbf{x} = \mathbf{0}$ is an unstable equilibrium of Eq. (24).

The direct method of Lyapunov is based on constructing an energylike function $v(\mathbf{x})$, known as a Lyapunov function, whose value decreases along the solution trajectories of Eq. (24). The actual details are somewhat complex, but the direct method provides a very powerful tool for analyzing nonlinear systems. *See* NONLINEAR CONTROL THEORY.

Advanced methods. The techniques described here have been extended to systems governed by partial or even functional differential equations, as well as to multidimensional systems (for example, two-dimensional digital filters). In addition, advanced mathematical techniques such as functional analysis and differential geometry have been applied to stability problems. *See* CONTROL SYSTEMS.

M. Vidyasagar

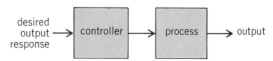

Fig. 1. Open-loop control system.

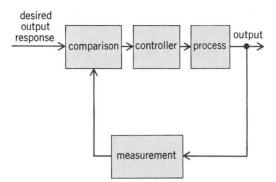

Fig. 2. Closed-loop control system.

and the reference input is amplified and used to control the process so that the difference is continually reduced. The feedback concept has been the foundation for control system analysis and design.

Because of the increasing complexity of the system under control and the interest in achieving optimum performance, the importance of control system engineering has grown over the years. Furthermore, as the systems become more complex, the interrelationship of many controlled variables must be considered in the control scheme.

History of automatic control. The first automatic feedback controller used in an industrial process was James Watt's flyball governor (1769) for controlling the speed of a steam engine. Prior to World War II, one main impetus for the use of feedback control in the United States was the development of the telephone system and electronic feedback amplifiers. The frequency domain was used primarily to describe the operation of the feedback amplifiers in terms of bandwidth and other frequency variables. In contrast, the Soviet theory, which was dominated by mathematicians, tended to utilize a time-domain formulation using differential equations.

A large impetus to the theory and practice of automatic control occurred during World War II, when it became necessary to design and construct automatic airplane pilots, gun-positioning systems, radar antenna control systems, and other military systems based on the feedback control approach. The complexity and expected performance of these military systems necessitated an extension of the available control techniques and fostered interest in control systems and the development of new insights and methods. *See Fire-control systems.*

Frequency-domain techniques continued to domi-

nate the field of control following World War II, with the increased use of the Laplace transform and the complex frequency plane. During the 1950s, the utilization of both analog and digital computers for control components became possible. These new controlling elements possessed an ability to calculate rapidly and accurately, which was formerly not available to the control engineer. Digital process control computers are employed especially for process control systems in which many variables are measured and controlled simultaneously by the computer. *See Analog computer; Digital computer; Process control.*

With the advent of the space age, it became necessary to design complex, highly accurate control systems for missiles and space probes. Furthermore, the necessity to minimize the weight of satellites and to control them very accurately spawned the important field of optimal control. It appears that control engineering must consider both the time-domain and the frequency-domain approaches simultaneously in the analysis and design of control systems. *See Guidance systems; Optimal control theory; Space navigation and guidance.*

Applications. Familiar control systems have the basic closed-loop configuration as shown earlier. For example, a refrigerator has a temperature setting for desired temperature, a thermostat to measure the actual temperature and the error, and a compressor motor for power amplification. Other examples in the home are the oven, furnace, and water heater. In industry, there are controls for speed, process temperature and pressure, position, thickness, composition, and quality, among many others.

In order to provide a mass transportation system for modern urban areas, a large, complex, high-speed system is necessary. Automatic control is necessary in order to maintain a constant flow of trains and to ensure comfortable deceleration and braking conditions at stations (**Fig. 3**). A measurement of the distance from the station and the speed of the train is used to determine the error signal and therefore the braking signal. *See Railroad control systems.*

The electric power industry is primarily interested in energy conversion, control, and distribution. It is critical that computer control be increasingly applied to the power industry in order to improve the efficiency of use of energy resources. Also, the control of power plants for minimum waste emission has become increasingly important. The modern large-capacity plants which exceed several hundred megawatts require automatic control systems which account for the interrelationship of the process variables and the optimum power production. It is common to have as many as 90 or more manipulated variables under coordinated control. Computer controls are also used to control energy use in industry and stabilize and connect loads evenly to gain fuel economy. *See Electric power systems engineering.*

Feedback control concepts have been applied to automatic warehousing and inventory control and to the automatic control of agriculture systems (farms). Automatic control of wind turbine generators, solar heating and cooling, and automobile engine performance are other important examples.

Also, there have been many applications of control system theory to biomedical experimentation, diagnosis, prosthetics, and biological control systems. The control systems under consideration range from the cellular level to the central nervous system and

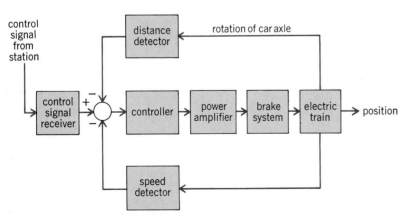

Fig. 3. Train-brake control system.

include temperature regulation and neurological, respiratory, and cardiovascular control. Most physiological control systems are closed-loop systems. However, they involve not one controller but rather control loop within control loop, forming a hierarchy of systems. The modeling of the structure of biological processes confronts the analyst with a high-order model and a complex structure. Prosthetic devices are designed to provide automatically controlled aids to the disabled. An artificial hand that uses force feedback signals and is controlled by the amputee's bioelectric control signals is an example. SEE MATHEMATICAL BIOLOGY; PROSTHESIS.

Finally, it has become of interest to attempt to model the feedback processes prevalent in the social, economic, and political spheres. This approach is undeveloped but appears to have a reasonable future. A simple lumped model of the national-income feedback control system is shown in **Fig. 4**. This type of model helps the analyst to understand the effects of government control—granted its existence—and the dynamic effects of government spending. Of course, many other loops not shown also exist. This type of political or social feedback model, while usually nonrigorous, does impart information and understanding. SEE SYSTEMS ANALYSIS; SYSTEMS ENGINEERING.

Analysis using the Laplace transform. In order to understand a complex control system, it is necessary to analyze the relationships between the system variables and to obtain a quantitative mathematical model of the system. Since the systems under consideration are dynamic in nature, the descriptive equations are usually differential equations. Furthermore, if these equations can be linearized, then the Laplace transform can be utilized in order to simplify the method of solution. In practice, the complexity of systems and incomplete knowledge of the relevant factors necessitate the introduction of assumptions concerning the system operation. Therefore, it is often useful to consider the physical system, delineate some necessary assumptions, and linearize the system. Then, by using the physical laws describing the linear equivalent system, one can obtain a set of linear differential equations. Finally, utilizing mathematical tools, such as the Laplace transform, a solution describing the operation of the system is obtained. In summary, the approach to dynamic system problems can be listed as follows: (1) define the system and its components; (2) formulate the mathematical model and list the necessary assumptions; (3) write the differential equations describing the model; (4) solve the equations for the desired output variables; (5) examine the solutions and the assumptions; and then (6) reanalyze or design. SEE DIFFERENTIAL EQUATION.

The Laplace transform of a function f of time t is given by Eq. (1), where s is a complex variable re-

$$F(s) = \mathscr{L}\{f(t)\} = \int_0^\infty f(t)\, e^{-st} dt \qquad (1)$$

lated to the frequency ω by the equations $s = \sigma + j\omega$ (where $j = \sqrt{-1}$). The analysis thus takes place in the frequency domain. The inverse Laplace transformation is usually obtained by a partial fraction expansion. SEE LAPLACE TRANSFORM.

The transfer function of a linear system is the ratio of the Laplace transform of the output variable to the Laplace transform of the input variable with all initial conditions assumed to be zero. The block diagram of

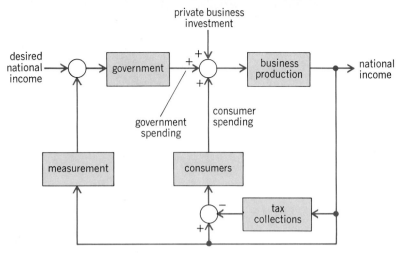

Fig. 4. Feedback control system model of the United States economy.

a negative-feedback control system is shown in **Fig. 5**. The transfer function $G(s)$ of the process is given by Eq. (2). The transfer function of the closed-loop

$$\frac{C(s)}{E_a(s)} = G(s) \qquad (2)$$

system can be obtained by the algebraic manipulation of the system equations. The error, $E_a(s)$, is given by Eq. (3), and the output is given by Eq. (4). Solving

$$\begin{aligned} E_a(s) &= R(s) - B(s) \\ &= R(s) - H(s)C(s) \end{aligned} \qquad (3)$$

$$\begin{aligned} C(s) &= G(s)E_a(s) \\ &= G(s)[R(s) - H(s)C(s)] \end{aligned} \qquad (4)$$

for $C(s)$, Eq. (5) is obtained, and therefore $C(s)$ is

$$C(s)[1 + G(s)H(s)] = G(s)R(s) \qquad (5)$$

given by Eq. (6). The closed-loop transfer function

$$C(s) = \frac{G(s)}{1 + G(s)H(s)} R(s) \qquad (6)$$

$T(s) = C(s)/R(s)$ is given by Eq. (7), and the error is

$$\frac{C(s)}{R(s)} = T(s) = \frac{G(s)}{1 + G(s)H(s)} \qquad (7)$$

then given by Eq. (8). It is clear that in order to re-

$$E_a(s) = \frac{1}{1 + G(s)H(s)} R(s) \qquad (8)$$

duce the error, the magnitude of $[1 + G(s)H(s)]$ must

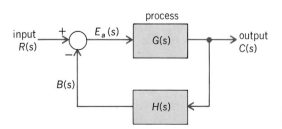

Fig. 5. Negative-feedback control system.

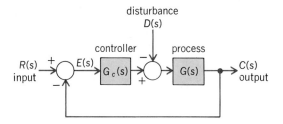

Fig. 6. Feedback control system with a disturbance.

be greater than one over the range of s under consideration.

Advantages of feedback control. The addition of feedback to a control system results in several important advantages.

Reduction in sensitivity. A process, represented by the transfer function $G(s)$, whatever its nature, is subject to a changing environment, aging, ignorance of the exact values of the process parameters, and other natural factors which affect a control process. In the open-loop system, all these errors and changes result in a changing and inaccurate output. However, a closed-loop system senses the change in the output due to the process changes and attempts to correct the output. The sensitivity of a control system to parameter variations is of prime importance. A primary advantage of a closed-loop feedback control system is its ability to reduce the system's sensitivity.

For the closed-loop case, if $G(s)H(s)$ is much greater than 1 for all complex frequencies of interest, then from Eq. (6) one obtains Eq. (9). That is, the

$$C(s) = \frac{1}{H(s)} R(s) \qquad (9)$$

output is affected only by $H(s)$, which may be constant. If $H(s) = 1$, the desired result is obtained; that is, the output is equal to the input. However, this approach cannot be used for all control systems because the requirement that $G(s)H(s)$ be much greater than 1 may cause the system response to be highly oscillatory and even unstable. But the fact that an increase in the magnitude of the loop transfer function $G(s)H(s)$ results in a reduction of the effect of $G(s)$ on the output is an exceedingly useful concept. Therefore, the first advantage of a feedback system is that the effect of the variation of the parameters of the process, $G(s)$, is reduced.

Control of transient response. One of the most important characteristics of control systems is their transient response, which often must be adjusted until it is satisfactory. If an open-loop control system does not provide a satisfactory response, then the process, $G(s)$, must be replaced with a suitable process. By contrast, a closed-loop system can often be adjusted to yield the desired response by adjusting the feedback loop parameters. It is often possible to alter the response of an open-loop system by inserting a suitable cascade filter, $G_1(s)$, preceding the process, $G(s)$. Then it is necessary to design the cascade transfer function $G_1(s)G(s)$ so that the resulting transfer function provides the desired transient response. Therefore, the second advantage of feedback control is control of the transient response of the system.

Reduction of effects of disturbance. The third most important effect of feedback in a control system is the control and partial elimination of the effect of disturbance signals. Many control systems are subject to

extraneous disturbance signals which cause the system to provide an inaccurate output. Electronic amplifiers have inherent noise generated within the integrated circuits or transistors, radar antennas are subjected to wind gusts, and many systems generate unwanted distortion signals due to nonlinear elements. Feedback systems have the beneficial aspect that the effect of distortion, noise, and unwanted disturbances can be effectively reduced.

As a specific example of a system with an unwanted disturbance, consider the speed control system for a steel rolling mill. Rolls passing steel through are subject to large load changes or disturbances. As a steel bar approaches the rolls, the rolls turn unloaded. However, when the bar engages in the rolls, the load on the rolls increases immediately to a large value. This loading effect can be approximated by a step change of disturbance torque.

The block diagram of a feedback control system with a disturbance, $D(s)$, is shown in **Fig. 6**, where $H(s) = 1$. Then the output, $C(s)$, may be shown to be given by Eq. (10) when the input, $R(s)$, is set to

$$C(s) = \frac{-G(s)D(s)}{1 + G_c(s)G(s)} \qquad (10)$$

zero. Then if $G_c(s)G(s)$ is much greater than one over the range of s, the approximate result given by Eq. (11) is obtained. Therefore, if $G_c(s)$ is made suffi-

$$C(s) \simeq \frac{-1}{G_c(s)} D(s) \qquad (11)$$

ciently large, the effect of the disturbance can be decreased by closed-loop feedback.

Reduction of steady-state error. The reduction of the steady-state error of a control system is another advantage of feedback control.

The error of the closed-loop system when $H(s) = 1$ is given by Eq. (12). The steady-state error

$$E(s) = \frac{1}{1 + G(s)} R(s) \qquad (12)$$

can be calculated by using the final-value theorem which is given by Eq. (13), where $e(t)$ is the error

$$\lim_{t \to \infty} e(t) = \lim_{s \to 0} sE(s) \qquad (13)$$

variable in the time domain. Therefore, using a unit step input as comparable input, Eq. (14) is obtained

$$e(\infty) = \lim_{s \to 0} s \left(\frac{1}{1 + G(s)} \right) \left(\frac{1}{s} \right)$$
$$= \frac{1}{1 + G(0)} \qquad (14)$$

for the closed-loop system. The value of $G(s)$ when $s = 0$ is often called the dc gain and is normally greater than one. Therefore, the open-loop system will usually have a steady-state error of significant magnitude. By contrast, the closed-loop system with a reasonably large dc-loop gain $G(0)H(0)$ will have a small steady-state error.

Costs of feedback control. While the addition of feedback to a control system results in the advantages outlined above, it is natural that these advantages have an attendant cost.

Increased complexity. The cost of feedback is first

manifested in the increased number of components and the complexity of the system. In order to add the feedback, it is necessary to consider several feedback components, of which the measurement component (sensor) is the key component. The sensor is often the most expensive component in a control system. Furthermore, the sensor introduces noise and inaccuracies into the system.

Loss of gain. The second cost of feedback is the loss of gain. For example, in a single-loop system, the open-loop gain is $G(s)$ and is reduced to $G(s)/[1 + G(s)]$ in a unity negative-feedback system. The reduction in closed-loop gain is $1/[1 + G(s)]$, which is exactly the factor that reduces the sensitivity of the system to parameter variations and disturbances. Usually, there is open-loop gain to spare, and one is more than willing to trade it for increased control of the system response.

However, it is the gain of the input-output transmittance that is reduced. The control system does possess a substantial power gain which is fully utilized in the closed-loop system.

Possibility of instability. Finally, a cost of feedback is the introduction of the possibility of instability. While the open-loop system is stable, the closed-loop system may not be always stable.

Necessity of considering problems. The addition of feedback to dynamic systems results in several additional problems for the designer. However, for most cases, the advantages far outweigh the disadvantages, and a feedback system is utilized. Therefore, it is necessary to consider the additional complexity and the problem of stability when designing a control system.

It is desired that the output of the system $C(s)$ equal the input $R(s)$. However, one might ask, "Why not simply set the transfer function $G(s) = C(s)/R(s)$ equal to one?" The answer to this question is that the process (or plant) $G(s)$ was necessary in order to provide the desired output; that is, the transfer function $G(s)$ represents a real process and possesses dynamics which cannot be neglected. To set $G(s)$ equal to one implies that the output is directly connected to the input. However, it must be recalled that a specific output, such as temperature, shaft rotation, or engine speed, is desired, while the input might be a potentiometer setting or a voltage. The process $G(s)$ is necessary in order to provide the physical process between $R(s)$ and $C(s)$. Therefore, a transfer function $G(s) = 1$ is unrealizable, and a practical transfer function must be adopted.

State variables and the time-domain. A useful approach to the analysis and design of feedback systems was outlined above. The Laplace transform was utilized to transform the differential equations representing the system into an algebraic equation expressed in terms of the complex variable, s. Utilizing this algebraic equation, a transfer function representation of the input-output relationship was obtained. The Laplace transform approach, carried out in terms of the complex variable s and accomplished in the frequency domain, is extremely useful; it is and will remain one of the primary tools of the control engineer. However, the limitations of the frequency-domain techniques and the attractiveness of the time-domain approach require a reconsideration of the time-domain formulation of the equations representing control systems.

The frequency-domain techniques are limited in applicability to linear, time-invariant systems. Furthermore, they are particularly limited in their usefulness for multivariable control systems because of the emphasis on the input-output relationship of transfer functions. By contrast, the time-domain techniques may be readily utilized for nonlinear, time-varying, and multivariable systems. A time-varying control system is a system for which one or more of the parameters of the system may vary as a function of time. For example, the mass of a missile varies as a function of time as the fuel is expended during flight. A multivariable system is a system with several input and output signals. Furthermore, the solution of a time-domain formulation of a control system problem is facilitated by the availability and ease of use of digital and analog computers.

Definition of state variables. The time-domain analysis and design of control systems utilizes the concept of the state of a system. The state of a system is a set of numbers such that the knowledge of these numbers and the input functions will, with the equations describing the dynamics, provide the future state and output of the system. For a dynamic system, the state of a system is described in terms of a set of state variables $[x_1(t), x_2(t), \ldots, x_n(t)]$. The state variables determine the future behavior of a system when the present state of the system and the excitation signals are known. A set of state variables $(x_1, x_2, \ldots, x_n)$ is a set such that knowledge of the initial values of the state variables $[x_1(t_0), x_2(t_0), \ldots, x_n(t_0)]$ at the initial time t_0, and of the input signals $u_1(t)$ and $u_2(t)$ for times t greater than t_0, suffices to determine the future values of the output and state variables. SEE *LINEAR SYSTEM ANALYSIS*.

A simple example of a state variable is the state of an on-off light switch. The switch can be in either the on or off position, and thus the state of the switch can assume one of two possible values. Thus if the present state (position) of the switch is known at t_0 and if an input is applied, the future value of the state of the element can be determined.

Example of dynamic system. The concept of a set of state variables which represent a dynamic system can be illustrated in terms of the spring-mass-damper system shown in **Fig. 7**. The number of state variables chosen to represent this system should be as few as possible in order to avoid redundant state variables. A set of state variables sufficient to describe this system is the position $y(t)$ and the velocity of the mass. Therefore, a set of state variables is defined as (x_1, x_2), given by Eqs. (15).

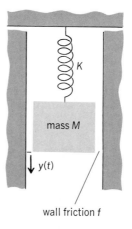

Fig. 7. Spring-and-mass system with wall friction.

mass M

$y(t)$

wall friction f

K

$$x_1(t) = y(t) \qquad x_2(t) = \frac{dy(t)}{dt} \tag{15}$$

The differential equation describing the behavior of the system is usually written as Eq. (16), where M is

$$M\frac{d^2y}{dt^2} + f\frac{dy}{dt} + Ky = u(t) \tag{16}$$

the mass, K is the force constant of the spring, and the coefficient f determines the frictional force exerted by the wall. In order to write Eq. (16) in terms of the state variables, one substitutes the definition of the state variables and obtains Eq. (17). Therefore, the

$$M\frac{dx_2}{dt} + fx_2 + Kx_1 = u(t) \tag{17}$$

differential equations which describe the behavior of the spring-mass-damper system may be written as a set of two first-order differential equations, Eqs. (18) and (19). This set of differential equations describes

$$\frac{dx_1}{dt} = x_2 \tag{18}$$

$$\frac{dx_2}{dt} = \frac{-f}{M}x_2 - \frac{K}{M}x_1 + \frac{1}{M}u \tag{19}$$

the behavior of the state of the system in terms of the rate of change of each state variable.

Matrix differential equation. The state of a system is described by the set of first-order differential equations written in terms of the state variables (x_1, x_2, . . . , x_n). These first-order differential equations may be written in general form as Eqs. (20), where

$$\begin{aligned}
\dot{x}_1 &= a_{11}x_1 + a_{12}x_2 + \ldots \\
&\quad + a_{1n}x_n + b_{11}u_1 + \cdots + b_{1m}u_m \\
\dot{x}_2 &= a_{21}x_1 + a_{22}x_2 + \cdots \\
&\quad + a_{2n}x_n + b_{21}u_1 + \cdots + b_{2m}u_m \\
&\;\;\vdots \\
\dot{x}_n &= a_{n1}x_1 + a_{n2}x_2 + \cdots \\
&\quad + a_{nn}x_n + b_{n1}u_1 + \cdots + b_{nn}u_m
\end{aligned} \tag{20}$$

$\dot{x} = dx/dt$. Thus this set of simultaneous differential equations may be written in matrix form. The column matrix consisting of the state variables is called the state vector and is written as Eq. (21), where the

$$\mathbf{x} = \begin{bmatrix} x_1 \\ x_2 \\ \cdot \\ \cdot \\ \cdot \\ x_n \end{bmatrix} \tag{21}$$

boldface indicates a matrix. The matrix of input signals is defined as $\mathbf{u}$. Then the system may be repre-

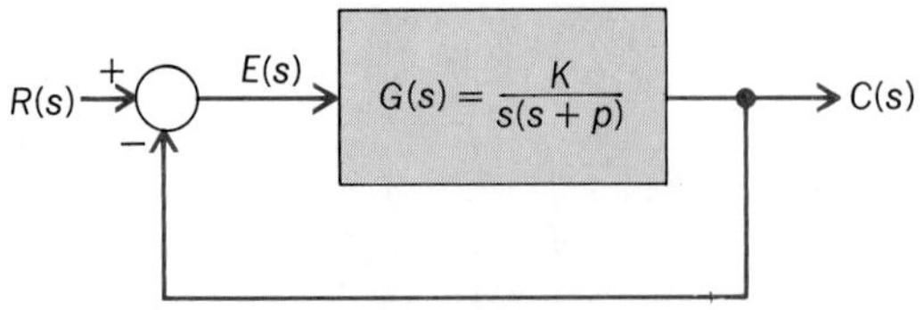

Fig. 8. Closed-loop second-order control system.

sented by the compact notation of the system vector differential equation as Eq. (22). The matrix $\mathbf{A}$ is an

$$\dot{\mathbf{x}} = \mathbf{Ax} + \mathbf{Bu} \tag{22}$$

$n \times n$ square matrix, and $\mathbf{B}$ is an $n \times m$ matrix. The vector matrix differential equation relates the rate of change of the state of the system to the state of the system and the input signals. In general, the outputs of a linear system may be related to the state variables and the input signals by the vector matrix equation (23), where $\mathbf{c}$ is the set of output signals expressed

$$\mathbf{c} = \mathbf{Dx} + \mathbf{Hu} \tag{23}$$

in column vector form. *See* MATRIX CALCULUS; MATRIX THEORY.

The solution of the matrix differential equation uses the matrix exponential given by Eq. (24). The matrix

$$\boldsymbol{\phi}(t) = e^{\mathbf{A}t} \tag{24}$$

exponential function describes the unforced response of the system and is called the fundamental or transition matrix $\boldsymbol{\phi}(t)$. Therefore, the solution may be written as Eq. (25).

$$\mathbf{x}(t) = \boldsymbol{\phi}(t)\mathbf{x}(0) + \int_0^t \boldsymbol{\phi}(t - \tau)\mathbf{Bu}(\tau)\,d\tau \tag{25}$$

The solution to the unforced system (that is, when $u = 0$) is given simply by Eq. (26). Hence, in order

$$\begin{bmatrix} x_1(t) \\ x_2(t) \\ \cdot \\ \cdot \\ \cdot \\ x_n(t) \end{bmatrix} = \begin{bmatrix} \phi_{11}(t) \cdots \phi_{1n}(t) \\ \phi_{21}(t) \cdots \phi_{2n}(t) \\ \cdot \qquad\quad \cdot \\ \cdot \qquad\quad \cdot \\ \cdot \qquad\quad \cdot \\ \phi_{n1}(t) \cdots \phi_{nn}(t) \end{bmatrix} \begin{bmatrix} x_1(0) \\ x_2(0) \\ \cdot \\ \cdot \\ \cdot \\ x_n(0) \end{bmatrix} \tag{26}$$

to determine the transition matrix, all initial conditions are set to zero except for one state variable, and the output of each state variable is evaluated. That is, the term $\phi_{ij}(t)$ is the response of the ith state variable due to an initial condition on the jth state variable when there are zero initial conditions on all the other states. The time-domain approach may be used to investigate the performance of a control system as well as to design improved systems.

Time response. The time response of a second-order feedback control system (**Fig. 8**) is illustrative of the normal range of response of a control system. The transfer function of the process of such a system has the general form $G(s) = K/s(s + p)$, where K is the gain of the process, and the transfer function contains a pole at the origin of the s-plane and one pole at $s = -p$. The response to a unit step input will be determined. The closed-loop output is given by Eq. (27). Utilizing a generalized notation, Eq. (27) may

$$C(s) = \frac{G(s)}{1 + G(s)} R(s)$$

$$= \frac{K}{s^2 + ps + K} R(s) \tag{27}$$

be rewritten as Eq. (28), where ω_n is the natural fre-

$$C(s) = \frac{\omega_n^2}{s^2 + 2\zeta\omega_n^2 + \omega_n^2} R(s) \tag{28}$$

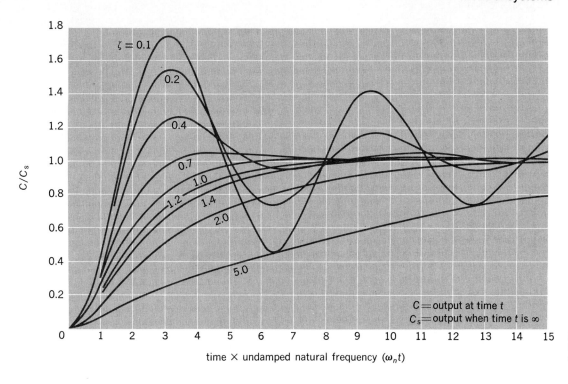

C = output at time t
C_s = output when time t is ∞

Fig. 9. Response of a second-order system to a step input.

quency and ζ is called the damping ratio. With a unit step input, this becomes Eq. (29), for which the tran-

$$C(s) = \frac{\omega_n^2}{s(s^2 + 2\zeta\omega_n s + \omega_n^2)} \qquad (29)$$

sient output is given by Eq. (30), where $\beta =$

$$c(t) = 1 - \frac{1}{\beta}e^{-\zeta\omega_n t} \sin(\omega_n t + \theta) \qquad (30)$$

$\sqrt{1 - \zeta^2}$ and $\phi = \tan^{-1} \beta/\zeta$. The transient response of this second-order system for various values of the damping ratio ζ is shown in **Fig. 9**. As ζ decreases, the overshoot of the response increases. It is usually desired to have a rapid response to a step input while minimizing the overshoot.

Standard performance measures are usually defined in terms of the step response of a system (**Fig. 10**). The swiftness of the response is measured by the rise time T_r and the peak time T_p. For underdamped systems with an overshoot, the 0–100% rise time is a useful index.

The similarity with which the actual response matches the step input is measured by the percent overshoot and settling time T_s. The percent overshoot, P.O., is defined by Eq. (31) for a unit step input,

$$\text{P.O.} = \frac{M_{p_t} - 1}{1} \times 100\% \qquad (31)$$

where M_{p_t} is the peak value of the time response. The settling time T_s is defined as the time required for the system to settle within a certain percentage δ of the input amplitude. This band of $\pm\delta$ is shown in Fig. 10. For the second-order system with a closed-loop damping constant ω_n, the response remains within 2% after four time constants τ, Eq. (32). Therefore, the

$$T_s = 4\tau = \frac{4}{\zeta\omega_n} \qquad (32)$$

settling time is defined as four time constants of the dominant response. Finally, the steady-state error of the system may be measured on the step response of the system as shown in Fig. 10. SEE TIME CONSTANT.

Stability of closed-loop systems. The transient response of a feedback control system is of primary interest and must be investigated. A very important characteristic of the transient performance of a system is the stability of the system. A stable system is defined as a system with a bounded system response. That is, if the system is subjected to a bounded input or disturbance and the response is bounded in magnitude, the system is said to be stable.

The concept of stability can be illustrated by con-

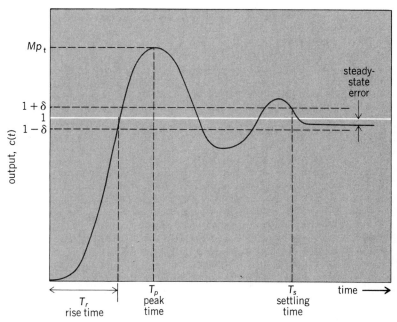

Fig. 10. Step response of a control system.

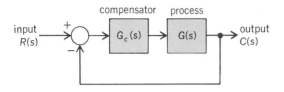

compensator   process

Fig. 11. Cascade compensator in a feedback system.

sidering a right circular cone placed on a plane horizontal surface. If the cone is resting on its base and is tipped slightly, it returns to its original equilibrium position. This position and response is said to be stable. If the cone rests on its side and is displaced slightly, it rolls with no tendency to leave the position on its side. This position is designated as neutral stability. On the other hand, if the cone is placed on its tip and released, it falls onto its side. This position is said to be unstable.

The stability of a dynamic system is defined in a similar manner. The response to a displacement, or initial condition, will result in either a decreasing, neutral, or increasing response. Specifically, it follows from the definition of stability that a linear system is stable if and only if the absolute value of its response to an impulse input $g(t)$, integrated over an infinite range, is finite.

There are three general approaches to determining the stability of a system: (1) the s-plane approach, (2) the frequency-plane ($j\omega$) approach, and (3) the time-domain approach. Only the first will be considered here.

The stability requirement of a linear system may be defined in terms of the location of the poles in the closed-loop system transfer function, which may be written as Eq. (33), where $q(s) = \Delta(s)$ is the char-

$$T(s) = \frac{p(s)}{q(s)} =$$

$$\frac{K\prod_{i=1}^{M}(s + z_i)}{S^N \prod_{k=1}^{Q}(s + \sigma_k)\prod_{m=1}^{R}[s^2 + 2\alpha_m s + (\alpha_m^2 + \omega_m^2)]}$$

(33)

acteristic polynomial whose roots are the poles of the closed-loop system. The output response for an impulse function input is then given by Eq. (34), when

$$c(t) = \sum_{k=1}^{Q} A_k e^{-\sigma_k t}$$
$$+ \sum_{m=1}^{R} B_m \left(\frac{1}{\omega_m}\right) e^{-\alpha_m t} \sin \omega_m^t \quad (34)$$

$N = 0$. It follows from Eq. (34) that the poles in the left-hand portion of the s plane result in a decreasing response for disturbance inputs. Similarly, poles on the imaginary axis (the $j\omega$ axis) and in the right-hand plane result in a neutral and an increasing response, respectively, for a disturbance input. Clearly, in order

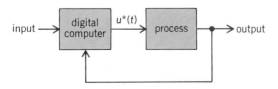

input → digital computer → $u^*(t)$ → process → output

Fig. 12. Digital computer control system.

to obtain a bounded response, the poles of the closed-loop system must be in the left-hand portion of the s plane. Thus, a necessary and sufficient condition that a feedback system be stable is that all the poles of the system transfer function have negative real parts.

Therefore, one can ascertain the stability of a feedback control system by determining the roots of the characteristic polynomial $q(s)$. A method for determining whether any of the roots of $q(s)$ lie in the right half of the s plane, and thereby ascertaining the stability of the system, was developed independently by A. Hurwitz and E. J. Routh, in the late 1800s. *SEE CONTROL SYSTEM STABILITY*.

Design and compensation. The performance of a feedback control system is of primary importance. A suitable control system is stable and results in an acceptable response to input commands, is less sensitive to the system parameter changes, results in a minimum steady-state error for input commands, and finally, is able to eliminate the effect of undesirable disturbances. A feedback control system that provides an optimum performance without any necessary adjustments is rare indeed. Usually one finds it necessary to compromise among the many conflicting and demanding specifications and to adjust the system parameters to provide a suitable and acceptable performance when it is not possible to obtain all the desired optimum specifications.

As discussed above, it is often possible to adjust the system parameters in order to provide the desired system response. However, it is often not possible to simply adjust a system parameter and thus obtain the desired performance. Rather, the scheme or plan of the system must be reexamined, and a new design or plan must be obtained which results in a suitable system. Thus, the design of a control system is concerned with the arrangement, or the plan, of the system structure and the selection of suitable components and parameters. For example, if one desires a set of performance measures to be less than some specified values, one often encounters a conflicting set of requirements. If these two performance requirements cannot be relaxed, the system must be altered in some way. The alteration or adjustment of a control system, in order to make up for deficiencies and inadequacies and provide a suitable performance, is called compensation.

In redesigning a control system in order to alter the system response, an additional component or device is inserted within the structure of the feedback system to equalize or compensate for the performance deficiency. The compensating device may be an electric, mechanical, hydraulic, pneumatic, or other type of device or network, and is often called a compensator. An electric circuit serves as a compensator in many control systems. The transfer function of the compensator is designated as $G_c(s) = E_{out}(s)/E_{in}(s)$, and the compensator may be placed in a suitable location within the structure of the system. The system when the compensator is placed in the forward path is illustrated in **Fig. 11**.

The performance of a system may be specified by requiring a certain peak time T_p, maximum overshoot, and settling time for a step input. Furthermore, it is usually necessary to specify the maximum allowable steady-state error for several test signal inputs and disturbance inputs. These performance specifications may be defined in terms of the desirable location of the poles and zeros of the closed-loop system transfer function, $T(s)$. Thus, the performance of a control system may be described in terms of frequency-domain performance measures.

The time-domain method, expressed in terms of state variables, may also be utilized to design a suitable compensation scheme for a control system. Typically, one is interested in controlling the system with a control signal $u(t)$, which is a function of several measurable state variables. Then one develops a state-variable controller which operates on the information available in measured form. This type of system compensation is quite useful for system optimization.

Quite often, in practice, the best and simplest way to improve the performance of a control system is to alter, if possible, the process itself. That is, if the system designer is able to specify and alter the design of the process which is represented by the transfer function $G(s)$, then the performance of the system may be readily improved. For example, in order to improve the transient behavior of a servomechanism position controller, one can often choose a better motor for the system. In the case of an airplane control system, one might be able to alter the aerodynamic design of the airplane and thus improve the flight transient characteristics. However, often the process is fixed and unalterable or has been altered as much as is possible and is still found to result in an unsatisfactory performance. Then the addition of compensation networks becomes useful for improving the performance of the system. In the following discussion, it will be assumed that the process has been improved as much as possible and that the $G(s)$ representing the process is unalterable.

Digital computer control systems. The use of a digital computer as a compensator device grew after 1970 as the price and reliability of digital computers improved dramatically. A block diagram of a single-loop digital control system is shown in **Fig. 12**. The digital computer in this system configuration receives the input and output and performs calculations in order to provide an output $u^*(t)$. It may be programmed to provide an output $u^*(t)$, so that the performance of the process is near or equal to that desired. Many computers are able to receive and manipulate several inputs, so a digital computer control system can often be a multivariable system. SEE DIGITAL COMPUTER; DIGITAL CONTROL.

Within a computer control system, the digital computer receives and operates on signals in digital (numerical) form, as contrasted to continuous signals. The measurement data are converted from analog form to digital form by means of a converter. After the digital computer has processed the inputs, it provides an output in digital form. This output is then converted to analog form through the use of a digital-to-analog converter. SEE ANALOG-TO-DIGITAL CONVERTER; DIGITAL-TO-ANALOG CONVERTER.

Automatic handling equipment for home, school, and industry is particularly useful for hazardous, repetitious, dull, or simple tasks. Machines that automatically load and unload, cut, weld, or cast are used by industry in order to obtain accuracy, safety, economy, and productivity.

Robots are programmable computers integrated with machines. They often substitute for human labor in specific repeated tasks. Some devices even have anthropomorphic mechanisms, including what might be recognized as mechanical arms, wrists, and hands. Robots may be used extensively in space exploration and assembly as well as for various industrial applications. SEE ROBOTICS.

Richard C. Dorf

Modeling and simulation. The design of feedback control systems generally assumes that a mathematical model of the system to be controlled is available. The model may have been obtained experimentally by applying test input signals and measuring the response, or analytically by utilizing the laws of physics to derive the equations representing the system. The model would generally be expressed either as a differential equation relating the system's input and output signals or as a transfer function in the complex Laplace variable $s = \sigma + j\omega$ for the special case where the system is linear. The values of the coefficients in the differential equation or transfer function would be functions of the physical quantities in the system. The form of the compensator used in the feedback control system to make it respond satisfactorily to inputs, as well as to possess sufficient stability, would depend on the model for the system being controlled.

At several stages in the control system design process, it is desirable to use a simulation of the control system. Simulation refers in general to the process of numerically representing the equations that describe the system on a digital computer. This allows the designer to analyze the preformance of the control system relative to the specifications, and indicates how the design should be modified in order to make its performance more acceptable. Analog computers, although still available for simulation, are now less used than digital computers. Simulation on a computer (analog or digital) permits the system to be analyzed without expending the time and the cost of constructing a physical model, or of testing the system itself. The risk of damage to the equipment by an incorrectly designed compensator is also reduced considerably.

System modeling. When models for physical systems are derived analytically, the equations are developed by applying the appropriate conservation laws. For an electrical system, Kirchhoff's current and voltage laws may be used to model the relationships between variables. For mechanical systems, Newton's second law of motion or Lagrange's equations may be used. An example is the spring-mass damper system of Fig. 7, and the variables and parameters of this simple example. The sum of the applied and reactive forces from the spring, mass, and damper is equal to zero. The reactive forces are given in Eqs. (35)–(37),

$$F_k = -K_y \qquad \text{for spring} \qquad (35)$$

$$F_m = M_{\ddot{y}} \qquad \text{for mass} \qquad (36)$$

$$F_f = -f_{\dot{y}} \qquad \text{for damper (wall friction)} \quad (37)$$

where the dots over the variables represent differentiation with respect to time. The model for this mechanical system, expressed as a second-order differential equation, becomes, through use of Newton's law, Eq. (16), where $u(t)$ is the total applied force.

Equation (16) is often not convenient in form for control system design. The model of this linear system may be expressed either in state-variable form or as a transfer function. In state-variable form, an nth-order differential equation is expressed as n first-order differential equations. For the model of Equation (16), the state variables $x_1 = y$ and $x_2 = \dot{y}$ and input variable u can be defined. The system model's state equations then become Eq. (38). The output equation for this model is Eq. (39).

$$\begin{bmatrix} \dot{x}_1 \\ \dot{x}_2 \end{bmatrix} = \begin{bmatrix} 0 & 1 \\ (K/M) & -(f/M) \end{bmatrix} \begin{bmatrix} x_1 \\ x_2 \end{bmatrix}$$
$$+ \begin{bmatrix} 0 \\ (1/M) \end{bmatrix} u \qquad (38)$$

$$y = \begin{bmatrix} 1 & 0 \end{bmatrix} \begin{bmatrix} x_1 \\ x_2 \end{bmatrix} + [0]u \qquad (39)$$

In order to develop the transfer-function model for this system, the Laplace transform is taken of Eq. (16) under the assumption that all initial conditions are zero. This yields Eq. (40). The transfer function

$$(Ms^2 + fs + K)Y(s) = U(s) \qquad (40)$$

is defined as the ratio of the Laplace transform of the output to the Laplace transform of the input with zero initial conditions. Equation (37) thus provides the transfer function, Eq. (41).

$$\frac{Y(s)}{U(s)} = \frac{(1/M)}{s^2 + (f/M)s + (K/M)} \qquad (41)$$

Comparison of Eqs. (16), (38), and (41) shows that the coefficients in the various model forms are related to the component parameters in the physical system. For simple systems, such as that in Fig. 7, the state equation or transfer-function models can be developed by inspecting the original differential equation.

In order to model a system experimentally, certain input test signals are chosen and applied to the system, and the output is measured. Perhaps the most common set of input signals for this purpose is composed of sinusoids of various frequencies. Each input signal is of the form given in Eq. (42) for some value

$$u(t) = \sin(\omega t) \qquad (42)$$

of angular frequency ω. The input signals are applied singly to the system until a steady-state condition is reached at the output, which will be of the form given in Eq. (43). Therefore, the ratio of the magnitudes of

$$y(t) = A(\omega) \sin[\omega t + \theta(\omega)] \qquad (43)$$

output to input signal is $A(\omega)$, and the phase shift of the output relative to the input is $\theta(\omega)$. The plots of $A(\omega)$ and $\theta(\omega)$ versus ω are called Bode plots. Generally, $20 \log_{10}[A(\omega)]$ is plotted rather than $A(\omega)$, thereby expressing the magnitude ratio in decibels (dB). From

known rules of constructing Bode plots, the transfer function for the system can be obtained from the experimental results. To obtain the best plots for developing the transfer function, the frequencies chosen for the input signals should be evenly spaced on a logarithmic scale. **Figure 13** shows the magnitude plot for a particular system. The circles are the values of magnitude ratio that would be determined experimentally if the corresponding sinusoidal frequencies were input to the system. The curve represents the actual magnitude. Equation (44) gives the transfer function for this system.

$$G(s) = \frac{500(s+1)(s+5)}{(s+0.2)(s^2+20s+2500)}$$
$$= \frac{5(1+s)(1+s/5)}{(1+s/0.2)[1+s/125+(s/50)^2]} \qquad (44)$$

The second form of $G(s)$ shows how the transfer function can be obtained from the plot. The dc (zero-frequency) gain in the plot is 14 dB = $20 \log_{10}(5)$. The plot starts to curve down near a pole of the transfer function and curves up near a zero. Underdamped complex poles cause an overshoot in the curve, with a sharper peak for less damping. The quadratic factor in the denominator of Eq. (44) corresponds to an undamped natural frequency $\omega_n = 50$ rad/s and a damping ratio $\zeta = 0.2$. The slopes of the Bode magnitude curve change by integral multiples of ± 20 dB per decade as poles or zeros are encountered.

Care must be taken in determining a system model experimentally. The amplitude of the input signal must be so chosen that the system response remains linear throughout the experiment. Feedback loops must not be broken if the open-loop system has unstable poles.

System simulation. Simulation is used to evaluate the performance of both open- and closed-loop systems. With simulation on a digital computer, either the continuous-time equations forming the mathematical model may be converted into a discrete-time state equation or difference equation, or they may be solved directly by numerical integration. In any case, a sampling period T must be chosen small enough for the important dynamics of the continuous-time system to be reproduced. In converting to a discrete-time system, a rule of thumb is to choose T so that the highest-frequency components of the original system are sampled five to ten times per period. In using numerical integration, T must be so chosen that the products of T with each of the poles of the system all lie inside the region of absolute stability for the particular integration operator used.

If the system model is given in the state-variable form shown in Eq. (45), then the discrete-time model

$$\dot{x}(t) = Ax(t) + Bu(t)$$
$$y(t) = Cx(t) + Du(t) \qquad (45)$$

which can be used for simulation is given by Eqs. (46) and (47). Here k takes on values $[0, 1, 2, \ldots]$

$$x[(k+1)T] = \Phi x(kT) + \Psi u(kT) \qquad (46)$$

$$y(kT) = Cx(kT) + Du(kT) \qquad (47)$$

and the matrices in Eq. (46) are defined by Eqs. (48)

$$\Phi = e^{AT} \qquad \psi = \int_0^T e^{A\tau}B \, d\tau \qquad (48)$$

assuming that $u(t)$ is constant in the interval from kT

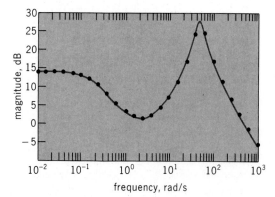

Fig. 13. Frequency domain magnitude for an experimentally determined model.

to $(k + 1)T$. Equations (46) and (47) can be solved for the output at each discrete point in time kT.

If the simulation is to be performed as an nth-order difference equation, the system model is expressed in the form of a discrete-time transfer function through the use of the Z transform, as shown in Eq. (49). In

$$\frac{Y(z)}{U(z)} = \frac{b_m z^m + b_{m-1} z^{m-1} + \cdots + b_1 z + b_0}{z^n + a_{n-1} z^{n-1} + \cdots + a_1 z + a_0} \quad (49)$$

this form, the sample period T is implicit in the values of the a_i and b_i coefficients. The solution for the output $y(kT)$ is given by Eq. (50). The output is gener-

$$\begin{aligned}
y(kT) = &-a_{n-1} y[(k - 1)T] - \cdots \\
&- a_1 y[(k - n + 1)T] - a_0 y[(k - n)T] \\
&+ b_m u[(k + m - n)T] + \cdots \\
&+ b_1 u[(k - n + 1)T] + b_0 u[(k - n]T] \quad (50)
\end{aligned}$$

ated recursively as a sum of previous values of the output and present (if $n = m$) and past values of the input.

If the simulation is to be performed by numerical integration, the continuous-time state model of Eq. (45) is used, along with the selected integration operator. Many computer installations have software which can provide highly acurate solutions to linear or nonlinear differential equations. SEE NUMERICAL ANALYSIS; SIMULATION.

Guy O. Beale

Bibliography. J. J. D'Azzo and C. H. Houpis, *Linear Control System Analysis and Design,* 3d ed., 1988; R. C. Dorf, *Modern Control Systems,* 4th ed., 1986; E. Eyman, *Modeling, Simulation, and Control,* 1988; B. C. Kuo, *Automatic Control Systems,* 5th ed., 1987; O. Mayr, *The Origins of Feedback Control,* 1970; C. L. Phillips and R. D. Harbor, *Feedback Control Systems,* 1988; J. M. Smith, *Mathematical Modeling and Digital Simulation for Engineers and Scientists,* 2d ed., 1987.

Controlled rectifier

A three-terminal semiconductor junction device with four regions of alternating conductivity type (*pnpn*), also called a thyristor. This switching device has a characteristic such that, once it conducts, the voltage in the circuit in which it conducts must drop below a threshold before the controlled rectifier regains control. Such devices are useful as high-current switches and may be used to drive electromagnets and relays.

The principle of operation can be understood by referring to the **illustration**. The central junction is reverse-biased (positive collector, grounded emitter). The wide n region between collector and base regions prevents holes injected at the collector junction from reaching the collector-to-base barrier by diffusion. The junction between emitter and base is the emitter. When operated as a normal transistor, this device shows a rapid increase of current gain of α (equal to I_c/I_e) with collector current. This effect may be due to a field-induced increase of transport efficiency across the floating n region, or to increased avalanching in the high-field barrier region, or to increased injection efficiency at the two forward-biased junctions, or to a combination of these phenomena.

With a floating-base region ($I_b = 0$), the device is a two-terminal device and collector current I_c must

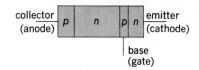

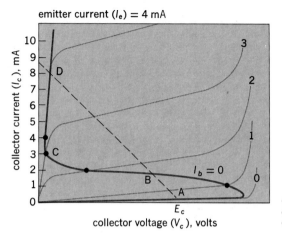

Controlled rectifier diagram and characteristics.

equal emitter current I_e.

By selecting the points on the illustrated characteristics where $I_c = I_e$, the characteristic for the base current $I_b = 0$ is shown as the heavy curve with the negative-resistance characteristic. This characteristic will be found in any transistor which shows an integrated α increasing from below unity at low collector currents to above unity at high collector currents.

If this device is operated as a three-terminal device, the switching between the nonconducting and conducting states can be controlled by the base. If in the grounded-emitter case the collector is biased to $+E_c$ as shown, the device will remain at point A until the base is pulsed positive by at least enough current to carry the emitter current to point B. At this point α exceeds unity and the device will spontaneously switch to point D in the conducting state. To reset the unit, either the emitter must be cut off or the collector voltage must be reduced so that the load line falls below the valley point C. The current of point C is called the holding current. Either of these results can be achieved by appropriate pulses on the base. Modern terminology usually refers to the rectifier terminals as anode and cathode and to the control terminal as the gate. SEE SEMICONDUCTOR RECTIFIER.

Overall current gain α may be maintained below unity for low anode-cathode currents by designing the junction between the anode and the floating n region and the junction between the gate and the cathode so each has a low-current injection efficiency below 0.5. For further discussion of four-layer devices SEE TRANSISTOR.

Lloyd P. Hunter

Bibliography. S. M. Sze, *Physics of Semiconductor Devices,* 2d ed., 1981; E. S. Yang, *Fundamentals of Semiconductor Devices,* 1978.

Conularida

A small group of extinct invertebrates showing a fourfold symmetry and a narrow pyramidal shape; the cross section commonly is square. Specimens are chitinous and frequently impregnated with calcium phos-

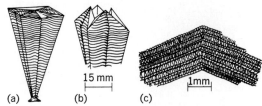

Concularida restorations. (*a*) Attachment disk. (*b*) Distal part of same individual with triangular flaps raised. (*c*) Part of exterior showing the ornamentation. (*After R. C. Moore, C. G. Lalicker, and A. G. Fischer, Invertebrate Fossils, McGraw-Hill, 1952*)

phate. The four side walls are characteristically ornamented by numerous transverse lines. The anterior part of each side wall is a triangular flap which may be bent over to cover the aperture, suggesting that the walls may have been flexible during life. The posterior tip may bear an attachment disk or may be broken off and sealed with a septum (see **illus.**). Internal structures are rare; the few known show fourfold symmetry.

Conularida are now regarded as a subclass of the Coelenterata because of the characteristic symmetry. In the past there has been uncertainty as to their affinities; they have been considered related to worms, gastropods, cephalopods, and bryozoans, among others. Several authorities still suggest that they are allied to the mollusks or to tube-building worms. Work by R. Kozlowski in 1968 suggested that Coelenterata affinities are not correct and that this minor group of fossils could represent an extinct phylum. *See* COELENTERATA.

Specimens are worldwide in occurrence and are known from rocks of Cambrian through Triassic age. They have been found in sandstone, shales, and limestones. Some may have been attached, but others are found in geologic settings suggestive of a floating mode of life. About 20 genera and 150 species are known. Except for a few local occurrences, Conularida are rare fossils.

Ellis L. Yochelson

Convection (heat)

The transfer of thermal energy by actual physical movement from one location to another of a substance in which thermal energy is stored. A familiar example is the free or forced movement of warm air throughout a room to provide heating. Technically, convection denotes the nonradiant heat exchange between a surface and a fluid flowing over it. Although heat flow by conduction also occurs in this process, the controlling feature is the energy transfer by flow of the fluid—hence the name convection. Convection is one of the three basic methods of heat transfer, the other two being conduction and radiation. *See* CONDUCTION (HEAT); HEAT RADIATION; HEAT TRANSFER.

Natural convection. This mode of energy transfer is exemplified by the cooling of a vertical surface in a large quiescent body of air of temperature t_∞. As shown in **Fig. 1a**, the lower-density air next to a hot vertical surface moves upward because of the buoyant force of the higher-density cool air farther away from the surface. At any arbitrary vertical location x, the actual velocity variation parallel to the surface will be similar to that sketched in Fig. 1b, increasing from

zero at the surface to a maximum, and then decreasing to zero as ambient surrounding conditions are reached. In contrast, the temperature of the air decreases from the heated wall value to the surrounding air temperature. These temperature and velocity distributions are clearly interrelated, and the distances from the wall through which they exist are coincident because, when the temperature approaches that of the surrounding air, the density difference causing the upward flow approaches zero.

The region in which these velocity and temperature changes occur is called the boundary layer. Because velocity and temperature gradients both approach zero at the outer edge, there will be no heat flow out of the boundary layer by conduction or convection. *See* BOUNDARY-LAYER FLOW.

An accounting of all of the energy streams entering and leaving a small volume in the boundary layer (Fig. 1c) during steady-state conditions yields Eq. (1), where κ, w, and c_p are the thermal conductivity,

$$\kappa\left(\frac{\partial^2 t}{\partial x^2} + \frac{\partial^2 t}{\partial y^2}\right) = wc_p\left(u\frac{\partial t}{\partial x} + v\frac{\partial t}{\partial y}\right) \quad (1)$$

the weight density, and specific heat at constant pressure of the air, and u and v are the velocity components in the x and y directions, respectively. Equation (1) states that the net energy conducted into the element equals the increase in energy of the fluid leaving (convected) over what it had entering.

At the surface, $u = v = 0$, and $\partial^2 t/\partial x^2 = \partial t/\partial x = 0$. Thus, $\kappa(\partial^2 t/\partial y^2) = 0$ and Eq. (2) is obtained,

$$\kappa\frac{\partial t}{\partial y} = \text{constant} = q/A \quad (2)$$

where q is the time rate of flow through an infinitesimally thin layer dx, and A is the cross-sectional area

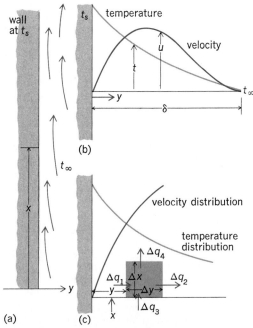

Fig. 1. Temperature and velocity distributions in air near a heated vertical surface. (*a*) Upward movement of hot air. (*b*) Distributions at arbitrary vertical location. The distance δ is that distance at which the velocity and the temperature reach ambient surrounding conditions. (*c*) Distributions in boundary layer.

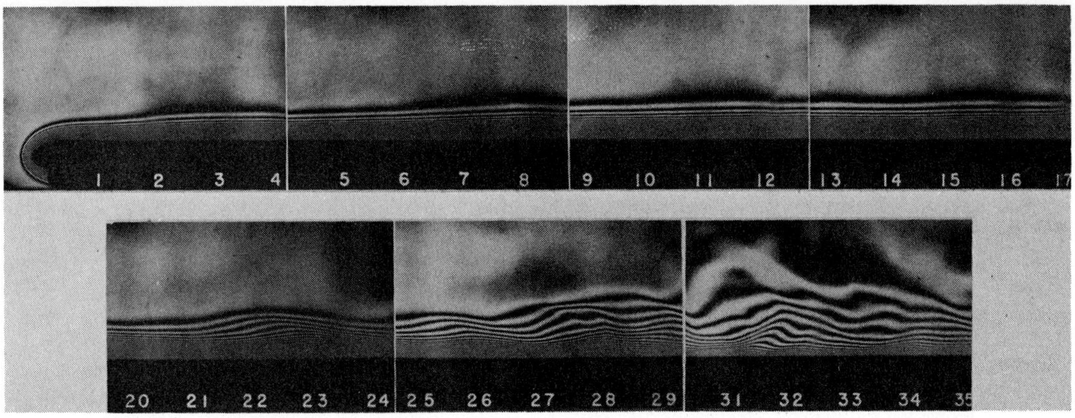

678425 26 27 28 29 31 32 33 34 35

Fig. 2. Laminar and turbulent natural convection flow along a vertical plate, as revealed by interference photographs. (*From E. R. G. Eckert and R. M. Drake, Jr., Heat and Mass Transfer, McGraw-Hill, 1959*)

normal to the direction of flow. This shows that the heat transfer in the immediate vicinity of the wall is by conduction through a thin layer of air which does not move relative to the surface. At a very small distance from the surface, the velocity becomes finite and some of the energy conducted normal to the surface is convected parallel to it. This process causes the temperature gradient to decrease, eventually to zero. The solution of a problem requires determination of the temperature distribution throughout the boundary layer. From this, the temperature gradient at the wall and the rate of heat flow can be computed.

The effect of energy leaving a surface and remaining in the boundary layer is (1) a gradual increase in temperature of the air in this layer as it moves upward, and (2) diffusion of energy farther from the surface, entraining more air in (thickening) the boundary layer. This is shown in **Fig. 2**, where the interference fringes indicate lines of constant temperature. **Figure 3** is a similar visualization of the free convection temperature field around a horizontal cylinder. The outer broad fringe indicates approximately the edge of the boundary layer which, by comparison with the cylinder (which measures 4 in. or 10 cm in diameter), is about ¾ in. (2 cm) thick.

Forced convection. The effect of blowing air across the cylinder is shown in **Fig. 4**. Here the boundary layer on the forward half of the cylinder has become so thin that it is not possible to resolve the isotherms within it. Although the natural convection forces are still present in this latter case, they are clearly negligible compared with the imposed forces. The process of energy transfer from the heated surface to the air is not, however, different from that described for natural convection. The major distinguishing feature is that the maximum fluid velocity is at the outer edge of the boundary layer, as is illustrated in **Fig. 5** for flow along a flat plate. This difference in velocity profile and the higher velocities provide more fluid near the surface to carry along the heat conducted normal to the surface. Consequently, boundary layers are very thin (greatly enlarged in Fig. 5 for clarity).

The properties of a fluid which influence its heat-convecting ability are the dynamic viscosity μ, heat capacity at constant pressure c_p, and thermal conductivity κ. These combine in a single significant property in the form $\mu c_p/\kappa = \mu/(\kappa/c_p)$, which is the ratio of the fluid viscosity to the quotient of its heat-conducting and -storage capacities. With proper units, this ratio is dimensionless and is called the Prandtl

number, N_{Pr}. Fluids such as liquid metals, having low values of N_{Pr} (**Fig. 6**), are particularly effective for convective heat-transfer applications. *SEE FLUIDS.*

Heat-transfer coefficient. The convective heat-transfer coefficient h is a unit conductance used for calculation of convection heat transfer. It was introduced by Isaac Newton and, until the mechanism of convection was properly interpreted, was thought to be a characteristic of the fluid flowing. To describe quantitatively the cooling of objects in air, Newton

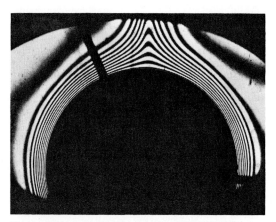

Fig. 3. Interference photograph of isotherms around heated horizontal cylinder. (*From D. L. Doughty and W. H. Giedt, Proc. Instrum. Soc. Amer., 7:115–119, 1952*)

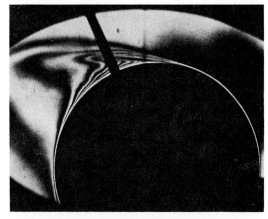

Fig. 4. Interference photograph showing isotherms around a heated cylinder normal to a 33 ft/s (10 m/s) airstream. (*From D. L. Doughty and W. H. Giedt, Proc. Instrum. Soc. Amer., 7:115–119, 1952*)

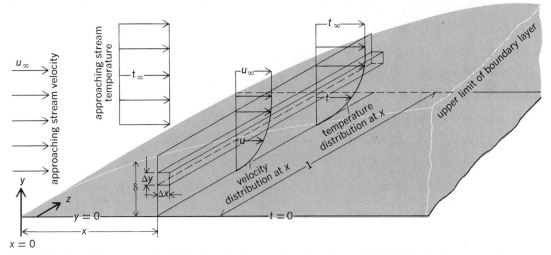

Fig. 5. Velocity and temperature distributions in a laminar boundary layer along a flat plate. The fluid is assumed to have a Prandtl number of 1 for which the velocity and temperature boundary layers will be of equal thicknesses. (*After W. H. Giedt, Principles of Engineering Heat Transfer, Van Nostrand, 1957*)

suggested Eq. (3). This equation, known as Newton's

$$q = hA(t_{surface} - t_{fluid}) \qquad (3)$$

law of cooling, is really a definition of h. This coefficient is determined by the slope of the fluid temperature distribution right at the surface, and the thermal conductivity κ of the fluid. Engineering units are Btu/(h)(ft^2)(°F).

Local and average coefficients. As the fluid in the heated or cooled boundary layer moves along an isothermal surface, it gradually approaches the temperature of the surface. This causes the temperature gradient in the fluid at the surface (and the rate of heat transfer) to decrease in the direction of flow. Taking, for example, an airstream at 80°F (27°C) moving at 50 ft/s (15 m/s) over a flat plate at 30°F (−1°C), the local heat-transfer coefficient h_x decreases in 1 ft (0.3 m) to about one-fifth of its leading-edge value. For practical calculations, an average heat-transfer coefficient h is more useful. This is obtained by integrating h_x over the heat-transfer surface and dividing by the surface area. For this system, h is a function of the Prandtl number N_{Pr}, κ, ν, and u_∞ of the fluid. For comparing geometrically similar systems involving different fluids, dimensional analysis shows that the specific properties can be combined into dimensionless parameters, conveniently reducing the number of independent variables. For the flat plate, these are the Nusselt number, $N_{Nu} = hl/\kappa$, the Reynolds number, $N_{Re} = lu_\infty/\nu$, and the Prandtl number. For example, it can be shown that in this case Eq. (4) holds. In the

$$N_{Nu} = 0.664 N_{Pr}^{1/3} \sqrt{N_{Re}} \qquad (4)$$

case of free convection, the Grashof number, $N_{Gr} = \beta g l^3 \, \Delta t/\nu^2$ (where β is the coefficient of thermal expansion), replaces the Reynolds number. Other dimensionless numbers pertinent to convection include the Stanton number, $N_{St} = h/\rho u c_p = N_{Nu}/N_{Re}N_{Pr}$, which is related to the skin-friction coefficient. *SEE DIMENSIONAL ANALYSIS; REYNOLDS NUMBER; SKIN FRICTION.*

Turbulent flow. Heat convection in turbulent flow is interpreted similarly to that in laminar flow, which has been implied in previous paragraphs. Rates of heat transfer are higher for comparable velocities, however, because the fluctuating velocity components of the fluid in a turbulent flow stream provide a macroscopic exchange mechanism which greatly increases the transport of energy normal to the main flow direction. Because of the complexity of this type of flow, most of the information regarding heat transfer has been obtained experimentally. Such results, combined with dimensional analysis, have yielded useful design equations, typical of which is Eq. (5), for predicting the heat transfer in a pipe, where d is the pipe diameter.

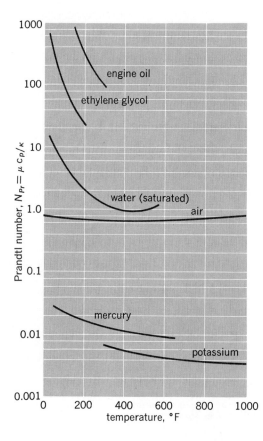

Fig. 6. Variation of Prandtl number with temperature. °C = (°F − 32)/1.8.

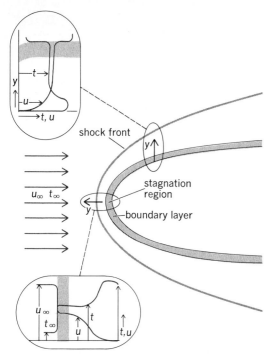

Fig. 7. Velocity and temperature distributions in the gas around a high-speed missile (thicknesses of shock wave and boundary layer exaggerated).

$$\frac{hd}{\kappa} = 0.023 \left(\frac{du}{\nu}\right)^{0.8} \left(\frac{\mu c_p}{\kappa}\right)^{0.4} \qquad (5)$$

SEE LAMINAR FLOW; TURBULENT FLOW.

Aerodynamic heating. Convection heat transfer which occurs during high-speed flight or high-velocity flow over a surface is known as aerodynamic heating. This heating effect results from the conversion of the kinetic energy of the fluid as it approaches a body to internal energy as it is slowed down next to the surface. In the case of a gas (**Fig. 7**), its temperature increases first, because of compression as it passes through a shock and approaches the stagnation region, and second, because of frictional dissipation of kinetic energy in the boundary layer along the surface. Typical velocity and temperature distributions near the surface of a missile are shown in Fig. 7. The maximum temperature possible would result from an adiabatic compression of the gas as it is slowed from the free stream velocity to zero. This maximum temperature depends on the square of the missile velocity, being, for example, around 1500°F (815°C) for a speed of 4000 ft/s or 1200 m/s (a Mach number of 4). Temperatures very close to the maximum may occur in the stagnation region; downstream maximum temperatures are lower because of decreasing pressure and heat convection toward the outer edge of the boundary layer.

Condensation and boiling. Condensation and boiling are important phase-change processes involving heat release or absorption. Because vapor and liquid movement are present, the energy transfer is basically by convection. Local and average heat-transfer coefficients are determined and used in the Newton cooling-law equation for calculating heat rates which include the effects of the latent heat of vaporization.

Condensation. Consider a saturated or superheated vapor of a single substance in some region. When it comes in contact with a surface maintained at a temperature lower than the saturation temperature, heat flow results from the vapor, releasing its latent heat and condensing on the surface. The condensation process may proceed in two more or less distinct ways. If there are no impurities in the vapor or on the surface (which need not be smooth), the condensate will form a continuous liquid film. If, however, such contaminants as fatty acids or mercaptans are present, the vapor will condense in small droplets. These increase in size until their weight causes them to run down the surface. In doing so, they sweep the surface free for formation of new droplets. For the same temperature difference between the vapor and surface, heat transfer with dropwise condensation may be 15–20 times greater than filmwise condensation. The dropwise type is therefore very desirable, but conditions under which it will occur are not predictable, and designs are limited to systems for which experimental results are available.

Boiling. In boiling, results indicate the existence of several regimes, as shown in **Fig. 8**. The important independent variable is the temperature difference between the hot surface and the fluid relatively far from the surface. For values of Δt up to approximately 10°F (A to B on the curve), the liquid is being superheated by natural convection, and q/A is proportional to $\Delta t^{5/4}$. With further increase in Δt (B to C), bubbles form at active nuclei on the heated surface. These bubbles break away and rise through the pool, their stirring action causing the heat transfer to be much above that due to natural convection. This phenomenon is called nucleate boiling, and q/A varies as Δt^3 to Δt^4. When the rate of bubble formation becomes so rapid that the bubbles cannot get away before they tend to merge, a vapor film begins to form, through which heat must flow by conduction. The rate of heating then decreases with Δt until complete film boiling is reached and heat flows by radiation and conduction through the film. Point F corresponds to the melting point of the wire.

The high heat rates which occur during boiling make it a very effective means of absorbing the energy capable of being released in furnaces and nuclear reactors. This is also one reason why the vapor power-generating cycles have been successful.

Mass and momentum transfer. In a gas, heat conduction occurs by transfer of kinetic energy from high-temperature to lower-temperature molecules. This process requires a change in location of molecules, called mass transfer. If, in addition to their ran-

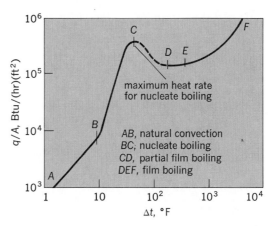

Fig. 8. Heat rate versus temperature difference during boiling of water at 212°F (100°C) on an electrically heated platinum wire Δt, °C = ⅝(Δt, °F).

Transfer equations

Phenomenon	Mass (diffusion)	Transfer of	
		Momentum (introducing shear stress)	Heat
By conduction	$\frac{W}{A} = D_{AB}\frac{\partial C}{\partial y}$	$\sigma = \frac{W}{A}\Delta u = \mu\frac{\partial u}{\partial y}$	$\frac{q}{A} = -\kappa\frac{\partial t}{\partial y}$
By convection	$\frac{W}{A} = (\rho v)\,C$	$\sigma = (\rho v)u$	$\frac{q}{A} = (\rho v)c_p t$

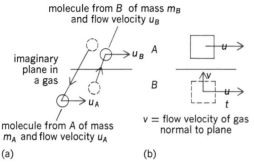

Fig. 9. Mass, momentum, and heat transfer. (a) By conduction: mass transfer = $m_B - m_A$; momentum transfer = $m_B u_B - m_A u_A$; heat transfer = $m_B c_B t_B - m_A c_A t_B$. (b) By convection: mass transfer (per unit time and area) = ρv; momentum transfer = $(\rho v)u$; heat transfer = $(\rho v)c_p t$.

dom velocities, the molecules have definite but different flow velocities, the molecular (mass) interchange will also result in a momentum transfer. This means that mass, momentum, and heat exchange are interrelated. This interrelationship exists in both conduction and convection phenomena (**Fig. 9**).

These three processes are described by similar equations, which are listed in the **table**. As mass conduction, W/A denotes the mass transfer per unit time and area of component A through B. The coefficient D_{AB} is a diffusion coefficient characterizing the diffusion of molecules of gas A through gas B, and C denotes the concentration of A (pounds per cubic foot) which is the driving potential for the process. The equation describing momentum transfer is basically an expression of Newton's second law of motion. The three coefficients, D_{AB}, μ, and κ, in these transport processes, are referred to as the transport properties. *See* Diffusion.

In the case of convection, the same mass flux ρv carries with it the concentration of C of a given species, the velocity u, and the enthalpy $c_p t$.

The similarity between the fundamental equations suggests that solutions for one process may be applicable to the others. The validity of this has been established for such cases as evaporation and condensation from a liquid surface into a gas, such as air, above it.

The applicability of these analogous solutions is, however, limited to conditions where one process can be regarded as independent of the others. When exchange rates are high and the properties of the species involved vary with temperature and pressure, it may be necessary to solve the differential equations governing each transport process simultaneously.

Warren H. Giedt

Bibliography. V. S. Arpaci and P. S. Larsen, *Convection Heater Transfer*, 1984; J. P. Holman, *Heat Transfer*, 5th ed., 1981; Y. Jaluria, *Natural Convection Heat and Mass Transfer*, 1980; W. M. Kays and M. Crawford, *Convective Heat and Mass Transfer*, 2d ed., 1980; W. M. Rohsenow and J. P. Hartnett (eds.), *Handbook of Heat Transfer*, 1973.

Convective instability

A state of fluid flow in which the distribution of body forces along the direction of the net body force is unstable and will thus break down. Fluid flows are subject to a variety of instabilities, which may be broadly viewed as the means by which relatively simple flows become more complex. Instabilities are an important step in the transition between smooth and turbulent flow, and in the atmosphere they are responsible for phenomena ranging from thunderstorms to low- and high-pressure systems. Meteorologists and oceanographers divide instabilities into two broads classes: convective and dynamic. *See* Dynamic instability.

In the broadest terms, convective instabilities arise when the displacement of a small parcel of fluid causes a force on that parcel which is in the same direction as the displacement. The parcel of fluid will then continue to accelerate away from its initial position, and the fluid is said to be unstable. In most geophysical flows, the convective motions that result from convective instabilities operate very quickly compared with the processes acting to destabilize the fluid; the result is that such fluids seem to be nearly neutrally stable to convection.

Types. The simplest type of convective instability arises when a fluid is heated from below or cooled from above. This heating and cooling result in a temperature distribution as shown in **Fig. 1**; here a small

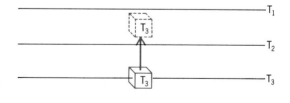

Fig. 1. Diagram showing temperature decreasing upward in a fluid and convective instability. When a small parcel of fluid is displaced upward from its initial position (solid-line box) to a new position (dashed-line box), it will be warmer than its environment and will continue to accelerate upward. T denotes temperature, with T_3 the highest and T_1 the lowest.

chunk of fluid at temperature T_3 is shown displaced upward. If it conserves its original temperature, it will be warmer than its environment at the same altitude and will accelerate upward, since warm air is less dense than cold air. The parcel is said to be positively buoyant. Warm air rises and cold air sinks; thus a fluid whose temperature decreases with altitude is convectively unstable, while one in which the temperature increases with height is convectively stable. A fluid at constant temperature is said to be convectively neutral. *See Archimedes' principle*.

The above description assumes that density depends on temperature alone and that density is conserved in parcels of fluid, so that when the parcels are displaced their density does not change. However, in the Earth's atmosphere, neither of these assumptions is true. In the first place, the density of air depends on pressure and on the amount of water vapor in the air, in addition to temperature. Second, the density will change when the parcel is displaced because both its pressure and its temperature will change. Because of these conditions it is convenient to define a quantity known as virtual potential temperature (θ_v) that both is conserved and reflects the actual density of air. This quantity is given by Eq. (1), where T is the

$$\theta_v \equiv T\left(\frac{1 + (r/0.622)}{1 + r}\right)\left(\frac{1000}{p}\right)^{0.287} \quad (1)$$

temperature in kelvins, p is the pressure in millibars, and r is the number of grams of water vapor in each gram of dry air. When θ_v decreases with height in the atmosphere, it is convectively unstable, while θ_v increasing with height denotes stability.

Over the tropical oceans, the θ_v of air adjacent to the sea surface is continually increased by addition of both heat and water vapor from the ocean, while at higher levels θ_v is decreased by cooling due to radiation to space. The result is that a layer roughly 1600 ft (500 m) deep is continually convecting. The convection is so efficient that this layer of the atmosphere is kept very close to a state of convective neutrality. Measurements show that θ_v is constant with height in this layer to within measurement error. A similar situation prevails over land during the day, when the land is heated by the Sun, but at night the land cools and the air becomes convectively stable (θ_v increases with height).

In the oceans, density is a function of pressure, temperature, and salinity; convection there is driven by cooling of the ocean surface by evaporation of water into the atmosphere and by direct loss of heat when the air is colder than the water. It is also driven by salinity changes resulting from precipitation and evaporation. In many regions of the ocean, a convectively driven layer exists near the surface in analogy with the atmospheric convective layer. This oceanic mixed layer is also nearly neutral to convection.

Convective instabilities are also responsible for convection in the Earth's mantle, which among other things drives the motion of the plates, and for many of the motions of gases within other planets and in stars.

Convective motions. Experiments have been performed to determine what kind of motions result from convective instability. Typically these experiments use a fluid contained between two parallel, horizontal plates maintained at constant temperature, with the lower plate warmer than the upper plate. If the temper-

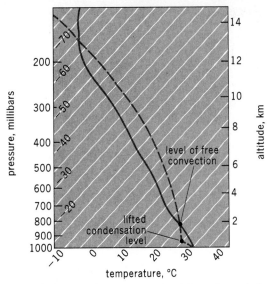

Fig. 2. Thermodynamic diagram with height on the vertical coordinate (with pressure decreasing upward), while temperature increases downward and to the right. The solid line shows an example of temperature measured at various levels in the atmosphere, typically by an ascending instrumented balloon. The broken line shows the temperature that a parcel would have if lifted from near the surface. At the lifted condensation level, water vapor begins to condense, and the parcel cools less rapidly above this point. At the level of free convection, it becomes warmer than its environment and would ascend freely above this point. 1 millibar = 10^2 pascals. 1 km = 0.6 mi. °F = (°C × 1.8) + 32.

ature difference is small, the viscosity of the fluid prevents convection from occurring. As the temperature difference is increased beyond a critical value, convection begins as steady overturning cells whose width is roughly equal to the distance between the plates. When the temperature difference is increased further, the cells begin to oscillate and eventually the flow becomes chaotic. Variations on this experiment include one in which the flux of heat, rather than the temperature, is fixed at the parallel plates. In this case, the cells tend to be much broader than they are deep.

Virtually all convection in nature is well into the chaotic regime determined by experiments, and yet natural convection is often organized in distinctive patterns. In the atmosphere, these include long rows of clouds known as cloud streets and regular hexagonal arrays of clouds whose width can be more than 50 times their depth. The reason for this is as yet unknown. Little is known about the organization of convecting fluid in the Earth's mantle, which determines the motion of the plates that make up the Earth's crust. Convection on the Sun is also known to be organized in clumps called granules; these are in turn organized in larger clumps, or supergranules.

Moist convection. In the atmosphere, convection is complicated by the phase changes of water substance. When water vapor condenses into the liquid waterdroplets that make up clouds, for example, the heat that was used to evaporate the water in the first place is released, making the air warmer than it would otherwise be. Conversely, when rain evaporates it takes up heat and makes the air cooler. In addition to these effects, suspended condensate (that is, cloud, rain, snow, hail, and so forth) adds directly to the effective density of the air. All of these effects make moist convection, which produces cumulus clouds and thunderstorms and is considerably more exotic than dry convection.

The criterion for convective instability of moist atmospheres has been the subject of much research. A basic tool for examining the stability of moist atmospheres is the thermodynamic diagram, an example of which is presented in **Fig. 2**. This diagram has pressure on the vertical coordinate, and temperature increases downward and to the right. (Since pressure decreases with height in the atmosphere, the vertical coordinate also represents height.) The temperature of air measured from an ascending balloon has been plotted as the solid line on this diagram. The thermodynamic diagram may contain various sets of background curves that permit a determination of the temperature of a parcel of air lifted from any point on the sounding. For example, the broken line shows the temperature a parcel of air would have if it were lifted from near the surface. At first, its temperature decreases rapidly with height, since the first law of thermodynamics states that temperature must fall with pressure. Then, when it reaches the lifted condensation level, water vapor begins to condense into cloud, releasing heat, and the temperature does not fall as fast with height.

In the example shown in Fig. 2, the lifted parcel has about the same temperature as the surrounding atmosphere up to its lifted condensation level, reflecting the neutral stability of the layer of air next to the surface. But if the parcel is displaced beyond its level of free convection, it will be warmer than its environment and will accelerate upward. This type of atmosphere is said to be conditionally unstable; that is, it is stable unless near-surface parcels are forcibly dis-

placed beyond their level of free convection. Such an atmosphere will not spontaneously convect, and thus the amount of conditional instability can build up to large values until something triggers convection. The result is severe thunderstorms and sometimes tornadoes. This situation is common in the plains of North America in spring. *See* THUNDERSTORM; TORNADO.

It was formerly believed that the tropical atmosphere is conditionally unstable, but convective stability that is assessed by using thermodynamic diagrams like the one in Fig. 2 omits an important contribution to density: the weight of suspended condensate such as rain, snow, and cloud. When this is included, the tropical atmosphere is nearly neutrally stable to parcels of air displaced upward from near the surface. It is maintained in such a state by the nearly continuous activity of moist convection, which takes the form of cumulus clouds and thunderstorms. In this sense, the whole depth of the tropical troposphere (a layer of air extending up to about 7 mi or 12 km altitude in the tropics) exists in a convectively adjusted state in analogy with the layer near the surface, which has constant θ_v. *See* TROPICAL METEOROLGY.

Convective instability may be driven by centrifugal as well as by gravitational forces. This is illustrated in **Fig. 3,** which shows the distribution of angular momentum per unit mass (M) in a rotating cylinder of fluid. This angular momentum is defined by expression (2), where r is the radius from the center and V

$$M \equiv rV \qquad (2)$$

is the velocity of the fluid rotating about the center. If a ring of fluid centered about the axis of the cylinder is displaced radially inward or outward, it conserves its angular momentum. For example, as the ring of fluid in Fig. 3 is displaced outward, its angular momentum is less than that of the surrounding fluid, and the laws of physics can be used to demonstrate that since the centrifugal force on that fluid is less than on the surrounding fluid, it will accelerate inward. The fluid is then said to be centrifugally stable. A corollary to Archimedes' law states that a fluid with low M tends to sink toward the axis of rotation, while fluid with high M rises away from the axis. A fluid whose angular momentum increases outward is centrifugally stable, while a fluid with M decreasing outward is unstable. The nature of this instability is much like that of thermal convection, and this, too, is a type of convective instability. *See* ANGULAR MOMENTUM.

In the atmosphere, it is possible to have a combination of centrifugal and buoyant convection. It turns out that the criterion for this to happen is that parcels displaced upward along angular momentum surfaces must become positively buoyant. These surfaces are usually sloped in the atmosphere because of the increase with height of atmospheric winds. Convection that results from this combined instability (which is also known as symmetric instability) is also generally sloped and is thus called slantwise convection. This form of convection results in bands of rain and snow embedded in winter storms; these bands may be hundreds of miles wide and as much as 600 mi (1000 km) long. The concept of convective neutrality applies here as well, as measurements made by flying instrumented aircraft along angular momentum surfaces show that parcels displaced along them are almost neutrally buoyant. *See* ATMOSPHERE; ATMOSPHERIC

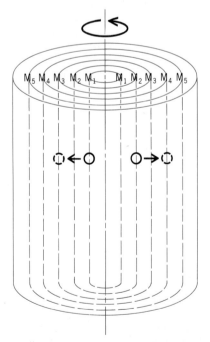

Fig. 3. Diagram of a cylinder of rotating fluid with an angular momentum per unit mass, *M*, which in this case increases outward from the axis (*M₁*, *M₂*, *M₃*, and so forth denote increasing values of *M*). This arrangement is centrifugally stable, since a ring of fluid (solid-line circle) displaced outward to a new position (broken-line circle) will have less centrifugal force acting on it than on its environment; it will therefore accelerate inward toward its initial position.

GENERAL CIRCULATION; FLUID FLOW; PRECIPITATION (METEOROLOGY); STORM.

Kerry A. Emanuel

Bibliography. A. K. Betts, Saturation point analysis of moist convective overturning, *J. Atmos. Sci.*, 39:1484–1505, 1982; S. Chandrasekhar, *Hydrodynamic and Hydromagnetic Stability*, 1961; K. A. Emanuel, On assessing local conditional symmetric instability from atmospheric soundings, *Mon. Weath. Rev.*, 111:2016–2033, 1983; D. K. Lilly, Severe storms and storm systems: Scientific background, methods and critical questions, *Pure Appl. Geophys.*, 113:713–734, 1975; J. S. Turner, *Buoyancy Effects in Fluids*, 1973.

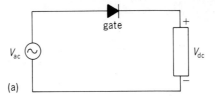

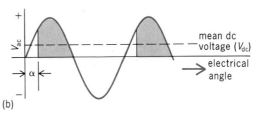

Fig. 1. Thyristor half-wave rectifier. (a) Circuit. (b) Voltage waveforms. Shaded areas indicate nonsmooth dc output waveform.

Converter

A device for processing alternating-current (ac) or direct-current (dc) power to provide a different electrical waveform. The term converter denotes a mechanism for either processing ac power into dc power (rectifier) or deriving power with an ac waveform from dc (inverter). Some converters serve both functions, others only one. SEE RECTIFIER.

Historically, converters were needed to accommodate various needs to match characteristics of the supply voltage. For example, dc motors, which in principle could be connected directly to early dc distribution systems, required a rectifier for operation with ac supplies. SEE DIRECT-CURRENT MOTOR.

While ac has been universally adopted for power distribution and consequently is the standard power supply for industrial, commercial, and domestic purposes, the range of demands for conversion far exceeds the occasional rectifier load. Converters are used for such applications as (1) rectification from ac to supply electrochemical processes with large controlled levels of direct current; (2) rectification of ac to dc followed by inversion to a controlled frequency of ac to supply variable-speed ac motors; (3) interfacing dc power sources (such as fuel cells and photoelectric devices) to ac distribution systems; (4) production of dc from ac power for subway and streetcar systems, and for controlled dc voltage for speed-control of dc motors in numerous industrial applications; and (5) transmission of dc electric power between rectifier stations and inverter stations within ac generation and transmission networks. SEE DIRECT-CURRENT TRANSMISSION; MOTOR SPEED CONTROL.

These are a few examples of the need for converters to be versatile, controllable, and able to handle a wide range of power levels. Furthermore, emphasis on long-term cost of power losses has drawn attention to the efficiency of converters and the ability to contribute to the overall efficiency of their loads. For example, it can be economically advantageous to recapture the kinetic energy of some mechanical loads, which are subject to frequent stops and starts, through a converter power reversal back to the supply (regeneration) rather than losing it as heat in friction braking. This increases operating efficiency in such applications as electric vehicles and steel rolling mills.

Converter development has been furthered by advances in power semiconductors and microprocessors for their control. Beyond the basic converter action, there may be control of voltage, current, power, reactive power, and frequency at the converter output terminals. SEE MICROPROCESSOR.

Types. Until the advent of power semiconductors, converter action was achieved either by rotary converters or mercury-arc valves. In a rotary converter, an electric motor drives a generator. For example, a dc motor drives an ac generator to provide ac voltage at a chosen frequency. The mercury-arc valve is a unidirectional switching device which relies on ionization between a pool of mercury and an anode. Later developments permitted the timing of the start of conduction to be controlled by a grid between the mercury and the anode. Some of the modern semiconductor circuits have evolved from mercury-arc technology.

The introduction of the thyristor (silicon-controlled rectifier) in the 1960s had an immediate effect on converter applications because of its ruggedness, reliability, and compactness. As a controlled, unidirectional, solid-state switch, it replaced mercury-arc devices with ratings in the kilowatt range and, through series–parallel combinations, at high power levels. Power semiconductor devices for converter circuits include (1) thyristors, controlled unidirectional switches that, once conducting, have no capability to suppress current; (2) triacs, thyristor devices with bidirectional control of conduction; (3) gate turn-off devices with the properties of thyristors and the further capability of suppressing current; and (4) power transistors, high-power transistors operating in the switching mode, somewhat similar in properties to gate turn-off devices.

Thyristors are available with ratings from a few watts up to the capability of withstanding several kilovolts and conducting several kiloamperes. All the devices incorporate one or more gates, that is, low-power input connections to permit point-on-wave control of turn-on relative to the waveform of applied voltage. SEE SEMICONDUCTOR RECTIFIER.

Circuits. Of the many different converter circuits, three types are described.

Half-wave rectifier. The thyristor in the controlled half-wave rectifier (**Fig. 1***a*) conducts on parts of the positive half-cycles of the ac supply to provide a mean dc voltage across the load. Conduction is delayed by the variable control angle α resulting in the nonsmooth dc waveform shown (Fig. 1*b*). In practice, the dc component in the supply current tends to cause saturation problems in supply transformers.

Full-wave circuit. A full-wave circuit (**Fig. 2***a*) has a bidirectional current in the ac supply transformer. Thyristors 1 and 2 are gated together and then, in

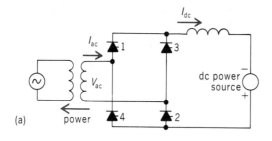

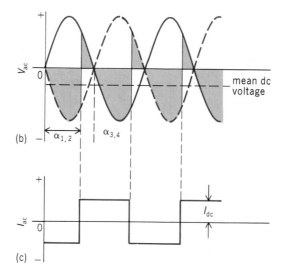

Fig. 2. Inverter operation by thyristor full-wave bridge. (a) Circuit. (b) Voltage waveforms. Shaded areas indicate nonsmooth dc waveform with negative mean value. (c) Current waveform.

The above examples are called bridge circuits. Inversion into an ac system that does not contain an ac voltage source requires a circuit that either develops internal reverse voltages to turn off thyristors or uses gate turn-off devices. Other types include the cyclo-converter which synthesizes a lower frequency from an ac waveform, and the pulse-width-modulated converter which processes a waveform into a series of segments that are recombined into the desired waveform. SEE PULSE MODULATION.

Three-phase bridge. Three-phase ac/dc converters are often used in industrial applications. Very high-power versions are used in power transmission systems.

Converters at the terminal stations of the dc transmission system either rectify three-phase ac voltages to dc or invert the dc voltage to feed power into the adjacent ac system. A three-phase converter bridge is made up of six thyristor valves conducting in sequence as numbered 1, 2, . . . , 6 in **Fig. 3a**. Bridges are usually connected in series pairs, as shown. The 30° ac phase difference between the converter-side voltages of the wye/wye and the wye/delta converter transformers permits a conduction sequence 1, 1′, 2, 2′, . . ., 6, 6′. This is called 12-pulse operation and gives a smoother dc waveform and less ac harmonic currents.

Each valve is made up of many thyristors in series together with additional components to aid in the voltage sharing between individual thyristors (Fig. 3b). Valves are often combined into single units, such as the quadrivalve in Fig. 3a.

Converters range from low-power applications, such as the power supply in a television, up to thousands of megawatts in dc power transmission. SEE ALTERNATING CURRENT; DIRECT CURRENT.

John Reeve

Bibliography. C. W. Lander, *Power Electronics*, 2d ed., 1987; M. A. Rashid, *Power Electronics*, 1988.

turn, dc-side current is transferred to thyristors 3 and 4 to provide the negative half-cycle of ac current. The inductor provides some smoothing of the dc waveform. By delaying gating by more than 90 electrical degrees while providing a dc power source (Fig. 2b), the sign of the mean dc voltage is reversed so that the circuit now inverts dc power into the ac supply. The ac system must provide an existing ac voltage waveform sufficient to naturally allow the thyristors to turn on and off in the conducting sequence. The ac current waveform is not sinusoidal (Fig. 2c); ac filters may be added to reduce the harmonic content. SEE COMMUTATION; ELECTRIC FILTER.

Convertiplane

An aircraft combining vertical takeoff and landing capabilities as in the helicopter with forward-flight effectiveness and high-speed potentials of the airplane. In forward flight a convertiplane relies, at least partially, on the fixed wing, while for vertical takeoffs, landings, and hovering, a separate vertical thrust generator is provided. Between vertical and forward-flight regimes, the aircraft goes through a conversion. SEE AIRPLANE; HELICOPTER; VERTICAL TAKEOFF AND LANDING (VTOL).

Of many systems suitable as vertical-thrust generators, those applied to practical convertiplanes will probably be either of the direct type, in which combustion products are used directly for thrust generation (turbojets and rockets), or of the indirect type, in which all or a part of the available combustion energy is used to drive a mechanical system, functioning as an actuator disk to accelerate the ambient air. The actuator disks may be either free (helicopter rotors and propellers) or shrouded (shrouded propellers and ducted fans).

The effectiveness of static thrust generators can be compared on the basis of energy expenditure per unit of time and per unit of thrust as power/thrust. The ideal power required per pound of thrust is proportional to the fully developed downwash velocity of

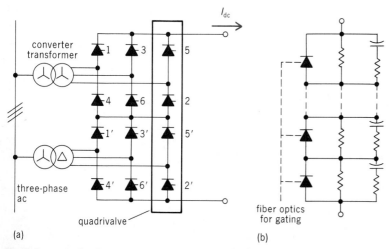

Fig. 3. Converter for dc power transmission. (a) Circuit of 12-pulse converter. (b) Thyristor valve.

the slipstream affected by the thrust generator. Actual shaft horsepower of an engine driving an actuator disk or the rate of release of thermal energy in a jet and a rocket will in turn be proportional to the ideal values.

Loading of rotors and propellers (thrust per square foot of the disk), as well as that of ducted fans, jets, and rockets (thrust per square foot of the jet exit area), is proportional to the square of the fully developed downwash velocity. Hence, lightly loaded static thrust generators will show a low ideal horsepower per pound of thrust and should be used, therefore, whenever a long time in hovering, or near hovering, is required. Such use will result in economy of power installed and fuel consumed in the aircraft.

Effectiveness of a lift generator in forward flight can be compared on the basis of the ideal energy expenditure (in Btu or joules) required per unit of lift (as 1 lbf or 1 newton) and a unit of distance flown (as 1 nautical mile or 1 km) at a given speed.

In that respect, the wing is by far the most efficient means of providing lift in horizontal flight. At slow speed the rotor is next, but it still requires at least about three times more energy per unit of lift and distance flown. At speeds exceeding 150–160 knots (170–180 mi/h or 77–82 m/s), the combined effect of the stall of the retreating blade and compressibility make the rotor even less effective.

In the convertiplane the fixed wing, optimized for forward flight, is combined with a vertical-thrust generator whose selection is governed by the desire of obtaining a well-balanced design. This usually means that the installed power, or thrust, established from hovering at a given altitude and ambient temperature, should differ as little as possible from that resulting from the requirements of the forward flight speed at a given altitude. In principle, the higher the speed requirements of the convertiplane, the more highly loaded the vertical-thrust generator that can be used. However, such operational requirements as limitations on downwash velocity in hovering may force a

Fig. 1. NASA-Army-Bell XV-15 in airplane configuration after conversion from helicopter mode. (*Bell Helicopter Textron*)

Fig. 2. Bell-Boeing V-22 Osprey tilt-rotor aircraft. (*Bell Helicopter Textron*)

Fig. 3. Harrier fighter aircraft, manufactured by Hawker Siddeley, now British Aerospace, and McDonnell Douglas. Close ground support and reconnaissance are provided through use of vectored-thrust turbofans.

deviation from this principle toward vertical-thrust generators more lightly loaded than the design optimum. For any one vehicle, the final design is the one that achieves optimum performance at design center and acceptable performance off center.

The principle of the convertiplane can be applied to many types of aircraft. The first practical attempts were directed toward rotary-wing aircraft. In the so-called compounds as much lift as possible is transferred in forward flight from the rotor (which is either put into autorotation, or slowed down) to more efficient fixed wings, while a propeller (or propellers) provides the forward thrust.

To achieve high speeds the helicopter-type rotor must be eliminated from the forward-flight configuration, which should be as close as possible to that of conventional aircraft. Configurations have been studied wherein rotors are stopped, folded, and retracted into the fuselage or into nacelles.

In the nonhelicopter-type convertiplane, the same device can be used both as a vertical-thrust generator and as a means of forward propulsion. Two basic solutions are technically feasible: (1) the wing remains fixed, while either the entire thrust generator or only the thrust vector tilts from the vertical to the horizontal position, and (2) both the wing and the thrust generator tilt as a unit.

In the fixed-wing, free-airscrew group, the NASA-Army-Bell XV-15, a flight research aircraft with 375 mi/h (167 m/s) speed capabilities, went through complete conversion in 1979 (**Fig. 1**). In the ensuing years of extensive flight testing, the XV-15 proved the feasibility of the tilt-rotor concept; thus paving the way for the development of an operational aircraft, the Bell-Boeing V-22 Osprey (**Fig. 2**), for the U.S. military forces, especially the Marine corps. Additional civil applications of the tilt-rotor are anticipated for a short-haul commercial transport, and studies have been undertaken for aircraft having up to 75-passenger-seat capacity.

The Harrier (manufactured by Hawker Siddeley, now British Aerospace, and McDonnell Douglas), with Pegasus vectored-thrust turbofan engines (manufactured by British Siddeley, now Rolls Royce), is a fighter that is produced for the military forces (including the U.S. Marines) of several Western countries and was an important factor for the British in the 1982 Falkland Islands conflict (**Fig. 3**). The Yakovlev, a Soviet single-seat, carrier-based combat aircraft, is similar in concept. Many studies in the United States and abroad are aimed toward development of more advanced fixed-wing and high-bypass–jet-based aircraft, especially for naval operations.

The LTV-Hiller-Ryan SC-142 VTOL is an example of the application of the movable-wing system to larger transports (gross weight in the 40,000-lb or 18,000-kg class), and the Canadair CL-84 to the utility-type transport.

Wieslaw Z. Stepniewski

Bibliography. J. P. Campbell, *Vertical Take-Off and Landing Aircraft*, 1962; B. W. McCormick, Jr., *Aerodynamics of V/STOL Flight*, 1967; Society of Automotice Engineers, *Proceedings of the Powered Lift Conference*, 1987; W. Z. Stepniewski and C. N. Keys, *Rotary-Wing Aerodynamics*, 1984; J. W. R. Taylor (ed.), *Jane's All the World's Aircraft*, revised periodically; E. Torrenbeek, *Synthesis of Subsonic Airplane Design*, 1982.

Conveyor

A horizontal, inclined, declined, or vertical machine for moving or transporting bulk materials, packages, or objects in a path predetermined by the design of the device and having points of loading and discharge fixed or selective. Included in this category are skip hoist and vertical reciprocating and inclined reciprocating conveyors; but in the strictest sense this category does not include those devices known as industrial trucks, tractors and trailers, cranes, hoists, monorail cranes, power and hand shovels or scoops, bucket drag lines, platform elevators, or highway or rail vehicles. The more usual basic types of conveyors and their normal, rather than exceptional, operating characteristics are shown in the **table**. *SEE BULK-HANDLING MACHINES.*

Gravity conveyors. The economical means for lowering articles and materials is by gravity conveyors. Chutes depend upon sliding friction to control the rate of descent; wheel and roller conveyors use rolling friction for this purpose (**Fig. 1**).

With a body resting on a declined plane, friction F opposes component P parallel to the surface of the plane of weight W of the body. When the angle of elevation equals the angle of repose, the body is just about to start sliding, or to express it another way, when the inclination of the plane equals the angle of repose, F equals P. In this position the tangent of the angle of repose equals the coefficient of friction. For metal on metal the coefficient of friction is 0.15 to 0.25, and for wood on metal it is 0.2 to 0.6, which is why steel tote boxes slide more readily on metal chutes than do wooden cases.

When the body is a smooth-surfaced container and the inclined plane is replaced by rollers, the rollers are spaced so that at least three rollers support the smallest container to be conveyed; then each roller supports one-third the weight of the load. Component P of the weight W which acts downward and parallel to the inclination of the plane produces a turning moment Pr about the roller's axle, where r is the radius of the roller. As the inclination of the conveyor is increased, P increases until, at the point where rolling is about to start, Pr equals Fr. In this case, however, the force of friction is made up of the friction between the contacting surfaces and the friction in the bearings of the rollers. The mass of a tubular roller is concentrated near its circumference; hence rollers have greater starting inertia than wheels, but after they have started rolling, they have a flywheel effect that tends to keep products in motion, especially when packages follow each other in close succession. Inclination (or declination) is expressed as the number of inches of rise or fall per foot of conveyor or per 5-ft (1.5-m) or 10-ft (3-m) length.

Gravity chutes. Gravity chutes may be made straight or curved and are fabricated from sheet metal or wood, the latter being sometimes covered with canvas to prevent slivering. The bed of the chute can be shaped to accommodate the products to be handled. In spiral chutes centrifugal force is the second controlling factor. When bodies on a spiral move too fast, they are thrown out toward a guard rail, and contact with the rail increases the friction thereby causing the bodies to slow down somewhat and settle back into the center of the runway and continue their descent at a controlled rate. Spirals with roller beds or wheels provide smooth descent of an article and tend to maintain the position of the article in its original starting position. Rollers may be of metal, wood, or plastic and can be arranged in an optimum position, depending upon the articles to be carried.

Changes in direction can be effected by turntables, ball transfer tables, or gradual S curves (**Fig. 2**). Generally speaking, wheel-type conveyors are less expensive than the roller varieties; however, the latter can withstand severer service. For successful use of wheeled conveyors, the conveyed article must have a smooth and firm riding surface.

Chain conveyors. There are three basic types of chain conveyors: (1) those that support the product being conveyed, (2) those that carry actuating elements between the two chains, and (3) those that operate overhead or are set into the floor (**Fig. 3**).

In the sliding chain conveyor either plain links or links with such special attachments as lugs are made up into the conveyors to handle cases, cans, pipes, and other similar products.

Slats or apron conveyors are fitted with slats of wood or metal, either flat or in special shapes, between two power chains. The slats handle such freight as barrels, drums, and crates. With the addition of cleats these conveyors carry articles up steep inclines. In this type of conveyor the article rides on the top strand of the slats.

Normal operating characteristics of typical conveyors

Conveyor Type	Paths						Typical products											Bulk materials							
	horizontal	declined	inclined	vertical	straight	curved	cases, boxes, etc.	barrels and kegs	drums	textile bags	paper bags	bottles and jars	cans	small parts	food products	lumber, pipe, etc.	towing trucks	free flowing	sluggish	dry	wet	cold	hot	nonabrasive	abrasive
GRAVITY CONVEYORS																									
Sliding friction																									
Skid		X			X		X	X																	
Chute		X		X	X	X				X	X				X			X		X		X	X	X	
Spiral chute		X			X	X				X	X				X										
Rolling friction																									
Wheel	X	X			X	X	X					X			X										
Roller	X	X			X	X	X	X	X						X										
Spiral wheel		X			X	X																			
Spiral roller		X			X	X	X	X																	
POWERED CONVEYORS																									
Continuous belt																									
Fabric	X	X	X		X		X	X	X	X	X	X	X	X	X	X		X	X	X	X	X	X	X	X
Flexible tube with zipper	X	X	X	X	X	X	X											X		X		X	X	X	X
Steel band	X	X	X		X		X	X	X	X	X	X	X	X	X	X		X	X	X	X	X	X	X	X
Woven or flat wire	X	X	X		X							X	X	X	X										
Linked rod	X	X	X		X	X						X	X	X	X										
Live roller (drive)																									
Flat belt	X	X	X		X	X	X	X	X	X	X	X	X	X	X	X									
V belt	X	X	X		X	X	X	X	X	X	X	X	X	X	X	X									
Sprocket	X	X	X		X	X	X	X	X	X	X	X	X	X	X	X									
Chain																									
Apron (slat)	X	X	X		X		X	X	X	X	X														
Free roller	X	X	X		X		X	X	X																
Pan	X	X	X		X							X	X	X											
Pusher bar	X	X	X		X		X																		
Hinged plate	X	X	X									X	X	X											
Flat top plate	X				X	X						X	X												
Bucket*	X	X	X	X	X													X		X	X	X	X	X	X
Flight	X	X	X	X	X													X		X		X	X	X	
Drag chain																									
Plain links with lugs	X	X	X		X	X	X		X						X		X								
Overhead trolley†	X	X	X		X	X											X								
Infloor trolley	X				X	X											X								
Cable																									
Overhead trolley†	X	X	X		X	X																			
Table-mounted trolley	X	X	X																						
Cableway‡	X	X	X		X	X																			
Spiral (screw)	X	X	X	X	X	X												X	X	X	X	X	X	X	X
Vertical																									
Rigid arm				X	X		X	X	X	X	X														
Pendant carriage				X	X		X	X	X	X	X														
Pneumatic																									
Tube (with carriers)	X	X	X	X	X	X									X										
Pressure or suction	X	X	X	X	X	X												X		X		X	X	X	X
Vibrating	X	X	X	X	X													X		X		X	X	X	X

*Some bucket conveyors may use endless rubber belts.
†Most products or bulk materials can be conveyed by selection of attachments to the trolleys.
‡Cableways or aerial tramways are used primarily for cross-country conveying of coal, ashes, lumber, concrete, and similar products used in construction.

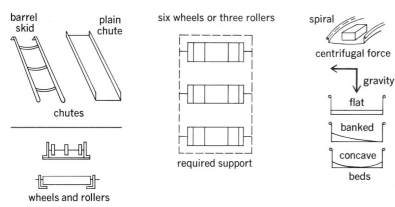

Fig. 1. Elements used in gravity conveyors.

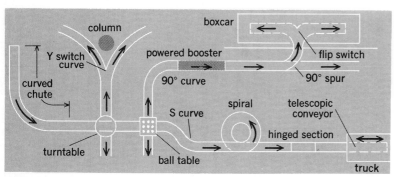

Fig. 2. Gravity components for conveyor lines.

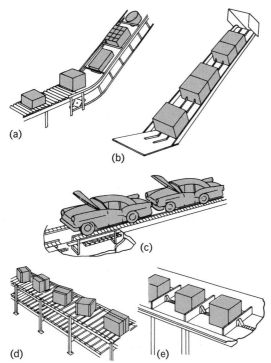

Fig. 3. Five examples of the basic types of chain conveyors. (a) Push-bar conveyor. (b) Twin chains carrying cartons. (c) Heavy-duty plates for assembly conveyor. (d) Apron or slat conveyor to carry miscellaneous articles. (e) Push-bar or cleat conveyor. (*After Conveyor Equipment Manufacturers Association, Conveyor Terms and Definitions, 3d ed., 1966*)

Push-bar conveyors are a variation of the slat conveyor, in which two endless chains are cross-connected at intervals by pusher bars which propel the load along a stationary bed or trough of the conveyor.

Powered conveyors. Gravity conveyors are limited to use in conditions where the material is to be lowered in elevation. The horizontal conveying distance capabilities of the gravity conveyor depend on the difference in elevation and the angle of sliding or rolling friction of the material to be conveyed.

To move loads on level or inclined paths, or declining paths that exceed the angle of sliding or rolling friction of the particular material to be conveyed, powered conveyors must be employed. The following are various types of powered conveyors.

Belt conveyors. Loads are moved on a level or inclined path by means of power-driven belts. Belt conveyors with rough-top belts make possible inclines up to 28°; cleated belts are limited on degree of incline only by the position of the center of gravity of the conveyed item.

Essential components of belt conveyors in addition to the belt itself are (1) a bed, which may be a combination bed and frame; (2) end rollers; (3) a take-up to adjust belt tension as it varies with age and atmospheric conditions; and (4) a power unit (**Fig. 4**). The bed of a fabric belt conveyor may be wood, or more usually flat or concave sheet metal; or it may be made up of rollers. The former are called slide beds, the latter roller beds. Slide beds provide quiet, smooth conveying, but they require more power than roller beds. The two end rollers are mounted in antifriction bearings with provision for lubrication.

Where the tail roller is mounted in adjustable end arms to provide take-up, the length of the conveyor is not constant. Where this feature is objectionable, automatic gravity, screw, or spring take-ups can be used. Another required adjustment is provided by mounting the rollers so that their bearings can be shifted slightly either forward or backward to train the belt to run true.

Belts can be powered in several ways. A gear-head motor can be connected directly to the head roller, or the drive may be through a chain and sprockets. These arrangements necessitate placing the drive outside the frame. When this is undesirable, one of two drives can be located under the frame. Power units usually provide fixed or variable speeds from a few feet per minute to over 50 ft/min (15 m/min); 50 ft/min is usual with package conveyors. Normally ¼-hp (190-W) motors power these units, but there are other, more powerful motors which are available for heavier duty.

Powered package conveyors can be combined in various ways to meet specific conditions. Powered curves are used, but wheel and roller sections are more usually employed to change direction. Inclined belt conveyors usually have a hump at the upper end to reduce the slapping by packages on the belt as they move from inclined to horizontal travel (**Fig. 5**). Cleats can be used, but they complicate the feeding of packages onto the line and require special construction features so that the cleats can pass around pulleys and be properly supported on the return portion of the conveying cycle.

Metal belts made of woven and flat wire, linked rods, and steel bands are engineered to make up conveyors, which are essentially the same as fabric belt conveyors. For example, woven wire belts can be driven by pulleys and these may be lagged, that is, covered with a material to improve traction. Contact with the drive pulley can be increased by snubbing with small-diameter idler pulleys which guide the belt to increase the angle of contact. Flat wire, linked rod, and woven wire belts with chain edges are driven positively by pulleys with sprocket teeth or by true sprockets. Steel band conveyors are driven by pulleys with unusually large diameters. Other components are similar to those used in fabric belt conveyors but are more rugged in construction to handle the heavier car-

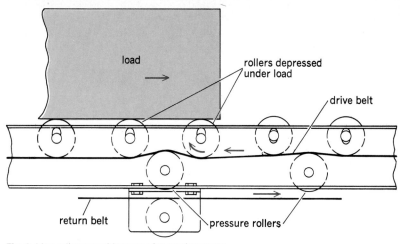

Fig. 6. Live rollers provide power for gravity curves.

riers and to meet the more exacting conditions under which metal belts operate. Metal belts are used, for example, in high-temperature operations such as baking, annealing, brazing, and heat-treating.

The choice of a belt for a given situation depends on the effect of the belt's surface on the product, amount of drainage required, and similar factors.

Live-roller conveyors. Objects are moved over series of rollers by the application of power to all or some of the rollers. The power-transmitting medium is usually belting or chain. In one arrangement a belt drive running under and in contact with the rollers turns them to propel the load forward (**Fig. 6**). In another variation small sprockets are attached to the ends of some or all of the roller shafts, the sprockets being powered by a chain drive.

Arthur M. Perrin

Vibrating conveyors. These mechanical devices are designed to move bulk materials along a horizontal, or almost horizontal, path in a controlled system (**Fig. 7**). They can be used to simply transport material from one point to another or to perform various functions en route, such as cooling, drying, blending, metering, spreading, and, by installing a screen, scalping or dedusting.

They can be fed by a belt conveyor or, more commonly, used to provide a precisely controlled outlet from a bin, hopper, or chute. In all cases they have the ability to take a vertical flow of material and change the direction of movement up to 90° with only a small change in the handling level.

A vibrating conveyor basically consists of a base or reaction mass and a driven mass connected to each other by springs. The base is excited by a power source such as an electromagnetic or electromechanical drive. A trough or pan forms part of the driven mass and serves as the material carrying surface. In operation, the trough is moved alternately forward and up, then down and back. This causes the conveyed material to move along the trough surface in a series of short hops.

Amplitude or frequency of vibration can be adjusted to control conveyor speed and output. Electromagnetic conveyors operating at 3600 vibrations per minute with a low amplitude [a displacement of 0.09 in. (0.2 cm), for example], provide a gentle motion to the conveyed product and are frequently used to handle fragile or friable materials. Electromechanical

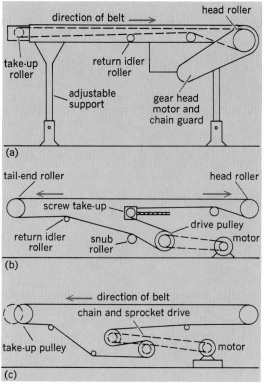

Fig. 4. Diagrams of three driven-belt conveyors. (*a*) Components of a fabric belt conveyor. (*b*) Wrap drive. (*c*) Tandem drive.

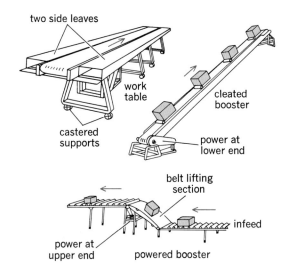

Fig. 5. Fabric belt conveyors.

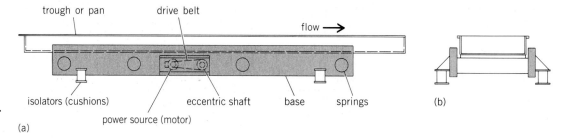

trough or pan   drive belt   flow ⟶

isolators (cushions)   eccentric shaft   base   springs   (b)

power source (motor)

Fig. 7. Vibrating conveyor. (a) Side view. (b) End view.   (a)

conveyors normally operate at a lower frequency but a much higher amplitude [a displacement of 0.5 in. (1.2 cm), for example], which results in rougher handling of the product but greatly increased capacity. The volume of material which can be handled ranges from a few pounds per hour up to thousands of pounds per hour. SEE VIBRATION MACHINE.

R. F. Merwin

Bibliography. E. A. Avallone and T. Baumeister III (eds.), *Marks' Standard Handbook for Mechanical Engineers*, 9th ed., 1987; Conveyor Equipment Manufacturers Association, *CEMA Standard No. 2: Conveyor Terms and Definitions*, 4th ed., 1982; J. Fruchtbaum, *Bulk Materials Handling Handbook*, 1986; T. Wireman, *Plant Layout and Material Handling*, 1984.

Cooling tower

A tower- or buildinglike device in which atmospheric air (the heat receiver) circulates in direct or indirect contact with warmer water (the heat source) and the water is thereby cooled. A cooling tower may serve as the heat sink in a conventional thermodynamic process, such as refrigeration or steam power generation, or it may be used in any process in which water is used as the vehicle for heat removal, and when it is convenient or desirable to make final heat rejection to atmospheric air. Water, acting as the heat-transfer fluid, gives up heat to atmospheric air, and thus cooled, is recirculated through the system, affording economical operation of the process.

Basic types. Two basic types of cooling towers are commonly used. One transfers the heat from warmer water to cooler air mainly by an evaporation heat-transfer process and is known as the evaporative or wet cooling tower. The other transfers the heat from warmer water to cooler air by a sensible heat-transfer process and is known as the nonevaporative or dry cooling tower. These two basic types are sometimes combined, with the two cooling processes generally used in parallel or separately, and are then known as wet-dry cooling towers.

Cooling process. With the evaporative process, the warmer water is brought into direct contact with the cooler air. When the air enters the cooling tower, its moisture content is generally less than saturation; it emerges at a higher temperature and with a moisture content at or approaching saturation. Evaporative cooling takes place even when the incoming air is saturated, because as the air temperature is increased in the process of absorbing sensible heat from the water, there is also an increase in its capacity for holding water, and evaporation continues. The evaporative process accounts for about 65–75% of the total heat

transferred; the remainder is transferred by the sensible heat-transfer process.

The wet-bulb temperature of the incoming air is the theoretical limit of cooling. Cooling the water to within 5 to 20°F (−15 to −6.7°C) above wet-bulb temperature represents good practice. The amount of water evaporated is relatively small. Approximately 1000 Btu (1055 kilojoules) is required to vaporize 1 lb (0.45 kg) of water at cooling tower operating temperatures. This represents a loss in water of approximately 0.75% of the water circulated for each 10°F (6°C) cooling, taking into account the normal proportions of cooling by the combined evaporative and sensible heat-transfer processes. Drift losses may be as low as 0.01–0.05% of the water flow to the tower (recent performance of 0.001% has been achieved) and must be added to the loss of water by evaporation and losses from blowdown to account for the water lost from the system. Blowdown quantity is a function of makeup water quality, but it may be determined by regulations concerning its disposal. Its quality is usually expressed in terms of the allowable concentration of dissolved solids in the circulating cooling water and may vary from two to six concentrations with respect to the dissolved solids content of the cooling water makeup.

With the nonevaporative process, the warmer water is separated from the cooler air by means of thin metal walls, usually tubes of circular cross section, but sometimes of elliptical cross section. Because of the low heat-transfer rates from a surface to air at atmospheric pressure, the air side of the tube is made with an extended surface in the form of fins of various geometries. The heat-transfer surface is usually arranged with two or more passes on the water side and a single pass, cross flow, on the air side. Sensible heat transfer through the tube walls and from the extended surface is responsible for all of the heat given up by the water and absorbed by the cooling air. The water temperature is reduced, and the air temperature increased. The nonevaporative cooling tower may also be used as an air-cooled vapor condenser and is commonly employed as such for condensing steam. The steam is condensed within the tubes at a substantially constant temperature, giving up its latent heat of vaporization to the cooling air, which in turn is increasing in temperature. The theoretical limit of cooling is the temperature of the incoming air. Good practice is to design nonevaporative cooling towers to cool the warm circulating water to within 25 to 35°F (14 to 20°C) of the entering air temperature or to condense steam at a similar temperature difference with respect to the incoming air. Makeup to the system is to compensate for leakage only, and there is no blowdown requirement or drift loss.

With the combined evaporative-nonevaporative

process, the heat-absorbing capacity of the system is divided between the two types of cooling towers, which are selected in some predetermined proportion and usually arranged so that adjustments can be made to suit operating conditions within definite limits. The two systems, evaporative and nonevaporative, are combined in a unit with the water flow arranged in a series relationship passing through the dry tower component first and the wet tower second. The airflow through the towers is in a parallel-flow relationship, with the discharge air from the two sections mixing before being expelled from the system. Since the evaporative process is employed as one portion of the cooling system, drift, makeup, and blowdown are characteristics of the combined evaporative-nonevaporative cooling tower system, generally to a lesser degree than in the conventional evaporative cooling towers.

Of the three general types of cooling towers, the evaporative tower as a heat sink has the greatest thermal efficiency but consumes the most water and has the largest visible vapor plume. When mechanical-draft cooling tower modules are arranged in a row, ground fogging can occur. This can be eliminated by using natural-draft towers, and can be significantly reduced with modularized mechanical-draft towers when they are arranged in circular fashion.

The nonevaporative cooling tower is the least efficient type, but it can operate with practically no consumption of water and can be located almost anywhere. It has no vapor plume.

The combined evaporative-nonevaporative cooling tower has a thermal efficiency somewhere between that of the evaporative and nonevaporative cooling towers. Most are of the mechanical-draft type, and the vapor plume is mitigated by mixing the dry warm air leaving the nonevaporative section of the tower with the warm saturated air leaving the evaporative section of the tower. This retards the cooling of the plume to atmospheric temperature; visible vapor is reduced and may be entirely eliminated. This tower has the advantage of flexibility in operation; it can accommodate variations in available makeup water or be adjusted to atmospheric conditions so that vapor plume formation and ground fogging can be reduced.

Evaporative cooling towers. Evaporative cooling towers are classified according to the means employed for producing air circulation through them: atmospheric, natural draft, and mechanical draft.

Atmospheric cooling. Some towers depend upon natural wind currents blowing through them in a substantially horizontal direction for their air supply. Louvers on all sides prevent water from being blown out of these atmospheric cooling towers, and allow air to enter and leave independently of wind direction. Generally, these towers are located broadside to prevailing winds for maximum sustained airflow.

Thermal performance varies greatly because it is a function of wind direction and velocity as well as wet- and dry-bulb temperatures. The normal loading of atmospheric towers is about 1–2 gal/min (3.7–7.5 liters/min) of cooling water per square foot of cross section. They require considerable unobstructed surrounding ground space in addition to their cross-sectional area to operate properly. Because they need more area per unit of cooling than other types of towers, they are usually limited to small sizes.

Natural draft. Other cooling towers depend for their air supply upon the natural convection of air flowing upward and in contact with the water to be cooled. Essentially, natural-draft cooling towers are chimney-like structures with a heat-transfer section installed in their lower portion, directly above an annular air inlet in a counterflow relationship with the cooling air (**Fig. 1**), or with the heat-transfer section circumscribing the base of the tower in a cross-flow relationship with the cooling airflow (**Fig. 2**). Sensible heat ab-

Fig. 1. Counterflow natural-draft cooling tower at Trojan Power Plant in Spokane, Washington. (*Research–Cottrell*)

Fig. 2. Cross-flow mechanical-draft cooling towers. (*Marley Co.*)

sorbed by the air in passing over the water to be cooled increases the air temperature and its vapor content and thereby reduces its density so that the air is forced upward and out of the tower by the surrounding heavier atmosphere. The flow of air through the tower varies according to the difference in specific weights of the ambient air and the air leaving the heat-transfer surfaces. Since the difference in specific weights generally increases in cold weather, the airflow through the cooling tower also increases, and the relative performance improves in reference to equivalent constant-airflow towers.

Normal loading of a natural-draft tower is 2–4 gal/(min)(ft^2) [1.4–2.7 liters/(s)(m^2)] of ground-level cross section. The natural-drafting cooling tower does not require as much unobstructed surrounding space as the atmospheric cooling tower does, and is generally suited for both medium and large installations. The natural-draft cooling tower was first commonly used in Europe. Then a number of large installations were built in the United States, with single units 385 ft (117 m) in diameter by 492 ft (150 m) high capable of absorbing the heat rejected from an 1100-MW light-water-reactor steam electric power plant.

Mechanical draft. In cooling towers that depend upon fans for their air supply, the fans may be arranged to produce either a forced or an induced draft.

Fig. 3. Mechanical-draft cooling towers. (a) Conventional rectangular cross-flow evaporative induced-draft type; (b) circular cross-flow evaporative induced-draft type. (*Marley Co.*)

Induced-draft designs are more commonly used than forced-draft designs because of lower initial cost, improved air-water contact, and less air recirculation (**Fig. 3**). With controlled airflow, the capacity of the mechanical-draft tower can be adjusted for economic operation in relation to heat load and in consideration of ambient conditions.

Normal loading of a mechanical-draft cooling tower is 2–6 gal/(min)(ft^2) [1.4–4.1 liters/(s)(m^2)] of cross section. The mechanical-draft tower requires less unobstructed surrounding space to obtain adequate air supply than the atmospheric cooling tower needs; however, it requires more surrounding space than the natural-draft towers do. This type of tower is suitable for both large and small installations.

Nonevaporative cooling towers. Nonevaporative cooling towers are classified as air-cooled condensers and as air-cooled heat exchangers, and are further classified by the means used for producing air circulation through them. *See Heat exchanger; Vapor condenser.*

Air-cooled condensers and heat exchangers. Two basic types of nonevaporative cooling towers are in general use for power plant or process cooling. One type uses an air-cooled steam surface condenser as the means for transferring the heat rejected from the cycle to atmospheric cooling air. The other uses an air-cooled heat exchanger for this purpose. Heat is transferred from the air-cooled condenser, or from the air-cooled heat exchanger to the cooling air, by convection as sensible heat.

Nonevaporative cooling towers have been used for cooling small steam electric power plants since the 1930s. They have been used for process cooling since 1940; a complete refinery was cooled by the process in 1958. The nonevaporative cooling tower was formerly used almost exclusively with large steam electric power plants in Europe. Interest in this type of tower gradually increased in the United States, and a 330-MW plant in Wyoming using nonevaporative cooling towers was completed in the late 1970s.

The primary advantage of nonevaporative cooling towers is that of flexibility of plant siting. There is seldom a direct economic advantage associated with the use of nonevaporative cooling tower systems in the normal context of power plant economics. They are the least efficient of the cooling systems used as heat sinks.

Cooling airflow. Each of the two basic nonevaporative cooling tower systems may be further classified with respect to type of cooling airflow. Both types of towers, the direct-condensing type and the heat-exchanger type, can be built as natural-draft or as mechanical-draft tower systems.

Design. The heat-transfer sections are constructed as tube bundles, with finned tubes arranged in banks two to five rows deep. The tubes are in a parallel relationship with each other and are spaced at a pitch slightly greater than the outside fin diameter, either in an in-line or a staggered pattern. For each section, two headers are used, with the tube ends secured in each. The headers may be of pipe or of a box-shaped cross section and are usually made of steel. The bundles are secured in an open metal frame.

The assembled tube bundles may be arranged in a V shape, with either horizontal or inclined tubes, or in an A shape, with the same tube arrangement. A similar arrangement may be used with vertical tube bundles. Generally, inclined tubes are shorter than horizontal ones. The inclined-tube arrangement is best suited to condensing vapor, the horizontal-tube arrangement best for heat-exchanger design. The A-shaped bundles are usually used with forced-draft airflow, the V-shaped bundles with induced-draft airflow. With natural-draft nonevaporative cooling towers there is no distinction made as to A- or V-shaped tube bundle arrangement; the bundles are arranged in a deck above the open circumference at the bottom of the tower, in a manner similar in principle to that used in the counterflow natural-draft evaporative cooling tower (Fig. 1). A natural-draft cooling tower with a vertical arrangement of tube bundles has been built, but this type of structure is not generally used.

Modularization of the bundle sections has become general practice with larger units; an installation of a circular module–type unit is shown in **Fig. 4**.

Combined evaporative-nonevaporative towers. The combination evaporative-nonevaporative cooling tower is arranged so that the water to be cooled first passes through a nonevaporative cooling section which is much the same as the tube bundle sections used with nonevaporative cooling towers of the heat-exchanger type. The hot water first passes through these heat exchangers, which are mounted directly above the evaporative cooling tower sections; the water leaving the nonevaporative section flows by force of gravity over the evaporative section.

The cooling air is divided into two parallel flow streams, one passing through the nonevaporative section, the other through the evaporative section to a common plenum chamber upstream of the induced-draft fans. There the two airflow streams are combined and discharged upward to the atmosphere by the fans. A typical cooling tower of this type is shown in **Fig. 5**.

In most applications of wet-dry cooling towers attempts are made to balance vapor plume suppression and the esthetics of a low silhouette in comparison with that of the natural-draft evaporative cooling tower. In some instances, these cooling towers are used in order to take advantage of the lower heat-sink temperature attainable with evaporative cooling when an adequate water supply is available, and to allow the plant to continue to operate, at reduced efficiency, when water for cooling is in short supply.

The more important components of evaporative cooling towers are the supporting structure, casing, cold-water basin, distribution system, drift eliminators, filling and louvers, discharge stack, and fans (mechanical-draft towers only). The counterflow type is shown in **Fig. 6**; the cross-flow type is shown in **Fig. 7**.

Treated wood, especially Douglas-fir, is the common material for atmospheric and small mechanical-draft cooling towers. It is used for structural framing, casing, louvers, and drift eliminators. Wood is commonly used as filler material for small towers. Plastics and asbestos cement are replacing wood to some degree in small towers and almost completely in large towers where fireproof materials are generally required. Framing and casings of reinforced concrete, and louvers of metal, usually aluminum, are generally used for large power plant installations. Fasteners for securing small parts are usually made of bronze, copper-nickel, stainless steel, and galvanized steel. Distribution systems may be in the form of piping, in galvanized steel, or fiber glass–reinforced plastics; and they may be equipped with spray nozzles of non-

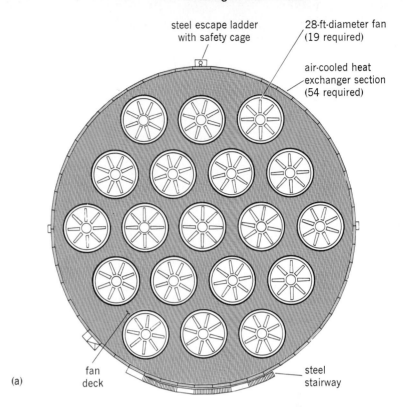

(a)

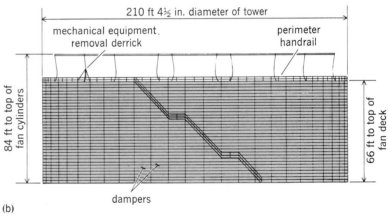

(b)

Fig. 4. Multifan circular nonevaporative tower. (*a*) **Plan.** (*b*) **Elevation. 1 ft = 0.3 m; 1 in. = 2.5 cm. (*Marley Co.*)**

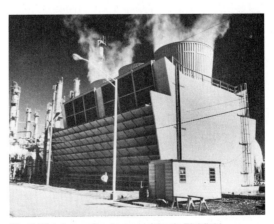

Fig. 5. Wet-dry cooling tower at Atlantic Richfield Company. (*Marley Co.*)

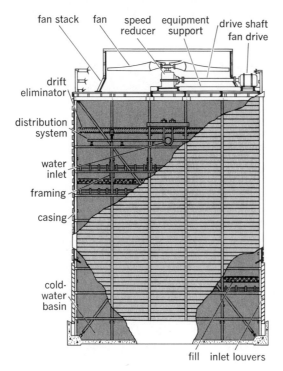

Fig. 6. Induced-draft counterflow cooling tower, showing the component parts.

corrosive material, in the form of troughs and weirs, or made of wood, plastic, or reinforced concrete. Structural framing may also be made of galvanized or plastic-coated structural steel shapes. Natural-draft cooling towers, especially in large sizes, are made of reinforced concrete. The cold-water basins for ground-mounted towers are usually made of concrete; wood or steel is usually used for roof-mounted towers. Fan blades are made of corrosion-resistant material such as monel, stainless steel, or aluminum; but most commonly fiber glass–reinforced plastic is used for fan blades.

The heat-transfer tubes used with nonevaporative cooling towers are of an extended-surface type usually with circumferential fins on the air side (outside). The tubes are usually circular in cross section, although elliptical tubes are sometimes used. Commonly used tube materials are galvanized carbon steel, ferritic stainless steel, and various copper alloys. They are usually made with wrapped aluminum fins, but steel fins are commonly used with carbon steel tubes and galvanized. Most designs employ a ratio of outside to inside surface of 20:25. Outside-diameter sizes range ¾–1½ in. (19–38 mm).

Tube bundles are made with tube banks two to five rows deep. The tubes are in parallel relationship with

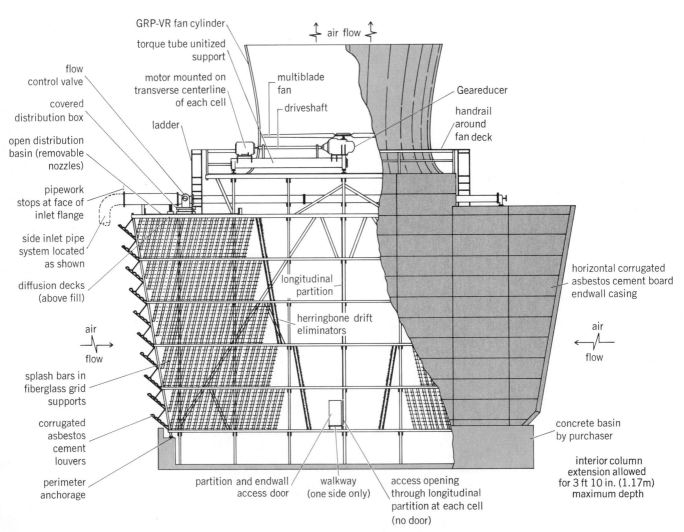

Fig. 7. Transverse cross section of a cross-flow evaporative cooling tower.

Fig. 8. Nonevaporative cooling tower, Utrillas, Spain. (*GEA*)

each other and secured in headers, either of steel pipe or of weld-fabricated steel box headers. The tube bundle assemblies are mounted in steel frames which are supported by structural steel framework. The bundle assemblies may be arranged in a V pattern, requiring fans of the induced-draft type, or in an A pattern, requiring fans of the forced-draft type (**Fig. 8**). Fans and louvers are similar to those described for evaporative cooling towers.

Nonevaporative cooling towers may also be used with natural airflow. In this case, the tube bundles are usually mounted on a deck within the tower and just above the top of the circumferential supporting structure for the tower. In this application, the tower has no cold-water basin, but otherwise it is identical with the natural-draft tower used for evaporative cooling with respect to materials of construction and design.

Performance. The performance of an evaporative cooling tower may be described by the generally accepted equation of F. Merkel, as shown below, where

$$\frac{KaV}{L} = \int_{T_2}^{T_1} \frac{dT}{h'' - h}$$

a = water-air contact area, ft²/ft³; h = enthalpy of entering air, Btu/lb; h'' = enthalpy of leaving air, Btu/lb; K = diffusion coefficient, lb/(ft²)(h); L = water flow rate, lb/(h)(ft²); T = water temperature, °F; T_1 = inlet water temperature, °F; T_2 = outlet water temperature, °F; and V = effective volume of tower, ft³/ft² of ground area. The Merkel equation is usually integrated graphically or by Simpson's rule. *SEE PERFORMANCE*.

The performance of a nonevaporative cooling tower may be described by the generally accepted equation

of Fourier for steady-state unidirectional heat transfer, using the classical summation of resistances formula with correction of the logarithmic temperature difference for cross-counterflow design in order to calculate the overall heat transfer. It is usual practice to reference the overall heat-transfer coefficient to the outside (finned) tube surface.

Evaluation of cooling tower performance is based on cooling of a specified quantity of water through a given range and to a specified temperature approach to the wet-bulb or dry-bulb temperature for which the tower is designed. Because exact design conditions are rarely experienced in operation, estimated performance curves are frequently prepared for a specific installation. These provide a means for comparing the measured performance with design conditions.

Joseph F. Sebald

Bibliography. D. R. Baker, *Cooling Tower Performance*, 1983; N. P. Cheremisinoff and P. N. Cheremisinoff, *Cooling Towers: Selection, Design, and Practice*, 1981; D. Q. Kern and A. D. Kraus, *Extended Surface Heat Transfer*, 1972; R. D. Landon and J. R. Houx, Jr., Plume abatement and water conservation with the wet-dry cooling towers, *Proc. Amer. Power Conf.*, Chicago, 35:726–742, 1973; J. I. Reisman and J. C. Ovard, Cooling towers and the environment: An overview, *Proc. Amer. Power Conf.*, Chicago, 35:713–725, 1973; J. F. Sebald, *Site and Design Temperature Related Economies of Nuclear Power Plants with Evaporative and Nonevaporative Cooling Tower Systems*, Energy Research and Development Administration, Division of Reactor Research and Development, C00-2392-1, January 1976; E. C. Smith and M. W. Larinoff, Power plant siting, performance and economies with dry cooling tower systems, *Proc. Amer. Power Conf.*, Chicago, 32:544–572, 1970.

Coordinate systems

Schemes for locating points in a given space by means of numerical quantities specified with respect to some frame of reference. These quantities are the coordinates of a point. To each set of coordinates there corresponds just one point in any coordinate system, but there are useful coordinate systems in which to a given point there may correspond more than one set of coordinates.

A coordinate system is a mathematical language that is used to describe geometrical objects analytically; that is, if the coordinates of a set of points are known, their relationships and the properties of figures determined by them can be obtained by numerical calculations instead of by other descriptions. It is the province of analytic geometry, aided chiefly by calculus, to investigate the means for these calculations.

The most familiar spaces are the plane and the three-dimensional euclidean space. In the latter a point P is determined by three coordinates (x,y,z). The totality of points for which x has a fixed value constitutes a surface. The same is true for y and z so that through P there are three coordinate surfaces. The totality of points for which x and y are fixed is a curve and through each point there are three coordinate lines. If these lines are all straight, the system of coordinates is said to be rectilinear. If some or all of the coordinate lines are not straight, the system is curvilinear. If the angles between the coordinate lines at each point are right angles, the system is rectangular.

Cartesian coordinate system. This is one of the simplest and most useful systems of coordinates. It is constructed by choosing a point O designated as the origin. Through it three intersecting directed lines OX, OY, OZ, the coordinate axes, are constructed. The coordinates of a point P are x, the distance of P from the plane YOZ measured parallel to OX, and y

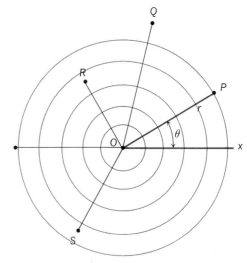

Fig. 2. Polar coordinate system.

and z, which are determined similarly (**Fig. 1**). In this system the coordinate lines through P are straight lines respectively parallel to the three coordinate axes. Usually the three axes are taken to be mutually perpendicular, in which case the system is a rectangular cartesian one. Obviously a similar construction can be made in the plane, in which case a point has two coordinates (x,y). It is this system that is used in the construction of graphs and charts of various data, whether observed or computed.

Polar coordinate system. This system is constructed in the plane by choosing a point O called the pole and through it a directed straight line, the initial line. A point P is located by specifying the directed distance OP and the angle through which the initial line must be turned to coincide with OP in position and direction (**Fig. 2**). The coordinates of P are (r,θ). The radius vector r is the directed line OP, and vectorial angle θ is the angle through which the initial line was turned, $+$ if turned counterclockwise, $-$ if clockwise. To each pair of values (r,θ) there corresponds just one point, but any point has an endless number of sets of coordinates. The coordinate lines in this case are radial lines through the pole (θ = constant) and concentric circles with center at the pole (r = constant). In spite of this lack of unique reciprocity between points and their coordinates, the polar system is useful in the study of spirals and rotations and in the investigation of motions under the action of central forces such as those of planets and comets.

Spherical coordinates. In three-dimensional euclidean space this system of coordinates is constructed by choosing a plane and in it constructing a polar coordinate system. At the pole O a polar axis OZ is constructed at right angles to the chosen plane. A point P, not on OZ, and OZ determine a plane. The spherical coordinates of P are then the directed distance OP denoted by ρ, the angle θ through which the initial line is turned to lie in ZOP and the angle $\phi = ZOP$ (**Fig. 3**). The coordinate lines are radial lines through O (θ and ϕ constant), meridian circles (ρ and ϕ constant), and circles of latitude (ρ and θ constant). This is an example of a curvilinear rectangle coordinate system. It is used in locating stars, in the study of spherical waves, and in problems in which there is spherical symmetry.

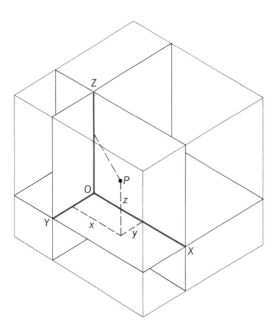

Fig. 1. Cartesian coordinate system.

Cylindrical coordinates. These are constructed by choosing a plane with a pole O, an initial line in it, and a polar axis OZ, as in spherical coordinates. A point P is projected onto the chosen plane. The cylindrical coordinates of P are (r,θ,z) where r and θ are the polar coordinates of Q and $z = QP$ (Fig. 3). This is also a curvilinear rectangular system, the coordinate lines being mutually perpendicular. This system is used in problems of fluid flow and in others in which there is axial symmetry.

What has been done above in the plane and in euclidean space of three dimensions can be extended to curved spaces of any number dimensions. For a space S of n dimensions with some known geometrical or physical properties, a coordinate system in which a point has coordinates $(u_1, u_2, \ldots , u_n)$ is chosen. The coordinate lines are curves along which only one coordinate varies. When the known properties of S are expressed in terms of these coordinates, it is then the province of differential geometry to investigate their consequences. SEE DIFFERENTIAL GEOMETRY.

Transformation of coordinates. By means of a system of equations the description of a geometrical object in one coordinate system may be translated into an equivalent description in another coordinate system. Thus in a given space of n dimensions if there is a coordinate system A in which a general point P has coordinates $(u_1, u_2, \ldots , u_n)$ and a coordinate system B in which P has the coordinates $(v_1, v_2, \ldots , v_n)$, the transformation from system A to system B is the set of equations $u_i = f_i(v_1, v_2, \ldots , v_n)$, for $i = 1, 2, \ldots , n$, which expresses each u in terms of the v's. These functions are obtainable from the relation between the two coordinate systems. They are not completely arbitrary, for they must be single-valued, independent, and such that if in the B system P has another set of coordinates, say $(v_1', v_2', \ldots , v_n')$, then $f_i(v_1, v_2, \ldots , v_n) = f_i(v_1', v_2', \ldots , v_n')$. Then if a geometrical locus in the space S is described by one or more equations of the form $F(u_1, u_2, \ldots , u_n) = 0$ in system A, the equivalent description in system B is one or more of the equations of the form

$$F[f_1(v), f_2(v), \ldots , f_n(v)] = 0$$

Other geometrical objects such as vectors and areas

have more complicated laws of transformation, but each such law is expressible in terms of the equations of transformation. As remarked above, a coordinate system is a mathematical language; a transformation of coordinates plays the role of a dictionary that translates from one language to another.

The most important transformations of coordinates are between rectangular cartesian coordinate systems. One such set of coordinate axes is obtainable from another by a translation (in the physical sense) and a rotation. Consider a coordinate system A with coordinate axes OX, OY, OZ, and coordinates of a point $P(x,y,z)$. System B is obtained by moving the axes without turning to a point O' with coordinates (a,b,c). The new axes are $O'X'$, $O'Y'$, $O'Z'$ and the coordinates of P in this system are (x', y', z'); then the equations of transformation are $x = x' + a$, $y = y' + b$, $z = z' + c$. If the new system is obtained from A by a rotation of the coordinate axes, let α_1, α_2, α_3; β_1, β_2, β_3; γ_1, γ_2, γ_3 be the angles which the new axes OX', OY', OZ' make with the original axes (only three of these angles are independent); then the equations of transformation are

$$x = x' \cos \alpha_1 + y' \cos \beta_1 + z' \cos \gamma_1$$
$$y = x' \cos \alpha_2 + y' \cos \beta_2 + z' \cos \gamma_2$$
$$z = x' \cos \alpha_3 + y' \cos \beta_3 + z' \cos \gamma_3$$

The equations of transformation may be differently interpreted. The equations of transformation make the point P of coordinates (x, y, z), in the same coordinate system, correspond to a point P' of coordinates (x', y', z'). This constitutes a mapping of the space upon itself which maps a figure into some other figure. In applications one usually seeks a mapping that carries some pertinent loci into simpler loci while preserving some relevant properties. SEE ANALYTIC GEOMETRY; BARYCENTRIC CALCULUS; CALCULUS; CONFORMAL MAPPING; CURVE FITTING; SPHERICAL HARMONICS.

Morris S. Knebelman

Coordination chemistry

A field which, in its broadest usage, is acid-base chemistry as defined by G. N. Lewis. However, the term coordination chemistry is generally used to describe the chemistry of metals and metal ions in their interactions with other molecules or ions. For example, reactions (1)–(3) show acid-base-type reactions;

$$Mg^{2+} + 6H_2O \rightarrow Mg(H_2O)_6{}^{2+} \tag{1}$$

$$Ni + 4CO \rightarrow Ni(CO)_4 \tag{2}$$

$$Fe^{2+} + 6CN^- \rightarrow Fe(CN)_6{}^{4-} \tag{3}$$

the products formed are coordination ions or compounds, and this area of chemistry is known as coordination chemistry.

Thus, it follows that coordination compounds are compounds that contain a central atom or ion and a group of ions or molecules surrounding it. Such a compound tends to retain its identity, even in solution, although partial dissociation may occur. The charge on the coordinated species may be positive, zero, or negative, depending on the charges carried by the central atom and the coordinated groups. These groups are called ligands, and the total number of at-

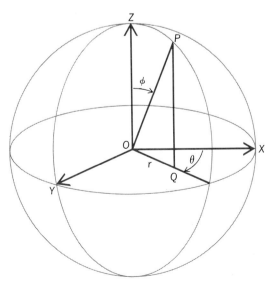

Fig. 3. Spherical coordinate system.

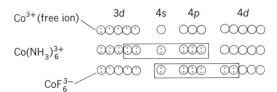

Fig. 1. Representation of the cobalt(III) complexes.

tachments to the central atom is called the coordination number. Other names commonly used for these compounds include complex compounds, complex ions, Werner complexes, coordinated complexes, chelate compounds, or simply complexes. SEE ACID AND BASE; CHELATION; COORDINATION COMPLEXES.

Experimental observations as early as the middle of the eighteenth century reported the isolation of coordination compounds. During that time and for the following 150 years, the valence theory could not adequately account for such materials. As a result, they were referred to as complex compounds, a term which is still in common usage, but not for the same reason. The correct interpretation of these compounds was finally given by Alfred Werner in 1893. He introduced the concept of residual or secondary valence, and suggested that elements have this type of valence in addition to their normal or primary valence. Thus, platinum(IV) has a normal valence of 4 but a secondary valence or coordination number of 6. This then led to the formulation of $PtCl_4 \cdot 6NH_3$ as $[Pt(NH_3)_6]^{4+}$, $4Cl^-$ and of $PtCl_4 \cdot 5NH_3$ as $[Pt(NH_3)_5Cl]^{3+}$, $3Cl^-$. The compound with five ammonias has only three ionic chlorides; the fourth is inside the coordination sphere, and therefore is not readily precipitated upon the addition of silver ion. Although the exact nature of the coordinate bond between metal and ligand remains the subject of considerable discussion, it is agreed that the formulations of Werner are essentially correct.

Coordinate bond. Three theories have been used to explain the nature of the coordinate bond. These are the valence bond theory, the electrostatic theory, including crystal field corrections, and the molecular orbital theory. Currently, the theory used almost exclusively is the molecular orbital theory. The valence bond theory for metal complexes was developed chiefly by Linus Pauling. This theory considers that the pair of electrons on the ligand enter the hybridized atomic orbitals of the metal and that the bond is either essentially covalent or essentially ionic. Several of the properties of these substances can be explained on the

basis of this theory. For example, cobalt(III) complexes are represented by **Fig. 1**. The orbital hybridization of $Co(NH_3)_6^{3+}$ is designated as d^2sp^3, and the complex is referred to as an inner orbital complex. Such a representation is consistent with the diamagnetic properties of this cation, for all electrons are paired. The orbital hybridization of CoF_6^{3-} is designated as sp^3d^2, and it is called an outer orbital complex. This ion is known to be paramagnetic, which is in keeping with the four unpaired electrons in the $3d$ orbitals. The term inner orbital is applied if the d orbital of a lower energy than the s and p is used in bonding, whereas outer orbital has reference to systems in which the d orbital used is at the same energy level as the s and p. SEE MOLECULAR ORBITAL THEORY.

The electrostatic theory, plus the crystal field theory for the transition metals, assumes that the metal-ligand bond is caused by electrostatic interactions between point charges and dipoles and that there is no sharing of electrons. Physicists have made good use of this theory to explain the properties of ionic crystalline solids, and it has now been extended to the metal complexes. For the nontransition metals, the parameters needed to determine the strength of the metal ligand bond are the charges and sizes of the central ions and the charges, dipole moments, polar-

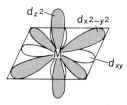

Fig. 3. Diagram of the spatial orientation of d orbitals in gaseous ion represented by M.

izabilities, and sizes of the ligands. In order to give an adequate explanation of the bonding for transition metals, it is also necessary to consider the orientation of the d orbitals in space. The five possible spatial configurations are shown in **Fig. 2**. For the gaseous ion M, all five of the d orbitals are of equal energy. However, as shown in **Fig. 3**, the $d_{x^2-y^2}$ and d_{z^2} orbitals are pointing directly toward the six ligands at the corners of an octahedron and, because of repulsive interaction with the ligands, are at a higher energy than the d_{xy}, d_{xz}, and d_{yz} orbitals which do not point toward the ligands. SEE CRYSTAL FIELD THEORY.

Thus, on the basis of the crystal field theory, the cobalt(III) complexes referred to above are designated as in **Fig. 4**. The $Co(NH_3)_6^{3+}$ is called a spin-paired complex, whereas CoF_6^{3-} is called spin-free. This refers to the fact that the electrons are all paired in the former because of the larger crystal field splitting, Δ_0, whereas in the latter, Δ_0 is small, and the electrons are not paired. In addition to explaining the structure and magnetic properties, this theory affords an adequate interpretation of the visible spectra of metal complexes.

The molecular orbital theory assumes that the electrons move in molecular orbitals which extend over all the nuclei of the metal-ligand system. In this manner, it serves to make use of both the valence bond theory and crystal field theory. The molecular orbital theory is therefore the best approximation to the nature of the coordinate bond because it is sufficiently

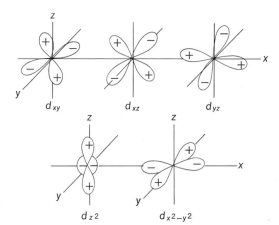

Fig. 2. Diagram orientation of d orbitals in space.

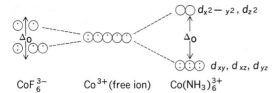

CoF$_6^{3-}$ Co^{3+}(free ion)   Co(NH$_3$)$_6^{3+}$

Fig. 4. Representation of the cobalt(III) complexes on the basis of crystal field theory.

flexible to permit both covalent and ionic bonding as well as the splitting of d orbitals into various energy levels. The complexes CoF$_6^{3-}$ and Co(NH$_3$)$_6^{3+}$ can be represented by the molecular orbital energy diagrams shown in **Fig. 5**.

It is apparent from these molecular orbital diagrams that this theory combines the desirable features of both the valence bond and the crystal field theories. The covalent bonding of the valence bond theory appears as the sigma (σ) bonded molecular orbitals, designated here as σ_s, σ_p, and σ_d. Likewise the crystal field splitting (Δ_0) of the crystal field theory now is the energy difference between the nonbonding d orbitals d_{xy}, d_{xz}, d_{yz} and the antibonding sigma orbital σ_d*. More complicated molecular orbital diagrams would include the contribution of π bonding in these systems.

Stabilities of complexes. The stability of metal complexes depends both on the metal ion and the ligand. There is a great deal of quantitative information on the stability of coordination compounds. These data show that in general the stability of metal complexes increases if the central ion increases in charge, decreases in size, and increases in electron affinity. Thus, alkali-metal ions have the least tendency to form complexes, and the highly polarizing transition-metal ions have the greatest tendency. However, even in the least favorable cases of the alkali-metal ions, there is ample evidence that coordination with certain ligands does occur. For the transition-metal ions, it also appears that the electronic configuration of the ion is significant with regard to the stability of its complexes. For example, regardless of the nature of the ligand, the stability of complexes of bivalent transition metals is Mn < Fe < Co < Ni < Cu > Zn. This so-called natural order of stability is explained in terms of the crystal field theory, which indicates that a maximum stabilization effect in spin-free complexes is realized with d^8 systems, such as exists in Ni^{2+}. The platinum metals, because of their large polarizing ability, generally form the most stable metal complexes.

Effect of ligand. Several characteristics of the ligand are known to influence the stability of complexes: (1) basicity of the ligand, (2) the number of metal-chelate rings per ligand, (3) the size of the chelate ring, (4) steric effects, (5) resonance effects, and (6) the ligand atom. Since coordination compounds are formed as a result of acid-base reactions where the metal ion is the acid and the ligand is the base, it follows that generally the more basic ligand will tend to form the more stable complex. Much of the available quantitative data on the formation constants of coordination compounds gives a good linear correlation with the base strength of the ligand. It is also known that a polydentate ligand, one attached to the metal ion at more than one point, forms more stable complexes than does an analogous monodentate ligand. For ex-

ample, ethylenediamine forms more stable complexes than does ammonia, as shown in the following inequality.

$$\left[\begin{array}{c} CH_2-NH_2 \\ | \\ CH_2-NH_2 \end{array} \!\! \begin{array}{c} NH_2-CH_2 \\ Zn \\ NH_2-CH_2 \end{array} \right]^{2+} > $$

$$\left[\begin{array}{c} H_3N \\ \\ H_3N \end{array} \!\! Zn \!\! \begin{array}{c} NH_3 \\ \\ NH_3 \end{array} \right]^{2+}$$

The ethylenediamine complex is referred to as a chelate complex. In general, an increase in the extent of chelation results in an increase in the stability of the complex. Ethylenediaminetetraacetate ion, a sexa-

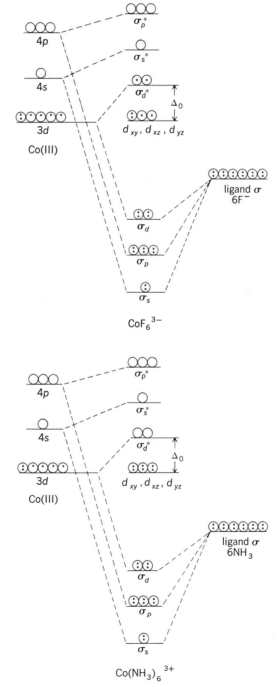

Fig. 5. Representation of the cobalt complexes CoF$_6^{3-}$ and Co(NH$_3$)$_6^{3+}$ according to molecular orbital theory.

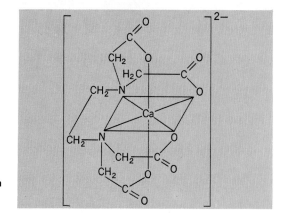

dentate ligand, is an excellent chelating agent and has found numerous applications as a sequestrant for metal ions. **Figure 6** is the structural formula of the chelate complex of calcium ion and ethylenediamine-tetraacetic acid.

The size of the chelate ring is likewise an important factor. For saturated ligands such as ethylenediamine, five-membered rings are the most stable, whereas six-membered rings are the most stable for chelates containing one or more double bonds, for example (I) and (II). Examples of smaller and of larger chelate

(I)

(II)

ring systems are known, but these are much less common and less stable. It has been observed that for analogous systems an increase in double bonding in the chelate ring increases the stability of the complex. For ligands of comparable base strengths, the acetyl-acetonate complexes of type (I) are more stable than the salicylaldehyde compounds of type (II). There are in effect two double bonds per chelate ring in (I) compared to one and a half in (II). This same phenomenon is believed to be responsible for the marked stability of the porphyrin complexes such as chlorophyll, hemoglobin, and the phthalocyanine dyes.

Steric factors often have a very large effect on the stability of metal complexes. This is most frequently observed with ligands having a large group attached to the ligand atom or near it. Thus complexes, of the type shown in **Fig. 7**, with alkyl groups R in the position designated are much less stable than the parent

Fig. 7. Structural formulas of metal complexes which are affected by steric factors.

complex where R = H. This results from the steric strain introduced by the size of the alkyl group on or adjacent to the ligand atom. In contrast to this, alkyl substitution at any other position results in the formation of more stable complexes because the ligand becomes more basic, and the bulky group is now removed from a position near the coordination site.

Finally, the ligand atom itself plays a significant role in controlling the stability of metal complexes. For most of the metal ions, the smallest ligand atom with the largest electron density will form the most stable complex. This means that the second period elements form more stable metal-ligand bonds than do other members of the same group, for example, N > P > As > Sb; O > S > Se > Te; F > Cl > Br > I. It is also known that for this same class of metal ions the stability order is N > O > F. These trends are in accord with the coordinate bond strengths expected on the basis of electrostatic interactions. These metal ions and the ligand atoms they prefer are not very polarizable, and are designated as hard Lewis acids and bases, respectively. A different trend in stability is found for a second class of metal ions, for example, Cu(I), Rh(I), Pd(II), Ag(I), Pt(II), and Hg(II). The stability trends observed for these metal ions are N ≫ P > As ≈ Sb, O ≪ S > Se ≈ Te, and F ≪ Cl < Br < I. These metal ions and the ligand atoms they prefer are polarizable, and are designated as soft Lewis acids and bases, respectively. Thus, Pearson has proposed the HASAB theory of acids and bases, which states that hard Lewis acids prefer hard bases and soft Lewis acids prefer soft bases. Since the soft metal ions also have a large number of d orbital electrons and the third-period ligand atoms have vacant d orbitals, the enhanced stability is attributed to the formation of dd-π bonding. This means that the filled d orbitals of the metal ion overlap with the empty d orbitals of the ligand atom. This is illustrated by **Fig. 8**. This same class of metal ions is the one that forms olefin complexes such as Zeise's salt, $K[Pt(CH_2\!\!=\!\!CH_2)Cl_3]$. The stability of these compounds is likewise attributed to π bonding.

Stability and reactivity. Often the most stable complex is also the least reactive or most inert. Therefore, the platinum metal complexes as well as cobalt(III) and chromium(III) complexes are usually very slow to react. However, there are exceptions to this, for instance, the rapid exchange of radiocyanide with the coordinated cyanide in the stable tetracyano complexes $Hg(CN)_4^{2-}$ and $Ni(CN)_4^{2-}$. Several factors, such as the electronic configuration of the central metal ion, its coordination number, and the extent of chelation, all have a marked effect on the rate of reaction of a given compound.

There is a sufficient data available to permit a fairly reliable classification for the relative reactivities of six-coordinated systems. Metal complexes are said to be labile if they have reacted completely at room temperature within the time of mixing, such as the instantaneous formation of the deep-blue color of $Cu(NH_3)_4^{2+}$ upon the addition of excess ammonia to $Cu(H_2O)_4^{2+}$. Complexes are designated as being inert if their reactions at room temperature proceed at a detectable rate and have half-lives longer than 2 min. On the basis of such a definition, it is possible to classify the labile complexes as being those of either the outer orbital type, such as CoF_6^{3-}, $Al(C_2O_4)_3^{3-}$, $SnCl_6^{2-}$, $Ni(NH_3)_6^{2+}$, or the inner orbital type with one or more vacant d orbital, such as TiF_6^{2-},

$V(CN)_6^{3-}$, $Mo(SCN)_6^{2-}$. The inert complexes are then the inner orbital type with no vacant d orbital, for example $Co(NH_3)_6^{3+}$, $Cr(H_2O)_6^{6+}$, $RhCl_6^{3-}$. A similar classification based on the crystal field theory is done in terms of the labile systems being those that are not stabilized greatly by the crystal field.

Reactions of complexes. The reactions referred to above are acid-base reactions, where one basic ligand replaces a less basic ligand coordinated to the acidic metal ion, for example, reaction (4). A reaction of

$$Co(NH_3)_5Cl^{2+} + H_2O \rightarrow Co(NH_3)_5H_2O^{3+} + Cl^- \quad (4)$$

this type is called a nucleophilic substitution reaction and is designated by the symbol S_N. There are at least two fundamentally different pathways conceivable for such a reaction. These are the dissociation and displacement mechanisms which are designated as S_N1 and S_N2, respectively. An S_N1 reaction goes by a two-step process, where the first step is a slow unimolecular heterolytic dissociation, reaction (5), followed by

$$Co(NH_3)_5Cl^{2+} \xrightarrow{slow} Co(NH_3)_5^{3+} + Cl^- \quad (5)$$

the rapid coordination of the entering group, reaction (6). An S_N2 reaction is one involving a bimolecular

$$Co(NH_3)_5^{3+} + H_2O \xrightarrow{fast} Co(NH_3)_5H_2O^{3+} \quad (6)$$

rate-determining step in which one nucleophilic reagent displaces another [reaction (7)]. Examples of

$$Co(NH_3)_5Cl^{2+} + H_2O \xrightarrow{slow}$$
$$H_2O \cdots Co(NH_3)_5 \cdots Cl^{2+} \xrightarrow{fast}$$
$$Co(NH_3)_5H_2O^{3+} + Cl^- \quad (7)$$

both types of substitution are known for the reaction of metal complexes. The specific reaction just cited, as well as analogous reactions of $Co(NH_3)_4Cl_2^+$ and $Co(NH_3)_4OHCl^+$, appears to proceed by an S_N1 mechanism. However, reactions of platinum(II) complexes are believed to involve an S_N2 process, for example, reaction (8).

$$Pt(NH_3)_3Cl^+ + NO_2^- \xrightarrow{slow}$$
$$O_2N \cdots Pt(NH_3)_3 \cdots Cl \xrightarrow{fast}$$
$$Pt(NH_3)_3NO_2^+ + Cl^- \quad (8)$$

Most recently, the symbols AID were introduced in order to better designate the mechanisms of these reactions. A = association and relates to S_N2, where the entering group becomes associated with the complex [slow step of Eq. (8)]; I = interchange and signifies that a group just outside the complex interchanges positions with a ligand on the metal; D = dissociation and relates to S_N1, where the leaving group dissociates away from the complex [Eq. (5)]. In general, I_d is similar to S_N1 and designates a dissociative interchange mechanism, whereas I_a is similar to S_N2 and designates an associative interchange mechanism.

Finally, coordination may have the effect of stabilizing an unusual oxidation state of the coordinated metal. For example, ligands such as CO, CN^-, CNR, 2,2'-bipyridine, and PR^3 are most effective in stabilizing low-valent states. Thus, compounds of the type

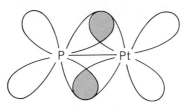

Fig. 8. Overlap of d orbitals of platinum with empty d orbitals of phosphorus ligand.

$Na_2[Fe(CO)_4]$, $Na[V(2,2'-bipyridine)_3]$, $K_4[Ni(CN)_4]$, and $Pt(PR_3)_4$ are known where the assigned oxidation states are Fe (2−), V (1−), Ni (0), and Pt (0). It is believed that the low oxidation states in these systems are stabilized by the formation of π-type molecular orbitals between the metal and ligand atoms. If the ligand orbitals are of low energy and vacant, they can accommodate the d-orbital electrons of the metal and permit the addition of other electrons leading to the reduction of the complex. In contrast, ligands such as O^{2-}, OH^-, and F^- tend to stabilize high-valence states giving complexes of the type K_3FeO_4, Na_3BiO_4, K_2NiF_6, and K_3CuF_6, where the oxidation states are Fe (6+), Bi (5+), Ni (4+), and Cu (3+). These ligands have no d orbitals but only filled p orbitals. One explanation for the stabilization of higher-valent states by such ligands is that in these systems the π-type molecular orbitals cannot be formed so the metal electrons are forced into higher-energy antibonding orbitals from which they may be readily removed. *See* CHEMICAL BONDING; MAGNETOCHEMISTRY; SOLID-STATE CHEMISTRY; STEREOCHEMISTRY: SUBSTITUTION REACTION.

Fred Basolo

Bibliography. J. C. Bailar, Jr. (ed.), *The Chemistry of the Coordination Compounds*, Amer. Chem. Soc. Monog. 131, 1956; F. Basolo and R. G. Pearson, *Mechanisms of Inorganic Reactions*, 2d ed., 1967; F. A. Cotton and G. Wilkinson, *Advanced Inorganic Chemistry: A Comprehensive Text*, 5th ed., 1988; D. Katakis and G. Gordon, *Mechanisms of Inorganic Reactions*, 1987; L. Pauling and Z. S. Herman, Valence-bond concepts in coordination chemistry and the nature of metal-metal bonds, *J. Chem. Educ.*, 61(7):582–587, July 1984; R. G. Pearson (ed.), *Hard and Soft Acids and Bases*, 1973.

Coordination complexes

A group of chemical compounds in which a part of the molecular bonding is of the coordinate covalent type.

This article summarizes the different types of compounds that are known and discusses their nomenclature, structure, stereochemistry, and synthesis. For a discussion of the nature of the coordinate bond and the stability and reactivity of complex compounds *SEE* CHELATION; COORDINATION CHEMISTRY.

Complex compounds contain a central atom or ion and a group of ions or molecules surrounding it. Many simple hydrates, such as $MgCl_2 \cdot 6H_2O$, are best formulated as $[Mg(H_2O)_6]Cl_2$ because it is known that the six molecules of water surround the central magnesium ion. Therefore, $[Mg(H_2O)_6]^{2+}$ is a complex ion, and $[Mg(H_2O)_6]Cl_2$ is a complex compound. The charge on this complex ion is +2, because this is the charge on the magnesium ion and the coordinated wa-

Table 1. Metal complexes with coordination numbers (C.N.) other than 4 or 6

C.N.	Complex	Structure
2	$[Ag(CN)_2]^-$ $[Ag(NH_3)_2]^+$	Linear
3	$[Cu(PR_3)_2I]$ $[Au(AsR_3)_2I]$	Planar
5	$[Fe(CO)_5]$ $[Zn(2,2',2''\text{-terpyridine})Cl_2]$	Trigonal bipyramid
	$[Ni(PR_3)_2Br_3]$ $\{Pd[o\text{-phenylene-bis-}$ $\text{(dimethylarsine)]}_2I\}I$	Tetragonal pyramid
7	ZrF_7^{3-} NbF_7^{2-} TaF_7^{2-}	Pentagonal bipyramid Face-centered trigonal prism
8	$Mo(CN)_8^{4-}$ TaF_8^{3-}	Dodecahedral Archimedean antiprism
9	$Nd(H_2O)_9^{3+}$	Face-centered trigonal prism
10	$Cd_2[Mo(CN)_8(H_2O)_2]$	Unknown structure

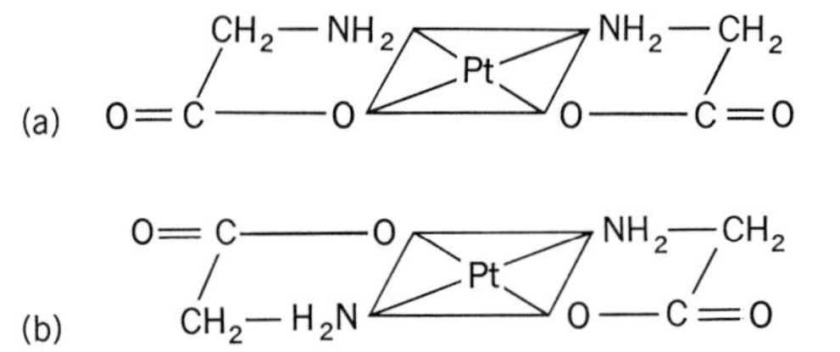

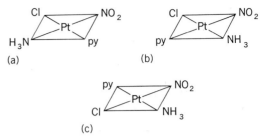

Fig. 2. Isomeric complexes. (a) *trans*-Chloropyridine-*trans*-nitroammineplatinum(II). (b) *trans*-Chloroammine-*trans*-nitropyridineplatinum(II). (c) *trans*-Chloronitro-*trans*-amminepyridineplatinum(II).

then the negative ligands with an ending of -o, and finally the metal followed by a Roman numeral in parentheses to designate its oxidation state. Neutral ligands are named as the molecule, except that H_2O is aquo and NH_3 is ammine. (3) The prefixes such as di, tri, and tetra are used before simple expressions such as chloro, aquo, and oxalato. Prefixes such as bis, tris, and tetrakis are used before complex expressions such as ethylenediamine, 2,2′-bipyridine, and trialkylphosphine. (4) Neutral complexes are named in the same way except that only one word is required. (5) Anionic complexes are also named according to these same rules except that an -ate ending is used. Additional rules are available to name the more complicated compounds, and some of these will be used in the discussion that follows.

Coordination numbers of 2–10 have been observed for different complex compounds. The most common

Fig. 1. Structure of (a) *cis*-diglycinatoplatinum(II) and (b) *trans*-diglycinatoplatinum(II).

ter molecules are neutral. However, if the coordinated groups are charged, then the charge on the complex is represented by the sum of the charge on the metal and that of the coordinated ions. See, for example, the progression of charges on the platinum(IV) complexes listed below. Thus the charge is $+4$ for $[Pt(NH_3)_6]^{4+}$ because Pt is $+4$ and NH_3 is neutral. But the charge is -2 for $[PtCl_6]^{2-}$ because of the 6 Cl^-, that is, $+4 - 6 = -2$.

Nomenclature. The metal complex may be a cation, have zero charge, or be an anion, as is exemplified by the following series of complexes:

$[Pt(NH_3)_6]Cl_4$	Hexaammineplatinum(IV) chloride
$[Pt(NH_3)_5Cl]Cl_3$	Chloropentaammine-platinum(IV) chloride
$[Pt(NH_3)_4Cl_2]Cl_2$	Dichlorotetraammine-platinum(IV) chloride
$[Pt(NH_3)_3Cl_3]Cl$	Trichlorotriammine-platinum(IV) chloride
$[Pt(NH_3)_2Cl_4]$	Tetrachlorodiammine platinum(IV)
$K[Pt(NH_3)Cl_5]$	Potassium pentachloro-ammineplatinate(IV)
$K_2[PtCl_6]$	Potassium hexachloro-platinate(IV)

These compounds are named according to rules set up by the Nomenclature Committee of the International Union of Pure and Applied Chemistry. Some of the rules are the following: (1) Name the cation first as one word followed by the anion as one word. (2) For a cationic complex, name the neutral ligands first,

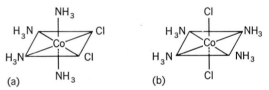

Fig. 3. Structure of (a) *cis*-dichlorotetraamminecobalt(III) ion and (b) *trans*-dichlorotetraamminecobalt(III) ion.

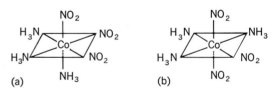

Fig. 4. Structure of (a) *cis*-(or facial-)trinitrotriammine-cobalt(III) and (b) *trans*-(or meridial-)trinitrotriammine-cobalt(III).

coordination numbers are 6 and 4. Complexes of a coordination number 6 generally have an octahedral structure, but may also be trigonal prismatic. Complexes of a coordination number 4 are either square planar or tetrahedral. **Table 1** gives examples of complexes having coordination numbers other than 4 or 6.

Structures. Metal complexes exhibit various types of isomerism. In many ways, inorganic stereochemistry is similar to that observed with organic compounds. Geometrical isomers are common among the inert complexes of coordination numbers 4 and 6. Square planar complexes of the type $Pt(NH_3)_2Cl_2$ exist in two forms. Likewise, cis-trans isomers of

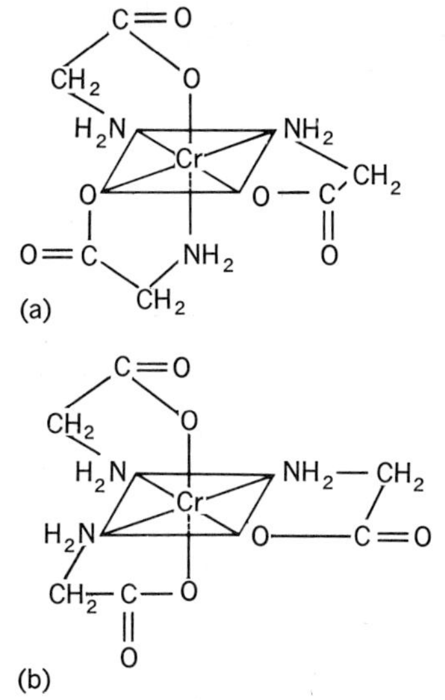

Fig. 5. Structure of (a) *fac*-triglycinatochromium(III) and (b) *mer*-triglycinatochromium(III).

Optical isomerism is also fairly common among these compounds. For example, 4-coordinate tetrahedral complexes containing unsymmetrical bidentate ligands, such as bis(benzoylpyruvato)beryllate(II) ion, have been resolved (**Fig. 6**). Optically active complexes of this type are likewise reported for boron(III) and for zinc(II). Most of the examples of optical activity occur with 6-coordinated systems containing three bidentate ligands, for example, trioxalatorhodiate(III) ion (**Fig. 7**). The resolution of 6-coordinated complexes of this type has been reported for the metal ions Al(III), As(V), Cd(II), Co(III), Cr(III), Ga(III), Ge(IV), Ir(III), Fe(II),(III), Ni(II), Os(II),(III), Pt(IV), Rh(III), Ru(II),(III), Ti(IV), and Zn(II).

The cis isomer of a complex containing two bidentate ligands and two monodentate ligands is asymmetric, and therefore exists in the form of mirror-im-

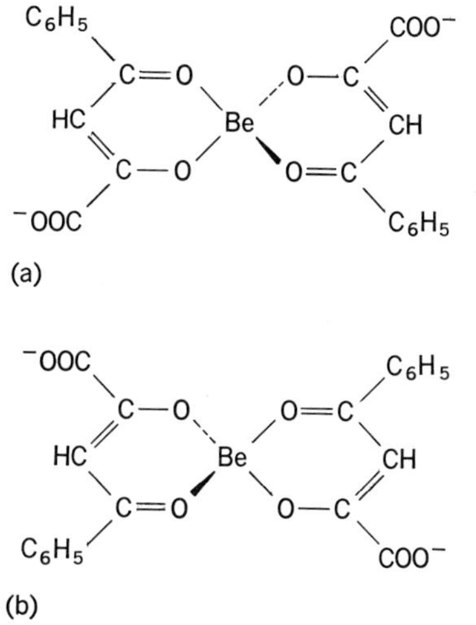

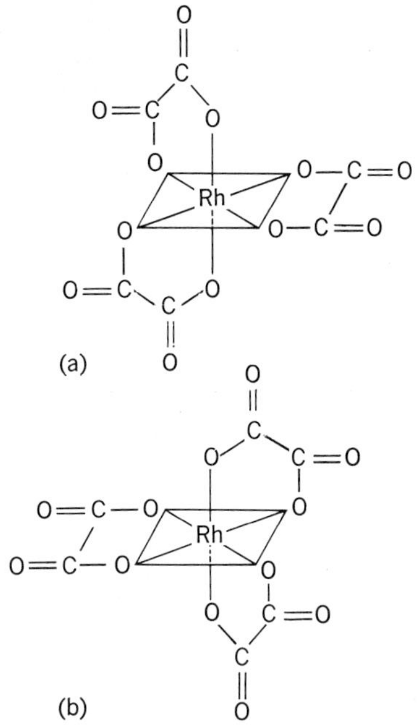

Fig. 6. Optical isomers of bis(benzoylpyruvato)beryllate(II) ion: (a) dextro and (b) levo.

Fig. 7. Optical isomers of trioxalatorhodiate(III) ion: (a) dextro and (b) levo.

Pt(NH$_2$CH$_2$COO)$_2$ type have been isolated (**Fig. 1**). It is apparent from the above examples that the isomer with the same ligands or ligand atoms in adjacent positions is called cis, whereas the trans isomer has its like groups in opposite positions. There are also examples of geometrical isomers of complexes containing four different ligands (**Fig. 2**). Although geometrical isomerism of square complexes is most common with platinum(II), it has also been observed with compounds of nickel(II), palladium(II), and gold(III). SEE STEREOCHEMISTRY.

There are many examples of geometrical isomers for 6-coordinated complexes of cobalt(III), chromium(III), and the platinum metals. Most of the examples are of the type [Co(NH$_3$)$_4$Cl$_2$]$^+$ (**Fig. 3**). If three of the ligands differ from the other three, then only two isomers are possible (**Fig. 4**). Similar isomers are obtained with an unsymmetrical bidentate ligand (**Fig. 5**). A complex with 6 different ligands, such as [Pt(NH$_3$)(py)(NH$_2$OH)(Cl)(Br)(NO$_2$)], can exist theoretically in 15 different geometrical forms. In practice, it has not been possible to isolate all of these 15 different isomers.

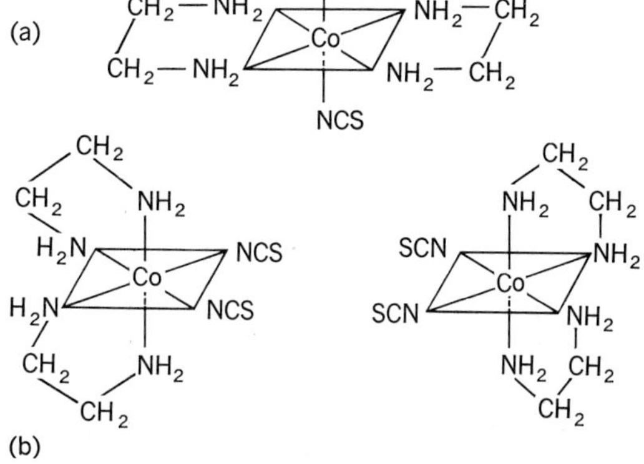

Fig. 8. Symmetric trans form of a complex and asymmetric cis structures. (a) *trans*-Isothiocyanatobis(ethylenediamine)cobalt(III) ion. (b) Asymmetric forms of *d,l-cis*-isothiocyanatobis(ethylene-diamine)cobalt(III) ion.

Table 2. Types of isomerism for complex compounds

Isomerism	Examples
Geometrical	cis- and trans-[Pt(NH$_3$)$_2$Br$_2$]
Optical	d- and l-[Ir(NH$_2$CH$_2$CH$_2$NH$_2$)$_3$]Cl$_3$
Coordination	[Co(NH$_3$)$_6$][Cr(CN)$_6$] and [Cr(NH$_3$)$_6$][Co(CN)$_6$]
Coordination position	[(R$_3$P)$_2$Pt⟨Cl⟩PtCl$_2$] and [Cl(R$_3$P)Pt⟨Cl⟩Pt(PR$_3$)Cl]
Hydrate	[Cr(H$_2$O)$_6$]Cl$_3$ and [Cr(H$_2$O)$_5$Cl]Cl$_2$ · H$_2$O
Ionization	[Pt(NH$_3$)$_4$Cl$_2$]SO$_4$ and [Pt(NH$_3$)$_4$SO$_4$]Cl$_2$
Linkage	[(NH$_3$)$_5$Co—NO$_2$]SO$_4$ and [(NH$_3$)$_5$Co—ONO]SO$_4$
Polymerization	[Co(NH$_3$)$_3$(NO$_2$)$_3$] and [Co(NH$_3$)$_6$][Co(NO$_2$)$_6$]
Conformation	[Ni(P(C$_6$H$_5$)$_2$CH$_2$C$_6$H$_5$)$_2$Br$_2$] square-planar and tetrahedral

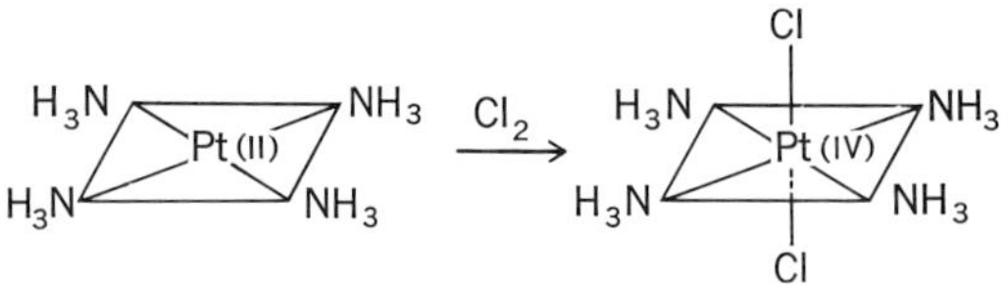

Fig. 9. The reaction in the trans isomer formation of platinum(IV) complexes.

age isomers, whereas the trans form is symmetrical and cannot be optically active (**Fig. 8**). Since the cis isomer of this type is optically active and the trans isomer is not, the successful resolution of one of the isomers is often used as a proof of its cis structure. Other types of isomerism are known for metal complexes (**Table 2**).

Synthesis. The synthesis of metal complexes containing only one kind of ligand generally involves just the reaction of the metal salt in aqueous solution with an excess of the ligand reagent, reaction (1).

$$[Ni(H_2O)_6](NO_3)_2 + 6NH_3 \rightarrow$$

$$[Ni(NH_3)_6](NO_3)_2 + 6H_2O \quad (1)$$

The desired complex salt can then be isolated by removal of water until it crystallizes, or by addition of a water-miscible organic solvent to cause it to separate. Many reactions, such as the one cited above, occur readily at room temperature. For the inert complexes (those slow to react), prolonged treatment at

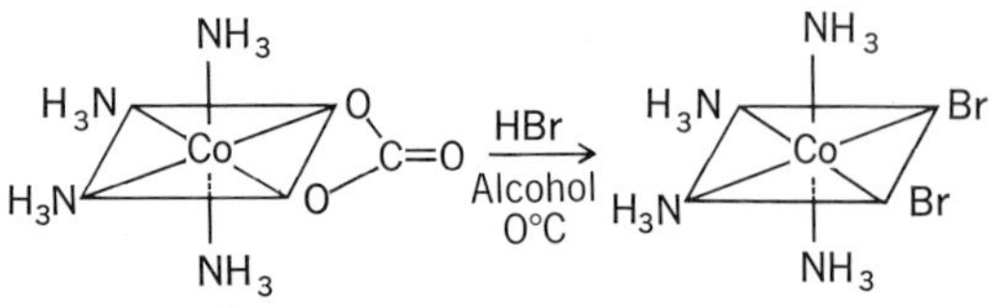

Fig. 10. Cis isomer formation of cobalt(III) complexes.

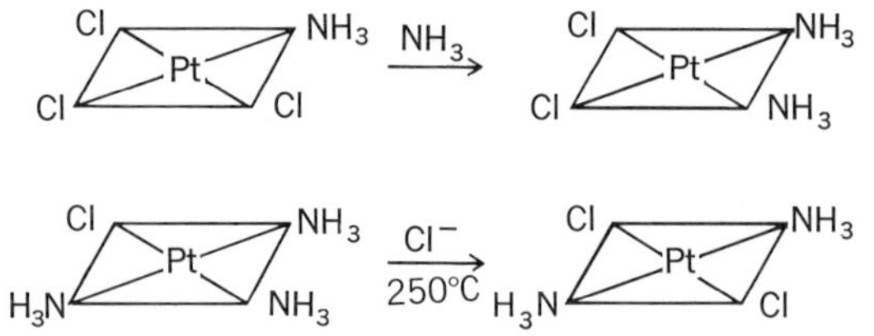

Fig. 11. Cis and trans isomer formation of platinum(II).

more drastic conditions is often necessary.

The preparation of geometrical isomers is much more difficult, and in most cases, the approach used is rather empirical. Generally, reactions yield a mixture of cis-trans products, and these are separated on the basis of their differences in solubility. The trans isomers of platinum(IV) complexes are prepared by the oxidation of the appropriate platinum(II) compound (**Fig. 9**). The cis isomers of cobalt(III) complexes can sometimes be prepared by the reaction of a carbonato complex with the desired acid (**Fig. 10**).

The easiest geometric isomers to prepare are those of platinum(II). For examples, the cis and the trans isomers of [Pt(NH$_3$)$_2$Cl$_2$] are prepared as shown in reactions (2). The second step in each of these reactions

$$K_2[PtCl_4] + 2NH_3 \rightarrow cis\text{-}[Pt(NH_3)_2Cl_2] + 2KCl$$

$$[Pt(NH_3)_4]Cl_2 \xrightarrow[\text{(392°F)}]{250°C} trans\text{-}[Pt(NH_3)_2Cl_2] + 2NH_3$$

(2)

results in the replacement of the ligand trans to a chloro group (**Fig. 11**). This phenomenon, that a negative ligand often has a greater labilizing effect on a group in the trans position than does a neutral group, for example, Cl$^-$ > NH$_3$, is called the trans effect. Extensive use has been made of this trans effect in the synthesis of desired platinum(II) complexes. The complex cis-[Pt(NG$_3$)$_2$Cl$_2$] (Fig. 11) is used as an antitumor drug for certain types of cancer.

Finally, the separation of optical isomers of metal complexes involves techniques similar to those used for organic compounds. The usual procedure is to convert the racemic mixture into diastereoisomers by means of an optically active resolving agent and then to separate the diastereoisomers by fractional crystallization. Nonionic complexes have been resolved by preferential adsorption on optically active quartz or sugars. SEE AMMINE; HYDRATE.

Fred Basolo

Bibliography. F. Basolo and R. G. Pearson, *Mechanisms of Inorganic Reactions*, 2d ed., 1967; F. A. Cotton and G. Wilkinson, *Advanced Inorganic Chemistry*, 5th ed., 1988; D. Katakis and G. Gordon, *Mechanisms of Inorganic Reactions*, 1987; A. E. Martel (ed.), *Coordination Chemistry*, vol. 1, 1971, vol. 2, 1978.

Coordination number

A count of the number of nearest-neighbor atoms around a reference atom; in effect, it is a census on the atomic scale. Determining coordination numbers for the same element, in compositionally different environments, is vital to understanding in a wide range of areas. Studies of the orbitals involved in chemical bonding, biological activity, and high-strength alloys represent a few of the fields that benefit from knowledge about the preferred neighbors of an atom, their number (the coordination number), and the geometry of their mutual arrangement, called the coordination polyhedron.

Determination. Knowledge of the composition and physical density, and a sample of material in crystalline form are required to decode the diffraction pattern experimentally recorded by monochromatic x-rays. Decoding is the procedure of making a mathematical model whose calculated diffraction pattern exactly matches the experimental one, and whose composition and density agree with the observed values. To construct this model, the material is conceived as being made of boxes that can be stacked together to fill three-dimensional space. The boxes may have unusual shapes, but they must all be identical and they must be stacked in the same orientation and have the correct composition.

The size and shape of the box determine the overall symmetry of the calculated diffraction pattern. The atoms within the box determine the density and the intensities of individual diffraction spots. Once a box is constructed, the atoms within it can be moved around until the model and experimental patterns match. A census is then taken to determine the coordination number and coordination polyhedron.

Examples. The **illustration** shows a cubic box. Cubic diffraction symmetry is common for many metals. Two examples will be considered. In the first, there are identical atoms at the corners and in the geometric center of the cube. There is no problem stacking cubes to fill three-dimensional space. Each box contains 2 atoms, 1 at the center and 1 at each of the 8 corners, each shared by 8 stacked boxes $[1 + (8/8) = 2]$. The coordination number is 8; the nearest neighbors of the center are at the corners, and the neighbors of a corner atom are at the centers of 8 boxes.

In the second example, the atoms at alternate corners (the ones surrounded by broken circles in the illustration) are eliminated (erased). This changes everything except the mutual arrangement of the center and 4 corner atoms. The coordination number becomes 4, and the coordination polyhedron becomes the inscribed tetrahedron. However, it is no longer possible to fill three-dimensional space with the box because translations do not generate identical boxes; under such translations, occupied corners become un-

occupied, and vice versa. The diffraction symmetry of this material is still cubic, but the correct model requires a much larger box to fill space.

Erasure simulates the coordination number and coordination polyhedron of silicon and diamond. Their boxes are larger cubes containing 8 atoms—4 enclosed in the box, 1 at each of the 6 faces, and 1 at each of the 8 corners $[4 + (6/2) + (8/8) = 8]$. Each atom is tetrahedrally coordinated as shown in the illustration. Silicon in all of its naturally known forms, such as silicon dioxide (SiO_2), silicates, glasses, and amorphous silicon, arranges 4 neighbors tetrahedrally. Clearly, coordination numbers and coordination polyhedra provide guidance in understanding interatomic forces and the architectures generated by them. An accurate map of these structures, in turn, enhances understanding and control of materials making transistors and other devices possible. SEE CRYSTAL STRUCTURE; CRYSTALLOGRAPHY.

Doris Evans

Bibliography. J. Donohue, *The Structures of the Elements*, 1974, reprint 1982; W. J. Moore, *Seven Solid States*, 1967; L. Pauling and R. Hayward, *The Architecture of Molecules*, 1964.

Copepoda

A subclass of Crustacea with eight orders. There are over 6000 species, most of which are free-living in aqueous environments, though many are parasitic or symbiotic with other aquatic animals. The free-living forms are the most abundant of all animals in the sea and directly constitute the main food for vast numbers of fishes, many invertebrates, and also at times for baleen whales.

Structure. The structure of copepods, although similar in basic pattern, varies greatly. In free-living forms the body is composed externally of a series of small chitinous cylinders, or segments, some of which may be more or less fused. The head bears two pairs of antennae. The first antennae are used for stabilizing, retarding sinking, olfactory functioning, and usually in the male for grasping the female. The second antennae are used for creating water currents for food straining and slow locomotion in the Calanoida, and for food grasping in the Cyclopoida and Harpacticoida. The feeding appendages are one pair of mandibles, two pairs of maxillae, and one pair of maxillipeds. During the nauplius larval stage, the second antennae are commonly provided with masticatory hooks for feeding. The thorax, consisting of six segments, bears the swimming feet. The genital aperture occurs on the sixth thoracic segment. The abdomen consists of one to five segments and bears no appendages except a pair of terminal processes known as caudal rami. The sexes are separate and sexual dimorphism is frequently evidenced, especially by modification in the male by one or both of the first antennae for clasping, and in many by marked specialization in the fifth pair of feet for the transfer of spermatophores, or packets of spermatozoa, to the female. Free-living species usually range in size from less than 0.02 in. (0.5 mm) to about 0.4 in. (10 mm) in length. A deep oceanic species, *Bathycalanus sverdrupi*, attains a length of 0.68 in. (17 mm). Some parasitic copepods, such as *Penella*, are greatly elongate, reaching up to 10 in. (250 mm) in length. SEE SEXUAL DIMORPHISM.

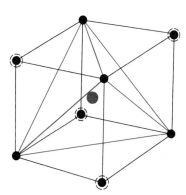

Crystal structures. In one example (see text), identical atoms occupy the corners and the geometric center of the cube. In a second example, the corner atoms surrounded by broken circles are eliminated, leaving the center and four remaining corner atoms, and the coordination polyhedron becomes the inscribed tetrahedron.

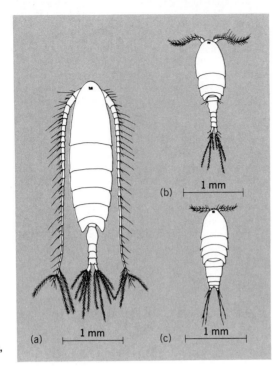

Free-living copepods. (a) *Calanus finmarchicus*, a calanoid. (b) *Cyclops*, a cyclopoid. (c) *Harpacticus*, a harpacticoid.

Classification. Seven orders of copepods are usually recognized. Three of these, Calanoida, Cyclopoida, and Harpacticoida, are wholly or primarily free-living and because of overwhelming numbers are of greatest general interest. The anatomical differences that set these three apart are, among others, the shape of the body and the position of the movable articulation separating the forebody (metasome) from the hindbody (urosome). In the calanoids and the cyclopoids the urosome is distinctly narrower than the metasome (**illus**. *a* and *b*). In the calanoids the movable articulation falls between the fifth and sixth thoracic segments, whereas in the cyclopoids, it is between the fourth and fifth segments. In the harpacticoids the urosome is commonly not narrow (illus. *c*), and the movable articulation falls between the fourth and fifth thoracic segments. In many harpacticoids there is considerable flexibility also between other segments. The length of the first antennae relative to body length is very useful in distinguishing these three orders. The first antennae in the Calanoida have 20–25 segments, usually extending to or beyond the caudal rami. In the Cyclopoida they have 6–17 segments, extending to the posterior border of the metasome or even the first body segment. In the Harpacticoida they have 5–9 segments, usually not extending to the posterior border of the first body segment. SEE CALANOIDA; CYCLOPOIDA; HARPACTICOIDA.

In the adult state the remaining orders (Poecilostomatoida, Siphonostomatoida, Monstrilloida, Misophrioida, and Mormonilliloida) are parasitic at least part of their lives. Many adult stages are so greatly modified that their crustacean relationship is obscure. Their numbers are relatively small, but they may be conspicuous and important as "fish lice," burrowing deep into the flesh or feeding externally upon the skin and gills. SEE MISOPHRIOIDA; MONSTRILLOIDA; MORMONILLOIDA; POECILOSTOMATOIDA; SIPHONOSTOMATOIDA.

Life history. The life histories of copepods differ in details, but in nearly all, development takes place from fertilized eggs which are either attached to the genital segment or are cast free in the water. The spermatozoa, transferred to the female genital segment in spermatophores, are stored in spermathecal sacs in the female. Usually one mating suffices for all subsequent eggs laid by the female. A nauplius larva hatches from the egg. This larva usually undergoes five molts in its growth. Most nauplii are self-sustaining, gathering their own food, but some nauplii hatched from yolky eggs are sustained by the yolk through one or more stages. The sixth naupliar stage metamorphoses into the first of six copepodid stages, the last of which is the adult. Only a few life histories have been studied completely. In *Calanus finmarchicus*, a calanoid, a summer brood in boreal waters requires about 2–3 months to reach maturity, but broods produced in the autumn live longer because they winter in deep water as subadults. The number of broods produced is related to latitude, fewer being produced in the far north.

In *Tisbe furcata*, a bottom-living harpacticoid, individual females may live for 70 days, producing 5–10 broods averaging 43 eggs per brood. The young may produce the first batch of eggs in 15 days. Because of their high fecundity and early sexual maturity certain harpacticoids have proved useful for genetic studies.

Habitats. Copepod habitats are extremely varied, but the vast majority of species live in the sea. They are mostly planktonic throughout their lives, some occurring only in coastal waters while others live in the open sea. Many frequent only the upper layers to about 650–1000 ft (200–300 m) depth. Others are bathypelagic at great depths. The harpacticoids, however, are primarily a benthic, or bottom-dwelling, group, though many are truly planktonic. A great many harpacticoid species live upon aquatic vegetation to which they cling with tiny claws. Others are found on the bottom feeding upon detritus. Some are especially adapted by slender wormlike bodies to burrow or creep into the bottom among the grains of sand. There are also many fresh-water copepods living in ponds, lakes, and streams in all parts of the world. Their wide dispersal to isolated bodies of water is probably brought about largely through transport on birds' feet. A few copepods live naturally in damp moss.

Economic biology. The role of copepods in the economy of the sea consists mainly of converting plant into animal substance. Many species are adapted to graze directly upon the chief synthesizers of organic material, the microscopic diatoms and dinoflagellates that constitute the great pasturage of the sea. Some species are predatory on other copepods, fish larvae, and other small animals, such as protozoa. Commonly copepods constitute about 70% of the zooplankton, and at times their numbers are so great as to impart a pinkish color to the sea. They are of sufficient size to be fed upon individually by young fishes and to be filtered from the water by the filtering apparatus (gill rakers) of such fishes as the herring, sardines, and others that require animal protein as food. At higher trophic levels these fishes in turn are, directly or indirectly, food for the more predatory fishes and mammals. SEE MARINE ECOLOGY.

Harry C. Yeatman

Bibliography. S. P. Parker (ed.), *Synopsis and Classification of Living Organisms*, 2 vols., 1982; C. B. Wilson, *Copepods of the Woods Hole Region*, U.S. Nat. Mus. Bull. 158, 1932.

measuring 1 cm in each direction, is 1.6730×10^{-6} ohm·cm at 20°C (68°F). Only silver has a greater volumetric conductivity than copper. On a relative basis in which silver is rated 100, copper is 94, aluminum 57, and iron 16.

The mass resistivity of pure copper for a length of 1 m weighing 1 g at 20°C (68°F) is 0.14983 ohm. The conductivity of copper on the mass basis is surpassed by several light metals, notably aluminum. The relative values are 100 for aluminum, 50 for copper, and 44 for silver.

By far the largest use of copper is in the electrical industry, and therefore high electrical conductivity is its most important single property, although for industrial use this property must be accompanied by suitable characteristics in other respects. *See* Conductor (electricity); Electrical conductivity of metals.

Chemistry. Copper chemistry involves several valence states.

Copper (I). Removal of the outer $4s$ valence electron from metallic copper yields the diamagnetic (no unpaired electrons) Cu(I) ion having the [argon]$3d^{10}$ electronic configuration. In contrast to Cu(II), Cu(I) compounds frequently are colorless because of the completely filled d shell. When color is present in Cu(I) compounds, it is due to visible light promoting electronic excitations from the filled Cu(I) d orbitals to empty ligand orbitals, or from ligands which themselves are colored. Cu(I) compounds (complexes) consist of two or more molecules or ions (ligands) bound to Cu(I) in a particular geometric arrangement (coordination geometry). The number of ligands (coordination number) most commonly is 4; 2- and 3-coordinations are less common, while 5-coordination is rare and 6-coordination is unknown. Idealized geometries for these coordination numbers are shown in **Fig. 1;** the geometries actually adopted by Cu(I) complexes usually are somewhat distorted from those shown. *See* Coordination number.

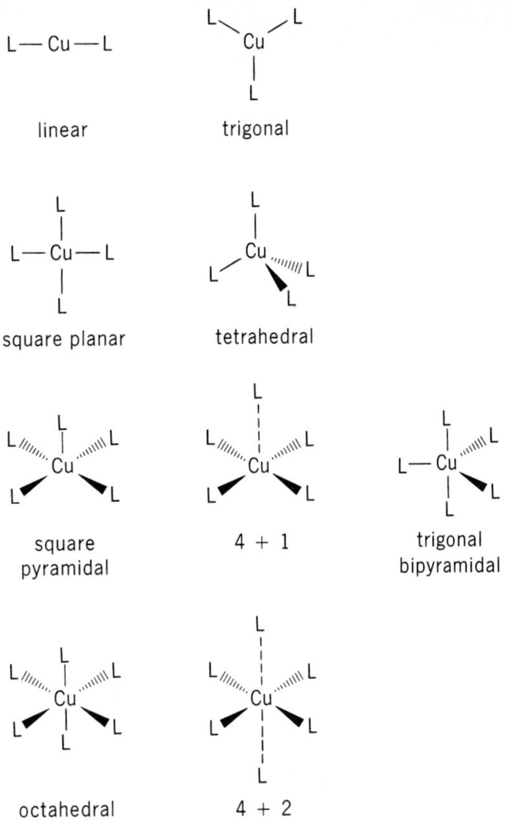

In addition to these mononuclear [one Cu(I)] complexes, a wide variety of polynuclear complexes are known that contain two or more Cu(I) ions (binuclear, trinuclear, and so forth). A simple example is the binuclear complex containing two trigonal Cu(I) ions, two bridging iodide ions, and two neutral terminal ligands (L), as shown in the following structure.

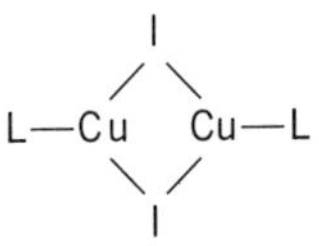

The simple binary compounds copper(I) chloride (CuCl) and copper(I) bromide (CuBr) are prepared by reacting solutions of Cu(II) in aqueous HCl or hydrogen bromide (HBr) with metallic copper [reaction(3)].

$$Cu(II)_{aq} + Cu + 2X^- \rightleftharpoons 2CuX \qquad (X = Cl, Br) \qquad (3)$$

Poorly soluble white CuCl and yellow CuBr precipitate when water is added to lower the halide ion (X^-) concentration and thus the concentration of the fairly soluble charged polyhalide complexes [reaction (4)].

$$CuX + nX^- \rightleftharpoons [CuX_n]^{n-} \qquad (n = 2, 3, 4) \qquad (4)$$

The direction of reaction (3) depends upon the natures of X and the other potential ligands that are

Fig. 1. Idealized coordination geometries of copper complexes.

present, sometimes including the solvent. This reaction proceeds as shown with Cl^-, Br^-, cyanide ion (CN^-), and a variety of inorganic and organic sulfur- and phosphorus-containing ligands such as thiourea [reaction (1)] that bind strongly to Cu(I) and stabilize it relative to Cu(II). The reverse reaction, disproportionation of Cu(I) to metallic copper and Cu(II), is promoted by ligands such as water, ammonia, amines, and acetate that preferentially bind and stabilize Cu(II). *See* Amine.

The binary compounds copper(I) oxide (Cu_2O) and copper(I) sulfide (Cu_2S) are more stable thermally than their Cu(II) analogs. Treatment of copper(II) oxide (CuO) and copper(II) sulfide (CuS) at high temperatures yields the Cu(I) compounds along with evolved O_2 and S, respectively.

A wide variety of mononuclear and polynuclear Cu(I) complexes can be prepared by reacting the appropriate ligands with a Cu(I) halide or by reducing a Cu(II) precursor in the presence of the ligand. *See* Coordination complexes.

In contrast to Cu(II), Cu(I) exhibits an extensive organometallic chemistry, that is, Cu(I) can bond to carbon. The class of compounds known as the lithium alkyl cuprates, $LiCuR_2$ (R = methyl, ethyl, or another alkyl group), are widely used in organic syntheses; carbon-carbon bonds may be formed between the R group and a carbon atom from a variety of organic substrates. Numerous Cu(I) complexes are known with alkene and acetylene ligands; some of these also are synthetically useful. The binding of ethylene ($H_2C=CH_2$), the simplest alkene, to an ill-defined Cu(I) substance present in plants such as bananas promotes rapid ripening. *See* Ethylene; Organometallic compound.

Copper(II). The incompletely filled d shell of Cu(II),

which has the [argon]$3d^9$ electronic configuration, results in Cu(II) complexes being colored and having interesting magnetic properties due to the unpaired electron; it also gives rise to conspicuous structural distortions. Known coordination numbers for Cu(II) complexes span the range 2–9, with the relative numbers of examples following the order $6 = 5 > 4 \gg 7 > 8 \gg 9 \approx 2$. Idealized structural types for the 4-, 5-, and 6-coordinate complexes are shown in Fig. 1. In contrast to Cu(I), Cu(II) does form 6-coordinate complexes, but these are distorted so that four coplanar bonds are shorter than the other two bonds (the 4 + 2 type shown in Fig. 1).

The shapes of the five d-valence orbitals are such that for a hypothetical octahedral complex having six Cu(II)-ligand bonds of equal length, three orbitals point between the ligands and are nonbonding, whereas two orbitals point directly at the ligands and participate in Cu(II)-ligand bonding. The energies of these two bonding orbitals are the same and exceed the energies of the other three d orbitals. Six of the nine d electrons fill these lower-energy orbitals, and the remaining three electrons are distributed among the two isoenergetic bonding d orbitals, either of which can be half-occupied. As explained by the Jahn-Teller theorem, this octahedral complex will distort in some way so that the two bonding orbitals will have different energies, and the d vacancy is then localized in the orbital with the highest energy. The form of this structural distortion frequently amounts to the lengthening of two opposite Cu(II)-ligand bonds (4 + 2 coordination). It also is common for one ligand to dissociate, completely leading to the 4 + 1 square pyramidal coordination; the loss of both opposed ligands results in planar 4-coordination. Thus electronic effects cause 4-coordinate Cu(II) complexes to be square planar with adjacent L—Cu—L angles of 90°. In contrast, 4-coordinate Cu(I) complexes typically are approximately tetrahedral, since in the absence of the electronic effects present in Cu(II) complexes, the larger bond angles of hte tetrahedron (109°) result in reduced ligand-ligand repulsions. *See* BOND ANGLE AND DISTANCE; JAHN-TELLER EFFECT.

Copper(II) complexes nearly always are colored. Some of the d-d excitations of electrons from the four filled d orbitals to the half-occupied d orbital are promoted by light in the visible spectral region. Many complexes absorb red light (the 600–750-nanometer region) and therefore are blue or green. Visible light also can promote electrons from filled orbitals of ligands like organic thiolates (the amino acid cysteine is an important example) into the half-occupied d orbital.

The divalent state of copper is by far the most common one, and a seemingly endless variety of mononuclear and polynuclear complexes are known. Simple Cu(II) salts are quite stable and soluble in water and in polar organic solvents such as methanol and acetonitrile. Complex formation is achieved upon the addition of the desired ligand. Copper(II) has a high affinity for ligands that possess one or more nitrogen-donor groups such as ethylenediamine [(II); reaction (5)] and bipyridine [(III); reaction (6)].

Oxygen-donor ligands such as ionized phenols and carboxylic acids also bind Cu(II) well, and complexes that contain one or more halide ions are common. The chemistry of Cu(II) complexation has been extensively developed and fine-tuned to the extent that sta-

$$Cu(II)_{aq} + 2N\,N \longrightarrow \tag{5}$$

$$\left[\begin{array}{c} OH_2 \\ N\text{---}Cu\text{---}N \\ N \qquad N \\ OH_2 \end{array} \right]^{2+} \qquad (N\,N = H_2NCH_2CH_2NH_2)$$
(II)

$$Cu(II)_{aq} + 2\,\widehat{N\,N} \xrightarrow[PF_6^-]{Cl^-} \tag{6}$$

$$\left[\begin{array}{c} N \\ N\text{---}Cu\text{----}Cl \\ N \qquad N \end{array} \right] PF_6 \qquad (\widehat{N\,N} = \text{(biphenyl structure)})$$
(III)

ble complexes having unnatural structures and ligation have been prepared. The binding of strongly reducing ligands such as thiolates and cyanide ion has been achieved without reduction of the Cu(II) to Cu(I), and rigid ligands have been designed that permit 4-coordinate Cu(II) to overcome the electronic factors favoring planar coordination and achieve nearly tetrahedral geometries.

Transition-metal ions having unpaired electrons act like little magnets whose effective strength (in units of the Bohr magneton) is given by $[n(n+2)]^{1/2}$, where n is the number of unpaired electrons. For mononuclear Cu(II) complexes, the occupancy of five d orbitals by nine electrons results in one unpaired electron and a predicted magnetic strength of 1.73 Bohr magnetons; the observed values span the range 1.7–2.0 Bohr magnetons. Binuclear complexes in which the Cu(II) units are bridged by one or two halide or hydroxide ions have interesting magnetic properties. For most of these complexes, the most energetically stable arrangement of these little magnets is such that their north-south directions are opposed (antiparallel). In this arrangement, the magnetic fields mutually cancel, and the binuclear units become diamagnetic even though the individual Cu(II) sites each have one unpaired electron. The parallel arrangement, in which the magnetic fields couple and add, occurs at a higher energy. In cases where this energy separation is fairly small, some of the binuclear units can be excited thermally to this higher-energy arrangement, and the observed average magnetic strength increases as the temperature is raised. *See* DIAMAGNETISM; MAGNETO-CHEMISTRY; MAGNETON.

Other valence states. The rare zero-valent state is restricted to the Cu_2, Cu_3 polyatomic species present in copper vapor, and several vaguely characterized reaction products of such species with carbon monoxide.

Trivalent copper has the [argon]$3d^8$ electronic configuration. Most Cu(III) complexes are 4-coordinate, square planar, and diamagnetic. The four most stable d orbitals are fully occupied, and the one at highest energy is empty. Simple purely inorganic complexes such as the one with cesium and fluorine ($CsCuF_4$) have been prepared by oxidizing Cu(II) with powerful chemical oxidants. Complexes of Cu(III) having organic ligands are best prepared by the electrochemical oxidation of the Cu(II) precursors.

Complexes in which Cu(III) is ligated either by four sulfur donors from organic dithiolates or by four nitrogen donors from amines or ionized amide groups of peptides have been the most fully studied. There is no unambiguous example of a naturally occurring Cu(III) compound; a suggestion that Cu(III) is present in an enzyme known as galactose oxidase remains the subject of controversy.

The chemistry of Cu(IV) has been restricted to a few fluoride and oxide complexes. Only those ligands that are most resistant to oxidation can be expected to survive exposure to such a powerful oxidant as Cu(IV). SEE COORDINATION CHEMISTRY.

Biological systems. Copper-containing proteins provide diverse biochemical functions, including copper uptake and transport (ceruloplasmin), copper storage (metallothionen), protective roles (superoxide dismutase), catalysis of substrate oxygenation (dopamine β-monooxygenase), biosynthesis of connective tissue (lysyl oxidase), terminal oxidases for oxygen metabolism (cytochrome *c* oxidase), oxygen transport (hemocyanin), and electron transfer in photosynthetic pathways (plastocyanin).

The first six classes of copper proteins are of particular importance to human biochemistry. It is the strategic variations in types of ligands, coordination geometries, valences, and numbers of copper ions present in the so-called active sites of these proteins that tune them for their vastly different chemical behavior. The chemically active sites in these proteins bind copper ions via the nitrogen-, sulfur-, and oxygen-donor ligands present as side chains of protein amino acid components such as histidine, cysteine, and glutamic acid. In contrast to iron, biological copper never is bound to an extrudable and isolable unit like a porphyrin; the active sites do not persist if the protein is destroyed. A healthy human body that weighs about 150 lb (70 kg) contains transition-metal ions in the approximate amounts of 0.004 oz (0.1 g) of copper, 0.07 oz (2 g) of zinc, and 0.2 oz (5 g) of iron; traces of manganese, molybdenum, chromium, cobalt, and nickel also are present. SEE AMINO ACIDS; PROTEIN.

It is useful to classify copper metalloproteins according to whether their active sites contain so-called blue copper, binuclear copper, conventional 4 + 1 or 4 + 2 copper, or a combination of these structural types.

Blue copper refers to the fact that the blue color of this Cu(II) site is considerably more intense (approximately 50 times) than the color of conventional Cu(II) complexes having coordination numbers of 4 to 6. Protein crystallographic studies have revealed the detailed structures of the blue sites present in a plant plastocyanin isolated from popular leaves and the bacterial azurin isolated from *Alcaligenes denitrificans* (**Fig. 2**). These sites share the feature of having approximately planar Cu(II)N$_2$S(cysteine) units additionally bound at a rather longer distance by methionine sulfur. Their intense color is due not to a *d-d* excitation but to the ligand-to-metal charge-transfer excitation of an electron from the tightly bonded S(cysteine) ligand into the single Cu(II) *d* vacancy. The Cu(I) forms of both metalloproteins are colorless. Both proteins participate in biological electron-transfer (redox) reactions whereby the valence at the active sites shuttles between the Cu(I) and Cu(II) states. SEE ELECTRON TRANSFER REACTION.

Binuclear copper sites are present in a wide variety of metalloproteins whose transport functions require the chemically reversible binding of O$_2$ or whose oxygenase functions require the binding of O$_2$ to acti-

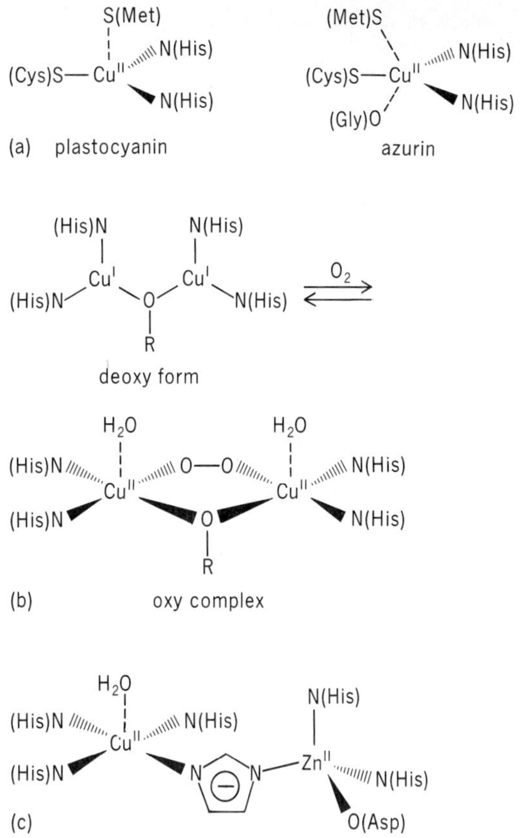

Fig. 2. Active site structures of copper metalloproteins. (*a*) Blue copper. (*b*) Hemocyanin and tyrosinase. (*c*) Superoxide dismutase. His = histidine. Met = methionine. Asp = aspartate.

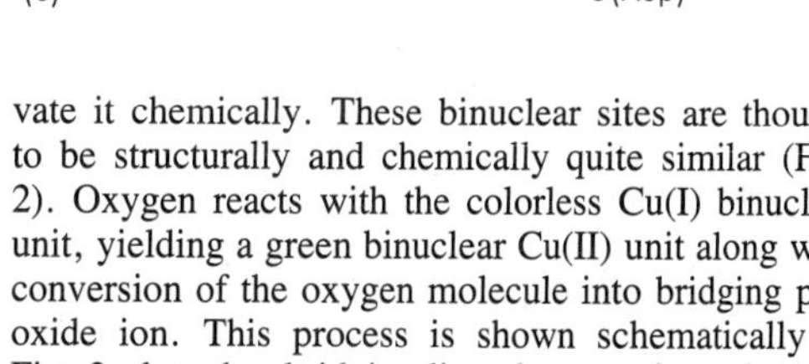

vate it chemically. These binuclear sites are thought to be structurally and chemically quite similar (Fig. 2). Oxygen reacts with the colorless Cu(I) binuclear unit, yielding a green binuclear Cu(II) unit along with conversion of the oxygen molecule into bridging peroxide ion. This process is shown schematically in Fig. 2; the other bridging ligand comes from the protein, and the N ligands represent imidazoles from histidine residues in the protein chains. The resulting oxy-complex is diamagnetic; the bridging ligands cause the two unpaired electrons on the CU(II) ions to adopt the antiparallel arrangement, and their magnetic fields mutually cancel. Units of this type are present in arthropods and mollusks; they bind oxygen and transport it to internal tissues where it is required for metabolism. Tyrosinases can also bind oxygen reversibly in the absence of oxidizable organic substrates. However, their main biochemical role in a wide variety of fungi, plants, and animals is to use the bound and chemically activated oxygen to oxidize phenols to catechols and quinones [reactions (7) and

$$\text{(phenol)} + O_2 + 2e^- + 2H^+ \longrightarrow \text{(catechol)} + H_2O \qquad (7)$$

$$\text{(catechol)} + \tfrac{1}{2} O_2 \longrightarrow \text{(quinone)} + H_2O \qquad (8)$$

(8)]. Tyrosinases in mushrooms are responsible for the dark brown colors that the fungi develop after being picked, and those in humans are responsible for the formation of the dark melanin skin pigments.

Small amounts of superoxide ion (O$_2^-$) are by-products of oxygen metabolism and are believed by some to be toxic. Protection against superoxide ion in red blood cells is achieved by the so-called dismutation reaction (9) that is efficiently catalyzed by the

$$2O_2^- + 2H^+ \rightarrow O_2 + H_2O_2 \qquad (9)$$

enzyme superoxide dismutase (SOD). Protein crystallographic studies have revealed that the active site contains a conventional 5-coordinate Cu(II) unit that is bridged to an approximately tetrahedral Zn(II) unit via a deprotonated imidazole group (Fig. 2). Metal-ion substitution studies indicate that reaction (9) occurs at the Cu(II) rather than the Zn(II) site. SEE OX-YGEN TOXICITY; SUPEROXIDE CHEMISTRY.

There are multicopper oxidases that contain all three types of copper active sites. Some of the best-studied examples include plant laccases and ascorbic acid oxidases. These enzymes catalyze the oxygenative dehydrogenation of organic substrates such as amines, phenols, and dehydroascorbic acid, as shown in reaction (10).

$$4AH + O_2 \rightarrow 4A + 2H_2O \qquad A = \overset{\diagdown}{\underset{\diagup}{\text{---}}} N, \overset{\diagdown}{\underset{\diagup}{\text{---}}} C \qquad (10)$$

SEE BIOINORGANIC CHEMISTRY; ENZYME.

Harvey Schugar

Bibliography. F. A. Cotton and G. Wilkinson, *Advanced Inorganic Chemistry: A Comprehensive Text*, 5th ed., 1988; L. Que, Jr. (ed.), *Metal Clusters in Proteins*, Amer. Chem. Soc. Symp. Ser. 372, 1988; G. Wilkinson (ed.), *Comprehensive Coordination Chemistry*, vol. 5: *Late Transition Elements*, 1987.

Copper alloys

Solid solutions of one or more metals in copper. Many metals, although not all, alloy with copper to form solid solutions. Some insoluble metals and nonmetals are intentionally added to copper alloy to enhance certain characteristics.

Copper alloys form a group of materials of major commercial importance because they are characterized by such useful mechanical properties as high ductility and formability and excellent corrosion resistance. Copper alloys are easily joined by soldering and brazing. Like gold alloys, copper alloys have decorative red, pink, yellow, and white colors. Copper has the second highest electrical and thermal conductivity of any metal. All these factors make copper alloys suitable for a wide variety of products.

Wrought alloys. Copper and zinc melted together in various proportions produce one of the most useful groups of copper alloys, known as the brasses. Six different phases are formed in the complete range of possible compositions. The relationship between composition and phases alpha, beta, gamma, delta, epsilon, and eta are graphically shown in the well-established constitution diagram for the copper-zinc system. Brasses containing 5–40% zinc constitute the largest volume of copper alloys. One important alloy, cartridge brass (70% copper, 30% zinc), has innumerable uses, including cartridge cases, automotive radiator cores and tanks, lighting fixtures, eyelets, rivets, screws, springs, and plumbing products. Tensile strength ranges from 45,000 lb/in.^{2} (310 megapascals) as annealed to 130,000 lb/in.^{2} (900 MPa) for spring temper wire. SEE BRASS; PHASE EQUILIBRIUM.

Lead is added to both copper and the brasses, forming an insoluble phase which improves machin-ability of the material. Free cutting brass (61% copper, 3% lead, 36% zinc) is the most important alloy in the group. In rod form it has a strength of 50,000–70,000 lb/in.^{2} (340–480 MPa), depending upon temper and size. It is machined into parts on high-speed (10,000 rpm) automatic screw machines for a multiplicity of uses.

Increased strength and corrosion resistance are obtained by adding up to 2% tin or aluminum to various brasses. Admiralty brass (70% copper, 2% aluminum, 0.023% arsenic, 21.97% zinc) are two useful condenser-tube alloys. The presence of phosphorus, antimony, or arsenic effectively inhibits these alloys from dezincification corrosion.

Alloys of copper, nickel, and zinc are called nickel silvers. Typical alloys contain 65% copper, 10–18% nickel, and the remainder zinc. Nickel is added to the copper-zinc alloys primarily because of its influence upon the color of the resulting alloys; color ranges from yellowish-white, to white with a yellowish tinge, to white. Because of their tarnish resistance, these alloys are used for table flatware, zippers, camera parts, costume jewelry, nameplates, and some electrical switch gear.

Copper forms a continuous series of alloys with nickel in all concentrations. The constitution diagram is a simple all alpha-phase system.

Nickel slightly hardens copper, increasing its strength without reducing its ductility. Copper with 10% nickel makes an alloy with a pink cast. More nickel makes the alloy appear white.

Three copper-base alloys containing 10%, 20%, and 30% nickel, with small amounts of manganese and iron added to enhance casting qualities and corrosion resistance, are important commercially. These alloys are known as cupronickels and are well suited for application in industrial and marine installations as condenser and heat-exchanger tubing because of their high corrosion resistance and particular resistance to impingement attack. Heat-exchanger tubes in desalinization plants use the cupronickel 10% alloy.

Copper-tin alloys (3–10% tin), deoxidized with phosphorus, form an important group known as phosphorus bronzes. Tin increases strength, hardness, and corrosion resistance, but at the expense of some workability. These alloys are widely used for springs and screens in papermaking machines.

Silicon (1.5–3.0%), plus smaller amounts of other elements, such as tin, iron, or zinc, increases the strength of copper, making alloys useful for hardware, screws, bolts, and welding rods.

Sulfur (0.35%) and tellurium (0.50%) form insoluble compounds when alloyed with copper, resulting in increased ease of machining.

Precipitation-hardenable alloys. Alloys of copper that can be precipitation-hardened have the common characteristic of a decreasing solid solubility of some phase or constituent with decreasing temperature. Precipitation is a decomposition of a solid solution leading to a new phase of different composition to be found in the matrix. In such alloy systems, cooling at the appropriate rapid rate (quenching) from an equilibrium temperature well within the all-alpha field will preserves the alloy as a single solid solution possessing relatively low hardness, strength, and electrical conductivity.

A second heat treatment (aging) at a lower temperature will cause precipitation of the unstable phase. The process is usually accompanied by an increase in hardness, strength, and electrical conductivity. Some

19 elements form copper-base binary alloys that can be age- or precipitation-hardened. *See Heat treatment (metallurgy)*.

Two commercially important precipitation-hardenable alloys are beryllium- and chromium-copper. Beryllium-copper (2.0–2.5% beryllium plus cobalt or nickel) can have a strength of 200,000 lb/in.^{2} (1400 MPa) and an electrical conductivity of less than 50% of IACS (International Annealed Copper Standard). Cobalt adds high-temperature stability, and nickel acts as a grain refiner. These alloys find use as springs, diaphragms, and bearing plates and in other applications requiring high strength and resistance to shock and fatigue.

Copper chromium (1% chromium) can have a strength of 80,000 lb/in.^{2} (550 MPa) and an electrical conductivity of 80%. Copper-chromium alloys are used to make resistance-welding electrodes, structural members for electrical switch gear, current-carrying members, and springs.

Copper-nickel, with silicon or phosphorus added, forms another series of precipitation-hardenable alloys. Typical composition is 2% nickel, 0.6% silicon. Strength of 120,000 lb/in.^{2} (830 MPa) can be obtained with high ductility and electrical conductivity of 32% IACS.

Zirconium-copper is included in this group because it responds to heat treatment, although its strength is primarily developed through application of cold deformation or work. Heat treatment restores high electrical conductivity and ductility and increases surface hardness. Tensile strength of 70,000 lb/in.^{2} (500 MPa) coupled with an electrical conductivity of 88% can be developed. Uses are resistance welding wheels and tips, stud bases for transistors and rectifiers, commutators, and electrical switch gear.

Fabrication. Copper alloys have a long history because of their ease of fabrication into utensils and products.

Melting. Alloys are initially prepared by melting the various elements involved in low-frequency induction furnaces. Usually the melt is protected from oxidation by covering it with charcoal. Copper-rich alloys are also deoxidized by adding elements such as phosphorus, lithium, or boron. The melt is usually poured at 100–200°F (38–93°C) above its melting point into oil- and graphite-dressed, water-cooled copper molds of suitable shape. The dressing protects the mold surfaces, produces a reducing gas that aids in the suppression of oxidation, and brings dirt and dross to the molten surface, preventing their inclusion in the casting.

Continuous casting of copper and its alloys is common practice and is accomplished by pouring the melt into the top of an open-end mold as the solidified casting is withdrawn from the bottom. Various mold designs are used, but all use water to cool the mold proper in addition to cooling sprays and tanks of water below the mold. Casting can be performed in either a vertical or a horizontal plane.

Mechanical properties of castings can be affected by such defects as large grain size, directional grain growth, and segregation. Either hot-working of the casting by rolling, forging, piercing, or extrusion, or cold-working by rolling or drawing with necessary intermediate annealing corrects these defects and markedly improves properties.

Processing. Processing varies with alloy, form, and size. Flat products are cast in heavy cakes for initial hot-rolling or in bars for cold-rolling. Copper and lead-free brasses are readily hot-rolled in a single heating from a 9-in. (23-cm) thick cake to a ½-in. (1.25-cm) thick slab. Further rolling is done cold with intermediate anneals. Leaded brasses, tin bronzes, and other hot short alloys (alloys that are brittle when heated above red heat) must be cold-rolled from the casting to final size, using intermediate anneals that heat the particular alloy above its recrystallization temperature.

Round copper rods and wire are started by hot-rolling castings (wire bar) in grooved rolls, but most copper alloys are extruded. Both are subsequently reduced in size by cold-rolling, drawing, or both. In extrusion, a billet placed in a press is forced through a lubricated die by a ram.

Tin and tin-lead bronzes which are hot short are continuously cast in rod form and are cold-rolled or drawn to finish sizes using intermediate anneals as necessary.

Copper and high-copper-content lead-free brass tubes are started by hot piercing wherein the solid billet is driven with a helical motion over a rotating mandrel to form a seamless tube. Tubes of most other alloys are formed by extrusion because the alloys do not stand piercing. In extrusion of tube, a mandrel is pushed through a billet forming the inside of the tube, and a ram pushes the billet over the mandrel and through a die to form a seamless tube.

Cast alloys. Copper alloy castings of irregular and complex external and internal shapes can be produced by various casting methods, making possible the use of shapes for superior corrosion resistance, electrical conductivity, good bearing quality, and other attractive properties. High-copper alloys with varying amounts of tin, lead, and zinc account for a large percentage of all copper alloys used in the cast form. Tensile strength ranges from 36,000 to 48,000 lb/in.^{2} (250–330 MPa) depending upon composition and size. Leaded tin-bronzes with 6–10% tin, about 1% lead, and 4% zinc are used for high-grade pressure castings, valve bodies, gears, and ornamental work. Bronzes high in lead and tin (7–25% lead and 5–10% tin) are mostly used for bearings. High tin content is preferred for heavy pressures or shock loading, but lower tin and higher lead for lighter loads, higher speeds, or where lubrication is less certain. A leaded red brass containing 85% copper and 5% each of tin, lead, and zinc is a popular alloy for general use.

High-strength brasses containing 57–63% copper, small percentages of aluminum and iron, and the balance zinc have tensile strengths from 70,000 to 120,000 lb/in.^{2} (490–830 MPa), high hardness, good ductility, and resistance to corrosion. They are used for valve stems and machinery parts requiring high strength and toughness.

Copper alloys containing less than 2% alloying elements are used when relatively high electrical conductivity is needed. Strength of these alloys is usually notably less than that of other cast alloys.

Aluminum bronzes containing 5–10.5% aluminum, small amounts of iron, and the balance copper have high strengths even at elevated temperature, high ductility, and excellent corrosion resistance. The higher-aluminum-content castings can be heat-treated, increasing their strength and hardness. These alloys are used for acid-resisting pump parts, pickling baskets, valve seats and guides, and marine propellers.

Additives impart special characteristics. Manganese is added as an alloying element for high-strength brasses where it forms intermediate compounds with other elements, such as iron and aluminum. Nickel additions refine cast structures and add toughness, strength, and corrosion resistance. Silicon added to copper forms copper-silicon alloys of high strength and high corrosion resistance in acid media. Beryllium or chromium added to copper forms a series of age- or precipitation-hardenable alloys. Copper-beryllium copper alloys are among the strongest of the copper-base cast materials. SEE ALLOY; COPPER; METAL CASTING; METAL FORMING.

Ralph E. Ricksecker

Bibliography. J. H. Mendenhall, *Understanding Copper Alloys*, 1977, reprint 1986; E. G. West, *Copper and Its Alloys*, 1982.

Copper loss

The power loss caused by the flow of current through copper conductors. When an electric current flows through a copper conductor (or any conductor), some energy is converted to heat. The heat, in turn, causes the operating temperature of the device to rise. This happens in transformers, generators, motors, relays, and transmission lines, and is a principal limitation on the conditions of operation of these devices. Excessive temperature rises lead to equipment failure. SEE CONDUCTOR (ELECTRICITY).

If the resistance R (in ohms) and the current I (in amperes) are both known, the copper loss in watts may be found from the equation $P = I^2R$. An alternative form is to know the voltage across the resistor V (in volts) and to determine the power loss by the equation $P = V^2/R$. V and I are effective values of the voltage and current, respectively. The resistance R varies with both frequency and temperature. SEE ELECTRICAL RESISTANCE.

In electric machines, there are typically constant losses, as well as those that vary with the square of the load, such as the copper loss. Maximum efficiency is obtained when the load is adjusted so as to make the constant and varying losses equal.

Since the discovery of superconductivity in 1911, much effort has been directed toward the development of devices based on superconducting principles, with superconductors replacing the copper coils in electrical devices and systems in order to reduce or eliminate copper loss. Since superconductors have traditionally operated at temperatures close to absolute zero (0 K or −459.67°F), such devices are practical only if the power required to operate the refrigeration equipment is less than the copper loss that is supplanted. In 1986, a ceramic superconductor, lanthanum barium copper oxide, was discovered whose superconducting temperature was 30 K (−406°F), and by 1988 another ceramic, thallium barium calcium copper oxide, was shown to have a superconducting temperature of 125 K (−235°F). The rapid increase of superconducting temperatures has given new hope to the idea of reducing copper loss. However, the process of going from a scientific discovery to technological realization requires solutions to many technical problems as well as complex problems involving economic, reliability, safety, and related considerations. SEE SUPERCONDUCTING DEVICES; WIRING.

Edwin C. Jones, Jr.

Bibliography. Electric Power Research Institute, *Cost Components of High-Capacity Transmission Options*, Rep. EL-1065, May 1979; K. Fitzgerald, Superconductivity: Fact versus fancy, *IEEE Spectrum*, 25(5):30–41, May 1988; W. Hively, Closer to room temperature, *Amer. Sci.*, 76(6):245–246, May–June 1988; J. L. Kirtley, Jr., and M. Furuyama, A design concept for large superconducting alternators, *IEEE Trans.*, PAS-94(4):1264–1269, July/August 1975.

Copper metallurgy

The economic production of pure copper metal, suitable for fabrication and use, from copper ores containing as little as 0.5% Cu. Over 90% of the consumption of primary copper in the Western world is produced from ores containing sulfide minerals (chalcopyrite, $CuFeS_2$; chalcocite, Cu_2S; and bornite, Cu_5FeS_4) that can be economically treated only by pyrometallurgical processes. SEE PYROMETALLURGY, NONFERROUS.

Production. The main processes used in the production of copper from sulfide ores are shown in **Fig. 1**. The mined ore (0.5–2.0% Cu) is finely ground, and then concentrated by flotation to form copper concentrates containing 20–30% Cu. The concentrates are then smelted at high temperatures (about 2280°F or 1250°C) to form a molten mixture of copper and iron sulfides called matte. The molten matte is converted to blister copper (about 99% Cu) by oxidizing the remaining iron and sulfur. After removing the residual sulfur and oxygen in an anode furnace, copper anodes are cast and then refined electrolytically to produce high-purity cathode copper (99.99% Cu), which is suitable for most uses. SEE ORE DRESSING.

Smelting processes. The reverberatory furnace (**Fig. 2**) is the oldest and most widely used smelting process. It consists of a refractory-lined chamber, typically 100 ft (30 m) long by 30 ft (10 m) wide, into which copper concentrates and silica flux are charged. Fuel-fired burners melt the charge, driving off the labile sulfur by reaction (1). Little iron sulfide

$$2CuFeS_2 \rightarrow Cu_2S \cdot 2FeS(l) + \frac{1}{2}S_2(g) \qquad (1)$$

is oxidized, so that fuel requirements are high, about 5.4×10^6 Btu per ton (6.3×10^6 kilojoules per metric ton) of Cu concentrate. Two molten layers are formed in the furnace: an upper layer slag of iron silicate with little copper ($\leq 0.5\%$ Cu); and a lower layer of matte (30–40% Cu), containing most of the original copper value. The slag and matte are drained separately from the furnace through tapholes into ladles. The slag is discarded, while the matte is transferred to the converting step. The sulfur evolved during smelting leaves the furnace in a 1–2% SO_2 gas stream, which is too dilute for economic treatment.

The reverberatory furnace cannot meet today's requirements for low-energy consumption and stringent environmental control. Therefore more efficient smelting processes have been developed since the 1960s. These processes use much less energy (typically 0.7–1.8×10^6 Btu per ton or 0.8–2.1×10^6 kJ per metric ton of Cu concentrate) and produce a strong SO_2 gas stream (10% SO_2) to reduce treatment costs. These processes use oxygen enrichment, and oxidize more iron sulfide to generate more heat and produce mattes with higher copper levels (50–75% Cu). The smelting reaction for these processes can be represented by reaction (2), where α is the fraction of FeS reacted (typically, in range $\alpha = 0.5$ to 0.9). Modern smelting processes fall into two categories:

$$2CuFeS_2 + 2\alpha \cdot SiO_2 + (1 + 3\alpha)O_2 \rightarrow$$
$$Cu_2S \cdot 2(1 - \alpha)FeS + 2\alpha \cdot FeO \cdot SiO_2$$
$$+ (1 + 2\alpha)SO_2 \quad (2)$$

flash smelting (Outokumpu flash smelting furnace) and bath smelting (Noranda process reactor). In flash smelting, dry concentrate is dispersed in an oxidizing gas stream, and the smelting reactions occur very rap- idly as the particles fall down a reaction shaft. The molten matte and slag are collected in a hearth, and the SO_2-containing gases exit via an uptake shaft. In bath smelting, moist concentrate is smelted continu- ously in a molten bath of matte and slag, which is vigorously stirred by the injection of air or oxygen- enriched air. In the Noranda process reactor, the air is injected through tuyeres into a vessel similar to an elongated converter.

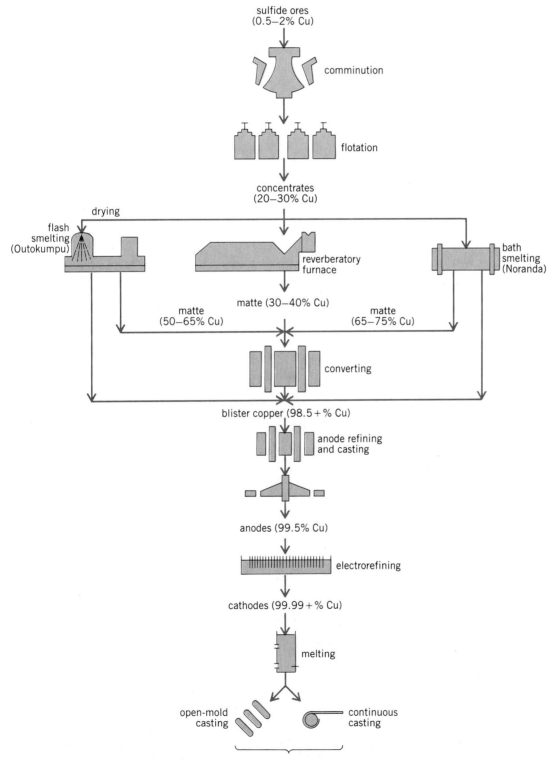

Fig. 1. Flow sheet of cop- per production from sul- fide minerals.

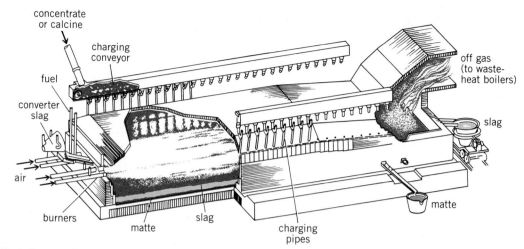

Fig. 2. Cutaway view of a reverberatory furnace for the production of copper matte from sulfide concentrates or roasted calcines. (*After A. K. Biswas and W. D. Davenport, Extractive Metallurgy of Copper, Pergamon Press, 1980*)

Matte converting. The molten matte is converted to blister in Peirce-Smith converters. The converter is a refractory-lined, cylindrical vessel, typically 30 ft (10 m) in length and 13 ft (4 m) in diameter (**Fig. 3**). The converter can be rotated about its axis, and is fitted on one side with a row of about 50 tuyeres through which air is injected. The top of the converter has a large mouth for charging molten matte and removing slag and product copper. Converting is a batch operation; initially several ladles of matte are charged, the air turned on, and the converter rotated until the tuyeres are submerged. The air, bubbling violently through the bath, gradually oxidizes the matte in two stages. In the first or slag-forming stage, iron sulfide is oxidized and fluxed with silica to form a fluid slag by reaction (3). The converter slag contains

$$2FeS(l) + 2SiO_2(s) + 3O_2(g) \rightarrow$$
$$2FeO \cdot SiO_2(l) + 2SO_2(g) \quad (3)$$

some copper (1–5% Cu), and is recycled to the smelting furnace or treated in a separate process. When all the iron has been removed, the remaining copper sulfide is further oxidized to blister copper by reaction (4). The converting process is sufficiently exothermic

$$Cu_2S(l) + O_2(g) \rightarrow 2Cu(l) + SO_2(g) \quad (4)$$

that no additional fuel is required. The blister copper from the converter is transferred by ladle to the anode furnace, where the residual sulfur and oxygen levels in the copper are reduced further. The copper is then cast into anodes for electrorefining.

Sulfur fixation. Smelting and converting a typical copper concentrate generates over 0.55 ton SO_2 per ton concentrate (0.50 metric ton SO_2 per metric ton concentrate), and the resulting SO_2 emissions must be controlled to meet local environmental standards. This is generally achieved by converting the SO_2 to

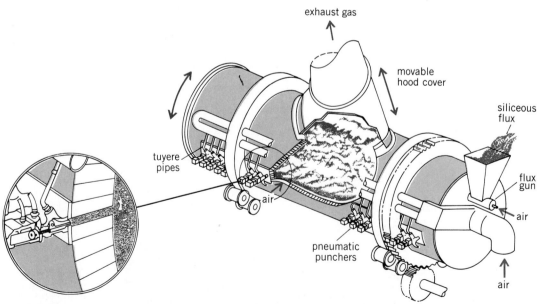

Fig. 3. Cutaway view of a horizontal side-blown Peirce-Smith converter for producing blister copper from matte. (*After A. K. Biswas and W. G. Davenport, Extractive Metallurgy of Copper, Pergamon Press, 1980*)

Copulatory organ 423

sulfuric acid in a contact acid plant, as long as the SO_2 concentration exceeds 4% and a viable market for acid exists. If local conditions are favorable, it is also possible to make liquid SO_2 or elemental sulfur from strong SO_2 gases. SEE SULFUR.

Electrorefining. The objective of electrorefining is to remove the remaining impurities in the anode copper (principally As, Bi, Ni, Pb, Sb, and Se) and produce a pure cathode copper (99.99 + % Cu). Also, many copper ores contain appreciable amounts of precious metals (Ag, Au, Pt, and so on), which are concentrated into the anode copper during smelting and are recovered as valuable by-products in electrorefining. The impure anodes are suspended alternately with pure copper cathodes in tanks through which an electrolyte of copper sulfate and free sulfuric acid is continuously circulated. When direct current is applied, the copper in the anodes is electrochemically dissolved and then plated as pure copper on the cathodes. Some of the anode impurities, such as arsenic and nickel, are less noble than copper and dissolve in the electrolyte, but they do not plate out at the cathode as long as their concentrations are controlled. The other impurities, such as silver, lead, and selenium, are virtually insoluble in the electrolyte and fall as slimes to the bottom of the tank. These slimes are recovered and processed for eventual recovery of selenium and the precious-metal values. SEE ELECTROCHEMICAL PROCESS; ELECTROMETALLURGY.

Hydrometallurgical processes. Oxidized copper ores are more effectively treated by hydrometallurgical processes. The ore is crushed, ground if necessary, and leached with dilute sulfuric acid, either by percolation through heaps of ore or by agitation in tanks. Copper is recovered from the resulting solution by either cementation or solvent extraction–electrowinning. In cementation, copper is precipitated by contact with scrap iron to form an impure cement copper, which is smelted, then refined. Solvent extraction–electrowinning has become the preferred process. In solvent extraction special organic reagents are used to selectively extract copper from solution. The resulting copper-containing organic phase is then stripped to give a pure and more concentrated aqueous copper solution for electrowinning. Electrowinning is similar to electrorefining except that an inert anode is used and more energy is required. Although electrowon cathode copper is generally not as pure as electrorefined copper, it is still suitable for many applications. SEE HYDROMETALLURGY; SOLVENT EXTRACTION.

Future trends. Considerable progress has been made in improving the efficiency of copper smelting since the 1960s through new process innovations and the use of oxygen in large tonnages. The trend toward the production of higher-matte grades (up to 75% Cu) is expected to continue in order to minimize energy requirements and SO_2-fixation costs. Pyrometallurgical processes will continue to dominate primary copper production, and it is anticipated that the ultimate goal of direct copper production from chalcopyrite concentrates in a single smelting and converting vessel will be achieved. SEE COPPER.

John G. Peacey

Bibliography. American Institute of Mining and Metallurgical Engineers (AIME), *Copper Smelting: An Update*, 1982; AIME, *Extractive Metallurgy of Copper*, vols. 1 and 2, 1976; A. K. Biswas and W. G. Davenport, *Extractive Metallurgy of Copper*, 1980; A. Butte, *Copper: The Science and Technology of the Metal, Its Alloys and Compounds*, 1954; E. G. West, *Copper and Its Alloys*, 1982.

Copulatory organ

An organ utilized by the male animal for insemination, that is, to deposit spermatozoa directly into the female reproductive tract. Various types of copulatory organs are found among the vertebrates, whereas cloacal apposition occurs in most other vertebrates which lack these structures. Amplexus, or false copulation, is peculiar to amphibians.

In fish, internal fertilization is restricted to certain groups. The pelvic fin of male elasmobranchs and holocephalians is modified for the transmission of sperm and is known as the clasper or clasping organ. A pair of anterior claspers (**Fig. 1**a) and a frontal

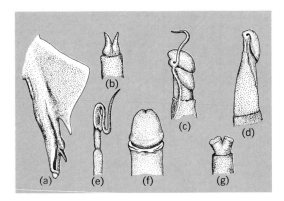

Fig. 1. Vertebrate copulatory organs. (a) Clasper of dogfish (*Squalus*). Glans penis of (b) opossum, (c) ram, (d) bull, (e) short-tailed shrew, (f) man, (g) Echidna.

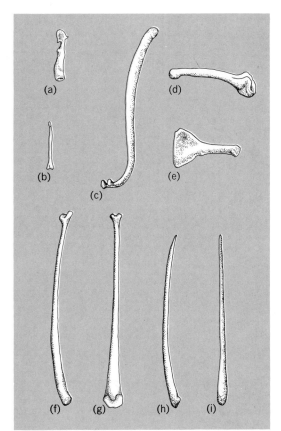

Fig. 2. Bacula of (a) fox squirrel, (b) cotton mouse, (c) otter, (d) *Neotoma lepida*, (e) *N. micropus*, (f, g) *Perognathus hispidus*, showing lateral and ventral views, (h, i) *P. baileyi*, showing lateral and ventral views.

clasper which protrudes from the head occur among holocephalians in addition to the modified pelvic clasper of elasmobranchs. The anal fin of teleosts, in which copulation occurs, may be elongated to form a gonopodium. Copulatory organs are lacking in amphibians, even in those in which internal fertilization occurs. Among caecelians, the cloaca of the male is muscular and protrusible; however, it is questionable that this structure is a true copulatory organ. Hemipenes which can be everted during copulation are common to both snakes and lizards; however, these structures lack erectile tissue. Turtles and crocodiles possess a single penis with associated erectile tissue, the corpora cavernosa, which becomes distended with blood. A few species of birds have a penis; among these are the ostriches and anseriforms. The penis is present in all mammals.

This copulatory organ (Fig. 1b–g) is variously modified morphologically. Among certain mammals such as rodents, bats, whales, some carnivores, and lower primates, a penis bone, known as the os priapi or baculum (**Fig. 2**), occurs, increasing the rigidity of the penis. *See Fertilization; Penis; Reproductive system.*

Charles B. Curtin

Coquina

A calcarenite or clastic limestone whose detrital particles are chiefly fossils, whole or fragmented. The term is most frequently used for an aggregate of large shells more or less cemented by calcite. If the rock consists of fine-sized shell debris, it is called a microcoquina. Encrinite is a microcoquina made up primarily of crinoid fragments (see **illus**.). Some coquinas

Coquinoid limestone, encrinite, showing many entire fossils in fine-grained matrix. Note lack of assortment and diversity of types. (*From F. J. Pettijohn, Sedimentary Rocks, 3d ed., Harper and Row, 1949, 1957*)

show little evidence of any transportation by currents; articulated bivalves are preserved in entirety and the shells are not broken or abraded. *See Calcarenite; Limestone.*

Raymond Siever

Coraciiformes

A diverse order of land birds that is found mainly in the tropics. The relationships of these families have been disputed, with no resolution; the closest affini-

ties may be with other land birds, such as the Piciformes and Passeriformes. *See Passeriformes; Piciformes.*

Classification The Coraciiformes are divided into four suborders and families: Alcedines, including the families Alcedinidae (kingfishers; 91 species), Todidae (todies; 5 species), and Momotidae (motmots; 9 species); Meropes, with the single family Meropidae (bee-eaters; 24 species), Coracii, containing the families Coraciidae (rollers; 11 species), Brachypteraciidae (ground-rollers; 5 species), Leptosomatidae (cuckoo-rollers; 1 species), Upupidae (hoopoes; 1 species), and Phoeniculidae (wood hoopoes; 8 species); and Bucerotes, with the single family Bucerotidae (hornbills; 45 species).

Several of these families, such as the different groups of rollers, are sometimes merged into a single family, and several families, such as the kingfishers and hornbills, are divided into subfamilies.

Fossil record. The fossil record of this group of basically forest birds is rather poor; however, there are a number of specimens known which appear to belong to the Coraciiformes. A better picture of the history and relationships of this order may appear once these fossils are better studied. Most interesting is a fossil tody (a group now restricted to the Greater Antilles) from the Oligocene of Wyoming, and a second fossil tody from the Eo-oligocene of France, and a fossil motmot (a group that is now restricted to the American tropics) from the Oligocene of Switzerland. These specimens suggest the possible earlier, broader distribution of several coraciiform families.

Characteristics. Coraciiform birds are characterized by a syndactyl foot in which the three anterior toes are joined together at the base, although the cuckoo rollers have a zygodactyle foot. Most groups are tropical or warm temperate and are brilliantly colored. Except for bee eaters and wood hoopoes, coraciiforms are generally solitary. Most temperate species are migratory; the tropical ones are permanent residents. Most coraciiforms feed on insects and other animal prey, including fish; hornbills are omnivorous and some feed mainly on fruit. All nest, either solitarily or in colonies in holes burrowed in earthern banks or in tree cavities. Wood hoopoes and bee eaters have young adults serving as helpers at the nest. The chicks are born naked and cared for by both parents until they can fly. The todies, motmots, and one small subfamily of kingfishers are found in the New World, but all other coraciiforms live in the Old World. The kingfishers are the largest and most diversified family, with many members catching their prey on dry land. The largest and most famous kingfisher is the kookaburra or laughing jackass of Australia. Rollers are widespread throughout the Old World tropics, but the ground rollers and the cuckoo rollers are restricted to Madagascar. Wood hoopoes are confined to the tropical forests of Africa, while the related hoopoes are widespread throughout Europe and Asia. The hornbills are the largest of the coraciiforms, feed in forests, and are excellent soarers. A number of species possess a large casque on the base of the upper jaw which may serve in courtship or species recognition. The male seals up the female with her egg in the nesting cavity in a large tree, and feeds her and the chick until the young bird is ready to fly. During this period, the female undergoes a complete molt. Some species of hornbills are threatened be-

(a)

Coralline algae on reefs. (*a*) Rongerik Atoll, Marshall Islands. (*b*) Near Ylig Point, Guam. (*c*) Stinging coral *Millepora* and coralline alga *Porolithon*, Bikini Atoll, Marshall Islands. (*d*) Closeup of same habitat, showing mostly *Porolithon*. (*e*) Coral, crustose coralline alga, and soft green algae, near Aga Point, Guam. In *a* and *b* the algal ridge near the reef edge is formed chiefly of *Porolithon* and *Lithophyllum*. (*Photographs by J. I. Tracey, Jr., USGS*)

(b)

(c)

(d)

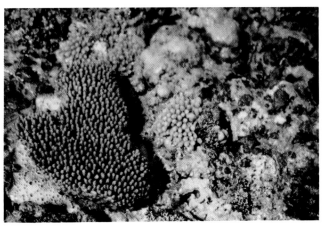

(e)

cause of habitat destruction, especially the loss of large trees with suitable nesting cavities. *See Aves*.

Walter J. Bock

Corallimorpharia

An order of the subclass Zoantharia or Hexacorallia. These sea anemones resemble corals in their weak musculature, capitate (spherically tipped) tentacles, and complex nematocysts. They do not possess a skeleton, however. The tentacles are arranged in radial rows, not cycles, on the disk, and the base has no musculature. The mesenterial filaments are without special ciliated tracts. The group occurs in shallow water from the temperate zone to the tropics. The species are solitary, although they may occur in large, asexually produced aggregations. Examples are *Corallimorphus* and *Corynactis*. *See Anthozoa*.

Cadet Hand

Coralline algae of the Pacific Ocean.

Corallinales

An order of red algae (Rhodophyceae), commonly called coralline algae. Only one family is recognized, the Corallinaceae, which formerly was assigned to the order Cryptonemiales. These algae are distinguished by the impregnation of cell walls with calcite, a form of calcium carbonate, which causes the thallus to be stony or brittle (see **illus.**). The few other red algae that are calcified are much softer than corallines and are impregnated with aragonite, another form of calcium carbonate. Many early biologists (including C. Linnaeus) thought that these plants were related to corals. *See Rhodophyceae*.

Coralline algae, comprising about 40 genera and 500 species, are widespread, abundant, and ecologically important. They are divisible into two groups on the basis of the presence or absence of uncalcified, moderately flexible joints (genicula) between calcified segments (intergenicula). Those with genicula are called articulated coralline algae, while those totally lacking genicula are called nonarticulated or, with reference to their usual habit, crustose coralline algae. The erect portions (fronds) of articulated corallines usually arise from a crustose base. They vary in height from a few millimeters to 30 cm (12 in.) and are pinnately, dichotomously, or proliferously branched. Segments are cylindrical or flattened.

Structure. The most simple nonarticulated coralline algae are individual crusts of varying extent and thickness (to 20 cm or 8 in. thick). These often become confluent and cover large expanses of substrate. Many crusts bear rounded or pointed, branched or unbranched protuberances that can break off and continue to grow as free nodules known collectively as maerl. In some species, only part of the crust is attached to the substrate, the main portion growing free and resembling lichens or bracket fungi. In three genera of nonarticulated corallines, the thallus is ribbonlike, the linear segments being dichotomously branched, dorsiventrally organized, and flexible.

Coralline algae are cytologically similar to other red algae except for calcification of the cell wall. The wall consists of three layers: a thin, uncalcified inner layer; a thick, calcified middle layer; and a thin, calcified outer layer or middle lamella. The calcite apparently is transported in vesicles from the cell lumen to the outer layers. Growth, which is restricted to uncalcified or decalcified areas or tissues, is usually very slow, especially at sites of low light intensity. Most corallines are perennial.

Reproduction. The life history follows the *Polysiphonia*-type, in which male and female plants alternate with a tetrasporophyte, all three phases being free-living and vegetatively identical or nearly so. Reproductive structures are produced in sunken or elevated chambers (conceptacles) with a roof penetrated by one or many pores. Tetraspores are produced by simultaneous zonate cleavage, a process unique to the order. They are relatively large and in most genera germinate in special patterns of cell division.

Habitat. Coralline algae are exclusively marine, although some species can tolerate a reduction in salinity to 13 parts per thousand. Some species thrive only where light is intense, as at the crest of a coral reef, while others grow only in shaded habitats or in deep water. Most species require constant immersion. Rock is the most common substrate in cold water, with some species preferring particular aspects, such as broad ledges, crevices, cobbles, or pebbles. Other hard substrates include shells, dead coral (the most common substrate in the tropics), and such artifacts as bottles and metal plates. Many corallines are obligate epiphytes, growing on other algae and on seagrasses, and a few overgrow such soft animals as sponges. An inconspicuous layer of corallines often underlies a turf of algae that at first glance appears to be growing directly on rock.

Unlike corals, coralline algae are equally abundant in cold or warm water, occurring from the Arctic to the Antarctic. At high latitudes, crustose corallines pave vast expanses of ocean floor. Along temperate shores, both crustose and articulated corallines are often abundant in tide pools. In areas of rough water, they typically form a broad band in the lowermost intertidal and uppermost subtidal zone. They may dominate deeper sites where herbivores consume fleshy algae. Stands of crustose corallines are made conspicuous by their bright pink or bluish-pink color.

The ecosystem in which coralline algae are most important is the coral reef, where they are primary producers, adding carbon to the ecosystem, adding new material to the reefs, and cementing together other calcareous organisms. They have been engaged in similar activities through the millennia, with modern genera recognizable in limestones at least as old as 150 million years (Jurassic).

Paul C. Silva; Richard L. Moe

Fossils. Coralline algae are important frame builders and sediment producers in tropical reefs throughout Cenozoic time. Although scleractinian corals are volumetrically more important in most living reefs, crustose coralline algae are essential binding and cementing agents. Prominent coralline algae of tropical reefs today are *Neogoniolithon*, *Porolithon*, and *Lithoporella*. *Mesophyllum* and *Lithothamnium* are mainly cold-water genera, but they do occur in low-latitude regions in deep-water environments.

Coralline algae have a rather complete fossil record consisting of numerous genera and species. The earliest definite representatives of the Corallinaceae appeared in Jurassic time, although fossil forms with similar morphologies are known from late Paleozoic rocks. Coralline algae are some of the more common and widespread skeletal constituents in Cretaceous and Cenozoic marine rocks. SEE ALGAE; ATOLL; REEF.

John L. Wray

Bibliography. H. W. Johansen, *Coralline Algae: A First Synthesis*, 1981; J. H. Johnson, *Limestone-Building Algae and Algal Limestones*, 1961; J. L. Wray, *Calcareous Algae*, 1977.

Corbino disk

A disk of conducting material with inner and outer concentric-ring electric contacts, named for O. M. Corbino, who first used this arrangement to study magnetoresistance in metals. When subjected to a magnetic field perpendicular to the plane of the disk, the current between the inner and outer ring contacts tends to spiral rather than flow straight out, thus introducing a considerable magnetoresistance. SEE MAGNETORESISTANCE.

Lloyd P. Hunter

Bibliography. D. A. Kleimann and A. L. Schawlow, Corbino disk, *J. Appl. Phys.*, 31:2176–2187, 1960.

Corbinotron

The combination of a corbino disk, made of high-mobility semiconductor material, and a coil arranged to produce a magnetic field perpendicular to the disk. This device can be used as a switch in which the current in the coil controls the resistance of the disk and hence the radial current flowing in the disk. SEE CORBINO DISK.

If the coil is connected in series with the disk, the resulting two-terminal device exhibits rectification. For one direction of the current, the self-generated magnetic field of the spiraling current in the disk opposes the magnetic field generated in the series coil, thus reducing the magnetoresistance of the disk. For the opposite direction of current, the self-generated magnetic field of the spiraling current reinforces that of the coil, and the magnetoresistance increases. SEE MAGNETORESISTANCE.

If the coil is connected in parallel with the disk, the device exhibits a negative-resistance characteristic for one direction of current.

Lloyd P. Hunter

Cordaitales

An extensive and for the most part natural grouping of forest trees of the late Paleozoic. With inclusion of certain presumed relatives, the stratigraphic range of the Cordaitales extends from the Upper Devonian to the Lower Triassic. The order may be divided into three families: Pityaceae, Cordaitaceae, and Poroxylaceae. The degree of relationship between the three families is quite uncertain, and the bulk of knowledge and significance of the group resides in the Cordaitaceae, abundant remains of which occur in Carboniferous sediments in most parts of the Earth.

Pityaceae. Fossil plants grouped in the family Pityaceae are known only as petrifactions of branches and wood, and the generic classification is based on anatomical features such as cortical structure, nodal vascularization, tracheary pitting, and other features of the secondary wood. The range of structural variation among the genera is so great that the "family" is patently an artificial grouping and yet collectively finds its closest affinity with the Cordaitaceae among known groups of vascular plants. The best-known and most widespread member of the Pityaceae is the genus *Callixylon* of Upper Devonian and lowermost Carboniferous (Mississippian) age. *Callixylon* was a forest tree with diameter up to 4 ft (1.2 m) or more. Stem fragments showing little taper have been found as much as 27 ft (8 m) in length. The wood of *Callixylon* has a pronounced, widely spaced, clustered arrangement of the bordered pit pairs on the tracheid walls (**Fig. 1**). The rays of the wood are commonly multiseriate and heterogeneous, differing from the wood rays of true cordaitean stems. Leaves and reproductive structures of *Callixylon* are now known to be *Archeopteris*. SEE PETRIFACTION.

Cordaitaceae. The genus *Cordaites* is the central taxonomic unit of the family Cordaitaceae. As is often the case with other groups of fossil plants, the various parts or organs of the plant are known under a series of separately designated generic categories and binomials. This cumbersome taxonomic proce-

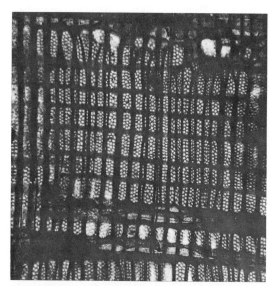

Fig. 1. Radial section through secondary wood of *Callixylon newberryi* showing alignment of pit groups on radial walls of tracheids. (*Photograph by C. A. Arnold*)

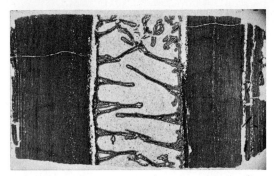

Fig. 2. Radial section through stem of *Cordaites transversa* showing pith septations. Specimen from Coal Measures, England. (*From C. A. Arnold, An Introduction to Paleobotany, McGraw-Hill, 1947*)

dure is a necessary consequence of the sedimentary process, whereby different parts of the plant body are separately and disjunctly preserved.

Plant tissues. The secondary wood of *Cordaites,* when it is preserved devoid of pith or cortex, is known as *Dadoxylon.* It is virtually indistinguishable in basic structure from the secondary wood of conifers of the extant family Araucariaceae. The trunks achieved a diameter of 3 ft (1 m) and the trees probably attained a height of 100 ft (30 m) or more. With preservation of other diagnostic features of the primary tissues such as pith, cortex, and leaf vascularization, cordaitean stems may be placed in the genera *Mesoxylon* or *Cordaites.* Pith casts, commonly 1.6–4 in. (4–10 cm) in diameter, are known as *Artisia* and often show external evidence of the septae or diaphragms which usually characterize the internal pith cavities of structurally preserved stems (**Fig. 2**). The leaves of cordaitean trees are all basically similar in design, though varying greatly in length from a few centimeters to over 40 in. (100 cm). Apices of the

leaves vary in acuteness but the leaves are characteristically linear in shape with numerous parallel veins and are devoid of a midrib (**Fig. 3**). Internal preservation indicates that the leaves were thick, coriaceous, and rather heavily cutinized. SEE PLANT TISSUE SYSTEMS.

Reproductive organs. The reproductive organs of the Cordaitaceae are quite thoroughly known and are of special interest in their bearing on the evolution of the reproductive structure of the great group of gymnosperms, the Coniferales. Both pollen-bearing and ovuliferous structures were produced in small multi-bracted florets, borne helically inserted in the axis of distinct bracts, on lax reproductive axes, or as inflorescences (Fig. 3). The individual florets, as well as the inflorescences as a whole, are unisexual. Both

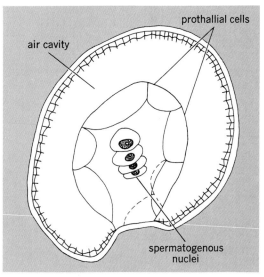

Fig. 4. Pollen grain of *Cordaites*.

male organs (stamens) and female organs (ovules) were borne terminally and grouped on the distal apex of the floret, surrounded below by the densely imbricated sterile bracts. Pollen grains are ellipsoidal with a circumferential air bladder; inside, and attached at the inner, proximal, surface, is a spherical grouping, one cell thick, of prothallial cells. A row of spermatogenous nuclei (**Fig. 4**) is occasionally shown within the prothallial mass. Histological details of the early stages of ovule development are not known, although it is known that the ovules were archegoniate. Structurally preserved seeds of cordaitean trees are 0.4–1.2 in. (1–3 cm) long, usually ovoid, and commonly winged. Internal organization closely resembles that of many seed-fern seeds of contemporary deposits, though less elaborate in structure. Coalified, compressed remains of cordaitean seeds often appear cordate in outline, with a conspicuous broad wing; if they are detached from vegetative organs, they cannot be distinguished from seeds of seed ferns. Cordaitean seeds have been described under the generic categories *Cardiocarpon,* *Mitrospermum,* *Cordaicarpus,* and *Samaropsis.*

Poroxylaceae. Remains of plants known from a limited locality in Europe under the generic name *Poroxylon* have been traditionally grouped in the

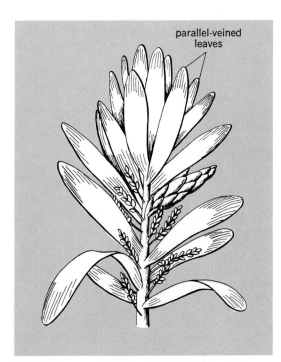

Fig. 3. Foliage, small florets, and inflorescence of *Cordaites.*

Cordaitales, perhaps most appropriately as the family Poroxylaceae. The status of the genus is unsatisfactory. It is probable that *Poroxylon* is a genus best classified among the seed ferns.

The evolutionary significance of the Cordaitales (the Cordaitaceae) lies in their importance in delineating early stages in the phylogeny of the Coniferales. This well-documented sequence in the evolution of vascular plants is described by Rudolf Florin. *See Paleobotany*.

Elso S. Barghoorn

Bibliography. H. N. Andrews, Jr., *Studies in Paleobotany*, 1961; C. H. Beck, The identity of *Archeopteris* and *Callixylon*, *Brittonia*, 12:351–368, 1960; T. Delevoryas, *Morphology and Evolution of Fossil Plants*, 1962; R. Florin, Evolution in cordaites and conifers, *Acta Horti Bergiani*, vol. 15, no. 10, 1951; R. Florin, Untersuchungen zur stammesgeschichte der coniferales och cordaitales, *Kgl. Svenska Vetenskapsakad. Handl.*, vol. (3)10, no. 1, 1931; D. H. Scott, *Studies in Fossil Botany*, 2 vols., 3d ed., 1962.

Cordierite

An orthorhombic magnesium aluminosilicate mineral of composition $Mg_2[Al_4Si_5O_{18}]$, $a = 1.710$ nanometers, $b = 0.973$ nm, $c = 0.936$ nm, $Z = 4$, space group *Cccm*. The crystal structure is related to beryl; magnesium is octahedrally coordinated by oxygen atoms, and the tetrahedrally coordinated aluminum and silicon atoms form a pseudohexagonal honeycomb, admitting limited amounts of K^+ and Na^+ ions and water molecules in the open channels. Limited amounts of Fe^{2+} may substitute for Mg^{2+}, and Fe^{3+} for Al^{3+}. The hardness is 7 (Mohs scale), specific gravity 2.6; luster vitreous; cleavage poor; and color greenish-blue, lilac blue, or dark blue, often strongly pleochroic colorless to deep blue (see **illus.**). Transparent pleochroic crystals are used as gem material. The disordered cordierite structure, $Mg_2[(Al,Si)_9O_{18}]$, is called indialite and is hexagonal, space group *P6/m cc*, isotypic with beryl. Osumilite, $KMg_2Al_3 \cdot [(Al,Si)_{12}O_{30}] \cdot H_2O$, is also related but has a structure built of double six-membered rings of tetrahedrons. Cordierite, indialite, and osumilite are difficult to distinguish. Pale colored varieties are often misidentified as quartz, since these minerals have many physical properties in common. *See Beryl; Silicate minerals*.

Cordierite possesses unusually low thermal expansion, and synthetic material has been applied to thermal-shock-resistant materials, such as insulators for spark plugs and low-expansion concrete. Cordierite is

an important phase in the system $MgO-Al_2O_3-SiO_2$. Between 800 and 900°C (1470–1650°F), glasses of cordierite composition crystallize. Below 500°C (930°F) at pressures up to 500 bars (50 megapascals), cordierite in the presence of water breaks down to form chlorite + pyrophyllite. Indeed, it is often found naturally altered to these minerals. Cordierite melts incongruently at 2000 bars (200 MPa) and 1125°C (2057°F) to mullite + spinel + liquid. Natural cordierties break down at temperatures of about 800°C (1470°F) and pressures above 8 kilobars (800 kilopascals) to form enstatite, sillimanite, and quartz.

Cordierite frequently occurs associated with thermally metamorphosed rocks derived from argilaceous sediments. A common reaction in regional metamorphism resulting in gneisses is garnet + muscovite → cordierite + biotite. It may occur in aluminous schists, gneisses, and granulites; though usually appearing in minor amounts, cordierite occurs at many localities throughout the world.

Paul B. Moore

Cordilleran belt

A mountain belt or chain which is an assemblage of individual mountain ranges and associated plateaus and intermontane lowlands. A cordillera is usually of continental extent and linear trend; component elements may trend at angles to its length or be nonlinear.

The term cordillera is most frequently used in reference to the mountainous regions of western South and North America, which lie between the Pacific Ocean and interior lowlands to the east (see **illus.**). The term was first applied to the Andes Mountains of South America in their entirety (Cordillera de los Andes), but individual mountain ranges within the Andean belt are now called cordilleras by some authorities. Farther north, the extensive and geologically diverse mountain terrane of western North America is formally known as the Cordilleran belt or orogen. This belt includes such contrasting elements within the United States as the Sierra Nevada, Central Valley of California, Cascade Range, Basin and Range Province, Colorado Plateau, and Rocky Mountains. *See Mountain systems*.

Cordilleras represent zones of intense deformation of the Earth's crust produced by the convergence and interaction of large, relatively stable areas known as plates. J. F. Dewey and J. M. Bird have analyzed mountain belts in terms of different modes of plate convergence. They contrast cordilleran-type mountain belts, such as the North American Cordillera, with collision-type belts, such as the Himalayas. The former develop during long-term convergence of an oceanic plate toward and beneath a continental plate, whereas the latter are produced by the convergence and collision of one continental plate with another or with an island arc. Characteristics of cordilleran-type mountain belts include their position along a continental margin, their widespread volcanic and plutonic igneous activity, and their tendency to be bordered on both sides by zones of low-angle thrust faulting directed away from the axis of the belt. The deformational, igneous, and metamorphic characteristics of cordilleran belts appear to be largely related to crustal compression, to melting of rocks at depth along the inclined plate boundary, and to attendant thermal ef-

Cordierite, from Tsilaizina, Madagascar. (*Specimen from Department of Geology, Bryn Mawr College*)

2 cm

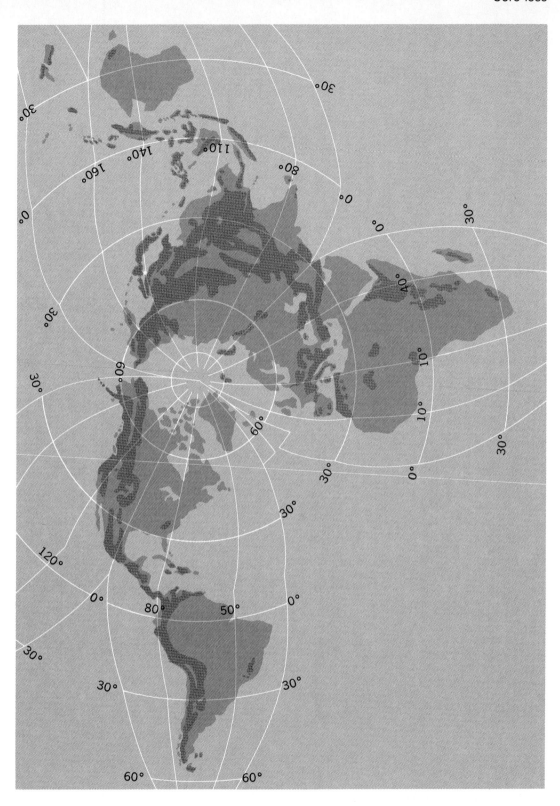

The arrangement of the mountains and the continents of the world. (*After P. E. James, An Outline of Geography, Ginn and Co., 1935*)

fects in the overlying continental plate. *See Orogeny.*
Gregory A. Davis

Bibliography. B. C. Burchfiel and G. A. Davis, Nature and controls of Cordilleran orogenesis, western United States: Extensions of an earlier synthesis, *Amer. J. Sci.,* Rodgers Volume, vol. 275A, 1975; J. F. Dewey and J. M. Bird, Mountain belts and the new global tectonics, *J. Geophys. Res.,* vol. 75, no. 14, 1970.

Core loss

The rate of energy conversion into heat in a magnetic material due to the presence of an alternating or pulsating magnetic field. It may be subdivided into two principal components, hysteresis loss and eddy-current loss. *See Eddy current; Magnetic hysteresis.*

Hysteresis loss. The energy consumed in magnetizing and demagnetizing magnetic material is called the hysteresis loss. It is proportional to the frequency and to the area inside the hysteresis loop for the material used. Hysteresis loss can be approximated empirically by using Eq. (1), where K_h is a constant

$$P_h = K_h f B_{max}^n \qquad (1)$$

characteristic of the material, f is the frequency, B_{max} is the maximum flux density, and n, called the Steinmetz coefficient, is often taken as 1.6 although it may vary from 1.5 to 2.5. *See MAGNETIZATION.*

Most rotating machines are stacked with silicon steel lamination, which have low hysteresis losses. The cores of large units are sometimes built up with cold-reduced, grain-oriented, silicon iron punchings having exceptionally low hysteresis loss, as well as high permeability when magnetized along the direction of rolling.

Eddy-current loss. Induced currents flow within the magnetic material because of variation in the flux. This eddy-current loss may be expressed to close approximation by using Eq. (2), where K_e is a constant

$$P_e = K_e (B_{max} f \tau)^2 \qquad (2)$$

depending on the volume and resistivity of the iron, B_{max} the maximum flux density, f the frequency, and τ the thickness of the lamination in the core. For 60-cycle rotating machines, core lamination of 0.014–0.018 in. (0.35–0.45 mm) are usually used to reduce eddy-current loss.

Measured core loss. The measured core loss in a rotating machine also includes eddy-current losses in solid structural parts, such as the frame, ventilating duct spacers, pole faces, and damper windings, as well as those due to burrs or other contacts between punchings. A number of precautions are taken to minimize these components, which become appreciable in large machines. Structural parts close to the core, including clamping plates, I-beam spacers, dovetail bars, and shields, may be made of nonmagnetic material. Some of these parts may be shielded from the variable flux by low-resistance plates or by flux traps of laminated, high-permeability, low-loss steel. The core punchings are deburred and are coated with a baked-on insulating varnish. The finished cores are often tested with an ac magnetizing coil to locate and correct hot spots caused by damage during assembly. Pole-face losses are reduced by avoiding excessive slot-width to air-gap ratios, by surface grooving of the poles, and by low-resistance amortisseur windings. *See ELECTRIC ROTATING MACHINERY.*

Leon T. Rosenberg

Coriander

A strong-scented annual herb. Coriander is cultivated in many places throughout the world for both seeds and leaves. The two forms are quite different in taste from one another, and both are used for flavor in a variety of foods. Only one species, *Coriandrum sativum,* is cultivated. Coriander is a member of the carrot family, Apiaceae (Umbelliferae), and is closely related to other spice seed plants such as cumin, caraway, anise, dill, and fennel. A number of distinct cultivars have been developed. Some, with longer maturity times and resulting higher leaf yield, are grown for cilantro, also called Chinese parsley. Other seed-producing types, having a more uniform seed set, and types that resist splitting, have been developed in the Soviet Union and India. *See ANISE; APIALES; CARAWAY; CUMIN; DILL; FENNEL.*

The growing of cilantro from seed to a harvestable size takes 40–60 days, and full maturity as coriander (dry ripe fruit) is reached in approximately 120 days. Coriander occurs as an erect herbaceous plant with one to many stems that reach about 4 ft (1.2 m) in height. When the plant is young, small deeply segmented leaves are formed, but as the plant ages the leaves become more and more divided until they become feathery in appearance, similar to dill.

The term coriander comes from the Greek *koris,* meaning bedbug, because the leaves and green fruit of this plant have an odor similar to bedbugs. Although some find it objectionable, it imparts a distinctive authentic taste to Mexican and Asian foods. Coriander seeds are used in pickling, curries, and for flavoring alcoholic beverages such as gin. The "seed," which is actually the whole fruit, is a two-seeded schizocarp borne in umbels and is approximately 0.1–0.3 in. (3–7 mm) in diameter. Longitudinal ridges run from top to bottom.

Cultivation is accomplished by planting seed in the spring at a rate of 9–14 lb/acre (10–15 kg/hectare). Pollination and weed control are similar to that of dill. *See APIALES; SPICE AND FLAVORING.*

Seth Kirby

Bibliography. S. Arctander, *Perfume and Flavor Materials of Natural Origin,* 1960; L. H. Bailey, *Manual of Cultivated Plants,* rev. ed. 1975; F. Rosengarten, *The Book of Spices,* 1969.

Coriolis acceleration

An acceleration which arises as a result of motion of a particle relative to a rotating system. Only the components of motion in a plane parallel to the equatorial plane are influenced. Coriolis accelerations are important to the circulation of planetary atmospheres, and also in ballistics. They are so called after G. de Coriolis, the French engineer and mathematician whose analysis of the phenomenon was published in 1835. *See ACCELERATION; BALLISTICS.*

Newton's second law of motion is valid only when the motions and accelerations are those observed in a coordinate system that is not itself accelerating, that is, an inertial reference frame. In order to utilize familiar concepts in mathematical treatment, the Earth is commonly treated as if it were fixed, as it appears to one observing from a point on the surface, and the Coriolis force is introduced to balance the acceleration observed by virtue of the observer's motion in the rotating frame. *See FRAME OF REFERENCE; NEWTON'S LAWS OF MOTION.*

As with the influenced components of motion, the Coriolis force is directed perpendicularly to the Earth's axis, that is, in a plane parallel to the equatorial plane. Since the direction of its action is also perpendicular to the particle velocity itself, the Coriolis force affects only the direction of motion, not the speed. This is the basis for referring to it as the deflecting force of the Earth's rotation. The magnitude of the Coriolis force is equal to the product of twice the projection of particle velocity into the equatorial plane, with the angular velocity of the Earth's rotation (7.292×10^{-5} s^{-1}).

The Coriolis force is directed to the right of the direction toward which the horizontal wind blows or the missile flies in the Earth's Northern Hemisphere, and to the left in the Southern. The horizontal component of the deflecting force is zero at the Equator, and a maximum at the poles. While negligible for the dynamics of small-scale phenomena such as cumulus convection and thunderstorms, Coriolis effects are critical to the dynamics of the Earth's large-scale wind currents. Indeed, the general circulations of planetary atmospheres owe their zonal nature and some other characteristics largely to the forces that arise by virtue of planetary rotation.

A simple illustration of a Coriolis effect in the Northern Hemisphere is afforded by a turntable in counterclockwise rotation, and an external observer who moves a marker steadily in a straight line from the axis to the rim of the turntable. The trace on the turntable is a right-turning curve, and obviously this is also the nature of the path apparent to an observer who rotates with the table. Now consider the contrasting case of an air parcel near the North Pole that moves directly south (away from the Earth's axis of rotation) so that its motion to an observer on the Earth is in a straight line. To a nonrotating observer in space, this same motion appears curved toward the east because of the increased linear velocity of the meridian at lower latitudes. The force necessary to produce the eastward acceleration in the inertial frame is equal to the Coriolis force and would be produced by a gradient of air pressure from west to east, and shown by north-south–oriented isobars. In the absence of the pressure gradient force, the Coriolis force would cause the trajectory of the southward-moving air to curve westward on the Earth's surface. Then an air parcel moving uniformly away from the North Pole along a line which appears straight to an observer in space would appear to earthbound observers to curve westward. SEE GEOSTROPHIC WIND; ISOBAR (METEOROLOGY).

Edwin Kessler

Bibliography. L. J. Battan, *Fundamentals of Meteorology*, 1979; J. M. Wallace and P. V. Hobbs, *Atmospheric Science, An Introductory Survey*, 1977.

Corn

Zea mays occupies a larger area than any other grain crop in the United States, where 60% of the world production is grown.

PRODUCTION

Production is widely distributed throughout the United States, but it reaches its greatest concentration in the states of Iowa, Illinois, Indiana, Ohio, Nebraska, Minnesota, Wisconsin, Michigan, and Missouri. This area, called the Corn Belt, is characterized by moderately high temperature, fertile, well-drained soils, and normally adequate rainfall. The Corn Belt accounts for approximately 80% of corn production in the United States. Although corn is grown in the United States primarily for livestock feed, about 10% is used for the manufacture of starch, sugar, corn meal, breakfast cereals, oil, alcohol, and several other specialized products. In many tropical countries, corn is used primarily for human consumption.

Origin and description. From its presumed origin in Mexico or Central America, corn has been intro-

duced into all the countries of the world that have suitable climatic conditions. However, the corn of today is far different from the primitive types found in prehistoric sites excavated in Arizona, Mexico, and Peru. The origin of corn is still unsettled, but the most widely held hypothesis assumes that corn developed from its wild relative teosinte (Z. *mexicana*) through a combination of favorable mutations, recognized and selectively propagated by early humans. Corn migrated from its center of origin and was being cultivated by the Indians as far north as New England when the first European colonists arrived, whose survival was due largely to the use of corn as food.

Botanically, corn is a member of the grass family. Certain seed types within the species Z. *mays* were given subspecific rank by early investigators. Among these types are the dents (*indentata*), flints (*indurata*), soft or flour (*amylaceae*), sweet (*saccharata*), pop (*everta*), and pod (*tunicata*). Each form (botanical variety) is conditioned by fairly few genetic differences, and each may exhibit the full range of differences in color, plant type, maturity, and so on, characteristic of the species. All types have the same number of chromosomes (10 pairs), and all may be intercrossed to produce fertile progeny. Dent corns are the most important in the United States. Kernel texture is less homogeneous in other areas of the world. Soft types tend to predominate at higher elevations, flint types at lower elevations. This pattern is due, in part, to differences in methods of food preparation, but it has also been influenced by the greater resistance of flint types to stored-grain insect pests, which are more prevalent at lower elevations. Sweet corn is grown more extensively in the United States than in any other country. It is eaten as fresh corn or canned or frozen. In other countries, flint, dent, or flour corns may be eaten fresh, but at a much more mature stage than the sweet corn eaten in the United States. The commercial production of popcorn is almost exclusively American. Pod corn is grown only as a genetic curiosity. SEE CYPERALES; GENETICS; REPRODUCTION (PLANT).

Varieties and types. Corn is a cross-pollinated plant; the staminate (male) and pistillate (female) inflorescences (flower clusters) are borne on separate parts of the same plant (**Fig. 1**). Plants of this type are called monoecious. The staminate inflorescence is the tassel; it produces pollen that is carried by the wind to the silks produced on the ears. Natural cross-pollination permits maintenance of a high degree of genetic variability. As corn was carried by the Indians to different environments in North, Central, and South America, a large array of varieties ultimately was developed. Many of these primitive varieties still exist today. Extensive efforts have been devoted to the collection, classification, and maintenance of this valuable reservoir of genetic variability. Settlers migrating westward took with them their corn seed (botanically fruits). In time, by both planned and natural selection, a large number of varieties and local strains became established and adapted to the new ecological conditions. These strains varied in color of grain, plant, and ear characteristics, and in length of time required to reach maturity. Because of the variability associated with cross-pollination, corn is a highly plastic species, and varieties have been developed that are well adapted to a frost-free growing season of less than 90 days or, in contrast, grow 12–15 ft (3.5–4.5 m) tall and require more than 200 days to reach

Fig. 1. A corn plant in full tassel and silk. The tassel produces pollen that is blown by wind to the silks. (*Courtesy of J. W. McManigal*)

maturity, as in certain areas in Guatemala. *See Flower; Inflorescence; Pollination.*

Hybrids. The development of varieties and strains of corn made possible the extension of its culture under diverse soil and climatic conditions. However, modern research methods, used for its further improvement, led to the present widespread use of hybrid corn. Essentially all of the corn grown in the United States is planted to hybrid seed, and this technology is being adopted in other areas of the world as rapidly as suitable hybrids are developed. *See Breeding (plant); Heterosis.*

Hybrid corn is the first generation of cross involving inbred lines. Inbred lines are developed by controlled self-pollination. When continued for several generations, self-pollination leads to reduction in

vigor but permits the isolation of types which are genetically pure or homozygous. Intense selection is practiced during the inbreeding phase to identify and maintain genotypes having the desired plant and ear type and maturity characteristics, and relative freedom from insect and disease attacks. After the inbred lines have become fairly pure, they are further evaluated in hybrid combinations. Crosses involving any two unrelated lines will exhibit heterosis, that is, yields above the means of the two parents. Only a very few combinations, however, exhibit sufficient heterosis to equal or surpass hybrids in current commercial use. Hence large-scale experimental testing is required to identify new useful combinations.

The first commercial hybrids were double crosses produced by crossing two single crosses, each, in turn, having been produced by the crossing of two inbred lines. Hybrid seed for commercial planting is produced in special crossing blocks (isolated fields). The field is planted with the parents (inbred lines or single crosses) in alternating groups of rows.

The plants in blocks to be used as female parents are detasseled before silk emerges. Thus all seed produced will be pollinated by the male parent. After drying, sizing, and treatment, the seed harvested from the female rows becomes the hybrid seed of commerce. Since the early 1960s the use of single-cross hybrids has been increasing, and in the Corn Belt today probably accounts for 80% or more of the seed used.

Several techniques have been used to control pollination. Initially, manual removal of the tassels was used exclusively. Then cytoplasmic male sterility was used. This type of male sterility is transmitted only through the cytoplasm of the female parent. Since no fertile pollen is produced, no detasseling is required. With this system, pollen production in the farmers' fields is provided either by blending (artificial mixing of sterile and fertile types) or by the use of genetic fertility-restoring factors introduced through the male parent. In 1970 possibly 90% of the hybrid seed used in the United States utilized a single type of cytoplasmic sterility: the Texas type. This cytoplasmic type proved to be susceptible to a new mutant race (race T) of one of the leaf-blight fungi, *Helminthosporium maydis*, and thus the use of this type of sterility was distincontinued. Investigations of other types of cytoplasmic sterility are under way. Machines which cut or pull tassels in the seed fields have been developed and are being used extensively. Other types of genetic control of pollen production have been developed, and some attention is being given to the possible use of chemical sterilants.

Companies specializing in hybrid seed production produce and market the entire seed supply. Gains from the use of hybrid seed have been spectacular. The shift from open-pollinated varieties to hybrids, beginning in the 1930s, accounted for a yield increase of 25%. In succeeding years these first hybrids have been replaced by newer, higher-yielding types which have greater disease and insect resistance. Hybrid succession is a continuing process; few hybrids have a commercial life of more than 5 years. The per-acre yield of corn has increased about threefold since the first introduction and use of hybrid seed. *See Agricultural science (plant).*

Planting. Planting dates depend upon temperature and soil conditions. Germination is very slow at soil temperatures of 50°F (10°C), and seedling growth is

Fig. 2. Two-row mounted picker for harvesting corn.

limited at temperatures of 60°F (16°C) or below. Because of these temperature relations, planting begins in Florida and southern Texas in early February and is completed in New York and New England by late May or early June. Planting rates are influenced by water supply, soil type, and fertility and by the maturity characteristics of the hybrid grown. Planting rates for full-season hybrids may vary from 8000 to 12,000 plants per acre (20,000 to 30,000 per hectare) in eastern Texas to 20,000 to 24,000 (50,000 to 59,000) in the central Corn Belt. Check-planting, with hills and rows spaced 38–42 in. (95–105 cm) apart, has been almost completely replaced by drilling. With planting rates above 16,000 plants per acre (40,000 per hectare), drilling in rows 24–36 in. (60–90 cm) apart has become common practice. The use of nitrogen fertilizer has increased greatly; lesser amounts of phosphorus and potash are applied as needed. Soil tests and experience provide useful guides to economic rates of fertilizer use under different environmental conditions.

There has been a marked trend toward reducing the amount of tillage in seedbed preparation. The variations in the planting systems used are influenced by soil type and topography, amount of crop residues on the soil surface, and the weed-control system to be employed. The use of the minimum tillage system, in which all soil preparation as well as planting is performed in one passage across the field, is increasing. This system, and less extreme modifications of the conventional system, reduces total energy costs.

Weed control. Weed control may be effected by cultivation or herbicides or both. With minimum tillage planting methods, weed control may be more difficult than with conventional seedbed preparation because of crop residues left on or near the surface. Weeds are easiest to control mechanically when they are small, preferably before they emerge. The most efficient and ecnomical tool for killing weeds at this stage is the rotary hoe. Unfortunately this or other mechanical cultivation systems are relatively ineffective during extended periods of rainy weather. Disk or shovel cultivators are used after the corn is 6–8 in. (15–20 cm) in height.

Increasing use has been made of selective herbicides. Most broadleaf and grassy weeds may be controlled without damage to corn. Herbicides may be applied either "preplant" or "preemergence." If applied prior to planting, herbicides may be left on the soil surface or incorporated into the soil. Preemergence applications may be made as the seed is planted, or at any time prior to crop emergence. The choice of the many herbicides available is dependent upon many factors, including the type of weeds to be controlled and the farming system used. Under unfavorable early-season weather conditions, neither mechanical nor chemical control nor their combination may be completely effective, and sizable yield reductions may result. SEE HERBICIDE.

Harvesting. Harvesting of corn has undergone a major revolution since 1930. In the 1930s most corn was husked by hand, and the ears were stored in slatted cribs. Harvesting began when the moisture content of the grain had been reduced, by natural drying, to 20–24%. The mechanical picker supplanted hand harvesting. The mechanical picker, in turn, has replaced by the picker-sheller or corn combine, which harvests the crop as shelled grain (**Fig. 2**). Energy requirements, however, have increased materially over the years. When harvested as shelled grain, at a relatively high moisture content (20–30%), the grain must be dried artificially for safe storage. High-moisture corn to be used for livestock feed may be stored in airtight silos or may be treated with certain chemical preservatives such as propionic acid. Corn stored under either of these systems is not suitable either for industrial processing or for seed. SEE AGRICULTURAL MACHINERY.

G. F. Sprague

Productivity. Corn is highly productive largely because it can use solar energy so efficiently. It is classed as a C-4 species, which means that its leaves use the more efficient of the two photosynthetic processes found in crop plants. This is the same photosynthetic system used by sugarcane, sorghum, other tropical grasses, and many persistent weeds. In addition to relatively efficient leaves, corn's growth habit allows planting densities thick enough to almost completely shade the soil. Thus, a well-managed cornfield receives almost all the solar energy the environment provides and uses it as efficiently as any plant known. SEE PHOTOSYNTHESIS.

The higher productivity of modern corn hybrids, as compared with earlier hybrids, when both are grown under disease-free conditions, results to a considerable degree from their ability to tolerate higher plant populations without excessive barrenness. This capability is largely due to the greater plant-to-plant uniformity introduced by single cross hybrids and to the development of varieties with more erect leaves. Such foliage minimizes mutual shading and permits more uniform illumination of all leaf surfaces.

The actual rate of dry matter production by an area of corn with ample water and fertility depends on the average amount of light and on the temperature. The highest value reported, 2 oz (52 g) of dry matter per 11 ft^2 of land area per day, was for high-population corn under cloudless conditions at Davis, California. A more representative range of values for the corn belt of the United States would be 0.9–1 oz (26–32 g) of dry matter per 11 ft^2 per day. Day-to-day variation is large because of differences in cloudiness and in temperatures. Such estimates suggest dry matter growth rates of 230–285 lb per acre per day (258 to 319 kg per hectare per day).

The corn plant grows vegetatively until about silking, after which all weight increase is in the form of grain. There is little transfer of earlier growth to grain, so almost the entire grain yield results from photosynthesis during the grain growth period, which runs from silking to maturity. Thus, grain yield results from a combination of the rate of dry matter production after silking and the duration of the grain growth period. Contrary to much popular opinion, grain yields are highest under cool conditions, when the lengthened grain growth period more than compensates for the slower growth rate. Thus, yields for single planting are much greater at higher altitudes in the tropics than at warmer, lower locations. Highest average grain yields are in favored areas of Washington and Oregon in the United States, in the North Island of New Zealand, and in cooler long-season locations in Europe. Consistently high yields in the Corn Belt of the United States result more from favorable soil moisture and excellent management than from climatic superiority.

Relationships among solar radiation, temperature, growing-season length, soil moisture, day length, soil fertility, and corn genotype in producing grain yields are complex and not well understood. Attempts to study the system as a whole, using simulation models on digital computers, may add considerably to knowledge of the subject.   William G. Duncan

DISEASES

More than 100 diseases of corn are known worldwide; yet damage or yield loss from disease seldom forces a grower out of production. World reduction in grain production due to disease is estimated to average 9% and in the Corn Belt to average 7–17%. The widespread practices of continuous cropping, increased fertilization, high plant populations, and relative genetic uniformity of hybrids have sometimes resulted in epidemics: Stewart's bacterial wilt, north-

Fig. 3. Corn smut on various parts of plant. (*University of Minnesota Agricultural Experimental Station*)

ern corn leaf blight epidemics in 1939–1943, the maize dwarf mosaic epidemics in the 1960s, and the southern corn leaf blight epidemic of 1970–1971.

Diseases are caused by bacteria, fungi, nematodes, viruses, several mycoplasmalike organisms, and one parasitic seed plant (*Striga* spp.). Pollutants, pesticides, and unfavorable climatic and soil factors are responsible for still other diseases.

Seedling diseases. Seed decay and seedling blight caused by seed- and soil-borne *Diplodia, Fusarium, Penicillium,* and *Pythium* species are troublesome when grain is planted in cool, moist soil. Planting sound, mature, and chemically treated seed reduces these problems.

Leaf diseases. Of approximately 25 leaf blights, only a few are economically important in the United States. Southern corn leaf blight, caused by *Helminthosporium maydis,* accounted for a 15% loss in yield in the 1970 epidemic in the nation. Yellow leaf blight, caused by *Phyllosticta maydis,* occurred about the same time, and both blights are found most commonly in hybrids with the Texas male-sterile cytoplasm. Northern corn leaf blight, caused by *Helminthosporium* (= *Exserohilum*) *turcicum,* has been destructive and can cause 50% loss in yield if the disease strikes before pollination. With new conservation tillage practices, other leaf diseases have become increasingly important due to fungi that overwinter on drop debris left on the ground. Resistant hybrids and, sometimes, crop rotation are generally used for control. Two of the bacterial leaf blights are of major concern. Stewart's wilt, caused by *Erwinia stewartii,* results in a severe wilt of sweet corn and a leaf blight of dent corn in eastern North America. Control is obtained by use of resistant hybrids or by insecticides that control the flea beetle vector. Goss' bacterial wilt and blight, or leaf freckles, occurs in Colorado, South Dakota, Nebraska, and Kansas, and is caused by *Corynebacterium nebraskense.* All types of corn are susceptible, and various degrees of blighting and stunting occur.

Smuts and rusts. Common or boil smut, caused by *Ustilago maydis,* is found wherever corn is grown, but losses in yield seldom exceed 2% over a wide area. Galls may appear on any part of the plant and are most prevalent in vigorously growing plants in soil high in organic matter and nitrogen, often under dry conditions, and increase with hail and other mechanical injuries (**Fig. 3**). Large galls on or above the ear are more destructive than galls below the ear. Control is to plant resistant hybrids. The small cinnamon brown to black pustules of common rust, caused by *Puccinia sorghi,* on leaves are found whever corn is grown. Most widely used hybrids have good field resistance. Southern rust, *P. polysora,* found mainly in the southeastern United States and other countries, appears to be spreading to northern states of the Corn Belt.

Stalk and root diseases. Stalk rots are universally important and among the most destructive diseases of corn in the world (**Fig. 4**). They are caused by *Fusarium moniliforme, F. roseum* ''Graminearum'' (*Gibberella zeae*), *Macrophomina phaseolina,* and to a lesser extent by *Diplodia zeae* and *Colletotrichum graminicola,* and several dozen other fungi and bacteria. Losses vary by season and region, but yield losses of 10–20% can occur among susceptible hybrids in the United States and 25–33% in other countries. Conditions favoring disease may cause plants to die before maturity. This results in poorly filled ears,

Fig. 4. Corn stalk rot, one of the most widespread corn diseases. (a) Corn lodged from stalk rot. (b) Bacterial stalk rot, with healthy corn stem on left. (*Ohio Agricultural Research and Development Center*)

but the greatest loss may come from stalk lodging that interferes with harvesting, or from ears fallen to the ground. Many factors affect stalk rot: unbalanced soil fertility, high plant populations, dry weather, leaf diseases, insect damage, hail, and, sometimes, continuous cropping to corn. The best control is to grow resistant hybrids and to follow good management practices. Root rots are caused usually by the same organisms that cause stalk rot, except that nematodes (eel- or roundworms) are sometimes involved. In fact, more than 40 species of nematodes have been reported to feed on corn roots. They can stunt plants, cause yellowing, and aid root-infecting fungi.

Ear diseases. Considerable damage to ears from ear and kernel infection occurs in humid regions where rainfall is above normal late in the season, and from insect and bird feeding, early frosts, and fallen stalks (**Fig. 5**). Infected grain reduces yield and lowers grain quality, and grain may contain mycotoxins (aflatoxins, estrogens, and vomitoxin) which makes animals ill when used as feed. Infections occur in the field from fungi such as *Diplodia, Fusarium (Gibberella), Helminthosporium,* and *Nigrospora,* or in storage from such fungi as *Aspergillus, Cladosporium,* and *Penicillium.* Storage fungi can be controlled if grain is stored at 13–15% moisture and at uniform temperatures between 39 and 50°F (4 and 10°C), with provision for adequate aeration.

Viral and mycoplasma diseases. Prior to 1963, viruses were not a factor in corn production except in localized areas outside the United States. That year, maize dwarf mosaic virus (MDMV) caused economic losses in the United States (**Fig. 6**). Almost complete destruction of susceptible corn occurred in some areas. MDMV has been identified in 15–20 states in the Corn Belt, as well as in the East and South. Aphids are vectors of MDMV, but leafhoppers are

Fig. 6. Workers examining corn plant infected with maize dwarf mosaic virus. (*Ohio Agricultural Research and Development Center*)

vectors for maize chlorotic dwarf virus (MCDV). Wheat streak mosaic virus (WSMV), widespread on certain inbred lines and hybrid corns in the Corn Belt, is transmitted by the mite *Eriophyes tulipae*. Other naturally occurring viruses of corn are sugarcane mosaic (a complex of Potyviruses that contain the MDMV subgroup), maize mosaic virus, maize chlorotic mottle virus (MCMV), and maize rayado fino virus. Corn lethal necrosis is a synergistic interaction between MCMV and either MDMV or WSMV, and is potentially destructive in the United States, but is currently confined to Kansas and Nebraska.

Some disease once thought to be caused by viruses are now known to be caused by a spiroplasma or a mycoplasmalike organism. Corn stunt, transmitted by a leafhopper, occurs in Texas and in the Rio Grande Valley, and is caused by a spiroplasma. The maize bushy stunt pathogen is a similar organism and has the approximate range of corn stunt in the United States and South America. The economic importance of virus diseases of corn has increasingly been recognized. SEE PLANT PATHOLOGY; PLANT VIRUSES AND VIROIDS.   *Thor Kommedahl*

PROCESSING

The high starch content and caloric value of corn have resulted in extensive production of animal feeds, human food, alcoholic beverages, and industrial products from the grain. Corn grain consists of three main parts: endosperm (horny and floury), germ, and hull (**Fig. 7**), which differ in composition (see **table**). The dry hard kernel of common dent corn is excellent for storage and handling but requires suitable processing for use. The whole grain may be ground, or alternatively, it may be degermed and dehulled during dry milling. It may also be wet-milled to further separate endosperm into starch and protein.

Fig. 5. Ear rots. (*a*) Storage rot. (*b*) Field rot. (c) Healthy ear. (*Ohio Agricultural Research and Development Center*)

Average composition, in percent, of whole corn and hand-dissected fractions (moisture-free basis)						
Fraction	% of kernel	Starch	Protein	Oil	Sugar	Minerals
Whole grain	—	71.5	10.3	4.8	2.0	1.4
Endosperm						
Horny	54.2	80.4	13.3	0.7	0.5	0.3
Floury	27.5	86.6	7.7	0.4	0.5	0.4
Germ	11.5	8.2	18.8	34.5	10.8	10.1
Bran	5.3	7.3	3.7	1.0	0.3	0.8
Tip cap	0.8	5.3	9.1	3.8	1.6	1.6

New World Indians discovered that cooking corn in solutions of lime or wood ashes (lye) rendered it soft and easier to grind or dehull. Lime-treated corn is used for preparation of tortillas and other popular Mexican-style foods, whereas lye-treated corn is widely sold as canned hominy. European immigrants introduced the water-powered stone grist mills to grind whole corn. Modern steel mills are used to grind most corn today.

Feed uses. About 90% of the corn produced in the United States is used for animal feeds. Most corn for feed is whole-ground by burr, roller, or hammer mills. Grinding facilitates mixing with other ingredients and pelleting, as well as improving feed conversion efficiency. Some steam flaking of corn is done to improve feed value. Because of its nutritional inadequacy, the grain is generally supplemented with protein sources such as soybean meal and with minerals and vitamins. *See Animal feeds.*

Dry milling. The high unsaturated oil content of corn germ limits the shelf life and restricts the use of whole ground corn. Of the corn dry-milled annually for food in the United States, only 15%, mostly white corn, is whole-ground by small mills. Some of this meal is bolted to remove bran hull particles.

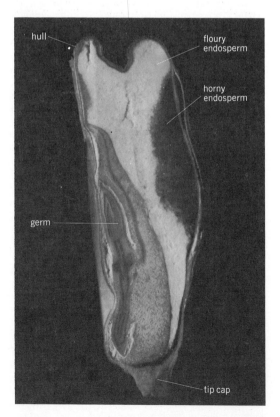

Fig. 7. Parts of the corn kernel shown in cross section. (*From G. E. Inglett, ed., Corn: Culture, Processing, Products, Avi Publishing Co., 1970*)

Most dry-milled corn is produced by degermer processes that fractionate grain into various-size endosperm particles, bran, and germ. After being cleaned, the grain is tempered by increasing its moisture content to 20–25%, often at an elevated temperature, to loosen and toughen the hull and germ. The grain next passes through degerminators, such as the Beall, that crack the endosperm and help free the germ and hull. The particle mixture is dried, then sent through a series of operations: aspiration to remove bran particles; roller milling to reduce endosperm particle size, release and flatten germ, and scrape endosperm from bran; and sieving to separate particles according to size. The germ and endosperm particles are separated by density differences on gravity tables.

Endosperm particles, after separation according to size, are classified into grits ($-3.5 + 28$ U.S. sieve), meal ($-28 + 75$), and flour (-75). Grits are used to prepare breakfast porridge, in corn flakes, and as a brewing substrate. Meal is a constituent of corn bread or muffin recipes and snack items. Corn flour is used in pancakes, soup thickeners, and dusting flours. Corn flour may be cooked on heated rolls or in extruders to pregelatinize its starch for use in baby foods and in industrial binders and adhesives.

Oil is expelled from the germ and refined to give 0.75 lb/bu (0.97 kg/m³) of corn. Residual germ meal, bran, and degermer fines are combined to produce hominy feed for cattle and swine. The germ fraction may also be highly purified and defatted by solvent extraction. This process yields additional oil and a germ flour food product containing 25% quality protein and ample vitamins and minerals.

Wet milling. Demand for purified starch and corn sugars has resulted in continued growth of the corn wet-milling industry in the United States.

In wet milling **(Fig. 8),** the cleaned corn is steeped in solutions of sulfur dioxide (0.15%) in large tanks 45 h at 120°F (50°C). Steeping softens the kernel and aids starch-protein separation. Fermentation of sugars to lactic acid retards other organisms in the steep media. The extracted solids in the steep liquor are then concentrated. The corn is next passed through an attrition mill that tears the kernels apart. Liquid cyclones are used to separate the lighter germ from endosperm particles. The particles are milled further in impact mills. Bran and other fragments are removed from starch and gluten (insoluble protein) by screening, and the starch and gluten are separated by centrifugation. After washing and purification, starch and gluten are dewatered and dried. Numerous cost- and energy-saving and environmental protection measures have been incorporated into modern corn wet-milling plants.

Operations in most modern plants are microprocessor-controlled, and production streams are monitored

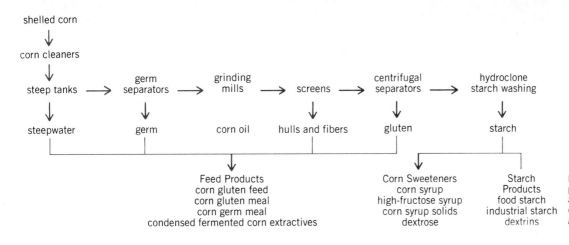

Fig. 8. Corn wet-milling process. (*After Nutritive Sweeteners from Corn, Corn Refiners Association, Inc., 1979*)

and regulated through central computers. In addition to use of more efficient boilers, heat recovered from steep liquor and syrup evaporators is used to dry feed. The water for washing starch and gluten is totally recycled for use during grinding grain, separating grain components, and steeping the grain. Mechanical vapor recompression evaporators used by some plants reduce energy requirements for concentrating steep water by almost half. In one plant, starch-gluten wash water is purified for reuse by reverse osmosis, with considerable energy savings.

Of the starch produced in the United States, 40% is used directly or chemically modified in food products and industrial uses. In addition to starch produced from normal dent corn, appreciable amounts of starches for special uses are prepared from two mutant corn hybrid lines. Starches from high-amylose corn (*ae*) and from waxy corn (*wx*) differ from normal corn starch in amounts of amylose and amylopectin components, and vary in solubility, gelation, viscosity, and other properties. *See* STARCH.

The remainder of the starch is dispersed in water, gelatinized by heat, and converted by acid or enzymic hydrolysis to corn syrup, which consists mainly of the sugar glucose (dextrose). Enzymic processes are finding increasing use, since they yield purer syrups. The starch is partially digested with α-amylase to yield soluble oligosaccharides, which are then cleaved to glucose with glucoamylase enzyme. Manufacture of crystalline dextrose requires nearly complete depolymerization of starch and crystallization from cooled 75% solutions. Corn syrups and dextrose are used extensively in foods and beverages. High-fructose corn syrup is manufactured by conversion of regular corn syrup with a purified isomerase enzyme derived from selected strains of species of *Streptomyces* bacteria. The enzyme is employed while immobilized on a solid support which facilitates reuse of continuous column operation. The produce, 42% fructose, 52% glucose, and 6% oligosaccharides, is sweeter than regular corn syrup. An 80–90% fructose syrup is produced commercially by fractionating the glucose and fructose on an ion-exchange column. This product is blended with 42% fructose syrup to yield a 55% fructose syrup that is widely used in popular beverages; it has sweetness comparable to sucrose solutions of equal concentration. *See* ENZYME; FRUCTOSE; GLUCOSE.

Oil is recovered from clean, dry, wet-milled germ by combined expeller-extraction operations at a yield of 1.25 lb/bu (1.61 kg/m^3). After further refinement, it is sold as table oil and used in margarine manufacture. Valuable protein-rich feed by-products are also obtained, including defatted corn germ meal and 60% protein gluten meal. Corn gluten feed consists of bran, residual starch and protein, and concentrated corn steep liquor. *See* FAT AND OIL (FOOD); MARGARINE.

Alcoholic beverages and alcohol. Whole ground corn is cooked, and the starch is hydrolyzed with malt enzymes to provide substrate for whiskey fermentation. Cooked grits treated with malt are used to produce beer by yeast fermentation. Corn sugar syrups are also widely used in brewing. Industrial ethyl alcohol may be distilled from the yeast fermentation broth of enzyme-treated cooked whole grain or wet-milling by-products. The need for gasoline has encouraged grain alcohol production for automotive fuel supplements. *See* ALCOHOL FUEL; DISTILLED SPIRITS; ETHYL ALCOHOL; FERMENTATION; FOOD ENGINEERING; FOOD MANUFACTURING; MALT BEVERAGE.

Joseph S. Wall

Bibliography. A. Cicuttini, W. A. Kollacks, and C. J. N. Rekers, Reverse osmosis saves energy and water in corn wet-milling, *Starke*, 35:149, 1983; W. C. Galinat, The origin of maize, *Annu. Rev. Genet.*, 5:447–478, 1971; G. E. Inglett (ed.), *Corn: Culture, Processing, Products*, 1970; G. E. Inglett (ed.), *Symposium on Sweeteners*, 1974; M. A. Joslyn and J. L. Heid, *Food Processing Operations*, vol. 3, 1964; R. H. Moll and C. W. Stuber, Quantitative genetics: Empirical results relevant to plant breeding, *Advan. Agron.*, 26:277–322, 1974; C. E. Morris, Huge plant for ethanol and HFCS, *Food Eng.*, pp. 111–112, June 1983; L. R. Nault et al., *Phytopathology*, vol. 57, 1967; M. C. Shurtleff (ed.), *Compendium of Corn Diseases*, 1980; G. F. Sprague (ed.), *Corn and Corn Improvement*, American Society of Agronomy, 1977; G. F. Sprague and W. E. Larson, *Corn Production*, USDA Handb. 322, 1975; W. N. Stoner and L. E. Williams, *47th Annual Report of the Southern Seedsmen Association*, 1965; A. J. Ullstrup, *USDA Agr. Handb.*, no. 199, 1978.

Cornales

An order of flowering plants, division Magnoliophyta (Angiospermae), in the subclass Rosidae of the class Magnoliopsida (dicotyledons). The order consists of 4 families and about 150 species. Within its subclass the order is marked by its usually woody habit, simple leaves, flowers with the perianth attached to the surface of the ovary and appearing to grow from the top of it (epigynous), usually well-developed endosperm, and fleshy fruits that remain closed at maturity (indehiscent). The various species of dogwood (*Cornus*) are well-known members of the order, as is the sour gum (*Nyssa sylvatica*, family Nyssaceae). *See* Dogwood; Magnoliophyta; Plant kingdom; Rosidae; Tupelo.

Arthur Cronquist; T. M. Barkley

Corner reflector antenna

A directional antenna consisting of the combination of a reflector comprising two conducting planes forming a dihedral angle and a driven radiator or dipole which usually is in the bisecting plane. It is widely used both singly and in arrays, gives good gain in comparison with cost, and covers a relatively wide band of frequency.

The **illustration** shows the general configuration

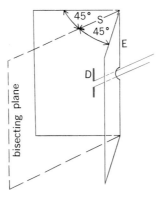

Diagram of corner reflector antenna.

for a 90° corner reflector. The distance S from the driven radiator D to edge E need not be critically chosen with respect to wavelength; for a 90° reflector D may lie between 0.25 and 0.7 wavelength. The overall dimensions of the reflector need not exceed 2 wavelengths in order to approximate the performance obtained with very large reflectors. Gain, as compared with an isotropic radiator, is about 12 dB.

The reflecting planes may be metal sheets or parallel wires separated by a small fraction of a wavelength and extending in the direction of current flow. *See* Antenna (electromagnetism).

J. C. Schelleng

Corona discharge

A type of electrical conduction that generally occurs at or near atmospheric pressure in gases. A relatively strong electric field is needed. External manifestations are the emission of light and a hissing sound. The particular characteristics of the discharge are determined by the shape of the electrodes, the polarity, the size of the gap, and the gas or gas mixture.

In some cases corona discharge may be desirable and useful, whereas in others it is harmful and attempts are made to minimize it. The effect is used for voltage division and control in direct-current nuclear particle accelerators. On the other hand, the corona

discharge that surrounds a high-potential power transmission line represents a power loss and limits the maximum potential which can be used. Because the power loss due to i^2r heating decreases as the potential difference is increased, it is desirable to use maximum possible voltage.

The shape of the electrodes has a very profound effect on the potential current characteristic. If the radius of curvature of the positive electrode is small compared to the gap between electrodes, the transition from the dark current region to the field-sustained discharge will be quite smooth. The effect here is to enable the free electrons to ionize by collision in the high field surrounding this electrode. One electron can produce an avalanche in such a field, because each ionization event releases an additional electron, which can then make further ionization. To sustain the discharge, it is necessary to collect the positive ions and to produce the primary electrons far enough from the positive electrode to permit the avalanche to develop. The positive ions are collected at the negative electrode, and it is their low mobility that limits the current in the discharge. The primary electrons are thought to be produced by photoionization (see **illus.**). For a discussion of ionic mobility *See* Electrical conduction in gases. *See also* Dark current.

The situation at the negative electrode is quite important. The efficiency for ionization by positive ions is much less than for electrons of the same energy. Most of the ionization occurs as the result of secondary electrons released at the negative electrode by positive-ion bombardment. These electrons produce ionization as they move from the strong field at the electrode out into the weak field. This, however, leaves a positive-ion space charge, which slows down the incoming ions. This has the effect of diminishing

Breakdown streamers in positive-point corona crossing from positive point to cathode below. (From L. B. Loeb, *Fundamental Processes of Electrical Discharge in Gases*, Wiley, 1939)

the secondary electron yield. Because the positive ion mobility is low, there is a time lag before the high field conditions can be restored. For that reason the discharge is somewhat unstable.

From the preceding it may be seen that the electrode shape is important. The dependence on the gas mixture is difficult to evaluate. Electronegative components will tend to reduce the current at a given voltage, because heavy negative ions have a low mobility and are inefficient ionization agents. The excess electron will not be tightly bound, however, and may be released in a collision. The overall effect is to reduce the number of electrons that can start avalanches near the positive electrode. Further, if a gas with low-lying energy states is present, the free electrons can lose energy in inelastic collisions. Thus it is more difficult for the electrons to acquire enough energy to ionize. In electrostatic high-voltage generators in pressurized tanks, it is quite common to use Freon and nitrogen to take advantage of this effect to reduce corona. In a pure monatomic gas such as argon, corona occurs at relatively low values of voltage.

In the potential current characteristic, the corona region is found above the dark current region and is field-sustained. Near the upper end it goes into either a glow discharge or a brush discharge, depending on the pressure. Higher pressure favors the brush discharge. See Brush discharge; Glow discharge.

For still higher potential difference, breakdown takes place and a continuously ionized path is formed. See Electric spark.

Glenn H. Miller

Bibliography. American Society for Testing and Materials, *Engineering Dielectrics*, vol. 1: *Corona Measurement and Interpretation*, 1979; L. B. Loeb, *Electrical Coronas*, 1965.

Fig. 2. Image of the solar corona as recorded in the green spectral line (530.3 nm) of Fe(XIV). The complex loop structures trace corresponding magnetic field configurations. The black central disk is an image of the occulting disk inside the coronagraph, slightly larger than the image of the Sun. The photograph was recorded with a 20-cm-aperture (8-in.) emission-line coronagraph. *(National Solar Observatory/Sacramento Peak Association of Universities for Research in Astronomy, Inc.)*

Coronagraph

A specialized astronomical telescope substantially free from instrumentally scattered light, used to observe the solar corona, the faint atmosphere surrounding the Sun. Coronagraphs can record the emission component of the corona (spectral lines emitted by high-temperature ions surrounding the Sun) and the white-light component (solar photospheric light scattered by free electrons surrounding the Sun) routinely from high mountain sites under clear sky conditions. The emission and white-light components are typically only a few millionths the brightness of the Sun itself. Hence, the corona is difficult to observe unless the direct solar light is completely rejected, and unless instrumentally diffracted and scattered light that reaches the final image plane of the coronagraph is small relative to the coronal light. See Solar corona; Sun.

The basic design (**Fig. 1**), as invented by B. Lyot, has an occulting disk in the primary image plane of the telescope to block the image of the Sun itself. In addition, the primary objective (a lens or superpolished mirror) is specially fabricated to minimize scattered light. Also, light diffracted by the objective rim must be suppressed. For this, an aperture (Lyot stop) is placed at an image of the objective as produced by a field lens, with the aperture diameter slightly smaller than that of the objective image. A camera lens behind the Lyot stop forms the coronal image at the final image plane. See Optical telescope.

To observe the emission corona, optical filters or spectrographs are used to isolate the wavelengths corresponding to the various coronal emission lines and thus minimize the contribution due to the sky (**Fig. 2**). To observe the white-light corona, polarization subtraction techniques are required to discriminate the coronal electron-scattered light (polarized with the electric vector tangential to the limb) from the variably polarized sky background. Coronagraphs designed for operation on satellites usually have a circular mask in front of the objective to completely shade it. This external occulting is more efficient, when no sky background is present, than if an internal occulting disk is used.

Raymond Smartt

Bibliography. D. E. Billings, *A Guide to the Solar Corona*, 1966; R. R. Fisher et al., New Mauna Loa coronagraph systems, *Appl. Opt.*, 20:1094–1101, 1981; R. M. MacQueen et al., The high altitude observatory coronagraph polarimeter on the Solar Maximum Mission, *Solar Phys.*, 65:91–107, 1980; R. N. Smartt, R. B. Dunn, and R. R. Fisher, Design and performance of a new emission-line coronagraph, *S.P.I.E.* (Society of Photo-optical Instrumentation Engineers), 288:395–402, 1981.

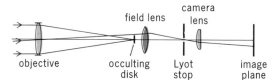

Fig. 1. Diagram showing the key features of a coronagraph used for observing the solar corona.

Coronatae

An order of the class Scyphozoa which includes mainly abyssal species. The exumbrella is divided by a circular or coronal furrow into two parts, an upper

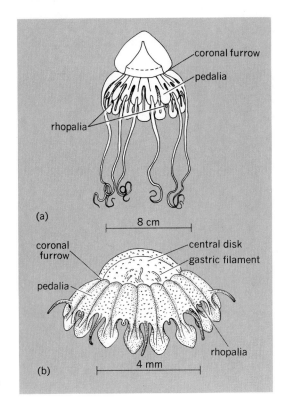

Coronatae. (a) *Periphylla*. (b) *Nausithoë*. (After L. H. Hyman, *The Invertebrates*, vol. 1, McGraw-Hill, 1940)

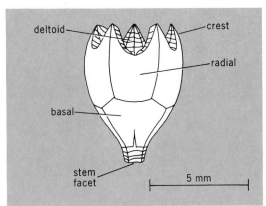

Stephanocrinus gemmiformis, Middle Silurian, Tennessee; side view of the theca.

central disk and a lower coronal part (see **illus.**). The central disk is usually domelike, but in *Periphylla* it often narrows toward the top. The coronal part has gelatinous thickenings, or pedalia, situated on the radii running from the center of the umbrella to the tentacles and the sensory organs. The pedalia are separated from one another by clefts, running down the radii, that are halfway between each tentacle and rhopalium (a sensory organ). The mouth has four lips. There are four groups of gastral filaments within the central stomach. Abyssal forms of this group are dark brown or reddish purple. The littoral species of *Nausithoë* is reported to show alternation of generations, but the life history of abyssal forms remains unknown. SEE SCYPHOZOA.

Tohru Uchida

Coronoidea

A small class of ''arm''- and brachiole-bearing, stemmed echinoderms in the subphylum Crinozoa, based on five genera known from the Middle Ordovician to Late Silurian of Europe and North America. Coronoids have a crested theca or body with well-developed pentameral symmetry and plate arrangement very similar to that found in blastoids (see **illus.**). Skeletal plates include three basals (two large and one smaller), five radials extending up to form the crests and bearing central notches for the ambulacra, five small plates supporting the arms, and four regular deltoids plus two anal deltoids in the fifth position around the mouth. The mouth is central on top of the theca and has five ambulacral grooves radiating from it. A coiled arm is attached to a mounting plate at the end of each ambulacral groove; each arm has biserial plating with smaller biserial branches (= brachioles) on both sides. The projecting crests have in-

ternal canals that connected with the body cavity and apparently served as respiratory structures. A thin stem supported the theca above the sea floor, allowing coronoids to live as attached, low- to medium-level suspension feeders. Coronoids had previously been assigned to the blastoids, eocrinoids, or crinoids by different researchers, but they have been elevated in rank to a separate class. Coronoids appear to be most closely related to the Blastoidea and may have been the ancestors of this class. SEE BLASTOIDEA; CRINOZOA; ECHINODERMATA; EOCRINOIDEA.

James Sprinkle

Bibliography. C. E. Brett et al., *J. Paleontol.*, 57:627–651, 1983.

Corrosion

In broad terms, the interaction between a material and its environment that results in a degradation of the physical, mechanical, or even esthetic properties of that material. More specifically, corrosion is usually associated with a change in the oxidation state of a metal, oxide, or semiconductor. For example, immersing a piece of sodium in water is an extreme example of corrosion; the vigorous reaction is the electrochemical oxidation of elemental sodium in the zero valence state to monovalent sodium ions (Na^+) dissolved in the water. Corrosion encompasses principles from diverse fields such as electrochemistry, metallurgy, physics, chemistry, and biology.

Economic impact. The cost of corrosion in the United States and other industrial countries has been estimated to be on the order of 4% of the gross national product. This cost represents both the replacement of structures or components and the protection against corrosion. Such direct losses are relatively easy to assess in comparison to indirect losses, which include cost of shutdown, loss of product or efficiency, and contamination. Another economic aspect is the source of corrosion-resistant metals. For example, chromium is considered to be of critical importance since it is responsible for the high-temperature oxidation resistance and low-temperature corrosion resistance of a wide range of iron- and nickel-based alloys. Much of the chromium being used is considered irreplaceable. Since chromium ores occur primarily in South Africa and the Soviet Union, concern is often expressed within the Western industrialized nations regarding a perceived vulnerability in supply. SEE CHROMIUM.

Engineering systems for service environments require materials that are chemically stable or that can be made chemically stable. Moreover, modern technology continues to demand materials that are capable of performing in increasingly hostile circumstances such as high-temperature batteries, deep sour-gas wells, magnetohydrodynamic channels, and fusion reactors. Additionally, as engineering systems continue to evolve, they will use new or advanced materials. Thin-film electronic, magnetic, and optical devices will require special attention in both civilian and military engineering systems. Likewise, it is necessary to deal with problems associated with more traditional systems, such as aging infrastructure, underground tanks and pipelines, water treatment plants, and power stations.

Reactions. Electrolytic corrosion consists of two partial processes: an anodic (oxidation) and cathodic (reduction) reaction (**Fig. 1**). In the absence of any external voltages, the rates of the anodic and cathodic reactions are equal, and there is no external flow of current. The loss of metal that is the usual manifestation of the corrosion process is a result of the anodic reaction, and can be represented by reaction (1). This

$$M \rightarrow M^{n+} + ne^- \qquad (1)$$

reaction represents the oxidation of a metal (M) from the elemental (zero valence) state to an oxidation state of M^{n+} with the generation of n moles of electrons (e^-). The anodic reaction may occur uniformly over a metal surface or may be localized to a specific area. If the dissolved metal ion can react with the solution to form an insoluble compound, then a buildup of corrosion products may accumulate at the anodic site. SEE OXIDATION-REDUCTION.

In the absence of any applied voltage, the electrons generated by the anodic reaction (1) are consumed by the cathodic reaction. For most practical situations, the cathodic reaction is either the hydrogen-evolution reaction or the oxygen-reduction reaction. The hydrogen-evolution reaction can be summarized as reaction (2). In this case, protons (H^+) combine with electrons

$$2H^+ + 2e^- \rightarrow H_2 \qquad (2)$$

to form molecules of hydrogen (H_2). This reaction is often the dominant cathodic reaction in systems at low pH. The hydrogen-evolution reaction can itself cause corrosion-related problems, since atomic hydrogen (H) may enter the metal, causing embrittlement, a phenomenon that results in an attenuation of the mechanical properties and can cause catastrophic failure. SEE EMBRITTLEMENT.

The second important cathodic reaction is the oxygen-reduction, given by reactions (3) and (4). These

$$O_2 + 4H^+ + 4e^- \rightarrow 2H_2O \qquad (3)$$

$$O_2 + 2H_2O + 4e^- \rightarrow 4OH^- \qquad (4)$$

represent the overall reactions in acidic and alkaline solutions, respectively. This cathodic reaction is usually dominant in solutions of neutral and alkaline pH. In order for this reaction to proceed, a supply of dissolved oxygen is necessary; hence the rate of this reaction is usually limited by the transport of oxygen to the metal surface.

Reactions (2)–(4) represent the overall reactions

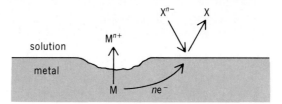

Fig. 1. Diagram of a corrosion cell showing the anodic and cathodic partial processes. X^{n-} = cathodic reactant, X = cathodic product, $x^{n-} + ne \rightarrow x$.

which, in practice, may occur by a sequence of reaction steps. In addition, the reaction sequence may be dependent upon the metal surface, resulting in significantly different rates of the overall reaction. The cathodic reactions are important to corrosion processes since many methods of corrosion control depend on altering the cathodic process. It should be pointed out that although the cathodic reactions may be related to corrosion processes which are usually unwanted, they are essential for many applications such as energy storage and generation.

Thermodynamic stability of metals. The equilibrium potential of a reaction is determined by a thermodynamic relationship known as the Nernst equation. By convention, all electrochemical reactions are written in terms of reduction [reaction (5)]. The

$$O + ne^- \rightleftharpoons R \text{ or } M^{n+} + ne \rightleftharpoons M \qquad (5)$$

Nernst equation is given by Eq. (6), where U_0 is the

$$U_0 = U^\theta - \frac{2.303RT}{nF} \log \frac{a_R}{a_O} \qquad (6)$$

equilibrium potential, U^θ is the equilibrium potential under standard conditions, a_R and a_O are the activities of the reduced and oxidized species, respectively, n is the number of moles of electrons transferred, F is Faraday's constant, R is the gas constant, and T is temperature. The activity coefficients are related to concentration, and in Eq. (6) a_R is the activity of the metal ions M^{n+} (for metal/ion electrodes this is usually assumed to be 10^{-6} mole/liter).

The standard reduction potentials for various metals are given in the **table**. Also included in this list are the standard potentials for the hydrogen evolution and oxygen reduction reactions. The equilibrium potential for the hydrogen evolution reaction at pH 0 is arbitrarily designated as zero volts, and all other potentials are referenced to this value.

This table can be used to predict the thermodynamic tendency of a metal to corrode by comparing the equilibrium potential of the metal to that of a cathodic reaction. If a metal has a more negative potential than a cathodic reaction, then it can be shown that the Gibbs free energy of the system is negative, and the metal will have a thermodynamic tendency to dissolve according to reaction (1). For example, the equilibrium potential for iron (Fe^{2+}/Fe) equilibrium is more negative than the equilibrium potentials for both cathodic reactions; iron will, therefore, dissolve to form ferrous ions in solution. Copper has an equilibrium potential more positive than that of the hydrogen reaction and, hence, will dissolve only with oxygen reduction.

A complete diagram of thermodynamic stability

Standard reduction potentials (U^θ)* for various metals

Metal	Electrode reaction	U^θ
Gold	$Au^{3+} + 3e^- = Au$	1.50
Oxygen reduction	$O_2 + 2H_2O + 4e^- = 4OH^-$	1.23
Palladium	$Pd^{2+} + 2e^- = Pd$	0.99
Silver	$Ag^+ + e^- = Ag$	0.80
Copper	$Cu^{2+} + 2e^- = Cu$	0.34
Hydrogen evolution	$2H^+ + 2e^- = H_2$	0.00
Lead	$Pb^{2+} + 2e^- = Pb$	−0.13
Tin	$Sn^{2+} + 2e^- = Sn$	−0.14
Nickel	$Ni^{2+} + 2e^- = Ni$	−0.25
Cobalt	$Co^{2+} + 2e^- = Co$	−0.28
Cadmium	$Cd^{2+} + 2e^- = Cd$	−0.40
Iron	$Fe^{2+} + 2e^- = Fe$	−0.44
Chromium	$Cr^{3+} + 3e^- = Cr$	−0.74
Zinc	$Zn^{2+} + 2e^- = Zn$	−0.76
Titanium	$Ti^{2+} + 2e^- = Ti$	−1.63
Aluminum	$Al^{3+} + 3e^- = Al$	−1.66
Magnesium	$Mg^{2+} + 2e^- = Mg$	−2.37
Sodium	$Na^+ + e^- = Na$	−2.71
Calcium	$Ca^{2+} + 2e^- = Ca$	−2.87
Potassium	$K^+ + e^- = K$	−2.93

*$T = 25°C$ (77°F), $a = 1$ mol/liter.

source of thermodynamic stability. However, it is noted that these diagrams are only thermodynamic predictions and cannot predict the kinetics, or rate, of a particular reaction. SEE ACTIVITY (THERMODYNAMICS); CHEMICAL THERMODYNAMICS.

Rate of corrosion. Corrosion rates are usually expressed in terms of loss of thickness per unit time. General corrosion rates may vary from on the order of centimeters per year [for example, iron in sulfuric acid (H_2SO_4)] to micrometers per year. Relatively large corrosion rates may be tolerated for some large structures, whereas for other structures small amounts of corrosion may result in catastrophic failure. For example, with the advent of technology for making extremely small devices, future generations of integrated circuits will contain components that are on the order of nanometers (10^{-9} m) in size, and even small amounts of corrosion could cause a device failure.

Many metals and alloys that exhibit a thermody-

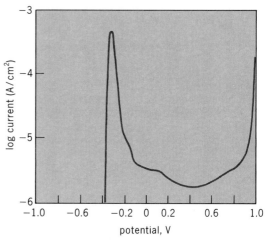

Fig. 3. Current–potential diagram for AISI 304 stainless steel in sulfuric acid (1 N H_2SO_4), $T = 25°C$ (77°F), scan rate = 0.3 mV/s.

can be constructed by considering the pH dependency of the system. For example, **Fig. 2** shows a potential–pH diagram for the iron–water system. Such diagrams are also known as Pourbaix diagrams. From this diagram it can be seen that species other than soluble metal ions exist at potentials more positive than the equilibrium potential. For example, in the region of stability of ferric [iron(III)] oxide (Fe_2O_3), iron is protected from corrosion by a thin film of this oxide; a phenomenon known as passivation. In regions where the ferrous ions (Fe^{2+}) or other soluble species are stable, iron will corrode.

Similar diagrams can be compiled for various metal–solution systems in order to provide a reference

namic tendency to corrode actually have remarkable corrosion resistance in many environments due to the formation of a passive film. Passivation is the term used to describe the formation of a thin protective oxide on the surface of a metal. Passive films are usually less than 10 nm in thickness but, in some cases, may consist of just a few atomic layers. Passive films are highly stable and can prevent severe corrosion of the underlying metal. For example, stainless steels such as Fe-18%Cr-8%Ni form a highly stable thin chromium oxide layer that is the main reason for the widespread application of stainless steels in industry as corrosion-resistant alloys.

The electrochemical characteristics of a metal or alloy are usually determined by a combination of experimental techniques of which the current–potential behavior is the most common. **Figure 3** shows a current–potential diagram for AISI 304 stainless steel (18–20% Cr, 8.0–10.5% Ni, <0.08% C) in 1 N H_2SO_4 solution at 25°C (77°F). (AISI is the acronym for the American Iron and Steel Institute.) The current is often plotted on a logarithmic scale, as in this figure, due to the exponential relationship between current and potential for many electrochemical reactions.

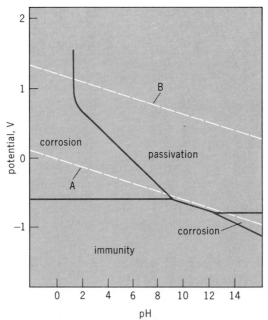

Fig. 2. Potential–pH diagram for the iron-water system: A = hydrogen equilibrium [see reaction (2) in text]; B = oxygen equilibrium [see reactions (3) and (4) in text]. (After M. Pourbaix, ed., Atlas of Electrochemical Equilibria in Aqueous Solutions, NACE, Houston, 1974)

In Fig. 3 it can be seen that the open circuit potential, or corrosion potential, is about -0.38 V; as the potential is increased, the current increases rapidly to a maximum. At this point a stable passive film is formed on the surface, and increasing the potential further results in a decrease in current to a low value. At more positive potentials, the current increases again, at about 1 V because of oxygen evolution from the surface.

The continuous stability of the passive film is critical for most corrosion resistance applications. A passive film can become unstable in an environment in which it can dissolve chemically or in the presence of certain aggressive anions. Halide ions, and in particular chloride ions, are the most common species that can cause passive film breakdown, and consequently many types of corrosion problems are related to exposure to seawater or other chloride-containing environments. This type of corrosion attack is usually characterized by localized breakdown of the film. For many common engineering alloys, the sites of localized attack are usually related to inhomogeneities such as precipitates or secondary phases.

Types. In some situations, corrosion may occur only at localized regions on a metal surface. This type of corrosion is characterized by regions of locally severe corrosion, although the general loss of thickness may be relatively small. Common types of localized corrosion are pitting corrosion and crevice corrosion. Other types of localized corrosion include a galvanic corrosion, stress corrosion cracking, and hydrogen embrittlement.

Pitting. This type of corrosion is usually associated with passive metals, although this is not always the case. Pit initiation is usually related to the local breakdown of a passive film and can often be related to the presence of halide ions in solution (**Fig. 4**). The local environment within a pit can be extremely acidic, resulting in a high rate of corrosion within this small region. For the common situation of a metal in salt water, the dissolution of metal within the pit results in the formation of metal chlorides (MCl). These salts are then hydrolyzed to form hydrochloric acid (HCl) within the pit, as shown in reaction (7).

$$\text{MCl} + \text{H}_2\text{O} \rightarrow \text{MOH} + \text{HCl} \qquad (7)$$

The decrease in pH within the pit caused by acid formation further increases the rate of metal dissolution, and such a reaction is often referred to as autocatalytic.

Crevice. This type of corrosion occurs in restricted or occluded regions, such as at a bolted joint, and is often associated with solutions that contain halide ions (**Fig. 5**). Crevice corrosion is initiated by a depletion of the dissolved oxygen in the restricted region. For this situation to occur, the dimension of the restricted region is typically less than 1 mm. As the supply of oxygen within the crevice is depleted, because of cathodic oxygen reduction, the metal surface within the crevice becomes activated, and the anodic current is balanced by cathodic oxygen reduction from the region adjacent to the crevice. The ensuing reactions within the crevice are the same as those described for pitting corrosion: halide ions migrate to the crevice, where they are then hydrolyzed to form metal hydroxides and hydrochloric acid. In neutral salt solutions the pH within a crevice may fall to a value as low as 2.

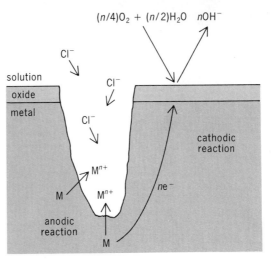

Fig. 4. Diagram of pitting corrosion. The diffusion of chloride ions into the pit to maintain electroneutrality is followed by hydrolysis of metal chlorides, resulting in the formation of hydrochloric acid (HCl).

Galvanic. Corrosion can also be accelerated in situations where two dissimilar metals are in contact in the same solution. This form of corrosion is known as galvanic corrosion. The metal with the more negative potential becomes the anode, while the metal with the more positive potential sustains the cathodic reaction (**Fig. 6**). In many cases the table of equilibrium potentials can be used to predict which metal of galvanic couple will corrode. For example, aluminum-graphite composites generally exhibit poor corrosion resistance since graphite has a positive potential and aluminum exhibits a highly negative potential. As a result, in corrosive environments the aluminum will tend to corrode while the graphite remains unaffected.

Tensile stress. Stress corrosion cracking and hydrogen embrittlement are corrosion-related phenomena associated with the presence of a tensile stress. Stress corrosion cracking results from a combination of stress and specific environmental conditions so that localized corrosion initiates cracks that propagate in the presence of stress (**Fig. 7**). Mild steels are susceptible to stress corrosion cracking in environments containing hydroxyl ions (OH^-; often called caustic cracking) or nitrate ions (NO_3^-). Austenitic stainless steels are susceptible in the presence of chloride ions

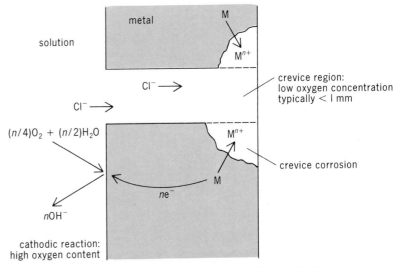

Fig. 5. Diagram of crevice corrosion, showing the mechanisms involved.

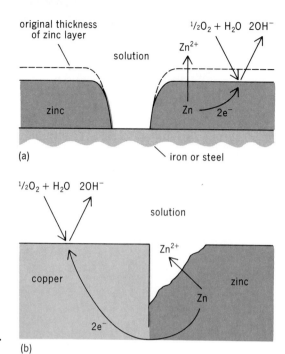

Fig. 6. Reactions of dissimilar metals in contact in the same solution. (a) Iron or steel protected from corrosion by a galvanic zinc coating. (b) Galvanic corrosion.

(Cl^-) and hydroxyl ions (OH^-). Other alloys that are susceptible under specific conditions include certain brasses, aluminum and titanium alloys.

Hydrogen embrittlement is caused by the entry of hydrogen atoms into a metal or alloy, resulting in a loss of ductility or cracking if the stress level is suf-

ficiently high. The source of the hydrogen is usually from corrosion (that is, cathodic hydrogen evolution) or from cathodic polarization. In these cases the presence of certain substances in the metal or electrolyte can enhance the amount of hydrogen entry into the alloy by poisoning the formation of molecular hydrogen. Metals and alloys that are susceptible to hydrogen embrittlement include certain carbon steels, high-strength steels, nickel-based alloys, titanium alloys, and some aluminum alloys. *SEE ALLOY; STAINLESS STEEL; STEEL.*

Other types. There are other types of corrosion reactions. Corrosion can be stimulated by biological organisms (microbially induced corrosion) and contaminants in the air (atmospheric corrosion). An example of the latter is corrosion caused by acid rain. *SEE ACID RAIN.*

Control. A reduction in the rate of corrosion is usually achieved through consideration of the materials or the environment. Materials selection is usually determined by economic constraints. For example, gold is resistant to corrosion in almost all solutions; however, it is not usually an economic choice. In addition, the corrosion resistance of a specific metal or alloy may be limited to a certain range of pH, potential, or anion concentration. As a result, replacement metal or alloy systems are usually selected on the basis of cost for an estimated service lifetime.

Control of the environment can be achieved in a number of ways. This may involve controlling the solution chemistry such as pH or ion concentration, or controlling the metal–solution interface. Control of this interface can be achieved through the addition of small amounts of chemicals, called inhibitors, that either form a barrierlike layer on the metal surface or stimulate film formation. Inhibitors can generally be classified as either inorganic salts, ions, or organic molecules. Examples of inorganic salts and ions that are commonly used as inhibitors are chromate ions (CrO_4^{2-}), nitrite ions (NO_2^-), and phosphate ions (PO_4^{3-}). Organic inhibitors are generally long-chain polar molecules with nitrogen, sulfur, oxygen, or hydroxyl groups. Organic coatings, inorganic coatings, and metallic coatings are all used to provide an interfacial barrier between the metal and solution. *SEE INHIBITOR (CHEMISTRY); METAL COATINGS.*

Electrochemical techniques can also be used for corrosion control. For example, the principle of cathodic protection is the application of a cathodic potential, or current, to a structure that prevents it from undergoing an anodic reaction. This can be achieved either by coupling the metal to be protected with a metal having a more negative open circuit potential (sacrificial anode) or by applying a cathodic current through an auxiliary electrode (impressed current).

Cathodic protection is often used on a small scale to prevent corrosion of domestic hot water tanks. In this application, a small magnesium electrode is connected to the tank electrically and immersed in the water. If corrosion does occur due to the presence of contaminant ions in the tank, the magnesium will preferentially corrode leaving the tank unaffected.

Cathodic protection is usually applied to a structure in conjunction with a coating, since the current required to protect a large uncoated structure would be uneconomical. Many underground pipes are coated and protected from corrosion by an impressed current system through a series of auxiliary anodes connected to the pipeline through a current source and embedded

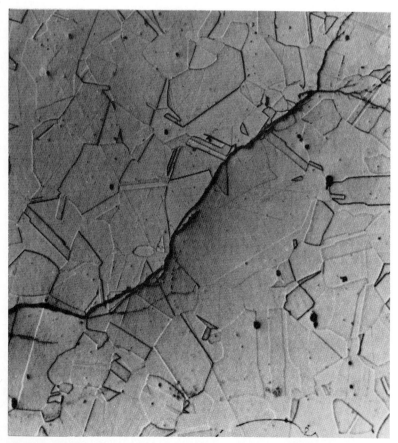

Fig. 7. Micrographic illustrating intergranular stress corrosion cracking of AISI 304 stainless steel exposed to boiling magnesium chloride at 90% of the yield stress.

close to the pipeline. Marine structures, such as oil drilling platforms, are frequently cathodically protected.

The future. Traditional corrosion engineering problems have included systems such as steam generators, heat exchangers, reaction vessels, bridges, oil platforms, and motor vehicles. These areas will continue to be important in terms of materials selection and corrosion control; however, with the search for new energy sources, the development of new materials, and the increase in microprocessor control in manufacturing, many new problems concerning the chemical stability and utilization of materials will have to be addressed.

The next generation of thin-film devices for integrated circuits and magnetic storage will be fabricated on a submicrometer scale ($<10^{-6}$ m), and a small amount of corrosion may cause a device failure. Microprocessor control is being utilized in increasingly aggressive environments, such as process plants and automobiles, where the importance of corrosion control in terms of device packaging will become critical. Advanced optical and magnetic devices will also require attention in order to avoid environmentally induced degradation.

Corrosion and corrosion control will also continue to be important in energy systems. For example, a major limitation to the development of solar cells has been the relatively poor corrosion stability of the more efficient semiconducting oxides. Other energy sources, such as geothermal energy and sour oil and gas wells, also require processing in highly corrosive environments. In addition, there is a continued search for more efficient materials for catalysis, fuel cells, batteries, photoelectrolysis, and other emerging technologies.

In summary, all engineering materials are susceptible to some form of environmental degradation. Whether traditional construction steels, ceramic superconductors with high critical temperatures, structural composites, optical fibers, and so forth, it is incumbent on users of such materials to be aware of the limits of chemical stability of these materials in service environments. SEE ELECTROCHEMISTRY.

R. M. Latanision; Peter C. Searson

Bibliography. *Agenda for Advancing Electrochemical Corrosion Science and Technology,* Nat. Mater. Adv. Bd. Rep. NMAB 438-2, 1987; M. G. Fontana, *Corrosion Engineering,* 1986; M. Pourbaix (ed.), *Atlas of Electrochemical Equilibria in Aqueous Solutions,* 1974; H. H. Uhlig and R. W. Revie, *Corrosion and Corrosion Control,* 1985; J. M. West, *Basic Corrosion and Oxidation,* 1980.

Cortex (plant)

The mass of primary tissue in roots and stems extending inward from the epidermis to the phloem. The cortex may consist of one or a combination of three major tissues: parenchyma, collenchyma, and sclerenchyma. In roots the cortex almost always consists of parenchyma, and is bounded, more or less distinctly, by the hypodermis (exodermis) on the periphery and by the endodermis on the inside.

Cortical parenchyma is composed of loosely arranged thin-walled living cells. Prominent intercellular spaces usually occur in this tissue. In stems the cells of the outer parenchyma may appear green due

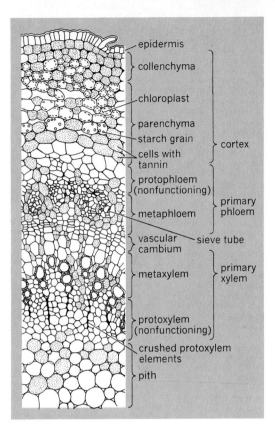

Fig. 1. Transverse section of the *Prunus* stem showing the cortex which is composed of collenchyma and parenchyma. The parenchyma has cells containing chloroplasts, starch grains, and tannin. (*After K. Esau, Plant Anatomy, 2d ed., John Wiley and Sons, 1967*)

to the presence of chloroplasts in the cells (**Fig. 1**). This green tissue is sometimes called chlorenchyma, and it is probable that photosynthesis takes place in it.

In some species the cells of the outer cortex are modified in aerial stems by deposition of hemicellulose as an additional wall substance, especially in the corners or angles of the cells. This tissue is called collenchyma, and the thickening of the cell walls gives mechanical support to the shoot.

The cortex makes up a considerable proportion of the volume of the root, particularly in young roots, where it functions in the transport of water and ions from the epidermis to the vascular (xylem and phloem) tissues (**Fig. 2**). In older roots it functions primarily as a storage tissue. The cells of the root cortex generally are similar in appearance and may occur in radial files, but histochemical tests reveal a variety of cell types in root cortex.

In addition to being supportive and protective, the cortex functions in the synthesis and localization of many chemical substances; it is one of the most fundamental storage tissues in the plant. The kinds of cortical cells specialized with regard to storage and synthesis are numerous.

Because the living protoplasts of the cortex are so highly specialized, patterns and gradients of many substances occur within the cortex, including starch, tannins, glucosides, organic acids, crystals of many kinds, and alkaloids. Oil cavities, resin ducts, and laticifers (latex ducts) are also common in the mid-cortex of many plants. SEE SECRETORY STRUCTURES (PLANT).

The young cortex in roots is functional in the selection and passage of solutes and ions into the plant. Experiments with radioactive ions indicate a pileup of solutes in the innermost cells of the cortex next to the

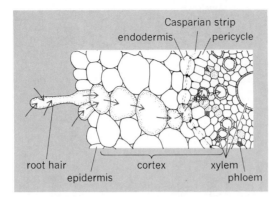

Fig. 2. Transverse section of wheat root, illustrating the kinds of cell that may be crossed by water and salts absorbed from the soil before they reach the tracheary elements of the xylem. The arrows indicate the direction of the movement through a selected series of cells. Among these, the living cells are partly stippled. The Casparian strip in the endodermis is shown as though exposed in surface views of the end walls. (After K. Esau, Plant Anatomy, 2d ed., John Wiley and Sons, 1967)

endodermis. Radioactive isotopes used in experimental work accumulate just outside the endodermis by a screening action of the endodermis and the innermost cells of the cortex. Cortical cells next to the endodermis apparently act as a selective barrier with the endodermis to the kinds of ions taken into the stele and vascular tissue.

There appears to be complete freedom of movement of water, solutes, and ions through the cortex from the epidermis, but the Casparian strips limit freedom of movement through the endodermis into the stele. The piling up of solutes in the inner cortex limits freedom of movement and may function in the selection of the kinds of ions that pass from the cortex through the endodermis into the stele. Numerous substance gradients, as well as gradients of oxidizing potential across the cortex, may be significant in the uptake of solutes in the active growing portion of young roots. The passage and transport of solutes and ions from the cortex into the vascular system is most pronounced in the cortex just back of the growing tip of the root. In this zone of nutrient absorption, it is believed that passage of solutes into the vascular tissue is effected by an active metabolic transport system in the inner cells of the cortex, with an active system of selection in the endodermis.

The gradient of decreasing oxygen and increasing carbon dioxide across the cortex and into the vascular tissue is associated with a low level of metabolic activity in the cells around the vascular xylem vessels. Cells around the xylem with low metabolic activity would be expected to lose salts, since energy is required to hold salts, into the open vessels of the xylem. Diffusion of solutes back into the cortex is prevented by the impervious Casparian band between all endodermal cells. A unidirectional movement of solutes into the xylem vessels of the plant is a result of the action of the cortex and endodermis. *See Cell walls (plant); Collenchyma; Endodermis; Hypodermis; Parenchyma; Root (botany); Sclerenchyma; Stem.*

D. S. Van Fleet

Bibliography. J. D. Mauseth, *Plant Anatomy*, 1988.

Corundum

A mineral with the ideal composition Al_2O_3. It is one of a large group of isostructural compounds including hematite (Fe_2O_3) and ilmenite ($FeTiO_3$), all of which crystallize in the hexagonal crystal system, trigonal subsystem. The structure is based on hexagonal closest packing of oxygen ions with aluminum in octahedrally coordinated sites. Corundum has the high hardness of 9 on Mohs scale and is therefore commonly used as an abrasive, either alone or in the form of the rock called emery, which consists principally of the minerals corundum and magnetite. Crystals occurring in igneous rocks usually have an elongated barrellike shape, while crystals from metamorphic rocks are generally tabular (see **illus.**). It is commonly complexly twinned. Such specimens have a pronounced parting along twin planes and when broken may have blocky shapes. There is no discernible cleavage. The specific gravity is approximately 3.98. *See Hematite; Ilmenite.*

Pure corundum is transparent and colorless, but most specimens contain some transition elements substituting for aluminum, resulting in the presence of color. Substitution of chromium results in a deep red color; such red corundum is known as ruby. The term "sapphire" is used in both a restricted sense for the "cornflower blue" variety containing iron and titanium, and in a general sense for gem-quality corundums of any color other than red. Green, yellow, gray, violet, and orange hues are not uncommon. Star ruby and star sapphire contain tiny needles of the mineral rutile. Reflection of light from these needles produces the six-rayed star in specimens which have been cut and polished en cabochon, with the cabochon base oriented perpendicular to the c axis. *See Ruby; Sapphire.*

Corundum occurs as a rock-forming mineral in both metamorphic and igneous rocks, but only in those which are relatively poor in silica, and never in association with free silica. Igneous rocks which most

(a)   (b)

Corundum. (a) Specimen taken from Steinkopf, South Africa (American Museum of Natural History). (b) Crystal habits (after C. S. Hurlbut, Jr., Dana's Manual of Mineralogy, 17th ed., John Wiley and Sons, 1959).

commonly contain corundum include syenites, nepheline syenites, and syenite pegmatites. It is also found as megacrysts in basaltic igneous rocks and in xenoliths contained in igneous rocks. Both contact and regionally metamorphosed silica-poor rocks may contain corundum. Original unmetamorphosed source rocks include bauxite and other aluminous sediments, as well as basic igneous rocks. Some metamorphic rocks with corundum are probably generated through the extraction of alkali and silica-rich magma. Corundum is chemically resistant to weathering processes and is therefore frequently found in alluvial deposits. Most of the gem-quality material is recovered from such sediments, as in the Mogok district of Burma (ruby), Thailand, Sri Lanka, and Australia. Deposits of gem-quality sapphire occur in Montana. SEE IG-NEOUS ROCKS; METAMORPHIC ROCKS.

Corundum is synthesized by a variety of techniques for use as synthetic gems of a variety of colors and as a laser source (ruby). The most important technique is the Verneuil flame-fusion method, but others include the flux-fusion and Czochralski "crystal-pulling" techniques. SEE GEM.

Donald R. Peacor

Cosmic background radiation

A nearly uniform flux of microwave radiation that is believed to permeate all of space. The discovery of this radiation in 1965 by A. Penzias and R. W. Wilson has had a profound impact on understanding the nature and history of the universe. The interpretation of this radiation as the remnant fireball from the big bang by P. J. E. Peebles, R. H. Dicke, R. G. Roll, and D. T. Wilkinson, and their correct prediction of the spectrum to be that of a blackbody, was one of the great triumphs of cosmological theory.

Origin of radiation. In the theory of the big bang, the universe began with an explosion 10–15×10^9 years ago. This big bang was not an explosion of matter into empty space, but an explosion of space itself. The early universe was filled with dense, hot, glowing matter; there was no region of space free of matter or radiation. (This is reflected in the present universe by the fact that space is more or less uniformly filled with galaxies; galaxy clusters and holes between clusters are believed to have grown from gravitational instabilities during the expansion.) The explosion of space increased the volume of the matter and radiation, and thus reduced the density and temperature. The initial temperature was so high that even for several hundred thousand years after the initial explosion the universe was still as hot as the surface of the present-day Sun. At this temperature the matter of the universe was in the form of a plasma of electrons, protons, alpha particles (helium nuclei), and photons. The photons were strongly absorbed and reemitted by the electrons, and their spectrum was similar to that of the Sun. About 500,000 years after the initial explosion, the expansion caused the temperature to drop enough that the electrons and protons recombined to form hydrogen atoms. Unlike the previous plasma, which was opaque to light, neutral hydrogen is transparent. From that time (called the time of the decoupling, or of recombination) until now, the cosmic photons have been traveling virtually unscattered, carrying information about the nature of the

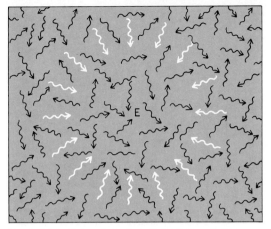

Fig. 1. Origin of cosmic background radiation. Arrows represent photons at a time when the universe became transparent, about 500,000 years after the big bang. Those photons which are reaching the Earth now are indicated by the white arrows. The position of the Earth is indicated by the letter E.

universe at the time of the decoupling (**Fig. 1**). SEE BIG BANG THEORY.

To an observer moving with the plasma, the photons have a blackbody spectrum with a characteristic temperature of a few thousand kelvins. (A blackbody spectrum is the characteristic emission from a perfectly absorbing object heated to the characteristic temperature. The orange glow emitted from a heated pan is approximately blackbody, as is the light emitted from the filament of a light bulb or from the surface of the Sun.) Although the glow from the plasma is in the visible region, as a result of the recessional velocity of the plasma from the Earth the radiation is redshifted from the visible into the microwave region, with a characteristic temperature of 3 K (5°F above absolute zero, $-459.67°F$). Detection of the radiation is really the observation of the shell of matter that last scattered the radiation. SEE HEAT RADIATION; REDSHIFT.

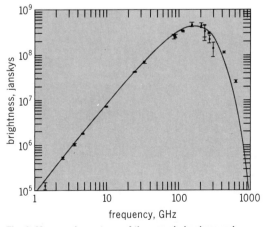

Fig. 2. Measured spectrum of the cosmic background radiation. Where they are not shown, the error bars are smaller than the point. The continuous line is the expected spectrum for a blackbody at a temperature of 2.813 K. Although the disagreement between theory and data at high frequencies is small, it is statistically significant. 1 jansky = 10^{-26} W · m^{-2} · Hz^{-1}. (*Steven M. Levin, University of California, Berkeley*)

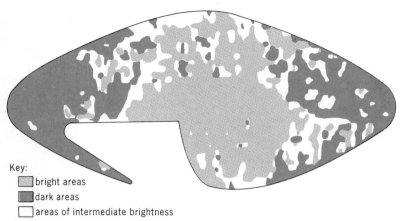

Key:
- bright areas
- dark areas
- areas of intermediate brightness

Fig. 3. Map of the cosmic microwave background measured at 3.3 mm wavelength from a balloon-borne package. The sky is bright in the direction toward which the Earth is moving, and dark in the opposite direction. No measurements were made in the lower left region. The small irregularities were caused by measurement uncertainty. (*Philip Lubin, University of California, Berkeley*)

The microwave radiation is coming from the most distant region of space ever observed, and was emitted earlier in time than any other cosmological signal. The radiation was originally termed cosmic background radiation because the discoverers foresaw that it would cause a background interference with satellite communications, but the term has taken on a vivid new meaning: the radiating shell of matter forms the spatial background in front of which all other astrophysical objects, such as quasars, lie. Until methods are devised to detect the neutrinos or gravity waves that were decoupled earlier, there will be no direct means of viewing beyond this background. *See* GRAVITATIONAL RADIATION DETECTORS; NEUTRINO.

Spectrum measurements. After the discovery of the radiation, the initial work was concerned primarily with the measurement of the radiation's color, that is, its spectrum of intensity at different frequencies.

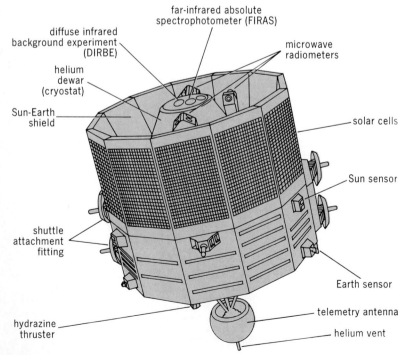

Fig. 4. COBE (Cosmic Background Explorer) satellite. (*NASA*)

far-infrared absolute spectrophotometer (FIRAS)

diffuse infrared background experiment (DIRBE)

helium dewar (cryostat)

Sun-Earth shield

microwave radiometers

solar cells

Sun sensor

shuttle attachment fitting

Earth sensor

hydrazine thruster

telemetry antenna

helium vent

Within a few years the low-frequency (long-wavelength) spectrum was known, but there were alternate theories that could explain the shape of the low-frequency end of the spectrum. Most astrophysicists reserved judgment on the cosmological interpretation until the spectrum was shown to have the characteristics expected of blackbody radiation at 3 K: a predicted peak in intensity at millimeter wavelengths, and the drop-off at shorter wavelengths known as the Wien tail. These were finally measured in a series of experiments carried out in a balloon gondola floating at an altitude above 99.9% of the Earth's atmosphere. The shape was essentially as predicted, and the big bang theory quickly became accepted as the standard model of cosmology.

Even in the big bang model, the spectrum is not expected to be exactly that of a blackbody. The universe is not completely transparent to the radiation, and scattering of photons from free electrons can increase the photons' energy, reducing the intensity of the radiation at low frequencies and increasing it at high frequencies. This process is known as the Sunyaev-Zel'dovich effect. Other distortions of the radiation arise if there is substantial dust formation in the early universe; or if large amounts of energy are released into the universe either before or after the decoupling, for example, from the formation of early stars or by the decay of unknown (but theoretically possible) particles such as massive neutrinos. There have been several measurements indicating small distortions in the radiation (**Fig. 2**), but these experiments are so difficult and the systematic errors so difficult to estimate that they are not yet widely accepted. If and when the measurements are verified, they can provide insight into some of the details of the early big bang.

Isotropy measurements. In describing their discovery, Penzias and Wilson said that the intensity of the radiation in different directions was isotropic (uniform in all directions) to better than 10%. With a few years, measurements showed it was isotropic to better than 1%. Anisotropy was first observed in the 1970s. The smoothness and large angular scale of the variation suggests that its cause is not cosmological but local: the motion of the Earth relative to the radiation. If the Earth is moving relative to the shell of matter that emitted the radiation, radiation coming from the direction in which the Earth is moving will be more intense, and radiation coming from the opposite direction will be weaker because of the Doppler effect. The intensity of the radiation due to the motion should be proportional to the cosine of the angle between the direction of motion and the direction of observation, and this angular dependence is exactly what is observed in the data (**Fig. 3**). From the maximum amplitude of the anisotropy (10^{-3} of the intensity of the 3 K signal) the velocity of the Earth can be deduced to be 10^{-3} of the speed of light, or about 180 mi/s (300 km/s). This velocity probably results from the gravitational acceleration of the Milky Way Galaxy toward the center of a large supercluster of galaxies (Local Supercluster). *See* DOPPLER EFFECT; GALAXY, EXTERNAL; MILKY WAY GALAXY.

Except for the cosine component, no other statistically significant deviation from the isotropy has been detected at the level of 1 part in 30,000. This uniformity has been one of the major puzzles of cosmology, since in the original big bang theory the radiation from different parts of space was decoupled before these regions had time to be in causal communication with each other. Many clever cosmological models were invented to solve this paradox, including the Mixmaster universe theory of J. Wheeler, but under close examination they all failed. Finally a possible solution was found in the variation of the big bang theory called the inflationary universe, elaborated by A. Guth and others. This theory takes into account recent discoveries in elementary particle theory. In the inflationary universe theory, a change in the state of the vacuum during the initial expansion of the universe (similar to the change in state of water as it cools from gas to liquid) explains how different regions of the universe came to thermal equilibrium with each other.

Even in the inflationary universe model, a small anisotropy is expected from early gravitational clumping of mass, the clumping that gave rise to galaxies and clusters of galaxies. Theories of galaxy formation must postulate large amounts of dark, unseen matter in order to provide the strong gravitational fields necessary for sufficient clumping. The nature of this matter is unknown, but it must be more massive than all the matter observed in stars and galaxies. Discovery and measurements of a cosmological anisotropy will make it possible to observe the beginning of galaxy formation and will provide critical information on the nature of the dark matter. A cosmological anisotropy is also expected if the universe is rotating, or if other exotic phenomena, such as cosmic stings or intense long-wavelength gravitational waves, play a role in the formation of the universe. See Cosmic string; Cosmology; Gravitation; Inflationary universe cosmology.

Advanced experiments. More sensitive experiments have been designed to continue to look for deviations from the blackbody spectrum and for cosmological anisotropy. The most sensitive of these, the NASA *COBE* (*Cosmic Background Explorer*) satellite (**Fig. 4**), has been designed to carry three instruments into an orbit 900 km (559 mi) high to study the background radiation over the range from 1 micrometer to 1 cm, improving on the sensitivity of previous experiments by a factor of 100. Its plan of operation calls for it to make a very sensitive measurement of the spectrum and intensity for every direction in the sky in the first 6 months of its flight and to repeat the measurement in the second 6 months.

Richard A. Muller

Bibliography. J. C. Mather, *Cosmic Background Explorer*, NASA, 1988; R. A. Muller, The cosmic background radiation and the new aether drift, *Sci. Amer.*, 238(5):64–74, 1978; R. A. Sunyaev and Ya. B. Zel'dovich, Cosmic background radiation as a probe of the contemporary structure and history of the universe, *Annu. Rev. Astron. Astrophys.*, 18:537–560, 1988; S. Weinberg, *The First Three Minutes*, 1977; R. Weiss, Measurements of the cosmic background radiation, *Annu. Rev. Astron. Astrophys.*, 18:489–536, 1988.

Cosmic electrodynamics

The science concerned with electromagnetic phenomena in ionized media encountered in interstellar space, in stars, and above the atmosphere. Because these ionized materials are excellent electrical conductors, they are strongly linked to magnetic-field lines; they can travel freely along but not across the field lines. Statistically this linkage tends to equalize the energies in the magnetic field and in the turbulent motion of the ionized material. Phenomena treated under cosmic electrodynamics include acceleration of charged particles to cosmic-ray energies, both in the galaxy and on the surface of the Sun; collisions between galaxies; the correlation of magnetic fields with galactic structure; sunspots and prominences; magnetic storms; aurora; and Van Allen radiation belts.

Alfvén waves—transverse waves which travel along the magnetic-field lines in a manner similar to the way waves travel along a stretched string—are often important. See Ionosphere; Magnetohydrodynamics.

Rolf K. M. Landshoff

Bibliography. H. Alfvén, *Cosmic Plasma*, 1981; J. H. Piddington, *Cosmic Electrodynamics*, 2d ed., 1981.

Cosmic rays

Electrons and the nuclei of atoms—largely hydrogen—that impinge upon Earth from all directions of space with nearly the speed of light. These nuclei with relativistic speeds are often referred to as primary cosmic rays, to distinguish them from the cascade of secondary particles generated by their impact against air nuclei at the top of the terrestrial atmosphere. The secondary particles shower down through the atmosphere and are found all the way to the ground and below.

Cosmic rays are studied for a variety of reasons, not the least of which is a general curiosity over the process by which nature can produce such energetic nuclei. Apart from this, the primary cosmic rays provide the only direct sample of matter from outside the solar system. Measurement of their composition can aid in understanding which aspects of the matter making up the solar system are typical of the Milky Way Galaxy as a whole and which may be so atypical as to yield specific clues to the origin of the solar system. Cosmic rays are electrically charged; hence they are deflected by the magnetic fields which are thought to exist throughout the Milky Way Galaxy, and may be used as probes to determine the nature of these fields far from Earth. Outside the solar system the energy contained in the cosmic rays is comparable to that of the magnetic field, so the cosmic rays probably play a major role in determining the structure of the field. Collisions between the cosmic rays and the nuclei of the atoms in the tenuous gas which permeates the Milky Way Galaxy change the cosmic-ray composition in a measurable way and produce gamma rays which can be detected at Earth, giving information on the distribution of this gas.

This modern understanding of cosmic rays has arisen by a process of discovery which at many times produced seemingly contradictory results, the ultimate resolution of which led to fundamental discoveries in other fields of physics, most notably high-energy particle physics. At the turn of the century several different types of radiation were being studied, and the different properties of each were being determined with precision. One result of many precise experiments was that an unknown source of radiation existed with properties that were difficult to characterize. In 1912 Viktor Hess made a definitive series of balloon flights which showed that this background radiation increased with altitude in a dramatic fashion. Far more penetrating then any other known at that time, this radiation had many other unusual properties and became known as cosmic radiation, because it clearly did not originate in the Earth or from any known properties of the atmosphere.

Unlike the properties of alpha-, beta-, gamma-, and x-radiation, the properties of cosmic radiation are not of any one type of particle, but are due to the interactions of a whole series of unstable particles, none of which was known at that time. The initial identification of the positron, the muon, the π meson or pion, and certain of the K mesons and hyperons were made from studies of cosmic rays.

Thus the term cosmic ray does not refer to a particular type of energetic particle, but to any energetic particle being considered in its astrophysical context. The effects of cosmic rays on living cells are discussed in a number of other articles: for example, SEE RADIATION INJURY (BIOLOGY). SEE ALSO ELEMENTARY PARTICLE.

Cosmic-ray detection. Cosmic rays are usually detected by instruments which classify each incident particle as to type, energy, and in some cases time and direction of arrival. A convenient unit for measuring cosmic-ray energy is the electronvolt which is the energy gained by a unit charge (such as an electron) accelerating freely across a potential of 1 V. One electronvolt equals about 1.6×10^{-19} joule. For nuclei it is usual to express the energy in terms of electronvolts per nucleon, since as a function of this variable the relative abundances of the different elements are nearly constant. Two nuclei with the same energy per nucleon have the same velocity.

Flux. The intensity of cosmic radiation is generally expressed as a flux by dividing the average number seen per second by the effective size or "geometry factor" of the measuring instrument. Calculation of the geometry factor requires knowledge of both the sensitive area (in square centimeters) and the angular acceptance (in steradians) of the detector, as the arrival directions of the cosmic rays are randomly distributed to within 1% in most cases. A flat detector of any shape but with area of 1 cm^2 has a geometry factor of π cm$^2\cdot$sr if it is sensitive to cosmic rays entering from one side only. The total flux of cosmic rays in the vicinity of the Earth but outside the atmosphere is about 0.3 nucleus/(cm$^2\cdot$s$\cdot$sr) [2 nuclei/(in.$^2\cdot$s$\cdot$sr)]. Thus a quarter dollar, with a surface area of 4.5 cm^2 (0.7 in.^{2}), lying flat on the surface of the Moon will be struck by $0.3 \times 4.5 \times 3.14 = 4.2$ cosmic rays per second.

Energy spectrum. The flux of cosmic rays varies as a function of energy. This type of function is called an energy spectrum, and may refer to all cosmic rays or to only a selected element or group of elements. Since cosmic rays are continuously distributed in energy, it is meaningless to attempt to specify the flux at any one exact energy. Normally an integral spectrum is used, in which the function gives the total flux of particles with energy greater than the specified energy [in particles/(cm$^2\cdot$s$\cdot$sr)], or a differential spectrum, in which the function provides the flux of particles in some energy interval (typically 1 MeV/nucleon wide) centered on the specified energy, in particles/[cm$^2\cdot$s$\cdot$sr$\cdot$(MeV/nucleon)]. The basic problem of cosmic-ray research is to measure the spectra of the different components of cosmic radiation and

to deduce from them and other observations the nature of the cosmic-ray sources and the details of where the particles travel on their way to Earth and what they encounter on their journey.

Types of detectors. All cosmic-ray detectors are sensitive only to moving electrical charges. Neutral cosmic rays (neutrons, gamma rays, and neutrinos) are studied by observing the charged particles produced in the collision of the neutral primary with some type of target. At low energies the ionization of the matter through which they pass is the principal means of detection. Such detectors include cloud chambers, ion chambers, spark chambers, Geiger counters, proportional counters, scintillation counters, solid-state detectors, photographic emulsions, and chemical etching of certain mineral crystals or plastics in which ionization damage is revealed. The amount of ionization produced by a particle is given by the square of its charge multiplied by a universal function of its velocity. A single measurement of the ionization produced by a particle is therefore usually not sufficient both to identify the particle and to determine its energy. However, since the ionization itself represents a significant energy loss to a low-energy particle, it is possible to design systems of detectors which trace the rate at which the particle slows down and thus to obtain unique identification and energy measurement. SEE CLOUD CHAMBER; GAMMA-RAY DETECTOR; GEIGER-MÜLLER COUNTER; IONIZATION CHAMBER; JUNCTION DETECTOR; PARTICLE TRACK ETCHING; PHOTOGRAPHIC MATERIALS; SCINTILLATION COUNTER; SPARK CHAMBER.

At energies above about 500 MeV/nucleon, almost all cosmic rays will suffer a catastrophic nuclear interaction before they slow appreciably. Some measurements are made using massive calorimeters which are designed to trap all of the energy from the cascade of particles which results from such an interaction. More commonly an ionization measurement is combined with measurements of physical effects which vary in a different way with mass, charge, and energy. Cerenkov detectors and the deflection of the particles in the field of large superconducting magnets or the magnetic field of the Earth itself provide the best means of studying energies up to a few hundred gigaelectronvolts per nucleon. Detectors employing the phenomenon of x-ray transition radiation promise to be useful for measuring composition at energies up to a few thousand GeV per nucleon. Transition radiation detectors have already been used to study electrons having energies of 10–200 GeV which, because of their lower rest mass, are already much more relativistic than protons of the same energies. SEE CERENKOV RADIATION; SUPERCONDUCTING DEVICES; TRANSITION RADIATION DETECTORS.

Above about 10^{14} eV, direct detection of individual particles is no longer possible since they are so rare. Such particles are studied by observing the large showers of secondaries they produce in Earth's atmosphere. These showers are detected either by counting the particles which survive to strike ground-level detectors or by looking at the flashes of light the showers produce in the atmosphere with special telescopes and photomultiplier tubes. It is not possible to directly determine what kind of particle produces any given shower. However, because of the extreme energies involved, which can be measured with fair accuracy and have been seen as high as 10^{20} eV (16 J), most of the collision products travel in the same direction as the primary and at essentially the speed of light. This center of intense activity has typical dimensions of only a few tens of meters, allowing it to be tracked (with sensitive instruments) like a miniature meteor across the sky before it hits the Earth at a well-defined location. In addition to allowing determination of the direction from which each particle came, the development of many such showers through the atmosphere may be studied statistically to gain an idea of whether the primaries are protons or heavier nuclei. Basically the idea behind these studies is that a heavy nucleus, in which the energy is initially shared among several neutrons and protons, will cause a shower that starts higher in the atmosphere and develops more regularly than a shower which has the same total energy but is caused by a single proton. SEE PARTICLE DETECTOR; PHOTOMULTIPLIER.

Atmospheric cosmic rays. The primary cosmic-ray particles coming into the top of the terrestrial atmosphere make inelastic collisions with nuclei in the atmosphere. The collision cross section is essentially the geometrical cross section of the nucleus, of the order of 10^{-26} cm^2 (10^{-27} in.^{2}). The mean free path for primary penetration into the atmosphere is given in **Table 1**. (Division by the atmospheric density in g/cm^3 gives the value of the mean free path in centimeters.)

When a high-energy nucleus collides with the nucleus of an air atom, a number of things usually occur. Rapid deceleration of the incoming nucleus leads to production of pions with positive, negative, or neutral charge; this meson production is closely analogous to the generation of x-rays, or bremsstrahlung, produced when a fast electron is deflected by impact with the atoms in a metal target. The mesons, like the bremsstrahlung, come off from the impact in a narrow cone in the forward direction. Anywhere from 0 to 30 or more pions may be produced, depending upon the energy of the incident nucleus. The ratio of neutral to charged pions is about 0.75. SEE BREMSSTRAHLUNG; MESON; NUCLEAR REACTION.

A few protons and neutrons (in about equal proportions) may be knocked out with energies up to a few GeV. They are called knock-on protons and neutrons.

A nucleus struck by a proton or neutron of the nucleonic component with an energy greater than approximately 300 MeV may have its internal forces momentarily disrupted so that some of its nucleons are free to leave with their original nuclear kinetic energies of about 10 MeV. The nucleons freed in this fashion appear as protons, deuterons, tritons, alpha particles, and even somewhat heavier clumps, radiating outward from the struck nucleus. In photographic emulsions the result is a number of short prongs radiating from the point of collision, and for this reason is called a nuclear star.

All these protons, neutrons, and pions generated by collision of the primary cosmic-ray nuclei with the nuclei of air atoms are the first stage in the development of the secondary cosmic-ray particles observed inside the atmosphere. Since several secondary particles are produced by each collision, the total number of energetic particles of cosmic-ray origin will at first increase with depth, even while the primary density is decreasing. Since electric charge must be conserved and the primaries are positively charged, the positive particles outnumber the negative particles in the secondary radiation by a factor of about 1.2. This factor is called the positive excess.

Table 1. Mean free paths for primary cosmic rays in the atmosphere

Charge of primary nucleus	Mean free path in air, g/cm^2*
$Z = 1$	60
$Z = 2$	44
$3 \leq Z \leq 5$	32
$6 \leq Z \leq 9$	27
$10 \leq Z \leq 29$	21

*1 g/cm^2 = 1.42 × 10^{-2} lb/in.^{2}

Electromagnetic cascade. The uncharged π^0 mesons decay into two gamma rays with a lifetime of about 9×10^{-17} s. The decay is so rapid that π^0 mesons are not directly observed among the secondary particles in the atmosphere. The two gamma rays, which together have the rest energy of the π^0, about 140 MeV, plus the π^0 kinetic energy, each produce a positron-electron pair. Upon passing sufficiently close to the nucleus of an air atom deeper in the atmosphere, the electrons and positrons convert their energy into bremsstrahlung. The bremsstrahlung in turn creates new positron-electron pairs, and so on. This cascade process continues until the energy of the initial π^0 has been dispersed into a shower of positrons, electrons, and photons with insufficient individual energies (≤ 1 MeV) to continue the pair production. The shower, then being unable to reproduce its numbers, is dissipated by ionization of the air atoms. The electrons and photons of such showers are referred to as the soft component of the atmospheric (secondary) cosmic rays, reaching a maximum intensity at an atmospheric depth of 150–200 g/cm^2 and then declining by a factor of about 10^2 down to sea level. SEE ELECTRON-POSITRON PAIR PRODUCTION.

Muons. The $\pi^\pm$ mesons produced by the primary collisions have a lifetime about 2.6×10^{-8} s before they decay into muons: $\pi^\pm \rightarrow \mu^\pm$ + neutrino. With a lifetime of this order a $\pi^\pm$ possessing enough energy (greater than 10 GeV) to experience significant relativistic time dilatation may exist long enough to interact with the nuclei of the air atoms. The cross section for $\pi^\pm$ nuclear interactions is approximately the geometrical cross section of the nucleus, and the result of such an interaction is essentially the same as for the primary cosmic-ray protons. Most low-energy $\pi^\pm$ decay into muons before they have time to undergo nuclear interactions.

Except at very high energy (above 500 GeV), muons interact relatively weakly with nuclei, and are too massive (207 electron masses) to produce bremsstrahlung. They lose energy mainly by the comparatively feeble process of ionizing an occasional air atom as they progress downward through the atmosphere. Because of this ability to penetrate matter, they are called the hard component. At rest their lifetime is 2×10^{-6} s before they decay into an electron or positron and two neutrinos, but with the relativistic time dilatation of their high energy, 5% of the muons reach the ground. Their interaction with matter is so weak that they penetrate deep into the ground, where they are the only charged particles of cosmic-ray origin to be found. At a depth equivalent of 300 m (990 ft) of water the muon intensity has decreased from that at ground level only by a factor of 20; at 1400 m (4620 ft) it has decreased by a factor of 10^3.

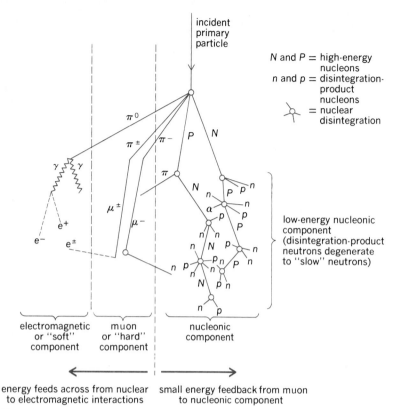

incident
primary
particle

N and P = high-energy
             nucleons
n and p = disintegration-
             product
             nucleons
       ⚛ = nuclear
           disintegration

low-energy nucleonic
component
(disintegration-product
neutrons degenerate
to "slow" neutrons)

electromagnetic
or "soft"
component

muon
or "hard"
component

nucleonic
component

←————————→
energy feeds across from nuclear
to electromagnetic interactions

————————→
small energy feedback from muon
to nucleonic component

Fig. 1. Cascade of secondary cosmic-ray particles in the terrestrial atmosphere.

cleonic component increases in intensity down to a depth of about 120 g/cm^2, and thereafter declines in intensity, with a mean absorption length of about 200 g/cm^2 (2.8 lb/in.^{2}).

The various cascades of secondary particles in the atmosphere are shown schematically in **Fig. 1**. Note that about 48% of the initial primary cosmic-ray energy goes into charged pions, 25% into neutral pions, 7% into the nucleonic component, and 20% into stars. The nucleonic component is produced principally by the lower-energy (about 5 GeV) primaries. Higher-energy primaries put their energy more into meson production. Hence in the lower atmosphere, a Geiger counter responds mainly to the higher-energy primaries (about 5 GeV) because it counts the muons and electrons, whereas a BF$_3$ counter detecting thermal neutrons responds more to the low-energy primaries.

Neutrinos. Cosmic neutrinos, detected for the first time from the explosion of the supernova 1987A, provide confirmation of theoretical calculations regarding the collapse of the cores of massive stars. Although neutrinos are produced in huge numbers (over 10^{15} passed through a typical human body from this supernova), they interact with matter only very weakly, necessitating a very large detector. Detectors consisting of huge tanks containing hundreds of tons of pure water located deep underground to reduce the background produced by other cosmic rays recorded less than two dozen neutrino events. Still larger detectors will permit observation of more distant supernovae and allow sensitive searches for point sources of high-energy neutrinos. Also, by measuring the fraction of non-neutrino-induced events containing multiple muons, these new detectors should determine the composition of cosmic rays at energies above 10^{15} eV. Some preliminary measurements indicate that these high-energy cosmic rays may consist primarily of iron nuclei rather than the protons that dominate at lower energies. Measurement of the flux of solar neutrinos, which is really quite a different problem, has begun to cause fundamental changes in thought about the physics of the Sun. *See* Neutrino; Solar neutrinos; Supernova.

Relation to particle physics. Investigations of cosmic rays continue to make fundamental contributions to particle physics. Neutrino detectors have set the best limit yet (about 10^{32} years) on the lifetime of the proton. These detectors are also able to study the physics of the neutrino and specifically to search for oscillations, or spontaneous conversions of one type of neutrino into another. Cosmic rays remain the only source of particles with energies above 1000 GeV. With the continued increase in the size and sensitivity of detectors, study of cosmic rays should continue to provide the first indications of new physics at ultrahigh energies. *See* Fundamental interactions; Proton.

Nucleonic component. The high-energy nucleons—the knock-on protons and neutrons—produced by the primary-particle collisions and a few pion collisions proceed on down into the atmosphere. They produce nuclear interactions of the same kind as the primary nuclei, though of course with diminished energies. This cascade process constitutes the nucleonic component of the secondary cosmic rays.

When the nucleon energy falls below about 100 MeV, stars and further knock-ons can no longer be produced. At the same time the protons are rapidly disappearing from the cascade because their ionization losses in the air slow them down before they can make a nuclear interaction. The neutrons are already dominant at 3500 m (11,550 ft), about 300 g/cm^2 (4.3 lb/in.^{2}) above sea level, where they outnumber the protons four to one. Thus the final stages in the lower atmosphere are given over almost entirely to neutrons in a sequence of low-energy interactions which convert them to thermal neutrons (neutrons of kinetic energy of about 0.025 eV) in a path of about 90 g/cm^2 (1.3 lb/in.^{2}). These thermal neutrons are readily detected in boron trifluoride (BF$_3$) counters. The nu-

Table 2. Properties of particles when all have a rigidity of 1 GV					
			Kinetic energy		Momentum,
Particle	Charge	Nucleons	MeV	MeV/nucleon	MeV/c
Electron	1	—	1000	—	1000
Proton	1	1	430	430	1000
^{3}He	2	3	640	213	2000
^{4}He	2	4	500	125	2000
^{16}O	8	16	2000	125	8000

Geomagnetic effects. The magnetic field of Earth is described approximately as that of a magnetic dipole of strength 8.1×10^{15} weber-meters (8.1×10^{25} gauss · cm^3) located near the geometric center of Earth. Near the Equator the field intensity is 3×10^{-5} tesla (0.3 gauss), falling off in space as the inverse cube of the distance to the Earth's center. In a magnetic field which does not vary in time, the path of a particle is determined entirely by its rigidity, or momentum per unit charge; the velocity simply determines how fast the particle will move along this path. Momentum is usually expressed in units of eV/c, where c is the velocity of light, because at high energies, energy and momentum are then numerically almost equal. By definition, momentum and rigidity are numerically equal for singly charged particles. The unit so defined is normally called the volt, but should not be confused with the standard and nonequivalent unit of the same name. **Table 2** gives examples of these units as applied to different particles with rigidity of 1 GV. This corresponds to an orbital radius in a typical interplanetary (10^{-9} tesla or 10^{-5} gauss) magnetic field of approximately 10 times the distance from the Earth to the Moon. *See Relativistic electrodynamics.*

The minimum rigidity of a particle able to reach the top of the atmosphere at a particular geomagnetic latitude is called the geomagnetic cutoff rigidity at that latitude, and its calculation is a complex numerical problem. Fortunately, for an observer near the ground, obliquely arriving secondary particles, produced by the oblique primaries, are so heavily attenuated by their longer path to the ground that it is usually sufficient to consider only the geomagnetic cutoff for vertically incident primaries, which is given in **Table 3**. Around the Equator, where a particle must come in perpendicular to the geomagnetic lines of force to reach Earth, particles with rigidity less than 10 GV are entirely excluded, though at higher latitudes where entry can be made more nearly along the lines of force, lower energies can reach Earth. Thus, the cosmic-ray intensity is a minimum at the Equator, and increases to its full value at either pole—this is the cosmic-ray latitude effect. Even deep in the atmosphere the variation with latitude is easily detected with BF$_3$ counters, as shown in **Fig. 2**. North of 45° the effect is slight because the additional primaries admitted are so low in energy that they produce few secondaries.

Accurate calculations of the geomagnetic cutoff must consider the deviations of the true field from that of a perfect dipole and the change with time of these deviations. Additionally the distortion of the field by the pressure of the solar wind must often be accounted for, particularly at high latitude. Such corrections vary rapidly with time because of sudden bursts of solar activity and because of the rotation of the Earth. Areas with cutoffs of 400 MV during the day may have no cutoff at all during the night. This day-night effect is confined to particles with energies so low that neither they nor their secondaries reach the ground, and is thus observed only on high-altitude balloons or satellites.

Since the geomagnetic field is directed from south to north above the surface of Earth, the incoming cosmic-ray nuclei are deflected toward the east. Hence an observer finds some 20% more particles incident from the west. This is known as the east-west effect. *See Geomagnetism.*

Table 3. Geomagnetic cutoff

Geomagnetic lat.	Vertical cutoff, GV
0°	15
±20°	11.5
+40°	5
±60°	1
±70°	0.2
±90°	0

Solar modulation. Figure 3 presents portions of the proton and alpha-particle spectra observed near the Earth but outside of the magnetosphere in 1973. Below 20 GeV/nucleon the cosmic-ray intensity varies markedly with time. S. Forbush was the first to show that the cosmic-ray intensity was low during the years of high solar activity and sunspot number, which follow an 11-year cycle. This effect is clearly seen in the data of Fig. 2 and has been extensively studied with ground-based and spacecraft instruments. While this so-called solar modulation is now understood in general terms, it has not been calculated in detail, in large part because of the lack of any direct measurements of conditions in the solar system out of the ecliptic plane, to which all present spacecraft are confined because of limitations on the power of rockets.

There is indirect evidence that the solar system is not at all spherically symmetric and that conditions near the ecliptic plane are quite special.

The primary cause of solar modulation is the solar wind, a highly ionized gas (plasma) which boils off the solar corona and propagates radially from the Sun at a velocity of about 400 km/s (250 mi/s). The wind is mostly hydrogen, with typical density of 5 protons/cm^3 (80 protons/in.^{3}). This density is too low for collisions with cosmic rays to be important. Rather, the high conductivity of the medium traps part of the solar magnetic field and carries it outward. The rotation of the Sun and the radial motion of the plasma combine to create the observed archimedean spiral pattern of the average interplanetary magnetic field. Turbulence in the solar wind creates fluctuations in the field which often locally obscure the average direction and intensity. This complex system of magnetic irregularities propagating outward from the Sun deflects and sweeps the low-rigidity cosmic rays out of the solar system.

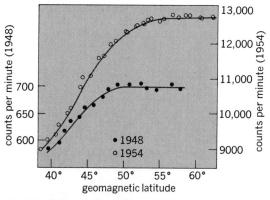

Fig. 2. Latitude variation of the neutron component of cosmic rays in 80°W longitude and at a height corresponding to an atmospheric pressure of 30 kPa (22.5 cm of mercury) in 1948, when the Sun was active, and 1954, when the Sun was deep in a sunspot minimum.

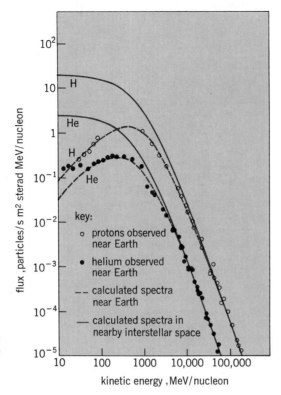

Fig. 3. Spectra of cosmic-ray protons and helium at Earth and in nearby interstellar space, showing the effect of solar modulation. Observations were made in 1973, when the Sun was quiet.

In addition to the bulk sweeping action, another effect of great importance occurs in the solar wind, adiabatic deceleration. Because the wind is blowing out, only those particles which chance to move upstream fast enough are able to reach Earth. However, because of the expansion of the wind, particles interacting with it lose energy. Thus, particles observed at Earth with 10 MeV/nucleon energy actually started out with several hundred megaelectronvolts per nucleon in nearby interstellar space, and those with only 100–200 MeV/nucleon initial energy probably never reach Earth at all. This is particularly unfortunate because at these lower energies the variation with energy of nuclear reaction probabilities would allow much more detailed investigation of cosmic-ray history. Changes in the modulation with solar activity are caused by the changes in the pattern of magnetic irregularities rather than by changes in the wind velocity, which are quite small. *See Magnetohydrodynamics; Plasma physics.*

There are several phenomenological classes of cosmic-ray variation besides the 11-year variation which are associated with the short-term variations of the solar wind and generally affect only low-energy particles.

A sudden outburst at the Sun, at the time of a large flare, yields a cosmic-ray decrease in space which extends to very high energies, 50 GeV or so. The magnetic fields carried in the blast wave from the flare sweep back the cosmic rays, causing a decrease in their intensity. This is termed the Forbush-type decrease, where the primary intensity around the world may drop in an irregular way as much as 20% in 15 h or 8% in 3 h, slowly recovering in the days or weeks that follow. Often, but not always, the Forbush decrease and geomagnetic storms accompany each other. Striking geographical variation is to be seen in the sharper fluctuations during the onset of a Forbush decrease.

The region in space where solar modulation is important is probably a sphere with a radius of 50–80 AU, although this is not at all certain. (1 AU, or astronomical unit, is the mean Earth-Sun separation, 1.49×10^8 km or 9.26×10^7 mi.) The *Pioneer 11* spacecraft did not find abrupt changes in intensity indicative of crossing any type of boundary out to a distance of 40 AU, although the fluxes detected clearly increased as it traveled outward. A distinct change in interplanetary phenomena seems to take place in the 10–15 AU range to a quieter, more azimuthally uniform, and less temporally varying situation. *See Solar magnetic field; Solar wind; Sun.*

Composition of cosmic rays. Nuclei ranging from protons to lead have been identified in the cosmic radiation. The relative abundances of the elements ranging up to nickel are shown in **Fig. 4**, together with the best estimate of the "universal abundances" obtained by combining measurements of solar spectra, lunar and terrestrial rocks, meteorites, and so forth. Most obvious is the similarity between these two distributions. However, a systematic deviation is quickly apparent: the elements lithium-boron and scandium-manganese as well as most of the odd-charged nuclei are vastly overabundant in the cosmic radiation. This effect has a simple explanation: the cosmic rays travel great distances in the Milky Way Galaxy and occasionally collide with atoms of interstellar gas—mostly hydrogen and helium—and fragment. This fragmentation, or spallation as it is called, produces lighter nuclei from heavier ones but does not change the energy/nucleon very much. Thus the energy spectra of the secondary elements are similar to those of the primaries.

Calculations involving reaction probabilities determined by nuclear physicists show that the overabundances of the secondary elements can be explained by assuming that cosmic rays pass through an average of about 5 g/cm² (0.07 lb/in.²) of material on their way to Earth. Although an average path length can be obtained, it is not possible to fit the data by saying that all particles of a given energy have exactly the same path length; furthermore, results indicate that higher-energy particles traverse less matter in reaching the solar system, although their original composition seems energy independent. *See Elements, cosmic abundance of.*

When spallation has been corrected for, differences between cosmic-ray abundances and solar-system or universal abundances still remain. The most important question is whether these differences are due to the cosmic rays having come from a special kind of material (such as would be produced in a supernova explosion), or simply to the fact that some atoms might be more easily accelerated than others. It is possible to rank almost all of the overabundances by considering the first ionization potential of the atom and the rigidity of the resulting ion, although this calculation gives no way of predicting the magnitude of the enhancement expected. It is also observed that the relative abundances of particles accelerated in solar flares are far from constant from one flare to the next. *See Ionization potential.*

Isotopes. The possibility of such preferential acceleration is one of the reasons why much cosmic-ray study is concentrated on determining the isotopic composition of each element, as this is much less likely to be changed by acceleration. It is apparent that the low-energy helium data in Fig. 3 do not fit the calculated values. Since it is known that this low-energy helium is nearly all ⁴He, whereas the higher-energy helium contains 10% ³He, one can be fairly certain that the deviation is due to a local source of energetic ⁴He within the solar system rather than a lack of understanding of the process of solar modulation. Similarly, a low-energy enhancement of nitrogen is pure ¹⁴N, whereas the higher-energy nitrogen is almost 50% ¹⁵N.

High-accuracy measurements of isotopic abundance ratios became available only in the late 1970s, but rapidly one puzzle emerged. The ratio of ²²Ne to ²⁰Ne in the cosmic-ray sources is estimated to be 0.37, while the accepted solar system value for this number is 0.12, which agrees well with the abundances measured in solar-flare particles. However, another direct sample of solar material—the solar wind—has a ratio of 0.08, indicating clearly that the isotopic composition of energetic particles need not reflect that of their source. Conclusions drawn from the observed difference in the solar and cosmic-ray values must be viewed as somewhat tentative until the cause of the variation in the solar material is well understood. *See Isotope.*

Electron abundance. Cosmic-ray electron measurements pose other problems of interpretation, partly because electrons are nearly 2000 times lighter than protons, the next lightest cosmic-ray component. Protons with kinetic energy above 1 GeV are about 100 times as numerous as electrons above the same energy, with the relative number of electrons decreasing slowly at higher energies. But it takes about 2000 GeV to give a proton the same velocity as a 1-GeV electron. Viewed in this way electrons are several thousand times more abundant than protons. (Electrical neutrality of the Milky Way Galaxy is maintained by lower-energy ions which are more numerous than cosmic rays although they do not carry much energy.) It is thus quite possible that cosmic electrons have a different source entirely from the nuclei. It is generally accepted that there must be direct acceleration of electrons, because calculations show that more positrons than negatrons should be produced in collisions of cosmic-ray nuclei with interstellar gas. Measurements show, however, that only 10% of the electrons are positrons. As the number of positrons seen agrees with the calculated secondary production, added confidence is gained in the result that there is indeed an excess of negatrons. *See Electron.*

Electrons are light enough to emit a significant amount of synchrotron radiation as they are deflected by the 10^{-10}-tesla (10^{-6}-gauss) galactic magnetic field. Measurement of this radiation by radio telescopes provides sufficient data for an approximate calculation of the average energy spectrum of electrons in interstellar space and other galaxies. Comparison of spectra of electrons and positrons measured at Earth with those calculated to exist in interstellar space provides the most direct measurement of the absolute amount of solar modulation. *See Radio astronomy; Synchrotron radiation.*

Properties of energy spectrum. At energies above 10^{10} eV, the energy spectra of almost all cosmic rays are approximated over many decades by functions in which the flux decreases as the energy raised to some negative, nonintegral power referred to as the spectral index. Such a power-law relationship is of course a straight line when plotted using logarithmic axes. A steep (that is, more rapidly falling with increasing en-

ergy) spectrum thus has a higher spectral index than a flat spectrum. The straight-line regions of the spectra in Fig. 3 correspond to a variation of flux with a spectral index of -2.7. A spectral index of -2.7 provides a good fit with the data up to 10^{15} eV total energy. Between 10^{15} and 10^{19} eV a steeper spectrum, with an index around -3.0, seems to be well established. Above 10^{19} eV the spectrum surprisingly flattens once more, returning to an index of about -2.7.

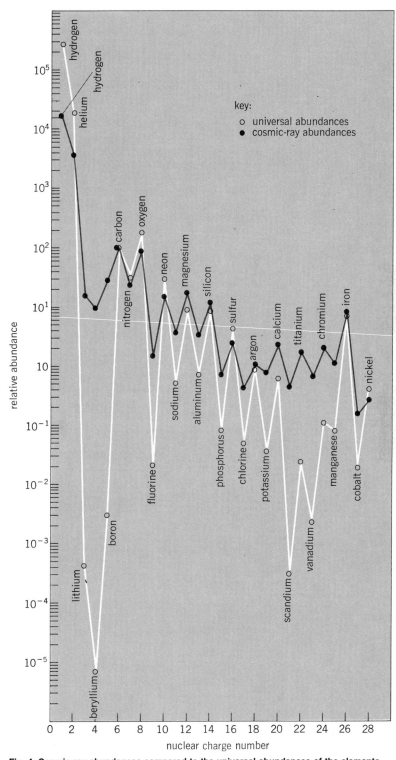

Fig. 4. Cosmic-ray abundances compared to the universal abundances of the elements. Carbon is set arbitrarily to an abundance of 100 in both cases.

The spectral index above 10^{20} eV has not been determined, because particles are so rare that they are almost never seen, even in detectors which cover several square kilometers and operate for many years. At such high energies, the individual particles are not identified, and changes in the measured-energy spectrum could be the result of composition changes. However, the evidence available indicates that the composition is essentially unchanged.

Age. Another important result which can be derived from detailed knowledge of cosmic-ray isotopic composition is the "age" of cosmic radiation. Certain isotopes are radioactive, such as beryllium-10 (^{10}Be) with a half-life of 1.6×10^6 years. Since beryllium is produced entirely by spallation, study of the relative abundance of ^{10}Be to the other beryllium isotopes, particularly as a function of energy to utilize the relativistic increase in this lifetime, will yield a number related to the average time since the last nuclear collision. Measurements show that ^{10}Be is nearly absent at low energies and yield an estimate of the age of the cosmic rays of approximately 10^7 years. An implication of this result is that the cosmic rays propagate in a region in space which has an average density of 0.1–0.2 atom/cm^3 (1.5–3 atoms/in.^{3}). This is consistent with some astronomical observations of the immediate solar neighborhood.

Very high-energy particles cannot travel long distances in the 2.7 K blackbody-radiation field which permeates the universe. Electrons of 15 GeV energy lose a good portion of their energy in 10^8 years by colliding with photons via the (inverse) Compton process, yet electrons are observed to energies of 100 GeV and over. A similar loss mechanism becomes effective at approximately 10^{20} eV for protons. These observations are of course not conclusive, but a safe statement is that a cosmic-ray age of 10^7 years is consistent with all currently available data. SEE COMPTON EFFECT; COSMIC BACKGROUND RADIATION.

Several attempts have been made to measure the constancy of the cosmic-ray flux in time. Variations in ^{14}C production, deduced from apparent deviations of the archeological carbon-dating scale from that derived from studies of tree rings, cover a period of about 10^3 years. Radioactive ^{10}Be in deep-sea sediments allows studies over 10^6 years, whereas etching of tracks left by cosmic rays in lunar minerals covers a period of 10^9 years. None of these methods has ever indicated a variation of more than a factor of 2 in average intensity. There are big differences in these time scales, and the apparent constancy of the flux could be due to averaging over variations which fall in the gaps as far as the time scales are concerned. Nevertheless, the simplest picture seems to be that the cosmic rays are constant in time at an intensity level which is due to a long-term balance between continuous production and escape from the Galaxy, with an average residence time of 10^7 years. SEE COSMOGENIC NUCLIDES; RADIOCARBON DATING.

Origin. Although study of cosmic rays has yielded valuable insight into the structure, operation, and history of the universe, their origin has not been determined. The problem is not so much to devise processes which might produce cosmic rays, but to decide which of many possible processes do in fact produce them.

In general, analysis of the problem of cosmic-ray origin is broken into two major parts: origin in the sense of where the sources are located, whatever they are, and origin in the sense of how the particles are accelerated to such high energies. Of course, these questions can never be separated completely.

Location of sources. It is thought that cosmic rays are produced by mechanisms operating within galaxies and are confined almost entirely to the galaxy of their production, trapped by the galactic magnetic field. The intensity in intergalactic space would only be a few percent of the typical galactic intensity, and would be the result of a slow leakage of the galactic particles out of the magnetic trap. It has not been possible to say much about where the cosmic rays come from by observing their arrival directions at Earth. At lower energies (up to 10^{15} eV) the anisotropies which have been observed can all be traced to the effects of the solar wind and interplanetary magnetic field. The magnetic field of the Milky Way Galaxy seems to be completely effective in scrambling the arrival directions of these particles.

Between 10^{15} and 10^{19} eV a smoothly rising anisotropy is measured, ranging from 0.1 to 10%, but the direction of the maximum intensity varies in a nonsystematic way with energy. At these energies, particles have a radius of curvature which is not negligible compared to galactic structures, and thus their arrival direction could be related to where they came from but in a complex way.

Above 10^{19} eV the radius of curvature in the galactic magnetic field becomes comparable to or larger than galactic dimensions, making containment of such particles in the Milky Way Galaxy impossible. Only a few hundred events greater than 10^{19} eV have been detected, but the directions from which they have come are plainly nonrandom. A clear minimum appears in the direction of the disk of the Milky Way Galaxy and a clear maximum near the north galactic pole, which is also the direction of the majority of the galaxies in the so-called local supercluster. The average distance of these galaxies is such that these particles in fact would be able to propagate to the Earth from them without losing their energy to photon collisions. This clearly defined anisotropy begins at the same energy at which the energy spectrum changes—further indication that these particles may have a different source from the low-energy particles. An alternative explanation, that particles are absorbed by material in the galactic plane, cannot at present be ruled out because no large detectors presently view the sky in the region of the south galactic pole. Thus it is not known whether the presence of the cluster of galaxies in the direction of the maximum flux is only coincidental.

Much effort has been devoted to construction of air-shower detectors with good directional resolution in order to search for point sources of gamma rays with energies greater than 10^{12} eV. A number of claims have been made regarding the detection of such high-energy gamma rays from the directions of well-known x-ray binary sources. It is generally believed that gamma rays of energy greater than 10^{12} eV can be produced only through interactions of high-energy protons or other hadrons. Unambiguous detection and study of these gamma rays will provide important information on the structure and operation of these exotic objects. SEE ASTROPHYSICS, HIGH-ENERGY; BINARY STAR; GAMMA-RAY ASTRONOMY; X-RAY STAR.

Direct detection of cosmic rays propagating in distant regions of the Milky Way Galaxy is possible by observing the electromagnetic radiation produced as they interact with other constituents of the Milky Way Galaxy. Measurement of the average electron spectrum using radio telescopes has already been mentioned. Proton intensities are mapped by studying the arrival directions of gamma rays (at about 50 MeV) produced as they collide with interstellar gas. Unfortunately, the amount of radiation in these processes depends upon both the cosmic-ray flux and the magnetic-field intensity or density of interstellar gas. Areas where cosmic rays are known to exist can be pointed out because the radiation is observed. But where no radiation is seen, it is not known whether its absence results from lack of cosmic rays or lack of anything for them to interact with. In particular, very little radiation is seen from outside the Milky Way Galaxy, but there is also very little gas or magnetic field there. There is therefore no direct evidence either for or against galactic containment.

A major difficulty with the concept of cosmic radiation filling the universe is the large amount of energy needed to maintain the observed intensity in the face of an expanding universe—probably more energy than is observed to be emitted in all other forms put together. *See Cosmology.*

Confinement mechanisms. Three possible models of cosmic-ray confinement are under investigation. All assume that cosmic rays are produced in sources, discrete or extended, scattered randomly through the galactic disk. Most popular is the "leaky box" model, which proposes that the particles diffuse about in the magnetic field for a few million years until they chance to get close to the edge of the Milky Way Galaxy and escape. This is a phenomenological model in that no mechanism is given by which either the confinement time or the escape probability as a function of energy can be calculated from independent observations of the galactic structure. Its virtue is that good fits to the observed abundances of spallation products are obtained by using only a few adjustable parameters. Variations of the model which mainly postulate boxes within boxes—ranging from little boxes surrounding sources to a giant box or static halo surrounding the whole Milky Way Galaxy—can be used to explain variations from the simple predictions. However, all attempts to calculate the details of the process have failed by many orders of magnitude, predicting ages which are either far older or far younger than the observed age.

A second model is that of the dynamical halo. Like the earlier static-halo model, it is assumed that cosmic rays propagate not only in the galactic disk but also throughout a larger region of space, possibly corresponding to the halo or roughly spherical but sparse distribution of material which typically surrounds a galaxy. This model is based on the observation that the energy density of the material which is supposed to be contained by the galactic magnetic field is comparable to that of the field itself. This can result in an unstable situation in which large quantities of galactic material stream out in a galactic wind similar in some respects to the solar wind. In this case the outward flow is a natural part of the theory, and calculations have predicted reasonable flow rates. In distinction to the solar wind, in which the cosmic rays contribute almost nothing to the total energy density, they may provide the dominant energy source in driving the galactic wind.

A third model assumes that there is almost no escape; that is, cosmic rays disappear by breaking up into protons which then lose energy by repeated collision with other protons. To accept this picture, one must consider the apparent 5-g/cm^2 (0.07-lb/in.^{2}) mean path length to be caused by a fortuitous combination of old distant sources and one or two close young ones. Basically, the objections to this model stem from the tendency of scientists to accept a simple theory over a more complex (in the sense of having many free parameters) or specific theory when both explain the data. *See Milky Way Galaxy.*

Acceleration mechanisms. Although the energies attained by cosmic-ray particles are extremely high by laboratory standards, their generation can probably be understood in terms of known astronomical objects and laws of physics. Even on Earth, ordinary thunderstorms generate potentials of millions of volts, which would accelerate particles to respectable cosmic-ray energies (a few gigaelectronvolts) if the atmosphere were less dense. Consequently, there are many theories of how the acceleration could take place, and it is quite possible that more than one type of source exists. Two major classes of theories may be identified—extended-acceleration regions and compact-acceleration regions.

Extended-acceleration regions. Acceleration in extended regions (in fact the Milky Way Galaxy as a whole) was first proposed by E. Fermi, who showed that charged particles could gain energy from repeated deflection by magnetic fields carried by the large clouds of gas which are known to be moving randomly about the Milky Way Galaxy. Many other models based on such statistical acceleration have since been proposed, the most recent of which postulates that particles bounce off shock waves traveling in the interstellar medium. Such shocks, supposed to be generated by supernova explosions, undoubtedly exist to some degree but have an unknown distribution in space and strength, leaving several free parameters which may be adjusted to fit the data.

Compact-acceleration regions. The basic theory in the compact-acceleration class is that particles are accelerated directly in the supernova explosions themselves. One reason for the popularity of this theory is that the energy generated by supernovas is of the same order of magnitude as that required to maintain the cosmic-ray intensity in the leaky box model.

However, present observations indicate that the acceleration could not take place in the initial explosion. Cosmic rays have a composition which is similar to that of ordinary matter and is different from the presumed composition of the matter which is involved in a supernova explosion. At least some mixing with the interstellar medium must take place. Another problem with an explosive origin is an effect which occurs when many fast particles try to move through the interstellar gas in the same direction: the particles interact with the gas through a magnetic field which they generate themselves, dragging the gas along and rapidly losing most of their energy. In more plausible theories of supernova acceleration, the particles are accelerated gradually by energy stored up in the remnant by the explosion or provided by the intense magnetic field of the rapidly rotating neutron star or pulsar which is formed in the explosion.

Such acceleration of high-energy particles is clearly observed in the Crab Nebula, the remnant of a supernova observed by Chinese astronomers in A.D. 1054. This nebula is populated by high-energy electrons which radiate a measurable amount of their energy as they spiral about in the magnetic field of the nebula. So much energy is released that the electrons would lose most of their energy in a century if it were not being continuously replenished. Pulses of gamma rays also show that bursts of high-energy particles are being produced by the neutron star—the gamma rays coming out when the particles interact with the atmosphere of the neutron star. Particles of cosmic-ray energy are certainly produced in this object, but it is not known whether the particles escape from the trapping magnetic fields in the nebula and join the freely propagating cosmic-ray population. SEE CRAB NEBULA; NEUTRON STARS; PULSAR; X-RAY ASTRONOMY.

Acceleration in the solar system. The study of energetic particle acceleration in the solar system is valuable in itself, and can give insight into the processes which produce galactic cosmic rays. Large solar flares, about one a year, produce particles with energies in the gigaelectronvolt range, which can be detected through their secondaries even at the surface of the Earth. It is not known if such high-energy particles are produced at the flare site itself or are accelerated by bouncing off the shock fronts which propagate from the flare site outward through the solar wind. Nuclei and electrons up to 100 MeV are regularly generated in smaller flares. In many events it is possible to measure gamma rays and neutrons produced as these particles interact with the solar atmosphere. X-ray, optical and radio mapping of these flares are also used to study the details of the acceleration process. By relating the arrival times and energies of these particles at detectors throughout the solar system to the observations of their production, the structure of the solar and interplanetary magnetic fields may be studied in detail.

In addition to the Sun, acceleration of charged particles has been observed in the vicinity of the Earth, Mercury, Jupiter, and Saturn—those planets which have significant magnetic fields. Again, the details of the acceleration mechanism are not understood, but certainly involve both the rotation of the magnetic fields and their interactions with the solar wind. Jupiter is such an intense source of electrons below 30 MeV that it dominates other sources at the Earth when the two planets lie along the same interplanetary magnetic field line of force. Although the origin of the enhanced flux of ^{4}He has not been identified with certainty, it may be generated by the interaction of the solar wind with interstellar gas in the regions of the outer solar system where the wind is dying out and can no longer flow smoothly. SEE JUPITER; MERCURY (PLANET); PLANETARY PHYSICS; SATURN.

Direct observation of conditions throughout most of the solar system will be possible in the next few decades, and with it should come a basic understanding of the production and propagation of energetic particles locally. This understanding will perhaps form the basis of an understanding of the problem of galactic cosmic rays, which will remain for a very long time the only direct sample of material from the objects of the universe outside the solar system.

Paul Evenson

Bibliography. A. M. Hillas, *Cosmic Rays*, 1972; F. K. Lamb (ed.), *High Energy Astrophysics*, 1985; M. S. Longair, *High Energy Astrophysics*, 1981; J. L. Osborne and A. W. Wolfendale (eds.), *Origin of Cosmic Rays: Proceedings of the NATO Advanced Study Institute*, 1975; M. A. Pomerantz, *Cosmic Rays*, 1971; G. Setti and G. Spada (eds.), *Origin of Cosmic Rays*, 1981; M. M. Shapiro (ed.), *Cosmic Radiation in Contemporary Astrophysics*, 1986.

Cosmic spherules

Solidified droplets of extraterrestrial materials that melted either during high-velocity entry into the atmosphere or during hypervelocity impact of large meteoroids onto the Earth's surface. Cosmic spherules are rounded particles that are millimeter to microscopic in size and that can be identified by unique physical properties. Although great quantities of the spheres exist on the Earth, they are ordinarily found only in special environments where they have concentrated and are least diluted by terrestrial particulates. SEE METEOR.

The most common spherules are ablation spheres produced by aerodynamic melting of meteoroids as they enter the atmosphere. Typical ablation spheres are produced by melting of submillimeter asteroidal and cometary fragments that enter the atmosphere at velocities ranging from 6.5 to 43 mi (11 to 72 km) per second. Approximately 10,000 tons of such particles collide with the Earth each year, and cosmic spheres in the size range of 0.004 to 0.04 in. (0.1 to 1.0 mm) are the most abundant form of this material that survives to reach the Earth's surface. The spheres are formed near 48 mi (80 km) altitude, where deceleration, intense frictional heating, melting, partial vaporization, and solidification all occur in only a few seconds time. During formation, the larger particles can be seen as luminous meteors or shooting stars. Impact spheres constitute a second and rarer class of particles that are produced when giant meteoroids impact the Earth's surface with sufficient velocity to produce explosion craters that eject molten droplets of both meteoroid and target materials.

Impact spheres. These are very abundant on the Moon, but they are rare on the Earth, and they have been found in only a few locations. Impacts large enough to produce explosion craters occur on the Earth every few tens of thousands of years, but the spheres and the craters themselves are rapidly degraded by weathering and geological processes. Meteoritic spherules have been found around a number of craters, including Meteor Crater in Arizona, Wabar in Saudi Arabia, Box Hole and Henbury in Australia, Lonar in India, and Morasko in Poland. They have also been found at the Sikhote-Alin meteorite shower site and at the location of the Tunguska explosion in the Soviet Union. Spheres from these craters include ablation spheres as well as true impact spheres produced either by shock melting of target and meteoroid or by condensation from impact-generated vapor. Impact spheres are also produced by the larger cratering events, some of which may have played roles in biological extinctions. Silica-rich glass spheroids (microtektites) are found in thin layers that are contemporaneous with the conventional tektites. Microtektites are believed to be shock-melted sedimentary materials that were ejected from large impact craters. They were ejected as plumes that covered substantial fractions of the surface of the Earth. The cumulative mass of microtektites in the 35-million-year-old North American tektite field is estimated to be equivalent in mass to 36 mi^3 (100 km^3) of solid rock. Microspher-

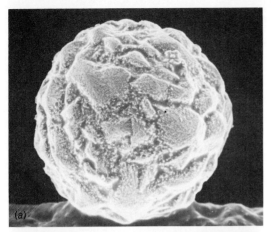

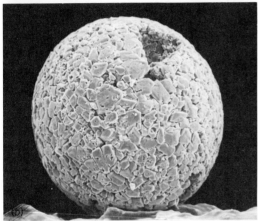

Scanning electron photomicrographs of stony cosmic spherules about 300 μm in diameter collected from the mid-Pacific ocean floor. (*a*) Sphere with so-called turtleback texture, indicative of rapid cooling from a very hot molten droplet. (*b*) Sphere with porphyritic texture, indicative of formation of a particle that was not so strongly heated during hypervelocity entry into the atmosphere as the sphere in *a*.

ules of a different composition have been found in the thin iridium-rich layer associated with the global mass extinctions at the Cretaceous–Tertiary boundary. *See Tektite*.

Ablation spheres. These fall to Earth at a rate of one 0.1-mm-diameter sphere per square meter per year, and every rooftop contains these particles. Unfortunately they are usually mixed in with vast quantities of terrestrial particulates, and they are very difficult to locate. They can, however, be easily found in special environments that do not contain high concentrations of terrestrial particles that could be confused with cosmic spheres larger than 0.004 in. (0.1 mm) in diameter. Such environments include the mid-Pacific ocean floor and certain ice deposits in Greenland and Antarctica. In mid-Pacific sediments the accumulation of terrestrial sediments is only a few meters per million years, and the spherules larger than 0.004 in. (0.1 mm) than are found in concentrations of roughly 10–100 per kilogram of sediment. Cosmic spheres are easily extracted from the sediment because most spherules are ferromagnetic and much larger than typical sediment particles. The highest ablation spherule concentrations on Earth (over one per gram of sediment) are found in the melt zones of

the Greenland ice cap, where melting ice leaves dust particles concentrated directly on the ice surface. *See Marine sediments*.

Cosmic ablation spherules can be grouped into two major types: type S (stony) and type I (iron). In polished sections it can be seen that the S spheres are composed of olivine, magnetite, and glass, with textures consistent with rapid crystallization from melt. Both of the type S spherules shown in the **illustration** have elemental compositions similar to those of stony meteorites. In general, types S spheres have compositions that are a close match with chondritic meteorites except for depletion of volatile elements such as sulfur and sodium that are lost during atmospheric entry. Many spheres are also depleted in nickel by a process that may also lead to the formation of the iron spheres. Typical type I spheres consist of iron oxide (magnetite and wustite) surrounding a core of either nickel-iron metal or a small nugget of platinum group elements. Nickel and the platinum group elements are concentrated during brief oxidation when the sphere is molten. Some of the iron spheres may be droplets of iron meteorites, but the most common ones appear to be droplets that separated from the stony spheres during atmospheric melting.

Identification. Cosmic spherules are of particular scientific interest because they provide information about the composition of comets and asteroids and also because they can be used as tracers to identify debris resulting from the impact of large extraterrestrial objects. The ablation spherules can be positively identified because of several unique properties, including distinctive oxygen isotope compositions and the presence of isotopes that are produced by cosmic rays such as aluminum-26 and manganese-53. Some spheres did not melt entirely, and they also retain high concentrations of noble gases implanted by the solar wind. In general, cosmic spherules can be confidently identified on the basis of their elemental and mineralogical compositions, which are radically different from nearly all spherical particles of terrestrial origin.

Don E. Brownlee, II

Bibliography. D. E. Brownlee, Cosmic dust: Collection and research, *Annu. Rev. Earth Planet. Sci.,* 13:147–173, 1985; C. Emiliani (ed.), *The Sea,* 1981; B. P. Glass et al., North American microtekitites from the Caribbean Sea and their fission track ages, *Earth Planet Sci. Lett.,* 19:184–192, 1973; P. W. Hodge, *Interplanetary Dust,* 1981; J. A. M. Mc-Donnel, *Cosmic Dust,* 1978.

Cosmic string

A hypothetical object that may account for the origin of large-scale structure in the universe. One of the most active areas of research in physics and astronomy is the search for a theory of the formation of such structure. The hot big bang model is highly successful, but cannot be the whole story. If, as is assumed, the universe was completely homogeneous and isotropic at early times, it would still be so now. But the universe is very lumpy, at least on scales up to clusters of galaxies. The simplest possibility is that the structure grew by the gravitational collapse of small initial fluctuations in the matter density. To produce the present structure, the density at early times

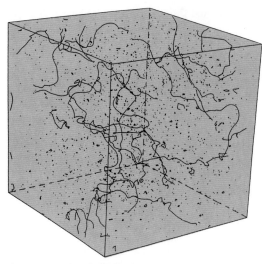

Computer model of a cubical box of cosmic strings. The strings oscillate at speeds close to the speed of light and reconnect if they cross. The length of an edge of the box is one-half the Hubble radius (approximately the distance traveled by light in the time since the big bang). (After A. Albrecht and N. Turok, Evolution of cosmic string networks, Phys. Rev. D, 40:973–988, 1989)

must have varied by approximately 1 part in 10^4. *See* Big bang theory.

Origin. An important idea, which emerged from particle physics in the early 1980s, was that the process that broke the symmetry between the particles and forces, which forms the basis for the standard model for particle physics, might break the spatial symmetry of the universe and produce the fluctuations. The process is analogous to the freezing of ordinary liquids.

Most liquids are homogeneous and isotropic. This is expected, because nothing in the description of atoms and their interactions singles out a particular direction or location in space. However, when a liquid freezes, the crystal structure of the solid picks a particular direction; but different directions are chosen in different regions. Thus the solid is full of defects where there is a mismatch between neighboring crystalline regions. *See* Crystal defects; Grain boundaries.

Cosmic strings are very similar to these defects. They are predicted to occur by some grand unification theories and superstring theories. In the very early universe, at high temperature the fields in these theories are random, just as the atoms of a liquid are. The density is quite uniform. As the universe cools below a certain temperature, some of the fields freeze. As they do so, defects are formed, just as in solids. *See* Elementary particle; Grand unification theories; Inflationary universe cosmology; Standard model; Superstring theory; Symmetry breaking.

Properties. In different theories, the defects may be at points, along lines, or in sheets. Cosmic strings are the linelike defects; they have special properties that make them ideally suited to forming structure later in the universe. For topological reasons they cannot have ends; they must form closed loops or continue on forever. They are very thin, about 10^{-15} the radius of a proton, but very massive. One meter of cosmic string weighs about 10^{20} kg.

Distribution. When cosmic strings form, the universe is filled with a random, tangled network of very long strings. The motion of the strings is quite complicated but, remarkably, is independent of the strings mass per unit length. Crudely speaking, the string's tension equals its mass per unit length times c^2, so that waves propagate on the string at the speed of light. This makes cosmic string theory very predictive: the distribution of the strings is fully specified at any time (given the initial conditions).

As the universe expands, the long strings chop themselves up into loops. This process is well described by a simple scaling theory. At any time after the network is produced, it has a characteristic scale, and the long strings look like random walks with steps of this scale. They continually chop off loops, also characterized by this scale. Furthermore, the scale grows linearly with time. The result is that, apart from a small correction, the total density in long string remains a fixed fraction (approximately 10^{-4}) of the total density in the universe. Detailed computer simulations have confirmed the scaling theory and have led to precise predictions of the distribution of the strings as the universe evolves (see **illus**.).

In the simplest picture of how structure forms around the strings, one loop of string is associated with one object. A loop 10 parsecs (2×10^{14} mi or 3×10^{14} km) long has a mass of 10^8 solar masses and, by now, accretes 10^{12} solar masses, a galaxy mass. Likewise, a loop of 10 kiloparsecs (2×10^{17} mi or 3×10^{17} km) accretes 10^{15} solar masses, the mass of a cluster of galaxies. While the mass of the accreted objects depends on the mass per unit length of string, the pattern in which the galaxies and clusters are laid down does not. Remarkably, the pattern in which cosmic string loops are produced closely matches the observed pattern of giant galaxy clusters. *See* Galaxy, External.

Direct observation. One direct way of observing the effects of cosmic strings is in the pattern of anisotropy produced in the observed microwave background radiation. A light ray that passes behind a moving string on its path to Earth is shifted toward the blue relative to a ray passing in front of the string. If observations of the microwave background radiation become sensitive to a part in 10^3, which they are close to doing, stripes should be observed in the temperature pattern on the sky. *See* Cosmic background radiation.

The second direct effect is gravitational lensing. Light is bent as it passes on either side of a string. Consequently, a galaxy observed behind a string would be seen as a double image. Several lensed galaxies have been observed, but there is no firm evidence that a cosmic string is responsible; a massive galaxy or dust cloud can produce the same effect. *See* Gravitational lens.

The third test is the gravity-wave background. Loops chopped off the network radiate away into gravity waves. The resulting background would be 10^{-3} of the density of the microwave background. There are limits on even this low a density in gravity waves, from accurate timing measurements on the millisecond pulsar. A gravity-wave background would lead to noise in the pulsar timing data due to Doppler shifts resulting from motion of the equipment from passing gravity waves. After corrections for the evolution of the pulsar, no such noise is observed. The simplest cosmic string theory is nearly ruled out by these measurements. *See* Gravitation; Pulsar.

However, cosmic strings may be more complex. In some grand unification theories the cosmic strings are superconducting and can carry enormous electric currents (10^{21} A) with zero resistance. Electric currents can build up on such strings via a dynamo effect, and the strings then build up a large magnetic field around them. This leads to a number of new ways for observing strings directly, and a variety of novel astrophysical phenomena. SEE COSMOLOGY; SUPERCONDUCTIVITY; UNIVERSE.

Neil Turok

Bibliography. A. Albrecht and N. Turok, Cosmic strings, *New Sci.*, 114:40–44, April 16, 1987; A. Albrecht and N. Turok, Evolution of cosmic string networks, *Phys. Rev. D*, 40:973–988, 1989; T. W. B. Kibble, Topology of cosmic domains, *J. Phys.*, A9:1387–1398, 1976. H. Mitter and F. Widder (ed.), *Particle Physics and Astrophysics: Current Viewpoints*, A. Vilenkin, Cosmic strings and domain walls, *Phys. Rep.*, 121:263–315, 1985.

Cosmocercoidea

A superfamily of nematodes parasitic in a variety of animals, including mollusks, amphibians, reptiles, and marsupial mammals. The weakly developed stoma is surrounded by either three or six lips. The esophagus is rhabditoid and divisible into corpus, isthmus, and valved posterior bulb, and has uninucleate glands. There are no intestinal or esophageal ceca. Males may possess a precloacal sucker. These nematodes are either oviparous or ovoviviparous. SEE NEMATA.

Armand R. Maggenti

Cosmochemistry

The science of the chemistry of the universe, particularly that beyond the Earth—that is, the abundance and distribution of elements, chemical compounds, and minerals; chemical processes, particularly in the formation of cosmic bodies; isotopic variations; radioactive transformations; and nuclear reactions, including those by which the elementary constituents were formed. Cosmochemistry is often functionally considered as a branch or extension of geochemistry. It is also closely related to astronomy and astrophysics. SEE GEOCHEMISTRY.

Techniques and sources of information. Many techniques of geochemistry can be applied to extraterrestrial samples which can be obtained for analysis in the laboratory. These include chemical analysis by a variety of methods, including neutron activation analysis for trace elements; electron-microprobe and ion-microprobe analysis of very small areas of polished surfaces; petrographic and x-ray-diffraction characterization of minerals and rocks; studies of magnetic properties and remanent magnetism; determination of relative abundances of isotopes of the elements, particularly by mass spectrometry; radiometric age determinations; and measurement of cosmic-ray-induced radioactivities and stable reaction products. SEE ACTIVATION ANALYSIS; COSMOGENIC NUCLIDE; ELECTRON PROBE ANALYSIS; ISOTOPE; MASS SPECTROMETRY; ROCK AGE DETERMINATION; ROCK MAGNETISM; SECONDARY ION MASS SPECTROMETRY (SIMS); X-RAY DIFFRACTION.

The most important of the available samples are meteorites, which are generally considered to be fragments of minor planets and possibly outgassed comets, and lunar samples that have been returned to Earth by spacecraft. Remote chemical studies have been made on the surfaces and in the atmospheres of planets and in the vicinity of comets. SEE METEORITE; SPACE PROBE.

For most of the objects of the universe, chemical information comes from analysis and theoretical interpretation of the radiations received from them on Earth and by artificial satellites. Spectral analysis of the light (visible, ultraviolet, infrared) received from the Sun, stars, and luminous nebulae give information on their elemental contents and on the few compounds that are stable in their exposed regions. Radio astronomy provides information on the interstellar medium, including a considerable number of molecular constituents. The interstellar medium and very high-temperature regions of stars, nebulae, and galaxies are studied in the short-wavelength ultraviolet and x-ray spectral regions by instruments aboard rockets and artificial satellites. SEE ASTRONOMICAL SPECTROSCOPY; INFRARED ASTRONOMY; RADIO ASTRONOMY; SATELLITE ASTRONOMY; ULTRAVIOLET ASTRONOMY; X-RAY ASTRONOMY.

Nonluminous bodies of the solar system can be characterized chemically to some extent by their reflection spectra—that is, albedo as a function of wavelength. Meteor spectra give information about these small fragments of asteroids and comets. Observations of the solar wind and energetic solar particles give information about the Sun's outer layers. Cosmic rays bring information about more distant regions of space. SEE ALBEDO; COSMIC RAYS; METEOR; SOLAR WIND.

Solar system. Study of the solar system involves the Sun, planets, asteroids, and comets.

Sun. The elemental composition of the Sun, which accounts for 99.9% of the mass of the solar system, is essentially that of the system as a whole, and undoubtedly also of the presumed precursor solar nebula. Lines of 74 elements have been identified in the solar spectrum, and it is probable that all of the 81 stable and 2 long-lived elements are present.

Many of the abundant elements can be determined quantitatively by analysis of their Fraunhofer lines. Helium can be detected, but its quantification in the Sun is difficult; the solar He:H atomic ratio is believed to be similar to that observed in other population I objects (hot stars, luminous nebulae), about 0.098 (ignoring the transformation of hydrogen to helium by thermonuclear reactions in the core, which does not mix with the surface). Carbon, nitrogen, and oxygen are determined relative to hydrogen in the solar photosphere. For most of the other elements, the relative abundances are assumed to be the same as in type I carbonaceous chondrite meteorites, which are believed to have condensed from the solar nebula with little or no fractionation. The relative abundances of the noble gases and mercury are deduced from population I objects or by use of nuclear systematics.

Carbon, nitrogen, and oxygen, like the noble-gas elements, are highly depleted in meteorites because of the volatility of the elements and their compounds. On the other hand, the observed solar values for lithium, beryllium, and boron are distinctly lower than those derived from meteorites, because those ele-

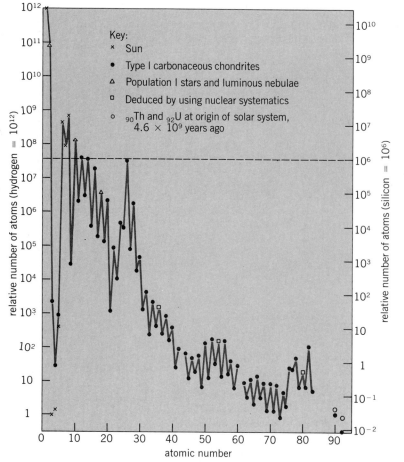

Fig. 1. Solar system elemental abundances. Meteoritic abundances relative to silicon have been normalized to solar abundances relative to hydrogen by using $^{14}Si/^1H = 3.58 \times 10^{-5}$. (*From data of E. Anders and N. Grevesse, Abundances of the elements: Meteoritic and Solar, Geochim. Cosmochim. Acta, 53:197–214, 1988*)

ments are being destroyed by thermonuclear reactions at the bottom of the Sun's outer convective layer. **Figure 1** shows the relative atomic abundances of the elements deduced for the solar system. The corresponding mass fractions are approximately 70.7% hydrogen, 27.4% helium, and 1.9% elements 3–92. *SEE ELEMENTS, COSMIC ABUNDANCES OF; SOLAR SYSTEM; SUN.*

Planets and asteroids. The terrestrial planets (Mercury, Venus, Earth, Mars), their satellites, and the asteroids consist largely of solids which have been separated from volatile elements. Their outer parts consist mainly of silicates, and each of these planets has a metallic core, the fraction of metal decreasing with increasing solar distance. The more massive Jovian planets (Jupiter, Saturn, Uranus, and Neptune) have retained large amounts of hydrogen, helium, other noble gases, and volatile compounds such as methane (CH_4), ammonia (NH_3), and water (H_2O). The overall compositions of these planets may be similar to that of the solar nebula, but the metallic and "rocky" elements are evidently concentrated in cores, and helium is probably partially concentrated in the interiors of Jupiter and Saturn since its abundances in these planets' atmospheres are less than the solar value. For a detailed discussion of the compositions of the planets *SEE PLANETARY PHYSICS.* For compositions of the individual planets *SEE EARTH; JUPITER; MARS; MERCURY (PLANET); NEPTUNE; PLUTO; SATURN;*

URANUS; VENUS. For discussions of planetary satellites *SEE* the articles on their planets and *MOON.*

Reflection spectroscopy of asteroids (minor planets) indicates considerable variety in surface compositions, many corresponding to known types of meteorites. The most abundant have low albedos (~2–5%) and spectra resembling carbonaceous chondrites. *SEE ASTEROID.*

Comets. The fluorescent emission spectra of the nearly straight type I tails of luminous comets show the presence of atoms, molecules, radicals, and ions consisting of hydrogen, carbon, nitrogen, oxygen, and sulfur. In the comas of bright comets, metallic elements are also seen. Shower meteors, which are outgassed fragments of comets, show spectral lines of common metallic elements. Huge atomic-hydrogen clouds have been observed extending from comets in the antisolar directions, and large amounts of hydroxyl radical (OH) are seen. Spacecraft measurements near Comet Halley indicated that, except for hydrogen and noble gases, its overall composition is similar to the Sun's, with minor fractionation of nonvolatile elements. These observations are consistent with the icy-conglomerate model, according to which comets were formed as masses of water, methane, ammonia, and carbon dioxide, in which are embedded organic compounds and metallic, siliceous, and carbonaceous particles. Comets, such as Halley, which have passed close to the Sun many times have dark mantles of devolatilized and partially sintered dust, and emissions are concentrated in localized areas. Dust particles entrained by the sublimating gases form curved type II reflecting tails. Photodecomposition of vaporized water produces the atomic hydrogen and hydroxyl clouds. *SEE COMET; INTERPLANETARY MATTER.*

Stars. Most stars in the Milky Way and other galaxies can be grouped into populations having various distinguishing characteristics, which are fundamentally related to their ages or times of formation. Within each population, most main-sequence stars and many of the evolved stars exhibit similar surface chemical compositions. Contents of heavier elements, such as carbon, nitrogen, oxygen, neon, and metallic elements, increase with decreasing population age. Regardless of age, the hydrogen-helium ratio seems to be roughly constant at ~0.1 on an atomic basis, and the relative amounts of the heavy elements to each other are roughly constant.

These observations can be explained by three sets of circumstances: (1) The interstellar medium, from which stars are formed, has gradually increased its content of heavy elements (thus also dust), rapidly in the early history of the Milky Way Galaxy and slowly subsequently, but the amount of helium resulting from hydrogen fusion and escaping from stars is small relative to the primordial amount. (2) The heavy elements ejected into the interstellar medium are produced by a variety of processes in many stars of a variety of types, and the mix of products is roughly constant. (3) All but the least luminous main-sequence stars and some moderately evolved stars have nonconvecting (radiative-energy-transporting) zones somewhere in their interiors, and so the products of nuclear transformations in their cores are never mixed into their surface layers; and low-luminosity stars have as yet transformed very little of their hydrogen into helium in their lifetimes. Therefore, all of these stars have preserved in their atmospheres the elementary composition of the media from which they were born, which has evolved over time.

Some main-sequence and slightly evolved stars and many highly evolved stars have spectra showing non-standard elemental contents which can be attributed to the presence in their surfaces of products of nuclear transformations in their atmospheres or interiors. Study of such stars helps scientists to understand the locales and mechanisms of energy generation and nucleosynthesis. SEE NUCLEOSYNTHESIS; STAR; STELLAR EVOLUTION.

Interstellar medium. This involves interstellar gas, dust, and molecules, as well as their chemistry.

Interstellar gas. Between the stars of the Milky Way are large quantities of gas and dust, with overall elemental composition similar to that of population I stars including the Sun, though definitely variable in elemental, chemical, and isotopic composition. Everywhere hydrogen dominates, and helium is the second most abundant element.

Neutral atomic hydrogen (H I) is detected and mapped by its characteristic 21-cm fine-structure radio emission (seen also in absorption in some cases). It is found to be concentrated in the spiral arms of the disk, mostly in diffuse clouds. Such clouds are typically ~30 parsecs (1 pc = 3.09 × 10^13 km = 1.92 × 10^13 mi = 3.26 light-years) in diameter with hydrogen concentration ~10–100 atoms/cm³ and temperature ~50–100 K (−370 to −280°F). They often show filamentary or shell-like structure, and amount to several percent of the mass of the stars.

Perhaps half of the interstellar matter is present as giant dense molecular clouds, of typical mass ~10^5 solar masses and dimension ~40 pc. These are most abundant in the inner galactic disk and central region. In these, molecular hydrogen (H₂) is undoubtedly the dominant constituent, although it cannot be observed there directly.

In less dense regions, molecular hydrogen can be detected by electronic absorption lines superposed on the ultraviolet continua of distant hot stars and extragalactic objects, and in somewhat warmer regions by infrared vibration-rotation emission lines. In the same regions, carbon monoxide can be detected by the 2.6-mm radio emissions of the abundant CO (that is, ¹²C¹⁶O) and its rarer isotopic forms ¹³CO and C¹⁸O. Radio emission from ¹³CO remains unsaturated in even the densest regions and thus serves as a tracer for molecular hydrogen and the giant molecular clouds. Molecular hydrogen is thereby found to have concentrations typically ~300 molecules/cm³ in the exteriors and up to ~10^7 molecules/cm³ in the densest cores of the giant cloud complexes.

More conspicuous though quantitatively less abundant are various kinds of luminous nebulae, collectively called H II regions, which owe their high temperatures (~10^4 K), ionization, and luminosity (fluorescence) to neighboring hot stars (both young and highly evolved). Typical H⁺ concentrations are ~10–10³/cm³. SEE NEBULA.

Interstellar dust. In cooler regions the gas is always accompanied by dust, whose mass is ~1% of that of the gas. The dust reveals itself mainly by absorption and scattering of starlight, this extinction being greater at shorter wavelengths and thus reddening the transmitted light. The spectral extinction indicates dominant particle sizes of ~0.01–0.1 micrometer. It

shows maxima at wavelengths attributed to graphite and/or amorphous carbon, amorphous solid water, and silicates. In addition, refractory oxides and metal (predominantly iron) may be present, and water, ammonia, and methane ices may mantle cores of refractory constituents. More complex molecules and organic polymers may also be present. SEE INTERSTELLAR EXTINCTION.

The *Infrared Astronomical Satellite* (*IRAS*) observed extended sources of infrared emission called infrared cirrus, often associated with H I clouds and molecular regions and correlated with extinction of starlight. The relative intensities at different wavelengths indicate frequent occurrence of "cold" dust (~30 K or −400°F) and "warm" dust (~300 K or 80°F). The latter contains a major component of very small (≤2 nanometers) grains, possibly carbonaceous, or very large molecules, which become temporarily heated internally by absorption of individual starlight photons.

Dust also produces dark nebulae seen in front of star fields of the Milky Way. Large clouds have a typical dimension of ~4 pc and mass of ~2000 solar masses, and smaller ones (Bok globules) are typically ~1 pc in diameter and have mass of ~60 solar masses. Frequently associated with H II regions are even smaller Bok globules (~0.1 pc or less in diameter and of ~1 solar mass). SEE GLOBULE.

In moderately transparent interstellar regions, absorption lines of heavy atoms and ions are observed. These elements are depleted relative to hydrogen as compared to population I stars; this is correlated positively with expected condensation temperature (**Fig. 2**). This strongly suggests that the atoms missing from the gas are bound in solid oxide or silicate grains formed at high temperatures. That is believed to occur in the expanding and cooling ejecta of giant stars with strong stellar winds, planetary nebulae, novae, and supernovae, and in interstellar shock fronts.

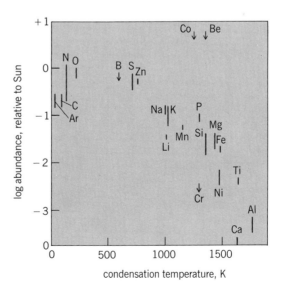

Fig. 2. Relative abundances of elements in a region of the interstellar medium versus temperature of condensation from gas of solar composition cooling in thermal and chemical equilibrium. Abundances are determined by absorption of ultraviolet light from the star Zeta Ophiuchi. Arrows indicate observational upper limits. (*After B. D. Savage and J. S. Mathis, Interstellar dust, Annu. Rev. Astron. Astrophys., 17:73, 1979*)

Interstellar molecules. Any neutral or charged aggregate of two or more atoms that exists in space is called a molecule; this includes many species that are known in chemistry as radicals and radical ions. Although very short-lived at the relatively high pressures of terrestrial environments, they can have long lifetimes in even dense interstellar regions because of the infrequency of collisions. *SEE FREE RADICAL; ION.*

A number of simple molecules are observed in diffuse clouds by visible and ultraviolet absorption and by infrared and radio-line emission. Many more molecules and several molecular ions are observed by radio-line emission in the giant clouds. Over 75 molecules and a half dozen molecular ions have been identified, and dozens of weak lines are not yet identified. Molecules containing up to 13 atoms have been observed. In diffuse clouds, photodissociation allows only relatively simple molecules to be built up to detectable concentrations. *SEE PHOTOCHEMISTRY.*

Interstellar chemistry. Two principal modes of formation of interstellar molecules have been proposed: grain-surface reactions (heterogeneous catalysis) and gas-phase reactions. Only reactions which are exoergic or very nearly so can occur by either mode at the low temperatures (below 100 K or $-280°F$) of molecular clouds.

A solid grain can catalyze the combination of two (or more) atoms or molecules by adsorbing both (or all) on its surface until they migrate into contact and combine, following which the product, generally a saturated molecule, escapes. Reactive centers on grain surfaces are probably rapidly covered by permanently bound atoms and molecules, and so the additional binding, effective in the process just described, is mainly by van der Waals forces. *SEE HETEROGENEOUS CATALYSIS.*

When a particular reaction can occur by both modes, the grain-surface mode should be more effective because of the much greater frequency of collisions involving the larger particles in spite of their low number abundance. A large fraction of low-temperature atom- and molecular-grain collisions results in sticking. Since hydrogen atoms are quite mobile on adsorbing surfaces and the adsorption energy for molecular hydrogen is small, the grain-catalyzed mechanism is undoubtedly the major one for the formation of molecular hydrogen. A few other small molecules may have appreciable formation rates on grains. However, the van der Waals binding energy increases with the number of atoms of a molecule in contact with the surface, inhibiting its release. Most heavy molecules formed on grain surfaces probably stay there. Thus it appears that most heavy molecules found in the gas phase must have been formed in that phase.

Most two-body reactions involving neutral molecules have activation energies, resulting in strong temperature dependence of reaction rates and effectively preventing low-temperature collisions from leading to reaction. However, exoergic reactions between positive ions and neutral molecules generally have zero activation energy, and thus are nearly insensitive to temperature and can take place at near-zero temperatures. The same is true for some reactions of free atoms or radicals with each other or molecules. Such gas-phase reactions account for most of the interstellar molecules heavier than molecular hydrogen in the cold molecular clouds.

The necessary ionization is provided by ultraviolet photons in the diffuse regions and by cosmic rays, which can penetrate even dense interstellar regions. Most of the initial ionization forms H_2^+, H^+, and

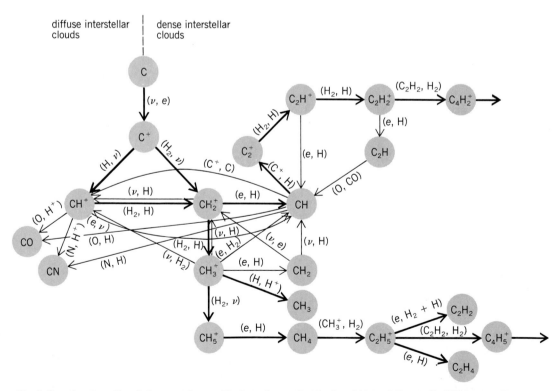

Fig. 3. Gas-phase reactions between carbon and hydrocarbon molecules in cold interstellar media. Within parentheses are the reactant and product not shown. ν = **photon (or, as reactant in dense clouds, cosmic-ray particle),** e = **negative electron. Heavy arrows show principal reaction paths leading to higher-order hydrocarbons. (After G. Winnewisser, The chemistry of interstellar molecules, Top. Curr. Chem., 99:39, 1981)**

He$^+$, which can extract electrons from heavier atoms and molecules to form principally C$^+$, O$^+$, N$^+$, and CO$^+$. Secondary reactions, especially chain reactions involving cations, then lead to a great diversity of products. Reactions involving the abundant molecular hydrogen and carbon monoxide are prominent.

The interaction of various reactions in cold interstellar regions is illustrated in **Fig. 3** for the simpler compounds of carbon. Similar schemes exist for other elements, especially oxygen and nitrogen.

In high-temperature regions, additional reactions, including many which have activation energies and some which are endoergic, can take place by ordinary thermal mechanisms. Elevated temperatures can be produced by interstellar shocks and by radiations from newly formed stars in collapsing clouds. SEE INTERSTELLAR MATTER.

Cosmic rays. Galactic cosmic rays (those not originating from the Sun) are highly energetic atomic nuclei, whose relative abundances reflect the composition of their source regions, but with considerable modification because of elemental differences in the acceleration processes and nuclear spallation-fragmentation reactions by collisions between heavy nuclei and hydrogen and helium nuclei in space. Modern balloon- and spacecraft-borne detectors can resolve individual masses up to the iron region, and the mass spectra are similar to the solar system abundances (Fig. 1). However, spallation and fragmentation greatly increase the relative abundances of lithium, beryllium, boron, and fluorine and the rarer isotopes of the other elements. Most of the solar system and stellar-atmosphere ^{6}Li, ^{9}Be, ^{10}B, and ^{11}B is believed to have been produced in this way. SEE COSMIC RAYS.

Galaxies. The mass fraction of heavy elements is enhanced in the central regions of the Milky Way and in similar spiral galaxies such as the Andromeda Galaxy (Messier 31), presumably because of the above-average rates of star formation and death of massive stars, which are chiefly responsible for dispersing products of nucleosynthesis into the interstellar medium.

The overall chemical composition of galaxies depends chiefly on the different stellar population distributions. Elliptical galaxies consist almost exclusively of population II stars, with very little interstellar gas and dust, and so virtually no formation of new stars occurs. Therefore they have relatively low contents of heavy elements. Spiral and undifferentiated irregular galaxies have high contents of gas and dust, resulting in high rates of star formation and nucleosynthesis, and thus considerable amounts of population I stars and relatively high contents of heavy elements. In some tidally interacting galaxies, induced star formation, called starburst, occurs, resulting in enhanced heavy-element abundances. SEE GALAXY, EXTERNAL; MILKY WAY GALAXY.

Cosmochronology. The radioactive dating methods based on primary natural radionuclides which are used in geochemistry are also applicable to cosmochemical samples in terrestrial laboratories. Methods based on short-lived extinct natural radionuclides are applicable to times very early in the solar system. Methods based on cosmic-ray-induced nuclides are especially useful for meteorites and lunar-surface materials, which have not been shielded by atmospheres.

Long-lived radionuclides. A number of meteoritic uranium,thorium-lead (U,Th-Pb), rubidium-strontium (Rb-Sr), and samarium-neodymium (Sm-Nd) ages have been obtained which are close to 4.56×10^9 years, and no reliable ages are older. This is regarded as the age of solid objects in the inner solar system. A few younger dates indicate subsequent disturbances due to igneous fractionation, metamorphism, or impact-induced shock, which "reset the clocks." The potassium-argon (K-Ar) and uranium, thorium-helium (U,Th-He) methods, which date retention of gaseous ^{40}Ar and ^{4}He, provide information about heating and shock events. SEE EARTH, AGE OF; GEOCHRONOMETRY.

Extinct natural radionuclides. Excess ^{129}Xe from the decay of now-extinct ^{129}I has been observed in a number of meteorites, especially chondrites. This indicates that they were formed as cool solids and incorporated live ^{129}I no later than about 10 times its half-life (1.59×10^7 years) following its last prior nucleosynthesis in a supernova. Differences in the predecay ^{129}I/^{127} ratio in different meteorites indicate a spread of at least 16 million years in their gas-retention-onset times.

Xenon isotopes attributable to spontaneous fission of ^{244}Pu (half-life 8.1×10^7 years) are also widely observed, especially in achondrites. Excess of fission-fragment radiation-damage tracks in some meteoritic minerals is also ascribed to this extinct radionuclide.

Aluminum-correlated ^{26}Mg excesses have been found in high-temperature condensates occurring as inclusions in carbonaceous chondrites, and can definitely be ascribed to the decay of extinct ^{26}Al (half-life 7.1×10^5 years). Similarly, ^{107}Ag excesses strongly correlated with Pd/Ag ratios in iron meteorites, which were differentiated in planetary bodies, require that the parent ^{107}Pd (half-life $\sim 6.5 \times 10^5$ years) was alive in the early solar system.

Decay products of other now-extinct radionuclides have been detected in meteorites as well, while traces of some nuclides with suitable half-lives have not been detected so far. A special case is ^{22}Na (2.6 years) $\rightarrow$ ^{22}Ne (Ne E), discussed below.

Nucleosynthesis of solar nebula material. The hydrogen and the bulk of the helium of the solar system undoubtedly originated in the cooling and expanding big bang fireball. The nuclides ^{129}I, ^{244}Pu, ^{235}U, ^{238}U, and ^{232}Th all result from the r-process of nucleosynthesis and were contributed by many supernovae to the interstellar medium from which the solar nebula formed. Analysis of the relative abundances of these radionuclides in meteorites at the time of their formation suggests that their nucleosynthesis began 8–15 $\times 10^9$ years ago and terminated about 1–2 $\times 10^8$ years before meteorite formation, with several percent of the r-products coming from the last contributing supernova. The ^{26}Al observations are interpreted by some as indicating a minor nucleosynthetic contribution much later. It is probable that the bulk of the stable elementary matter was synthesized on a similar time scale. SEE SUPERNOVA.

Ages of bodies in solar system. Dating of lunar rocks and soil indicates an age of the Moon as a body of about 4.4–4.6 $\times 10^9$ years, about the same as that of the meteorites, and it is believed that all of the planets were formed at close to the same time. Dating of lunar lava basalts indicates that igneous processes were active on the Moon as recently as 3.1×10^9 years ago. It seems likely that most meteorites are fragments of asteroids, some of which must have been strongly heated and then cooled rather rapidly following their formation. However, a few rare basaltic-type meteorites, the SNC (Shergotty-Nakhla-Chassigny) group, have crystallization ages of about 1.3 $\times$ 10^9 years, and it is widely postulated that they are ejecta from Mars, which must have been volcanically active then, although it is no longer so.

Measurements of stable and radioactive cosmogenic (cosmic-ray-produced) nuclides in meteorites have shown that most iron meteorites have existed as small bodies in space, produced by collisional fragmentation from larger bodies, for less than 1 $\times$ 10^9 years, and most stony meteorites, which are more fragile, for less than 2 $\times$ 10^7 years.

Isotopes. Limited information is available on the isotopic composition of matter outside the solar system. In some cool giant stars, carbon and oxygen isotopes can be detected in optical and infrared spectra, and $^{13}C/^{12}C$, $^{17}O/^{16}O$, $^{18}O/^{16}O$, and $^{17}O/^{18}O$ ratios are generally greater than in the terrestrial elements. This can be attributed in part to variable participation of the carbon-nitrogen-oxygen thermonuclear-energy cycle in different stars. *See* Carbon-nitrogen-oxygen cycles.

In the interstellar medium, isotope ratios of several elements can be determined from ultraviolet absorption and microwave molecular-emission spectra. The D/^{1}H ratio in atomic hydrogen, $\sim 1.8 \times 10^{-5}$, is considerably lower than the terrestrial value ($\sim 1.5 \times 10^{-4}$, which is elevated because of selective escape of ^{1}H from the atmosphere), but it is greatly enhanced in molecules because of mass-dependent chemical fractionation in ion–molecule reactions. The $^{13}C/^{12}C$, $^{18}O/^{16}O$, $^{17}O/^{16}O$, and $^{15}N/^{14}N$ ratios also differ markedly from their solar system values and vary significantly from the galactic center to the disk. These observations are clues to past stellar activity in the Milky Way Galaxy.

Studies of meteorites and the Moon have shown that for the most part there is a close similarity between the isotopic composition of meteoritic and lunar elements and their terrestrial counterparts. This is evidence that the bulk of the matter of the solar nebula, which was presumably derived from a number of different nucleosynthetic sources, was well mixed both physically and chemically before the formation of planetary bodies.

However, careful examination of many meteorite types and particularly of selected small phases of meteoritic matter has shown that there are numerous cases of differences. Those which cannot be accounted for by radioactive and cosmic-ray-induced nuclear transformations within the samples are referred to collectively as isotopic anomalies, and can be attributed to isotopic fractionation and to variable nuclear processes acting on matter which was not thoroughly mixed in the solar nebula.

The D/^{1}H ratio is variable in meteorites, greater than 30–fold enhancements being found in some organic polymer fractions, which are presumed to be derived from relict interstellar molecules. The $^{13}C/^{12}C$ ratio is close to the terrestrial mean (~ 0.011) in major meteorite phases, but is greatly enhanced in some minor carbonaceous particles in some primitive chondrites. Considerable enhancements of $^{15}N/^{14}N$ have been observed in minor nonmetallic phases of some iron meteorites. These are probably derived from surviving interstellar phases of anomalous composition.

Oxygen is the principal major meteoritic element to exhibit significant isotopic variations. Because it has three stable isotopes, fractionation and mixing of isotopically different components can be distinguished. **Figure 4** illustrates schematically many of the observations. (The complete range of variations is considerably greater than shown here, $\delta^{18}O$ ranging from about $-41‰$ to about $+35‰$ and $\delta^{17}O$ from about $-42‰$ to about $+17‰$) Mass-dependent fractionation yields points on a line of slope 0.5 on this diagram, as illustrated by the terrestrial fractionation line. The Earth-Moon field refers to estimated bulk compositions. Meteorites having similar compositions are inferred to have formed in the inner part of the solar nebula. Objects away from this line must have been derived partly or wholly from matter of different oxygen composition. The array near a line of slope close to 1.0 suggests mixing of two end-member reservoirs, one near the terrestrial fractionation line and one very rich in ^{16}O but with about the same $^{18}O/^{17}O$ ratio ($\delta^{18}O$ about $-40‰$, $\delta^{17}O$ about $-42‰$). The array with slope ~ 1.0 above the terrestrial fractionation line implies at least one more ^{16}O-poor reservoir, and combined fractionation and mixing effects are also indicated. It is believed that the ^{16}O-rich and ^{16}O-poor oxygen resided in dust components that were not completely vaporized and mixed in the solar nebula, and the intermediate fraction, which accounted for the majority of planetary-object oxygen, resided in the gas.

The next most abundant meteoritic element showing large anomalies is magnesium. It also has three stable isotopes, but because one of them is partly radiogenic (^{26}Mg from extinct ^{26}Al) distinction between fractionation and nucleosynthetic anomalies is complicated. The larger anomalies imply an incompletely

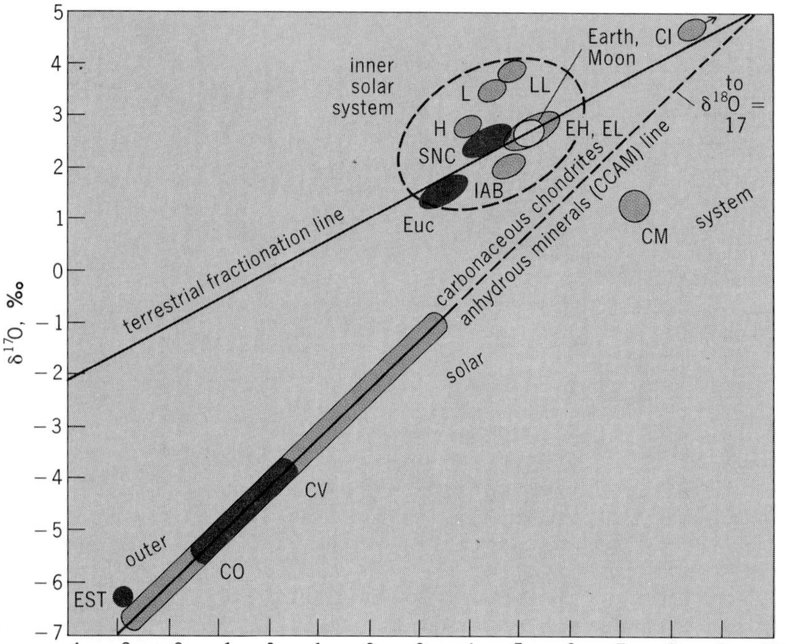

Fig. 4. Oxygen-isotope composition of solar system objects. $\delta^{17}O$ and $\delta^{18}O$ are the relative changes in the ratios $^{17}O/^{16}O$ and $^{18}O/^{16}O$ from their values in standard mean ocean water. CI, CM, CV, CO: types of carbonaceous chondrites. H, L, LL: types of ordinary chondrites. EH, EL: types of enstatite chondrites. Euc: eucrite, howardite, and diogenite achondrites, mesosiderites, and pallasites. SNC: Shergotty-Nakhla-Chassigny group of achondrites. IAB: iron meteorites of groups IA and IB. EST: Eagle Station trio of pallasites. (*After J. T. Wasson, Meteorites: Their Record of Early Solar-System History, W. H. Freeman, 1985*)

mixed interstellar dust component. Smaller anomalies are due to mass fractionation, as is also the case for silicon.

Calcium, titanium, and chromium exhibit anomalies, principally variable excesses of their heaviest isotopes ^{48}Ca, ^{50}Ti, and ^{54}Cr. These are attributed to an incompletely mixed dust component containing matter nucleosynthesized in a neutron-rich environment.

The noble gases in meteorites, being highly depleted, show strong effects of cosmic-ray irradiation, which enhances principally the lighter isotopes. The most pronounced anomaly is that of ^{22}Ne, which is strongly enhanced in some differential-heating release fractions. Indications are of the presence of a component of possibly pure ^{22}Ne (called NE E), which is attributed to the decay of short-lived ^{22}Na in dust particles which incorporated that radionuclide close to sites of its production, most likely in novae. Krypton and xenon show components with the isotopic signature of the s-process of nucleosynthesis, which occurs in red giant stars. These must have been mixed to the surfaces and incorporated into stellar winds, or expelled in planetary nebulae, and trapped or implanted in circumstellar dust particles. Noble-gas anomalies often occur in particles showing anomalies also of carbon, oxygen, and other mineral-forming elements. *See Nova; Planetary nebula.*

Chemistry in cosmic evolution. According to current orthodoxy, the matter of the universe was created in an expanding "primordial fireball" (following the big bang) about $1-2 \times 10^{10}$ years ago, mostly in the form of ^{1}H and ^{4}He in close to the present proportions, about 4:1 on a mass basis. Small amounts of ^{2}H, ^{3}He, and ^{7}Li were also formed then. *See Big bang theory; Cosmology.*

Galactic evolution. Successive condensations in the expanding universe produced galaxy clusters, galaxies, star clusters, and individual stars. The most massive early stars generated heavy elements in their interiors and ejected them into the interstellar medium mainly through stellar winds and supernova eruptions. Later-formed stars in the Milky Way Galaxy, formed before it had collapsed to a disk, thus inherited small amounts of heavy elements. These are the population II stars, including those in globular clusters, which have typically 0.01–0.1% of elements heavier than helium. Subsequent star formation occurred mainly in the disk, with a rather rapid buildup of heavy elements, reflected in the initial composition of successive generations of stars. Most population I stars, of which the Sun is a member, have about 1–3% of heavy elements.

Solar system formation. About 4.6×10^9 years ago a fragment of a giant molecular gas-dust cloud in a spiral arm of the Milky Way collapsed gravitationally to form the solar nebula. The collapse may have been initiated or accelerated by a nearby supernova, some of whose newly formed stable and radioactive nuclides may have been incorporated into the nebula. In the center of the nebula, accelerated contraction produced the Sun, initially much more luminous than now, probably in a T Tauri phase characterized by strong stellar winds. The outer parts collapsed into a disk, and the inner regions at least were heated by gravitational compression, solar radiation, and electromagnetic disturbances to temperatures sufficient to vaporize most of the dust grains.

As the Sun decreased in luminosity and became a

Stability fields of equilibrium solar nebula condensates*

Phase	Formula	Temperature limits, K[†]	
		Upper[‡]	Lower[¶]
Corundum	Al_2O_3	1758	1513
Perovskite	$CaTiO_3$	1647	1393
Melilite	$Ca_2Al_2SiO_7$–$Ca_2MgSi_2O_7$	1625	1450
Spinel	$MgAl_2O_4$	1513	1362
Metallic iron	(Fe,Ni)	1473	
Diopside	$CaMgSi_2O_6$	1450	
Forsterite	Mg_2SiO_4	1444	
	Ti_3O_5	1393	1125
Anorthite	$CaAl_2Si_2O_8$	1362	
Enstatite	$MgSiO_3$	1349	
Eskolaite	Cr_2O_3	1294	
Metallic cobalt	Co	1274	
Alabandite	MnS	1139	
Rutile	TiO_2	1125	
Alkali feldspar	$(Na,K)AlSi_3O_8$	~1000	
Troilite	FeS	700	
Magnetite	Fe_3O_4	405	
Water ice	H_2O	≤200	

*For total pressure of 10^{-3} atm (100 pascals).
[†]°F = (K × 1.8) − 459.67.
[‡]For condensation with decreasing temperature or disappearance with rising temperature.
[¶]For conversion to or from forms stable at lower temperatures.
SOURCE: L. Grossman, Condensation in the primitive solar nebula, *Geochim. Cosmochim. Acta*, 36:597–619, 1972.

stable main-sequence star, the nebula cooled while maintaining a distribution of temperature decreasing with increasing distance from the Sun. As the temperature at a particular radial distance decreased, a series of solids condensed from the gas. The **table**, based on thermochemical calculations, indicates the probable order of condensation of minerals, many of which are now found in meteorites. There is evidence, particularly from isotope fractionations, that some of the matter was subjected to repeated temperature decreases and increases with condensation and partial or complete revolatilization

When the temperature had dropped to somewhat below the condensation temperature, metal, sulfide, and siliceous grains initially of equilibrium composition adhered and accreted into small, loosely bound aggregates. Many of these were flash-heated by an uncertain process or processes so that they were instantly melted and rapidly cooled to form spherical objects. These comprise at least some of the abundant millimeter-sized chondrules now observed in chondritic meteorites. Further aggregation of dust and particles produced larger objects called planetesimals, perhaps meter- and kilometer-sized and even larger. The increased transparency of the nebula allowed solar radiation pressure and a strong solar wind in the inner regions to drive away the remaining gases. The planetesimals coalesced further to form the terrestrial planets, fortuitously large ones sweeping up all of the others at about their solar distances. The proportion of low-temperature condensates, including most oxides and silicates, relative to iron, a high-temperature condensate, increased with increasing distance from the Sun, and this is reflected in the densities and core sizes of the terrestrial planets. *See Earth interior.*

In the outer part of this region, conditions were such that a single large planet did not form; but instead a number of minor planets, the progenitors of the asteroids, resulted. Gravitational perturbations by nearby massive Jupiter may have played a role. Reflection spectroscopy of the asteroids indicates a variety of chemical and mineralogical types, with some correlation with mean solar distance. In particular, in the outer part of the asteroid belt there is a high proportion of apparently carbonaceous bodies, indicating lower temperatures of formation than in the inner part of the belt. For the minor planets represented by the chondritic meteorites, there is evidence based mainly on trace-element analyses that the temperature was decreasing during their accretion, so that a layered structure resulted.

The formation of Earth's Moon is a special case, but whatever process was responsible created a fractionation of refractory elements, which are enhanced in the Moon, from more volatile elements, which are enriched in the Earth.

Still farther out, where enormous amounts of gases and solids were present, gravitational attraction became important, and uncondensed gases and unaccreted grains were also swept up by the giant planets which formed there. Each of these planets may have accreted nearly all of the hydrogen and helium that were associated with the heavier elements in its region of the nebula; the differing proportions of hydrogen, helium, and heavier elements in the planets' visible parts may reflect different fractionation and redistribution of elements in their interiors. Pluto's low density indicates that it is composed largely of ices, probably mostly water, though spectroscopy indicates solid and gaseous methane on its surface. This reflects both its low gravity and the extremely low temperature far out in the solar nebula.

The formation of the large satellites of the Jovian planets, which mostly have direct low-eccentricity orbits nearly in their parents' equatorial planes, presumably followed a pattern similar to the formation of the Sun's satellites, but with the difference that the temperature was much lower. Consequently they have low densities and consist largely of ices, predominantly water, surrounding small rocky cores.

At much greater distances, innumerable comets formed as aggregates of dust grains and ices. Those with the longest periods have randomly distributed orbital orientations and make up the enormous Oort cloud, which formed before the collapse of the solar nebula to a disk. Those of somewhat shorter periods have predominantly prograde orbits and make up the Kuiper belt beyond the orbit of Neptune, evidently formed after the collapse and shrinkage of the disk but before appreciable heating and evaporation of ices.

Planetary evolution. After and even during accretion, evolutionary changes within the planetary bodies and their atmospheres began to take place. Most chondrites show evidence that each was at one time a part of the regolith of a bombarded planetary body, probably still in the accretionary stage, and that it was later subjected to higher temperatures and pressures in the interior, resulting in lithification and varying degrees of metamorphism.

The major heat sources were probably gravitation, radioactive disintegration, and electromagnetic energy propagated in the solar wind. The last might have been important for asteroidal and smaller bodies. If appreciable amounts of intermediate-lived and now-extinct radionuclides (particularly ^{26}Al and ^{60}Fe) were present, they could have had important heating effects

in asteroidal bodies. Relatively strong short-lived heat sources in the meteorite parents, which lose heat rapidly by surface radiation, were necessary to account for formation of the achondrite and iron meteorites, which are igneous rocks, followed by rapid cooling. Gravitational energy of accretion becomes important for the larger asteroids and the planets, which retain heat better, and long-lived radionuclides (^{235}U, ^{40}K, ^{238}U, and ^{232}Th) are effective in large bodies on long time scales. Their interiors rise to high temperatures, and when melting occurs, mass redistribution releases still more gravitational energy, especially if an iron or troilite core is formed. *See Earth, heat flow in.*

Melting, gravitational redistribution, recrystallization, and outgassing produced stratified interiors and crusts, hydrospheres and atmospheres, volcanism, other geologic phenomena, and igneous rocks, both plutonic and extrusive, in the terrestrial planets. Any original atmospheres of the latter were probably swept away by the solar wind, and the present atmospheres of Venus, Earth, and Mars presumably resulted from interior outgassing. High temperatures and solar wind have stripped Mercury's atmosphere, and thermal upper-atmosphere evaporation has caused loss of much hydrogen from the other planets' atmospheres and most of the other light gases from Mars. Conversion of the presumably carbon dioxide (CO_2)-rich early atmosphere of the Earth to an oxygen (O_2)-rich atmosphere is probably the result of its biosphere's activity. *See Atmosphere, evolution of; Geophysics; Planetary physics; Solar wind.*

Truman P. Kohman

Bibliography. E. Anders and N. Grevesse, Abundances of the elements: Meteoritic and solar, *Geochim. Cosmochim. Acta*, 53:197–214, 1989; J. F. Kerridge and M. S. Mathews (eds.), *Meteorites and the Early Solar System*, 1988; J. M. Moran and P. T. P. Ho (eds.), *Interstellar Matter*, 1988; S. K. Runcorn, G. Turner, and M. M. Woolfson (eds.), *The Solar System: Chemistry as a Key to Its Origin*, 1988; J. W. Truran, Nucleosynthesis, *Annu. Rev. Nucl. Part. Phys.*, 34:53–97, 1984; G. Turner and C. T. Pillinger (eds.), *Diffuse Matter in the Solar System: Comet Halley and Other Studies*, 1987.

Cosmogenic nuclides

Nuclides produced by the interaction of cosmic rays with matter. Cosmic rays are composed of high-energy particles of nuclear matter; they fragment target nuclear material and produce a variety of stable and radioactive nuclides known as cosmogenic nuclides. Typical energies of nuclei in the primary cosmic-ray particles lie in the range 1–10 GeV; the total energy range is, however, much wider, 0.1–10^{10} GeV. Protons (p) are the most abundant nuclei in the cosmic radiation, constituting more than 90% of all nuclei above a given kinetic energy per nucleon. Helium (He) and heavier nuclei in roughly the same proportion as in the solar system material account for the rest of the nuclei. Since the binding energy of a nucleon in nuclear matter is approximately 10 MeV, primary cosmic rays and even the secondary particles produced in their interactions are very effective in breaking up nuclear matter. *See Cosmic rays; Cosmochemistry; Proton; Solar system.*

The isotopic composition of any matter exposed to cosmic radiation thus gets altered because of nuclear

Table 1. Often-studied cosmogenic nuclides with half-lives exceeding 2 weeks

Isotopes	Half-life*	Main targets†
^{3}H‡	12.3 years	O, Mg, Si, Fe (N, O)
^{3}He, ^{4}He	S	O, Mg, Si, Fe (N, O)
^{7}Be‡	53 days	O, Mg, Si, Fe (N, O)
^{10}Be‡	1.6×10^6 years	O, Mg, Si, Fe (N, O)
^{14}C‡	5730 years	O, Mg, Si, Fe (N)
^{20}Ne, ^{21}Ne, ^{22}Ne	S	Mg, Al, Si, Fe
^{22}Na‡	2.6 years	Mg, Al, Si, Fe (Ar)
^{26}Al	7.1×10^5 years	Si, Al, Fe (Ar)
^{32}Si‡	100–200 years	(Ar)
^{35}S‡	87 days	Fe, Ca, K, Cl (Ar)
^{36}Cl‡	3.0×10^5 years	Fe, Ca, K, Cl (Ar)
^{36}Ar, ^{38}Ar	S	Fe, Ca, K
^{37}Ar‡	35 days	Fe, Ca, K (Ar)
^{39}Ar‡	269 years	Fe, Ca, K (Ar)
^{40}K	1.3×10^9 years	Fe,
^{39}K, ^{41}K	S	Fe
^{41}Ca	1.0×10^5 years	Ca, Fe
^{46}Sc	84 days	Fe
^{48}V	16 days	Fe, Ti
^{53}Mn	3.7×10^6 years	Fe
^{54}Mn	312 days	Fe
^{55}Fe	2.7 years	Fe
^{56}Co	79 days	Fe
^{59}Ni	7.6×10^4 years	Ni, Fe
^{60}Fe	1.5×10^6 years	Ni
^{60}Co	5.27 years	Co, Ni
^{81}Kr‡	2.1×10^5 years	Rb, Sr, Zr (Kr)
^{78}Kr, ^{80}Kr, ^{82}Kr, ^{83}Kr	S	Rb, Sr, Zr
^{129}I‡	1.6×10^7 years	Te, Ba, La, Ce (Xe)
$^{124-132}$Xe	S	Te, Ba, La, Ce, I

*S = stable.
†Elements from which most production occurs; those in parentheses are for the Earth's atmosphere.
‡Atmospheric cosmogenic isotopes.

interactions; in this process, the composition of the cosmic-ray beam itself also gets altered. Cosmogenic nuclides have been studied in diverse forms of extraterrestrial and terrestrial matter and in the cosmic-ray beam with a view toward understanding the evolutionary history of matter. Studies of cosmogenic nuclides in the cosmic-ray beam itself give information on the characteristics of propagation of cosmic-ray particles in the galaxy (for example, the amount of matter traversed and the travel time). The information thus obtained is generally not available from other studies. *See* Milky Way Galaxy; Isotope.

Cosmogenic nuclides have received increasing emphasis and importance since higher-sensitivity techniques for detection of nuclides have become available. This aspect is of great significance in cosmic-ray studies since the cosmic-ray flux is small; the time-averaged cosmic-ray flux in the solar system at 1–10 astronomical units (AU) is about 0.3 particle/$(cm^2 \cdot s \cdot sterad)$. The cosmic-ray flux incident at the top of the Earth's atmosphere varies between about $0.02/(cm^2 \cdot s)$ at the Equator and about $1/(cm^2 \cdot s)$ at the poles; the latitude effect is caused by the screening of charged particles by the Earth's geomagnetic field. *See* Astronomical unit; Geomagnetism.

The discovery of cosmogenic carbon-14 (^{14}C) primarily produced in the Earth's atmosphere by capture of thermal neutrons (n) with nitrogen-14 (^{14}N; n + ^{14}N = ^{14}C + p + 0.625 MeV) by W. F. Libby and colleagues in 1947 triggered searches and studies of cosmogenic nuclides in a wide variety of extraterrestrial and terrestrial samples. These studies were suc-

Table 2. Principal "atmospheric" cosmic-ray-produced isotopes and their applications

Nuclides*	Applications
Short half-lives (½ hr–2.6 years) ^{34m}Cl, ^{38}Cl, ^{39}Cl, ^{18}F, ^{31}Si, ^{38}S, ^{24}Na, and ^{28}Mg	Cloud physics
Half-lives of days–years ^{32}P, ^{33}P, ^{7}Be, ^{35}S, and ^{22}Na	Atmospheric structure, large-scale air circulation, and scavenging
Half-lives of more than 10 years ^{14}C	Archeology and paleobotany
^{3}H, ^{32}Si, ^{39}Ar, ^{14}C, ^{36}Cl, ^{10}Be, and ^{129}I	Air-sea exchange, geochemical and biological cycles, paleomagnetic reversal records, and cosmic-ray prehistory
^{32}Si, ^{39}Ar, ^{14}C, ^{81}Kr, ^{36}Cl, ^{81}Kr, and ^{10}Be	Hydrology and glaciology: chronology of groundwaters, lacustrine sediments, and glaciers

*Grouped in order of increasing half-lives in each category.

cessful, and further investigations led to opening of new frontiers in solar system astrophysics and in earth sciences, as well as in cosmic-ray physics. A novel technique, accelerator mass spectrometry, was introduced in the late 1970s. Accelerator mass spectrometry allows identification and counting of individual atoms; thus it is very sensitive, especially for long-lived cosmogenic radionuclides (which were identified earlier on the basis of detection of their characteristic radiation). It has become possible to measure a large number of radionuclides of half-lives 10^3–10^6 years at levels of 10^6 atoms. This corresponds to detection that is about 10^4–10^6 times more sensitive than the earlier techniques based on detecting the radiations emitted by these nuclei. This has dramatically changed the scope of application of cosmogenic nuclides. SEE ACCELERATOR MASS SPECTROMETRY; NEUTRON.

Earlier work. The first studies and applications of cosmogenic nuclides occurred during the period 1947–1977. The detection of natural cosmogenic ^{14}C (half-life 5730 years) in methane extracted from sewage gas was a major task, considering the isotope measurement techniques then available. The ability to detect the feeble radioactivity arising from nuclear interactions of cosmic rays led to an increased interest on the part of nuclear physicists and geophysicists. The Libby group also accomplished another difficult feat by detecting cosmogenic hydrogen-3 (^3H; half-life 12.3 years). The ratio of cosmogenic ^3H to ^1H atoms in natural water is of the order of 10^{-18}.

Following its discovery, ^{14}C was used for archeological dating and for studying a wide variety of geophysical problems such as air-sea exchange, biological cycles, and large-scale ocean circulation. During 1947–1955 the widespread occurrence of cosmogenic nuclides in meteorites and in terrestrial samples was realized, and extensive efforts were made in several laboratories to detect them. These studies, carried out principally during the 1950s, established that the concentrations of a wide variety of nuclides in meteorites and in terrestrial samples (in the atmosphere, in wet precipitation, in sediments, and in a great variety of samples from the upper layers of the Earth) could be measured, and that it was also possible to elucidate the physical and chemical processes governing their accumulations in these samples. **Table 1** lists some 30 nuclides of half-lives exceeding 2 weeks that are often studied. Atmospheric cosmogenic nuclides are

primarily those for which the target nuclei are nitrogen (N), oxygen (O), and argon (Ar); the low abundances of krypton (Kr) and xenon (Xe) limit the studies of their nuclear products to a few nuclides. The greater variety of cosmogenic nuclides which can be studied in extraterrestrial samples is due to the fact that they contain magnesium (Mg), aluminum (Al), silicon (Si), iron (Fe), and heavier target nuclei in appreciable concentrations and that they are exposed to higher fluxes of cosmic radiation. SEE MARITIME METEOROLOGY; OCEAN CIRCULATION; RADIOCARBON DATING; SEAWATER.

Cosmogenic nuclides in meteorites have proven useful principally for the study of (1) exposure ages of meteorites, (2) temporal variations in the flux of galactic cosmic radiation during the past 10^9 years, and (3) the characteristics of low-energy charged particles (mostly protons) accelerated by the Sun (the solar cosmic radiation) during the recent and early history of the solar system. The latter records are found in grains within the gas-rich meteorites. Availability of documented lunar samples since 1969 has made it possible to study in detail the temporal variations in the long-term average flux of solar cosmic-ray protons during the past 5 million years. These studies indicate that the constancy of the time-averaged solar cosmic-ray flux in the 10–100-MeV kinetic energy interval has remained the same within a factor of 3 during this period. The long-term averaged fluxes of galactic cosmic-ray particles at the meteoritic orbits (1–3 AU) have been determined to be the same as those observed near Earth since 1960, within a factor of 2. SEE METEORITE; MOON.

On the other hand, studies of atmospheric cosmogenic nuclides make possible study of the dynamical transport and mixing processes responsible for their global dispersion on the Earth. **Table 2** lists the principal applications of radionuclides in geophysical studies; included are seven short-lived radionuclides with half-lives of 0.5 h–1 day not listed in Table 1. These nuclides find applications in studies of short-term processes in the atmosphere, for example, cloud formation, and scavenging of aerosols by wet precipitations.

The estimated global average production rates of cosmogenic nuclides of half-lives exceeding 1 week are listed in **Table 3**, separately for 1 cm^2 columns in the troposphere and in the total atmosphere. The cor-

Table 3. Production rates of cosmic-ray-produced isotopes in the atmosphere*

Radioisotope	Half-life	Production rate, atoms/($cm^{-2} \cdot s^{-1}$)		Global inventory, g
		Troposphere	Total atmosphere	
^3He	Stable	6.7×10^{-2}	0.2	3.2×10^9
^{10}Be	1.5×10^6 years	1.5×10^{-2}	4.5×10^{-2}	2.6×10^8
^{26}Al	7.1×10^5 years	3.8×10^{-5}	1.4×10^{-4}	1.1×10^6
^{36}Cl	3.0×10^5 years	4.0×10^{-4}	1.1×10^{-3}	1.5×10^7
^{81}Kr	2.1×10^5 years	$\sim1.8 \times 10^{-7}$	$\sim4 \times 10^{-7}$	2.6×10^3
^{14}C	5730 years	1.1	2.5	7.7×10^7
^{32}Si	~200 years	5.4×10^{-5}	1.6×10^{-4}	4×10^2
^{39}Ar	269 years	2.0×10^{-3}	5.6×10^{-3}	2.3×10^4
^3H	12.3 years	8.4×10^{-2}	0.25	3.5×10^3
^{22}Na	2.6 years	2.4×10^{-5}	8.6×10^{-5}	1.9
^{35}S	87 days	4.9×10^{-4}	1.4×10^{-3}	4.5
^7Be	53.3 days	2.7×10^{-2}	8.1×10^{-2}	32
^{37}Ar	35 days	2.8×10^{-4}	8.3×10^{-4}	1.1
^{33}P	25.3 days	2.2×10^{-4}	6.8×10^{-4}	0.6
^{32}P	14.3 days	2.7×10^{-4}	8.1×10^{-4}	0.4

*After D. Lal and B. Peters, Cosmic-ray-produced radioactivity on the Earth, *Handbook of Physics*, vol. 46/2, pp. 551–612, 1967.

Table 4. Principal *in situ* terrestrial cosmic-ray-produced isotopes and their applications

Samples	Isotopes*	Applications
Rocks	^{39}Ar, ^{14}C, ^{36}Cl, ^{26}Al, ^{10}Be, ^{21}Ne, ^{3}He	Weathering and erosion processes, uplift rates, subsidence, glacial history, eruption ages, cosmic-ray prehistory
Sands, soils	^{39}Ar, ^{14}C, ^{36}Cl, ^{26}Al, ^{10}Be	Transportation and burial histories: depositional and erosional rates, rates of movements of sand dunes
Glaciers	^{14}C, ^{36}Cl, ^{10}Be	Accumulation and ablation rates, chronology, climatic changes
Tree rings	^{14}C, ^{10}Be	Temporal changes in solar activity, geomagnetic field, and climate

*Grouped in order of increasing half-lives.

responding terrestrial inventories are also listed in Table 3.

Accelerator mass spectrometry. Until 1960 the field of atmospheric cosmogenic isotopes continued to open up new frontiers by providing so-called tracers for the study of the exposure history of meteorites and of a variety of geophysical and geochemical processes. Work in the field then slowed down during the 1960s, both in meteoritics and in terrestrial studies, with no new directions. The development of accelerator mass spectrometry made it possible to detect most of the long-lived radionuclides at levels of approximately 10^6 atoms, corresponding to a gain in sensitivity of detection by a factor of about 10^6 over the earlier methods, which were based on detection of the decay product. The gain in sensitivity was particularly important for studies of both meteorites and rare terrestrial material, for example, samples of old ice. The immediate result of increased detection sensitivity was the development of more comprehensive databases and longer records in the case of terrestrial samples, and a widening of the scope of application of cosmogenic nuclides in general. Advances occurred principally in the applications of beryllium-10 (^{10}Be) and ^{14}C. There is now available a long time series of ^{10}Be in polar ice cores, useful for delineating secular variations in solar-terrestrial relationships, for example, cosmic-ray 'modulation by solar plasma and climatic changes influencing the fallout of ^{10}Be. Accelerator mass spectrometry has made it possible to study the ^{14}C present in 1 liter of seawater and in a tiny sample of foraminaferan shells (composed of calcium carbonate; $CaCO_3$). These studies are providing detailed insight into changes in the ocean circulation during the past 20,000 years, for example, the temporal and spatial changes in deep-water formation.

The cosmogenic nuclide iodine-129 (^{129}I; half-life 1.6×10^7 years) was detected in iron meteorites by using accelerator mass spectrometry. This nuclide fills in nicely the large gap in half-lives of nuclides studied in meteorites, manganese-53 (^{53}Mn; half-life 3.7×10^6 years) and potassium-40 (^{40}K; half-life 1.3×10^8 years), and it can provide a higher resolution study of the temporal change in the flux of galactic cosmic rays.

In the field of terrestrial studies, use of accelerator mass spectrometry was extended to include studies of nuclides produced *in situ* in solids, for example, rocks, tree rings, and ice. These studies have made it possible to consider a wide range of geophysical processes which could not be studied earlier (see **Table 4**). The basis of working with *in situ* cosmogenic nuclides in terrestrial solids is the fact that the isotope

production rates are very sensitive to depth. Thus processes such as erosion and sedimentation can be modeled based on observations of isotopes of different half-lives. *See* Radioisotope (geochemistry).

The nuclide production rates in a given sample depend on the altitude and latitude of the location where it is exposed. The surface production rates can be estimated fairly well from the **illustration,** which gives the rate of production of all nuclear disintegrations in the atmosphere as a function of latitude and altitude. Typical yields of different nuclides in nuclear disintegrations in different target elements (Table 1) lie between 1 and 10%.

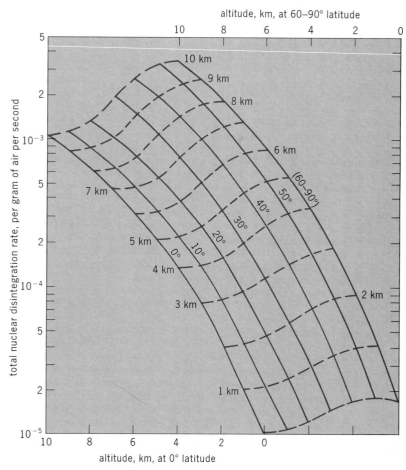

Total rate of nuclear disintegrations in the atmosphere (with energy release greater than 40 MeV) is plotted as a function of altitude and geomagnetic latitude, 0–90°, in steps of 10°. Curves for different latitudes have been successively displaced along the abscissa by 1 km (0.6 mi). (*After D. Lal and B. Peters, Cosmic-ray-produced radioactivity on the Earth, Handbook of Physics, vol. 46/2, pp. 551–612, Springer-Verlag, Berlin, 1967*)

Table 5. Global cosmic-ray source function

Characteristic	Value
Flux of hydrogen nuclei	7.8×10^{25}/year (~1/cm^2 · s)
Mass flux of hydrogen nuclei	130 g/year
Mass flux of all nuclei	170 g/year
(H, He, C, N, O, and so forth)	
Mean energy of a cosmic-ray particle	3.6×10^9 eV/nucleon (~4 × 10^{13} K)
Incident energy flux	3.7×10^{35} eV/year (1.4 × 10^{16} cal/year or 5.9 × 10^{16} J/year)

In terrestrial samples, the isotopic changes produced by cosmic rays can be studied up to depths of about 10^6 g · cm^{-2} below sea level, deeper than in the case of meteorites (about 10^3 g · cm^{-2}). The reason for this is that pi mesons produced in the Earth's atmosphere decay to penetrating muons. The penetrating component has been used as cosmic "x-rays" in a variety of applications—for example, to determine the geometry of pyramids and, in civil engineering, to determine the overburden in different directions. SEE MESON.

The incident flux of cosmic rays on the earth is small (**Table 5**) compared to the solar energy flux (5.2×10^{24} cal/year or 2.2×10^{25} joules/year) or heat flow from the interior of the earth (2.4×10^{20} cal/year or 1×10^{21} J/year), but it is important because of its high specific energy. SEE EARTH, HEAT FLOW IN; INSOLATION.

D. Lal

Bibliography. E. C. Anderson et al., Natural radiocarbon from cosmic radiation, *Phys. Rev.*, 72:931–936, 1947; V. Faltings and P. Harteck, The tritium concentration in the atmosphere, *Z. Naturforsch.*, 5A:438, 1950; M. Honda and J. R. Arnold, Effects of cosmic rays on meteorites, *Handbook of Physics*, Vol. 46/2, pp. 613–632, 1967; D. Lal, In situ produced cosmogenic isotopes in terrestrial rocks, *Annu. Rev. Earth Planet. Sci.*, 16:355–388, 1988; D. Lal and B. Peters, Cosmic-ray produced radioactivity on the Earth, *Handbook of Physics*, vol. 46/2, pp. 551–612, 1967; W. F. Libby, *Radiocarbon Dating*, 2d ed., 1955; A. E. Litherland, Accelerator mass spectrometry, *Nucl. Instrum. Meth. Phys. Res.*, 233 (B5):100–108, 1984; G. M. Raisbeck and F. Yiou, Production of long-lived cosmogenic nuclei and their applications, *Nucl. Instrum. Meth. Phys. Res.*, 233 (B5):91–99, 1984; R. C. Reedy, J. R. Arnold, and D. Lal, Cosmic ray record in solar system, *Science*, 219:127–135, 1983; J. A. Simpson, Elemental and isotopic composition of the galactic cosmic rays, *Annu. Rev. Nucl. Part. Sci.*, 33:323–381, 1983.

Cosmology

The study of the large-scale structure and the evolution of the universe, and including, in its modern connotation, that part of cosmogony that deals with the origin of the universe and of the chemical elements. SEE UNIVERSE.

Structure of the universe. Beginning about 1924, E. P. Hubble developed ways of estimating distances to remote galaxies, and also studied their distribution in space. From counts of galaxy images on photographs of 1283 sample regions of the sky, Hubble concluded that on the large scale the distribution of galaxies is homogeneous and isotropic. Hubble's finding was the first observational evidence for the cosmological principle, which states that at any instant of cosmic time the universe is, on the large scale, the same as seen by all hypothetical observers; that is, it is the same everywhere (is isotropic and homogeneous). SEE GALAXY, EXTERNAL.

Galaxy clusters. On the small scale, however, Hubble found a tendency for galaxies to cluster. A few dozen individual clusters of galaxies were known when Hubble carried out his survey, and he selected the sample fields to avoid these known clusters. There were, however, too many fields with too few galaxies, and too many fields with too many for their distribution to be completely random in space. Hubble suggested that all galaxies either are members of, or were formed in, groups and clusters. The Milky Way Galaxy, in fact, is a member of a system of about two dozen galaxies, most of them much smaller than the Milky Way, spread over a region of space with a linear diameter of about 3×10^6 light-years (2×10^{19} mi or 3×10^{19} km); the system is called the Local Group. SEE LOCAL GROUP.

The general tendency of galaxies to cluster was confirmed by the analysis of the galaxy distribution in two post–World War II photographic surveys of the sky: the Lick Astrographic survey with the 20-in. (50-cm) astrographic camera, and the Palomar Sky Survey with the 48-in. (1.2-m) telescope. Distribution of galaxy images on the Lick photographs was found to be compatible with a statistical model of complete clustering. Occasional clusters, the great clusters, contain thousands of member galaxies each, and have radii of 5×10^6 to 2×10^7 light-years (3×10^{19} to 1.2×10^{20} mi or 5×10^{19} to 2×10^{20} km).

Superclusters. G. O. Abell cataloged 2712 of these great clusters, and defined a homogeneous statistical sample of about 1800. On the large scale, the clusters in the statistical sample confirm Hubble's finding of homogeneity and isotropy—further observational evidence for the cosmological principle. On the small scale, however, even the clusters are clumped into larger aggregates, and Abell identified 17 obvious superclusters. Subsequently, statistical studies of catalogs of galaxies and clusters of galaxies confirmed that the spatial distribution of galaxies is correlated over scales that are large compared to sizes of individual clusters.

The Local Group appears to be a member of such a larger aggregate, the Local Supercluster, whose center may lie in the general direction of the nearest fairly rich cluster of galaxies, the Virgo cluster, some $30–50 \times 10^6$ light-years (2–3 × 10^{20} mi or 3–5 × 10^{20} km) distant. Several other individual superclusters have been studied in some detail, and evidence on their structures is rapidly accumulating. Typical superclusters are 1–3×10^8 light-years (0.6–1.8×10^{21} mi or 1–3×10^{21} km) across and contain one or two great clusters, although some contain 10 or more. They also contain dozens of lesser clusters, and hundreds of groups like the Local Group, as well as individual galaxies. There may also be significant quantities of ionized intercluster gas, although at present there is firm evidence for intergalactic gas only in certain rich clusters; this gas has a temperature of 10^8 K (10^8°F) and is detected by its thermal emission of x-rays.

There is an additional dark component, whose existence is inferred from mass estimates of clusters by using virial analysis, which expresses a balance between kinetic and potential energy. This dark matter probably comprises more than 90% of the mass density of the universe, and its nature is very mysterious. Dark matter is associated with the outer regions of spiral galaxies, as is evidenced by the flat rotation curves of spirals. At present it is unknown whether all the dark matter is clustered like the galaxies on the length scales of group and superclusters. It is possible that the large voids between superclusters contain significant amounts of ionized gas and dark matter. *See* X-RAY ASTRONOMY.

Large-scale homogeneity. Present observations suggest that superclusters are the end of the hierarchy. One way to sample the distribution of matter to various depths is to count galaxies to successively fainter limits. If counts are extended to a limit of faintness such that the line of sight from the Earth encounters only one or two superclusters, the presence of those aggregates is clearly apparent in the irregular way that the counts increase with faintness. But if the counts are continued to very faint limits, they always smooth out, consistent with a large-scale homogeneous distribution of matter, washing out all traces of superclustering. In 1977 G. Rainey made such counts in three widely separated directions in the sky and found that he could definitely rule out general clustering of matter on scales of 10^9 light-years (6×10^{21} mi or 10^{22} km) or more. He also found a remarkably high degree of isotropy in those different directions. Other similar studies, some to much fainter limits, are in agreement with Rainey's and further strengthen observational support for the cosmological principle.

In summary, the galaxy distribution in the universe is observed to be very lumpy on scales of up to a few hundred million light-years, but on larger scales appears to be the same everywhere. The cosmological principle, therefore, is a starting assumption for nearly all theories of cosmology, and for all that are widely considered today. There is, of course, no guarantee that the observed large-scale homogeneity extends throughout the entire universe, but without that assumption there is no basis for formulating a general cosmological model.

Gravitation in the universe. The present-day density of matter in the universe exceeds the mass equivalent of the energy of radiation by at least a factor of 10^2; thus the universe today is dominated by matter. Moreover, bulk matter is electrically neutral to a very high degree. Hence the only known long-range force that can be important in influencing the dynamical evolution of the universe is gravitation.

In the seventeenth century Newton considered the effects of gravitation on the universe at large, and concluded that in an infinite universe the potential due to gravitation would everywhere be the same, and hence an infinite universe could be stable against gravitational collapse (except locally, where irregularities in the density could cause local collapses that might become stars).

On the other hand, the new theory of gravitation embodied in Einstein's general theory of relativity leads to field equations that, in their simplest form, do not permit a static stable universe even if it is infinite. In general relativity the field equations describe the curvature of space-time by matter; it is in curved space-time that all objects, material bodies as well as photons of light, move along unaccelerated paths (geodesics). Imbued with the idea that the universe should be static, Einstein, in his effort to apply the field equations to cosmology, modified them with the introduction of a term now called the cosmological constant. The cosmological constant, if positive, implies a repulsive force that is greater in proportion to distance. By choosing just the correct value for the constant, the force can be made to balance gravitation on the large scale, and permit a static universe. *See* GRAVITATION; RELATIVITY; SPACE-TIME.

The cosmological constant does not violate the assumptions on which the field equations are derived, for it enters as a constant of integration, and in the most general mathematical case it should be present. But there is no analog for such a repulsion in newtonian theory, and there is no observational evidence for its existence; just as one usually evaluates a constant of integration from the boundary conditions of the physical problem involved, so one would normally assign to the cosmological constant the value zero. Its introduction here with just the correct value to provide a static universe seemed an ad hoc procedure that was adopted for the sole purpose of saving the theory.

Expanding universe. In the 1920s it occurred to others that the universe need not necessarily be static, and that the field equations in their simplest form (zero cosmological constant) can apply perfectly well to a uniformly expanding or contracting universe. In particular, the Soviet mathematician A. Friedmann in 1922, and independently the Belgian cosmologist G. Lemaître in 1927, proposed general relativistic models of an expanding universe. Indeed, Lemaître suggested observations that could test the hypothesis.

These were observations begun about 1912 by V. M. Slipher of the spectra of the "nebulae" that were later found to be galaxies. Slipher found the features in the spectra of the objects to be displaced from their normal wavelengths by the Doppler effect, which indicated that the nebulae are moving away from the Earth at high speeds. *See* DOPPLER EFFECT; REDSHIFT.

Now, if the cosmological principle is applied to an expanding universe, the universe must expand uniformly; otherwise, irregularities would be generated that would make the universe inhomogeneous. In a uniform expansion, the separations of all pairs of objects must increase by the same factor in the same time, which means that widely separated objects, having farther to move apart, must separate at a greater speed than objects close together, which have less far to move to increase their separations by the same relative amount. In other words, every observer, everywhere in an expanding universe, will see every other object moving away from him or her at a speed that is proportional to its distance; it is a necessary and unique condition for a uniform expansion. Of course, since all observers see the same effect, this fact gives no information about the observer's location in the universe, nor of any "center." In an infinite expanding universe, the concept of "center" has no meaning.

By 1929 Hubble had found approximate distances to more than 40 of the galaxies whose radial velocities (speeds of recession) has been measured by Slipher. Hubble's analysis showed there to be a significant correlation between velocity and distance, with velocities ranging up to nearly 1200 mi/s (2000 km/s). Meanwhile, M. L. Humason had begun observing the spectra of still more distant objects with the 100-in. (2.5-m) telescope, and in 1931 Hubble and Humason jointly published a comparison of the distances and radial velocities of galaxies receding at speeds up to 15 times those of the most remote galaxies observed by Slipher. The velocities and distances were in direct proportion: the expansion of the universe had been firmly established. The observed relation between the velocities and distances of galaxies is now called the Hubble law.

As noted above, the cosmological principle applies only to the universe on the large scale. Similarly, the uniform expansion applies only to the large-scale universe—not to individual galaxies and clusters of galaxies. Like the Earth, solar system, and stars, galaxies are bound tightly together with their own mutual gravitation; they are gravitationally stable and do not expand with the universe—a point often misunderstood. At least the rich clusters of galaxies also appear to be gravitationally bound, and gravitation must play some role in the dynamics of superclusters, although just how much is not yet known.

The mutual gravitational attractions of the member galaxies in a cluster accelerate them to high speed (often 1.6×10^3 mi/s or 10^3 km/s or more) in their more or less random paths back and forth through the cluster. When the radial velocity of a galaxy in a group or cluster is observed, therefore, the velocity of recession of the cluster itself caused by the expansion of the universe always has superimposed on it the line-of-sight component of the galaxy's velocity due to the local gravitational accelerations by other cluster members. In a few nearby clusters the speeds of individual galaxies within them are great enough that a few galaxies are approaching the Earth, even though the clusters to which they belong are receding.

Cosmological models. The Hubble law alleviates the need for the positive cosmological constant introduced by Einstein to allow a static universe. Nevertheless, theoreticians have carefully considered the much larger range of cosmological models that are possible with both positive and negative values of the cosmological constant, and it cannot be guaranteed that future observations will never require such nonzero values. The cosmological constant is not, however, needed at present, and the only models for the universe that receive the wide attention of both theoreticians and observers are those based on general relativity with zero cosmological constant. These models will now be described qualitatively.

Because the cosmological principle requires that the universe expand uniformly, its evolution can be described uniquely with a simple change in scale.

Thus the actual distance $r(t)$ between two remote objects in space at time t can be expressed as the product of a suitably normalized time-varying scale factor, usually denoted as $R(t)$, and the fixed distance u between those objects at some arbitrarily chosen time; thus, Eqs. (1) hold. The radial velocity V_r of a galaxy

$$r(t) = R(t)u \qquad (1a)$$

$$u = r(t)/R(t) \qquad (1b)$$

is simply the rate of change of $r(t)$. Since u is constant, Eq. (2) holds. According to the Hubble law,

$$V_r = \frac{dr(t)}{dt} = \frac{dR(t)}{dt} u = \frac{1}{R(t)} \frac{dR(t)}{dt} r(t) \qquad (2)$$

Eq. (3) is valid, where the constant of proportionality,

$$V_r = Hr(t) \qquad (3)$$

H, is called the Hubble constant. Note that the Hubble law does not imply that a given galaxy speeds up as it recedes to greater distance; rather at any given time the galaxies that happen to be more remote must recede faster. *See* Hubble constant.

Comparison of Eqs. (2) and (3) shows that Eq. (4)

$$H = \frac{1}{R(t)} \frac{dR(t)}{dt} \qquad (4)$$

holds, so the Hubble "constant" is not really constant over time, but has a value that changes as the universe evolves and the scale $R(t)$ changes with time. Even if all galaxies maintained their present speeds, H would diminish as the scale grows. In fact, however, the mutual gravitation of the objects in the universe must slow its expansion, so that galaxies, as they recede from each other, are gradually reducing their relative speeds. A major goal of observational cosmology is to determine how strongly the gravitation of the universe decelerates its expansion.

Closed, open, and flat universes. In general relativity, photons and material objects flow unaccelerated along geodesics in space-time, and that space-time is curved by the presence of matter. In a random distribution of massive objects the curvature can be very complex, but for the universe at large, adoption of the cosmological principle results in an enormous simplification. Except for small-scale local effects, the curvature is the same everywhere. If the mean universal density of matter and energy is high enough, the curvature is positive, in which case the universe is unbounded but finite, for information can never be received from beyond a particular volume of space. Such a universe is said to be closed, and the geometry of space-time is elliptical (in which the circumference c of a circle of radius r is less then $2\pi r$). If the universal density is below a critical value, however, the curvature is negative, the universe is open and infinite, and the geometry of space-time is hyperbolic (in which c is greater than $2\pi r$). The dividing line between these two classes of space-time, requiring the critical density that does not quite close the universe, is flat space-time, in which the universe is infinite and the geometry is euclidean.

A closed universe is closed not only to radiation but to matter, for it implies that the geodesics of particles of matter and of photons are closed: such a universe will eventually cease expanding and start to contract. The contraction will continue until all of the matter and radiation approaches a singularity at the origin of a universal black hole—the "big crunch," as J. A. Wheeler described it. It has been speculated that such a universe may undergo another "big bang" to begin a new cycle, and thenceforth oscillate between successive expansions and contractions, each contraction followed by a new big bang. There is, however, no physical theory to account for future big bangs and subsequent expansions, so the "oscillating" universe is at present no more than a speculative extension of the closed universe.

Alternatively, the open universe will expand forever into greater and greater infinities; neither radiation nor material objects will ever come together again. Closed and open universes are analogous to rockets fired from the surface of the Earth at respectively less than and greater than the velocity of escape. A rocket launched with greater than the escape velocity is always slowed by the Earth's gravitation, but it has enough energy that it never stops; similarly, in the open universe the mean density of matter and radiation is too low to provide enough gravitation to ever stop the expansion.

The flat universe has the critical density to barely stop the expansion, analogous to a rocket launched with precisely the escape velocity. In this case the scale $R(t)$ increases as $t^{2/3}$. It is mathematically equivalent to a model by Einstein later modified by the astronomer W. de Sitter, and so is also called the Einstein–de Sitter universe. **Figure 1** shows the scale of the universe R plotted against time in various cosmological models.

Age of universe. All general relativistic models of the universe with zero cosmological constant have a unique beginning, usually called the big bang, and a finite age. The maximum possible age for any of these models is that of the open universe with so low a mean density that gravitation can be ignored, so that the galaxies continue to recede at constant speed. In this case, since the time it takes a given galaxy to reach its present distance is that distance divided by its speed, the time since that galaxy began receding with zero distance, or the age of the expansion, is simply the reciprocal of the Hubble constant. With the range of current estimates for the Hubble constant, that maximum possible age lies in the range $1\text{–}2 \times 10^{10}$ years. The age of the Einstein–de Sitter universe is just two-thirds of the reciprocal of the Hubble constant; open universes have greater ages than the Einstein–de Sitter universe, and closed universes have smaller ages.

Selecting between cosmological models. The future of the universe, that is, whether it is open or closed, depends on how rapidly the expansion is being slowed by gravitation. A measure of this slowing is called the deceleration parameter, denoted q_o. (The formal definition of q_o is $-\ddot{R}_0 R_0 / \dot{R}_0^2$, where the dots indicate time derivatives and the zero subscripts the present-day value of the scale, R.) The value of q_o is proportional to the mean density of the universe, and so is always positive; that is, any density produces gravitation that results in a deceleration. For the Einstein–de Sitter universe, q_o has the value ½; it is less than ½ for the open and greater than ½ for the closed universes. It is possible to find q_o directly from observations, but those observations are difficult to make and the present-day evaluations of q_o are highly uncertain. Some of the procedures are discussed in the following.

Precise form of the Hubble law. As explained above,

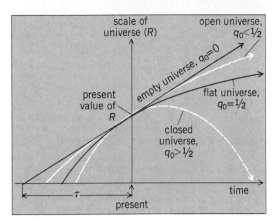

Fig. 1. Scale of the universe R plotted against time in various cosmological models; τ is the maximum possible age for universes with cosmological constant equal to 0.

any uniform expansion results in a linear relation between velocity and distance (so long as the velocities are small compared to that of light; at very great distances the velocities become very large, and special relativity requires that they asymptotically approach the speed of light). Instantaneous velocities and distances are not observed, however, because of the finite speed of light. The velocity observed for a galaxy 10^8 light-years away is the speed of recession it had 10^8 years ago when the light, by which it is now observed, was emitted. Both the distance and speed of recession of the galaxy have subsequently changed, and how much they have changed depends in a calculable way on q_o—that is, how rapidly the expansion is slowing. A plot of the Hubble law therefore shows deviations from linearity for objects at large distances (**Fig. 2**), and comparison of the observed deviations from those predicted for various values of q_o can, in principle, determine q_o. Velocities of galaxies can be measured with sufficient precision by measuring the Doppler shifts of their spectral lines, but relative distances cannot be determined accurately enough to make the test. The main problem is that distances of galaxies must be estimated from their apparent brightnesses and their assumed intrinsic luminosities. The luminosity of a bright galaxy is expected to evolve as its member stars gradually burn out, dimming the galaxy and changing its color toward the red. Additionally, bright galaxies can merge with their neighbors, or cannibalize smaller satellite galaxies, increasing their luminosity with time. The resulting net evolutionary correction is quite uncertain. Estimates of those corrections lead to an interpretation of the ob-

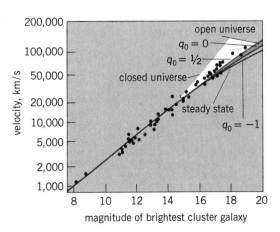

Fig. 2. Hubble law curves for various cosmological models. Data points indicate observed values.

served Hubble law that is compatible with either an open or a closed universe. Angular sizes of galaxies or a cluster of galaxies, again in principle, can also be used as distance indicators when proper allowance is taken of photometric and special relativistic effects, but the present precision of measurements is not good enough to improve the discrimination between open and closed models. Again, uncertain evolutionary corrections cloud the interpretation of this test.

Counts of galaxies and other objects. In a static homogeneous universe the numbers of objects should increase as the cube of the distance to which they are counted, but in an expanding universe the density decreases with time. Thus when one looks out to great distance, one also looks back in time to a denser universe; the counts should increase at a greater rate than with the cube of the distance, and how much greater depends on q_o. The difficulty in applying this test lies in the problem of knowing the distances of the counted objects, particularly because their intrinsic luminosites can evolve in time. The quasars, for example, have very large radial velocities indicating enormous distances; quasars must therefore be objects that existed in the remote past but no longer today. *See Quasar.*

Deuterium-hydrogen ratio. In the first few minutes of the big bang, temperatures were high enough to support nuclear fusion. Calculations show that deuterium was formed, but would have fused almost completely into helium unless the density of the matter were relatively low; for appreciable deuterium to survive the big bang, the density then would have to have been low enough that the present-day density would be far below the critical value to close the universe. Observations of the present ratio of deuterium to hydrogen indicate a value in the range 10^{-5} to 10^{-4}, high enough to imply an open universe. This conclusion of an open universe must be regarded with caution, however, for it is not certain that the observed deuterium is primordial. *See Big bang theory.*

Mean matter density. Several other tests for determining q_o from the observations have been attempted, but the one with the most significant (although not definitive) result is that of the mean matter density in the universe. The critical density to close the universe is of order 10^{-30} to 10^{-29} g/cm^3 (depending on the value of the Hubble constant). The expansion rate of the universe is given by the Hubble constant; if the mean mass density exceeds the critical value, the expansion rate will be insufficient to allow the matter in the universe to escape its own self-gravity and the universe would then be closed. The ratio of the average mass density $\bar{\rho}$ to this critical density ρ_c is defined as $\Omega \equiv \bar{\rho}/\rho_c$. In a matter-dominated universe $\Omega = 2q_o$, so $\Omega = 1$ corresponds to an Einstein–de Sitter model.

To measure $\bar{\rho}$ one can simply multiply the mean number of density of galaxies in the universe by the mass per galaxy. The mean number density of galaxies is well determined from observations, but the mass per galaxy is not. Including only the well-observed inner regions of galaxies, where dark matter is not in evidence, Ω is found to be less than < 0.01, indicating a universe open by a large margin. On the scale of clusters, however, virial estimates give a much higher mass per galaxy, consistent with $\Omega \sim 0.1$–0.2. The discrepancy simply reflects the fact that the mysterious dark matter is not tightly clustered on the scale size of galaxies, and that galaxies do not precisely trace the underlying mass density. If $\Omega = 1$, then the dark matter must be considerably less clustered than the galaxies on all scales. Until the nature of the dark matter and the physics of galaxy formation are understood, the estimates of Ω will remain inconsistent and subject to large systematic error.

Gravitational instability and dark matter. The observed small-scale inhomogeneity of the universe has presumably evolved from an earlier, more homogenous state through the well-known gravitational instability of cold matter, first discussed in application to the expanding universe by E. M. Lifshitz (1946). Small positive density fluctuations in the early universe would decelerate more rapidly than lower density regions, resulting in a density contrast that grows with time. In the linear perturbation limit with $\Omega = 1$, the density contrasts grow in proportion to $R(t)$ until the contrasts become large, at which time these lumps fragment out of the expansion of the universe, and collapse roughly by a factor of 2 to become virialized objects. These objects can be either galaxies, clusters, or superclusters. Thus a hierarchy of structure can evolve in the universe from small initial fluctuations. The natural expectation is for small structures to separate out of the expansion first and then to clump further into large associations, much as observed for galaxies, clusters, and superclusters.

The order and timing of the hierarchical clustering depends on the nature of the initial spectrum of small density fluctuations established in the very early universe, and this in turn depends on the nature of the dark matter. If $\Omega = 1$, the dark matter cannot be composed of ordinary matter (protons, electrons, and so forth), as this would violate observed constraints on helium and deuterium. Possible candidates for the dark matter are neutrinos with nonzero rest mass, or some other weakly interacting, but heavier, elementary particle. *See Neutrino.*

The standard model of the big bang predicts a neutrino abundance of approximately 10^2/cm^3, and if the neutrinos have a rest mass that exceeds by 6×10^{-5} the electron rest mass, they would close the universe. If neutrinos do dominate the mass density of the universe, the initial density fluctuation spectrum would have been severely damped on scales smaller than supercluster scales. In this event, superclusters would have formed prior to galaxies, which could form only by fragmentation after the large superclusters had collapsed. Detailed numerical simulations suggest that the observed superclusters have formed very recently, leaving no time for galaxy formation. Observations strongly suggest that galaxy formation did not occur in the recent past, thus contradicting the expectations of a neutrino-dominated universe.

Alternative candidates for the dark matter are either primordial black holes or some undiscovered stable elementary particle expected from theories of supersymmetry or grand unified physics. These candidate particles are expected to be much more massive than neutrinos, and there would be no damping of density fluctuations on a small scale in the early universe. In this situation, the hierarchical clustering process will result in the formation of galaxies before the collapse of clusters and superclusters. Detailed numerical simulations show that this model is consistent with $\Omega = 1$ only if galaxies are poor traces of the mass distribution in the universe, so that the voids between superclusters are more deficient of galaxies than they are of matter. Whether this is reasonable depends on complex details of galaxy formation. *See Elementary particle; Grand unification theories*.

Early universe. In 1931 Lemaître proposed that the universe had an explosive beginning with the fission of a primeval atom, which ultimately disintegrated to produce the atoms that make up the universe today. In 1948 G. Gamow, R. A. Alpher, and R. C. Herman proposed that the present elements of the universe were formed by fusion in the big bang. Modern calculations show that elements heavier than helium cannot be produced in material with the combination of temperature and density expected in the big bang, and the current view is that the heavy elements are produced in the cores of stars advanced in their evolution and in supernovae explosions. But otherwise the fusion hypothesis of Gamow and his associates is now accepted. The best current model, called the standard model of the big bang, is that of R. A. Wagoner, W. A. Fowler, and F. Hoyle. *See Nucleosynthesis*.

At an age of 1 s, the temperature was about 10^{10} K (10^{10} °F), and matter and radiation were in equilibrium. In the early universe the temperature varied with time as $t^{-1/2}$, and after a few seconds pair production and annihilation ceased, and the present-day neutrinos were released. After a few minutes fusion ceased, and the only nuclei to survive were those of hydrogen (about 75% by mass), helium, and a trace of deuterium. After a million years, when the temperature had dropped to about 3000 K (5000°F) and the density of protons to 10^3/cm^3, protons and electrons combined to make neutral hydrogen, and the universe became transparent to visible radiation. It is presumed that stars and galaxies formed during the next 10^9 years, although present understanding of the galaxy formation process is poor.

Whereas the formation and early evolution of matter in the universe is very uncertain, and is the subject of considerable theoretical controversy, the general scenario of the evolution of the universe from a hot, dense state has strong support from three very different kinds of observations:

1. The present expansion of the universe, verified by Hubble and Humason in 1931.

2. The present relative abundance of hydrogen and helium. The standard model of the big bang predicts that the ratio by mass should be 3:1, independent of the value of q_o. Within observational uncertainty this is just the ratio observed today, except for small local enhancements of helium that are understood in terms of fusion in stellar interiors.

3. Microwave radiation arriving isotropically from space, and with the spectrum characteristic of that from a black body of temperature 2.7 K (-454.8°F). This cosmic background radiation is described next.

Cosmic background radiation. Imagine a box filled with radiation. If this box expands adiabatically and has size $R(t)$, then the radiation within the box cools with $T(t)$ proportional to $R^{-1}(t)$. This analogy of an adiabatic expansion applies to the universe, even though it has no walls. The observed temperature of the radiation is today 2.7 K (-454.8°F), and when the universe was 1000 times smaller than its present size, the temperature was 2700 K (4400°F). At this temperature and above, the matter would be ionized, and the free electrons would provide enormous opacity for scattering the radiation. Below this temperature, the matter recombines to become neutral, and the opacity drops very suddenly to a negligible value. Thus the microwave radiation detected today was last scattered at the epoch of recombination, when the universe was less than 10^6 years old. *See Heat radiation*.

Cosmic background radiation was predicted independently by a number of workers from before 1950; it was discovered accidentally in 1965 by A. A. Penzias and R. W. Wilson. Penzias and Wilson were attempting to use a large horn antenna, designed for satellite communication, for the absolute calibration of galactic radio sources when they encountered the unexpected background noise. Many subsequent observations at many wavelengths, from high-altitude balloons and a U-2 reconnaissance aircraft (converted by NASA for space research), as well as from the ground, have confirmed the blackbody spectrum of the background radiation.

It is important to understand that there is no "site" for the big bang; it was everywhere. The Earth is in the universe now and has always been surrounded by it; consequently the background radiation comes from all directions in space, and, of course, gives no indication of any center or preferred point of observation. The background radiation is also the present "horizon" of the observable universe. To look farther into space, it would be necessary to look farther back into time, and hence into the opaque fireball which vision cannot penetrate. The background radio radiation is from that last radiating surface before the universe became transparent, and that surface serves as an opaque curtain completely surrounding the Earth. That horizon recedes in time, however, for in the remote future it would be possible to look a farther distance into the past, and thus farther into space.

Steady-state universe. The microwave background is the best evidence that the universe has evolved from a hot dense state, and is generally regarded as specifically ruling out an alternative cosmology advanced by Hoyle, H. Bondi, and T. Gold in 1948. Their steady-state universe was based on the perfect cosmological principle, namely that the universe is the same (on the large scale) not only everywhere, but for all time. As the universe expands, new matter would have to be created to maintain a constant density, and at just the right rate to maintain the same ratio of young and old galaxies. The steady-state universe was eternal, with no specific origin, and could show no hot opaque radiating surface in the past.

Anisotropy. On the small scale (a few minutes of arc) the background radiation is exceedingly isotropic—to better than 3 parts in 10^6—which places a corresponding upper limit on the temperature variations in the emitting matter. Density fluctuations in that material, which must certainly have been present in order to provide the gravitational instabilities required to condense galaxies and clusters, would produce temperature variations within a factor of 30 of this limit, and so it is anticipated that small-scale inhomogeneities will be found when observations of higher sensitivity become possible. *See Anisotropy (physics)*.

A large-scale anisotropy in the background radiation has, however, been observed independently by a group headed by D. T. Wilkinson, who have sent a horn antenna to an altitude of 17 mi (27 km) in a balloon, and by a group headed by G. Smoot, who have flown a similar antenna on a U-2 aircraft at an altitude of 12 mi (20 km). The anisotropy is compatible with a dipole of amplitude 0.003 K (0.005°F) superimposed on an isotropic background, and is generally interpreted as being due to a peculiar motion of the Sun with respect to the background itself of about 200 mi/s (300 km/s). The radiation from the direction in the sky toward which the Sun approaches is Doppler-shifted to slightly shorter wavelengths, relative to the average, and appears as a very slightly hotter blackbody, and that from the opposite direction in the sky is shifted the opposite way, and mimics a slightly cooler blackbody. When account is taken of the rotation of the Galaxy and of its motion in the Local Group, it is found that the Local Group appears to be moving at about 300 mi/s (500 km/s) with respect to the background toward a direction about 40° south of the Virgo cluster. One interpretation is that the Local Supercluster has enough gravitation to slow its expansion somewhat with respect to that of the universe as a whole, so that the Local Group is moving away from the center of the supercluster with a speed a few hundred kilometers per second less than it would if there were no gravitational drag and the Local Group were moving with the general expansion of the universe. *See* Cosmic background radiation.

Inflationary universe. The large-scale homogeneity of the universe is deeply mysterious in the standard big bang, since regions on opposite sides of the sky have never been in causal contact with each other (that is, the light travel time from one point to the other is longer than the age of the universe). Without causal contact there is no physical process by which these regions would be expected to have the same temperature and density, as observed; in the standard model, one must therefore postulate that the initial conditions of the universe were very smooth.

A revolutionary consmological model, termed the inflationary universe, was suggested by A. Guth to be a result of symmetry breaking of grand unification physics in a very early phase of the universe, at an age of 10^{-35} s. The universe could have supercooled while in a false-vacuum state, with its energy dominated by the false vacuum, which acts as a large positive cosmological constant. During this phase, the size of the universe would increase exponentially with time, according to Einstein's general relativity theory. Eventually the universe would make the transition from the false vacuum to the true vacuum, and all the latent heat of the supercooled state would reheat the universe to very high temperatures, creating all the particles present today. In this model, a single vacuum fluctuation inflates to become far larger than the entire observed universe. The universe is homogeneous on a large scale because it was causally connected at a very early time. The curvature of the universe should be very small, and $\Omega = 1$ and $q_o = \frac{1}{2}$ should be measured to high precision. This model unambiguously predicts the spectrum of initial density fluctuations that later grow to become galaxies and clusters of galaxies. The observations at present are consistent with this prediction.

George O. Abell; Marc Davis

Bibliography. J. D. Barrow and J. Silk, *The Left Hand of Creation: The Origin and Evolution of the Expanding Universe*, 1986; J. Bernstein and G. Feinberg, *Cosmological Constants: Papers in Modern Cosmology*, 1986; G. Contopoulos and D. Kotsakis, *Cosmology*, 1987; A. H. Guth and P. J. Steinhardt, The inflationary universe, *Sci. Amer.*, 250(5):116–128, May 1984; P. J. E. Peebles, *Physical Cosmology*, 1972; D. W. Sciama, *Modern Cosmology*, 1982; J. Silk, *The Big Bang*, 1980; S. Weinberg, *The First Three Minutes*, 1977; S. Weinberg, *Gravitation and Cosmology*, 1972.

Cotter pin

A split pin (see **illus.**) formed by doubling a piece of wire semicircular in cross section to form a loop at one end. After insertion in a hole or a slot in a nut and through a mating crosswise hole in a bolt, the ends of the pin are separated and bent to hold it in

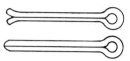

Two of the common forms of cotter pin.

place. This fastener is not adaptable to quick assembly but is widely used in locations where service inspection is difficult and where failure would be disastrous. Cotter pins are similarly used in many other applications to prevent relative rotation or sliding. *See* Nut (engineering).

Paul H. Black

Cotton

A fiber obtained from the cotton plant *Gossypium*, of the order Malvales. It has been used for more than 3000 years. It is the most widely used natural fiber, because of its versatility and comparatively low cost to produce and to manufacture into finished products. Cotton traditionally has been used alone in textile products, but blends with artificial fibers have become increasingly important. *See* Malvales; Natural fiber; Textile.

Unlike most other fibers obtained from plants, the cotton fiber is a single elongated cell. Under the microscope, it resembles a flattened, spirally twisted tube with a rough surface. The fiber cell has many convolutions, with a collapsed inner canal (lumen). Chemically, cotton is essentially pure cellulose. In its original state, cotton contains 3–5% natural waxes

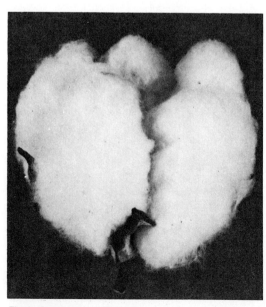

Fig. 1. Open cotton boll ready to harvest.

Fig. 2. Productive field of open cotton. (*K. Bilbrey, County Agent, Mississippi County, Arkansas*)

and gums, mostly in the outer wall of the cells. The natural waxes on the fiber surface act as a finish which facilitates spinning. Cotton is hygroscopic and contains 6–8% moisture under normal atmospheric conditions. *See Cellulose.*

Cultivation and harvesting. To mature, cotton requires about 180 days of continuously warm weather with adequate moisture and sunlight; frost is harmful and may kill the plant. The ground must be thoroughly plowed and the soil pulverized. In the United States, usually in March or April, carefully selected seeds are planted in rows. The plants require systematic fertilization. When they are about 3 in. (7.5 cm) high, they are weeded and thinned. The plants begin to bloom in June or July, about 4 months after planting. Creamy white flowers appear first, change to a reddish-purple in about 2 days, and then fall off, leaving seed pods that grow to full size by August or September. The cotton capsules or bolls must be protected against the boll weevil or other insects. *See Entomology, economic; Fertilizer; Fertilizing; Flower; Fruit; Seed.*

When fully grown, the cotton plant may be 3–6 ft (1–2 m) in height. Its wide green leaves partially conceal some of the bolls until they burst and expose the fleecy white fiber, which indicates that the cotton is ready for harvesting (**Fig. 1**). Not all cotton bolls open at the same time, but the ripening period has been shortened and pickings have been reduced to one or two. When the raw cotton is harvested, it contains seeds, leaf fragments, and dirt that must be removed before baling. The cotton seeds alone constitute approximately two-thirds of the weight of the picked cotton.

Products and processing. When the bolls open, the fiber and seed, or "seed cotton," is harvested mostly by machines. The fiber, or lint as it is then called, is separated from the seed by gins. The lint is compressed into bales, covered with jute bagging, and bound with metal bands for ease of handling. Bales weigh about 500 lb (225 kg) each. The seed (except that portion needed for planting the next crop and for other farm uses) is transferred from the gins, usually by trucks, to oil mills for processing. After baling, the lint is sampled, graded, and sold. The bales then pass through commercial channels and eventually reach cotton mills, where they are broken open and the lint is blended, cleaned, carded, and spun into yarns for many uses. In the oil mill processing industry, the cottonseed is separated into fuzz

or linters, hulls, oil, and protein cake. Each of these products is converted to several subproducts. The oil and cake are the most valuable products. The oil subproducts are largely used as human food; the protein cake, either as cracked cake or ground meal, is used as livestock feed. Hulls are also fed to livestock, and the linters are converted into chemical cellulose. *See Animal feeds; Fat and oil (food).*

Distribution and production. The cotton plant is one of the world's major agricultural crops. It is grown in all countries lying within a wide band around the Earth. The limits of this band in the New World are at about 37°N latitude and at about 32°S latitude. In the Old World the band spreads northward to 47°N in the Ukraine of the Soviet Union, but only to about 30°S in Africa and Australia. In addition to the effects of latitude, suitability of climate for the growth of cotton is also regulated by elevation, wind belts, ocean currents, and mountain ranges. As a result of the effects of these climatic factors, the topography of the land, the nature of the soil, and the availability of irrigation water when needed, cotton culture actually is carried out in an irregular and greatly extended world pattern. The regions of more intensive culture comprise the Cotton Belt of the United States, the northern valleys of Mexico, India, West Pakistan, eastern China, Central Asia, Australia, the Nile River Valley, East Africa, South Africa, northeastern and southern Brazil, northern Argentina, and Peru.

Stimulants to production. The United States has made three major contributions to world cotton production. These were the development of Upland cotton, the invention of the saw gin, and the development of knowledge of cotton culture.

Upland cotton arose from an annual form of *G. hirsutum* native to the plateaus of southern Mexico. This species was introduced into what is now southern United States at different times from the colonial period to the first decade of the twentieth century. The name Upland is derived from the upland country in southeastern United States, where this stock was first grown commercially. This cotton was shipped to England under the name of American Upland. From the time of its early development as a crop in the American Cotton Belt, it has been very hardy and productive, and versatile in its many uses (**Fig. 2**).

Upland production was handicapped at first by the

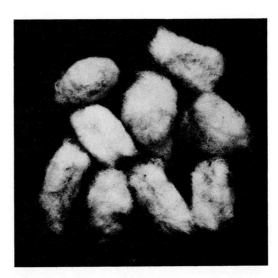

Fig. 3. Ginned seed of Upland cotton, showing dense and tight covering of short fuzz hairs on seed coat. (*J. O. Ware, Journal of Heredity*)

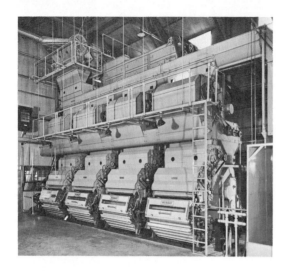

Fig. 4. Modern cotton gin, showing the four stands and the accessory equipment for cleaning and conveying the cotton. (*Murray Company of Texas*)

difficulty of ginning. The seed coat has a tight covering of short fuzz hairs (**Fig. 3**). Only sawlike teeth can penetrate and remove the longer coat of lint from the seed and leave the fuzz behind with the seed coat undamaged. The saw gin, incorporating this principle, was gradually developed into the modern gin plant (**Fig. 4**).

In the history of cotton culture, simple but efficient farm tools were developed and the mule was generally the source of farm power. The cotton fields were well tilled, the weeds controlled, the cotton carefully picked by hand, and the best seed selected for the next crop. However, when mechanical power began to replace animal power, the United States quickly mechanized both the cultural and harvesting phases of cotton production. (**Figs. 5** and **6**). Great progress in research, plant breeding, technology, and mechanics paved the way for advancements when economic conditions were ready for their adoption. Upland cotton and some of the improvements made in handling this crop spread to most cotton-growing countries as they began commercial production. Upland cotton provides about 85% of the world supply of raw cotton for factory use. SEE AGRICULTURAL SCIENCE (PLANT); BREEDING (PLANT).

The other 15% of world production is supplied from three other cultivated species: *G. barbadense, G. arboreum,* and *G. herbaceum.* The extra-long staple type includes Egyptian, Tanguis, American-Pima, and Sea Island. Egyptian is grown in Egypt and the Sudan, Tanguis in Peru, American-Pima in southwestern United States, and Sea Island in the West Indies. Sea Island cotton formerly was grown on islands and coastal areas of the southeastern United States.

Fig. 5. Planting cotton with a four-row tractor-propelled planter. (*John Deere Co.*)

Gossypium arboreum and *G. herbaceum* are commonly referred to as Asiatic cottons and include the shortest-staple types of cultivated cottons. The growth of the former is largely confined to India and China and of the latter to Central Asia, western India, and the Near East. The saw gins usually used for Upland cotton in other countries are simpler and carry less cleaning equipment than those used in the United States. Most other countries still employ much handwork in cotton culture; most of the crop is harvested by hand.

Most commercial cottons are now annuals. However, in the tropics some races of the regular cultivated species are perennial. All cottons at one time were tropical and perennial.

Wild cotton. Besides the four cultivated species, *Gossypium* includes about 30 wild and lintless species that are widely scattered and occur mostly in sparsely covered desert and tropical areas of the world. All of

Fig. 6. Two-row, self-propelled cotton picker in operation. (*International Harvester Co.*)

the wild species may ultimately be of plant breeding interest. Some of them already have been used in crosses to improve quality values in cultivated cottons.

Elton G. Nelson

Diseases. Pathogens may attack or invade the cotton plant through the roots or aboveground parts. Diseases reduce the yield potential 15 to 20%, and also have undesirable effects on quality of fiber and seed.

Root diseases are caused by soil-borne organisms. Severity of disease is proportional to the number of infective pathogen propagules in soil and to the degree of susceptibility in varieties. Minimizing losses involves using resistant varieties and practices that reduce pathogen numbers in the soil.

The seedling disease complex occurs wherever cotton is grown. Symptoms are seed rot, preemergence damping-off, postemergence damping-off, and damage to roots of plants that survive. The primary organisms are the fungi *Rhizoctonia solani, Pythium* spp., *Thielaviopsis basicola,* and *Glomerella gossypii* (**Fig. 7**), and the root-knot nematode *Meloidogyne incognita.*

Using high-quality seed treated with protectant fungicides is essential to the control of seedling diseases. Application of soil fungicides at planting is an additional practice that gives more effective control. Cotton varieties with resistance to seedling pathogens, preservation of seed quality, and with improved performance under early-season cool-wet conditions are now available.

Although the wilt-inducing fungus *Fusarium oxysporum* f. *vasinfectum* or the root-knot nematode *M. incognita* (Fig. 9*c*) can damage cotton in single infections, losses due to double infections are greater. Management practices effective against both organisms must therefore be used. The most economical practice is the use of resistant varieties, but it is best to use resistant varieties and rotations with root-knot–resistant crops such as small grains, corn, sorghum, or some grasses.

Verticillium wilt, caused by *Verticillium alboatrum*, occurs in the cooler regions of the world (**Fig. 8**). Effective control can be achieved with integrated management. High populations of 25 to 30 plants per meter of row help reduce losses. Too much nitrogen causes wilt to be more severe. Serious losses are usually limited to irrigated cotton, cotton in high-rainfall regions, and in poorly drained areas of fields. Reducing the amount of irrigation water by one-third will often give higher yields. The wilt fungus survives as

Fig. 8. Verticillium wilt of young cotton plant.

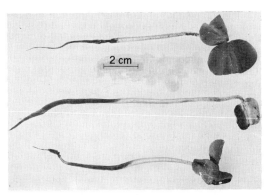

Fig. 7. Diseased cotton seedlings which have been affected by *Glomerella gossypii*.

microsclerotia in the soil and in undecomposed plant debris. Resistant crops such as barley and sorghum in rotation with cotton have been effective in reducing the inoculum potential. Many resistant varieties are available, and their use, along with other practices that reduce disease severity, gives effective control.

Root rot, caused by *Phymatotrichum omnivorum*, occurs in alkaline soils in the southwestern and western regions of the United States and in Mexico. The soil in infested areas has more sclerotia, lower bacterial populations, less organic matter, and less sodium than other soils. Control depends on reducing sclerotia production and preventing survival of the fungus between cropping seasons.

Burial of barnyard manure or residues of such legumes as Hubam clover and winter peas helps control the disease. Turning the upper 8–12 in. (20 to 30 cm) of soil with a moldboard plow a few days after harvesting a crop of cotton reduces sclerotia production. Rotation of cotton with grain crops also helps to keep numbers of sclerotia low.

Planting early in the season with fast-maturing varieties permits early boll set and maturation of the crop before soil temperatures reach levels for the pathogen to become highly active. Restoration of sodium to a level equal to that in noninfested soil pre-

vents production of new sclerotia, and is now being used for control. Best control comes from using short-season management with practices that reduce sclerotia production and survival.

Diseases of aboveground plant parts may be caused by bacteria or fungi. These organisms may survive in undecomposed plant debris, on alternative hosts, or in and on seed. Sanitation of planting-seed and shredding and burial of residues from a cotton crop reduce the ability of foliage pathogens to survive the winter.

Bacterial blight, caused by *Xanthomonas malvacearum*, affects all plant parts and occurs wherever cotton is grown (**Fig. 9***a* and *b*). The bacterium sur-

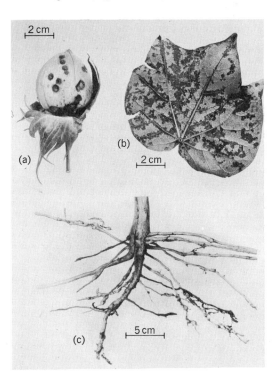

Fig. 9. Bacterial blight on cotton plant. (*a*) Bacterial blight lesions on bolls. (*b*) Bacterial blight lesions on leaves. (*c*) Root-knot galls on cotton roots.

vives in and on seed and as dried exudates on unde-composed plant debris. Chemical delinting removes the bacterium from the seed surface, but no treatment is available for eliminating internal transmission. Using seed from disease-free fields ensures no transmission. Management to ensure debris decomposition eliminates crop residue as a means of transmission. Varieties with high horizontal resistance to the 18 races of the pathogen are available for many cotton-growing regions of the world.

Boll rot may be caused by a number of fungi, bacteria, or yeasts. Some, such as *G. gossypii* and *X. malvacearum*, are primary pathogens. Many gain entrance through natural openings such as boll sutures and nectaries; others enter through wounds caused by insects or primary pathogens.

Control hinges on using practices that reduce inoculum densities and promote dryness within the leaf canopy. Varieties with okra-shaped leaves and frego bracts (which are strap-shaped leaves) are significant in reducing boll rot. Sanitation should be practiced to reduce transmission on seed and survival of organisms on debris. See PLANT PATHOLOGY.

Luther S. Bird

Bibliography. M. E. Selsom, *Cotton*, 1982; G. M. Watkins, *A Compendium of Cotton Diseases*, American Phytopathology Society, 1981.

Cotton effect

The characteristic wavelength dependence of the optical rotatory dispersion curve or the circular dichroism curve or both in the vicinity of an absorption band.

When an initially plane-polarized light wave traverses an optically active medium, two principal effects are manifested: a change from planar to elliptic polarization, and a rotation of the major axis of the ellipse through an angle relative to the initial direction of polarization. Both effects are wavelength dependent. The first effect is known as circular dichroism, and a plot of its wavelength (or frequency) dependence is referred to as a circular dichroism (CD) curve. The second effect is called optical rotation and, when plotted as a function of wavelength, is known as an optical rotatory dispersion (ORD) curve. In the vicinity of absorption bands, both curves take on characteristic shapes, and this behavior is known as the Cotton effect, which may be either positive or negative (**Fig. 1**). There is a Cotton effect associated with each absorption process, and hence a partial CD curve or partial ORD curve is associated with each

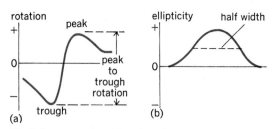

Fig. 2. Curves used to determine relative rotatory intensities. (a) Partial ORD curve. (b) Partial CD curve.

particular absorption band or process. See POLARIZED LIGHT; OPTICAL ROTATORY DISPERSION.

Measurements. Experimental results are commonly reported in either of two sets of units, termed specific and molar (or molecular). The specific rotation $[\alpha]$ is the rotation in degrees produced by a 1-decimeter path length of material containing 1 g/ml of optically active substance, and the specific ellipticity θ is the ellipticity in degrees for the same path length and same concentration. Molar rotation $[\varphi]$ (sometimes $[M]$) and molar ellipticity $[\theta]$ are defined by Eqs. (1) and (2). For comparisons among different

$$[\varphi] = [\alpha]M/100 \qquad (1)$$

$$[\theta] = \theta M/100 \qquad (2)$$

compounds, the molar quantities are more useful, since they allow direct comparison on a mole-for-mole basis.

The ratio of the area under the associated partial CD curve to the wavelength of the CD maximum is a measure of the rotatory intensity of the absorption process. Moreover, for bands appearing in roughly the same spectral region and having roughly the same half-width (**Fig. 2**), the peak-to-trough rotation of the partial ORD curve is roughly proportional to the wavelength-weighted area under the corresponding partial CD curve. In other words, relative rotatory intensities can be gaged from either the pertinent partial ORD curves or pertinent partial CD curves. A convenient quantitative measure of the rotatory intensity of an absorption process is the rotational strength. The rotational strength R_i of the ith transition, whose partial molar CD curve is $[\theta_i(\lambda)]$, is given by relation (3).

$$R_i \approx 6.96 \times 10^{-43} \int_0^\infty \frac{[\theta_i(\lambda)]}{\lambda} \, d\lambda \qquad (3)$$

Molecular structure. The rotational strengths actually observed in practice vary over quite a few orders of magnitude, from $\sim 10^{-38}$ down to 10^{-42} cgs and less; this variation in magnitude is amenable to stereochemical interpretation. In this connection it is useful to classify optically active chromophores, which are necessarily dissymmetric, in terms of two limiting types: the inherently dissymmetric chromophore, and the inherently symmetric but dissymmetrically perturbed chromophore. See OPTICAL ACTIVITY.

A symmetric chromophore is one whose inherent geometry has sufficiently high symmetry so that the isolated chromophoric group is superimposable on its mirror image, for example, the carbonyl group $\mathrm{>C{=}O}$. The transitions of such a chromophore can become optically active, that is, exhibit a Cotton effect, only when placed in a dissymmetric molecular

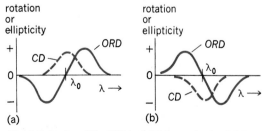

Fig. 1. Behavior of the ORD and CD curves in the vicinity of an absorption band at wavelength λ_0 (idealized). (a) Positive Cotton effect. (b) Negative Cotton effect.

environment. Thus, in symmetrical formaldehyde, $H_2C=O$, the carbonyl transitions are optically inactive; in ketosteroids, where the extrachromophoric portion of the molecule is dissymmetrically disposed relative to the symmetry planes of the $>C=O$ group, the transitions of the carbonyl group exhibit Cotton effects. In such instances the signed magnitude of the rotational strength will depend both upon the chemical nature of the extrachromophoric perturbing atoms and their geometry relative to that of the inherently symmetric chromophore. In a sense, the chromophore functions as a molecular probe for searching out the chemical dissymmetries in the extrachromophoric portion of the molecule.

The type of optical activity just described is associated with the presence of an asymmetric carbon (or other) atom in a molecule. The asymmetric atom serves notice to the effect that, if an inherently symmetric chromophore is present in the molecule, it is almost assuredly in a dissymmetric environment, and hence it may be anticipated that its erstwhile optically inactive transitions will exhibit Cotton effects. Moreover, the signed magnitude of the associated rotational strengths may be interpreted in terms of the stereochemistry of the extrachromophoric environment, as compared with that of the chromophore. But an asymmetric atom is not essential for the appearance of optical activity. The inherent geometry of the chromophore may be of sufficiently low symmetry so that the isolated chromophore itself is chiral, that is, not superimposable on its mirror image, for example, in hexahelicene.

In such instances the transitions of the chromophore can manifest optical activity even in the absence of a dissymmetric environment. In addition, it is very often true that the magnitudes of the rotational strengths associated with inherently dissymmetric chromophores will be one or more orders of magnitude greater ($\sim 10^{-38}$ cgs, as opposed to $< 10^{-39}$ cgs) than those associated with inherently symmetric chromophores. In the spectral regions of the transitions of the inherently dissymmetric chromophore, it will be the sense of handedness of the chromophore itself that will determine the sign of the rotational strength, rather than perturbations due to any dissymmetric environment in which the inherently dissymmetric chromophore may be situated.

The sense of handedness of an inherently dissymmetric chromophore may be of considerable significance in determining the absolute configuration or conformations of the entire molecule containing that chromophore. Accordingly, the absolute configuration or conformation can often be found by focusing attention solely on the handedness of the chromophore itself. For example, in the chiral molecule shown in **Fig. 3** there is a one-to-one correspondence between the sense of helicity of the nonplanar diene chromophore present and the absolute configuration at the asymmetric carbon atoms. Hence there exists a one-

Fig. 3. Structural formula of (+)-*trans*-9-methyl-1,4,9,10-tetrahydronaphthalene.

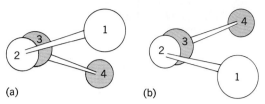

Fig. 4. Schematic representation of the twisted diene chromophore showing the two possible handednesses; the numbering is as indicated in Fig. 3. (a) Right-handed. (b) Left-handed.

to-one correspondence between the handedness of the diene and the absolute configuration of the molecule. Since it is known that a right-handed diene helix (**Fig. 4**) associates a positive rotational strength with the lowest diene singlet transition in the vicinity of 260 nanometers, by examination of the pertinent experimental Cotton effect (positive), the absolute configuration of the molecule is concluded to be as shown.

Other examples of inherently dissymmetric chromophores are provided by the helical secondary structures of proteins and polypeptides. Here the inherent dissymmetry of the chromophoric system arises through a coupling of the inherently symmetric monomers, which are held in a comparatively fixed dissymmetric disposition relative to each other through internal hydrogen bonding. The sense of helicity is then related to the signs of the rotational strengths of the coupled chromophoric system. The destruction of the hydrogen bonding destroys the ordered dissymmetric secondary structure, and there is a concomitant decrease in the magnitude of the observed rotational strengths.

Albert Moscowitz

Bibliography. L. D. Barron, *Molecular Light Scattering and Optical Activity*, 1983; S. F. Mason, *Molecular Optical Activity and the Chiral Discriminations*, 1982; *Proc. Roy. Soc. London*, ser. A, 297:1–172, 1976; L. Velluz et al., *Optical Circular Dichroism: Principles, Measurements, and Applications*, 1969.

Coulomb excitation

A process in which two atomic nuclei that approach each other undergo transitions to excited states that are caused by the long-range Coulomb forces acting between the nuclei. The process can occur even when the nuclei do not come sufficiently close to allow the strong short-range nuclear forces to act.

That nuclei can be excited by this process is not so important in itself. What has proved to be enormously fruitful is the fact that experimental results based on this process have provided accurate values for electrical or electromagnetic properties of nuclei. To take an easily visualized example, Coulomb excitation measurements were instrumental in showing that many nuclear species are not spherical but are, instead, shaped like a football. *See* COULOMB'S LAW; NUCLEAR STRUCTURE.

Experiments. The field of experimental Coulomb excitation began in the mid-1950s. Early experiments involved bombarding targets with beams of protons or alpha particles obtained from accelerators such as Van de Graaffs, cyclotrons, or linacs. Nuclei are made of protons and neutrons; the positively charged protons cause an intense electrostatic (Coulomb) field to surround nuclei. Positively charged bombarding particles, such as protons or alpha particles, must have a large kinetic energy in order to overcome the strong Coulomb repulsion of nuclei and thereby reach the close distances required for the attractive nuclear forces to act.

Coulomb excitation was discovered when protons of much too small an energy to overcome this Coulomb repulsion of the target nuclei were observed to still cause nuclear excitation. The occurrence of nuclear excitation was confirmed by detecting the gamma rays emitted from the target when the nuclei decayed back down to the ground state. SEE ALPHA PARTICLES; PARTICLE ACCELERATOR; PROTON.

Although the first Coulomb excitation measurements were carried out with beams of protons or alpha particles, it was soon realized that the use of beams of more highly charged projectiles would dramatically increase the probabilities for Coulomb excitation. Experiments have been done with beams of ^{16}O, ^{20}Ne, ^{40}Ar, ^{84}Kr, and ^{136}Xe ions. Coulomb excitation with these heavy ions is characterized by the term "multiple excitation." Many more excited states of nuclei are appreciably populated with these heavy-ion beams.

Theory and interpretations. The theory of pure Coulomb excitation is well understood, but mathematically complicated. The modern approach uses computer programs to generate specific theoretical results applicable to a particular experimental situation.

In order to interpret Coulomb excitation results precisely, it is important that the influence of nuclear forces be negligible so that a "pure" Coulomb excitation situation exists. Of course, at higher bombarding energies, where nuclear forces are an important part of the reaction, Coulomb excitation is still present; and, in fact, the combination of Coulomb and nuclear forces produces an interesting reaction situation which has been the subject of much research. SEE NUCLEAR REACTION; SCATTERING EXPERIMENTS (NUCLEI).

Measurement and applications. Coulomb excitation can be measured either by the direct analysis of the energy spectrum of the scattered projectiles or by the analysis of the gamma rays subsequently emitted by the excited nuclei. Both methods are used extensively. The chief advantage of a direct measurement of the spectrum of the scattered projectiles is the good accuracy with which excitation probabilities can be determined. A high degree of accuracy is valuable in determining such nuclear properties as static quadrupole moments of excited states, the existence and magnitude of hexadecapole or E4 moments, and the possibility of centrifugal stretching of nuclei.

The chief advantages offered by the detection of gamma rays following Coulomb excitation are (1) the excellent energy resolution (approximately 2 keV) of Ge(Li) gamma ray detectors can be used to study the excited nuclear states; (2) good statistical accuracy (high counting rates) is achieved in much shorter times on the accelerator; and (3) more extensive information on the properties of the nuclear states can be extracted from gamma ray measurements. From measurements of angular distributions of the gamma rays, spin-parity values can be assigned to excited states, and information about the strength of magnetic dipole transitions in nuclei can be gained. SEE GAMMA RAYS; NUCLEAR SPECTRA.

Heavy-ion collisions. One of the important frontiers of nuclear physics research is the use of heavy-ion projectiles; producing collisions between large complex nuclei allows interesting questions to be asked about the behavior and structure of nuclei. In such collisions, the Coulomb interaction is always important. Thus it is likely that the Coulomb excitation process will continue to play a vital role in nuclear physics research.

Paul H. Stelson

Bibliography. B. I. Bleaney and B. Bleaney, *Electricity and Magnetism*, 3d ed., 1989; Coulomb's Law Committee, *Amer. J. Phys.*, 18:6–11, 1950; E. M. Purcell, *Electricity and Magnetism*, Berkeley Physics Course, vol. 2, 1965; E. M. Pugh and E. W. Pugh, *Principles of Electricity and Magnetism*, 2d ed., 1976.

Coulomb explosion

A process in which a molecule moving with high velocity strikes a solid and the electrons that bond the molecule are torn off rapidly in violent collisions with the electrons of the solid; as a result, the molecule is suddenly transformed into a cluster of charged atomic constituents that then separate under the influence of their mutual Coulomb repulsion. The initial velocity of the molecule is typically greater than 3×10^6 ft/s (10^6 m/s), and it takes on the order of 10^{-17} s for electrons to be torn off the molecule. Typically, it takes about 10^{-15} s for the initial Coulomb potential energy of the cluster to be converted into kinetic energy as the charged fragments recede from one another. SEE COULOMB'S LAW.

Coulomb explosions are most commonly studied using a particle accelerator, normally employed in nuclear physics research (Van de Graaff generator, cyclotron, and so forth), to produce a beam of fast molecular ions that are directed onto a solid-foil target. The Coulomb explosion of the molecular projectiles begins within the first few tenths of a nanometer of penetration into the foil, continues during passage of the projectiles through the foil, and runs to completion after emergence of the projectiles into the vacuum downstream from the foil. Detectors located downstream make precise measurements of the energies and charges of the molecular fragments together with their angles of emission relative to the beam direction. The Coulomb explosion causes the fragment velocities to be shifted in both magnitude and direction from the beam velocity. The corresponding shifts in energy and angle are small, but if the foil target is thin (approximately 10 nm) and of light material (for example, carbon), the blurring effects of energy-loss straggling and multiple scattering in the foil can be kept small relative to the Coulomb explosion effects. SEE CHARGED-PARTICLE BEAMS; PARTICLE ACCELERATOR.

Consider a beam of 3-MeV HeH$^+$ ions incident on a 10-nm-thick carbon foil. Upon striking the foil, each projectile produces a 2.4-MeV alpha particle and a 600-keV proton separated by about 80 picometers (the bond length for HeH$^+$). The Coulomb explosion causes the separation to grow to about 100 pm during traversal of the foil. Downstream, the fragments achieve asymptotic shifts in energy and angle that are determined by the initial orientation of the molecule. The maximum energy shift (± 8.4 keV) is obtained when the internuclear vector in the projectile

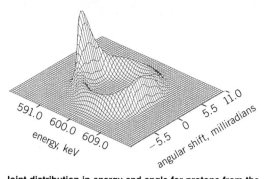

Joint distribution in energy and angle for protons from the dissociation of 3-MeV HeH$^+$ ions incident on a 19.5-nm-thick carbon foil.

is parallel to the beam direction. There is then no angular shift. The sign of the energy shift for a given fragment depends on whether it is leading or trailing its partner. If the internuclear vector is perpendicular to the beam, the maximum angular shift (0.4° for the proton, 0.1° for the alpha particle) is achieved, and there is no shift in energy. A joint distribution in energy and angle of protons from similar dissociations (with a 19.5-nm-thick foil) is shown in the **illustration**. The width of the "rim" reflects variations in the internuclear distance of the incident molecular ions due to their vibrational motion.

Coulomb explosion experiments of this type serve two main purposes. First, they yield valuable information on the interactions of fast ions with solids. For example, it is known that a fast ion generates a polarization wake that trails behind it as it traverses a solid. This wake can be studied in detail by using diatomic molecular-ion beams, since the motion of a trailing fragment is influenced not only by the Coulomb explosion but also by the wake of its partner (rather like the way a boat is affected by the wake of another in front of it). The nonuniform distribution of intensity around the ring in the illustration can be accounted for quantitatively in terms of forces due to the polarization wakes. Second, Coulomb-explosion techniques can be used to determine the stereochemical structures of molecular-ion projectiles. For example, with this method it was demonstrated experimentally for the first time that the H$_3^+$ molecule is equilateral-triangular. SEE ELECTRON WAKE; MOLECULAR STRUCTURE AND SPECTRA; STEREOCHEMISTRY.

Donald S. Gemmell

Bibliography. D. S. Gemmell, Determining the stereochemical structures of molecular ions by "Coulomb-explosion" techniques with fast (MeV) molecular ion beams, *Chem. Rev.,* 80:301–311, 1980; Z. Vager and D. S. Gemmell, Polarization induced in a solid by the passage of fast charged particles, *Phys. Rev. Lett.,* 37:1352–1354, 1976.

System (SI) of units, $k_0 = 1/(4\pi\epsilon_0)$, where ϵ_0 is called the permittivity of empty space and has the value $\epsilon_0 = 8.85 \times 10^{-12}$ farad/m. Thus, Coulomb's law is as in the equation below, where q_1 and q_2 are expressed in

$$ F = \frac{1}{4\pi\epsilon_0} \frac{q_1 q_2}{r^2} $$

coulombs, r is expressed in meters, and F is given in newtons. SEE ELECTRICAL UNITS AND STANDARDS.

The direction of F is along the line of centers of the point charges q_1 and q_2, and is one of attraction if the charges have the same sign. For a statement of Coulomb's law as applied to point magnet poles SEE MAGNETOSTATICS.

Experiments have shown that the exponent of r in the equation is very accurately the number 2. Ernest (Lord) Rutherford's experiments, in which he scattered alpha particles by atomic nuclei, showed that the equation is valid for charged particles of nuclear dimensions down to separations of about 10^{-12} cm. Nuclear experiments have shown that the forces between charged particles do not obey the equation for separations smaller than this.

The direct force that one charged particle exerts on another is unaffected by the presence of additional charge and, in any electrostatic system, the equation gives this direct force between q_1 and q_2 under any conditions of charge configuration, including that in which intervening and surrounding matter is present and the molecules of the matter are polarized so that their charges contribute to this configuration. The total force on any one charge, say q_1, is the vector sum of the separate direct forces on q_1 due to q_2, q_3, q_4, and so on, each force computed separately by use of the equation as if all other charges were absent.

The permittivity ϵ of a medium is defined by $\epsilon = \epsilon_r \epsilon_0$, where ϵ_r is the relative permittivity of the medium. It was formerly known also as the relative dielectric constant or specific inductive capacity. SEE PERMITTIVITY.

If two free point charges q_1 and q_2 are immersed in an infinite homogeneous isotropic dielectric, the total force on one of them, say q_1, is given by $F = q_1 q_2 / 4\pi\epsilon r^2$ and the use of ϵ (in place of ϵ_0) takes proper account of the forces on q_1 due to the polarization charges of the dielectric molecules. It is only in a few special cases that this latter formulation of Coulomb's law is valid. SEE ELECTRIC CHARGE; ELECTROSTATICS.

Ralph P. Winch

Bibliography. B. I. Bleaney and B. Bleaney, *Electricity and Magnetism*, 3d ed., 1989; Coulomb's Law Committee, *Amer. J. Phys.*, 18:6–11, 1950; E. M. Purcell, *Electricity and Magnetism*, Berkeley Physics Course, vol. 2, 2d ed., 1985; E. M. Pugh and E. W. Pugh, *Principles of Electricity and Magnetism*, 2d ed., 1976.

Coulomb's law

For electrostatics, Coulomb's law states that the direct force F of point charge q_1 on point charge q_2, when the charges are separated by a distance r in free space, is given by $F = k_0 q_1 q_2 / r^2$, where k_0 is a constant of proportionality whose value depends on the units used for measuring F, q, and r. It is the basic quantitative law of electrostatics. In the International

Coulometer

Electrolysis cell in which a product is obtained with 100% efficiency as a result of an electrochemical reaction. The quantity of electricity, that is, the number of coulombs of electricity (Q), can be determined very accurately by weighing the product that is deposited on an electrode in the course of the electrochemical reaction. The relationship between the

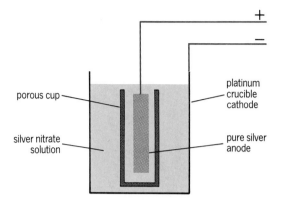

Fig. 1. Silver coulometer. The porous cup catches particles of silver that fall off the anode.

weight of the product formed in the coulometer and the quantity of electricity used is given by Faraday's laws of electrolysis. When a constant current of i amperes flows through the electrolyte in the coulometer for t seconds, the number of coulombs passed is given by Eq. (1). If the current varies in the course of the

$$Q = it \tag{1}$$

electrolysis, the simple current-time product in Eq. (1) is replaced by the current-time integral, Eq. (2).

$$Q = \int_0^t i\, dt \tag{2}$$

When Q coulombs of electricity are passed through the electrolyte, the weight in grams of the material that is deposited on the electrode (w) is given by Eq. (3), where n is the number of electrons transferred per

$$w = \frac{QM}{Fn} \tag{3}$$

mole of material deposited, M is its molecular weight, and F is the Faraday constant, $96{,}487 \pm 1.6$ coulombs.

Equation (3) is fundamental in coulometry and is a mathematical statement of Faraday's laws. This equation is used for the accurate determination of Q, the current-time integral, by weighing or measuring a product that is formed at an electrode by an electrochemical reaction that occurs with 100% current efficiency. The electrolysis cell that is used for this purpose is a coulometer.

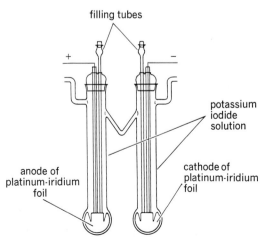

Fig. 2. Iodine coulometer.

Only a few electrode reactions proceed with the 100% current efficiency that is required for the use of Eq. (3). The deposition of silver or copper (in a silver or copper coulometer), the evolution of oxygen and hydrogen (in a gas coulometer), and the oxidation of iodide to iodine (in an iodine coulometer) are examples of electrode reactions that have been successfully employed. One coulomb of electricity will deposit 1.1180 mg of silver at the cathode in a silver coulometer (**Fig. 1**) or liberate 1.315 mg of iodine at the anode in an iodine coulometer (**Fig. 2**). Although these classical chemical coulometers are capable of measuring the quantity of electricity with high precision and accuracy, their use is time-consuming and inconvenient; and they have been largely replaced by operational amplifier integrator circuits or digital circuits that display in a direct readout the number of coulombs passed during electrolysis. See Coulometric analysis; Electrochemical equivalent; Electrolysis.

Quintus Fernando

Bibliography. S. A. Borman (ed.), *Instrumentation in Analytical Chemistry*, 1982; G. W. Milner and G. Phillips, *Coulometry in Analytical Chemistry*, 1967.

Coulometric analysis

An electroanalytical method that is based on the accurate measurement of the quantity of electricity that passes through a solution during the course of an oxidation or reduction reaction occurring at an electrode. The quantity of electricity or the net charge transferred (Q) can be calculated from Faraday's laws of electrolysis and is related to the amount of material that is produced or consumed at an electrode. See Coulometer; Electrochemical techniques; Electrolysis.

In Eq. (1), Q is given by the integral of the current

$$Q = \int_0^t i\, dt = \frac{nwF}{M} \tag{1}$$

(i, amperes) that flows through the electrolytic cell over the time of electrolysis (t, seconds); w is the weight in grams of a species, having molecular weight M, that is produced or consumed during the electrolysis; n is the number of electrons transferred per mole; and F is the Faraday constant equal to $96{,}487 \pm 1.6$ coulombs. If Q is measured accurately, then w can be determined directly from Eq. (1), provided that the electrolysis reaction occurs with a known stoichiometry and 100% current efficiency. See Electrochemical equivalent; Stoichiometry.

If the species of interest is either oxidized or reduced at an electrode during the electrolysis, the electroanalytical technique is a primary coulometric analysis; if the material that is electrolytically generated reacts quantitatively with the species of interest in solution, the technique is a secondary coulometric analysis, often referred to as a coulometric titration. All coulometric analyses, however, can be considered to be coulometric titrations in which the titrants are electrons. Coulometric titrations can be used for analytical determinations in which acid-base, precipitation, complexometric, and redox reactions are employed.

Coulometric titrations have a number of important advantages over conventional titrations in which a standard solution of a titrant is dispensed from a buret. A few micrograms of a compound can be determined coulometrically with great accuracy because the quantity of electricity (Q) that corresponds to 1 microgram can be measured easily, whereas conventional titrimetry would require the use of ultramicro volumetric techniques. It is not necessary to prepare and store standard solutions or use primary standards. There are no dilution effects in coulometric titrations. Volatile or gaseous titrants such as chlorine, bromine, and iodine or unstable titrants such as copper(I) or chromium(II) can be readily generated at an electrode. Coulometric titrations are often used for the determination of hazardous materials because they are much simpler to automate and perform by remote control than conventional titrations. *SEE TITRATION*.

Methods. Coulometric analyses may be divided into two groups: controlled-potential coulometry, in which the applied potential is controlled, and controlled-current coulometry (also known as constant-current coulometry), in which the applied current is controlled and maintained at a constant value. In controlled-current coulometry, the quantity of electricity or the net charge transferred (Q) can be evaluated if the current and the elapsed time for the electrolysis are known. Both these quantities can be measured very accurately. It is difficult, however, to ensure that a single reaction occurs with 100% current efficiency during the course of the electrolysis. As the electrode reaction proceeds, the concentration of the electroactive species decreases; eventually, the migration of this species to the electrode is insufficient to maintain the required current, and some other electrolytic reaction occurs to maintain the current at a constant value. The current efficiency, therefore, decreases significantly from 100%. For this reason the controlled-current method is seldom used in primary coulometric analysis; the controlled-potential method is preferred.

Controlled-potential analysis. In controlled-potential coulometry, the species that is being determined is reduced or oxidized at the working electrode. The potential of the working electrode is maintained at a desired value with the aid of a potentiostat, and extraneous electrode reactions are thereby prevented from occurring. As the desired electrode reaction proceeds at the working electrode, the current decreases exponentially, from a relatively high initial value to a very low value at the end of the reaction. Therefore the value of the net charge transferred (Q) has to be obtained by an integration method. Modern electronic current-time integrators with an accuracy of at least 0.01% are used for this purpose.

The design of the electrolysis cell in which the controlled-potential coulometric analysis is carried out is optimized to facilitate the electrode reaction. Three electrodes are used in the electrolysis cell: a working electrode with a large area, a counter electrode to complete the electrolysis circuit, and a reference electrode for the measurement of the potential for the working electrode (**Fig. 1**). The solution volume in the cell is small, and it is stirred steadily to replenish the material that is oxidized or reduced at the electrode surface. In controlled-potential coulometry, no end-point detection system is required; the electrolysis is terminated when the current has decreased to about 0.1% of the initial current. The sensitivity of the method is limited by the background current that is observed in the presence of the supporting electrolyte alone. A correction for the background current can be applied by carrying out a blank coulometric analysis.

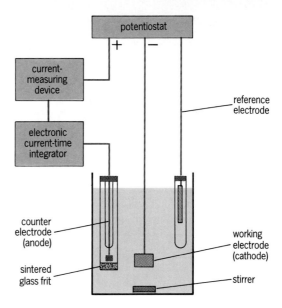

Fig. 1. Schematic representation of the apparatus for controlled-potential coulometry.

Controlled-potential coulometry has all the advantages of electrodeposition methods. Moreover, it is not necessary that a weighable solid product be deposited at an electrode. The reaction product formed at the electrode can be soluble, for example, the reduction of iron(III) to iron(II); or it can be gaseous, for example, the oxidation of hydrazine to nitrogen. *SEE ELECTRODEPOSITION ANALYSIS*.

Controlled-current analysis. Controlled-current coulometry has been used extensively in secondary coulometric analyses in which an intermediate species is generated electroanalytically and allowed to react quantitatively with a species of interest in solution. Stable constant-current sources are relatively easy to construct, and the quantity of electricity (Q) can be determined accurately by measurement of the current and time of electrolysis. The equipment commonly used in controlled-current coulometry is shown in **Fig. 2**. The platinum generator electrode with a large surface area (0.8–1.6 in.^{2} or 5–10 cm^2) and the isolated auxiliary electrode are connected to a constant-current source. The end point is detected potentiometrically or amperometrically with a pair of electrodes or by any other appropriate end-point detection method.

An example of this type of coulometry is the reduction of cerium(IV) to cerium(III) in a sulfuric acid solution under controlled-current conditions. As the electrolysis reaction proceeds, the cerium(IV) is con-

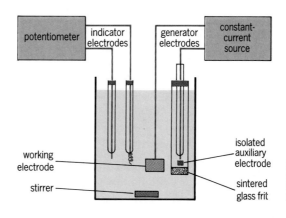

Fig. 2. Schematic representation of the apparatus for controlled-current coulometry.

tinuously depleted; at some point, another reduction process must occur to maintain the current at the controlled value. The reaction that takes place at the electrode in the presence of acid is the reduction of hydrogen ions to hydrogen gas. When this occurs electricity is consumed, and the relationship between Q and the number of moles of cerium(IV) in solution is no longer valid. If, however, some iron(III) is added to the solution containing the cerium(IV), the reduction of iron(III) to iron(II) will occur in preference to the reduction of hydrogen ions. The iron(II) that is formed will be instantaneously oxidized to iron(III) by the cerium(IV); and the titration efficiency for the reduction of cerium(IV) to cerium(III) is 100%, although only a part of the electricity consumed is used in generating the intermediate titrant. In many instances, all of the electricity is used to generate the intermediate titrant, for example, the oxidation of bromide ions to bromine, which then reacts quantitatively with a species that is to be determined in solution.

Virtually all conventional types of titrations have been performed coulometrically. Acid-base titrations can be carried out by the generation of hydrogen ions and hydroxyl ions. The electrolytic generation of hydroxyl ions in a carbonate-free solution is advantageous for the determination of dilute acid solutions. The electrogeneration of halogens, especially bromine, has been widely used for the analysis of organic compounds. Titrants that are difficult to prepare, such as copper(I), silver(II), and uranium(IV), can be readily electrogenerated. Most of these coulometric determinations can be performed with samples containing 0.01 to 0.1 mg of reactive species with an accuracy of about 0.2%.

Reaction kinetics. Controlled-current coulometry has been used for the determination of the rates of chemical reactions in solution. The reactant, for example a halogen, is electrogenerated in solution at a constant predetermined rate. There it reacts with an organic compound of known concentration, and the concentration of the unreacted halogen is monitored as a function of time. The rate of reaction of the halogen with the organic compound can be evaluated from this information. Alternatively, the rate of generation can be adjusted until it is equal to the rate of depletion of the halogen by the chemical reaction; hence, the rate of the chemical reaction can be calculated. Controlled-current coulometry is a useful technique for the addition of small concentrations of a reactant to a solution for the investigation of the kinetics of relatively rapid reactions. *See* CHEMICAL DYNAMICS.

Quintus Fernando

Bibliography. A. J. Bard and L. R. Faulkner, *Electrochemical Methods*, 1980; S. A. Borman (ed.), *Instrumentation in Analytical Chemistry*, 1982; P. T. Kissinger and W. R. Heinemen (eds.), *Laboratory Techniques in Electroanalytical Chemistry*, 1984.

Countercurrent exchange (biology)

Engineers have known for decades that efficient, almost complete heat or other exchange could be achieved between two fluids flowing in opposite directions in separate tubes. Such countercurrent systems have been "invented" numerous times by living organisms for all types of exchange function. They are most commonly found in the circulatory, respiratory, and excretory (kidney) systems, serving in heat, oxygen, and ion exchange. Biological countercurrent systems can be classified into two main types: downhill exchanges and hairpin multipliers. In both cases, the basic mechanism is the same—exchange of substance between fluids flowing in opposite directions—but the consequences are very different.

Downhill exchanges. These countercurrent systems are commonest in the circulatory system where their morphological structure is a rete mirabile (a wonderful net) of closely oppressed sets of small arteries and veins. They are also found in gills of fish and in the minute air tubules of the avian lung. The principle of downhill exchanges is simple, as shown in **Fig. 1**. Fluids flow in opposite directions in separate tubes with the possibility of exchange, for example, heat flow or diffusion of oxygen, between them. The fluid entering one tube is warmest at that end, while that entering the second tube is coolest at the other end. Heat flows from higher to lower temperature. As heat flows from the warmer to the cooler tube, the fluid in the warmer tube cools down slightly and moves down along the tube. But as the slightly cooler fluid has moved further, it comes into contact with still cooler fluid in the second tube, and additional heat flow can occur. Thus, as the warmer fluid flows down the tube, it constantly loses heat, but always comes into contact with even cooler fluid in the second tube; the reverse is true for the initially cool fluid as it receives heat and warms up. Although the temperature differential between the two fluids is small at any point along the length of the countercurrent system, almost all the heat contained in the warmer tube is transferred to the cooler tube. Exchange of heat or oxygen occurs by passive diffusion. Most of the heat that entered the countercurrent system at one end leaves the system at the same end.

Rete of blood vessels thus serve as thermal isolating mechanisms within the body. They are found in appendages of mammals and birds (for example, whale flippers, the tail of beavers, and legs of gulls) to prevent excessive heat loss from these uninsulated parts. In reverse, masses of warm muscles in rapidly

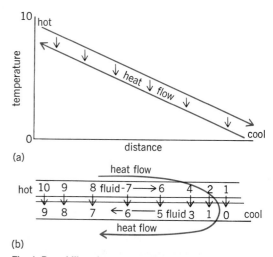

Fig. 1. Downhill exchange countercurrent system. (*a*) Graph showing the relationship between distance and temperature in a downhill system. (*b*) Schematic showing the two tubes carrying hot fluid and cool fluid and the mechanism by which heat exchange takes place.

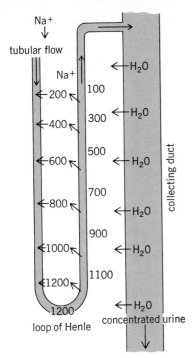

Fig. 2. Hairpin multiplier countercurrent system (loop of Henle of nephron).

swimming fish, such as mackerel, tuna, and the mako shark, are isolated from the rest of the body and the gills, where heat loss would occur, by sets of rete. In desert mammals a rete located between the veins draining evaporating (that is, cooling) surfaces and the carotid artery cools the arterial blood before it reaches the brain.

Downhill exchange systems in the gills of fish and in the air tubules of birds permit maximum exchange of oxygen from the environment into the blood. Blood in respiratory capillaries flows against the water or air current and thus can pick up most of the oxygen contained in the external fluid. The advantage of downhill exchangers is that they achieve greater efficiency without extra energy cost simply by arranging flow in a countercurrent rather than in a concurrent fashion.

Hairpin multipliers. These exchange systems take their name from the structure of the tubes, which have a hairpin turn between the afferent (descending) and the efferent (ascending) limbs. Hairpin countercurrent systems are found in the nephron (the loop of Henle) of the kidney and in the capillary system of the gas gland in the swim bladder of many fish. In contrast to downhill systems, which operate by passive transport, hairpin multipliers must employ active transport of materials. These are always materials pumped out of the efferent limb of the system.

Operation of a hairpin multiplier, such as the loop of Henle, is shown in **Fig. 2**. As the kidney filtrate flows down the afferent limb and up the efferent limb, sodium ions (Na^+) are actively transported out of the efferent limb. The sodium diffuses back into the afferent limb and is carried once again around the hairpin turn, together with additional sodium that constantly enters the system via the afferent limb. Continual active transport of sodium along the entire length of the efferent limb and its diffusion back into the afferent limb will result in the accumulation of

sodium at the bottom of the loop and in the surrounding interstitial fluid, and in the establishment of a steep osmotic gradient from the top to the bottom of the hairpin system. Yet at no point within this multiplier mechanism are individual cells exposed to excessive osmotic pressure. The steep gradient of osmotic pressure is used for final concentration of the urine as it passes down the collecting ducts. As water is drawn from the urine by the osmotic pressure of the interstitial fluid, the more concentrated urine passes into an area of even greater osmotic pressure—an example of downhill exchange. Thus, both types of countercurrent systems operate in the kidney to recover water from the urine. Desert mammals improve their water-conserving abilities by simply increasing the length of the loop of Henle; hence the strength of the osmotic pressure at the lower end of the hairpin is also increased.

The hairpin multiplier in the gas gland of the swim bladder of deep-sea marine fish serves to concentrate oxygen. Lactic acid is produced in the gas gland located at the bend of the hairpin capillaries and secreted into the blood. Lactic acid drives oxygen from hemoglobin faster than it can recombine with the hemoglobin molecule. The oxygen diffuses out of the efferent limb and into the afferent limb. With the constant addition of oxygen by arterial blood in the efferent limb, a very steep concentration gradient increase of over 1000-fold is achieved, thereby filling the swim bladder with gaseous oxygen against the great water pressures at depth of 330 ft (1000 m) or more. SEE CIRCULATORY SYSTEM; KIDNEY; RESPIRATORY SYSTEM; SWIM BLADDER.

Walter J. Bock

Bibliography. K. Schmidt-Nielsen, *Animal Physiology*, 2d ed., 1981; K. Schmidt-Nielsen, *How Animals Work*, 1972; K. Schmidt-Nielsen et al. (eds.), *Primitive Mammals*, 1980.

Countercurrent transfer operations

Industrial processes in chemical engineering or laboratory operations in which heat or mass or both are transferred from one fluid to another, with the fluids moving continuously in very nearly steady state or constant manner and in opposite directions through the unit. Other geometrical arrangements for transfer operations are the parallel or concurrent flow, where the two fluids enter at the same end of the apparatus and flow in the same direction to the other end, and the cross-flow apparatus, where the two fluids flow at right angles to each other through the apparatus. These two arrangements are ordinarily not as efficient as countercurrent flow, but do find certain applications in industry and the laboratory.

Heat transfer. In heat transfer there can be almost complete transfer in countercurrent operations. The limit is reached when the temperature of the colder fluid becomes equal to that of the hotter fluid at some point in the apparatus. At this condition the heat transfer is zero between the two fluids. The amount of actual heat transfer is determined by economical design, that is, by comparing the value of the transferred heat with the cost of the heat exchange equipment. The economically optimum heat transfer has been studied for many years in engineering and is changing constantly as the costs of basic forms of energy increase.

Most heat transfer equipment has a solid wall between the hot fluid and the cold fluid, so the fluids do not mix. Heat is transferred from the hot fluid through the wall into the cold fluid. Another type of equipment, fewer in number but significant in size and use, does use direct contact between the two fluids—for example, the cooling towers used to remove heat from a circulating water stream. Cooling towers are of the countercurrent type and the cross-flow type. For more discussion of heat transfer SEE COOLING TOWER; HEAT BALANCE; HEAT EXCHANGER; HEAT TRANSFER.

Mass transfer. This process involves the changing compositions of mixtures, and is done usually by physical means instead of chemical reactions because of the lower costs. Even as heat is transferred from a region of high temperature to one of lower temperature, a material is transferred within a single phase from a region of high concentration to one of lower concentration by processes of molecular diffusion and eddy diffusion. In typical mass transfer processes, at least two phases are in direct contact in some state of dispersion, and mass (of one or more substances) is transferred from one phase across the interface into the second phase. Similar to heat transfer, mass transfer takes place between two immiscible phases until equilibrium between the two phases is attained. In heat transfer, equilibrium denotes an equality of temperature in the two phases, but in mass transfer there is seldom an equality of concentration in the two equilibrium phases. This means that a component may be transferred from a phase at low concentration (but at a concentration higher than that at equilibrium) to a second phase of greater concentration. The approach to equilibrium is controlled by diffusion transport across phase boundaries. Because this is a relatively slow process, the transport rate is increased by increasing the total interfacial area, by decreasing the thickness of the near-stagnant films adjacent to the interface, and by more frequent renewals of the interfacial films.

Although the two phases may be in concurrent flow or cross-flow, usual arrangements have the phases moving in countercurrent directions. The more dense phase enters near the top of a vertical cylinder and moves downward under the influence of gravity. The less dense phase enters near the bottom of the cylinder and moves upward under the influence of a small pressure gradient. In some cases, centrifugal force is used instead of gravity to provide phase separations.

For discussions of specific mass transfer operations SEE ADSORPTION; CHEMICAL SEPARATION TECHNIQUES; CRYSTALLIZATION; DISTILLATION; DRYING; ELECTROPHORESIS; EXTRACTION; LEACHING.

Frank J. Lockhart

Bibliography. R. H. Perry and D. W. Green (eds.), *Perry's Chemical Engineers' Handbook*, 6th ed., 1984; R. E. Treybal, *Mass Transfer Operations*, 3d ed., 1980.

Couple

A system of two parallel forces of equal magnitude and opposite sense (**Fig. 1**). Forces P at normal distance p constitute a counterclockwise couple C_z (viewed from $+Z$) in the OXY plane; forces Q at arm q, constitute a clockwise couple C_x (from $+X$) in OYZ.

Under a couple's action a rigid body tends only to

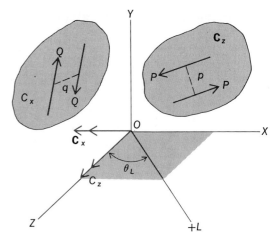

Fig. 1. Vector properties of a couple.

rotate about a line normal to the couple's plane. This tendency reflects the vector properties of a couple. SEE RIGID-BODY DYNAMICS.

Vector properties. The total force of a couple is zero. The total moment $\mathbf{C}$ of a couple is identical about any point. Accordingly, $\mathbf{C}$ is the moment of either force about a point on the other and is perpendicular to the couple's plane. SEE RESULTANT OF FORCES; STATICS.

In Fig. 1, $\mathbf{C}_z$, at the origin for convenience, is the moment of couple C_z. Its magnitude $|C_z| = +P_p$. Its sense, by the convention of moment, is $+Z$. Also $\mathbf{C}_x$, of sense $-X$, represents couple C_x; $|C_x| = +Q_q$.

Scalar moment. The moment of a couple about a directed line is the component of its total moment in the line's direction. For example, the moment of couple C_z about line L is $M_L = |C_z| \cos \theta_L = +P_p \cos \theta_L$. Also $M_x = 0$, $M_y = 0$, and $M_z = +P_p \cos 0° = +P_p$.

Equivalent couples. Couples are equivalent whose total moments are equal. In **Fig. 2**, the paired linear forces and the counterclockwise curl represent counterclockwise couple C_1, C_2, and C_3 in parallel planes OXY, AXY, and BXY. When their total moments are

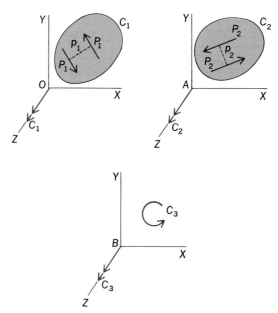

Fig. 2. Equivalent couples.

directed the same ($+Z$), and their magnitudes $|C_1| = P_1 p_1$, $|C_2| = P_2 p_2$, and $|C_3|$ are equal, then $\mathbf{C}_1 = \mathbf{C}_2 = \mathbf{C}_3$ and these couples are equivalent. Thus $\mathbf{C}_1$ can represent these or any number of other equivalent couples.

Nelson S. Fisk

Coupled circuits

Two or more electric circuits are said to be coupled if energy can transfer electrically or magnetically from one to another. If electric charge, or current, or rate of change of current in one circuit produces electromotive force or affects the voltage between nodes in another circuit, the two circuits are coupled.

Between coupled circuits there is mutual inductance, resistance, or capacitance, or some combination of these. The concept of a mutual parameter is based on the loop method of analysis. A mutual parameter can be one that carries two or more loop currents; such a network has conductive coupling because electricity can flow from one circuit to the other. SEE NETWORK THEORY.

Also, there can be purely inductive coupling, which appears if the magnetic field produced by current in one circuit links the other circuit. A two-winding transformer is an application of inductive coupling, with energy transferred through the magnetic field only.

It is also possible to have mutual capacitance, with energy transferred through the electric field only. Examples are the mutual capacitance between grid and plate circuits of a vacuum tube, or the capacitive interference between two transmission lines, as a power line and a telephone line, that run for a considerable distance side by side.

Polarity. With inductive coupling, polarity may need to be known, particularly if two circuits are coupled in more than one way. Do two kinds of coupling produce voltages that add or subtract?

There are several ways to show the relative polarities of inductive coupling. **Figure 1** shows, somewhat pictorially, two coils wound on the same core. Current flowing into the upper end of coil 1 would produce magnetic flux upward in the core, and so also would current flowing into the upper end of coil 2. For this reason the upper ends of the two coils are said to be corresponding ends.

If a wiring diagram is drawn with some such semipictorial sketch of the coils, it is not difficult to determine which are corresponding ends. However it is easier, both in representation and in interpretation, to indicate the corresponding ends symbolically. For this purpose dots are placed on a diagram at the corresponding ends of coupled coils. Such dots are shown in Fig. 1, though they are not needed there. Dots are also shown in **Fig. 2**, where they give the only means of identifying corresponding ends of the coils.

Note that the bottom ends of the coils shown in Fig. 1 are also corresponding ends, and that the two lower ends might have been dotted instead of the two upper ends; it makes no difference. But if the sense of winding of either coil 1 or coil 2 in Fig. 1 were reversed, then one dot (either one) would have to be shifted correspondingly.

Voltage equations. In addition to showing corresponding ends of two coils, Fig. 2 indicates that the two coils couple to each other. This coupling is iden-

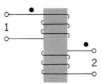

Fig. 1. Two coils wound on a single core, illustrating polarity.

tified as L_{12}. There is voltage from a to b in Fig. 2 if current i_1 is changing through the self-inductance L_{11} of the coil in circuit 1. There is additional voltage from a to b if current i_2 is changing in circuit 2 because of the mutual inductance L_{12} between the circuits. Thus, with circuit 1 coupled to circuit 2 and to any number of other circuits, Eqs. (1) and (2) hold.

$$v_{ab} = \left(L_{11} \frac{di_1}{dt} \pm L_{12} \frac{di_2}{dt} \pm \cdots \right) \qquad (1)$$

$$v_{cd} = \left(\pm L_{21} \frac{di_1}{dt} + L_{22} \frac{di_2}{dt} \pm \cdots \right) \qquad (2)$$

Here L_{11} and L_{22}, the self-inductances of the circuits, are positive numbers (inductances measured in henrys). With regard to the mutual terms, two questions of polarity must be asked. Are the mutual inductances such as L_{12} and L_{21} always positive numbers? And are the signs before the mutual terms, shown as $\pm$ in Eq. (1), actually $+$ or are they $-$? Answers to these questions are not always the same, and may be different as given by different authors, but the most usual simple procedure to answer them is as follows:

1. Draw a circuit diagram, such as Fig. 2, with nominal positive directions of currents shown by arrows and with dots at corresponding ends of coupled coils.

2. Let each mutual inductance such as L_{12} be a positive number.

3. Write equations such as Eqs. (1) and (2) with the following signs: First, if both arrows enter dotted ends of a pair of coupled coils, or if both arrows enter undotted ends, use $+$ before the corresponding mutual-impedance term. Second, if one arrow enters a dotted end and the other an undotted end, use $-$ before the corresponding mutual-impedance term.

However, this simple procedure sometimes fails, for it may be impossible to dot corresponding ends of all pairs of coupled coils in a network if there are three or more coils. A more general method follows:

1. Draw a circuit diagram with nominal positive directions of currents shown by arrows; place dots at coil ends arbitrarily, as may be convenient.

2. Determine corresponding ends of pairs of coupled coils, considering the coils two at a time. For the mutual inductance between a pair of coils with cor-

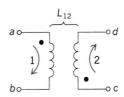

Fig. 2. Mutual inductance. Dots identify corresponding ends of coils.

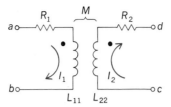

Fig. 3. A pair of coupled circuits.

responding ends dotted, use a positive number, such as $L_{12} = 5$. For mutual inductance between a pair of coils with noncorresponding ends dotted, use a negative number, such as $L_{12} = -5$.

3. Write equations according to rule 3, above.

Steady-state equations. The differential Eqs. (1) and (2) are quite general. For steady-state operation at a single frequency, it is often simpler to have phasor or transform equations. Such equations can be in terms of reactances instead of inductances ($X = \omega L = 2\pi f L$ where f is frequency in cycles per second, or hertz). With the usual interpretation of phasor or transform equations, the steady-state relations corresponding to Eqs. (1) and (2) are shown in Eqs. (3) and (4). The rules in the foregoing paragraphs deter-

$$V_{ab} = j\omega(L_{11}I_1 \pm L_{12}I_2 \pm \cdots)$$
$$= j(X_{11}I_1 \pm X_{12}I_2 \pm \cdots) \quad (3)$$

$$V_{cd} = j\omega(\pm L_{21}I_1 + L_{22}I_2 \pm \cdots)$$
$$= j(\pm X_{21}I_1 + X_{22}I_2 \pm \cdots) \quad (4)$$

mine the choice of $+$ or $-$ in these equations; also, a mutual reactance such as X_{12} is a positive number (of ohms) if the corresponding mutual inductance such as L_{12} is a positive number (of henrys), as determined by the above rules. *See Alternating-current circuit theory*.

Equality of mutual inductance. Because of this equality it is not uncommon, if there are only two coupled coils in a network, to use the letter M in place of both L_{12} and L_{21}. This use of M will be adopted in the following paragraphs.

Coefficient of coupling. The coefficient of coupling between two circuits is, by definition, Eq. (5). If the

$$k = M/\sqrt{L_{11}L_{22}} \quad (5)$$

circuits are far apart or, because of orientation, have little mutual magnetic flux, they are loosely coupled and k is a small number. Values of k for circuits with loose coupling may be in the range between 0.01 and 0.10. For closely (or tightly) coupled circuits with air-core coils, k may be around 0.5. In a transformer with a ferromagnetic core, k is very nearly 1.00.

Ideal transformers. An ideal transformer is defined as one in which primary and secondary currents are related inversely as the number of turns in the windings; this is shown in Eq. (6). Voltages across the

$$I_1/I_2 = N_2/N_1 \quad (6)$$

primary and secondary windings are in direct proportion to the numbers of turns, as shown in Eq. (7).

$$V_1/V_2 = N_1/N_2 \quad (7)$$

An actual transformer may be almost ideal or not

at all ideal, depending on how it is made, and construction in turn depends on the purpose for which it is to be used. In an ideal transformer $k = 1$; in any actual transformer k is less than 1. In an actual transformer there is magnetizing current, so Eq. (6) is not exact. Also, an actual transformer has resistance and leakage reactance, so Eq. (7) is not exact. Nevertheless, the relations of ideal transformers are closely approximated by 60-Hz power transformers, and these relations are more or less close for other transformers that have ferromagnetic cores. *See Transformer*.

Whereas coupling k is desirably as close as possible to unity in a transformer that couples power from one circuit to another, k may purposefully be considerably less than unity in a transformer used for another purpose. In an oscillator, coupling need only be sufficient to sustain oscillation. In a band-pass amplifier, coupling is determined by bandwidth requirements. *See Oscillator; Resonance (alternating-current circuits)*.

Equivalent circuits. It is often convenient to substitute into a network, in place of a pair of inductively coupled circuits, an equivalent pair of conductively coupled circuits. Circuits so substituted are equivalent if the network exterior to the coupled circuits is unaffected by the change; in many cases this requirement implies that input current and voltage and output current and voltage are unaffected by the change.

Voltages and currents at the terminals of the coupled circuits of **Fig. 3** are related by Eqs. (8) and (9).

$$(R_1 + j\omega L_{11})I_1 - j\omega MI_2 = V_{ab} \quad (8)$$

$$(R_2 + j\omega L_{22})I_2 - j\omega MI_1 = V_{cd} \quad (9)$$

Voltages and currents at the terminals of the network of **Fig. 4** are also related by Eqs. (8) and (9), so the network of Fig. 4 is equivalent to the coupled circuits of Fig. 3. This is not immediately obvious, but it appears when loop equations for the network are simplified to a form that shows the requirement that $L_{11} = L_1 + aM$ and $L_{22} = L_2 + M/a$ and that $V_{c'd'} = aV_{cd}$, where a is the turn ratio of the ideal transformer shown, equal to N_1/N_2 in Eqs. (6) and (7).

It seems at first that this substitution, resulting in the network of Fig. 4, has produced something more complicated than the circuit of Fig. 3, but the value of substituting will now be shown.

Mathematically, a could be any number, but practically there are two particularly advantageous values for a. When the coupling between coils is loose so that k, the coefficient of coupling, is small, and if the coils have somewhere near the same number of turns, it is advantageous to let $a = 1$. With $a = 1$ the

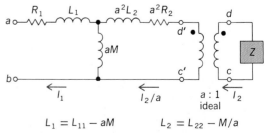

Fig. 4. A network equivalent to coupled circuits, feeding a load of impedance Z.

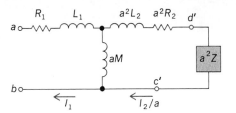

network of Fig. 4 is simplified to **Fig. 5**, leaving only the conductively coupled circuits with parameters R_1 and L_1, M, and L_2 and R_2, where $L_1 = L_{11} - M$, and $L_2 = L_{22} - M$. Many loosely coupled circuits particularly in radio circuits, are conveniently represented by this equivalent network with $a = 1$.

On the other hand, if coupling is close and especially when the two coupled coils have widely different numbers of turns (a situation that is typical of transformers), it is more convenient to let a equal the actual turn ratio of the coils. If, for example, there are 10 times as many turns in the primary winding of a transformer as there are in the secondary winding, it is well to let $a = 10$. (Letting $a = 1$ in such a transformer would result in a negative value for L_2 which, though correct for analysis, is difficult to visualize.)

With a equal to the actual transformer turn ratio, the following interpretation of Fig. 4 is usual and convenient. All causes of power loss and voltage drop have been put into the equivalent T network. Only the turn ratio, the actual transforming function, remains in the ideal transformer. It now becomes possible to consider, to study, and even to design the separate functions independently.

Transformers. The equivalent circuit of Fig. 4 is so commonly used in transformer work that a number of the quantities have been given special names. With the concept that power is supplied to one winding of the transformer, this is called the primary winding; the other, the secondary winding, provides power to a load. There may be a third, or tertiary, winding, perhaps providing power to another load at a different voltage or with a different connection, and even other windings, on the same transformer core.

However, it is only when a transformer has become part of a system that there is any meaning in designating the windings as primary and secondary, for a two-winding transformer can be used to carry power in either direction. The terms "high-voltage side" and "low-voltage side" are preferable for a transformer that is not part of a system. In the following discussion the words primary and secondary really mean nothing more than the windings of the transformer that are designated by the subscripts 1 and 2.

Speaking of a transformer with N_1 primary turns and N_2 secondary turns, let a be the turn ratio N_1/N_2. Then, referring to Fig. 4, R_1 is primary resistance; L_1 is primary leakage inductance; aM is primary exciting inductance; a^2L_2 is secondary leakage inductance referred to the primary side; and a^2R_2 is secondary resistance referred to the primary side. Usually, operation of power transformers is described in terms of these leakage inductances (or leakage reactances) and in terms of resistances, inductances, or reactances all referred to one side of the transformer or the other.

Transformation of impedance. Figure 6 is the same as Fig. 4 except that the ideal transformer with

turn ratio of a:1 has been eliminated and the impedance of the load has been changed from Z to a^2Z. This conversion makes the network of Fig. 6 equivalent to that of Fig. 4 at the input terminals, and it suggests a concept that is quite useful in communications.

If a load with impedance Z is preceded by an ideal transformer with turn ratio a, the input impedance to the transformer is a^2Z. Thus a load of impedance Z can be made to act like a load with impedance a^2Z by using an ideal transformer with turn ratio a. A useful application is to connect a load of one impedance to an incoming line that has a different terminal impedance through an impedance-matching transformer. By this means a transformer with turn ratio a will match a load with resistance R to a source with resistance a^2R, thereby providing for maximum power transfer to the load.

The preceding paragraph assumes an ideal transformer to provide a perfect impedance match. Figure 6 shows the T network of resistances and inductances that an actual transformer introduces into the network. In practice, the transformer used must be good enough so that these resistances and inductances are acceptable.

Core loss. Neither the equations nor the circuits of this article have taken into account the power loss in a ferromagnetic core. Both eddy-current loss and hysteresis loss may be appreciable; these iron losses are commonly great enough to affect the economics of power transformers, and even to prohibit the use of metal cores with ratio-frequency current. See CORE LOSS.

The relations of core loss are not linear and cannot accurately be included in linear equations, but a satisfactory approximation is used with power transformers for which frequency and applied voltage change little if at all. With these restrictions, core loss can be represented by the loss in a resistance shunted around mutual inductance aM of Fig. 4. This approximation of loss is much better than neglecting core loss entirely, and is usual in work with power transformers. See CIRCUIT (ELECTRICITY).

Hugh H. Skilling

Bibliography. L. P. Huelsman, *Basic Circuit Theory*, 2d ed., 1984; D. E. Johnson, *Basic Electric Circuit Analysis*, 2d ed., 1984; R. E. Risdale, *Electric Circuits*, 2d ed., 1983.

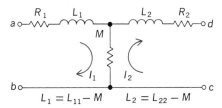

Fig. 5. A conductive network equivalent to the coupled circuits of Fig. 3 ($a = 1$).

Coupling

A mechanical fastening device for connecting the ends of two shafts together. There are three major coupling types: rigid, flexible, and fluid.

Rigid coupling. This connection is used only on shafts that are perfectly aligned. The flanged-face coupling (**Fig. 1a**) is the simplest of these. For this type of coupling the flanges must be keyed to the shafts. The clamp, or keyless compression, coupling

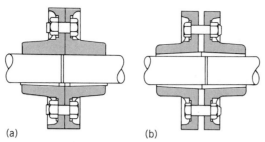

(a)   (b)

Fig. 1. Rigid coupling. (a) Flanged-face coupling. (b) Clamp coupling. (After E. A. Avallone and T. Baumeister III, eds., Marks' Standard Handbook for Mechanical Engineers, 9th ed., McGraw-Hill, 1987)

(Fig. 1b) has split cylindrical elements which clamp the shaft ends together by direct compression, through bolts, and by the wedge action of conical sections. This coupling is generally used on line shafting to transmit medium or light loads. *See Machine key.*

Flexible coupling. This connection is used for shafts which are misaligned either laterally or angularly. It also absorbs the impact due to fluctuations in shaft torque or angular speed. The Oldham, or double-slider, coupling (**Fig. 2a**) may be used to connect shafts that have only lateral misalignment. Because the tongues move about in the slots, the coupling must be well lubricated and can be used only at low speeds. The geared "fast" flexible coupling (Fig. 2b) uses two interior hubs on the shafts with circumferential gear teeth surrounded by a casing having internal gear teeth to mesh and connect the two hubs. Considerable misalignment can be tolerated because the teeth inherently have little interference. This completely enclosed coupling provides means for better lubrication, and is thus applicable for higher speeds. The rubber flexible coupling (**Fig. 3a**) is used to transmit the torque through a comparatively soft rubber section acting in shear. The type shown in Fig. 3b loads the intermediate rubber member in compression. Both types are recommended for light loads only.

Universal joint. This is a flexible coupling for connecting shafts with much larger values of misalign-

ment than can be tolerated by the other types of flexible coupling. Shaft angles up to 30° may be used. The initial universal joint, credited to Robert Hooke (**Fig. 4**), was a swivel arrangement by which two pins at right angles allowed complete angular freedom between two connected shafts. However, it suffers a loss in efficiency with increasing angles. *See Universal joint.*

Fluid coupling. This type has two basic parts: the input member, or impeller, and the output member, or runner (**Fig. 5**). There is no mechanical connection between the two shafts, power being transmitted by

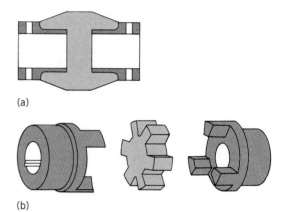

(a)

(b)

Fig. 3. Rubber flexible coupling. (a) Shear type. (b) Compression type. (After E. A. Avallone and T. Baumeister III, eds., Marks' Standard Handbook for Mechanical Engineers, 9th ed., McGraw-Hill, 1987)

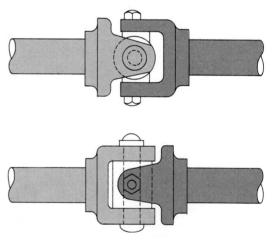

Fig. 4. Hooke's universal joint. (After E. A. Avallone and T. Baumeister III, eds., Marks' Standard Handbook for Mechanical Engineers, 9th ed., McGraw-Hill, 1987)

(a)

(b)

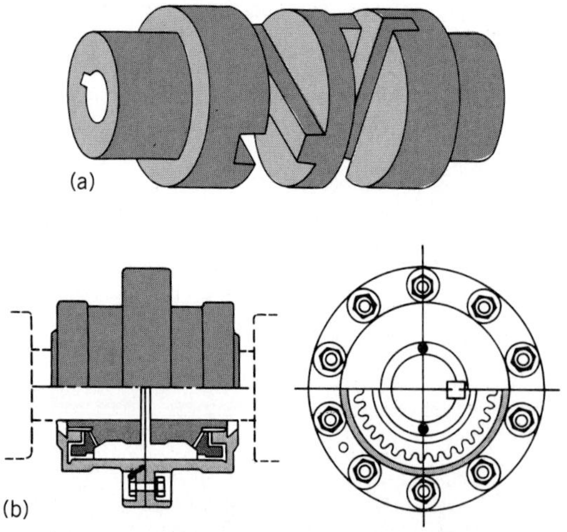

Fig. 2. Flexible coupling. (a) Oldham (double-slider) coupling. (b) Geared "fast" flexible coupling. (After E. A. Avallone and T. Baumeister III, eds., Marks' Standard Handbook for Mechanical Engineers, 9th ed., McGraw-Hill, 1987)

kinetic energy in the operating fluid. The impeller is fastened to the flywheel and turns at engine speed. As this speed increases, fluid within the impeller moves toward the periphery because of centrifugal force. The circular shape of the impeller directs the fluid toward the runner, where its kinetic energy is absorbed as torque delivered by shaft. The positive pressure behind the fluid causes flow to continue toward the hub and back through the impeller. The to-

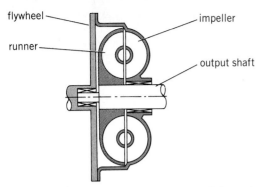

Fig. 5. Fluid coupling. (*After E. A. Avallone and T. Baumeister III, eds., Marks' Standard Handbook for Mechanical Engineers, 9th ed., McGraw-Hill, 1987*)

roidal space in both the impeller and runner is divided into compartments by a series of flat radial vanes. SEE FLUID COUPLING; SHAFTING.

Y. S. Shin

Bibliography. E. A. Avallone and T. Baumeister III (eds.), *Marks' Standard Handbook for Mechanical Engineers*, 9th ed., 1987.

Covellite

A mineral having composition CuS and crystallizing in the hexagonal system. Tabular hexagonal crystals of covellite are rare (see **illus.**); it is usually massive or occurs in disseminations through other copper min-

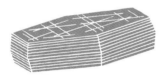

Covellite crystal habit. (*After L. G. Berry and B. Mason, Mineralogy, Freeman, 1959*)

erals. The luster is metallic and the color indigo blue. There is perfect basal cleavage yielding flexible plates; the hardness is 1.5 (Mohs scale), the specific gravity 4.7. Covellite is a common though not abundant mineral in most copper deposits. It is a supergene mineral and is thus found in the zone of sulfide enrichment associated with other copper minerals, principally chalcocite, chalcopyrite, bornite, and enargite, and is derived by their alteration. SEE BORNITE; CHALCOCITE; CHALCOPYRITE; COPPER; ENARGITE.

Cornelius S. Hurlbut, Jr.

Cover crops

Crops grown for the express purpose of preventing and protecting a bare soil surface. Many crops serve this purpose, and some give a high direct economic return. Other crops of lower return value are mainly important as a part of a system of good soil and water conservation and soil productivity maintenance, and it is these that are ordinarily considered cover crops.

Basic benefits of cover crops are in the prevention of soil detachment by wind, rainfall, and splash ero-

sion (**Figs. 1** and **2**) and in the return of cover crop roots and tops to replenish and add to the supply of soil organic matter.

Permanent cover. Cover crops can be readily grown in humid climates and in arid and semiarid climates where irrigation provides sufficient water so that cover crops do not rob the soil of needed moisture. Reseeded rangeland is an outstanding example of the use of cover crops. Long-lived grasses such as blue gramma (*Bouteloua gracilis*), buffalo grass (*Buchloë dactyloides*), and the blue-stems (*Andropogon* sp.) have been widely seeded upon many different soil conditions in the Great Plains. When properly managed, these grasses can supply excellent ground cover and grazing for livestock. SEE GRASS CROPS.

Many special areas such as road banks, ditch banks, and recreation areas need the protection of permanent vegetative cover of high quality. A variety of grasses and legumes may be used, depending upon soil and climatic conditions. Crown vetch (*Coronilla varia*) provides a luxurious cover for many road bank conditions. Reed canary grass (*Phalaris arundinacea*) and tall fescue (*Festuca elator*) have been widely and successfully utilized to stabilize ditch bands and waterways. In the northern United States, Kentucky bluegrass (*Poa pratensis*) provides cover for many recreation areas, while in the southern United States Bermuda grass (*Cynodon dactylon*) serves the same purpose.

Management considerations. With the development of low-cost nitrogen fertilizer and more efficient herbicides and insecticides, cover crops now require special management. Low-cost commercial nitrogen has tended to replace the dependence on the nitrogen produced by leguminous cover crops. Efficient herbicides, such as atrazine, when sprayed over a row crop area will destroy not only all broadleaf and grass weeds but also most seeded cover crops, making it

Fig. 1. Soil splash container in a representative location of domestic ryegrass (mixture of *Lolium perenne* and *L. multiflorum*) cover located in cabbage field. This level of cover almost eliminates soil splashing by raindrops.

Fig. 2. Alfalfa (*Medicago sativa*) being used as a cover crop. Less effective in preventing splash erosion than domestic ryegrass, it does fix nitrogen from the air.

impossible to use domestic ryegrass or other cover crops between crop rows. The use of herbicide in bands, directly over the rows of the main crop, enables the grower to cultivate the row crop and to seed the cover crop between the rows. *See* Fertilizer; Herbicide.

Annual winter legumes. A wide range of possibilities exist for the use of cover crops, especially where growers wish to maintain and improve soil productivity. In some areas the so-called naturally reseeding annual winter legumes can be utilized in the row crop each year. Once these plants have matured in a field, there are usually enough hard-coated seeds to establish in the autumn after having remained dormant during the dry summer period. Thus, provided enough seed is scattered to maintain the hard seed supply, one can obtain seed, grazing plants, or hay from a cultivated field plus a normal row crop such as corn or cotton. With this system soil fertility has been known to improve year after year under a system of dual harvests.

In general, these kinds of cover crops are restricted to the southeastern United States, where the growing season is long and the winter is mild. Varieties of vetch, most of which are annual legumes, are utilized, including hairy vetch (*Vicia villosa*) and smooth vetch (*V. villosa* var. *glabrescens*), the two commonly grown. Crimson clover (*Trifolium incarnatum*), rough pea (*Lathyrus hirsutus*), button clover (*Medicago orbicularis*), bur clover (*M. arabica*), and black medic (*M. lupulina*) are also used as naturally reseeding legumes.

Fig. 3. Early spring view of rolling corn field protected by stalk residues and domestic ryegrass cover.

Spring crops. Many ordinary spring-sown crops can be advantageously used as cover crops. Spring oats (*Avena sativa*) sown in the fall are killed by cold weather but provide excellent ground cover during the fall and winter. Buckwheat (*Fagopyrum esculentum*) can be used as a cover crop to protect the soil and as a smother crop to control weeds, and at the same time can be turned under for soil improvement. Cereal rye (*Secale cereale*) is widely seeded as a cover crop following the harvest of the main crop.

Advantages. In northern areas it is not unusual for a cover crop like domestic ryegrass to return 4 or more tons of roots and tops to the soil (Fig. 1 and **Fig. 3**). In warmer climates more is produced. Where a legume is utilized, the added benefit of nitrogen fixation is also obtained (Fig. 2). As compared to an annual splash loss of 5 tons of soil per acre per year (11 metric tons per hectare per year) on bare land, cover-cropped areas suffer little or no splash erosion. They also provide better support and traction for harvesting equipment. *See* Agricultural soil and crop practices; Nitrogen fixation; Soil conservation.

Paul J. Zwerman

Bibliography. C. H. Bornman et al., *Research, Breeding and Production of Crop Plants*, 1985; S. J. Goldman et al., *Erosion and Sediment Control Handbook*, 1986; J. F. Power (ed.), *The Role of Legumes in Conservation Tillage Systems*, 1987.

Cowpea

Vigna unguiculata ssp. *unguiculata*, a legume of such ancient cultivation that its country of origin, either in Asia or Africa, is uncertain. Large-scale production occurs mainly in the southern United States, Africa, the Mediterranean region, and Australia. It was cultivated as early as 1714 in North Carolina. Cowpeas are grown primarily in the United States as a vegetable crop and are also known as southern peas and blackeye peas, the latter referring to a group of varieties. It is also grown as a hay and forage crop but to a much less extent than in the past. In the tropics and subtropics, cowpeas provide food for millions of people and feed for a large number of livestock. It is cultivated extensively in western Africa and is the principal source of dietary protein in Nigeria.

Varieties. There are at least 200 varieties in the United States, but only about 20 of these are of commercial and home garden importance. All varieties have been arranged in groups based on seed size, shape, color, and color pattern, and on pod color and plant type, either upright or vining.

Cultivation. Cowpeas are grown commercially in the southern United States for fresh market, canning, freezing, and also in home gardens. Soil type is not usually a limiting factor as they grow well on most soils, require very little fertilizer compared with most vegetable crops, and withstand short periods of drought. Seeds are planted when the danger of a killing frost is over. Plantings are made in rows 3 to 3½ ft (0.9 to 1 m) apart with seeds spaced 3 to 4 in. (8 to 10 cm) apart. Two cultivations are generally required during the season.

Harvesting. Depending on variety, from 60 to 75 days are required to reach maturity. Pods are harvested before the seeds become dry, although some blackeye varieties are marketed as dry, mature seeds.

Almost all of the commercial acreage for processing is harvested with machines that remove the pods from the vines and shell them in one operation. With other machines, the pods are harvested and shelled in the processing plant. The use of cowpeas as a vegetable crop has increased steadily. Plant breeders continue to develop more productive, disease-resistant varieties with improved nutritive values. SEE LEGUME FORAGES.

Blake B. Brantley, Jr.

Diseases. Diseases caused by fungi and viruses are primary constraints in cowpea production. Roots can be destroyed by the fungi *Rhizoctonia solani, Sclerotium rolfsii,* and *Pythium* sp. *Cercospora cruenta* and *C. canescens* cause leaf spots, which sometimes result in defoliation of mature cowpea plants. Stem anthracnose and pod spot are caused by *Colletotrichum* sp. *Fusarium oxysporum* infects vascular tissue and causes wilting.

Virus diseases of cowpea are characterized by leaf discolorations (mottle, mosaic, chlorosis), stunted growth, and reduced yields. Malformed leaves and distorted growth are caused by cowpea mosaic and southern bean mosaic viruses. Although cucumber mosaic virux causes a very mild disease in cowpea, it frequently occurs in a mixed infection with blackeye cowpea mosaic virus and results in a synergistic interaction with severly stunted plants (see **illus.**). SEE PLANT VIRUSES AND VIROIDS.

The bacterium *Xanthomonas vignicola* can cause both leaf blight and stem canker. In sandy soils a root-knot disease caused by nematodes (*Meloidogyne* sp.) can be particularly injurious to general plant growth.

The most common method to control cowpea diseases is biological. Resistant cultivars are known for many of the diseases, and chemical control usually is

Severe mosaic and stunting of cowpea caused by a mixed infection of cucumber mosaic virus and blackeye cowpea mosaic virus. (From G. Pio-Ribeiro, S. D. Wyatt, and C. W. Kuhn, Cowpea stunt: A disease caused by synergistic interaction of two viruses, Phytopathology, 68:1260–1265, 1978)

not applied. Breeding programs have attempted to incorporate resistance to the major diseases into single cowpea lines. SEE BREEDING (PLANT); PLANT PATHOLOGY.

Cedric W. Kuhn

Bibliography. G. Pio-Ribeiro, S. D. Wyatt, and C. W. Kuhn, Cowpea stunt: A disease caused by a synergistic interaction of two viruses, *Phytopathology*, 68:1260–1265, 1978; K. O. Rachie, Improvement of food legumes in Tropical Africa, *Nutritional Improvement of Food Legumes by Breeding*, Symposium of the Protein Advisory Group, Food and Agriculture Organization, Rome, pp. 83–92, 1973; R. J. Williams, Diseases of cowpea [*Vigna unguiculata* (L.) Walp.] in Nigeria, *PANS*, 21:253–267, 1975.

Coxsackievirus

A large subgroup of the genus *Enterovirus* in the family Picornaviridae. The coxsackieviruses produce various human illnesses, including aseptic meningitis, herpangina, pleurodynia, and encephalomyocarditis of newborn infants. SEE ENTEROVIRUS; PICORNAVIRIDAE.

Coxsackieviruses measure about 28 nanometers in diameter; they resemble other enteroviruses in many biological properties, but differ in their high pathogenicity for newborn mice. Inapparent infections can be induced in adult mice, chimpanzees, and cynomolgus monkeys. Some types, especially A7, produce severe central nervous system lesions in monkeys. Coxsackieviruses of group A and group B differ in their effects on suckling mice; those of group A produce widespread myositis in skeletal muscle, whereas group B viruses produce necrosis of embryonic fat pads (brown fat). At least 23 antigenically distinct types in group A are now recognized, and 6 in group B. Several types (B1–6, A9) grow well in tissue culture. SEE TISSUE CULTURE.

After incubation for 2–9 days, during which the virus multiplies in the enteric tract, clinical manifestations appear which vary widely.

Diagnosis is by isolation of virus in tissue culture or infant mice. Stools are the richest source of virus. Neutralizing and complement-fixing antibodies form during convalescence and are also useful in diagnosis. SEE ANTIBODY; COMPLEMENT-FIXATION TEST; NEUTRALIZING ANTIBODY.

The coxsackieviruses have worldwide distribution. Infections occur chiefly during summer and early fall, often in epidemic proportions. Spread of virus, like that of other enteroviruses, is associated with family contact and contacts among young children. SEE ANIMAL VIRUS; VIRUS CLASSIFICATION.

Jospeh L. Melnick; M. E. Reichmann

Bibliography. D. O. White and F. Fenner, *Medical Virology*, 3d ed., 1986.

CPT theorem

A fundamental ingredient in quantum field theories, which dictates that all interactions in nature, all the force laws, are unchanged (invariant) on being subjected to the combined operations of particle-antiparticle interchange (so-called charge conjugation, C), reflection of the coordinate system through the origin (parity, P), and reversal of time, T. In other words, the CPT operator commutes with the hamiltonian.

The operations may be performed in any order; *TCP*, *TPC*, and so forth, are entirely equivalent. If an interaction is not invariant under any one of the operations, its effect must be compensated by the other two, either singly or combined, in order to satisfy the requirements of the theorem. *SEE NONRELATIVISTIC QUANTUM THEORY; QUANTUM FIELD THEORY*.

The *CPT* theorem appears implicitly in work by J. Schwinger in 1951 to prove the connection between spin and statistics. Subsequently, G. Lüders and W. Pauli derived more explicit proofs, and it is sometimes known as the Lüders-Pauli theorem. The proof is based on little more than the validity of special relativity and local interactions of the fields. The theorem is intrinsic in the structure of all the successful field theories. *SEE QUANTUM STATISTICS; RELATIVITY; SPIN (QUANTUM MECHANICS)*.

Significance. *CPT* assumed paramount importance in 1957, with the discovery that the weak interactions were not invariant under the parity operation. Almost immediately afterward, it was found that the failure of *P* was attended by a compensating failure of *C* invariance. Initially, it appeared that *CP* invariance was preserved and, with the application of the *CPT* theorem, invariance under time reversal. Then, in 1964 an unmistakable violation of *CP* was discovered in the system of neutral *K* mesons. *SEE PARITY (QUANTUM MECHANICS)*.

One question immediately posed by the failure of parity and charge conjugation invariance is why, as one example, the π^+ and π^- mesons, which decay through the weak interactions, have the same lifetime and the same mass. It turns out that the equality of particle-antiparticle masses and lifetimes is a consequence of *CPT* invariance and not *C* invariance alone. A casual proof can be obtained by applying the *CPT* operation to a free particle. This results in an antiparticle with precisely the same momentum and total energy and therefore, from special relativity, the same rest mass as the original particle. This obvious result can be generalized heuristically to include equality of lifetimes by including an imaginary component in the mass.

Experimental tests. The validity of symmetry principles such as the *CPT* theorem should not be assumed beyond the limits of the experimental tests. In that the *CPT* theorem predicts the equality of particle-antiparticle properties such as mass, lifetime, and the magnitude of the magnetic moment, measurements which compare these quantities constitute tests of the theorem. For example, it is known from studies of the neutral *K*-meson system that the mass of K^0 is the same as the $\overline{K^0}$ to within 1 part in 10^{18}. The magnitude of the magnetic moment of the electron is known to be the same as the positron to 1 part in 10^{10}. The lifetime of the π^+ and π^- mesons is the same to better than 0.1%, and so with the μ^+ and μ^- mesons.

The discovery of *CP* violations in 1964 in the system of neutral *K* mesons reopened the question of *CPT* invariance. That system has been analyzed in great detail, and it is known now that at least 90% of the observed *CP* violation is compensated by a violation of time-reversal invariance. That is, not more than 10% of the observed *CP* violation can be credited to a violation of *CPT*. These are all experimental limits and do not in any way suggest a violation of *CPT*.

The remarkable sensitivity of the *K* mesons to pos-

sible departures from *CPT* invariance, as demonstrated by the precision with which the masses of the K^0 and $\overline{K^0}$ are known to be the same, suggests that this system is perhaps the best one for further experimental tests of the validity of the *CPT* theorem. *SEE ELEMENTARY PARTICLE; MESON; SYMMETRY LAWS (PHYSICS)*.

Val L. Fitch

Crab

The name applied to arthropods in sections Anomura and Brachyura of the reptantian suborder of the order Decapoda, class Crustacea. The term crab is also sometimes used for two species of sucking lice (order Anoplura) which prey upon humans. *Phthirus pubis* is the crab louse which inhabits the pubic region.

Section Anomura includes over 1400 species in 12 different families. Commonly called hermit, king, sand, or mole crabs, the anomurans have lobsterlike abdomens which bend beneath the cephalothorax in crablike manner, setting them apart in a somewhat ill-defined group.

Section Brachyura encompasses the so-called true crabs, with reduced abdomens folded snugly beneath the cephalothorax. More than 4500 species making up 28 families have been grouped into subsections. In addition to subsections Dromiacea, Gymnopleura, and Oxystomata, the Hapalocarcinidea has one family of the same name containing three genera of small crabs with elongated abdomens. Young coral gall crab females settle on coral and become enveloped by growing coral. Small openings in the coral, maintained by the crab's water currents, are large enough to permit passage of miniature males (less than 0.08 in. or 2 mm long) and larvae; the female is stuck there for life.

Subsection Brachygnatha is the largest group of brachyurans and includes the most typical crabs. It has 20 families encompassing over 3700 species which occupy a variety of habitats, undergo different types of development, and exhibit contrasting patterns of behavior. Family Cancridae includes popular edible species of the genus *Cancer*. The Portunidae are the swimming crabs, most of which have paddle-shaped modifications of the terminal segments of their fifth pereopods. This structure, advantageous in swimming, is disadvantageous to the blue crab, *Cal-*

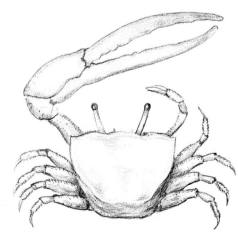

Male fiddler crab (*Uca pugnax*).

linectes sapidus, in that the outline of the new exo-skeleton in the thin paddle allows identification of premolt crabs. Such crabs, called "peelers," are kept in cypress wood floats until they molt, then sold as soft-shelled delicacies; the meat of hard-shelled crabs is also delectable. Blue crab fishing and marketing is one of the largest of the shellfish industries on the Atlantic coast of the United States.

Families Potomidae, Pseudothelphusidae, and Trichodactylidae include fresh-water species. These crabs are unique in that, like crayfish, the hatching young resemble adults; there is no series of larval stages culminating in metamorphosis. The only other fresh-water crabs are four species in the family Grapsidae. These crabs hatch as zoeae which complete their non-feeding larval development rapidly, using yolk retained from the eggs for nutrition. One Jamaican species, Metapaulias depressus, lives in water contained in bromeliad plants. Other grapsid crabs are marine or semiterrestrial forms which invade fresh waters but must undergo postembryonic development in the sea.

The Gecarcinidae, or land crabs, live in humid, tropical regions. Cardisoma and Gecarcinus may migrate several kilometers from the sea to dig burrows in moist areas. Although the land crabs retain gills, portions of the moist inner surface of the branchial chamber have become highly vascularized to allow respiration in air. A female returns to the sea only to release larvae as they hatch from the fertilized egg mass she carries.

Most Pinnotheridae are tiny, therefore called pea crabs, and live commensally with other organisms. Different species live in the mantle cavities of oysters or other mollusks, in polychaete tubes or burrows, in branchial cavities of ascidians, in cloacas of sea cucumbers, or in the burrows of other crustaceans. After establishing a commensal habitat, many undergo a molt, after which the exoskeleton remains soft.

Other crabs commonly seen at the coast belong to the family Xanthidae (the mud crabs) or the family Ocypodidae. These are amphibious burrowers. Uca spp. (Ocypodidae) are called fiddler crabs because males have one enlarged cheliped, the weight of which may be equal that of the rest of the animal (see **illus.**). In different species, this claw is used to beat the ground, creating vibrations to threaten, or to wave, an activity which may be a part of courtship. Ocypode, a ghost crab, does not have an enlarged cheliped but makes a variety of sounds by stridulation, rubbing the ridged inner surface of the claw against a tubercle on the base of the same cheliped.

Many of the other brachygnathans are marine and inhabit deeper waters. The best known of these are the Majidae, decorator or spider crabs. The Japanese spider crab, Macrocheira kaempferi, can have a span up to 11.9 ft (3.6 m) [body width of 18 in. or 45 cm] between the tips of its chelipeds; it is the largest living crab. SEE ARTHROPODA; CRUSTACEA; DECAPODA (CRUSTACEA); MEROSTOMATA.

Marilyn S. Kerr

Crab Nebula

The Crab Nebula in Taurus (see **illus.**) is the most remarkable known gaseous nebula. Observed optically, it consists of an amorphous mass that radiates a continuous spectrum and is involved in a mesh of delicate filaments which radiate a bright line spectrum

Crab Nebula, in the constellation Taurus, emitter of strong radio waves and of x-rays.

characteristic of typical gaseous nebulae.

This nebula has been identified as the expanding remnant of a supernova that appeared in A.D. 1054, reached an apparent magnitude of -5, corresponding to an absolute magnitude of about -18 (roughly 1.6 $\times 10^9$ times as bright as the Sun). From a comparison of the angular expansion rate of the elliptical nebular shell with the expansion velocity in kilometers per second, measured by the Doppler shift, distance estimates ranging from 1300 to 2000 parsecs (1 parsec = 1.9×10^{13} mi = 3.1×10^{13} km) have been obtained. If a distance of 1800 parsecs is adopted, the nebular size is 3.1×2.1 parsecs. The total mass of the nebula is comparable to that of the Sun. Presumably it comprises only material from the star not yet mixed with the interstellar medium. SEE DOPPLER EFFECT; SUPERNOVA.

The Crab Nebula also radiates strongly in the radio range, and the infrared, ultraviolet, x-ray, and gamma-ray spectral regions. Furthermore, the radiation of the central amorphous mass is strongly polarized, even in the x-ray region—indicating that the source of the radiation must be the synchrotron emission of electrons accelerated in a field on the order of 10^{-8} tesla. Energy must be supplied to these electrons at at a rate equivalent to about 30,000 times the solar power output. SEE SYNCHROTRON RADIATION.

The source of energy for the Crab Nebula appears to be a remarkable, rapidly spinning object known as a pulsar, or neutron star, which is presumed to be the residue of the supernova that created the nebula. Its period of variability (which is associated with its rotation period) is 0.033 s. The period increases uniformly with time except for sudden spin-ups or glitches, when the rotation suddenly speeds up. These glitches are interpreted as readjustments in a crust of a neutron star, that is, as "star-quakes." Energy output variations are observed in the radio, infrared, optical, and x-ray range. The total amount of electromagnetic energy emitted by the star is about 10^{29} W (10^{36} ergs/s), but this is only a small fraction of the amount supplied to high-energy particles. The complex pulses observed both optically and at radio frequencies can be interpreted by an oblique magnetized rotator, but the mechanisms responsible for accelerations of particles to high energies are not fully understood. The ultimate source of the energy of the Crab Nebula is the rotational energy of the neutron star; in the period 1054–1975 the pulsar lost about 10^{43} joules (10^{50} ergs), mostly in particle energy. At present, the pulsar is losing about 10^{31} W (10^{38} ergs/

s), but the rate of energy loss may have been much higher at an earlier epoch. The Crab Nebula may be regarded as the Rosetta Stone of high-energy astrophysics. SEE ASTROPHYSICS, HIGH-ENERGY; NEUTRON STAR; PULSAR.

One might anticipate that a search of the sky would find some object roughly similar to the Crab Nebula. Supernova remnants (for example, the Veil Nebula in Cygnus) abound in the Milky Way Galaxy, in the Magellanic Clouds, and in the Triangulum Spiral, M33, but in many instances what is observed is material thrown off by the supernova mixed with the neighboring interstellar medium. There is no trace of the residual star. Only the Vela supernova pulsar has been found optically, a twenty-sixth-magnitude star detected with the Anglo-Australian telescope. It is the faintest star ever observed. Remote pulsars cannot be detected, but it would seem that in some instances the star simply shattered completely or the residue disappeared as a black hole. SEE BLACK HOLE; NEBULA.

Lawrence H. Aller

Bibliography. *Crab Nebula: 46th International Astronomical Union Symposium*, Manchester, 1971; S. Mitton, *The Crab Nebula*, 1978.

Crabapple

A fruit, represented commercially by such varieties as Martha, Hyslop, and Transcendent, comprising hybrids between *Malus baccata* (Siberian crabapple) and *M. domestica* (cultivated apple).

Trees may grow to 40 ft (1.2 m) and are very hardy. At one time the fruit of the crabapple was esteemed because of its high pectin content and its usefulness in jam and jelly manufacture. With the introduction of commercial pectin preparations, demand for the crabapple declined sharply. Except for its use as a pickled product, there is little commercial interest in this fruit. SEE PECTIN.

Twenty or more species of Oriental flowering crabapples, some of which produce edible fruit of indifferent quality, are listed. SEE APPLE; FRUIT; FRUIT, TREE; ROSALES.

Harold B. Tukey

Cracking

A process used in the petroleum industry to reduce the molecular weight of hydrocarbons by breaking molecular bonds. Cracking is carried out by thermal, catalytic, or hydrocracking methods. Increasing demand for gasoline and other middle distillates relative to demand for heavier fractions makes cracking processes important in balancing the supply of petroleum products.

Thermal cracking depends on a free-radical mechanism to cause scission of hydrocarbon carbon-carbon bonds and a reduction in molecular size, with the formation of olefins, paraffins, and some aromatics. Side reactions such as radical saturation and polymerization are controlled by regulating reaction conditions. In catalytic cracking, carbonium ions are formed on a catalyst surface, where bond scissions, isomerizations, hydrogen exchange, and so on, yield lower olefins, isoparaffins, isoolefins, and aromatics. Hydrocracking, a relative newcomer to the industry, is based on catalytic formation of hydrogen radicals to break carbon-carbon bonds and saturate olefinic bonds. Hydrocracking converts intermediate- and high-boiling distillates to middle distillates, high in paraffins and low in cyclics and olefins. Hydrocracking also causes hydrodealkylation of alkyl-aryl components in heavy reformate to produce benzene and naphthalene. SEE HYDROCRACKING.

Thermal cracking. This is a process in which carbon-to-carbon bonds are severed by the action of heat alone. It consists essentially in the heating of any fraction of petroleum to a temperature at which substantial thermal decomposition takes place through a thermal free-radical mechanism followed by cooling, condensation, and physical separation of the reaction products. SEE FREE RADICAL.

There are a number of refinery processes based primarily upon the thermal cracking reaction. They differ primarily in the intensity of the thermal conditions and the feedstock handled.

Visbreaking is a mild thermal cracking operation (850–950°F or 454–510°C) where only 20–25% of the residuum feed is converted to mid-distillate and lighter material. It is practiced to reduce the volume of heavy residuum which must be blended with low-grade fuel oils.

Thermal gas-oil or naphtha cracking is a more severe thermal operation (950–1100°F or 510–593°C) where 45% or more of the feed is converted to lower molecular weight. Attempts to crack residua under these conditions would coke the furnace tubes.

Steam cracking is an extremely severe thermal cracking operation (1100–1400°F or 593–760°C) in which steam is used as a diluent to achieve a very low hydrocarbon partial pressure. Primary products desired are olefins such as ethylene, propylene, and butadiene.

Fluid coking is a thermal operation where the residuum is converted fully to gas-oil products boiling lower than 950°F (510°C) and coke. The thermal conversion is carried out on the surface of a fluidized bed of coke particles.

The fluid coking process has been extended to include gasification of the product coke in a vessel that is heat-integrated into the plant. Low-Btu gas which can be desulfurized is produced rather than a product coke stream.

Delayed coking is a thermal cracking operation wherein a residuum is heated and sent to a coke drum, where the liquid has an infinite residence time to convert to lower-molecular-weight hydrocarbons which distill out of the coke drum, and to coke which remains in the drum and must be periodically removed.

In fluid coking and delayed coking, there is total conversion of the very heavy high-boiling end of the residuum feed.

Although there are many variations of visbreaking and thermal cracking, most commonly a feedstock that boils at higher temperatures than gasoline is pumped at inlet pressures of 75–1000 pounds per square inch gage (0.52–6.9 megapascals) through steel tubes so placed in a furnace as to allow gradual heating of the coil to temperatures in the range 850–1100°F (454–593°C). The flow rate is controlled to provide sufficient time for the required cracking to lighter products; the time may be extended by subsequently passing the hot products through a reaction chamber that is maintained at a high temperature. To achieve optimum process efficiency, part of the overhead product ordinarily is returned to the cracking unit for further cracking (**Fig. 1**).

Crude oils differ in their compositions, both in molecular weight and molecular type of hydrocarbon. Since refiners must make products in harmony with market demand, they often need to alter the molecular structure of the hydrocarbons. The cracking of heavy distillates and residual oils increases the yields of gasoline and the light intermediate distillates used primarily as domestic heating oils, as well as providing low-molecular-weight olefins needed for the manufacture of chemicals and polymers.

Beginning in 1912, thermal cracking proved for many years to be eminently suitable for this purpose. During the period 1920–1940, more efficient automobile engines of higher compression ratios were developed. These engines required higher-octane-number gasolines, and thermal cracking operations in the United States were expanded to meet this need. Advantageously, thermal cracking reactions produce olefins and aromatics, leading to gasolines generally of higher octane number than those obtained by simple distillation of the same crude oils. The general nature of the hydrocarbon products and the basic mechanism of thermal cracking is well described by the free-radical theory of the pyrolysis of hydrocarbons. *SEE PYROLYSIS.*

In the early 1930s, the petrochemical industry began its growth. Olefinic gases from thermal cracking operations, especially propylene and butylenes, were used as the chief raw materials for the production of aliphatic chemicals. Stimultaneously, practical catalytic processes were invented for polymerizing propylene and butylenes to gasoline components, and for dimerizing and hydrogenating isobutylene to isooctane (2,2,4-trimethylpentane), the prototype 100-octane fuel. Just prior to World War II, the alkylation of light olefins with isoparaffins to produce unusually high-octane gasoline components was discovered and extensively applied for military aviation use. These developments resulted in intense engineering efforts to bring thermal cracking to maximum efficiency, as exemplified by a number of commercial processes made available to the industry.

Since World War II, thermal cracking has been largely supplanted by catalytic cracking, both for the manufacture of high-octane gasoline and as a source of light olefins. It is, however, still used widely for the mild cracking of heavy residues to reduce their viscosities and for the final cracking of low-grade residuum.

Since carbon-to-hydrogen bonds are also severed in the course of thermal cracking, two hydrogen atoms can be removed from adjacent carbon atoms in a saturated hydrocarbon, producing molecular hydrogen and an olefin. This reaction prevails in ethane and propane cracking to yield ethylene and propylene, respectively. Methane cracking is a unique case wherein molecular hydrogen is obtained as a primary product and carbon as a coproduct. These processes generally operate at low pressures and high temperatures and in some cases utilize regenerative heating chambers lined with firebrick, or equipment through which preheated refractory pebbles continuously flow. Such conditions also favor the production of aromatics and diolefins from normally liquid feedstocks and are applied commercially to a limited extent despite relatively low yields of the desired products.

Catalytic cracking. This is the major process used throughout most of the world oil industry for the production of high-octane quality gasoline by the conversion of intermediate- and high-boiling petroleum dis-

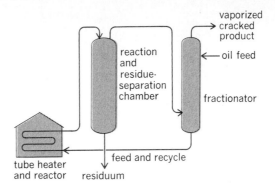

Fig. 1. Thermal cracking unit.

tillates to lower-molecular-weight products. Many aspects involve proprietary technology. In general, oil heated to within the lower range of thermal cracking temperatures (850–1025°F or 454–551°C) reacts in the presence of an acidic inorganic catalyst, typically a silica-alumina zeolite, under low pressures (10–35 psig or 70–240 kilopascals). Gasoline of much higher octane number is obtained than from thermal cracking, a principal reason for the widespread adoption of catalytic cracking. All nonvolatile carbonaceous materials are deposited on the catalyst as coke and are burned off during catalyst regeneration. *SEE ZEOLITE.*

In contrast to thermal cracking, low-grade residual oils are not generally processed, because excessive amounts of coke are deposited on the catalyst, and inorganic components of these oils contaminate the catalyst. Preferential feedstocks are heavier-boiling distillate gas oils. However, in recent years with catalyst improvement, residual stocks can also be processed. These stocks are generally good-quality residuals in low concentration as they yield 2–3 barrels of gas oil feed when processed in a full unit.

Many processes have become available to improve catalytic cracking feedstocks. These include hydrogenation, deasphalting (including the Rose process), deep-vacuum distillation, and separations or low-conversion processes such as the Art process. These processes generally reject the poorer components of the feed, such as Conradson carbon (a residual carbon), metallic contaminants, and possibly heavy aromatics. Hydrogenation can upgrade feedstocks by saturation or aromatics, and remove metals to considerably improve catalytic cracker product selectivity.

Catalytic cracking, as conceived by E. J. Houdry in France, reached commercial status in 1936 after extensive engineering development by American oil companies. In its first form, the process used a series of fixed beds of catalyst in large steel cases. Each of these alternated between oil cracking and catalyst-coke burning at intervals of about 10 min and provided for heat and temperature control.

Successful operation led to major engineering improvements, and the goals of much improved efficiency, enlarged capacity, and ease of operation were achieved by two different systems. One employs a moving bed of small pellets or beads of catalyst traveling continuously through the oil-cracking vessel and subsequently through a regeneration kiln. The beads are lifted mechanically or by air to the top of the structure to flow down through the vessels again. This process has two commercially engineered embodiments, the Thermofor (or TCC) and the Houdriflow processes, which are similar in general process arrangements.

These moving-bed processes are limited in size and

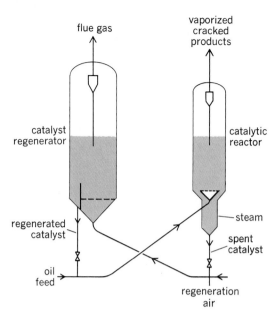

flue gas

vaporized cracked products

catalyst regenerator

catalytic reactor

regenerated catalyst

steam

spent catalyst

oil feed

regeneration air

Fig. 2. Fluid catalytic cracking unit.

became technically obsolete. They generally have been replaced by another type of unit, the fluid-solids, as dictated by economic considerations, and no new moving-bed units are being constructed. In the fluid-solids unit, a finely divided powdered catalyst is transported between oil-cracking and air-regeneration vessels in a fluidized state by gaseous streams in a continuous cycle. This system employs the principle of balanced hydrostatic heads of fluidized catalyst between the two vessels. Catalyst is moved by injecting heated oil vapors into the transport line from the regenerator to the reactor, and by injecting air into the transport line from the reactor stripper to the regenerator. Large amounts of catalyst can be moved rapidly; cracking units of total oil intake as great as 180,000 bbl/day (28,800 m³/day) are in operation (**Fig. 2**).

In both the moving-bed and fluidized systems, the circulating catalyst provides the cracking heat. Coke deposited on the catalyst during cracking is burned at controlled air rates during regeneration; heat of combustion is converted largely to sensible heat of the catalyst, which supplies the endothermic heat of cracking in the reaction vessel.

Gasoline of 90–95 research octane number without tetraethyllead is rather uniformly produced by catalytic cracking of fractions from a wide variety of crude oils, compared with 65–80 research octane number via thermal cracking, the latter figures varying with crude oil source. *See* OCTANE NUMBER.

Although the primary objective of catalytic cracking is the production of maximum yields of gasoline concordant with efficient operation of the process,

large amounts of normally gaseous hydrocarbons are produced at the same time. The gaseous hydrocarbons include propylene and butylenes, which are in great demand for chemical manufacture. Isobutane and isopentane are also produced in large quantities and are valuable for the alkylation of olefins, as well as for directly blending into gasoline as high-octane components.

The other chief product is the material boiling above gasoline, designated as catalytically cracked gas oil. It contains hydrocarbons relatively resistant to further cracking, particularly polycyclic aromatics. The ligher portion may be used directly or blended with straight-run and thermally cracked distillates of the same boiling range for use as a diesel oil component and heating oils. Part of the heavier portion can be recycled with fresh feedstock to obtain additional conversion to lighter products. The remainder is withdrawn for blending with residual oils to reduce the viscosity of heavy fuel, or else subjected to a final step of thermal cracking.

Thus, the catalytic cracking process is used in refineries to shift the production of products to match swings in market demand. It can process a wide variety of feeds to different product compositions. For example, light gases, gasoline, or diesel oil can be emphasized by varying process conditions, feedstocks, and boiling range of products (see **table**).

To account for the difference between the product compositions obtained by catalytic and thermal cracking, the mechanism of cracking over acidic catalysts has been investigated intensively. In thermal cracking, free radicals are reaction intermediates, and the products are determined by their specific decomposition patterns. In contrast, catalytic cracking takes place through ionic intermediates, designated as carbonium ions (positively charged free radicals) generated at the catalyst surface. Although there is a certain parallelism between the modes of cracking of free radicals and carbonium ions, the latter undergo rapid intramolecular rearrangement reactions prior to cracking. This leads to more highly branched hydrocarbon structures than those from thermal cracking, and to important differences in the molecular weight distribution of the cracked products. Furthermore, the cracked products undergo much more extensive secondary reactions with the catalyst. *See* REACTIVE INTERMEDIATES.

The catalytic cracking mechanism also favors the production of aromatics in the gasoline boiling range; these reach quite high concentrations in the higher-boiling portion. This characteristic, together with the copious production of branched aliphatic hydrocarbons especially in the lower-boiling portion, is largely responsible for the high octane rating of catalytically cracked gasoline.

Representative yield structures for three different processing objectives in catalytic cracking

Process variables	Light gases	Gasoline	Light cycle oil
Feed	Light gas oil	Gas oil	Gas oil
Reactor temperature, °F (°C)	990 (532)	950–990 (510–532)	890–900 (477–482)
Light gases, wt %	4.5	2.8	1.6
Propane/propylene, vol %	15	10.0	7.5
Butane/butylene, vol %	22	16.4	11.2
Gasoline, vol %	46	69.5	32.6
Catalytic diesel oil, vol %	18	10	43.6
Bottom, vol %	5	5	5

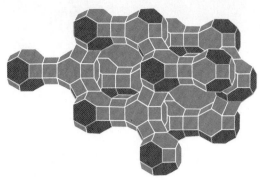

Fig. 3. Model of zeolite type Y.

Cracking catalysts must have two essential properties: (1) a chemical composition capable of maintaining a high degree of acidity, preferably as readily available hydrogen ions (protons); and (2) a physical structure of high porosity so that these active sites are available for cracking. Mechanical durability is also necessary for industrial use.

Cracking catalysts are essentially silica-alumina compositions. A dramatic improvement in catalytic unit performance occurred with the switch from acid-treated clays (montmorillonite or kaolinite) to synthetic silica-aluminas. After 1960, a new group of aluminosilicates, molecular sieve zeolites, were introduced into the catalyst formulation. These crystalline materials (**Fig. 3**) have cracking activity 50 to 100 times the previous amorphous catalyst. They permit cracking to greater conversion levels, producing more gasoline, less coke, and less gas. *SEE ZEOLITE.*

As the catalyst particles pass through the reactor regenerator system every 3 to 15 min they are gradually deactivated through loss of active surface area by the effect of heat and steam and through contamination by the trace metallic components on the feedstocks, mainly sodium and vanadium. The catalyst particles also undergo mechanical attrition, and fines are lost in the reactor and regenerator gases. To compensate, fresh catalyst is added.

Cracking catalysts have evolved considerably since the mid-1960s, when zeolite crystals were incorporated into the matrix. The sodium and hydrogen ions in the crystals have been replaced with rare-earth elements to provide thermal stability. Surface area of the catalysts has been reduced to provide increased resistance to contaminant metals (Ni and V) and to reduce coke-make of the reaction.

With increased levels of metals on the catalysts, processes have been developed to passivate the effect of the metals. One process involves addition of antimony to the catalytic feedstock. Antimony thus incorporated into the catalyst can reduce hydrogen and carbon yield attributable to the metals.

The new zeolite catalysts have resulted in considerable change in the process itself. The catalysts are made more resistant to thermal degradation primarily by the exchange of sodium and hydrogen ions in the zeolite with rare-earth elements. Regenerator temperatures can be safely raised to the 1350°F (732°C) or greater level. The carbon on regenerated catalyst is reduced to the 0.05 wt % level resulting in improved gasoline yields. In addition, all the carbon monoxide produced at lower generation temperatures (10% concentration in the regenerator flue gas) can be combusted in the regenerator if desired, making for a

more efficient recovery of combustion heat and reduced atmospheric pollution. The effluent CO concentration can be reduced to less than 0.05 vol%, if desired.

Additives have been developed to improve the efficiency of regeneration. Combustion promoters can ensure complete combustion of the flue gas to CO_2 even at relatively low temperatures of 1250°F (695°C). These promoters generally contain an element from the platinum group in extremely low concentrations, the ppm range. They may be added as separate powders or in admixtures with the catalyst. Other additives are being produced which are claimed to reduce sulfur oxides in the flue gas.

The high-activity zeolite catalyst has permitted units to be designed with all riser cracking (**Fig. 4**), wherein all the cracking reaction takes place in a relatively dilute (less than 2 lb/ft³ or 3.2 kg/m³) catalyst suspension in a 2- or 5-s residence time. No dense-bed (10–15 lb/ft³ or 16–24 kg/m³) cracking exists in these units.

Many old units are being converted to riser crackers, and virtually all new units feature riser cracking. Some riser crackers are also provided with small dense beds to achieve an optimum yield pattern. Some have riser-type regenerators directed to improve regenerator efficiency.

Over the years, there have been continued improvements to the process in many areas. Feeds available for processing have become much heavier, and include stocks such as vacuum residua containing contaminant metal up to 30–40 ppm. Many catalyst formulations have been developed to increase catalytic activity and to resist the high level of contaminant metals on the catalyst. Levels as high as 10,000 ppm of nickel and vanadium on catalyst have been run commercially. Nickel, vanadium, and copper catalyze detrimental competing dehydrogenation reactions that produce coke and light gases (hydrogen and methane). These heavier stocks have produced larger

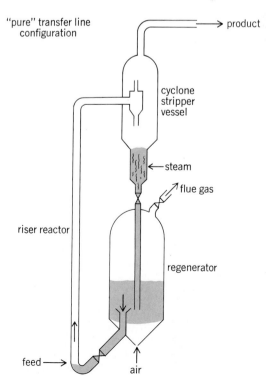

Fig. 4. Riser catalytic cracker.

amounts of coke which must be burned in the regenerator, causing regenerator temperatures to rise. Staged regeneration is reported to permit temperatures as high as 1500°F (815°C), since water of combustion is liberated in the lower-temperature first stage, and thus hydrothermal deactivation is not as much a factor in the second stage. Additionally, facilities have been provided internally and externally to the regenerator to remove excess heat of combustion.

With more emphasis on environmental protection, complex facilities are provided to remove pollutants from the effluent regeneration gases. Third-stage cyclone collectors, electrostatic precipitators, and scrubbers are used commercially to meet regulations.

To conserve energy in the process, flue gas expanders are used in the flue gas system to provide more than enough energy to compress the regeneration air.

With the continuing emphasis on conservation of petroleum resources, catalytic cracking assumed a more important role. The heavier products from a refinery could be replaced by coal-, shale-, or tar-sand-derived products. This creates a need for greater conversion of petroleum to gasoline, with emphasis on catalytic cracking since it is the cheapest major conversion process. *SEE ALKYLATION (PETROLEUM); AROMATIZATION; COAL LIQUEFACTION; HETEROGENEOUS CATALYSIS; ISOMERIZATION; PETROLEUM PROCESSING; TUBE-STILL HEATER.*

Edward Luckenbach

Bibliography. F. H. Blanding, Reaction rates in catalytic cracking of petroleum, *I and EC*, 45:1185–1197, June 1953; G. D. Hobson, *Modern Petroleum Technology*, 5th ed., 1983; W. L. Nelson, *Petroleum Refinery Engineering*, 1958; E. I. Shaheen, *Catalytic Processing in Petroleum Refining*, 1983; P. B. Venuto and E. T. Habib, Jr., *Fluid Catalytic Cracking with Zeolite Catalysts*, 1979.

Cranberry

The large-fruited American cranberry, *Vaccinium macrocarpon*, a member of the heath family, Ericaceae, is a native plant of open, acid peat bogs in northeastern North America. Selections from the wild have been cultivated since the early nineteenth century. It is an evergreen perennial vine producing runners and upright branches with conspicuous terminal flower buds (**Fig. 1**). *SEE ERICALES.*

Cultivation. A cranberry bog is made by removing the vegetation on a maple, cedar, or brown-brush swamp, draining it by cutting "shore," "lateral," and "main" ditches, spreading 2 or 3 in. (5 or 8 cm) of sand over the leveled peat, and inserting cuttings of the selected variety of vines through the sand into the peat. Rooting occurs readily within a month, but 3 or 4 years of growth and care are required before the first commercial crop is produced.

Care includes frost protection in spring and fall, now largely provided by solid-set, low-gallonage sprinkler systems which effect protection much quicker and with only a tenth of the water formerly required with flood frost protection. Flooding is used in winter to protect the cranberry vines, not from cold but from desiccation when the root zone is frozen. Early-season floods are used for drowning pest insects.

Because well-tended cranberry bogs continue to produce annual crops for a century or more without replanting, a half-inch (1.3 cm) layer of sand is spread every 3 or 4 years over the vines (by spreading

Fig. 1. Cranberry uprights in full bloom.

on winter ice when thick enough) to cover the accumulating cranberry leaves on the soil surface where the cranberry girdler insect would otherwise breed and multiply.

Harvesting. Commercial cranberry growing is confined to Massachusetts, New Jersey, Wisconsin, Washington, and Oregon and to several provinces in Canada. Only about 20% of the cranberries are sold as fresh fruit, and most of these are dry-harvested by machines. Most of the processing fruit is harvested in flood waters, either by machines which pick and deliver the fruit into towed plastic boats or by water reels which detach the berries to float and be driven by wind to shore where they are elevated into bulk trucks. Berries for juice manufacture are usually vine-ripened and deep-frozen for a month or more prior to thawing and extraction. The older form of hand harvest with wood-toothed scoops is now gone, the scoops sometimes being used for ditch-edge picking.

Good-quality fresh cranberries can be stored for several months, refrigerated, with very little loss to decay. Good-quality cranberries can be kept in deep-freeze storage for several years with only minor moisture loss. Frozen berries on thawing are soft and juicy, unlike the firm fresh berry, and must be utilized promptly.

Cranberry bogs in full bloom in early July are a sight to remember because they have 40,000,000 to 50,000,000 white or pink flowers per acre (100,000,000 to 125,000,000 per hectare), each of which must be visited by pollinating honeybees or bumblebees to set the berries. The control of frost and winter injury, the control of weeds, insects, and diseases, and the development of more efficient harvest techniques have seen productivity rise. Supplies of the fruit are ample to satisfy both domestic and export demand.

The special requirements of the cranberry plant for low fertility, acid soil, and winter protection make it a poor choice for home garden cultivation. *SEE FRUIT GROWING, SMALL.*

Chester E. Cross

Diseases. Fruit rots are the most economically important diseases of cranberry. Eight species of fungi cause almost all of the loss attributed to these diseases.

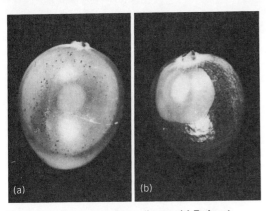

Fig. 2. Symptoms of cranberry disease. (*a*) Early rot symptoms on fruit; black specks in the affected area are the fruiting bodies of the fungus. (*b*) Early symptoms of blotch rot on fruit; all fungal fruit rots appear identical at this stage.

the host plant are all dependent on the presence of adequate moisture. Bruising during harvest, sorting, or packing significantly increases the amount of post-harvest fruit rot.

Infection before harvest is common in the production areas with warmer climates and higher rainfall. In such areas fungicide sprays, at midbloom and every 14 days until all blossoms have fallen, have provided excellent control.

Practices that reduce fruit rots include rapid drying of the fruit, reducing fruit temperatures to 36–40°F (2–4°C), and reducing the time the fruit remains in storage. Dry harvest results in less disease than wet harvest. SEE PLANT PATHOLOGY.

John K. Springer

Bibliography. H. F. Bergmann, Disorders of cranberries, *Yearbook of Agriculture*, pp. 789–796, 1953; E. G. Hall and K. J. Scott, *Storage and Market Diseases of Fruit*, 1982.

Symptoms of all fruit rots are almost identical, and include the development of a discolored spot on the surface of the fruit; the spot enlarges rapidly until the entire berry is affected (**Fig. 2**). Later the infected berry shrivels into a raisinlike mummy.

Several factors influence the amount of fruit rot that occurs. Higher temperatures during the growing season increase fungal activity and the level of disease. High moisture levels also increase disease, because spore release, spore germination, and penetration of

Cranial nerve

Any peripheral nerve which has its central nervous system connection with the brain, as opposed to the spinal cord, and reaches the brain through a hole (foramen) in the skull. Nerve fibers are sensory if they carry information from the periphery to the brain, and motor if they carry information from the brain to the periphery. Sensory fibers are classified as somatic sensory if they come from the skin or muscle sense

Cranial nerves of vertebrates				
Number	Name	Fiber types	Peripheral origin or destination	Vertebrates possessing this nerve
—	Terminal	Somatic sensory	Anterior nasal epithelium	Almost all
I	Olfactory	Special sensory	Olfactory mucosa	All
—	Vomeronasal	Special sensory	Vomeronasal mucosa	Almost all
II	Optic	Special sensory	Retina of eye	All
III	Oculomotor	Somatic motor	Four extrinsic eye muscles	All
IV	Trochlear	Somatic motor	One extrinsic eye muscle	All
V	Trigeminal	Special visceral motor	Muscles of mandibular arch derivative	All
		Somatic sensory	Most of head	All
VI	Abducens	Somatic motor	One extrinsic eye muscle	All
—	Anterior lateral line	Special sensory	Lateral line organs of head	Fish and larval amphibians
VII	Facial	Special visceral motor	Muscles of hyoid arch derivative	All
		General visceral motor	Salivary glands	All
		Somatic sensory	Small part of head	All
		Visceral sensory	Anterior pharynx	All
		Special sensory	Taste, anterior tongue	All
VIII	Vestibulocochlear	Special sensory	Inner ear	All
	Posterior lateral line	Special sensory	Lateral line organs of trunk	Fish and larval amphibians
IX	Glossopharyngeal	Special visceral motor	Muscles of third branchial arch	All
		General visceral motor	Salivary gland	All
		Somatic sensory	Skin near ear	All
		Visceral sensory	Part of pharynx	All
		Special sensory	Taste, posterior tongue	All
X	Vagus	Special visceral motor	Muscles of arches 4–6	All
		General visceral motor	Most viscera of entire trunk	All
		Visceral sensory	Larynx and part of pharynx	All
		Special sensory	Taste, pharynx	All
XI	Spinal accessory	Special visceral motor	Some muscles of arches 4–6	Reptiles, birds, mammals
XII	Hypoglossal	Somatic motor	Muscles of tongue and anterior throat	Reptiles, birds, mammals

organs, visceral sensory if they come from the viscera, and special sensory if they come from special sense organs such as the eye and ear. Motor fibers are classified as somatic motor if they carry information to somatic striated muscles; general visceral motor if they carry information to glands, smooth muscle, or cardiac muscle; and special visceral motor if they carry information to visceral striated muscle. A cranial nerve may have only one fiber type or several; cranial nerves with several fiber types are called mixed nerves.

In mammals, 12 pairs of cranial nerves, numbered I through XII, are usually described (see **table**). However, mammals also have an anterior unnumbered nerve, the terminal nerve. Many vertebrates also have two pairs of lateral line nerves (unnumbered), and lack discrete nerves XI and XII. *See* Brain; Nervous system (vertebrate).

Douglas B. Webster

Craniata

The major subphylum of the phylum Chordata comprising the vertebrates, or the cyclostomes to mammals. The subphylum is also known as the Vertebrata. Their most distinctive characteristic is the cranium or braincase. A column of articulated vertebrae extends down the body from the cranium to the end of the tail in all forms except some cyclostomes (the column is absent in myxinoids and very small in petromyzontids); it protects the nerve cord and supports the body. Other distinctive vertebrate features are bony tissue, a kidney with the nephron as the basic unit, a heart, red and white blood cells, a liver and a pancreas, specialized sense organs such as a distinctive eye and a lateral line organ plus ear, several unique endocrine organs such as the pituitary and thyroid, a true skin, and the neural crest—a unique by-product of neural tube formation that gives rise to a series of features from gill cartilages and pigment cells to the adrenal medulla.

The earliest vertebrates—agnathan fish—lacked jaws and teeth, paired fins, and a lung, all of which originated in the higher groups, the gnathostomes. Vertebrates originated in fresh waters, with the kidney being a water balance organ, and heavy dermal bone being protection against larger predatory invertebrates. They were initially slow-moving bottom-dwelling forms, become free-swimming pelagic later, with several evolutionary lines reinvading the oceans. The main line of evolution leading to the tetrapods remained always in fresh waters. *See* Vertebrata.

Walter J. Bock

Crank

In a mechanical linkage or mechanism, a link that can turn about a center of rotation. The crank's center of rotation is in the pivot, usually the axis of a crankshaft, that connects the crank to an adjacent link. A crank is arranged for complete rotation (360°) about its center; however, it may only oscillate or have intermittent motion. A bell crank is frequently used to change direction of motion in a linkage (see **illus.**). *See* Linkage (mechanism).

In mechanisms where energy input fluctuates, it is sometimes useful to design a crank stoutly, as a form

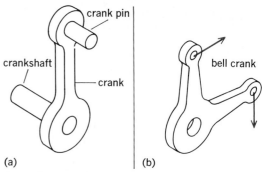

Cranks (a) for changing radius of rotation, and (b) for changing direction of translation.

of flywheel, so that considerable rotational energy is stored in it, tending thereby to make the output smoother inasmuch as the inertia of such a crank absorbs peak input and releases energy more smoothly to output.

Input and output cranks of a four-bar linkage may express in their relative angular placement a relation $\beta = \beta(\alpha)$, which is excellent in control operations where electric current computers would be vulnerable. *See* Four-bar linkage.

Douglas P. Adams

Creep (materials)

The time-dependent strain occurring when solids are subjected to an applied stress. *See* Stress and strain.

Kinds of phenomena. Some of the different kinds of creep phenomena that can be exhibited by materials are shown in the **illustration**. The strain $\epsilon = \Delta L/L_0$, in which L_0 is the initial length of a body and ΔL is its increase in length, is plotted against the time t for which it is subjected to an applied stress. The most common kind of creep response is represented by the curve A. Following the loading strain ϵ_0, the creep rate, as indicated by the slope of the curve, is high but decreases as the material deforms during the primary creep stage. At sufficiently large strains, the material creeps at a constant rate. This is called the secondary or steady-state creep stage. Ordinarily this is the most important stage of creep since the time to failure t_f is determined primarily by the secondary creep rate $\dot{\epsilon}_s$. In the case of tension creep, the secondary creep stage is eventually interrupted by the onset of tertiary creep, which is characterized by internal fracturing of the material, creep acceleration, and finally failure. The creep rate is usually very temperature-dependent. At low temperatures or applied stresses the time scale can be thousands of years or longer. At high temperatures the entire creep process can occur in a matter of seconds. Another kind of creep response is shown by curve B. This is the sort of strain-time behavior observed when the applied stress is partially or completely removed in the course of creep. This results in time-dependent or anelastic strain recovery.

Examples. Creep of materials often limits their use in engineering structures. The centrifugal forces acting on turbine blades cause them to extend by creep. In nuclear reactors the metal tubes that contain the fuel undergo creep in response to the pressures and forces exerted on them. In these examples the occur-

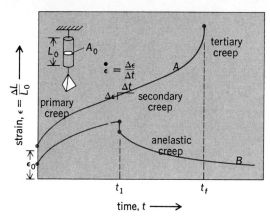

Typical creep curves for materials.

rence of creep is brought about by the need to operate these systems at the highest possible temperatures. Creep also occurs in ordinary structures. For example, when a bolt is tightened to fasten two parts together, the tension stress in the bolt can be quite high. Over a long period of time the bolt creeps, with the result that the force holding the two parts together diminishes. Another example of stress relaxation caused by creep is found in prestressed concrete beams, which are held in compression by steel rods that extend through them. Creep and stress relaxation in the steel rods eventually leads to a reduction of the compression force acting in the beam, and this can result in failure. *See* Prestressed concrete.

There are many examples of creep that are not related to high technology. The window panes in very old homes or cathedrals are often thicker at the bottom than they are at the top due to creep under the force of gravity. Similarly, when a glacier flows down the side of a mountain the ice undergoes extensive creep deformation.

Mechanism. The mechanism of creep invariably involves the sliding motion of atoms or molecules past each other. In amorphous materials such as glasses, almost any atom or molecule within the material is free to slide past its neighbor in response to a shear stress. In plastics, the long molecular chains can slide past each other only to a limited extent. Such materials typically show large anelastic creep effects (curve B in the illustration).

For crystalline materials, creep deformation also involves the sliding of atoms past each other, but here the sliding can occur only within the cores of crystal dislocations. Thus, creep of metals and ceramics is usually governed by the motion of dislocations. In addition to gliding along their slip planes, the dislocations can climb out of their slip planes by the absorption or emission of lattice vacancies. This process is controlled by the rate at which lattice vacancies diffuse to or from climbing dislocations and gives rise to the strong temperature dependence of creep.

Materials with creep resistance. Understanding of the mechanisms that control creep deformation makes it possible to design materials with superior creep resistance. When solute atoms are added to metals, they are attracted to the strain fields of the dislocations. There they inhibit dislocation motion and in this way improve the creep resistance. Many of the aluminum alloys used for aircraft structures are strengthened in this way. The addition of second-phase particles to alloys is another way to improve

the creep resistance. The most effective strengthening phases are oxides, carbides, or intermetallic phases, because they are usually much stronger than the host metal and therefore create strong obstacles to dislocation motion. Materials containing finely dispersed, strong particles of a stable phase are usually very creep-resistant. Nickel-based superalloys, used in gas-turbine engines, derive their creep resistance from these effects. *See* Aluminum alloys; Crystal defects; Diffusion; High-temperature materials; Metal, mechanical properties of; Plasticity; Rheology.

William D. Nix; Jeffery C. Gibeling; Kevin J. Hemker

Bibliography. M. F. Ashby and L. M. Brown (eds.), *Perspectives in Creep Fracture*, 1983; A. C. Cocks (ed.), *Mechanics of Creep Brittle Materials*, 1988; H. E. Evans, *Mechanisms of Creep Fracture*, 1984; J. H. Gittus, *Creep, Viscoelasticity and Creep Fracture in Solids*, 1975; H. Kraus, *Creep Analysis*, 1980; H. J. McQueen and W. J. McGregor Tegart, The deformation of metals at high temperatures, *Sci. Amer.*, 232(4):116–124, April 1975; J. P. Poirier, *Creep of Crystals*, 1985; H. Ruesch et al., *Creep and Shrinkage: Their Effect on the Behavior of Concrete Structures*, 1983.

Creeping flow

Fluid flow in which the velocity of the flow is very small. For creeping flow, the Reynolds number $R_e = Ud/v$ is small (less than unity). Here U is the reference velocity, d is the reference length, and v is the kinematic viscosity of the fluid. For low Reynolds number, the inertial force is negligible and the nonlinear terms in Navier-Stokes equations are neglected. The resultant flow is known as Stokes flow. One of the important applications of creeping flow is the motion of a tiny particle in a viscous flow, which is important in two-phase flow. If the particle is a sphere, its drag coefficient is $C_D = 24/R_e$ if R_e is less than 1, where the reference length d is the diameter of the sphere. When R_e is greater than 1, but not too large, the drag coefficient is a little larger than $24/R_e$. Such a flow is known as Oseen's flow, and may be considered as an upper limit of creeping flow. Another application of creeping flow is the lubrication problem which was initiated by O. Reynolds, who showed that two parallel or near-parallel surfaces can slide one over the other with only slight frictional resistance, even under great normal pressure, provided that a film of viscous flow is maintained. *See* Fluid flow; Fluid-flow principles; Navier-Stokes equations; Reynolds number; Viscosity.

S. I. Pai

Bibliography. J. Happel and H. Brenner, *Low Reynolds Number Hydrodynamics*, 1965; W. F. Hughes, *Introduction to Viscous Flow*, 1979; S. I. Pai, *Two-Phase Flows*, 1977; S. I. Pai, *Viscous Flow Theory*, vol. 1: *Laminar Flow*, 1956; F. M. White, *Viscous Fluid Flow*, 1974.

Creosote

Either of two products of tar distillation: commercial creosote is a toxic substance obtained by high-temperature carbonization of soft coal, and pharmaceutical creosote is an aromatic mixture obtained by redistillation of wood tar.

Commercial creosote is obtained from creosote oil, which amounts to about 30–35% of coal tar. It contains numerous tar acids such as cresols, xylenols, naphthols, and higher homologs, as well as various tar bases such as picolines and collidines, and hydrocarbons. The boiling range varies from 200 to 350°C (390–660°F), and commercial creosote has a specific gravity greater than water. It is usually prepared according to standards established by the American Wood-Preservers Association and is used for wood treatment. Small amounts are consumed in the manufacture of lampblack, in disinfectant sprays for fruit trees, and as melting auxiliaries for pitch and bitumen.

Hardwood creosote is prepared from settled hardwood tar, chiefly derived from beechwood. The tar is dried and distilled at 140–150°C (280–300°F), the pyroligneous acid separated and the acidic residue dissolved in caustic. The alkaline solution is steam distilled to remove tar bases and hydrocarbons; then the crude creosote is fractionally precipitated with a dilute mineral acid such as sulfuric acid. The precipitated oils are dried and redistilled to furnish pharmaceutical-grade creosote, formerly used in medicine for treatment of bronchitis and as an antiseptic. See COAL CHEMICALS; CRESOL; WOOD PRESERVATION.

Frank Wagner

Cresol

Any one of the three methylphenols. The name is also often used to describe a mixture of the three. The structural formulas for the individual isomers are shown in the **illustration**.

Properties. At room temperature the ortho and para isomers are colorless solids when highly purified, whereas the meta isomer is a liquid, as is a mixture of the three. The cresols are moderately soluble in water and readily soluble in alcohol and ether. They are only slightly less acidic than phenol, and they also resemble the parent substance in toxicity.

As typical phenols the cresols can be converted into ethers and esters, in which the hydrogen of the hydroxyl group is replaced by an alkyl or acyl group, respectively. The compounds are quite susceptible to oxidation, which accounts for the dark color which liquid cresol acquires upon aging. In the presence of appropriate catalysts, cresols combine with three molecular equivalents of hydrogen to form methylcyclohexanols. All three isomers are quite reactive in aromatic substitution reactions, such as the reaction with bromine, which is shown below for *para*-cresol.

Note that the hydroxyl group dominates the path of substitution, so that the bromine is introduced preferentially in the positions indicated rather than in the alternative positions.

Production. In industry the cresols are usually associated with products known as cresylic acids, which also contain varying amounts of xylenols and other high-boiling phenolic substances. The traditional

Structural formulas for the three isomeric methylphenols: (a) *ortho*-cresol, (b) *meta*-cresol, (c) *para*-cresol.

sources are the by-product tar acids from coke ovens and the petroleum refinery caustic-wash extracts. The production of cresols by synthetic methods has grown steadily, particularly those methods that produce single isomers needed for many modern uses. The syntheses in use parallel some of those used in phenol production. Cresols may also be produced by the alkylation of phenol.

Uses. Solutions of cresol have long been used as disinfectants. Large quantities are utilized in the manufacture of synthetic resins, antioxidants, surface-active agents, plasticizers, and antibacterial agents. The phosphate ester of cresol is used as an additive for fuel and lubricating oils. See PHENOL.

Martin Stiles

Cress

A prostrate hardy perennial crucifer of European origin belonging to the plant order Capparales. Watercress (*Nasturtium officinale*) is generally grown in flooded soil beds and used for salads and garnishing. Propagation is by seed or stem cuttings. High soil moisture is necessary. Leafy stems are cut usually 180 days after planting. Virginia is an important producing state.

Garden cress (*Lepidium sativum*) is a cool-season annual crucifer of western Asian origin grown for its flavorful leaves. Propagation is by seed, and leaves are harvested approximately 2 months after planting.

Upland or spring cress (*Barbarea verna*) is a biennial crucifer of European origin, grown and harvested similarly to garden cress but of lesser commercial importance. See CAPPARALES; VEGETABLE GROWING.

H. John Carew

Cretaceous

The latest system of rocks or period of the Mesozoic Era. It includes that part of the Earth's history from about 127,000,000 to 64,000,000 years ago—the time of the greatest flooding of the Earth's surface during the Mesozoic Era. It was also the time of extensive chalk deposition over much of the Northern Hemisphere; hence the origin of the term *terrains crétacés* by d'Omalius d'Halloy in 1822.

The Cretaceous, bounded below by the Jurassic and above by the Tertiary, marks the rapid rise and spread of deciduous trees over the landscape. Its close was marked by the withdrawal of the seas from the continents, and by the extinction of the dinosaurs, flying reptiles and huge marine reptiles, and the ammonites and certain other groups of marine mollusks that had been so conspicuous throughout the Mesozoic Era. See JURASSIC; TERTIARY.

Paleogeography. Much of the Earth's present land surface was submerged by shallow seas during the

| CENOZOIC | QUATERNARY |
| | TERTIARY |

(geologic time scale chart)

- CENOZOIC: QUATERNARY, TERTIARY
- MESOZOIC: CRETACEOUS, JURASSIC, TRIASSIC
- PALEOZOIC: PERMIAN, CARBONIFEROUS (PENNSYLVANIAN, MISSISSIPPIAN), DEVONIAN, SILURIAN, ORDOVICIAN, CAMBRIAN
- PRECAMBRIAN

rican part of Gondwana and India. Another broader arm of the Pacific Ocean crossed present-day Central America and formed a North Atlantic Ocean to the present British Isles. SEE CONTINENTS, EVOLUTION OF.

A rift began splitting South America from Africa near the close of the Jurassic. This rift began in the south and, by the beginning of the Cretaceous, had worked its way north as far as present-day Nigeria. Lacustrine sediments formed at first in the rift, and later beds of salt, anhydrite, and black shales accumulated. By the end of the Early Cretaceous, the rift had widened and deepened enough to allow marine water to advance as far north as Nigeria. A rift had also developed trending northwest from Nigeria to the North Atlantic Ocean. Inasmuch as North America was drifting westward and Africa northward, the continents separated completely in the early part of the Late Cretaceous. A rift that had formed earlier in the middle of the North Atlantic Ocean eventually split Greenland from Eurasia as North America drifted westward. By the close of the Cretaceous, Madagascar had rifted from Africa, and India had drifted northward toward its present position.

Broad climatic belts were present in the Cretaceous. Most workers recognize two belts, a southern Mediterranean or Tethyan belt and a northern or boreal belt. The Mediterranean belt, extending from Central America eastward across southern Europe and northern Africa to India and the East Indies, was characterized by much deposition of limestone and other calcareous rocks. Characteristic fossils include large thick-shelled gastropods, certain ammonites, and sessile bivalves known as rudistids which formed reefs. The boreal belt was characterized by an extinct group of squids known as belemnites, certain bivalves (such as aucellas), and numerous ammonites. Temperatures were warmer than at present and probably more uniform. Palms grew as far north as Alaska and Greenland.

The Cretaceous is generally divided into two epochs, Lower Cretaceous and Upper Cretaceous. These are subdivided into stages whose names derive from localities in western Europe, where the Cretaceous was first studied intensively (see **table**). The Senonian is commonly divided into three substages: Campanian (Champagne, France), Santonian (Saintonge, France), and Coniacian (Cognac, France). The Neocomian is often divided into four substages: Barremian (Barrème, France), Hauterivian (Hauterive, France), Valanginian (Valangin, France), and Berriasian (Berrias, France).

Lower Cretaceous. During Early Cretaceous time three seas—Tethys, Northern, and Russian—covered much of Europe. The Tethys submerged the northern

Cretaceous. Areas that were most extensively and frequently flooded include Europe, northernmost Africa, Madagascar, northern India, Japan, the western margin of North and South America, Mexico and the Gulf coastal plain of the United States, and the western interior of the United States and Canada. Two seas covered much of Europe most of the time. A Northern sea extended from the British Isles eastward across Germany and Poland to Russia. A southern sea, Tethys, covered most of the present Mediterranean Sea and the surrounding parts of southern Europe and northernmost Africa. The Tethys sea also crossed Asia as a belt through Turkey, the Persian Gulf area, West Pakistan, the Himalayan region, Assam, Burma, and Sumatra. The Tethys is believed to have been moderately deep, whereas the Northern sea and the vast Western Interior sea of North America were much shallower.

According to concepts of continental drift, at the beginning of the Cretaceous the Earth consisted of four closely spaced landmasses and a vast Pacific Ocean. The landmasses consisted of South America united to Africa, an India landmass lying to the east, North America united to Europe by way of Greenland, and Antarctica united to Australia. The northern landmass (North America–Eurasia) is known as Laurasia, and the rest of the landmasses are known collectively as Gondwana. Laurasia and Gondwana almost touched in one place, at what is now Spain and Morocco. Eastward from this point, a narrow seaway (Tethys Sea) extended to the Pacific Ocean and separated the Eurasian part of Laurasia from the north Af-

Stages of the Cretaceous

Epoch	Stage	Source of name
Upper Cretaceous	Maestrichtian	Maastricht, Netherlands
	Senonian	Sens, France
	Turonian	Touraine, France
	Cenomanian	Cenomanum, Latin for Le Mans, France
Lower Cretaceous	Albian	Aube, France
	Aptian	Apt, France
	Neocomian	Neocomum, Latin for Neuchâtel, Switzerland

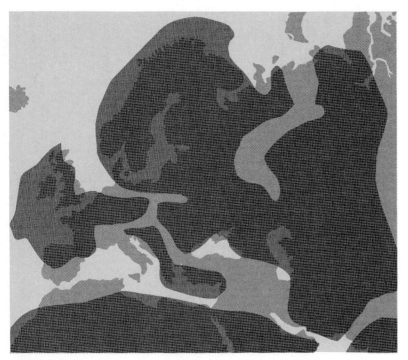

Fig. 1. Land and sea in Europe and adjoining areas during the Neocomian. Darker shading over contemporary land and sea indicates Neocomian land.

part of Morocco, Algeria, and Tunisia, southern and eastern Spain, southern France, Italy, Albania, western Greece and Yugoslavia, and parts of Switzerland, Austria, Hungary, Romania, Bulgaria, and Turkey. The Northern sea, an arm of the Arctic or Boreal Ocean, extended through the present North Sea area and flooded the eastern part of England and much of northern Germany and Poland. The Russian sea was first an arm of the Boreal Ocean and later an extension of the Tethys sea. Limestone and marl were the dominant sediments formed in the Tethys whereas marl, clay, shale, and sandstone are characteristic sediments of the Northern and Russian seas.

The fauna of the warm Tethys sea differed considerably from that of the cooler Northern and Russian seas. In southeastern France thick masses of limestone (Urgonian limestone) were formed by the accumulation of large foraminifers (orbitolines), corals, stromatoporids, bryozoans, rudistids, and large gastropods (nerineas). The cephalopods in the deeper parts of the Tethys were dominantly the rather smooth ammonites (phylloceratids, lytoceratids) and flat belemnites. The shallow Northern sea contained a more varied pelecypod fauna and very ornate ammonites. Aucellas and distinctive forms of ammonites were featured in the cool Russian sea.

During the Neocomian, the Tethys and Northern seas may have been connected by a strait through Poland. West of this possible strait was a land mass covering the southern half of Germany, western Czechoslovakia, Belgium, northern and western France, northern Spain, and most of the eastern edge of the British Isles. In some areas along the margins of this land mass deltaic deposits of sandstone and clay accumulated. The largest area of deltaic sediments was in a lowland covering part of Belgium, northern France, and southern England. Here these nonmarine rocks, known as the Wealden Beds, contain fossil plants, fresh-water gastropods and pelecypods, and

dinosaurs known as iguanodons. During the Aptian, the Wealden area sank below sea level and the Tethys and Northern seas were connected by a strait through northern France and southeastern England. The land mass centering in southern Germany was reduced to an island known as the mid-European island, which persisted throughout Cretaceous time (**Fig. 1**).

The Russian sea was a south-trending gulf of the Boreal Ocean during the early part of the Neocomian. Later in the Neocomian the sea extended southward and merged with the Tethys. During the Aptian the sea was closed off from the Boreal Ocean and became a narrow northeastern arm of the Tethys sea.

The Albian seas were very extensive in Europe. The Northern sea spread eastward across southern Russia and merged with the Russian sea. The mid-European island was reduced in size. The Northern sea transgressed westward too and submerged most of England, where the Gault clay, famous for its abundant and exceptionally well-preserved fossils, was deposited. Glauconitic clay, marl, and sandstone characterize the rocks of Albian age.

Africa. In Africa marine waters invaded Angola and Nigeria while sandstones of nonmarine origin accumulated in northern Algeria. During the Early Cretaceous a seaway extended northeastward through the present Mozambique Channel to the Persian Gulf. This sea encroached upon the western part of Madagascar and the eastern margin of Africa. In southeastern Saudi Arabia this sea merged with the Tethys.

Asia. The Tethys seaway through the Himalayan area of northern India seems to have persisted throughout the Early Cretaceous. Nonmarine strata totaling as much as 12,000 ft (3700 m) thick were deposited in northwestern and west-central China during the Neocomian. Nonmarine rocks also characterize the older part of the Neocomian of Japan. The Aptian there is chiefly marine sandstone and shale, but includes some reef-limestone containing fossils reminiscent of the Urgonian reef-limestones of the European Tethys. The Albian is represented by conglomerates, sandstones, and shales and marks an extensive transgression of marine waters.

Australia. Australia was above the sea during the Neocomian. The east-central part was a lowland in which fluviatile and lacustrine deposition took place. In the Aptian and Albian this lowland area was invaded by shallow marine waters which submerged about a third of the continent and divided it into two or three large islands.

North and South America. Along the Pacific Coast of North America many thousands of feet of shale, sandstone, and conglomerate of marine origin reveal the encroachment of the Pacific Ocean into the continent. The rich ammonite faunas of California show the presence of all the stages of the Lower Cretaceous. Pacific waters also transgressed upon the western part of South America. Marine Aptian limestone, shale, and sandstone in northern Colombia and Venezuela suggest a strait connecting the Pacific and Atlantic Oceans. The area of the present Caribbean Sea was probably land that contained numerous volcanoes.

Mexico and much of the Gulf Coastal area of the United States were flooded by a gulf of the Atlantic Ocean. During the Neocomian a peninsula extended from Arizona south through Sonora and thence southeast along the western edge of Mexico as far as the latitude of Mexico City. A much smaller peninsula extended from Texas south into Coahuila. Central

America and the southwestern part of Mexico were above sea level. The Pacific and Atlantic Oceans were united by means of the narrow passage west of Mexico City. Near the shores coarse clastic sediments were formed, whereas farther east away from the shore limestone and marl accumulated. During the Aptian and Albian the Mexican sea spread widely, crossed the Coahuila peninsula, and submerged the Gulf Coastal area of the United States as well as southwestern Texas and southern New Mexico and Arizona. Limestone was formed in central and eastern Mexico, but near the shore in Sonora a thick sequence of limestone, shale, agglomerate, and lava was deposited.

Neocomian rocks are not known with certainty in the western interior of the United States. In Alberta and British Columbia nonmarine deposits with many important beds of coal formed during this time. Nonmarine rocks of Aptian and early Albian age are widely distributed over the western interior of the United States and Canada. During middle Albian time an arm of the Arctic Ocean covered northern Alaska and the Mackenzie River area and advanced southeastward across northeastern British Columbia, most of Alberta and Saskatchewan, and southwestern Manitoba. About the same time marine waters spread north from the Gulf Coastal area and inundated Kansas, Nebraska, and the Dakotas, and the eastern halves of New Mexico, Colorado, Wyoming, and Montana, to merge with the Canadian sea. This vast seaway thus split North America into two land masses. Sandstone and dark noncalcareous shale characterize the sediments.

Nonmarine deposition occurred in Virginia, Maryland, and Delaware during the Neocomian and Albian. In west Greenland nonmarine rocks also accumulated during the Early Cretaceous. East Greenland was partly submerged in the Neocomian and Albian.

Upper Cretaceous. The great transgression of the seas in Europe, which began in the Aptian and Albian, reached its peak during the Late Cretaceous. The mid-European island was further reduced in size and most of the British Isles was submerged. An arm of the Northern sea extended eastward across Poland to Russia. The Tethys sea still covered the Mediterranean area but spread widely and merged with the Northern sea through wide straits in France, Romania, and in the region of the Caspian Sea.

Cenomanian deposits of the Northern sea are typically glauconitic clay, sandstone, and sandy marl near the shores, and glauconitic chalk, marly chalk, and marl farther from the shores. Sediments formed in the Tethys sea area ranged from sandstones and conglomerates near the shores to rudistid limestones or shales and sandstones with oysters. During the Turonian, marly chalk was deposited in much of the Northern sea, and marls, sandstones, and rudistid limestones in the Tethys. The Senonian of the Northern sea area is characterized by white chalk with flint nodules in France, the British Isles, Denmark, and northernmost Germany. Farther south in Germany sandy marl and sandstone were deposited near the mid-European island. In the northern part of the Tethys sea the Senonian rocks display many facies. In southwestern France marly limestone, rudistid limestone, and sandy limestone are typical. In southeastern France the upper part of the Senonian consists of brackish-water, lacustrine, and fluviatile beds. In Italy, Yugoslavia, Albania, and Greece, rudistid limestones mark the

Senonian. Rocks of Maestrichtian age are not as widespread as those of the Senonian. The beds are largely tuffs in the Netherlands and Belgium, chalks and greensands in southwestern France, chalks in northern Germany, and marly and sandy beds farther south near the mid-European island. In southeastern France the Maestrichtian is represented by a continental facies of lacustrine marls, lignites, and sandstones. The Danian also has a very restricted distribution. The type Danian of Denmark is limestone with abundant hydrocorals, bryozoans, and mollusks. The Danian is represented by marine limestone in southwestern France and by nonmarine rocks in southeastern France. Continental formations also feature the Danian of Spain and Portugal.

Africa. During the Cenomanian the Tethys sea spread southward over the Sahara and deposited marls and marly limestones with oysters, echinoids, and some ammonites. Sandstone, shale, and limestone of marine origin are present in Nigeria and north of there in the eastern part of French West Africa. This suggests the probability that the Tethys sea and the Atlantic Ocean were connected through the Sahara, thus dividing Africa into an eastern and a western land mass. A seaway still existed through the Mozambique Channel to West Pakistan, and Cenomanian sediments were deposited in western Madagascar and in places along the east coast of Africa. Limy sediments continued to form in North Africa during the Turonian. Nigeria, the eastern part of French West Africa, and the Atlantic coastal part of French Equatorial Africa were submerged, as well as the western part of Madagascar and the eastern part of South Africa. The Tethys sea continued to cover the northern edge of Africa from Morocco to Egypt during the Maestrichtian. Chalk and marl were deposited in Morocco and Algeria. Shale, sandstone, and coal beds formed in Nigeria and, in Angola, marls and limestones. In North Africa the Danian is represented by marine beds with numerous nautiloids.

Australia and Asia. Most of Australia was above the sea during Late Cretaceous time. Chalk, chalky clay, and greensand of Senonian age are present along the westernmost part of the continent.

The Himalayan seaway across the northern border of India persisted through the Late Cretaceous. In addition the east coastal area of India was submerged.

Cenomanian and Turonian times were marked in Japan by crustal unrest and by a regression of marine waters from parts of the area. Coarse sediments are common. The Senonian was a time of marine transgression, and shale and fine-grained sandstone were the dominant rocks formed. In the Maestrichtian a regression of the seas occurred and coarse clastic sediments were again common.

North and South America. North America was extensively flooded during the Late Cretaceous. The pattern of the seaways established by the great Albian transgression was followed throughout most of the later Cretaceous. During the Cenomanian, Turonian, and Senonian stages Pacific waters invaded much of Alaska, Oregon, and California, and deposited shales, sandstones, and conglomerates. Marine rocks of Senonian or Maestrichtian age are found along the west coast of parts of British Columbia. The Pacific Ocean also encroached upon the western margin of South America and laid down mostly shales in the Cenomanian, limy beds in the Turonian, and shale and sandstone in the Senonian. Part of Central America

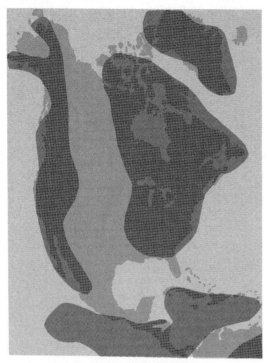

Fig. 2. Land and sea in North America and Central America during the Turonian. Darker shading over contemporary land and sea indicates Turonian land.

and much of the present area of the Caribbean Sea were probably land with many active volcanoes. An arm of the Pacific Ocean crossed northern Colombia and Venezuela and merged with the Atlantic. The Gulf Coastal region of the United States and most of Mexico were under water during the Cenomanian. Only the Sonoran peninsula and the westernmost edge of southern Mexico were emergent. Limestone and marl were the commonest rocks formed. Carbonate deposition continued through the Turonian.

In the early Cenomanian the sea seems to have withdrawn from the great seaway established during the Albian through the western interior of the United States and Canada. In the late Cenomanian, marine waters again filled this low area and united the Gulf of Mexico with the Arctic Ocean. The seaway spread somewhat during the early Turonian, and at one time its width reached from central Utah to western Iowa. Limestones and marls were formed over most of this region in the early Turonian, but sandstones and non-calcareous shales in the later Turonian reflected rising mountains in western Utah (**Fig. 2**).

During the early Senonian the Mexican–Western Interior sea was gradually reduced in size as mountains formed to the west. Near the western shore thick deposits of shale and sandstone accumulated with thick coal in some areas. Chalky beds were deposited farther east. Late in the Senonian, carbonate deposition almost ceased in the western interior of the United States and sandstone and shale was the rule. By the late Maestrichtian, the sea had completely withdrawn from the region and fluviatile deposition occurred to the end of the Cretaceous. These latest Cretaceus rocks, usually assigned to the Danian, have yielded remarkable horned dinosaurs.

Along the Atlantic Coastal Plain of the United States, nonmarine deposition occurred during the Cenomanian and in the early Senonian. Later in the Senonian and in the Maestrichtian, Atlantic waters

transgressed upon the continent, depositing marl, clay, and sand from central South Carolina northeastward through New Jersey.

In Greenland the Cenomanian is known from a very small area near the center of the east coast, and the Turonian is known from small areas near the center of the east and west coasts. The Senonian rocks are more widespread; they are represented mainly by marine shales on both coasts. SEE MESOZOIC.

William A. Cobban

Bibliography. D. Eicher and L. McAlester, *History of the Earth*, 1980; W. B. Harland, A. G. Smith, and B. Wilcock (eds.), The Phanerozoic time-scale, *J. Geol. Soc. London*, vol. 1205, 1964; F. Keller et al., *Introduction to Geology*, 1979; J. M. Poort, *Historical Geology: Interpretations and Applications*, 3d ed., 1980; J. B. Reeside, Jr., Paleoecology of the Cretaceous seas of the western interior of the United States, in H. S. Ladd (ed.), *Treatise on Marine Ecology and Paleoecology*, Geol. Soc. Amer. Mem. 67, vol. 2, 1957; F. B. Van Houten, *Ancient Continental Deposits*, 1977.

Criconematoidea

A superfamily of ectoparasitic nematodes of plants ranging from migratory to highly specialized sessile forms that emulate endoparasites. In all known species the males have atrophied mouthparts and do not feed. The procorpus and valved metacorpus of the characteristic esophagus are merged so that no clear distinction exists between them. Following the enlarged corpus is a short, narrow isthmus that ends in a slight terminal swelling, and a postcorpus of the same length. The body shape varies from stout and wormlike with heavy annulation, often highly ornamented with fringes, to obese with very fine annulations and no ornamentation. Females always have one anteriorly directed ovary.

Sedentary root parasitism is accomplished in several ways. In some species the spear or feeding stylet is nearly one-half the length of the body. This feature allows the nematode to stay in one place and feed on several cells by probing deeply and widely. Other species induce terminal root galls that are attractive to other individuals of the same species. The most sophisticated sessile forms embed the anterior portion of the body into the root; their feeding induces the production of giant nutritive plant cells.

Two important species are *Criconemoides xenoplax* and *Tylenchulus semipenetrans*. The former, through excessive root damage, predisposes peach trees to bacterial canker, an aerial disease caused by *Pseudomonas syringae*. *Tylenchulus semipenetrans* is commonly known as citrus nematode and is the most important citrus parasite. The females lay eggs in a gelatin to which dirt adheres. In the field, dirty roots, from which the mud cannot be washed, are symptomatic of citrus nematode. SEE NEMATA; PLANT PATHOLOGY.

Armand R. Maggenti

Crinoidea

A class of radially symmetrical Crinozoa in which the body is flower-shaped and either is carried on top of a long, anchored stem (**Fig. 1**), as in sea lilies, or is stemless and free to move about, as in comatulids or feather stars. The flowerlike appearance is the result of a circlet of arms that range like petals around the edge of a cup-shaped plated structure called the calyx. SEE BLASTOIDEA; CYSTOIDEA; PELMATOZOA.

About 690 living species are known, and of these, 80 are stemmed forms. More than 5500 fossil species, representing about 800 genera, have been described. All living crinoids are referred to one subclass, the Articulata, the main subject of this article. The three other subclasses of crinoids, the Inadunata, Camerata, and Flexibilia, are known only from fossils.

Stem joints of fossil crinoids were first described by Georgius Agricola (George Bauer) in 1546 and first illustrated by Conrad Gesner in 1565. The initial discovery of a living stemless comatulid was made by an Italian, Fabio Columna, in 1592, who thought that the specimen was a 10-armed starfish. It was not until 1761 that a living stemmed crinoid, from Martinique, was described by Jean Guettard. The relationship between fossil and living crinoids was first deduced by Edward Lloyd in 1699; but it was not until 1821 that J. S. Miller united all these forms into one group, which he named the Crinoidea, or "lily-shaped animals."

No crinoids are of direct economic importance to humans because they are not used for food, nor do they appear to be eaten by other animals. None are venomous.

Ecology. Crinoids occur in all seas except the Black and the Baltic. Shallow-water comatulids are most abundant and varied around tropical coral reefs. Many are nocturnal, coming out of crevices in the reefs to perch on coral heads at night to feed. The arms are arrayed in a fan-shaped filtration net to catch suspended food which consists mainly of zooplankton less than 0.02 in. (0.5 mm) in size. Tintinnids and dinoflagellates compose more than 50% of their diet. Crinoids prefer areas with moderate currents so that food can be carried through the feeding fan. Stemless comatulids with more than 10 arms are sluggish crawlers and are most abundant at depths of about 50 ft (15 m). Delicate 10-armed varieties, most common in 50 to 100 ft (15 to 30 m) of water, can swim gracefully.

All living stemmed crinoids inhabit water more than 330 ft (100 m) deep, where they and some comatulids are widespread. The deepest-water crinoid, *Bathycrinus*, lives down to 32,000 ft (9700 m) in the Kuril Trench and to 27,000 ft (8210 m) in the Kermadec Trench. Sea lilies use the cirri on the lower end of the stem to anchor themselves on sea-floor sediments; and comatulids, which also have cirri on the centrodorsal plate, use them for temporary attachment.

Crinoids are infested with a host of endoparasites and ectoparasites. The most common are peculiar polychaete worms, *Myzostoma*, that typically feed from the ambulacral grooves of the arms and tegmen. They have a rounded body bordered by a fringe of limblike organs, and a protrusible proboscis for feeding. Among other crinoid parasites are small snails that drill into the arms of the host, flatworms, copepods, crabs, shrimps, ostracods, ophiuroids, and one type of clingfish.

Morphology. The arms of small crinoid specimens may be less than 1.4 in. (3 cm) from tip to tip; those of the largest known living forms (from the Okhotsk Sea) can reach 3 ft (1 m) from tip to tip. Shallow-water crinoids, all comatulids, are brilliantly colored because of the presence of conjugated carotenoid pig-

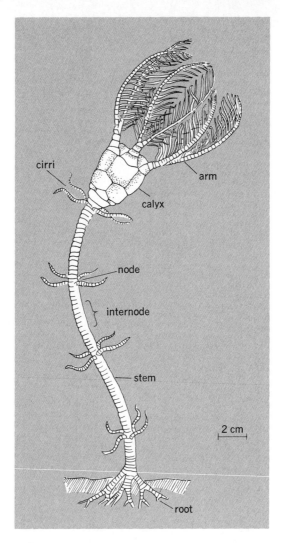

Fig. 1. Schematic diagram of a crinoid.

ments, which may be yellow, green, purple, or red. The arms may be banded with concentric zones of color, or contrasted black melanoid pigments may occur together with white or yellow zones. Sea lilies and deep-water comatulids have more subdued colors, often white, yellow, or green. SEE CAROTENOID.

Crinoid larval development is known exclusively from the comatulids. Upon rupture of the egg, a free-swimming, bilaterally symmetrical larva, termed the doliolaria, is formed. This larva swims about, using four or five rings of cilia, for a few hours or days, and then settles and attaches to the sea floor by the anterior end. During this time the rudimentary skeletal plates of the calyx begin to form. A short stem develops for attachment, and the developing viscera undergo rotation and metamorphosis, so that the mouth and anus both open on the upper side. The young comatulids break loose from the stem after about 6 weeks, assuming a free-living mode of existence. Sea lilies remain attached to the stem as adults.

The skeleton of the calyx is first formed of two circlets of five plates each. Those of the lower circlet are called basals; those of the upper, orals. The orals either develop into five guard plates surrounding the mouth or dwindle and disappear in the adult. Later, another circlet of five plates that support the arms, termed radials, appears between the basals and orals. A calyx with only basals and radials is termed monocyclic (**Fig. 2a**). If a further circlet of plates develops below the basals, the plates are termed infrabasals,

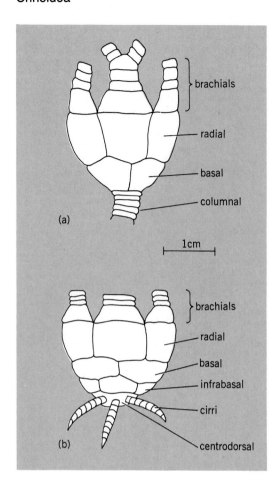

Fig. 2. Types of calyces. (a) Monocyclic sea lily calyx with stem. (b) Dicyclic comatulid calyx with infrabasals and centrodorsal plate.

and the calyx is termed dicyclic (Fig. 2b). All living crinoids are judged to have evolved from a dicyclic ancestor, although the infrabasals may not appear in the adult. The upper surface of the calyx is termed the tegmen. The mouth is situated near the center of the tegmen, and the anal opening lies to one side, between two of the ambulacral grooves (**Fig. 3**). The free-moving arms, usually five in number, occur around the upper margin of the calyx. From the five radial corners of the mouth the five open ambulacral grooves extend across the tegmen up onto each of the five arms.

Each of the five arms is usually divided into one or more branches, for a total of 10 to 30 branches; in extreme cases, as many as 200 branches occur. The

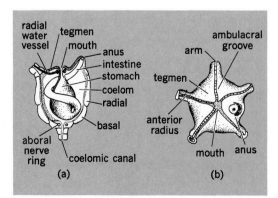

Fig. 3. Calyx anatomy. (a) Dissection in lateral aspect. (b) Calyx from adoral aspect.

arms are built of brachial plates arranged in a single series (uniserial arm; see **Fig. 4a**). In some fossil crinoids the brachial plates are arranged in a double series (biserial arms; see Fig. 4b). In living crinoids, and in some fossil crinoids, the arms carry small, uniserial, lateral branches called pinnules. These provide the meshwork of the filtration fan for feeding, and bear many tiny triplets of tube feet, or podia, that take up oxygen and produce mucous strands for capturing food. The tube feet also gather up food-laden mucous strands and, along with cilia, pass them down the ambulacral grooves to the mouth. Pinnules near the tegmen are commonly specialized for cleaning it and warding off ectoparasites. These pinnules may lack ambulacral grooves.

The stem is attached to the aboral pole of the calyx. It may be long and slender with a rootlike organ attaching it to the sea floor, or it may carry regular whorls of appendages, called cirri, arranged in groups of five at points on the stem termed nodes (Fig. 4c). The cirri on the distal part of the stem also serve to anchor the crinoid on the sea floor. Comatulids have a stem only during the very early growth stages. They retain the uppermost plate of the stem, termed the centrodorsal, which becomes enlarged and bears a dense cluster of cirri that serve for temporary attachment.

Muscular system. True muscle fibers are confined to intersegmental bundles that unite the plates of the arms and pinnules, which are capable of rapid and prolonged movements. The plates of the stem and cirri are joined together by elastic ligament fibers that penetrate deeply into each plate (Fig. 4d). The stem and cirri can be moved only quite slowly. Ligament fibers are also present in the arms and pinnules, where they oppose contraction by the muscles.

Coelom. A perivisceral coelom forms the body cavity in the calyx (Fig. 3a). The viscera are suspended in it. Aging individuals develop calcified sheets of mesoderm—trabeculae—between the gut and the body wall. The coelom extends into the stem and arms as a canal.

Water-vascular system. The water-vascular system has the general structure for the phylum, namely, a ring canal and radial water vessels. The radial vessels extend up each arm and send branches into each pinnule and into each group of tube feet. The water-vascular system of crinoids differs from that of other living echinoderms in that there is no direct connection of the ring canal with the exterior. On the tegmen there are numerous small ciliated funnels (hydropores), situated interradially, that open internally into the coelom. The ring canal replenishes its fluid by numerous short tubes (stone canals) that also open into the coelom. Some fossil crinoids have a distinct madreporite, or sieve plate, that is an enlarged posterior oral.

A poorly developed system of irregular blood sinuses, known as the hemal system, may occur.

Digestive system. The mouth opens into a short esophagus that leads into an enlarged midgut where digestion takes place. The midgut may have several lateral outpocketings, termed diverticulae. Sacculated digestive glands occur in the intestinal wall. The hindgut may be extended into a short anal sac or tube that can be moved and is capable of strong rhythmical contractions. From the mouth to the rectum, the gut makes a complete 360° coil within the calyx, so that the mouth and anal opening are situated close to each other on the tegmen. Some comatulids have an excentric, rather than a central, mouth, and in these crinoids the gut makes four complete coils internally.

Nervous system. The principal nervous system of crinoids is aboral, with a central nerve mass situated at the aboral pole of the calyx. From this center, branches extend to the cirri of comatulids, or down the center of the stem in sea lilies. Other branches extend through the brachial plates of the arms, where small branches innervate the muscles and tube feet. Two other nervous systems in crinoids, the oral and deep oral systems, are centered around the esophagus and serve to innervate the tegmen and floor of the ambulacral grooves. Crinoids have no special sensory organs, but they are sensitive to sunlight, water currents, and the presence of food.

Reproduction. The sexes are separate but cannot be distinguished by external examination. The small ovaries and testes are situated on proximal pinnules in many comatulids, but they may also be housed directly on the arms or, rarely, within the calyx. The eggs remain on the pinnules or arms until the doliolarian larva is formed; then, when the eggs rupture, the free-swimming larvae are released. A few Antarctic comatulids, such as *Isometra vivipara*, are known to be viviparous; the young are retained in special brood pouches. Under unfavorable conditions or during handling, crinoids may cast off some of their arms, tegmen, or viscera. Regeneration occurs within a few weeks unless the central aboral nerve mass is affected, in which case regeneration fails.

Geologic history. The class Crinoidea has an extensive fossil record. The oldest known crinoid is an enigmatic form from Middle Cambrian rocks of British Columbia, Canada. Crinoids became increasingly common during Ordovician time, represented especially by monocyclic inadunates, the Disparida, and by small camerates. Most early forms have small crowns, simple arms, and high, conical cups. The flexible crinoids evolved from the inadunates during the Ordovician. *SEE CAMBRIAN; ORDOVICAN.*

In Silurian time, crinoids began to inhabit coral reefs. Individuals became more robust and thick-plated, with short, stout stem and arms. Devonian rocks contain the oldest advanced inadunates, the poteriocrinoids, with pinnulate arms and brachial muscular articulations.

The Mississippian (early Carboniferous) Period is sometimes called the Age of Crinoids. More genera and species are known from this interval than any other time in Earth history. The advanced monocyclic camerates flourished, especially in the early Mississippian Burlington Limestone of Iowa, Illinois, and Missouri. Many of these crinoids became extinct in middle Mississippian time, and by the close of the period the poteriocrinoids had become the dominant crinoid stock. *SEE MISSISSIPPIAN.*

Late Paleozoic rocks are dominated by advanced inadunate crinoids. These developed a low, bowl-shaped cup; biserial arms; large, porous anal sacs; and other advanced features. There was a tremendous evolutionary diversification at the familial and generic levels. The most diverse and one of the youngest Paleozoic faunas is known from Permian rocks of the island of Timor, Indonesia, where many bizarre, endemic forms occur. The flexible and camerate subclasses both became extinct at the close of Permian time. The inadunates suffered a tremendous decline in diversity, with only one genus, *Encrinus,* being

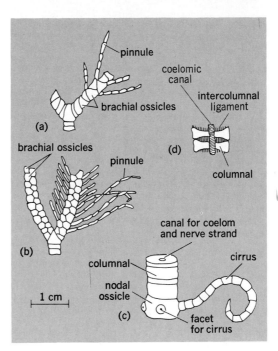

Fig. 4. Stem and arm. (a) Uniserial arm. (b) Biserial arm. (c) Stem ossicles. (d) Stem in vertical section.

known from Middle Triassic rocks of Europe. Prior to their extinction, the inadunates gave rise to the subclass Articulata, the only surviving group of crinoids. *SEE PALEOZOIC; PERMIAN; TRIASSIC.*

Lower Triassic rocks have yielded abundant disarticulated crinoid plates, but complete specimens have never been described. By later Triassic time, both stalked crinoids and small, stemless, pelagic crinoids, called roveacrinoids had evolved. The roveacrinoids proliferated druing the Mesozoic Era and finally became extinct at the close of the Cretaceous. In the Jurassic Period, the stemless comatulid crinoids, so dominant today, first appeared, although their ancestry is poorly understood. Stalked crinoids continued to be common in shallow-water Mesozoic rocks but during the Cenozoic Era became restricted to bathyal and abyssal depths, where they still live today. The comatulids have thrived in shallow, tropical waters, although deep-water and high-latitude species are known. *SEE ARTICULATA (ECHINODERMATA); CAMERATA; CENOZOIC; CRETACEOUS; CRINOZOA; ECHINODERMATA; FLEXIBILIA; INADUNATA; JURASSIC; MESOZOIC; REGENERATION (BIOLOGY).*

N. Gary Lane

Bibliography. J. Binyon, *Physiology of Echinoderms,* 1972; L. Fishelson, Ecology of the northern Red Sea crinoids and their epi- and endozoic fauna, *Mar. Biol.,* 26:183–192, 1974; D. B. Macurda, Jr., and D. L. Meyer, Feeding posture of modern stalked crinoids, *Nature,* 247(5440):394–396, 1974; D. L. Meyer, Distribution and living habits of comatulid crinoids near Discovery Bay, Jamaica, *Bull. Mar. Sci.,* 23(2):244–259, 1973; D. L. Meyer, Feeding behavior and ecology of shallow-water unstalked crinoids (Echinodermata) in the Caribbean Sea, *Mar. Biol.,* 22(2):105–129, 1973; R. C. Moore and C. Teichert (eds.), *Echinodermata 2: Crinoidea,* pt. T of *Treatise on Invertebrate Paleontology,* 3 vols., 1978; S. P. Parker (ed.), *Synopsis and Classification of Living Organisms,* 2 vols., 1982; C. R. C. Paul and A. B. Smith (eds.), *Echinoderm Phylogeny and Evolutionary Biology,* 1988; J. Rutman and L. Fishelson,

Food composition and feeding behavior of shallow-water crinoids at Eilat (Red Sea), *Mar. Biol.*, 3:46–57, 1969.

Crinozoa

A subphylum of the Echinodermata including radially symmetrical echinoderms showing a partly meriodional pattern of growth. This tends to produce an aboral cup-shaped theca and a partly radially divergent pattern of growth, forming appendages, called brachioles or arms and bearing ambulacral grooves. Included in the subphylum are the blastoids, coronoids, crinoids, cystoids, edrioblastoids, eocrinoids, parablastoids, and paracrinoids. SEE BLASTOIDEA; CORONOIDEA; CRINOIDEA; ECHINODERMATA; EDRIOBLASTOIDEA; EOCRINOIDEA; PARABLASTOIDEA; PARACRINOIDEA.

Howard B. Fell

Critical care medicine

The treatment of acute, life-threatening disorders, usually in intensive care units. Critical care medicine has been practiced informally for many decades in trauma centers, postanesthesia recovery rooms, coronary care units, delivery rooms, emergency rooms, and postoperative areas. The facilities and trained personnel available in the intensive care unit (ICU) permit extensive monitoring of physiological variables, organization of complex, multidisciplinary diagnostic and therapeutic plans, administration of therapy to predetermined goals, and expert nursing care.

Critical care thus runs counter to the traditional division of specialities by organ or organ system. Specialists in critical care undergo training beyond a primary qualification (internal medicine, surgery, anesthesia, or pediatrics), and must be able to manage acute respiratory, cardiovascular, metabolic, cerebral, and renal problems, as well as infections. The patients may be newborns, children, or adults suffering from trauma or acute life-threatening disease. Patients having failure of multiple organs, complicated medical problems, disorders falling into several medical specialities, or a need for 24-h care often become the responsibility of the critical care specialist.

The intensive care unit is the most labor-intensive, technically complex, and expensive part of hospital care. Of the average daily cost of a bed, about 70% is for personnel. The intensive care unit, however, may be crucial to the patient's survival.

Historical background. Although there were probably precedents, it was during the Crimean War in 1856 that Florence Nightingale placed the most serious casualties nearest the nursing station for the sake of better and more efficient care. Operational efficiency became important in the deployment of scarce medical resources and personnel, particularly in combat. In the civilian area, a special room was used for the postoperative care of neurosurgical patients at Johns Hopkins Hospital in 1923. In 1930 a postanesthesia recovery room was installed at Tübingen, Germany, as well as a unit that applied new knowledge and techniques to postoperative needs. Casualty facilities and trauma hospitals were developed in Europe around World War II. Shock wards were established in field hospitals to resuscitate battle casualties before and after their operations; the importance of blood transfusions, early operation, and postoperative care were recognized. Subsequently, trauma units, coronary care units, and postoperative care were recognized. Trauma units, coronary care units, and neonatal intensive care units were established in the United States to promote expediency and efficiency. In community hospitals the need for intensive care units was mainly the result of the shortage of nurses during and after World War II. It became essential to group patients according to the severity of their illness.

A historical example illustrates these developments. In the poliomyelitis epidemic of 1952 in Copenhagen, an estimated 5000 persons were afflicted; about 10% of these had respiratory paralysis from bulbar polio. Tubes placed through surgically prepared openings in the trachea and manual ventilation were instituted for patients with respiratory paralysis. To assist patients who could not breathe, medical teaching and research were suspended and the students worked day and night in shifts to ventilate patients by hand. This extraordinary effort reduced the mortality rate from 85 to 40%. This episode stimulated clinical research toward the development of positive-pressure mechanical ventilation. An early model of the Engstrom respirator proved to be reliable; through subsequent developments the iron lung, with its negative-pressure whole-body ventilation, was superseded. Mechanical ventilation was used for patients having crushed chest injuries and was employed during thoracic surgery and abdominal surgery. Respiratory and multidisciplinary intensive care units were rapidly developed for patients whose respiration was inadequate for any of various reasons.

Internists became involved in critical care when it was discovered that apparent death from heart attack could be treated by electric shock: dysrhythmias (abnormal heart rhythms) were found to be the most common cause of death from myocardial infarction in the first 12 h after the onset of symptoms. Resuscitation was found to be possible only if a well-equipped, trained team capable of inserting the tracheal tube and performing defibrillation arrived within minutes of cardiac arrest. This finding gave impetus to the development of paramedic transport of patients (prehospital emergency care) and in-hospital coronary care units.

In the late 1950s, modern emergency resuscitation developed. It was based on the observation that the expired breath of the rescuer could be used to breathe for the patient with ventilatory arrest. Mouth-to-mouth breathing and external cardiac massage led to the technique of cardiopulmonary resuscitation (CPR) and subsequently to protocols for basic life support and advanced life support.

The introduction of cardiac catheterization, its application to the diagnosis and management of congenital and acquired heart disease, and its later application to cardiovascular monitoring have been basic to managing circulatory problems in patients in the intensive care unit. The advent of plastic catheters facilitated such procedures as intravascular blood sampling, infusion of fluids and titration of drugs, and circulatory measurements. In the early 1970s, the flow-directed, balloon-tip pulmonary artery catheter was introduced, which in combination with a thermodilution system permits estimation of cardiac output.

Although renal failure had been managed with peritoneal dialysis beginning in 1923, hemodialysis (dialysis of blood through an external membranous coil) came into wide clinical use only after World War II. Today, in addition to being used in outpatient and other inpatient settings, hemodialysis is employed in intensive care units, for example, to treat patients who have multiple organ failure.

Many advances in medicine have increased the need for critical care. Among them are heart operations, transplantation of various organs, and cancer chemotherapy.

Life-support systems. Perhaps the single most useful function of the intensive care unit is to provide life-support systems for desperately ill patients who would not survive without them. Such systems are briefly discussed below.

Mechanical ventilation. A mechanical ventilator automatically delivers oxygen-enriched air at positive pressure through an endotracheal tube. The machine may be driven by pressure (pressure-regulated) or volume (volume-regulated). Adjustments may be made to control rate of ventilation, maximum airway pressure, peak flow, ratio of time of inspiration to that of expiration, and oxygen concentration. Ventilator modes determine how much effort the patient makes and how much assistance the ventilator provides; the ventilator can entirely control ventilation. Different clinical conditions require somewhat different ventilator modes.

Cardiopulmonary resuscitation. Resuscitation for cardiopulmonary arrest must be begun at the site and time of arrest whether in the intensive care unit, elsewhere in the hospital, or in the community. Arrest may occur more frequently in the intensive care unit than elsewhere because the sickest patients are usually there. Out in the field, without equipment, cardiopulmonary resuscitation consists of (a) airway control (positioning of head, inflation of lungs, manual clearing of mouth and throat), (b) breathing support [mouth-to-mouth (or nose) ventilation], and (c) circulation support (control of external hemorrhage, mechanical chest compressions if the patient is pulseless, and circulatory support with intravenous fluids if available). When equipment is available, cardiopulmonary resuscitation consists of the above plus (a) pharyngeal suctioning, tracheal intubation, or, if this is not immediately successful, creation of an opening into the trachea in the neck; (b) ventilation manually (using a bag and mask) or mechanically, with oxygen added; and (c) mechanical chest compressions, or if blood pressure and pulse do not respond immediately, open-chest massage with direct cardiac compressions.

Advanced life support consists of the use of cardiac drugs, fluids, electrocardiographic monitoring for cardiac dysrhythmias, defibrillation by electrical shock with external direct current of 200–400 joules, and management of subsequent complications and multiple organ support in an intensive care unit.

Peritoneal dialysis and hemodialysis. In the intensive care unit, peritoneal dialysis and hemodialysis are used to help treat such challenging conditions as renal failure, multiple organ failure, drug overdose, sepsis (widespread infection), posttrauma conditions, and postoperative renal failure. Blood is withdrawn through intravascular catheters and pumped through a dialysis chamber and back to the patient's vascular system. Among items removed by dialysis are the waste products, urea and creatinine; excess water, potassium, and sodium; drugs; and organic acids and other noxious metabolic end-products.

Circulatory support with intraaortic balloon pumping. The intraaortic balloon pump is the temporary device that is most widely used to assist the failing heart. It is employed when the ventricle of the heart cannot pump enough blood to meet the minimum needs of the body. The balloon, which is placed in the descending aorta, is rapidly inflated in diastole (the time of cardiac relaxation) and rapidly deflated at the onset of systole (the time of cardiac contraction). When the pump is properly synchronized with the heart, the balloon deflation produces a vacuumlike effect that helps the left ventricle to contract. The device also augments diastolic pressure, which improves blood flow and oxygen supply to the heart. Combining the mechanical assistance with the ventricle's own limited ability can permit adequate circulation.

Extracorporeal membrane oxygenation. Extracorporeal membrane oxygenation equipment is a modification of the apparatus that is routinely used to oxygenate the blood outside the body during open heart surgery. It is used mainly in near-term infants who do not respond to maximum ventilatory and medical support.

Life-sustaining therapy. Another aspect of critical care is the provision of life-sustaining therapy. Components include administration in intravenous fluids, provision of nutritional support, and control of infections.

Fluid and electrolyte administration. Fluids are given mainly to replace fluid losses and to provide circulatory support in patients with conditions such as shock or circulatory deficiencies. Intravenous and intraarterial catheters are placed to allow the monitoring of arterial and central venous pressures and the rapid administration of fluids, such as blood, blood components, and various solutions.

Drugs often are used in conjunction with fluid therapy. Such drugs include agents that stimulate cardiac contraction, agents that increase blood pressure, and agents that reduce vascular resistance to cardiac output.

Nutritional support. In life-threatening crises, nutritional requirements must be maintained. A patient who cannot take nourishment by mouth receives it either via a tube introduced through the nose into the stomach or duodenum, or parenterally (through a vein). Glucose, fat, amino acids, and other substances are given. Total parenteral nutrition can sustain life indefinitely in patients who are unable to eat and have higher than normal nutritional requirements because of sepsis, injury, or prior malnutrition.

Control of infection. A large percentage of patients in intensive care units have infections that are part of their primary disease or were acquired in the hospital. The latter infections are particularly difficult to eradicate because they are usually resistant to most antibiotics. Infectious agents are identified by culturing bronchial secretions and the fluids from body cavities, wounds, and elsewhere. The sensitivity of each bacterial agent to an array of antibiotics must be tested.

The control of the environment in the intensive care unit is a major problem. Cross contamination between various patients and between patients and personnel must be constantly monitored to prevent outbreaks of infection. SEE HOSPITAL INFECTIONS.

Other acute crises. Various types of crises are managed by critical care specialists. The following are some examples. Upper gastrointestinal hemorrhage may occur because of bleeding duodenal ulcers, esophagitis, gastritis, perforations of the stomach or esophagus, or other conditions. Lower gastrointestinal bleeding may arise from conditions such as colon tumors, colon diverticula, and hemorrhoids. Therapy usually consists of rapid blood replacement and then, once the patient's condition is stable, operative control of the bleeding source. SEE HEMORRHAGE.

Acute abdominal crises. Acute catastrophic abdominal conditions can include acute pancreatitis, intestinal obstruction, perforation of a duodenal ulcer, stones of the kidney or ureter, and many other conditions. Many patients with these conditions are admitted to the intensive care unit during their preoperative diagnostic work-up and for stabilization of their cardiorespiratory and fluid and electrolyte status; they may be readmitted postoperatively for management.

Cardiac emergencies. Among the cardiac emergencies sometimes treated in intensive care units are serious abnormalities of heart rate and rhythm, toxicity from drugs used to treat heart disease, and cardiac tamponade (compression of the heart by blood or fluid in the sac surrounding it). For some conditions treatment can include emergency placement of a pacemaker.

Coma management. The sudden onset of altered mental status or coma may result from head injury, cerebrovascular accidents (stroke), drug overdose, postoperative conditions after neurosurgery, meningitis, brain tumors, and various other causes. Patients with altered mental status or coma require special nursing to prevent complications. A major consideration, especially after neurosurgical operations and head injuries, is to monitor the level of consciousness; a change in consciousness is usually the first warning of increasing intracranial pressure, which if not promptly corrected may lead to irreversible brain damage.

Acute endocrine disorders. Critical care is used in treating a number of acute endocrine catastrophies that are uncommon but potentially fatal. Examples include severe conditions caused by excessive or deficient thyroid hormone secretion, hypoparathyroidism after thyroid surgery, and adrenocortical insufficiency (Addison's disease). SEE ENDOCRINE MECHANISMS.

Injuries. Trauma from automobile accidents, penetrating injuries from violent crimes, and blunt trauma from falls or assaults pose serious technical and logistic problems that must be rapidly addressed. The injured patient must promptly receive appropriate diagnostic tests, monitoring, and therapy. Sometimes surgery must be undertaken while resuscitation efforts are still under way.

Hepatic failure and metabolic problems. Critical care also is used in managing hepatic failure and metabolic problems. Hepatic insufficiency or failure can result from conditions such as alcoholic cirrhosis, viral hepatitis, drug reactions, and poisoning. Often infection is present, as immunocompetence is impaired in liver failure. SEE CIRRHOSIS; LIVER.

Pulmonary problems. Patients whose breathing is seriously impaired by conditions such as pneumonia, asthma, pulmonary embolism (blood clot in the lung), smoke inhalation, and near-drowning may receive critical care. Measures available to assist such patients include use of mechanical ventilators (discussed above), administration of humidified supplemental oxygen through a mask or nasal prongs, and use of bronchodilator drugs.

William C. Shoemaker

Bibliography. W. C. Shoemaker et al. (eds.), *Textbook of Critical Care*, 2d ed., 1989.

Critical mass

That amount of fissile material (^{233}U, ^{235}U, or ^{239}Pu) which permits a self-sustaining chain reaction. The critical mass ranges from about 2.1 lb (950 g) of ^{235}U or a smaller amount of ^{239}Pu for dissolved compounds through 35 lb (16 kg) for a solid metallic sphere of ^{235}U, and up to hundreds of tons for some power reactors. It is increased by the presence of such neutron absorptive materials as admixed ^{238}U, aluminum pipes for flow of coolant, and boron or cadmium control rods. It is reduced by a moderator, such as graphite or heavy water, which slows down the neutrons, inhibits their escape, and indirectly increases their chance to produce fission. SEE CHAIN REACTION (PHYSICS); NUCLEAR FISSION; REACTOR PHYSICS.

John A. Wheeler

Critical path method (CPM)

A diagrammatic network-based technique, similar to the program evaluation and review technique (PERT), that is used as an aid in the systematic management of complex projects. The technique is useful in organizing and planning; analyzing and comprehending; problem detecting and defining; alternative action simulating; improving (replanning); time and cost estimating; budgeting and scheduling; and coordinating and controlling. It has its greatest value in complex projects which involve many interrelated events, activities, and resources (time, money, equipment, and personnel) which can be allocated or assigned in a variety of ways to achieve a desired objective. It can be used to complete a multifaceted program faster or with better utilization of resources by reassignments, trade-offs, and judiciously using more or less assets for certain of the individual activities composing the overall project. SEE PERT.

CPM was introduced in 1957–1958 by E. I. du Pont de Nemours & Company. Present applications include research and development programs; new product introductions; facilities planning and designs; plant layouts and relocations; construction projects; equipment installations and start-ups; major maintenance programs; medical and scientific researches; weapons systems developments; and other programs in which cost reduction, progress control, and time management are important. It is best suited to large, complex, one-time or first-time projects rather than to repetitive, routine jobs.

A simplified example will show the features of the critical path method and the steps involved in constructing and using the network-based system:

Project: Design personal home-use computer.

Objective: Complete project as soon as possible.

Procedure: The CPM procedure for this project involves eight steps:

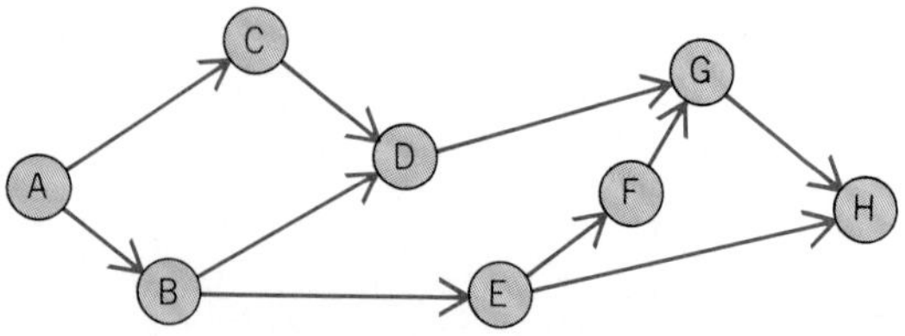

Fig. 1. Network of relationship between events (circles) and their activities (arrows).

Table 1. CPM events table

Prior event	Event	Following event
—	A = Authorization to start received	B and C
A	B = Computing circuits designed	D and E
A	C = Video circuits designed	D
B and C	D = Keyboard and cabinets designed	G
B	E = Programming completed	F and H
E	F = Operating systems completed	G
D and F	G = Testing and debugging finished	H
E and G	H = Design specifications and programs finalized	—

Step 1: List required events; arrange as to best guess of their sequence; assign a code letter or number to each. (See center column of **Table 1**.) [Events are accomplished portions or phases of the project. Activities are the resources applied to progress from one event to the next. This example has been greatly simplified for illustrative purposes. A real case could have several hundred events and activities and would be a candidate for computer analysis, for which programs are available.]

Step 2: Determine which event(s) must precede each event before it can be done and which event(s) must await completion of that event before it can be done. Construct a table to show these relationships. (See left and right columns of Table 1.)

Step 3: Draw the network, which is a diagram of the relationship of the events and their required activities that satisfies the dictates of Table 1 (see **Fig. 1**).

Step 4: List the required activities and enter the best estimate available of the time required to complete each activity. In PERT, three time estimates are entered: the most optimistic, the most likely, and the most pessimistic. This is the basic way in which PERT differs from CPM. Some users of CPM enter a second time estimate for each activity, the first being the normal time and the second being the crash (or fast at any price) time. **Table 2** shows the list of activities (AB designates the activity required in going from event A to event B), the expected normal time, the maximum number of weeks that each activity can be shortened under a crash program, and the cost per week to shorten, if it can be shortened.

Step 5: Write the normal expected time to complete each activity on the diagram (**Fig. 2**). Note that the relative lengths of the arrows (activities) have no relationship to their time magnitudes.

Step 6: Find the critical path, that is, the longest route from event A to event H. The longest path is the total time that the project can be expected to take

unless additional resources are added or resources from some of the other paths in which there may be slack can be reallocated to activities which lie in the critical path. In the example shown in Fig. 2, the initial critical path is A–B–D–G–H, or 5 + 12 + 18 + 5 = 40 weeks. If the critical path time works out to be less or more than the target time to complete the project, there is then a positive or negative float.

Step 7: Shorten the critical path. The stated objective for this example was to complete the project as soon as possible. Other projects might have different objectives. An examination of Table 2 and Fig. 2 suggests that the activities lying along the critical path can be shortened for the costs shown in **Table 3**. The total expected time to complete the project can be shortened from 40 weeks to 30 weeks—a 25% improvement. A judgment can be made by those responsible for the project as to whether the 10-week time savings is worth the additional cost of $7000. In some cases, the critical path is so shortened that it is no longer the longest path, and another path becomes the critical one. Only reductions in the critical path time reduce the total project time, and activities in noncritical paths may run over or late to the limits of their floats without causing the project to be delayed. The floats are the differences between the total expected time required for each possible path in the network

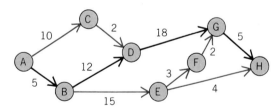

Fig. 2. Network diagram modified to show time to complete activities. The critical path is indicated by the black arrows.

Table 2. Time and cost factors for activities

Activity	Expected time to complete, weeks	Maximum possible time reduction, weeks	Cost (per week) to reduce time
AB	5	1	$1000
AC	10	3	500
BD	12	3	1000
BE	15	2	500
CD	2	0	—
DG	18	4	500
EF	3	0	—
EH	4	1	2000
FG	2	0	—
GH	5	2	500

and the total time for the critical path, and are measures of spare time available.

Step 8: Construct a schedule of the project showing the earliest and latest permissible start and completion dates for each activity. To do this, the individual floats are calculated. With the slack in the critical path set at zero at the onset of the project's implementation, any slippage in the completion of any event along that path will result in a delay in the completion of the whole project, unless made up later.

The use of CPM adds another step to a project, and it requires continual updating and reanalysis as conditions change, but experience has shown that the effort can be a good investment in completing a project in less time, with less resources, with more control, and with a greater chance of on-time, within-budget completion. SEE INDUSTRIAL ENGINEERING.

Vincent M. Altamuro

Critical phenomena

The unusual physical properties displayed by substances near their critical points. The study of critical phenomena of different substances is directed toward a common theory.

Critical points. Ideally, if a certain amount of water (H_2O) is sealed inside a transparent cell and heated to a high temperature T, for instance, $T > 647$ K (374°C or 705°F), the enclosed water exists as a transparent homogeneous substance. When the cell is allowed to cool down gradually and reaches a particular temperature, namely the boiling point, the enclosed water will go through a phase transition and separate into liquid and vapor phases. The liquid phase, being more dense, will settle into the bottom

half of the cell. This sequence of events takes place for water at most moderate densities. However, if the enclosed water is at a density close to 322.2 kg · m⁻³, rather extraordinary phenomena will be observed. As the cell is cooled toward 647 K (374°C or 705°F), the originally transparent water will become increasingly turbid and milky, indicating that visible light is being strongly scattered. Upon slight additional cooling, the turbidity disappears and two clear phases, water and vapor, are found. This phenomenon is called the critical opalescence, and the water sample is said to have gone through the critical phase transition. The density, temperature, and pressure at which this transition happens determine the critical point and are called respectively the critical density ρ_c, the critical temperature T_c, and the critical pressure P_c. For water, $\rho_c =$ 322.2 kg · m⁻³, $T_c = 647$ K (374°C or 705°F), and $P_c = 2.21 \times 10^7$ pascals. SEE OPALESCENCE.

Different fluids, as expected, have different critical points. Although the critical point is the end point of the vapor pressure curve on the pressure-temperature (P-T) plane (**Fig. 1**), the critical phase transition is qualitatively different from that of the ordinary boiling phenomenon that happens along the vapor pressure curve. In addition to the critical opalescence, there are other highly unusual phenomena that are manifested near the critical point; for example, both the isothermal compressibility and heat capacity diverge to infinity as the fluid approaches T_c. SEE THERMODYNAMIC PROCESSES.

Many other systems, for example, ferromagnetic materials such as iron and nickel, also have critical points. The ferromagnetic critical point is also known as the Curie point. As in the case of fluids, a number of unusual phenomena take place near the critical point of ferromagnets, including singular heat capacity and divergent magnetic susceptibility. The study of critical phenomena is directed toward describing the various anomalous and interesting types of behavior near the critical points of these diverse and different systems with a single common theory. SEE CURIE TEMPERATURE; FERROMAGNETISM.

Order parameters. One common feature of all critical phase transitions is the existence of a quantity called the order parameter. The net magnetization M is the order parameter for the ferromagnetic system. At temperatures T above T_c and under no external field, there is no net magnetization. However, as the temperature of the system is cooled slowly through T_c, a net magnetization M will appear precipitously. The increase in M is more gradual as temperature is reduced further. The nonzero magnetization is due to the partial alignment of the individual spins in the ferromagnetic substance. M is called the order parameter since the state with partial alignment of spins is more ordered than that with no alignment. The density difference in the liquid and vapor phases, $(\rho_l - \rho_v)/\rho_c$, is the proper order parameter in the fluid

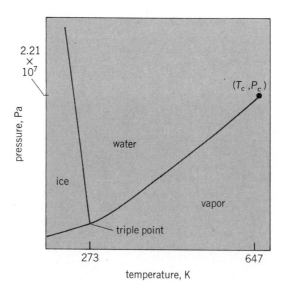

Fig. 1. Phase diagram of water (H_2O) on pressure-temperature (P-T) plane.

Table 1. Order parameters, theoretical models, and classification according to universality class of various physical systems*

Universality class	Theoretical model	Physical system	Order parameter
$d = 2$ $n = 1$	Ising model in two dimensions	Adsorbed films	Surface density
$n = 2$	XY model in two dimensions	Helium-4 films	Amplitude of superfluid phase
$n = 3$	Heisenberg model in two dimensions		Magnetization
$d > 2$ $n = \infty$	Spherical model	None	
$d = 3$ $n = 0$	Self-avoiding random walk	Conformation of long-chain polymers	Density of chain ends
$n = 1$	Ising model in three dimensions	Uniaxial ferromagnet	Magnetization
		Fluid near a critical point	Density difference between phases
		Mixture of liquids near consolute point	Concentration difference
		Alloy near order-disorder transition	Concentration difference
$n = 2$	XY model in three dimensions	Planar ferromagnet	Magnetization
		Helium 4 near superfluid transition	Amplitude of superfluid phase
$n = 3$	Heisenberg model in three dimensions	Isotropic ferromagnet	Magnetization
$d \leqslant 4$ $n = -2$		None	
$n = 32$	Quantum chromodynamics	Quarks bound in protons, neutrons, etc.	

*From K. Wilson, Problems in physics with many scales of length, *Sci. Amer.*, 241(2):158–179, August 1979.

system; ρ_c in the denominator is the critical density. This order parameter has temperature dependence similar to that for the net magnetization M of a ferromagnetic system. For $T > T_c$ the order parameter is equal to zero, since there is only one homogeneous phase in the fluid. As the system is cooled through T_c, the fluid system phase separates with a precipitous increase in the difference of the liquid and vapor densities. A number of critical systems and their respective order parameters are listed in **Table 1**.

The order parameters assume power law behaviors at temperatures just below T_c. In the fluid and ferromagnetic systems, for $t < 0$, they follow Eqs. (1) and (2), and for $t > 0$ they obey Eqs. (3). In these equations, $t \equiv (T - T_c)/T_c$ is the reduced temperature, and

$$M = B(-t)^\beta \qquad (1)$$

$$\frac{\rho_l - \rho_v}{\rho_c} = B(-t)^\beta \qquad (2)$$

$$M = 0 \qquad \frac{\rho_l - \rho_v}{\rho_c} = 0 \qquad (3)$$

tions, $t \equiv (T - T_c)/T_c$ is the reduced temperature, and β and B are respectively the critical exponent and amplitude for order parameter.

Measurement of the order parameter of fluid neon near the critical temperature [$T_c = 44.48$ K ($-228.57°C$ or $-379.43°F$), $\rho_c = 484$ kg · m^{-3}, $P_c = 2.72 \times 10^6$ Pa] is shown in **Fig. 2**. The densities of the liquid (upper branch) and vapor (lower branch) phases as a function of temperature are deduced by measuring the dielectric constant of neon in the bottom and top halves of a sample cell. Data at reduced temperatures t between -4×10^{-4} and -2×10^{-3} are shown as broken lines. Careful data analysis shows that the simple power function given in Eq. (2) is not adequate to describe the data over the entire temperature range. This is the case because even in the range of reduced temperatures t between -4×10^{-4} and -4×10^{-2} the fluid is not yet inside the asymptotic or true critical region, and cor-

rection terms are needed. When the density data are analyzed according to the predicted functional form given by Eq. (4) the exponent is found to be 0.327

$$\frac{\rho_l - \rho_v}{\rho_c} = B(-t)^\beta \, [1 + B_1 |t|^{0.5}$$
$$+ \, B_2|t| + \cdots] \qquad (4)$$

$\pm$ 0.002, in excellent agreement with the theoretically predicted value of 0.325 as discussed below. The value of β found for other simple fluids also converges toward the theoretical value.

Power law behavior of other quantities. The anomalous behavior of other thermodynamic quantities near the critical point can also be expressed by power laws. These can be characterized by a set of critical exponents, labeled (besides β) α, γ, δ, ν, and

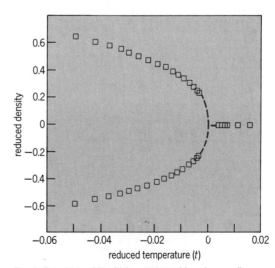

Fig. 2. Densities of liquid (upper branch) and vapor (lower branch) neon near its critical temperature. Reduced density is defined as $(\rho - \rho_c)/\rho_c$ and reduced temperature is defined as $t = (T - T_c)/T_c$. (After M. Pestak and M. H. W. Chan, Equation of state of N$_2$ and Ne near their critical points: Scaling, correction to scaling, and amplitude ratios, *Phys. Rev.*, B30:274–288, 1984)

Table 2. Critical exponents and power laws in pure fluids and ferromagnets*

Thermodynamic quantity	Fluid	Ferromagnet	Power law
Specific heat	C_v	C_H	$\sim (t)^{-\alpha}$ for $t > 0$ $\sim (-t)^{-\alpha'}$ for $t < 0$
Order parameter $= S$	$(\rho_l - \rho_v)\rho_c^{-1}$	M	$= 0$ for $t > 0$ $\sim (-t)^{\beta}$ for $t < 0$
Response function	$K_t = \dfrac{1}{\rho}\left(\dfrac{\partial \rho}{\partial P}\right)_T$	$X = \left(\dfrac{\partial M}{\partial H}\right)_T$	$\sim (t)^{-\gamma}$ for $t > 0$ $\sim (-t)^{-\gamma'}$ for $t < 0$
Critical isotherm	$P - P_c$	H	$\sim |S|^{\delta} \cdot$ sign of S
Correlation length	ξ	ξ	$\sim (t)^{-\nu}$ for $t > 0$ $\sim (-t)^{-\nu'}$ for $t < 0$
Critical correlation function of fluctuation	$G(r)$	$G(r)$	$\sim (r)^{-(d-2+\eta)}$

*d = spatial dimensionality of the critical system; H = magnetic field strength; C_v = specific heat at constant volume; C_H = specific heat at constant magnetic field strength; K_H = isothermal compressibility; X = susceptibility.

η: α characterizes the divergent heat capacity; γ the susceptibility (of magnets) and isothermal compressibility (of fluids); δ the critical isotherm; ν the correlation length; and η the critical correlation function. The functional forms of these power laws for the fluid and magnet systems are shown in **Table 2**. The critical opalescence phenomenon discussed above is closely related to the strong density fluctuations induced by the divergent isothermal compressibility near T_c. When the density fluctuations are correlated at lengths comparable to the wavelength of visible light, intense scattering occurs.

Mean field theories. The earliest attempts to understand the critical behavior were the van der Waals model for fluids (1873) and the Weiss model for ferromagnets (1907). These are mean field theories in the sense that the state of any particular particle in the system is assumed to be determined by the average properties of the system as a whole. In these models, all particles can be considered to contribute equally to the potential at each site. Therefore, the mean field theory essentially assumes the intermolecular interaction to be of infinite range at all temperatures. The mean field theories are qualitatively quite successful in that they predict the existence of critical points and power law dependence of the various thermodynamic quantities near the critical point. They are not quantitatively correct because the predicted values for the various exponents are not in agreement with exact

model calculations or with experimental results (**Table 3**).

Scaling hypothesis. Theoretical efforts in the study of critical phenomena have been centered on predicting correctly the value of these critical exponents. One of the most important developments is the hypothesis of scaling. This hypothesis is model-independent and applicable to all critical systems. The underlying assumption is that the long-range correlation of the order parameter, such as the spin fluctuation in the ferromagnetic system and the density fluctuation in the fluid system near T_c, is responsible for all singular behavior. This assumption leads to a particular functional form for the equation of state near the critical point. With this simple assumption, it has been demonstrated that a number of inequalities among critical exponents that can be proved rigorously by thermodynamic arguments are changed into equalities. These equalities, or scaling laws, show that there are only two independent critical exponents; once two exponents in a system are given, all other exponents can be determined. What is truly impressive about this simple hypothesis is that the scaling laws, Eqs. (5)–(9), have been shown to be correct in almost all

$$\alpha = \alpha', \quad \gamma = \gamma', \quad \nu = \nu' \tag{5}$$

$$2 = \alpha + \gamma + 2\beta \tag{6}$$

$$2 = \alpha + 2\beta\delta - \gamma \tag{7}$$

Table 3. Values of critical exponents

Systems	α	β	γ	δ	ν	η
Mean field model	0 (discontinuity)	½	1	3	½	0
Ising model (exact) ($d = 2, n = 1$)	0 (logarithmic discontinuity)	⅛	7/4	15	1	0.25
Ising model (approx.) ($d = 3, n = 1$)	0.110	0.325	1.24	4.82	0.63	0.03
Heisenberg model (approx.) ($d = 3, n = 3$)	−0.10	0.36	1.38	4.80	0.705	0.03
Fluids*						
Xe	0.08 ± .02	0.325 − 0.337	1.23 ± 0.03	4.40		
SF_6		0.321 − 0.339	1.25 ± 0.03			
Ne		0.327 ± 0.002	1.25 ± 0.01			
N_2		0.327 ± 0.002	1.233 ± 0.010			
Ferromagnets* (isotropic)						
Iron, Fe	−0.09 ± .01	0.34 ± .02	1.33 ± 0.02			0.07 ± 0.07
Nickel, Ni	−0.09 ± .03	0.37 ± .03	1.34 ± 0.02	4.2 ± 0.1		

*Experimental data are the averaged measured values of a number of experiments.

$$\nu d = 2 - \alpha \qquad (8)$$

$$\gamma = \nu(2 - \eta) \qquad (9)$$

real and theoretical model critical systems.

The meaning of these exponents is explained above and also in Table 2; d is not an exponent but the spatial dimensionality; the primed and unprimed exponents represent the value below and above T_c, respectively. A large number of theoretical and experimental activities are concerned with putting the scaling hypothesis on a firm fundamental basis and understanding its universality and limitations.

Model systems. A great deal of insight has been gained by the construction and solution of model systems that can be solved exactly. The most famous one is the two-dimensional ($d = 2$) Ising model solved by L. Onsager in 1944. In this model, spins (little magnets) on a lattice are allowed to point in either the up or down directions, and it is assumed that only pairs of nearest-neighboring spins can interact. Onsager found a critical point for this system and calculated the values of the various critical exponents. The solution of this model is important because this is one of the very few model systems with exact solutions, and it is often used to check the validity of approximation techniques. There are many other model systems similar to the Ising model. They are distinguished from each other by the spatial (d) and spin (n) dimensionality. The spin can be oriented along an axis ($n = 1$, Ising model), or in any direction on a plane ($n = 2$, XY model), or in any direction in space ($n = 3$, Heisenberg model). SEE ISING MODEL.

These models are essentially simplified versions of real physical systems. The three-dimensional Heisenberg model ($d = 3$, $n = 3$), for example, clearly resembles the isotropic ferromagnets; the three-dimensional Ising model ($d = 3$, $n = 1$) can be found to correspond to the pure fluid system. The correspondence can be shown if the space accessible to the fluid is divided into lattice sites, and at each site the spin parameter is considered to be up if the site is occupied and down if it is not. There are other physical systems beside pure fluids that resemble the three-dimensional Ising model, for example, the binary fluid near its consolute mixing point, the uniaxial ferromagnet, and an alloy near the order-disorder transition. A great deal of effort and ingenious mathematical techniques have been employed to obtain approximate solutions to these model systems.

Universality hypothesis. It has been observed that the measured values for the critical exponents are rather close to the calculated ones of the corresponding model system. This observation leads to the hypothesis of critical universality. According to this hypothesis, the details of the particle-particle interaction potential in the vicinity of the critical point are not important, and the critical behavior is determined entirely by the spatial dimensionality d and the spin dimensionality n. All systems, both model and real, that have the same value of d and n are said to be in the same universality class and have the same critical exponents. The hypotheses of scaling and universality are closely related: since the length of correlation between density or spin fluctuation diverges as one approaches the critical point, and any interaction potential between particles is finite in range, the details of interparticle potential are expected to be increasingly less important as one approaches the critical point. It has been shown that scaling laws can be derived from the universality hypothesis. Classification of model and physical systems according to universality classes is shown in Table 1.

Renormalization group. The renormalization group (RG) method, originally used in quantum field theory, has been introduced in the study of critical phenomena. By employing a set of symmetry transformations, the ideas contained in the principle of universality and in the scaling hypothesis can be reformulated and incorporated much more economically. As a result, a fully operational formalism is obtained from which critical exponents can be calculated explicitly. Beside the success in critical phenomena, the renormalization group method is also found to be a very useful technique in many diverse areas of theoretical physics. SEE QUANTUM FIELD THEORY.

The validity and indeed the elegance of the concept of universality and the theory of critical phenomena are borne out by experiments. As stated above, the order parameter exponents for simple fluids were found in a number of experiments to be equal to 0.327 ± 0.002. This is in excellent agreement with the value of 0.325 calculated for the three-dimensional Ising model ($n = 1$, $d = 3$) by renormalization-group and high-temperature series expansion methods. **Figure 3** shows the measured densities of the liquid and vapor phases of a two-dimensional fluid. The two-dimensional fluid is achieved by adsorbing methane molecules on an atomically flat graphite surface at monolayer and submonolayer coverages. The curve in Fig. 3, which traces the boundary separating the liquid-vapor coexistence region from the hypercritical fluid region, also reflects the liquid (upper half) and vapor (lower half) densities. It is qualitatively different from that for a bulk or three-dimensional fluid as shown in Fig. 2. When the curve is subjected to an analysis according to the power law given in Eq. (2), the resulting exponent β is found to be 0.127 ± 0.020, in excellent agreement with the value of $\frac{1}{8}$ as predicted by Onsager for the two-dimensional Ising model ($d = 2$, $n = 1$).

Dynamical effects. Besides the static critical phenomena, there are many interesting dynamical effects near the critical point, including critical slowing down, the dynamics of density and spin fluctuations,

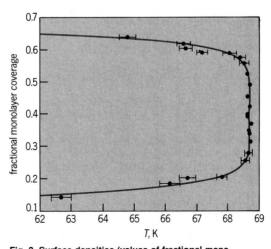

Fig. 3. Surface densities (values of fractional monolayer coverage) of monolayer methane adsorbed on graphite near its two-dimensional critical temperature. (*After H. K. Kim and M. H. W. Chan, An experimental determination of a two-dimensional liquid-vapor critical point exponent, Phys. Rev. Lett., 53:170–173, 1984*)

thermal and mass transport, and propagation and attenuation of sound. Understanding of these effects is far from complete.

Most critical phenomena experiments have been performed on three-dimensional systems, but a number have been done in two-dimensional and quasi-two-dimensional systems. The experiments include the order-disorder transition and the continuous melting transition of gases bound to a graphite surface and the superfluid transition of ^{4}He films on substrates. There has also been considerable interest in the influence of disorder on fluid and magnetic systems. These experiments provide interesting physical realizations of the various model systems in two dimensions. SEE PHASE TRANSITIONS; STATISTICAL MECHANICS.

Moses H. W. Chan

Bibliography. C. S. Domb and M. S. Green (eds.), *Phase Transitions and Critical Phenomena*, vols. 1–12, 1972–1988; S. K. Ma, *Modern Theory of Critical Phenomena*, 1976; P. Pfeuty and G. Toulouse, *Introduction to the Renormalization Group and to Critical Phenomena*, 1977; H. E. Stanley, *Introduction to Phase Transition and Critical Phenomena*, 1971; K. Wilson, Problems in physics with many scales of length, *Sci. Amer.*, 241:(2)158–179, August 1979.

Crocco's equation

A relationship between vorticity and entropy gradient for the steady flow of an inviscid compressible fluid. Crocco's equation, given below, pertains to isoener-

$$\mathbf{v} \times \boldsymbol{\omega} = -T \operatorname{grad} s$$

getic flow, which is a common type of flow where the total energy, or stagnation enthalpy, of the fluid per unit mass is constant throughout the fluid. In the equation $\mathbf{v}$ is the fluid velocity vector, $\boldsymbol{\omega} = \operatorname{curl} \mathbf{v}$ is the vorticity vector, T is the fluid temperature, and s is the entropy per unit mass of the fluid. (The equation in this form is applicable in SI units or any other coherent system of units.) Thus entropy gradients can occur only at right angles to $\mathbf{v}$ and $\boldsymbol{\omega}$. An irrotational flow is one where $\boldsymbol{\omega} = 0$ throughout the flow, and it follows from Crocco's equation that it must also be an isentropic flow, that is, one with constant entropy throughout. An example of a steady, inviscid, isoenergetic flow that is rotational, and hence nonisentropic, is that behind the curved shock wave at the nose of a blunt body traveling at supersonic speeds; the flow is uniform and isentropic ahead of the shock wave, but because the entropy rise through the shock depends on the inclination of the shock front to the oncoming stream, the flow is nonisentropic behind the curved shock. SEE FLUID-FLOW PRINCIPLES; ISENTROPIC FLOW; LAPLACE'S IRROTATIONAL MOTION; UNITS OF MEASUREMENT.

Arthur E. Bryson, Jr.

Crocodile

The common name used for 14 species of large reptiles included in the family Crocodylidae, one of the three families of the order Crocodilia, which also includes the alligators, caimans, and the gharial (also known as the gavial). Like all crocodilians, the croc-

A typical crocodile, with short, powerful legs and a long, narrow head.

odiles are primarily distributed throughout the tropical regions of the world. Species occur in both saltwater and fresh-water habitats. Crocodiles are generally omnivorous, feeding on invertebrates, fish, other reptiles and amphibians, birds, and mammals—practically any animal they can overpower. A few, very narrow-snouted species are believed to subsist primarily on fish. Crocodiles are primarily aquatic and nocturnal, leaving the water only to bask by day or to build their nests. Some species construct burrows into the banks of rivers or lakes where they spend part of their time. SEE CROCODILIA; GAVIAL.

Anatomy. These animals are powerful predators with large teeth and strong jaws. Large adults of some species may exceed 20 ft (6 m) in length and are capable of overpowering and eating large grazing mammals such as deer and cattle and even, occasionally, humans. The webbed feet, flattened tail, and placement of the nostrils, eyes, and ears on raised areas of the head are adaptations for an aquatic existence (see **illus.**). The raised nostrils, eyes, and ears allow the animals to float almost completely submerged while still monitoring their environment.

The body, particularly the back, is covered with a series of bony plates which are free and unfused and not connected with the skeletal system proper. The crocodile does not molt the epidermis in a single unit or in large pieces, as do most reptiles.

Reproduction. Reproduction in crocociles, and in all crocodilians, is the most elaborate of the reptiles. Courtship and mating occur in the water. The female digs a hole in the soil for the 30 or so eggs, or, in the case of several species, constructs, near the water, an elevated nest of available soil and vegetation. Nests may be as large as 12 ft (3.5 m) across and 4 ft (1.2 m) high and are often guarded by the female, which may also construct wallows adjacent to the nest where she remains for much or all of the incubation period. Incubation takes approximately 2 months. In some species, at least, the hatchlings croak or grunt from within the nest, and the female opens the nest to assist their emergence. The young may remain together as a pod with the female for a year or more. Hearing and vocal communication are well developed in the crocodiles, and a variety of bellows, snarls, and grunts are utilized in their elaborate social behavior.

Conservation and ecology. Crocodiles are considered very valuable for the leather obtained from their hides, and all species are becoming very rare owing to hunting and to the loss of their habitats through land development for other uses. Most countries with native crocodile populations are implementing conservation measures, and international efforts are being made to regulate trade in crocodile products. Papua–New Guinea has developed a conservation-farming program for its species which, if successful, could serve as a model for the conservation of crocodiles and all crocodilians throughout the world.

The population biology of crocodiles is poorly known. Historically, they existed in many areas in large numbers and were evidently of considerable importance in the energetics and nutriment cycling in their environments. The present sparse populations make studies of their ecological importance difficult.

Size. There is considerable controversy over just how large crocodiles may grow. The saltwater crocodile (*Crocodylus porosus*) of the Indo-Pacific and Australia; the Nile crocodile (*C. niloticus*) of Africa; the American crocodile (*C. acutus*) of Central and South America, many Caribbean islands, and Florida; and the Orinoco crocodile (*C. intermedius*) of the Orinoco drainage in South America are reliably reported to approach or slightly exceed 20 ft (6 m) in total length. Johnson's crocodile (*C. johnsoni*) of Australia, the Mugger crocodile (*C. palustris*) of India and Sri Lanka (Ceylon), the Siamese crocodile (*C. siamensis*) of Malasia and Indonesia, the New Guinea and Philippine crocodiles (*C. novaeguineae* and *C. novaeguinea mindorensis*) of Papua–New Guinea and the Philippines, Morelet's crocodile (*C. moreletti*) of Mexico, Guatemala, and Belize (British Honduras), the Cuban crocodile (*C. rhombifer*) of Cuba, and the narrow-snouted crocodile (*C. cataphractus*) of Africa have a maximum length of between 9 and 14 ft (3 and 4 m). The false gavial (*Tomistoma schlegeli*) of Malaysia, Borneo, and Sumatra may occasionally reach 16 ft (5 m) in length. The broad-snouted crocodiles (*Osteolaemus tetraspis*) of west-central Africa seldom exceed 6 ft (2 m) in total length. SEE ALLIGATOR; ARCHOSAURIA; CARDIOVASCULAR SYSTEM; DENTITION; REPTILIA; THECODONTIA.

Howard W. Campbell

Bibliography. P. Brazaitis, *The Identification of Living Crocodilians*, 1974; C. A. W. Guggisberg, *Crocodiles: Their Natural History, Folklore, and Conservation*, 1972; E. Jarvik, *Basic Structure and Evolution of Vertebrates*, 2 vols., 1981; W. T. Neill, *The Last of the Ruling Reptiles: Alligators, Crocodiles, and Their Kin*, 1971.

Crocodylia

An order of the class Reptilia (infraclass Archosauria) which is composed of large, voracious, aquatic species which include the alligators, caimans, crocodiles, and gavials. The group has a long fossil history from Late Triassic times and its members are the closest living relatives of the extinct dinosaurs and the birds. The 21 or 22 living species are found in tropic areas of Africa, Asia, Austria, and the Americas. One form, the saltwater crocodile (*Crocodylus porosus*), has traversed oceanic barriers from the East Indies as far east as the Fiji Islands. SEE ARCHOSAURIA.

Morphology. The order is distinguished from other living reptiles in that it has two temporal foramina, an immovable quadrate, a bony secondary palate, no shell, a single median penis in males, socketed teeth, a four-chambered heart, and an oblique septum that

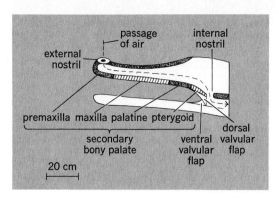

Fig. 1. Sagittal section of crocodylian head to show secondary palate and air passage.

completely separates the lung cavities from the peritoneal region.

Certain of these unique features and other salient characteristics of the Crocodylia are intimately associated with their aquatic life. The secondary palate, composed of medial expansions of premaxillary, maxillary, palatine, and pterygoid bones, divides the mouth cavity into two separate regions. The area above the bony palate forms an air passage extending from the external nostrils at the tip of the snout posteriorly to near the orbital region. The lower space retains the usual functions of the mouth, and is bordered above by the secondary palate, and below by the lower jaw. A special pair of fleshy flaps are found at the posterior end of the mouth cavity and form a valvular mechanism which separates the mouth from the region where the air passage opens into the throat (**Fig. 1**). This complex arrangement allows crocodylians to breathe even though most of the head is under water, or the mouth is open holding prey or full of water. Inasmuch as these great reptiles are extremely active when submerged and may remain under water for considerable periods, the increased efficiency of respiration provided by the oblique septum is an obvious adaptation. The significance of the four-chambered heart also lies in its contribution to rapid cellular respiration through an increased efficiency of circulation. Interestingly, all three of these structural characteristics are also found in mammals, although differing in detail. The Crocodylia, however, are not closely related to any of the reptiles that ultimately gave rise to the mammal stock. Other aquatic adaptations include valvular external nostrils that are closed during immersion, a recessed eardrum that can be covered by a skin flap under water, webbed toes,

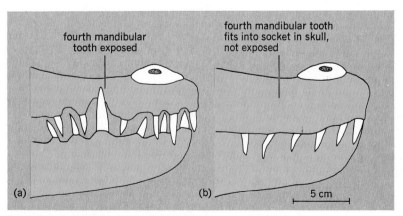

Fig. 2. Comparative lateral view of crocodylian snouts, showing tooth arrangement. (a) Gavialidae and Crocodylinae. (b) Alligatorinae.

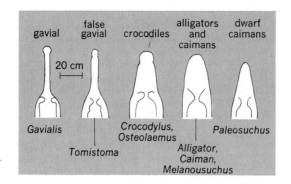

Fig. 3. Dorsal view of crocodylian heads.

and a long, compressed, muscular tail that propels the crocodylian through the water with strong lateral thrusts. In addition, the eyes, adapted for nocturnal activity by a vertically elliptical pupil, and the nostrils are mounted in raised areas of the head so that the animal may see and breathe without exposing much of its body above water.

Feeding. Crocodylians are carnivorous and prey upon insects, mollusks, fishes, and mammals. Usually only the larger adults attack terrestrial prey, which is usually caught at the water's edge and after being grasped in the powerful jaws is dragged down to be drowned. Crocodylians are extremely vicious reptiles and are as dangerous as sharks among aquatic organisms. The jaws, numerous sharp teeth, and the habit of twisting and rolling under water after the prey is seized, all contribute to their bad reputation. Even on land they are imposing creatures, especially when utilizing the muscular tail as a weapon.

Reproduction. During the breeding season male crocodilians set up territories on land which they defend against intruders of the same species. During this period their loud roars are frequently heard at night. Fertilization is internal and the hard-shelled eggs are deposited in excavations in the sand or in large nests of decaying vegetation, depending upon the species. In some forms the female guards the nest and several females may take turns in protecting a communal nest. It has been reported that when the young hatch the female may liberate them from the nest and lead them to the nearest water. The method of locomotion on land is by progression on all fours, with the belly and head held off the ground, and the tail dragging behind. In the water, movement is produced by lateral undulations of the tail and the forelimbs are held flat against the sides of the body.

Classification. The living species are placed in two families and eight genera. The family Crocodylidae contains two subgroups: the true crocodiles, Crocodylinae, including the genera *Crocodylus* found in all tropic areas, *Osteolaemus* in central Africa, and the false gavial (*Tomistoma*) in Malaya and the East Indies; the alligators and caimans, Alligatorinae, including the genera *Alligator* of the southeastern United States and near Shanghai, China, the *Caiman* from Central and South America, and *Melanosuchus* and *Paleosuchus* of South America. The gavial (*Gavialis gangeticus*) of India and north Burma is the only living member of the family Gavialidae. Crocodiles differ most obviously from alligators and caimans in head shape and in the position of the teeth, although other technical details also separate them (**Figs. 2** and **3**). The gavial differs from other living forms in its extremely long and narrow snout, with 27 or more

teeth on the upper jaw and 24 or more on the lower. Even the false gavial (*Tomistoma*), which resembles the gavial in head shape, has no more than 21 teeth on the upper jaw and 20 on the lower. SEE ALLIGATOR; CROCODILE; GAVIAL.

Jay M. Savage

Fossils. Crocodylians first appear in deposits of Late Triassic or possibly Early Jurassic age in North America and South Africa. *Protosuchus* from Arizona was a small armored reptile 32 in. (80 cm) long, not greatly different from Triassic thecodonts. Its flattened skull roof, slightly elongate coracoid, rodlike proximal carpals, and pubis forming only a small portion of the border of the acetabulum indicate that it was a true crocodile. *Nothochampsa*, from South Africa, had a similar skull, and *Erythrochampsa*, from slightly later deposits of that country, had an equally characteristic pelvis. These genera are commonly placed in a primitive suborder, Protosuchia. The extent of their secondary palates is unknown.

Most Mesozoic crocodylians differ from surviving members of the order in the less extensive development of the secondary palate and in having platycoelous vertebrae (in which the centra are shallowly concave at each end). The internal nares or choanae are elongate openings between the palatines and pterygoids (**Fig. 4**). The best-known fossils are remains of specialized marine crocodiles. Teleosaurs such as *Mystriosaurus* and *Steneosaurus* (**Fig. 5**) were long-snouted, completely armored animals with webbed toes. These were confined to the Jurassic. Smooth-skinned *Metriorhynchus* and *Geosaurus* of the Jurassic and Early Cretaceous had lost the armor and de-

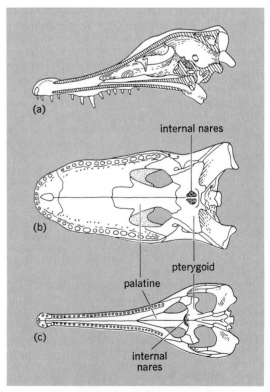

Fig. 4. Crocodylian skulls, showing position of nasal openings. (*a*) Longitudinal cross section of a crocodile skull to show nasal passages. (*b*) Palate of *Alligator* skull; nasal opening at rear of pterygoids. (*c*) Palate of a mesosuchian, *Steneosaurus*; narial opening between palatines.

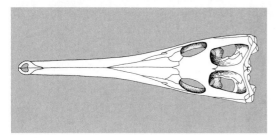

Fig. 5. Skull of teleosaur.

veloped paddlelike feet and even a small tail fin. *Hyposaurus* of the Cretaceous foreshadowed the gavials.

Fresh-water deposits of Late Jurassic and Cretaceous age have yielded remains of primitive crocodiles such as *Goniopholis* with broader, flatter heads which may include ancestors of both alligators and crocodiles. Many genera have been named, often from fragments.

In South America the mesosuchian crocodiles developed a side branch with narrow, deep skulls and compressed, bladelike teeth suggestive of the carnivorous dinosaurs. It is thought that the cretaceous *Baurosuchus* and Eocene *Sebecus* (**Fig. 6**) were more terrestrial and predaceous than typical crocodilians; they lack the flattened heads and usual aquatic specializations of the order.

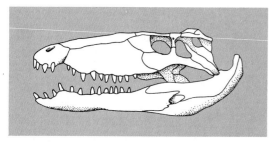

Fig. 6. Skull of *Sebecus*, Eocene mesosuchian from South America.

Eusuchia, or modern crocodiles, with a fully developed secondary plate and procoelous vertebrae (centra concave anteriorly and convex posteriorly) appear late in the Cretaceous and have continued to the present day in the tropics. The largest known crocodile, *Phobosuchus,* whose skull was 6.5 ft (2 m) long and whose body length has been estimated at 50 ft (15 m), lived during the Cretaceous and may have preyed on dinosaurs. *Thoracosaurus* was a long-snouted forerunner of the false gavial (*Tomistoma*) in the Late Cretaceous and Eocene of both North America and Europe. Ancestors of the alligators, *Allognathosuchus* and *Diplocynodon*, are known from the Paleocene and Eocene. Specimens indistinguishable from *Crocodylus* abound in the Eocene, and are foreshadowed by the Cretaceous *Leidyosuchus*.

Crocodylians formerly were much more widely distributed in temperate latitude than today; abundant paleobotanical evidence confirms a warmer climate in mid-latitudes at this time. It seems probable that the restriction of crocodylians to the tropics was a direct result of cooling climate in the late Cenozoic. *See* REPTILIA.

Joseph T. Gregory

Bibliography. E. H. Colbert, *The Age of Reptiles*, 1965; E. Jarvik, *Basic Structure and Evolution of Vertebrates*, 2 vols., 1981; S. P. Parker (ed.), *Synopsis and Classification of Living Organisms*, 2 vols., 1982; A. S. Romer, *Vertebrate Paleontology*, 3d ed., 1966; K. P. Schmidt and R. F. Inger, *Living Reptiles of the World*, 1957; S. W. Williston, *Water Reptiles of the Past and Present*, 1914.

Crocoite

A mineral with the chemical composition $PbCrO_4$. Crocoite occurs in yellow to orange or hyacinth red, monoclinic, prismatic crystals with adamantine to vitreous luster (see **illus.**); it is also massive granular.

Needlelike crocoite crystals in association with cerussite and dundasite, Dundas, Tasmania. (*Specimen courtesy of Department of Geology, Bryn Mawr College*)

Fracture is conchoidal to uneven. Hardness is 2.5–3 on Mohs scale and specific gravity is 6.0. Streak, or color of the mineral powder, is orangish-yellow. It fuses easily.

Crocoite is a secondary mineral associated with other secondary minerals of lead such as pyromorphite and of zinc such as cerussite. It has been found in mines in California and Colorado. *See* LEAD.

Edward C. T. Chao

Cro-Magnon people

The earliest representatives of the modern humans, *Homo sapiens*; a people who were biologically and behaviorally similar to modern human hunters and gatherers. The name Cro-Magnon is commonly used to refer to the earliest group of modern humans in western Europe, but it is frequently taken to include all of the early modern humans across the Old World. In that sense, Cro-Magnon is synonymous with early modern humans.

The name Cro-Magnon derives from the human skeletal remains found in 1868 in the Cro-Magnon rock shelter in Les Eyzies, France (see **illus.**). They were the first fossil humans to be found securely associated with the remains of extinct animals and Paleolithic stone tools, and the name of this site has therefore been used for human fossils of the same time period. *See* PALEOLITHIC.

Early modern humans emerged between 50,000 and 100,000 years ago from more archaic human ancestors in sub-Saharan Africa. They subsequently spread northward and into Eurasia, absorbing the lo-

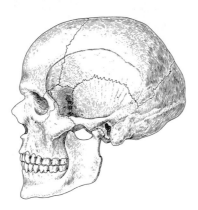

Skull of a Cro-Magnon male. *(After M. F. Ashley Montagu, An Introduction to Physical Anthropology, 2d ed., Charles C. Thomas, 1951)*

cal groups of late archaic humans. They had spread into northern Africa and the Near East by 40,000–45,000 years ago, into central Europe by 36,000 years ago, into western Europe by 32,000 years ago, and eastward to eastern Asia and Australia by about 35,000 years ago. The resultant populations of early modern humans, including the Cro-Magnon people proper in western Europe, therefore included among their ancestors both local populations of late archaic humans and early modern humans spreading out of sub-Saharan Africa and through the Near East. *See Neandertals.*

The Cro-Magnon people and other geographical groups of early modern humans lived during the last glacial period of the Pleistocene Epoch, eventually giving rise to later modern human populations, which were the aboriginal populations of the last 20,000 years. *See Glacial epoch.*

The Cro-Magnon people were the first humans that were biologically the same as modern humans and essentially would blend in with living peoples. They differed from modern humans primarily in their tendency to have a rugged, athletic build. This is evident in the enlarged attachment areas for many of their limb muscles and in the large dimensions of their limb bones. In addition, they had teeth that were slightly larger than those of most modern humans and heavily built faces. Early modern people and the European Cro-Magnon people in particular were relatively tall, being about 5 ft 10 in. (178 cm) in height on the average. In fact, until the twentieth century, they were virtually the only group of humans to reach their full biological growth potential. As a result of their large statures and muscularity, and hence large body masses, they had brains that were, on the average, larger than those of most modern humans (90 in.^{3} or 1500 cm^3, as opposed to 78–84 in.^{3} or 1300–1400 cm^3 averages for modern humans).

These early modern humans were successful hunters and gatherers, occupying almost all of the inhabitable regions of the Old World. Although frequently portrayed as big-game hunters, they in fact lived by hunting small to medium-size animals, especially various deer, goats, and antelopes, and by gathering wild plants for food. They were sufficiently successful that the levels of stress on their bodies, as reflected in lesions on their skeletons and teeth, are among the lowest known for prehistoric humans.

Their effectiveness as hunters and gatherers was due in part to their technology. They developed elaborate stone tool technologies, producing long blades that became blanks for tools with replaceable cutting

edges and points. This advance was made possible by use of bone and wood for carefully made hafts and handles for the sharp stone edges. Yet, their ability to effectively live a subsistence lifestyle largely depended on their extensive knowledge of the environment, so that they could harvest wild game and plants, rather than seek them opportunistically as did their predecessors. This knowledge was communicated between groups through the first symbolic systems known, which consisted of various geometric notational systems and forms of representational art. They were also the first humans to wear jewelry, and hence the first to modify their personal social images, suggesting more complex social roles than had previously existed.

The Cro-Magnon people and other early modern humans of the Upper Pleistocene were therefore the first humans to be biologically the same as modern humans and to use elaborate technology and information processing to successfully exploit all of the inhabitable environments of the Old World. *See Fossil human.*

Erik Trinkaus

Bibliography. F. H. Smith, *The Origins of Modern Humans*, 1984; E. Trinkaus (ed.), *The Biocultural Emergence of Modern Humans in the Late Pleistocene*, 1989; M. H. Wolpoff, *Paleontology*, 1980.

Cromwell Current

The strong subsurface current that flows eastward along the Equator in the Pacific Ocean. The Cromwell Current, also known as the Pacific Equatorial Undercurrent, flows from north of New Guinea in the western Pacific to the Galápagos Islands in the eastern Pacific. Its length, about 8700 mi (14,000 km), is larger than one-third the circumference of the Earth. The current is typically 120–240 mi (200–400 km) wide and 300–900 ft (100–300 m) thick. Its core of highest speed, usually 1.5–4.5 ft/s (0.5–1.5 m/s), is found at depths of 150–600 ft (50–200 m) [see **illus.** *a*]. *See Pacific Ocean.*

Observations. The Cromwell Current was discovered in the central Pacific in 1952, when researchers measured eastward speeds of over 1 knot (0.5 m/s) by tracking buoys tethered to subsurface drogues. Eastward drifts of long-line fishing gear, occasional eastward ship drifts, and a few direct measurements of eastward currents near the Equator had been reported previously, but had not been recognized by oceanographers as signs of a major ocean current. Extensive measurements of the Cromwell Current have been made, but there are still large areas with few measurements, and many aspects of the dynamics are yet to be understood.

The Cromwell Current is usually a continuous flow along the Equator from about 140°E to 91°W. It may at times originate further to the west, and its terminus in the east sometimes extends to the South American coast, where it turns south into the Peru-Chile Undercurrent. The high-speed core is found in the tropical thermocline, typically near 68°F (20°C), and rises with the thermocline from 600 ft (200 m) in the western Pacific to 150 ft (50 m) or less in the east. In the western Pacific there is sometimes a second core of eastward flow in the mixed layer. *See Thermocline.*

The average transport of the Cromwell Current in

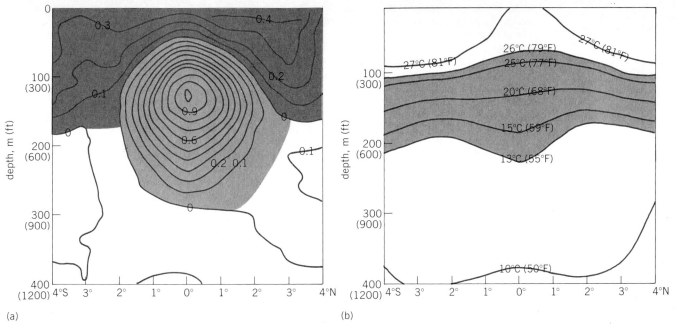

(a)   (b)

Cromwell Current contours of (a) zonal velocity component (m/s) and (b) temperature contours averaged over one year in the central Pacific. (After R. Lukas and E. Firing, The geostrophic balance of the Pacific Equatorial Undercurrent, Deep-Sea Res., 31(1):61–66, 1984)

the central Pacific is $1-1.4 \times 10^9$ ft³ ($3-4 \times 10^7$ m³) per second, similar to the North Equatorial Countercurrent and comparable to the Kuroshio and the Gulf Stream. Large changes in transport occur with a variety of time scales, from days to years. Eastward-traveling, equatorially trapped internal waves (called Kelvin waves) can cause the transport to nearly double for a week or two at a time. In the central and eastern Pacific, stronger easterly winds during September through January strengthen the South Equatorial Current and weaken the upper half of the Cromwell Current, which flows east below the westward South Equatorial Current. The extensive westerly winds associated with the El Niño phenomenon can destroy the east-west slope of the thermocline, and with it the Cromwell Current. This was observed for the first time during the exceptionally strong 1982–1983 El Niño. Previously the Cromwell Current had been thought to be a permanent feature of the circulation. *See* GULF STREAM; OCEAN WAVES.

Dynamics. The Cromwell Current is indirectly forced by the southeast tradewinds that blow westward along the Equator over most of the Pacific. The winds blow the surface waters to the west as the South Equatorial Current. Since the ocean basin is bounded, the wind stress maintains a slope of the sea surface, with sea level in the western Pacific about 1.5 ft (0.5 m) higher than in the east. The resulting eastward pressure-gradient force drives the Cromwell Current. Currents are relatively weak below the thermocline; there the pressure force due to the sea-level slope is compensated by an opposite force due to the thermocline slope down to the west. *See* WIND STRESS.

Because the vertical component of the Coriolis force is zero at the Equator, the existence of the Cromwell Current can be explained without reference to the Earth's rotation. However, the Coriolis force is important in the dynamics and structure of the current. The time-averaged eastward current is nearly in geostrophic balance, consistent with the spreading of the thermocline at the Equator (illus. *b*). A few degrees away from the Equator, the easterly tradewinds cause a poleward Ekman flux in the upper few tens of yards, while the eastward pressure force is geostrophically balanced by an equatorward flow in and above the thermocline. The surface divergence and subsurface convergence cause upwelling in and above the core of the Cromwell Current. This circulation in the vertical-meridional plane advects momentum and heat, contributing significant nonlinearities to the dynamics of the current. Equatorial upwelling also has major biological, geological, and climatic consequences. *See* CORIOLIS ACCELERATION; MARITIME METEOROLOGY; OCEAN CIRCULATION; TROPICAL METEOROLOGY.

Eric Firing

Bibliography. J. McCreary, A linear stratified ocean model of the equatorial undercurrent, *Phil. Trans. Roy. Soc. London*, A298(1444):603–635, 1981; S. G. H. Philander, Equatorial undercurrent: Measurements and theories, *Rev. Geophys. Space Phys.*, 11(3):513–570, 1973; S. G. H. Philander and R. C. Pacanowski, The generation of equatorial currents, *J. Geophys. Res.*, 85(C2):1123–1136, 1980; B. A. Warren and C. Wunsch (eds.), *Evolution of Physical Oceanography*, 1981.

Crop micrometeorology

Crop micrometeorology deals with the interaction of crops and their immediate physical environment. Especially, it seeks to measure and explain net photosynthesis (photosynthesis minus respiration) and water use (transpiration plus evaporation from the soil) of crops as a function of meteorological, crop, and soil moisture conditions. These studies are complex because the intricate array of leaves, stems, and fruits modifies the local environment and because the processes of energy transfers and conversions are inter-related. As a basic science, crop micrometeorology is related to plant anatomy, plant physiology, meteorology, and hydrology. Expertise in radiation exchange theory, boundary-layer and diffusion processes, and turbulence theory is needed in basic crop micrometeorological studies. A practical goal is to provide improved plant designs and cropping patterns for light interception, for reducing infestations of diseases, pests, and weeds, and for increasing crop water-use efficiency. Shelter belts are modifications that have been used in arid or windy areas to protect crops and seedlings from a harsh environment. *SEE AGRICULTURAL METEOROLOGY; AGRICULTURAL SCIENCE (PLANT); MICROMETEOROLOGY.*

Unifying concepts. Conservation laws for energy and matter are central to crop micrometeorology. Energy fluxes involved are solar wavelength radiation, consisting of photosynthetically active radiation (0.4–0.7 micrometer) and near-infrared radiation (0.7–3 μm); far-infrared radiation (3–100 μm); convection in the air; molecular heat conduction in and near the plant parts and in the soil; and the latent heat carried by water vapor. The main material substances transported to and from crop and soil surfaces are water vapor and carbon dioxide. However, fluxes of ammonia, sulfur dioxide, pesticides, and other gases or pollutants to or from crop or soil surfaces have been measured. These entities move by molecular diffusion near the leaves and soil, but by convection (usually turbulent) in the airflow. During the daytime generally, and sometimes at night, airflow among and above crops is strongly turbulent. However, often at night a stable air layer forms because of surface cooling caused by emission of far-infrared radiation back to space, and the air flow becomes nonturbulent. Fog or radiation frosts may result. The aerodynamic drag and thermal (heat-absorbing) effects of plants contribute to the pattern of air movement and influence the efficiency of turbulent transfer.

Both field studies and mathematical simulation models have dealt mostly with tall, close-growing crops, such as maize and wheat, which can be treated statistically as composed of infinite horizontal layers. Downward-moving direct-beam solar and diffuse sky radiation are partly absorbed, partly reflected, and partly transmitted by each layer. Less photosynthetically active radiation than near-infrared radiation is transmitted to ground level and reflected from the crop canopy because photosynthetically active radiation is strongly absorbed by the photosynthetic pigments (chlorophyll, carotenoids, and so on) and near-infrared radiation is only weakly absorbed. The plants act as good emitters and absorbers of far-infrared radiation. Transfers of momentum, heat, water vapor, and carbon dioxide can be considered as composed of two parts; a leaf-to-air transfer and a turbulent vertical transfer. As a bare minimum, two mean or representative temperatures are needed for each layer: an average air temperature and a representative plant surface temperature. Because some leaves are in direct sunlight and some are shaded, a representative temperature is difficult to obtain. Under clear conditions, traversing solar radiation sensors show a bimodal frequency distribution of irradiances in most crop communities; that is, most points in space and time are exposed to either high irradiances of direct-beam radiation or low irradiances characteristic of shaded conditions. Models of radiation interception have been developed which predict irradiance on both shaded leaves and on exposed leaves, depending on the leaf inclination angle with respect to the rays. The central concept of both experimental studies and simulation models is that radiant energy fluxes, sensible heat fluxes, and latent heat fluxes are coupled physically and can be expressed mathematically. This interdependence applies to a complex crop system as well as to a single leaf.

Photosynthesis. Studies of photosynthesis of crops using micrometeorological techniques do not consider the submicroscopic physics and chemistry of photosynthesis and respiration, but consider processes on a microscopic and macroscopic scale. The most important factors are the transport and diffusion of carbon dioxide in air to the leaves and through small ports called stomata to the internal air spaces. Thence it diffuses in the liquid phase of cells to chloroplasts, where carboxylation enzymes speed the first step in the conversion of carbon dioxide into organic plant materials. Solar radiation provides the photosynthetically active radiant energy to drive this biochemical conversion of carbon dioxide. Progress has been made in understanding the transport processes in the bulk atmosphere, across the leaf boundary layer, through the stomata, through the cells, and eventually to the sites of carboxylation. Transport resistances have been identified for this catenary process: bulk aerodynamic resistance, boundary-layer resistance, stomatal diffusion resistance, mesophyll resistance, and carboxylation resistance. All these resistances are plant factors which control the rate of carbon dioxide uptake by leaves of a crop; however, boundary-layer resistance and especially bulk aerodynamic resistance are determined also by the external wind flow.

Carbon dioxide concentration and photosynthetically active radiation are two other factors which control the rate of crop photosynthesis. Carbon dioxide concentration does not vary widely from about 315 microliters per liter (0.0315 vol %). Experiments have revealed that it is not practical to enrich the air with carbon dioxide on a field scale because the carbon dioxide is rapidly dispersed by turbulence. Therefore carbon dioxide concentration can be dismissed as a practical variable. Solar radiation varies widely in quantity and source distribution (direct-beam or diffuse sky or cloud sources). Many species of crop plants have leaves which can utilize solar radiation having flux densities greater than full sunlight (tropical grasses such as maize, sugarcane, and Burmuda grass, which fix carbon dioxide through the enzyme phosphoenolpyruvate carboxylase). Other species have leaves which may give maximum photosynthesis rates by individual leaves at less than full sunlight (such as soybean, sugarbeet, and wheat, which fix carbon dioxide through the enzyme ribulose 1,5-diphosphate carboxylase). However, in general, most crops show increasing photosynthesis rates with increasing irradiances for two reasons. First, more solar energy would become available to the shaded and partly shaded leaves deep in the crop canopy. Second, many of the well-exposed leaves at the top of a crop canopy are exposed to solar rays at wide angles of inclination so that they do not receive the full solar flux density. These leaves will respond to increasing irradiance also. Furthermore, increased diffuse to direct-beam ratios of irradiance (which could be caused by haze or thin clouds) may increase the irradiance on shaded leaves and hence increase overall crop photosynthesis.

If crop plants lack available soil water, the stomata may close and restrict the rate of carbon dioxide uptake by crops. Stomatal closure will protect plants against excessive dehydration, but will at the same time decrease photosynthesis by restricting the diffusion of carbon dioxide into the leaves. SEE PHOTOSYNTHESIS.

Transpiration and heat exchange. Transpiration involves the transport of water vapor from inside leaves to the bulk atmosphere. The path of flow of water vapor is from the surfaces of cells inside the leaf through the stomata, through the leaf aerodynamic boundary layer, and from the boundary layer to the bulk atmosphere. Sensible heat is exchanged by convection directly from plant surfaces; therefore there is no stomatal diffusion resistance associated with this exchange. Stomatal diffusion resistance does affect heat exchange from leaves, however, because when stomata are open wide (low resistance) much of the heat exchanged from leaves is in the form of latent heat of evaporation of water involved in transpiration.

Small leaves, such as needles, convect heat much more rapidly than large leaves, such as banana leaves. Engineering boundary-layer theory suggests that boundary-layer resistance should be proportional to the square root of a characteristic dimension of a leaf and inversely proportional to the square root of the airflow rate past a leaf. Experiments support these relationships.

Under high-irradiance conditions, low air humidity, high air temperature, and low stomatal diffusion resistance will favor high transpiration, whereas high air humidity, low air temperature, and high stomatal diffusion resistance will favor sensible heat exchange from leaves. The function of wind is chiefly to enhance the transport rather than determine which form of convected energy will be most prominent. In arid environments, the latent energy of transpiration from crops may exceed the net radiant energy available, because heat from the dry air may actually be conducted to crops which will cause transpiration to increase. In those areas, crop temperature is lower than air temperature. SEE LEAF.

Flux methods. At least three general methods have been employed to measure flux density of carbon dioxide, water vapor, and heat to and from crop surfaces on a field scale. These methods are restricted to use in the crop boundary layer immediately above the crop surface, and they require a sufficient upwind fetch of a uniform crop surface free of obstructions. Flux densities obtained by these methods will reflect the more detailed interactions of crop and environment, but will not explain them.

The principle of the energy balance methods is to partition the net incoming radiant energy into energy associated with latent heat of transpiration and evaporation, sensible heat, photochemical energy involved in photosynthesis, heat flux into the soil, and heat stored in the crop. Measurement of net input of radiation to drive these processes is obtained from net radiometers, which measure the total incoming minus the total outgoing radiation. The most important components—latent heat, sensible heat, and photochemical energy—are determined by average vertical gradients of water vapor concentration (or vapor pressure), air temperature, and carbon dioxide concentration.

The principle of the bulk aerodynamic methods is to relate the vertical concentration gradients of those transported entities to the vertical gradient of horizontal wind speed. The transports are assumed to be related to the aerodynamic drag (or transport of momentum) of the crop surface. Corrections are required for thermal instability or stability of the air near the crop surface.

The eddy correlation methods are direct methods which correlate the instantaneous vertical components of wind (updrafts or downdrafts) to the instantaneous values of carbon dioxide concentration, water vapor concentration, or air temperature. Under daytime conditions, turbulent eddies, or whorls, transport air from the crop in updrafts, which are slightly depleted in carbon dioxide, and conversely, turbulence transports air to the crop in downdrafts which are representative of the atmospheric content of these entities. More basic and applied research is being done on eddy correlation methods because they measure transports through direct transport processes.

Plant parameters. The stomata are the most important single factor in interactions of plant and environment because they are the gateways for gaseous exchange. Soil-to-air transfers are also very important while crops are in the seedling stage until a large degree of ground cover is attained. Coefficients of absorption, transmission, and reflection by leaves of photosynthetically active, near-infrared, and long-wavelength infrared radiation are not very different among crop species, but the geometric arrangement and stage of growth of plants in a crop may affect radiation exchange greatly. The crop geometry also interacts with radiation-source geometry (diffuse to direct-beam irradiance, solar elevation angle). Crop micrometeorology attempts to show how the plant parameters interact with the environmental factors in crop production and water requirements of crops under field conditions. SEE PHYSIOLOGICAL ECOLOGY (PLANT).

L. H. Allen, Jr.

Bibliography. J. Goudriaan, *Crop Micrometeorology*, 1977; E. Lemon, D. W. Stewart, and R. W. Shawcroft, The sun's work in a cornfield, *Science*, 174:371–378, 1971; J. L. Monteith (ed.), *Vegetation and the Atmosphere*, vol. 1, 1975; vol. 2, 1976; R. E. Munn, *Biometeorological Methods*, 1971; N. J. Rosenberg and B. L. Blad, *Microclimate: The Biological Environment*, 2d ed., 1983; W. D. Sellars, *Physical Climatology*, 1965; L. P. Smith (ed.), *The Application of Micrometeorology to Agricultural Problems*, 1972; O. G. Sutton, *Micrometeorology*, 1977.

Crossing-over (genetics)

The process whereby one or more gene alleles present in one chromosome may be exchanged with their alternative alleles on a homologous chromosome to produce a recombinant (crossover) chromosome which contains a combination of the alleles originally present on the two parental chromosomes. Genes which occur on the same chromosome are said to be linked, and together they are said to compose a linkage group. In eukaryotes, crossing-over may occur during both meiosis and mitosis, but the frequency of meiotic crossing-over is much higher. This article is concerned primarily with meiotic crossing-over. *See Allele; Chromosome; Gene; Linkage (genetics).*

Crossing-over is a reciprocal recombination event which involves breakage and exchange between two nonsister chromatids of the four homologous chromatids present at prophase I of meiosis; that is, crossing-over occurs after the replication of chromosomes which has occurred in premeiotic interphase. The result is that half of the meiotic products will be recombinants, and half will have the parental gene combinations (**Fig. 1**). Using maize chromosomes which carried both cytological and genetical markers, H. Creighton and B. McClintock showed in 1931 that genetic crossing-over between linked genes was accompanied by exchange of microscopically visible chromosome markers. *See Recombination (genetics).*

During meiosis, crossing-over occurs at the pachytene stage, when homologous chromosomes are completely paired. At diplotene, when homologs separate, the sites of crossing-over become visible as chiasmata, which hold the two homologs of a bivalent together until segregation at anaphase I. Each metaphase I bivalent will necessarily have at least one chiasma. In favorable material, such as grasshopper spermatocytes, it is possible to observe that each diplotene chiasma involves a crossover of two of the four chromatids at one site.

Where two or more crossovers occur in one biva-

lent, they usually do not cluster together but are widely separated; this is known as chiasma interference. The occurrence of one crossover event appears to preclude the occurrence of a second crossover in the immediate vicinity. In addition, the distribution of occurrence of chiasmata along a chromosome may be localized; the probability that a crossover will occur is higher in some chromosome segments and lower in other segments.

In general, the closer two genes are on a chromosome, that is, the more closely linked they are, the less likely it is that crossing-over will occur between them. Thus, the frequency of crossing-over between different genes on a chromosome can be used to produce an estimate of their order and distances apart; this is known as a linkage map. *See Genetic mapping.*

Molecular mechanisms. Since each chromatid is composed of a single deoxyribonucleic acid (DNA) duplex, the process of crossing-over involves the breakage and rejoining of DNA molecules. Although the precise molecular mechanisms have not been determined, it is generally agreed that the following events are necessary: (1) breaking (nicking) of one of the two strands of one or both nonsister DNA molecules; (2) heteroduplex (hybrid DNA) formation between single strands from the nonsister DNA molecules; (3) formation of a half chiasma, which is resolved by more single-strand breakages to result in either a reciprocal crossover, a noncrossover, or a nonreciprocal crossover (conversion event).

Two molecular models of recombination which

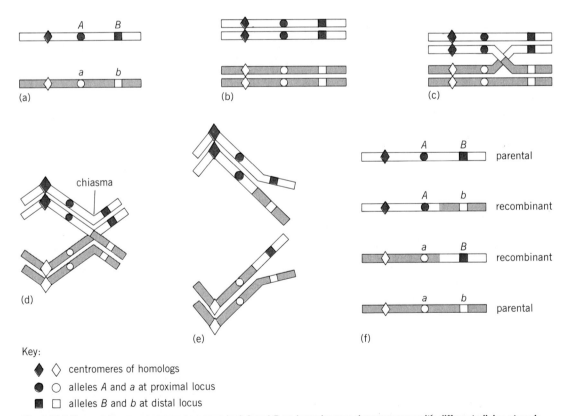

Fig. 1. Meiotic crossing-over between two gene loci *A* and *B* on homologous chromosomes with different alleles at each locus. (*a*) Prior to replication. (*b*) After replication in premeiotic interphase. (*c*) Crossing-over at pachytene between nonsister chromatids of paired homologs (bivalent). (*d*) Diplotene bivalent held together by a chiasma at crossover point. (*e*) Anaphase I segregation of chromosomes in bivalent. (*f*) Four meiotic products after anaphase II segregation. In general, centromeres and loci proximal to the chiasma (crossover) segregate at first division, while loci distal to the chiasma segregate at second division.

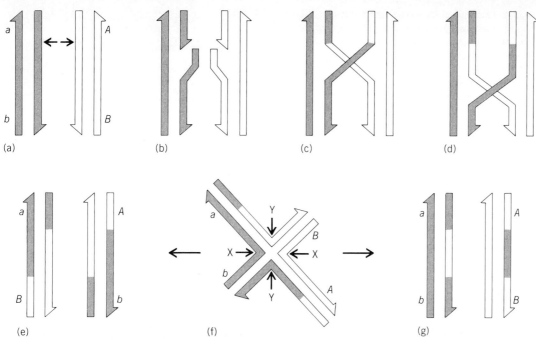

Fig. 2. Molecular model of recombination, based on that of R. Holliday. Only the two recombinant chromatids are shown. (*a*) Two homologous nonsister chromatid DNA molecules, with gene loci *ab* and *AB,* respectively; arrows are paired nicking sites. (*b*) After single-strand nicks at equivalent sites in both chromatids, nicked strands separate. (*c*) Nicked strands are displaced and reanneal to opposite duplex forming a half chiasma. (*d*) Migration of half chiasma increases length of heteroduplex DNA. (*e*) Isomerization of structure in *d* by rotation gives open half-chiasma form; paired nicks occur in two strands. (*f*) Nicks at X,X resolve the half chiasma into reciprocal crossover chromatids *aB* and *Ab,* with heteroduplexes in middle regions; or (*g*) nicks at Y,Y resolve the half chiasma into noncrossover chromatids *ab* and *AB,* with heteroduplexes in middle regions.

have gained credence are those of R. Holliday and of M. Meselson and C. Radding. Holliday's model postulates nicks in both chromatids at the initiation of crossing-over (**Fig. 2**). Meselson and Radding postulate an initial single-strand cut in only one DNA strand. Repair synthesis displaces this strand, which pairs with its complement on the other chromatid, thereby displacing and breaking the other strand of that DNA molecule. Following pairing and ligation of the two remaining broken ends, a half chiasma is formed. Other models have been postulated in which recombination is initiated by a double-stranded break in one chromatid. In all the above models, gene conversion can occur in the middle region of the molecules (with or without outside marker crossing-over) by mismatch repair of heteroduplex DNA.

Ultrastructural cytology. Pachytene, the meiotic stage at which crossing-over is considered to occur, corresponds with the period of close pairing or synapsis of homologous chromosomes. Electron microscopy has revealed that proteinaceous structures, the synaptonemal complexes (**Fig. 3**), are involved in the synapsis of chromosomes. A synaptonemal complex forms during zygotene by pairing of axial elements from two homologous chromosomes. It is present along the whole length of each pachytene bivalent and disappears at diplotene. Evidence from inhibitor studies and mutant stocks shows that the synaptonemal complex is necessary for meiotic crossing-over to occur. However, in cases such as desynaptic mutants, some hybrids, and the female silkworm, complete pachytene synaptonemal complexes have been observed, but no crossing-over occurs, showing that the synaptonemal complex alone is not sufficient to cause crossing-over.

In *Drosophila melanogaster* oocytes, the occurrence at pachytene of dense spherical bodies bridging the central region of the synaptonemal complex has been described. These bodies coincided in number and position with expected crossover events, and therefore were named recombination nodules. A variety of oval and bar-shaped recombination nodules (Fig. 3) have also been found in organisms as diverse as fungi, humans, rat, silkworm, and maize. In many cases their number correlates with crossover frequency. It has been suggested that recombination nodules are prerequisites for crossing-over. If this is

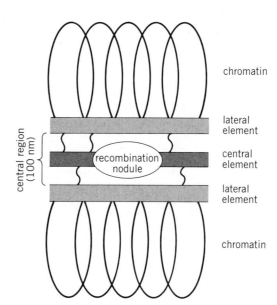

Fig. 3. Longitudinal section of a synaptonemal complex with a recombination nodule.

so, the recombination nodule may represent a complex of enzymes involved in the early events of recombination (nicking, strand separation, repair synthesis).

DNA repair synthesis has been observed during pachytene in lily microsporocytes, and has been shown to be reduced in an achiasmatic mutant. Prophase I of lilies is characterized by the presence of several proteins which could have a role in crossing-over, for example, DNA binding protein, endonucleases, ligases, and kinase. Inhibition of protein synthesis at zygotene-pachytene results in failure of crossing-over. Thus both DNA synthesis and protein synthesis appear necessary for meiotic crossing-over in lily. SEE MEIOSIS; MITOSIS.

Sex chromosomes. The differentiated X and Y sex chromosomes in human males and many animals (Z and W chromosomes in female birds) have small regions near one tip which undergo pairing and crossing-over at meiotic prophase I. Electron microscopy of the pachytene XY reveals the formation of a short synaptonemal complex segment with a recombination nodule in the majority of cases; the presence of a chiasma between the X and Y at metaphase I indicates the occurrence of crossing-over. An obligatory crossover in the XY bivalent is necessary to ensure regular segregation of X and Y to opposite poles at anaphase I. The pairing region contains a few gene loci on both X and Y chromosomes which exhibit an autosomallike inheritance pattern. Recombination between genes and DNA sequences in this pseudoautosomal region confirms the occurrence of obligatory crossing-over. The rare occurrence of XX males in some cases is accounted for by abnormal recombination events outside the pseudoautosomal region which have transferred the male sex-determining gene from the Y to the X chromosome. SEE SEX DETERMINATION; SEX-LINKED INHERITANCE.

C. B. Gillies

Bibliography. D. G. Catcheside, *The Genetics of Recombination*, 1977; C. B. Gillies (ed.), *Fertility and Chromosome Pairing: Recent Studies in Plants and Animals*, 1988; P. B. Holm and S. W. Rasmussen, Human meiosis, VI. Crossing-over in human spermatocytes, *Carlsberg Res. Commun.*, 48:385–413, 1983; B. Lewin, *Genes*, 3d ed., 1987; P. B. Moens (ed.), *Meiosis*, 1987; F. Rouyer et al., A gradient of sex linkage in the pseudoautosomal region of the human sex chromosomes, *Nature*, 319:291–295, 1986.

Crossopterygii

An infraclass of the bony fishes (class Osteichthyes), also known as fringe-finned fishes, that forms one of the two major divisions of the lobe-finned fishes (Sarcopterygii). The group first appeared as fossils in the Early Devonian; in the Paleozoic they were mostly small to medium-sized carnivorous fish [to 12 in. (30 cm), rarely to 6.5 ft (2 m)] living in shallow tropical seas, estuaries, and fresh waters. There were two principal groups: a diverse set of fishes termed Rhipidista and the Coelacanthini. Their principal radiations were in the Devonian, and by the Mississippian they were in sharp decline. The Rhipidista were wholly extinct by the Middle Permian, but the coelacanths underwent a second, smaller, Mesozoic radiation and managed to survive to the present day as the

Living coelacanth, *Latimeria chalumnae*; 5 ft (1.5 m). (*From P. P. Grassé, ed., Traité de Zoologie, tome 13, fasc. 3, 1958*)

most famous lobe-fin of all, the living species *Latimeria chalumnae* (see **illus**.). Crossopterygii are characterized by a unique hinge in the skull that allowed the front portion to be raised and lowered during feeding and respiratory movements. SEE SARCOPTERYGII.

Rhipidistia. Members of the order Rhipidistia were principally fusiform, fast-swimming, carnivores that flourished in the rivers and lakes of the Late Devonian. They could breathe air, and the use of lungs as well as gills gave them an advantage in warm, shallow-water environments where dissolved oxygen was often low. The group includes at least three distinct lineages, and is especially important because one subgroup (Osteolepiformes) has specialized paired fins and a skull structure, including internal nostrils or choanae, that are shared only with the first land vertebrates, which also appeared in the Late Devonian. It is widely held, therefore, that the ancestor of all the tetrapod vertebrates was a rhipidistian. The rival theory finds ancestry among the Dipnoi (lungfishes). SEE DIPNOI.

Coelacanthini. Members of the order Coelacanthini are characterized by a special trifid tail and scales ornamented with tubercles. They are not thought to be close to the ancestors of tetrapods. The group underwent a modest Devonian radiation in parallel with the Rhipidistia and Dipnoi, and then a second Mesozoic radiation, but was universally thought to have become extinct at the end of the Cretaceous, when the last fossils are found. However, in December 1938, a living species, *Latimeria chalumnae*, was caught by a trawler off the South African coast and identified. Subsequently, some 150 more specimens have been caught by fishing people of the Comoro Islands (northwest of Madagascar). In their skeletal evolution, coelacanths have been extremely conservative. *Latimeria* is so like its Mesozoic and Paleozoic predecessors that it is called a living fossil, and it is being actively studied for clues to the biology of all ancient lobe-finned fishes. SEE LIVING FOSSIL.

The coelacanths are principally marine fishes, although there are fossil fresh- and brackish-water forms. *Latimeria* has a specialized blood chemistry, similar to that of sharks, involving urea retention that allows it to live in salt water. It is large (to 6.5 ft or 2 m), but most of the fossil taxa were smaller. In all coelacanths the lungs have been reduced to a single, so-called "swim bladder" that in *Latimeria* is filled with fat. Off the Comoro Islands, *Latimeria* lives along the steep, rocky submarine coasts, probably at depths of no more than 1900 ft (600 m). Its distribution and general biology are still poorly known, but the blue color, characteristics of the eye, stomach contents, and the composition of the fats in the tissues all suggest that it is not a genuinely deep-sea fish.

Most catches have been made between December and April, and all at night. Since 1938, it has not been found anywhere except the Comoro Islands, and the first specimen is often thought to have been a stray. Alternatively, populations may occur more widely but sparsely throughout the western Indian Ocean.

Studies of fresh and frozen tissues suggest that the fish is rather sedentary. Underwater photographs have shown that the fish remains stationary in a current by sculling actions of the mobile paired and median fins. The tail is used only for short, sharp bursts of speed in pursuit of prey such as fishes and cephalopods. *Latimeria* has a unique sensory chamber in the snout which may be electroreceptive.

Latimeria, as well as at least some of the fossil taxa, is ovovivparous. Females contain 12–18 grapefruit-sized eggs that are fertilized internally and retained until hatching. Some one to three *Latimeria* are caught each year, and there is considerable debate as to whether this is depleting the stocks. When brought to shore, specimens have lived up to 24 h, giving hope that eventually one might be studied in the laboratory. SEE OSTEICHTHYES.

Keith Thomson

Bibliography. E. Jarvik, *Basic Structure and Evolution of Vertebrates,* 2 vols., 1980; N. A. Lockett, Some advances in coelacanth biology, *Proc. Roy. Soc. London,* B208:265–307, 1980; J. A. Moy-Thomas and R. S. Miles, *Palaeozoic Fishes,* 1971.

Crosstalk

Interference in a communications channel (disturbed channel) caused by activity in other communications channels (disturbing channels). The term was originally used to denote the presence in a telephone receiver of unwanted speech signals from other conversations, but its scope has been extended by common usage to include like effects in other types of communications.

Causes. The cause of crosstalk is some form of coupling mechanism between the disturbed channel and the disturbing channels. Communications channels are normally segregated by space, frequency, or time division, or by some combination of the three, to avoid such coupling, but economic and other constraints often preclude complete segregation.

Space division. In space-division segregation, each communication channel is assigned its own transmission medium, for example, a pair of wires is a multipair cable (see **illus.**) or a separate radio propagation path. Coupling between channels is caused by the physical proximity and relative orientation of the transmission media. The coupling is usually electromagnetic, a linear phenomenon that is independent of signal level in the channels. SEE ELECTROMAGNETIC COMPATIBILITY.

Frequency division. In frequency-division segregation, a single, common transmission medium (for example, a coaxial cable system) is used to provide many communications channels, and each channel is assigned a separate frequency band within the medium. A signal is placed in its channel by amplitude modulation (AM) or angle modulation [frequency modulation (FM) or phase modulation (PM)]. With AM systems, nonlinearities in the transfer characteristics of system elements create harmonic, or intermodulation, distortion. Typically, nonlinearities are

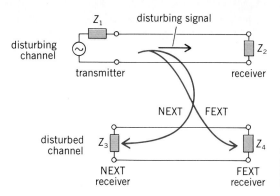

Crosstalk in a space-division system. Z_1, Z_2, Z_3, and Z_4 designate impedances.

introduced by system amplifiers that are used to compensate for loss in the medium. Intermodulation products (distortions) have frequencies that are combinations (sums, differences, multiples) of the frequencies of the signals being carried by the system. Thus, nonlinearities spawn many coupling paths among the channels of the system. The coupling is nonlinear, that is, the losses of the paths are functions of signal amplitudes. In angle-modulation systems, the principal cause of intermodulation distortion is frequency dependency in the amplitude and delay characteristics of the transmission medium (including amplifiers), but variation of the delay characteristic of the medium as a function of signal level (AM to PM conversion) can also cause intermodulation distortion. The resulting coupling paths among the channels of the system are similar to those in an AM system. SEE AMPLITUDE MODULATION; DISTORTION (ELECTRONIC CIRCUITS); FREQUENCY MODULATION; MODULATION; PHASE MODULATION.

Time division. In time-division segregation, a single common transmission medium is time-shared to provide many communications channels, with each channel allotted a separate time interval. These may be assigned on the basis of relative activity among the channels, or each channel may be periodically assigned a like interval regardless of activity, which is usually the case with digital systems. Interchannel coupling is the result of the presence of vestigial energy in the transmission medium from previous channel assignments. The coupling tends to be linear (the loss tends to be independent of signal amplitude), and often involves an impedance that is common to both the disturbing and disturbed channels. SEE PULSE MODULATION.

Classification. Crosstalk is classified in a variety of ways. The type of coupling (electromagnetic, intermodulation, or common impedance) indicates the mechanism. The terms near-end crosstalk (NEXT) and far-end crosstalk (FEXT) indicate the relative directions of signal propagation in the disturbed and disturbing channels (see illus.). The terms direct crosstalk, where the disturbing channel couples directly to the disturbed channel, and indirect crosstalk, where the coupling path between disturbing and disturbed channels involves a third, or tertiary, channel, are often used to further describe crosstalk caused by electromagnetic coupling. Interaction crosstalk (IXT) is a term used to further describe indirect crosstalk that couples from the disturbing channel to the tertiary channel at one place, propagates along the tertiary channel, and subsequently couples into the disturbed channel at another place. Transverse crosstalk is a term that includes all direct and indirect crosstalk that is not interaction crosstalk.

Intelligible crosstalk is understood by the receiver of the disturbed channel, whereas nonintelligible crosstalk is not. Intelligible crosstalk most often occurs between channels carrying similar analog signals, such as analog speech signals. Intelligible crosstalk between channels, one of which is carrying a digital signal, or between channels carrying different types of analog signals (a voice signal and a video signal) is less likely to occur because the receiver does not comprehend the information in the disturbing signal. Nonintelligible crosstalk often affects the receiver in much the same way as noise, but in some cases it may be nearly intelligible, for example, crosstalk from one analog speech channel to another where intermodulation coupling has inverted the frequency spectrum without masking the syllabic nature of the disturbing speech signal. *See* ELECTRICAL NOISE.

Quantitative measures. The magnitude of crosstalk that becomes disruptive to a communications channel depends on many factors: the type of signals involved, the intelligibility of the crosstalk, the acuity of the receiver, the activity factors for the channels, and so on. In some cases crosstalk affects the disturbed channel in much the same way as noise, and signal-to-noise ratios are applicable measures. Digital channels tend to be of this nature. Typically, binary (two-level) digital signals require peak signal–to–root-mean-square noise ratios equal to, or better than, about 20 dB (voltage or current ratios of 10:1 or better). Crosstalk from one analog video channel to another produces an effect much different from noise, but a signal-to-interference ratio is an appropriate measure because the primary signal is usually present. Peak signal-to-peak interference ratios of 60 dB or better (voltage or current ratios of 1000:1 or better) are generally required for good reception of such signals. *See* SIGNAL-TO-NOISE RATIO.

In other cases, such as intelligible crosstalk in telephony, signal-to-crosstalk ratios are not applicable measures because the primary signal is not always present to mask the interference. In telephony, detection of intelligible crosstalk by the receiver is particularly disruptive—it does not really distort the primary signal, but it represents a breach of privacy—and the detection of nonintelligible crosstalk that seems to be intelligible (of a syllabic nature) is almost as disruptive. Ever-present noise in the disturbed channel tends to mask crosstalk detection. To prevent detection, the noise power must exceed the crosstalk power by about 10 dB (noise-to-crosstalk power ratio of 10:1). When a disturbed telephony channel has very little noise, the typical receiver can detect crosstalk levels of about -85 dBm (-85 decibels below 1 milliwatt, or 3×10^{-9} mW). For comparison, a typical level for the primary signal at the receiver is about -25 dBm (3×10^{-3} mW). Objectives for the crosstalk performance of a telephony network are often expressed by crosstalk index, a measure of the probability of the detection of intelligible crosstalk during the course of a telephone call. Values for crosstalk index are expressed in percentages, and typical values are less than 1%. *See* DECIBEL.

Remedies. Most remedies for reducing crosstalk entail some technique for decreasing coupling among the communications channels involved. The use of twisted pairs in multipair cables, of shielding for each pair, or of coaxial conductors and optical fibers in place of pairs are common techniques for reducing electromagnetic coupling where space division alone is not adequate. Improved control of signal levels and improved linearity in amplifiers are effective for frequency-division systems. Often such improvements are made possible by advances in technology. Selection of the appropriate type of modulation (AM, FM, PM) is also important. Crosstalk within multichannel digital systems that transport digital versions of analog signals can be reduced with the help of a separate coder-decoder for each analog channel in place of a common, time-shared coder-decoder for all channels. *See* ELECTRICAL SHIELDING; OPTICAL FIBERS.

Signal processing in the disturbed channel can sometimes be effective in reducing crosstalk. Syllabic compandors, which are sometimes used with telephony channels on analog carrier systems to reduce noise in silent intervals during conversations, are effective in reducing crosstalk as well. Pulse shaping (modification of the signal spectrum) can be effective in reducing crosstalk between digital transmission systems and in reducing intersymbol interference within a digital transmission system. One of the most important tools for control of crosstalk is spectrum management, a systematic approach to the assignment of frequency bands, signal levels, and the like in such way as to offer the most efficient use of the transmission media. *See* ELECTRICAL COMMUNICATIONS; ELECTRICAL INTERFERENCE; RADIO SPECTRUM ALLOCATIONS; TELEPHONE SERVICE.

Jonathan W. Smith

Bibliography. AT&T, Bell Telephone Companies, and Bell Telephone Laboratories, *Telecommunications Transmission Engineering*, vol. 1: *Principles*, 2d ed., 1977; Bell Telephone Laboratories, *Transmission Systems for Communications*, 5th ed., 1982; K. I. Park, Intelligible crosstalk performance of voice-frequency customer loops, *Bell Sys. Tech. J.*, 57:3001–3029, 1978; T. K. Sen, Masking of crosstalk by speech and noise, *Bell Sys. Tech. J.*, vol. 49:561–584, 1970.

Crown ethers

Macrocyclic organic compounds generally composed of repeating ethylene (CH_2CH_2) units separated by noncarbon atoms such as oxygen, nitrogen, sulfur, phosphorus, or silicon. Cryptands are similar to crown ethers but have a third organic chain attached at two points on the crown ring, usually trivalent nitrogen. Although other carbon-containing subunits such as propylene ($CH_2CH_2CH_2$) or methylene (CH_2) may be included, these subunits are less common. The conformations of the ethylene unit are more favorable than those of propylene, and the ethylene unit is more stable to hydrolysis than is the O—CH_2—O linkage (an acetal). Organic units sterically equivalent to the ethylene unit, such as 1,2-benzo, have been incorporated, as have units sterically equivalent to two ethyleneoxy units, such as 2,6-bis(hydroxymethyl)pyridine. Replacement of an OCH_2CH_2O unit by the 1,2-dihydroxyethyl unit of tartaric acid (HOOC—CHOH—CHOH—COOH) or by the adjacent hydroxyl groups of carbohydrates (sugars) alters hydrophobicity and introduces chirality. By far, the most common heteroatom present in the macrorings of crowns [X in $(XCH_2CH_2)_n$] is oxygen; but as more intricate structures are prepared, nitrogen, sulfur, phosphorus, silicon, or siloxy residues are becoming much more common. *See* STEREOCHEMISTRY.

The name crown ether is informal, developed to bypass the complex systematic names that these compounds are given under traditional nomenclature rules. A formal system of nomenclature for compounds of this type has yet to be developed, although several systems have been suggested. Of these, the name "coronand" for a cycle and "coronates" for its complex have been most widely adopted. By tradition, however, crown ethers are named according to the total number of atoms composing the macrocyclic ring and the number of heteroatoms contained within it. A crown composed of five ethyleneoxy [CH$_2$CH$_2$O] units would be called 15-crown-5. Substitution of one ethyleneoxy unit by a 1,2-benzo unit would lead to benzo-15-crown-5. Substitution of a nitrogen atom for an oxygen would transform 15-crown-5 into aza-15-crown-5. Structures of 12-crown-4 and dibenzo-18-crown-6 are shown below.

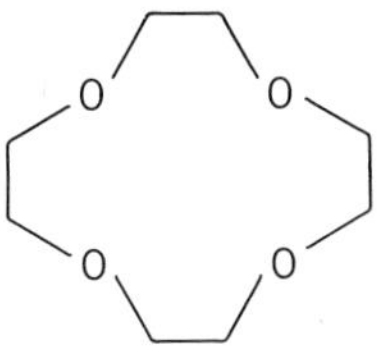

12-Crown-4

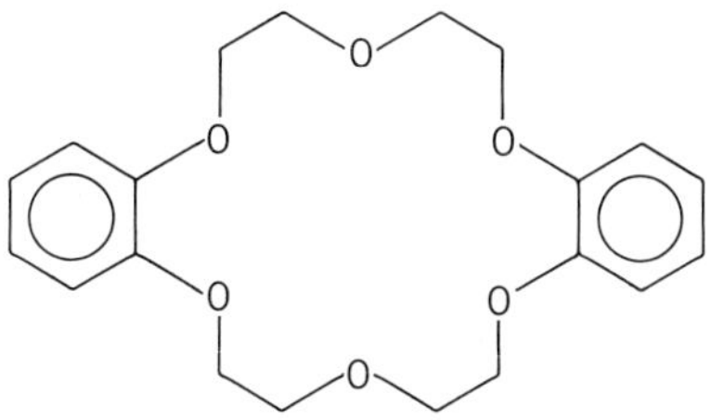

Dibenzo-18-crown-6

The crown ethers have been known for many years. Certain species of them were prepared in Germany in the 1930s and in England in the 1950s. Their modern history dates from their preparation by C. J. Pedersen, who synthesized a large number of such compounds and recognized that they have the ability to complex a variety of cations, such as Na$^+$, K$^+$, Ca^{2+}, Ag$^+$, and ammonium (NH$_4^+$).

Classes of macrocycles. Several classes of macrocyclic polyethers are known. Substitution of oxygen by nitrogen, sulfur, phosphorus, or silicon leads respectively to azacrowns, thiacrowns, phosphacrowns, or silacrowns. Likewise, when the ether link (—O—) is replaced by an ester link (—CO—O—), macrocyclic ester (lactone) analogs are formed. More than one ester unit and mixtures of esters, amides, small heterocycles, and heteroatoms are common. *SEE ESTER; ETHER; LACTAM; LACTONE.*

Cryptands. J. M. Lehn added a third strand to the simple macrocyclic polyethers and formed three-dimensional compounds based on the crown framework. Typically, two of the oxygen atoms across the ring from each other are replaced by nitrogens, and a third ethyleneoxy chain is attached to them. Known as cryptands, these structures completely encapsulate cations smaller than their internal cavities and strongly bind the most similar in size. Systematic nomenclature of cryptands is complicated. A cryptand is usually presumed to be composed of ethyleneoxy chains attached to a nitrogen atom at either end. A typical cryptand

would be N[(CH$_2$CH$_2$O$_2$)CH$_2$CH$_2$]$_3$N. Since each chain contains two oxygen heteroatoms, it is referred to as [2.2.2]-cryptand, or simply [2.2.2]. A cryptand composed of chains having two, three, and four ethyleneoxy units tied to nitrogen atoms at either end has one, two, and three oxygens respectively in the chains and is called [3.2.1] cryptand.

Lariat ethers. This name has been given to compounds which, like the cryptands, have a ring and a third strand, but the additional chain is attached at only one point. This permits three-dimensional encapsulation of a cation while retaining a high degree of flexibility. Two-armed lariat ethers are referred to as bibracchial lariat ethers or BiBLEs. Three-armed systems are TriBLEs, and so forth. A typical lariat ether, *N*-(2-methoxyethyl)-monoaza-15-crown-5, and three typical cryptands are shown below.

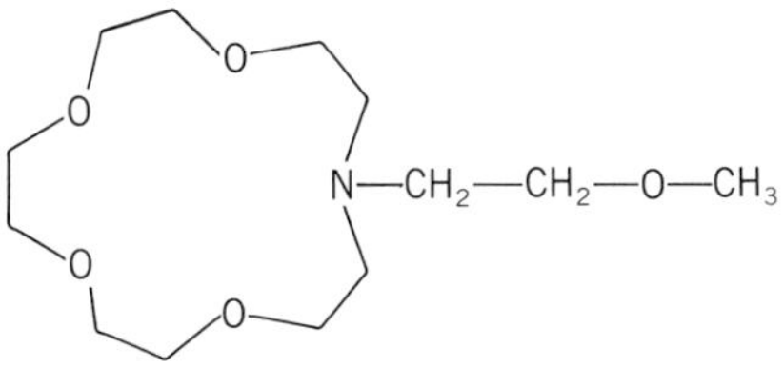

N-(2-methoxyethyl)monoaza-15-crown-5

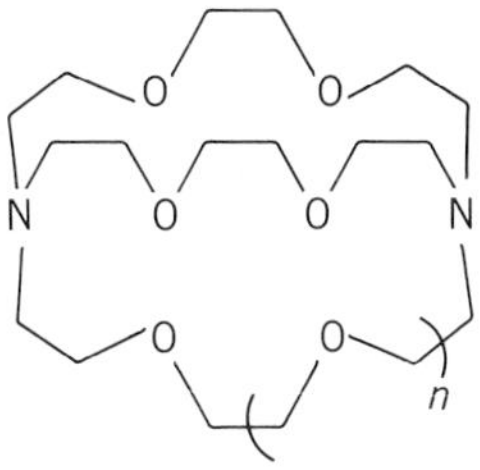

$n = 1$: [2.2.2] cryptand
$n = 2$: [3.2.2] cryptand
$n = 3$: [4.2.2] cryptand

Spherands. Compounds other than simple cryptands have been developed which are capable of either partially (cavitands) or completely (spherands) enveloping cations. The donor atoms (O, N, S) are arranged so that they provide a solvation sphere to the encapsulated cation. The earliest example of a spherand, Lehn's all-aliphatic spherand, is shown below along with an all-aromatic spherand, developed by D. J. Cram. Macrotricyclic compounds have properties similar to those of spherands. They consist of two crown ether–type structures bridged by two ethyleneoxy or diethyleneoxy strands. The structures are roughly tube-shaped. An even more elaborate and enveloping complexing agent has been dubbed a carcerand.

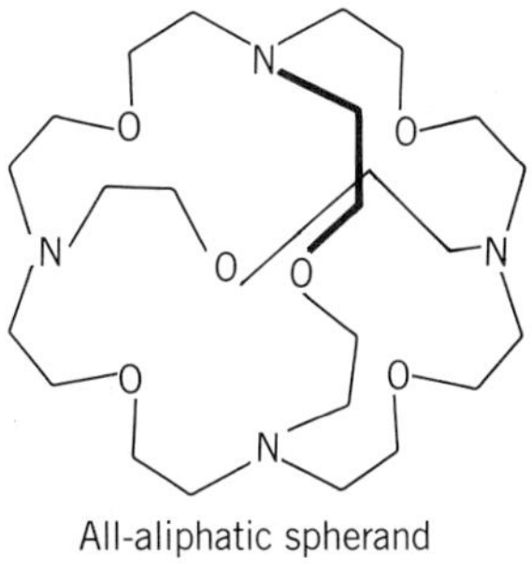

All-aliphatic spherand

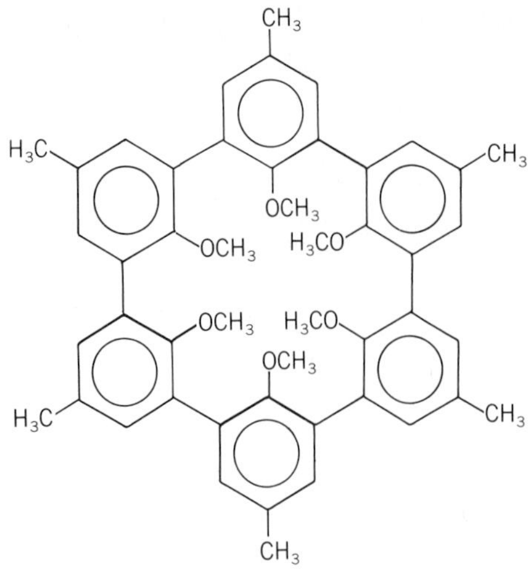

All-aromatic spherand

Complexation phenomena. It is the ability of these so-called host compounds to complex a variety of guest species which makes this family of structures interesting. A crown ether can be described as a doughnut which has an electron-rich and highly polar hole and a greasy or lipophilic (hydrophobic) exterior. As a result, these compounds are usually quite soluble in organic solvents but accommodate positively charged species in their holes. The complexation process shown in reaction (1) is usually characterized by

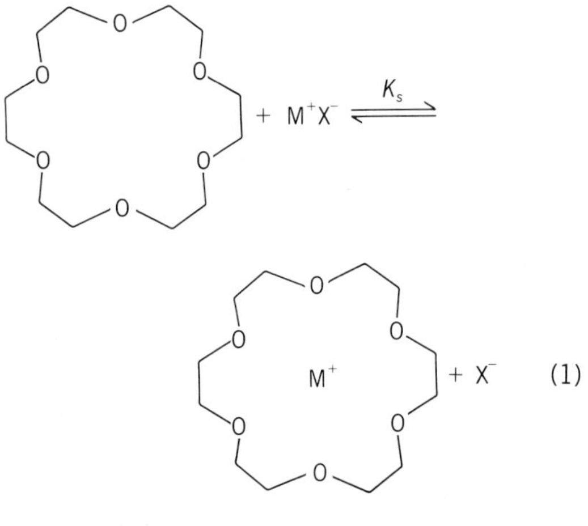

a binding or stability constant (K_s) of the form $K_s = k_c/k_d$, where k_c represents the rate constant for complex formation ($k_{complex}$) and k_d represents the rate of decomplexation ($k_{decomplex}$ or $k_{release}$).

The two important components of complexation are the magnitude of the equilibrium constant and the kinetics of the process. For simple crown ethers, k_c is usually very large and k_d is also large. These rate constants are solvent-dependent, but k_d decreases much more in nonpolar solvents than does k_c. Thus, K_s is smaller in polar solvents and larger in nonpolar solvents. Both k_c and k_d are lower for cryptands and spherands than for crowns, but k_d is lowered much more. As a result, binding constants (K_s) for the three-dimensional hosts are generally much higher (they bind cations much more strongly) than for simple two-dimensional crowns.

Binding constants. Binding or stability constants can be assessed by several methods. One approach is to

dissolve in water a cation such as Li$^+$ or Na$^+$ paired with a colored organic anion such as picrate. Shaking with an equal volume of organic solvent does not extract the salt. When a crown ether is added, the cation is complexed and transported into the organic phase. The crown cation–complex is accompanied by the colored anion, so the extent of extraction can be determined by colorimetric analysis. The extraction constants are expressed numerically as the percent of total salt extracted. Such experiments suggest that crown ethers coordinate best with metallic cations whose diameters are similar in size to the crown hole. For example, 15-crown-5 binds Na$^+$ more strongly than K$^+$, and 18-crown-6 binds K$^+$ more strongly than either Na$^+$ or Li$^+$. The hole of 15-crown-5 is estimated to be 0.20 nanometer across, and the cation diameter of Na$^+$ is about the same. The cation diameter of 18-crown-6 is 0.27–0.30 nm, and the diameter of K$^+$ is near 0.27 nm. This hole-size relationship is often observed when binding is assessed by the picrate extraction technique, and clearly obtains in general for the less flexible ligands (cryptands, spherands, and so forth). It is not strictly obeyed by flexible crowns and lariat ethers, especially when binding is assessed by homogeneous binding-constant measurements. SEE COLORIMETRY.

Homogeneous binding constants are determined for a ligand and a cation present simultaneously in solution. These K_s values may be determined by a variety of techniques (nuclear magnetic resonance, conductometric methods, calorimetry), but are most often obtained from either calorimetric or ion-selective electrode studies. In the series of simple crowns from 12-crown-4 to 24-crown-8, homogeneous binding constants for K$^+$ are higher than for Na$^+$, Ca^{2+}, or ammonium, and 18-crown-6 is the best cation binder in the series. Homogeneous binding constants correlate well with extraction constants for less flexible ligands. It should also be noted that values for K_s are highest (strongest association of ligand and cation) in the least polar solvents and lowest in solvents such as water.

Complexing species. A variety of organic cations have been found to complex with crown ethers and related hosts. It has been suggested that for a host-guest interaction to occur, the host must have convergent binding sites and the guest must have divergent sites. This is illustrated by the interaction between optically active dibinaphtho-22-crown-6 and optically active phenethylammonium chloride, as shown below. The crown ether oxygen atoms converge to the center of a hole and the ammonium hydrogens diverge from nitrogen. Three complementary O—H—N hydrogen bonds stabilize the complex. In this particular case, different steric interactions between the optically active crown and the enantiomers of the complex permit resolution of the salt.

Other organic cations have also been complexed,

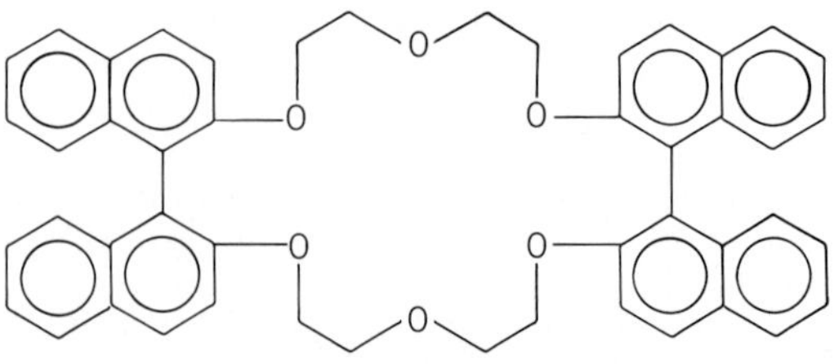

Dibinaphtho-22-crown-6

R,R-Dibinaphtho-22-crown-6 complex of
phenethylammonium chloride

either by insertion of the charged function in the crown's polar hole or by less distinct interactions observed in the solid state. Arenediazonium cations, for example, are complexed in solution and in the solid state by various crown ethers. Also, a compound such as 18-crown-6 traps acetonitrile in its matrix when it crystallizes. Other noncharged, molecular guests such as nitromethane, thiourea, and dimethyl acetylenedicarboxylate have been trapped in crown crystals as well. *See Coordination chemistry; Coordination complexes.*

Applications. Crown ethers and related species have numerous interesting properties that have led to many applications.

Cation complexation. The striking ability of neutral macrocyclic polyethers to complex with alkali and alkaline-earth cations as well as a variety of other species has proved of considerable interest to the chemistry community. Crown ethers may complex the cation associated with an organic salt and cause separation of the ions. In the absence of cations to neutralize them, many anions show considerably enhanced reactivity. For example, decarboxylations, certain anionic rearrangements, and nucleophilic substitutions all proceed at enhanced rates in the presence of crowns. *See Organic reaction mechanism.*

By far, the most common effect of crown or cryptand compounds is to enhance the reactivity of counteranions by complexation of cations. There are two notable exceptions. First, when complexed by crowns, the normally unstable or explosive arenediazonium cations exhibit remarkably enhanced stability. More dramatic is the complete isolation of sodium cation, which allowed the preparation of Na^- for the first time. The crystals of cryptated Na^+ in the presence of Na^- are deep gold in color. This ability of cryptands to surround and protect the cation has also allowed the preparation of the first electride, a cation salt of a lone electron. *See Alkali metals; Alkaline-earth metals.*

Phase-transfer catalysis. One of the important modern developments in synthetic chemistry was the use of the phase-transfer technique. Nucleophiles such as cyanide are often insoluble in media that dissolve organic compounds with which they react. Thus 1-bromooctane may be heated in the presence of sodium cyanide for days with no product formation. When a crown ether is added, two things change. First, solubility is enhanced because the crown wraps about the cation, making it more lipophilic. This, in turn, makes the entire salt more lipophilic. Second, by solvating the cation, the association between cation and anion and the interactions with solvent are weakened, thus activating the anion for reaction. This approach has been used to assist the dissolution of $KMnO_4$ in benzene (to give so-called purple benzene) in which solvent permanganate is a powerful oxidizing agent.

One striking example of solubilization is the displacement of chloride by fluoride in dimethyl 2-chloroethylene-1,1-dicarboxylate by using the KF complex of dicyclohexano-18-crown-6, reaction (2).

$$ClHC=C(COOCH_3)_2 \longrightarrow FHC=C(COOCH_3)_2 \quad (2)$$

In reaction (2), a crown provides solubility for an otherwise insoluble or marginally soluble salt. Use of crowns to transfer a salt from the solid phase into an organic phase is often referred to as solid-liquid phase-transfer catalysis. Such reactions may also be conducted by using liquid-liquid conditions. Generally, the most concentrated possible aqueous solution is contacted with an organic solvent containing the substrate. An example is the reaction between concentrated aqueous potassium cyanide and 1-bromooctane. The crown ether catalyst complexes the cation, transferring it from the aqueous to the organic phase. The cyanide anion accompanies the cation into the organic phase, where it is poorly solvated and therefore highly reactive. The cyanide ion and the organic bromide react to give 1-cyanooctane in high yield, reaction (3).

$$KCN + Br-C_8H_{17} \xrightarrow{crown} NC-C_8H_{17} + KBr \quad (3)$$

See Catalysis; Phase-transfer catalysis.

Cation transport. Because of the ability of crown ethers to complex cations and because of their obvious similarity to such macrocyclic ionophores as valinomycin, extensive studies of crowns as transport agents have been undertaken. Many such studies have been conducted in U-tube devices or variants thereof. In this experiment, a U-shaped tube is partly filled with chloroform or other organic solvent. The sidearms are then filled with water. A salt or mixture of salts is dissolved in one aqueous phase (source phase). The crown, cryptand, or other ionophore is dissolved in the chloroform, which serves as a liquid membrane. The crown complexes the cation and conducts it across the membrane to the other aqueous phase (receiving phase). Such experiments are used to learn about the selectivities of ligands for cations and to understand the behavior of membrane systems. It is hoped that these membranes will be of commercial importance, perhaps in separating valuable metals from common ones.

Indicators and sensors. Since crown ethers and related species complex cations selectively, they can be used as sensors. Crowns have been incorporated into electrodes for this purpose, and crowns having various appended chromophores have been prepared. When a cation is bound within the macroring, a change in electron density is felt in the chromophore. The chromophores are often nitroaromatic residues and therefore highly colored. The color change which accompanies complexation can be easily detected and quantitated. It seems likely that such species will find future application in the rapid or automated analysis of various biological fluids. *See Bioelectronics; Ion-selective membranes and electrodes.*

Toxicity. Some crowns and cryptands, like many other organic compounds, are irritants or toxic. 15-Crown-5 has produced a slight redness in skin tests and is reported to be toxic in mice at a level of about 0.01 oz/lb (800 mg/kg) of body weight. This means that a 150-lb (70-kg) human would have to ingest about 2 oz (56 g) of this compound to be in danger. The toxicity of simple aspirin is about half this (that is, twice as much aspirin, 0.0267 oz/lb or 1.75 g/kg, is required for the same toxic effect), so the toxicity level of this crown is not high. Of course, not all crowns have been tested, and some may be dangerous. See Toxicology.

George W. Gokel

Bibliography. D. J. Cram, Preorganization—from solvents to spherands, *Angewandte Chemie International Edition*, 25:1039, 1986; F. de Jong and D. N. Reinhoudt, *Stability and Reactivity of Crown Ether Complexes*, 1981; G. W. Gokel et al., Clarification of the hole-size cation diameter relationship in crown ethers and a new method for determining calcium cation homogeneous equilibrium binding constants, *J. Amer. Chem. Soc.*, 105:6786, 1983; G. W. Gokel and S. H. Korzeniowski, *Macrocyclic Polyether Syntheses*, 1982; M. Hiraoka, *Crown Compounds, Their Characteristics and Applications*, 1982; R. M. Izatt and J. J. Christensen (eds.), *Synthetic Multidentate Macrocyclic Compounds*, 1978; J. M. Lehn, Supramolecular chemistry—Scope, perspectives, molecules, supermolecules, and molecular devices, *Angewandte Chemie International Edition*, 27:89, 1988; G. A. Melson (ed.), *Coordination Chemistry of Macrocyclic Compounds*, 1978; W. P. Weber and G. W. Gokel, *Phase Transfer Catalysis in Organic Synthesis*, 1977.

Crown gall

A neoplastic disease of primarily woody plants, although the disease can be reproduced in species representing more than 90 plant families. The disease results from infection of wounds by the free-living soil bacterium *Agrobacterium tumefaciens* which is commonly associated with the roots of plants.

The first step in the infection process is the site-specific attachment of the bacteria to the plant host. Up to half of the bacteria become attached to host cells after 2 h. At 1 or 2 weeks after infection, swellings and overgrowths take place in tissue surrounding the site of infection, and with time these tissues proliferate into large tumors (see **illus.**). If infection takes place around the main stem or trunk of woody hosts, continued tumor proliferation will cause girdling and may eventually kill the host. Crown gall is therefore economically important, particularly in nurseries where plant material for commercial use is propagated and disseminated.

Crown gall tumors can be cultured free of bacteria on synthetic, defined media and, under proper conditions, can reinitiate tumors when grafted back onto healthy plants. Unlike healthy normal cells, crown gall tumor cells do not require an exogenous source of phytohormones (auxins and cytokinin) for growth in culture because they readily synthesize more than sufficient quantities for their own growth. The tumor cells also synthesize basic amino acids, each conjugated with an organic acid, called opines. Octopine and nopaline are opines that are condensation products of arginine and pyruvic acid, and arginine and 2-keto glutaric acid, respectively. The tumor cells also grow about four times faster and are more permeable to metabolites than normal cells.

These cellular alterations, such as the synthesis of opines and phytohormone regulation, result from bacterial genes introduced into host plant cells by *A. tumefaciens* during infection. Although it is not under-

stood how these genes are introduced into the plant cell, the genes for the utilization of these opines and for regulating phytohormone production have been found to be situated on an extrachromosomal element called the pTi plasmid. This plasmid, harbored in all tumor-causing *Agrobacterium* species, also carries the necessary genetic information for conferring the tumor-inducing and host-recognition properties of the bacterium. It is thus possible to convert avirulent strains of *A. tumefaciens* and related species to virulent forms by introducing the pTi plasmid into them through standard microbial genetic procedures. Also, the host range of a given *Agrobacterium* can be modified by introducing a pTi plasmid from a bacterial strain having a different host range.

Analyses made of crown gall nuclear genetic material (DNA) have revealed the presence of a specific segment of pTi plasmid DNA known as the T-DNA. Several distinct genes such as chloramphenicol acetyltransferase and β-galactosidase have been inserted into the T-DNA by recombinant DNA methods. These genes can be introduced into plants through infection by *A. tumefaciens* carrying a pTi plasmid with one of these genes inserted in an appropriate location within the T-DNA. These molecular and genetic manipulations have made it feasible to genetically modify plants for desirable traits, a procedure popularly called genetic engineering.

Transfer of the T-DNA requires a set of genes that are clustered in another region of the pTi plasmid, known as the *vir* region. *Vir* region genes are essential for virulence of *A. tumefaciens*, and seem to respond to signals from wounded plant tissue. Although these signals remain to be identified, the interactions with the *vir* genes suggest a long-time association and dependency of *A. tumefaciens* with its host plants.

Crown gall on peach.

Consequently, crown gall is a result of this unique bacteria-plant interaction, whereby *A. tumefaciens* genetically engineers its host to produce undifferentiated growth in the form of a large tumor, in which there is the synthesis of a unique food source in the form of an opine for specific use by the bacterial pathogen. SEE BACTERIAL GENETICS; GENETIC ENGINEERING; PLANT HORMONES; PLANT PATHOLOGY.

Clarence J. Kado

Crushing and pulverizing

The reduction of materials such as stone, coal, or slag to a suitable size for their intended uses such as road building, concrete aggregate, or furnace firing. Crushing and pulverizing are processes in ore dressing needed to reduce valuable ores to the fine size at which the valueless gangue can be separated from the ore. These processes are also used to reduce cement rock to the fine powder required for burning, to reduce cement clinker to the very fine size of portland cement, to reduce coal to the size suitable for burning in pulverized form, and to prepare bulk materials for handling in many processes. SEE MATERIALS-HANDLING EQUIPMENT; ORE DRESSING.

Equipment suitable for crushing large lumps as they come from the quarry or mine cannot be used to pulverize to fine powder, so the operation is carried on in three or more stages called primary crushing, secondary crushing, and pulverizing (see **table**). The three stages are characterized by the size of the feed material, the size of the output product, and the resulting reduction ratio of the material. The crushing-stage output may be screened for greater uniformity of product size.

Reduction in size is accomplished by five principal methods: (1) crushing, a slow application of a large force; (2) impact, a rapid hard blow as by a hammer; (3) attrition, a rubbing or abrasion; (4) sudden release of internal pressure; and (5) ultrasonic forces. The last two methods are not in common use.

Crushers. All the crushers used in primary and secondary crushing operate by crushing as defined above except the hammer mill, which is largely impact with some attrition.

Primary crushers. The Blake jaw crusher using a double toggle to move the swinging jaw is built in a variety of sizes from laboratory units to large sizes having a feed inlet 84 by 120 in. (213 by 305 cm; **Fig. 1**). Large units have a capacity of over 1000 tons/h (900 metric tons/h) and produce a 9-in. (23-cm) product. The Dodge jaw crusher uses a single toggle or eccentric and is generally built in smaller sizes.

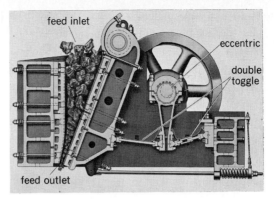

Fig. 1. Blake-type jaw crusher. (*Allis Chalmers Co.*)

The Gates gyratory crusher has a cone or mantle that does not rotate but is moved eccentrically by the lower bearing sleeve (**Fig. 2**). A 42- by 134-in. (107- by 340-cm) crusher has a capacity of 850 tons/h (770 metric tons/h), crushing rock from 27 to 8 in. (69 to 20 cm).

The Symons cone crusher also has a gyratory motion, but has a much flatter mantle or cone than does the gyratory crusher (**Fig. 3**). The top bowl is spring-mounted. It is used as a primary or secondary crusher.

Angle of nip is the largest angle that will just grip a lump between the gyratory mantle and ring, be-

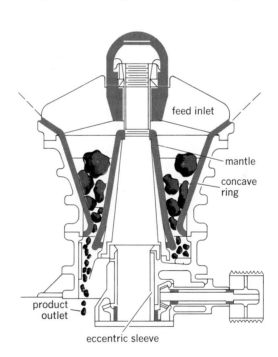

Fig. 2. Gates-type gyratory crusher. (*Allis Chalmers Co.*)

Crushing specifications				
Category	Feed	Product	Reduction ratio	Equipment used
Primary crushing	27–12 in. (69–30 cm)	9–4 in. (23–10 cm)	3:1	Jaw, gyratory, cone
Secondary crushing (one or two stages)	9–4 in. (23–10 cm)	1–½ in. (25–13 mm)	9:1	Hammer mill, jaw, gyratory, cone, smooth roll, and toothed rolls
Pulverizing	1–½ in. (25–13 mm)	60–325 mesh	60:1	Ball and tube, rod, hammer, attrition, ball race, and roller mills

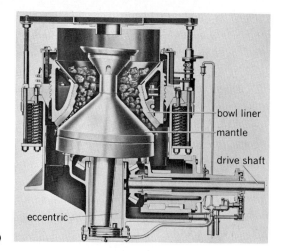

Fig. 3. Symons cone crusher. (*Nordberg Co.*)

tween the smooth jaws of a jaw crusher, or between the pair of smooth rolls. Depending on the material, it is 8–48°, but 32° is commonly used.

Secondary crushers. The single-roll crusher and the double-roll crusher have teeth on the roll surface and are used mainly for coal (**Fig. 4**). Single-roll crushers 36 in. (91 cm) in diameter by 54 in. (137 cm) long have a capacity of 275 tons/h (250 metric tons/h) crushing run-of-mine coal to 1¼-in. (32-mm) ring size. Smooth rolls without teeth are sometimes used for crushing ores and rocks.

The hammer crusher is the type of secondary crusher most generally used for ore, rock, and coal (**Fig. 5**). The reversible hammer mill can run alternately in either direction, thus wearing both sides of the hammers (**Fig. 6***a*). A hammer mill 42 in. (107 cm) in diameter by 82 in. (208 cm) long crushes 300 tons/h (270 metric tons/h) of coal to ¼ in. (6 mm) for coke oven feed. The ring coal crusher is a modification of the hammer mill using rings instead of hammers (Fig. 6*b*).

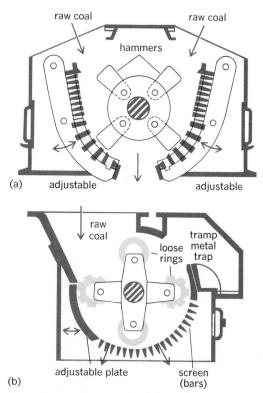

Fig. 5. Hammer crusher, a secondary crusher for ore, rock, and coal. (*Jeffrey Co.*)

Pulverizers. In open-circuit pulverizing, the material passes through the pulverizer once with no removal of fines or recirculation.

In closed-circuit pulverizing, the material discharged from the pulverizer is passed through an external classifier where the finished product is removed

Fig. 4. Types of roll crusher. (*a*) Single-roll. (*b*) Double-roll. (*Link Belt Co.*)

Fig. 6. Two examples of coal crusher. (*a*) Hammer mill type. (*b*) Ring type. (*Penna Crusher*)

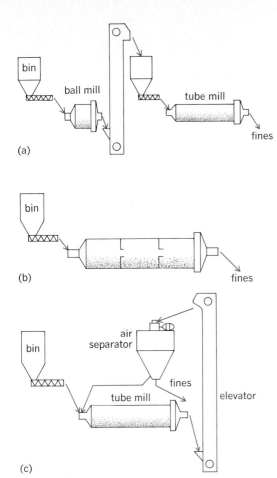

Fig. 7. Pulverizers. (*a*) Open-circuit type with preliminary ball mill stage followed by tube mill. (*b*) Open-circuit three-compartment tube mill. (*c*) Closed-circuit tube mill with an air separator.

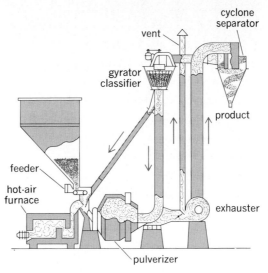

Fig. 9. Conical ball mill pulverizer in closed circuit with classifier. (*Hardinge Co., Inc.*)

of tumbling pulverizer. Both consist of a horizontal rotating cylinder containing a charge of tumbling or cascading steel balls, pebbles, or rods. If the length of the cylinder is less than 2–3 diameters, it is a ball mill. If the length is greater than 3 diameters, it is a tube mill. Both types are extensively used in ore and rock pulverizing where the material is abrasive, as it is easy to add balls during operation. *See Tumbling mill.*

Ball mills charged with large steel balls 2–3 in. (5–8 cm) in diameter are often used as preliminary pulverizers in series with a tube mill in open circuit (**Fig. 7***a*). The Krupp mill is a preliminary ball mill which

and the oversize is returned to the pulverizer for further grinding.

Ball and tube mills, rod mills, hammer mills, and attrition mills are pulverizers operating by impact and attrition. In ball race and roller pulverizers, crushing and attrition are used.

Buhrstone pulverizer. The Buhrstone pulverizer is one of the oldest forms of pulverizers; a few still operate in flour and feed mills. The pulverizing is done between a rotating lower stone and a floating upper stone, which is restrained from rotating. Modern pulverizers use tumbling or rolling action instead of the grinding action of such earlier mills. For example, smooth double rolls are used for pulverizing flour and other materials. *See Buhrstone mill.*

Tumbling pulverizers. There are two principal classes

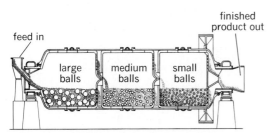

Fig. 8. Three-compartment tube mill pulverizer, containing three sizes of balls. (*Hardinge Co., Inc.*)

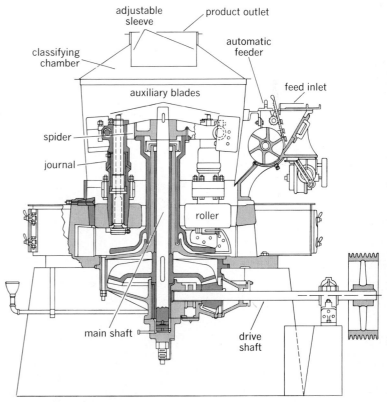

Fig. 10. A pulverizer of the airswept pendulum type. (*Raymond Pulverizer Division, Combustion Engineering Co.*)

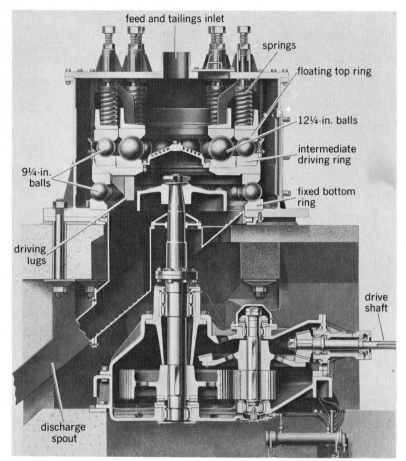

Fig. 11. Ball race rock pulverizer for use in closed circuit. (*Babcock and Wilcox Co.*)

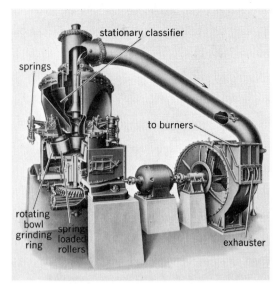

Fig. 13. Medium-speed bowl-mill coal pulverizer, which feeds burners. (*Combustion Engineering Co.*)

discharges the material through a screen which forms the cylindrical surface.

Pebble mills use balls and liners of quartz for application where contamination of the product with iron would be detrimental. For products coarser than 6 mesh, steel rods are sometimes used instead of balls. *SEE PEBBLE MILL*.

In some cement plants and many mining operations, ball, tube, rod, and pebble mills operate wet, water being added with the feed and the product being discharged as a slurry. Rake or cone type classifiers are used in closed circuits that employ wet grinding.

Tube mills for fine pulverization are used extensively in open circuit, or as compartment mills, or as closed-circuit mills (**Fig. 7**). A three-compartment, grate-discharge tube mill has larger balls in the first, medium-size balls in the second, and smaller balls in the third compartment (**Fig. 8**).

Cascade pulverizers are another form of tumbling pulverizer; they use the large lumps to do the pulverizing. The Aerfall pulverizer is built in diameters up to 16 ft (5 m). Feed is 12- to 18-in. (30- to 46-cm) lumps and the unit is airswept with an external classifier. A few steel balls are sometimes used but not over 2½% of the volume.

The airswept conical ball mill is used in closed circuit with an external air classifier (**Fig. 9**). The conical shape of the mill classifies the balls, keeping the

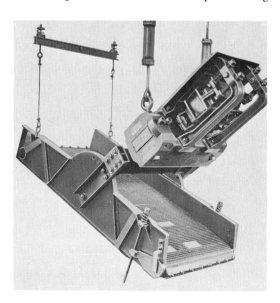

Fig. 12. Vibrating screen classifier. (*Jeffrey Co.*)

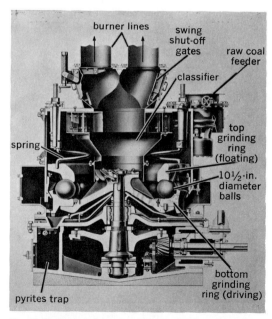

Fig. 14. Medium-speed ball race coal pulverizer, operated under pressure. 10½ in. = 27 cm. (*Babcock and Wilcox Co.*)

large balls at the feed end, where they are needed to crush the larger lumps, and the smaller balls at the small discharge end where the material is finer. This conical ball mill is also used extensively in open circuit in rocks and ores.

Roller pulverizers. The airswept pendulum roller pulverizer has rollers pivoted from a rotating spider on a vertical shaft; the centrifugal force of the rollers supplies the pressure between rollers and ring. Hot air enters below the rollers and picks up the fines. The rotating rollers and spider provide a centrifugal classifying action which returns the oversize to the pulverizing zone. For higher fineness a rotating classifier is added in the top housing (**Fig. 10**).

The ball race pulverizer uses large steel balls between races (**Fig. 11**). The bottom race is fixed, the intermediate race is rotated through driving lugs, and the top races are restrained from rotating by stops and are loaded by the springs. The material flows by centrifugal force from the center outward through the 12¼-in. (31-cm) and 9½-in. (23.5-cm) balls. A bail race unit in closed circuit with an air separator is used for pulverizing cement rock.

Screens. Material is separated into desired size ranges by screening. For products 2 in. (5 cm) or larger, an open-end perforated cylindrical rotary screen is used. It is on a slight inclination, and the undersize goes through the holes and the oversize passes out through the open lower end.

For finer products from 2 in. (5 cm) down to fine powder passing through a 200-mesh sieve, shaking or vibrating screens of woven wire are used.

Shaking screens have either an eccentric drive or an unbalanced rotating weight to produce the shaking. Vibrating screens have electric vibrators (**Fig. 12**). They are often hung by rods and springs from overhead steel, giving complete freedom to vibrate. In some applications the screen is double decked, thus producing three sizes of product.

Direct-fired pulverizers for coal. Most modern large coal-fired furnaces are fed directly by the pulverizer, the coal rate being regulated by the rate of pulverizing. Air at 300–600°F (150–315°C) is supplied to the pulverizer to dry and preheat the coal. In a bowl-type medium-speed pulverizer, the springs press the pivoted stationary rolls against the rotating

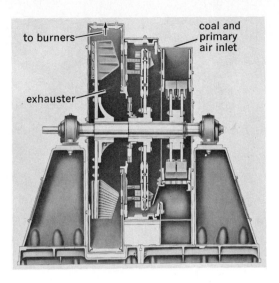

Fig. 16. High-speed coal pulverizer. (*Riley Stoker Corp.*)

bowl grinding ring, crushing the coal between them (**Fig. 13**). The self-contained stationary cone classifier returns the oversize to the pulverizing zone.

The medium-speed pulverizer of the ball race type uses closely spaced 18-in.-diameter (46-cm) balls between a lower rotating race and a floating top race (**Fig. 14**). The single coil springs restrain the top race and also apply the pressure needed. A stationary cone classifier is used. This pulverizer operates under pressure, using a blower on the clean inlet air instead of an exhauster on the coal-laden outlet air.

Slow-speed pulverizers of the ball mill type have classifiers and return the oversize for regrinding on both ends (**Fig. 15**).

High-speed pulverizers usually incorporate multiple stages such as a preliminary hammer mill and a secondary pulverizing stage. A built-in exhaust fan exhausts the fuel and air from the unit (**Fig. 16**). SEE BALL-AND-RACE-TYPE PULVERIZER.

Ralph M. Hardgrove

Bibliography. R. H. Perry and D. W. Green (eds.), *Perry's Chemical Engineers' Handbook*, 6th ed., 1984; C. L. Prasher, *Crushing and Grinding Process Handbook*, 1987.

Crustacea

The hierarchical rank of the Crustacea is a matter of continuing debate among carcinologists. They are regarded as a phylum, distinct from other arthropodous taxa by proponents of the concept of polyphyly in the Arthropoda. Alternatively, they are given the rank of subphylum or superclass by those who view the Arthropoda as a monophyletic taxon. Regardless of which of the three alternative ranks is assigned, the Crustacea are a monophyletic assemblage.

Species of Crustacea such as the shrimp, prawn, crab, or lobster (**Fig. 1**) are familiar. However, there are many more with less common vernacular names such as the water fleas, beach fleas, sand hoppers, fish lice, wood lice, sow bugs, pill bugs, barnacles, scuds, slaters, and krill or whale food.

The Crustacea are one of the most difficult animal groups to define because of their great diversity of structure, habit, habitat, and development. No one character or generalization will apply equally well to all. The adults of some aberrant or highly specialized

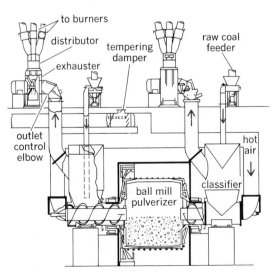

Fig. 15. Slow-speed coal pulverizer of the ball mill type; classifier sorts out oversize. (*Foster Wheeler Corp.*)

parasitic species are not recognizable as crustaceans, and evidence of their kinship rests on a knowledge of their life histories. Virtually every form of marine animal life as well as a number of fresh-water animals have their crustacean parasites.

Characterization. Crustaceans have segmented, chitin-encased bodies; articulated appendages; mouthparts known as mandibles during some stage of their life, however modified they may be for cutting,

chewing, piercing, sucking, or licking; and two pairs of accessory feeding organs, the maxillules and maxillae. One or the other pair is sometimes vestigial or may be lacking.

The Crustacea are unique in having two pairs of antennae: the first pair, or antennules, and the second pair, the antennae proper. The latter, moreover, are almost always functional at some stage of every crustacean's life, although they disappear in adult barna-

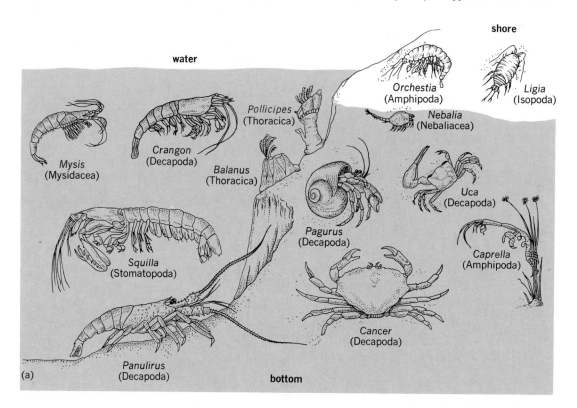

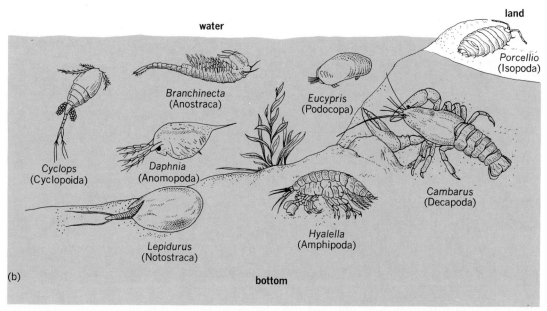

Fig. 1. Some forms of Crustacea in their respective habitats, mostly reduced but not to same scale. Orders are given in parentheses. (*a*) Marine forms, and common names: *Pollicipes*, goose barnacle; *Balanus*, acorn barnacle; *Mysis*, opossum shrimp; *Orchestia*, sand hopper; *Caprella*, "skeleton shrimp"; *Crangon*, shrimp; *Panulirus*, spiny rock lobster; *Pagurus*, hermit crab; *Cancer*, edible crab; *Uca*, fiddler crab; *Squilla*, mantis shrimp. (*b*) Fresh-water forms, and common names if recognized. *Branchinecta*, fairy shrimp; *Daphnia*, water flea; *Eucypris*, seed or mussel shrimp; *Lepidurus*, tadpole shrimp; *Cyclops*, copepod; *Porcellio*, sow bug (on land); *Cambarus*, crayfish. (*After T. I. Storer et al., General Zoology, 6th ed., McGraw-Hill, 1979*)

cles, in some parasites and in some lower shrimplike groups. In some of the same forms the antennules may also become reduced or vestigial.

TAXONOMY

Not only is there a lack of agreement on the rank of the Crustacea per se, but there is no consensus on hierarchial levels of subordinate taxa. The classification presented here is restricted to extant taxa and is, at best, a compromise among opposing opinions.

Superclass Crustacea
  Class Cephalocarida
  Class Branchiopoda
    Order: Anostraca
        Spinicaudata
        Laevicaudata
        Ctenopoda
        Anomopoda
        Onychopoda
        Haplopoda
        Notostraca
  Class Remipedia
  Class Maxillopoda
    Subclass Mystacocarida
    Subclass Cirripedia
      Order: Ascothoracica
          Thoracica
          Acrothoracica
          Rhizocephala
    Subclass Copepoda
      Order: Calanoida
          Harpacticoida
          Cyclopoida
          Poecilostomatoida
          Siphonostomatoida
          Monstrilloida
          Misophrioida
          Mormonilloida
    Subclass Branchiura
    Subclass Tantulocarida
  Class Ostracoda
    Subclass Myodocopa
      Order: Myodocopida
          Halocyprida
    Subclass Podocopa
      Order: Platyocopida
          Podocopida
  Class Malacostraca
    Subclass Phyllocarida
      Order Leptostraca
    Subclass Hoplocarida
      Order Stomatopoda
    Subclass Eumalacostraca
      Superorder Syncarida
        Order: Bathynellacea
           Anaspidacea
      Superorder Pancarida
        Order Thermosbaenacea
      Superorder Peracarida
        Order: Spelaeogriphacea
           Mysidacea
           Mictacea
           Amphipoda
           Isopoda
           Tanaidacea
           Cumacea
      Superorder Eucarida

Order: Euphausiacea
       Amphionidacea
       Decapoda

GENERAL MORPHOLOGY

The true body segments, the somites or metameres, are usually somewhat compressed or depressed (**Fig. 2**). Rarely are they circular rings. The somite's dorsum, or upper side, is the tergum or tergite; the lower or ventral surface, the sternum or sternite; and either lateral portion, the pleuron or epimeron. When prolonged or extended downward to overhang the attachment of the biramous appendages, of which one pair is a part of every typical crustacean somite, the structure is termed the epimeron. Between the embryonic anterior or prostomial element and the telson at the posterior end, the linear series of somites making up the body of a crustacean are more or less distinctly organized into three regions or tagmata, the head, thorax, and abdomen. Where regional organization of the postcephalic somites is not clearly marked, they

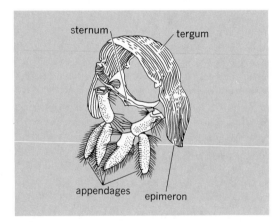

Fig. 2. Diagrammatic representation of an abdominal somite of a lobster with its appendages and skeletal elements.

collectively form the trunk. The somites are variously fused with one another in diagnostic combinations in different groups of the Crustacea.

The body. The head region of the simplest type is found in the characteristic crustacean larva, the nauplius. When the nauplius hatches, it has only three pairs of appendages, the antennules, antennae, and mandibles, which correspond to the three fused somites of the primary head of the lower Crustacea. In more advanced forms, and at later stages, two additional somites bearing the maxillules and maxillae become coalesced with this protocephalon to form the most frequently encountered crustacean head, or cephalon. Thoracic somites frequently are fused with the cephalon to form a cephalothorax. A dorsal shield or carapace of variable length arises from the dorsum of the third cephalic somite and covers the cephalon and cephalothorax to varying extent (**Fig. 3**). In some crustaceans it is in the form of a bivalved shell; in the barnacles it is modified as a fleshy mantle reinforced by calcareous plates. The carapace reaches its greatest development in the malacostracan Decapoda where, including the head, it fuses dorsally with the eight thoracic somites and extends down laterally over the attachment of the appendages to enclose the branchial chamber on either side of the cephalothorax.

There is a single median "nauplius" eye on the

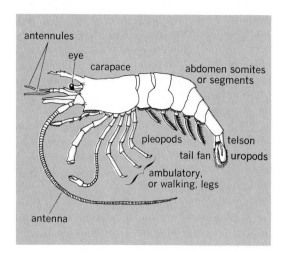

Fig. 3. External morphology of a shrimp.

nauplius head. This persists to become the sole visual organ in many of the lower Crustacea. In some it disappears; occasionally it remains as a vestige in the species with paired compound eyes.

The chitinous cuticle covering the crustacean body is its external skeleton (exoskeleton). The apodemes develop from this as ingrowths for muscle attachment. Further development of the apodemes results in the formation of the endophragmal skeleton in the higher Crustacea. The chitin is flexible at the joints, in foliaceous appendages, and throughout the exoskeletons of many small and soft-bodied species, but

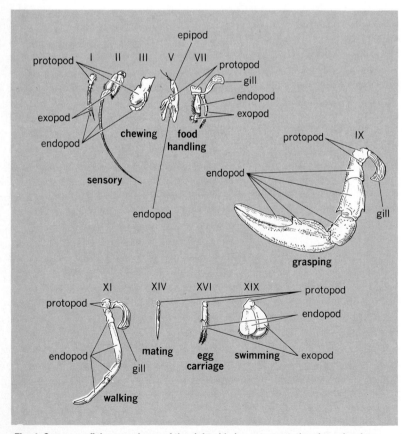

Fig. 4. Some crayfish appendages of the right side in ventroposterior view, showing structural differentiation to serve various functions. (*After T. I. Storer et al., General Zoology, 6th ed., McGraw-Hill, 1979*)

it is often thickened and stiff in others. It becomes calcified in many species as a result of the deposition of lime salts, and figuratively can be as hard as rock in some crabs.

Appendages. The paired appendages (**Fig. 4**) are typically biramous and consist of two branches, the endopod and exopod. These arise from the distal end of a penduncle, the protopod, which is composed of two segments: the coxopod (coxa), articulated to the body, and the basipod (basis). Either the endopod or exopod or both may be reduced or suppressed, or both may be coalesced as in the foliaceous limbs of lower Crustacea. When the subdivisions of the rami of such an appendage do not clearly reveal their relation to the appendage as a whole, they are labeled or counted as endites and exites. The endopod is definitely segmented in the higher Crustacea. The segments, which are usually present in full complement from the peduncular basipod outward, are the ischiopodite (ischium), meropodite (merus), carpopodite (carpus), propodite (propodus), and dactylopodite (dactyl). The endopods are variously modified to serve a variety of functions and needs such as sensory perception, respiration, locomotion, prehension and comminution of food, cleansing, defense, offense, reproduction, and sex recognition and attraction. If retained in the adult, the exopod may remain leaf- or paddlelike, or become flagellated structures, facilitating swimming or aiding respiration.

Digestive system. The digestive system is a relatively straight tube that curves dorsally from the mouth, which is situated on the underside of the head between the mandibles. In a few small species it may be coiled. The anterior margin of the mouth, or upper lip, is often a fleshy labrum; the hinder margin or lower lip is the metastoma or hypostoma. When the metastoma is bilobed, the lobes are called paragnatha, each of which may have a movable distal portion, but in no sense do these or other armatures of the paragnatha represent appendages. In the alimentary tract three regions are recognizable, the foregut, midgut, and hindgut. The anterior part of the foregut often is esophageal in nature, whereas the posterior and greater part usually consists of two parts. In the anterior portion in the Eucarida, a complex and elaborate grinding mechanism, the gastric mill, is developed. The posterior chamber is divided into dorsal and ventral filtering compartments for the straining of food. The midgut in most cases is provided with several ceca or diverticula which produce digestive secretions or serve as organs for the absorption of food. When these ceca are present in considerable number, they become organized into a sizable gland, the midgut gland of higher Crustacea. The hindgut is typically short and terminates, with few exceptions, in a muscular anus on the underside of the telson.

Circulatory system. Most crustaceans have a heart perforated by openings, or ostia, which admit venous blood from the pericardial sinus in which it is located. The heart may be elongated and tubular and extend through the greater part of the body, but generally it is a more compact organ. It pumps blood through an arterial system or through connecting sinuses or lacunae within the body tissues. Some lower Crustacea do not have a heart, and the muscular movements of the animal or its alimentary tract circulate the blood through the body cavities or sinuses. The blood of the great majority of the crustaceans is bluish because it contains the respiratory pigment hemocyanin. A few crustaceans have red blood as a result of the presence of erythrocrurion. *See Respiratory pigments (inverte-brate).*

Respiratory system. Crustacea take up oxygen by means of gills, the general body surface, or special areas of it. Some of the few species that have become more or less terrestrial in their habits have developed modifications of their branchial mechanism such as villi, ridges, or water-retaining recesses which when sufficiently moist enable them to breathe air. Some sow or pill bugs have special tracheal developments in their abdominal appendages for the same purpose.

Nervous system. In the crustacean nervous system a supraesophageal ganglion somewhat larger than the other ganglia is considered to be the brain. It is connected by circumesophageal commissures to a double ventral nerve cord with segmentally arranged ganglia. These become reduced in number and, in the brachyuran crabs, form a large ganglionic mass centered in the thorax.

Organs of special sense. The organs of special sense are the eyes, the antennules, and the antennae. Crustacean eyes (photoreceptors) are of two types: the median (nauplius eye) and simple eyes (frontal organs); and compound eyes. Simple and median eyes and lateral eyes consist only of light sensory cells. In contrast, compound eyes are composed of many subunits (ommatidia), each having separate optical elements. Compound eyes, therefore, provide varying amounts of actual vision, and in at least some species color differentiation. The antennules and antennae are provided with a variety of sensory structures for the reception of chemical and mechanical stimuli. Specific structures on the antennules called aesthetascs act as organs of smell. Taste chemoreceptors are usually found on the mouthparts and dactyls of the pereopods. Many of the hairs (setae) and bristles found on the crustacean body and appendages act as mechanoreceptors. Organs of balance (statocysts) are also present on basal segments of the antennules in many crustaceans, but on the uropods in most mysids. Although crustaceans may not hear in the accepted sense, they are sensitive to sound waves and vibrations, and many have stridulating or sound-producing mechanisms on appendages or on somites overlapping one another on the carapace, which produce snapping, clocking, rasping, or rubbing noises. Crustacea have setae variously located on their bodies which are sensitive to various external stimuli. *See Chemoreception; Eye (invertebrate); Photoreception.*

Glands and glandular systems. Those glands recognized as definite excretory organs are the maxillary and antennal glands in the adult crustaceans. They are located in the cephalon and rarely are both present at the same time. The antennal gland is best developed in the Malacostraca. It is often called the green gland and opens to the exterior on the underside of the coxal segment of the antennal peduncle. Many crustaceans, especially deep-sea forms, have phosphorescent or luminous organs. The barnacles and some other crustaceans have cement glands. This may also be true of tube-building species which construct their homes of extraneous materials. Species that produce encysted or drought-resistant eggs have other glands of special secretion. The most studied crustacean gland is the sinus gland of the eyestalks. This gland is part of a neurosecretory system producing hormones that con-

trol color change and pattern, molting cycle, oogenesis, and egg development within the ovary. *See Endocrine system (invertebrate); Neurosecretion.*

Reproductive system. The sexes are separate in most Crustacea and usually can be differentiated from each other by secondary sex characters. Chief among these characters are the size and shape of the body, appendages, or both, and placement of the genital apertures. Hermaphroditism is the rule in the Cephalocarida, Remipedia, some ostracods, sessile Cirripedia (barnacles), in isolated cases in other crustaceans, and in certain parasitic forms. Parthenogenesis (eggs developing and hatching without prior fertilization) occurs frequently in some of the lower crustaceans such as the Branchiopoda and Ostracoda, which have what might be called an alternation of generations. The parthenogenetic generations (and there are usually very many of them in succession) alternate with a generation produced by fertilized eggs. *See Parthenogenesis.*

Protandric hermaphroditism has been observed in some decapods, and protogyny has been reported in

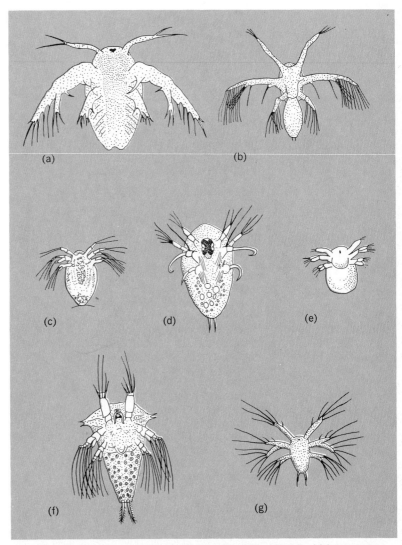

Fig. 5. Nauplii of (a) *Apus cancriformis*; (b) *Branchinecta occidentalis*; (c) *Lernaeocera branchialis*; (d) *Haemocera danae*; (e) *Meganyctiphanes norvegica*; (f) *Sacculina carcini*; (g) *Penaeus setiferus*. Enlarged but not to same scale. (*After R. E. Snodgrass, Crustacean Metamorphoses, vol. 131, no. 10, Smithson. Inst. Publ. 4260, 1956*)

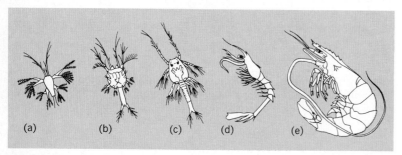

Fig. 6. Life history of *Penaeus*. Some stages enlarged, others reduced but not to same scale. (*a*) Nauplius. (*b*) Protozoea. (*c*) Zoea. (*d*) Mysis. (*e*) Adult. (*After T. I. Storer et al., General Zoology, 6th ed., McGraw-Hill, 1979*)

tanaids and some carideans. In the former circumstance, individuals reach maturity as males but subsequently beome functional females. In the latter condition the development is reversed. Sexual differentiation is under control of the androgenic gland. This is a reversal of sex in the individual or in the species that reach maturity as males but become female in their third year and function as females.

The eggs of most crustaceans are carried attached to the female until hatched. Some females develop brood pouches in which the young are retained for a time. A nutrient secretion which sustains the young until they are released is produced in some species having a brood chamber. Penaeid shrimp and a few of the lower Crustacea deposit their eggs in the medium in which they live, in some cases attaching them to aquatic vegetation.

Development. The nauplius larva is characteristic of Crustacea (**Fig. 5**). Its oval, unsegmented body has three pairs of appendages indicative of three embryonic somites. These are the antennules which are uniramus in the nauplius, the biramous antennae, and mandibles. Besides subserving other functions such as sensory and feeding, all three are the organs of locomotion of this free-swimming larva. Their basal portions, the gnathobases, are modified to serve as masticatory organs and for passing food to the mouth with its relatively large labrum. There is a single dorsal, median, nauplius eye composed of three closely united, similar parts which form a relatively simple compound eye and a frontal organ, probably sensory, evidenced externally by a very minute pair of projections. This frontal organ seems to persist in the form of anterior sensory cells in the Branchiopoda, as setae in the Copepoda, and as papillae even in the late larval stages of some Malacostraca, although these do not appear to be homologous in all groups.

The integument of the upper surface of the nauplius body has the appearance of a dorsal shield, as in most crustaceans. Typical as this first larval stage may be, it does not appear in all crustaceans. It is common in the lower forms, but in many of the higher forms it occurs during development in the egg, and the young are hatched as a different and more advanced larva or, as in many Malacostraca, in a form similar to the adult. Life histories vary from the simple to the complex within the different groups of Crustacea. Only the euphausiid and penaeid shrimps among the Malacostraca hatch as nauplii (**Fig. 6**). Following the naupliar stage, whether it is passed in the egg or not, very remarkable larval stages follow in some groups: the free-swimming cypris larvae of barnacles; the phyllosomae of the spiny lobsters; the erichthus and

alima larvae of the mantis shrimps (Stomatopoda); and the zoeae and megalopa of crabs.

PHYSIOLOGICAL MECHANISMS

Because of its tough, chitinous, or calcified exoskeleton, a crustacean does not usually grow without molting. Crustaceans have also developed the remarkable feature of autotomy and regeneration to minimize injury and hemorrhage from loss of limb through accident or to an enemy.

Molting (ecdysis). This process involves several steps: (1) preparation, which includes some degree of resorption of the old cuticle; (2) the formation of a new, temporarily soft and thin one within it; (3) the accumulation and storing of calcium in the midgut gland or as lenticular deposits (gastroliths), one of which occurs on either side of the foregut in certain forms such as crayfishes and lobsters. The preparatory period is less complicated in the thinly chitinous forms. The actual molt follows. The old shell or cuticle splits at strategic and predetermined places, permitting the crustacean within, already enclosed in new but still soft exoskeleton, to withdraw. There is little question but that a temporary absorption of water enables the animal to split or crack his housing. Upon withdrawal of the entire animal, absorption of water again rapidly takes place with a pronounced increase in body size. The chitinous lining of the foregut and hindgut as well as the endophragmal skeleton are shed along with the exoskeleton. In the immediate postmolt period the crustacean is quite helpless and, if unsheltered, is at the mercy of its enemies. It is in the soft-shelled stage immediately after molting that the East Coast blue crab (*Callinectes*) becomes a much-sought seafood delicacy. The tender new cuticle is reinforced rapidly by the resorbed chitin, and hardened by whatever reserves of calcium the animal may have stored, supplemented and extended by the far more plentiful supplies in solution in the sea which may be absorbed or ingested by the growing crustacean. Molting takes place quite frequently in the larval stages when growth is rapid, but becomes less frequent as the animal ages. In many species there is a terminal molt at maturity. Hormones from the sinus gland play an important role in both initiating and inhibiting molting. *SEE GASTROLITH.*

Autotomy and regeneration. The mechanisms of autotomy and regeneration are developed in the crustaceans to minimize injury or loss to an enemy. When an appendage is broken, it is cast off or broken at the fracture or breaking plane. This sacrifice often enables the victim to escape. This plane runs through the basis of the appendage in all crabs and all but a few lobster species, and in some other Malacostraca. It is thought by many, however, to represent the line of juncture of the basis and ischium in the pereopods and chelipeds which in these forms are usually fused. The break in other crustaceans with completely jointed pereopods occurs between the unfused basis and ischium. At the fracture plane there is an infolding of epidermal tissue forming a double-walled diaphragm with a small medial opening for the passage of nerves and blood vessels. This is so designed that severance of the limb will occur at that place. The proximal wall of that diaphragm prevents undue loss of body fluids or laceration of tissue, and facilitates the speedy closure of its foramen with quickly clotting blood.

Even more remarkable is the fact that crustaceans,

by voluntary muscular contraction, can part with a limb which may be injured.

Crustacea also have the ability to regenerate lost parts. Although the regenerated parts are not always the same size as the original in the first molt after injury, increase in size in successive molts soon restores a lost limb to virtually its former appearance. During the regenerative stages among alpheid shrimps and lobsters which generally exhibit marked dimorphism in the chelipeds when the major chela is lost, the minor one often may take on more or less the character of the former major chela. Then, the chela replacing the lost one assumes the habitus of the former minor chela during successive molts. *See Regeneration (biology)*.

PHYLOGENY

The geologic history of crustaceans dates from the earliest fossilferous rock formations, the Cambrian, estimated to be about 350 million years of age. Many lower forms of crustaceans have been discovered in those strata, including some thought to be distantly related, morphologically, to present-day Malacostraca. Until the discovery of phosphatized fossils, preservation, especially of the appendages, left much to be desired. It is not unlikely that because of their small size and fragility many crustaceans of bygone ages never did become recognizably fossilized. Interest in the hypothetical ancestral crustacean has been greatly stimulated by the discovery of the small forms for which the classes Cephalocarida and Remipedia were established (**Fig. 7**). *See Cephalocarida*.

An intensive study of the crustaceans, which are quite diverse both structurally and biologically, raises a much discussed question as to the monophyletic or polyphyletic origin of the subclasses.

BIONOMICS AND ECONOMICS

Crustacea are ubiquitous. They live at almost all depths and levels of the sea, in fresh waters at elevations up to 12,000 ft (3658 m), in melted snow water, in the deepest of the sea's abysses more than 6 mi (9 km) down, and in waters of 0°C (32°F) temperature. They live on land, trees, and mountains, although most of these species must descend to salt water areas again to spawn their young. Some live in strongly alkaline waters and others in salt water which is at the saturation point, still others in hot springs and hydrothermal vents with temperatures in excess of 55°C (131°F).

All crustaceans of sufficient size are eaten by people somewhere in the world. Very few properly pre-

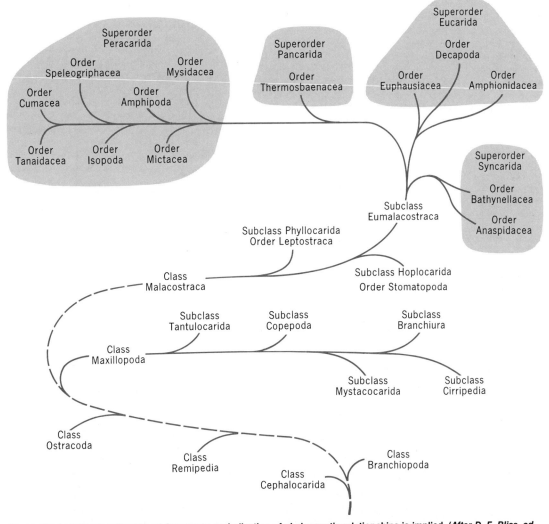

Fig. 7. Chart of the classification of Crustacea; no indication of phylogenetic relationships is implied. (*After D. E. Bliss, ed., The Biology of Crustacea, vols. 1–10, Academic Press, 1982–1985***)**

pared have proved to be poisonous so far as is known. Myriads of small fry are the sustenance of larger aquatic animals, chiefly fish and whales.

Crustacea are of all sizes, ranging from copepods 1/100 in. (0.25 mm) long to huge spider crabs of Japan, which span 12 ft (3.7 m) from tip to tip of the laterally extended legs. Individuals of the giant xanthid crab of Australia have been recorded weighing 20–30 lb (9–14 kg), and the American lobster, the heaviest so far known, tops all crustaceans at 44½ lb (20 kg).

Most crustaceans are omnivorous and essentially scavengers. Many are filter feeders and screen particulate life, plankton, and organic detritus from the waters in which they live; others are largely carnivorous, still others vegetarian. Among the vegetarians are the grazers of the ocean meadows which convert the microscopic plant life (diatoms) into flesh and food for larger animals which in turn are harvested as food for humans.

Waldo L. Schmitt; Patsy A. McLaughlin

Bibliography. H. Balss, *Decapoda and Stomatopoda*, in H. G. Bronn (ed.), *Klassen und Ordnungen des Tierreichs*, vol. 5, 1940–1957; D. E. Bliss (ed.), *The Biology of Crustacea*, vols. 1–10, 1982–1985; W. T. Calman, *Crustacea*, in E. R. Lankester (ed.), *Treatise on Zoology*, pt. 7, 1909; A. P. Gupta (ed.), *Arthropod Phylogeny*, 1979; K. J. Muller and D. Walossek, External morphology and larval development of the Upper Cambrian maxillopod *Bredocaris admiralis*, *Fossils and Strata*, vols. 23, 1988; S. P. Parker (ed.), *Synopsis and Classification of Living Organisms*, 2 vols., 1982; W. L. Schmitt, *Crustaceans*, 1964.

Crux

The Southern Cross, in astronomy, the most celebrated of the constellations of the far south. The four principal bright stars of the group α, γ, β, and δ form the figure of a cross, giving the constellation its name, in contradistinction to the Northern Cross of Cygnus in the north, which is larger and more distinct, having a bright star to mark its center (see **illus.**). The brightest star in Crux is Alpha Crucis at

Line pattern of the constellation Crux. The grid lines represent the coordinates of the sky. The apparent brightness, or magnitude, of the stars is shown by the sizes of dots, graded by appropriate numbers as indicated.

the foot of the cross, which has received the artificial name, Acrux. This is a navigational star. The star at the top of the cross is Gamma Crucis, sometimes called Gacrux. The line joining these two stars points approximately toward the South Celestial Pole. SEE CONSTELLATION; CYGNUS.

Ching-Sung Yu

Cryobiology

The use of low-temperature environments in the study of living plants and animals. The principal effects of cold on living tissue are destruction of life and preservation of life at a reduced level of activity. Both of these effects are demonstrated in nature. Death by freezing is a relatively common occurrence in severe winter storms. Among cold-blooded animals winter weather usually results in a comalike sleep that may last for a considerable time. SEE HIBERNATION.

While these natural occurrences suggest the desired ends of cryobiology, activities in this field differ in that much lower temperatures are employed than are present in natural environments. The extreme cold of liquid nitrogen (boiling at $-320°F$ or $-196°C$) can cause living tissue to be destroyed in a matter of seconds or to be preserved for an indefinite time, certainly for years and possibly for centuries, with essentially no detectable biochemical activity.

The result achieved when heat is withdrawn from living tissue depends on processes occurring in the individual cells. Basic knowledge of the causes of cell death, especially during the process of freezing, and the discovery of methods which circumvent these causes have led to practical applications both for long-term storage of living cells or tissue (cryopreservation) and for calculated and selective destruction of tissue (cryosurgery). SEE CELL SENESCENCE AND DEATH.

Fundamental processes. The biochemical constituents of a cell are either dissolved or suspended in water. During the physical process of freezing, water tends to crystallize in pure form, while the dissolved or suspended materials concentrate in the remaining liquid. This principle has been suggested as one means for producing fresh water from seawater. In the living cell, however, this process is quite destructive. SEE SALINE WATER RECLAMATION.

In a relatively slow freezing process ice first begins to form in the fluid surrounding the cells, and the concentration of dissolved materials in the remaining liquid increases. The equilibrium between the fluid inside the cell and the fluid surrounding the cell is upset. A concentration gradient is established across the cell wall, and water moves out of the cell in response to the osmotic force. As freezing continues, the cell becomes quite dehydrated. Salts may concentrate to extremely high levels. For example, sodium chloride may reach 200 g/ liter (7 oz/qt) concentration in the final stages of freezing. In a similar manner the acid-base ratio of the solution may be altered during the concentration process. Dehydration can affect the gross organization of the cell and also the molecular relationships, some of which depend on the presence of water at particular sites. Cellular collapse resulting from loss of water may bring in contact intracellular components normally separated to prevent destructive interaction. Finally, as the ice crystals grow in size, the cell walls may be ruptured by the ice crystals themselves or by the high concentration gradients which are imposed upon them.

By speeding the freezing process to the point that temperature drop is measured in degrees per second, some of these destructive events can be modified. For example, ice crystals will form within the cells so that less water will be lost, and the size of each crystal will be minimized, tending to reduce incidence of destruction of the cell walls. However, most of the previously mentioned destructive processes will prevail.

To prevent dehydration, steps must be taken to stop the separation of water in the form of pure ice so that all of the cell fluids can solidify together. The chief tools used to accomplish this are agents that lower the freezing point of the water. Typical materials used for industrial antifreeze agents are alcohols or glycols (polyalcohols). Glycerol, a polyalcohol which is compatible with other biochemical materials in living cells, is frequently used in cell preservation.

Besides the antifreeze additive, refrigeration procedures are designed to control the rate of decline in temperature to the freezing point, through the liquid-solid transition, and below, to very low temperatures. As a result of these procedures, the materials in the cell and in the surrounding fluid solidify together in a state more nearly resembling a glass than a crystalline material. In this form cells can be preserved apparently for as long as one may desire and will regain all their normal functions when properly thawed.

In the case of preservation of sizable aggregates of cells in the form of whole organs, the problems are significantly increased. With a large mass, control of the cooling rate is much more difficult. Additives may not reach cells inside a large organ in sufficient quantity to depress the freezing point adequately. The organ may suffer from oxygen deprivation or from other interruption of its normal environment quite apart from the freezing process itself. While there are signs of progress, whole-organ preservation is not yet a normally successful procedure.

Uses of cryopreservation. The earliest commercial application of cryopreservation was in the storage of animal sperm cells for use in artificial insemination. In the early post–World War II period artificial insemination of dairy herds was practiced widely. However, sperm were viable for only a few days when cooled but not frozen. The development of techniques for sperm freezing, storage, and transport using liquid nitrogen refrigeration, which occurred in the mid-1950s, greatly increased the use of artificial insemination, both for breeding stock and for human reproduction. In the frozen state sperm may be transported great distances. However, of even greater value is the fact that sperm cells can be stored for many years. *See* Breeding (animal).

In an entirely similar manner the microorganisms used in cheese production can be frozen, stored, and transported without loss of lactic acid–producing activity. Pollen from various plants can be frozen for storage and transport. Normally this is of little value since fertile seeds can also be conveniently stored and transported without special preservation techniques. However, in plant-breeding experiments it is sometimes desired to crossbreed plants that do not grow in similar environments. In such instances the transport of frozen pollen can be utilized.

In biological research the growth of a pure strain or culture of a particular species is a necessity. For long-term studies these cultures must be continually renewed by growth on a suitable medium. Given enough time, these cultures undergo mutations; therefore, the species characteristics may change during the course of the investigation. By preparing a sufficiently large culture at the start of the investigation and freezing most of it, the investigator is assured of a continuous supply of unaltered microorganisms for his study no matter how long it may take. Banks of carefully established cell lines are stored in frozen condition for use by future generations of biologists.

Another application of cryopreservation is the storage of whole blood or separated blood cells. Frozen storage permits the buildup of the large stocks of blood required in a major catastrophe. It is also useful in storing the less common types of blood. Through frozen blood storage a person can actually develop a stock of their own blood for autotransfusion, a practice which is almost essential for those individuals with rare blood types. *See* Blood.

Freezing and preservation equipment. While other refrigerant materials or mechanical refrigeration systems can be used, most cryobiology apparatus has been designed to use liquid nitrogen as the refrigerant because it is convenient, biologically inert, relatively cheap, and plentiful. *See* Nitrogen.

When a controlled-rate freezer is used, the material to be frozen is placed in a well-insulated chamber. Liquid nitrogen is admitted to another portion of the chamber at a rate controlled by an electrically operated valve. The liquid nitrogen is immediately vaporized by the air that is circulated through the chamber by a motor-driven fan. A thermocouple inserted in a sample of the material to be frozen is connected to the controller. Using the temperature of the thermocouple as a guide, the controller adjusts the flow of liquid nitrogen and the output of an electrical heater to obtain just the cooling rate desired. This procedure is outlined in **Fig. 1**. Working storage of sperm for artificial insemination or for small laboratories employs specially designed liquid nitrogen containers as illustrated in **Fig. 2**. A large opening (2.5-in. or 6.3-

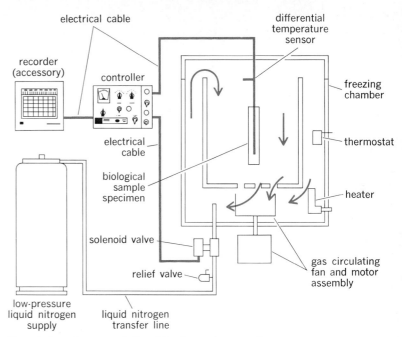

Fig. 1. Schematic representation of freezing system operation.

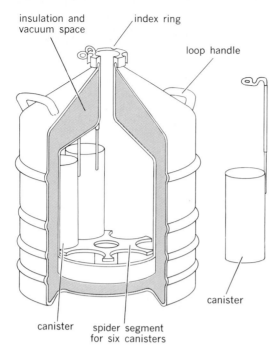

Specifications:

height = 23 3/8 in.
diameter = 17 3/4 in.
neck tube diameter = 2 1/2 in.
ampule capacity = 360 (1.2-cc ampules, 6/rack)
= 484 (1.0-cc ampules, 8/rack)
maximum liquid
nitrogen capacity = 30 liters
working holding time = 60 days

Fig. 2. Liquid nitrogen container. 1 in. = 2.5 cm. 1 liter = 1.06 qt.

cm neck-tube diameter in diagram) is provided so that materials can be inserted and removed. Since the container is designed to allow very little leakage of heat into the inner container, one charge of liquid nitrogen can last for 60 days in active use. The frozen material is normally stored in small glass ampules. Six or eight ampules are clipped onto a vertical rod, and ten rods are mounted in a canister 2 in. (5 cm) in diameter. A typical storage container holds six canisters, or 360–480 ampules. A porous plug fits into the neck tube to hold the canister handles in place and to reduce the leakage of heat into the container.

While containers of the type just mentioned can be used to ship frozen biologicals by all normal means of transport, there are materials that would not conveniently fit into a container with such a small opening. Furthermore, some types of shipment require that the container be able to assume any position during transport. For these applications the equipment illustrated in **Fig. 3** has been developed. The entire top of this equipment is removable, leaving an opening as large as 14 in. (35 cm) in diameter. The liquid nitrogen refrigerant is absorbed in disks of porous material. The porous material can easily be shaped to conform to the contours of the object to be stored. Because the refrigerant liquid is absorbed in the porous packing disks, the container can be shipped in any attitude, although somewhat better results are obtained in the upright position.

There is a container made for the stationary storage of large amounts of frozen material equipment. While this unit has much the same general appearance as a chest-type frozen food storage refrigerator, it is far

superior in insulation performance. The unit can be left untended for a week without loss of effectiveness. The capacity of a large refrigerator is about 90,000 ampules (250 times the capacity of the portable container in Fig. 2), and several hundred units of frozen blood can be stored in it. Each item to be stored is mounted in a vertical rack. Stored material is retrieved by lifting the rack high enough to reach the desired package and then returning it to position. The brief exposure to ambient temperatures does not warm the other packages more than 1 or 2 degrees above their storage temperature of −320°F (−196°C).

Cryosurgery. As was previously mentioned, the causes of cellular destruction from freezing can be effected deliberately to destroy tissue as a surgical procedure. A technique for controlled destruction of tissue in the basal ganglia of the brain was developed for relief from tremor and rigidity in Parkinson's disease.

One significant advantage of cryosurgery is that the apparatus can be employed to cool the tissue to the extent that the normal or the aberrant function is suppressed; yet at this stage the procedure can be reversed without permanent effect. When the surgeon is completely satisfied that the exact spot to be destroyed has been located, the temperature can be low-

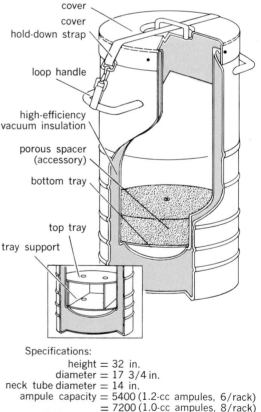

Specifications:

height = 32 in.
diameter = 17 3/4 in.
neck tube diameter = 14 in.
ampule capacity = 5400 (1.2-cc ampules, 6/rack)
= 7200 (1.0-cc ampules, 8/rack)
working liquid
nitrogen capacity = 40.5 liters
working holding time = 19 days
bulk storage capacity = 1.6 ft
LD-40 used with accessory porous spacers
spacer thickness = 3 in.
spacer diameter = 13 1/2 in.
liquid nitrogen capacity
per spacer = 6.3 liters

Fig. 3. Liquid nitrogen container which is used for shipment of large samples. 1 in. = 2.5 cm. 1 ft = 0.3 m. 1 liter = 1.06 qt.

ered enough to produce irreversible destruction. This procedure is of particular assistance in neurosurgery.

A second major advantage of cryosurgery is that the advancing front of reduced temperatures tends to cause the removal of blood and the constriction of blood vessels in the affected area. This means that little or no bleeding results from cryosurgical procedures.

A third major advantage of cryosurgery is that the equipment employs a freezing apparatus (which is about the size of a large knitting needle) that can be placed in contact with the area to be destroyed with a minimum incision to expose the affected area.

The equipment used in cryosurgery utilizes the same operating principles as the controlled-rate freezer. The flow of liquid nitrogen is controlled by an electrically operated valve whose function is directed by a controller that is based on the temperature sensed by a thermocouple located at the freezing element. Since it is desirable to freeze only the area immediately surrounding the tip of the instrument, the liquid nitrogen lead, the gaseous nitrogen return, the thermocouple leads, and the electrical supply for a small heater are all encased in a vacuum-insulated jacket.

Other uses. Besides the preservation or destruction of life, low temperatures can be employed for other purposes. For example, it may be desirable in the formulation of drugs to release antibodies or other active agents from their cells. One method by which this can be achieved is by grinding the cells while they are frozen. In frozen form the cells are hard and brittle and can be fractured easily without resulting in any chemical changes in the enzyme or antibody that is being released. SEE CRYOGENIC ENGINEERING.

A. W. Francis

Bibliography. *CIBA Foundation Symposium 52: The Freezing of Mammalian Embryos*, 1978; P. Dowzou, *Cryobiochemistry: An Introduction*, 1977; L. K. Lozina-Lozinskii, *Studies in Cryobiology*, 1974; J. E. Rash and C. S. Hudson, *Freeze-Fracture: Methods, Artifacts, and Interpretations*, 1979.

Cryogenic engineering

A branch of engineering specializing in technical operations at low temperatures, arbitrarily defined here as below 150 K ($-190°F$). Cryogenic engineers are competent in low-temperature physics and chemistry and in conventional branches of engineering, such as mechanical, chemical, electrical, and nuclear.

Production and handling of cryogens. Cryogens are liquefied gases used to maintain low-temperature environments. Such gases and others used in the chemical industry are usually isolated from mixtures such as the atmosphere, natural gas, or a chemical by-product stream. The desired components are separated and purified by cryogenics. Depending upon the mixture, one or more of the following low-temperature techniques must be used: fractional distillation, scrubbing, and selective adsorption. The last technique is particularly applicable for the separation and purification of the noble, or inert, gases. SEE LIQUEFACTION OF GASES.

Both oxygen and nitrogen are obtained by the fractional distillation of air. Most hydrogen gas is obtained from the catalytic cracking (reformation) of hydrocarbons; some is obtained by the more expensive electrolysis of water. The rare gases argon, neon, and krypton, which are becoming increasingly useful in industry, are obtained as a mixture from air-separation plants and purified by low-temperature adsorption. SEE ARGON; ATMOSPHERIC GAS PRODUCTION; HYDROGEN; KRYPTON; NEON; NITROGEN; OXYGEN.

Gases with critical temperatures below ambient temperature, and therefore not liquefiable by pressure alone, have ordinarily been transported and stored in heavy-walled steel cylinders at high pressure. In the case of hydrogen, for example, about 100 lb of steel must be used to contain each pound of gas (100 kg of steel for each kilogram of gas) at high pressure. In current cryogenic technology, on the other hand, when hydrogen is stored as a liquid, the corresponding weight ratio is only 3. Comparable savings justify the storage of nitrogen and oxygen as liquids in almost all large-scale operations.

Liquid helium and hydrogen. Helium is recovered most economically by a cryogenic condensation method from several helium-rich (more than 0.3% helium) natural gas sources. With the depletion of the existing natural gas sources, a helium shortage was expected by the end of the 1980s but did not materialize. A conservation program in effect from 1961 to 1973 stockpiled some 3.8×10^{10} ft^3 (1.08×10^9 m^3), but after 1973 helium was discarded from gas streams. Ultimately, helium will have to be recovered from the atmosphere, where its concentration is 0.0005% by volume. Such recovery is expensive and extremely energy-intensive: it would require the energy equivalent of one-half barrel of oil to extract 1 ft^3 of helium at standard pressure and temperature (STP), or 2.8 m^3 of oil to extract 1 STP m^3. Helium can be obtained as a by-product of air-separation plants at lower cost and energy consumption, but the air-liquefaction industry is too small to supply more than a few percent of present requirements. Some current uses of helium can be met by other gases, for example, argon in heliarc welding. However, helium is unique in low-temperature applications, particularly those involving superconductivity, but the discovery of high-temperature conductors offers some hope that the higher-boiling cryogens can also be used for applications of superconductivity. However, helium is an important refrigerating medium for very low-temperature applications, and because some of these uses may require substantial quantities, it is important to conserve as much of the present helium supply as possible. SEE HELIUM; SUPERCONDUCTING DEVICES.

Large-scale hydrogen-liquefaction plants are in operation throughout the United States; many helium liquefiers ranging in capacity from 2.6 to 925 gal/h (10 to 3500 liters/h; the largest is at Fermi National Accelerator Laboratory) provide widely distributed sources of liquid helium. Thermal insulation reduces losses to almost negligible proportions, so that liquid hydrogen and helium can be transported over long distances with very small losses. The containers vary in size up to 10,000 gal (38 m^3) for helium. Hydrogen is shipped by truck, rail tank car, and barge.

Liquefied natural gas. Liquefied natural gas (LNG), essentially liquid methane, is shipped in insulated tankers with capacities up to 790,000 bbl (125,000 m^3).

Natural gas is liquefied at various locations in the United States and throughout the world to serve as storage to meet peak utility requirements. At use centers, storage vessels have been built in sizes up to about 1×10^6 bbl (1.6×10^5 m^3). In the past, the liquefied natural gas has been warmed in atmospheric heat exchangers, wasting the available refrigeration. Newer installations have been built in conjunction with air-separation facilities, so that refrigeration made available by the gasification of the liquefied natural gas can reduce the energy requirements of air liquefaction. See Liquefied natural gas (LNG).

Metering. The greatly expanded traffic in cryogenic liquids has increased the importance of accurate metering of quantities delivered or used. Metering is complicated by losses arising from various sources. Despite numerous attempts to develop mass flow meters, the transfer of cryogenic fluids is still usually evaluated by liquid-level measurements and sometimes by volume-flow measurements. Both methods require additional measurement of pressure and temperature to determine liquid density and mass.

Associated pumps and vaporizers. Because the stored liquid is usually converted by the consumer to a gas at ambient temperature and sometimes at high pressure, exploitation of liquid storage has necessitated the development of cryogenic pumps and vaporizers. This in itself was a difficult engineering feat, posing numerous low-temperature bearing, seal, and lubrication problems.

Efficiency of refrigeration. The cost of the energy needed to refrigerate and liquefy gases is of greatest concern to the cryogenics industry. The theoretical maximum refrigeration efficiency, that is, the smallest amount of input energy required to refrigerate at a lower temperature, is given by the Carnot cycle; actual cryogenic gas-expansion refrigerators are, however, much less efficient. A detailed examination of existing refrigerators and liquefiers has shown that the degree to which Carnot efficiency is approached is a function of refrigeration capacity but not of temperature, and the larger the size, the more efficient the refrigerator. Very small refrigerators may operate with 1% (or even less) of the theoretical efficiency, whereas the largest refrigerators seem to approach a limit of about 35% of the Carnot efficiency. Computer analysis has been applied to refrigeration cycles to allow optimization of such parameters as flow rates and temperatures at various points in the process. Actual efficiencies of components, for example, expanders and compressors, are included, and the irreversibilities arising in the refrigeration process are examined. Such analyses suggest that the ultimate efficiency to be expected from existing cryogenic refrigeration techniques is about 40% of Carnot. For this reason, it is of interest to investigate other methods. See Carnot cycle; Refrigeration.

Applications. Oxygen is used in oxidation processes to make chemicals, such as methanol and ethylene oxide. Nitrogen is used in the production of ammonia and fertilizers, for protecting hazardous chemicals with an inert atmosphere, and in liquid form for the purification of gases with extremely low boiling points such as hydrogen.

Metal industry. Metallurgical applications for pure gases are increasing as new alloy steels are created and as the production of hitherto rare or seldom-refined metals is undertaken in order to meet the demands of the nuclear and space age. Oxygen, nitrogen, argon, and hydrogen, all produced cryogenically, are commonly used. In order to meet the growing demand for oxygen economically, on-site generators have been developed to supply the gas from plants constructed adjacent to the user and piped "over the fence" in quantities up to 500 tons (450 metric tons) per day. Previously, liquid oxygen was delivered to the consumer from a large central air-separation plant. Some of the smaller plants are automatically controlled, requiring only periodic visits for maintenance. The liquid produced is stored in large insulated tanks, from which it is withdrawn as needed. See Steel manufacture.

Liquefied cryogens are used in several metal fabrication processes: in the precipitation hardening of steel at 180 K ($-136°F$) with liquid nitrogen, in the fabrication of vessels by cold stretching with high-pressure nitrogen at 140 K ($-208°F$), and in the quenching of complex metal parts with liquid nitrogen.

Food preservation. Large quantities of liquid nitrogen are employed in the preservation of food by rapid freezing. This is carried out by direct immersion in liquid nitrogen or by contact with liquid nitrogen spray or cold vapor. Among the advantages are low cost, rapid freezing, and low water loss. Liquid nitrogen is also used to maintain low temperatures during the transportation of frozen food. From the thermodynamic standpoint, the use of such low temperatures is inefficient. However, these refrigeration systems are maintenance-free, simple, and cheap to operate. See Food preservation.

Cryopump. A high-speed vacuum pump has been developed capable of producing an extremely low ultimate vacuum and requiring less power than conventional diffusion pumps. The principle is well known: lowering the pressure within a gas-tight enclosure by freezing out condensable gases on a cold surface. In the cryopump a low-temperature refrigeration unit circulates helium at 20 K through coils in plate condensers upon which the gas to be pumped condenses. The cryopump is employed in large environmental simulation facilities such as high-altitude wind tunnels. Small commercial cryopump units are in common use. See Vacuum pump.

Transportation. Cryogenic engineering is applied in several modes of transportation. Superconducting magnets are used, notably in Japan in magnetically levitated trains with speeds of 300 mi/h (500 km/h) or more.

The use of liquid hydrogen fuel appears promising for both jet aircraft and land transportation, especially automotive. At 13.8 K ($-434.8°F$) solid hydrogen can exist in equilibrium with the liquid. Such mixtures, called slush hydrogen, can flow like a liquid. Because the density of the solid is about 12% higher than that of the coexisting liquid, slush hydrogen can be useful as a fuel in air or space applications requiring smaller and lighter storage tanks. Hydrogen fuel offers the advantages of cleanliness and high efficiency, producing no solid or carbon monoxide emissions to contribute to smog and no carbon dioxide to promote the so-called greenhouse effect, a cause of considerable concern. However, hydrogen does not occur naturally in sufficient amounts to be a primary source of energy; it can be considered only as an energy carrier. Because energy must be expended to produce hydrogen, other savings must be found before it can achieve economic parity with fossil fuels. Furthermore, procedures for its safe handling must be established. See Aircraft fuel.

Electric power systems. The application of superconductivity and associated cryogenic systems to electric power generation and transmission appears promising. Superconducting magnets are being developed to provide the magnetic fields needed for the generation of electric power by magnetohydrodynamics (MHD).

They will also be needed for the magnetic confinement of plasmas in most designs now under consideration for thermonuclear (nuclear fusion) power generation. Because energy can be stored efficiently in large magnetic fields, superconducting magnets are being considered for energy storage in electric power transmission systems. A superconducting magnetic energy storage (SMES) system was used to analyze the dynamic response characteristics of the northwest power grid of the United States; and a much larger system has been proposed for electric power load leveling and peak shaving. The application of superconductivity to electric power transmission has been investigated for ac and dc systems, both of which appear to be technically and economically feasible for transporting large blocks of power (5000 MW or more). All of these systems require large-scale, sophisticated, reliable cooling systems. *See Energy storage; Magnetohydrodynamic power generator; Nuclear fusion*.

Solid-state electronics. Several elementary superconducting (Josephson junction) circuits have been developed to produce binary memory elements, switching devices, and multivibrators for high-speed computers. Although the small size, negligible power inputs, increased speed, and high reliability of these units promise significant advantages over other types of computer components, performance has not yet met commercial requirements. *See Cryotron; Josephson effect*.

Devices with molecular high-gain, low-noise microwave amplifiers are called masers. A weak incident signal of the appropriate resonance frequency can release stored energy from a paramagnetic crystal cooled below about 4.2 K. A crystal transmitting more energy than it receives from the input signal operates as a microwave amplifier. When the sensitivity of an electromagnetic detection system is limited by thermal noise, for example, in radio astronomy, the maser has proved particularly significant. *See Maser; Radio telescope*.

Materials research. Efforts have been made to develop new and better superconductors with improved mechanical properties and the ability to carry higher currents at higher temperatures. The development of niobium-germanium (Nb_3Ge) with a transition temperature of over 23 K ($-418°F$) created the possibility of superconducting devices operating with liquid hydrogen coolant. The discovery that certain metal oxide ceramics become superconducting above 120 K ($-244°F$) suggests that liquid nitrogen might be used as coolant. Research has been extensive into the low-temperature mechanical and thermal properties of structural materials, including epoxy resins and composites. *See Superconductivity*.

Isotope separation. The increasing demands of atomic energy programs, both fission and fusion, for large quantities of separated hydrogen isotopes led to renewed interest in low-temperature distillation for producing pure deuterium. Major production facilities have been constructed in the Soviet Union, Germany, France, and Switzerland. Besides being the most economical, the distillation method requires the least time to reach equilibrium, involves the smallest capital investment, and extracts the most deuterium from the feed. *See Deuterium; Isotope (stable) separation*.

Physics research. Cryogenic engineering is necessary in low-temperature research, which has contributed significantly to the understanding of supercon-ducting and superfluid helium systems, where quantum-mechanical effects are exhibited on a macroscopic scale. Of similar importance are measurements of thermal, electric, magnetic, optical, and structural properties of materials to temperatures into the nanokelvin range. To establish such extreme temperature environments and make physical measurements requires the most advanced cryoengineering skills. *See Liquid helium; Low-temperature physics; Superfluidity*.

Superconducting magnets are frequently used in beam focusing and beam bending in high-energy particle accelerators. For example, the Energy Doubler at Fermi National Accelerator Laboratory uses 800 dipole and 216 quadrupole superconducting magnets arranged in a ring 4 mi (6 km) in circumference. Steering the proton beam requires highly sensitive magnet alignment and stability. The next-generation proton collider, the superconducting supercollider, is planned to be over 10 times larger in circumference and, if the project proceeds, will involve the largest cryogenic engineering task to date. *See Particle accelerator*.

Medicine and biology. Low temperatures have found increasing use in biological studies and medical procedures. Since biological activity is greatly reduced as the temperature is lowered, reactions can be stopped and studied at each step. In addition, a means is provided for the preservation of biological material. The liquid-nitrogen freezing and preservation of human blood and of cattle sperm cells are examples. Cryogenic surgical procedures include operations for Parkinson's disease and the removal of cataracts and skin cancers. *See Cryobiology*.

The first significant widespread use of superconductive devices has been the application of nuclear magnetic resonance for tomographic examination of the human body. Magnetic resonance imaging devices analyze soft, primarily hydrogenic body tissue and thereby complement x-ray tomography, which is better adapted to reveal structure in hard body tissue (bone). Magnetic resonance imaging requires magnetic fields of up to 50,000 gauss (5 teslas) or more over volumes that can encompass the human body, requiring superconducting solenoids. The method is noninvasive, and the magnetic fields have little or no biological effect. *See Medical imaging*.

Nuclear, military, and space applications. An important military application of cryogenics has been the development of oxygen-breathing apparatus for crews of high-altitude, long-range aircraft. By storing oxygen as a dense liquid in lightweight, well-insulated containers from which it can be vaporized as needed, significant weight reductions are effected over the older method of high-pressure, low-density gas storage in heavy steel cylinders. The need for liquid oxygen to fill these containers has led in turn to the development for the military services of portable oxygen generators in a wide range of sizes for air bases and ships.

Cryogenic engineers have also been active in the aerospace field. Because the specific impulse of a rocket is inversely proportional to the square root of the molecular weight of the exhaust gases, low-molecular-weight fuels such as hydrogen are highly desirable. These are best handled and stored as cryogenic liquids. Improvement in rocket performance has required large-scale gas-liquefaction plants for hydrogen, oxygen, and other gases; large-scale storage and transport facilities; high-capacity, high-pressure pumps; more efficient thermal insulation; lightweight fuel tanks; and specialized valves and fittings for cryogenic fluids. *See Rocket propulsion*.

Space satellite operations are often dependent upon the refrigeration of instrumentation, especially radiation detection devices. In many cases, these missions are limited (and thereby defined) by the extent to which such cooling can be provided. For example, it is desirable to launch observation satellites with active lives of 3–5 years. This, however, is not yet possible because cryogenic refrigerators have not been devised to operate unattended for such long periods. *See Cryogenics; Low-temperature thermometry*.

F. J. Edeskuty; K. D. Williamson, Jr.; W. E. Keller
Bibliography. *Advances in Cryogenic Engineering*, annually; R. F. Barron, *Cryogenic Systems*, 2d ed., 1985; C. F. Gottzmann, *Cryogenic Processes and Equipment in Energy Systems*, 1980; B. A. Hands (ed.), *Cryogenic Engineering*, 1986; G. G. Haselden (ed.), *Cryogenic Fundamentals*, 1971; R. B. Scott, *Cryogenic Engineering*, 1959; G. Walker, *Cryocoolers*, vols. 1 and 2, 1983.

Cryogenics

The science of producing and maintaining very low temperatures. Before World War II, research at very low temperatures was carried out at only a few laboratories because of the large amount of time and money required for the construction, operation, and maintenance of gas liquefaction and related facilities. Beginning in the prewar years, however, recognition of the value of applied cryogenics in certain industrial processes led to a rapid expansion of scientific and industrial low-temperature technology. In 1939, liquid helium was available in fewer than a dozen laboratories throughout the world, whereas now it is available in almost any large research institute. Similarly, many prewar experimental applications of cryogenic engineering, such as plants to produce pure oxygen in tonnage amounts for metallurgical processes, have come into common use. In the vanguard of industrial cryogenics are plants capable of producing large quantities of liquefied hydrogen and helium for industrial and military use. *See Atmospheric gas production; Liquid helium*.

Production of low temperatures. From a practical standpoint, any object may be cooled to and maintained at a low temperature simply by placing it in contact with a suitable liquefied gas held at a constant pressure. The temperatures afforded by baths of liquefied gases are given in the **table.** In principle, a liquefied gas can be used to provide constant temperatures from its triple point to its critical point. Temperatures are varied by changing the pressure above the liquid, and, within the liquid range, any desired temperature can be maintained by removing precisely the amount of gas vaporized by heat passage into the bath liquid. Heat cannot be excluded entirely, and baths eventually boil away unless replenished.

Liquefaction of gases. In order to cool and liquefy a gas, various procedures may be employed. The most frequently used are (1) successive steps of compression followed by cooling and expansion through a throttling valve—this so-called Joule-Thomson expansion lowers the gas temperature until liquid is formed; (2) cooling the gas at constant pressure with a series of progressively colder refrigerator reservoirs; (3) allowing the gas itself to constitute the refrigerator working substance until it has cooled sufficiently to

liquefy partially—the process is continued with the addition of makeup gas and withdrawal of the liquid formed in each cycle. *See Liquefaction of gases*.

As may be seen from the table, liquefied gases do not exist over the entire low-temperature range. The lowest temperature at which it is practical to use a liquid bath is about 0.3 K; gaps exist between 5 and 14 K and between 44 and 55 K. If it is necessary to extend the accessible temperature range or reach certain ranges more conveniently, other methods of cold production are available.

Refrigerators. Refrigerators can, in principle, be designed to produce and maintain any predetermined temperature. For refrigerators operating cyclically, the low temperature produced is not strictly constant because of the periodic withdrawal of heat from the low-temperature reservoir, but the temperature variation can be large or extremely small, depending on the cycle characteristics. If, however, the refrigeration process is noncyclical, for example, where the working substance changes phase in the low-temperature reservoir, the reservoir temperature remains constant. Solid-to-liquid and liquid-to-gas phase changes can be used, although more often the low-temperature reservoir is simply a liquid bath in which the liquid boiling off is continuously replenished. *See Refrigeration cycle*.

Refrigerator cycles operating above about 1 K generally involve compression and expansion of appropriately chosen gases. Because it is impractical or impossible to use gases at lower temperatures, liquid or solid working materials are employed with different techniques. One such method is adiabatic demagnetization of paramagnetic ions in solids, involving electron-spin paramagnets, for use between 0.003 and 1 K. This method can be extended to lower temperatures (10^{-7} K or slightly lower) through the adiabatic demagnetization of nuclear-spin paramagnets, and to higher temperatures (300 K or higher) by using a paramagnetic or ferromagnetic working material in continuous magnetic refrigeration cycles. Other examples include the helium-3/helium-4 dilution refrigerator (0.003–0.3 K); adiabatic compression of a liquid-solid mixture of helium-3, made possible because of

Physical constants of low-temperature liquids

Liquid	Triple point Temp., K	Triple point Pressure, atm[†]	Normal boiling point, K	Critical point Temp., K	Critical point Pressure, atm[†]
Helium-3			3.2	3.3	1.2
Helium-4			4.2	5.2	2.3
n-Hydrogen	14.0	0.07	20.4	33.2	13.0
n-Deuterium	18.7	0.17	23.7	38.4	16.4
n-Tritium	20.6	0.21	25.0	42.5*	19.4*
Neon	24.6	0.43	27.1	44.7	26.9
Fluorine	53.5	<0.01	85.0	144.0	55.0
Oxygen	54.4	<0.01	90.2	154.8	50.1
Nitrogen	63.2	0.12	77.4	126.1	33.5
Carbon monoxide	68.1	0.15	81.6	132.9	34.5
Argon	83.8	0.68	87.3	150.7	48.0
Propane	85.5	<0.01	231.1	370.0	42.0
Propene	87.9	<0.01	226.1	365.0	45.6
Ethane	89.9	<0.01	184.5	305.4	48.2
Methane	90.7	0.12	111.7	191.1	45.8

*Calculated.
[†]1 atm = 1.013 × 10^5 pascals.

a slope reversal in the helium-3 melting curve, known as Pomeranchuk cooling (0.002–0.05 K); and hyperfine enhanced nuclear refrigeration, which exploits certain ions with magnetic moments that increase greatly with application of an external magnetic field (0.001–0.05 K). *SEE ADIABATIC DEMAGNETIZATION.*

Principles of cold production. The same principles of refrigeration may be used to explain gas liquefaction and the periodic temperature reduction of the refrigerator working substance. Cold production is related directly to the concept of temperature, which in turn is intimately connected with entropy S. Because S is not easily defined directly, it is more convenient to define increases in S of a system as the heat absorbed by the system isothermally and reversibly divided by the temperature at which it is absorbed; that is, $dS = dq/T$. Entropy is a measure of the intrinsic thermal disorder associated with a substance, and, at the absolute zero of temperature, the entropy of every substance in internal equilibrium is zero. Entropy is increased by the introduction of thermal energy or by varying some other parameter upon which S depends. A conjugate relation exists so that, provided all other parameters are held constant, a change in T results in a change of S of the same sign. The significance of S for refrigeration cycles stems from the concept of isentropic (or constant S) processes, which result in the maximum possible decrease in T for a given spontaneous parametric change. Such processes, being both adiabatic (thermally isolated) and reversible, represent the ideal case and can be only approximated in practice. Nevertheless, the isentropic process provides a criterion against which the efficiency of any actual process may be gauged. In **Fig. 1**, the variation of S and T for an idealized substance is shown at two different constant values of a thermodynamic parameter X, such as pressure or magnetization. It is seen that an isentropic parameter change from A to B produces the maximum reduction in T. (Spontaneous processes resulting in a decrease in S with a change in X are impossible.) *SEE ENTROPY; THERMODYNAMIC PRINCIPLES.*

On the other hand, because all real cooling processes are to some extent irreversible, the temperature drop for such a process resulting from a change in X is always less than that obtained from a reversible isentropic process for the same change in X. The cooling achievable by an irreversible process is represented in Fig. 1 by the path $A \rightarrow C$. (Under certain conditions, it is even possible in an irreversible process to produce, for the same parameter change, heating instead of cooling, as shown by the path $A \rightarrow C'$.)

Cooling by refrigeration. If a suitable working substance is repeatedly subjected to one of the cooling procedures described in Fig. 1, a temperature lower than ambient may be maintained indefinitely. Machines capable of carrying out such processes (refrigerators) operate over a closed sequence of states through which the working substance passes (the refrigeration cycle). All reversible refrigeration cycles have the same efficiency, the hypothetical maximum possible. One such cycle, the Carnot cycle, is illustrated in **Fig. 2**. First a suitable substance, showing a large change in S with both T and some other parameter such as pressure, is cooled isentropically ($A \rightarrow B$) from a hot reservoir at temperature T_H to a cold reservoir at T_C. It is then allowed to absorb heat isothermally ($B \rightarrow C$) from the cold reservoir by further changing the parameter as required to maintain constant temperature. Following this, a second isentropic

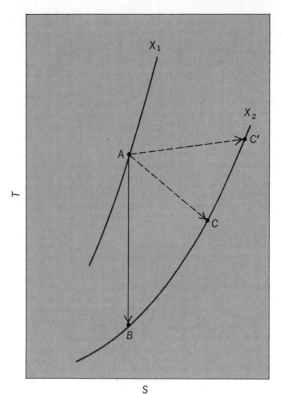

Fig. 1. Variation of entropy with temperature for two different constant values of thermodynamic parameter X.

process ($C \rightarrow D$) is carried out by thermally isolating the working substance and reversing the parametric change adiabatically. During this step, the working substance heats until it reaches T_H, whereupon thermal contact is made between the working substance and the hot reservoir. The net heat Q_c extracted from the low-temperature reservoir and the work W done in completing the cycle are then transferred ($D \rightarrow A$) isothermally to the hot reservoir by further parameter changes until the working substance has returned to its original state and the cycle is ready to be repeated. As seen in Fig. 2, this sequence is represented by a rectangle on a T-S diagram. The efficiency of the Carnot cycle is $W/Q_c = (T_H/T_C) - 1$; Q_c is also called the refrigeration power. Actual refrigerators always operate at a lower efficiency than that of a Carnot cycle, again because truly isentropic processes are not achievable in practice. *SEE CARNOT CYCLE.*

Cooling of a gas. In applying the above principles to a gas, parameters X_1 and X_2 are the constant pressures

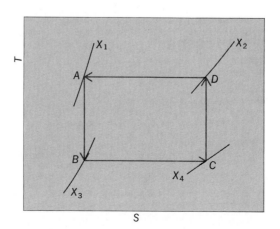

Fig. 2. An ideal refrigeration cycle on a termperature entropy (T-S) diagram.

P_1 and P_2, with $P_2 < P_1$. If the gas were expanded isentropically with the performance of external work (as in the driving of a turbine or piston expansion engine), maximum cooling would be obtained as shown by the path $A \rightarrow B$ in Fig. 1. Actually, the temperature drop is somewhat less than ideal because of frictional losses. An irreversible expansion of the gas from P_1 to P_2, for example, through a throttle valve, would result in a smaller temperature drop, such as that given by the path $A \rightarrow C$.

When a gas is used as the refrigerator working substance, the cycle is located between two isobars on the T-S diagram, with the cooling phase occurring during the expansion of the compressed gas. The closeness of approach by a gas refrigerator to the performance of a Carnot machine depends on how much importance the designer places in such requirements as reliability, simplicity of operation, and efficiency. In practice, however, the maximum efficiency for gas refrigerators is about 35% of Carnot efficiency for the largest machines; for smaller refrigerators the figure drops to 1% or less. The achievable percentage of Carnot efficiency has empirically been found to be independent of both the species of working gas used (helium, hydrogen, nitrogen, and so forth) and its temperature.

For certain applications in which the refrigeration must be supplied at a constant low temperature, a portion of the gas is allowed to liquefy in the cold reservoir. The absorption of heat then occurs by evaporation of the liquid at constant pressure and hence constant temperature. If the heat absorption can occur over a range of low temperatures, however, the refrigeration is usually supplied directly by the expanded cold gas.

Adiabatic demagnetization.. If in Fig. 1 the parameter curves are assumed to represent two different but constant magnetizations of a paramagnetic or ferromagnetic material, M_1 and M_2 with $M_2 < M_1$, and if initially the specimen is magnetized by an external magnetic field to a value M_1 at a temperature corresponding to that given by point A, then an isentropic demagnetization of the specimen to M_2 results in the temperature drop $A \rightarrow B$. In practice, this is accomplished by thermally isolating the magnetized material before reducing the magnetic field. After the low temperature T_B has been reached, experiments may be carried out on the magnetic material itself, or a second substance can be thermally connected with the magnetic material and cooled by it to some intermediate temperature dependent on the relative heat capacities of the magnetic material and sample.

In a magnetic refrigerator designed to operate in the millikelvin range, the high-temperature reservoir could consist of a liquid helium bath held at approximately 1.2 K (helium-4), 0.3 K (helium-3), or 0.020 K (helium-3–helium-4 dilution refrigerator), depending on what final temperature is required. The full cycle, shown in Fig. 2, carries the magnetic material through a sequence of isentropic and isothermal changes. However, experiments at these low temperatures usually take advantage only of the legs $D \rightarrow A$ (magnetization) and $A \rightarrow B$ (demagnetization) in a so-called one-shot cooling. In order to carry out quasi-adiabatic and isothermal changes at will, a heat switch is required between the working substance and the heat reservoirs. In some magnetic refrigerators the heat switch may be a lead wire, the thermal conductivity of which can be changed by one or two factors of 10 when it is caused to pass from its superconducting to its normal state.

Continuously operating magnetic refrigerators do, however, work repeatedly through the full cycle. For such machines, currently under development, specific magnetic materials are chosen by matching the thermomagnetic properties (for example, entropy as functions of temperature and magnetic field) to the desired temperature range to be spanned. For example, gallium gadolinium garnet is a good working material to cover the range 2–20 K.

In both continuous and one-shot magnetic refrigerators, the degree of cooling achieved is dependent on the strength of the magnetic field applied. For this reason, superconducting magnets are required to produce the best results. Because the magnetic fields can be applied fairly slowly compared with the time constant for thermal equilibrium of the magnetic spins of the working material, the magnetization process ($D \rightarrow A$ in Fig. 2) can be carried out very nearly reversibly. Therefore the magnetic-cooling processes can be considerably more efficient than gas-cooling processes, where the corresponding $D \rightarrow A$ step (compression) is in fact highly nonisothermal and a major source of the low efficiency of the gas machine. *See* Magnet.

Helium-3–helium-4 dilution refrigerator. Another parameter X that may be varied to obtain cooling is the relative concentration c of two chemical species that form a nonideal solution. In such a solution, the mixing of the two components is accompanied by a heat of mixing ΔH_m and an excess entropy of mixing ΔS_m. Surprisingly, the liquid solutions of the two helium isotopes, helium-3 and helium-4, are nonideal and, below about 0.86 K, separate into two phases, depending on c. In the region of miscibility, the mixing process absorbs heat from the surroundings and thereby forms the basis for a refrigeration system. Solutions with less than approximately 6% helium-3 are miscible at all temperatures, so that very low temperatures can be obtained.

In practice, a continuously operating dilution refrigerator has as its principal element a mixing chamber containing a phase-separated mixture, with nearly pure helium-3 (the light phase) on top and a dilute mixture of helium-3 ($c = 6$–7%) in the bottom phase. Circulated by a pump at ambient temperature, helium-3 is continuously introduced into the upper phase, forced across the phase boundary into the phase of low helium-3 concentration (where cooling occurs), and then withdrawn from the latter phase to continue the cycle. An object to be cooled can be attached to the bottom of the mixing chamber.

The dilution refrigerator can be used to provide stable temperature environments from 0.3 K (helium-3 liquid bath) down to about 0.01 K, delivering sufficient refrigeration power to be of considerable use in a wide variety of experiments. The low-temperature limitation near 0.01 K arises from the frictional heat produced by the flowing helium-3 (inasmuch as in this temperature range the viscosity of the fluid rises rapidly with decreasing temperature), which reduces the net refrigeration power to zero.

Other refrigeration cycles. Although the above discussion is based upon Carnot refrigeration cycles, it should be remembered that any heat engine run in reverse constitutes a refrigerator. Correlative to this is the existence of a large number of other cycles, each with specific characteristics which make them more

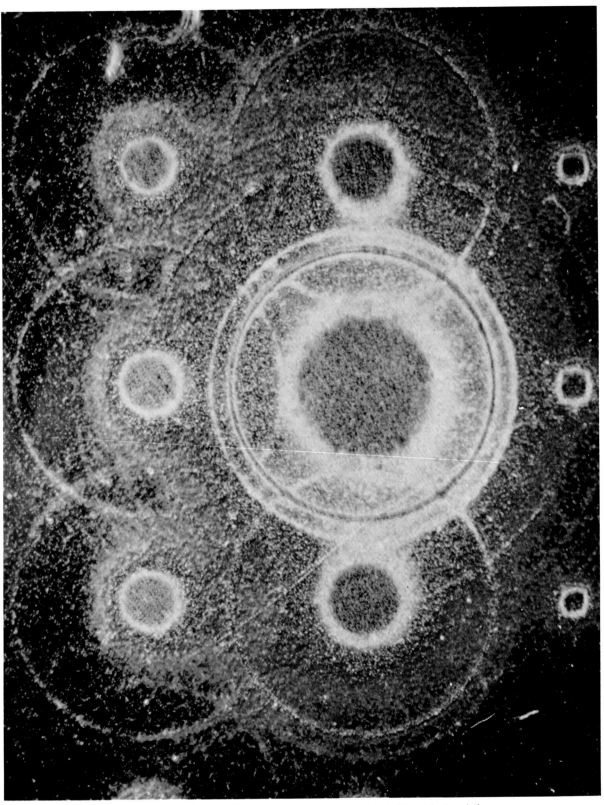

High-resolution electron diffraction patterns recorded at 1.8 K with a color translation technique which permits quantitative evaluation of characteristic parameters of single-crystal films and deposited thin layers at liquid helium temperatures. This illustrates the potential of cryoelectron microscopy and diffraction in the study of novel electron optical phenomena, including direct observation of Josephson junction devices and other critical components of superconducting computers under cryogenic operating conditions. (*Courtesy of Humberto Fernández-Morán*)

or less suitable for certain applications. An example of a refrigerator utilizing a cycle different from the one discussed above is the Stirling-cycle air liquefier developed in 1954. In this device, heat is absorbed and rejected to the heat reservoirs isothermally during expansion and compression of a gaseous working substance (hydrogen or helium), whereas the cooling and heating phases of the cycle can be considered to occur at either constant pressure or constant volume with the aid of a regenerator or heat accumulator. *See Stirling engine*.

Maintenance of low temperatures. Once a temperature lower than ambient has been produced, it may be maintained indefinitely (provided the region itself contains no heat sources) if heat flow to the cooled region can be prevented. Because no perfect thermal insulators exist, the problem becomes one of minimizing the heat flow to the refrigerated region. Devices designed to provide low-temperature environments in which operations may be carried out under controlled conditions are called cryostats.

Heat flows irreversibly from a high- to a low-temperature zone by three different mechanisms: radiation, convection, and conduction. Radiative heat losses are proportional to $(T_H^4 - T_C^4)$ where T_H and T_C are the temperatures of the zones, but this very strong transfer process can be minimized (the proportionality constant reduced) by coating the cold region with a material having the lowest possible emissivity. Because emissivities of 0.02 are practically attainable, a coated surface can be made to reflect 98% of the radiant energy incident upon it. Radiation heat leak may be further diminished by the use of similarly coated radiation shields maintained at temperatures intermediate between the hot and cold surfaces. Finally, the temperature of the hot radiating surface may often be reduced by cooling it to some temperature intermediate between ambient and the low temperature in question by some relatively inexpensive refrigerant. By the use of one or more of the above techniques, the heat leak from radiation can almost always be reduced to acceptable proportions. *See Emissivity; Heat radiation; Heat transfer*.

Heat leak from convection is also relatively easy to eliminate, either by a very high vacuum in the space between the hot and cold surfaces or by packing this space with an insulating material such as a foamed plastic or a powder. The gas is thus effectively displaced or localized sufficiently to prevent convection. In addition, the efficiency of a powder insulation is often enhanced by evacuating it to a pressure of a few millitorrs (1 millitorr = 0.13 pascal). *See Convection (heat)*.

Conduction along the solid supports, piping, instrument lead wires, openings, and the insulation itself constitutes the remaining mechanism by which heat can flow to the refrigerated region. As a general rule, all connecting material between hot and cold surfaces should have the lowest possible thermal conductivity, the smallest cross-sectional area, and the longest length consistent with mechanical stability, operating, and design requirements. *See Conduction (heat)*.

These three heat-leak mechanisms are usually satisfactorily suppressed in a device known as a dewar (thermos bottle), a double-walled vessel with evacuated annular space and with walls fabricated from thin material of low thermal conductivity (glass or stainless steel) and coated with a low-emissivity surface (silver or polished metal). Cryostats incorporate one or more (nested) dewars to contain cryogenic liquid baths; for example, a liquid nitrogen bath is usually used to cool the outside of the dewar containing a liquid helium bath. *See Dewar flask*.

Finding the best solution to any problem requires an analysis of all the factors contributing to the heat leak. Beginning with the largest, these leaks are successively reduced. Under these conditions, it frequently happens that a suboptimal but less expensive material can be used for a component when its contribution to the total heat leak is small. *See Cryogenic engineering; Heat insulation; Low-temperature physics; Refrigeration*.

F. J. Edeskuty; K. D. Williamson, Jr.; W. E. Keller

Bibliography. *Advances in Cryogenic Engineering*, annually; A. Arkharov, *Theory and Design of Cryogenic Systems*, 1981; C. A. Bailey, *Advanced Cryogenics*, 1971; D. S. Betts, *Refrigeration and Thermometry below One Kelvin*, 1976; S. Fluegge (ed.), *Handbuch der Physik*, vols. 14 and 15, 1956; W. E. Keller, *Helium-3 and Helium-4*, 1969; B. Law (ed.), *Cryogenics Handbook*, 1981; O. V. Lounasmaa, *Experimental Principles and Methods below 1 K*, 1974; S. W. Van Sciver, *Helium Cryogenics*, 1986.

Cryolite

A mineral with chemical composition Na_3AlF_6. Although it crystallizes in the monoclinic system, cryolite has three directions of parting at nearly 90°, giving it a pseudocubic aspect. Hardness is 2½ on Mohs scale and the specific gravity is 2.95. Crystals are usually snow-white but may be colorless and more rarely brownish, reddish, or even black (see **illus.**).

White translucent cryolite in association with siderite and galena, Ivigtut, Greenland. (*Specimen courtesy of Department of Geology, Bryn Mawr College*)

The mean refraction index is 1.338, approximately that of water, and thus fragments become invisible when immersed in water. Cryolite, associated with siderite, galena, and chalcopyrite, was discovered at Ivigtut, Greenland, in 1794. This locality remains the only important occurrence.

Cryolite was once used as a source of metallic sodium and aluminum, but now is used chiefly as a flux in the electrolytic process in the production of aluminum from bauxite. *See Aluminum; Bauxite*.

Cornelius S. Hurlbut, Jr.

Cryotron

A current-controlled switching device based on super-conductivity for use primarily in computer circuits. The early version has been superseded by the tunneling cryotron, which consists basically of a Josephson junction. In its simplest form (see **illus.**) the device has two electrodes of a superconducting material (for example, lead) which are separated by an insulating film only about 10 atomic layers thick. For the elec-

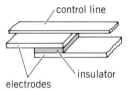

control line

insulator

electrodes

Diagram of the tunneling cryotron.

trodes to become superconducting, the device has to be cooled to a few degrees above absolute zero. The tunneling cryotron has two states, characterized by the presence or absence of an electrical resistance. They can be considered as the "on" and "off" states of the switch, respectively. Switching from on to off is accomplished by a magnetic field generated by sending a current through the control line on top of the junction. The device can switch in a few pico-seconds and has a power consumption of only some microwatts. These properties make it an attractive switching device for computers, promising perfor-mance levels probably unattainable with other de-vices. SEE JOSEPHSON EFFECT; SUPERCONDUCTING DE-VICES; SUPERCONDUCTIVITY.

P. Wolf

Bibliography. W. Anacker, Computing at 4 degrees Kelvin, *IEEE Spectrum*, 16:26–37, May 1979.

Cryptobiosis

A state in which the metabolic rate of an organism is reduced to an imperceptible level. The several kinds of cryptobiosis (hidden life) include anhydrobiosis (life without water), cryobiosis (life at low tempera-tures), and anoxybiosis (life without oxygen). Of these, most is known about anhydrobiosis; therefore, the discussion will be restricted to that type.

States of anhydrobiosis occur only in the early de-velopmental stages of various organisms, including plant seeds, bacterial and fungal spores, cysts of cer-tain crustaceans, and certain insect larvae. They occur in both developmental and adult stages of certain soil-dwelling invertebrates (rotifers, tardigrades, and nem-atodes), mosses, lichens, and certain ferns.

Induction of anhydrobiosis. Many organisms that are capable of entering an anhydrobiotic state require a period of induction of the morphological and bio-chemical alterations. For example, rotifers, tardi-grades, and nematodes include species that live in the water associated with soil. When this water dries up, the animals dry up also, but are not killed by the de-hydration. When they come in contact with water again, they rapidly swell and assume active life, often within minutes. In order to enter this remarkable state the animals require slow dehydration, during which they undergo pronounced morphological changes.

Tardigrades and rotifers contract longitudinally, as-suming a so-called tun configuration (**Fig. 1**). Nema-todes typically coil into tight spirals (**Fig. 2**). In all these animals, the morphological changes character-istic of cryptobiosis are thought to reduce the rate of evaporative water loss by reducing the animal's sur-face area accessible to the surrounding air. In addi-tion, contraction of the body is thought to entail or-dered packing of organ systems and intracellular contents, thereby obviating mechanical damage.

Electron microscopy studies on dry organisms lend support to these suppositions. During the slow dehy-dration the animals show changes in their chemical composition. The nematode *Aphelenchus avenae*, for which there is the most information, synthesizes two carbohydrates, free glycerol and trehalose, that are thought to be important to its survival of dehydration. It appears that these molecules are made from storage lipids by way of the glyoxylate cycle, a biochemical pathway that is unusual among animals. Once the an-imals have produced these compounds, they can be dehydrated to less than 2% water content and can be kept in this dehydrated state for years. Workers in this field originally believed that both these compounds might be involved in stabilizing dry cells, but other studies have shown that while trehalose appears to be present at high concentrations in all animals that sur-vive extreme dehydration, glycerol is not always present. Trehalose is also found at high concentra-tions in lower plants, such as yeast cells and the des-ert resurrection plant, that survive dehydration, but

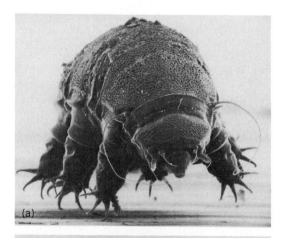

(a)

(b)

Fig. 1. Scanning electron micrographs of tardigrades: (a) active and (b) anhydrobiotic. (From J. H. Crowe and A. F. Cooper, Jr., Cryptobiosis, Sci. Amer., 225:30–36, 1971)

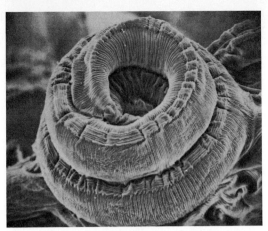

Fig. 2. Scanning electron micrograph of anhydrobiotic nematode. (*From J. H. Crowe and K. A. C. Maden, Anhydrobiosis in nematodes: Evaporative water loss and survival, J. Exp. Zool., 193:323–334, 1975*)

the analog of trehalose in higher plants appears to be sucrose. Thus, the data available from comparative biochemistry strongly suggest that the disaccharides trehalose and sucrose are involved in protecting dry organisms from damage due to dehydration.

In contrast with the above examples, cysts of crustaceans like the brine shrimp *Artemia* and seeds of plants seem to require no inductive period prior to anhydrobiosis. An *Artemia* cyst, for example, is an embryo covered with a capsule that is produced by the mother. The encapsulated embryo is released from the mother in a partially dehydrated state, and may afterward be rapidly reduced to low water content. Female *Artemia* also produce embryos without capsules that proceed with normal development without an intervening period of anhydrobiosis. In fact, these nonencapsulated embryos appear incapable of surviving dehydration. Encapsulated embryos contain large amounts of free glycerol and trehalose, similar to the amounts found in anhydrobiotic nematodes. Embryos produced without capsules contain insignificant amounts of glycerol and far less trehalose. Thus, induction of anhydrobiosis certainly occurs in *Artemia*, but the inductive mechanisms, which are poorly understood, probably act on the parent rather than directly on the embryo. Similar arguments can be made concerning the induction of anhydrobiosis in the seeds of higher plants and bacterial and fungal spores.

Biochemical adaptation and anhydrobiosis. Aside from gross mechanical stresses, dehydration results in a variety of other stresses that the chemical composition of anhydrobiotes may inhibit or obviate altogether. For example, it is well established that most macromolecules and cell membranes are maintained in a functional state by hydrophobic interactions. When water is removed, molecules and membranes lose their functionality; in proteins hydrophobic residues are exposed, and membranes undergo phase transitions that are likely to result in displacement of integral proteins from the membrane. A considerable body of evidence is accumulating which suggests that disaccharide compounds like trehalose may interact with macromolecules and membranes, stabilizing them at low water activities. For example, it has been possible to dry membranes in the presence of trehalose and to obtain membranes that show normal bio-

logical activity and structure upon rehydration. By contrast, when the same membranes are dried without trehalose, extensive structural damage occurs and all biological activity is lost. Similarly, proteins that are usually labile to drying can be dried in the presence of trehalose without loss of activity. SEE CELL MEMBRANES.

Another possible consequence of dehydration is that molecules, normally separated by bulk water in hydrated cells, would come into proximity, potentially resulting in chemical reactions between these molecules. Trehalose could serve as a substitute for bulk water, preventing such reactions. Yet another possible consequence of dehydration arises from the observation that oxygen is deleterious to anhydrobiotes. The organisms survive much longer in the dry state if oxygen is excluded.

It has been shown that damage to the dry organisms is probably due to direct oxidation of their chemical constituents, resulting in the formation of highly reactive free radicals; and that free glycerol inhibits such oxidation, which may explain the presence of glycerol in many dry but viable organisms. Finally, the reaction between reducing sugars and protein in the dry state, which is known as the browning reaction, could have serious consequences for dry tissues. This reaction involves the formation of a covalent bond between the protein and sugar, resulting in an insoluble product and denatured protein. It is possible that glycerol inhibits this reaction. It is not known whether sugars like trehalose show a browning reaction with dry proteins, since it is a nonreducing disaccharide of glucose. Thus, it would be an advantage to the anhydrobiote to store low-molecular-weight carbohydrates in this form.

Longevity of anhydrobiotes. It is not known just how long anhydrobiotes can survive. Some nematodes have been revived after being kept dry for 39 years. A museum specimen of moss that had been kept dry for over 120 years yielded a number of viable rotifers and tardigrades. Cysts of *Artemia* survive under appropriate conditions (under vacuum and at temperatures below 32°F or 0°C) for at least 10 years and probably considerably longer. Possibly the most celebrated (and most debated) case of extreme longevity of anhydrobiotes is that of the viable lotus seeds recovered from a dry peat bed in Manchuria by Professor I. Ohga in 1927. These seeds were initially reported to be about 400 years old, but radiocarbon dating suggested the seeds were no more than 150 years old. Other, similar reports have appeared, but the greatest authenticated longevity for a seed was obtained from the accidental hydration and subsequent germination of a seed in the British Museum in the course of an air raid in 1942. The seed had been mounted on a herbarium sheet in 1702. Thus, though there is no definitive data on the limit of longevity of anhydrobiotes, it seems likely to be impressive—on the order of centuries.

A factor that may be related to the longevity of anhydrobiotes is their phenomenal resistance to environmental extremes. In the 1920s researchers found that anhydrobiotic rotifers and tardigrades could be exposed to temperatures as high as 304°F (151°C) and as low as −328°F (−200°C), provided their water content was low. It has also been shown that anhydrobiotic rotifers and tardigrades survived exposure to a temperature at the brink of absolute zero (0.008 K). The reported LD_{50} (median lethal dose) for x-ray exposure of anhydrobiotic tardigrades is 570,000 roentgens; by contrast, the LD_{50} for humans is about 500 roentgens. Tardigrades and nematodes will survive exposure to high vacuum (less than 10^{-6} mmHg or 10^{-4}Pa). In fact, the tardigrade shown in Fig. 1b and the nematode in Fig. 2 were each kept at such a vacuum and were bombarded with electrons for 15 min or longer in a scanning electron microscope. When the animals were moistened after removal from the microscope they both recovered, although the tardigrade died shortly afterward. Similar observations have been made on cysts of *Artemia*, seeds, and spores.

The resistance of anhydrobiotes to environmental extremes has important economic consequences. As a result, considerable research has been devoted to killing anhydrobiotes if they are detrimental to humans and their various agricultural endeavors or to preserving the anhydrobiotes if they are beneficial. For example, many soil-dwelling nematodes that exhibit anhydrobiosis are important agricultural pests. In the anhydrobiotic state they are remarkably resistant to chemical control measures. *Artemia* cysts and seeds, on the other hand, are beneficial to humans, and the fact that they can be stored and transported dry is particularly convenient. *Artemia* cysts have become vital links in aquaculture systems. The cysts are hydrated and hatched, and the young brine shrimp are fed to fish, crustaceans, and other aquatic animals maintained in culture.

Other applications of anhydrobiosis. The discovery that biological membranes and proteins can be dried in the presence of trehalose without loss of activity has led to applications that may be important in human welfare. For instance, liposomes are artificial membranes enclosing an aqueous compartment. These structures are being used for delivery of water-soluble drugs to cells. The liposomes are injected intravenously and are transported through the circulatory system. They fuse with cell membranes, depositing the drug in the cytoplasm of the cell. Ultimately, it may be possible to target specific cells with liposomes. One difficulty with this application is that liposomes are not stable structures. However, it is now possible to dry the liposomes with trehalose and to store them in this state. When the liposomes are needed, water can be added and they are ready for immediate use. There are similar applications in other areas of the pharmaceutical industry, medicine, and agriculture.

Metabolism of anhydrobiotes. A central question in the study of anhydrobiosis has been whether metabolism actually ceases. Most investigations in this area have involved attempts at measuring the metabolic rate of the dry organism directly. Increasingly sophisticated measurements have been made, using, for example, oil-filled cartesian divers to record oxygen uptake, radiotracers to study metabolism of radiolabeled metabolic intermediates, and gas chromatography to determine emission of carbon dioxide from the anhydrobiote. In all these studies the rate of metabolism recorded is low or undetectable. However, such observations cannot be construed to imply that metabolism is lacking altogether. The argument can always be made that metabolism does proceed in the anhydrobiote but the techniques used are not sufficiently sensitive to measure it. The most convincing argument is the one based on the hydration properties of proteins in laboratory cultures: If all the water con-

tained by an anhydrobiotic organism were associated with protein, the protein would still exist in a semicrystalline state, and enzymes would not possess catalytic activity. Thus, enzyme-mediated metabolism must almost certainly cease at the low water contents that these organisms can possess. However, some chemical reactions do proceed in anhydrobiotes; mostly they are deleterious ones, such as the browning reaction. Such reactions could be construed as metabolism. But would it then follow that chemical reactions in rocks, air, and dead animals are metabolism, or even that a cosmic metabolism comprises the total chemical reactions of the universe.

Thus, the available evidence strongly suggests that dry anhydrobiotes are ametabolic. If that is the case, a philosophical question immediately arises concerning the nature of life. If metabolism is absent, the organism is generally referred to as dead. By this line of reasoning, anhydrobiotes would therefore be dead, returning to life when they are rehydrated. But it is also known that some anhydrobiotes die while in the dry state, in the sense that an increasing proportion of the population fails to resume activity upon rehydration. It follows, then, that they must have "died" while they were "dead."

This philosophical quandary can be avoided by applying the definition of life adopted by most students of anhydrobiosis: An organism is alive, provided its structural integrity is maintained. When that integrity is violated, it is dead. SEE METABOLISM; PLANT METABOLISM.

John H. Crowe; Lois M. Crowe

Bibliography. J. H. Crowe et al., *Biochim. Biophys. Acta Membranes Rev.*, 947:367–384, 1988; J. H. Crowe and J. S. Clegg, *Anhydrobiosis*, 1973; J. H. Crowe and J. S. Clegg, *Dry Biological Systems*, 1978; C. Leopold, *Membranes, Metabolism and Dry Organisms*, 1986; S. Weisburd, Death-defying dehydration, *Sci. News*, 133:107–110, February 13, 1988.

Cryptography

The various methods for writing in secret code or cipher. As society becomes increasingly dependent upon computers, the vast amounts of data communicated, processed, and stored within computer systems and networks often have to be protected, and cryptography is a means of achieving this protection. It is the only practical method for protecting information transmitted through accessible communication networks such as telephone lines, satellites, or microwave systems. Furthermore, in certain cases, it may be the most economical way to protect stored data. Cryptographic procedures can also be used for message authentication, personal identification, and digital signature verification for electronic funds transfer and credit card transactions. SEE COMPUTER SECURITY; DATA COMMUNICATIONS; DATABASE MANAGEMENT SYSTEMS; DIGITAL COMPUTER; ELECTRICAL COMMUNICATIONS.

Cryptographic algorithms. Cryptography must resist decoding or deciphering by unauthorized personnel; that is, messages (plaintext) transformed into cryptograms (codetext or ciphertext) have to be able to withstand intense cryptanalysis. Transformations can be done by using either code or cipher systems. Code systems rely on code books to transform the plaintext words, phrases, and sentences into codetext or code groups. To prevent cryptanalysis, there must be a great number of plaintext passages in the code book and the code group equivalents must be kept secret. To isolate users from each other, different codes must be used, making it difficult to utilize code books in electronic data-processing systems.

Cipher systems are more versatile. Messages are transformed through the use of two basic elements: a set of unchanging rules or steps called a cryptographic algorithm, and a set of variable cryptographic keys. The algorithm is composed of enciphering ($\mathbf{E}$) and deciphering ($\mathbf{D}$) procedures which usually are identical or simply consist of the same steps performed in reverse order, but which can be dissimilar. The keys, selected by the user, consist of a sequence of numbers or characters. An enciphering key (Ke) is used to encipher plaintext (X) into ciphertext (Y) as in Eq. (1),

$$\mathbf{E}_{Ke}(X) = Y \qquad (1)$$

and a deciphering key (Kd) is used to decipher ciphertext (Y) into plaintext (X) as in Eq. (2).

$$\mathbf{D}_{Kd}[\mathbf{E}_{Ke}(X)] = \mathbf{D}_{Kd}(Y) = X \qquad (2)$$

Algorithms are of two types—conventional and public-key (also referred to as symmetric and asymmetric). The enciphering and deciphering keys in a conventional algorithm either may be easily computed from each other or may be identical [Ke = Kd = K, denoting $\mathbf{E}_K(X) = Y$ for encipherment and $\mathbf{D}_K(Y) = X$ for decipherment]. In a public-key algorithm, one key (usually the enciphering key) is made public, and a different key (usually the deciphering key) is kept private. In such an approach it must not be possible to deduce the private key from the public key.

When an algorithm is made public, for example, as a published encryption standard, cryptographic security completely depends on protecting those cryptographic keys specified as secret. *SEE ALGORITHM*.

Unbreakable ciphers. Unbreakable ciphers are possible. But the key must be randomly selected and used only once, and its length must be equal to or greater than that of the plaintext to be enciphered. Therefore such long keys, called one-time tapes, are not practical in data-processing applications.

To work well, a key must be of fixed length, relatively short, and capable of being repeatedly used without compromising security. In theory, any algorithm that uses such a finite key can be analyzed; in practice, the effort and resources necessary to break the algorithm would be unjustified.

Strong algorithms. Fortunately, to achieve effective data security, construction of an unbreakable algorithm is not necessary. However, the work factor (a measure, under a given set of assumptions, of the requirements necessary for a specific analysis or attack against a cryptographic algorithm) required to break the algorithm must be sufficiently great. Included in the set of assumptions is the type of information expected to be available for cryptanalysis. For example, this could be ciphertext only; plaintext (not chosen) and corresponding ciphertext; chosen plaintext and corresponding ciphertext; or chosen ciphertext and corresponding recovered plaintext.

A strong cryptographic algorithm must satisfy the following conditions: (1) The algorithm's mathematical complexity prevents, for all practical purposes, solution through analytical methods. (2) The cost or time necessary to unravel the message or key is too great when mathematically less complicated methods are used, because either too many computational steps are involved (for example, in trying one key after another) or because too much storage space is required (for example, in an analysis requiring data accumulations such as dictionaries and statistical tables).

To be strong, the algorithm must satisfy the above conditions even when the analyst has the following advantages: (1) Relatively large amounts of plaintext (specified by the analyst, if so desired) and corresponding ciphertext are available. (2) Relatively large amounts of ciphertext (specified by the analyst, if so desired) and corresponding recovered plaintext are available. (3) All details of the algorithm are available to the analyst; that is, cryptographic strength cannot depend on the algorithm remaining secret. (4) Large high-speed computers are available for cryptanalysis.

In summary, even with an unlimited amount of computational power, data storage, and calendar time, the message or key in an unbreakable algorithm cannot be obtained through cryptanalysis. On the other hand, although a strong algorithm may be breakable in theory, in practice it is not.

Computational complexity. The strength of a cryptographic scheme can be measured by the computational complexity of the task of cryptanalysis. The term complexity, when referring to a program or algorithm to accomplish a given task, means the number of elementary operations used by this program. The complexity of a task is the least possible number of elementary operations used by any program to accomplish this task. This is directly related to sequential time, or the time used by a conventional sequential computer. Of course, the time used by a faster computer will be less than that used by a slower computer. Other measures of importance are storage and parallel time, or the time used on a highly parallelized computer.

Given a particular algorithm (computer program) for solving a problem, analysis of the algorithm can involve probability theory, detailed knowledge of the problem at hand, and other disciplines, but meaningful estimates of the resource consumption of the algorithm can usually be provided. This gives an upper bound to the complexity of the given problem.

However, nontrivial lower bounds are very hard to obtain. This is a fundamental problem in the design of cryptographic systems: it is very difficult to ensure that a system is sufficiently hard to crack. Without a good lower bound, the possibility that someone will find a fast algorithm for cryptanalyzing a given scheme must always be anticipated.

Problem classes P and NP. An important direction of theoretical work concerns the consideration (from computer science) of P versus NP. The class P consists of those problems which can be solved in polynomial time. That is, there are constants c and k such that, if the input to the problem can be specified in N bits, the problem can be solved on a sequential machine in time $c \times N^k$. Roughly speaking, these are the tractable problems. They include multiplication of two large numbers, exponentiation modulo a large prime, running the Data Encryption Standard (discussed below), and roughly any straightforward problem which does not involve searching.

The class NP (nondeterministic polynomial time) consists of problems which can be solved by searching. Roughly speaking, a possible solution to a problem in NP is to guess in turn each of 2^N possible values of some N-bit quantity, do some polynomial-time work related to each guess, and if some guess turns out to be correct, report the answer.

An example of a problem in NP is the knapsack problem: Given a set of integers $\{A_1, A_2, \ldots, A_n\}$ and a target integer B, can a subset of the A_i be selected without repetition (say $\{A_1, A_3, A_8\}$) such that their sum $(A_1 + A_3 + A_8)$ is the target B? One algorithm for solution is to try all possible subsets and just see whether any has the desired property. This algorithm requires exponential time, so called because the size of the input (n) occurs in the exponent of the formula expressing the running time (in this case, roughly 2^n). In fact, all known algorithms for this problem require exponential time. But there may be an unknown algorithm which runs in polynomial time; no proof prohibiting this is currently known.

Certainly, any problem in P is also in NP. A major outstanding question in computer science is whether P equals NP or whether NP is strictly larger.

There is a particular collection of problems in NP, including the knapsack problem, which are termed NP-complete. These can be thought of as the hardest problems in NP. More precisely, either P = NP or P ≠ NP. If P = NP, then any problem in NP is also in P. If P ≠ NP, then there are problems that cannot be done in polynomial time on a conventional sequential computer but can be done in polynomial time on a nondeterministic computer. An important mathematical result states: if there are any such problems in NP but not in P, then each NP-complete problem is also in NP but not in P. This class has particular significance for cryptography.

In a good cryptographic system, certain operations (like decryption) are easy for those in possession of the key and difficult for those without the key. (In some public-key applications, encryption should be easy and decryption should be difficult.) The legitimate user, in possession of the key, should be able to easily decrypt messages, and this task should be polynomial-time on a conventional sequential machine. The cryptanalyst who could first guess and verify the correct key would be able to decrypt easily (in polynomial time). This could be done by searching over the space of possible keys and attempting to verify each one in polynomial time. Since the problem can be solved by searching, decryption is in NP.

If P = NP, then decryption would also be in P, and a good cryptographic system would most likely be difficult to design. Even if no way was seen to easily decrypt without the key, P = NP would guarantee the existence of an algorithm whereby the cryptanalyst could "easily" decrypt any message. Of course, this is not a proof, merely an intuitive argument. In particular, "easy" and "polynomial-time" are not exactly the same thing.

If P ≠ NP, then the NP-complete problems might form a good starting point for cryptographic system design. They are in NP, so that a machine endowed with fortunate guesses or inside information (that is, the key) can easily solve the problem (decrypt). But they are not in P, so that machines without such inside information would require time larger than any fixed polynomial (that is, the cryptanalyst's job would be "hard").

Unfortunately, mathematicians and cryptographers have not yet learned how to transform an NP-complete problem into a secure cryptographic system. In one attempt to do so, R. Merkle and M. Hellman devised a public-key scheme for encryption based on the

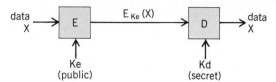

Fig. 1. Public-key cryptographic system used for privacy only.

knapsack problem. They showed how to choose a secret key K and generate from that key a set $\{A_i\}$ so that the legitimate user, knowing K, could easily solve a knapsack problem based on $\{A_i\}$, while the opponent, without K, would presumably have a difficult time. However, it turned out that the particular choice of $\{A_i\}$ prescribed in this scheme, chosen for ease of decoding by the legitimate user in possession of the key, rendered the scheme liable to attack by integer programming. Although some instances of knapsack problems are difficult, the special instances used by Merkle and Hellman were shown to be easy.

Even if an NP-complete problem is eventually transformed successfully into a cryptographic system, a proof of the difficulty of cryptanalysis of that system (or any other) can be expected to be taxing. Such a proof would probably also prove that P ≠ NP, and this conjecture has eluded computer scientists for several years. Thus, for now, the designers of cryptographic systems must rely on experience rather than rigorous proof of the difficulty of cryptanalysis.

Examples. Examples of the most efficient known attacks on several popular cryptographic systems are discussed below. The Data Encryption Standard (DES) apparently requires key exhaustion in order to break it. That is, no known method is faster than trying in turn each of the $2^{56} \simeq 10^{17}$ keys, a task which would strain the largest computing facilities for years to come. An attack has been devised that allows this to be done in 2^{56} computations once and for all, storing a table of $2^{40} \simeq 10^{12}$ words, and using this table with a relatively modest amount of computation (2^{40} steps) to break a given key, under the assumption of chosen plaintext or chosen ciphertext. This attack fails when chaining techniques are employed [such as cipher block chaining (CBC) and cipher feedback (CFB) modes of encryption], so that DES is still safe. The attack also fails against repeated encipherment with three independent keys. There is no guarantee against a more efficient, analytic attack, although none is known at present.

The RSA algorithm, also discussed below, is based on the difficulty of factoring large numbers. However, a family of algorithms has been developed for factoring such numbers. Fifty-digit numbers are routinely factored in a matter of hours, and sixty- or seventy-digit numbers will be handled in a matter of days or months, respectively, before long. If the modulus

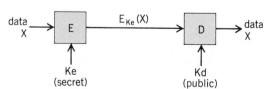

Fig. 2. Public-key cryptographic system used for message authentication only.

involved has, say, 200 digits, then RSA should be secure in the foreseeable future.

Another proposed number-theoretic cryptographic scheme is based on the difficulty of solving the discrete logarithm problem. This problem involves modular arithmetic. Two integers a and b are congruent for the modulus m if their difference $(a - b)$ is divisible by the integer m. This is expressed by the symbolic statement $a \equiv b \pmod{m}$, where mod is short for modulus. If P is a prime and E is an integer, E is a primitive element when $E^0 \bmod P$, $E^1 \bmod P$, $E^2 \bmod P, \ldots, E^{P-2} \bmod P$ are all different, and take on each nonzero value mod P exactly once. Equivalently, if m is the least positive integer such that $E^m = 1 \bmod P$, E is primitive when $m = P - 1$. SEE NUMBER THEORY.

If P is a prime and E a primitive element mod P, then given X it is straightforward to compute Y such that Eq. (3) is satisfied, while it is more difficult to

$$E^X \equiv Y \bmod P \tag{3}$$

recover X given Y, E, and P. But a modification of the factoring algorithm discussed above handles this discrete logarithm problem as well. This algorithm recovers X in a time given by expression (4) for some

$$e^{c\sqrt{(\log P \ \log \log P)}} \tag{4}$$

small constant c; this is eventually smaller than any fixed fractional power of P such as $P^{1/6}$. Thus, this scheme also requires a large modulus P for security.

In summary, an outstanding problem in the field of computational complexity is to devise a provably secure cryptographic system. A second, perhaps easier, problem is to devise a cryptographic system which is provably at least NP-hard.

Privacy and authentication. Anyone can encipher data in a public-key cryptographic system (**Fig. 1**) by using the public enciphering key, but only the authorized user can decipher the data through possession of the secret deciphering key. Since anyone can encipher data, message authentication is necessary in order to identify a message's sender.

A message authentication procedure can be devised (**Fig. 2**) by keeping the enciphering key secret and making the deciphering key public, provided that the enciphering key cannot be obtained from the deciphering key. This makes it impossible for nondesignated personnel to encipher messages, that is, to produce $\mathbf{E}_{Ke}(X)$. By inserting prearranged information in all messages, such as originator identification, recipient identification, and message sequence number, the messages can be checked to determine if they are genuine. However, because the contents of the messages are available to anyone having the public deciphering key, privacy cannot be attained.

A public-key algorithm provides privacy as well as authentication (**Fig. 3**) if encipherment followed by decipherment, and decipherment followed by encipherment, produce the original plaintext, as in Eq. (5). A message to be authenticated is first deciphered

$$\mathbf{D}_{Kd}[\mathbf{E}_{Ke}(X)] = \mathbf{E}_{Ke}[\mathbf{D}_{Kd}(X)] = X \tag{5}$$

by the sender (A) with a secret deciphering key (KAd). Privacy is ensured by enciphering the result with the receiver's (B's) public enciphering key (KBe).

Effective data security with public-key algorithms demands that the correct public key be used, since

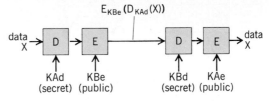

Fig. 3. Public-key cryptographic system used for both message authentication and privacy. KAe and KAd are enciphering and deciphering keys of the sender (A). KBe and KBd are enciphering and deciphering keys of the receiver (B).

otherwise the system is exposed to attack. For example, if A can be tricked into using C's instead of B's public key, C can decipher the secret communications sent from A to B and can transmit messages to A pretending to be B. Thus key secrecy and key integrity are two distinct and very important attributes of cryptographic keys. While the requirement for key secrecy is relaxed for one of the keys in a public-key algorithm, the requirement for key integrity is not.

In a conventional cryptographic system, data are effectively protected because only the sender and receiver of the message share a common secret key. Such a system automatically provides both privacy and authentication (**Fig. 4**).

Digital signatures. Digital signatures authenticate messages by ensuring that the sender cannot later disavow messages; the receiver cannot forge messages or signatures; and the receiver can prove to others that the contents of a message are genuine and that the message originated with that particular sender. The digital signature is a function of the message, a secret key or keys possessed by the sender and sometimes data that are nonsecret or that may become nonsecret as part of the procedure (such as a secret key that is later made public).

Digital signatures are more easily obtained with public-key then with conventional algorithms. When a message is enciphered with a private key (known only to the originator), anyone deciphering the message with the public key can identify the originator. The originator cannot later deny having sent the message. Receivers cannot forge messages and signatures, since they do not possess the private key.

Since enciphering and deciphering keys are identical in a conventional algorithm, digital signatures must be obtained in some other manner. One method is to use a set of keys to produce the signature. Some of the keys are made known to the receiver to permit signature verification, and the rest of the keys are retained by the originator in order to prevent forgery.

Data Encryption Standard. Regardless of the application, a cryptographic system must be based on a cryptographic algorithm of validated strength if it is

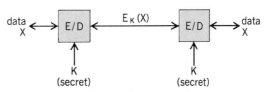

Fig. 4. Conventional cryptographic system in which message authentication and message privacy are provided simultaneously. K represents a common secret key.

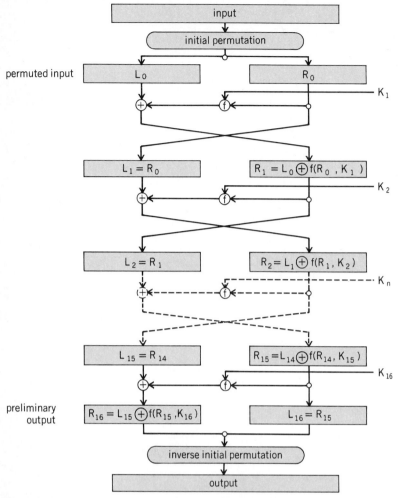

Fig. 5. Enciphering computation in the Data Encryption Standard. (*After Data Encryption Standard, FIPS Publ. 46, National Bureau of Standards, 1977***)**

Table 1. Modulo 2 addition		
A	B	A $\oplus$ B
0	0	0
0	1	1
1	0	1
1	1	0

The interaction of data, cryptographic key (K), and f is shown in **Fig. 5**. The externally supplied key consists of 64 bits (56 bits are used by the algorithm, and up to 8 bits may be used for parity checking). By shifting the original 56-bit key, a different subset of 48 key bits is selected for use in each round. These key bits are labeled K_1, K_2, . . ., K_{16}. To decipher, the keys are used in reverse order (K_{16} is used in round one, K_{15} in round two, and so on).

At each round (either encipherment or decipherment), the input is split into a left half (designated L) and a right half (designated R) [Fig. 5]. R is transformed with f, and the result is combined, using modulo 2 addition (also called the EXCLUSIVE OR operation; see **Table 1**), with L. This approach, as discussed below, ensures that encipher and decipher operations can be designed regardless of how f is defined.

Consider the steps that occur during one round of encipherment (**Fig. 6**). Let the input block (X) be denoted $X = (L_0, R_0)$, where L_0 and R_0 are the left and right halves of X, respectively. Function f transforms R_0 into $f_{K_1}(R_0)$ under control of cipher key K_1. L_0 is then added (modulo 2) to $f_{K_1}(R_0)$ to obtain R_1, as in Eq. (6). The round is then completed by setting L_1 equal to R_0.

$$L_0 \oplus f_{K_1}(R_0) = R_1 \qquad (6)$$

The above steps are reversible without introducing any new parameters or requiring that f be a one-to-one function. The ciphertext contains L_1, which equals R_0, and therefore half of the original plaintext is immediately recovered (**Fig. 7**). The remaining half, L_0, is recovered by recreating $f_{K_1}(R_0)$ from $R_0 = L_1$ and adding it (modulo 2) to R_1, as in Eq. (7). However, to use the procedure in Fig. 6 for en-

$$R_1 \oplus f_{K_1}(R_0) = [L_0 \oplus f_{K_1}(R_0)] \oplus f_{K_1}(R_0) = L_0 \qquad (7)$$

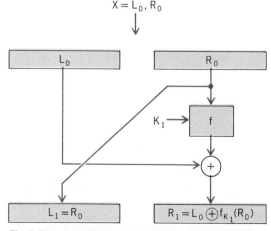

Fig. 6. Transformation of input block (L_0, R_0). (*After C. H. Meyer and S. M. Matyas, Cryptography: A New Dimension in Computer Data Security, John Wiley and Sons, 1980***)**

to be acceptable. The DES is such a validated conventional algorithm already in the public domain. (Since public-key algorithms are relatively recent, their strength has yet to be validated.)

During 1968–1975, IBM developed a cryptographic procedure that enciphers a 64-bit block of plaintext into a 64-bit block of ciphertext under the control of a 56-bit key. The National Bureau of Standards accepted this algorithm as a standard, and it became effective on July 15, 1977.

Conceptually, the DES can be thought of as a huge key-controlled substitution box (S-box) with a 64-bit input and output. With such an S-box, 2^{64} different transformations or functions from plaintext to ciphertext are possible. The 56-bit key used with DES thus limits the number of usable functions to 2^{56}.

A single huge S-box is impossible to construct. Therefore DES is implemented by using several smaller S-boxes (with a 6-bit input and a 4-bit output) and permuting their concatenated outputs. By repeating the substitution and permutation process several times, cryptographic strength "builds up." The DES encryption process consists of 16 iterations, called rounds. At each round a cipher function (f) is used with a 48-bit key. The function comprises the substitution and permutation. The 48-bit key, which is different for each round, is a subset of the bits in the externally supplied key.

 cipherment as well as decipherment, the left and right halves of the output must be interchanged; that is, the ciphertext (Y) is defined by Eq. (8). This modified

$$Y = [L_0 \oplus f_{K_1}(R_0)], R_0 \qquad (8)$$

procedure easily extends to n rounds, where the keys used for deciphering are $K_n, K_{n-1}, \ldots, K_1$.

RSA public-key algorithm. The RSA algorithm (named for the algorithm's inventors, R. L. Rivest, A. Shamir, and L. Adleman) is based on the fact that factoring large composite numbers into their prime factors involves an overwhelming amount of computation. (A prime number is an integer that is divisible only by 1 and itself. Otherwise, the number is said to be composite. Every composite number can be factored uniquely into prime factors. For example, the composite number 999,999 is factored by the primes 3, 7, 11, 13, and 37; that is, $999,999 = 3^3 \cdot 7 \cdot 11 \cdot 13 \cdot 37$).

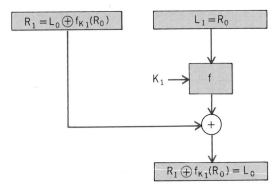

Fig. 7. Recovery of L_0. (*After C. H. Meyer and S. M. Matyas, Cryptography: A New Dimension in Computer Data Security, John Wiley and Sons, 1980*)

To describe the RSA algorithm, the following quantities are defined:

p and q are primes	(secret)
$r = p \cdot q$	(nonsecret)
$\phi(r) = (p - 1)(q - 1)$	(secret)
Kd is the private key	(secret)
Ke is the public key	(nonsecret)
X is the plaintext	(secret)
Y is the ciphertext	(nonsecret)

Based on an extension of Euler's theorem, Eq. (9),

$$X^{m\phi(r)+1} \equiv X \;(\text{mod } r) \qquad (9)$$

the algorithm's public and private keys (Ke and Kd) are chosen so that Eq. (10) or, equivalently, Eq. (11)

$$Kd \cdot Ke = m\phi(r) + 1 \qquad (10)$$

$$Kd \cdot Ke \equiv 1[\text{mod } \phi(r)] \qquad (11)$$

is satisfied. By selecting two secret prime numbers p and q, the user can calculate $r = p \cdot q$, which is made public, and $\phi(r) = (p - 1)(q - 1)$, which remains secret and is used to solve Eq. (11). (Tests are available to determine with a high level of confidence if a number is prime or not.) To obtain a

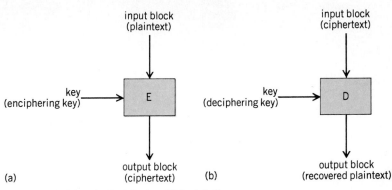

Fig. 8. Block cipher. (*a*) Enciphering. (*b*) Deciphering.

unique solution for the public key (Ke), a random number, or secret key (Kd), is selected that is relatively prime to $\phi(r)$. (Integers a and b are relatively prime if their greatest common divisor is 1.) Ke is the multiplicative inverse of Kd, modulo $\phi(r)$, and Ke can be calculated from Kd and $\phi(r)$ by using Euclid's algorithm. Equation (9) can therefore be rewritten as Eq. (12), which holds true for any plaintext (X).

$$X^{Kd \cdot Ke} \equiv X \;(\text{mod } r) \qquad (12)$$

Encipherment and decipherment can now be interpreted as in Eqs. (13) and (14). Moreover, because

$$\mathbf{E}_{Ke}(X) = Y \equiv X^{Ke} \;(\text{mod } r) \qquad (13)$$

$$\begin{aligned}\mathbf{D}_{Kd}(Y) &\equiv Y^{Kd} \;(\text{mod } r) \\ &\equiv X^{Ke \cdot Kd} \;(\text{mod } r) \equiv X \;(\text{mod } r)\end{aligned} \qquad (14)$$

multiplication is a commutative operation (Ke $\cdot$ Kd = Kd $\cdot$ Ke), encipherment followed by decipherment is the same as decipherment followed by encipherment [Eq. (5)]. Thus the RSA algorithm can be used for both privacy and digital signatures.

Finally, since $X^{Ke}\,(\text{mod } r) \equiv (X + mr)^{Ke}\,(\text{mod } r)$ for any integer m, $\mathbf{E}_{Ke}(X) = \mathbf{E}_{Ke}(X + mr)$. Thus the transformation from plaintext to ciphertext, which is many-to-one, is made one-to-one by restricting X to the set $\{0, 1, \ldots, r - 1\}$.

Block ciphers. A block cipher (**Fig. 8**) transforms a string of input bits of fixed length (termed an input

Table 2. Hexadecimal and binary notation	
Hexadecimal digit	Binary digits
0	0000
1	0001
2	0010
3	0011
4	0100
5	0101
6	0110
7	0111
8	1000
9	1001
A	1010
B	1011
C	1100
D	1101
E	1110
F	1111

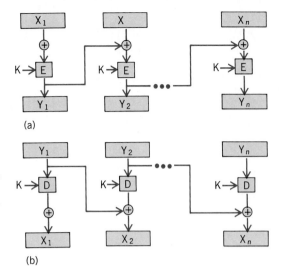

Fig. 9. Block chaining with ciphertext feedback. (a) Enciphering. (b) Deciphering. (After C. H. Meyer and S. M. Matyas, Cryptography: A New Dimension in Computer Data Security, John Wiley and Sons, 1980)

block) into a string of output bits of fixed length (termed an output block). In a strong block cipher, the enciphering and deciphering functions are such that every bit in the output block jointly depends on every bit in the input block and on every bit in the key. This property is termed intersymbol dependence.

The following example (using DES) illustrates the marked change produced in a recovered plaintext when only one bit is changed in the ciphertext or key. Hexadecimal notation (**Table 2**) is used. If the plaintext 1000000000000001 is enciphered with a (56-bit) key 30000000000000, then the ciphertext 958E6E627A05557B is produced. The original plaintext is recovered if 958E6E627A05557B is now deciphered with 30000000000000. However, if the leading 9 in the ciphertext is changed to 8 (a one-bit change) and the ciphertext 858E6E627A05557B is now deciphered with key 30000000000000, the recovered plaintext is 8D4893C2966CC211, not 1000000000000001. On the other hand, if the leading 3 in the key is changed to 1 (another one-bit change) and the ciphertext 958E6E627A05557B is now deciphered with key 10000000000000, the recovered plaintext is 6D4B945376725395. (The same effect is also observed during encipherment.)

In the most basic implementation of DES, called block encryption or electronic codebook mode (ECB), each 64-bit block of data is enciphered and deciphered separately. Every bit in a given output block depends on every bit in its respective input block and on every bit in the key, but on no other bits.

As a rule, block encryption is used to protect keys. A different method, called chained block encryption,

is used to protect data. In chaining, the process of enciphering and deciphering is made dependent on other (prior) data, plaintext, ciphertext, and the like, also available at the time enciphering and deciphering takes place. Thus every bit in a given output block depends not only on every bit in its respective input block and every bit in the key, but also on any or all prior data bits, either inputted to, or produced during, the enciphering or deciphering process.

Sometimes data to be enciphered contain patterns that extend beyond the cipher's block size. These patterns in the plaintext can then result in similar patterns in the ciphertext, which would indicate to a cryptanalyst something about the nature of the plaintext. Thus, chaining is useful because it significantly reduces the presence of repetitive patterns in the ciphertext. This is because two equal input blocks encipher into unequal output blocks.

A recommended technique for block chaining, referred to as cipher block chaining (CBC), uses a ciphertext feedback (**Fig. 9**). Let X_1, X_2, . . . , X_n denote blocks of plaintext to be chained using key K; let Y_0 be a nonsecret quantity defined as the initializing vector; and let Y_1, Y_2, . . . , Y_n denote the blocks of ciphertext produced. The ith block of ciphertext (Y_i) is produced by EXCLUSIVE ORing Y_{i-1} with X_i and enciphering the result with K, as in Eq. (15), where $\oplus$ denotes the EXCLUSIVE OR operation,

$$E_K(X_i \oplus Y_{i-1}) = Y_i \quad i \geq 1 \quad (15)$$

or modulo 2 addition. Since every bit in Y_i depends on every bit in X_1 through X_i, patterns in the plaintext are not reflected in the ciphertext.

The ith block of plaintext (X_i) is recovered by deciphering Y_i with K and EXCLUSIVE ORing the result with Y_{i-1}, as in Eq. (16). Since the recovered plain-

$$D_K(Y_i) \oplus Y_{i-1} = X_1 \quad i \geq 1 \quad (16)$$

text X_i depends only on Y_i and Y_{i-1}, an error occurring in ciphertext Y_j affects only two blocks of recovered plaintext (X_j and X_{j+1}).

Stream ciphers. A stream cipher (**Fig. 10**) employs a bit-stream generator to produce a stream of binary digits (0's and 1's) called a cryptographic bit stream, which is then combined either with plaintext (via the $\boxplus$ operator) to produce ciphertext or with ciphertext (via the $\boxplus^{-1}$ operator) to recover plaintext. (Traditionally, the term key stream has been used to denote the bit stream produced by the bit-stream generator. The term cryptographic bit stream is used here to avoid possible confusion with a fixed-length cryptographic key in cases where a cryptographic algorithm is used as the bit-stream generator.)

Historically, G. S. Vernam was the first to recognize the merit of a cipher in which ciphertext (Y) was produced from plaintext (X) by combining it with a secret bit stream (R) via a simple and efficient operation. In his cipher, Vernam used an EXCLUSIVE OR operation, or modulo 2 addition (Table 1), to combine the respective bit streams. Thus encipherment and decipherment are defined by $X \oplus R = Y$ and $Y \oplus R = X$, respectively. Therefore $\boxplus = \boxplus^{-1} = \oplus$. Modulo 2 addition is the combining operation used in most stream ciphers, and for this reason it is used in the following discussion.

If the bit-stream generator were truly random, an unbreakable cipher could be obtained by EXCLUSIVE ORing the plaintext and cryptographic bit stream. The cryptographic bit stream would be used directly as the

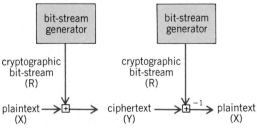

Fig. 10. Stream cipher concept. (After C. H. Meyer and S. M. Matyas, Cryptography: A New Dimension in Computer Data Security, John Wiley and Sons, 1980)

key, and would be equal in length to the message. But in that case the cryptographic bit stream must be provided in advance to the communicants via some independent and secure channel. This introduces insurmountable logistic problems for heavy data traffic. Hence, for practical reasons, the bit-stream generator must be implemented as an algorithmic procedure. Then both communicants can generate the same cryptographic bit stream—provided that their algorithms are identically initialized. **Figure 11** illustrates a cryptographic bit stream produced with a key-controlled algorithm.

When modulo 2 addition is used as the combining operation, each bit in the output ciphertext (recovered plaintext) is dependent only upon its corresponding bit in the input plaintext (ciphertext). This is in marked contrast to the block cipher which exhibits a much more complex relationship between bits in the plaintext (ciphertext) and bits in the ciphertext (recovered plaintext). Both approaches, however, have comparable strength.

In a stream cipher the algorithm may generate its bit stream on a bit-by-bit basis, or in blocks of bits. This is of no real consequence. All such systems are stream ciphers, or variations thereof. Moreover, since bit streams can be generated in blocks, it is always possible for a block cipher to be used to obtain a stream cipher. However, because both the sender and receiver must produce cryptographic bit streams that are equal and secret, their keys must also be equal and secret. Therefore public keys in confirmation with a public-key algorithm cannot be used in a stream-cipher mode of operation.

For security purposes, a stream cipher must never predictably start from the same initial condition, thereby producing the same cryptographic bit stream. This can be avoided by making the cryptographic bit stream dependent on a nonsecret quantity Z (known as seed, initializing vector, or fill), which is used as an input parameter to the ciphering algorithm (Fig. 11).

In a stream cipher, Z provides cryptographic strength and establishes synchronization between communicating cryptographic devices—it assures that the same cryptographic bit streams are generated for the sender and the receiver. Initialization may be accomplished by generating Z at the sending device and transmitting it in clear (plaintext) form to the receiver.

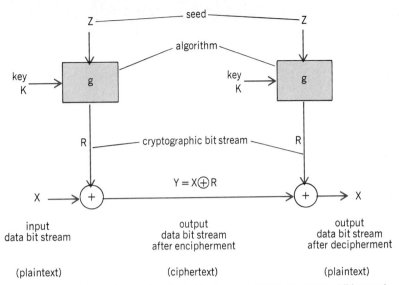

Fig. 11. Stream cipher using an algorithmic bit stream generator, modulo 2 addition, and seed. (*After C. H. Meyer and S. M. Matyas, Cryptography: A New Dimension in Computer Data Security, John Wiley and Sons, 1980*)

Cipher feedback. A general approach to producing cryptographic bit streams is the automatic modification of the algorithm's input using feedback methods. In a key auto-key cipher, the cryptographic bit stream generated at time $t = \tau$ is determined by the cryptographic bit stream generated at time $t < \tau$. In a ciphertext auto-key cipher the cryptographic bit stream generated at time $t = \tau$ is determined by the ciphertext generated at time $t < \tau$. A particular implementation of a ciphertext auto-key cipher, recommended by the National Bureau of Standards, is called cipher feedback (**Fig. 12**).

In cipher feedback, the leftmost n bits of the DES output are EXCLUSIVE ORed with n bits of plaintext to produce n bits of ciphertext, where n is the number of bits enciphered at one time ($1 \le n \le 64$). These n bits of ciphertext are fed back into the algorithm by first shifting the current DES input n bits to the left, and then appending the n bits of ciphertext to the right-hand side of the shifted input to thus produce a new DES input used to the next iteration of the algorithm.

A seed value, which must be the same for both

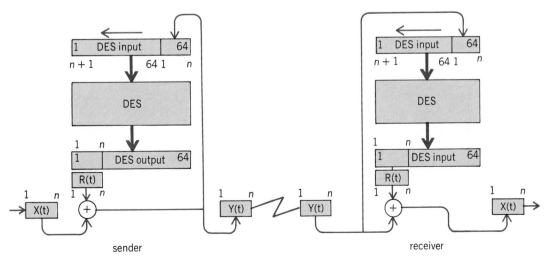

sender   receiver

Fig. 12. Cipher feedback. (*After C. H. Meyer and S. M. Matyas, Cryptography: A New Dimension in Computer Data Security, John Wiley and Sons, 1980*)

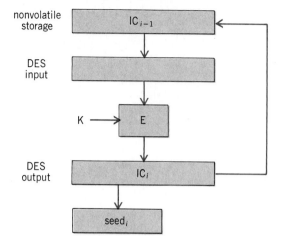

Fig. 13. Seed generation using the DES algorithm. (*After C. H. Meyer and S. M. Matyas, Cryptography: A New Dimension in Computer Data Security, John Wiley and Sons, 1980*)

sender and receiver, is used as an initial input to the DES in order to generate the cryptographic bit stream. Federal Standard 1026 defines a 48-bit seed for all cipher feedback implementations, thus ensuring compatibility among users. The communicating nodes are synchronized by right-justifying the seed in the input to the DES and setting the remaining bits equal to 0.

One method to generate seed values with the DES is illustrated in **Fig. 13**. IC_0 (for initial condition) is a starting value supplied by the user and is placed in nonvolatile storage (where data remain permanent). IC_1 is produced by enciphering IC_0 with the (stored) cryptographic key, IC_2 is produced by enciphering IC_1, and so forth. At each iteration, IC_i replaces IC_{i-1}, and $seed_i$ is the left-most m bits ($m \leq 64$) of IC_i.

The cipher feedback approach is self-synchronizing, since any bit changes occurring in the ciphertext during transmission get shifted out of the DES input after 64 additional ciphertext bits are sent and received. If, for example, 8 bits are enciphered at one time, as shown in Fig. 12, and a bit is altered in $Y(t_1)$ changing it to $Y^*(t_1)$, then the sender's and receiver's inputs are as shown in **Fig. 14**, where the 40-bit seed is defined as S1, S2, . . . , S5. In this case, the blocks of ciphertext, given by $Y^*(t_1)$, $Y(t_2)$, . . . , $Y(t_8)$, can be correctly deciphered only at the receiver by chance since the DES input is incorrect in each case. After eight blocks of uncorrupted ciphertext have been received, given by $Y(t_2)$, $Y(t_3)$, . . . , $Y(t_9)$, both the sender's and receiver's cryptographic devices will have equal DES inputs again.

In general, any bit changes in an n-bit block of ciphertext can cause a change in any of the corresponding n bits of recovered plaintext and in any of the 64 bits of recovered plaintext immediately following. However, a permanent "out-of-sync" condition will result if a ciphertext bit is added or dropped, since the integrity of the block boundary is lost. To recover from such an error, the sender and receiver would have to establish the beginning and ending of blocks of bits that are enciphered at one time ($n = 8$ bits in the given example).

On the other hand, if enciphering takes place on a bit-by-bit basis ($n = 1$), then the property of self-synchronization is maintained even when bits are lost or added. This is because blocks are bits, and therefore the block boundary cannot be disturbed. SEE IN-FORMATION THEORY.

Carl H. Meyer; Stephen M. Matyas; Don Coppersmith

Bibliography. H. J. Becker and F. C. Piper (eds.), *Cryptography and Coding*, 1989; *Data Encryption Standard*, FIPS Publ. 46, National Bureau of Standards, January 1977; W. Diffie and M. Hellman, New directions in cryptography, *IEEE Trans. Inform. Theory*, IT-22:644–654, November 1976; M. R. Garey and D. S. Johnson, *Computers and Intractability: A Guide to the Theory of NP-Completeness*, 1979; C. H. Meyer and S. M. Matyas, *Cryptography: A New Dimension in Computer Data Security*, 1982; R. L. Rivest, A. Shamir, and L. Adleman, Method for obtaining digital signatures and public-key cryptosystems, *Comm. ACM*, 21(2):120–126, February 1978; J. Seberry and J. Pieprzyk, *Cryptography: An Introduction to Computer Security*, 1989; C. E. Shannon, Communication theory of secrecy systems, *Bell Syst. Tech. J.*, 28:656–715, 1949.

Cryptomonadida

An order of the class Phytamastigophorea, also known as the Cryptomonadina. Cryptomonads are considered to be protozoa by zoologists and algae by botanists. Most species occur in fresh water. They are olive-green, blue, red, brown, or colorless. All flagellated members have two subequal flagella inserted

Fig. 14. Self-synchronizing feature in cipher feedback. (*After C. H. Meyer and S. M. Matyas, Cryptography: A New Dimension in Computer Data Security, John Wiley and Sons, 1980*)

iteration	DES input at sender								DES input at receiver							
0	0	0	0	S1	S2	S3	S4	S5	0	0	0	S1	S2	S3	S4	S5
1	0	0	S1	S2	S3	S4	S5	$Y(t_1)$	0	0	S1	S2	S3	S4	S5	$Y(t_1)$
2	0	S1	S2	S3	S4	S5	$Y(t_1)$	$Y(t_2)$	0	S1	S2	S3	S4	S5	$Y(t_1)$	$Y(t_2)$
3	S1	S2	S3	S4	S5	$Y(t_1)$	$Y(t_2)$	$Y(t_3)$	S1	S2	S3	S4	S5	$Y(t_1)$	$Y(t_2)$	$Y(t_3)$
4	S2	S3	S4	S5	$Y(t_1)$	$Y(t_2)$	$Y(t_3)$	$Y(t_4)$	S2	S3	S4	S5	$Y(t_1)$	$Y(t_2)$	$Y(t_3)$	$Y(t_4)$
5	S3	S4	S5	$Y(t_1)$	$Y(t_2)$	$Y(t_3)$	$Y(t_4)$	$Y(t_5)$	S3	S4	S5	$Y(t_1)$	$Y(t_2)$	$Y(t_3)$	$Y(t_4)$	$Y(t_5)$
6	S4	S5	$Y(t_1)$	$Y(t_2)$	$Y(t_3)$	$Y(t_4)$	$Y(t_5)$	$Y(t_6)$	S4	S5	$Y(t_1)$	$Y(t_2)$	$Y(t_3)$	$Y(t_4)$	$Y(t_5)$	$Y(t_6)$
7	S5	$Y(t_1)$	$Y(t_2)$	$Y(t_3)$	$Y(t_4)$	$Y(t_5)$	$Y(t_6)$	$Y(t_7)$	S5	$Y(t_1)$	$Y(t_2)$	$Y(t_3)$	$Y(t_4)$	$Y(t_5)$	$Y(t_6)$	$Y(t_7)$
8	$Y(t_1)$	$Y(t_2)$	$Y(t_3)$	$Y(t_4)$	$Y(t_5)$	$Y(t_6)$	$Y(t_7)$	$Y(t_8)$	$Y(t_1)$	$Y(t_2)$	$Y(t_3)$	$Y(t_4)$	$Y(t_5)$	$Y(t_6)$	$Y(t_7)$	$Y(t_8)$
9	$Y(t_2)$	$Y(t_3)$	$Y(t_4)$	$Y(t_5)$	$Y(t_6)$	$Y(t_7)$	$Y(t_8)$	$Y(t_9)$	$Y(t_2)$	$Y(t_3)$	$Y(t_4)$	$Y(t_5)$	$Y(t_6)$	$Y(t_7)$	$Y(t_8)$	$Y(t_9)$

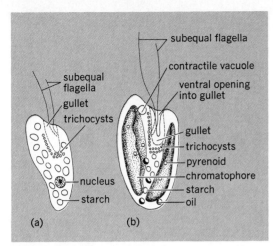

Examples of cryptomonads. (a) *Chilomonas paramecium.* (b) *Cryptomonas erosa.*

in a gullet opening through an obliquely truncate anterior end (see **illus.**). In *Nephroselmis* and *Protochrysis* the gullet is laterally displaced and a contractile vacuole opens into it. The nucleus is posterior. Oil and starch or a related compound are formed and pyrenoidlike bodies are present. *Cyathomonas* and *Chilomonas* are colorless; the latter has been widely used in nutritional studies, and extensive studies have been made of its cytology, demonstrating trichocysts around the gullet and muciferous bodies under the pellicle.

Phaeoplax is a palmelloid type with nonmotile cells embedded in a gelatinous matrix. The reproductive cells are flagellated. Radiolarians have symbiotic yellow cells, zooxanthellae, which are cryptomonads. Although few in genera and species, Cryptomonadida form extensive marine and fresh-water blooms. *See Cilia and flagella; Phytamastigophorea; Protozoa.*

James B. Lackey

Cryptophyceae

A small class of biflagellate unicellular algae (cryptomonads) in the chlorophyll *a–c* phyletic line (Chromophycota). In protozoological classification, these organisms constitute an order, Cryptomonadida, of the class Phytomastigophora. Cryptomonads are 4–80 micrometers long, ovoid or bean-shaped, and dorsoventrally flattened. The cell is bounded by a moderately flexible periplast comprising the plasmalemma and underlying rectangular or polygonal proteinaceous plates. A tubular invagination (gullet, groove, or furrow) traverses the ventral cytoplasm and opens just below the apex of the cell. A pair of subequal flagella, which are covered with hairs and small scales, arise from the center or apical end of the gullet.

Cryptomonads may be photosynthetic, osmotrophic, or phagotrophic. In photosynthetic species, chloroplasts occur singly or in pairs. The cryptomonad that is symbiotic in the ciliate *Mesodinium* is exceptional in having numerous chloroplasts. The photosynthetic lamellae, which are composed of loosely appressed pairs of thylakoids, are surrounded by four membranes. As in other members of the Chromophycota, the inner pair constitutes the chloroplast envelope while the outer pair represents endoplasmic reticulum confluent with the outer membrane of the nuclear envelope. The periplastidial compartment (space between the pairs of membranes on one side of the chloroplast) is more extensive than in other chromophytes. In other algae the compartment contains a dense network of tubules and vesicles, but in cryptomonads it contains starch grains, ribosomes, and a unique structure called a nucleomorph, in addition to scattered tubules and vesicles. The nucleomorph, which is smaller than the chloroplast, is bounded by a double membrane with numerous pores. It contains fibrillogranular inclusions. One nucleomorph, which always divides before the cell enters mitotic prophase, is closely associated with each chloroplast. If it can be shown that the nucleomorph is a vestigial nucleus, as it seems to be, the hypothesis that cryptomonad chloroplasts are evolutionarily derived from an endosymbiont will be supported.

Photosynthetic pigments include chlorophyll *a* and *c*, α-carotene and β-carotene (the former being predominant, an unusual ratio for algae), and red and blue phycobiliproteins closely related to those in red algae but occurring in the intrathylakoidal space rather than forming phycobilisomes. The color of the chloroplasts, and thus the color of the cryptomonad, depends upon pigment composition and may be green, olive, brown, yellow, red, or blue. Pigment composition is affected by environmental conditions. Thus, red cryptomonads at a low nitrate level gradually and reversibly bleach to green. Pyrenoids, usually capped with starch, are present in the chloroplasts of most species while eyespots are present in a few.

A single wormlike or reticulate mitochondrion, differing from those of other chromophytes in having flattened rather than tubular or inflated cristae, is present in each cell. Most cryptomonads have two sizes of ejectile organelles (trichocysts or ejectosomes)—those large enough to be seen with the compound microscope form rows along the gullet, and those much smaller are situated at corners of the periplast plates. Discharge of trichocysts imparts a characteristic darting movement to the cell. Contractile vacuoles may be present in fresh-water species.

Reproduction is by longitudinal binary fission. Sexual reproduction has not been confirmed.

Cryptomonads are found in fresh, brackish, marine, and hypersaline bodies of water, sometimes in great abundance. Several species are endosymbionts of marine ciliates and invertebrates, and of fresh-water dinoflagellates. They do not appear to be closely related to any other group of algae and thus are sometimes assigned to their own division. In addition to typical cryptomonads (order Cryptomonadales), of which about 200 species have been described, there are two monospecific genera with a life history that includes only a brief motile phase. In these genera, which constitute the order Tetragonidiales, the cells usually are nonflagellate and sedentary and are said to have a cellulose wall. They form zoospores, however, which are typically cryptomonad in structure. *See Algae; Chromophycota; Cryptomonadida.*

Paul C. Silva; Richard L. Moe

Bibliography. R. W. Butcher, *An Introductory Account of the Smaller Algae of British Coastal Waters*, pt. 4: *Cryptophyceae*, 1967; R. E. Norris, Cryptophyceae, in S. P. Parker (ed.), *Synopsis and Classification of Living Organisms*, 2 vols., 1982.

Cryptostomata

An extinct order of bryozoans in the class Stenolaemata. Cryptostome colonies have delicate, flattened branches of bifoliate construction in which the short, tubular, solid-walled, moderately complex zooecia are tightly packed back to back and open out onto the two opposite sides of the branches. *See* Bryozoa; Stenolaemata.

Morphology. Mostly small, erect, and twiglike or frondlike, cryptostome branches are distinctly regionated, showing an abrupt transition between endozone and exozone, and with the endozonal portion of each zooecium quite short relative to the portion lying in the exozone. Some colonies bear mesopores and a few acanthrods and monticules or maculae; but ovicells, stolons, nanozooecia, vibracula, avicularia, and coelomic chambers are lacking.

Individual cryptostome zooecia are markedly curved or bent; adjacent ones are fused together. Zooecial walls, thin in the endozone but thick in the exozone, have both laminated and finely granular portions. In many species the zooecia contain one or two hemisepta at the endozone-exozone transition; a few forms also posseses diaphragms. Round in outline and lacking an operculum, the zooecial aperture in many species is surrounded by a raised rim (peristome).

History and classification. Possibly evolving from early cystoporates or from an ancestor they shared with them, cryptostomes first appeared in the sea at the start of the Middle Ordovician. They became moderately common immediately, and remained so, until their extinction during latest Permian time.

Bifoliate cryptostomes are also termed ptilodictyoids or timanodictyoids. The order Cryptostomata was much more broadly defined in the past, and included the Fenestrata and Rhabdomesonata (the latter, also twiglike, were occasionally grouped with the bifoliates as the Habrovirgatina). Moreover, although formerly suggested as possibly ancestral to cheilostomes, the cryptostomes may have died out without modern descendants. *See* Cheilostomata; Cystoporata; Fenestrata; Rhabdomesonata.

Roger J. Cuffey

Cryptozoon

A type of stromatolite; a sedimentary organic structure formed by blue-green algae, sedimentary particles, and fine biochemical calcareous material, resulting from the activity of the algae. The organisms develop rounded masses shaped like a cabbage or flat-

Fig. 2. Weathered slab of Lower Ordovician limestone containing concentrically banded *Cryptozoon*, Saratoga Springs, New York. (*N.Y. State Museum and Science Service*)

tened cabbage which commonly range in size from 6 in. (15 cm) to more than 2 ft (0.6 m) across. Structurally they are composed of thin laminae, but they show little or no microstructure (**Fig. 1**).

Cryptozoon colonies developed in such numbers as to be important builders of limestone (**Fig. 2**). Probably the best-known occurrence is at Saratoga Springs, New York. However, they are found in all continents in rocks ranging in age from Precambrian to Permian. They were especially abundant during the late Proterozoic, Cambrian, and Early Ordovician. *See* Algae; Stromatolite.

J. Harlan Johnson

Bibliography. W. Goldring, *Algal Barrier Reefs in the Lower Ozarkian of New York*, N.Y. State Mus. Bull. 315, 1938; J. H. Johnson, *Limestone-Building Algae and Algal Limestones*, Colo. Sch. Mines Spec. Publ., 1961; J. H. Johnson, *Review of the Ordovician Algae*, 1961; B. W. Logan, R. Rezak, and R. N. Ginsburg, Classification and environmental significance of algal stromatolites, *J. Geol.*, 72(1):68–83, 1964; C. R. Stauffer, Cryptozoons of the Shakopee Dolomite, *Paleontology*, 19(4):376–379, 1945.

Fig. 1. *Cryptozoon* structure in the Lower Magnesian dolostone (Ordovician) of western Wisconsin. Oolites are common in the depressions between adjacent masses. Successive layers are characteristically convex upward. Two-thirds natural size. (*After R. R. Shrock, Sequence in Layered Rocks, McGraw-Hill, 1948*)

Crystal

This term, as used in science and technology, usually denotes a single crystal. A single crystal is a solid throughout which the atoms or molecules are arranged in a regularly repeating pattern. In electronics the term crystal is usually restricted to mean a single crystal which is piezoelectric. Single crystals include most gems, piezoelectric quartz crystals used in controlling the frequencies of radio transmitters, and single crystals of galena (lead sulfide) used in crystal radios. *See* Single crystal.

Most crystalline solids are made up of millions of tiny single crystals called grains and are said to be polycrystalline. These grains are oriented randomly with respect to each other. Any single crystal, however, no matter how large, is a single grain. Single crystals of metals many cubic centimeters in volume are relatively easy to prepare in the laboratory.

Single crystals differ from polycrystalline and amorphous substances in that their properties are anisotropic. Young's modulus, for example, is different for different directions in the crystal. Anisotropy is responsible for the fact that crystals will cleave (split) along very flat planes which are characteristic of the atomic stacking pattern. SEE CRYSTAL DEFECTS; CRYSTAL GROWTH; CRYSTAL OPTICS; CRYSTAL STRUCTURE; CRYSTALLOGRAPHY.

Herman H. Hobbs

Crystal absorption spectra

The wavelength or energy dependence of the attenuation of electromagnetic radiation as it passes through a crystal, due to its conversion to other forms of energy in the crystal. When atoms are grouped into an ordered array to form a crystal, their interaction with electromagnetic radiation is greatly modified. Free atoms absorb electromagnetic radiation by transitions between a few electronic states of well-defined energies, leading to absorption spectra consisting of sharp lines. In a crystal, these states are broadened into bands, and the cores of the atoms are constrained to vibrate about equilibrium positions. The ability of electromagnetic radiation to transfer energy to bands and ionic vibrations leads to broad absorption spectra that bear little resemblance to those of the free parent atoms. SEE ABSORPTION OF ELECTROMAGNETIC RADIATION; ATOMIC STRUCTURE AND SPECTRA; BAND THEORY OF SOLIDS; LATTICE VIBRATIONS.

The absorption spectrum of a crystal includes a number of distinct features. These are shown in the **illustration** for a semiconducting crystal, gallium arsenide. Absorption spectra of insulating crystals are similar except that electronic absorption is shifted to higher energy. Intrinsic absorption and extrinsic absorption differ according to whether energy would be extracted from electromagnetic radiation even if the crystal were perfect, or if defects or impurities are required. Intrinsic absorption includes lattice absorption in the infrared, electronic absorption by transitions from bonding to antibonding states (valence and conduction bands) in the visible and ultraviolet, and electronic absorption by transitions from core levels in the x-ray region. Extrinsic absorption is much weaker and is generally observable only in the region of transparency of semiconductors and insulators. Free-carrier absorption may be either intrinsic or extrinsic: it is intrinsic to metal and semimetal crystals with their partially filled bands, and extrinsic to semiconductors and insulators with their normally filled or empty bands. Free-carrier processes dominate the absorption spectra of metals from the lowest frequencies throughout the ultraviolet, with lattice and band-to-band absorption processes being much less pronounced. SEE ELECTRIC INSULATOR; METAL; SEMICONDUCTOR.

Properties. Absorption is measured in terms of the absorption coefficient, which is defined as the relative rate by which electromagnetic radiation is attenuated per unit length. The absorption coefficient is typically expressed in units of cm^{-1}. Absorption coefficients of solids range from less than $2 \times 10^{-7} \; cm^{-1}$ in the exceedingly highly refined glasses used in fiber-optic communications systems to higher than $10^6 \; cm^{-1}$ for band-to-band electronic transitions in semiconductors and insulators. Absorption is related to reflection through a more fundamental quantity called the dielectric function. Absorption is also related to the index of refraction by a relationship called the Kramers-Kronig transformation, which can be evaluated if the spectral dependence of the absorption coefficient is known over a sufficiently wide frequency range. SEE DISPERSION RELATIONS; REFLECTION OF ELECTROMAGNETIC RADIATION.

Because energy is extracted from electromagnetic radiation by work (force × distance) performed on the crystal, considerable insight into absorption processes can be obtained from simple mechanical considerations. Although force is just equal to the electric field times the charge, the distance that the charge moves depends inversely on mass and frequency and on the bonding configuration. The mass dependence suggests that absorption associated with the heavy ions of the lattice is much weaker than that associated with the light electrons that hold the lattice together, as can immediately be verified from the illustration. The bonding dependence suggests that absorption associated with the free electrons of a metal such as chromium is different from that associated with the tightly bound electrons of an insulator such as glass, as is immediately verified from observation.

Lattice absorption. Lattice absorption arises from the excitation of phonons, which are collective vibrations of the atomic cores about their equilibrium lattice positions. Lattice absorption occurs in the infrared and is responsible for the intrinsic structure in the absorption spectra of crystals below about 0.1 eV (wavelengths above about 10 micrometers). In a diatomic, partially ionic crystal such as gallium arsenide, an electric field that forces the gallium cores one way simultaneously forces the arsenic cores the other way, leading to the generation of transverse optic phonons. The process becomes very efficient at the natural mechanical resonance of the lattice at $8.8 \times 10^{12} \; s^{-1}$ (energy of 36 meV, wavelength of 34 μm), and causes the strong reststrahl absorption peak in the spectrum of illus. *b*. Since the lattice-restoring forces are anharmonic, it is also possible for electromagnetic radiation to create several phonons simultaneously at higher multiples of the reststrahl frequency. These multiphonon processes are responsible for the structure shown from 10 to 30 μm in **illus.** *b*. SEE IONIC CRYSTALS; MOLECULAR STRUCTURE AND SPECTRA; PHONON.

Bonding-to-antibonding transitions. Electronic transitions between bonding (filled valence) bands and antibonding (empty conduction) bands yield the crystal analog of the line absorption spectra of free atoms. These intrinsic absorption processes dominate the optical behavior of semiconductors and insulators in the visible and ultraviolet regions of the electromagnetic spectrum, as binding energies of most materials are of the order of magnitude of electronvolts per atom. Although analogous transitions also occur in metals, these are generally overwhelmed by free-carrier effects with the spectacular exceptions of copper, silver, and gold, where free-carrier and band-to-band absorption processes are of comparable importance.

The energy of an electron within a band is not constant but is a function of the velocity or momentum of the electron relative to the crystal lattice. This dependence on momentum causes optical structure within a bonding-to-antibonding absorption band. The most readily apparent such feature is the fundamental

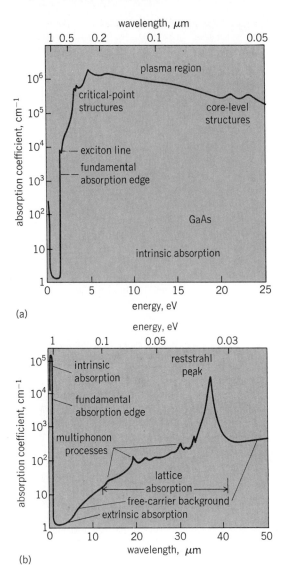

wavelength, μm

energy, eV

(a)

energy, eV

wavelength, μm

(b)

Absorption spectrum of GaAs. The region in which each type of absorption process is important is indicated. (a) Spectrum at higher energies. (b) Spectrum at lower energies.

about 100–1000 times less probable than direct transitions. As absorption coefficients for indirect transitions are correspondingly reduced relative to those for direct transitions, it is not surprising that indirect absorption is typically seen only in semiconducting crystals such as germanium, silicon, gallium phosphide, and aluminum arsenide where the highest valence and lowest conduction band states occur at different momenta. In other semiconducting crystals, for example, indium antimonide, gallium arsenide, and zinc oxide, the highest valence and lowest conduction-band states occur at the same momentum and the fundamental absorption edge is direct.

The detailed dependence of absorption on frequency in the vicinity of the fundamental absorption edge depends not only on the variation of electron energy with momentum but also on many-body effects, that is, interactions between the electron excited into the conduction band and the vacancy, or hole, left behind. Because electrons and holes are charges of opposite sign, there is a strong tendency for an electron and hole to bind together in a state like a hydrogen atom. This state is called an exciton. As a result of the positive binding energy of the exciton, the energy needed to create it is less (by the binding energy) than that of the forbidden gap itself. Consequently, electromagnetic radiation is absorbed at energies less than what would be expected in the absence of electron-hole interactions. The lowest possible exciton absorption process produces the single absorption line seen at the fundamental edges of most semiconductors and insulators at low enough temperatures, and is analogous to the lowest absorption line of a hydrogen atom. Excitonic effects also substantially enhance optical absorption at energies immediately above the fundamental edge. *See* Exciton.

Direct thresholds analogous to the fundamental absorption edge also occur at higher energies, and cause additional structure in bonding-to-antibonding absorption spectra. These features are generally described in terms of critical points. In three-dimensional energy bands, four types of critical points occur. Local or absolute minima, including the fundamental absorption edge itself and other thresholds where transitions between new band pairs become possible, are of type M_0. Saddle points where the interband energy reaches local minima in two dimensions and a local maximum in the third are of type M_1. Saddle points where the interband energy reaches a local minimum in one dimension and maxima in the other two are of type M_2. Points where the interband energy reaches a local maximum in all three dimensions are of type M_3.

Critical point structure begins to die out at energies greater than 10 eV, as the crystal potential in which the excited electrons propagate becomes small compared to their kinetic energy. This is the plasma region shown in illus. *a*. Although small groups of absorption features typically appear at higher energies (at 20 eV in illus. *a*), these high-energy structures no longer involve bonding states but are associated with the electronic levels of the atomic cores. Being highly localized, core electrons are virtually unaffected by the crystal and (apart from minor shifts in electrostatic potential) can be viewed as being the same as in the free parent atoms. The core-level features of gallium arsenide seen near 20 eV in illus. *a* originate from the gallium 3*d* core electrons; the 3*d* core electrons of arsenic give rise to similar features (not shown) near 45 eV. Deeper-lying core levels give corresponding

absorption edge of semiconductors and insulators, which marks the boundary between the range of transparency at lower energies and the strong electronic absorption that occurs at higher energies. The energy of the fundamental absorption edge is determined by the forbidden gap, which is the energy difference between the highest filled valence and lowest empty conduction level of the crystal. The forbidden gap may be nonexistent, as in metals such as iron, lead, and white tin, and in semimetals such as arsenic, antimony, and bismuth; zero, as in the semiconductors gray tin and one of the mercury-cadmium-telluride alloys; in the infrared, as in the semiconductors germanium, silicon, gallium arsenide, and lead sulfide; in the visible or near-ultraviolet, as in the semiconductors gallium phosphide, cadmium sulfide, and gallium nitride; or in the ultraviolet, as in insulators such as diamond, strontium titanate, sodium chloride, silicon dioxide, and lithium fluoride.

Bonding-to-antibonding absorption processes are of two types, direct or indirect, according to whether the initial and final electronic states have the same momentum or whether the electron must absorb or emit a phonon to conserve momentum in the process. The need to involve phonons makes indirect transitions absorption features at higher energies. Core-level features in both crystalline and free-atom absorption spectra tend to be broad owing to the short lifetimes of core holes created by the absorption process.

Extrinsic absorption. Extrinsic absorption processes involve states associated with deviations from crystal perfection, such as vacancies, interstitials, and deliberately or inadvertently introduced impurities. Since the concentration of imperfections is typically low relative to the concentration of host atoms (10^{12}–10^{19} cm^{-3} compared to 10^{23} cm^{-3}), extrinsic absorption is weak relative to intrinsic absorption. Consequently, as with indirect absorption, extrinsic absorption is typically seen only in spectral regions where the crystal is otherwise transparent. However, as electromagnetic radiation can interact with many absorbing centers in passing through macroscopic lengths of ordinarily transparent material, extrinsic absorption is an important phenomenon. Extrinsic processes are responsible for the colors of most gemstones, for example, the red of ruby and blue of sapphire, and also for the poor optical quality of industrial diamonds. Extrinsic processes are also vitally important in luminescence, which depends almost entirely on the types and concentrations of impurities in a crystal. *SEE LUMINESCENCE.*

Free-carrier absorption. Unbound charges such as conduction electrons and valence holes can interact with electromagnetic radiation at all frequencies. However, the absorption of electromagnetic radiation requires that the motion induced by the electromagnetic field be coupled to dissipative mechanisms involving charge-charge or charge-lattice collisions. Thus, although all common metals have large densities of free carriers (of the order of magnitude of 10^{23} cm^{-3}), it is the existence of highly efficient loss mechanisms in transition metals such as iron, nickel, chromium, platinum, and palladium that gives them their characteristic neutral gray appearance. In contrast, the loss mechanisms in aluminum and in the noble metals copper, silver, and gold are much less efficient, leading among other effects to substantially higher reflectances for these materials.

Free-carrier effects are typically described phenomenologically by the simple Drude model with electron density and scattering lifetime as the only two parameters. At frequencies below the reciprocal of the scattering lifetime, absorption is roughly independent of frequency and proportional to the direct-current conductance. Above this value, absorption drops roughly as the square of the frequency. The intrinsic absorption properties of electrons in transition metals and the extrinsic absorption properties of free carriers in semiconductors can both be described approximately by the Drude model. However, the noble metals copper and gold cannot be so described, as a result of the strong interband thresholds in the visible that also give these metals their characteristic colors. Free-electron absorption in high-temperature superconductors may be affected similarly by a strongly frequency-dependent loss mechanism, although the detailed nature of this mechanism is still uncertain. *SEE FREE-ELECTRON THEORY OF METALS; SUPERCONDUCTIVITY.*

David E. Aspnes

Bibliography. J. N. Hodgson, *Optical Absorption and Dispersion in Solids*, 1970; T. S. Moss et al., *Semiconductor Optoelectronics*, 1973; J. I. Pankove, *Optical Processes in Semiconductors*, 2d ed., 1976; R. K. Willardson and A. C. Beer (eds.), *Semiconductors and Semimetals*, vol. 3: *Optical Properties of III–V Compounds*, 1967; J. M. Ziman, *Principles of the Theory of Solids*, 2d ed., 1972.

Crystal counter

A device, more correctly described as a crystal detector, that detects ionizing radiations of all types and is adaptable to measuring neutrons. The sensitive element is a single crystal with a dc resistance normally higher than 10^{12} ohms. The crystals are small and are cut or grown to volumes ranging from less than 1 mm^3 (6×10^{-5} in.^{3}) to approximately 200 mm^3 (1.2×10^{-2} in.^{3}).

Crystal detectors fall into two categories: Certain crystals act as thermoluminescent detectors, of which lithium fluoride (LiF), lithium borate ($Li_2B_4O_7$), and calcium sulfate ($CaSO_4$) are among the best known. Other crystals, for example, cadmium telluride (CdTe) and mercury iodide (HgI_2), act as ionization chambers that deliver either pulses or a dc signal, depending upon the associated electronic circuitry. *SEE IONIZATION CHAMBER; THERMOLUMINESCENCE.*

Diamond deserves special mention. It is a unique crystal that functions as a thermoluminescent detector or, if suitable contacts are made, as an ionization chamber. The efficiency of the diamond detector in the thermoluminescent or ionization chamber mode is strongly dependent on the impurity atoms included within the crystal lattice, with nitrogen and boron playing dominant roles. Not all diamonds are good detectors; only the rare and expensive natural types IB or IIA are appropriate. Besides being stable and nontoxic, diamond has an additional attractive feature as a detector. As an allotrope of carbon, it has the atomic number $Z = 6$. Human soft tissue has an effective $Z = 7.4$, so that diamond is a close tissue-equivalent material, an essential characteristic for biological dosimetry, for example, in measurements in living organisms. Because of diamond's desirable properties as a detector, economical techniques for the synthesis of diamonds with the correct types and concentrations of impurity atoms are being actively sought for both thermoluminescent and ionization chamber modes. *SEE DIAMOND.*

Good crystal detectors are insulators and therefore have significant band-gap energies. A large band gap impedes the spontaneous excitation of charge carriers between the valence and conduction bands, thus lowering leakage currents and movement of charge carriers to trapping centers. Room temperature devices are consequently possible. Band gaps for some typical crystal detectors are listed in the **table.** *SEE BAND THEORY OF SOLIDS; ELECTRIC INSULATOR; TRAPS IN SOLIDS.*

Thermoluminescence. Thermoluminescence can be explained in terms of the band theory of solids. It

Band gaps of crystal detectors		
Crystal	Band gap, eV	Mode
Diamond	5.4	Thermoluminescence and conduction
LiF	11.0	Thermoluminescence
CdTe	1.45	Conduction
HgI$_2$	2.15	Conduction

is to be understood within the framework of the distinctions between fluorescence, which is luminosity continuing only as long as the radiation is continued, and is essentially temperature-independent, and phosphorescence, which is observable luminosity even after the removal of the radiation source. Phosphorescence has an intensity decay time which is highly sensitive to the crystal temperature. This situation can arise when the atom or molecule is excited by ionizing radiation—for example, from its ground state E_0 to a metastable state E_m, from which the immediate return to the ground state is prohibited by selection rules (**Fig. 1a**). If, however, energy $E_c - E_m$ can be supplied to the system, then it can be further excited to a state E_c from which it can return to the ground state directly with the emission of a photon. *See FLU-ORESCENCE; LUMINESCENCE; PHOSPHORESCENCE.*

Crystals always contain defects in the lattice. Some, such as impurity atoms, can be introduced intentionally or unintentionally during the growing phase. Others are distortions of the lattice. These de-

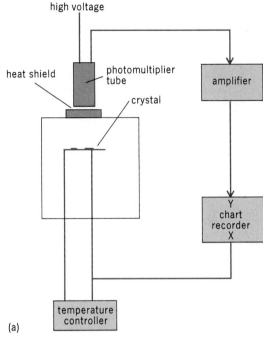

(a)

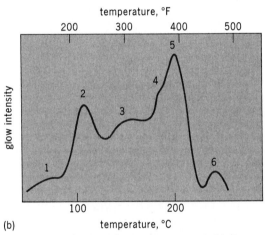

(b)

Fig. 2. Thermoluminescence: (a) measurement; (b) glow curve.

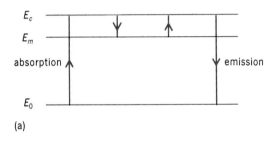

(a)

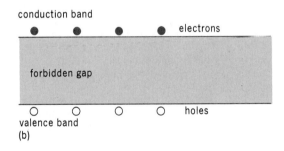

(b)

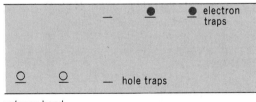

(c)

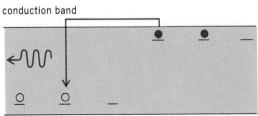

(d)

Fig. 1. Crystal band theory of luminescence: (a) energy levels; (b) conduction and valence bands; (c) hole and electron traps; (d) photon emission.

fects perturb the periodicity of the lattice and constitute localized energy levels within the band gap between the valence and conduction bands. After irradiation and excitation of the crystal, the carriers moving through their respective bands (electrons through the conduction band and holes through the valence band; Fig. 1b) can become trapped at these defect lattice sites (Fig. 1c). Upon heating of the crystal, the trapped electrons (holes) may gain sufficient energy to escape from the trapping centers to the conduction (valence) band, from which they may make a direct transition back to the valence (conduction) band, become retrapped, or combine with a trapped hole (electron). The last alternative, called recombination, can lead to energy release, one form of which is the emission of a photon (Fig. 1d). The trapped carriers are known as recombination centers if the emission of photons dominates the energy release of the recombination. The degree of heating required to empty the traps is a function of their depth, and the rate at which they are emptied is a function

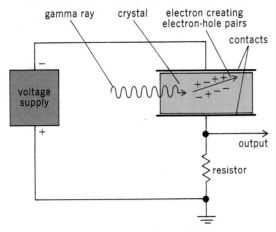

Fig. 3. Crystal detector in conduction mode.

of the temperature. SEE CRYSTAL DEFECTS; ELECTRON-HOLE RECOMBINATION; HOLE STATES IN SOLIDS.

The crystal is heated at a controlled rate on a metal tray by means of an electric current. The photon emission from the crystal is monitored by a photomultiplier the output of which is directed toward an appropriate recording device (**Fig. 2a**). The result is a "glow curve" (Fig. 2b), the area of which correlates with the number of traps depopulated. These traps were, of course, the traps which were populated during irradiation, their number being directly related to the radiation-field intensity. The integrated light output therefore becomes a direct measure of the total radiation dose. The peaks 1–6 (Fig. 2b) are indicative of the various trapping levels within the crystal. SEE DOSIMETER; PHOTOMULTIPLIER.

Conduction mode. An essential feature of the crystal operating as a detector in the conduction mode is the forming of the contacts. The contacts must be ohmic and should not form a Schottky barrier, which would severely impede the charge signal from the crystal. A polarizing voltage is applied, usually to opposite faces of the crystal, and a field of 10^4–10^5 V/cm is used to collect the charges. (**Fig. 3**).

A charged particle entering the crystal transfers its kinetic energy to the bulk of the crystal by creating charge carriers (electron-hole pairs). A photon of sufficient energy interacts with the crystal atoms, losing all or part of its energy through the photoelectric effect, the Compton effect, or pair production. In each of these processes, electrons are either liberated or created, and they in turn have their energy dissipated in the bulk of the crystal by creating charge carriers. SEE COMPTON EFFECT; ELECTRON-POSITRON PAIR PRODUCTION; GAMMA-RAY DETECTORS; PHOTOEMISSION.

When the carrier pair is created, the individual carriers move under the influence of the electric field toward the oppositely charged contacts. On arriving at the contacts, the charges can be measured at the output point either as a dc or as a pulse signal, depending upon the circuitry. It is, however, necessary for full efficiency of the counting system that both types of carriers are collected equally. This makes the property of ion mobility within the crystal very important, and both types should have equal or nearly equal mobility. A condition called hole trapping arises if this criterion is not satisfied. It leads to poor resolution of the overall detecting system, which is also a consequence of high leakage currents or electronic noise within the circuit elements. Because of the high resis-

tance of the crystal, the length of the leads and the screening of all connections must be carefully controlled to reduce the inherent input capacitance of the front end of the system. Resolution can also be improved by cooling the crystal.

Sensitivity and resolution. In the thermoluminescent mode the crystals measure the total dose of the applied radiation, whereas in the conduction mode they measure the instantaneous dose rate; in both cases it is ultimately the crystal itself that limits the sensitivity and resolution of the system. Present methods of synthesis for crystals permit the detection of radiation fields down to nearly background values (0.1 microgray/h or 10^{-5} rad/h) even with crystals as small as 1 mm^3 (6×10^{-5} in.^{3}). Small crystals make detectors possible that are capable of very high spatial resolution. This feature is important in electron radiation therapy, for example, where the isodose curves lie very close together. SEE PARTICLE DETECTOR; RADIATION BIOLOGY; RADIOLOGY.

R. J. Keddy

Bibliography. R. K. Bull, Nuclear tracks radiation measurement, *Int. J. Rad. Appl. Instrum.* part D, 11 (1/2):105–113, 1986; R. J. Keddy, T. L. Nam, and R. C. Burns, The detection of ionizing radiations by natural and synthetic diamond crystals and their application as dosimeters in biological environments, *Carbon*, 26(3):345–356, 1988; R. C. Whited and M. M. Schieber, Detectors in nuclear science, *Nucl. Instrum. Meth.*, 162:113–123, 1979.

Crystal defects

Departures of a crystalline solid from a regular array of atoms or ions. A "perfect" crystal of sodium chloride (NaCl), for example, would consist of alternating Na$^+$ and Cl$^-$ ions on an infinite three-dimensional simple cubic lattice, and a simple defect (a vacancy) would be a missing Na$^+$ or Cl$^-$ ion. There are many other kinds of possible defects, ranging from simple and microscopic, such as the vacancy and other structures illustrated in **Fig. 1**, to complex and macroscopic, such as the inclusion of another material, or a surface.

Natural crystals always contain defects, often in abundance, due to the uncontrolled conditions under which they were formed. The presence of defects which affect the color can make these crystals valuable as gems, as in ruby [chromium replacing a small

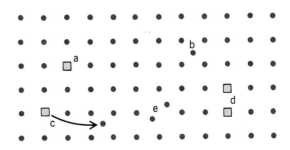

key:
a = vacancy (Schottky defect)
b = interstitial
c = vacancy—interstitial pair (Frenkel defect)
d = divacancy
e = split interstitial
▢ = vacant site

Fig. 1. Some simple defects in a lattice.

fraction of the aluminum in aluminum oxide (Al_2O_3)]. Crystals prepared in the laboratory will also always contain defects, although considerable control may be exercised over their type, concentration, and distribution.

The importance of defects depends upon the material, type of defect, and properties which are being considered. Some properties, such as density and elastic constants, are proportional to the concentration of defects, and so a small defect concentration will have a very small effect on these. Other properties, such as the color of an insulating crystal or the conductivity of a semiconductor crystal, may be much more sensitive to the presence of small numbers of defects. Indeed, while the term defect carries with it the connotation of undesirable qualities, defects are responsible for many of the important properties of materials, and much of solid-state physics and chemistry and materials science involves the study and engineering of defects so that solids will have desired properties. A defect-free silicon crystal would be of little use in modern electronics; the use of silicon in devices is dependent upon small concentrations of chemical impurities such as phosphorus and arsenic which give it desired electronic properties.

Surfaces and interfaces. The importance of surfaces depends very much upon the properties which are being considered and upon the geometry of the specimen. In a specimen no more than a few hundred atoms thick, a significant fraction of the atoms is close to the surface, while in a sample 0.4 in. (10^{-2} m) thick or greater, a much smaller fraction of the atoms is close to the surface. In the latter case, it makes sense to consider bulk properties which are surface-independent and shape-independent. Even in this case, surface effects may be important, as in metals which reflect visible light from their surfaces, but they may generally be separated from bulk effects.

Surfaces are of great importance in determining the chemical properties of solids, since they are in contact with the environment. Such properties include chemical reactions involving the surface, such as corrosion; they also include the role of solids as catalysts in chemical reaction. It is often found that surface atoms reconstruct; that is, they displace and form bonds different from those existing in the bulk. Experimental studies of surfaces are difficult, because of the high vacuum which must be maintained and the relatively small number of atoms which are involved. SEE CATALYSIS; CORROSION; SURFACE PHYSICS.

An interface is the boundary between two dissimilar solids; it is in a sense an interior surface. For example, a simple metal-oxide-semiconductor (MOS) electronic device has two interfaces, one between metal and semiconductor, and the other between semiconductor and oxide. (In this case the oxide is not crystalline.) Interfaces tend to be chemically and electrically active and to have large internal strains associated with structural mismatch.

Chemical impurities. An important class of crystal defect is the chemical impurity. The simplest case is the substitutional impurity, for example, a zinc atom in place of a copper atom in metallic copper. Impurities may also be interstitial; that is, they may be located where atoms or ions normally do not exist.

In metals, impurities usually lead to an increase in the electrical resistivity. Electrons which would travel as unscattered waves in the perfect crystal are scattered by impurities. Thus zinc or phosphorus impurities in copper decrease the conductivity. Impurities can play an important role in determining the mechanical properties of solids. As discussed below, an impurity atom can interact with an extended structural defect called a dislocation to increase the strength of a solid; hydrogen, on the other hand, can lead to brittle fracture of some metals. SEE ELECTRICAL CONDUCTIVITY OF METALS; ELECTRICAL RESISTIVITY.

Impurities in semiconductors are responsible for the important electrical properties which lead to their widespread use. The electronic states of all solids fall into energy bands, quasicontinuous in energy. In nonmetals there is a band gap, a region of energy in which no states exist, between filled (valence) and empty (conduction) bands. Because electrons in filled bands do not conduct electricity, perfect nonmetals act as insulators. Impurities (and other defects) often introduce energy levels within the forbidden gap (**Fig. 2**). If these are close in energy to the valence or conduction band, electrons may be thermally excited from a nearby filled level (donor) into the conduction band, or from the valence band into a nearby empty level (acceptor), in the latter case leaving an unoccupied state or hole in the valence band. In both cases, conduction will occur in an applied electric field. Semiconductor devices depend on the deliberate addition (doping) of appropriate impurities. Other impurities may have energy levels which are not close in energy to either valence or conduction band (so-called deep traps). These may be undesirable if they trap the conducting particles. Thus careful purification is an important part of semiconductor technology. SEE ACCEPTOR ATOM; BAND THEORY OF SOLIDS; DONOR ATOM; HOLE STATES IN SOLIDS; SEMICONDUCTOR; TRAPS IN SOLIDS; ZONE REFINING.

The energy levels associated with impurities and other defects in nonmetals may also lead to optical absorption in interesting regions of the spectrum. Ruby and other gems are examples of this; in addition, the phenomenon of light emission or luminescence is often impurity-related. Solid-state lasers generally involve impurity ions. SEE CRYSTAL ABSORPTION SPECTRA; LASER; LUMINESCENCE.

Structural defects. Even in a chemically pure crystal, structural defects will occur. Thermal equilibrium exists when the Gibbs free energy G is given by Eq. (1), plus a pressure-times-volume term which can of-

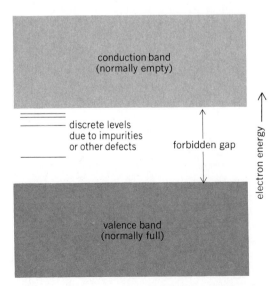

Fig. 2. Band structure of an imperfect nonmetallic crystal, including defect states which occur in the normally forbidden energy gap.

conduction band
(normally empty)

discrete levels due to impurities or other defects

forbidden gap

electron energy ⟶

valence band
(normally full)

$$G = E - TS \qquad (1)$$

ten be neglected. Here E is the energy, T is the absolute temperature, and S is the entropy. A perfect crystal has a lower energy E than one containing structural defects, but the presence of defects increases the entropy S. Thus at $T = 0$ K, a perfect crystal is stable, but at higher values of T, the quantity G has its lowest value when structural defects exist. SEE FREE ENERGY.

Crystals may be grown by solidifying the molten substance. If a crystal is cooled slowly (annealed), the number of defects will decrease. However, as the temperature decreases it becomes more difficult for atoms to move in such a way as to "heal" the defects, and in practice there are often more defects than would be expected in thermal equilibrium.

In thermal equilibrium the number of defects at temperatures of interest may be nonnegligible. Application of the methods of statistical mechanics yields Eq. (2). Here n is the number of defects, N is the

$$\frac{n}{N} = Ae^{-BE/kT} \qquad (2)$$

total number of possible atomic sites, E is the energy required to create a defect, and k is Boltzmann's constant. The quantities A and B are dimensionless constants, generally of order 1, whose actual value depends on the type of defect being considered. For example, for vacancies in monatomic crystals such as silicon or copper, $A = B = 1$. E is typically of order 1 eV; since $k = 8.62 \times 10^{-5}$ eV/K, at $T = 1000$ K (1340°F), $n/N = \exp\left[-1/(8.62 \times 10^{-5} \times 1000)\right] \simeq 10^{-5}$, or 10 parts per million. For many purposes this fraction would be intolerably large, although as mentioned this number may be reduced by slowly cooling the sample. SEE STATISTICAL MECHANICS.

Simple defects. Besides the vacancy, other types of structural defects exist (Fig. 1). The atom which left a normal site to create a vacancy may end up in an interstitial position, a location not normally occupied. Or it may form a bond with a normal atom in such a way that neither atom is on the normal site but the two are symmetrically displaced from it. This is called a split interstitial. The name Frenkel defect is given to a vacancy-interstitial pair, whereas an isolated vacancy is a Schottky defect. (In the latter case the missing atom may be thought of as sitting at a normal site at the surface.)

Simple defects are often combined. For example, when a small amount of alkaline-earth halide is melted with an alkali halide and the resulting mixture is recrystallized, the mixed solid is found to have a large number of alkali-ion vacancies, approximately one for each alkaline-earth ion. Since the alkaline-earth ions are divalent, the alkali-ion vacancies act as "charge compensators" to make the crystal neutral. Such a crystal has a relatively high electrical conductivity associated with the movement of positive ions into the vacancies (or, alternatively, movement of the vacancies through the crystal). This general phenomenon of ionic conductivity is of practical interest, for example, in attempts to develop solid-state batteries. SEE IONIC CRYSTALS; SOLID-STATE BATTERIES.

Simple defects tend to aggregate if thermally or optically stimulated. The F-center in an alkali halide is an electron trapped at a halogen-ion vacancy. Under suitable conditions of optical or thermal stimulation, it is found that F-centers come together to form F_2-

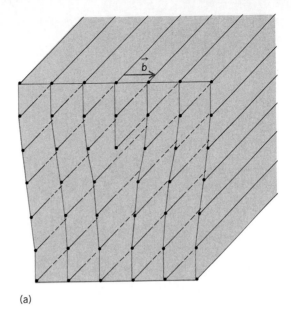

(a)

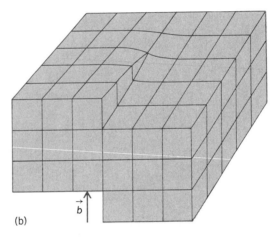

(b)

Fig. 3. Extended structural defects. (a) Edge dislocation. (b) Screw dislocation. In each case the Burgers vector is denoted by $\vec{b}$.

or M-centers (two nearest F-centers), F_3- or R-centers (three nearest F-centers), and so on. SEE COLOR CENTERS.

Extended defects and mechanical properties. The simplest extended structural defect is the dislocation. An edge dislocation is a line defect which may be thought of as the result of adding or subtracting a half-plane of atoms (**Fig. 3a**). A screw dislocation is a line defect which can be thought of as the result of cutting partway through the crystal and displacing it parallel to the edge of the cut (Fig. 3b). The lattice displacement associated with a dislocation is called the Burgers vector $\vec{b}$.

Dislocations generally have some edge and some screw character. Since dislocations cannot terminate inside a crystal, they either intersect the surface or an internal interface (called a grain boundary) or form a closed loop within the crystal.

Dislocations are of great importance in determining the mechanical properties of crystals. A dislocation-free crystal is resistant to shear, because atoms must be displaced over high-potential-energy barriers from one equilibrium position to another. It takes relatively little energy to move a dislocation (and thereby shear the crystal), because the atoms at the dislocation are barely in stable equilibrium. Such plastic deformation is known as slip. SEE PLASTIC DEFORMATION OF METAL.

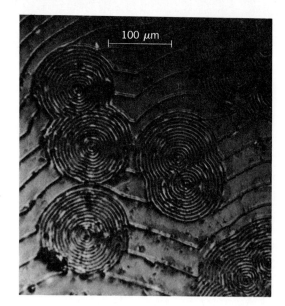

Fig. 4. Crystal-growth spirals on the surface of a silicon carbide crystal. Each spiral step originates in a screw dislocation defect. From the side the surface looks like an ascending ramp wound around a flat cone. (*General Electric Co.*)

of iron carbide, whose major function is dislocation blocking.

A large, randomly oriented collection of dislocations can tangle and interfere with their respective motions. This is the principle behind work-hardening.

Dislocations have other important properties. The growth of crystals from the vapor phase tends to occur where screw dislocations intersect the surface, for in these regions there are edges rather than planes, and the attractive forces for atoms will be greater. **Figure 4** shows spiral growth patterns of silicon carbide (SiC) derived from a screw dislocation. SEE CRYSTAL GROWTH.

A common planar structural defect is the stacking fault. This can be thought of as the result of slicing a crystal in half, twisting the two pieces with respect to each other, and then reattaching them. Another planar defect, the grain boundary, is the boundary between two crystallites. In some cases it can be described in terms of a parallel set of edge dislocations. SEE GRAIN BOUNDARIES; TWINNING (CRYSTALLOGRAPHY).

Alloys. A solid containing more than one element is called an alloy, especially when the solid is metallic. There are a vast number of possible alloys, and the present discussion will treat two-component or binary systems as examples.

The term stoichiometry is used to describe the chemical composition of compounds and the resulting crystal structure. The alkali halides are particularly simple, forming compounds with well-defined compositions (for example, NaCl). The situation becomes more complicated for compounds in which one or more components can take on more than one valence state. In particular, in the transition-metal oxides different compositions can occur, with different crystal structures and defects. For example, titanium dioxide (TiO_2) represents a well-defined perfect crystal, whereas TiO_{2-x} would have an oxygen deficiency whose value is described by the subscript x. In this case the deficit oxygens are associated with neither oxygen vacancies nor titanium interstitials, but rather with defect aggregates called shear planes.

If two elements are mixed uniformly throughout the solid, the atomic arrangement may be described as ordered or disordered. For example, an alloy composed of equal numbers of two metal atoms may form an ordered lattice with a regular periodic arrangement of the two atoms with respect to each other, or a disordered lattice in which some of the atoms are on the "wrong" site (**Fig. 5**). The ordered arrangement is most stable at very low temperatures. The temperature at which the structure becomes disordered is called the transition temperature of the order-disorder transformation. SEE ALLOY STRUCTURES.

Defect chemistry. Most of the preceding discussion has treated the simpler types of defects and their static properties, that is, their properties under conditions such that the nature and location of the defects do not change with time. Situations have also been alluded to in which the number, locations, or types of defects change with time, either because of thermal effects (for example, slowly cooling a hot sample and annealing vacancies) or because of external irradiation or defect interactions (causing, for example, the aggregation of alkali-halide F-centers).

For both scientific and practical reasons, much of the research on crystal defects is directed toward the dynamic properties of defects under particular conditions, or defect chemistry. Much of the motivation for

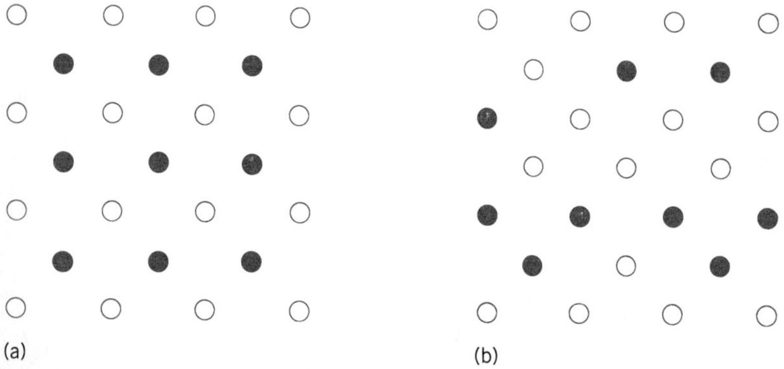

To extend the earlier argument about the difficulty in preparing perfect crystals, it should be anticipated that most crystals will contain dislocations in ample numbers and that special care would have to be taken to prepare a dislocation-free crystal. The latter is in most cases impractical (although in fact some electronic materials are dislocation-free). Although practical metallurgy developed centuries before dislocations were known to exist, most strengthening processes are methods for dealing with dislocations.

Dislocations will not be mechanically deleterious if they cannot move. Three major ways of hindering dislocation motion are pinning by dissolved impurities, blocking by inclusions, and tangling. Each of these is described briefly below.

Impurities tend to collect near dislocations if the solid is sufficiently hot that the impurities can move. The presence of these impurities along a dislocation removes the energetic equivalence of the dislocation in different positions; now it requires energy to move the dislocations unless the cloud of dissolved impurities can move as well; and to move them will involve overcoming a large energy barrier. This approach's success is lessened at higher temperatures, when the impurities tend to migrate away.

Small particles of a second substance can "block" dislocation motion. Steel is iron with small inclusions

(a)   (b)

Fig. 5. Binary alloy. (a) Ordered lattice. (b) Disordered lattice.

this arises from the often undesirable effects of external influences on material properties, and a desire to minimize these effects. Examples of defect chemistry abound, including one as familiar as the photographic process, in which incident photons cause defect modifications in silver halides or other materials. Properties of materials in nuclear reactors is another important case. A few examples will be considered here.

Thermal effects. Thermal effects are the easiest to understand from a microscopic point of view: raising the temperature increases the amplitude of atomic vibrations, thereby making it easier for atoms to "hop" from place to place over potential-energy barriers. Almost all dynamic processes are temperature-dependent, occurring more readily at high temperatures. *SEE DIFFUSION; LATTICE VIBRATIONS.*

Defect creation by irradiation. Radiation can have profound effects on materials in both direct and indirect ways. For example, a beam of high-energy electrons, neutrons, or atomic nuclei (such as alpha particles) can create defects by simple collision processes: the incident particle imparts a portion of its momentum and energy to a lattice atom, thereby releasing the atom from its normal postion. If the released atom has enough energy, it may collide with and release a second lattice atom, thereby creating a second vacancy, and so on; thus one incident particle may lead to the production of a number of defects. Other processes also exist. *SEE CHARGED-PARTICLE BEAMS; RADIATION DAMAGE TO MATERIALS.*

A beam of photons may also create defects through a variety of processes. If the photon energy equals or exceeds the band gap, ionization can occur; that is, electrons can be excited from core or valence bands into the conduction bands as the photons are absorbed. Very energetic photons (gamma rays) may be absorbed or scattered, generating fast electrons, which in turn eject atoms through collision processes such as those occurring with incident particles. *SEE GAMMA RAYS.*

Lower-energy photons (x-rays, ultraviolet, visible, and so on) have insufficient energy to create defects by such processes. In some solids simple ionization results in defect creation; for example, in alkali halides excitation of an electron on a halogen ion leads after atomic rearrangements to the formation of an excited halogen molecule-ion which is unstable with respect to a translation along its axis. This leads to the creation of a halogen ion vacancy plus an electron (*F*-center) next to a molecule-ion which is crowded into a normal halogen ion position (*H*-center). A replacement sequence can occur (**Fig. 6**) in which each halogen displaces its neighbor in the indicated direction. Eventually this replacement sequence ends, and the result is a pair of isolated *F*- and *H*-centers. The first stages of this process can occur in very short times (on the order of 10^{-12} s).

Laser irradiation is often observed to result in crystal destruction, even when the crystal is an insulator and the photon energy is well below the band gap. One way this can occur is first by ionization of defects by the laser, producing a few conduction electrons. These conduction electrons can in turn be accelerated, by the large electric field present in the laser beam, to sufficiently high energy to ionize normal atoms; thus more electrons are available to be accelerated, and a type of cascade process occurs which leads to destruction of the crystal. *SEE LASER-SOLID INTERACTIONS.*

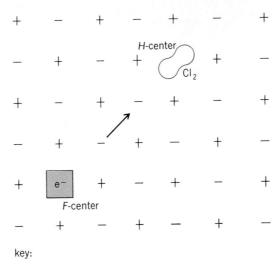

key:

+ = K⁺

− = Cl⁻

Fig. 6. *F*- and *H*-center formation in KCl after a replacement sequence of halogen-halogen collisions along the direction indicated by the arrow.

Effects of radiation on preexisting defects. More subtle effects must often be considered when studying the effects of radiation on device performance. This involves the interaction of radiation with existing (and desirable) defects. Aggregation of *F*-centers occurs when alkali halides are irradiated in the *F*-absorption band. The process involves several steps: (1) The excited *F*-center loses its electron to the conduction band, leaving behind a positively charged vacancy. (2) The conduction electron is subsequently trapped at another *F*-center, forming a (negative) *F'*-center. (3) Under the attractive Coulomb force, the vacancy moves toward the *F*-center. (4) The extra electron of the *F'*-center returns to the vacancy either by ionization or by tunneling, leading to two *F*-centers closer than normal. (5) The process repeats, until the two *F*-centers are adjacent.

Externally activated defect processes are also found in semiconductors, typically (but not solely) associated with irradiation. Charge-state processes involve the motion of preexisting defects following a change in the charge of the defect. For example, band-gap light generates electrons and holes in silicon, which can be trapped, for example, at a vacancy V in silicon. The doubly negative vacancy (V^{2-}) has a considerably smaller activation energy for motion than less negatively charged vacancies, and will consequently migrate more readily. Presumably differences in locations of the neighboring atoms are responsible for the different activation energies of motion.

Recombination processes may also occur. Here an electron and hole recombine at a defect and transfer their (electronic) energy to vibrational energy of the defect, leading to its motion. This effect has been observed, for example, at a recombination junction in gallium arsenide (GaAs), where defects introduced by proton bombardment were destroyed under charge injection. *SEE ELECTRON-HOLE RECOMBINATION.*

A third possibility is excited-state effects. In this case a defect in an excited electronic state and its surroundings relax into a configuration in which defect motion is easier; in some instances, there may be a zero barrier energy to overcome. One of the best examples of this is the type II *F_A*-center in alkali halide crystals, an *F*-center with an alkali impurity (usually Li) as one of its six nearest alkali neighbors (**Fig. 7**).

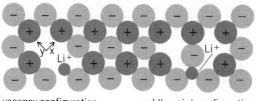

vacancy configuration   saddle-point configuration

(a)

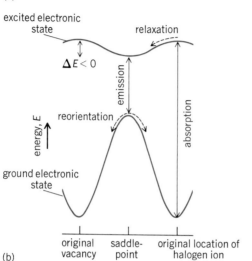

(b)

Fig. 7. Reorientation of the type II F_A-center in alkali halides. (a) Vacancy and saddle-point configurations. (b) Schematic potential energy functions for motion of the neighboring (halogen)$^-$ ion. (After W. Beall Fowler, ed., *Physics of Color Centers*, Academic Press, 1968)

After a photon is absorbed, the system relaxes into a "double-well" configuration with a Cl$^-$ which was initially adjacent to both the Li$^+$ and the vacancy moving into a "saddle-point" position half way to the vacancy. The excited electron is then shared by the two small equivalent "wells." When the system returns to the ground state, there is equal probability of the Cl$^-$ returning to its original site or to the original vacancy; in the latter case the vacancy will have moved. Similar processes are thought to occur in semiconductors.

Bistability and metastability. Many semiconductor defects are found to exhibit bistability or metastability. An example is the defect formed in silicon from an interstitial carbon–substitutional carbon pair. This defect has three stable charge states: it can be neutral, it can trap an electron (−), or it can give up an electron (+). Furthermore, it is found to have two distinctly different atomic configurations (whose structures have been proposed but are not known with certainty). Configuration A is stable when the defect is in either the (+) or (−) charge state, while configuration B is stable when the defect is neutral. This phenomenon is called bistability. In addition, in each case the other configuration also has a potential-energy minimum but at a slightly higher energy; this is an example of metastability. Thus when A is stable, B is metastable, and vice versa. At low temperature a metastable state can have a long lifetime if the energy barrier to atomic rearrangement is sufficiently high.

Negative U. Some semiconductor defects possess a characteristic called negative U. This term comes about because the defect behaves as if the electron–electron interaction (U) were attractive. An example is interstitial boron in silicon: this defect can exist in either (−) or (+) charge states, but the neutral state is not stable. This means that more energy is gained

when a neutral boron traps an electron than was lost when that electron was removed from another neutral boron: the total energy of the system is lower if half the defects are (+) and half (−) than if all are neutral. In chemistry this phenomenon is known as disproportionation. The occurrence of negative U in semiconductors involves several factors: first, the net Coulomb repulsion between defect electrons is reduced by the interactions with other electrons in the solid (polarization and hybridization). Second, the energies gained from atomic rearrangements can sometimes exceed the remaining Coulomb repulsions, thereby favoring the more ionic charge states.

W. Beall Fowler

Topological defects. Defects such as vacancies and interstitials have been known for many years, as have topological defects such as dislocations. The study of two-dimensional systems and their associated defects has led to the discovery of a new phase of matter which has no crystalline positional order and behaves like a fluid under shear, but has well-defined axes like a crystal. An example is the layered liquid-crystal compound 650 BC, which has properties intermediate between a liquid and a crystal.

Order in crystals. To understand topological defects, it is first necessary to understand the nature of order in crystals that have both translational and bond-orientational order. Translational order means that the atoms of the crystal form a well-defined periodic lattice, while orientational order is weaker and means that the directions of the bonds between neighboring atoms are the same throughout the system. **Figure 8** shows that a translation of a patch of crystal by $\vec{u}$ from its ideal position will spoil translational but not orientational order (Fig. 8a), whereas a rotation will disrupt both (Fig. 8b). In principle, orientational order may exist in the absence of translational order, but not vice versa, and a system of point atoms or spherical molecules can exist in three phases: a fully ordered crystal, an intermediate anisotropic fluid with orientational order only, and an isotropic fluid with neither. Systems of nonspherical molecules such as liquid crystals may have many more complex phases. In the great majority of real three-dimensional systems, only the first and last possibilities are realized, with an abrupt melting or freezing transition between states of complete order and complete disorder. However, the intermediate phase with long-range orientational but short-range translational order corresponds to the hexatic phase observed in various liquid-crystal systems and also in xenon adsorbed on graphite. The name hexatic comes from the underlying hexagonal structure of the parent crystal. The three phases can

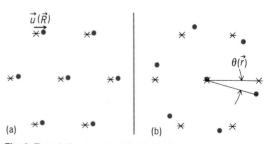

(a)   (b)

Fig. 8. Translations and rotations of a crystal. The ideal positions are denoted by crosses. (a) A patch of crystal translated by $\vec{u}$ from its ideal position with no change of the crystal axes. (b) A patch rotated by an angle θ relative to the perfect crystal.

be distinguished by diffraction measurements.

Dislocations and disclinations. A topological defect is one that disrupts a particular order. A crystal, which has two types of order, has two types of topological defect. A translational defect is a dislocation (Fig. 3), and an orientational defect is a disclination. A dislocation may be regarded as a disclination dipole as shown in **Fig. 9**, which is drawn as a square lattice for simplicity. Circling around the $+\pi/2$ $(-\pi/2)$ disclination requires making five (three) right-angle turns and the orientation of the axes is quickly lost. Circling round the dipole involves four right angles, and so this does not disrupt orientational order. However, an attempt to go around the dipole while keeping count of the number of lattice spacings on a path which would close if the lattice were perfect results in failure to return to the starting point by the Burgers vector $\vec{b}$. In other words, the dislocation disrupts translational order only, while a disclination disrupts both translational and orientational order.

Two-dimensional systems. Topological defects do not occur to any significant extent in most bulk crystals in thermal equilibrium. They are present in general as frozen-in metastable states and have a dramatic effect on mechanical properties. In films a few atoms thick and in three-dimensional systems of very weakly coupled layers, they are, however, present in thermal equilibrium and lead to some unusual effects. A graphic way of visualizing a two-dimensional crystal and its excitations and defects is to imagine the crystal lattice drawn on an elastic sheet. This can be distorted in a number of ways, all of which are present to some extent depending on the energy involved, as in Eq. (2). There are the obvious smooth distortions created by stretching and compressing different regions without cutting. Interstitials and vacancies result from the additional or removal of small pieces containing a single atom and regluing the edges. Dislocations are introduced by cutting the sheet, displacing the edges by one lattice spacing, and rejoining the edges, after adding or removing some material as necessary. In a real system this requires diffusing in interstitials or vacancies. Disclinations involve the addition or removal of an angular segment contained between two crystal axes and clearly require a huge amount of energy in the crystalline state.

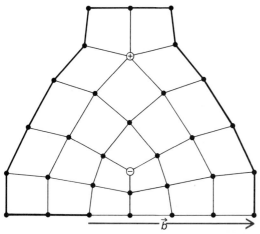

Fig. 9. Disclination dipole in a square lattice. The atom denoted by ⊕ (⊖) has five (three) neighbors and corresponds to a $+\pi/2$ $(-\pi/2)$ disclination. The heavy line is a perfect lattice closed path.

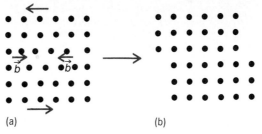

(a)   (b)

Fig. 10. Defect unbinding transition in a two-dimensional crystal. (*a*) A dislocation pair of equal and opposite Burgers vectors $\vec{b}$ in a square lattice. Above the melting temperature, an infinitesimal shear stress, indicated by the arrows above and below the lattice, will cause the dislocations to move and the crystal to distort. (*b*) Result of this motion.

The equilibrium crystalline state is one with some smooth distortions, some vacancies and interstitials, and a small density of dislocation pairs of equal and opposite Burgers vectors (**Fig. 10***a*). The essential ideas of the role played by topological defects can be understood by focusing on the members of a single bound pair. As the temperature is raised, the average spacing between them increases until it equals the spacing between pairs. The density of dislocations remains small. At this stage, the dislocations become independent of each other, positional order is lost, and no force is required to move them about the crystal. The system will behave exactly like a very viscous fluid under a shear stress. The result of this motion is shown in Fig. 10*b*.

Provided the density of dislocations and point defects is sufficiently low, the resulting fluid has orientational order. As the temperature is further increased, pairs of equal and opposite disclinations appear in the fluid and their separation will eventually diverge. Above this temperature, all orientational order is lost and the fluid is isotropic.

This discussion assumes that the density of defects remains quite low; in that case the melting of the crystal is via two stages of defect unbinding transitions of the Kosterlitz-Thouless type. Should the density of local defects, such as vacancies, interstitials, and closely bound dislocation pairs, be large, then the crystal melts by an explosive proliferation of these directly to the isotropic fluid. This is the route taken by the great majority of melting transitions both in the bulk and in films, but the defect unbinding mode has been observed experimentally. The theoretical discovery of the hexatic phase led to the experimental identification of the bulk hexatic B and smectic I and F liquid-crystal phases. *See* CRYSTAL STRUCTURE; CRYSTALLOGRAPHY; LIQUID CRYSTALS.

John M. Kosterlitz

Bibliography. J. H. Albany (ed.), *Defects and Radiation Effects in Semiconductors, 1978*, 1979; J. H. Crawford, Jr., and L. M. Slifkin (eds.), *Point Defects in Solids*, vols. 1, 2, and 3, 1972, 1975, 1978; C. Domb and J. L. Lebowitz (eds.), *Phase Transitions and Critical Phenomena*, 1983; N. B. Hannay (ed.), *Treatise on Solid State Chemistry*, vol. 2: *Defects in Solids*, 1975; W. Hayes and A. M. Stoneham, *Defects and Defect Processes in Nonmetallic Solids*, 1985; C. Kittel, *Introduction to Solid State Physics*, 6th ed., 1986; J. M. Kosterlitz and D. J. Thouless, Metastability and phase transitions in two dimensional systems, *J. Phys.*, C6:1181–1203, 1973; S. Mrowec,

Defects and Diffusion in Solids: An Introduction, 1980; T. Mura, *Micromechanics of Defects in Solids,* 1982; F. R. N. Nabarro, *Theory of Dislocations,* 1967; P. S. Pershan, *Structure of Liquid Crystal Phases,* 1988; K. J. Standburg, Two dimensional melting, *Rev. Mod. Phys.,* 60:161–207, 1988.

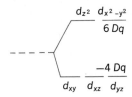

Fig. 2. The splitting of the five *d* orbitals because of an octahedral crystal field.

Crystal field theory

An essentially ionic approach to chemical bonding which is often used with coordination compounds. These compounds consist of a central transition-metal ion that is surrounded by a regular array of coordinated atoms or ligands. Accordingly, the ligands are assumed to be sources of negative charge which perturb the energy levels of the central metal ion. In this respect the ligands subject the metal ion to an electric field which is analogous to the electric or crystal field produced by the regular distribution of nearest neighbors within an ionic crystalline lattice. For example, the crystal field produced by the Cl ion ligand in octahedral $TiCl_6^{3-}$ is considered to be similar to that produced by the octahedral array of the six Cl ions about each Na ion in NaCl. The Na ion with its rare-gas configuration has an electronic charge distribution which is spherically symmetric both within and without the crystal field. The paramagnetic Ti(III) ion, which possesses one 3d electron (d^1), has a spherically symmetric charge distribution only in the absence of the crystal field produced by the ligands. The presence of the ligands destroys the spherical symmetry and produces a more complex set of energy levels within the central metal ion. The crystal field theory allows the energy levels to be calculated and related to experimental observation. See Coordination chemistry.

To illustrate the results of a typical crystal field calculation, assume that the single *d* electron in the Ti(III) ion will experience a coulombic repulsion with each of the six nearest neighbor Cl ions which are taken as point negative charges. This model for coulombic repulsion may be described mathematically as the summation $eq\Sigma r^{-1}$, where *e* and *q* are the electronic and ligand charges, respectively, and *r* is the distance between the electron and the ligand. The summation extends over all the ligands. A detailed quantum mechanical calculation can then be made. Fortunately, it is possible to arrive at identical results by a very qualitative procedure. This method considers the spatial orientation of *d* orbitals with relation to the ligands when both are viewed within the same coordinate system, as shown in **Fig. 1**.

The d_{z^2} orbital is oriented along the *z* axis with a ring in the *xy* plane, while the $d_{x^2-y^2}$ orbital is oriented only along the *x* and *y* axes. Although it is not visually obvious, both are equivalent. The d_{xy}, d_{xz}, and d_{yz} orbitals are oriented between the *x*, *y*, and *z* axes and are geometrically equivalent. The ligands are placed along the coordinate *x*, *y*, and *z* axes. An electron in any of the five orbitals will experience a coulombic repulsion due to the crystal field of the ligands. However, since the d_{z^2} and $d_{x^2-y^2}$ orbitals are directed toward the ligands, an electron in these orbitals will undergo far more repulsion than one in either of the three other orbitals. The imposition of six point negative charges, then, will not allow the five *d* orbitals to be equally energetic (degenerate) as they are in the bare Ti(III) ion, but causes three of these orbitals to be more stable than the remaining two. The difference in energy is termed $10Dq$. Thus, in the ground state of $TiCl_6^{3-}$, the single *d* electron will be found in the lower set of orbitals, whose energy is $-4Dq$. The crystal field stabilization energy (CFSE) is then said to be $4Dq$ (**Fig. 2**).

Coordination compounds are often colored. Crystal field theory suggests that in the case of $TiCl_6^{3-}$ its color is a result of an electronic excitation of the electron from the threefold set of orbitals into the twofold set. In the spectrum of $TiCl_6^{3-}$, an absorption band maximum is found at $13,000\ cm^{-1}$, which is then the energy associated with the transition, or $10Dq$. See Molecular structure and spectra.

Several electrons. When more than one *d* electron is present, the spectroscopic evaluation of *Dq* is not as simple, but can generally be accomplished. Nevertheless, the CFSE may be easily and formally obtained for these cases, as shown in the **table**, since this energy is simply the number of electrons occupying the orbital multiplied by the orbital energy. Thus, the CFSE amounts to $8Dq$ in an octahedral d^2 complex, and $(6 \times 4) - (2 \times 6) = 12Dq$ in a similar d^8 complex. With d^4, d^5, d^6, and d^7, however, two possibilities exist. If the crystal field is sufficiently strong to overcome the repulsion energy which will result from pairing the electrons in the lower set

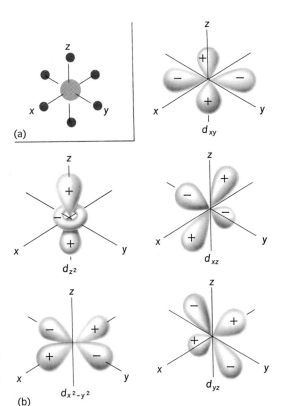

Fig. 1. Spatial orientation of *d* orbitals with relation to ligands in $TiCl_6^{3-}$. (*a*) The coordinate system for the octahedral $TiCl_6^{3-}$ ion. (*b*) The five *d* orbitals. The d_{z^2} orbital is symmetric with respect to the *z* axis.

Crystal field stabilization energy

	Octahedral CFSE (Dq)	
d^n	Weak	Strong
1	4	4
2	8	8
3	12	12
4	6	16
5	0	20
6	4	24
7	8	18
8	12	12
9	6	6

of orbitals, the maximum number of electrons will be found in the lower set. This situation is termed the strong-field case. In the weak-field case, the electron-pairing (repulsive) energy is greater than the crystal field (attractive) energy and the maximum number of unpaired electrons will result. Thus, in a strong crystal field a d^5 ion should have only one unpaired electron and a CFSE of $20Dq$. This is found in the $Mn(CN)_6^{4-}$ ion, which has been shown experimentally to possess only one unpaired electron. In most complexes of Mn(II), such as $Mn(H_2O)_6^{2+}$, the crystal field is weak so that five unpaired electrons result and the CFSE is zero.

Simple considerations such as these have enabled inorganic chemists to understand why certain coordination compounds containing a given metal ion may exhibit full paramagnetism, while others containing the metal in exactly the same formal oxidation state may show either a much weaker paramagnetism or none at all. A striking example of this is provided by paramagnetic CoF_6^{3-}, which possesses six unpaired d electrons and the diamagnetic $Co(NH_3)_6^{3+}$, in which none of the d electrons are unpaired. Both contain Co(III).

Tetrahedral array. Tetrahedral complexes may be treated in a similar fashion. The results indicate that the only difference, when compared to the octahedral complex, is in the orbital splitting pattern and the relative magnitude of Dq. The tetrahedral array of ligands causes an inversion of the pattern such that the d_{z^2} and $d_{x^2-y^2}$ orbitals lie lowest. The difference in energy between the two sets of orbitals is now found to be 4/9 of that in an octahedral complex, or Dq(tetrahedral) = $(4/9)Dq$(octahedral). This particularly simple result has led to the understanding of many stereochemical phenomena. An important early example was found in the cation distribution of normal and inverse spinels. The former are double oxides having the general formula $M(II)[M'(III)]_2O_4$ in which the oxygens lie in a close-packed system. The divalent metal ions, M(II), occupy one-eighth of the tetrahedral holes. In the inverse spinel $M'(III)[M(II)M'(III)]O_4$ the divalent metal ions have changed places with one-half of the trivalent ions.

Experimentally, it has been found that $MnCr_2O_4$ [containing Mn(II) with five unpaired electrons] and $NiFe_2O_4$ [containing Fe(III) with five unpaired electrons] have the normal and inverse structures, respectively. A simple application of crystal field theory results in exact agreement with experiment. In $MnCr_2O_4$ the ion at each octahedral site has a stability of $12Dq$, while the tetrahedral site has no CFSE. The total CFSE is then $24Dq$. If the structure were in-

verse, the total CFSE would be only $(4/9) \times 12 + 12 = 17.3Dq$. With $NiFe_2O_4$ the CFSE of the normal spinel is only $5.3Dq$, but in the inverse structure the CFSE increases to $12Dq$. In general, agreement between predicted cation distribution and that experimentally observed is good.

The application of these methods to conventional coordination compounds also meets with a fair amount of success. For example, with only one or two exceptions octahedral coordination of Cr(III) and diamagnetic Co(III) prevails in all compounds, as one would predict from crystal field theory. Important exceptions to these rules do exist and point out that other factors, such as ligand-ligand repulsions, sometimes outweigh the CFSE. Octahedral coordination of Ni(II) is favored over the tetrahedral arrangement insofar as CFSE is concerned, yet the tetrahedral $NiCl_4^{2-}$, $NiBr_4^{2-}$, and NiI_4^{2-} ions are well known.

Anomalous effects. Heats of hydration, lattice energies, crystal radii, and oxidation potentials of transition-metal ions and their complexes contain apparent anomalies which are best explained in terms of an effect due to the crystal field. As an example, when the heats of hydration of the ions of the first transition series are plotted in **Fig. 3** with respect to atomic number, a peculiar double-humped curve is obtained. The results for those ions not possessing any CFSE, that is, Ca, Mn, and Zn, lie nearly on a straight line. In the absence of any other effect, it might be expected that with the successive addition of each nuclear charge, a monotonic increase in the heat of hydration would occur. Instead, it was found that the heat of hydration of those metal ions possessing CFSE is far more exothermic than would be predicted by arguments pertaining only to the successive increase in atomic number. In fact, the excess heat is best accounted for in terms of the CFSE. The total heat of hydration may be written as $\Delta H = \Delta H° + $ CFSE, where $\Delta H°$ is the heat of hydration that would be expected for a hypothetical metal ion which ignored the crystal field. The change in ΔH^0 with atomic number would then be expected to be monotonic. When the observed heats of hydration are corrected for the CFSE obtained from spectroscopic data for the resulting $M(H_2O)_6^{2+}$ complexes, the expected

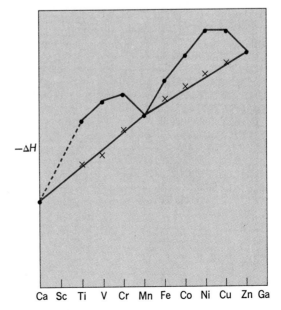

Fig. 3. Variation of heat of hydration $-\Delta H$ of divalent ions of first transition series. Experimental values are given by the dots, and the corresponding values, after correction for CFSE, are given by the x's.

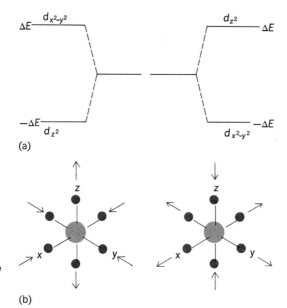

Fig. 4. The splitting of the $d_{x^2-y^2}$ and d_{z^2} orbitals by an axial distortion of the octahedron. (a) Alternate possiblities for writing the electronic configuration. (b) The two modes of distortion.

monotonic increase is observed. Similar double-humped curves occur with the lattice energies and crystal radii and can be explained by including the effects due to the crystal field.

Stereochemical anomalies can also often be explained through the judicious use of simple arguments. A particularly important example is found in complexes of Cu(II). X-ray crystallography has established that most "octahedral" complexes containing that ion are in fact elongated along one axis. According to a theorem due to H. A. Jahn and E. Teller, this behavior is not unexpected. The theorem states that a system possessing a degenerate ground state will distort in some unspecified manner to remove the degeneracy.

The degeneracy in a regular octahedral complex of Cu(II) is easily illustrated by the possibility of writing the electronic configuration in two distinct, but equally energetic, ways:

$$(d_{xy})^2(d_{xz})^2(d_{yz})^2(d_{z^2})^2(d_{x^2-y^2})^1$$

or $$(d_{xy})^2(d_{xz})^2(d_{yz})^2(d_{x^2-y^2})^2(d_{z^2})^1$$

For each, the CFSE is $6Dq$. In addition to the twofold degeneracy, there are two separate means by which the degeneracy may be removed. If the ligands along the z axis of the octahedron move away from the metal ion while those in the xy plane move toward the center of the octahedron, then according to simple crystal field arguments, this movement will result in stabilizing the d_{z^2} orbital with respect to the $d_{x^2-y^2}$ orbital. Alternatively, the completely opposite movement will stabilize the $d_{x^2-y^2}$ orbital with respect to the d_{z^2} orbital (**Fig. 4**). In either case, an additional increment of energy is added to the CFSE of the Cu(II) complex: $6Dq + \Delta E$. Thus, the driving force for the distortion is the additional stabilization energy ΔE. Crystal field theory is unable to predict which mode of distortion will occur, but it clearly predicts that a distortion should occur. From structure determinations through the use of x-rays, it is found that elongation along one axis is by far the most predominant mode.

Complete theory. The many successes of crystal field theory in the interpretation of the natural phe-

nomena associated with transition-metal compounds should not lead one to the conclusion that the bonding within these compounds can be truthfully represented by a strictly ionic model. Crystal field theory is essentially a very specialized form of the more complete molecular orbital theory. The need for the more complete theory becomes obvious when an attempt is made to rationalize the absolute value of Dq, the observation of ligand-to-metal electronic transitions, and certain observations in both nuclear magnetic resonance and electron spin resonance experiments, which can only be interpreted in terms of some covalent bonding. SEE MOLECULAR ORBITAL THEORY.

R. A. D. Wentworth

Bibliography. R. L. DeKock and H. B. Gray, *Chemical Structure and Bonding*, 1980; T. M. Dunn, D. S. McClure, and R. G. Pearson, *Some Aspects of Crystal Field Theory*, 1965; M. Gerloch and R. C. Slade, *Ligand-Field Parameters*, 1973; H. L. Schlafer and G. Glieman, *Basic Principles of Ligand Field Theory*, 1969.

Crystal growth

The growth of crystals, of which all crystalline solids are composed. This article discusses the processes by which crystal growth occurs, and methods of artificially growing crystals.

GROWTH PROCESSES

Crystal growth generally comes about by means of the following processes occurring in series: (1) diffusion of the atoms (or molecules, in the case of molecular crystals such as hydrocarbons or polymers) of the crystallizing substance through the surrounding environment (or solution) to the surface of the crystal, (2) diffusion of these atoms over the surface of the crystal to special sites on the surface, (3) incorporation of atoms into the crystal at these sites, and (4) diffusion of the heat of crystallization away from the crystal surface. The rate of crystal growth may be limited by any of these four steps. The initial formation of the centers from which crystal growth proceeds is known as nucleation. SEE CRYSTALLIZATION; NUCLEATION.

Increasing the supersaturation of the crystallizing component or increasing the temperature independently increases the rate of crystal growth. However, in many physical situations the supersaturation is increased by decreasing the temperature. In these cir-

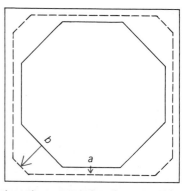

Fig. 1. Schematic representation of cross section of crystal at three stages of regular growth.

cumstances the rate of crystal growth increases with decreasing temperature at first, goes through a maximum, and then decreases. Often the growth is greatly retarded by traces of certain impurities.

If nucleation occurs in several locations in the medium, the crystals grow isolated from one another for a time. However, if several differently oriented crystals are growing, they may finally impinge on one another, and intercrystalline (grain) boundaries will be formed. At relatively high temperatures, the average grain size in these polycrystalline aggregates increases with time by a process called grain growth, whereby the larger grains grow at the expense of the smaller. *See Grain boundaries.*

Regular growth. During its growth into a fluid phase, a crystal often develops and maintains a definite polyhedral form which may reflect the characteristic symmetry of the microscopic pattern of atomic arrangement in the crystal. The bounding faces of this form are those which are perpendicular to the directions of slowest growth. How this comes about is illustrated in **Fig. 1**, in which it is seen that the faces *b*, normal to the faster-growing direction, disappear, and the faces *a*, normal to the slower-growing directions, become predominant. Growth forms, like that shown in the figure, are not necessarily equilibrium forms, but they are likely to be most regular when the departures from equilibrium are not large. *See Crystal structure.*

The atomic binding sites on the surface of a crystal can be of several kinds. Thus an atom must be more weakly bound on a perfectly developed plane of atoms at the crystal surface (site A in **Fig. 2**) than at a ledge formed by an incomplete plane one atom thick (site B). Atom A binds with only three neighboring atoms, whereas atom B binds to five neighbors. (An atom in the middle of the island monolayer has bonds with nine neighbors.) Therefore, the binding of atoms in an island monolayer on the crystal surface will be less per atom than it would be within a completed surface layer.

The potential energy of a crystal is most likely to be minimum in forms containing the fewest possible ledge sites. This means that, in a regime of regular crystal growth, dilute fluid, and moderate departure from equilibrium, the crystal faces of the growth form are likely to be densely packed and atomically smooth. There will be a critical size of monolayer, which will be a decreasing function of supersaturation, such that all monolayers smaller than the critical

Fig. 3. A dendrite. Growth direction is upward for primary protuberances; side branches grow perpendicularly; side branches on the side branches are also seen growing perpendicularly. (*After W. Kurz and D. J. Fisher, Fundamentals of Solidification, Trans Tech Publications, 1986***)**

size tend to shrink out of existence, and those which are larger will grow to a complete layer. The critical monolayers form by a fluctuation process. Kinetic analyses indicate that the probability of critical fluctuations is so small that in finite systems perfect crystals will not grow, except at substantial departures from equilibrium. That, in ordinary experience, finite crystals do grow in a regular regime only at infinitesimal departures from equilibrium is explained by the screw dislocation theory. According to this theory, growth is sustained by indestructible surface ledges which result from the emergence of screw dislocations in the crystal face. *See Crystal defects.*

Irregular growth. When the departures from equilibrium (supersaturation or undercooling) are sufficiently large, the more regular growth shapes become unstable and cellular (grasslike) or dendritic (treelike) morphologies develop. Essentially, the development of protuberances on an initially regular crystal permits more efficient removal of latent heat or of impurities, but at the cost of higher interfacial area and the associated excess surface energy. In the cellular structure, the protuberances align, forming an advancing front and rejecting impurities laterally into the spaces between protuberances. In the dendritic morphology, these protuberances become narrower and develop side branches, which are other protuberances growing laterally (**Fig. 3**). When the supersaturation becomes so great that the energy associated with the increase in interfacial area is unimportant, protuberances proliferate and the crystal grows in a multibranched form that is even more complicated; its shape is characterized by fractal geometry. This limit, in which atoms stick permanently wherever they land, is called diffusion-limited aggregation because steps 2–4 in the series of processes outlined above occur very rapidly, and the rate of crystal growth is limited by diffusion (step 1). *See Fractals.*   Michael J. Aziz; David Turnbull

GROWTH METHODS

The advent of semiconductor-based technology generated a demand for large, high-quality single crystals, not only of semiconductors but also of associated electronic materials. With increasing sophistication of semiconductor devices, an added degree of freedom in materials properties was obtained by vary-

Fig. 2. A model of a crystal surface showing the inequivalent binding sites at which the atoms labeled A and B are located. (*General Electric Co.***)**

ing the composition of major components of the semiconductor crystal over very short distances. Thin, multilayered single-crystal structures, and even structures that vary in composition both normal and lateral to the growth direction, are often required.

Bulk single-crystal growth. Bulk single crystals are usually grown from a liquid phase. The liquid may have approximately the same composition as the solid; it may be a solution consisting primarily of one component of the crystal; or it may be a solution whose solvent constitutes at most a minor fraction of the crystal's composition. **Figure 4** shows the phase diagram and corresponding component partial pressures for a hypothetical binary compound. The stoichiometry range of Fig. 4a results from the deviation from the 1-to-1 composition caused by point defects in the crystal structure. Nonstoichiometry, high partial pressures, and relatively high point-defect concentrations compared to growth at lower temperatures are all disadvantages of melt growth. Nevertheless, melt growth is used for virtually all bulk semiconductor crystals and for many other industrial crystals. *See* Nonstoichiometric compounds; Phase equilibrium.

Czochralski method. The most important bulk crystal

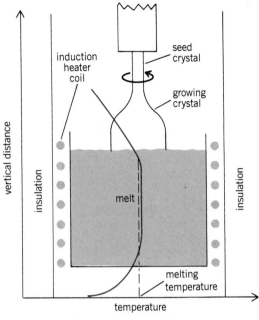

Fig. 5. Czochralski crystal growth and temperature distribution.

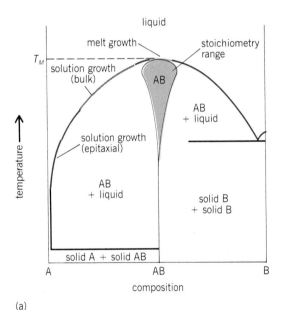

(a)

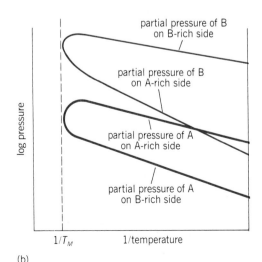

(b)

Fig. 4. Hypothetical binary system AB having simple compound formation. (a) Phase diagram. (b) Corresponding partial pressure curves. The extent of the stoichiometry range is greatly exaggerated for illustration.

growth technique is the crystal-pulling or Czochralski method, in which a rotating seed crystal is dipped into the melt (**Fig. 5**). Rotation reduces radial temperature gradients, and slow withdrawal of the rotating seed results in growth of a cylinder of single-crystal material. The conditions for optimum growth vary widely, and pulling rates range up to a few inches per hour. Crystal diameter and length depend upon the details of the temperature and pulling rate, and the dimensions of the melt container. Crystal quality depends very critically upon minimization of temperature gradients that enhance the formation of dislocations. Pulled silicon crystals 6 in. (15 cm) in diameter are important for the semiconductor industry. Ruby, sapphire, and group III–V compound semiconductor crystals are among the many crystals that are routinely grown by the Czochralski technique.

Melt-growth variations. There are several other variations of the melt-growth technique. One of them is the zone-refining method, in which a narrow molten region is passed through a solid rod of single-crystal or polycrystalline material. Another is the Bridgman technique, in which single-crystal growth is achieved for some materials by moving the melt through a temperature gradient (**Fig. 6**). Generally the container has a restricted region where the first solidification takes place as the liquidus temperature moves past it. An initially formed crystallite in that region propagates with the moving liquidus front, and a single crystal consisting of the entire volume of the melt can result. For some variants of this technique, the liquid is rich in the less volatile component of the solid compound. Growth is driven by maintaining the partial pressure of the more volatile component as required to maintain the liquidus composition. *See* Zone refining.

Solvent methods. Other solution growth methods utilize a solvent in which the crystal components may have extensive solubility, but which itself is not usually appreciably soluble in the crystal. Often, small seed crystals are used to initiate growth which is maintained by subsequent cooling of a saturated so-

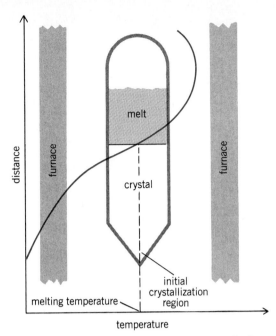

Fig. 6. Vertical Bridgman crystal and temperature distribution.

lution or diffusion of components through a temperature gradient from a source nutrient. Garnets and other magnetic oxides are often grown from molten salts; quartz crystals are routinely grown from aqueous solution at high temperature and pressure. *SEE CRYSTAL WHISKERS; FERRIMAGNETIC GARNETS; SINGLE CRYSTAL.*

Epitaxial growth. The evolution of methods for the growth of very thin but very high-quality epitaxial layers has resulted largely from the need for such layers of semiconductors and magnetic garnets.

Liquid-phase epitaxy. The technique most closely related to the methods used for bulk crystal growth is liquid-phase epitaxy. For a typical binary semiconductor, growth is done onto a substrate single-crystal wafer from a solution rich in the component with the lowest partial pressure (Fig. 4). For a binary compound, the grown layers may differ only in impurity concentrations to modify their electrical characteristics. More often, multilayered structures with layers differing in major component composition but having the same crystal structure and lattice parameter are required. The simplest example of liquid-phase epitaxy with major composition changes in layers is the growth of layers of aluminum gallium arsenide ($Al_xGa_{1-x}As$; $1 > x > 0$) on a gallium arsenide (GaAs) substrate. In this example, for all x, the solid solution $Al_xGa_{1-x}As$ has essentially the same lattice parameter, so that this is a naturally lattice-matched system. An isotherm of the ternary phase diagram for this system is shown in **Fig. 7**. A typical range of growth compositions can yield layers with very different solid compositions.

Growth by liquid-phase epitaxy is done in an apparatus in which the substrate wafer is sequentially brought into contact with solutions that are at the desired compositions and may be supersaturated or cooled to achieve growth. For crystalline solid solutions other than $Al_xGa_{1-x}As$, very precise control over solution compositions is required to achieve a lattice match. Typically, structures grown by liquid-

phase epitaxy have four to six layers ranging widely in composition and having thickness from 10^{-7} to 10^{-6} m.

Molecular-beam epitaxy. The desirability of highly reproducible growth and even thinner epitaxial layers of III–V compounds on large wafer areas has led to the development of molecular-beam epitaxy and several forms of chemical-vapor deposition. Molecular-beam epitaxy is an ultrahigh-vacuum technique in which beams of atoms or molecules of the constituent elements of the crystal provided by heated effusion ovens, impinge upon a heated substrate crystal (**Fig. 8**). It has been used for epitaxial layers as thin as 0.5 nanometer (**Fig. 9**). Molecular-beam epitaxy has also been used for group II–VI and VI–VI compounds and for silicon. *SEE ARTIFICIALLY LAYERED STRUCTURES; MOLECULAR BEAMS.*

The ability to achieve epitaxy of compounds under vacuum conditions requires that it be possible to obtain a nominally stoichiometric crystal from impingement of a flux of component atoms and molecules whose net composition is far from that at stoichiometry. It is also required that surface adsorption and mobility of the impinging components be sufficient to ensure that atoms finally reside in their most stable lattice sites. How this can happen is best illustrated with gallium arsenide. Growth of gallium arsenide is achieved by impinging Ga atoms and As_2 or As_4 molecules onto a heated GaAs wafer. The Ga atoms essentially all stick to the surface because of the relatively low Ga vapor pressure at the growth temperature of 900–1300°F (500–700°C). Only those As atoms that interact with a Ga atom stay on the surface, so that the growing layer is stoichiometric provided the As-to-Ga ratio striking the surface is somewhat greater than unity. Crystalline solid solutions such as $Al_xGa_{1-x}As$ are obtained by adding aluminum (Al) to the beams with an Al-containing effusion oven (Fig. 8). Shutters are used to block the beams and permit changes in the beam composition for multilayer growth. *SEE GALLIUM.*

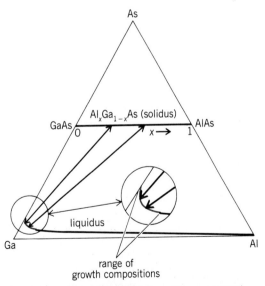

Fig. 7. An isotherm in the aluminum-gallium-arsenic (Al-Ga-As) ternary phase diagram. Each corner represents a pure element and each point within the diagram a unique composition, all at a given temperature. Each composition on the liquidus is in equilibrium with one composition on the solidus. The tie lines (lines with arrowheads) show two such sets of points.

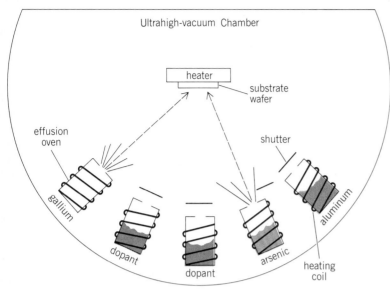

Fig. 8. Molecular-beam epitaxy apparatus.

Chemical-vapor deposition. Chemical-vapor deposition of epitaxial layers is also often used for semiconductors. Again, the group III–V compounds are of major importance. The growth of GaAs will be used for illustration. One of the most common chemical-vapor deposition methods is a hot-wall procedure in which the growing layer is generated by reaction (1).

$$As_4 + 4GaCl + 2H_2 \leftrightarrows 4GaAs + 4HCl \qquad (1)$$

The As_4 is generated either by the thermal decomposition of arsine, AsH_3, as in reaction (2), or by reaction (3). The gallium chloride (GaCl) is generated by

$$4AsH_3 \longrightarrow 6H_2 + As_4 \qquad (2)$$

$$4AsCl_3 + 6H_2 \longrightarrow As_4 + 12HCl \qquad (3)$$

reaction (4). The reactants are introduced into a

$$2HCl + 2Ga \longrightarrow 2GaCl + H_2 \qquad (4)$$

heated fused-silica reaction tube with a carrier gas, usually molecular hydrogen (H_2) or nitrogen (N_2). A temperature gradient is maintained so that the substrate seed is downstream of the decomposition region for reactions (1), (2), and (3), and is at a temperature that favors growth of GaAs by reaction (1). Crystalline solid solutions of a variety of III–V compounds can be grown by this process. Layers of $Al_xGa_{1-x}As$ are not generally grown this way because of thermodynamic limitations on composition control, and the very high reactivity of Al-containing vapor species with most hot-wall reactor materials.

Metal organic chemical-vapor deposition. An alternate process is the metal organic (MOCVD) process, in which metal organic compounds are decomposed near the surface of the heated substrate wafer. This method depends upon the fact that a very large variety of MR_x compounds (where M is a metal and R is an organic radical) such as $Ga(CH_3)_3$, $As(CH_3)_4$, and $Zn(CH_3)_2$ are available. The organic radical may be a methyl, ethyl, or more complex group. Generally, a compound is available as a liquid with a convenient vapor pressure so that it can be transported by saturating a stream of H_2. Most often, group V elements are transported as the hydrides. Thus, typically, indium phosphide (InP) is grown by reaction (5).

$$In(C_2H_5)_3 + PH_3 \longrightarrow InP + 3C_2H_6 \qquad (5)$$

Growth by MOCVD is most often done in a cold-wall chamber containing an induction-heated platform for the substrate wafer. Decomposition of the reactants apparently occurs in a boundary-layer region near the crystal surface. The MOCVD technique has proved highly successful for the growth of GaAs, $Al_xGa_{1-x}As$, InP, and $Ga_xIn_{1-x}P_yAs_{1-y}$, among others, and its versatility for multilayer growth of very thin, very high-quality layers is comparable to that of molecular-beam epitaxy. *See* SEMICONDUCTOR; SEMICONDUCTOR HETEROSTRUCTURES.

Morton B. Panish

Bibliography. H. Arend and J. Hulliger (eds.), *Crystal Growth in Science and Technology*, 1990; W. K. Burton et al., The growth of crystals and the equilibrium structure of their surfaces, *Phil. Trans. Roy. Soc. London*, 243A:299–358, 1951; J. C. Brice, *Crystal Growth Processes*, 1986; C. H. L. Goodman, *Crystal Growth*, vol. 1, 1974, vol. 2, 1978; N. B. Hannay (ed.), *Treatise on Solid State Chemistry*, vol. 5: *Changes of State*, 1975; J. W. Mathews (ed.), *Epitaxial Growth*, vols. A and B, 1975; L. M. Sander, Fractal growth processes, *Nature*, 322:789–793, 1986; J. P. van der Eerden, *Fundamentals of Crystal Growth*, 1990; A. W. Vere, *Crystal Growth: Principles and Progress*, 1987.

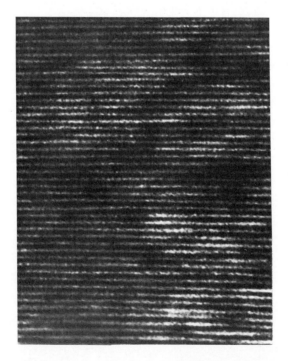

Fig. 9. Cross section of a gallium arsenide–aluminum arsenide (GaAs–AlAs) single crystal with individual gallium arsenide and aluminum arsenide layers each about 5 × 10^{-10} m thick.

Crystal optics

The study of the propagation of light, and associated phenomena, in crystalline solids. The propagation of light in crystals can actually be so complicated that not all the different phenomena are yet completely understood, and not all theoretically predicted phenomena have been demonstrated experimentally.

For a simple cubic crystal the atomic arrangement is such that in each direction through the crystal the crystal presents the same optical appearance. The atoms in anisotropic crystals are closer together in some planes through the material than in others. In anisotropic crystals the optical characteristics are different in different directions. In classical physics the progress of an electromagnetic wave through a material involves the periodic displacement of electrons. In anisotropic substances the forces resisting these displacements depend on the displacement direction. Thus the velocity of a light wave is different in different directions and for different states of polarization. The absorption of the wave may also be different in different directions. *See Dichroism; Trichroism.*

In an isotropic medium the light from a point source spreads out in a spherical shell. The light from a point source embedded in an anisotropic crystal spreads out in two wave surfaces, one of which travels at a faster rate than the other. The polarization of the light varies from point to point over each wave surface, and in any particular direction from the source the polarization of the two surfaces is opposite. The characteristics of these surfaces is opposite. The characteristics of these surfaces can be determined experimentally by making measurements on a given crystal.

For a transparent crystal the theoretical optical behavior is well enough understood so that only a few measurements need to be made in order to predict the behavior of a light beam passing through the crystal in any direction. It is important to remember that the velocity through a crystal is not a function of position in the crystal but only of the direction through the lattice. For information closely related to the ensuing discussion *See Polarized light. See also Crystal structure; Refraction of waves.*

Index ellipsoid. In the most general case of a transparent anisotropic medium, the dielectric constant is different along each of three orthogonal axes. This means that when the light vector is oriented along each direction, the velocity of light is different. One method for calculating the behavior of a transparent anisotropic material is through the use of the index ellipsoid, also called the reciprocal ellipsoid, optical indicatrix, or ellipsoid of wave normals. This is the surface obtained by plotting the value of the refractive index in each principal direction for a linearly polarized light vector lying in that direction (**Fig. 1**). The different indices of refraction, or wave velocities associated with a given propagation direction, are then given by sections through the origin of the coordinates in which the index ellipsoid is drawn. These sections are ellipses, and the major and minor axes of the ellipse represent the fast and slow axes for light proceeding along the normal to the plane of the ellipse. The length of the axes represents the refractive indices for the fast and slow wave, respectively. The most asymmetric type of ellipsoid has three unequal axes. It is a general rule in crystallography that no property of a crystal will have less symmetry than the class in which the crystal belongs. In other words, if a property of the crystal had lower symmetry, the crystal would belong in a different class. Accordingly, there are many crystals which, for example, have four- or sixfold rotation symmetry about an axis, and for these the index ellipsoid cannot have three unequal axes but is an ellipsoid of revolution. In such a crystal, light will be propagated along this axis as though the crystal were isotropic, and the velocity of propagation will be independent of the state of polarization. The section of the index ellipsoid at right an-

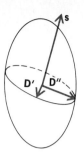

Fig. 1. Index ellipsoid, showing construction of directions of vibrations of D vectors belonging to a wave normal s. (*After M. Born and E. Wolf, Principles of Optics, 6th ed., Pergamon Press, 1980*)

gles to this direction is a circle. Such crystals are called uniaxial and the mathematics of their optical behavior is relatively straightforward. *See Crystallography.*

Ray ellipsoid. The normal to a plane wavefront moves with the phase velocity. The Huygens wavelet, which is the light moving out from a point disturbance, will propagate with a ray velocity. Just as the index ellipsoid can be used to compute the phase or wave velocity, so can a ray ellipsoid be used to calculate the ray velocity. The length of the axes of this ellipsoid is given by the velocity of the linearly polarized ray whose electric vector lies in the axis direction. *See Phase velocity.*

The ray ellipsoid in the general case for anisotropic crystal is given by Eq. (1), where α, β, and γ are the

$$\alpha^2 x^2 + \beta^2 y^2 + \gamma^2 z^2 = 1 \qquad (1)$$

three principal indices of refraction and where the velocity of light in a vacuum is taken to be unity. From this ellipsoid the ray velocity surfaces or Huygens wavelets can be calculated as just described. These surfaces are of the fourth degree and are given by Eq. (2). In the uniaxial case $\alpha = \beta$ and Eq. (2) factors

$$\left(x^2 + y^2 + z^2 \right)\left(\frac{x^2}{\alpha^2} + \frac{y^2}{\beta^2} + \frac{z^2}{\gamma^2} \right)$$
$$- \frac{1}{\alpha^2}\left(\frac{1}{\beta^2} + \frac{1}{\gamma^2} \right)x^2 - \frac{1}{\beta^2}\left(\frac{1}{\gamma^2} + \frac{1}{\alpha^2} \right)y^2$$
$$- \frac{1}{\gamma^2}\left(\frac{1}{\alpha^2} + \frac{1}{\beta^2} \right)z^2 + \frac{1}{\alpha^2\beta^2\gamma^2} = 0 \qquad (2)$$

into Eqs. (3). These are the equations of a sphere and

$$x^2 + y^2 + z^2 = \frac{1}{\alpha^2}$$
$$\gamma^2(x^2 + y^2) + \alpha^2 z^2 = 1 \qquad (3)$$

an ellipsoid. The z axis of the ellipsoid is the optic axis of the crystal.

The refraction of a light ray on passing through the surface of an anisotropic uniaxial crystal can be calculated with Huygens wavelets in the same manner as in an isotropic material. For the ellipsoidal wavelet this results in an optical behavior which is completely different from that normally associated with refraction. The ray associated with this behavior is termed the extraordinary ray. At a crystal surface where the optic axis is inclined at an angle, a ray of unpolarized light incident normally on the surface is split into two beams: the ordinary ray, which proceeds through the

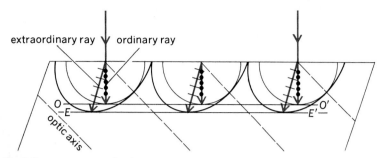

Fig. 2. Huygens construction for a plane wave incident normally on transparent calcite. If one proceeds to find the common tangents to the secondary wavelets shown, the two plane waves labeled *OO'* and *EE'* are obtained. (*After F. A. Jenkins and H. E. White, Fundamentals of Optics, 4th ed., McGraw-Hill, 1976*)

surface without deviation; and the extraordinary ray, which is deviated by an angle determined by a line drawn from the center of one of the Huygens ellipsoidal wavelets to the point at which the ellipsoid is tangent to a line parallel to the surface. The construction is shown in **Fig. 2**. The two beams are oppositely linearly polarized. When the incident beam is inclined at an angle ϕ to the normal, the ordinary ray is deviated by an amount determined by Snell's law of refraction, the extraordinary ray by an amount which can be determined in a manner similar to that of the normal incidence case already described. The plane wavefront in the crystal is first found by constructing Huygens wavelets as shown in **Fig. 3**. The line from the center of the wavelet to the point of tangency gives the ray direction and velocity.

The relationship between the normal to the wavefront and the ray direction can be calculated algebraically in a relatively straightforward fashion. The extraordinary wave surface, or Huygens wavelet, is given by Eq. (3), which can be rewritten as Eq. (4),

$$\epsilon^2(x^2 + y^2) + \omega^2 z^2 = 1 \tag{4}$$

where ϵ is the extraordinary index of refraction and ω the ordinary index. A line from the center of this ellipsoid to a point (x_1, y_1, z_1) on the surface gives the velocity of a ray in this direction. The wave normal corresponding to this ray is found by dropping a perpendicular from the center of the ellipsoid to the plane tangent at the point (x_1, y_1, z_1). For simplicity consider the point in the plane $y = 0$. The tangent at the point $x_1 z_1$ is given by Eq. (5). The slope of the normal to

$$\epsilon^2 x x_1 + \omega^2 z z_1 = 1 \tag{5}$$

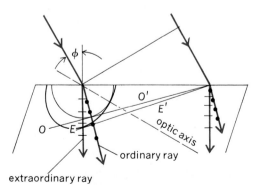

Fig. 3. Huygens construction when the optic axis lies in the plane of incidence. (*After F. A. Jenkins and H. E. White, Fundamentals of Optics, 4th ed., McGraw-Hill, 1976*)

this line is given by Eq. (6). The tangent of the angle

$$\frac{z}{x} = \frac{\omega^2 z_1}{\epsilon^2 x_1} \tag{6}$$

between the optic axis and the wave normal is the reciprocal of this number, as shown in Eq. (7), where

$$\tan \varphi = \frac{\epsilon^2 x_1}{\omega^2 z_1} = \frac{\epsilon^2}{\omega^2} \tan \psi \tag{7}$$

ψ is the angle between the ray and the optic axis. The difference between these two angles, π, can be calculated from Eq. (8). This quantity is a maximum

$$\tan \tau = \frac{\tan \varphi - \tan \psi}{1 + \tan \varphi \tan \psi} \tag{8}$$

when Eq. (9) holds.

$$\tan \varphi = \pm \frac{\epsilon}{\omega} \tag{9}$$

One of the first doubly refracting crystals to be discovered was a transparent variety of calcite called Iceland spar. This uniaxial crystal cleaves into slabs in which the optic axis makes an angle of 45° with one pair of surfaces. An object in contact with or a few inches from such a slab will thus appear to be doubled. If the slab is rotated about a normal to the surface, one image rotates about the other. For the sodium D lines at 589 mμ the indices for calcite are given by Eqs. (10). From these, the maximum angle

$$\epsilon = 1.486 \qquad \omega = 1.659 \tag{10}$$

τ_{max} between the wave normal and the ray direction is computed to be 6°16'. The wave normal at this value makes an angle of 41° 52' with the optic axis. This is about equal to the angle which the axis makes with the surface in a cleaved slab. Accordingly, the natural rhomb gives nearly the extreme departure of the ray direction from the wave normal.

Interference in polarized light. One of the most interesting properties of plates of crystals is their appearance in convergent light between pairs of linear, circular, or elliptical polarizers. An examination of crystals in this fashion offers a means of rapid identification. It can be done with extremely small crystals by the use of a microscope in which the illuminating and viewing optical systems are equipped with polarizers. Such a polarizing microscope is a common tool for the mineralogist and the organic chemist.

In convergent light the retardation through a birefringent plate is different for each direction. The slow and fast axes are also inclined at a different angle for each direction. Between crossed linear polarizers the plate will appear opaque at those angles for which the retardation is an integral number of waves. The locus of such points will be a series of curves which represent the characteristic interference pattern of the material and the angle at which the plate is cut. In order to calculate the interference pattern, it is necessary to know the two indices associated with different angles of plane wave propagation through the plate. This can be computed from the index ellipsoid. SEE INTERFERENCE OF WAVES.

Ordinary and extraordinary indices. For the uniaxial crystal there is one linear polarization direction for which the wave velocity is always the same. The wave propagated in this direction is called the ordinary wave. This can be seen directly by inspection of the index ellipsoid. Since it is an ellipsoid of revolution, each plane passing through the center will produce an ellipse which has one axis equal to the axis of the ellipsoid. The direction of polarization will always be at right angles to the plane of incidence and the refractive index will be constant. This constant index is called the ordinary index. The extraordinary index will be given by the other axis of the ellipse and will depend on the propagation direction. When the direction is along the axis of the ellipsoid, the extraordinary index will equal the ordinary index.

From the equation of the ellipsoid, Eq. (11), one

$$\frac{x^2}{\omega^2} + \frac{y^2}{\omega^2} + \frac{z^2}{\epsilon^2} = 1 \qquad (11)$$

can derive the expression for the ellipse and in turn Eq. (12) for the extraordinary index n_e as a function

$$n_e = \frac{\omega\epsilon}{(\epsilon^2 \cos^2 r + \omega^2 \sin^2 r)^{1/2}} \qquad (12)$$

of propagation direction. Since the ellipsoid is symmetrical it is necessary to define this direction only with respect to the ellipsoid axis. Here r is the angle in the material between the normal to the wavefront and the axis of the ellipsoid, n_e is the extraordinary index associated with this direction, ω is the ordinary index, and ϵ is the maximum value of the extraordinary index (usually referred to simply as the extraordinary index).

A slab cut normal to the optic axis is termed a Z-cut or C-cut plate (**Fig. 4**). When such a plate is placed between crossed linear polarizers, such as Nicol prisms, it gives a pattern in monochromatic light shown in **Fig. 5**. The explanation of this pattern is obtained from the equations given for the indices of refraction. To a first approximation the retardation for light of wavelength λ passing through the plate at an angle r can be written as Eq. (13), where d is the

$$\Gamma = \frac{(n_e - \omega)d}{\lambda \cos r} \qquad (13)$$

thickness of the plate. The axis of the equivalent retardation plate at an angle r will be in a plane containing the optic axis of the plate and the direction of light propagation. Wherever Γ is a whole number of waves, the light leaving the plate will have the same polarization as the incident light. The ordinary index is constant. The extraordinary index is a function of r alone, as seen from Eq. (12). Accordingly, the locus of direction of constant whole wave retardation will be a series of cones. If a uniform white light background is observed through such a plate, a series of rings of constant whole wave retardation will be seen. Since the retardation is a function of wavelength, the rings will appear colored. The innermost ring will have the least amount of color. The outermost ring will begin to overlap and disappear as the blue end of the spectrum for one ring covers the red end of its neighbor. The crystal plate has no effect along the axes of the polarizers since here the light is polarized along the axis of the equivalent retardation plate. The family of circles is thus bisected by a dark cross. When the crystal is mounted between like circular polarizers, the dark cross is not present and the center of the system of rings is clear. During World War II crystal plates were used in this fashion as gunsights. The rings appear at infinity even when the plate is

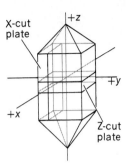

Fig. 4. X-cut and Z-cut plates in $NH_4H_2PO_4$ crystal.

close to the eye. Lateral movement of the plate causes no angular motion of the ring system. The crystal plate could be rigidly fastened to a gun mount and adjusted so as to show at all times the direction in space in which the gun was pointing.

Uniaxial crystal plates in which the axis lies in the plane of the plate are termed X-cut or Y-cut. When such plates are observed between crossed polarizers, a pattern of hyperbolas is observed.

When crystal plates are combined in series, the patterns become much more complex. In addition to the fringes resulting from the individual plates, a system of so-called moiré fringes appears.

Negative and positive crystals. In calcite the extraordinary wave travels faster than the ordinary wave. Calcite and other materials in which this occurs are termed negative crystals. In positive crystals the extraordinary wave travels slower than the ordinary wave. In the identification of uniaxial minerals one of the steps is the determination of sign. This is most easily demonstrated in a section cut perpendicular to the optic axis. Between crossed linear polarizers the pattern in convergent light, as already mentioned, is a series of concentric circles which are bisected by a dark cross. When a quarter-wave plate is inserted between one polarizer and the crystal, with its axis at 45° to the polarizing axis, the dark rings are displaced outward or inward in alternate quadrants. If the rings are displaced outward along the slow axis of the quarter-wave plate, the crystal is positive. If the rings are displaced inward along the axis, then the crystal is negative.

Fig. 5. Interference figure from fluorspar that has been cut perpendicular to the optic axis and placed between crossed Nicol prisms. (*After M. Born and E. Wolf, Principles of Optics, 6th ed., Pergamon Press, 1980*)

Fig. 6. Interference figure from Brazil topaz. (*After M. Born and E. Wolf, Principles of Optics, 6th ed., Pergamon Press, 1980*)

If a Z-cut plate of a positive uniaxial crystal is put in series with a similar plate of a negative crystal, it is possible to cancel the birefringence so that the combination appears isotropic.

Biaxial crystals. In crystals of low symmetry the index ellipsoid has three unequal axes. These crystals are termed biaxial and have two directions along which the wave velocity is independent of the polarization direction. These correspond to the two sections of the ellipsoid which are circular. These sections are inclined at equal angles to the major axes of the ellipsoid, and their normals lie in a plane containing the major and intermediate axes. In convergent light between polarizers, these crystals show a pattern which is quite different from that which appears with uniaxial crystals. A plate cut normal to the major axis of the index ellipsoid has a pattern of a series of lemniscates and ovals. The directions corresponding to the optic axes appear as black spots between crossed circular polarizers. The interference pattern between crossed linear polarizers is shown in **Fig. 6.**

Angle between optic axes. One of the quantities used to describe a biaxial crystal or to identify a biaxial mineral is the angle between the optic axes. This can be calculated directly from the equation for the index ellipsoid, Eq. (14), where α, β, and γ are the three

$$\frac{x^2}{\alpha^2} + \frac{y^2}{\beta^2} + \frac{z^2}{\gamma^2} = 1 \qquad (14)$$

indices of refraction of the material. The circular sections of the ellipsoid must have the intermediate index as a radius. If the relative sizes of the indices are so related that $\alpha > \beta > \gamma$, the circular sections will have β as a radius and the normal to these sections will lie in the xz plane. The section of the ellipsoid cut by this plane will be the ellipse of Eq. (15). The radius

$$\frac{x^2}{\alpha^2} + \frac{z^2}{\gamma^2} = 1 \qquad (15)$$

of length β will intersect this ellipse at a point $x_1 z_1$ where Eq. (16) holds. The solution of these two equa-

$$x_1^2 + z_1^2 = \beta^2 \qquad (16)$$

tions gives for the points $x_1 z_1$ Eq. (17). These points

$$x_1 = \pm \sqrt{\frac{\alpha^2(\gamma^2 - \beta^2)}{\gamma^2 - \alpha^2}}$$

$$z_1 = \pm \sqrt{\frac{\gamma^2(\alpha^2 - \beta^2)}{\alpha^2 - \gamma^2}} \qquad (17)$$

and the origin define the lines in the xz plane which are also in the planes of the circular sections of the ellipsoid. Perpendiculars to these lines will define the optic axes. The angle Ω between the axes and the z direction will be given by Eq. (18).

$$\tan^2 \Omega = \frac{z_1^2}{x_1^2} = \frac{(1/\alpha^2) - (1/\beta^2)}{(1/\beta^2) - (1/\gamma^2)} \qquad (18)$$

The polarization direction for light passing through a biaxial crystal is computed in the same way as for a uniaxial crystal. The section of the index normal to the propagation direction is ordinarily an ellipse. The directions of the axes of the ellipse represent the vibration direction and the lengths of the half axes represent the indices of refraction. One polarization direction will lie in a plane which bisects the angle made by the plane containing the propagation direction and one optic axis and the plane containing the propagation direction and the other optic axis. The other polarization direction will be at right angles to the first. When the two optic axes coincide, this situation reduces to that which was demonstrated earlier for the case of uniaxial crystals.

Conical refraction. In biaxial crystals there occurs a set of phenomena which have long been a classical example of the theoretical prediction of a physical characteristic before its experimental discovery. These are the phenomena of internal and external conical refraction. They were predicted theoretically in 1832 by William Hamilton and experimentally demonstrated in 1833 by H. Lloyd.

As shown earlier, the Huygens wavelets in a biaxial crystal consist of two surfaces. One of these has its major axis at right angles to the major axis of the other. The two thus intersect. In making the geometrical construction to determine the ray direction, two directions are found where the two wavefronts coincide and the points of tangency on the two ellipsoids are multiple. In fact, the plane which represents the wavefront is found to be tangent to a circle which lies partly on one surface and partly on the other. The directions in which the wavefronts coincide are the optic axes. A ray incident on the surface of a biaxial crystal in such a direction that the wavefront propagates along an axis will split into a family of rays which lie on a cone. If the crystal is a plane parallel slab, these rays will be refracted at the second surface and transmitted as a hollow cylinder parallel to the original ray direction. Similarly, if a ray is incident on the surface of a biaxial crystal plate at such an angle that it passes along the axes of equal ray velocity, it will leave the plate as a family of rays lying on the surface of a cone. The first of these phenomena is internal conical refraction; the second is external conical refraction. Equation (19) gives the half angle ψ

$$\tan \psi = \frac{\beta}{\sqrt{\alpha\gamma}} \sqrt{(\beta - \alpha)(\gamma - \beta)} \qquad (19)$$

of the external cone.

Bruce H. Billings

Bibliography. M. Born and E. Wolf, *Principles of Optics*, 6th ed., 1980; F. A. Jenkins and H. E. White, *Fundamentals of Optics*, 4th ed., 1976; E. E. Wahlstrom, *Optical Crystallography*, 5th ed., 1979; E. A. Wood, *Crystals and Light*, 1977; P. Yeh, *Optical Waves in Layered Media*, 1988.

Crystal structure

The arrangement of atoms, ions, or molecules in a crystal. Crystals are defined as solids having, in all three dimensions of space, a regular repeating internal unit of structure. This definition, which dates from early studies of crystals, expresses very concisely what crystals are.

The interior of crystals has been studied by the use of x-rays, which excite signals from the atoms that inhabit the crystal. The signals are of different strengths and depend on the electron density distribution about atomic cores. Thus, there is no probe analogous to a constant light that illuminates a landscape uniformly. Light atoms give weaker signals and hydrogen is invisible to x-rays. However, the mutual atomic arrangements that are called crystal structures can be derived once the chemical formulas and physical densities of solids are known, based on the knowledge that atomic positions are not arbitrary but are dictated by crystal symmetry, and that the diffraction signals received are the result of systematic constructive interference between the scatterers within the regularly repeating internal unit of pattern.

This article discusses fundamental concepts, the dimensionality dependence of crystal structure determination, the structures found for elements and chemically simple compounds, and the conclusions drawn. The relationship between low coordination and structural variability is illustrated, as is a noncrystallographic packing. For related information *SEE* C OORDI - NATION NUMBER ; C RYSTALLOGRAPHY ; P OLYMORPHISM ; X- RAY CRYSTALLOGRAPHY ; X- RAY DIFFRACTION .

F UNDAMENTAL C ONCEPTS

Crystals should be imagined as three-dimensional solids. They have fundamental units of structure—atoms, ions, or molecules—mutually arranged to form a geometrically regular unit of pattern, or motif, that continuously repeats itself by translations in three dimensions. Crystals are thus defined in terms of space, population, and mutual arrangement. Crystal space is represented as an indefinitely extended lattice of periodically repeating points. The periodicity of the lattice is defined by the lengths and mutual orientations of three lattice vectors that enclose the pattern. Population is defined as the total number and kind of fundamental units of structure that form the pattern.

Ordering and states of matter. It is the chemical ordering to make a pattern, and the geometric regularity with which that pattern is repeated in space, that distinguishes among the three classical states of matter—gas, liquid, and solid—and also distinguishes between the crystals and noncrystalline solids. Gases are freely moving atoms or molecules that have no interaction that maintains a contact between them. They are totally disordered. Liquids of various viscosities are essentially incompressible. This means that there are fundamental units of structure that maintain a contact distance between them as they move freely in space. This distance is called the short-range spatial order of the liquid. Short-range order is similar to the order in a bag of marbles. *SEE* G AS ; L IQUID .

Glasses are similar to liquids but have rigid structures because contact between fundamental units is maintained. Their geometric regularity of repetition extends to a few diameters of the fundamental unit. This is an intermediate range of spatial order, larger than that of a liquid. *SEE* A MORPHOUS SOLID ; G LASS .

The order and periodicity of crystals must extend to about 100 nanometers in all three dimensions of space to give the sharply defined diffraction signals required for mapping structural details by x-rays. Intermediate states of order are seen in liquid crystals, which have long molecules as fundamental units of structure. These are arranged with their lengths parallel to each other, but without periodicity, in the nematic state. In the smectic state there is orientation in equally spaced planes but no sideways periodicity, like traffic moving freely on a multilane highway. *SEE* L IQUID CRYSTALS .

Requirement for three-dimensional order. The necessity of meeting the order and periodicity requirements in three dimensions for exact mapping of crystal structure is well illustrated by elemental carbon. In the graphite structure (**Fig. 1**) the atoms are arranged in six-member, planar, hexagonal rings, which form two-dimensional extensive and periodic layers by sharing their edges. Every carbon atom is trigonally bonded to three others in the same sheet, with separations of 0.142 nm. Normal to this sheet, the layer spacing is 0.342 nm, but this does not define a period. There are no binding forces between carbon atoms in different sheets, except for the weak van der Waals attractive forces; the layers therefore slide over each other like cards in a deck. There is therefore no lattice periodicity in the third dimension. The structure is fundamentally two-dimensional, and was more difficult to define than the three-dimensional, tetrahedrally bonded diamond structure, where the carbon-carbon separation is 0.154 nm. There are no sharp diffraction signals from which the interlayer spacing could be determined precisely because, lattice periodicity being absent in this direction, there is no fixed carbon-carbon separation. *SEE* G RAPHITE .

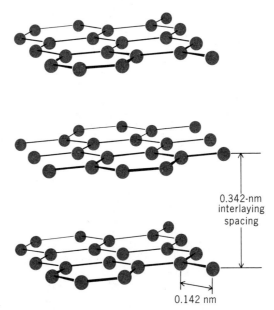

0.342-nm interlaying spacing

0.142 nm

Fig. 1. Crystal structure of graphite.

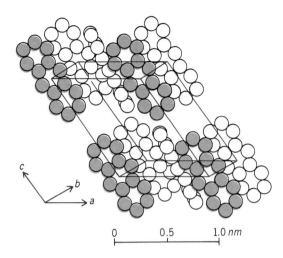

crystallizing and the medium that provides the components of growth and the growth environment. The process is reflected within the structures formed as an assemblage of atoms is collected and made relatively immobile by releasing the energy known as the heat of crystallization. The resulting crystal lattices resemble a mosaic of slightly misaligned adjacent regions. This is actually fortunate for research in x-ray crystallography. Perfect alignment would result in subtraction of energy by interference with the primary beam, due to a 180° phase reversal of the reflected beam (primary extinction). Internally diffracted beams would also be attenuated by internal reflection from regions above them (secondary extinction). *See* C*rystal growth.*

Detailed structure and advanced probes. Each of the spatially misaligned mosaic blocks of a single crystal is assumed to maintain lattice periodicity within it. This assumption is confirmed by the sharp diffraction patterns observed. There are some "wrong" atoms, vacant lattice sites, trapped gas atoms, and so forth, and the atomic occupants jiggle about while also vibrating cooperatively and synchronously in complex internal modes of motion. Intricate patterns of electron exchange are enacted, and systematic changes in spin orientations can occur for an atom with a magnetic moment. *See* C*rystal defects;* E*xchange interaction;* L*attice vibrations.*

Details like these are important for understanding the relationships between structure determination on the atomic and molecular levels and the cooperative behavior that determines bulk properties and functions. They are explored by methods other than conventional x-ray crystallography. For example, the power and tunability of synchrotron x-radiation makes it possible to discriminate against a specific atom by not exciting it to generate a signal. In effect, it is possible to sequentially "light up" different atoms and resolve specific details by parts. Intricate structures with very large atomic populations thereby yield maps with details resolved. *See* S*ynchrotron radiation.*

Arrangement of atoms and molecules.

Arrangement of atoms and molecules. Figure 2 illustrates the lattice symmetry and the fundamental unit of pattern and symmetry operations of a crystal structure that generate clones of the pattern in different orientations. The link between a crystal structure and either physical properties or biological function requires examination of the mutual arrangement of all the atoms and molecules, the distances that separate atom centers, and the distribution of empty space.

Figure 2 shows the crystal structure of anthracene ($C_{14}H_{10}$), an organic dye with blue fluorescence. The crystal symmetry is monoclinic with three unequal lattice vectors ($a = 0.856$ nm, $b = 0.604$ nm, $c = 1.116$ nm), and one angle, $\beta \neq 90°$ ($\beta = 124.7°$). There are 14 carbon atoms, arranged as three fused hexagonal rings in each unit of pattern and two units of pattern in each unit cell: eight at the cell vertices, each shared by eight cells (8/8 = 1), and two halves within each cell. The long axes of the molecules are parallel, inclined at about 9° from the c axis of the crystal. The 10 hydrogen atoms, invisible to x-rays but seen by neutron diffraction since they have a magnetic moment, symmetrically surround each unit of pattern. *See* A*nthracene;* N*eutron diffraction.*

The cohesive forces that cause these nonpolar molecules to come together and pack themselves into a crystal are explained by quantum mechanics as instantaneously induced dipole moments of one molecule on another, due to the noncoincidence of the centers of charge of an atom's nucleus and the electron cloud. Their average over time is zero, so that they are not observed as a net dipole moment. They provide the small cohesive energy of the crystal and the rapidly varying, minute locations of lines of communication within the crystal. *See* C*hemical bonding;* C*ohesion (physics);* I*ntermolecular forces.*

Crystal imperfection. The realities of crystal space are, in general, not as they are found in perfect diamonds, nor as they are modeled to be, even in crystals that appear to be freely formed, as in the case of crystallization from supersaturated solutions or from a melt. If scaled up, these crystals would win no prizes in crafting; not in quiltmaking for accurate replication of details within a unit of pattern; nor in wall building for precise translations of identical unit-cell bricks; nor in architecture for the exact symmetries of their extensive designs. The growth process is characterized by constraints and turbulences, and by the dynamic interaction between the substance that is

Space Lattices and Symmetry Operations

Figure 3 shows a rectangular space lattice with two possible cells outlined. These have the same cell volumes but different symmetries. Since crystallographic unit cells are completely defined by three lattice vectors, the crystal symmetry referenced to this lattice

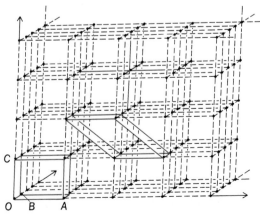

Fig. 3. A space lattice, two possible unit cells, and the environment of a point.

can be no higher than orthorhombic: $a \neq b \neq c$ ($OA \neq OB \neq OC$), and all angles equal to 90°. This and a possible monoclinic cell, with the same vectors a and b (OA and OB) and one angle not equal to 90°, are outlined. If the OAB plane is rotated and the vector a (OA) is extended to terminate at the next lattice point, then all angles differ from 90° and the crystal symmetry represented becomes triclinic. The mutual arrangement and atom coordinates of the cell population must be such that the environment, seen from every point of the space lattice, remains the same.

Screw axes. These combine the rotation of an ordinary symmetry axis with a translation parallel to it and equal to a fraction of the unit distance in this direction. **Figure 4** illustrates such an operation, which is symbolically denoted 3_1 and 3_2. The translation is respectively 1/3 and 2/3. The helices are added to help the visualization, and it is seen that they are respectively right- and left-handed. The projection on a plane perpendicular to the axis shows that the relationship about the axis remains in spite of the displacement. A similar type of arrangement can be considered around the other symmetry axes, and the following possibilities arise: 2_1, 3_1, 3_2, 4_1, 4_2, 4_3, 6_1, 6_2, 6_3, 6_4, and 6_5.

If screw axes are present in crystals, it is clear that the displacements involved are of the order of a few tenths of nanometer and that they cannot be distinguished macroscopically from ordinary symmetry axes. The same is true for glide mirror planes, which combine the mirror image with a translation parallel to the mirror plane over a distance that is half the unit distance in the glide direction.

The handedness of screw axes is a very important feature of many biological and mineral structures. They come in enantiomorphic pairs, with one member of the pair, the left-handed one, preferred, in the biological molecules found on Earth. In concentrated sugar solutions this screw symmetry operation is apparently maintained in some fashion since different directions for rotation of the plane of polarized light are observed. *See* Optical activity; Stereochemistry.

The helices of deoxyribonucleic acid (DNA) and of quartz are alike in that they contain what is essentially a structural backbone that can both flex and change partners with molecules that are not in the structure while maintaining structural continuity. In quartz, this is reflected by the direction of negative expansion when heated. The helical chain of $(SiO_4)^{4-}$ vertex-linked tetrahedra changes shape as two linked tetrahedra rotate about their common oxygen vertex. The overall configurational change is like that of a squashed mattress coil when it is sat upon. *See* Deoxyribonucleic acid (DNA); Silica minerals.

Space groups. These are indefinitely extended arrays of symmetry elements disposed on a space lattice. A space group acts as a three-dimensional kaleidoscope: An object submitted to its symmetry operations is multiplied and periodically repeated in such a way that it generates a number of interpenetrating identical space lattices. The fact that 230 space groups are possible means, of course, 230 kinds of periodic arrangement of objects in space. When only two dimensions are considered, 17 space groups are possible.

Space groups are denoted by the Hermann-Mauguin notation preceded by a letter indicating the Bravais lattice on which it is based. For example, P $2_1 2_1 2_1$ is an orthorhombic space group; the cell is

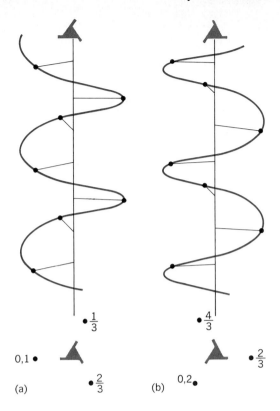

Fig. 4. Screw axes (a) 3_1 and (b) 3_2. Lower parts of figure are projections on a plane perpendicular to the axis. Numbers indicate heights of points above that plane.

primitive and three mutually perpendicular screw axes are the symmetry elements. J. D. H. Donnay and D. Harker have shown that it is possible to deduce the space group from a detailed study of the external morphology of crystals.

COMMON STRUCTURES

A discussion of some common structural systems found in metals, crystalline compounds, and minerals is given below.

Metals. In general, metallic structures are relatively simple, characterized by a close packing and a high degree of symmetry. Manganese, gallium, mercury, and one form of tungsten are exceptions. Metallic elements situated in the subgroups of the periodic table gradually lose their metallic character and simple structure as the number of the subgroup increases. A characteristic of metallic structures is the frequent occurrence of allotropic forms; that is, the same metal can have two or more different structures which are most frequently stable in a different temperature range.

The forces which link the atoms together in metallic crystals are nondirectional. This means that each atom tends to surround itself by as many others as possible. This results in a close packing, similar to that of spheres of equal radius, and yields three distinct systems: close-packed (face-centered) cubic, hexagonal close-packed, and body-centered cubic.

Close packing. For spheres of equal radius, close packing is interesting to consider in detail with respect to metal structures. Close packing is a way of arranging spheres of equal radius in such a manner that the volume of the interstices between the spheres is minimal. The problem has an infinity of solutions. The manner in which the spheres can be most closely packed in a plane A is shown in **Fig. 5**. Each sphere is in contact with six others; the centers form a regu-

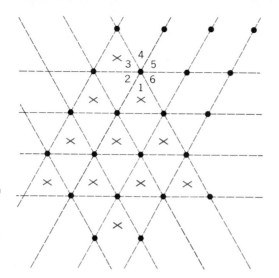

Fig. 5. Close packing of spheres of equal radius in a plane. The centers for the A, B, and C layers are indicated by a dot, a cross, and an open circle, respectively. Cavities between spheres are numbered.

lar pattern of equilateral triangles. The cavities between the spheres are numbered. A second, similar plane can be positioned in such a way that its spheres rest in the cavities 1, 3, 5 between those of the layer A. The new layer, B, has an arrangement similar to that of A but is shifted with respect to A. Two possibilities exist for adding a third layer. Its spheres can be put exactly above those of layer A (an assembly ABA is then formed), or they may come above the interstices 2, 4, 6. In the latter case the new layer is shifted with respect to A and B and is called C. For each further layer two possibilities exist, and any sequence such as ABCBABCACBA . . . in which two successive layers have not the same denomination is a solution of the problem. In the vast majority of cases, periodic assemblies with a very short repeat distance occur.

Face-centered cubic structure. This utilizes close packing characterized by the regular repetition of the sequence ABC. The centers of the spheres form a cubic lattice F, as shown in **Fig. 6a**. This form contains four sets of planes. The densely packed planes of the type A, B, C are perpendicular to the threefold axis and can therefore be written as {111}. Since these are the closest packed planes of the structure, d_{111} is greater than any other d_{hkl} of the lattice. The densest rows in these planes are $\langle 110 \rangle$.

It is relatively easy to calculate the percentage of the volume occupied by the spheres. The unit cell has a volume a^3 and contains four spheres of radius R, their volume being $16\pi R^3/3$. The spheres touch each other along the face diagonal (Fig. 6a); $a\sqrt{2}$ is therefore equal to $4R$, or $R = a\sqrt{2}/4$. Substitution gives for the volume of the spheres $\pi\sqrt{2}a^3/6$, which

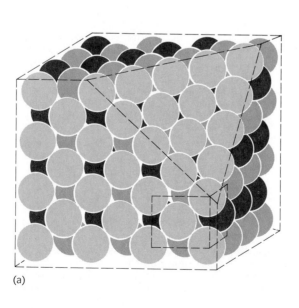

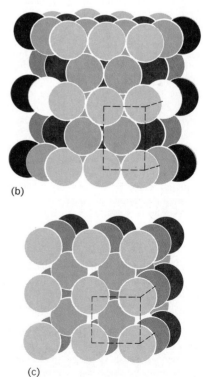

Fig. 6. Close packing of spheres in space. (a) Cubic close packing (face-centered cubic). One set of close-packed {111} planes (A, B, or C) is shown. (b) Hexagonal close packing. (c) Body-centered cubic arrangement.

(a)

(b)

(c)

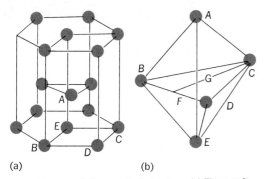

(a)   (b)

Fig. 7. Hexagonal close-packed structure. (a) Three unit cells, showing how the hexagonal axis results. One of the cells is fully outlined. (b) Calculation of the ratio c/a. Distance AE is equal to height of cell.

is 73% of the volume of the cube. It is clear that the same percentage of the unit volume is filled in all other close-packed assemblies, that is, assemblies corresponding to other alternations of the planes A, B, C.

Hexagonal close-packed structure. This is a close packing characterized by the regular alternation of two layers, or $ABAB$ . . . . The assembly has hexagonal symmetry (**Fig. 7**). Six spheres are at the corners of an orthogonal parallelepiped having a parallelogram as its base; another atom has as coordinates (1/3, 1/3, 1/2). The ratio c/a is easily calculated. The length $BD = a$, the edge of the unit cell. The height of the cell is $AE = 2AG = c$, defined in Eqs. (1) and (2).

$$AG = \sqrt{a^2 - \left(\frac{2}{3}\frac{a\sqrt{3}}{2}\right)^2} = a\sqrt{2/3} \qquad (1)$$

$$c = 2AG = a\sqrt{8/3} \quad c/a = \sqrt{8/3} = 1.633 \qquad (2)$$

This latter value is important, for it permits determination of how closely an actual hexagonal structure approaches ideal close packing.

Body-centered cubic structure. This is an assembly of spheres in which each one is in contact with eight others (Fig. 6c).

The spheres of radius R touch each other along the diagonal of the cube, so that measuring from the centers of the two corner cubes, the length of the cube

diagonal is $4R$. The length of the diagonal is also equal to $a\sqrt{3}$, if a is the length of the cube edge, and thus $R = a\sqrt{3/4}$. The unit cell contains two spheres (one in the center and 1/8 in each corner) so that the total volume of the spheres in each cube is $8\pi R^3/3$. Substituting $R = a\sqrt{3/4}$ and dividing by a^3, the total volume of the cube, gives the percentage of filled space as 67%. Thus the structure is less dense than the two preceding cases.

The closest-packed planes are {110}; this form contains six planes. They are, however, not as dense as the A, B, C planes considered in the preceding structures. The densest rows have the four $\langle 111 \rangle$ directions.

Tabulation of structures. The structures of various metals are listed in **Table 1**, in which the abbreviations fcc, hcp, and bcc, respectively, stand for face-centered cubic, hexagonal close-packed, and body-centered cubic. The structures listed hcp are only roughly so; only magnesium has a c/a ratio (equal to 1.62) that is very nearly equal to the ratio (1.63) calculated above for the ideal hcp structure. For cadmium and zinc the ratios are respectively 1.89 and 1.86, which are significantly larger than 1.63. Strictly speaking, each zinc or cadmium atom has therefore not twelve nearest neighbors but only six. This departure from the ideal case for subgroup metals follows a general empirical rule, formulated by W. Hume-Rothery and known as the 8-N rule. It states that a subgroup metal has a structure in which each atom has 8-N nearest neighbors, N being the number of the subgroup.

Table 2 compares metallic (elemental) and ionic

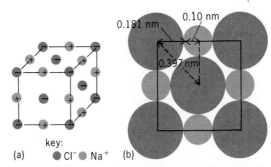

key:
(a)   ● Cl^- ● Na^+ (b)

Fig. 8. Structure of sodium chloride. (a) Cubic unit cell. (b) Basal plane of the unit cell. Effective sizes of ions are shown.

Table 1. Metal structures

Metal	Modification	Stability range	Structure
Cadmium	α	To melting point	hcp
Chromium	α	To melting point	bcc
	β	Electrolytic form	Hexagonal
Cobalt	α	To 788°F (420°C)	hcp mixed with fcc
	β	788°F (420°C) to melting point	fcc
Gold	α	To melting point	fcc
Iron	α	To 1668°F (909°C)	bcc
	γ	1668–2557°F (909–1403°C)	fcc
	δ	2557°F (1403°C) to melting point	bcc
Magnesium	α	To melting point	Nearly hcp
Nickel	α	To melting point	fcc
	β	Electrolytic form	Hexagonal
Silver	α	To melting point	fcc
Zinc	α	To melting point	hcp
Zirconium	α	To 1584°F (862°C)	hcp
	β	1584°F (862°C) to melting point	bcc

Table 2. Metallic and ionic radii in crystals

Period	Element	Atomic number Z	Metallic structure		Ionic structure	
			Structure type	Radius, nm	Radius, nm	Charge
First short	Lithium (Li)	3	bcc	0.152	0.07	1$^+$
	Berylium (Be)	4	hcp	0.111	0.035	2$^+$
Second short	Sodium (Na)	11	bcc	0.186	0.10	1$^+$
	Magnesium (Mg)	12	hcp	0.160	0.07	2$^+$
First long	Potassium (K)	19	bcc	0.231	0.133	1$^+$
	Calcium (Ca)	20	fcc	0.197	0.100	2$^+$
Second long	Rubidium (Rb)	37	bcc	0.240	0.150	1$^+$
	Strontium (Sr)	38	fcc	0.210	0.110	2$^+$
Third long	Cesium (Cs)	55	bcc	0.266	0.170	1$^+$
	Barium (Ba)	56	bcc	0.217	0.135	2$^+$

radii of the same elements. The first and second elements of their respective periods are listed. Both radii are determined from families of compounds. Metallic radii are always larger—for the light metals, very much larger—than the ionic radii. Thus, although the metallic structures are close-packed arrangements, the atoms obviously do not touch each other as they do in the sphere-packing models.

Crystalline compounds. Simple crystal structures are usually named after the compounds in which they were first discovered (diamond or zinc sulfide, cesium chloride, sodium chloride, and calcium fluoride). Many compounds of the types A^+X^- and $A^{2+}X_2^-$ have such structures. They are highly symmetrical, the unit cell is cubic, and the atoms or ions are disposed at the corners of the unit cell and at points having coordinates that are combinations of 0, 1, 1/2, or 1/4.

Sodium chloride structure. This is an arrangement in which each positive ion is surrounded by six negative ions, and vice versa. The arrangement is expressed by stating that the coordination is 6/6. The centers of the positive and the negative ions each form a face-centered cubic lattice. They are shifted one with respect to the other over a distance $a/2$, where a is the repeat distance (**Fig. 8a**). Systematic study of the dimensions of the unit cells of compounds having this structure has revealed that:

1. Each ion can be assigned a definite radius. A positive ion is smaller than the corresponding atom and a negative ion is larger. Figure 8b shows the effective sizes of ions in sodium chloride.

2. Each ion tends to surround itself by as many

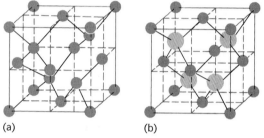

(a)   (b)

Fig. 10. Tetrahedral crystal compound structures. (a) Diamond. (b) Zinc blende.

others as possible of the opposite sign because the binding forces are nondirectional.

On this basis the structure is determined by two factors, a geometrical factor involving the size of the two ions which behave in first approximation as hard spheres, and an energetical one involving electrical neutrality in the smallest possible volume. In the ideal case all ions will touch each other; therefore, if r_A and r_X are the radii of the ions, $4r_X = a\sqrt{2}$ and $2(r_A + r_X) = a$. Expressing a as a function of r_X gives $r_A/r_X = \sqrt{2} - 1 = 0.41$. When r_A/r_X becomes smaller than 0.41, the positive and negative ions are no longer in contact and the structure becomes unstable. When r_A/r_X is greater than 0.41, the positive ions are no longer in contact, but ions of different sign still touch each other. The structure is stable up to $r_A/r_X = 0.73$, which occurs in the cesium chloride structure. *See Ionic crystals.*

Cesium chloride structure. This is characterized by a coordination 8/8 (**Fig. 9a**). Each of the centers of the positive and negative ions forms a primitive cubic lattice; the centers are mutually shifted over a distance $a\sqrt{3}/2$. The stability condition for this structure can be calculated as in the preceding case. Contact of the ions of opposite sign here is along the cube diagonal (Fig. 9b).

Diamond structure. In this arrangement each atom is in the center of a tetrahedron formed by its nearest neighbors. The 4-coordination follows from the well-known bonds of the carbon atoms. This structure is illustrated in **Fig. 10a**. The atoms are at the corners of the unit cell, the centers of the faces, and at four points having as coordinates (1/4, 1/4, 1/4), (3/4, 3/4, 1/4), (3/4, 1/4, 3/4), (1/4, 3/4, 3/4). The atoms can

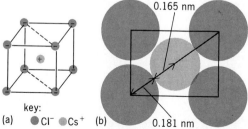

0.165 nm

key:

(a)   ● Cl$^-$ ● Cs$^+$ (b)   0.181 nm

Fig. 9. Structure of cesium chloride. (a) Cubic unit cell. (b) Section of the cell through a diagonal plane, showing the close packing of Cs$^+$ and Cl$^-$ ions, and effective sizes of ions.

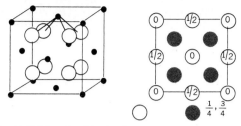

Fig. 11. Calcium fluoride structure.

be divided into two groups, each forming a face-centered cubic lattice; the mutual shift is $a\sqrt{3/4}$.

Zinc blende structure. This structure, shown in Fig. 10*b*, has coordination 4/4 and is basically similar to the diamond structure. Each zinc atom (small circles in Fig. 10*b*) is in the center of a tetrahedron formed by sulfur atoms (large circles), and vice versa. The zinc atoms form a face-centered cubic lattice, as do the sulfur atoms.

Calcium fluoride structure. **Figure 11** shows the calcium fluoride structure. If the unit cell is divided into eight equal cubelets, calcium ions are situated at corners and centers of the faces of the cell. The fluorine ions are at the centers of the eight cubelets. There exist three interpenetrating face-centered cubic lattices, one formed by the calcium ion and two by the fluorine ions, the mutual shifts being (0, 0, 0), (1/4, 1/4, 1/4), (3/4, 3/4, 3/4).

Minerals. Silica (SiO_2) is the most abundant material in the Earth's crust. It is surpassed only by carbon in variety of architectural structures. As in carbon, these structures also have widely ranging, dynamic properties. **Figure 12** indicates schematically how this abundance of structures can evolve from the 4:2 coordination of silica where each silicon atom has four tetrahedrally disposed oxygen atoms about it. Figure 12*a* represents an isolated tetrahedron with unsatisfied charge about each oxygen atom. Joining tetrahedra to make a single chain (Fig. 12*b*) in effect subtracts oxygen atoms and lowers the charge. Joining two such chains (Fig. 12*c*) extends the structure and also reduces the charge since another oxygen is subtracted. It is readily seen how the single chain of Fig. 12*b* could be wrapped as a helix, thereby extending the charged regions into the third dimension. *See* SILICATE MINERALS.

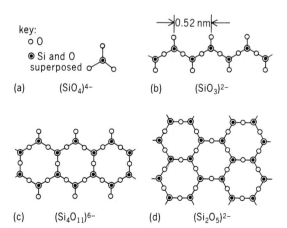

Fig. 12. Structures of silicates. (*a*) Fundamental SiO_4 tetrahedron. (*b*) Linear chain of tetrahedra. (*c*) Band formed by two linked linear chains. (*d*) Plane of linked tetrahedra.

Figure 13 shows a polyhedral structural model of the zeolite faujasite. The composition is $NaCa_{0.5}(Al_2Si_5O_{14}) \cdot 10H_2O$. Every vertex of the truncated octahedra with hexagonal faces is the location of a tetrahedrally coordinated silicon or aluminum atom. The structure continues to fill space indefinitely. However, while model space is filled according to lattice requirements, physical space is quite empty. This is a charged framework structure in which mobile species,

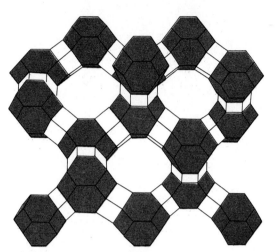

Fig. 13. Crystal structure of faujasite. The silicon and aluminum atoms are situated at the apices of the truncated octahedra, which are joined to form a three-dimensional framework.

such as water (H_2O) molecules and cations, can freely flow through the large pipes that are an integral part of the structure. Cation exchange occurs with this flow. *See* MOLECULAR SIEVE; ZEOLITE.

STRUCTURES AND PERIODICITY

The requirement that structures have three-dimensional periodicity that can be referenced to a Bravais lattice of mathematical points with three lattice vectors, that every point of the Bravais lattice have the same environment about it, and therefore that identical unit cells repeat periodically by translations in all three directions, is a requirement for accurate crystal structure determination by x-rays. As was demonstrated by the example of graphite, it is not a requirement imposed upon physical reality. Compositionally modulated structures exist; indeed, they are to be expected. As growth occurs, local composition changes, the heat of crystallization is released, and growing surfaces become available as templates for fresh deposition. Nor does the fact that crystals, as defined, give sharp x-ray diffraction patterns imply that all diffraction patterns come from crystals, as defined.

Finally, the random packing of spheres in contact, resembling a bag of marbles, is very dense, while close-packed arrangements need not be physically dense. **Figure 14** shows how a dense-sphere packing, with one-dimensional periodicity only, along a five-fold symmetry axis, will fill space. A two-layer construction is shown in which the first layer forms concentric pentagons with odd numbers of spheres on each side, and a second layer is added in which the corresponding numbers of spheres are even. The compact two-layer constructions extend indefinitely in two dimensions and can be stacked to form an infinite

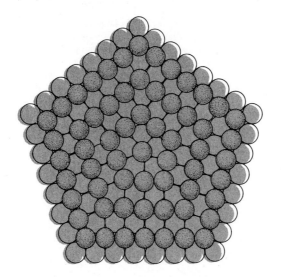

Fig. 14. Sphere packing with fivefold symmetry. (*After A. F. Wells, Models in Structural Inorganic Chemistry, Oxford University Press, 1970*)

packing along the fivefold axis. The density is 0.724, which is denser than the body-centered cubic close packing of spheres.

Doris Evans

Bibliography. C. S. Barrett and T. B. Massalski, *Structure of Metals*, 3d ed., 1980; J. D. Brock, Liquids, crystals and liquid crystals, *Phys. Today*, 42(7):52–59, July 1989; A. Guinier, *The Structure of Matter*, 1984; M. La Brecque, Quasicrystals, *Mosaic* (National Science Foundation), 18(4):2–20, Winter 1987–1988; W. J. Moore, *Seven Solid States*, 1966; A. F. Wells, *Structural Inorganic Chemistry*, 5th ed., 1983.

Crystal whiskers

Single crystals that have grown in filamentary form. Such filamentary growths have been known for centuries and can form by a variety of mechanisms. Some grow from their base: either these are extruded to relieve pressure in the base material or they grow as a result of a chemical reaction at the base. In both cases, the growth occurs at a singularity in the base material. Other crystal whiskers grow at their tip, into a supersaturated medium, and the filamentary growth results from a strong anisotropy in the growth rate. *See Single crystal; Supersaturation.*

The growth of extruded metal whiskers was studied

Crystal whiskers grown by the vapor-liquid-solid process. (*AT&T Bell Laboratories*)

in the 1950s, when it was discovered that tin whiskers growing from solder were causing electrical short circuits. These whiskers were typically a few micrometers in diameter and a few millimeters in length. This shorting problem was solved by adding more lead to the solder. Great interest in the whiskers developed after it was discovered that the strength exhibited by some whiskers approached that expected theoretically for perfect crystals. This great strength results from the internal and surface perfection of the whiskers, whereas the strength of most materials is limited by defects. The interest in the high strength of the whiskers centered on the possibility of using them in composites to increase the strength of more ductile matrix materials. Fibers of silica, boron, and carbon, which are much easier to fabricate in large quantity than whiskers, also exhibit similarly high strengths, and are now used in composites. *See Composite material; Crystal defects.*

The controlled growth of whiskers of a wide variety of materials is possible by means of the vapor-liquid-solid mechanism (see **illus.**). Here, a liquid droplet containing impurities rides on the tip of the whisker, providing a preferential site for condensation. Many electronic materials are prone to whisker formation by the vapor-liquid-solid process during chemical vapor deposition, so that control of the deposition to eliminate whisker formation is essential. Whiskers of many substances have also been grown in supersaturated solutions, but the mechanism of growth is not understood in detail. *See Crystal growth; Vapor deposition.*

Kenneth A. Jackson; Richard S. Wagner

Bibliography. S. S. Brenner, Growth and properties of "whiskers," *Science*, 128:569–575, 1958; A. P. Levitt, *Whisker Technology*, 1970.

Crystallization

The formation of a solid from a solution, melt, vapor, or a different solid phase. Crystallization from solution is an important industrial operation because of the large number of materials marketed as crystalline particles. Fractional crystallization is one of the most widely used methods of separating and purifying chemicals. This article discusses crystallization of substances from solutions and melts. Not discussed is biological crystallization, involving formation of teeth and bone, otoconia in the inner ear, renal calculi (kidney stones), biliary calculi (gallstones), crystals in some forms of arthritis, and dental plaque. For crystallization in glass, a supercooled melt, *see Glass.* Polymer crystals obtained from solutions are used to study the properties of these crystals, while crystallization of polymer melts dramatically influences polymer properties; *see Polymer.* For methods of preparing large crystals *see Single crystal.* For crystallization from vapors *see Sublimation.* For solubility and other relationships between solid and liquid phases *see Phase equilibrium; Solution.*

Solutions. In order for crystals to nucleate and grow, the solution must be supersaturated; that is, the solute must be present in solution at a concentration above its solubility. Different methods may be used for creating a supersaturated solution from one which is initially undersaturated. The possible methods depend on how the solubility varies with temperature. Two examples of solubility behavior are shown in the

illustration. Either evaporation of water or cooling may be used to crystallize potassium nitrate (KNO_3), while only evaporation would be effective for NaCl. An alternative is to add a solvent such as ethanol which greatly lowers the solubility of the salt, or to add a reactant which produces an insoluble product. This causes a rapid crystallization perhaps more properly known as precipitation. *See Precipitation (chemistry)*.

Nucleation. The formation of new crystals is called nucleation. At the extremely high supersaturations produced by addition of a reactant or a lower-solubility solvent, this nucleation may take place in the bulk of the solution in the absence of any solid surface. This is known as homogeneous nucleation. At more moderate supersaturations new crystals form on solid particles or surfaces already present in the solution (dust, motes, nucleation catalysts, and so on). This is called heterogeneous nucleation. When solutions are well agitated, nucleation is primarily secondary, that is, from crystals already present. Probably this is usually due to minute pieces breaking off the crystals by impact with other crystals, with the impeller, or with the walls of the vessel. *See Nucleation*.

Crystal size. Generally, large crystals are considered more desirable than small crystals, probably in the belief that they are more pure. However, crystals sometimes trap (occlude) more solvent as inclusions when they grow larger. Thus there may be an optimal size. The size distribution of crystals is influenced primarily by the supersaturation, the amount of agitation, and the growth time. Generally, the nucleation rate increases faster with increasing supersaturation than does the growth rate. Thus, lower supersaturations, gentle stirring, and long times usually favor large crystals. Low supersaturations require slow evaporation and cooling rates.

Crystal habit. Often the habit (shape) of crystals is also an important commercial characteristic. In growth from solutions, crystals usually display facets along well-defined crystallographic planes, determined by growth kinetics rather than by equilibrium considerations. The slowest-growing facets are the ones that survive and are seen. While habit depends somewhat on the supersaturation during growth, very dramatic changes in habit are usually brought about by additives to the solution. Strong habit modifiers are usually incorporated into the crystal, sometimes preferentially. Thus the additive may finally be an impurity in the crystals.

Fractional crystallization. In fractional crystallization it is desired to separate several solutes present in the same solution. This is generally done by picking crystallization temperatures and solvents such that only one solute is supersaturated and crystallizes out. By changing conditions, other solutes may be crystallized subsequently. Occasionally the solution may be supersaturated with respect to more than one solute, and yet only one may crystallize out because the others do not nucleate. Preferential nucleation inhibitors for the other solutes or seeding with crystals of the desired solute may be helpful. Since solid solubilities are frequently very small, it is often possible to achieve almost complete separation in one step. But for optimal separation it is necessary to remove the impure mother liquor from the crystals. Rinsing of solution trapped between crystals is more effective for large-faceted crystals. Internally occluded solution cannot be removed by rinsing. Even high temperatures may

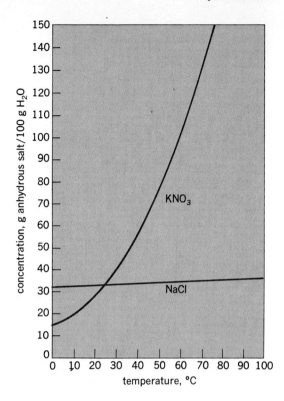

Temperature-solubility curves for two salts. °F = (°C × 1.8) + 32.

not burst these inclusions from the crystal. Repeated crystallizations are necessary to achieve desired purities when many inclusions are present or when the solid solubility of other solutes is significant.

Melts. If a solid is melted without adding a solvent, it is called a melt, even though it may be a mixture of many substances. That is the only real distinction between a melt and a solution. Crystallization of a melt is often called solidification, particularly if the process is controlled by heat transfer so as to produce a relatively sharp boundary between the solid and the melt. It is then possible to slowly solidify the melt and bring about a separation, as indicated by the phase diagram. The melt may be stirred to enhance the separation. The resulting solid is usually cut into sections, and the purest portion is subjected to additional fractional solidifications. Alternatively the melt may be poured off after part of it has solidified. Zone melting was invented to permit multiple fractional solidifications without the necessity for handling between each step. Fractionation by the above techniques appears to be limited to purification of small batches of materials already fairly pure, say above 95%. *See Zone refining*.

Fractional crystallization from the melt is also being used for large-scale commercial separation and purification of organic chemicals. It has also been tested for desalination of water. Rather than imposing a sharp temperature gradient, as in the above processes, the melt is relatively isothermal. Small, discrete crystals are formed and forced to move countercurrent to the melt. At the end from which the crystals exit, all or part of the crystals are melted. This melt then flows countercurrent to the crystals, thereby rinsing them of the adhering mother liquor. *See Crystal growth; Crystal structure; Crystallography*.

William R. Wilcox

Bibliography. S. J. Jancic and P. A. Grootscholten, *Industrial Crystallization*, 1984; J. W. Mullin (ed.), *Industrial Crystallization*, 1976; J. Nývlt, *Industrial Crystallization: The Present State of the Art*, 2d ed., 1983; J. Nývlt and O. Sohnel, *The Kinetics of Industrial Crystallization*, 1985; A. D. Randolph and M. A. Larson, *Theory of Particulate Processes: Analysis and Techniques of Continuous Crystallization*, 1971.

Crystallography

The branch of science that deals with the geometric forms of crystals. The field of crystallography originated in the prehistoric observation of geometric forms of minerals, such as a collection of quartz (SiO_2) crystals **(Fig. 1).** These crystals grow as hexagonal pyramids, with growth being terminated by the coming together of external faces to form a pinnacle.

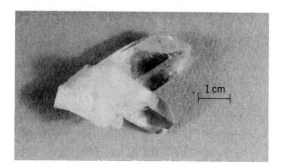

Fig. 1. Natural, intergrown quartz crystals.

Equally beautiful forms can be made by very simple procedures. How to describe, classify, and measure such forms are the first questions of crystallography. Revealing the forces that made them and the activities within them are the modern directions of the field. This article deals with the rigor, scope, and impact of crystallography, its interdisciplinary nature, and the key results and successes that have made it historically continuous and vital. For other aspects of crystallography *See* COORDINATION NUMBER; CRYSTAL STRUCTURE; POLYMORPHISM.

Crystallography is essential to progress in the applied sciences and technology and developments in all materials areas, including metals and alloys, ceramics, glasses, and polymers, as well as drug design. It is equally vital to progress in fundamental physics and chemistry, mineralogy and geology, and computer science, and to understanding of the dynamics and processes of living systems. This vitality and power are reflected in the rapid exploitation and commercial development of its methods and the support given to data files that record and make available its results.

CLASSICAL FUNDAMENTALS

Understanding of crystals with well-developed faces can be obtained with visible light, a simple mechanical device, mathematics, and the imagination as the only tools.

Faces and zones. Slow changes in the composition of the in-feeding substances and in the environment of growth (temperature, pressure, and atmospheric composition) yield many minerals that have large

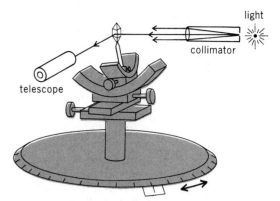

Fig. 2. Reflecting goniometer.

crystals that are optically transparent and chemically well defined, with well-defined shapes. The external morphology reflects growth rates in different directions. These directions remain constant during the course of the growth process, and are represented mathematically as the normals to sets of parallel planes that are imagined as being added on as growth proceeds. The faces that meet and define an edge belong to a zone, a zone being a set of planes that share one common direction, the zone axis. The invariance of interfacial angles, measured by rotation about an axis that is defined by the zone direction, was discovered in the seventeenth century. *See* CRYSTAL GROWTH.

Goniometer. Figures 2 and **3** show the optical goniometer and the method of tagging and measuring morphological geometries. Translation in two orthogonal directions aligns a face of the crystal so that it reflects light into a telescope when a zone axis is vertical, and the position is recorded. The crystal is then reoriented by rotation about the goniometer axis and two calibrated circular arcs, horizontal or equatorial and vertical or azimuthal, to reflect light from another face into the telescope.

Spherical projection. Interfacial angles are calculated from spherical geometry. Figure 3 illustrates the procedure for a crystal having well-developed faces of which three are mutually perpendicular. The normals to these faces are the natural directions for con-

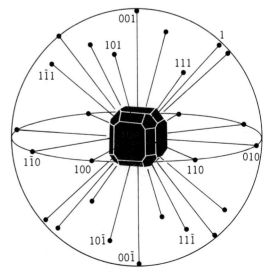

Fig. 3. Spherical projection of normals to crystal faces.

structing an orthogonal frame of reference for measurement. The crystal is imagined to be shrunk and placed at the center of a sphere with coordinates (0,0,0). The face normals, labeled [100], [010], and [001], define the directions of an orthogonal reference system. Normals to the same set of planes, but oppositely directed, are labeled [$\bar{1}$00], [0$\bar{1}$0], and [00$\bar{1}$]. The reversal of sign indicates that the crystal must be rotated 180° to obtain the same view. Rotation about the [001] direction interchanges the positions of [100] and [$\bar{1}$00] faces and their bounding edges. Rotation about [010] turns these faces upside down. Correct designations for group movements and symmetry operations are clearly essential for establishing and maintaining orientation in crystal space. The directions of face normals determine points at which the imagined sphere is pierced. The solid angles between an array of such points, all lying on the same great circle of the sphere, belong to a zone.

Bravais lattices. Optical measurements and stereographic projections established the constancy of interfacial angles, independent of how well developed the faces are. Such properties as the cleavage of large rhombohedral crystals of calcite ($CaCO_3$) into little rhombs suggested that the large crystal could be rep-

resented by geometrically identical smaller units stacked together, by translation, to fill space. The 14 lattices of Bravais (**Fig. 4**) enlarged the seven crystal systems of optical mineralogy by adding centering points to them: body (I), face (F), and base (C) centers. The 14 lattices define three-dimensional distributions of mathematical points such that the environments of all points of the lattice are identical. They also define the symmetries of frameworks for constructing mathematical models to represent the observed and measured realities—models made from cells of the smallest volume, but also highest symmetry, that stack together by translation to fill space.

Stacking of model cells does not imply that a crystal grows by stacking identical bricks; a lattice of identically surrounded mathematical points does not imply that any real objects, atoms or molecules, are located at the points; and filling space by translation of identical cells does not imply that the space defined by the cells is filled. Rather, the Bravais lattices are a formalism for representing observed geometries and symmetries of real crystals by three-dimensional lattices of identically surrounded points. **Figure 5** shows a two-dimensional square array of points with imaginary cells outlined. The four different shapes with one

Bravais lattice cells	Axes and interaxial angles	Examples
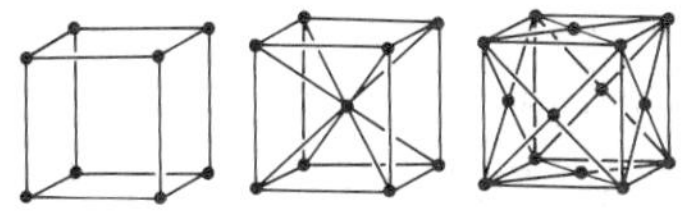 Cubic P Cubic I Cubic F	Three axes at right angles; all equal: $a = b = c$; $\alpha = \beta = \gamma = 90°$	Copper (Cu), silver (Ag), sodium chloride (NaCl)
 Tetragonal P Tetragonal I	Three axes at right angles; two equal: $a = b \neq c$; $\alpha = \beta = \gamma = 90°$	White tin (Sn), rutile (TiO_2), β-spodumene ($LiAlSi_2O_6$)
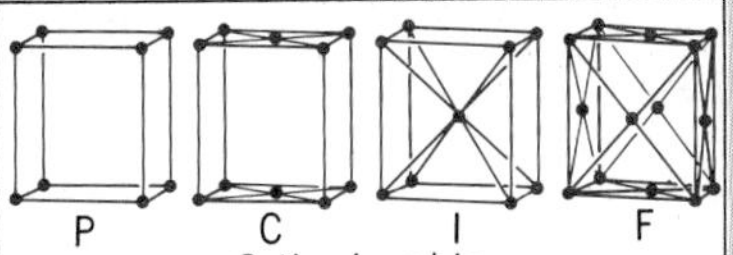 P C I F Orthorhombic	Three axes at right angles; all unequal: $a \neq b \neq c$; $\alpha = \beta = \gamma = 90°$	Gallium (Ga), perovskite ($CaTiO_3$)
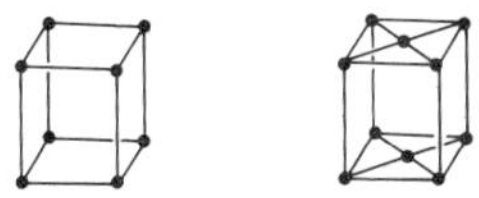 Monoclinic P Monoclinic C	Three axes, one pair not at right angles, of any lengths: $a \neq b \neq c$; $\alpha = \gamma = 90° \neq \beta$	Gypsum ($CaSO_4 \cdot 2H_2O$)
 Triclinic P	Three axes not at right angles, of any lengths: $a \neq b \neq c$; $\alpha \neq \beta \neq \gamma \neq 90°$	Potassium chromate (K_2CrO_7)
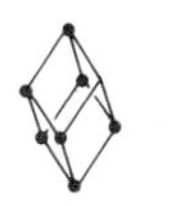 Trigonal R (rhombohedral)	Rhombohedral: three axes equally inclined, not at right angles; all equal: $a = b = c$; $\alpha = \beta = \gamma \neq 90°$	Calcite ($CaCO_3$), arsenic (As), bismuth (Bi)
 Trigonal and hexagonal C (or P)	Hexagonal: three equal axes coplanar at 120°, fourth axis at right angles to these: $a_1 = a_2 = a_3 \neq c$; $\alpha = \beta = 90°$, $\gamma = 120°$	Zinc (Zn), cadmium (Cd), quartz (SiO_2) [P]

Fig. 4. The 14 Bravais lattices, derived by centering of the seven crystal classes (P and R) defined by symmetry operators.

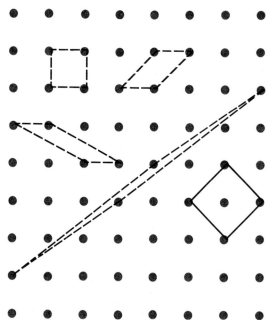

Fig. 5. Two-dimensional lattice of points. All cells fill space by translation; that is, the surroundings of all lattice points are the same. The four cells indicated by broken lines are P, with one unit of pattern and the same areas. The cell indicated by solid lines is C, with two units of pattern and twice the area of the others.

lattice point per cell have the same area and one elementary motif per cell (P in Fig. 4). The square with two units of motif, one within it and the four that are common to each of four cells (4/4 = 1), has twice the area of the other shapes; it is centered. The lengths of three edges, and the angles between them, represent crystal symmetry with reference to three-dimensional lattices of such points.

Miller indices. The lattices also provide the means to identify imaginary planes within the cell; these are called Miller indices (h,k,l). They consist of small whole numbers. For example, each of the six faces of a simple cube, with the origin of a coordinate frame of reference at the cube body center, is normal to one of the reference axes and parallel to the plane defined by the two others. The six faces are indexed as their normals in Fig. 3—(100), ($\bar{1}$00), (010), (0$\bar{1}$0), (001), and (00$\bar{1}$)—to represent a face that intercepts the $x,\bar{x}$ axis but not the y and z; the $y,\bar{y}$ axis but not the x and z; and so forth. Hypothetical parallel planes with ½

the interplanar spacing are represented as (200), ($\bar{2}$00), (020), and so forth.

Figure 6 shows the indexing of faces for an octahedron—the solid obtained by truncating the eight vertices of a cube until all the original faces disappear and a new regular solid with six vertices and eight faces, the dual of the cube with eight vertices and six faces, emerges. The truncation simulates the relative change in growth rates that exposes interior planes as bounding faces, thereby creating the architecture of crystals. Nothing has changed except the faces exposed. The octahedron has full cubic symmetry. The eight bounding faces have intercepts (111), ($\bar{1}$11), (1$\bar{1}$1), (11$\bar{1}$), ($\bar{1}\bar{1}$1), (1$\bar{1}\bar{1}$), ($\bar{1}$1$\bar{1}$), and ($\bar{1}\bar{1}\bar{1}$).

Symmetry operations and space groups. A complete mathematical formalism for modeling an external morphological form, and the symmetry relations between imagined units of structure within it, was in place in the nineteenth century. The symmetry operators include rotation axes, glide and mirror planes, and left- and right-handed screw axes which will simultaneously rotate and translate a three-dimensional object to create its clone in a different spatial position and orientation. The operators minimize the detail required to specify the spatial arrangements of patterns and objects that fill two-dimensional and three-dimensional space.

Symmetry is easily detected in common objects like a car wheel, where one nut and either a five- or sixfold rotation axis will generate other nuts holding the wheel in place. Wallpaper and quilts are useful for determining the minimal unit of pattern that must be specified within a cell, the combination of symmetry operations required to generate the complete pattern within a cell, and different shapes to outline two-dimensional cells that can be moved to fill the plane by translation. Quilts have well-defined symmetry operations and cells and identical shapes of nonidentical clones, if made from different fabrics. These are analogous to the so-called color space groups of crystallography, space groups that greatly increase the number of distinguishably different symmetries beyond the classical 230 by adding a fourth coordinate to the three space coordinates. This is done to encode a real difference that will be manifested in some property. The different directions of the magnetic moments of chemically identical atoms of an element such as iron provide an example of the need for representing a difference on the atomic level between cells that are otherwise identical.

Properties and models. Many physical and chemical properties were known, and models that accounted for them were proposed, before there was a means of getting inside crystals and mapping the components and mutual arrangements of components. Physical properties include the opacity and ductility of metals; the flakiness of mica; the hardness of diamond; the lubricity of graphite; double refraction from calcite, which gives two images of an object; pyroelectricity, where a thermal gradient builds up opposite charges on different faces; the piezoelectric effect in quartz; and the inherent anisotropy of crystals and the handedness of quartz, a handedness seen as either a counterclockwise or clockwise rotation of the plane of polarization of light. Polymorphs were known, as were chemical isomers, chemical compositions with the same molecular formula but different properties, such as ammonium cyanate (NH_4CNO) and urea [$CO(NH_2)_2$]. *See Birefringence; Crystal op-*

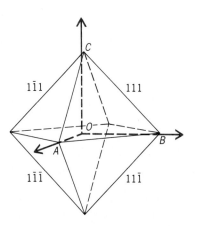

Fig. 6. Indexing of the four front faces of an octahedron.

TICS; *Diamond; Graphite; Isomorphism (crystallography); Metal, mechanical properties of; Piezoelectricity; Polarized light; Pyroelectricity; Quartz.*

Chemically identical but conformationally different (chair and boat) geometries had been proposed for cyclohexane. Steric hindrance was a concept, and structure models had been proposed and built to explain observed properties. Several were later confirmed by x-ray crystallography. Models based on the packing of ion spheres for sodium chloride (NaCl) and potassium chloride (KCl) were particularly insightful. Compositions, densities, and refractive indices in different directions had been measured, molecular formulas conjectured, and magnetic effects explored. A mathematically rigorous framework for mapping the internal mutual arrangements existed before there was a probe to enter into materials with light (radiation) that would send comprehensible signals revealing the species that built the structures, how these species were arranged, and what the arrangements implied regarding the forces between atoms. *See Conformational analysis; Stereochemistry.*

Modern Crystallography

A suitable light or radiation to serve as a probe came with W. C. Roentgen's discovery of x-rays in 1895. The nature of x-radiation, whether waves or particles, was not immediately known, but the wavelength expected if x-rays were waves could be calculated to be of the order of 0.1 nanometer, comparable to the dimensions of atoms and 10^3–10^4 shorter than visible light. *See Electromagnetic radiation; X-rays.*

Laue theory. The great insight of M. von Laue followed, leading to quantum physics and the reconciliation of the dual wave–particle nature of electromagnetic radiation, and also providing the means to get within the structures that chemists and mineralogists were modeling. Laue asked what would happen if x-rays really were waves; a crystal was really a regular, periodic, three-dimensional array of atoms; and the two entities were to meet. He concluded that the atoms would be excited to vibrate cooperatively, generating interference phenomena analogous to that of an optical diffraction grating. The same phenomenon can be observed by looking at a distant light through the mesh of a fine fabric or sieve. Spots are seen, which get larger as the mesh spacing gets smaller, regularly arranged about the direction of the incident beam.

Equations (1)–(3) must be satisfied to give constructive interference from a regular three-dimensional array of scattering lattice points; here a, b, c

$$a(\alpha - \alpha_0) = h \qquad (1)$$

$$b(\beta - \beta_0) = k \qquad (2)$$

$$c(\gamma - \gamma_0) = l \qquad (3)$$

are the repeat distances of the space lattice; α and α_0, β and β_0, and γ and γ_0 are the direction cosines of the diffracted and incident beams; and h, k, and l are integers defining the order of a particular diffracted beam. Experiment showed all speculations to be correct: scattered x-rays combine with each other as waves do, crystals are three-dimensional arrays of regularly spaced scatterers, and the atoms and molecules of chemical models are real scatterers with ~0.1-nm separation between their centers. Broadened diffraction spots reveal considerable distortion from

lattice ideality, while very sharp spots come from crystals that are perfect within the resolution of the beam. *See X-ray crystallography; X-ray diffraction.*

Bragg spectrometer and modeling. The Bragg x-ray spectrometer uses the goniometer of Fig. 2 to study the x-ray spectrum, with the light replaced by x-rays, the collimator by a slit system, and the telescope by an ionization chamber. With mica and other crystals, the beam from a platinum target was shown to consist of continuous radiation, analogous to white light, and superimposed monochromatic radiation of various wavelengths that results from specific transitions between electron shells. The simple three-dimensional cubic chessboard model for structures of alkali halides (sodium chloride and potassium chloride), proposed many years before by optical crystallographers, was confirmed by W. L. Bragg by modeling all the spots found in Laue patterns, recorded on film by a stationary crystal, as well as the intensity differences seen when potassium chloride is substituted for sodium chloride. The following assumptions were used: (1) the mathematical planes of the lattice can act as x-ray mirrors; (2) the x-ray beam is heterogeneous; and (3) the scattering power of an atom is dependent upon its atomic weight.

With a two-dimensional chessboard lattice arrangement, Avogadro's number, and the experimental densities, the distance between the sodium and chlorine atoms in a base-centered square cell containing two atoms (sodium and chlorine) per cell can be calculated. This distance is the sum of the radii of hypothetical atoms of sodium and chlorine. The required probe thus exists for entering structures and obtaining interpretable signals from atoms. The intensities of these signals can be understood in terms of interference from atoms whose distribution in space is modeled by idealized lattice and symmetry theory. The apparent sizes of scatterers—atoms, ions, or molecules—can be measured by assuming spheres in contact. The physical density and molecular formula determine the cell contents. Systematic work recording patterns for the same element in different chemical combinations and decoding the patterns by matching them with hypothetical models of scatterers, whose mutual arrangements are assigned from existing lattice and symmetry theory, gives values for the radii of spherical atoms.

Figure 7 shows the Laue pattern of calcium fluoride (CaF_2), recorded on flat photographic film. The spots are indexed by the projection method of Fig. 3. The distribution of spots reveals the geometric regularity and lattice symmetry of arrangement. The pattern has a threefold rotation axis (spots repeat every $360/3 = 120°$) and a vertical mirror plane, normal to the film, which in effect reflects the spots. Spots on the left are seen on the right as if reflected by a mirror.

Bragg equation. The Bragg equation simplifies the Laue conditions to be met for constructive interference by using what had already been surmised for minerals: that crystals grow as if parallel planes were being added on. Equation (4) makes use of this to

$$\lambda = 2d_1 \sin \theta \qquad (4)$$

decode observed patterns as if they originated from scattering by imaginary planes reflecting the x-ray light as if the planes were mirrors within a unit cell. Here, λ is the x-ray wavelength, $d' = d/n$, where d

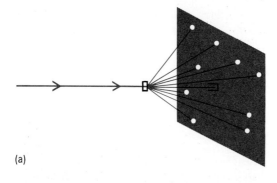

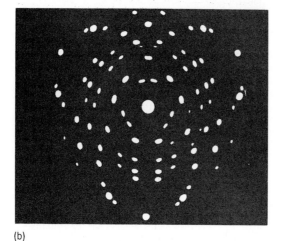

Fig. 7. Laue pattern of calcium fluoride (CaF_2), which has cubic symmetry. The x-ray beam is along a body diagonal (111). (*a*) Experimental setup for generating the pattern. (*b*) Pattern on photographic film. (***Photographic Library, Palais de la Découverte***)

(a)

(b)

is the true spacing of lattice planes and n is the number of imaginary planes between them, and θ is the reflection angle.

For orthogonal crystals, Eq. (5) relates the Miller

$$\frac{1}{d^2} = \frac{h^2}{a^2} + \frac{k^2}{b^2} + \frac{l^2}{c^2} \qquad (5)$$

indices of a plane, h, k, and l, the spacing d, and the lattice unit cell dimensions, a, b, and c. For the cubic system this simplifies to Eq. (6). More complex

$$\frac{1}{d^2} = \frac{h^2 + k^2 + l^2}{a^2} \qquad (6)$$

equations, involving angles, apply to lower symmetry.

X-ray spectra and synchrotron radiation. X-ray spectra have two origins. The beam ordinarily used in x-ray crystallography consists of sharp line spectra emitted by a target material when target atoms are ionized by removing electrons from innermost orbits. A limited heterogeneous spectrum of continuous (white) radiation is emitted during the sudden reduction of energy of the bombarding electrons. This is the radiation used in x-ray spectroscopy. The same phenomenon, reduction of electron beam energy, but with enormous energies and highly parallel beams, is the source of synchrotron radiation. Synchrotron beams add an extremely powerful probe to x-ray crystallography. They are tunable x-ray sources; therefore, some atoms in a structure can often, in effect, be switched off by using wavelengths that will not excite them. In addition, the high energies give dif-

fraction patterns in very short times. Therefore, the problem of degradation of biologically active materials during long x-ray exposures is eliminated. *See* Bremsstrahlung; Synchrotron radiation; X-ray spectroscopy.

Technological and Scientific Applications

The sharp x-ray line spectra, characteristic of the bombarded element, are the primary probes for determining interior structural detail.

Techniques. Following the introduction of beams of this radiation, experimental and analytical techniques developed rapidly. Cameras with cylindrical film and enclosed powdered samples record all diffraction lines as arcs of concentric circles. This fundamental powder method has endured since 1917 and is now employed with improved beam purity and optics, improved diffractometers which couple sample and detector rotation, electronic detection, rapid sequential recording, and computer indexing programs that provide patterns of compounds, mixtures of phases, and dynamic changes that occur when crystals are subjected and respond to external stress. The method is applied to single crystals, polycrystalline aggregates, and multiple-phase mixtures, randomly disposed either in space or in geometrically designed composite materials. *See* X-ray powder methods.

Data collection. The x-ray patterns of virtually every material that can be handled have been recorded and indexed. Comprehensive collections of experimental and model calculation results for metals, minerals, and inorganic compounds are kept current and accessible through the Joint Committee for Powder Diffraction Study (JCPDS). More sharply focused data files for structures of organic and biological interest are also extensive and report molecular configurations.

Basic structures and properties. Counting the number of neighbors about an atom and measuring the distances between their centers from the diffraction patterns of elements and compounds yield the basic elements of structure: atoms, molecules, ions, and the modern concept of the chemical bond. These procedures also yield explanations for some of the properties of materials.

Many metals fill space with a close-packed periodic arrangement of spheres in contact, but the space is really quite empty. The positively charged atom cores repel each other electrostatically and maintain large equilibrium separations through a sea of mobile electrons, which is the source of metallic conductivity. The mobility of the electron sea, the emptiness of the structures, and the mutual repulsions of the atom cores explain metallic ductility. The effective radii of atoms of the elementary metals are seen to be much greater than those found when the metals are in combination with other elements. Effective radii are also different in compounds having different numbers of neighbors around the same element, smaller for fewer near neighbors, larger for more. L. Pauling's theory of the chemical bond is based upon such crystallographic data. *See* Chemical bonding; Electrical conductivity of metals; Free-electron theory of metals.

Relation of structure and composition. Decoded patterns of materials of very different compositions and even from widely different origins show surprising similarities when viewed from within the framework of their structural architectures. For example, crystallographic patterns reveal much similarity of structural architecture between fibrous protein materials, such as wool, hair, and the myosin of muscle, which have very different compositions. SEE FIBROUS PROTEIN.

An example of materials of different origins is provided by silica (SiO_2) and ice (H_2O), which are extraordinarily alike in their polymorphic forms, the geometry of their building units, and the connectivity schemes that link the units to form the solids. There are four tetrahedrally arranged oxygen atoms about a silicon atom in silica and four hydrogen atoms about an oxygen atom in all the forms of ice. The variety of possible forms is seen as the many spatial arrangements that can be made by connecting each tetrahedron at its corners to other tetrahedra. SEE SILICA MINERALS; WATER.

In contrast to these architectural similarities between chemically different substances, other patterns show different structural architectures for identical compositions. Furthermore, fundamental differences in apparent size (as mentioned above) and mutual arrangements are observed among atoms of the same element. These differences depend upon the composition that the atoms are part of and also on the conditions of formation. The division between atomic, ionic, and covalently bonded arrangements is established by systematic measurements on the periodic lattice that models the observed diffraction patterns.

Studies of response to change. Modeling crystallographic results with space-filling unit cells of the proper density, independent of the mutual arrangement of atoms or molecules within the cell, provides the means for following, in the laboratory, the degrees of freedom of the overall structure, which are the types of adjustment that the structure makes to accommodate change without disruption. These are determined in several ways. Static determinations, which involve varying the composition systematically and comparing the patterns and densities of a family of compositions, reveal solid solutions and phase relations important to materials development. Dynamic determinations involve changing the temperature or pressure between a sequence of pattern recordings. Thermal expansion and compressibility are important both in developing new materials and properties and in understanding geological processes.

Dynamic studies. The foundation of space-filling lattice theory allows internal changes to be studied under laboratory conditions. Even when the details of mutual rearrangements are not known exactly, the size and shape of unit cells with the proper physical density can often be obtained from indexed patterns, density measurements, and a correct chemical formula. Subjecting a polycrystalline powder to external sources of stress is a dynamic way of focusing on structural details, such as the directions of strongest bonding, that are strongly implied but not directly measured. Heating, cooling, and the application of pressure cause changes in the size and shape of the unit cell as responses to the stress applied.

Changes in relative diffracted intensities indicate that the phase relationships that build up the total intensity are changing. This is often a signal that new modes of interatomic motion have been generated. Dynamic applied crystallography reveals many fundamental phenomena. A solution of elements that were initially uniformly distributed can coalesce as one-dimensional strings or two-dimensional platelets in specific crystallographic directions within the host lattice. A lively pattern of cooperative movements at the atomic level of structure can be observed crystallographically.

Crystals always sense and respond to environmentally imposed stress. They strive to equilibrate their structures to the new environment. Permanent and irreversible processes, as well as reversible ones, occur frequently. Detecting these is critical to the fundamental understanding of the solid state as well as the development of compositions and processes that produce materials that survive under conditions and environments of use.

Structure of glass. The surprising observation that there are true glasses that fill space uniformly with nonperiodic arrangements stimulated a reexamination of scattering theory and improvements in the detection of weak signals. The atomic structure of glass is explained by radial density distributions of scatterers in physical space such that all identical atoms in the structure have the same number and relative arrangement of their nearest neighbors. A geometrically irregular network that distributes spherical concentric shells of appropriate scatterers at larger distances accounts for the background pattern of diffuse scattering. A physical model of identical tetrahedral units, completely connected at their vertices and randomly arranged in space, fits the experimental scattering data. SEE AMORPHOUS SOLID; GLASS.

Living systems and other challenges. The dynamics of living systems, the difficulties in distinguishing light elements, and the inherent ambiguities of measuring, decoding, and mapping are continuing challenges. Major achievements of crystallography include the determination of the structures of deoxyribonucleic acid (DNA), proteins, other biological compounds, and boranes; the development of direct methods of phase determination; and the determination of the structure and mechanism of a photosynthetic center. SEE BORANE; DEOXYRIBONUCLEIC ACID (DNA); PHOTOSYNTHESIS; PROTEIN.

The problems encountered in the study of biological systems include the large atomic populations of these systems and the small scale of resolution needed to observe how components arrange themselves and how their configurations readjust, in response to subtle signals, with great speed and over extensive distances. **Figure 8** illustrates an ongoing challenge in the study of living systems: producing good crystals from material that does not ordinarily crystallize, and could not perform its biological function if it did crystallize. Although the crystal of porcine elastomer is considered large, it is 0.5 mm in the largest dimension, about 1/100 the size of the crystals in Fig. 1. It grew aboard Spacelab in 1988. Since convection forces on Earth disrupt the growth process, a gravity-free environment is required to grow a crystal of this protein, see inside it, and map its internal architecture. Only then can the rapid communication and readjustments that take place in living systems be modeled and understood. SEE SPACE PROCESSING.

Advanced probes and detailed modeling. Many kinds of radiation are used to enter within and excite signals from a crystal structure. All are employed to determine detail and fix positions for molecules and atoms within a space that is modeled as the real interior. Success in modeling depends upon extracting regularities and symmetries from signals that originate in the inner space of a material. These regularities are used to reconstruct both the summation of many am-

Fig. 8. Large porcine elastomer crystal grown in Spacelab (1988). (*Charles Bugg, University of Alabama*)

plitudes that yields the observed intensities, and the kind and mutual arrangement of the scatterers that locate these intensities in their observed positions. Relating these arrangements to internal activities is the goal of all crystallography. SEE ELECTRON DIFFRACTION; NEUTRON DIFFRACTION.

Icosahedral crystals. Conditions of formation in some three-component metal melts produce very small and morphologically well-defined icosahedral crystals. The idealized interior space of these crystals may not be correctly mapped in terms of diffraction patterns referenced to the three-vectors of a Bravais lattice, but require so-called higher-dimensional spaces. SEE QUASICRYSTAL.

Doris Evans

Bibliography. A. Guinier and R. Jullien, *The Solid State: From Superconductors to Superalloys*, 1989; T. Hahn (ed.), *International Union of Crystallography: Space-Group Symmetry*, 1985; C. Hammond, *Introduction to Crystallography*, 1990; A. Holden and P. Morrison, *Crystals and Crystal Growing*, 1982; J. Karle, Macromolecular structure from anomalous dispersion, *Phys. Today*, 42(6):22–29, June 1989; C. H. MacGillavary, *Symmetry Aspects of M. C. Escher's Periodic Drawings*, 1965; A. F. Wells, *Three Dimensional Nets and Polyhedra*, 1977.

Ctenophora

A phylum of exclusively marine organisms, formerly included in the jellyfish and polyps as coelenterates. These animals, the so-called comb jellies, possess a biradial symmetry of organization. They are characterized by having eight rows of comblike plates (*ctenos* = comb) as the main locomotory structures. Most of these animals are pelagic, but a few genera are creeping. Many are transparent and colorless; others are faintly to brightly colored. Many of these organisms are hermaphroditic. Development is biradially symmetrical, with a cydippid larval stage. Five orders constitute this phylum as follows:

Phylum Ctenophora
Order: Cydippida
Lobata
Cestida
Platyctenida
Beroida

Morphology. The body is gelatinous and extremely fragile; its form may be globular, pyriform, or bell- or helmet-shaped. Some species resemble a ribbon. Their size ranges from ⅛ to 20 in. (3 mm to 50 cm). All are biradially symmetrical. The axis of symmetry is determined by the mouth and the organ of equilibrium, or statocyst (**Fig. 1**). The mouth leads into the flattened, elongated pharynx. The sagittal plane is thus referred to as the stomodeal plane. The other plane of symmetry is perpendicular to the sagittal plane and is marked (Beroida excepted) by tentacles. This is known as the tentacular, transverse, or equatorial plane. Eight meridional comb-plate rows or ribs stretch from the aboral pole on the surface of the body.

Comb plates. These are the most characteristic structure of the ctenophores, possessed by all members of the phylum. Each plate consists of a great number of very long related cilia. Successive plates are arranged as meridional ribs or costae which extend from the aboral pole toward the mouth. Each rib is connected to the statocyst by a thin ciliary furrow. Of the eight ribs, four are stomodeal and are located on both sides of the stomodeal plane. The remaining four are subtentacular ribs and are situated on both sides of the tentacular plane. Comb rows are secondarily modified in some specialized ctenophores, and ribs may be entirely absent in the adult stages of a few genera of Platyctenida.

Tentacles. With the exception of Beroida, all ctenophores possess two highly extensile tentacles bearing many lateral branches (tentilla). The base of each tentacle is anchored on the bottom of a sheath which is formed by a deep depression in the body surface, and into which the tentacular apparatus may be completely retracted. Tentacles and tentilla are thickly covered with adhesive cells (colloblasts) which are exclusive to the phylum. Each colloblast is connected by the nervous system. A colloblast consists of a cell

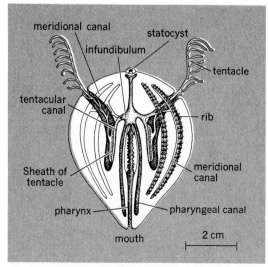

Fig. 1. Structure of a cydippid ctenophore.

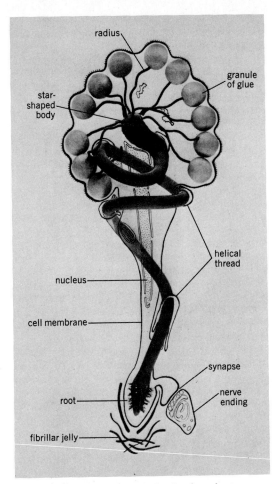

Fig. 2. Colloblast from the tentacle of a ctenophore.

iated cells. The peculiarity of these structures lies in a contractile diaphragm which allows complete opening or closing between the mesoglea and the lumen of the gastrovascular canals. The cell rosettes may be considered peculiar excretory or hydrostatic organs, since reversible liquid flows were observed through them with changes in seawater salinity.

Mesoglea. The maintenance of shape in the ctenophores depends on the mesoglea. This inner jelly, very rich in water (95%) is mainly made of protein-carbohydrate complexes. Collagen was biochemically detected, but has not been related to definite mesogleal structures. The mesoglea is crossed by numerous smooth muscle cells which are among the longest (2 in. or 5 cm) of the animal kingdom. Nerve processes connecting the muscle cells have been observed in the mesoglea, where scattered mesenchymal cells are also found.

Nervous system and sensory organs. Ctenophores possess a nervous system organized in a subectodermic synaptic network, which is more condensed along the ribs, under the statocyst, and in the tentacles. There is some evidence for the existence of a cholinergic system together with a monoaminergic one. Although intramesogleal neurons exist, innervation of the endoderm has not been observed. Sensory cells sensitive to water movement are located over the entire surface of the body and are connected to the nervous network. They are most abundant along the tentacles, the lips, and at the aboral pole. The nerve net propagates a localized excitation over the entire body of the animal and complex behavior patterns may take place (feeding behavior, for example). The statocyst is situated in a hollow of the aboral pole. The main feature of this equilibrium organ (**Fig. 3**) is the statolith, a mass of cells containing calcareous granules, held by and linked to four S-shaped tufts of cilia, the balancers. Each balancer, interradially located, is connected with the subpharyngeal and subtentacular costae of the same quadrant by the two corresponding ciliary furrows. The floor of the hollow contains secretory cells involved in the formation of the statolith and sensory cells, presumably photoreceptors and baroreceptors. A fence of incurved cilia forms a transparent dome over this aboral sensory area. Paired structures called the polar fields lie close to the statocyst, in the pharyngeal plane, and consist of linear or crescentic ciliated areas, the function of which is unknown.

Gonads. Many ctenophores are hermaphroditic. Gonads of endodermal origin develop in the outer side of the wall of the meridional canals or their branches. Both the ovaries and spermaries generally develop as continuous tracts of germ cells. The ovaries occupy the perradial and the spermaries the interradial side of the wall. Contrary to the spermatozoa of Porifera and

which looks like a mushroom (**Fig. 2**), the foot of which is firmly anchored in the fibrillar jelly surrounding the neuromuscular axis of the tentillum. This foot contains a helicoidal twisted thread which reaches the expanded head of the colloblast. Here, the helical thread divides into a great number of radii. Each radius terminates by a granule of glue. When a prey collides with a tentillum, some colloblasts burst and their granules stick the prey to the helical threads. These threads apparently act as shock absorbers of the captured prey's movements rather than as springs projecting the head. Tentacles are organs in continuous growth. Their formative tissue is located at their base, in the bottom of the tentacular sheath. In the adult stages of the Lobata and Cestida, the main tentacles, as described above, are reduced, and a series of tentilla with colloblasts develop on the margin of the mouth.

Gastrovascular system. The pharynx is connected aborally with a small chamber, the stomach, or infundibulum, from which a canal system common to all ctenophores ramifies throughout the mesoglea. The most important of the canals are the meridional canals which underlie the ribs, the pharyngeal (stomodeal) canals which run along the pharynx, and the tentacular canals (when tentacles exist) which go to the base of the tentacles. While the pharynx has an ectodermal lining, the stomach and the canal system are lined with endodermal epithelium which includes digestive cells. The peripheral canals bear special structures, the cell rosettes, consisting of the superposition of two circles of eight cil-

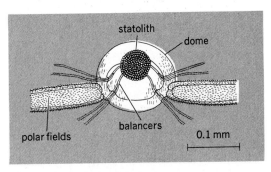

Fig. 3. Features of a statocyst.

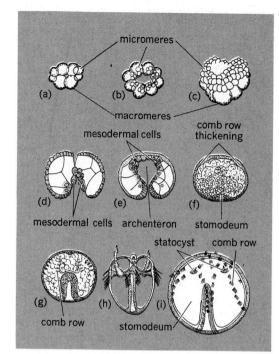

micromeres

(a) (b) (c)

macromeres

comb row
thickening

mesodermal cells

(d) (e) (f)

mesodermal cells   archenteron   stomodeum

statocyst   comb row

(g) (h) (i)

comb row   stomodeum

Fig. 4. Embryology. (*a*) Eight-cell stage. (*b*) Micromeres, viewed from above. (*c*) Later stage in division of micromeres. (*d*) Gastrula, section. (*e*) Invagination. (*f*) Stomodeum and thickenings for comb rows. (*g*) Stomodeum invaginated; comb rows differentiated. (*h*) Cydippid larva. (*i*) Embryo with statocysts and comb rows.

Cnidaria, the ctenophore spermatozoon possesses a well-differentiated acrosome. Gonducts seem to be of general occurrence. Functional gonads in the larval stage occur throughout the phylum.

Embryology. Fertilization usually occurs in seawater. Pelagic ctenophores are self-fertile, but cross-fertilization might also take place inside a swarm of ctenophores. Ctenophore eggs provide the unique example of a biradial pattern of cleavage (**Fig. 4**). The blastomeres are arranged in respect to the future two main planes of symmetry of the animal. Gastrulation is first epibolic (the macromeres are gradually covered by the multiplication of the micromeres), but ends with invagination of the last micromeres produced by macromeres. Development of the embryo is determinate. The claimed endodermal formation of a true mesoderm, giving rise to muscular and mesenchymal cells, is still debatable. The earliest larva formed in the development of all types of ctenophores is the cydippid, with a globular body, a pair of primary tentacles (or rudiments), and eight costae of comb plates. In the Lobata, Cestida, and Platyctenida, the larva undergoes secondary changes in later development.

Physiology. The ctenophores move through the sea with the oral pole forward, using comb plates as oars. The beating of successive plates occurs along each costa by metachronal waves and is initiated for the two costae of the same quadrant by the corresponding balancer of the statocyst. Balancer and comb-row activities and some muscular contractions are integrated by the nervous system.

Almost all ctenophores are luminescent. This capacity, linked to a calcium-dependent photoprotein, is assumed by specialized cells underlying the comb plates.

The ctenophores feed on zooplankton. Prey are trapped with the tentacles (Cydippida, Cestida) or oral extensions of the body (Lobata), or are engulfed directly through the widely opened mouth (Beroida). Extracellular digestion begins in the pharynx, where gland cells secrete enzymes. The partially digested material is then carried throughout the gastrovascular

system, where it is phagocytized by digestive cells.

Ctenophores have high powers of regeneration. Asexual reproduction (in a few platyctenid genera) is by regeneration of an entire organism from a small piece of the adult body.

Ecology. Ctenophores are themselves important plankton organisms. They are quite common and have a worldwide distribution in all seas, where they can appear in enormous swarms. Some genera stand great changes in the seawater salinity. Because of their voracity as predators of zooplankton, they play an important role in plankton equilibrium and in fisheries.

Phylogeny. The Ctenophora constitute a well-defined phylum. Among its orders, the Cydippida are undoubtedly the most primitive, with radiating evolution of the other orders. The ancestor of the phylum is possibly to be found among the Trachylina (Cnidaria, Hydrozoa). However, because of such forms as *Coeloplana* (Platyctenida), ctenophores have also been suspected of affinity with the polyclad Turbellaria. SEE POLYCLADIDA; TRACHYLINA.

J.-M. Franc

Bibliography. G. R. Harbison, L. P. Madin, and N. R. Swanberg, On the natural history and distribution of oceanic ctenophores, *Deep-Sea Research*, 25:233–256, 1978; L. H. Hyman, *The Invertebrates: Protozoa through Ctenophora*, 1940; A. G. Mayer, *Ctenophores of the Atlantic Coast of North America*, Carnegie Institution of Washington, Publ. 162, 1912; S. C. Morris et al. (eds.), *The Origins and Relationships of Lower Invertebrates*, 1985; S. P. Parker (ed.), *Synopsis and Classification of Living Organisms*, 2 vols., 1982.

Ctenopoda

An order of branchiopod crustaceans formerly included in the order Cladocera. The body is up to about 4 mm (0.2 in.) in length. Although superficially similar to the large order Anomopoda in the protection of the trunk and its limbs by a functionally bivalved carapace, in the nature of the eye, in the use of antennae for swimming, and in the possession of a somewhat similar postabdomen, ctenopods differ in various ways. SEE ANOMOPODA.

The six trunk limbs are all constructed on a similar plan but differ in detail among themselves. They beat with a metachronal rhythmn and by so doing draw in suspended food particles. Some species have become planktonic; *Holopedium* surrounds itself by a sphere of jelly to assist flotation. Others are sedentary; *Sida* attaches itself by a sucker at the back of the head.

Reproduction is mostly by parthenogenesis. Eggs and young are carried in a dorsal brood pouch beneath the carapace. Resistant eggs are produced, usually in autumn in temperate zone species, and are shed freely. Most species occur in fresh water, where they are widely distributed, but *Penilia* is marine. SEE BRANCHIOPODA.

Geoffrey Fryer

Ctenostomata

An order of bryozoans in the class Gymnolaemata. Ctenostomes have inconspicuous delicate colonies made up of relatively isolated, simple, short, straight, vase-shaped zooecia with solid chitinous walls. SEE GYMNOLAEMATA.

Morphology. Most ctenostome colonies are encrusting threadlike networks of stolons or proximal portions of autozooecia, with the distal portions of the autozooecia arising erect from the substrate at regular intervals. The network usually lies upon the surface of the substrate, but sometimes in a hollow excavated within the substrate; comparably shaped minute perforations found in dead or fossil shells are apparently traces of similar boring ctenostomes. A few other ctenostome colonies are tuftlike or are gelatinous masses. Though sometimes having stolons or rhizoids, ctenostome colonies lack ovicells or heterozooids (avicularia, vibracula, and so on). The colonies cannot be divided into distinct endozone and exozone regions.

The walls of individual zooecia may be thin and membranous or moderately thick and firm; they are mostly chitinous, but occasionally those portions encrusting on the substrate may be also weakly calcified. Transverse partitions, such as diaphragms, are absent within each zooecium. Round in outline, and of the same diameter as the zooecium, the zooecial aperture lacks an operculum, and instead is closed by a flexible body-wall collar which folds into pleats resembling a comblike row of tiny spines.

Life cycle. Each ctenostome zygote develops into only one larva. In some species, the zygotes are shed directly into the water and there develop into larvae (cyphonautes) which swim freely for up to several weeks. In other species, the zygotes are sheltered within the body of the parent zooid until fully developed into larvae; these swim freely for less than a day. After attaching to a substrate, each larva metamorphoses (by complete degeneration of its larval structures) into an ancestrular zooid. Additionally, the few fresh-water (but not marine) ctenostomes produce irregularly shaped resistant resting bodies (hibernacula) by asexual budding when unfavorable environmental conditions threaten.

History and classification. First appearing early in the Early Ordovician, ctenostomes are the oldest known bryozoans and may possibly have been the stem group from which the other ectoproct groups evolved, the stenolaemates early in the Paleozoic and the cheilostomes and possibly phylactolaemates late in the Mesozoic. Although always (even today) a relatively minor group among the bryozoans, the ctenostomes include both many marine and a few fresh-water species. Ctenostomes have diversified into two major groups; in stoloniferous ctenostomes, the zooids of a colony are budded off only from stolons, but in carnose ctenostomes, the zooids are budded off directly from older zooids. *See* Bryozoa; Cheilostomata; Phylactolaemata; Stenolaemata.

Roger J. Cuffey

Ctenothrissiformes

A small order of teleostean fishes that are of particular interest because they seem to be on or near the main evolutionary line leading from the generalized, soft-rayed salmoniforms to the spiny-rayed beryciforms and perciforms, dominant groups among higher bony fishes.

Ctenothrissiformes lack fin spines but they share the following characters with primitive beryciforms: pelvic fins that are thoracic, with seven or eight rays, and pectoral fins that are placed well up on the flank (see **illus.**); a subocular shelf; a scaly opercle; a me-

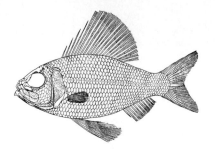

Cretaceous ctenothrissid, *Ctenothrissus radians*; length 10–12 in. (25–30 cm). (*After C. Patterson, Phil. Trans. Roy Soc. London, Ser. B, Biol. Sci., no. 739, vol. 247, 1964*)

socoracoid lacking; a similar premaxilla; maxillary teeth; an orbitosphenoid; and 17 branched caudal rays. Except for the last three these features persist in generalized Perciformes and indicate a major evolutionary trend away from typical soft-rayed fishes. The scales are ctenoid or cycloid, and there are nine branchiostegals. *See* Scale (zoology).

The few fishes confidently assigned to the Ctenothrissiformes, *Ctenothrissus* (Ctenothrissidae) and *Aulolepis* and *Pateroperca* (Aulolepidae), are known as Upper Cretaceous marine fossils from Europe and southwestern Asia. Two little-known genera of Recent oceanic fishes, *Macristium* and *Macristiella* (Macristiidae), are assigned by some authors to the order but some doubt attends this placement. *See* Beryciformes; Perciformes; Salmoniformes.

Reeve M. Bailey

Bibliography. C. Patterson, A review of Mesozoic Acanthopterygian fishes, with special reference to those of the English Chalk, *Phil. Trans. Roy. Soc. London*, Ser. B, Biol. Sci., no. 739, vol. 247, 1964.

Cube

A parallelepiped whose six faces are all squares. The cube (see **illus.**) is one of the five regular solids known to the ancient Greeks, who proposed the famous problem (now proved to be unsolvable) of con-

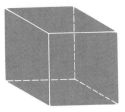

A cube, a type of parallelepiped.

structing, with the use of the compass and unmarked straightedge alone, a cube whose volume is twice that of a given cube. The edge of the desired cube may be found by use of conic sections. A cube has eight vertices and twelve edges. Each vertex is on three edges and three faces, each edge is on two vertices and two faces, and each plane is on four vertices and four edges. *See* Parabola; Parallelepiped; Polyhedron; Regular polytopes.

Leonard M. Blumenthal

Cubeb

The dried, nearly ripe fruit (berries) of a climbing vine, *Piper cubeba*, of the pepper family (Piperaceae). This species is native to eastern India and Indomalaysia. The crushed berries are smoked, and the inhaled smoke produces a soothing effect in certain respiratory ailments. Cubebs are used in medicine as a stimulant, expectorant, and diuretic. Medicinal properties of cubeb are due to the presence of a volatile oil which formerly was thought to stimulate healing of mucous membranes. SEE *PIPERALES*.

P. D. Strausbaugh/Earl L. Core

Cubomedusae

An order of the class Scyphozoa. They are good swimmers and their distribution is mostly tropical. The umbrella is cubic, and a pedalium, a gelatinous leaf-shaped body, is present at the base of each ridge of the exumbrella (see **illus.**). A well-developed ten-

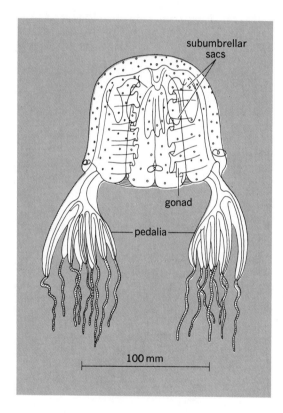

subumbrellar sacs

gonad

pedalia

100 mm

Cubomedusan, *Chiropsalmus quadrumanni*. (*After L. Hyman, The Invertebrates, vol. 1, McGraw-Hill, 1940*)

tacle, or a cluster of tentacles, arises from each pedalium. The sensory organs are on the perradii. They are the most highly developed in the Coelenterata, each one having six ocelli and a statocyst. The manubrium, a peduncle connecting the mouth with the stomach, has four lips. The cubic stomach connects with four radial pockets, and between these pockets are the interradial septa. The leaf-shaped gonads develop from the septa and extend into the pockets. No connecting stage between polyp and medusa is known. SEE *SCYPHOZOA*.

Tohru Uchida

Cuculiformes

An order of birds with zygodactyl feet in which the outer fourth toe is reversed. The relationship of the touracos to the cuckoos is debated: some workers

have argued that the peculiar hoatzin of South America is closely related to the cuckoos, but this conclusion is controversial; the hoatzin will be placed in its own order next to the Cuculiformes. SEE *GALLIFORMES; OPISTHOCOMIFORMES*.

Classification. The order Cuculiformes is divided into the suborder Musophagi, containing the single family Musophagidae (touracos; 18 species), and the suborder Cuculi, with the lone family Cuculidae (cuckoos; 129 species). The Cuculidae are divided into six distinct subfamilies, some of which have a limited distribution, such as the couas (Couinae) which are confined to Madagascar.

Fossil record. The fossil record of the Cuculiformes is poor and comprises only scattered finds. Several fossils of the Musophagidae, today restricted to Africa, have been found in the Oligocene and Miocene of Europe, indicating that this family once had a wider distribution.

Characteristics. Touracos are African woodland birds with a loose green, gray, or brown plumage with patches of bright yellow, red, or violet. The green pigment (turacoverdin) is unique and is the only true green pigment known in birds. The red pigment (turacin) is a water-soluble copper-based pigment unique to birds. These feathers must be protected by overlying feathers. The nest is a bulky platform placed in a tree, with incubation and care of the chicks shared by the parents. The young leave the nest at an early age (about 10 days) and crawl about the trees before being able to fly.

Cuckoos are a diverse group, as indicated by the six subfamilies recognized. They are mostly arboreal, but some are terrestrial, and some, such as the American roadrunners (*Geococcyx*) are fast runners with top speeds of about 15 mi/h (25 km/h) and fly rarely. Most cuckoos are secretive birds, sulking in heavy vegetation. They are strong fliers but do not fly often. Temperate species are migratory. Cuckoos feed on insects, small vertebrates, and other animals, and rarely on fruit. Most are solitary, but the anis (Crotophaginae) live in flocks and build a cooperative nest housing several females. Many species of Old World cuckoos (Cuculinae) are nest parasites. Females of these species belong to different "egg races" in which the eggs resemble those of different hosts. In a host nest the female lays an egg, which is small compared to the size of the cuckoo. Incubation is short, and the young cuckoo pushes the other eggs or young out of the nest. The host parents feed the cuckoo, even when it dwarfs them, until it is fledged. Many groups of cuckoos, including the American species, build their own nests and care for their own young. SEE *AVES*.

Walter J. Bock

Cucumber

A warm-season annual cucurbit (*Cucumis sativus*) native to India and belonging to the plant order Campanulales. SEE *CAMPANULALES*.

Production. The cucumber is grown for its immature fleshy fruit which is used primarily for pickling and for slicing as a salad (**Fig. 1**). The development of gynoecious (bearing only female flowers) and hybrid varieties (cultivars) with increased disease resistance is resulting in a continuing change in the varieties of both pickling and slicing cucumbers. In addition, the advent of mechanical harvesting and chemical weed control for pickling cucumbers is altering traditional methods of culture, for example, higher plant populations per acre and single harvests instead of multiple pickings. Important cucumber-producing states are Florida and South Carolina for fresh market and Michigan and Wisconsin for pickling. SEE SQUASH; VEGETABLE GROWING.

H. John Carew

Diseases. Diseases of cucumbers reduce plant growth, fruit yield, and product quality in both field and greenhouse plantings. Because cucumbers are grown all months of the year, it is necessary to guard against seeding in unfavorable environments and to control seed decay and seedling blight.

Bacterial diseases. Bacterial wilt is a serious disease in the midwestern, north-central, and northeastern regions of the United States, but it is not important in the South or West. Plants infected with the causal bacterium *Erwinia tracheiphila* first show a wilting of one or a few leaves which remain green, but later all the leaves wilt and the plant dies. A white, stringy, viscid bacterial ooze shows in freshly cut wilting stems. Bacteria-infested 12-spotted and striped cucumber beetles infect the plant while feeding upon it. The bacteria multiply in the plant and become distributed throughout the vascular system. The bacteria overwinter in the body of the insect. Because the only way in which the plants can become infected is through the feeding of the cucumber beetles, control depends upon early and prompt destruction of these pests by insecticides. Antibiotics have sometimes proved useful against the bacterium.

Angular leaf spot, incited by the bacterium *Pseudomonas lachrymans*, causes water-soaked spots on leaves and bordered, irregular lesions on fruits. The lesions later develop into a brown firm rot extending into the flesh (**Fig. 2**). In moist conditions, bacteria ooze from the lesion in "tears" which dry into a whitish residue. The causal bacterium overwinters on infected vine refuse, is seed-borne, and is spread by rain and surface water. Control consists of using California seed, which is low in infection, soaking it for 5 min in a mercuric chloride solution (1 part in 1000 parts of water), rinsing the seed in water, and planting promptly. Fixed copper sprays can achieve control, but must be used carefully as they can be phytotoxic

(a)

(b)

Fig. 2. Angular leaf spot disease of cucumber caused by bacterium *Pseudomonas lachrymans.* (a) Leaf spots. (b) Fruit lesions. (*Courtesy of J. C. Walker*)

to young plants. There are some resistant varieties available.

Fungal diseases. Downy mildew is a destructive disease in the eastern and southern states where the weather is warm and moist, is less damaging in the north-central states, and rarely occurs in the Southwest. It is incited by the leaf-inhabiting fungus *Pseudoperonospora cubensis*, which causes angular, yellowish spots; the spots later turn brown when the leaves shrivel (**Fig. 3**). In moist conditions, the lower leaf surface is covered with the fruiting layer of the fungus. Entire vines may be killed. The loss of foliage interferes with normal flower set and fruit development. Maturing fruits fail to color properly, are tasteless, and are usually sunburned. Planting of resistant varieties, such as Ashley, Palmetto, Palomar, Santee, and Stono, and treating plants with copper fungicides will provide maximum protection from downy mildew.

Scab is a serious disease in the northern and northeastern states, especially during cool moist weather. The causal fungus, *Cladosporium cucumerinum*, produces small, circular, halo-bordered, water-soaked lesions on leaves and stems. The greatest damage is to the fruit and appears as sunken, dark-brown spots on immature fruit and as rough corky lesions on mature fruit. The fungus survives in plant refuse and on the seed. Two-year crop rotation, seed treatment, and well-drained soil are important in control of scab. Control is achieved by planting resistant varieties, such as the slicing variety Highmoor or the pickling varieties Wisconsin SR 6 and SMR 12.

Viral diseases. Cucumber mosaic is the most destructive disease caused by a virus. The pathogen causes mottling of terminal leaves and dwarfing of vines, as well as mottling and varying malformations of the fruits. Plants can be infected at any growth

Fig. 1. Cucumber fruits. (*Asgrow Seed Co.; Upjohn Co.*)

Fig. 3. Downy mildew of cucumber leaf showing yellow angular blotches induced by *Pseudoperonospora cubensis*. (*Courtesy of D. E. Ellis*)

stage. The virus is harbored in a number of perennial hosts and is spread by aphids. The best means of control is provided by planting resistant varieties, such as the slicing varieties Burpee Hybrid, Niagara, Ohio MR 200, Sensation Hybrid, Shamrock, and Surecrop Hybrid, and pickling varieties Ohio MR 17, Ohio MR 25, Wisconsin SMR 12, and Yorkstate. *See Aphid; Hemiptera; Plant pathology; Plant viruses and viroids.*

Frank L. Caruso

Culture

The cultivation of cells in the laboratory. Bacteria and yeasts may be grown suspended in a liquid medium or as colonies on a solid medium (usually solidified by 1.5–2.5% agar); molds grow on moist surfaces; and animal and plant cells (tissue cultures) usually adhere to the glass or plastic beneath a liquid medium. Cultures must provide sources of energy and raw material for biosynthesis, as well as a suitable physical environment.

The materials supplied determine which organisms can grow out from a mixed inoculum. Some bacteria (prototrophic) can produce all their constituents from a single organic carbon source; hence they can grow on a simple medium (including K^+, NH_3, Mg^{2+}, Fe^{3+}, PO_4^{3-}, and SO_4^{2-}). Other cells (auxotrophic) lack various biosynthetic pathways and hence require various amino acids, nuclei acid bases, and vitamins. Obligatory or facultative anaerobes grow in the absence of O_2; many cells require elevated CO_2. Cultures isolated from nature are usually mixed; pure cultures are best obtained by subculturing single colonies. Viruses are often grown in cultures of a host cell, and may be isolated as plaques in a continuous lawn of those cells.

In diagnostic bacteriology, species are ordinarily identified by their ability to grow on various selective media and by the characteristic appearance of their colonies on test media. These tests include hemolytic reactions with sheep or other red cells embedded in the solid medium. Indicator dyes in the medium are used to detect colonies that ferment various sugars.

In ordinary cultures the cells are at all possible stages in their division cycle, and the composition of the medium changes continually as a result of their metabolism (until growth ceases, in the stationary phase of the culture). On transfer of an inoculum to fresh medium, there may be a lag phase, without multiplication, followed by a phase of exponential growth. Synchronous cultures are achieved by blocking growth or harvesting cells at a specific stage; the cells then divide in synchrony for several generations. In continuous cultures, fresh medium flows into the vessel and full-grown culture flows at the same rate (such as in a chemostat); the cells are therefore harvested from a medium of constant composition. *See Bacterial growth; Chemostat.*

Laboratory cultures are often made in small flasks, test tubes, or covered flat dishes (petri dishes). Industrial cultures for antibiotics or other microbial products are usually in fermentors of 10,000 gallons (37,850 liters) or more. The cells may be separated from the culture fluid by centrifugation or filtration.

Bernard D. Davis

Technique. Specific procedures are employed for isolation, cultivation, and manipulation of microorganisms, including viruses and rickettsia, and for propagation of plant and animal cells and tissues. A relatively minute number of cells, the inoculum, is introduced into a sterilized nutrient environment, the medium. The culture medium in a suitable vessel is protected by cotton plugs or loose-fitting covers with overlapping edges so as to allow diffusion of air, yet prevent access of contaminating organisms from the air or from unsterilized surfaces. The transfer, or inoculation, usually is done with the end of a flamed, then cooled, platinum wire. Sterile swabs may also be used and, in the case of liquid inoculum, sterile pipets.

The aqueous solution of nutrients may be left as a liquid medium or may be solidified by incorporation of a nutritionally inert substance, most commonly agar or silica gel. Special gas requirements may be provided in culture vessels closed to the atmosphere, as for anaerobic organisms. Inoculated vessels are held at a desired constant temperature in an incubator or water bath. Liquid culture media may be mechanically agitated during incubation. Maximal growth, which is visible as a turbidity or as masses of cells, is usually attained within a few days, although some organisms may require weeks to reach this stage. *See Chemostat; Embryonated egg culture; Tissue culture.*

Media. Combinations of many nutrients are used for the growth of microorganisms as well as plant and animal tissues. Water is the chief constituent of most media. A medium may consist of natural materials, such as enzymatic digests or extracts of yeast or beef. Milk, potato slices, and chick embryos are other common examples. Artificial media are prepared by mixing various ingredients according to particular formulas. A complex medium contains at least one crude ingredient derived from a natural material, hence of unknown chemical composition. A chemically defined, or synthetic, medium is one in which the chemical structure and amount of each component are known.

Copolymer

A macromolecule in which two or more different species of monomer are incorporated into a polymer chain. The properties of copolymers depend on both the nature of the monomers and their distribution in the polymer. Thus, monomers A and B can polymerize randomly to form ABBAABA; they can alternate to give ABAB; they can form blocks AAABBB; or one monomer can be grafted onto a polymer of the other:

AAAA
B
B

Copolymers can be prepared by all the known methods of polymerization: addition polymerization of vinyl monomers (by free-radical, anionic, cationic, or coordination catalysis), ring-opening polymerization, or condensation polymerization.

Reactivity ratios. In the polymerization of two monomers, M_1 and M_2, the monomers can add to a growing chain ending in either monomer, designated as M_1^* and M_2^*. To a first approximation, only the terminal group of the growing chain is important, so two reactivity ratios can be defined: r_1 is the relative reactivity of chain end M_1^* to monomers M_1 and M_2, and r_2 is the relative reactivity of chain end M_2^* to monomers M_2 and M_1. The reactivity ratios can be determined experimentally by analyzing polymer compositions at different monomer feeds. If $r_1 \approx r_2 \approx 1$ ($r_1 r_2 = 1$), the so-called ideal copolymerization, the growing chain ends show no preference for either monomer. Therefore, the copolymer will show a random distribution of monomers, and the composition of the copolymer will be determined by the composition of the monomer feed. If r_1 and $r_2 < 1$ ($r_1 r_2 < 1$), each growing chain end prefers the opposite monomer, and the copolymer will tend to be alternating. If r_1 and $r_2 > 1$ ($r_1 r_2 > 1$), each growing chain prefers its own monomer, and block copolymers or a mixture of homopolymers will form. If $r_1 \gg 1$ and $r_2 \ll 1$, copolymerization is very difficult, whereas if r_1 and $r_2 \approx 0$ ($r_1 r_2 = 0$), only alternating copolymers will form, since neither growing end will add its own monomer.

Hundreds of reactivity ratios have been determined. For radical-initiated polymerizations, reactivities have been correlated with structural effects by the Q-e scheme, Q being a measure of resonance and e of polar effects. From this scheme, reasonable predictions can be made how two monomers will behave in copolymerization.

Random copolymers. Styrene and butadiene, polymerized by free radicals in emulsion, form a random copolymer ($r_1 r_2 = 1.02$). Styrene-butadiene rubber (SBR), containing the two monomers in approximately a 1:3 ratio and constituting the largest-volume synthetic rubber manufactured, is used primarily in tires. Butadiene-acrylonitrile, prepared in the same manner, is a highly solvent-resistant rubber. Styrene-acrylonitrile is a much tougher and more solvent-resistant plastic than polystyrene, and is used in molded appliances and automotive parts. *SEE RUBBER; STYRENE.*

Polyethylene and polypropylene are both crystalline plastics, whereas an ethylene-propylene random copolymer is a useful elastomer. Even though the reactivity ratios are very unfavorable (ethylene = 18 and propylene = 0.06), copolymers can be prepared by adjusting the monomer concentrations. To make the rubber sulfur-vulcanizable, a small amount of a diene, in which one double bond is polymerizable and the other is not, is added; the copolymer is known as EPR and the terpolymer as EPDM. These polymers are known for their resistance to oxidation, since they contain little or no unsaturation to react with oxygen or ozone in the air.

Poly(vinyl chloride) is a very useful plastic, but it is very difficult to process because it decomposes near its molding temperature. A copolymer containing a small amount of vinyl acetate is more easily fabricated, yet it contains most of the good properties of the homopolymer (a polymer made from a single monomer species). Major uses of poly(vinyl chloride) are in floor coverings, film, and phonograph records. *SEE POLYVINYL RESINS.*

Alternating copolymers. Maleic anhydride and vinyl ethers homopolymerize with great difficulty by using radical initiators, yet form perfectly alternating copolymers ($r_1 = r_2 = 0$). Maleic anhydride also alternates with styrene. These resins have many uses, including adhesives, latex paints, and floor polishes.

Block copolymers. Although styrene-butadiene random copolymer is an important sulfur-vulcanizable rubber, a styrene-butadiene-styrene block copolymer (SBS) has very different properties. It is prepared with an alkyllithium catalyst, and a block copolymer is made possible because "living" chains are involved. First, a polystyrene chain is prepared. Since the chain end is still an alkyllithium group, it polymerizes butadiene, which is added next, attached to the polystyrene chain. The process is then repeated to add another block of polystyrene to form the SBS triblock copolymer. This copolymer is important because it need not be vulcanized to make a useful rubber. At elevated temperature a uniform melt is formed, and the material can be readily extruded or injection-molded, processes which are much more economical than those needed to form sulfur-vulcanizable rubbers. As the material cools, however, the polystyrene blocks solidify and form a separate phase. These tiny balls of plastic suspended in a rubbery matrix serve the same function as the sulfur cross-links in a vulcanized rubber. This inexpensive fabrication method has led to very rapid growth in areas where high temperatures are not encountered, such as in shoe soles. For some purposes isoprene is substituted for butadiene, and hydrogenation of the double bonds leads to much better stability.

Elastomeric fibers can also be prepared from block copolymers. Here, the soft, or rubbery, segments can be a polyether, prepared by ring-opening polymerization, or a polyester, prepared by condensation, and the hard segments are urethanes. The resulting $(AB)_n$ multiblock copolymers form highly elastic fibers, which are used to make swimsuits, hosiery, and foundation garments.

Graft copolymers. Acrylonitrile-butadiene-styrene terpolymers (ABS resins) are important thermoplastic structural plastics. The terpolymer is prepared by copolymerizing acrylonitrile and styrene dissolved in polybutadiene rubber. The chains of copolymer grow attached to the rubber by chemical bonds. Properties can be varied by varying the polybutadiene molecular weight and the amounts of styrene and acrylonitrile. The resulting product is extremely tough (while polystyrene is brittle) and opaque, indicating that separate phases are present. SEE ACRYLONITRILE; POLYACRYLONI-TRILE RESINS; POLYMER; POLMERIZATION.

David S. Breslow

Bibliography. T. Alfrey, Jr., J. J. Bohrer, and H. Mark, *Copolymerization*, 1952; J. Brandrup and E. H. Immergut, *Polymer Handbook*, 3d ed., vol. 1, 1988; R. J. Ceresa, *Block and Graft Copolymers*, 1962; G. E. Ham, *Copolymerization*, 1964; *Kirk-Othmer Encyclopedia of Chemical Technology*, 3d ed., vol. 6, 1979.

Copper

A chemical element, symbol Cu, atomic number 29, atomic weight 63.546. Copper, a nonferrous metal, is the twentieth most abundant element present in the Earth's crust, at an average level of 68 parts per million (0.22 lb/ton or 0.11 kg/metric ton). Copper metal

and copper alloys have considerable technological importance due to their combined electrical, mechanical, and physical properties. The discoveries that mixed-valence Cu(II)/Cu(III) oxides exhibit superconductivity (zero electrical resistance) at temperatures as high as 125 K ($-234°F$; liquid nitrogen, a cheap coolant, boils at 90 K or $-297°F$) have generated intense international competition to understand these new materials and to develop technological applications. Although some pure copper metal is present in nature, commercial copper is obtained by reduction of the copper compounds in ores followed by electrolytic refining. The rich chemistry of copper is restricted mostly to the valence states Cu(I) and Cu(II); compounds containing Cu(0), Cu(III), and Cu(IV) are uncommon. Soluble copper salts are potent bactericides and algicides at low levels and toxic to humans in large doses. Yet copper is an essential trace element that is present in various metalloproteins required for the survival of plants and animals. SEE COP-PER ALLOYS; COPPER METALLURGY.

Chemical properties. Copper is located in the periodic table between nickel and zinc in the first row of transition elements and in the same subgroup as the other so-called coinage metals, silver and gold. The electronic configuration of elemental copper is $[1s^2 2s^2 2p^6 3s^2] 3d^{10} 4s^1$ or $[argon] 3d^{10} 4s^1$. At first glance, the sole $4s$ electron might suggest chemical similarity to potassium, which has the $[argon] 4s^1$ configuration. However, metallic copper, in sharp contrast to metallic potassium, is relatively unreactive. The higher nuclear charge of copper relative to that of potassium is not fully shielded by the 10 ad-

ditional d electrons, with the result that the copper $4s$ electron has a higher ionization potential than that of potassium (745.5 versus 418.9 kilojoules/mole, respectively). Moreover, the second and third ionization potentials of copper (1958.1 and 3554 J/mole, respectively) are considerably lower than those of potassium, and account for the higher valence-state accessibility associated with transition-metal chemistry as opposed to alkali-metal chemistry. SEE ELECTRON CONFIGURATION; TRANSITION ELEMENTS; VALENCE.

Copper is obtained as the nonradioactive isotopes ^{63}Cu and ^{65}Cu that are present in natural abundance 69.9% and 30.91%, respectively. The nucleus of ^{63}Cu contains 34 neutrons and that of ^{65}Cu contains 36 neutrons; both contain 29 protons. Bombardment of zinc with high-energy deuterons in a nuclear reactor yields nine short-lived radioactive isotopes of copper that may be separated chemically from the excess zinc and the radioactive gallium that also is produced. These isotopes have mass numbers of 58, 59, 60, 61, 62, 64, 66, 67, and 68, with respective half-lives of 3.2 s, 81 s, 24 min, 3.33 h, 9.9 min, 12.9 h, 5.1 min, 61 h, and 30 s. Copper-64 has been useful for tracer studies of copper metabolism in humans and animals.

Metallic copper is fairly unreactive. It will not dissolve in an aqueous solution of a nonoxidizing acid such as hydrochloric acid (HCl). The oxidation potential for the reaction $Cu \rightarrow Cu(II) + 2e^-$ is -0.34 V in normal ionic solution at 25°C (77°F). Copper dissolves in the presence of an oxidizing acid such as nitric acid or in that of reagents such as thiourea $[=C(NH_2)_2]$ that strongly bind and stabilize Cu(I) ions [reaction (1)]. Dissolution and formation of the

$$Cu + 3S=C(NH_2) + HCl \rightarrow \\ Cu[S=C(NH_2)_2]_3Cl + \tfrac{1}{2}H_2 \quad (1)$$

Cu(II) state is promoted by the presence of dioxygen (O_2) and reagents, for example, ammonia (NH_3), that bind Cu(II) well [reaction (2)]. Copper and copper-

$$Cu + 4NH_3 + \tfrac{1}{2}O_2 + H_2O \rightarrow \\ [Cu(NH_3)_4]^{2+} + 2OH^- \quad (2)$$

rich alloys that are used in applications such as sheathing roofs or casting statues oxidize slowly in air to yield a thin, green surface coating composed of an insoluble mixture of hydroxo-carbonate and hydroxo-sulfate Cu(II) salts.

Physical properties. Copper is a comparatively heavy metal. The density of the pure solid is 8.96 g/cm^3 (5.18 oz/in.^{3}) at 20°C (68°F). The density of commercial copper varies with method of manufacture, averaging 8.90–8.92 g/cm^3 (5.14–5.16 oz/in.^{3}) in cast refinery shapes, 8.93 g/cm^3 (5.16 oz/in.^{3}) for annealed tough-pitch copper, and 8.94 g/cm^3 (517 oz/in.^{3}) for oxygen-free copper. The density of liquid copper is 8.22 g/cm^3 (4.75 oz/in.^{3}) near the freezing point.

The melting point of copper is $1083.0\mp0.1°C$ ($1981.4\mp0.2°F$). Its normal boiling point is 2595°C (4703°F).

The coefficient of linear expansion of copper is $1.65 \times 10^{-5}/°C$ at 20°C.

The specific heat of the solid is 0.092 cal/g at 20°C (68°F). The specific heat of liquid copper is 0.112 cal/g, and of copper in the vapor state about 0.08 cal/g.

Electrical properties. The electrical resistivity of copper in the usual volumetric unit, that of a cube

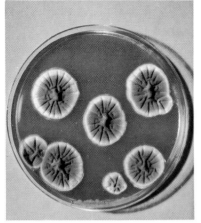

Penicillium, penicillin.

Streptomyces aureofaciens,
chlortetracycline (Aureomycin).

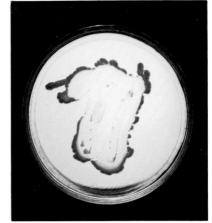

Streptomyces californicus,
viomycin (Viocin).

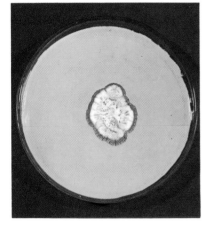

Streptomyces griseus, streptomycin.

Bacillus subtilis, bacitracin.

Bacillus polymyxa, polymyxin.

Streptomyces halstedii, carbomycin.

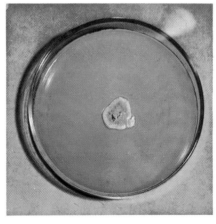

Streptomyces albus, actinomycetin.

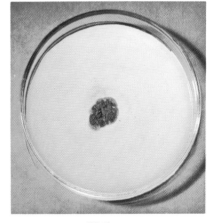

Streptomyces antibioticus,
oleandomycin (Matromycin).

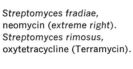

Streptomyces fradiae,
neomycin (*extreme right*).
Streptomyces rimosus,
oxytetracycline (Terramycin).

Microorganisms and the antibiotics they
produce. *Penicillium chrysogenum* (mold)
grown on potato-dextrose agar; all others
(bacteria) grown on Pridham's yeast-ex-
tract agar. (*Chas. Pfizer and Co., Inc.*)

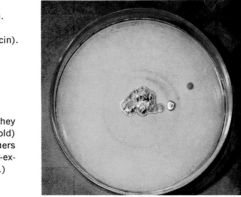

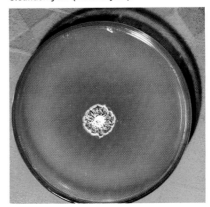

All media furnish minerals, sources of nitrogen, carbon, and energy, and sometimes vitamins or other growth factors. An ingredient furnishing carbon—usually the chief source—is called the substrate for the utilizing organism.

Selective medium. Usually this medium contains an individual organic compound as the sole source of carbon, nitrogen, or sulfur for growth of an organism. The organism oxidizes the carbon compound, thereby obtaining energy for vital processes. Strictly inorganic media are utilized by some bacteria, the chemoautotrophs. They obtain carbon exclusively from carbon dioxide and energy from the oxidation of ammonia, nitrate, hydrogen gas, or reduced, inorganic, sulfur compounds. Algae and a few bacterial types are photoautotrophs, securing their energy from light.

An agar medium of a given composition permits selection, through development of colonies, or organisms capable of utilizing the specific sources of carbon, energy, nitrogen, or sulfur available. A colony on such a medium is prima-facie evidence that such organisms have been selected, but this must be confirmed by restreaking (subculturing on agar medium) on a medium of the same composition.

Enrichment media. These are liquid media of a given composition which permit preferential emergence of certain organisms that initially may have made up a relatively minute proportion of a mixed inoculum. This enrichment of the population in the desired type expedites isolation by direct plating on the corresponding selective solid medium. Because of the numerous subtle, competitive forces, enrichment cultures yield only a relatively small proportion, that is, the fastest growers, of all the organisms potentially capable of developing in a given enrichment medium.

Indicator media. These media, usually solid, contain substances capable of changing color in the vicinity of the colony which has effected a particular chemical change, such as fermenting a certain sugar. Thus, organisms with distinctive types of metabolism may be readily detected by the color change in the medium, even when the surface of the medium contains many other colonies. Indicators may be used in liquid media to characterize a pure culture of organisms.

Elective culture. No single culture medium or environment will support growth of more than a small proportion of the known types of microorganisms. Each organism has characteristic nutritional and environmental requirements. Thus, in a culture composed of more than one type of organism, the medium and conditions are selective for only one type of organism present and the other types are suppressed. These selective, or elective, culture principles and techniques are invaluable for the detection and isolation of specified organisms from natural sources. Elective culture techniques are more effective for obtaining bacteria from natural sources than for obtaining fungi, actinomycetes, and other groups; the faster growth rates, larger numbers, and greater variety of bacteria usually confer a competitive advantage upon them. In general, elective culture procedures are applicable to organisms capable of utilizing particular sources of energy, carbon, nitrogen, and other nutrients in a given physicochemical environment. There is no principle whereby elective cultures may be employed to obtain selectively organisms capable of producing particular compounds. Basically, two procedures are available, direct isolation and liquid enrichment.

Direct isolation. This requires a solid medium, usually agar, in petri plates. The surface is lightly sprinkled with finely divided particles of soil or other inoculum material. Only organisms favored by that medium and those conditions will produce colonies of appreciable size; these may then be purified. Literally thousands of other kinds of bacteria fail to develop, although they might develop in other selective media. Thus, the direct isolation technique is a means of selecting certain minority members from natural, mixed microbial populations. Since most of the colonies are on a solid medium, develop independently, and are physically separate, competition rarely is a factor determining the emergence of the various colonies. Although certain fast-growing and characteristically spreading colonies may tend to overgrow neighboring colonies, the chief advantage of the procedure is that slow-growing bacteria have a chance to reveal themselves. This technique is, therefore, the best for obtaining the largest variety of bacteria capable of developing on a given medium. Colonies from a natural inoculum appearing under any one environmental condition are a small representation of a wider spectrum of bacteria capable of developing. For example, two sets of plates incubated, respectively, in the presence and absence of oxygen will yield different organisms, as will plates adjusted to different hydrogen ion concentrations (pH) and plates incubated at various temperatures.

Elective media containing common organic compounds, such as carbohydrates or amino acids, will yield large numbers of different organisms. Similarly, most such organisms are capable of utilizing a number of different but related compounds. However, certain substrates and conditions are highly specific for a few organisms or even for a single kind of organism. Occasionally, growth of a colony occurs at the expense of degradation of biosynthetic products or accessory growth factors which are liberated from one or more adjacent colonies and diffuse into the medium.

Although the direct-isolation method does not permit preferential development of colonies capable of forming specific products, such colonies as may be desired may be detected among those on an agar plate. Thus, acid producers may be distinguished by a zone of color change of an indicator in the medium or by dissolution of powdered calcium carbonate suspended in the medium. Clear zones result from the destruction of other insoluble substrates, such as starch, proteins, and lipids, by colonies producing the corresponding enzymes. Other metabolic products may be detected by flooding or spraying colony-containing plates with chemical or biological reagents, giving a distinctive reaction with the desired product.

The specificity can be enhanced by incorporation of substances that selectively inhibit growth of certain kinds of organisms, especially of those in the majority. For example, in a medium adjusted to pH 4.0–4.5 most bacteria and actinomycetes are inhibited, and thus filamentous fungi and yeasts may be selectively obtained. Other chemicals are used to inhibit undesirable organisms in specimens of body fluids, thereby facilitating isolation of the pathogenic bacteria in the minority.

Liquid enrichment. A small amount of soil or other natural material is added to a prescribed fluid medium which then is incubated. After appreciable growth of bacteria occurs, a drop of the culture is used to inoculate a second flask of the same medium. Organisms originally representing small minorities are thus preferentially selected by the medium and the conditions, and by the time turbidity develops in the secondary culture, they represent practically the majority populations. The chief disadvantage of this procedure is that only a small proportion of all the bacterial types potentially capable of developing in a given medium will be found. Potentially competent organisms are in direct competition, and the specific conditions usually select relatively few of the fastest growers. Liquid-enrichment cultures also contain organisms unable to attack the original substrate; these scavenge products of degradation and biosynthesis, as well as living and dead cells, of the primary organisms.

Pure culture. A pure culture contains cells of one kind, all progeny of a single cell. The concept of a pure culture is an important one in microbiology because most considerations of microorganisms require dealing with only one kind of a cell at a time. Many techniques have been evolved for the isolation and maintenance of pure cultures.

Isolation of aerobic bacteria. Isolation of pure cultures may be achieved by a variety of methods, the main differences being in isolation of aerobic versus anaerobic bacteria.

1. Pour-plate procedures. These involve a thorough distribution of a microbial mixture in cooled, melted agar medium. After solidification of the medium, individual cells are embedded and proliferate to form a colony. Purity of the colony cannot be taken for granted since it may have arisen from two adjacent cells (mixed colony); so the colony is dispersed in sterile water and the pour-plate procedure repeated. Well-isolated colonies in the second plating ordinarily can be shown to be pure. This method is based on the use of a homogeneous suspension of individual cells. The great majority of the colonies on secondary plates should be similar and, seen microscopically, the cells of such colonies should be one kind. However, certain organisms mutate characteristically, producing different colony types (dissociation).

2. Streak plates. The surface of the solid medium in a petri dish is streaked with an inoculating wire or glass rod in a continuous movement so that most of the surface is covered. The later parts of the streak should not intersect with the earlier parts. In later portions of the streak, the separate colonies are the out-

growth of individual cells. The chief advantage of this method is isolation of majority members of a mixed population.

3. Roll-tube technique. This technique, employed chiefly in tissue culture, achieves a homogeneous gaseous and nutrient environment by the elimination of diffusion gradients. During incubation the tubes are held in a wheellike instrument at an angle of about 15° from the horizontal. The wheel rotates vertically about once every 2 min. Constant liquid film formation over the inner surface of the tube facilitates oxygen transfer to the medium.

4. Submerged culture. In this method the microorganisms are grown in a liquid medium subjected to continuous, vigorous agitation. A homogeneous cellular suspension results and all cells in the medium are simultaneously provided with nearly optimal rates of diffusion of oxygen and nutrients. The mechanical agitation may be achieved by continuous bubbling of sterile air through the medium, by a propeller system, by combination of both, or by continuous rotary or reciprocating shaking machines.

Isolation of anaerobic bacteria. Special cultural conditions may be provided for anaerobes, organisms that do not require oxygen or that grow only in the complete absence of oxygen. Fluid thioglycollate medium, the Brewer anaerobic petri dish, and the Brewer anaerobic jar are some of the methods used (see **illus.**). Fluid thioglycollate medium contains sodium thioglycollate, a reducing compound which lowers the oxygen content of the medium. In the anaerobic petri dish, the thioglycollate agar medium and the petri dish cover, with its restricted air space and close fit to the medium surface, give anaerobic conditions. The anaerobic jar is operated by replacing the air in the jar with hydrogen or illuminating gas. A platinized catalyst heated by electricity causes the oxygen remaining in the jar to combine with hydrogen to form water or, in the case of illuminating gas, water and carbon dioxide. In this way free oxygen is removed from the jar.

1. Shake culture. This is a method for isolating anaerobic bacteria. Various dilutions of an inoculum are throughly mixed with cooled, melted agar medium occupying about two-thirds the volume of test tubes. After solidification a 1-in. (2.5-cm) layer of melted vaspar, a 50:50 mixture of paraffin wax and petrolatum, is added. When the mixture has hardened, the culture is sealed against diffusion of oxygen during incubation. In tubes containing higher dilutions of inoculum, separated discrete colonies develop in different positions in the agar. With soft glass tubes, the glass is cut and the agar column broken in the region of a desired colony, which is then isolated. Sometimes the seal is first removed by warming that area of the tube; the column of agar is then slowly ejected into a sterile container by compressed air carried to the bottom of the tube through a fine capillary tube introduced between the agar and the glass. Colonies are reached by slicing the agar with a sterile scalpel.

2. Anaerobic plates. Pour or streak plates are incubated in a closed vessel in which the air is replaced by an inert oxygen-free gas such as helium or nitrogen. Alternatively, the oxygen in a closed vessel is absorbed chemically by mixing pyrogallol and alkali. Specially designed petri dishes permit oxygen absorption in individual dishes.

Single-cell isolation. Single-cell isolation employs microscopic techniques to verify the presence of a single cell, which is then used to initiate a culture.

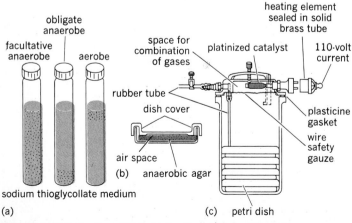

Some techniques for the cultivation of anaerobic bacteria. (a) Fluid thioglycollate medium showing aerobic, anaerobic, and facultative anaerobic growth. (b) Brewer anaerobic petri dish. (c) Brewer anaerobic jar. (*Baltimore Biological Laboratory, Inc.*)

1. Little-drop technique. Minute drops of dilutions of a cellular suspension are examined microscopically on a glass slide. A drop containing a single cell is transferred with a capillary pipet to an appropriate culture medium. This procedure is best suited for isolation of large cells, such as yeasts and mold spores, which can be located quickly with the high-dry lens of the compound microscope. Evaporation problems and the difficulty of obtaining drops small enough to examine with the oil immersion lens limit this procedure for bacteria. A variation of this procedure was originated by J. Lederberg. Little drops of a diluted cellular suspension are deposited on a slide covered with a thin film of oil. The Lederberg technique, by permitting rapid isolation of single bacterial cells, obviates to some extent the problems enumerated before.

2. Micromanipulation. The disadvantages of the little-drop technique for the isolation of single bacterial cells are eliminated with this technique. A micromanipulator permits study of drops in a moist chamber on the microscope stage, reducing evaporation difficulties and enabling the use of the oil immersion lens. The manipulations are performed by remotely controlled capillary micropipets activated by air pressure. *SEE MICROMANIPULATION.*

Maintenance. This is concerned with the preservation of culture viability and is accomplished by practices that reduce the basal metabolism of the cells, thereby enhancing longevity and eliminating the need for frequent transfers.

1. Lyophilization. This procedure, also called freeze-drying, is the best method for preserving culture viability. Freshly grown cells are suspended, in a glass tube, in a small volume of a sterile fluid, such as blood serum or skim milk. A small amount of a reducing agent such as ascorbic acid may help prolong the viability of the culture. The suspension is quickly frozen at approximately $-94°F$ ($-70°C$) by immersion in a dry ice–acetone mixture. The frozen material is then dehydrated, without melting, under high vacuum, or the gas space may be filled with nitrogen or helium prior to sealing. Storage in the absence of oxygen may preserve the viability of a culture for years, even without refrigeration. The mortality, during lyophilization and storage, varies with the culture and the suspending fluid. Lyophilized cells are rejuvenated by inoculation into fresh nutrient medium.

2. Slants. Slants are the most convenient method of maintenance for periods of 1–2 months. Cotton-plugged tubes containing an agar medium are slanted after sterilization until the agar solidifies. The inoculum is streaked on the agar surface. After maximal growth at the appropriate temperature, the slant culture is stored in a refrigerator and used repeatedly as inoculum. With long storage periods, the tube should be sealed to retard desiccation. Viability of slant cultures not used frequently may be prolonged by filling the tube with sterile mineral oil to about 1 in. (2.5 cm) above the agar. This retards water loss from and oxygen access to the culture.

3. Stabs. Stabs are used to maintain anaerobic bacteria. A tube about two-thirds filled with agar medium is pierced vertically for about 2 in. (5 cm) with an inoculating wire. Growth develops in the anaerobic depths of the agar along the stab.

4. Strict anaerobe maintenance. Strict anaerobes, in liquid or solid media, may be maintained in an atmosphere from which the oxygen has been absorbed by alkaline pyrogallol. Also, the air may be replaced by an inert gas, such as helium or nitrogen. The most convenient procedure employs liquid media containing approximately 0.1% of a reducing agent, such as sodium thioglycollate, to keep the oxidation-reduction potential low enough for growth of anaerobes and to protect it from oxygen diffusing from the air. Special vessels or atmospheres are not required, but fully developed cultures are sealed against oxygen and refrigerated.

5. Use of sterile soil. Sterile soil may be used to carry stock cultures of sporeforming organisms. Autoclaved soil supports limited growth of the organism, which then sporulates. The spores remain viable indefinitely when kept refrigerated. *SEE MICROBIOLOGICAL METHODS.*

Jackson W. Foster/R. E. Kallio

Cumacea

An order of the class Crustacea, subclass Malacostraca. All have a well-developed carapace which is fused dorsally with at least the first three thoracic somites and which overhangs the sides to enclose a branchial cavity. The telson may be free or may appear lacking because it is fused with the last abdominal somite. Eyes are sessile and, except for a few genera, are coalesced in a single median organ or wholly wanting. There are eight pairs of thoracic appendages, the first three modified as maxillipeds and the last five as peraeopods. The first maxillipeds bear an epipodite forming a respiratory organ with a forward-directed siphon and a posterior part carrying branchial lobes. Natatory exopodites may be present on the third maxillipeds and the first four pairs of peraeopods or some of them. The uropods are styliform. There is no free larval stage. In the few species whose life history has been investigated there are five to seven molts after release from the brood pouch.

There is always some sexual dimorphism in adults, but immature males resemble females until the later stages. The second antennae are always vestigial in females and nearly always well developed with a long flagellum in males (see **illus.**). Females often have a reduced number of thoracic exopodites. Pleopods are never present in the female but there may be one to five pairs in the male. The form and sculpturing of the body and the number of spines and setae on the appendages often differ between the sexes. Oostegites are present on the third to sixth thoracic limbs of the female.

All known species have a characteristic shape owing to the inflated carapace and slender abdomen. Most are small, 0.04–0.48 in. (1–12 mm) in length; the largest reach 1.4 in. (35 mm). The order is rather uniform and classification is based on trivial characters. Seven families are recognized. Affinities are

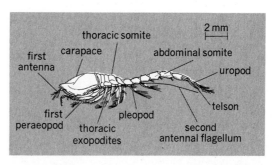

Typical adult male cumacean.

probably closer to the Mysidacea than to the Tanaidacea. A few fossils are now known.

Cumacea are found in all seas and in some estuaries and areas of brackish water, including the Caspian Sea. Most inhabit water shallower than 650 ft (200 m), but many occur in great depths. They normally burrow in sand or mud with the front protruding. Coastal species, and especially the more active adult males, may often be captured above the bottom. They feed on organic detritus or very small organisms and may themselves form part of the diet of various fishes. About 740 species have been described and placed in 82 genera. Many more remain to be found, as they have been poorly investigated in most areas. *See* MALACOSTRACA; SEXUAL DIMORPHISM.

Norman S. Jones

Bibliography. H. G. Bronn, *Klassen and Ordnungen des Tierreichs*, vol. 5, bk. 4, 1941; N. S. Jones, *Galathea Reports*, vol. 10, 1969; R. Lankester, *A Treatise on Zoology*, pt. 7, fasc. 3, 1909; S. P. Parker (ed.), *Synopsis and Classification of Living Organisms*, 2 vols., 1982; T. H. Waterman (ed.), *The Physiology of Crustacea*, 2 vols., 1960, 1961.

Cumin

Cuminum cyminum, a plant for which the whole or ground dried ripe fruit, commonly called seed, is a popular spice. It is a major ingredient in both chili powder and curry, and is added to meat sauces, rice, bread, pickles, soups, and other foods.

Roman caraway is another common name for this member of the parsley family (Apiaceae). The only species in its genus, cumin exhibits a variety of plant types depending on the seed source. A small annual herb about 1–2 ft (0.3–0.6 m) tall, cumin grows upright as a single slender stem with many branches. The deep green leaves are 0.5–2 in. (1.3–5 cm) long, thin, and finely divided. *See* APIALES.

The strong, pungent green-spicy odor and flavor of cumin is attributable largely to cuminaldehyde, the main constituent of the essential oil, and other aldehydes.

This herb is native to the Mediterranean region. Presently cumin is commercially grown in Iran, southern Russia, China, India, Morocco, and Turkey. The three major types of cumin seed, Iranian, Indian, and Middle Eastern, vary in seed color, essential oil quantity, and flavor. Cumin does best in a mild growing climate, free from hot dry periods, on well-drained sandy loam soil. In 90–120 days the plant matures; the small fragile plant is generally hand-harvested. Commonly cumin is planted at 30–35 lb/acre (33.7–39.3 kg/ha), and yields 400–500 lb seed/acre (450–560 kg/ha). Upon distillation cumin seed yields 2.5–5% essential oil, used in both perfumery and flavoring liqueurs. Cumin is also used medicinally. *See* SPICE AND FLAVORING.

Seth Kirby

Cummingtonite

A group of Ca-free to low-Ca amphiboles that crystallize with monoclinic symmetry and have the idealized chemical formula $(Mg,Fe)_7Si_8O_{22}(OH)_2$. The name cummingtonite refers to those varieties containing between 30 and 70% of the Mg end member. More-Fe-rich varieties are called grunerite; more-Mg-

rich varieties, although relatively rare in nature, are called magnesio-cummingtonite. Anthophyllite, the orthorhombic amphibole with the same idealized chemical formula, is generally restricted to more-Mg-rich compositions, but has a compositional range that partially overlaps that of cummingtonite. The conditions that favor the formation of anthophyllite over cummingtonite are poorly understood. The two amphiboles are known to coexist in the same rock. *See* ANTHOPHYLLITE.

Cummingtonite commonly occurs as pale- to dark-brown or green aggregates of fibrous crystals, often in radiating clusters. Hardness is 5–6 on Mohs scale; density is 3.1–3.6 g/cm² (1.8–2.1 oz/in.³), increasing with increasing Fe content. Cummingtonite has distinct, prismatic (110) cleavage that produces typical amphibole cross sections with approximately 124° and 56° angles.

Cummingtonite is generally considered to be a metamorphic mineral, but it has been reported from silicic volcanic rocks and, rarely, plutonic igneous rocks. It occurs in a variety of metamorphic rock types (amphibolite, schist, gneiss, granulite) that have undergone medium- to high-grade metamorphism. It most commonly occurs in metamorphosed iron formation, but can also be a constituent of metamorphosed mafic and ultramafic igneous rocks. *See* AMPHIBOLE; AMPHIBOLITE; GNEISS; GRANULITE; METAMORPHISM; SCHIST.

Robert K. Popp

Cuprite

A mineral having composition Cu_2O and crystallizing in the isometric system. Cuprite is commonly in crystals showing the cube, octahedron, and dodecahedron

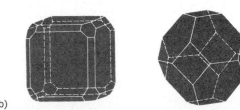

Cuprite. (a) Crystals and veins in limonite-bearing rock, Globe, Arizona (specimen courtesy of Department of Geology, Bryn Mawr College). (b) Crystal habits (after C. S. Hurlbut, Jr., Dana's Manual of Mineralogy, 17th ed., John Wiley and Sons, 1959).

(see **illus**.). More rarely, it is found in long capillary crystals known as chalcotrichite. It is also in fine-grained aggregate or massive.

Cuprite is various shades of red and a find ruby-red in transparent crystals which have a metallic to ada-mantine luster. The hardness is 3.5–4 (Mohs scale) and the specific gravity is 6.1.

Cuprite is a widespread supergene copper ore found in the upper oxidized portions of copper deposits in which it is associated with limonite, malachite, azur-ite, native copper, and chrysocolla. Fine crystals have been found at Cornwall, England, and Chessy, France. In the early mining operations of many cop-per deposits cuprite has been an important ore mineral but gives way to the primary ores in depth. It has served as an ore in the Congo, Chile, Bolivia, and Australia. In the United States it has been found at Clifton, Morenci, Globe, and Bisbee, all in Arizona. SEE COPPER.

Cornelius S. Hurlbut, Jr.

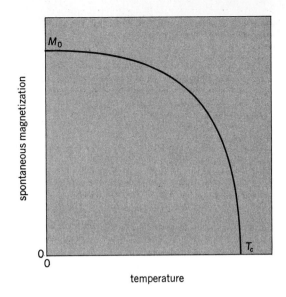

Variation of spontaneous magnetization of a ferromagnetic material with temperature.

Curie temperature

The critical or ordering temperature for a ferromag-netic or a ferrimagnetic material. The Curie tempera-ture T_C is the temperature below which there is a spontaneous magnetization M in the absence of an ex-ternally applied magnetic field, and above which the material is paramagnetic. In the disordered state above the Curie temperature, thermal energy over-rides any interactions between the local magnetic mo-ments of ions. Below the Curie temperature, these in-teractions are predominant and cause the local moments to order or align so that there is a net spon-taneous magnetization.

In the ferromagnetic case, as temperature T in-creases from absolute zero, the spontaneous magne-tization decreases from M_0, its value at $T = 0$. At first this occurs gradually, then with increasing rapid-ity until the magnetization disappears at the Curie temperature (see **illus**.). Many physical properties show an anomaly or a change in behavior at the Curie temperature. There is a peak in specific heat and in magnetic permeability, and there are changes in the behavior of such properties as electrical resistivity, elastic properties, and thermal expansivity. In ferri-magnetic materials the course of magnetization with temperature may be more complicated, but the spon-taneous magnetization disappears at the Curie temper-ature. The Curie temperature can be determined from magnetization versus temperature measurements or from the related anomalies in other physical proper-ties. The Curie temperature can be altered by changes in composition, pressure, and other thermodynamic parameters.

Well above the Curie temperature the magnetic susceptibility follows the Curie-Weiss law, and in fact the characteristic temperature θ in that the law is called the paramagnetic Curie temperature. Generally θ is slightly above T_C. The susceptibility deviates from Curie-Weiss behavior as the temperature ap-proaches T_C and θ. SEE CURIE-WEISS LAW; FERRIMAGNE-TISM; FERROMAGNETISM; MAGNETIC SUSCEPTIBILITY; MAG-NETIZATION; PARAMAGNETISM.

In antiferromagnetic materials the corresponding ordering temperature is termed the Néel temperature (T_N). Below the Néel temperature the magnetic sub-lattices have a spontaneous magnetization, though the net magnetization of the material is zero. Above the Néel temperature the material is paramagnetic. SEE ANTIFERROMAGNETISM.

The ordering temperatures for magnetic materials vary widely (see **table**). The ordering temperature for ferroelectrics is also termed the Curie temperature, below which the material shows a spontaneous elec-tric moment. SEE FERROELECTRICS; PYROELECTRICITY.

J. F. Dillon, Jr.

Bibliography. N. W. Ashcroft and N. D. Mermin, *Solid State Physics*, 1976; S. Chikazumi and S. H. Charap, *Physics of Magnetism*, 1964, reprint 1978; A. H. Morrish, *Physical Principles of Magnetism*, 1966, reprint 1980.

Curie-Weiss law

A relation between magnetic or electric susceptibili-ties and the absolute temperature which is followed by ferromagnets, antiferromagnets, nonpolar ferro-electrics and antiferroelectrics, and some paramag-nets. The Curie-Weiss law is usually written as the following equation, where χ is the susceptibility, C is

Ordering temperatures of magnetic materials*

Ferromagnets	Curie temperature (T_C), °F (K)	Ferrimagnets	Curie temperature (T_C), °F (K)	Antiferromagnets	Néel temperature (T_N), °F (K)
Co	2039 (1388)	Fe_3O_4	1085 (858)	NiO	620 (600)
Fe	1418 (1043)	$MnFe_2O_4$	572 (573)	Cr	100 (311)
$CrBr_3$	−393 (37)	$Y_3Fe_5O_{12}$	548 (560)	$FeCl_2$	−417.0 (23.7)
$GdCl_3$	−455.7 (2.2)				

*From F. Keffer, *Handbuch der Physik*, vol. 18, pt. 2, Springer, 1966.

$$\chi = C/(T - \theta)$$

a constant for each material, T is the absolute temperature, and θ is called the Curie temperature. Antiferromagnets and antiferroelectrics have a negative Curie temperature. The Curie-Weiss law refers to magnetic and electric behavior above the transition temperature of the material in question. It is not always precisely followed, and it breaks down in the region very close to the transition temperature. Often the susceptibility will behave according to a Curie-Weiss law in different temperature ranges with different values of C and θ. See Curie temperature; Electric susceptibility; Magnetic susceptibility.

The Curie-Weiss behavior is usually a result of an interaction between neighboring atoms which tends to make the permanent atomic magnetic (or induced electric) dipoles point in the same direction. The strength of this interaction determines the Curie temperature θ. In the magnetic case the interaction is caused by Heisenberg exchange coupling.

In many paramagnetic salts, especially those of the iron group, the Curie-Weiss behavior is due to crystalline distortions and their effect upon the atomic orbital angular momenta which contribute to the magnetization. For further discussion and derivation of the Curie-Weiss law see Ferromagnetism. See also Antiferromagnetism; Ferroelectrics.

Elihu Abrahams; Frederic Keffer

Curium

A chemical element, symbol Cm, in the actinide series, with an atomic number of 96. The electronic configuration of the neutral atom (beyond the radon core) is $5f^7 6d 7s^2$.

Curium does not exist in the terrestrial environ-

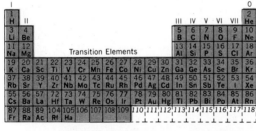

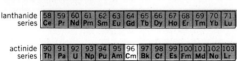

ment, but may be produced artificially, as was first done in 1944 by G. T. Seaborg, R. A. James, and Albert Ghiorso, who synthesized and identified the isotope ^{242}Cm. The synthesis was effected by the bombardment of ^{239}Pu (plutonium-239) with helium ions of about 32 MeV energy, $^{239}_{94}\text{Pu}(\alpha,n)^{242}_{96}\text{Cm}$.

Larger amounts of ^{242}Cm, produced by the prolonged neutron irradiation of ^{241}Am, Eq. (1), pro-

$$^{241}_{95}\text{Am}(n,\gamma)^{242\text{m}}_{95}\text{Am} \xrightarrow{\beta^-} {}^{242}_{96}\text{Cm} \qquad (1)$$

vided material for the first isolation of visible amounts of curium by L. B. Werner and I. Perlman in 1947.

Isotopes of curium

Mass no.	Half-life	Mode of disintegration
238	2.5 h	Orbital electron capture, <90%; alpha-particle emission, >10%
239	2.9 h	Orbital electron capture, >99.9%
240	28 d	Alpha particle emission
	1.9×10^6 y	Spontaneous fission
241	35 d	Orbital electron capture, 99+%; alpha-particle emission, 0.96%
242	162.5 d	Alpha particle emission
	6.09×10^6 y	Spontaneous fission
243	29 y	Alpha particle emission, >99%; orbital electron capture, 0.26%
244	18.12 y	Alpha particle emission
	1.35×10^7 y	Spontaneous fission
245	8540 y	Alpha particle emission
246	4716 y	Alpha particle emission
	1.85×10^7 y	Spontaneous fission
247	1.56×10^7 y	Alpha particle emission
248	3.70×10^5 y	Alpha particle emission
	4.11×10^6 y	Spontaneous fission
249	65 m	Beta particle (electron) emission
250	1.13×10^4 y	Spontaneous fission

Higher mass isotopes—^{244}Cm, ^{247}Cm, ^{248}Cm—are now the principal sources of the element.

The known isotopes of curium, along with their half-lives and modes of decay, are listed in the **table**. Where, for a given mass, separate half-lives and modes of disintegration are listed, the decay processes occur simultaneously.

The isotope ^{244}Cm is of particular interest because of its potential use as a compact, thermoelectric power source, through conversion to electrical power of the heat generated by nuclear decay.

Metallic curium may be produced by the reduction of curium trifluoride with barium vapor at 1350°C (2462°F). The metal has a silvery luster, tarnishes in air, and has a specific gravity of 13.5. Its room-temperature structure is double-hexagonal close-packed, similar to that of lanthanum. A face-centered cubic structure has been found at higher temperatures. The melting point has been determined as 1340 ± 40°C (2440 ± 72°F). The metal dissolves readily in common mineral acids with the formation of the tripositive ion, Eq. (2). In this process 142 kilocalories of

$$\text{Cm} + 3\text{H}^+ = \text{Cm}^{3+} + 1\tfrac{1}{2}\text{H}_2(g) \qquad (2)$$

heat are evolved per mole of curium dissolved.

A number of solid compounds of curium have been prepared and their structures determined by x-ray diffraction. These include CmF_4, CmF_3, CmCl_3, CmBr_3, CmI_3, Cm_2O_3, and CmO_2. In the trihalides the ionic radius of Cm^{3+} is 0.094 nanometer. Isostructural analogs of the compounds of curium are observed in the lanthanide series of elements. See X-ray diffraction.

Curium trichloride may be prepared by treating the sesquioxide with $\text{HCl}(g)$ at 500°C (930°F). Synthesis of the tribromide and triiodide may be effected similarly by using $\text{HBr}(g)$ and $\text{HI}(g)$. The trifluoride is readily prepared by precipitation of aqueous Cm^{3+} with hydrofluoric acid and drying of the precipitate. This reacts at about 300°C (570°F) with elemental fluorine to yield the tetrafluoride. The dioxide is obtained by treating curium(III) oxalate with ozone and oxygen at about 300°C (570°F). The black oxide has the fluorite structure with crystal lattice parameter, $a = 0.537$ nm. The dioxide is relatively unstable, and may be decomposed in vacuum at 600°C (1100°F) to yield the cubic sesquioxide with $a_0 = 0.550$ nm.

In aqueous solution the tetrapositive ion is highly unstable. It may be prepared, however, as a fluoro-complex ion by dissolving solid curium tetrafluoride in 15 M CsF at 0°C. The light-yellow solution shows sharp absorption maxima, the strongest being at 451.5 and 864 nm. Complete reduction to Cm(III) occurs in about 20 min at room temperature. Cm(III) in aqueous solutions shows little absorption in the visible region of the spectrum, but weak lines are found in the region 370–400 nm. Strong absorption occurs farther into the ultraviolet, particularly near 280 nm.

The magnetic susceptibility of CmF_3, near room temperature, gives a calculated effective magnetic moment of 7.9 Bohr magnetons, similar to the value of 7.94 for Gd^{3+}, the lanthanide analog of curium. The spectroscopic g value for Cm^{3+} is 1.99 as compared to 1.93 for Gd^{3+}. The difference is well accounted for by the greater tendency toward j-j coupling in the $5f$ as compared with the $4f$ transition series.

Curium is the seventh member of the actinide series of elements. Its half-completed f subshell ($5f^7 6d 7s^2$) is analogous to that of gadolinium in the lanthanide series. The chemical properties of curium are so similar to those of the typical rare earths that, if it were not for its radioactivity, it might easily be mistaken for one of these elements. SEE ACTINIDE ELEMENTS; TRANSURANIUM ELEMENTS.

Glenn T. Seaborg

Bibliography. *Gmelins Handbuch der anorganischen Chemie: Transurane*, vol. 4, 1972, vols. 7a and 8, 1973; E. K. Hyde, I. Perlman, and G. T. Seaborg, *The Nuclear Properties of the Heavy Elements*, vols. 1 and 2, 1964; J. Katz and G. T. Seaborg, *The Chemistry of the Transuranium Elements*, vol. 3 of *Kernchemie in Einzeldarstellungen*, 1971; G. T. Seaborg, *Man-Made Transuranium Elements*, 1963.

Currant

A fruit (berry) in the genus *Ribes* in the family Saxifragaceae. Cultivated black and red currants and gooseberries all belong to this genus. *Ribes* species having prickles or spines are called gooseberries, and those that do not are called currants. The berries are produced in clusters on bushes, and cultivars ripen in midsummer (**Fig. 1**).

Varieties. *Ribes* are not widely grown commercially for fruit in the United States, but red currants and gooseberries are popular in home gardens for use in jellies and pies. Several *Ribes* species, particularly

Fig. 2. Anthracnose disease on (*a*) gooseberry (small leaves) and (*b*) currant (large leaves).

golden currant (*R. aureum*) and fuchsia-flowered gooseberry (*R. speciosum*), are used as shrubs in landscaping. In the United States important red currant cultivars include Cherry, Minnesota 71, Perfection, Red Lake, White Imperial, and Wilder; popular gooseberry cultivars include Fredonia, Pixwell, Poorman, Red Jacket, and Welcome. In Europe cultivars of the black currant (*R. nigrum*) are extensively grown commercially for juice concentrate. Wild *Ribes* species occur widely in the United States, and desirable, edible types are sometimes gathered during the season when the fruit is locally abundant. SEE FRUIT GROWING, SMALL; GOOSEBERRY.

Currant and gooseberry diseases. Many of the same diseases affect both currants and gooseberries. Several virus diseases occur in Europe in these crops, reducing yield and fruit quality, often without showing symptoms on the leaves. In Great Britain and several other European countries where *Ribes* species are important crops, local growers can purchase virus-tested, certified plants from nurseries, thus reducing the likelihood of planting virus-infected stock. Tomato ringspot virus, causing leaf-spotting patterns and reduced vigor in red currants in the United States, is spread in the soil by the American dagger nematode (*Xiphinema* sp.), and can be controlled by appropriate soil fumigation prior to planting.

Anthracnose is a leaf- and fruit-spotting fungus disease caused by *Pseudopeziza ribis* (**Fig. 2**). It may be serious in wet seasons, causing leaves to yellow and drop by midsummer. Control involves elimination of overwintered leaves that harbor the fungus; pruning to allow rapid drying of interior foliage; and use of approved fungicides, usually fixed coppers, at bloom and harvest.

Powdery mildew (caused by *Sphaerotheca morsuvae*) is a fungus disease that affects cultivated *Ribes*, particularly gooseberries, producing white, powdery coatings that later darken on young shoots, leaves, and fruits. Powdery mildew often stunts infected plants and renders the fruit visually unappealing. The disease is hard to control, but sanitation and spraying with approved fungicides just as the buds open in spring and during the growing season often help.

Other important fungus diseases of *Ribes* include white pine blister rust (caused by *Cronartium ribicola*) on black currant, leaf spot (caused by *Mycosphaerella ribis*), and cane blight (caused by *Botryosphaeria ribis*). SEE ROSALES; PLANT PATHOLOGY.

Richard H. Converse

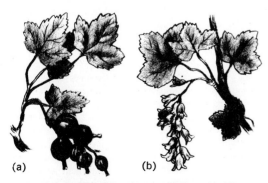

Fig. 1. Black-currant (*Ribes nigrum*). (*a*) Branch with cluster of fruits. (*b*) Branch with flowers.

Bibliography. H. W. Anderson, *Diseases of Fruit Crops*, 1956; N. W. Frazier (ed.), *Virus Diseases of Small Fruits and Grapevines*, 1970; J. S. Shoemaker, *Small Fruit Culture*, 1975.

Current balance

An apparatus with which force is measured between current-carrying conductors. The purpose of the measurement is to establish the value of the ampere; in terms of its International System (SI) definition; therefore the relative locations of the conductors are prescribed so that the force between them can be computed from their geometry and the current they carry. Experimentally this force is measured in terms of the gravitational force on a known mass; thus the ampere may be said to be weighed.

The determination of the ampere is one of the basic measurements used to assign values to electrical units in terms of mechanical units of length, mass, and time. By this assignment the electrical units of power and energy are made to have values identical with the corresponding mechanical units. The current balance is employed at the U.S. National Bureau of Standards and at the national laboratories of other countries as one of the steps in establishing the values of the absolute electrical units used in science and technology throughout the world. *See* ELECTRICAL UNITS AND STANDARDS; PHYSICAL MEASUREMENT.

Operation. In a current balance, force is measured between fixed and movable coils (see **illus.**). This force can be expressed as the product of the currents in the coils by the rate of change of their mutual inductance with displacement. If the coils carry the same current, Eq. (1) holds for a linear displacement,

$$F = I^2 \frac{\partial M}{\partial x} \qquad (1)$$

or Eq. (2) for an angular displacement, where F is the

$$F = I^2 \frac{\partial M}{\partial \theta} \qquad (2)$$

mechanical force, M is the coefficient of mutual inductance, and I is the current being measured. The coil construction must be such that $\partial M/\partial x$ or $\partial M/\partial \theta$ can be computed from dimensional measurements.

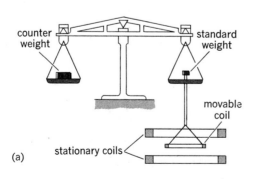

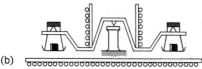

Current balances. (a) Rayleigh. (b) Pellat.

The movable coil is attached to an equal-arm weighing balance, and the force between the coils is made equal to a weight, which is the gravitational force on a known mass.

Types. Three types of current balance have been used in ampere determinations. In two of these, the Rayleigh and the Ayrton-Jones balances, the fixed and movable coils are coaxial, and the motion is linear along the coil axis. In the third type, the Pellat balance, the axes of the fixed and movable coils are at right angles, and the movable coil turns about its center, the pivot point of the weighing balance.

In the Rayleigh balance the coils are circular and have a square winding section. The fixed coils are located above and below, but coaxial with, the smaller, movable coil. The coil spacing is such that the force on the movable coil is maximum. Because of the small winding section, the ratio of radii of fixed and movable coils is the most important geometrical constant needed, and this ratio can be measured by electrical means. The radius of the movable coil is about half that of the fixed coils because for this ratio the exact dimensions of the winding cross section are of reduced importance. The value of current is determined from two weighings. With the standard weight removed from the right-hand balance pan, the current is given such direction that the force is downward. The counterweight is of such value that the beam is balanced. The current is then reversed in the fixed coils so that the force is upward, and the standard weight is simultaneously placed on the pan to restore balance. If the two rest points of the balance are the same, the standard weight is just double the force between fixed and movable coils.

In the Ayrton-Jones balance, and in a modification of this balance used at the National Bureau of Standards, the fixed and movable coils are single-layer solenoids. Here the diameter and winding pitch of the coils are of critical importance in computing force, and the coil construction is such that these dimensions can be measured accurately. The balance arrangement and the weighing procedure are the same as in the Rayleigh balance.

Single-layer solenoids, constructed so that diameter and pitch can be accurately determined, are also used in the Pellat balance. The coil forms are of fused silica or low-expansion glass for dimensional stability, and the winding is laid in a spiral groove, cut in the outer face of the cylinder, so that the pitch is uniform. The balance beam is of fused quartz. The weighing procedure is similar to that described above, and the torque required for balance is the product of the weight by its lever arm.

The electromotive force (emf) of a standard cell is also assigned in an absolute ampere determination. The current being measured is passed through a standard resistor, and the resulting voltage drop is compared to the emf of the cell. The resistor and cell are stable reference standards and are used to preserve the values of the electrical units in the intervals between absolute ampere and ohm determinations. *See* CURRENT MEASUREMENT.

Forest K. Harris

Bibliography. R. L. Driscoll, Measurement of current with a Pellat-type electrodynamometer, *J. Res. Nat. Bur. Stand.*, 60(4):287–296, 1958; R. L. Driscoll and R. D. Cutkosky, Measurement of current with National Bureau of Standards current balance, *J. Res. Nat. Bur. Stand.*, 60(4):297–305, 1958; F. K. Harris, *Electrical Measurements*, 1952, reprint 1975;

P. Kantrowitz et al., *Electronic Measurements*, 1979; P. Vigoureux, A determination of the ampere, *Metrologia*, 1(1):3–7, 1965.

Current comparator

An instrument for determining the ratio of two currents, based on Ampère's laws. Many electrical measurements are made by using voltage ratio transformers to compare the unknown to a standard. The current comparator provides the means for similar measurements to be performed with current ratio techniques. Moreover, the application of the current comparator, unlike the voltage ratio transformer, is not limited to alternating currents but can be used with direct currents as well. SEE INSTRUMENT TRANSFORMER.

The current comparator is based on Ampère's circuital law, which states that the integral of the magnetizing force around a closed path is equal to the sum of the currents which are linked with that path. Thus, if two currents are passed through a toroid by two windings of known numbers of turns, and the integral of the magnetizing force around the toroid is equal to zero, the current ratio is exactly equal to the inverse of the turns ratio. SEE AMPÈRE'S LAW; MAGNETIC FIELD.

For alternating-current comparators, the ampere-turn unbalance is given by the voltage at the terminals of a uniformly distributed detection winding on a magnetic core. Direct-current comparators use two magnetic cores which are modulated by alternating current in such a way that the dc ampere-turn unbalance is indicated by the presence of even harmonics in the voltage. However, neither method is an exact measure of the integral of the magnetizing force, and various design features are added to overcome this deficiency. The most important of these are the magnetic shields, which are configured as hollow toroids. They protect the magnetic core and detection winding from the leakage fluxes of the current-carrying windings and ambient magnetic fields, and they are responsible for an improvement in accuracy of about three orders of magnitude. The copper shields supplement this action at higher frequencies and also provide mechanical protection for the magnetic steels.

A cutaway view of the construction of an ac comparator is shown in the **illustration**. The compensation winding provides the means for magnetizing the outer magnetic shield without affecting overall ampere-turn balance, thus enabling energy transfer between the ratio windings. It has the same number of turns as one of the ratio windings so that they may be connected together in parallel.

Applications of the ac comparator include current and voltage transformer calibration, measurement of losses in large capacitors, inductive reactors and power transformers, resistance measurement, and the calibration of active and reactive power and energy meters.

In dc applications, the current comparator provides the means to resolve the first three or four most significant digits of a balance by turns-count, thus eliminating problems associated with switch contact resistance and thermal electromotive forces. Its uses include eight-decade resistance and thermometry bridges, a seven-decade potentiometer, and high current ratio standards. SEE CURRENT MEASUREMENT; ELECTRICAL MEASUREMENTS.

W. J. M. Moore; N. L. Kusters

Bibliography. W. J. M. Moore and P. N. Miljanic, *The Current Comparator*, 1988.

Current density

A vector quantity equal in magnitude to the instantaneous rate of flow of electric charge, perpendicular to a chosen surface, per unit area per unit time. If a wire of cross-sectional area A carries a current I, the current density J is I/A. The units of J in the International System (SI) are amperes per square meter.

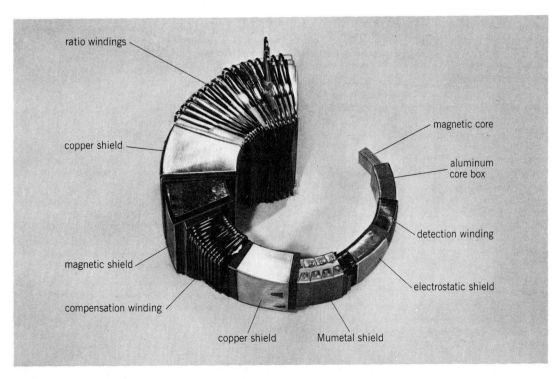

Cutaway view of an audiofrequency ac comparator. (*From W. J. M. Moore, Current comparator applications in precision measurements, ISA Advances in Instrumentation, vol. 25, pt. 3, pp. 1–8, Instrument Society of America, 1970*)

The concept of current density is useful in treating the flow of electricity through a conductor of nonuniform cross section, and in electromagnetic theory. In terms of current density Ohm's law can be written $J = E/\rho$, where E is electric field strength and ρ is resistivity. Current density is related to the number per unit volume n, the charge e, and the average velocity v of the effective charge carriers in a conductor by the formula $J = nev$.

John W. Stewart

Current measurement

The measurement of the rate of passage of electric charges in a circuit. The unit of measurement, the ampere (A), is one of the base units of the International System of Units (SI). It is defined as that constant current which, if maintained in two straight parallel conductors of infinite length, of negligible circular cross section, and placed 1 m apart in vacuum, would produce between these conductors a force equal to 2×10^{-7} newton per meter of length.

Determination of the ampere. In order to establish an electrical unit in accordance with the SI definition, it is necessary to carry out an experimental determination. Obviously the ampere cannot be realized exactly as defined. Electromagnetic theory has to be used to relate a practical experiment to the definition. For many years the current balance was used, in which two cylindrical solenoids are suspended from the opposite arms of a conventional balance. Fixed coils are so placed around the suspended coils that the magnetic interaction between the currents and coils causes an upward force on one side of the balance and a downward force on the other. The sum of these forces is measured by balancing against the force of gravity acting on an accurately known mass. A force equivalent to about 0.03 oz (1 g) weight has to be measured in the presence of a standing load of the order of 2 lb (1 kg). The accuracy of the experiment is limited to about 10 parts per million by the difficulties of the mechanical measurements to define the coils and convection currents. *See Current balance.*

An alternative determination of the SI unit of current was devised by B. P. Kibble and carried out at the national standards institutions of the United Kingdom and the United States. It depends on the availability of a satisfactory SI standard of resistance. The unit of resistance can be set up through the calculable capacitor, which is employed in several national standards laboratories. Through the concept of frequency (the reciprocal of time, another SI base quantity), the unit of impedance obtained from a calculable capacitor can be used to calibrate a resistor in SI units. In Kibble's virtual power balance, a coil is suspended partly in a strong permanent magnetic field. The experiment has two parts. First, a current I is passed through the coil and the resulting force F is measured by weighing. The current and force are proportional and related by a factor depending on the geometry of the apparatus. Second, the coil is moved at a controlled speed and the voltage induced in the coil is measured. The voltage V and the rate of displacement (dx/dt) are also proportional and depend on the same geometrical factor. The results of the two operations can be combined to eliminate the geometrical factor, giving the result shown in the equation below, where electric power = mechanical power. The result of the

$$V \times I = F \times \frac{dx}{dt}$$

experiment, when combined with the SI unit of resistance (V/I), allows the ampere to be determined and all the electrical units to be set up in accordance with the SI definitions. *See Resistance measurement.*

In practice, current or virtual power balances are not convenient as working standards, and so the unit of electric current is actually maintained as voltage through the alternating-current (ac) Josephson effect and resistance through the quantum Hall effect or calculable capacitor. *See Electrical units and standards; Hall effect; Josephson effect.*

Direct and low-frequency currents. The moving-coil (d'Arsonval) meter measures direct currents (dc) from 10 microamperes to several amperes. The accuracy is likely to be a few percent of the full-scale indication, although precision instruments can achieve 0.1% or even better. Above 1 milliampere a shunt usually carries the major part of the current; only a small fraction is used to deflect the meter. Since the direction of deflection depends on the direction of the current, the d'Arsonval movement is suitable for use only with unidirectional currents. Rectifiers are used to obtain dc and drive the meter from an ac signal. The resulting combination is sensitive to the rectified mean value of the ac waveform.

In the moving-iron meter, two pieces of soft magnetic material, one fixed and one movable, are situated inside a single coil. When current flows, both pieces become magnetized in the same direction and accordingly repel each other. The moving piece is deflected against a spring or gravity restoring force, the displacement being indicated by a pointer. As the repulsive force is independent of current direction, the instrument responds to low-frequency ac as well as dc. The natural response of such a movement is to the root-mean-square (rms) value of the current. The accuracy of moving-iron meters is less than that of moving-coil types. *See Ammeter.*

Radio-frequency currents. For radio-frequency applications it is essential that the sensing element be small and simple to minimize inductive and capacitive effects. In a thermocouple meter the temperature rise of a short, straight heater wire is measured by a thermocouple and the corresponding current is indicated by a d'Arsonval movement. In a hot-wire ammeter the thermal expansion of a wire heated by the current is mechanically enhanced and used to deflect a pointer. Both instruments, based on heating effects, respond to the rms value of the current. Above 100 MHz, current measurements are not made directly, as the value of current is likely to change with position owing to reflections and standing waves. *See Microwave measurements; Thermocouple.*

Large currents. Above 50 A the design of shunts becomes difficult. For ac, current transformers can be used to reduce the current to a level convenient for measurement. At the highest acuracy, current comparators may be used in which flux balance is detected when the magnetizing ampere-turns from two signals are equal and opposite.

Direct-current comparators are available in which dc flux balance is maintained and any unbalance is used to servo a second, or slave, current signal. For the highest accuracy, second-harmonic modulators are used, and for lower precision, Hall effect sensors. Electronically balanced ac and dc current comparators

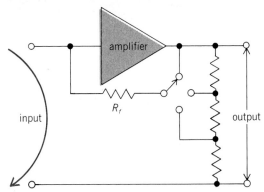

Measurement of very small currents by using an operational amplifier.

make clip-around ammeters possible, in which an openable magnetic core can be closed around a current-carrying conductor. This allows the meter to be connected into and removed from the circuit without breaking it or interrupting the current. SEE CURRENT COMPARATOR; INSTRUMENT TRANSFORMER.

Very small currents. The obvious method for measuring a very small current is to determine the voltage drop across a large resistor. A sensitive voltage detector having very low offset current is required, for example, an electrometer. Electrometers based on MOSFET (metal-oxide-semiconductor field-effect transistor) devices have overtaken other designs where the very highest resolution is required, as they can have offset current drifts less than 10^{-16} A. In order to provide a low impedance to the measured current, it is preferable to use the electrometer device in an operational-amplifier configuration (see **illus.**). The input becomes a virtual ground, and so stray capacitance across the input connection does not degrade the rate of response of the circuit as seriously as in the simple connection. SEE TRANSISTOR.

The operational-amplifier configuration can be extended to provide a variety of ranges from each high-value resistor. Driving the resistor from the amplifier output through a voltage divider of ratio N gives a similar result to the use of a resistor N times higher in value. This also has the advantage that extremely high-value resistors can be avoided.

Operational amplifiers suitable for this application are readily available, based not only on field-effect transistors but also on varactors. The latter are semiconductor equivalents to the mechanical vibrating capacitor electrometers. SEE AMPLIFIER; ELECTRICAL MEASUREMENTS; ELECTROMETER; OPERATIONAL AMPLIFIER; VARACTOR; VOLTAGE MEASUREMENT.

R. B. D. Knight

Bibliography. W. D. Copper and A. D. Helfrick, *Electronic Instrumentation and Measurement Techniques,* 3d ed., 1985; R. D. Cutkosky, New NBS measurement of the absolute farad and ohm, *IEEE Trans. Instrum. Meas.,* IM-23:305–309, 1974; S. Geczy, *Basic Electrical Measurements,* 1984; F. K. Harris, *Electrical Measurements,* 1975; B. P. Kibble et al., The NPL moving coil ampere determination, *IEEE Trans. Instrum. Meas.,* IM-32(1):141–143, March 1983; U.S. Department of Commerce, National Bureau of Standards, *Precision Measurement and Calibration, Electricity—Low Frequency,* Spec. Publ. 300, vol. 3., 1968; P. Vigoureux, Determination of the ampere, *Metrologia,* 1:3–7, 1965.

Current sources and mirrors

A current source is an electronic circuit that generates a constant direct current which flows into or out of a high-impedance output node. A current mirror is an electronic circuit that generates a current which flows into or out of a high-impedance output node, which is a scaled replica of an input current, flowing into or out of a different node.

Basic circuit configurations. Most specifications of analog integrated circuits depend almost uniquely on the technological parameters of the devices, and on the direct or alternating current that flows through them. The voltage drop over the devices has much less impact on performance, as long as it keeps the devices in the appropriate mode of operation (linear or saturation). High-performance analog integrated-circuit signal processing requires that currents be generated and replicated (mirrored) in an accurate way, independent of supply voltage and of those device parameters that are least predictable (for example, current gain β in a bipolar transistor). Hence, current sources and mirrors occupy a large portion of the total die area of any analog integrated circuit. They are also used, but less often, in discrete analog circuits.

Two basic circuit configurations are shown in **Fig. 1**. The circuit in Fig. 1a is suitable for implementation in bipolar integrated-circuit technology, or in complementary metal-oxide-semiconductor (CMOS) integrated-circuit technology if Q_1 and Q_2 are replaced by enhancement metal-oxide-semiconductor field-effect transistors (MOSFETs). Formulas will be derived here only for the bipolar configuration, but can be extended easily to MOS circuit configurations. The collector-base short circuit guarantees that transistor Q_1 will be always in the linear range. (If it is assumed, for the moment, that the current gain β and the output impedance r_o of the bipolar transistors are infinite), then the currents are determined by Eqs. (1)–(3). Here, I_{C1} is the collector current of Q_1; V_{BE1}

$$I_{C1} = I_{S1} \cdot \left(\exp \frac{V_{BE1}}{U_T} - 1 \right)$$

$$= \frac{V_{DD} - V_{SS} - V_{BE1}}{R} \tag{1}$$

$$I_{out} = I_{S2} \cdot \left(\exp \frac{V_{BE2}}{U_T} - 1 \right) \tag{2}$$

$$V_{BE1} = V_{BE2} = 0.7 \text{ volt} \tag{3}$$

and V_{BE2} are the base-emitter voltages of Q_1 and Q_2; V_{DD} and V_{SS} are the drain and source supply voltages; I_{S1} and I_{S2} are the saturation currents of Q_1 and Q_2, which are proportional to the emitter areas A_{E1} and A_{E2}; and U_T is the thermal voltage kT/q, where k is Boltzmann's constant, T is the absolute temperature, and q is the electron charge. (At room temperature, $U_T \simeq 16$ mV.) For two transistors, side by side on the same integrated circuit, the ratio of the two saturation currents I_{S1}/I_{S2} can be realized with an error of less than 1%; Eqs. (1)–(3) can be rewritten as Eq. (4). This equation implies that the output current will

$$I_{out} = \frac{A_{E2}}{A_{E1}} \cdot \frac{V_{DD} - V_{SS} - 0.7 \text{ V}}{R}$$

$$= \frac{A_{E2}/A_{E1}}{I_{ref}} \tag{4}$$

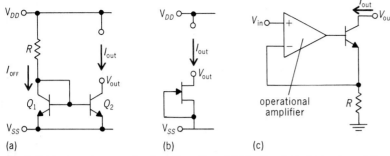

(a) (b) (c)

Fig. 1. Basic current-source circuit configurations. (a) Circuit suitable for implementation in bipolar or enhancement metal-oxide-semiconductor (MOS) integrated-circuit technology, for example, complementary MOS (CMOS) technology. (b) Circuit suitable for implementation in depletion field-effect transistor (FET) integrated-circuit technology, for example, negative MOS (NMOS), bipolar-junction FET (BiFET), or gallium arsenide metal-semiconductor FET (MESFET) technologies. (c) More general active circuit (both integrated-circuit and discrete applications).

be independent of the technological parameters of the bipolar transistors but dependent on the value of resistor R and on the supply voltages. Even if resistor R or the supply voltages are varied, I_{C1} and I_{C2} always remain equal: I_{C1} is mirrored in I_{C2}. Equations (1)–(3) demonstrate a very important principle in analog integrated-circuit design: replica biasing—I_{C1} is forced to produce a base-emitter voltage V_{BE} on transistor Q_1. This voltage creates exactly the right current in Q_2, under the assumption that both halves of the current mirror are exact replicas. The circuit of Fig. 1a thus acts as a current source characterized by Eq. (4) if the register R is present, and as a current mirror also characterized by Eq. (4) if the resistor is replaced by an input current source of value I_{ref}.

A simple and elegant one-transistor current source can be realized with a junction field-effect transistor (JFET) or depletion MOSFET (Fig. 1b). Indeed, the source-to-gate connection guarantees a constant and

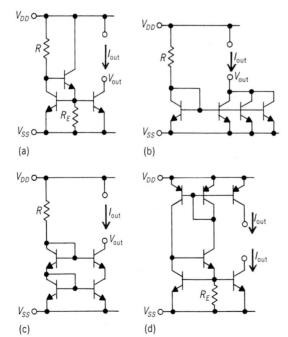

(a) (b)

(c) (d)

Fig. 2. Improved current-source configurations. (a) Circuit for reduction of effect of finite β. (b) Replica biasing for integer current mirror ratio. (c) Cascoded current mirror for enhanced output impedance. (d) Power-supply-independent current source.

predictable transistor biasing, since the threshold voltage V_T is negative, as in Eq. (5). The output current

$$V_{GS} - V_T = -V_T > 0 \qquad (5)$$

realized in this way is often denoted by I_{DSS}.

Finally, a more general circuit configuration (Fig. 1c) employs an operational amplifier to create a constant current given by Eq. (6). This circuit has been

$$I_{out} = V_{in}/R \qquad (6)$$

successfully applied in both integrated and discrete circuits to create constant or voltage-dependent current sources.

Variations. Although the circuit in Fig. 1a is the most common circuit configuration, with applications in bipolar and CMOS technology [with bipolars replaced by negative MOS (NMOS) transistors], it has some serious drawbacks that are addressed in the improved circuits in **Fig. 2**.

The output current I_{out} in the circuit of Fig. 1 is affected by the finite current gain β in the bipolar transistors as given by Eq. (7). The circuit in Fig. 2a

$$\frac{\Delta I_{out}}{I_{out}} = \frac{1}{1 + (2/\beta)} \qquad (7)$$

has a considerably lower error due to finite β.

Current mirror ratios I_{S2}/I_{S1} different from unity can be realized by transistor pairs with unequal emitter areas; however, manufacturing tolerances can now lead to significant systematic errors. Integer mirror ratios can be realized without accuracy degradation, if arrays of identical unit transistors are used (3 to 1 in Fig. 2b).

The basic current source or current mirror of Fig. 1 has an output impedance equal to the dynamic output resistance r_o of the transistors. Cascodes (Fig. 2c) enhance the output resistance of the current source according to Eq. (8), where G_m is the transconduct-

$$R_{out,source} = r_o \cdot (G_m \cdot r_o) \qquad (8)$$

ance of the cascode transistor. The increase in R_{out} is usually by a factor of about 100.

It is often very important that analog circuit specifications be independent of the applied supply voltages. Figure 2d shows a simple circuit in which the current is independent of the supply voltages. The realized current I_{out} can be found by solving transcendental equation (9).

$$I_{out} = V_T \cdot \frac{\log_e(I_{out}/I_S)}{R_E} \qquad (9)$$

This circuit has two operating points: I_{out} given by Eq. (9) or I_{out} equal to zero. To avoid having the circuit lock into the currentless mode during power-up, a startup circuit must be added. SEE INTEGRATED CIRCUITS; TRANSISTOR.

Peter M. VanPeteghem

Bibliography. P. R. Gray and R. G. Meyer, *Analysis and Design of Analog Integrated Circuits*, 1984; R. Gregorian and G. C. Temes, *Analog MOS Integrated Circuits for Signal Processing*, 1986.

Curry

A mixture of plant spices used with meats and fish and other seafoods, each requiring its own specific combination. One recipe calls for a mixture of turmeric, coriander, cinnamon, cumin, ginger, cardamon, cayenne pepper, pimiento, black pepper, long pepper, cloves, and nutmeg. See separate articles on each of these spices. *SEE SPICE AND FLAVORING.*

Perry D. Strausbaugh/Earl L. Core

Curve fitting

A procedure in which the basic problem is to pass a curve through a set of points, representing experimental data, in such a way that the curve shows as well as possible the relationship between the two quantities plotted. It is always possible to pass some smooth curve through all the points plotted, but since there is assumed to be some experimental error present, such a procedure would ordinarily not be desirable. *SEE IN-TERPOLATION.*

The first task in curve fitting is to decide how many degrees of freedom (number of unspecified parameters, or independent variables) should be allowed in fitting the points and what the general nature of the curve should be. Since there is no known way of answering these questions in a completely objective way, curve fitting remains something of an art. It is clear, however, that one must make good use of any background knowledge of the quantities plotted if the above two questions are to be answered. Therefore, if it was known that a discontinuity might occur at some value of the abscissa, one would try to fit the points above and below that value by separate curves.

Against this background knowledge of what the curve should be expected to look like, one may observe the way the points fall on the paper. It may even be advantageous to make a few rough attempts to draw a reasonable curve "through" the points.

A knowledge of the accuracy of the data is needed to help answer the question of the number of degrees of freedom to be permitted in fitting the data. If the data are very accurate, one may use as many degrees of freedom as there are points. The curve can then be made to pass through all the points, and it serves only the function of interpolation. At the opposite extreme when the data are very rough, one may attempt to fit the data by a straight line representing a linear relation, Eq. (1), between y and x. Using the above information and one's knowledge of the functions that have been found useful in fitting various types of experimental curves, one selects a suitable function and tries to determine the parameters left unspecified. At this point there are certain techniques that have been worked out to choose the optimum value of the parameters.

$$y = ax + b \qquad (1)$$

One of the most general methods used for this purpose is the method of least squares. In this method one chooses the parameters in such a way as to minimize the sum, Eq. (2), where y_i is the ordinate of ith

$$S = \sum_{i=1}^{n} [y_i - f(x_i)]^2 \qquad (2)$$

point and $f(x_i)$ the ordinate of the point on the curve having the same abscissa x_i as this point.

Method of least squares. If one attempts to fit the data with a straight line, one lets y equal the values shown in Eq. (3), and Eq. (2) becomes (4).

$$y = f(x) = ax + b \qquad (3)$$

$$S = \sum_{i=1}^{n} [y_i - ax_i - b]^2 \qquad (4)$$

Since S is to be a minimum,

$$\frac{\partial S}{\partial a} = \frac{\partial S}{\partial b} = 0$$

and therefore Eqs. (5) hold.

$$a \sum_i x_i^2 + b \sum_i x_i = \sum_i x_i y_i$$
$$a \sum_i x_i + bn = \sum_i y_i \qquad (5)$$

The solution of these equations is shown as Eqs. (6), where the term Δ is defined in Eq. (7). These

$$a = \frac{1}{\Delta}\left(\sum_i x_i \sum_i y_i - n \sum_i x_i y_i\right)$$

$$b = \frac{1}{\Delta}\left(\sum_i x_i \sum_i x_i y_i - \sum_i x_i^2 \sum_i y_i\right) \qquad (6)$$

$$\Delta = \left(\sum_i x_i\right)^2 - n \sum_i x_i^2 \qquad (7)$$

values of a and b are those that minimize S and therefore represent the proper choice of the parameters of a straight line according to the criterion of least squares. *SEE LEAST-SQUARES METHOD.*

Method of averages. This method has less theoretical justification, but it is easier to apply than the method of least squares.

If all the points (x_i, y_i) for $i = 1, 2, \ldots, n$ representing the data were to lie on a straight line, one could require that a and b satisfy all n of Eqs. (8).

$$y_i = ax_i + b \qquad i = 1, 2, \ldots, n \qquad (8)$$

Any two of these equations could then be used to determine a and b. One could also add these equations together in two groups to obtain Eqs. (9).

$$\sum_{i=1}^{k} y_i = a \sum_{i=1}^{k} x_i + bk$$

$$\sum_{i=k+1}^{n} y_i = a \sum_{i=k+1}^{n} x_i + b(n - k) \qquad (9)$$

Suppose now that the points do not lie on a straight line, then Eqs. (8) are no longer valid. One may, however, assume that a best line through the points is given by Eq. (3) with a and b determined by requiring that Eqs. (9) be satisfied. There is some arbitrariness in the choice of k, but it is usually taken to be a whole number near $n/2$.

Consider the data below:

x	y
0.5	0.4
1.0	1.3
1.5	1.6
2.0	2.3
2.5	3.2
3.0	3.6
3.5	4.2
4.0	5.2

Letting $k = 4$, one obtains the following values for Eqs. (9):

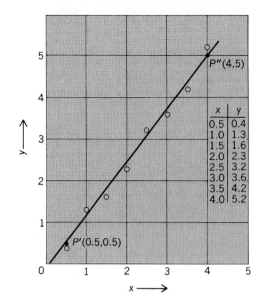

x	y
0.5	0.4
1.0	1.3
1.5	1.6
2.0	2.3
2.5	3.2
3.0	3.6
3.5	4.2
4.0	5.2

Graphical method of curve fitting.

$$5a + 4b = 5.6$$
$$13a + 4b = 16.2$$

Subtracting, one has $8a = 10.6$ or $a = 1.325$, and therefore $b = -0.256$; thus, the best straight line is given by the equation

$$y = 1.325x - 0.256$$

Graphical method. In this method one first plots the points (x_i, y_i) on coordinate paper using suitable scales along the X and Y axes (see **illus.**). One then takes a transparent straightedge and adjusts it so the points lie as near the edge as possible with about the same numbers above as there are below. The straight line corresponding to the position of the edge is drawn in, and the coordinates of two points (x', y') and (x'', y'') near each end of the line are read off. The straight line drawn is then given by Eq. (3), with a and b determined by the simultaneous equations labeled (10).

$$\begin{aligned} y' &= ax' + b \\ y'' &= ax'' + b \end{aligned} \qquad (10)$$

The coordinates of points P' and P'' in the illustration are, respectively, (0.5,0.5) and (4,5); therefore, $\alpha = 1.286$ and $b = -0.144$.

Use of nonlinear scales. Suppose one tries to fit the data points (x_i, y_i) with any general functional relation $y = f(x, \alpha, \beta)$ having two undetermined parameters α and β. It often happens that it is possible to introduce two new variables, Eqs. (11), that will satisfy linear Eq. (12). From the given data points (x_i, y_i)

$$X = F(x) \qquad Y = G(y) \qquad (11)$$

isfy linear Eq. (12). From the given data points (x_i, y_i)

$$Y = aX + b \qquad (12)$$

one can obtain a new set of data points (X_i, Y_i) by the use of Eqs. (11). One may then fit these points by a straight line in the XY plane, given by Eq. (12), using any of the techniques described above. The required equation in x and y is then

$$G(y) = aF(x) + b$$

which will determine the original constants α and β.

For example, suppose one assumes that Eq. (13)

$$y = \beta e^{\alpha x} \qquad (13)$$

holds. Then, by taking the natural logarithm of both sides, one has Eq. (14). Thus a linear equation as

$$\log y = ax + \log \beta \qquad (14)$$

given in Eq. (12) exists, as shown in Eqs. (15) and

$$X = x \qquad Y = \log y \qquad (15)$$

(16). A graphical solution involves use of semilog pa-

$$a = \alpha \qquad b = \log \beta \qquad (16)$$

per in which the vertical displacement of points is proportional to $\log y$. On such paper, therefore, Eq. (14), or its equivalent, Eq. (13), plots as a straight line.

Consider next fitting data by Eq. (17), which may

$$y = \beta x^{\alpha} \qquad (17)$$

also be written

$$\log y = \alpha \log x + \log \beta$$

Thus again the linear equation labeled (12) connects, as shown in Eqs. (18), and Eqs. (16) again hold for

$$X = \log x \qquad Y = \log y \qquad (18)$$

the constants. Logarithmic paper makes it simple to give the data points the rectangular coordinate values X and Y. The best line through these points then determines a and b and, therefore, also α and β.

If the plotting on logarithmic paper seems to lie about a curved line, one may try an equation of the form of Eq. (19), which may be written as Eq. (20).

$$y = \beta x^{\alpha} + \gamma \qquad (19)$$
$$\log (y - \gamma) = \alpha \log x + \log \beta \qquad (20)$$

One therefore works with a linear relation between $\log x$ and $\log (y - \gamma)$.

To obtain γ one draws a smooth curve through the points (x_i, y_i) and picks out two points (x', y') and (x'', y'') near the ends of this curve. Next one chooses $x''' = \frac{1}{2}(x'x'')$ and finds the point on the curve (x''', y''') with this value of x. It can be shown quite simply that Eq. (21) holds. The values of a and b and therefore of α and β are found as before.

$$\gamma = \frac{y'y'' - y'''}{y' + y'' - 2y'''} \qquad (21)$$

If one decides to fit the data by a parabolic curve represented by Eq. (22), one may first draw a smooth

$$y = \alpha + \beta x + \gamma x^2 \qquad (22)$$

curve through the points and select some point $(x'y')$ on this curve. The parabola will pass through this point if Eq. (23) holds. From Eqs. (22) and (23) one

$$y' = \alpha + \beta x' + \gamma x'^2 \qquad (23)$$

has Eq. (24), which shows that $(y - y')/(x - x')$ is

$$\frac{y - y'}{x - x'} = \beta + \gamma(x + x') \qquad (24)$$

a linear function of x. Therefore, for each data point (x_i, y_i) one may plot a new point (X_i, Y_i), as shown in Eqs. (25). The best straight line through these points

$$X_i = x_i \qquad Y_i = \frac{y_i - y'}{x_i - x'} \qquad (25)$$

given by Eq. (12) can be determined as before and this determines β and γ in Eq. (24). The value of α is then determined by use of Eq. (23). *See Coordinate systems; Extrapolation; Graphic methods*.

Kaiser S. Kunz

Bibliography. P. Lancaster and K. Salkauskas, *Curve and Surface Fitting*, 1986; E. M. Mikhail, *Observations and Least Squares*, 1982.

Cutaneous sensation

The sensory quality of skin. The skin consists of two main layers: the epidermis, which is the outermost protective layer, and the dermis, which consists of a superficial layer called the papillary dermis and a deeper layer called the reticular dermis (**Fig. 1**). Beneath the dermis is a layer of loose connective tissue, the subcutaneous tissue, which attaches the skin to underlying structures. *See Skin*.

Sensory Receptors

Sensory receptors in or beneath the skin are peripheral nerve-fiber endings that are differentially sensitive to one or more forms of energy. The sensory endings can be loosely categorized into three morphological groups (**Fig. 2**): endings with expanded tips, such as Merkel's disks found at the base of the epidermis; encapsulated endings, such as Meissner's corpuscles (particularly plentiful in the dermal papillae), and other organs located in the dermis or subcutaneous tissue, such as Ruffini endings, Pacinian corpuscles, Golgi-Mazzoni corpuscles, and Krause's end bulb; and bare nerve endings that are found in all layers of the skin (some of these nerve endings are found near or around the base of hair follicles).

Available evidence indicates that the encapsulated and expanded tip endings and the hair follicle endings are mechanoreceptors and that the specialized end organs serve to influence the dynamic range of mechanical stimuli to which the receptor responds, but not the type of energy to which it is most sensitive (for example, mechanical as opposed to thermal). The response specificity of a receptor is determined by some as yet unidentified property of the membrane of the nerve ending. Thus, receptors having bare nerve endings still differ in the specificities of their responses to certain stimuli.

Specificities receptors. There is a remarkable relationship between the response specificities of cutaneous receptors and five primary qualities of cutaneous sensation, the latter commonly described as touch-pressure, cold, warmth, pain, and itch. Each quality is served by a specific set of cutaneous peripheral nerve fibers. More complex sensations must result from an integration within the central nervous system of information from these sets of nerve fibers. For example, tickle sensations may result for the temporal and spatial sequence of stimulating a succession of touch receptors with a slowly moving stimulus lightly applied to the skin. Exploration of the skin surface with a rounded metal point reveals that there exist local sensory spots on the skin, stimulation of which (by any type of energy and by any temporal pattern of stimulation) evokes only one of the five qualities of sensation. Thus may be plotted maps of pressure, warm, cold, pain, or itch spots. *See Pain*.

There is also a correlation between different qualities of sensation and the sizes (diameters) of the sensory peripheral nerve fibers serving them. The larger-diameter nerve fibers of 6–12 micrometers serve the sense of touch-pressure (and such temporal variations as vibratory sensations or tickle sensations). The sense of coolness and the sense of pricking pain are served by different sets of small myelinated fibers (1–5 μm diameter); sensations of warmth, burning or aching pain, or itch are each mediated by different sets of unmyelinated nerve fibers (0.2–1.5 μm diameter). Since the conduction velocity of nerve impulses along a peripheral nerve is inversely related to the fiber's diameter, there may be a temporal sequence of sensations evoked by an abruptly delivered stimulus. For example, stubbing the toe or pricking the skin of the foot or hand with a pin may elicit first a sense of touch (mediated by the faster-conducting myelinated fibers), followed almost immediately by a sharp pain of short duration (smaller myelinated fibers) and, after as long as a second or so, a more agonizing pain of longer duration (mediated by the slowly conducting unmyelinated fibers).

Local anesthetic infiltrated around a peripheral nerve first blocks conduction in unmyelinated fibers (with a corresponding loss of the capacity to sense warmth, burning pain, and itch), followed by a block of myelinated fibers in ascending order of their diameters with concomitant losses of the senses of coolness, pricking pain, tickle, and touch-pressure. The exact opposite sequence of sensation loss occurs when a nerve is blocked by asphyxia produced by a pressure cuff around the arm, with the larger-diameter fibers blocked first and the unmyelinated fibers last.

The best evidence for specificity of sensory peripheral nerve fibers and their receptor endings is provided by experiments in humans and animals in which

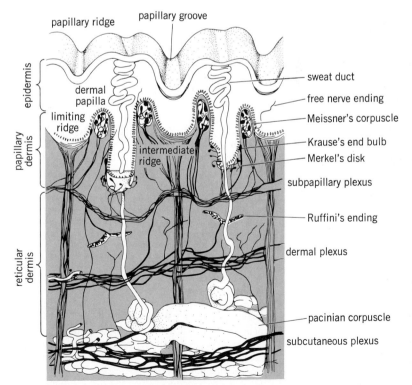

Fig. 1. A section of glabrous (nonhairy) human skin cut vertically through the papillary ridges to show locations of sensory receptors. (*After V. B. Mountcastle, ed., Medical Physiology, vol. 1, 14th ed., C. V. Mosby, 1980*)

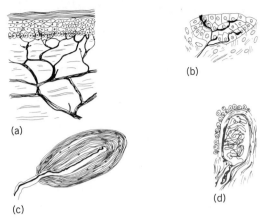

Fig. 2. Types of receptor endings in the skin: (a) free nerve endings; (b) Merkel's disk; (c) Pacinian corpuscle; (d) Meissner's encapsulated ending. (After A. Brodal, Neurological Anatomy in Relation to Chemical Medicine, Oxford University Press, 1969)

a recording electrode is brought in contact with a portion of the nerve (such as following microdissection of the nerve or by inserting the electrode through the skin and into the nerve), and the electrical signs of impulse discharges of single nerve fibers are recorded in response to cutaneous stimulation. Evidence is that each of the elementary qualities of sensation is served by a different set of peripheral nerve fibers.

Mechanoreceptors. The sense of touch-pressure is mediated by large, myelinated fibers. One type, believed to terminate in Merkel's disks, responds with increasing frequency of discharge to increases in pressure applied to the skin. The frequency of discharge is linearly related to the amount of skin indentation produced by a punctate stimulus and linearly related to numerical estimates, made by human observers, of the perceived magnitude of pressure sensations. Furthermore, the capacity of this mechanoreceptor to respond differentially to gaps of different sizes in a smooth surface applied to the passive skin can account for the human capacity for spatial resolution. The perceptual threshold (minimal separation) for spatially resolving gaps applied to the passive fingertip is a gap of 0.8 mm. Spatial resolution is very good on the fingertip, where the density of mechanoreceptors is high, and relatively poor on the back, where receptor density is sparse.

Other types of mechanoreceptors with large diameter myelinated fibers respond to mechanical movement of the skin and can be considered as velocity detectors. One type ends in Meissner corpuscles and responds best (that is, requires the least amplitude of skin displacement) to mechanical oscillations of 30 Hz. Another type ends in the Pacinian corpuscle and is even more sensitive, but only at higher frequencies (best frequency is about 225 Hz). The combined response properties of receptors with Meissner's or Pacinian corpuscles provide the necessary information required to account, at the level of the receptor, for the capacities of humans to detect and discriminate between mechanical oscillatory stimuli of different frequencies and amplitudes applied to the skin. Such receptors probably have an important role in conveying information about surface texture when the finger is moved back and forth over the surface of an object. *See* Mechanoreceptors.

Thermoreceptors. There are two types of thermoreceptors responsive to temperature changes in human skin. Warmth is served by a set of unmyelinated nerve fibers with bare nerve endings. These fibers have an ongoing discharge (in warm skin) that is eliminated by gentle cooling. Their frequencies of discharge increase nearly linearly with increases in stimulus intensity over the range of temperatures described as increasingly warm by human observers. The sense of coolness is served by a set of thinly myelinated nerve fibers with bare nerve endings. These have an ongoing discharge (in cool skin) that is eliminated by gentle warming. They increase their rates of discharge nearly linearly with decreases in temperatures described as increasingly cool by human observers. Neither type of thermoreceptor responds to mechanical stimulation. The response properties of these types of receptors have been shown to provide the information required to account for human capacities to discriminate small differences in the temperatures of stimuli applied to the skin.

Nociceptors. These receptors have bare nerve endings in or beneath the skin and are selectively responsive to noxious stimuli described as painful when applied to human skin. There are different sets of thinly myelinated nerve fibers and unmyelinated fibers whose endings are differentially responsive to only noxious thermal stimuli (hot or cold, or both) or only to noxious mechanical stimuli (such as pricking or pinching the skin). Other types of nociceptors are less specific, and the most common of these is one which responds to any noxious stimulus, whether it be chemical, thermal, or mechanical, and thus is called the polymodal nociceptor. It typically has an unmyelinated nerve fiber, but those with thinly myelinated fibers have been found as well. The minimal intensities of heat required to evoke a response in the typical polymodal nociceptor with an unmyelinated fiber has been shown to elicit minimal sensations of pain in humans. Furthermore, the frequencies of discharge in these receptors in response to heat stimuli increase as a slightly positively accelerating function of increasing temperature over the range of approximately 109–124°F (43 to 51°C) [higher intensities are not commonly tested], as do numerical estimates by humans of the magnitude of pain evoked by such stimuli. Following injury to the skin, such as that produced by a superficial burn, the threshold responses to mechanical or thermal stimuli of many nociceptors may be greatly lowered. This is believed to account, at least in part, for the increased painfulness of stimuli applied to injured skin. In such a state of heightened sensitivity to pain, termed hyperalgesia, normally innocuous stimulation such as gentle rubbing or warming can evoke a sensation of pain.

Itch. Itch sensations are believed to be mediated by a set of unmyelinated nerve fibers different from those serving pain. Itch can be evoked by mechanical, electrical, and chemical stimuli. Certain chemical stimuli are the most effective in producing itch, and these may be externally applied to the skin or released within the skin under certain conditions. For example, enzymes called proteinases may be found in certain plants and, when delivered to the skin via spicules, are powerful itch-producing agents. A naturally occurring substance is histamine, which is released from mast cells in the skin following local injury or in response to foreign substances that trigger an allergic reaction. Some chemicals produce itch by causing histamine to be released in the skin, while others act independently to evoke itch and can do so in regions of skin previously depleted of histamine.

Robert Lamotte

TOUCH SENSATIONS

The hands and the oral cavity, including the lips, are the body structures primarily used to tactilely explore small objects in the environment. Touch sensations can easily be aroused from other areas of the body, for example, the stimulation of body hairs, but the resolution of fine detail by these areas is not as great as that of the hands or the oral cavity.

Stimulus characteristics. A mechanical event applied to the skin may be divided into static and dynamic components. During indentation, mechanical energy is expended to move the skin from one position to another. The quantity of energy can be expressed in terms of the applied force per unit dimension, for example, tension (g/mm) or pressure (g/unit area), or in terms of work (force × indentation depth). Work, compared to pressure or tension, has proved to be a better stimulus measurement to account for variations in touch sensations that are the result of variations in the mechanical properties of the skin. The depth of skin indentation produced by a given force is related to the elastic properties of the skin, which are, in turn, related to age, site of stimulation, and the underlying tissue. If the force is made large and indentation depth controlled, much of the confounding that results from the varying mechanical properties of the skin can be avoided. In this way indentation depth becomes a convenient measurement by which to express the intensities of cutaneous mechanical stimuli.

The velocity, or rate, of indentation has also been shown to be related to threshold tactile sensations. The rate of indentation has little effect on the absolute tactile threshold when the rate of indentation is greater than about 0.3 mm per second. At slower rates of indentation, however, the depth of indentation required to produce a just-detectable tactile sensation increases dramatically. Rate of indentation also has a small but consistent effect on the judged intensity of tactile sensations. Acceleration and deceleration in indentation rate have so far not been implicated as stimulus variables for tactile sensations.

Absolute sensitivity. Sensitivity to tactile stimuli varies from one part of the body to another. The nose, lips, cheek, and forehead are the most sensitive parts, followed closely by the fingertips and palms, but even here there are large differences. Single tactile points on the palmar aspects of the fingers are highly sensitive with a median indentation depth threshold of 0.011 mm, whereas the tactile points in the center of the palm and the lateral aspects of the fingers have a median indentation depth threshold of 0.036 mm.

Intensity of sensations. As the intensity of a touch stimulus (for example, depth of indentation) increases, the intensity of the resulting tactile sensation increases. However, the intensity of the tactile sensation appears to increase in two stages, at least when rates of indentation of 0.4 mm per second or greater are used. Tactile sensations have been judged to be just about the detection level for indentation depths of 0.005 mm and rates of 1 or 10 mm per second. As

indentation depth increases, the resulting sensation is judged to increase slowly, but when indentation depths greater than about 0.5 mm are used, the sensation increases considerably more with each stimulus intensity increment. This two-stage response probably results from the cooperation of two different receptor mechanisms to produce the touch sensations.

Spatial discriminations. There are two traditional means of assessing the passive, tactile, spatial, discriminatory capabilites of the skin. These are point localization, the ability to locate the point on the skin that has been touched, and the two-point threshold, the distance separating two pointed stimulators to experience two rather than one point of stimulation. Measurements of the accuracy of point localization show the index finger to have the smallest error (1.4 mm) while the back shows the largest error (12.5 mm). The two-point threshold has been found to be smallest on the middle-finger pad (2.5 mm, the approximate separation of braille dots). The next smallest two-point threshold was on the index-finger pad (3.0 mm), and largest on the calf of the leg (47.0 mm). This means that the finger pads can detect much smaller (finer) features of an object than can other skin areas.

Temporal discriminations. When rapidly vibrating mechanical stimuli are applied to the skin, another stimulus dimension is added—the frequency of repetition. When applied to the index-finger pad, the resulting sensation, for repetition frequencies up to about 40 Hz, may best be described as a flutter. At higher frequencies the flutter changes to a buzzing sensation.

The lower limit to which the skin is sensitive is defined as that frequency at which the individual pulses fuse to form a single continuous sensation. This occurs, as it does in audition, at about 15 Hz. The upper limit is defined as that frequency at which the buzzing sensation disappears and the sensation becomes that of steady skin indentation, about 640 Hz.

Thresholds of vibratory sensations vary with the frequency of the vibration. There is little change in sensitivity for frequencies up to about 40 Hz. As the frequency of vibration increases beyond 40 Hz, the threshold decreases; that is, the receptors are more sensitive to the higher frequencies, until they reach their peak sensitivity at about 250 Hz, beyond which their sensitivity decreases. Sensitivity to increased frequency once again shows two stages. Two separate receptor processes, each with different rules of operation, combine to form the range of response.

For many years scientists and engineers have sought ways in which the skin might be used as a substitute means of communication for the blind or deaf. There are difficulties, such as cost, ease of learning, and the cosmetic appearance, in the development of a suitable device. In addition, the skin presents its own difficulties in terms of the amount of information that it can convey to the brain. Three information channels are possible: body site of stimulation, frequency, and intensity of the vibratory stimuli. The reliability of point localization is poor except on the fingertips and face, and varies widely with body site. Frequency requires discrimination by the skin—vibrations of 40 and 44 or 200 and 220 Hz applied to the finger pad can be discriminated about

75% of the time, but like point localization, this varies with the site of stimulation. The range of useful frequencies is also severely limited—up to about 600 Hz. Discrimination between intensities of vibration not only varies with the site of stimulation but also varies over a wide range with the frequency of the vibration.

Temperospatial sensations. When a light tactile stimulus, such as the tip of a feather, is drawn across the skin so as to stimulate successive touch points in temporal succession, a tickle sensation is often experienced. This is a different sensation than that of itch, although the response elicited may be the same—rubbing or scratching. Little is known of the necessary and sufficient stimulus conditions to arouse a tickle.

Active touch. The limits of fine discrimination (two-point discrimination) discussed above involved the finger pads resting passively on the feature to be detected. Much finer features (texture) can be detected, however, if the finger pads are actively moved over the surface. Scanning an object with the finger increases the information flow in a tactile sensation. It is not the movement of the finger per se over the object, for just as fine a discrimination can be made if the object is moved under the stationary finger. Perhaps small vibrations, shearing forces on the skin, the temporal pattern of tactile receptor stimulation, or some similar dimensions have been added by the movement.

Another, more complex aspect of active touch is object recognition, or stereognosis. Here, not only are the fine features of the object distinguished, but so is the object's overall form. Kinesthesis, the muscle and joint sense, is used.

THERMAL SENSATIONS

The thermal sense is unique among the senses in that an absence of sensation occurs at about 86–97°F (30–36°C). This represents a considerable amount of thermal energy, but this is the temperature domain in which homeothermic systems operate. Changes in temperature imposed from outside change the thermal environment of the (warm and cold) receptors, presumably producing changes in their metabolism and hence their response rates, without an exchange of energy between the stimulus temperature (change) and the receptor. In other sensory systems the absence of the appropriate energy form leads to the absence of sensation. In these systems stimulation is the result of an exchange of energy between the stimulus and the receptor.

The thermal sense is unique in another way. Its primary function—an important input to the body temperature–regulating system—does not reach awareness. Only relatively innocuous sensations of warm and cool are perceived, and only when the skin temperature drops below about 64°F or 18°C (cold-pain) or rises above 113–115°F or 45–46°C (heat-pain) does the sensation become noxious. This represents engagement of a different system—the nocifensor system—and is not properly a part of the thermal sensing system.

Stimulus characteristics. There are six principal stimulus variables that alter the responsiveness of the human temperature-sensing system: (1) the temperature to which the skin has been adapted; (2) the direction of the temperature change; (3) the intensity of the temperature change; (4) the rate of the temperature change; (5) the area of the skin stimulated; and (6) the site of application of the thermal stimuli.

The band of skin temperatures between about 86 and 97°F (30 and 36°C) is called the zone of physiological zero (**Fig. 3**). When skin temperatures do not exceed the limits of this band, given time, thermal sensations adapt completely; that is, they disappear. At lower skin temperatures a cool sensation persists no matter how long that temperature is maintained—the stimulus adapts but the sensation never disappears completely. With further reductions in skin temperature the sensation would be described as cold, and finally at about 64°F (18°C) is replaced by a dull, aching sensation—cold-pain. When the skin temperature is raised above the zone of physiological zero, much the same series of sensations occur except that their qualities are warm to hot, and finally at about 113°F (45°C) the hot is replaced by a stinging sensation—heat-pain.

The direction of a temperature change, when skin temperature is within the zone of physiological zero, produces thermal sensations consistent with the direction of the change—an increase produces a warm sensation and a decrease produces a cool sensation. A small increase in temperature from a skin temperature below physiological zero produces a decrease in the persisting cool sensation, not a warm sensation. Similarly, a small decrease in temperature from a skin temperature above physiological zero produces a decrease in the persisting warm sensation, not a cool sensation.

The intensity of a temperature change, when skin temperature is within the zone of physiological zero, produces sensations of proportional intensity. When skin temperature is below the zone of physiological zero, sensitivity to decreases in temperature remains unchanged from that in the zone of physiological zero. Sensitivity to temperature increases is reduced, and the farther the skin temperature is below the lower limit of physiological zero, the less sensitive it becomes to temperature increases. A similar circumstance occurs for temperature decreases when the skin temperature is above the upper limit of physiological zero.

When the rate of a temperature change is less than about 0.05°F (0.03°C) per second, sensitivity to that change decreases. At the extreme, skin temperature can be changed from one to the other limit of physiological zero without producing a thermal sensation if done sufficiently slowly, about 0.013°F (0.007°C) per second. Sensitivity to temperature changes does not increase at rates of change faster than 0.05°F (0.03°C) per second, at least up to rates of 2.7°F (1.5°C) per second.

The size of the area of the skin stimulated by tem-

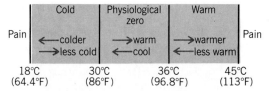

Fig. 3. The relationship between skin temperature, physiological zero, and the direction of temperature change as they affect thermal sensations. (*From D. R. Kenshalo, Sr., Biophysics and psychophysics of feeling, in E. C. Carterette and M. P. Friedman (eds.), Handbook of Perception, vol. 6B, chap. 2, Academic Press, 1978*)

perature changes markedly influences sensitivity. For temperature increases, areal summation is almost complete for areas up to about 30 in.^{2} (200 cm^2). Thus, there is almost complete reciprocity between the area stimulated and the intensity of the stimulus, so that if area is doubled the stimulus intensity required to reach detectability will be halved. This means that the whole palm of the hand is a much more sensitive heat detector than just the fingertip. Areal summation for temperature decreases is somewhat less than that for temperature increases but is still considerable. Such generous summation is not without a price. The thermal sensing system is almost devoid of any ability for spatial localization. Subjects are able to judge only with 80% accuracy whether radiant heat stimuli were delivered to the back or abdomen. This is easy to understand when one considers that the primary function of the cutaneous temperature-sensing system is to inform the central temperature-regulating system of the heat load experienced at the body surface rather than to specify the spatial location of warm and cold objects.

Various parts of the body differ in their sensitivity to temperature increases and decreases. For small temperature increases the forehead is the most sensitive, followed by the chest, forearm, calf, thigh, and abdomen. For small temperature decreases the body trunk is the most sensitive, followed by the limbs and finally the face. With relatively large changes in skin temperature (both increases and decreases) such regional differences in sensitivity are markedly reduced.

Paradoxical thermal sensations. Some cold points, when touched by a hot thermal probe (113°F or 45°C), will yield a cold sensation. The phenomenon of a hot probe giving rise to a cold sensation is called paradoxical cold. Paradoxical cold sensations do not occur, however, when a hot stimulus is applied over a large area. The necessary and sufficient conditions for paradoxical warmth (arousal of a warm sensation by application of a cold probe to a warm point) have not been established.

Heat sensations. The sensation of heat has been held to be uniquely different from either warm or heat-pain sensations. Its threshold ranges from 104–115°F (40 to 46°C) with a mean at about 108–109°F (42–43°C).

Two theories dealing with the mechanisms of heat sensations have been advanced. The first maintains that heat sensations are synthesized from the simultaneous stimulation of warm and paradoxically stimulated cold receptors. The second theory maintains that heat sensations arise from their own specifically sensitive receptors. Neither theory has been satisfactorily verified or refuted. *See Nervous system (vertebrate); Sensation; Somesthesis.*

Dan R. Kenshalo, Sr.

Bibliography. E. C. Carterette and M. P. Friedman (eds.), *Handbook of Perception*, vol. 6, 1978; V. B. Mountcastle (ed.), *Medical Physiology*, vol. 1, 14th ed., 1980.

Cyanamide

A term used to refer to the free acid, H_2NCN, or commonly to the calcium salt of this acid, $CaCN_2$, which is properly called calcium cyanamide. Calcium cyanamide is manufactured by the cyanamide process, in which nitrogen gas is passed through finely divided calcium carbide at a temperature of 1000°C (1830°F).

Most plants are equipped to produce the calcium cyanamide from the basic raw materials of air, limestone, and carbon according to reactions (1)–(3).

$$CaCO_3 \overset{\Delta}{\rightarrow} CaO + CO_2 \qquad (1)$$

$$CaO + 3C \rightarrow CaC_2 + CO \qquad (2)$$

$$CaC_2 + N_2 \rightleftharpoons CaCN_2 + C \qquad (3)$$

Most calcium cyanamide is used in agriculture as a fertilizer; some is used as a weed killer, as a pesticide, and as a cotton defoliant.

Cyanamide, H_2NCN, is prepared from calcium cyanamide by treating the salt with acid. It is used in the manufacture of dicyandiamide and thiourea. *See Carbon; Nitrogen.*

E. Eugene Weaver

Cyanate

A compound containing the —OCN group and derived from cyanic acid, HOCN. Cyanates are isomeric with fulminates which have the same atoms arranged thus, —ONC.

Like hydrogen cyanide and thiocyanic acid, cyanic acid may exist in two forms, $HO—C{\equiv}N$ and $H—N{=}C{=}O$, the latter of which is called isocyanic acid. The alkali and alkaline-earth metal cyanates contain an anion which may also be written in two forms: $[N{\equiv}C—O]^-$ and $[O{=}C{=}N]^-$. These salts are water-soluble.

The heavy metal cyanates are more covalent in nature and are water-insoluble. The main use of cyanates is in the synthesis of organic compounds.

Ammonium cyanate rearranges easily to urea: $NH_4OCN \rightleftharpoons NH_2CONH_2$.

The alkali metal cyanates may be prepared by the direct oxidation of the cyanide with oxygen or another suitable oxidizing agent under controlled conditions. *See Carbon; Cyanide; Nitrogen; Thiocyanate.*

E. Eugene Weaver

Cyanide

A compound containing the —CN group, for example, potassium cyanide, KCN; calcium cyanide, $Ca(CN)_2$; and hydrocyanic (or prussic) acid, HCN. Chemically, the simple inorganic cyanides resemble chlorides in many ways. Organic compounds containing this group are called nitriles. Acrylonitrile, CH_2CHCN, is an important starting material in the manufacture of fabrics, plastics, and synthetic rubber.

HCN is a weak acid, having an ionization constant of 1.3×10^{-9} at 18°C (64°F). In the pure state, it is a highly volatile liquid, boiling at 26°C (79°F).

HCN and the cyanides are highly toxic to animals, a lethal dose to humans usually considered to be 100–200 mg (0.0035–0.007 oz). Death has been attributed, however, to as low a dose as 0.57 mg of HCN per kilogram (9×10^{-6} oz per pound) of body weight. As poisons, they are rapid in their action. When these compounds are inhaled or ingested, their physiological effect involves an inactivation of the cytochrome respiratory enzymes, preventing tissue utilization of the oxygen carried by the blood.

The cyanide ion forms a variety of coordination complexes with transition-metal ions. Representative complexes are those of gold $[Au(CN)_2]^-$, silver $[Ag(CN)_2]^-$, and iron $[Fe(CN)_6]^{4-}$. This property is responsible for several of the commercial uses of cyanides. In the cyanide process, the most widely used method for extracting gold and silver from the ores, the finely divided ore is contacted with a dilute solution of sodium or potassium cyanide. The metal is solubilized as the complex ion, and, following extraction, the pure metal is recovered by reduction with zinc dust. In silver-plating, a smooth adherent deposit is obtained on a metal cathode when electrolysis is carried out in the presence of an excess of cyanide ion. SEE ELECTROPLATING OF METALS; GOLD METALLURGY; SILVER METALLURGY.

$Ca(CN)_2$ is extensively used in pest control and as a fumigant in the storage of grain. In finely divided form, it reacts slowly with the moisture in the air to liberate HCN.

In case hardening of metals, an iron or steel article is immersed in a bath of molten sodium or potassium cyanide containing sodium chloride or carbonate. At temperatures of 750°C (1380°F) and above, the cyanide decomposes at the surface, forming a deposit of carbon which combines with and penetrates the metal. The nitrogen also contributes to the increased hardness at the surface by forming nitrides with iron and alloying metals. SEE SURFACE HARDENING OF STEEL.

NaCN and KCN are produced commercially by neutralization of hydrogen cyanide with sodium or potassium hydroxide. The neutralized solution must then be evaporated to dryness in a strictly controlled manner to avoid undue losses. $Ca(CN)_2$ is produced primarily from calcium cyanamide by reaction with carbon in the presence of sodium chloride at 1000°C (1830°F). SEE COORDINATION CHEMISTRY; FERRICYANIDE; FERROCYANIDE; TOXICOLOGY.

Francis J. Johnston

Bibliography. R. H. Dreisbach, *Handbook of Poisoning: Prevention, Diagnosis, and Treatment*, 10th ed., 1982; M. J. Sienko and R. A. Plane, *Chemistry: Principles and Applications*, 1979.

Cyanobacteria

A large and heterogeneous group of photosynthetic microorganisms, formerly referred to as blue-green algae. They had been classified with the algae because their mechanism of photosynthesis is similar to that of algal and plant chloroplasts; however, the cells are prokaryotic, whereas the cells of algae and plants are eukaryotic. The name cyanobacteria is now used to emphasize the similarity in cell structure to other prokaryotic organisms. SEE ALGAE; CELL PLASTIDS.

General characteristics. All cyanobacteria can grow with light as an energy source, generating energy and reducing power through oxygen-evolving photosynthesis; carbon dioxide (CO_2) is fixed into organic compounds via the Calvin cycle, the same mechanism used in green plants. Thus, all species will grow in the absence of organic nutrients. However, some species will assimilate organic compounds into cell material if light is available, and a few isolates are capable of growth in the dark by using organic compounds as carbon and energy sources. Some cyanobacteria can shift to a different mode of photosynthesis, in which hydrogen sulfide rather than water serves as the electron donor. Molecular oxygen is not evolved during this process, which is similar to that in purple and green photosynthetic sulfur bacteria. SEE PHOTOSYNTHESIS.

The photosynthetic pigments of cyanobacteria include chlorophyll *a* (also found in algae and plants) and phycobiliproteins. One of these latter pigments, *c*-phycocyanin, is blue, and led to the name cyanobacteria (Greek cyano-, blue), and in combination with the green chlorophyll pigment gives cells a blue-green coloration. However, other species also contain a red phycobiliprotein, phycoerythrin, and these are red or brown in color. The coloration of the organisms can also be affected by the presence of orange or yellow carotenoid pigments. SEE CAROTENOID; CHLOROPHYLL.

Form and structure. Cyanobacteria have limited metabolic versatility in comparison with other bacteria, but they are extremely diverse morphologically. Species may be unicellular or filamentous (see **illus.**). Both types may aggregate to form macroscopically visible colonies. The cells range in size from those typical of bacteria (0.5–1 micrometer in diameter) to 60 μm in *Oscillatoria princeps*. The latter is among the largest-sized cells known in prokaryotes.

When examined by electron microscopy, the cells of cyanobacteria appear similar to those of gram-negative bacteria. The cell wall contains an outer lipid membrane and a layer of peptidoglycan. Many species produce extracellular mucilage or sheaths that promote the aggregation of cells or filaments into colonies.

The photosynthetic machinery is located on internal membrane foldings called thylakoids. Chlorophyll *a* and the electron transport proteins necessary for photosynthesis are located in these lipid membranes, whereas the water-soluble phycobiliprotein pigments are arranged in particles called phycobilisomes which are attached to the lipid membrane.

Several other types of intracellular structures are found in some cyanobacteria. Gas vesicles are rigid shells of protein constructed in a way to exclude water from the interior. The hollow structures, which have a diameter of about 70 nanometers and lengths up to 1000 nm, have densities much less than that of water and thus may confer buoyancy on the organisms. They are often found in cyanobacteria that grow in the open waters of lakes. Polyhedral bodies, also known as carboxysomes, contain large amounts of ribulose bisphosphate carboxylase, the key enzyme of CO_2 fixation via the Calvin cycle. Several types of storage granules may be found; glycogen and polyphosphate granules are also found in other organisms, but cyanophycin granules, which consist of polymers of the amino acids aspartic acid and arginine, are unique to cyanobacteria. These storage granules serve as reserves of carbon and energy, phosphorus, and nitrogen, respectively. They are synthesized when the particular nutrient is present in excess in the environment, and the reserves are utilized when external sources of the nutrient are unavailable.

Specialized cells. Some filamentous cyanobacteria can undergo a differentiation process in which some of the normal vegetative cells become heterocysts. The process occurs when combined nitrogen (ammonia and nitrate) is depleted in the environment. Heterocysts contain the enzyme nitrogenase, which catalyzes the conversion of nitrogen gas to ammonia. Nitrogen fixed in the heterocysts is translocated to the vegetative cells in the filament, to allow their continued growth in the absence of external ammonia or nitrate. Other changes occur during the differentiation into heterocysts. The cell wall becomes thicker, and the heterocyst develops intracellular connections with adjacent vegetative cells. In addition, the heterocysts lose the portion of the photosynthetic apparatus responsible for oxygen evolution. These changes are necessary for nitrogen fixation to occur, because nitrogenase is inactivated by oxygen. Thus, the heterocyst becomes an anaerobic site for nitrogen fixation, and depends upon translocation of organic compounds from vegetative cells to provide the reducing power to drive nitrogen fixation. *See Nitrogen fixation*.

In some filamentous species, resting spores called akinetes are formed. These cell forms are more resistant to periods of drying, freezing, or darkness than are the normal vegetative cells.

Occurrence. Cyanobacteria can be found in a wide variety of fresh-water, marine, and soil environments. They are more tolerant of environmental extremes than are eukaryotic algae. For example, they are the dominant oxygenic phototrophs in hot springs (at

temperatures up to 72°C or 176°F) and in hypersaline habitats such as may occur in marine intertidal zones. In these areas, they are often components of microbial mats, layers of microbial protoplasm up to several centimeters in thickness. In addition to cyanobacteria, other phototrophic bacteria and heterotrophic bacteria proliferate in these mats.

Cyanobacteria are often the dominant members of the phytoplankton in fresh-water lakes that have been enriched with inorganic nutrients such as phosphate. Up until the 1970s, cyanobacteria were not thought to be important members of the oceanic plankton, but it is now known that high population densities of small, single-celled cyanobacteria occur in the oceans, and that these are responsible for 30–50% of the CO_2 fixed into organic matter in these environments. *See Phytoplankton*.

Some cyanobacteria form symbiotic associations with eukaryotic photosynthetic organisms. These associations are based on the capacity of heterocystous cyanobacteria to fix nitrogen. *Nostoc* species reside in specialized cavities on the undersurface of liverworts, such as *Anthoceros*. In this environment, 30–40% of

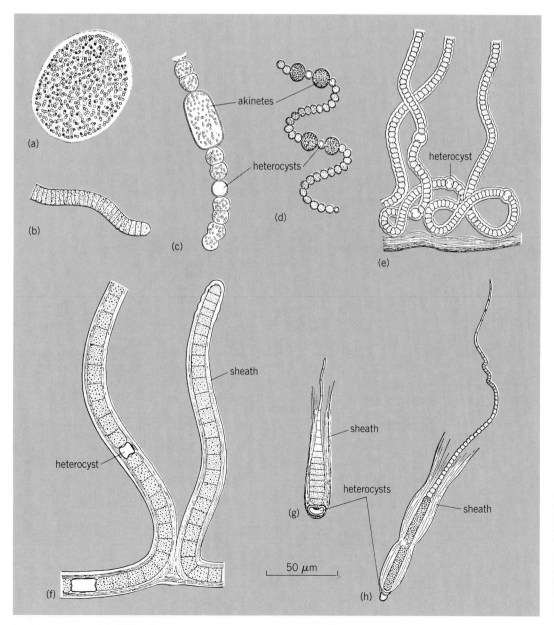

Morphological diversity of cyanobacteria. (a) *Microcystis* sp.; (b) *Oscillatoria* sp.; (c, d) *Anabaena* sp.; (e) *Nostoc* sp. with gelatinous capsule; (f) *Scytonema* sp.; (g) *Calothrix* sp.; (h) *Gleotrichia* sp. (After L. H. Tiffany and M. E. Britton, The Algae of Illinois, University of Chicago Press, 1952)

the *Nostoc* cells become heterocysts (a frequency tenfold higher than in free-living species), and 90% of the nitrogen fixed in the heterocysts is assimilated by the liverwort. Associations also occur with the water fern *Azolla*, some cycads, and the angiosperm *Gunnera*.

Lichens are symbiotic associations between a fungus and a photosynthetic microorganism. About 8% of the lichens involve a cyanobacterium, which can provide both fixed nitrogen and fixed carbon to the fungal partner. *SEE LICHENES.*

Dispersal. Cyanobacteria in open waters or in microbial mats can adjust their vertical position in the environment. Planktonic cyanobacteria do so by regulating the buoyancy conferred by gas vesicles. These cells can lose buoyancy by accumulating dense polymers such as glycogen. Filamentous cyanobacteria in microbial mats are often capable of gliding motility, a process in which cells lacking flagella can move in contact with surfaces. In both instances, the organisms adjust their vertical positions to avoid high, damaging light irradiances. However, their behavior is also modulated by the presence of inorganic nutrients or toxic substances such as hydrogen sulfide.

Evolution. Cyanobacteria are thought to be the first oxygen-evolving photosynthetic organisms to develop on the Earth, and hence responsible for the conversion of the Earth's atmosphere from anaerobic to aerobic about 2 billion years ago. This development permitted the evolution of aerobic bacteria, plants, and animals. *SEE BACTERIA; CYANOPHYCEAE; LIFE, ORIGIN OF.*

Allan Konopka

Bibliography. Y. Cohen, R. W. Castenholtz, and H. O. Halvorson (eds.), *Microbial Mats: Stromatolites,* 1984; W. M. Darley, *Algal Biology: A Physiological Approach,* 1982; T. Platt and W. K. W. Li (eds.), *Photosynthetic Picoplankton,* 1987; B. A. Whitton and N. G. Carr (eds.), *The Biology of Cyanobacteria,* 1982.

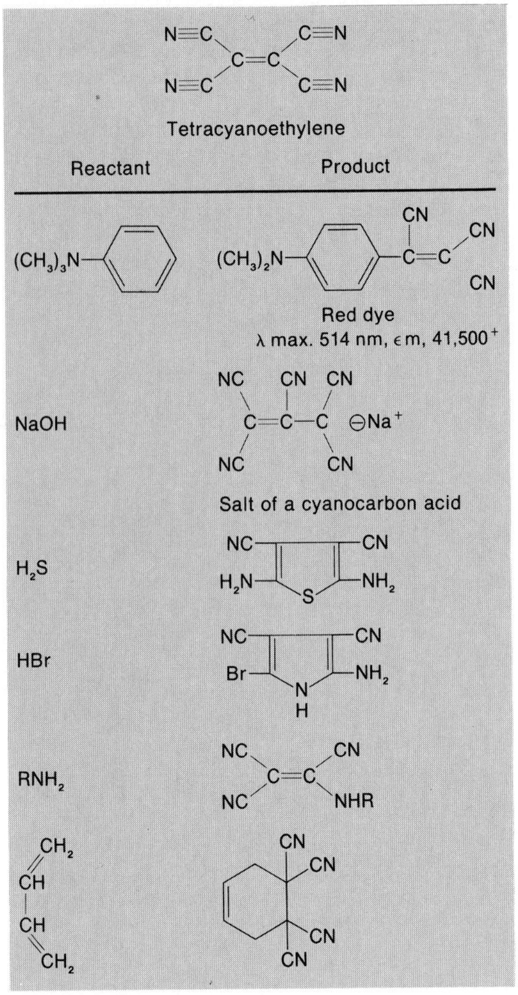

Reactions of tetracyanoethylene.

Cyanocarbon

A derivative of hydrocarbon in which all of the hydrogen atoms are replaced by the $-C\equiv N$ group. Only two cyanocarbons, dicyanoacetylene and dicyanobutadiyne, were known before 1958. Since then, tetracyanomethane, hexacyanoethane, tetracyanoethylene, hexacyanobutadiene, and hexacyanobenzene were synthesized.

The term cyanocarbon has been applied to compounds which do not strictly follow the above definition: tetracyanoethylene oxide, tetracyanothiophene, tetracyanofuran, tetracyanopyrrole, tetracyanobenzoquinone, tetracyanoquinodimethane, tetracyanodithiin, pentacyanopyridine, diazomalononitrile, and diazotetracyanocyclopentadiene.

Tetracyanoethylene, the simplest olefinic cyanocarbon, is a colorless, thermally stable solid, having a melting range of 198–200°C (388–392°F). It is a strong π-acid, readily forming stable complexes with most aromatic systems. These complexes absorb radiation in or near the visible spectrum.

Charge-transfer salts derived from tetracyanoquinodimethane and tetrathiafulvalene are highly conducting and have attracted much interest as possible replacements for metal conductors.

Cyanocarbon acids are among the strongest protonic organic acids known and are usually isolated only as the cyanocarbon anion salt. These salts are usually colored and stabilized by resonance of the type indicated in the reaction below. Twenty-five similar forms

$$\text{NC}-\underset{\ominus}{\overset{\overset{\displaystyle CN}{|}}{C}}-CN \leftrightarrow NC-\overset{\overset{\displaystyle CN}{|}}{C}=C=N\ominus$$

can be written for the 2-dicyanomethylene-1,1,3,3-tetracyanopropanediide ion, $(NC)_2C=C[C(CN)_2]_2$, which accounts for the similarity between the acid strength of the free acid and that of sulfuric acid.

The reactivity of cyanocarbons is illustrated by the facile replacement of one of the cyano groups in tetracyanoethylene by nucleophilic attack under very mild conditions and by its addition to dienes. Examples of these reactions are shown in the **illustration.** *SEE ORGANIC CHEMISTRY.*

Owen W. Webster

Bibliography. N. L. Allinger et al., *Organic Chemistry,* 2d ed., 1976; M. F. Ashworth, *Analytical Methods for Organic Cyano Groups,* 1971; T. L. Cairns et al., Cyanocarbon chemistry, *J. Amer. Chem. Soc.,* 80:2775–2844, 1958; R. T. Morrison and R. N. Boyd, *Organic Chemistry,* 5th ed., 1987; Z. Rappaport (ed.), *The Chemistry of the Cyano Group,* 1970, and Suppl. C ed. by S. Patoi and Z. Rappaport, 1983.

Cyanoethylation

A chemical reaction involving the addition of acrylo-nitrile (CH_2=CH—CN) to compounds carrying a reactive hydrogen, to introduce the β-cyanoethyl grouping (—CH_2CH_2CN). The reaction is represented by reaction (1), where Z is CN or a C=O group.

$$ZH + CH_2=CH—CN \rightleftharpoons ZCH_2CH_2CN \qquad (1)$$

Cyanoethylation can be effected with HBr, HCl, HCN, and with compounds containing the functional groups

$$\overset{|}{—AsH}, \overset{|}{—BH}, \overset{|}{—CH} \text{ (when activated by adjacent electron-withdrawing groups),}$$

$$\overset{|}{—NH}, —OH, \overset{|}{—PH}, —SH, \overset{|}{—SiH}, \overset{|}{—SnH}$$

The products can be subjected further to the usual nitrile reactions, such as hydration, hydrolysis, and reduction as in reactions (2)–(4).

$$ZCH_2CH_2CN \rightarrow ZCH_2CH_2CONH_2 \qquad (2)$$

$$ZCH_2CH_2CN \rightarrow ZCH_2CH_2COOH \qquad (3)$$

$$ZCH_2CH_2CN \rightarrow ZCH_2CH_2CH_2NH_2 \qquad (4)$$

Because of its versatility, cyanoethylation has assumed great importance in synthesis, as shown by the fact that over 1200 compounds have been used as reactants. Introduction of the cyanoethyl group tends to increase hydrophobic character and resistance to biological attack. Natural products, notably cellulose, have been cyanoethylated to take advantage of these properties.

Substrates for cyanoethylation. The nature of the catalyst is used to classify substrates into three groups.

(1) No catalyst:

$$\overset{|}{—AsH}, \overset{|}{—BH}, \overset{|}{—NH} \text{ (aliphatic),}$$

$$\overset{|}{—PH}, \overset{|}{—SnH}, HBr, HCl$$

(2) Basic catalyst (for example, NaOH, $NaOCH_3$, benzyltrimethylammonium hydroxide):

$$\overset{|}{—CH}, \overset{|}{—NH} \text{ (amide and certain heterocycles),}$$

$$—OH, \overset{|}{—PH}, \overset{|}{—P(O)H}, —SH, \text{ and HCN}$$

(3) Acid catalyst (for example, acetic acid with or without a copper salt):

$$\overset{|}{—NH} \text{ (aromatic)}, —NH_2 \text{ (t-carbinamines)}$$

Yields in cyanoethylation are generally high. The reaction is strongly exothermic and is usually carried out at moderate temperatures in a solvent such as dioxane or t-butanol. Acrylonitrile itself has been used as a diluent, but this method is not widely applicable in base-catalyzed cyanoethylation because of competing anionic polymerization of acrylonitrile. High temperatures and high concentrations of base tend to reverse the process.

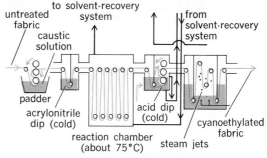

Fig. 1. Flow diagram for cyanoethylation. 75°C = 157°F. (Cyanoethylation of Cotton, American Cyanamid Co., Monsanto Chemical Co., and the Institute of Textile Technology, September 1956)

Cyanoethylation of natural materials. Since the extent of cyanoethylation can be varied by choice of conditions, a variety of products can be obtained from a given material. These may be further modified by reactions of the cyanoethyl group, for example, to give polyelectrolytes.

Partially cyanoethylated cellulose, in which one-sixth of the hydroxyl groups have reacted, is rot- and termiteproof; it has improved resistance to degradation by heat and by acids, as well as altered dyeing characteristics and greater hydrophobicity. Crystallinity and morphology of the cellulose are retained. Utility is in marine nets and ropes, electrical insulation, industrial fabrics, structural wood, and dimensionally stable paper.

In the cyanoethylation of cotton the fabric is first padded with 6% sodium hydroxide solution and with acrylonitrile, and then passed continuously into a chamber heated to about 75°C (157°F) by acrylonitrile vapor (**Fig. 1**). After a residence time of about 3.5 min, the caustic-impregnated cotton is neutralized, washed, and dried (**Fig. 2**). Unreacted acrylonitrile is recovered and recycled.

Highly cyanoethylated cellulose (two-thirds or more of hydroxyl groups reacted) is an amorphous, thermoplastic material which is soluble in acrylonitrile, dimethylformamide, and pyridine. Unhindered rotation of the polar cyanoethyl segments imparts an unusually high dielectric constant (13.3 at 60 cycles and 25°C or 77°F), a property useful in capacitors and in matrices for electroluminescent phosphors. Starch, cyanoethylated to retain water solubility, has the abil-

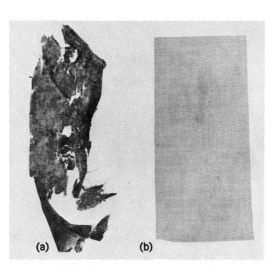

Fig. 2. Appearance of cotton after burial in soil for 2 weeks. (a) Untreated. (b) Partially cyanoethylated.

ity to prevent soil from redepositing on washed fabrics. More highly cyanoethylated starch is soluble in organic solvents. Cyanoethylsucrose is used as a dielectric fluid.

In the cyanoethylation of proteins the free amino and possibly the amide groups are attacked. Complete resistance to putrefaction results when casein is cyanoethylated with as little as a 5–10% by weight of acrylonitrile. Cyanoethylated wool had increased affinity for both cationic dyes and direct dyes. *See Textile Chemistry*.

Other uses. Cyanoethylation is also used in the preparation of amino acids, chemotherapeutic agents, dyes, blowing agents, long-chain diamines, special-purpose silicones, food stabilizers, and liquids to effect separations by gas chromatography. *See Acrylonitrile*.

Norbert M. Bikales

Bibliography. *Kirk-Othmer Encyclopedia of Chemical Technology*, 3d ed., vol. 7, 1979; J. March, *Advanced Organic Chemistry*, 3d ed., 1985; R. T. Morrison and R. N. Boyd, *Organic Chemistry*, 5th ed., 1987; J. D. Roberts and M. C. Caserio, *Basic Principles of Organic Chemistry*, 2d ed., 1977.

Cyanogen

A colorless, highly toxic gas having the molecular formula C_2N_2. Cyanogen belongs to a class of compounds known as pseudohalogens, because of the similarity of their chemical behavior to that of the halogens. Liquid cyanogen boils at $-21.17°C$ ($-6.11°F$) and freezes at $-27.9°C$ ($18.2°F$) at 1 atm (10^5 Pa). The density of the liquid is 0.954 g/ml at the boiling point.

Cyanogen may be prepared by prolonged heating of mercuric cyanide at 400°C (750°F), reaction (1), or

$$Hg(CN)_2 \rightarrow Hg + C_2N_2 \qquad (1)$$

by allowing a solution of copper sulfate to flow slowly into a solution of potassium cyanide, reaction (2). The copper(II) cyanide is unstable, decomposing

$$2KCN + CuSO_4 \rightarrow Cu(CN)_2 + K_2SO_4 \qquad (2)$$

to give cyanogen and copper(I) cyanide, reaction (3).

$$2Cu(CN)_2 \rightarrow 2CuCN + C_2N_2 \qquad (3)$$

Cyanogen is also formed by the reaction in the gas phase of hydrogen cyanide and chlorine at 400–700°C (750–1300°F) in the presence of a catalytic agent such as activated carbon, reaction (4). Cyanogen chloride,

$$2HCN + Cl_2 \rightarrow C_2N_2 + 2HCl \qquad (4)$$

ClCN, is an intermediate in this reaction. When heated to 400°C (750°F), cyanogen gas polymerizes to a white solid, paracyanogen $(CN)_x$.

Cyanogen reacts with hydrogen at elevated temperatures in a manner analogous to the halogens, forming hydrogen cyanide, reaction (5). With hydrogen sul-

$$C_2N_2 + H_2 \rightleftharpoons 2HCN \qquad (5)$$

fide, H_2S, cyanogen forms thiocyanoformamide or dithiooxamide. These reactions are shown by (6) and (7).

$$C_2N_2 + H_2S \rightarrow NC\overset{\overset{\displaystyle S}{\|}}{-}C-NH_2 \qquad (6)$$

$$C_2N_2 + 2H_2S \rightarrow H_2N\overset{\overset{\displaystyle S}{\|}}{C}-\overset{\overset{\displaystyle S}{\|}}{C}NH_2 \qquad (7)$$

Cyanogen burns in oxygen, producing one of the hottest flames known from a chemical reaction. It is considered to be a promising component of high-energy fuels.

Structurally, cyanogen is written $N\equiv C-C\equiv N$. *See Cyanide*.

Francis J. Johnston

Bibliography. W. H. Brown and E. P. Rogers, *General, Organic and Biochemistry*, 2d ed., 1983; V. Migrdichian, *The Chemistry of Organic Cyanogen Compounds*, 1947; Z. Rappaport (ed.), *The Chemistry of the Cyano Group*, 1971.

Cyanophyceae

A class of prokaryotic organisms coextensive with the division Cyanophycota of the kingdom Monera. Because these organisms have chlorophyll *a* and carry out oxygen-evolving photosynthesis, they have traditionally been aligned with algae and, with regard for their characteristic color, called blue-green algae. Microbiologists have emphasized the prokaryotic structure of these organisms and aligned them with bacteria, as the Cyanobacteria. Other names applied to these organisms include Cyanophyta at the level of division and Myxophyceae or Schizophyceae at the level of class. Blue-greens range in form from unicells 1–2 micrometers in diameter to filaments 10 cm (4 in.) long. *See Cyanobacteria*.

Classification. The taxonomy of the blue-greens is in an especially unsteady state, polarized by two viewpoints. At one pole (Geitlerian system) nearly every morphological variant encountered in nature is accorded taxonomic recognition. At the other pole (Drouetian system) great morphological plasticity is attributed to each species. Geitler recognized 4 orders, 22 families, 140 genera, and more than 1200 species, while Drouet recognized 2 orders, 6 families, 24 genera, and 61 species. Most specialists of Cyanophyceae steer a middle course. A further major consideration has recently been introduced by microbiologists, who have proposed their own system of classification based solely on characters revealed in pure cultures.

Unicellular forms, which may aggregate in colonies or loosely constructed filaments and which reproduce by binary fission or spores, constitute the order Chroococcales, from which two additional orders, Chamaesiphonales and Pleurocapsales, are sometimes segregated. Filamentous forms, which reproduce by hormogonia, constitute the order Nostocales (= Hormogonales or Oscillatoriales). The Nostocales may be restricted to unbranched or falsely branched forms (in which the ends of a trichome adjacent to a rupture grow out as a pair of branches), while those forms in which cells divide in more than one plane (true branching) constitute the order Stigonematales.

Habitat. Geographically and ecologically, blue-green algae are nearly as ubiquitous as bacteria. They are especially abundant in the plankton of neutral or alkaline eutrophic fresh waters and tropical seas, often forming blooms. Habitats for benthic forms include hot springs, snow and ice, soil, rocks, tree trunks, and buildings. Cyanophyceae live symbiotically with a large variety of animals and plants, including sponges, diatoms, liverworts, cycads, the aquatic fern *Azolla*, and the flowering plant *Gunnera*. They constitute the phycobiont of many lichens.

In addition to contributing to food chains, blue-green algae play specific beneficial roles. Nitrogen-fixing forms greatly enrich rice paddies. *Spirulina*, a traditional food in parts of Mexico and central Africa, is grown commercially and marketed widely as a high-protein dietary supplement. On the other hand, blue-green algae are often a nuisance. They clog filters, impart undesirable tastes and odors to domestic water supplies, and make unusable or at least unattractive many swimming pools, aquariums, and fountains. Cyanophycean blooms are often toxic to fish, birds, and livestock.

Blue-green algae were pioneers on Earth and are known from rocks at least as old as 2.3 billion years. They are believed to have been responsible for the accumulation of oxygen in the primeval atmosphere and to have been involved in the formation of laminated reeflike structures called stromatolites. To judge from similar modern structures (as at Shark Bay, Western Australia), stromatolites were produced by algal mats that entrapped detrital sediments and sometimes deposited calcium carbonate. SEE ALGAE; CRYPTOZOON; STROMATOLITE.

Paul C. Silva; Richard L. Moe

Bibliography. H. C. Bold and M. J. Wynne, *Introduction to the Algae: Structure and Reproduction*, 1985; F. Drouet, Revision of the Stigonemataceae with a summary of the classification of the blue-green algae, *Beih. Nova Hedwigia*, Heft 66, 1981; S. P. Parker (ed.), *Synopsis and Classification of Living Organisms*, 2 vols., 1982.

Cyatholaimoidea

A superfamily of nematodes. It is distinguished by the tightly coiled multispiral amphids located a short distance posterior to the cephalic sensilla. The vestibule of the stoma is lined with 12 riblike rugae. The esophastome is armed with a strong dorsal tooth or minute denticles. These nematodes are found in marine, terrestrial, and fresh-water environments. SEE NEMATA.

Armand R. Maggenti

Cybernetics

The study of communication and control within and between humans, machines, organizations, and society. This is a modern definition of the term cybernetics, which was first utilized by N. Wiener in 1948 to designate a broad subject area he defined as "control and communication in the animal and the machine." A distinguishing feature of this broad field is the use of feedback information to adapt or steer the entity toward a goal. When this feedback signal is such as to cause changes in the structure or parameters of the system itself, it appears to be self-organizing. SEE ADAPTIVE CONTROL.

Wiener's thinking was influenced by his work during World War II on the aiming of an antiaircraft gun

ahead of its target in order to achieve a successful intercept. He was also familiar with the work of physiologists and believed that similar mechanisms are exhibited in biological systems to provide for goal-directed behavior. He developed the statistical methods of autocorrelation, prediction, and filtering of time-series data to provide a mathematical description of both biological and physical phenomena. The use of filtering to remove unwanted information or noise from the feedback signal mimics the selectivity shown in biological systems in which imprecise information from a diversity of sensors can be accommodated so that the goal can still be reached. SEE ESTIMATION THEORY; FIRE-CONTROL SYSTEMS; HOMEOSTASIS; STOCHASTIC CONTROL THEORY.

This similarity between physical and biological systems was also recognized by a number of scientists and engineers in the 1940s, 1950s, and 1960s, such as W. McCulloch, who described the brain as a digital computer. Computing systems that exhibit behavior somewhat akin to that of humans are said to possess artificial intelligence. The field of neural networks is a subset of artificial intelligence in that intelligent behavior can be realized with a collection of computing elements (like biological neurons) connected together to merge various inputs together into a single signal, generally through use of a weighted sum of the individual signals. Usually, a threshold is set and, if this weighted sum is above the set threshold, a neuron fires, indicating that something of importance has happened. Neural networks that adjust the weights by using internal feedback appear to be self-organizing, and are thus consistent with Wiener's view of cybernetic systems. SEE ARTIFICIAL INTELLIGENCE; CONTROL SYSTEMS; ELECTRICAL COMMUNICATIONS; HUMAN-FACTORS ENGINEERING; HUMAN-MACHINE SYSTEMS; NEURAL NETWORK; SYSTEMS ENGINEERING.

Donald W. Bouldin

Bibliography. T. Kohonen, *Self-Organization and Associative Memory*, 1984; W. McCulloch, *Embodiments of Mind*, 1965; M. Singh (ed.), *Systems and Control Encyclopedia*, 1987; N. Wiener, *Cybernetics*, 1948; N. Wiener, *Extrapolation, Interpolation and Smoothing of Stationary Time Series with Engineering Applications*, 1949.

Cycadales

An order of the class Cycadopsida of the division Pinophyta (gymnosperms) consisting of four families, Cycadaceae, Stangeriaceae, Zamiaceae, and Boweniaceae, with perhaps 100 species. The order dates from the upper Carboniferous and has few living representatives. The cycads were distributed worldwide in the Mesozoic, but today are restricted to subtropical and tropical regions, with the plants occurring in small colonies; few have broad distributions.

The cycads, often incorrectly referred to as palms, range from a few inches (*Zamia*) to 65 ft (20 m; *Macrozamia*) tall. The stems are cylindrical and often branched; in some the stem is subterranean, in others it is mainly above the ground. The pinnate (or bipinnate in *Bowenia*) leaves are borne at the apex of the stems. Microsporophylls and megasporophylls are borne in cones of highly varied appearance. Male and female cones appear on separate plants (dioecious). SEE CYCADOPSIDA; PINOPHYTA; PLANT KINGDOM.

Thomas A. Zanoni

Cycadeoidales

An order of extinct plants that formed a conspicuous part of the landscape during the Triassic, Jurassic, and Cretaceous periods. The Cycadeoidales (or Bennettitales) had unbranched or sparsely branched trunks with a terminal crown of leaves (family Cycadeoidaceae), or they were branched, at times profusely (family Williamsoniaceae).

The squat, barrel-shaped trunks of members of the Cycadeoidaceae were covered with a dense armor of persistent leaf bases, and were terminated by a crown of pinnately divided leaves (see **illus.**).

Reconstruction of a plant of the genus *Cycadeoidea*, showing terminal crown of leaves, persistent leaf bases, and positions of the fruiting structures. (*Courtesy of T. Delevoryas*)

On the trunk surface, among the leaf bases, were borne complex fruiting structures, each with a central seed-bearing receptacle, surrounded by a whorl of pollen-bearing organs that were derived from modified leaves. Dicotyledonous embryos have been observed in the seeds.

In members of the Williamsoniaceae, fruiting structures were often stalked, and were flowerlike in appearance. They had both pollen-bearing and seed-bearing structures in the same "flower," or these organs may have been borne in separate "flowers." Seeds were borne on the surface of a domelike receptacle, while the pollen-bearing organs were fused at the base to form a whorl of simple or pinnate structures. Foliage was either pinnately divided or entire. No known relatives of the Cycadeoidales exist at the present time, although members of the order Cycadales show some resemblances to the extinct group. SEE CYCADALES; EMBRYOBIONTA; PALEOBOTANY; PLANT KINGDOM.

Theodore Delevoryas

Cycadopsida

A class of the division Pinophyta (gymnosperms), consisting of six orders: Lagenostomales (=Lyginopteridales, pteridosperms or seed ferns), Trigonocarpales (= Medullosales), Cycadales (cycads), Bennettitales (cycadeoids or bennettites), Gnetales (gnetum),

and Welwitschiales (welwitschia). The Cycadales, Gnetales, and Welwitschiales are the only orders with living species. The Cycadales date back to the Triassic; the Bennettitales lived in the Triassic to the Cretaceous; and the Lagenostomales and Trigonocarpales can be traced to the Carboniferous to the Permian. Fossil Gnetales and Welwitschiales are only known from pollen finds. SEE CYCADALES; GNETALES; WELWITSCHIALES.

The Cycadopsida are gymnosperms with radially symmetrical seeds and bilateral or radial cupules. Stems are simple (unbranched) to pinnate or dichotomous-branched. Leaves are usually compound (except in Gnetales and Welwitschiales). The seeds are borne on leaves (seed ferns), or on modified leaves (megasporophylls) that may be aggregated into a simple strobilus (cone) at the end of the stem or in clusters of strobili (Welwitschiales) or on linear axes (Gnetales). SEE PINOPHYTA; PLANT KINGDOM.

Thomas A. Zanoni

Cyclanthales

An order of flowering plants, division Magnoliophyta (Angiospermae), sometimes called Synanthae or Synanthales, in the subclass Arecidae of the class Liliopsida (monocotyledons). The order consists of the single tropical American family Cyclanthaceae, with about 180 species. They are herbs or, seldom, more or less woody plants, with characteristic leaves that have a basal sheath, a petiole, and an expanded, usually bifid (cleft into two equal parts) blade which is often folded lengthwise (plicate) like that of a palm leaf. The numerous, small, unisexual flowers are crowded into a spadix which serves as a sort of pseudanthium. One of the species, *Carludovica palmata*, is a principal source of fiber for panama hats. SEE ARECIDAE; LILIOPSIDA; MAGNOLIOPHYTA; PLANT KINGDOM.

Arthur Cronquist

Cyclic nucleotides

Mononucleotides in which the phosphate moiety is attached to both the third and the fifth carbon atoms of the ribose component of the molecule, thus forming a cyclic diester. They are more correctly referred to as 3',5'-mononucleotides, but the adjective "cyclic" was applied to them early, and this usage has by now become very common. The only cyclic nucleotides which have been unequivocally shown to exist in living tissues are 3',5'-adenylic acid and 3',5'-guanylic acid, also known as cyclic AMP and cyclic GMP, respectively. They are important because they influence a large variety of essential biological processes, and sometimes the two nucleotides may act in opposition to each other. Cyclic AMP was discovered first; thus more is known about its role as a regulatory agent. However, cyclic GMP may ultimately prove to be equally important. A third cyclic nucleotide, 3',5'-cytidylic acid (cyclic CMP), has been reported to exist in some cells, but its role is poorly understood. SEE NUCLEIC ACID.

G. Alan Robison

Bibliography. H. Cramer and J. Schultz (eds.), *Cyclic Nucleotides: Mechanisms of Action*, 1977; N. D. Goldberg et al., in *Advances in Cyclic Nucleo-*

tide Research, vol. 5, pp. 307–330, 1975; J. G. Hardman, G. A. Robison, and E. W. Sutherland, *Annu. Rev. Physiol.*, 33:311, 1971.

Cycloamylose

Any of a group of cyclic oligomers of glucose in the normal C-1 conformation in which the individual glucose units are connected by 1,4 bonds. They are called Schardinger dextrins (after the discoverer), cyclodextrins (to emphasize their cyclic character), or cycloamyloses (to emphasize both their cyclic character and their amylose origin). The most common ones are cyclohexaamylose (with six glucose units in a cyclic array) and cycloheptaamylose (with seven glucose units in a cyclic array). The toroidal shape of these molecules has been determined by x-ray analysis (**Fig. 1**). The molecular weights are around 1000.

In water or in a mixture of water and dimethyl sulfoxide, cycloamyloses form 1:1 complexes with many organic molecules and most benzenoid derivatives. X-ray and spectroscopic evidence indicates that the guest molecule (usually a small organic molecule) is bound tightly in the cavity of the host molecule (the cycloamylose) just like a key fitting into a lock. The result is an inclusion complex, which is what is formed by enzymes when they first bind the molecules whose reactions they subsequently catalyze. *SEE CATALYSIS; CLATHRATE COMPOUNDS.*

Inclusion complex formation is probably the most important similarity between cycloamyloses and enzymes. In the cycloheptaamylose-catalyzed hydrolysis of *p*-nitrophenyl acetate, a nonlinear increase in the rate of reaction is obtained when one increases the concentration of cycloamylose. However, a linear increase is obtained when the same data are plotted in a double reciprocal manner. These two concentration dependencies are exactly those seen in enzyme-catalyzed reactions, and show that enzymes and cycloamyloses increase the rate of reaction in the same way.

Specificity is one of the most important aspects of enzyme action. A particular enzyme can catalyze a given reaction but cannot catalyze a closely related reaction. The cycloamyloses provide the answer to how an enzyme recognizes a given compound but ignores a closely related compound.

The importance of the stereochemistry of binding may be seen easily in the models of complexes of

(a) (b)

Fig. 2. Molecular models of cycloamylose complexes. (a) Cyclohexaamylose-*m-t*-butylphenyl acetate complex. (b) Cyclohexaamylose-*p-t*-butylphenyl acetate complex. (From R. L. Van Etten et al., *Acceleration of phenyl ester cleavage by cycloamyloses: A model for enzymatic specificity*, J. Amer. Chem. Soc., 89:3249, 1967)

corresponding meta- and para-substituted esters with cyclohexaamylose: cyclohexaamylose *m-t*-butylphenyl acetate and cyclohexaamylose *p-t*-butylphenyl acetate. In the former complex, a glucosidic hydroxyl group is practically adjacent to the ester carbonyl group with which it is to react, leading to a facile reaction (**Fig. 2***a*). In the latter complex, the glucosidic hydroxyl group is several tenths of a nanometer away from the ester carbonyl group with which it must react, leading to a slow reaction (Fig. 2*b*).

Cycloamyloses have been used to mimic many enzymes, such as ribonuclease, transaminase, and carbonic anhydrase. The catalysis of the hydrolysis of phenyl acetates discussed above is a mimicry of the enzyme chymotrypsin by a cycloamylose. Both cycloamyloses and chymotrypsin catalyze via three steps: binding; acylation of the catalyst; and deacylation of the acyl-catalyst. The mechanisms by which cycloamyloses and chymotrypsin catalyze this reaction are identical: cycloamyloses form a noncovalent complex followed by the formation of an acyl-cycloamylose intermediate, while chymotrypsin forms a noncovalent enzyme-substrate complex followed by an acyl-chymotrypsin intermediate. Cycloamyloses, like chymotrypsin, catalyze the hydrolysis of both esters and amides.

Cycloamyloses have also been used to effect selectivity in several chemical reactions. Anisole can be chlorinated by HOCl in the presence of the cyclohexaamylose exclusively in the position para to the methoxy group, although the absence of cyclohexaamylose gives the usual mixture of ortho and para chlorinated products. The ortho position of anisole is buried inside the cavity when the anisole is bound in the cyclohexaamylose, and so reaction takes place in the para position only, although the methoxy group is an ortho, para director in aromatic electrophilic substitution. In a similar sense, it was found that the introduction of aldehyde groups into many phenols by chloroform and sodium hydroxide leads solely to the para-substituted benzaldehyde because the dichlorocarbene attacking agent is prevented from approaching the ortho position by the cycloheptaamylose.

Cycloamyloses are the premier example of chemical compounds that act like enzymes in inclusion complex formation followed by reaction. *SEE ENZYME; STEREOCHEMISTRY; STEREOSPECIFIC CATALYST; STERIC EFFECT.*

Myron L. Bender

Bibliography. M. L. Bender et al., *The Bioinorganic*

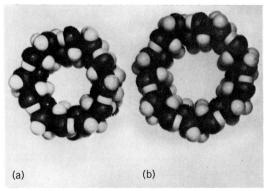

(a) (b)

Fig. 1. Molecular models of cycloamyloses viewed from the secondary hydroxyl side of the torus. (a) Cyclohexaamylose. (b) Cycloheptaamylose. (From M. L. Bender and M. Komiyama, Cyclodextrin Chemistry, Springer, 1978)

Chemistry of Enzymatic Catalysis, 1984; M. L. Bender and M. Komiyama, *Cyclodextrin Chemistry*, 1978; R. Breslow, Artificial enzymes, *Science*, 218:532–537, 1982; D. W. Griffiths and M. L. Bender, Cycloamyloses as catalysts, *Adv. Catalysis*, 23:209–261, 1973.

Cyclocystoidea

A class of small, disk-shaped echinozoans in which the lower surface of the body probably consisted of a suction cap for adhering to substrate, and the upper surface was covered by separate plates arranged in concentric rings. The mouth lay at the center of the upper surface, and the anus, also on the upper surface, lay some distance between the margin and the mouth. Little is known of the habits of cyclocystoids. They occur in rocks of middle Ordovician to middle Devonian age in Europe and North America. SEE ECHINODERMATA; ECHINOZOA.

Howard B. Fell

Bibliography. R. C. Moore (ed.), *Treatise on Invertebrate Paleontology*, pt. U, 3(1):188–210, 1966.

Cycloid

A curve traced in the plane by a point on a circle that rolls, without slipping, on a line. If the line is the x axis of a rectangular coordinate system, at whose origin O the moving point P touches the axis, parametric equations of the cycloid are $x = a(\theta - \sin\theta)$, $y = a(1 - \cos\theta)$, when a is the radius of the rolling circle, and the parameter θ is the variable angle through which the circle rolls (see **illus.**). One arch is

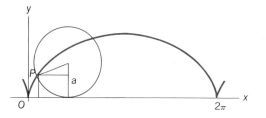

Diagram of cycloid; symbols explained in text.

obtained where θ assumes all values from 0 to 2π. The length of an arch is $8a$, and the area bounded by an arch and the x axis is $3\pi a^2$ (three times the area of the rolling circle).

Two important properties of the cycloid are the following: (1) It is isochronous, that is, if the line is horizontal, then the time of descent of a particle starting from rest at a point of a cycloidal runway and falling to its lowest point is independent of the starting point. (2) It is a brachistochrone; that is, if P_1, P_2 are two points in a vertical plane, but not in a vertical line, the curve down which a particle will fall from P_1 to P_2 in the shortest time is an arc of a cycloid. SEE ANALYTIC GEOMETRY.

Leonard M. Blumenthal

Cyclone

A vortex in which the sense of rotation of the wind about the local vertical is the same as that of the Earth's rotation. Thus, a cyclone rotates clockwise in

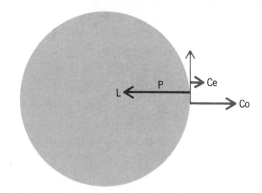

Fig. 1. Force balance for gradient flow about a cyclone in the Northern Hemisphere. Thin arrow indicates wind velocity; L = low-pressure center; Co = Coriolis force; Ce = centrifugal force; and P = pressure gradient force.

the Southern Hemisphere and counterclockwise in the Northern Hemisphere. In meteorology the term cyclone is reserved for circulation systems with horizontal dimensions of hundreds of kilometers (tropical cyclones) or thousands of kilometers (extratropical cyclones). For such systems the Coriolis force due to the Earth's rotation and the centrifugal force due to flow curvature are both directed to the right of the direction of the flow. Away from the surface of the Earth these two forces are approximately balanced by the pressure gradient force, which is directed toward low pressure (see **Fig. 1**), so that there must be a pressure minimum at the center of the cyclone, and cyclones are sometimes simply called lows.

This three-way balance is referred to as the gradient wind. In large-scale extratropical cyclones the centrifugal force is of secondary importance. The balance is then primarily between the Coriolis force and the pressure gradient force. In this case the flow, which is parallel to the isobars with high pressure to the right of the direction of motion (in the Northern Hemisphere), is said to be in geostrophic balance. SEE AIR PRESSURE; CORIOLIS ACCELERATION; GRADIENT WIND; VORTEX.

Extratropical cyclones. These are the common weather disturbances which travel around the world from west to east in mid-latitudes. They are generally associated with fronts, which are zones of rapid transition in temperature. In a typical mid-latitude cyclone (see **Fig. 2**) the air poleward of the frontal boundary is cool and dry, while the air equatorward is warm and moist. The poleward and upward motion of warm moist air south of the warm front, which extends eastward from the low-pressure center, is responsible for much of the precipitation associated with cyclonic storms. The equatorward and downward motion of cool dry air behind the cold front, which extends southwest of the low-pressure center, is responsible for the fine weather which generally follows the passage of a mid-latitude cyclone.

Formation of extratropical cyclones is referred to as cyclogenesis. In the classic Norwegian cyclone model, which is favored by many weather forecasters, cyclogenesis begins with the development of a wavelike disturbance on a preexisting frontal discontinuity separating a cold polar air mass from a warm subtropical air mass (**Fig. 3a**). The circulation associated with the wave then distorts the frontal surface to produce a closed vortical circulation with a warm front

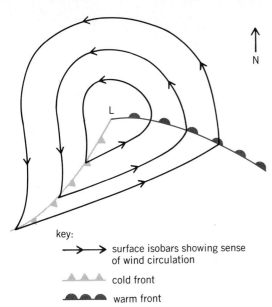

key:

⟶ surface isobars showing sense
     of wind circulation

▲▲▲ cold front

●●● warm front

Fig. 2. Plane view of an extratropical cyclone in the Northern Hemisphere as depicted on a surface weather map. The L designates a pressure low.

ahead of the vortex and a trailing cold front (Fig. 3b). In the later stages the cold front may overtake the warm front to produce an occlusion, a region where at the surface the warm air is limited to a narrow band (Fig. 3c).

A more modern theoretical model of cyclogenesis, on the other hand, suggests that most extratropical cyclones develop as dynamical instabilities in broad regions of enhanced horizontal temperature gradients and vertical shear of the geostrophic wind associated with the subtropical jet stream. In this instability process, which is referred to as baroclinic instability, the cyclone grows by converting potential energy associated with the horizontal temperature gradient in the mean flow into kinetic energy. As baroclinically unstable systems develop, the disturbance motion tends to intensify temperature contrasts near the surface so that warm and cold fronts develop. Thus, in this theory, fronts are considered to be secondary features produced by cyclogenesis, rather than acting as causes.

Cyclogenesis occurs over very broad areas of the Earth. However, it is most frequent over the western oceans (such as in the regions east of Japan and off the east coast of North America) and in the lee of

major mountain barriers (such as the Rockies and the Alps).

Occasionally, over maritime regions during the cold season, extratropical cyclones can develop rapidly into dangerously intense storms with winds comparable to those found in hurricanes. If decreases in surface pressure occur at a rate exceeding 1 millibar (100 pascals) per hour for a period of greater than 24 h, the development is called explosive cyclogenesis. Explosive cyclogenesis tends to occur in strong baroclinic zones when a weak surface disturbance receives reinforcement from a cyclonic disturbance aloft, and the static stability of the atmosphere (that is, the resistance to vertical motions) is weak so that strong cumulus convection occurs. SEE BAROCLINIC FIELD; FRONT; GEOSTROPHIC WIND; JET STREAM.

Tropical cyclones. By contrast, tropical cyclones, derive their energy from the release of latent heat of condensation in precipitating cumulus clouds. The moisture necessary to sustain this precipitation is supplied by the frictionally driven flow of moist air toward low pressure near the Earth's surface and its subsequent ascent into the cumulus clouds. Over the tropical oceans, where moisture is plentiful, tropical cyclones can develop into intense vortical storms (hurricanes and typhoons), which can have wind speeds in excess of 200 mi/h (100 m·s⁻¹). SEE HURRICANE; PRECIPITATION (METEOROLOGY); STORM; TROPICAL METEOROLOGY; WIND.

J. R. Holton

Bibliography. J. R. Holton, *An Introduction to Dynamic Meteorology*, 2d ed., 1979; B. J. Hoskins, *Dynamical processes in the atmosphere and the use of models*, *Quart. J. Roy. Meteorol. Soc.*, 109:1–21, 1983; R. J. Reed and M. D. Albright, A case study of explosive cyclogenesis in the Eastern Pacific, *Month. Weath. Rev.*, 114:2297–2319, 1986; J. M. Wallace and P. V. Hobbs, *Atmospheric Science: An Introductory Survey*, 1977.

Cyclone furnace

A water-cooled horizontal cylinder in which fuel is fired and heat is released at extremely high rates. When firing coal, the crushed coal, approximately 95% of which is sized at ¼ in. (0.6 cm) or less, is introduced tangentially into the burner at the front end of the cyclone (see **illus.**). About 15% of the combustion air is used as primary and tertiary air to impart a whirling motion to the particles of coal. The whirling, or centrifugal, action on the fuel is further increased by the tangential admission of high-velocity secondary air into the cyclone. SEE FURNACE CONSTRUCTION.

The combustible is burned from the fuel in the cyclone furnace, and the resulting high-temperature gases melt the ash into liquid slag, a thin layer of which adheres to the walls of the cyclone. The incoming fuel particles, except those fines burned in suspension, are thrown to the walls by centrifugal force and caught in the running slag. The secondary air sweeps past the coal particles embedded in the slag surface at high speed. Thus, the air required to burn the coal is quickly supplied and the products of combustion are rapidly removed.

The products of combustion are discharged through a water-cooled reentrant throat at the rear of the cyclone into the boiler furnace. The part of the molten

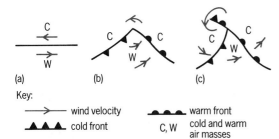

(a)   (b)   (c)

Key:

⟶ wind velocity

▲▲▲ cold front

●●● warm front

C, W cold and warm air masses

Fig. 3. Classic Norwegian cyclone model showing three stages in the life of a polar front cyclone. (a) Frontal discontinuity, (b) distorted frontal surface, (c) occlusion. (After B. J. Hoskins, Dynamical processes in the atmosphere and the use of models, Quart. J. Roy. Meteorol. Soc., 109:1–21, 1983)

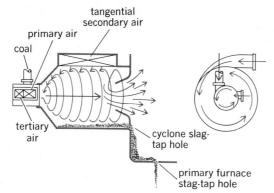

coal

primary air

tangential secondary air

tertiary air

cyclone slag-tap hole

primary furnace stag-tap hole

Schematic diagram of cyclone furnace. (*After E. A. Avallone and T. Baumeister III, eds., Marks' Standard Handbook for Mechanical Engineers, 9th ed., McGraw-Hill, 1987*)

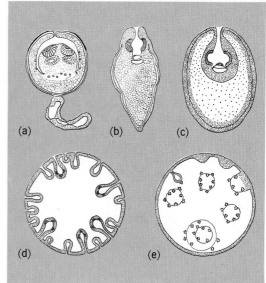

(a)   (b)   (c)

(d)   (e)

Fig. 2. Immature stage, the metacestode, of tapeworms. (*a*) Cysticercoid. (*b*) Plerocercoid. (*c*) Cysticercus. (*d*) Coenurus. (*e*) Hydatid cyst.

slag that does not adhere to the cyclone walls also flows toward the rear and is discharged through a taphole into the boiler furnace.

Essentially, the fundamental difference between cyclone furnaces and pulverized coal–fired furnaces is the manner in which combustion takes place. In pulverized coal–fired furnaces, particles of coal move along with the gas stream; consequently, relatively large furnaces are required to complete the combustion of the suspended fuel. With cyclonic firing, the coal is held in the cyclone and the air is passed over the fuel. Thus, large quantities of fuel can be fired and combustion completed in a relatively small volume, and the boiler furnace is used to cool the products of combustion.

Gas and oil can also be burned in cyclone furnaces at ratings and with performances equaling those of coal firing. When oil is the fuel, it is injected adjacent to the secondary air ports and directed downward into the secondary airstream. The oil is picked up and sufficiently atomized by the high-velocity air. Gas is fired similarly through flat, open-ended ports located in the secondary air entrance. SEE BOILER; STEAM-GEN-ERATING FURNACE; STEAM-GENERATING UNIT.

G. W. Kessler

Cyclophyllidea

The order which includes most tapeworms that inhabit the gut of warm-blooded vertebrates. They are frequently referred to as the Taenioidea. Each worm has a head, or scolex, and a segmented body, called the strobila. The scolex typically has an apical rostellum, or muscular pad and hooks, and two pairs of lateral suckers (**Fig. 1***a*). New segments are produced immediately posterior to the scolex, so that the oldest

segment is at the hind end. As a segment is pushed further from the scolex, the male and female reproductive organs mature and open on the lateral margin (Fig. 1*b*). Ova, fertilized by sperm from the same or another segment, gather in a uterus. The ripe terminal segments containing infectious eggs are shed into gut contents of the host (Fig. 1*c*).

If an intermediate host, such as an invertebrate, or less often a vertebrate, eats the eggs, an immature stage, the metacestode, develops. Commonly it grows in the hemocoel, or blood cavity, of invertebrates or in the liver of vertebrates, but depending on the tapeworm, it may occur on or in almost any organ. Generally the metacestode has one or more scoleces characteristic of the adult worm, but the remainder of the body may have one of several forms. The cysticercoid, usually small and found in invertebrate hosts, has the scolex retracted but not invaginated, and frequently there is a taillike appendage (**Fig. 2***a*). The remaining types of metacestodes of this group lack the tail appendage and have the scolex invaginated, that is, pulled in on itself, as when the finger of a glove is pushed outside in. A plerocercoid, which occurs in both invertebrates and vertebrates, has a solid body (Fig. 2*b*), whereas the other types of metacestodes, all of which occur only in vertebrates, have a

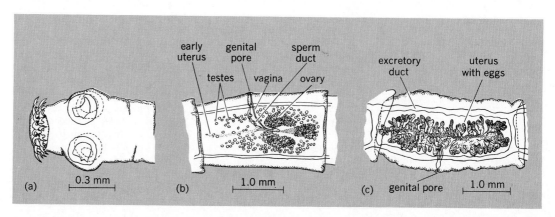

early uterus   genital pore   sperm duct

testes   vagina   ovary

excretory duct   uterus with eggs

(a)   0.3 mm   (b)   1.0 mm   (c)   genital pore   1.0 mm

Fig. 1. Tapeworm. (*a*) Scolex. (*b*) Mature segment. (*c*) Ripe segment.

Some cyclophyllidean tapeworms, their hosts, and immature forms

Name	Definitive host	Intermediate host	Type of metacestode
Dipylidium caninum (double-pored tapeworm of dogs and cats)	Dogs and cats, occasionally children	Fleas and biting lice	Cysticercoid
Echinococcus granulosus (hydatid tapeworm)	Dogs, other canines	Many vertebrates, including humans	Hydatid
Hymenolepis nana (dwarf tapeworm)	Humans, mice, rats	None, or fleas and beetles	Cysticercoid
Moniezia expansa (double-pored ruminant tapeworm)	Sheep, other ruminants	Oribatid mites	Cysticercoid
Taenia saginata (beef tapeworm)	Humans	Bovine	Cysticercus
Taenia solium (pork tapeworm)	Humans	Swine and humans	Cysticercus
Multiceps multiceps (gid tapeworm)	Dogs, other canines	Sheep, other vertebrates, including humans	Coenurus

body formed into a bladder. If a single invaginated scolex is present, the bladder is called a cysticercus (Fig. 2*c*). When many scoleces, up to a hundred or so, protrude inward from the common bladder wall, this is called a coenurus (Fig. 2*d*). The hydatid cyst is a modified bladder containing large numbers of attached and free-floating brood capsules, each with one or more scoleces (Fig. 2*e*). The metacestode usually remains quiescent until the intermediate host is eaten by a definitive host and then the tapeworm matures in the gut, thus completing the cycle.

A number of tapeworms occur in humans and domestic animals, either as adults or metacestodes, but rarely in both stages (see **table**). Symptoms caused by the adult worm vary from no visible effect to diarrhea, severe abdominal pain, convulsions, and occasionally an increased appetite. Diagnosis of tapeworms in the gut is based on finding eggs or segments in the feces, whereas diagnosis of metacestodes usually requires x-ray pictures and various skin tests. Certain drugs remove tapeworms from the gut, but removal of the metacestodes usually requires surgery. SEE COENUROSIS; ECHINOCOCCOSIS; PLATYHELMINTHES; TAPEWORM DISEASE.

Reino S. Freeman

Cyclopoida

An order of small copepod Crustacea. The abundant fresh-water and salt-water species form an important link in the food chains of aquatic life, consuming tiny plants, animals, and detritus and, in turn, furnishing food for small fish, some large fish, and other organisms. Some species are important intermediate hosts for human parasites. Females of the family Ergasilidae are parasitic on aquatic animals, while the males are free-swimming. SEE FOOD WEB.

The front part of the body is oval and sharply separated from the tubular hind end, which bears two caudal rami with distinctly unequal setae (**Fig. 1**). Usually 10 body segments are present in the male, and 9 in the female because of fusion of two to form a genital segment. Appendages (**Fig. 2**) are the first and second antennae, mandibles, first and second maxillae, maxillipeds, and five pairs of swimming legs, the last of which are alike and rudimentary. Males usually have rudimentary sixth legs. Appendages are usually biramous, with a 1- or 2-segmented basal portion. One branch is missing from the first antennae, and usually from the second antennae, the mandibles, and fifth legs. The first antennae are never much longer than the anterior portion of the body and consist of 6–17 segments. Usually there are either 12 or 17. Both first antennae of the males are modified for grasping females. The digestive tract consists of a narrow, dilatable esophagus, an expanded stomach, a long intestine, and a short rectum opening on the dorsal side of the last body segment.

A heart is absent. Colorless blood containing ameboid corpuscles is circulated by general body movements and rhythmic movements of the digestive tract. External respiration is through the body surface. Striated muscles are highly developed. The genital systems are shown in Fig. 1. In mating, the male grasps the female with the first antennae and by

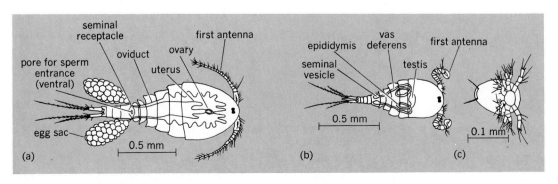

Fig. 1. *Cyclops vernalis.*
(*a*) Dorsal view of female.
(*b*) Dorsal view of male.
(*c*) Ventral view of the first of the six nauplius stages.

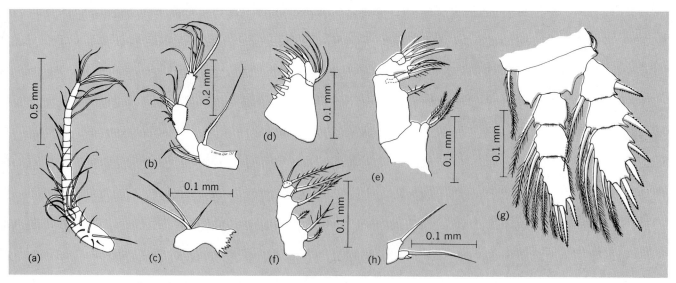

Fig. 2. Representative appendages of *Cyclops*. (*a*) First antenna of female. (*b*) Second antenna. (*c*) Mandible. (*d*) First maxilla. (*e*) Second maxilla. (*f*) Maxilliped. (*g*) Typical swimming leg (right third leg). (*h*) Fifth leg.

means of the swimming legs places two kidney-shaped spermatophores at the opening of the female's seminal receptacle. Egg sacs are paired and lateral. Rarely, they are subdorsally located. There are six nauplius and six copepodid stages, the last being the adult. Some species encyst in the copepodid stages to survive adverse conditions. A small brain in the head is connected, around each side of the esophagus, to a ventral chain of ganglia. There is a median dorsal eye. Cyclopoids are intermediate hosts of the parasitic guinea worm (*Dracunculus medinensis*), and sometimes the fish tapeworm (*Dibothriocephalus latus*). Most fresh-water species live in shallow water, swimming from plant to plant, but salt-water species are generally water-treaders. Food is not secured by filtration, but is seized and eaten with the biting mouthparts. Many species have a worldwide distribution. SEE COPEPODA.

Harry C. Yeatman

Bibliography. B. Dussart, *Les Copepodes des Eaux Continentales*, vol. 2, 1969; R. Gurney, *British Fresh-water Copepoda*, vol. 3, 1933; F. Kiefer, *Ruderfuss-Krebse (Copepoden)*, 1960; S. P. Parker (ed.), *Synopsis and Classification of Living Organisms*, 2 vols., 1982; G. O. Sars, *Crustacea of Norway*, vol. 6, 1918.

Cyclostomata (Bryozoa)

An order of bryozoans in the class Stenolaemata. Cyclostomes tend to have delicate colonies composed of relatively isolated, loosely bundled, or tightly packed, comparatively simple, long, slender, and tubular zooecia, with thin, highly porous, calcareous walls. When regionated, their colonies display a very gradual transition between endozone and exozone. SEE BRYOZOA; STENOLAEMATA.

Morphology. Cyclostome colonies most commonly are small and delicate, but some are moderately large and massive. They may be inconspicuous encrusting threadlike networks, thin encrusting sheets, nodular masses, or erect tuftlike, twiglike, or frondlike growths. Most cyclostome colonies are not obviously regionated, but some exhibit indistinct endozone and

exozone regions of quite variable relative proportions. Lacking cystopores, stolons, acanthorods, monticules, maculae, vibracula, and avicularia, cyclostome colonies may sometimes bear rhizoids, nanozooecia, mesopores, and various small coelomic chambers; many colonies also display well-developed ovicells (brood chambers) which have evolved by complete modification of initially normal-appearing zooids (gonozooids).

Individual cyclostome zooecia are straight to slightly curved long tubes. The walls of adjacent zooecia may be either distinctly separate or fused together. The zooecial walls, generally thin throughout the colony and of indistinct or variable microstructure, consist of calcareous material perforated by many tiny pores which are filled in life with soft tissues (pseudopores). In addition, some species may possess a chitinous cuticle covering the external surface of the outermost zooecial walls in the colony. The space within each zooecium usually lacks transverse calcareous partitions, but sometimes is crossed by a few diaphragms. Round or circular in outline, the zooecial aperture lacks an operculum, but in some forms is surrounded by a raised rim (peristome). Cyclostome zooids lack epistomes, have few to moderate numbers of tentacles, and (like the colony overall) can be either hermaphroditic or dioecious.

Life cycle. Immediately after fertilization, the cyclostome zygote enters an ovicell and divides into as many as 100 embryos (polyembryony), each of which then develops into a larva. After escaping from the ovicell, the larva swims freely for less than a day before settling down to the bottom and attaching to a substrate. There, the larva metamorphoses (by means of complete degeneration of its internal structures) into an ancestrular zooid from which repeated asexual budding produces the rest of the colony (but no resistant resting bodies).

History and classification. Exclusively marine, cyclostomes (occasionally termed Stenostomata) are known first from the Early Ordovician, when they may possibly have evolved from cystoporates (which were formerly included in the Cyclostomata). Rare and inconspicuous throughout the rest of the Paleozoic and Triassic, cyclostomes became moderately

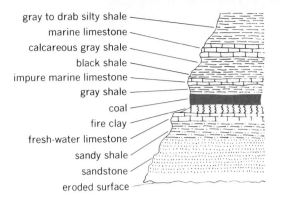

gray to drab silty shale
marine limestone
calcareous gray shale
black shale
impure marine limestone
gray shale
coal
fire clay
fresh-water limestone
sandy shale
sandstone
eroded surface

Cyclothem showing the complete succession of members that are recognized in Illinois.

Cyclostomata (Chordata)

The simplest and most primitive of living vertebrates. Cyclostomes, also known as Cyclostomi and Marsipobranchii, are the modern representatives of the class Agnatha, whose other members are not known after the Paleozoic. Since soft-bodied animals with naked skin and a cartilaginous skeleton are not easily preserved, they have been unknown as fossils until the recent fortunate discovery of a few in the Pennsylvanian rocks of Illinois. The Cyclostomata are descended at least in part (lampreys—Petromyzonida or Petromyzontiformes) from the subclass Cephalaspidomorpha (Monorhina). Whether the hagfishes (Myxiniformes) are similarly related or derived from the Pteraspidomorpha (Diplorhina) is debated by paleontologists. SEE CEPHALASPIDOMORPHA; FOSSIL; PTERASPIDOMORPHA.

Cyclostomes differ notably from all other Recent fishes. The absence of jaws and the presence of a single median nostril are distinctive. In addition, they have an uncalcified cartilaginous skeleton and lack true teeth and paired fins and girdles; a complex branchial basket is present and consists of 5–15 pairs of spherical pouches that open internally and have 1–15 pairs of porelike external apertures. Only one or two semicircular canals occur; ribs, spleen, and sympathetic nervous system are lacking, and the heart has no conus arteriosus. There is no cloaca. Cyclostomes are eellike in form and have tough, more or less slimy skin. They are oviparous and may be either bisexual or hermaphroditic. Development may involve a protracted larval stage (ammocoetes) followed by profound structural change at metamorphosis, as in lampreys, or it may be direct, as in hagfishes. SEE AGNATHA; MYXINIFORMES; PETROMYZONTIDA.

Robert H. Denison

Bibliography. D. Bardack and E. S. Richardson, New Agnathous fishes from the Pennsylvanian of Illinois, *Fieldiana, Geol.,* 33:489–510, 1977.

Cyclothem

A specific sequence of different kinds of rocks, one of which is generally coal. These sequences, each on the order of tens of feet in thickness, occur repeatedly, one above the other, in coal-bearing strata. Cyclothems have been recognized mainly in rocks of late Paleozoic age in the eastern United States but also have been described in late Mesozoic coal-bearing strata of the Rocky Mountain region. The **illustration** shows a composite cyclothem of western Illinois. The diagram is idealized in the sense that not all of the different kinds of rocks shown occur in every cyclothem of that region. Cyclothems in other areas depart significantly from those in Illinois. In Iowa, Kansas, and Nebraska, sandstones and coals are thinner and, in many places, absent, but limestones and shales containing marine fossils are abundant. In the Appalachian region sandstones and coal beds are thicker and more numerous than in Illinois, whereas

marine limestones or even shales containing marine fossils are relatively rare.

The lateral extent and boundaries of cyclothems are a subject of debate. Some geologists claim that cyclothems are very widespread, extending across the entire eastern half of the United States; others, while admitting lateral continuity of some rock units in smaller areas, believe that many cyclothems are mainly local deposits. Similarly, some authorities state that the boundary of the cyclothem should be placed at the base of the sandstone shown in the diagram because they believe that that sandstone rests on an eroded surface of great lateral extent. Others think that the eroded surface at the base of the sandstone is only a local phenomenon and that the boundary should be placed at the coal bed or at the top of the fireclay.

Because there has been no real agreement about factual knowledge concerning cyclothems, there is even less about their origin. Those who advocate the notion of widespread cyclothems limited at top and bottom by regional erosion surfaces are divided in opinion concerning their origin. Some believe that the eroded surface represents an episode of uplift of the land, and the succeeding deposits a period of subsidence, resulting first in nonmarine deposition and then in marine deposition. Others believe that the land surface may have remained stable but that the sea level rose and fell in response to the growth and melting of continental ice sheets in the Southern Hemisphere. Some support for this notion is found in the late Paleozoic glacial deposits recognized in portions of South Africa and Australia. At the time when cyclothems were being formed, these now isolated areas are believed to have been joined in a single continent designated Gondwana.

Geologists who do not agree that cyclothems are deposits of regional extent are inclined to favor combinations of (1) local subsidence influenced by differential compaction of underlying sediments, (2) regional periodic subsidence, and (3) differential rates of local sediment influx due to shifting of stream patterns such as are observed on modern alluvial plains and coastal regions. SEE SEDIMENTOLOGY; STRATIGRAPHY.

John C. Ferm

Bibliography. D. V. Ager, *Nature of the Stratigraphical Record,* 1981; P. McL. Duff et al., *Cyclic Sedimentation,* 1967; J. C. Ferm, in E. D. McKee et al. (eds.), *Paleotectonic Investigations of the Pennsylvanian System in the United States,* U.S. Geol. Surv. Prof. Pap. 853, 1975; D. F. Merriam (ed.), *Symposium on Cyclic Sedimentation,* Kans. Geol. Surv. Bull. 169, 1964; H. R. Wanless, Late Paleozoic cycles of sedimentation in the United States, *18th International Geological Congress Report*, pt. 4, pp. 17–28, 1950; J. M. Weller, Cyclic sedimentation of the Pennsylvanian period and its significance, *J. Geol.*, 38:97–135, 1930.

Cyclotron resonance experiments

The measurement of charge-to-mass ratios of electrically charged particles from the frequency of their helical motion in a magnetic field. Such experiments are particularly useful in the case of conducting crystals, such as semiconductors and metals, in which the motions of electrons and holes are strongly influenced by the periodic potential of the lattice through which they move. Under such circumstances the electrical carriers often have "effective masses" which differ greatly from the mass in free space; the effective mass is often different for motion in different directions in the crystal. Cyclotron resonance is also observed in gaseous plasma discharges and is the basis for a class of particle accelerators. *See* BAND THEORY OF SOLIDS; PARTICLE ACCELERATOR.

The experiment is typically performed by placing the conducting medium in a uniform magnetic field H and irradiating it with electromagnetic waves of frequency ν. Selective absorption of the radiation is observed when the resonance condition $\nu = qH/2\pi m^*c$ is fulfilled, that is, when the radiation frequency equals the orbital frequency of motion of the particles of charge q and effective mass m^* (c is the velocity of light). The absorption results from the acceleration of the orbital motion by means of the electric field of the radiation. If circularly polarized radiation is used, the sign of the electric charge may be determined, a point of interest in crystals in which conduction may occur by means of either negatively charged electrons or positively charged holes. *See* HOLE STATES IN SOLIDS.

For the resonance to be well defined, it is necessary that the mobile carriers complete at least $1/2\pi$ cycle of their cyclotron motion before being scattered from impurities or thermal vibrations of the crystal lattice. In practice, the experiment is usually performed in magnetic fields of 1000 to 100,000 oersteds (1 oersted = 79.6 amperes per meter) in order to make the cyclotron motion quite rapid ($\nu \approx 10-100$ gigahertz, that is, microwave and millimeter-wave ranges). Nevertheless, crystals with impurity concentrations of a few parts per million or less are required and must be observed at temperatures as low as 1 K in order to detect sharp and intense cyclotron resonances.

The resonance process manifests itself rather differently in semiconductors than in metals. Pure, very cold semiconductors have very few charge carriers; thus the microwave radiation penetrates the sample uniformly. The mobile charges are thus exposed to radiation over their entire orbits, and the resonance is a simple symmetrical absorption peak.

In metals, however, the very high density of conduction electrons present at all temperatures prevents penetration of the electromagnetic energy except for a thin surface region, the skin depth, where intense shielding currents flow. Cyclotron resonance is then observed most readily when the magnetic field is accurately parallel to the flat surface of the metal. Those conduction electrons (or holes) whose orbits pass through the skin depth without colliding with the surface receive a succession of pulsed excitations, like those produced in a particle accelerator. Under these curcumstances cyclotron resonance consists of a series of resonances $n\nu = qH/2\pi m^*c$ ($n = 1, 2, 3, \ldots$) whose actual shapes may be quite complicated. The resonance can, however, also be observed with the magnetic field normal to the metal surface; it is in this geometry that circularly polarized exciting radiation can be applied to charge carriers even in a metal.

Cyclotron resonance is most easily understood as the response of an individual charged particle; but, in practice, the phenomenon involves excitation of large numbers of such particles. Their net response to the electromagnetic radiation may significantly affect the overall dielectric behavior of the material in which they move. Thus, a variety of new wave propagation mechanisms may be observed which are associated with the cyclotron motion, in which electromagnetic energy is carried through the solid by the spiraling carriers. These collective excitations are generally referred to as plasma waves. In general, for a fixed input frequency, the plasma waves are observed to travel through the conducting solid at magnetic fields higher than those required for cyclotron resonance. The most easily observed of these excitations is a circularly polarized wave, known as a helicon, which travels along the magnetic field lines. It has an analog in the ionospheric plasma, known as the whistler mode and frequently detected as radio interference. There is, in fact, a fairly complete correspondence between the resonances and waves observed in conducting solids and in gaseous plasmas. Cyclotron resonance is more easily observed in such low-density systems since collisions are much less frequent there than in solids. In such systems the resonance process offers a means of transferring large amounts of energy to the mobile ions, a necessary condition if nuclear fusion reactions are to occur. *See* NUCLEAR FUSION; PLASMA PHYSICS; WAVES AND INSTABILITIES IN PLASMAS.

Walter M. Walsh

Bibliography. C. Kittel, *Introduction to Solid State Physics*, 5th ed., 1976; P. M. Platzman and P. A. Wolff, *Waves and Interactions in Solid State Plasmas*, 1973.

Cydippida

The largest order of the phylum Ctenophora, comprising 5 families (Bathyctenidae, Haeckelidae, Lampeidae, Mertensidae, and Pleurobrachiidae) and 11 genera. The body of cydippids is usually globular or cylindrical in shape (see **illus**.); sizes range from a few millimeters to about 15 cm (6 in.). Most cydippids are colorless and transparent, but some deep-sea species are pigmented dark red. The comb rows are of equal length, and the eight meridional canals lying beneath them end blindly at the oral end. Adult cydippids retain the larval morphology common to all ctenophores (except Beroida) and are usually thought to be the most primitive order. All forms have two main tentacles, which have side branches (tentilla) in some species. Cydippids catch prey on the outstretched tentacles, contracting them to bring it to the mouth. In three families (Bathyctenidae, Haeckelidae, and Lampeidae) the tentacles emerge from the body near the oral end; these cydippids have extensile mouths and catch large gelatinous animals. In the Pleurobrachiidae and Mertensidae, the tentacles exit near the aboral end, have numerous side branches, and catch mostly small crustacean prey. The well-

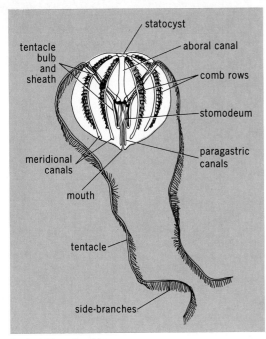

Typical *Pleurobrachia* sp.

known genus *Pleurobrachia* is widely distributed in temperate coastal waters and estuaries, where it can be a major predator of copepods and larval fish. Many other species of cydippids occur in oceanic waters, from the surface to mesopelagic depths. *See Beroida; Ctenophora.*

Laurence P. Madin

Cygnus

The Swan, in astronomy, is a conspicuous northern summer constellation. The five major stars of the group, α, γ, β, ε, and δ, are arranged in the form of a cross (see **illus.**). Hence Cygnus is often called the

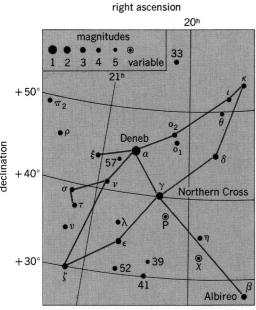

Line pattern of constellation Cygnus. Grid lines represent coordinates of sky. Apparent brightness, or magnitudes, of stars is shown by sizes of "dots," which are graded by appropriate numbers as indicated.

Northern Cross, to distinguish it from the Southern Cross of the constellation Crux. The constellation is represented by a swan with widespread wings flying southward. The bright star Deneb, signifying tail, is the tail of the swan; it lies at the head of the cross. Albireo, a beautiful double star of contrasting orange and blue colors, is the head of the swan. The whole constellation lies in, and parallel to, the path of the Milky Way. It contains several splendid star fields. *See Constellation; Crux.*

Ching-Sung Yu

Cylinder

The solid of revolution obtained by revolving a rectangle about a side is called a cylinder, or more precisely a right circular cylinder. More generally the word cylinder is used in solid geometry to describe any solid bounded by two parallel planes and a portion of a closed cylindrical surface. In analytic geometry, however, the word cylinder refers not to a solid but to a cylindrical surface (see **illus.**). This is a surface generated by a straight line which moves so that it always intersects a given plane curve called the directrix, and remains parallel to a fixed line that intersects the plane of the directrix. (Part or all of the

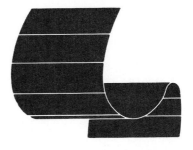

Diagram of cylindrical surface.

directrix may consist of straight-line segments.) The generating line in each of its positions is called an element of the cylinder, and a section of the cylinder made by a plane perpendicular to the elements is called a right section. All right sections of a cylinder are congruent. A circular cylinder (cylindrical surface) is one whose right sections are circles. Its other sections made by planes cutting all the elements are ellipses. Cylinders whose right sections are ellipses, parabolas, or hyperbolas are called elliptic cylinders, parabolic cylinders, or hyperbolic cylinders, respectively. All these are quadric cylinders. The lateral surface of a prism is a special type of cylindrical surface whose directrix and right sections are polygons.

The volume of a solid cylinder is $V = Bh$, where B denotes the area of the base, and h its altitude (height) measured perpendicular to the base. For a right circular cylinder whose base is a circle of radius r, the volume is $V = \pi r^2 h$, and the total surface area is $S = 2\pi r^2 + 2\pi rh$. *See Euclidean geometry.*

J. Sutherland Frame

Cylindrocorporoidea

A superfamily of nematodes with both free-living forms and intestinal parasites of amphibians, reptiles, and certain mammals. The lips surrounding the oral opening are well developed, and the lateral lips bear the small amphids. The stoma is very elongate, measuring one-fourth or more of the esophageal length. The corpus of the esophagus is enlarged into a cylindroid muscular complex followed by a glandular postcorpus. Males always have rudimentary caudal alae, and the spicules are elongate and thin. *See Nemata*.

Armand R. Maggenti

Cyperales

An order of flowering plants, division Magnoliophyta (Angiospermae), in the subclass Commelinidae of the class Liliopsida (monocotyledons). The names Glumiflorae, Graminales, and Poales have also been used for this order. There are only two families, the Poaceae (Gramineae), with about 8000 species, and the Cyperaceae, with nearly 4000. The Cyperales are Commelinidae with reduced, mostly wind-pollinated or self-pollinated flowers that have a unilocular, two- or three-carpellate ovary bearing a single ovule. The flowers are arranged in characteristic spikes or spikelets representing reduced inflorescences. The perianth is represented only by a set of bristles or tiny scales, or is completely missing. The leaves generally have a well-defined sheath and a narrow blade, often with a small adaxial appendage (the ligule) at the junction of the two. The stomates have two supporting cells and are arranged in straight files or rows of one or two, all oriented in the same direction. The pollen is uniformly trinucleate, and vessels are present in all vegetative organs.

The Cyperaceae, or sedge family, includes the bulrushes (*Scirpus*) and the papyrus (*Cyperus papyrus*) of Egypt, as well as the sedges (*Carex*). The Poaceae, or grass family, embraces all true grasses, including bamboos and the cereal grains such as wheat, maize, oats, and rye. The two families differ in a number of more or less consistent technical characters of the inflorescence, fruits, stems, and leaves. The well-developed intercalary meristem at the base of the blade in the grass leaf permits the plants to withstand grazing, and may be largely responsible for the dominance of grasses in large areas of the world with a suitable climate (seasonal rainfall, with the dry season too severe to permit forest growth). *See Bamboo; Barley; Cereal; Commelinidae; Corn; Flower; Grass crops; Liliopsida; Magnoliophyta; Millet; Oats; Plant kingdom; Rice; Rye; Sorghum; Sugarcane; Timothy; Wheat*.

Arthur Cronquist; T. M. Barkley

Cypress

The true cypress (*Cupressus*), which is very close botanically to the cedars (*Chamaecyparis*). The principal differences are that in *Cupressus* the cones are generally larger and the cone scales bear many seeds, whereas in *Chamaecyparis* each cone scale bears only two seeds; and in *Cupressus* the branchlets are not flattened as in *Chamaecyparis*. *See Cedar; Pinales*.

All of the species of *Cupressus* in the United States are western and are found from Oregon to Mexico. The Arizona cypress (*Cupressus arizonica*) found in the southwestern United States and the Monterey cypress (*Cupressus macrocarpa*) of California are medium-sized trees and are chiefly of ornamental value. The Italian cypress (*Cupressus sempervirens*) and its

varieties are handsome ornamentals, but usually do well only in the southern parts of the United States. Other trees are also called cypress, such as the Port Orford cedar (*Chamaecyparis lawsoniana*) known also as the Lawson cypress, and the Alaska cedar (*Chamaecyparis nootkatensis*), known also as the Nootka cypress or cedar.

The bald cypress (*Taxodium distichum*) is an entirely different tree that grows 120 ft (36 m) tall and 3–5 ft (1–1.5 m) in diameter. It is found in the swamps of the South Atlantic and Gulf coastal plains and in the lower Mississippi Valley. The soft needle-like leaves and short branches are deciduous; hence, they drop off in winter and give the tree its common name. Also known as the southern or red cypress, this tree yields a valuable decay-resistant wood used principally for building construction, especially for exposed parts or where a high degree of resistance to decay is required as in ships, boats, greenhouses, railway cars, and railroad ties. *See Forest and forestry; Tree*.

Arthur H. Graves/Kenneth P. Davis

Cypriniformes

An order of actinopterygian fishes which ranks second in size only to the Perciformes. The Cypriniformes together with the order Siluriformes (catfishes) compose the superorder Ostariophysi. Although the perciforms dominate modern seas, the ostariophysans command preeminent faunal importance in fresh waters. At least one-half of the fresh-water fishes of each continent, with the exception of Australia, belong to this group. *See Perciformes*.

Ostariophysi. The characters of the Ostariophysi agree in the main with the Salmoniformes, the generalized basal order from which most teleosts have evolved, but with one notable difference: The anterior four or five vertebrae are much modified as a protective encasement for a series of bony ossicles that form a chain connecting the swim bladder with the inner ear, the so-called Weberian apparatus. Other characters include an abdominal pelvic fin, if present, usually with many rays; a pelvic girdle which does not have contact with the cleithra; and a more or less horizontal pectoral fin placed low on the side. Usually there are no fin spines, but one or two spines derived from consolidated soft rays may be found in dorsal, anal, or pectoral fins, and a spinelike structure is present in the adipose fin of certain catfishes. The upper jaw is often bordered by premaxillae and maxillae, the latter commonly toothless and rudimentary. An orbitosphenoid is present, and a mesocoracoid is usually present. There is a swim bladder, usually with a duct. Cycloid scales are present, are modified into bony plates, or are absent, and branchiostegal rays are variable in number and arrangement. The superorder Ostariophysi consists of two well-marked orders, about 34 families, and approximately 5000 species. *See Ear; Salmoniformes; Swim bladder*.

The order Cypriniformes (or Eventognathi) includes the characins, minnows, and their allies (see **illus.**). The Weberian apparatus of the Cypriniformes involves only the first four vertebrae. The body is typically invested in cycloid scales, although these are sometimes lost. In further contrast to catfishes, the cypriniforms have parietal, suboperculum, symplectic, and intercalar bones; they lack a pectoral spine (present in most catfishes); and they have intermuscular bones. *See Scale (zoology)*.

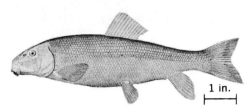

Northern redhorse (*Moxostoma macrolepidotum*), of the order Cypriniformes (1 in. = 2.5 cm). (*After G. B. Goode, Fishery Industries of the United States, 1884*)

Suborders. The eight families of the Cypriniformes, numbering about 3000 species, may be arranged in three suborders: Characoidei, Gymnotoidei, and Cyprinoidei.

Characoidei. This suborder includes the single family Characidae (split by some authors into several smaller families) or characins, characterized by usually toothed jaws, a maxilla which enters the gape, the usual presence of an adipose fin, and opposed upper and lower pharyngeal bones. The characins are primarily fresh-water fishes that abound in South America, where they have undergone an extensive adaptive radiation resulting in perhaps 850 species. A few have spread north into Central America and one into the southern United States. In African fresh waters there is less diversity of form, but there are about 150 species. The group appears to date from the Cretaceous. Tropical aquarium fishes include many characins, some of which, like the neon tetra, are extremely colorful. The notoriously predaceous piranha (*Serrasalmus*) of South American rivers belongs in this group.

Gymnotoidei. The suborder Gymnotoidei includes a single family, Gymnotidae, consisting of about 16 genera and perhaps 40 or 45 species that inhabit fresh waters of South and Central America. They are not known as fossils. These are eel-shaped fishes with numerous vertebrae, absence of pelvic and developed dorsal fins, and an anus which is located far forward; all species, so far as is known, can produce an electric shock. The dreaded electric eel, well known to visitors of public aquariums, is the largest species. SEE ELECTRIC ORGAN (BIOLOGY).

Cyprinoidei. The suborder Cyprinoidei includes about 6 families of primarily fresh-water fishes. They have toothless jaws, no adipose fin, and falciform lower pharyngeal bones, usually with teeth, which are opposed to a pair of padlike processes from the basioccipital bone. The Cyprinidae, including minnows and carps, are the largest family of fishes, with about 275 genera and more than 1500 species. This group is known from the Paleocene of Europe and the Eocene of Asia, but did not reach North America until the Oligocene. The center of abundance and diversity of form is in southeastern Asia. The family is well represented in Africa, Europe, Asia, and North America (about 230 species), but there are none in Central or South America, Madagascar, Australia, or in that part of the East Indies lying east of Lombok. Most species are small, but many are large enough to be of major importance in the fisheries, especially in eastern and southern Asia.

The carp (*Cyprinus carpio*), one of the hardiest of all fishes, has been widely disseminated as a food fish, and is being actively cultivated in many countries. The ornamental goldfish (*Carassius auratus*) is very familiar and is a commonly used experimental animal. The suckers (family Catostomidae) comprise 12 genera and about 65 species living in North America and eastern Asia. They probably originated in Asia from cyprinid ancestry. The loaches (family Cobitidae) include about 20 genera and 150 species that live in Asia, Africa, and Europe. Most are small and many are more or less eel-shaped.

In the mountainous regions of southeastern Asia are found three other small families of the Cyprinoidei: the Homalopteridae, with 12 genera and 51 species; the Gastromyzontidae (not separated from the preceding by some authors) with 16 genera and 30 species; and the Gyrinocheilidae, with 1 genus and 3 species. All are adapted to life in torrential streams. Another southeastern Asiatic genus with 2 species is recognized by some as a family, Psilorhynchidae. SEE ACTINOPTERYGII; CARP; EEL; SILURIFORMES.

Reeve M. Bailey

Cyrtosoma

A subphylum of the Mollusca comprising members of the classes Monoplacophora, Gastropoda, and Cephalopoda. The use by paleontologists of subphyla, including Cyrtosoma and Diasoma, in the classification of the phylum Mollusca has not been generally adopted by students of living forms. There is no disagreement over the further division of the phylum into eight clearly distinguishable classes.

Primitive members of the Cyrtosoma have, or had, a conical univalved shell, often twisted into a spiral. The relatively small single-shell aperture forced the anus to lie close to the mouth, either behind it (Monoplacophora and Cephalopoda) or ahead of it in those forms which have undergone torsion (Gastropoda; **illus**. *f*). Studies of the comparative anatomy and history of living and fossil mollusks indicate that monoplacophorans are the most primitive cyrtosomes and that the other two classes were derived independently from the Monoplacophora (see illus.). SEE CEPHALOPODA; GASTROPODA; MONOPLACOPHORA.

Although both gastropods and monoplacophorans are found in earliest Cambrian rocks in Siberia and China, intermediate forms which link the two classes (*Pelagiella*; illus. *e*) survived through the Cambrian. Apparently asymmetric coiling of the shell of some early monoplacophorans caused the body to begin to rotate with respect to the shell, thus initiating torsion. When this process was completed, the shell was carried so that it coiled posteriorly, and the first gastropods had evolved. Other early monoplacophorans, which remained bilaterally symmetrical, gave rise to animals with a variety of symmetrical shell forms, including those of a diverse extinct group, the Bellerophontida. The bellerophonts have usually been placed in the Gastropoda, but there is mounting evidence that they were untorted and hence monoplacophorans.

The third cyrtosome class, the Cephalopoda, did not appear until the end of the Cambrian. Its first members (illus. *h*) had small conical shells with numerous calcareous partitions (septa) at the apical end. Similar septate monoplacophorans are known from slightly older rocks, but these lacked a siphuncle, used for buoyancy control. It was the appearance of the siphuncle which enabled the first cephalopods to swim, and it has been responsible for the great post-Cambrian success of that class.

Each cyrtosome class has, in turn, been the most diverse molluscan group. In the Ordovician the ce-

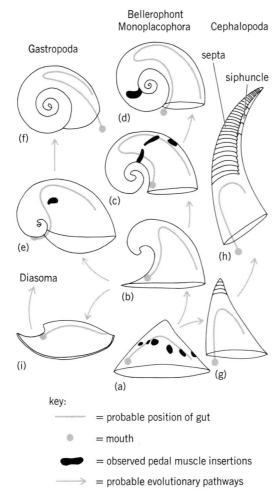

Gastropoda

Bellerophont
Monoplacophora   Cephalopoda

septa

siphuncle

(d)

(c)

(f)

(e)

Diasoma

(h)

(b)

(i)

(g)

(a)

key:

——— = probable position of gut

● = mouth

🖤 = observed pedal muscle insertions

——→ = probable evolutionary pathways

Evolution of the Cyrtosoma. (a) Lenaella. (b) Latouchella. (c) Cyrtolites. (d) Bucania. (e) Pelagiella. (f) Aldanella. (g) Knightoconus. (h) Plectronoceras. (i) Mellopegma. Organisms a–e, g, and i are monoplacophorans, f is an early gastropod, and h is the oldest cephalopod.

phalopods succeeded the monoplacophorans as the most diverse molluscan class, and they held this position until the close of the Mesozoic. They were followed by the gastropods, which by then had diversified into a great variety of aquatic and terrestrial habitats. It has been estimated that about 80% of the species of living mollusks are gastropods.

Although primitive cyrtosomes, including the only living shelled cephalopod, *Nautilus*, are shell-bearing animals, many advanced cyrtosomes have either lost or greatly reduced their shells (examples include slugs, nudibranchs, squids, and octopuses). One rare family of gastropods has even evolved a truly bivalved shell. Consequently, it is not any particular shell form which unites all members of the Cyrtosoma; rather, it is a knowledge of their evolutionary history. SEE DIASOMA; MOLLUSCA.

Bruce Runnegar

Bibliography. J. Pojeta and B. Runnegar, *The Paleontology of Rostroconch Mollusks and the Early History of the Phylum Mollusca*, U. S. Geol. Surv. Prof. Pap. 968, 1976.

Cystoporata

An extinct order of bryozoans in the class Stenolaemata. Cystoporates tend to have robust colonies composed of relatively simple, long, slender, and tubular zooecia (with thin, somewhat porous calcareous walls), separated by blisterlike vesicles (cystopores)

stacked upon one another. Their colonies display a very gradual transition between comparatively indistinct endozone and exozone regions. SEE BRYOZOA; STENOLAEMATA.

Morphology. Cystoporate colonies range from small and delicate to large and massive; they can be thin encrusting sheets, tabular or nodular masses, or erect twiglike or frondlike growths. Some cystoporate colonies are not regionated; others display indistinct endozone and exozone regions. Lacking known ovicells, stolons, acanthorods, vibracula, and avicularia, cystoporate colonies may sometimes include mesopores in addition to the ubiquitous cystopores. Many colonies exhibit monticules or maculae or both.

Individual cystoporate zooecia are straight to markedly curved long tubes. Because the zooecia are generally separated from one another by cystopores or vesicles, the walls of nearby zooecia remain distinctly separate and thin throughout the colony. The zooecial walls are calcareous, finely granular in microstructure, and perforated in some species by a few minute pores. The space within each zooecium may be crossed by diaphragms or may lack transverse calcareous partitions. Round or oval in outline, the zooecial aperture lacks an operculum, but often is surrounded by a raised rim (peristome) or partially covered over by a hoodlike lunarium.

History and classification. The cystoporates may possibly have evolved from the earliest ctenostomes. Cystoporates first appeared late in the Early Ordovician, somewhat earlier than representatives of other stenolaemate orders; thus, the cystoporates may possibly have in turn given rise to those orders, although all these forms may simply share a common ancestor further back in time. Apparently marine, the cystoporates (or ceramoporoids or fistuliporoids; formerly included in the Cyclostomata) became moderately common by Mid-Ordovician time—contributing to building small reefs, as well as to level-bottom communities—and remained so until they died out in the latest Permian. SEE CTENOSTOMATA; CYCLOSTOMATA (BRYOZOA).

Roger J. Cuffey

Cytochalasin

Any one of a class of hydrophobic antibiotics produced by fungi. Examples are cytochalasins A, B, C, D, E, G, and H, which possess varying specificities. The molecular weight of a commonly used form, cytochalasin B (CB), is 479.6. The cytochalasins (from Greek *cytos,* cell; *chalasis,* relaxation) elicit a diversity of responses. Examples of cytochalasin effects, usually reversible on removal of the antibiotic, are given in the **table**. The availability of radiolabeled cytochalasins greatly enhances the analytical power of this family of research tools and opens the way toward elucidation of its modes of action. Major targets for cytochalasin action are certain proteins associated with the plasma membrane, and actin microfilaments.

Plasma membrane. There is evidence that cytochalasins bind to plasma membrane proteins, including some involved in the transport of hexose sugars. Labeled cytochalasins, applied to a variety of animal cells, attain maximal binding within 1 to 5 min. Dissociation on transfer to cytochalasin-free medium is equally rapid, but as much as half the cytochalasin

Cytochalasin effects attributed to binding either to cell membranes or to actin

Target	Effect	Comments
Plant and animal cells	Inhibits uptake of small molecules (such as glucose, deoxyglucose, thymidine)	Not always accompanied by change in cellular cyclic AMP
Yeast cells	Cytochalasin A inhibits growth and sugar uptake	Acts as a sulfhydryl agent
Achyla	Cytochalasain A inhibits growth	Acts as a sulfhydryl agent
Animal and fungal cells	Inhibits secretion of vesicle-packaged enzymes	Acts on an event in secretion subsequent to Ca^{2+} influx
Animal cells	(a) Modifies lateral migration of integral membrane proteins during immune responses, (b) reduces electrophoretic mobility, and (c) exhibits specific steps in fertilization	(a) An example is the inhibition of clustering or "capping" of these proteins
Pancreas cells	Cytochalasin B inhibits glucose-induced electrical potential	Occurs when secretion of insulin is not inhibited
Fucoid brown algae	Inhibits establishment of cell polarity	Associated with polar flow of Ca^{2+}

remains associated with the cells. Both whole cells and fractionated cell components have been used to establish cytochalasin binding to plasma membranes. Red blood cells, for instance, each have 3×10^5 high-affinity cytochalasin binding sites, all in the plasma membrane. Binding is competitively inhibited by D-glucose; treatments which inactivate glucose-transport-related binding sites also eliminate cytochalasin binding; the estimated number of glucose binding sites, $2-5 \times 10^5$, corresponds well with the number of CB-binding sites. Most of these CB-binding sites appear therefore to be glucose transport sites. Similar results from other mammalian cells help to account for the potent inhibition of glucose uptake often observed with cytochalasins. Does the resulting glucose starvation explain the rapid "freezing" effect on cell motion that is the most universal of cytochalasin responses? Apparently not. Cellular motion is arrested long before glucose starvation occurs, and "freezing" occurs even in cells whose glucose uptake machinery is immune to cytochalasin. SEE CELL MEMBRANES.

The potent competitive inhibition of glucose uptake by CB has been attributed to the fact that these molecules can behave as structural analogs. The oxygen atoms on glucose, believed to react, through hydrogen bonds, with receptor sites (A, B, C, and D in **illus.** *a*) on a glucose carrier protein bound to the plasma membrane, correspond closely with the distribution of oxygen atoms on carbons 1, 19, 18, and 4 of CB (illus. *b*). This structural comparison has proved successful in predicting whether or not other molecules will inhibit glucose transport, and promises to be valuable in probing the structure of the glucose-binding site of the carrier protein.

CB inhibits membrane differentiation; membranes of cells preparing for mitosis fail to acquire the capacity to transport the DNA precursor thymidine.

Cytochalasins modify the infectivity of viruses applied to mammalian cells; for example, cytochalasin D increases the infectivity of polio virus.

Actin microfilaments. Evidence, accumulated using electron-microscopic, biochemical, and biophysical techniques, point to the conclusion that actin microfilaments have binding sites for cytochalasins. Microfilaments (6–8 nm in diameter) are associated with protoplasmic movements in plant and animal cells and are constructed of actin protein subunits, closely related to muscle actin. During the growth and division cycle of a cell, microfilaments undergo dramatic changes in distribution, which is probably achieved through reversible disassembly followed by

reassembly of the microfilaments from free subunits of actin. Actin microfilament bundles form the skeletons of microvilli, fingerlike outgrowths of the cell surface, particularly abundant in cancer cells and in normal, actively dividing cells. Microfilaments, like actin filaments from muscle, can be "decorated" along their length with "arrowheads" of myosin fragments, visible by electron microscopy. Arrowheads on a microfilament all point in the same direction, indicating filament polarity. Polarity can be described in terms of how a microfilament would be "decorated" with arrowheads, yielding a "pointed" and a "blunt" end. Cytochalasins bind near the blunt end of microfilaments, as do proteins such as vinculin,

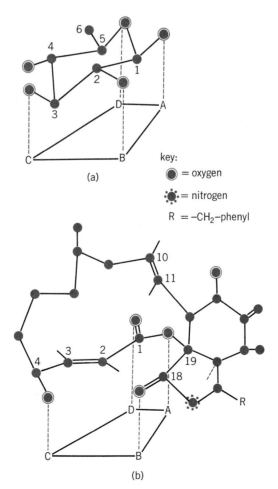

Structural models of (a) glucose and (b) cytochalasin B, demonstrating the close structural correspondence in the positions of four oxygen atoms believed to interact, via hydrogen bonds, with a membrane-bound glucose carrier protein bearing receptor sites A, B, C, and D.

which, like cytochalasins, can inhibit microfilament elongation. In cell regions where microfilaments appear attached to the cell membrane (such as at the tips of intestinal epithelial microvilli), all arrowheads of the microfilament point away from the membrane, just as the arrowheads in muscle actin point away from the attachment zone, the Z band. Since muscle contraction is achieved by an energy-dependent ratchetlike interaction between parallel actin and myosin filaments, it is assumed that actin microfilaments in nonmuscle cells may generate streaming by interaction with myosin. *See Muscle proteins*.

Myosins are more difficult to study than actins— more difficult to recognize using electron microscopy, and more difficult to characterize biochemically, since they are larger molecules and seem to be far more heterogeneous. Indeed this variation in myosins may dictate the type of movement generated by different actin-myosin systems. The presence of myosinlike proteins on cell organelles and vesicles suggests that movement of such cell inclusions in contact with microfilaments may be due to musclelike shearing forces generated between actin and myosin. Cytoplasmic streaming is frequently arrested by cytochalasins.

Electron-microscopic evidence documenting cytochalasin effects is varied. Different workers describe modification of microfilaments in terms of disruption, hypercondensation, or detachment from an anchoring site. Others find no evidence of microfilament changes, and studies using isolated actin and myosin fragments revealed no inhibition of arrowhead formation. However, CB did decrease the viscosity of an actin-myosin mixture and inhibited the ATPase of the myosin moiety. In addition, cytochalasin inhibits the microfilament polymerization from actin subunits which occurs in the presence of phalloidin, an alkaloid from a poisonous mushroom. *See Muscle*.

Locomotion and cell shape. In animals, cytochalasins inhibit cytokinesis, cell movement, and embryonic morphogenesis, as well as intracellular movement such as the transport of melanin granules. Chemotaxis in leukocytes may be stimulated or inhibited depending on antibiotic concentration. Microvilli of many cell types are diminished in length and number by cytochalasins, although some microvilli appear to be resistant. *See Cell motility*.

In plants, intracellular movements such as cytoplasmic streaming and chloroplast movements are inhibited. Tip growth in root hairs, pollen tubes, and fungi is also inhibited. Mitotic rate in root tip is slowed, without affecting phragmoplast formation.

Miscellaneous actions. In the animal cell, nuclear extrusion is induced; lymphocyte-mediated destruction of target cells is inhibited; and there is selective "pulverization" of certain chromosomes, which are converted to the unraveled, interphase form, while the other chromosomes in the cell remain in the condensed, metaphase form. This phenomenon may account for the selective disappearance of specific chromosomes in certain cell lines cultured in cytochalasins. In plants, cytochalasins inhibit root growth and water uptake in onion seedlings; cells become more spherical in shape.

Applications. The great value of the cytochalasins as research tools is that they appear to achieve their reversible impact on cell behavior with a minimum of undesirable side effects such as inhibition of respiration or protein synthesis. Cytochalasin is extensively applied as a chemical "scalpel" to enucleate mammalian cells rapidly, precisely, and efficiently in studies of nuclear-cytoplasmic relations and in cell hybridization and nuclear transplant work. Another major application is in examining the consequences of arrested cytoplasmic movement. For instance, plant cells with complete CB arrest of streaming have been found to show only mild inhibition of transport of the hormone auxin, supporting the conclusion that auxin transport within a cell depends more on diffusion than on cytoplasmic streaming.

Challenges for the future. The research potential of natural cytochalasins and of synthetic analogs remains to be explored. They will be valuable for studying the intricate cellular scaffolding known as the cytoskeleton, and also for labeling certain membrane proteins. Radiolabelled cytochalasins bound to target sites can be irradiated with ultraviolet light to yield a label covalently attached to the receptor protein. Cytochalasins have paved the way for work on proteins, such as vinculin and villin, which demonstrates cytochalasinlike activity. Such proteins may play key roles in directing the intracellular "traffic" in action. *See Antibiotic*.

Donovan Des S. Thomas

Bibliography. S. Lin, D. V. Santi, and J. A. Spudich, *J. Biol. Chem.*, 249:2268, 1974; B. K. Mookerjee et al., *Biol. Chem.*, 263:1290, 1981; E. Puszkin, *J. Biol. Chem.*, 248:7754, 1974; G. Schatten and H. Schatten, *Exp. Cell Res.*, 135:311, 1981; N. F. Taylor and G. L. Gagneja, *Can. J. Biochem.*, 53:1078, 1975; J. A. Wilkins and S. Lin, *Cell*, 28:83, 1982.

Cytochemistry

The science concerned with the chemistry of cells, mainly with determining the location of chemical constituents in cell organelles or other sites.

Methods. Cytochemical methods are designed to produce a visible end product or an increase in density that can be identified in cells by using the microscope. Staining reactions that yield colored end products can be studied with the light microscope, while reactions based on the use of fluorescent dyes such as fluorescein or rhodamine can be examined with the fluorescence microscope. Many cytochemical procedures have been adapted for use with the transmission electron microscope, an instrument that has a much greater resolving power than the light microscope. The scanning electron microscope, which has a resolving power intermediate between these two instruments, is especially useful for mapping chemical components on the surface of cells. Other microscopes with more limited applications to cytochemistry include the phase-contrast, interference, polarizing, and ultraviolet microscopes. *See Electron microscope; Fluorescence microscope; Interference microscope; Phase-contrast microscope; Scanning electron microscope*.

Specimen preparation. A variety of cell preparations are used in cytochemical studies. They include cell smears, sections of tissue, and cells that have been cultured in suspension or as monolayers on glass, plastic, or other substrates. The preparation of cells is an important factor in obtaining a satisfactory cytochemical localization. For light microscopy, cells are generally fixed in reagents such as formaldehyde or glutaraldehyde and then dehydrated in alcohol or a similar reagent. Solid tissues are fixed, dehydrated, and embedded in paraffin or a similar material. Sections are then prepared on a microtome and mounted on glass slides. Chemical constituents that are sensitive to these procedures can be localized in unfixed, frozen cells or tissue sections. Frozen specimens are also commonly used for fluorescence microscopy. *See Microtechnique*.

For transmission electron microscopy, cell specimens are fixed, dehydrated, and embedded in a special plastic. Cytochemical reactions are performed on the specimen after it is fixed but before it is dehydrated and embedded (pre-embedding method). Alternatively, the reaction may be carried out directly on a thin section prepared from the embedded material (postembedding method). Chemical constituents that might be lost or inactivated because of these procedures can be localized in thin sections prepared from unfixed, frozen (cryofixed) specimens. The sections are then placed in cytochemical media, fixed, and examined in the electron microscope. Components that are located within membranes such as the plasma membrane can be identified by using freeze fracture, a technique in which rapidly frozen cells are ruptured (fractured) so as to split open the membrane and expose the interior surfaces. After the cytochemical reaction is complete, a thin layer of carbon together with a metal, usually platinum, is evaporated over the specimen to produce a replica of the surface, which is removed and examined in the electron microscope. A number of cytochemical reactions can be performed on freeze-fractured specimens, especially if they involve the use of gold-labeled reagents which can be identified in the electron microscope.

Autoradiography. The sites within cells where macromolecules are synthesized can be identified by autoradiography. In this method, cells or tissue sections are exposed to small precursor molecules that have been labeled with a radioactive isotope, such as tritium (^{3}H). The specimens are coated with a photographic emulsion and placed in the dark. When the emulsion is developed, silver grains can be detected over cellular structures where the radioactive precursor has been incorporated. By using a fine-grain emulsion, the method can be adapted for use with the electron microscope. Autoradiography is also useful for identifying receptors for hormones or other molecules located on the cell surface. *See Autoradiography*.

Chemical components. The chemical constituents that are commonly identified by cytochemical methods are proteins, enzymes, carbohydrates and lipids, and various ions.

Proteins. Many proteins can be localized in cells with a high degree of specificity by using immunocytochemical methods. In this procedure, a purified protein (antigen) obtained from one species of animal is injected into another animal species to produce a specific antibody. After isolation and purification, the antibody is applied to cells, where it binds specifically to the antigen. A second antibody, prepared in similar fashion but labeled with a marker molecule, usually fluorescein or peroxidase, is then applied to the cell. Fluorescein-labeled antigen-antibody complexes can be detected by fluorescence microscopy (see **illus.**). Those labeled with peroxidase are first incubated for peroxidase activity with diaminobenzidine as substrate. The end product, oxidized diaminobenzidine, is visible by light microscopy. After exposure to os-

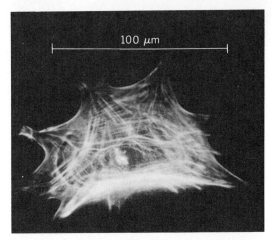

Fluorescence micrograph of a cultured endothelial cell from the cornea of a rabbit. After fixation in methanol and extraction with acetone, the cell was reacted with antiserum against the contractile protein, actin, followed by a fluorescein-labeled second antibody. The crisscrossing white lines represent actin-containing stress fibers within the cell. (*Courtesy of Dr. Sheldon R. Gordon*).

mium tetroxide, the product is rendered electron-dense and can thus be used in the pre-embedding method for electron microscopy. *See Immunochemistry; Immunofluorescence*.

Proteins can also be localized directly on thin, plastic sections (postembedding method) by using a secondary antibody labeled with gold particles that can be identified in the electron microscope. This procedure has also been adapted for light microscopy by using silver enhancement, technique in which the colloidal gold deposit is enlarged manyfold and converted into a visible, black precipitate by depositing silver over it.

Antigens that are present in very low concentrations can be detected by using biotin and avidin, two proteins that have a remarkably high affinity for each other. In this procedure, the specimen is exposed sequentially to primary antibody, secondary antibody, coupled to biotin, and avidin coupled to fluorescein, peroxidase, or gold particles. The high affinity between biotin and avidin ensures rapid formation of a stable complex between the biotin-labeled protein and the avidin-labeled marker. The specificity of the immunocytochemical reaction generally depends on demonstrating that normal, nonimmune serum or antiserum that has been precipitated with its specific antigen is unreactive. *See Protein*.

Enzymes. Most enzymes are free to diffuse in cells and are therefore best localized after fixation. Although many enzymes are markedly inhibited by fixatives, they often retain sufficient activity to be demonstrated by cytochemical means. More sensitive enzymes such as the dehydrogenases can be localized by using unfixed, frozen cells or tissue sections. Staining reactions are available for localizing various enzymes, such as the phosphatases, glycosidases, carboxylic esterases, and peptidases. They generally depend on the use of dyes such as naphthol derivatives, indoxyl and indoxyl amine derivatives, or tetrazolium salts, which are converted by the enzyme in the cell to an insoluble end product.

The most reliable methods are those characterized by rapid coupling of the intermediate product, formation of a highly insoluble end product, and minimal diffusion of both the enzyme and the end product. Procedures that demonstrate peroxidase activities using diaminobenzidene as substrate are among the most reliable because they possess these characteristics. In contrast to diaminobenzidine, the end products of most dye reactions lack sufficient density to be detected in the electron microscope. Reactions for demonstrating phosphatase activities using lead as a "trapping" agent have been especially useful since the end products, lead sulfide and lead phosphate, are readily visible by light and electron microscopy, respectfully. Since enzymes are proteins, many are potentially demonstrable by immunocytochemical methods, provided they are sufficiently antigenic to elicit an antibody response when injected into animals. *See* ENZYME.

Carbohydrates and lipids. Various types of carbohydrates, including those linked to proteins or lipids (glycoconjugates), can be identified with the periodic acid–Schiff reaction, in which aldehyde groups released by oxidation react to form a reddish end product. Acid mucopolysaccharides (glycosaminoglycans) can be visualized with cationic (positively charged) dyes, such as ruthenium red and Alcian blue, which bind to negatively charged sulfate or carboxyl groups. Sugar residues in various polysaccharides can be identified with high specificity by using lectins, which are proteins, mainly of plant origin, that bind specifically to sugar groups such as glucose, mannose, and galactose. By using lectins coupled to fluorescein, peroxidase, or gold particles, these sugars can be identified by fluorescence, light, or electron microscopy. The procedure is especially useful for mapping carbohydrate residues on the cell surface. Low concentrations of sugar residues can also be identified by using the biotin-avidin system described above. The specificity of lectin reactions is demonstrated by showing that a lectin is not reactive if it is first complexed with its specific sugar. *See* CARBOHYDRATE.

Lipids can be identified by using dyes such as oil red O or Sudan black B, which are soluble only in a specific type of lipid. Staining reactions for some of the complex lipids, such as phospholipids, are also available. *See* LIPID.

Nucleic acids. Deoxyribonucleic acid (DNA) can be demonstrated by the classical Feulgen reaction or with the dye methyl green–pyronin, which demonstrates both DNA and ribonucleic acid (RNA). Specific DNA or RNA sequences in cells, chromosomes, or other subcellular organelles can be detected by in-situ hybridization, a highly sensitive method that is based on the ability of single strands of nucleic acid to recognize and base-pair with strands bearing a complementary sequence. For light microscopy, hybridization is accomplished by applying a highly radioactive nucleotide to fixed cells. Reactive sites are detected by autoradiography. Nucleotide probes can also be labeled with nonisotopic markers such as fluorescein or peroxidase or with biotin for use with the biotin-avidin system. Probes labeled with gold particles can be used to localize nucleotide sequences with the electron microscope. *See* NUCLEIC ACID.

Ions. Various ions, such as sodium, potassium, magnesium, calcium, and zinc, can be identified in cells by using specific fluorescent indicators. These reagents bind with high affinity to specific ions, producing shifts in emission spectra that can be detected by fluorescence microscopy. *See* CELL (BIOLOGY); CELL, SPECTRAL ANALYSIS OF.

Edward Stanley Essner

Bibliography. J. M. Polak and S. Van Noorden, *Immunocytochemistry: Practical Applications in Pathology and Biology*, 2d ed., 1986; E. J. Sanders (ed.), *Cytochemistry of the Cell Surface and Extracellular Matrix during Early Embryonic Development*, 1986; L. A. Sternberger, *Immunocytochemistry* 3d ed., 1985.

Cytochrome

Any of a group of proteins that carry as prosthetic groups various iron porphyrins called hemes. Hemes also constitute prosthetic groups for other proteins (such as hemoglobin, whose function is to bind oxygen reversibly; catalase, which decomposes hydrogen peroxide; and peroxidase, which utilizes hydrogen peroxide to oxidize various substrates), but the function of prosthetic groups in the cytochromes is largely restricted to oxidation to the ferric heme, with the iron in the 3 + valence state, and reduction to ferrous heme with a 2 + iron. Thus, by alternate oxidation and reduction the cytochromes can transfer electrons to and from each other and other substances, and can operate in the oxidation of substrates. The energy released in their oxidation reactions is conserved by using it to drive the formation of the energy-rich compound adenosinetriphosphate (ATP) from adenosinediphosphate (ADP) and inorganic phosphate. This process of coupling the oxidation of substrates to phosphorylation of ADP is called oxidative phosphorylation. In cells of eukaryotic organisms, the cytochromes have rather uniform properties; they are part of the respiratory chain and are located in the mitochondria. In contrast, prokaryotes exhibit much more varied cytochromes. They have numerous a-, b-, and c-type cytochromes, as well as many others given different designations. Cytochromes are found even in metabolic pathways that employ oxidants other than oxygen. *See* ADENOSINEDIPHOSPHATE (ADP); ADENOSINETRIPHOSPHATE (ATP); MITOCHONDRIA; PROTEIN.

Respiratory chain. There are four cytochromes in the respiratory chain of eukaryotes, termed respectively aa_3, b, c, and c_1. There are two main sources of electrons for the respiratory chain, succinate dehydrogenase, the enzyme that oxidizes the major metabolic intermediate succinate to fumarate, and NADH dehydrogenase, the enzyme that oxidizes the reduced form of the coenzyme nicotinamide adenine dinucleotide (NADH) to its oxidized form (**Fig. 1**). Both of these substrates are products of the major energy-yielding metabolic pathways that are localized in the interior structure (matrix) of mitochondria. Electrons coming from two other mitochondrial enzymes also serve to reduce coenzyme Q; these enzymes are glycerol phosphate dehydrogenase and fatty acyl–coenzyme A dehydrogenase. The latter is a member of the fatty acid oxidation system, and fatty acid oxidation is an efficient and important source of metabolic energy. Cytochrome aa_3, also called cytochrome oxidase, functions by oxidizing reduced cytochrome c (ferrocytochrome c) to the ferric form. It then transfers the reducing equivalents acquired in this reaction to molecular oxygen, reducing it to water. This is a very unusual reaction in that most biological reductions of oxygen take it only to partially reduced forms such as hydrogen peroxide (H_2O_2) or to free-radical forms such as the superoxide radical (O_2^-), which are both toxic to cells and have to be destroyed. The cytochrome oxidase reaction is also probably the most important reaction in biology since it drives the entire respiratory chain and takes up over 95% of the oxygen employed by organisms, thus providing nearly all of the energy needed for living processes. Life, in its myriad forms that utilize oxygen as terminal metabolic oxidant, could never have evolved without a system capable of reducing oxygen to a nontoxic end product, water. *SEE LIFE, ORIGIN OF; RESPIRATION.*

Cytochrome oxidase. The cytochrome oxidase of eukaryotes is a very complex protein assembly containing from 8 to 13 polypeptide subunits, two hemes, *a* and a_3, and two atoms of copper. The two hemes are chemically identical but are placed in different protein environments, so that heme *a* can accept an electron from cytochrome *c* and heme a_3 can react with oxygen or substances that inhibit its reaction with oxygen, such as carbon monoxide and cyanide. When cytochrome oxidase has accepted four electrons, one from each of four molecules of reduced cytochrome *c*, both its hemes and both its copper atoms are in the reduced form, and it can transfer the electrons in a series of reactions to a molecule of oxygen to yield two molecules of water.

Cytochrome oxidase is a transmembrane protein in that it straddles the inner membrane of mitochondria, part of it appearing on the matrix side, part being within the membrane, and part on the outer surface or cytochrome *c* side of the inner membrane (Fig. 1). It is closely associated with membrane phospholipids, and this organization is apparently essential for carrying out the second aspect of respiratory chain function, that of conserving the energy of oxidation. *SEE CELL MEMBRANES.*

Chemiosmosis. The energy released during oxidation is utilized to actively pump protons (H^+) from the matrix of the mitochondrion through the inner membrane into the intermembrane space. This creates a proton gradient across the membrane, with the matrix space having a lower proton concentration and the outside having a higher proton concentration. This chemical and potential gradient can be released by allowing protons to flow down the gradient and back into the mitochondrial matrix, thereby driving the formation of ATP.

The enzyme complex involved in this process, ATP synthase, consists of a transmembrane protein assembly termed F_0, which serves as a channel for protons flowing from their region of higher concentration in the intermembrane space, a stalk connecting F_0 and F_1, the assembly carrying out ATP synthesis (originally named coupling factor 1). F_1 contains five types of polypeptides and appears as spheres or knobs tightly aligned along the mitochondrial matrix surface of the inner mitochondrial membrane. There are three sites in the respiratory chain at which protons are pumped into the intermembrane space, namely, at the NADH dehydrogenase, the cytochrome *c* reductase, and the cytochrome *c* oxidase. A pair of electrons flowing down the respiratory chain yields three molecules of ATP, a remarkable feat of energy conservation. This is called the chemiosmotic mechanism of oxidative phosphorylation, which is generally considered a true picture of respiratory chain function.

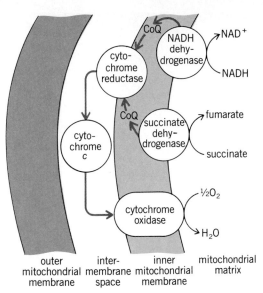

Fig. 1. The respiratory chain and its relation to the mitochondrial membranes. The thick colored arrows indicate the direction of movement of electrons (reducing equivalents); the thin black arrows, the direction of oxidation of the substrates (succinate and NADH) to the products (fumarate and NAD^+) and of the reduction of oxygen to water. The thicknesses of the outer membrane, the intermembrane space, and the inner membrane are all about the same at 7 nm, while that of cytochrome c is about 3.5 nm. The various components of the respiratory chain are indicated. CoQ refers to coenzyme Q.

Normally, the inner mitochondrial membrane is impermeable to protons, which allows the establishment of the crucial gradient. However, a large variety of substances can render it permeable, making it impossible to build up the outside concentration of protons, since they flow back in as soon as they are transported out. Such substances are termed uncouplers, since in their presence the oxidation of substrates no longer results in the formation of ATP; oxidative phosphorylation is uncoupled from respiration, and all of the energy of the process is released as heat. This is what happens when the thyroid gland produces too much of its normal hormones, thyroxine and triiodothyronine; among the many symptoms of the disease thyrotoxicosis are weight loss and fever, both directly related to uncoupling of oxidative phosphorylation. *SEE CELL PERMEABILITY; CHEMIOSMOSIS.*

Cytochrome c. Unlike the inner membrane, the outer mitochondrial membrane is highly permeable to molecules up to molecular weights possibly as high as 10,000. This allows free passage of all substrates, cofactors, and small ions, but keeps within the intermembrane space the several proteins that reside there, including cytochrome *c*, at a molecular weight of 12,500 (Fig. 1). Cytochrome *c* is the only protein member of the respiratory chain that is freely mobile in the mitochondrial intermembrane space. It is a small protein consisting of a single polypeptide chain of 104 to 112 amino acid residues, wrapped around a single heme prosthetic group. The cytochromes *c* of eukaryotes are all positively charged proteins, with strong dipoles, while the systems from which cytochrome *c* accepts electrons, cytochrome reductase, and to which cytochrome *c* delivers electrons, cytochrome oxidase, are negatively charged. There is good evidence that this electrostatic arrangement cor-

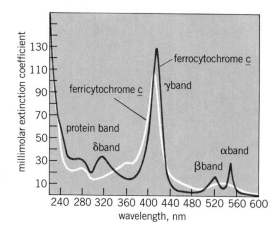

Fig. 2. Optical absorption spectrum of cytochrome c in the reduced form (ferrocytochrome c) and in the oxidized form (ferricytochrome c).

rectly orients cytochrome c as it approaches the reductase or the oxidase, so that electron transfer can take place very efficiently, even though the surface area at which the reaction occurs is less than 1% of the total surface of the protein.

The amino acid sequences of the cytochromes c of eukaryotes have been determined for well over 100 different species, from yeast to humans, and have provided some very interesting correlations between protein structure and the evolutionary relatedness of different taxonomic groups. The yeast protein differs from human cytochrome c at 39 amino acids out of 104, leaving 65 identical amino acids. In the same type of comparison, the human protein differs from chimpanzee cytochrome c by no residues, the rhesus monkey cytochrome c by 1, the horse protein by 11, the fruit fly protein by 26, and the wheat protein by 35. The extensive degree of similarity over the entire range of extant organisms has been taken as evidence for the hypothesis that all living forms are descendants from a single event. Interestingly, the arrangement in space of the polypeptide chain of cytochrome c, often termed the cytochrome c fold, not only is conserved in all the cytochromes c of eukaryotes, from fungi and plants to humans, but is also evident in various types of bacterial cytochromes c which have very different functional activities and amino acid sequences. It is obvious that this is an ancient structure, developed long before the divergence of plants and animals, which in the course of its evolutionary descent has been adapted to serve a variety of electron transfer functions in different organisms. *See* PROTEIN, EVOLUTION OF.

Cytochrome reductase. Like cytochrome oxidase, the cytochrome reductase complex (Fig. 1) is an integral membrane protein system. There are numerous subunits, consisting of two molecules of cytochrome b, one molecule of a nonheme iron protein, and one molecule of cytochrome c_1. Like the hemes of cytochrome oxidase, the hemes of cytochrome b are not covalently bound to the protein chain; they can be separated from the protein without breaking it up. The heme of cytochrome c_1 is covalently bound to the polypeptide chain. Again, as in the case of the oxidase, the two cytochrome b hemes are chemically identical, but are present in somewhat different protein environments and can be distinguished by their spectra. The reductase complex is reduced by reaction with the reduced form of the fat-soluble coenzyme Q, dissolved within the inner mitochondrial membrane, which is itself reduced by the succinate dehydroge-

nase, the NADH dehydrogenase and other systems. *See* COENZYME.

Absorption spectra. The hemes give characteristic spectra to the various cytochromes (**Fig. 2**). In the reduced forms, there are three major heme bands, termed, respectively, from longer to shorter wavelengths, the α, β, and γ bands. The last is also called the Soret band after its discoverer. There is also an absorption band farther in the ultraviolet region of the spectrum which is not related to the heme but is due to the aromatic amino acid residues of the protein, namely, tryptophan, tyrosine, and phenylalanine. The wavelength maximum of the reduced α band is characteristic for each cytochrome. In eukaryotes it is at or very near 605 nanometers for cytochrome oxidase, 562 and 566 nm for the two cytochromes b, 553 nm for cytochrome c_1, and 550 nm for cytochrome c.

It is by following the changes in these absorption bands with purified preparations of the various cytochromes, partially purified preparations, or even intact mitochondria and whole cells, that many of the aspects of respiratory chain function described above were discovered.

Other cytochromes. In addition to the mitochondrial respiratory chain cytochromes, animals have a heme protein, termed cytochrome P450 (because the carbon monoxide complex has its spectral absorption maximum at 450 nm), located in the liver and adrenal gland cortex. In the liver it is part of a mono-oxygenase system that can utilize oxygen and the reduced coenzyme NADPH, to hydroxylate a large variety of foreign substances and drugs and thus detoxify them; in the adrenal it functions in the hydroxylation of steroid precursors in the normal biosynthesis of adrenocortical hormones. *See* ADRENAL GLAND; LIVER.

Two varieties of cytochrome b, termed b_{563} (or earlier b_6) and b_{559}, and one of cytochrome c, c_{552} (earlier called cytochrome f), are involved in the photosynthetic systems of plants. Other plant cytochromes occur in specialized tissues and certain species. *See* BIOLOGICAL OXIDATION; PHOTOSYNTHESIS.

E. Margoliash

Bibliography. R. E. Dickerson, The structure and history of an ancient protein, *Sci. Amer.*, 226(4):58–72, 1972; L. Ernster (ed.), *Bioenergetics*, 1984; T. E. King et al. (eds.), *Oxidases and Related Redox Systems*, 1988; G. W. Pettigrew and G. R. Moore, *Cytochromes c: Biological Aspects*, 1987; M. Wikström et al, *Cytochrome Oxidase*, 1981.

Cytokinesis

The physical partitioning of a plant or animal cell into two daughter cells during cell reproduction. There are two modes of cytokinesis: by a constriction (the cleavage furrow in animal cells and some plant cells; see **illus.**) or from within by an expanding cell plate (the phragmoplast of many plant cells). In either mode, cytokinesis requires only a few minutes, beginning at variable times after the segregation of chromosomes during mitosis (nuclear division). In the vast majority of cases the resulting daughter cells are completely separated. Since they are necessarily smaller cells as a result of cytokinesis, most cells grow in volume between divisions.

Occasionally, cytokinesis is only partial, permitting nutrients and metabolites to be shared between cells through open intercellular channels. Should cytokinesis fail to occur at all, mitosis may cause more than one nucleus to accumulate. Such a cell is a syncytium and demonstrates that mitosis and cytokinesis are separate events. Some tissues normally contain syncytia, for example, binucleate cells in the liver and multinucleate plant endosperm. Some whole organisms such as slime molds are syncytial.

Relations to mitosis. Cytokinesis is precisely and indispensably linked to mitosis, yet the actual mechanisms are entirely distinct. The timing of cleavage furrow or phragmoplast formation with respect to mitotic anaphase has already been mentioned. The plane of cell partitioning is perpendicular to the axis of mitosis and coincides with the plane previously occupied by the chromosomes at metaphase. Despite the reliability of this correlation, the chromosomes themselves are not essential for cytokinesis. Experiments performed upon living cells have shown that it is the cell's machinery for chromosome separation, the mitotic apparatus, that provides the essential positional signal to other parts of the cytoplasm which initiates cytokinesis. Subsequently, the mitotic apparatus is no longer involved in cytokinesis and can be destroyed or even sucked out without affecting cytokinesis. Neither the origin, nature, or immediate target of this signal is known with certainty. *See Chromosome; Mitosis.*

Mechanisms. A cleavage furrow develops by circumferential contraction of the peripheral cytoplasm, usually at the cell's equator. In many cells the cleavage furrow is a completely encircling constriction, but in others, especially certain dividing egg cells, the cleavage furrow forms as a crease at one point on the surface and progresses through the cell from one side to the other. In the former, the mitotic apparatus is centrally located; in the latter, it is excentric, so it signals the initiation of furrowing at one point. This difference in pattern is minor; the actual mechanism of furrowing is very similar among a wide diversity of cell types in lower and higher animals and certain plants.

The physical forces of contraction exhibited by a cleavage furrow are evidently greater than the forces of resistance elsewhere. By inserting extremely delicate flexible needles in the path of a constricting cleavage furrow, the actual force of contraction has been measured. Electron microscopic analysis of the peripheral cytoplasm beneath the cleavage furrow consistently reveals a specialization called the contractile ring. This transient cell organelle is composed of numerous long, thin protein fibers oriented circumferentially within the plane of furrowing. These microfilaments are about 5 nanometers in thickness, appear to attach to the cell membrane, and are known to be composed of actin intermixed with myosin. Both of these proteins are intimately involved in force generation in muscle cells. Thus, the present theory of cytokinesis by furrowing implicates the contractile ring as a transient, localized intracellular "muscle" that squeezes the cell in two. *See Muscle proteins.*

The discovery of the role of the contractile ring in altering cell shape during cytokinesis has been a keystone for understanding the control of cell shape and cell movement in general. It represents one clear example in an expanding list of cellular events in which actin is employed by cells other than muscle cells to determine specific cell behavior.

In plant cells the dominant mode of cytokinesis involves a phragmoplast, a structure composed of fibrous and vesicular elements that resemble parts of the mitotic apparatus. Microtubules (the fibers) appear to convey a stream of small membranous vesicles toward the midline where they fuse into a pair of partitioning cell membranes. Cellulose cell walls are subsequently secreted between these membranes to solidify the separation between daughter cells. This mode of cytokinesis is well suited to plant cells whose stiff cell walls cannot participate in furrowing. Surprisingly, however, there are instances among the algae where cleavage furrows are the normal mode of cytokinesis. Occasionally, both cleavage furrows and phragmoplasts are employed in the same cell. *See Cell walls (plant); Plant cell.*

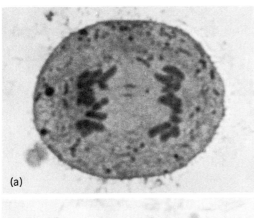

(a)

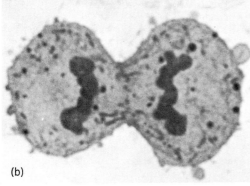

(b)

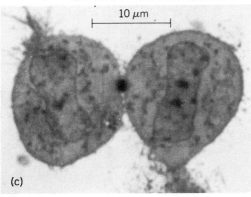

10 μm

(c)

Human cells at three stages of cytokinesis by furrowing, as seen through the light microscope. (*a*) After the end of anaphase, condensed chromosomes (the prominent dark bodies in two groups) have separated about 10 μm, and a barrel-shaped cell outline marks the onset of cleavage furrow formation. (*b*) A few minutes later the cleavage furrow constricts the cell halfway; coincidentally, chromosomes have fused but remain condensed. (*c*) At the end of furrowing, daughter cells are briefly connected by a dense bridge containing remnants of the mitotic apparatus; chromosomes have decondensed, and each daughter cell contains a reformed nucleus.

As in the case of the contractile ring, the exact mechanisms of phragmoplast function are unresolved. Moreover, the ways in which the mitotic apparatus sets up these structures are topics that require further investigation.

Functional significance. Cytokinesis is the terminal step in cell reproduction and offers a variety of benefits to the tissue or organism in which it occurs. Cell propagation is needed to counteract cell attrition through cell death. In growing tissues, cytokinesis increases the number and total surface area of living units, thereby facilitating access to nutrients for enlargement. SEE CELL SENESCENCE AND DEATH.

In early embryonic development, where cell division is not followed by cell growth, cytokinesis has the effect of subdividing the voluminous cytoplasm of the egg. Specialized cytoplasmic substances in eggs are sometimes segregated into particular cells by cytokinesis, thereby permanently channeling their development. In embryos the increase in cell number and in the complexity of positional relationships among cells permits genetically controlled cell differences to emerge. This process of cell differentiation during embryogenesis would not be possible without the metabolic isolation established by cytokinesis. SEE CELL DIFFERENTIATION; EMBRYOGENESIS.

Cytokinesis is sometimes used to sacrifice unneeded nuclei; during meiotic cell division in the ovary, nearly all of the cytoplasm is reserved for one of the daughter cells destined to be the egg. By a process of highly asymmetric cytokinesis, the smaller daughter cell is partitioned off with so little cytoplasm that it cannot survive. Thus, cytokinesis is a fundamental aspect of the biology of cells and achieves a level of diversity among cells that nuclear division alone cannot. SEE CELL (BIOLOGY); CELL DIVISION; MEIOSIS.

Thomas. E. Schroeder

Bibliography. H. W. Beams and R. G. Kessel, Cytokinesis: A comparative study of cytoplasmic division in animal cells, *Amer. Sci.*, 64:279–290, 1976; S. Inoue and R. E. Stephens (eds.), *Molecules and Cell Movement*, 1975; J. Lash and J. R. Whittaker (eds.), *Concepts of Development*, 1974; A. M. Zimmerman and A. Forer (eds.), *Mitosis/Cytokinesis*, 1981.

Cytokinins

A group of plant hormones (phytohormones) that, together with other plant hormones, induces plant growth and development. The presence of an unknown compound in the phloem that stimulated cell proliferation of potato tubers was demonstrated in the early 1900s by G. Haberlandt. In the 1940s J. van Overbeek found that the milky endosperm from immature coconuts is rich in compounds that promote cell proliferation. Then, in 1954, during investigations of the nutritional requirements of tobacco callus tissue cultures, F. Skoog and C. O. Miller discovered a very active substance that stimulated cell division and was formed by the partial breakdown of aged or autoclaved herring sperm deoxyribonucleic acid (DNA). The compound was named kinetin because it induced cytokinesis of cultured tobacco cells. Kinetin has not been found in plants, and it can be formed from a mixture of autoclaved furfuryl alcohol and ad-

enine. In 1961 Miller obtained an active substance that stimulates cell division from the milky endosperm of corn (*Zea mays*) seeds; in 1964 D. S. Letham, identified this active compound and named it zeatin. Cytokinin is now the accepted term for those naturally occurring and synthetic compounds that promote the cell division (cytokinesis) and growth of various plant tissues.

Analogs. The discovery of kinetin stimulated the synthesis of over 150 different species of kinetin analogs. Tobacco pith tissue culture bioassays indicate that most of the highly active cytokinins are N6-substituted adenines containing 5-C alkyl side chains (**Fig. 1**). Two of the most potent synthetic cytokinins are kinetin and N6-benzyl adenine. N,N'-diphenylurea and its derivatives are also considered cytokinins. These compounds induce cell division and growth but are yet to be confirmed as naturally occurring cytokinins.

Since the isolation of the first naturally occurring cytokinin, zeatin, more than 25 different cytokinins have been isolated from plants. A few cytokinin bases such as isopentenyladenine and *cis*-zeatin have been found in transfer ribonucleic acid (tRNA) of various organisms. The most active form of cytokinin is *trans*-zeatin isomer; the *cis* isomer of zeatin, on the other hand, shows little cytokinin activity.

Each cytokinin can exist as a free base, that is, a nucleoside in which a ribose group is attached to the nitrogen atom of position 9 or a nucleotide in which phosphate is esterified to the 5′ carbon of ribose. In most cases, bases have been shown to be more active than their corresponding nucleosides or nucleotides in bioassays. However, it is still not clear what the number of active forms of cytokinins may be, because cytokinin bases can be converted to the corresponding nucleosides and nucleotides in plant cells within a short period of time. SEE BIOASSAY.

A group of cytokinin analogs has been reported to be highly potent inhibitors of the cytokinin-requiring callus tissues. On the basis that growth inhibition was abolished or reduced in the presence of higher concentrations of cytokinin, these analogs are named anticytokinins or cytokinin antagonists. One of the most potent anticytokinins is 3-methyl-7-Nn-pentylamino-pyrazol[4,3-*d*]-pyrimidine (**Fig. 2**).

Biosynthesis and metabolism. Since certain tRNA species contain cytokinins, it has been suggested that turnover of tRNA serves as a source of free cytokinins, but it is now generally accepted that tRNA turnover is not a major source of free cytokinin under normal conditions. Indeed, it has been demonstrated that plant tissues contain enzymes that are capable of the direct synthesis of free cytokinins. The enzyme Δ^2-isopentenylpyrophosphate: AMP-Δ^2-isopentenyltransferase, or isopentenyltransferase, which has been isolated from various organisms, catalyzes the formation of Δ^2-isopentenyladenosine-5′-monophosphate, a cytokinin nucleotide, from Δ^2-isopentenylpyrophosphate and adenosine-5′-monophosphate. Once the cytokinin nucleotide is formed, other enzymes in plant cells convert it to the corresponding nucleoside and base. Oversynthesis of cytokinins in plant tissue causes abnormal growth: crown gall tumor disease caused by the bacterium *Agrobacterium tumefaciens* is an example of excessive production of cytokinins in local tumor tissue. Tissue from crown gall tumors can grow on a simple medium lacking plant hormones because the tumor tissue overproduces both cytokinin and auxin. This is due to the insertion of a piece of bacterial plasmid DNA into the plant nuclear genomes causing activation of the gene responsible for the regulation of cytokinin production. Roots have been shown to be the major site of cytokinin biosynthesis, but stems and leaves are also capable of synthesizing cytokinins. It is possible that all actively dividing cells are capable of cytokinin biosynthesis. *See* Crown gall.

Cytokinins applied to plant tissues are rapidly converted to a variety of metabolites by enzymes which catalyze side-chain modifications, side-chain cleavage, or conjugation of cytokinins with sugars or amino acids. The formation of glycosides and amino acid conjugates are very common metabolic responses of plant tissues to externally applied cytokinins.

Detection. The qualitative and quantitative aspects of cytokinin extracted from plant tissues vary with the types of extraction solvent and the procedure employed, but, in general, the purification and identification of naturally occurring cytokinins has been facilitated by the development of high-performance liquid chromatography, gas chromatography–mass spectrometry, and monoclonal antibodies. The levels of the major cytokinins in plant tissues range from 0.01 to 150 nanomoles per 100 g of fresh tissues. However, most of the reported cytokinin levels in plant tissues are rough estimates, because the various forms of cytokinins are not accurately measured. Certain cytokinins are also present in tRNAs of plants, animals, and microorganisms.

Roles of the cytokinins. Cytokinins exhibit a wide range of physiological effects when applied externally to plant tissues, organs, and whole plants. Exogenous applications of this hormone induce cell division in tissue culture in the presence of auxin. The formation of roots or shoots depends on the relative concentrations of auxin and cytokinin added to the culture medium. High auxin and low cytokinin concentrations lead to root formation, while low auxin and high cytokinin concentrations give shoots. Tissue culture techniques have been employed by plant biotechnologists to grow genetically engineered plant cells into whole plants. Cytokinins appear to be necessary for the correlated phenomena of mitosis and nucleic acid synthesis. Cytokinins delay the aging process in detached leaves by slowing the loss of chlorophyll. Retention of chlorophyll is accompanied by chlorophyll synthesis. Exogenously applied cytokinins are effective on fresh flowers, vegetables, and fruits as postharvest preservatives; however, the use of cytokinins for the preservation of plant materials is not practiced widely.

Cytokinin effects also include breaking of dormancy, promotion of seed germination, stimulation and nutrient mobilization, enhanced anthocyanin and flavanoid synthesis, increased resistance to disease, and stimulation of the opening of stomates. Under some conditions, cytokinin enhances shoot branching and the expansion of cells in leaf disks and cotyledons. In a mammalian system, cytokinin nucleotides have been shown to suppress leukemic cell growth, but the mechanism by which this occurs is unknown. Certain cytokinins located adjacent to the anticodon of tRNA have been shown to influence the attachment of amino acyl tRNA to the ribosome–messenger ribonocleic acid (mRNA) complex and may be in-

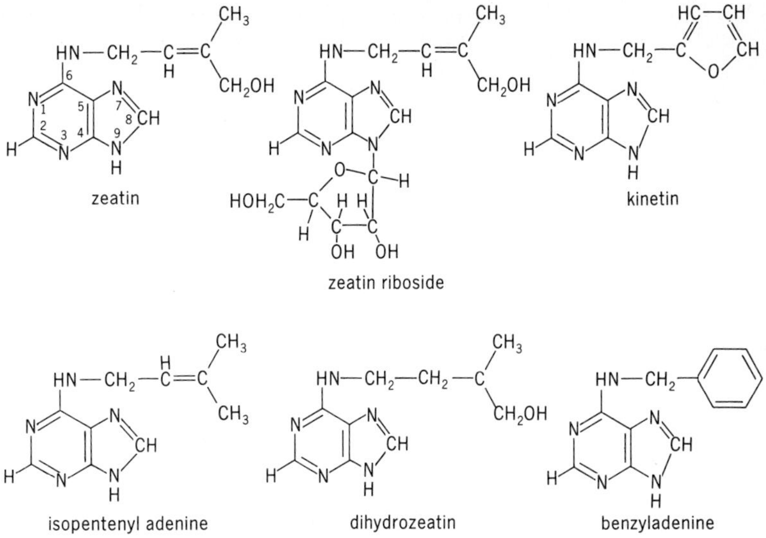

Fig. 1. Structures of naturally occurring (zeatin, zeatin riboside, isopentenyl adenine, and dihydrozeatin) and synthetic (kinetin and benzyladenine) cytokinins. All of these compounds are adenine derivations in which the purine ring is numbered as shown for zeatin. Zeatin can exist in the trans (as shown) or cis (with CH_3 and CH_2OH groups interchanged) configuration.

volved in protein synthesis. *See* Auxin; Dormancy; Plant physiology; Tissue culture.

Cytokinin-binding molecules. It is a widely held view that there are macromolecules such as proteins which interact between cytokinin molecules and the cytokinin-responsive genes. According to this hypothesis, the initial event of hormone action involves the binding of a hormone to a receptor, with the hormone–receptor complex then binding to a specific region of gene. There have been numerous papers reporting the isolation of cytokinin-specific and relatively high-affinity cytokinin-binding proteins. However, the physiological role of any of the reported cytokinin-binding proteins is unknown, and these proteins cannot yet be considered to be cytokinin receptors.

Cytokinin-modulated gene expression. The mechanism of action of cytokinin on plant growth and development is poorly understood. It has been demonstrated that specific proteins are induced, enhanced, reduced, or suppressed by the hormone. Some of the enzymes or proteins regulated by cytokinins have been identified. The activities of acid phosphatase, ribulose 1,5-bisphosphate carboxylase, invertase, light-harvesting chlorophyll *a/b*-binding protein, and hydroxypyruvate reductase are all enhanced by the hormone, but the details of how these enzymes are

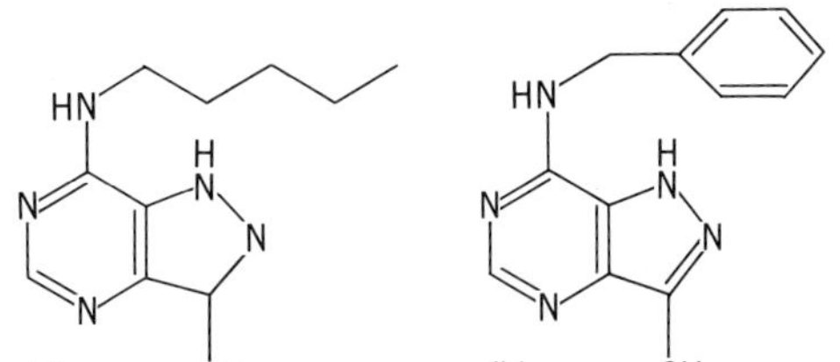

Fig. 2. Structures of two synthetic cytokinin antagonists or anticytokinins: (*a*) 3-methyl-7(pentylamino)pyrazolo(4,3-*d*)pyrimidine and (*b*) 3-methyl-7(benzylamino) pyrazolo(4,3-*d*)pyrimidine.

stimulated by cytokinins remain to be studied. Applications of cytokinin to cytokinin-responsive tissues also regulate the synthesis of specific mRNAs; some of the mRNAS are induced and some are suppressed. By using modern techniques, plant molecular biologists are able to show that mRNA changes occur within 60 min of cytokinin application to plant cells. Initial evidence suggests that cytokinins regulate, at least in part, the transcriptional process of gene expression by turning on or off specific genes and stimulating or suppressing the synthesis of specific mRNAs. Another possible action of cytokinins is the regulation of a posttranscriptional process such as stabilization of mRNA. Scientists also demonstrated that cytokinins specifically increase the rate of protein synthesis and the effect seems to be on the translation of mRNA into proteins. SEE PLANT GROWTH; PLANT HORMONES.

Chong-maw Chen

Bibliography. S. D. Kung and C. J. Arntzen (eds.), *Plant Biotechnology*, 1988; D. S. Lethan and L. M. S. Palni, The biosynthesis and metabolism of cytokinins, *Annu. Rev. Plant Physiol.*, 34:163–197, 1983; H. F. Linskens and J. F. Jackson (eds.), *Modern Methods of Plant Analysis, New Series*, vol. 5, 1987; C. O. Morris, Genes specifying auxin and cytokinin biosynthesis in phytopathogens, *Annu. Rev. Plant Physiol.*, 37:509–538, 1986; F. Skoog and D. J. Armstrong, Cytokinins, *Annu. Rev. Plant Physiol.*, 21:359–384, 1970; M. B. Wilkins (eds.), *Advanced Plant Physiology*, 1984.

Cytolysis

An important immune function involving the dissolution of certain cells. There are a number of different cytolytic cells within the immune system that are capable of lysing a broad range of target cells. The most thoroughly studied of these cells are the cytotoxic lymphocytes, which appear to be derived from a wide range of cells and may employ a variety of different lytic mechanisms. Cytotoxic cells are believed to be essential for the elimination of oncogenically or virally altered cells, but they can also play a detrimental role by mediating graft rejection. There are two issues regarding cytotoxic lymphocytes that are of concern: one is the target structure that is being recognized on the target cell, that is, the cell that is killed, which triggers the response; and the other is the lytic mechanism. SEE CELLULAR IMMUNOLOGY.

Cytotoxic lymphocytes. When freshly isolated, large granular lymphocytes from peripheral blood are tested in cytotoxicity assays, they spontaneously lyse certain tumor cells. These cytotoxic cells are called natural killer cells and are probably essential for immune surveillance. That is, these cells may be responsible for lysing spontaneously occurring tumors in the body. They are unique in that no previous sensitization is required for them to kill. It is not known what structure on the target cell must be recognized in order for the lytic event to be triggered, but a number of different molecules have been implicated.

Another killer cell, called the lymphokine-activated killer cell, is a product of culturing large granular lymphocytes in relatively high concentrations of the thymus-derived lymphocyte (T-cell) growth factor. It appears as though these cells might be derived from natural killer cells. They lyse any target cell, including cells from freshly isolated tumors, and are being employed in cancer therapy. Lymphokine-activated killer cells may also be important in mounting a vigorous response under conditions of extreme immunological stress. Very little is known about the mechanisms by which these cells recognize and lyse the target cell.

The last group of cytotoxic cells is the cytotoxic T lymphocyte. These are T cells that can lyse any target cell in an antigen-specific fashion. That is, as a population they are capable of lysing a wide range of target cells, but an individual cytotoxic T lymphocyte capable of lysing only those target cells which bear the appropriate antigen. In general cytotoxic T lymphocytes recognize the major histocompatibility complex class I antigens alone, for example, in graft rejection, or foreign antigens, for example, viruses in the context of self class I molecules. These are truly immune cells in that they require prior sensitization in order to function. These cells are thought to mediate graft rejection, mount responses against viral infections, and play a major role in tumor destruction. SEE ANTIGEN; HISTOCOMPATIBILITY.

Degranulation model. Very little is known about how cytotoxic cells kill their target cells, but many models have been proposed to explain the lytic mechanism. The dominant model is that cytotoxic cells release the contents of cytotoxic granules after specific interaction with the target cell (**Fig. 1**). After target cell binding, the Golgi apparatus, which is an organelle that prepares proteins for secretion, and the granules containing the cytolytic molecules orient toward the target cell in preparation for secretion. Vectorially oriented granule exocytosis then occurs, with the release of cytolytic molecules into the intercellular space between the killer and target cells. This results in the leakage of salts, nucleotides, and proteins from the target cell, leading to cell death. SEE GOLGI APPARATUS.

The best characterized of these lytic molecules is variously called perforin, cytolysin, or pore-forming protein. This molecule lyses a variety of tumor cells

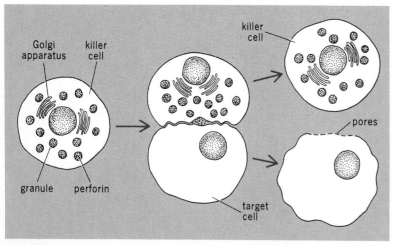

Fig. 1. Degranulation model of killing. The killer cell binds to the appropriate target cell, after which the Golgi apparatus and the cytolytic granules migrate toward the target cell. The cytolytic granules then fuse with the external membrane, releasing the pore-forming molecules into the intercellular space. The pores polymerize in the target cell membrane, resulting in osmotic lysis of the target cell. The killer cell goes on to repeat the entire process. (*After J. D.-E. Young and Z. A. Cohn, Cell-mediated killing: A common mechanism, Cell, 46:641–642, 1986*)

as well as erythrocytes. It is somewhat similar to the lytic pore-forming C9 component of the complement system and has an apparent molecular mass of about 70,000 daltons. In the presence of the calcium ion (Ca^{2+}), this protein polymerizes into pores or ring structures, which can be visualized by electron microscopy of the target cell membranes. Calcium ions appear to be required for both insertion of the pore-forming protein into membranes and its polymerization into pores.

Another molecule that has been found, at least in cytotoxic T lymphocytes, is similar to the tumor necrosis factor, a cytolytic molecule of macrophages that lyses cells in a Ca^{2+}-independent way although at a relatively slow rate. It is found within the cytolytic granules, but it is not clear what role, if any, this molecule plays in target cell lysis.

Several unique serine esterases have also been identified in cytotoxic T lymphocytes, both at the protein level and the gene level. These enzymes occur in the cytolytic granules, presumably along with the pore-forming proteins. They are released when a cytotoxic cell recognizes its specific target cell and can serve as a convenient measure of degranulation. The function of these serine esterases is unknown, but they may be involved in the activation or polymerization of the pore-forming protein much like similar enzymes that have been implicated in the activation of the complement cascade.

Overall, there are a number of observations that support the degranulation model. Cytotoxic lymphocytes contain numerous, electron-dense secretory granules in their cytoplasm. These granules and the Golgi apparatus reorient toward the target cell upon contact, after which the granules appear to be released. In addition, granules isolated from cytotoxic clones contain cytolytic molecules as determined by their lytic capacity and by the lesions they form on the target cell membrane, and are able to lyse tumor targets. Target cells killed by direct cytotoxic T lymphocyte–mediated lysis have also been claimed to have circular pores in their membranes, although reports are conflicting.

The degranulation model has been accepted because it explains how the target cell is lysed, it incorporates antigen specificity, and it is similar to the mechanism employed in other systems, including those used by bacterial and yeast toxins. There are, however, a number of problems with the secretion model, at least in cytotoxic T lymphocytes. For instance, some researchers have been unable to detect perforin or serine esterase activity in primary cytotoxic T lymphocytes in the living organism. In addition, a number of cytotoxic T-lymphocyte cloned cell lines have granules that contain serine esterase, but do not have significant levels of pore-forming molecules. If pore formation is the primary mechanism of killing, it is not clear why these cytotoxic T lymphocytes do not contain detectable levels of the activity. It could be argued that very little is required, so that even though there is no detectable activity, there may be sufficient levels to mediate lysis of a single target cell.

It has been observed that when a target cell dies after cytotoxic T-lymphocyte interaction, nuclear damage with rapid deoxyribonucleic acid (DNA) fragmentation precedes detectable plasma membrane damage. If perforin were the primary mechanism, plasma membrane damage should be the first detect-

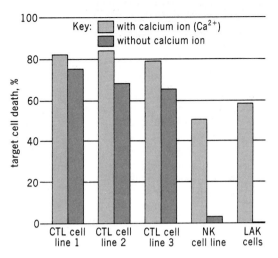

Fig. 2. Cytotoxic T lymphocyte–mediated killing but not natural killer– or lymphokine-activated killer mediated killing in the absence of extracellular calcium ion (Ca^{2+}).

able event in the target cell. Purified granule extracts are able to lyse target cells, but they do not cause DNA fragmentation within the cell. It has also been demonstrated that a mutation in a cytotoxic T lymphocyte–sensitive target cell can render it resistant to lysis, even though the cell is still recognized by the cytotoxic T lymphocyte and can still be lysed by the granule contents. A revertant could be isolated that could again be lysed by the cytotoxic T lymphocyte. These results suggest that there is a molecule on target cells that controls the ability of the cells to undergo lysis but that has not been accounted for in the lytic mechanism described so far.

Probably the most convincing argument against degranulation as the sole lytic mechanism, at least in cytotoxic T lymphocytes, is the observation that killing by cytotoxic T lymphocytes can occur in the complete absence of extracellular calcium ion (**Fig. 2**). This is difficult to reconcile with the degranulation model because calcium ion is required for both the actual degranulation event and the polymerization of the pore-forming proteins in the target cell membrane. Taken together, these observations bring into question the role of these pore-forming cytolytic molecules in cytotoxic T lymphocyte–mediated lysis.

Programmed cell death model. If primary cytotoxic T lymphocytes do not use the degranulation mechanism, they must have another way of killing. A number of models have been put forth, but one of the most attractive is the programmed cell death model. In this model, a cytotoxic T lymphocyte merely triggers an endogenous suicide pathway within the pathogenically altered cell so that it goes on to kill itself by a mechanism that produces nuclear disintegration and DNA fragmentation prior to plasma membrane damage. The target cell plays an active role in its own demise, which is consistent with much of the data available on cytotoxic T lymphocyte–mediated killing. It would make sense for all of the cells in the body to have a built-in self-destruct mechanism so that if a cell became pathogenically altered, it could be easily triggered to die. SEE IMMUNOLOGY.

H. L. Ostergaard

Bibliography. G. Moller (ed.), Molecular mechanisms of T cell–mediated lysis, *Immunol. Rev.*, 103:38–51, 1988; H. L. Ostergaard et al., Cytotoxic T lymphocyte mediated lysis can occur in the absence of serine esterase release, *Nature*, 330:71–72, 1987;

J. H. Russell, International disintegration model of cytotoxic lymphocyte–induced target damage, *Immunol. Rev.*, 72:97–117, 1983; J. D.-E. Young and Z. A. Cohn, How killer cells kill, *Sci. Amer.*, 258:38–44, 1988.

Cytomegalovirus infection

An illness also known as cytomegalic inclusion disease. In young infants clinical cases are characterized by jaundice, enlargement of liver and spleen, and blood and circulatory disturbances. Cytomegalovirus infection appears in its most severe, and sometimes fatal, form in infants under 2 years of age, who became infected before birth or when newborn. In the congenital infection the virus may persist even in the presence of antibody, an unusual situation similar to that found in congenital rubella infection. Mild or inapparent infection is common during late childhood and adolescence; it is recognized by the detection of the virus, which may be excreted in the urine for months. The virus may remain hidden for years, but its reappearance is brought on by immunosuppressive drugs, such as those given to persons receiving organ transplants. SEE ANTIBODY; RUBELLA; TRANSPLANTATION BIOLOGY.

The virus, a member of the herpesvirus group, produces enlargement of cells in the affected organs or in tissue cultures, as well as large, intranuclear inclusion bodies and sometimes intracytoplasmic inclusions. SEE HERPES; VIRAL INCLUSION BODIES.

Several animal species (monkeys, mice, guinea pigs) also have species-specific cytomegaloviruses. The virus of the monkey is related to that of humans. SEE ANIMAL VIRUS.

Joseph L. Melnick

Bibliography. M. Benyesh-Melnick, Cytomegaloviruses, in N. J. Schmidt and E. H. Lennette (eds.), *Diagnostic Procedures for Viral and Richettsial Diseases*, 4th ed., 1969; H. Frankel-Conrat and P. Kimball, *Virology*, 1982.

Cytoplasm

That portion of living cells bordered externally by the plasma membrane (cell membrane) and internally by the nuclear envelope. In the terminology of classical cytology, the substance in living cells and in living organisms not compartmentalized into cells was called protoplasm. It was assumed at the time that the protoplasm of various cells was similar in structure and chemistry. Results of research on cell chemistry and ultrastructure after about 1960 showed that each cell type had a recognizably different "protoplasm." Primarily for that reason, the term protoplasm gradually fell into disuse in contemporary biology. The terms cytoplasm and nucleoplasm have been retained and are used descriptively; they are used almost synonymously with the terms cytosome (body of cytoplasm) and nucleus, respectively.

Observation by optical microscopy. Optical microscopes are limited in resolving power at the level of about 0.2 micrometer. Therefore only the larger structures in the cell could be observed and named by early cytologists. Structures visible in the light microscope include the nucleus, with its chromosomes and nucleolus, and the cytoplasm, with its larger organelles such as chloroplasts (in plants only), mitochondria, Golgi bodies, and centrosome (containing two centrioles), as well as inclusions such as lipid droplets, pigment granules, yolk (in eggs), starch grains, and vacuoles. The main impression that early microscopists gained from observations of cells with an ordinary bright-field microscope was that cells were transparent and replete with particles of different types, and that many exhibited motion of some kind, usually Brownian motion (thermally induced collisions between molecules and granules in the cell).

Modern optical microscopes operate in modes that produce vastly improved contrast. Dark-field and phase-contrast microscopy greatly increases the visibility of cytoplasmic particles of all kinds. Interference microscopes show the shapes of cells by representing differences in optical path (refractive index difference times the thickness) in brightness contrast or color contrast. Differential interference microscopes render all cytoplasmic surfaces shadowcast, and provide a more detailed view of the various interfaces in cells caused by the existence of organelles and inclusions, vacuolar membrane systems, and various kinds of filamentous structures. Polarizing microscopes reveal ordered or crystalline structures (such as the mitotic spindle, stress fibers, and myofibrils) when they are doubly refracting, or birefringent. Fluorescence microscopes reveal naturally fluorescent substances (fluorochromes) such as chlorophyll in plants, or fluorochrome-labeled antibodies to specific protein molecules. These optical probes have made it possible to discover the location of enzymes, hormones, and structural proteins in the cytoplasm, sometimes even while cells are alive. SEE MICROSCOPE.

Regional differentiation. Many cells, especially the single-celled organisms or protistans, have regional cytoplasmic differentiation. The outer region is the cortex or ectoplasm, and the inner region is the endoplasm. In some echinoderm eggs, for example, endoplasmic organelles and inclusions can be sedimented at moderate centrifugal accelerations, while cortical granules that are denser than most other particles remain fixed in the cortex. In many cases the cortical layer is a gel made up of a meshwork of cytoskeletal fibers, especially F-actin. In many ciliates the cortical layer contains organelles peculiar to that region. For example, the cortex of *Paramecium* contains the basal bodies of cilia and a system of fibrils associated with them, as well as rows of trichocysts which are fired like tiny javelins when the *Paramecium* is injured. In amebae, movement requires that the cell maintain an outer ectoplasmic gel surrounding the more fluid flowing endoplasm, which nevertheless also has some gellike properties.

Apparently the cytoplasm of nearly all cells is contractile when stimulated. In some cells the machinery of contraction is localized in the cortex (for example, the ciliates *Stentor*, *Spirastomum*, and *Vorticella;* the flagellates *Euglena* and *Peranema;* the testacean *Difflugia;* and many sporozoans).

Centrifugal sedimentation of cytoplasm. The organelles and inclusions of some cells can be caused to sediment into layers based on their density by the application of centrifugal acceleration. Particles with high density are sedimented to the centrifugal pole, while less dense (buoyant) particles are moved to the centripetal pole. In classical experiments by E. N. Harvey and E. B. Harvey in the 1930s, it was shown that eggs of the sea urchin *Arbacia* could be stratified while under observation on a cushion of sucrose solution in a special microscope-centrifuge, actually a centrifuge with the microscope built into it. If the eggs were centrifuged while in a sucrose density gradient, they were pulled into two halves by the force caused by the centrifugal acceleration acting upon the density difference between the upper and lower halves of the egg and the local density in the gradient.

Centrifugation experiments clearly showed that organelles and inclusions fall into classes differentiated by their density, and focused attention on the properties of the ground cytoplasm (or hyaloplasm) in which the organelles and inclusions were bathed. The early centrifugation experiments led to later efforts directed at measuring the consistency of the ground cytoplasm, and to isolation and purification of fractions of cytoplasmic particles. *See Centrifugation*.

Observation by electron microscopes. The fine structure of cells could not be studied until the development of electron microscopes and a technique for cutting very thin sections (slices) of cells embedded in a suitable plastic. The first ultramicrotome for slicing cells embedded in plastic was developed by K. R. Porter. The scanning electron microscope (SEM) permits a detailed view of the cell surface or into a cell that has been opened or rendered partially transparent by treatment with electrically conductive fixative compounds. The resolution of the SEM is limited at between 2 and 5 nanometers, depending on instrument design. *See Scanning electron microscope*.

The conventional transmission electron microscope (CTEM or TEM) is used to produce images and electron micrographs of thin sections (about 40–60 nm thick). The resolving power of the TEM is about 0.2 nm, about a thousand times greater than any light microscope.

The high-voltage electron microscope (HVEM) is similar in design to the TEM but operates with an electron beam of $1-3 \times 10^6$ eV, about 10 times greater voltage than the CTEM. The higher penetrating power and reduced specimen damage at higher voltages make it possible to obtain stereo images of thick sections or whole cells up to 2–5 micrometers in thickness. The HVEM has been very important in obtaining a comprehensive view of the relationships of parts of cells not appreciated from thin sections. *See Electron microscope*.

Cytoplasmic ultrastructure. Electron microscopy revolutionized concepts of cell structure and function. Organelles too small to be seen clearly with light microscopes were discovered, named, and characterized after 1955. Organelles previously known were characterized as to their detailed inner morphology.

Centrioles. The two centrioles within the centrosome or central body of the cytoplasm appear as barely visible spherical dots in the optical microscope. In the electron microscope the two centrioles can be seen to be short bundles of 27 tubules arranged in a characteristic pattern, each bundle perpendicular to the other. Centrioles are present in most animal cells and absent in most plant cells, and are organizing centers for the microtubular structures in the mitotic spindles of dividing animal cells and in cilia and flagella. The basal bodies of the latter are related to centrioles. *See Centriole*.

Endoplasmic reticulum. In cells known to synthesize proteins for secretion (for example, cells of the pancreas), the cytoplasm just outside the nucleus was known from optical microscopy to be basophilic. This region of the cell, called the ergastoplasm, has an affinity for basic stains long known to mark the presence of nucleic acids. In the electron microscope, this region of the cell was found to contain massive amounts of rough endoplasmic reticulum consisting of stacks of flat membranous cisternae covered with ribosomes, particles of ribonucleoprotein which play an important role in protein synthesis. Face-on views of the endoplasmic reticulum show the ribosomes in lines connected by delicate strands of messenger ribonucleic acid, macromolecules that serve as an intermediate carrier of genetic information for the synthesis of specific protein molecules. The rough endoplasmic reticulum has been shown to be a temporary repository for proteins to be exported from the cell or incorporated into membranes. Smooth endoplasmic reticulum is more likely to be tubular than flat, and lacks attached ribosomes; it is involved in the synthesis and storage of other macromolecules. *See Endoplasmic reticulum; Ribosomes*.

Golgi complex. The Golgi complex is another class of organelles consisting of stacks of membranous sacs or cisternae that are slightly curving in shape. The Golgi complex serves to package proteins that the cell will secrete. The edges of the Golgi cisternae break off and form Golgi vesicles containing the secretions. *See Golgi apparatus*.

Mitochondria. Mitochondria, usually rod-shaped particles 0.5–3 μm long, consist of two membranes, the inner one thrown into folds called christae, on which are located assemblies of respiratory pigments and enzymes. Mitochondria are often spoken of as the powerhouses of the cell. Their matrix surrounding the inner membrane folds contains the enzymes that catalyze the oxidative phase of glycolysis (carbohydrate breakdown). Mitochondria each contain a circular strand of deoxyribonucleic acid (DNA), which contains the genetic information coding for some, but not all, of the mitochondrial enzymes and structural proteins. *See Mitochondria*.

Lysosomes. Lysosomes are cytoplasmic particles in the same size range as mitochondria, and can be recognized by an electron-dense matrix with no folded inner membrane. Lysosomes differ in density from mitochondria, and thus can be separated from them by density gradient centrifugation. Lysosomes contain several hydrolytic enzymes that can catalyze the breakdown of various classes of macromolecules. When cells ingest foreign material (endocytosis), lysosomes fuse with the vacuole containing the ingested material and supply enzymes required for its intracellular digestion. *See Endocytosis; Lysosomes*.

Microbodies. Another class of cytoplasmic particles are the microbodies, consisting of peroxisomes and glyoxisomes, particles that contain different groups of oxidative enzymes.

Chemical deposits. Cells contain inclusions that are not functional entities but, rather, deposits of certain chemicals. Many cells contain small droplets of fat or oil that are buoyant and can be displaced by centrifugation to the centripetal pole of the cell. Certain large amebas contain bipyrimidal or flat crystals of triuret, a nitrogen excretion product of purine and pyrimidine metabolism. In the same cells are many so-called refractile bodies containing calcium and magnesium polyphosphate compounds. Plant cells contain starch particles which can be broken down to glucose to obtain chemical energy; many animal cells contain glycogen deposits that play a similar role. The inclusions found in various cell types are more numerous and varied than the organelles. They can be removed by centrifugation and microsurgery without harming the cell, and are therefore nonessential to life.

Role in cell division. When the cell divides, a new organelle, the mitotic spindle, is formed, which divides the replicated chromosomes into equal groupings for the daughter nuclei to be formed. In some cells the mitotic spindle forms, at least in part, while the chromosomes are still within the nuclear envelope. In other cells the spindle itself is entirely intranuclear.

Cytoplasmic division (cytokinesis) in animal cells and many protists is accomplished by a contractile ring of microfilaments composed of the protein F-actin. Contraction involves the interaction of F-actin with another protein, myosin. In plant cells cytokinesis is accomplished by the transport of Golgi vesicles containing precursors of the new cell plate along microtubules of the phragmoplast, a reorganized remnant of the mitotic spindle. SEE CELL DIVISION.

Cytochemistry. Cytoplasm contains mostly water, from 80 to 97% in different cells, except for spores and other inactive forms of living material, in which water may be present in lesser amounts. The dry mass of cells consists mainly of macromolecules: proteins, carbohydrates, nucleic acids, and lipids associated with membranes. The small molecules present in cells are mainly metabolites or metabolic intermediates. The principal ions other than the hydrogen and hydroxyl ions of water are the cations of potassium, sodium, magnesium, and calcium, and the anions chloride and bicarbonate. The concentrations of these ions are maintained within narrow physiological limits by ion-pumping mechanisms and the presence of natural chelators or ion-binding substances. Therefore analyses of whole cells for calcium ion concentration often give values two to three orders of magnitude higher than the concentration of free calcium ions in cytoplasm.

Many other elements are present in cytoplasm in smaller amounts. Iron is found in cytochrome pigments in mitochondria; magnesium is present in chlorophyll in chloroplasts; copper, zinc, iodine, bromine, and several other elements are present in trace quantities.

The presence of certain chemical compounds in the cell can be measured quantitatively by spectroscopic techniques applied either to living cells or to fixed cells subjected to chemical reactions in place. Cytochrome pigments, for example, can be demonstrated in living mitochondria by their absorption bands in the blue region of the spectrum. The same holds for different forms of chlorophyll and accessory photosynthetic pigments in plant cells. Nucleic acid absorption can be detected in the ultraviolet region of the spectrum.

By carrying out chemical reactions in fixed cells, it is possible to localize proteins, carbohydrates, and the two basic types of nucleic acids, DNA and ribonucleic acid (RNA). Enzymes can be localized by chemical reactions that demonstrate the product of the reactions they catalyze.

One of the most powerful techniques in cytochemistry is cell fractionation. Cells are first homogenized or otherwise disrupted in a medium that keeps organelles and inclusions intact and, if possible, functional. The resulting suspension of cell organelles and inclusions is centrifuged first at low accelerations to remove nuclei and other heavy particles, and then the homogenate is centrifuged at a higher acceleration in a manner designed to separate particle classes on the basis of density. In this way it is possible to obtain a nuclear fraction, a mitochondrial fraction, and so on. The purity of each fraction is monitored by optical or electron microscopy, and when the desired criterion of purity has been met, fractions are submitted to chemical analysis. In this way it has been possible to construct lists of substances, especially enzymes, known to be present in different organelles or cell compartments.

The many proteins that lack enzymatic function may be regulatory or structural in function. The best way to identify and localize these is by fluorescent-antibody techniques. A given protein is first isolated and purified. Repeated injections of the purified protein (an antigen) into rabbits causes the production of a rabbit antibody to that protein. Antibody isolated from the blood serum of the rabbit and subsequently purified is then used as a probe to look for the antigenic protein by using fluorescent antibodies to rabbit serum proteins produced by some other animal, such as a goat or sheep. This technique is known as the indirect immunofluorescent technique or the "sandwich" technique. SEE CYTOCHEMISTRY; IMMUNOFLUORESCENCE.

Cytoplasmic rheology. In the first two decades of this century a controversy arose as to whether cytoplasm was fluid or gel. Efforts to resolve the controversy by objective physical experiments were not at first successful, because there was insufficient knowledge of the complexity of cytoplasmic organization, and because certain "simplifying" assumptions turned out to be misleading.

Fluids in which the rate of shear (flow) is proportional to the applied shear stress (force) are called newtonian fluids. Therefore fluids with a nonlinear force-flow relationship are designated non-newtonian. The slope of the force-flow relationship for a fluid is defined as its fluidity, while the reciprocal of the slope is viscosity, a measure of fluid resistance. Fluids with nonlinear force-flow relations obviously exhibit different slopes and therefore different viscosities at different rates of shear; that is, they exhibit a different fluid resistance at different rates of flow. Thixotropy is a reduction in viscosity at higher flow rates, and indicates a breakage of affinities or bonds contributing to fluid resistance. Dilatancy is the opposite effect: an increase in resistance with increased rate of flow. SEE BIORHEOLOGY.

The first objective methods used in an attempt to gain quantitative information about the consistency of cytoplasm relied on the assumption, which seemed reasonable at the time, that cytoplasm might be a newtonian fluid. Since this assumption disregarded the possibility that cytoplasm might be structured, the fact that the methods used applied forces that destroyed cytoplasmic structure was overlooked.

It is now clear that cytoplasm is highly nonlinear in rheological behavior due to the presence of cytoskeletal proteins in the form of microfilaments, microtubules, intermediate filaments, and various other cytoskeletal proteins in dynamic association with one another. These skeletal elements can confer sufficient structure on cytoplasm in some cells so that a measurable yield force is required to cause any flow at all. The presence of a yield point classifies any fluid as a gel. Some gels can be strong, as in the case of gelatin or agar, while other cytoplasmic gels are weak in comparison. Even weak gel structure is important, because the forces acting upon the cytoplasm are also weak. Cytoplasm in some cells also shows the property of viscoelasticity. Viscoelastic fluids have a "memory" for their original shape, and when deformed tend to recoil toward their original shape.

It is very difficult to characterize the rheological properties of cytoplasm quantitatively, because they depend on the force applied in order to make the measurement. Reliable viscosity values have been found for cytoplasm in the range of 5 centipoise to 10 poise or 0.005 to 1.0 pascal · second (water has a viscosity of about 1 centipoise or 0.001 Pa · s).

In some cells it has been possible to differentiate between the macroviscosity, the viscosity of the cytosol and cytoskeleton, and the microviscosity, the viscosity of the cytosol alone. The former is always an order of magnitude (or more) more viscous than the latter.

Important experiments carried out by Y. Hiramoto showed that the yield point and viscoelastic constants of cytoplasm could be measured by the displacement of an iron sphere inside a cell by an electromagnet. These studies showed the complex nature of the rheological properties of cytoplasm, and the fact that these properties can change with the life cycle of the cell.

Ground cytoplasm: cytosol and cytoskeleton. Sedimentation of cells by centrifugation shows that organelles and inclusions can be separated from the ground cytoplasm, the fluid phase of the cytoplasm in which they are suspended. The ground cytoplasm in turn has been shown to consist of a cytoskeletal network and the cytosol, the fluid in which the cytoskeleton is bathed.

The cytoskeleton consists of several biopolymers of wide distribution in cells. Microtubules have been observed in electron micrographs of a vast number of different cell types. They are about 25 nm in diameter and very long. They consist of the protein tubulin, and are frequently covered by a fuzzy layer of microtubule-associated proteins. Microfilaments of more than one type have been observed. The most common and ubiquitous are those composed of F-actin, one of the most common proteins in muscle of all kinds. Microfilaments are usually 6–7 nm in diameter and several micrometers long. If composed of F-actin, they can be decorated in a characteristic arrowhead pattern by the S-1 subfragment of skeletal muscle myosin to reveal their polarity. Intermediate filaments of several types are known. Most have a diameter of about 10 nm, and are sometimes called 10-nm filaments. At least some of these intermediate filaments are related to the protein keratin, which is present in skin, nails, hair, and horns.

Many proteins have been found that appear to have either a regulatory or cross-linking (gel-forming) function. They bind to one or more of the cytoskeletal proteins and act to control contractility, motility, or cytoplasmic consistency. The patterns of distribution of cytoskeletal proteins have been demonstrated by fluorescent antibody probes which show the presence of a given specific protein by labeling it with a fluorescent marker using the specificity of an antigen-antibody reaction. SEE CYTOSKELETON.

Cytoplasmic motility. In most cells the smaller particles exhibit brownian motion due to thermal agitation. In some cells lacking extensive cytoskeletal structure, particles can be moved freely around the cell by brownian motion. In others they are restricted by the cytoskeletal elements. SEE BROWNIAN MOVEMENT.

Particles of various types may also undergo saltatory motions which carry them farther than brownian motion possibly could. Such excursions are usually either linear or curvilinear, and result from the interaction of a particle with an element of the cytoskeleton such as one or more microtubules or microfilaments. Single particles or trains of particles may exhibit saltation, a jumping motion.

When masses of particles move together, they are usually accompanied by a flow of the cytoplasm around them called cytoplasmic streaming. Streaming of cytoplasm is best seen in cells of *Nitessa* or *Chara* (green algae), in amebas and paramecia (protists), in cellular slime molds (fungi), and in many cells (for example, eggs and leukocytes) of higher animals. There is evidence to indicate that cytoplasmic streaming involves the interaction of the proteins F-actin (microfilaments) and myosin. However, the manner of interaction and the cellular regulatory systems appear to differ widely. The movement of cells of both protists and higher organisms involves contractile, cytoskeletal, and regulatory proteins which exhibit differences from species to species and from cell to cell. SEE CELL MOBILITY.

Perhaps the most intricate and important form of motility is that which occurs in neurons (nerve cells) of all types: axoplasmic transport. Neurons synthesize transmitter substances near their nuclei in the cell body. Vesicles containing these transmitter substances must be transported long distances (up to 6 ft or 2 m in humans) in order to be released at the synapses where one nerve cell excites the next. Anterograde axoplasmic transport brings the vesicles to the nerve endings, whereas retrograde transport carries chemical messages back to the nerve cell body. Failure of this transport mechanism in certain diseases (for example, Alzheimer's disease and senility of old age) is due to the failure of this important process. SEE ALZHEIMER'S DISEASE.

Muscular movement is made possible by virtue of paracrystalline interdigitating arrays of F-actin thin filaments and myosin thick filaments in the myofibrils in the cytoplasm of muscle fibers. Muscle contraction is a highly specialized form of contractility, which is one of the basic properties of cytoplasm. SEE CELL BIOLOGY; MUSCLE; MUSCLE PROTEINS.

Robert Day Allen

Bibliography. B. R. Brinkly and K. R. Parta, *International Cell Biology*, 1977; C. De Duve, *A Guided Tour of the Living Cell*, 1985; E. D. P. De Robertis and E. M. De Robertis, Jr., *Cell and Molecular Biology*, 8th ed., 1987; G. Karp, *Cell Biology*; 2d ed., 1984; S. Seno and Y. Okada, *International Cell Biology 1984*, 1987.

Cytoplasmic inheritance

The control of genetic differences by hereditary units other than those carried in the nucleus; also known as extrachromosomal inheritance. SEE GENETICS.

Nonmendelian patterns. Detection of cytoplasmic control requires the demonstration of inheritance not following the chromosomal or mendelian rules.

Maternal influence. Transmission of a character difference only through the maternal line is exemplified by the uniparental transmission of chloroplast deficiency in certain white-green variegated mutants in flowering plants and by the inheritance of carbon dioxide sensitivity in *Drosophila* through the female. Cases such as these have led to the conclusion that cytoplasmic genetic factors are carried in the egg cell and that pollen grains on the one hand and spermatozoa on the other contribute mainly or solely a nucleus at fertilization. Transmission through the female line has also been described in mammals, including humans, mainly with respect to certain myopathies (muscle diseases). *See* FERTILIZATION; MATERNAL INFLUENCE; MENDELISM.

Many examples of cytoplasmic inheritance occur in microorganisms, the best known being the respiratory-deficient mutants *poky* in the fungus *Neurospora* and *petite* in the yeast *Saccharomyces*. These mutants are unable to synthesize cytochromes *a* and *b* and certain other mitochondrial enzymes. In *Neurospora*, where cross-fertilization is effected by sowing the protoperithecium (female) of one parental strain with the conidia (male) of the other, it is the mitochondrial character of the protoperithecial parent that is transmitted to the sexual progeny. Thus, if the female parent is *poky* and conidia come from a normal strain, all the progeny from the cross are *poky*. Since the conidia of normal *Neurospora* contain the cytoplasmic information for the synthesis of functional mitochondria, as can be seen when conidia are used as vegetative propagants, it must be concluded either that this information is not transmitted at fertilization or that it is rejected in the zygote.

Mitochondria. In the unicellular yeast, distinction into male and female parental strains is not possible, and fertilization involves the fusion of two haploid cells of opposite mating type to give a diploid zygote cell. This can give rise by budding to a clone of diploid vegetative cells, any one of which can be induced to go through meiosis to give a tetrad of haploid sexual spores. When cells of the mitochondrial mutant *petite* are crossed with cells of normal strain, diploid clones usually segregate into *petite* and normal. Sexual progeny, obtainable only from normal segregants, are all normal. In this case, it is generally thought that there is a mixture of cytoplasmic mitochondrial factors of the two parental strains in the zygotes and that either mitochondrial type may be transmitted to the daughter cells. Cytoplasmic mitochondrial factors for resistance to certain antibiotics have been identified and, in crosses between different resistant strains, a mitochondrial recombinational process occurs during clonal development from zygotes. Thus in a cross between two strains resistant to different antibiotics, recombinant diploid cells resistant to both or resistant to neither can arise during vegetative growth from zygotes. Physical evidence of recombination has been demonstrated in crosses between yeast strains in which different mitochondrial deoxyribonucleic acid (DNA) [see below] endonuclease cleavage sites recombine to give new fragment patterns in diploid segregants. Mitochondrial recombination has been demonstrated in higher plants but not clearly in higher animals. *See* MITOCHONDRIA; RECOMBINATION (GENETICS).

Chloroplasts. In the unicellular alga *Chlamydomonas*, cytoplasmic inheritance of certain chloroplast characters has been demonstrated. In these cases the departure from mendelian inheritance is seen in postmeiotic segregation and recombination; that is, these genetic factors segregate and recombine in the haploid clones that develop from individual sexual progeny.

Again, in the asexual *Euglena* the presence of cytoplasmic genetic elements controlling chloroplast synthesis has long been postulated. This was inferred mainly from the results of experimental procedures which block the synthesis of the chloroplast, leading to the loss of the organelle. Cells once having lost the chloroplast are apparently unable to regenerate this structure.

Kinetoplasts. Similarly, genetic autonomy of membrane-limited organelles called kinetoplasts (parabasal bodies) in trypanosomes is inferred from the inability of akinetoplastic mutant cells and their descendants to resynthesize the structure.

Genetic factors. Claims to having demonstrated genetic elements in the cytoplasm have been supported, at least in the case of mitochondria, chloroplasts, and kinetoplasts, by the findings that each of these organelles has its own complement of DNA, the genetic substance. Is the autonomy of these membrane-bound bodies attributable to the fact that each one carries the complete genetic blueprint for its own synthesis? From the estimates of the many macromolecular components of these complex bodies and particularly the various protein species, it is clear, certainly in the case of organelles, that the amount of organelle DNA is not sufficient to code for all of its components. Other genetic information, presumably of chromosomal origin, would seem to be required. The impression then is that the biosynthesis of these cytoplasmic bodies involves a high degree of coordination between cytoplasmic and nuclear genetic factors and that organelles have conditional autonomy.

Aspects of the self-reproducing capacity of chloroplasts and mitochondria have been studied in some detail since it is possible to obtain these organelles free from the cell by differential centrifugation of cell homogenates. Mitochondria, for example, have been shown, by the use of radioactive labeling of molecules, to incorporate exogenous supplies of amino acids into their proteins, and DNA and ribonucleic acid (RNA) precursors into DNA and RNA, respectively. These findings show that mitochondria have an intrinsic system for protein synthesis based on a DNA-RNA mechanism. A comparable system has also been demonstrated for chloroplasts; in both cases the physical properties of the DNA and the RNA species of the organelles are recognizably different from the properties of the general protein-synthesizing system of the cell.

Another important difference is that antibiotics which inhibit protein synthesis in bacteria specifically inhibit protein synthesis in chloroplasts and mitochondria, indicating affinities with the bacterial system. Other similarities are the requirement for *N*-formylmethionyl-tRNA, and the fact that DNA is a nonchromatin, circular molecule. Such evidence has led to the hypothesis that these organelles were once symbiotic bacteria that became incorporated into the general cell system and subject to nuclear control in the course of evolution while retaining some of the heritable properties of the bacterial system. However, a unique feature of these organelles is a departure from the universal genetic code in some codons: for example, UGA (stop codon) specifies tryptophan in mitochondria. *See* GENETIC CODE.

The complete nucleotide sequence of mitochondrial DNA and the gene content of this genome have been determined for the following organisms: human, cow, mouse, frog, fruit fly, a nematode, and yeast. Coding sequences comprise some 16 to 17 kilobasepairs, and the products of these genes are remarkably similar in all eukaryotic organisms, although the order and organization of genes within the mitochondrial genome may vary considerably between higher and lower orders. The genetic complement in all metazoan mitochondria, for example, specifies the RNA components of the organelle's protein-synthesizing system (2 ribosomal ribonucleic acids and 22 transfer ribonucleic acids), subunits I, II, and III of the polymeric cytochrome c oxidase, subunits 6 and 8 of mitochondrial adenosinetriphosphatase (ATPase), cytochrome b and 7 components of the respiratory-chain nicotinamide adenine dinucleotide (NADH) dehydrogenase. The specificity of few reading frames have yet to be identified.

Fragment patterns generated by endonuclease digestion of mitochondrial DNA can reveal differences between individuals at the nucleotide level (cleavage sites). Such differences between male and female parents in mammals, including humans, have been followed in the progeny, and the results clearly show that the mitochondrial chromosome is transmitted through the female parent. These findings substantiate the genetic evidence of maternal inheritance.

It is only in the case of organelles that cytoplasmic inheritance begins to be understood because of the presence and replication of DNA in these bodies. Other cases of nonmendelian inheritance being attributed to cytoplasmic genetic units are difficult to substantiate. SEE DEOXYRIBONUCLEIC ACID (DNA); RIBONUCLEIC ACID (RNA).

David Wilkie

Human disease. Two neurological diseases are transmitted by maternal inheritance, Leber's optic atrophy and a type of encephalomyopathy labeled MERRF (myoclonus epilepsy with ragged red fibers). Leber's optic atrophy is characterized by subacute bilateral visual loss usually affecting young, otherwise healthy men, and leading to permanent blindness. A point mutation of mitochondrial DNA has been documented in 9 of 11 families. The main symptoms and signs of MERRF are myoclonus, incoordination, dementia, deafness, lactic acidosis, and presence of abnormal mitochondria in muscle fibers (these fibers are called ragged red). Although the mutation of mitochondrial DNA has not been identified, there are partial defects of oxidative phosphorylation suggesting involvement of two complexes of the respiratory chain.

Major deletions of mitochondrial DNA have been found in patients with weakness of extraocular muscles (progressive external ophthalmoplegia) or Kearns-Sayre syndrome, a multisystem neurological disorder dominated by progressive external ophthalmoplegia, retinal degeneration, heart block, incoordination, and dementia. Because most cases are sporadic, the deletions must have occurred in the ovum or in the zygote. The precise relationship between mitochondrial DNA deletions and biochemical and clinical phenotypes remains to be clarified.

Salvatore Di Mauro

Bibliography. G. Bernardi, The *petite* mutation in yeast, *Trends Biochem. Sci.*, 4:197–221, 1979; R. E. Giles, Maternal inheritance of human mitochondrial DNA, *Proc. Nat. Acad. Sci.*, 77:6715–6719, 1980; L. A. Grivell, Mitochondrial DNA, *Sci. Amer.*, 248:78–89, March 1983; D. R. Newth and M. Balls (eds.), *Maternal Effects in Development*, 1979; E. K. Nikoskelainen, Leber's hereditary optic neuropathy: A hereditary disease, *Acta. Neurol. Scand.*, 69 (suppl. 98):172–173, 1984; R. Sager, *Cytoplasmic Genes and Organelles*, 1972; D. Wilkie, Cytoplasmic genetic systems of eukaryotic cells, *Brit. Med. Bull.*, 29:263–270, 1973; D. R. Wolstenholme, DNA sequences of mitochondrial genomes of nematode worms, *Proc. Nat. Acad. Sci.*, 84:1324–1328, 1987.

Cytoskeleton

A system of filaments found in the cytoplasm of cells which is responsible for cell shape, cell locomotion and elasticity, interconnection of the major cytoplasmic organelles, cell division, chromosome organization and movement, and the adhesion of a cell to a surface or to other cells.

Since this filamentous system is too small to be resolved by light microscopy, its several subcomponents, structural features, and cytoplasmic associations were not discovered until the electron microscope was developed. With this instrument, which is capable of magnifying a section of a cell many thousands of times, three major classes of filaments could be resolved on the basis of their diameter and cytoplasmic distribution: actin filaments (or microfilaments) each with an average diameter of 6 nanometers, microtubules with an average diameter of 25 nm, and intermediate filaments whose diameter of 10 nm is intermediate to that of the other two classes. In muscle cells, an additional class of thick filaments, known as myosin filaments, is found whose function is to interact with actin filaments and generate the force necessary for muscle contraction. The presence of this system of filaments in all cells, as well as their diversity in structure and cytoplasmic distribution, has been recognized only in the modern period of biology.

A technique that has greatly facilitated the visualization of these filaments as well as the analysis of their chemical composition is immunofluorescence applied to cells grown in tissue culture. In this technique, the protein that is to be localized is injected into an animal to elicit an antibody response. The resulting antibody is highly specific for the protein. Following several purification steps, the antibody is chemically attached to a fluorescent dye. The cells, which are usually grown on a glass cover slip, are fixed with a cross-linking agent such as formaldehyde in order to stabilize their cytoskeletons. They are subsequently rendered permeable to the fluorescent antibody by exposure to a detergent or to a polar solvent such as ethanol or acetone. The cells are then treated with the fluorescent antibody solution for a few minutes; unbound antibodies are washed away, and the cover slip is examined under a regular light microscope equipped with fluorescence optics. The fluorescent antibody binds to the protein it is specific for, and upon excitation of the fluorescent dye, the location of that protein is revealed. Extensions of this technique involve the use of antibodies and electron microscopy. In this technique, known as immunoelectron microscopy, the resolution of subcellular localization is increased appreciably. Furthermore, it can be used to detect the binding site of one protein to another, thereby expanding the detail at which cytoskeletal molecular interactions among each other and in the cytoplasm can be understood. *SEE IMMUNO- FLUORESCENCE.*

Actin filaments. Actin, a protein highly conserved in evolution, is the main structural component of actin filaments in all cell types, both muscle and nonmuscle. As revealed by immunofluorescence and electron microscopy, actin filaments assume a variety of configurations depending on the type of cell and the state

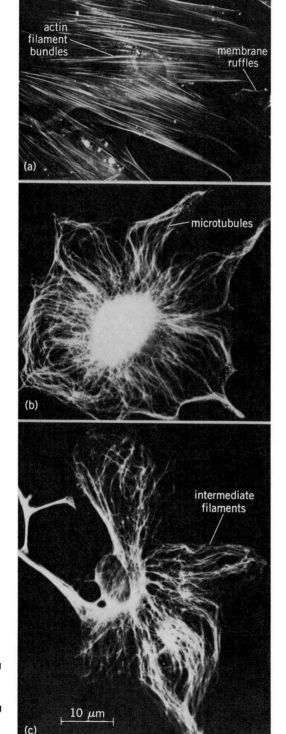

Fig. 1. Immunofluorescence with human cells grown in tissue culture. (*a*) Distribution of actin in filament bundles and membrane ruffles using fluorescent antibodies to muscle actin. (*b*) Distribution of microtubules using fluorescent anitbodies to their major subunit tubulin. (c) Distribution of intermediate filaments using fluorescent antibodies to their major subunit from fibroblasts.

it is in. They extend a considerable distance through the cytoplasm in the form of bundles (**Fig. 1***a*). These bundles are also known as stress fibers since they are important in determining the elongated shape of the cell and in enabling the cell to adhere to the substrate and spread out on it. In a single migratory or resting cell, most of the actin filament bundles are located just below the cell membrane and are frequently in close association with it. The sites where the bundles make contact with the cell membrane, known as adhesion plaques, correspond to the sites of cell-substratum contact. It has been estimated from time-lapse motion pictures that most of the actin filament bundles are oriented parallel to the long axis of the cell and shift their orientation as the cell changes direction; the bundles reorient themselves over a period of 15 to 60 min.

Actin filaments can exist in forms other than straight bundles. In rounded cells that do not adhere strongly to the substrate (such as dividing cells and cancer cells), the filaments form an amorphous meshwork that is quite distinct from the highly organized bundles. This filamentous meshwork is usually located in the cell cortex, which is the area just beneath the cell membrane. It is also found in areas of membrane ruffling in motile cells. Membrane ruffles are sheetlike extensions of the cytoplasm which function as the main locomotory organ of the cells (**Fig. 2**). The membrane ruffle forms transient contacts with the substrate on which the cell rests, stretching the rest of the cell among several temporary adhesions. When the cell is spread out to its fullest extent, it maintains one or two primary ruffles, each of which tends to lead the cell in a particular direction. When the transient contacts of a membrane ruffle are converted to permanent contacts, they become adhesion plaques and function to organize long actin filament bundles. Thus, the two filamentous states, actin filament bundles and actin filament meshworks, are interconvertible polymeric states of the same molecule. They appear to be the extremes of a gradient, and the ratio between them depends on a variety of factors, including the extent of cell motility, the age of the cell, the degree of adhesion to the substrate, and the stage of the cell cycle. Bundles give the cell its tensile strength, adhesive capability, and structural support, while meshworks provide elastic support and force for cell locomotion.

Actin filaments are involved in two main cellular functions: cell motility, and cell shape and elasticity (Fig. 1*a*). These broad types of function are mediated by specific accessory proteins that interact with actin to change the properties, functional associations, or the polymeric state of actin filaments. Thus, the cell has molecules which regulate the polymerization of actin to single filaments, molecules which regulate the assembly of single actin filaments to filament bundles, molecules which interconvert actin filament bundles and actin filament meshworks, molecules which mediate the interaction of actin filaments with the plasma membrane, and finally molecules, such as myosin, which enable actin filaments to move the cell as a whole or mediate specific motile activities to the cytoplasm, such as cell cleavage. In turn, the spatial and temporal interaction of these proteins with actin is regulated, in part, by the cytoplasmic hydrogen ion concentration (pH) and the calcium ion concentration. In the cytoplasm these regulatory mechanisms coexist in a dynamic equilibrium, many of them operating

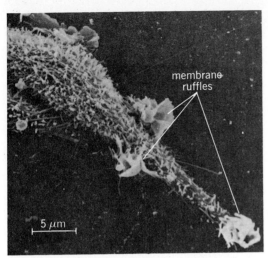

membrane ruffles

5 μm

Fig. 2. Scanning electron micrograph of the surface of a human fibroblast grown in tissue culture. The small protrusions on the cell surface are known as microvilli or microspikes and are composed of a core of 20–30 actin filaments.

simultaneously, to govern cell motility and maintenance of cytoplasmic structure.

The importance of accessory actin-binding proteins in the maintenance of cell shape and elasticity has been demonstrated through the study of certain severe human hemolytic anemias, such as hereditary spherocytosis, pyropoikilolytosis, and elliptocytosis. In these diseases one or more of the actin-binding proteins are defective. Normally, this battery of proteins binds to and crosslinks actin filaments, forming a pliable and deformable network under the plasma membrane of the red blood cell. Mutations in any one of these accessory proteins cause structural alterations in this skeletal network, resulting in the fragmentation of the plasma membrane as the cells deform their shapes during circulation. The consequence is cell lysis and hemolysis. A number of these accessory proteins are now known to be common to red blood cells, lens cells, muscle cells, and nerve cells, among others, in association with actin.

A main function of actin filaments is to interact with myosin filaments to generate the force that allows muscles to contract, and many nonmuscle cells to move. The interaction of these two filament systems is extremely well coordinated by a different set of proteins that interact with either actin filaments or myosin filaments. The interaction of these proteins with their respective filaments is regulated by Ca^{2+} ions, by the H^+ ion concentration in the cytoplasm, and by direct covalent modifications such as phosphorylation. SEE MUSCLE PROTEINS.

Microtubules. These slender cylindrical structures exhibit a cytoplasmic distribution distinct from actin filaments. Microtubules originate in structures that are closely associated with the outside surface of the nucleus known as centrioles. Fluorescent antibodies specific to the major structural protein of these filaments, known as tubulin, reveal a wavy system of filaments

originating in the centriole area and extending to the cell's outer membrane (Fig. 1b). Unlike the other two classes of filaments, microtubules are highly unstable structures and appear to be in a constant state of polymerization-depolymerization. During mitosis, they rapidly disaggregate and reorganize around the mitotic spindle, where they function in moving the chromosomes to the two newly forming daughter cells. In the resting cell, microtubules are involved in the determination of a cell's shape and in cytoplasmic transport of vesicles and other small organelles. In conjunction with the centrioles, they also function as the "information processing center" of the cell since they translate external stimuli into decisions about the direction of locomotion of a cell. In more specialized cells, like the sperm cell, they function to propagate the cell through body fluids. SEE CENTRIOLE; MITOSIS.

Intermediate filaments. These filaments function as the true cytoskeleton. Unlike microtubules and actin filaments, intermediate filaments are very stable structures. If a cell is exposed to a mild nonionic detergent, most of the microtubules, actin filaments, and various cytoplasmic organelles are rapidly removed, but intermediate filaments and nuclei are left insoluble. In the intact cell, they anchor the nucleus, positioning it within the cytoplasmic space. During mitosis, they form a filamentous cage around the mitotic spindle in a fixed place during chromosome movement. As seen using immunofluorescence (Fig. 1c), intermediate filaments have a cytoplasmic distribution independent of actin filaments and microtubules. However, their distribution in the cytoplasm depends very much on the type of cell and its stage in differentiation. For example, in undifffferentiated muscle cells these filaments are found scattered in the cytoplasm, but in mature, fully differentiated cells they wrap around the individual contractile units, the myofibrils, and function to interconnect them to the plasma membrane. In chicken red blood cells, which contain a nucleus, these filaments function to anchor the centrally located nucleus to the plasma membrane. In mammalian red blood cells, however, which lose their nuclei, these filaments are removed prior to enucleation. In all epithelial cells, and epidermal cells in particular, a special class of intermediate filaments, known as keratin filaments, interconnect the nuclei to special areas of cell-cell contact at the plasma membrane known as desmosomes. In this fashion, an epithelial layer is integrated mechanically both intercellularly and intracellularly to withstand tension and possess elastic properties. Therefore, intermediate filaments are a fundamental component of the morphogenesis of certain differentiating cells. SEE MUSCLE.

Elias Lazarides

Bibliography. A. D. Bershadsky and J. M. Vasiliev, *Cytoskeleton*, 1988; G. G. Borisy et al. (eds.), *Molecular Biology of the Cytoskeleton*, 1984; A. B. Fulton, *The Cytoskeleton: Cellular Architecture and Choreography*, 1984; L. Goldstein and D. M. Prescott (eds.), *Cell Biology: A Comprehensive Treatise*, vol. 4, 1980; D. Marme (ed.), *Calcium and Cell Physiology*, 1985; J. W. Shay (ed.), *Cell and Molecular Biology of the Cytoskeleton*, 1986; P. Traub, *Intermediate Filaments*, 1985.